About the Authors

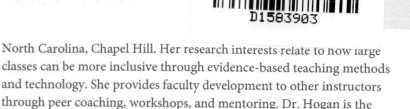

ERIC J. SIMON

is a professor in the Department of Biology and Health Science at New England College (Henniker, New Hampshire). He teaches introductory biology to science majors and nonscience majors, as well as upper-level courses in tropical marine biology and careers in science. Dr. Simon received a B.A. in biology and computer science and an M.A. in biology from Wesleyan University, and a Ph.D. in biochemistry from Harvard University. His research focuses on innovative ways to use technology to increase active learning in the science classroom, particularly for nonscience majors. Dr. Simon is also the author of the introductory biology textbook *Biology: The Core* and a coauthor of *Campbell Biology: Concepts & Connections*, 8th Edition.

To Muriel, my wonderful mother, who always supported my efforts with love, compassion, great empathy, and an unwavering belief in me

JEAN L. DICKEY

is Professor Emerita of Biological Sciences at Clemson University (Clemson, South Carolina). After receiving her B.S. in biology from Kent State University, she went on to earn a Ph.D. in ecology and evolution from Purdue University. In 1984, Dr. Dickey joined the faculty at Clemson, where she devoted her career to teaching biology to nonscience majors in a variety of courses. In addition to creating content-based instructional materials, she developed many activities to engage lecture and laboratory students in discussion, critical thinking, and writing, and implemented an investigative laboratory curriculum in general biology. Dr. Dickey is the author of *Laboratory Investigations for Biology*, 2nd Edition, and is a coauthor of *Campbell Biology: Concepts & Connections*, 8th Edition.

To my mother, who taught me to love learning, and to my daughters, Katherine and Jessie, the twin delights of my life

KELLY A. HOGAN

is a faculty member in the Department of Biology and the Director of Instructional Innovation at the University of North Carolina at Chapel Hill, teaching introductory biology and introductory genetics to science majors. Dr. Hogan teaches hundreds of students at a time, using active-learning methods that incorporate technology such as cell phones as clickers, online homework, and peer evaluation tools. Dr. Hogan received her B.S. in biology at the College of New Jersey and her Ph.D. in pathology at the University of North Carolina, Chapel Hill. Her research interests relate to now large classes can be more inclusive through evidence-based teaching methods and technology. She provides faculty development to other instructors through peer coaching, workshops, and mentoring. Dr. Hogan is the author of *Stem Cells and Cloning*, 2nd Edition, and is lead moderator of the *Instructor Exchange*, a site within MasteringBiology® for instructors to exchange classroom materials and ideas. She is also a coauthor of *Campbell Biology: Concepts & Connections*, 8th Edition.

To the good-looking boy I met in my introductory biology course many moons ago—and to our two children, Jake and Lexi, who are everyday reminders of what matters most in life

JANE B. REECE

has worked in biology publishing since 1978, when she joined the editorial staff of Benjamin Cummings. Her education includes an A.B. in biology from Harvard University (where she was initially a philosophy major), an M.S. in microbiology from Rutgers University, and a Ph.D. in bacteriology from the University of California, Berkeley. At UC Berkeley, and later as a postdoctoral fellow in genetics at Stanford University, her research focused on genetic recombination in bacteria. Dr. Reece taught biology at Middlesex County College (New Jersey) and Queensborough Community College (New York). During her 12 years as an editor at Benjamin Cummings, she played a major role in a number of successful textbooks. She is the lead author of *Campbell Biology*, 10th Edition, and *Campbell Biology: Concepts & Connections*, 8th Edition.

To my wonderful coauthors, who have made working on our books a pleasure

NEIL A. CAMPBELL

(1946–2004) combined the inquiring nature of a research scientist with the soul of a caring teacher. Over his 30 years of teaching introductory biology to both science majors and nonscience majors, many thousands of students had the opportunity to learn from him and be stimulated by his enthusiasm for the study of life. While he is greatly missed by his many friends in the biology community, his coauthors remain inspired by his visionary dedication to education and are committed to searching for ever-better ways to engage students in the wonders of biology.

Detailed Contents

1 Introduction: Biology Today

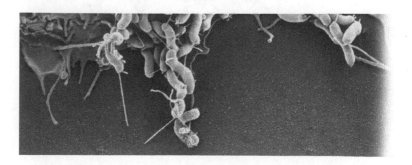

5 The Working Cell 108
▶ CHAPTER THREAD: NANOTECHNOLOGY

6 Cellular Respiration: Obtaining Energy from Food 124
▶ CHAPTER THREAD: EXERCISE SCIENCE

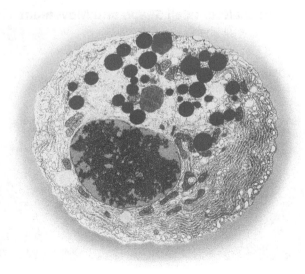

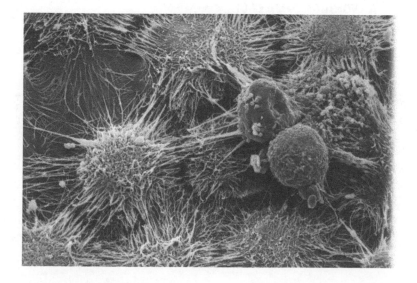

UNIT 3 Evolution and Diversity

16 The Evolution of Plants and Fungi — 348

▶ CHAPTER THREAD: PLANT-FUNGUS INTERACTIONS

17 The Evolution of Animals — 370

▶ CHAPTER THREAD: HUMAN EVOLUTION

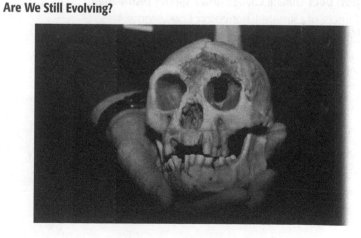

UNIT 4 Ecology

UNIT 5 Animal Structure and Function

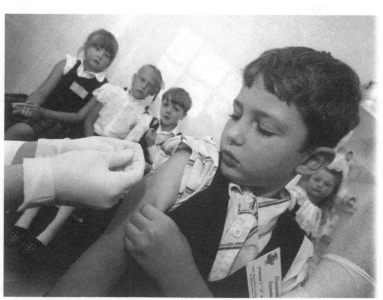

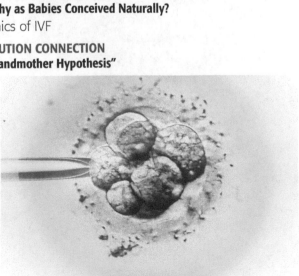

UNIT 6 Plant Structure and Function

Appendices

Discover Why Biology *Matters*

Campbell Essential Biology highlights how the concepts that you learn in your biology class are relevant to your everyday life.

- **NEW! Why Biology Matters Photo Essays** use dynamic photographs and intriguing scientific observations to introduce each chapter. Each scientific tidbit is revisited in the chapter.

15 The Evolution of Microbial Life

Why Microorganisms Matter

If your family took a vacation in which you traveled 1 mile for every million years in the history of life, you'd still be asking, "Are we there yet?" after driving from Miami to Seattle.

According to a recent study, infection by the parasite *Toxoplasma* makes mice lose their fear of cats.

▲ Seaweeds aren't just used for wrapping sushi—they're in your ice cream, too.

▲ You have microorganisms to thank for the clean water you drink every day.

326

MasteringBiology®

NEW! Everyday Biology Videos briefly explore interesting and relevant biology topics that relate to concepts that students are learning in class. These 20 videos can be assigned in MasteringBiology with assessment questions.

● **UPDATED! Chapter Threads** weave a single
compelling topic throughout the chapter. In
Chapter 15, human microbiota are explored.

Biology and Society essays
relate biology to your life and
interests. This example discusses
the microorganisms that live in
your own body.

 Human Microbiota BIOLOGY AND SOCIETY

Our Invisible Inhabitants

You probably know that your body contains trillions of individual cells, but did you know that they aren't
all "you"? In fact, microorganisms residing in and on your body outnumber your own cells by 10 to 1. That
means 100 trillion bacteria, archaea, and protists call your body home. Your skin, mouth, and nasal passages
and your digestive and urogenital tracts are prime
real estate for these microorganisms. Although each
individual is so tiny that it would have to be magnified hundreds of times for you to see it, the weight of
your microbial residents totals two to five pounds.

We acquire our microbial communities during the
first two years of life, and they remain fairly stable
thereafter. However, modern life is taking a toll on
that stability. We alter the balance of these communities by taking antibiotics, purifying our water,
sterilizing our food, attempting to germproof our
surroundings, and scrubbing our skin and teeth.
Scientists hypothesize that disrupting our microbial
communities may increase our susceptibility to infectious diseases, predispose us to certain cancers,
and contribute to conditions such as asthma and
other allergies, irritable bowel syndrome, Crohn's
disease, and autism. Researchers are even investigating whether having the wrong microbial community
could make us fat. In addition, scientists are studying

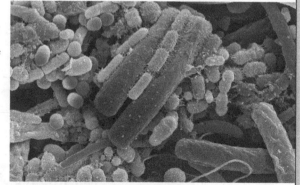

Colorized scanning electron micrograph of bacteria on a human tongue (14,500×).

how our microbial communities have evolved over the course of human history. As you'll discover in the
Evolution Connection section at the end of this chapter, for example, dietary changes invited decay-causing
bacteria to make themselves at home on our teeth.

Throughout this chapter, you will learn about the benefits and drawbacks of human-microbe interactions.
You will also sample a bit of the remarkable diversity of prokaryotes and protists. This chapter is the first of
three that explore the magnificent diversity of life. And so it is fitting that we begin with the prokaryotes,
Earth's first life-form, and the protists, the bridge between unicellular eukaryotes and multicellular plants,
fungi, and animals.

Process of Science explorations
give you real-world examples of how the
scientific method is applied. Chapter 15
explores a recent investigation into the
possible role of microbiota in obesity.

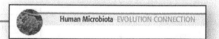

Evolution Connection essays
conclude each chapter by
demonstrating how the theme of
evolution runs throughout all of
biology. The example in Chapter
15 discusses how changes in
the typical human diet over
generations is linked to bacteria
that cause tooth decay.

● **Additional updated Chapter Threads and essays** include
radioactivity in Chapter 2, muscle performance in Chapter 6,
and theft of used cooking oil for biofuel recycling in Chapter 7.

Identify "Big Picture" Themes

Examples of major themes in biology are highlighted throughout the text to help you see how overarching biology concepts are interconnected.

- **NEW!** **Important Themes in Biology** are introduced in Chapter 1 to underscore unifying principles that run throughout biology.

MAJOR THEMES IN BIOLOGY				
Evolution	**Structure/Function**	**Information Flow**	**Energy Transformations**	**Interconnections within Systems**
Evolution by natural selection is biology's core unifying theme and can be seen at every level in the hierarchy of life.	The structure of an object, such as a molecule or a body part, provides insight into its function, and vice versa.	Within biological systems, information stored in DNA is transmitted and expressed.	All biological systems depend on obtaining, converting, and releasing energy and matter.	All biological systems, from molecules to ecosystems, depend on interactions between components.

- These themes—Evolution, Structure/Function, Information Flow, Energy Transformations, and Interconnections within Systems—are **signaled with icons** throughout the text to help you notice the reoccurring examples of the major themes.

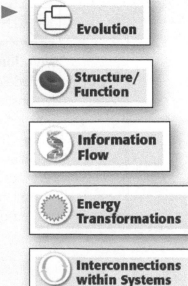

- Evolution
- Structure/ Function
- Information Flow
- Energy Transformations
- Interconnections within Systems

 Human Microbiota EVOLUTION CONNECTION

- The role of evolution throughout all of biology is further explored in depth at the end of each chapter in **Evolution Connection** discussions.

Recognize Analogies and Applications

Analogies and applications to everyday life make unfamiliar biology concepts easier to visualize and understand.

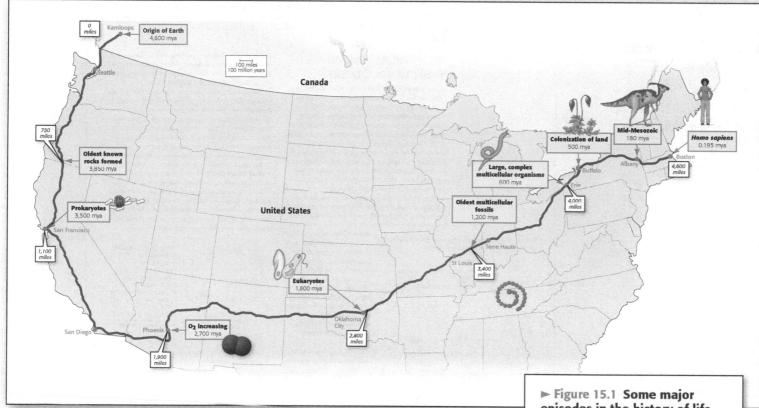

▶ **Figure 15.1 Some major episodes in the history of life.** On this 4,600-mile metaphorical road trip, each mile equals 1 million years in Earth's history.

If your family took a vacation in which you traveled 1 mile for every million years in the history of life, you'd still be asking, "Are we there yet?" after driving from Miami to Seattle.

● **NEW analogies and applications** have been added throughout the prose and the illustrations, making it easier to learn and remember key concepts for the first time. Examples include:

● comparing the significant differences between prokaryotic and eukaryotic cells to the differences between a bicycle and an SUV (Chapter 4)

● comparing the process of DNA winding into chromosomes with the act of winding yarn into a skein (Chapter 10)

● comparing a 4,600-mile road trip that describes the scale of biological evolution on Earth (Chapter 15)

● comparing signal transduction to email communication (Chapter 27*)

● comparing how dominoes relate to an action potential moving along an axon (Chapter 27*)

* Chapters 21–29 are included in the expanded version of the text that includes coverage of animal and plant anatomy and physiology.

Boost Your Scientific Literacy

A wide variety of exercises and assignments can help you move beyond memorization and think like a scientist.

● **UPDATED! Process of Science essays** appear in every chapter and walk through each step of the scientific method as it applies to a specific research question. ▶

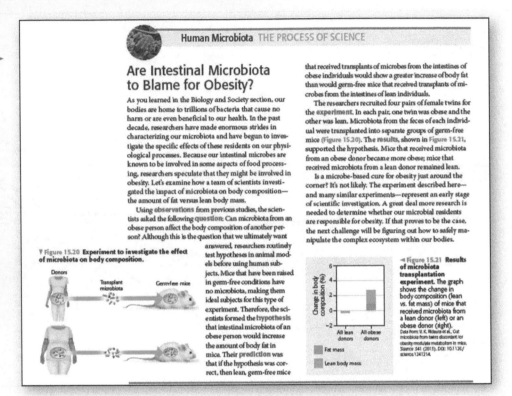

Human Microbiota THE PROCESS OF SCIENCE

Are Intestinal Microbiota to Blame for Obesity?

As you learned in the Biology and Society section, our bodies are home to trillions of bacteria that cause no harm or are even beneficial to our health. In the past decade, researchers have made enormous strides in characterizing our microbiota and have begun to investigate the specific effects of these residents on our physiological processes. Because our intestinal microbes are known to be involved in some aspects of food processing, researchers speculate that they might be involved in obesity. Let's examine how a team of scientists investigated the impact of microbiota on body composition—the amount of fat versus lean body mass.

Using observations from previous studies, the scientists asked the following question: Can microbiota from an obese person affect the body composition of another person? Although this is the question that we ultimately want answered, researchers routinely test hypotheses in animal models before using human subjects. Mice that have been raised in germ-free conditions have no microbiota, making them ideal subjects for this type of experiment. Therefore, the scientists formed the hypothesis that intestinal microbiota of an obese person would increase the amount of body fat in mice. Their prediction was that if the hypothesis was correct, then lean, germ-free mice

that received transplants of microbes from the intestines of obese individuals would show a greater increase of body fat than would germ-free mice that received transplants of microbes from the intestines of lean individuals.

The researchers recruited four pairs of female twins for the experiment. In each pair, one twin was obese and the other was lean. Microbiota from the feces of each individual were transplanted into separate groups of germ-free mice (Figure 15.20). The results, shown in Figure 15.21, supported the hypothesis. Mice that received microbiota from an obese donor became more obese; mice that received microbiota from a lean donor remained lean.

Is a microbe-based cure for obesity just around the corner? It's not likely. The experiment described here—and many similar experiments—represent an early stage of scientific investigation. A great deal more research is needed to determine whether our microbial residents are responsible for obesity. If that proves to be the case, the next challenge will be figuring out how to safely manipulate the complex ecosystem within our bodies.

▼ Figure 15.20 **Experiment to investigate the effect of microbiota on body composition.**

Donors / Transplant microbiota / Germ-free mice

◄ Figure 15.21 **Results of microbiota transplantation experiment.** The graph shows the change in body composition (lean vs. fat mass) of mice that received microbiota from a lean donor (left) or an obese donor (right). Data from: V. K. Ridaura et al., Gut microbiota from twins discordant for obesity modulate metabolism in mice. *Science* 341 (2013). DOI: 10.1126/science.1241214.

Fat mass
Lean body mass

MasteringBiology®

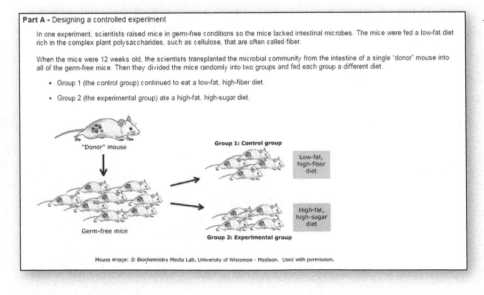

Part A - Designing a controlled experiment

In one experiment, scientists raised mice in germ-free conditions so the mice lacked intestinal microbes. The mice were fed a low-fat diet rich in the complex plant polysaccharides, such as cellulose, that are often called fiber.

When the mice were 12 weeks old, the scientists transplanted the microbial community from the intestine of a single "donor" mouse into all of the germ-free mice. Then they divided the mice randomly into two groups and fed each group a different diet.

- Group 1 (the control group) continued to eat a low-fat, high-fiber diet.
- Group 2 (the experimental group) ate a high-fat, high-sugar diet.

"Donor" mouse

Germ-free mice

Group 1: Control group
Low-fat, high-fiber diet

Group 2: Experimental group
High-fat, high-sugar diet

Mouse image: © Biochemistry Media Lab, University of Wisconsin · Madison. Used with permission.

◄ **NEW! Scientific Thinking Activities** are designed to help you develop an understanding of how scientific research is conducted.

NEW! Evaluating Science in the Media Activities challenge you to recognize validity, bias, purpose, and authority in everyday sources of information.

Learn to Interpret Data

Data interpretation is important for understanding biology and for making many important decisions in everyday life. Exercises in the text and online will help you develop this important skill.

● **NEW! Interpreting Data end-of-chapter questions** help you learn to use quantitative material by analyzing graphs and data. This example from Chapter 10 invites you to examine historical data of flu mortality. Other examples include:

- Chapter 13: Learn how markings on snail shells affect predation rates in an environment

- Chapter 15: Calculate how quickly bacteria can multiply on unrefrigerated food

14. Interpreting Data The graph below summarizes the number of children who died of all strains of flu from 2007 until 2013. Each bar represents the number of child deaths occurring in one week. Why does the graph have the shape it does, with a series of peaks and valleys? Looking over the Biology and Society section at the start of the chapter, why does the graph reach its highest points near the middle? Based on these data, when does flu season begin and end in a typical year?

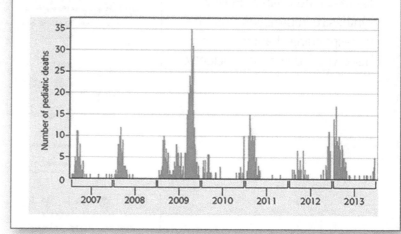

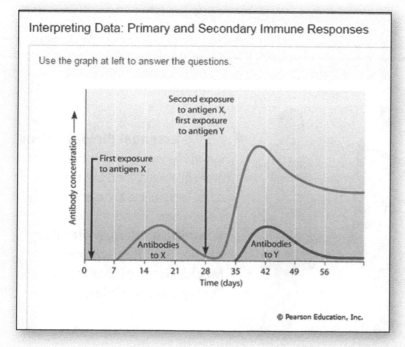

Interpreting Data: Primary and Secondary Immune Responses

Use the graph at left to answer the questions.

MasteringBiology®

◀ **NEW! Interpreting Data Activities** help you build and practice data analysis skills.

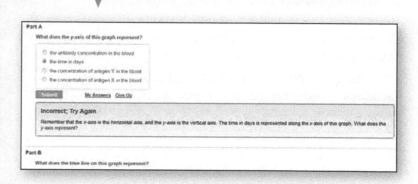

Maximize Your Study Time

Campbell Essential Biology and the **MasteringBiology** homework, tutorial, and assessment program work hand-in-hand to help students succeed in introductory biology.

● **The Chapter Review** offers ▶ a built-in study guide that combines words with images to help you organize the key concepts. The unique figures in the Chapter Review synthesize information from the corresponding chapter, which helps you study more efficiently.

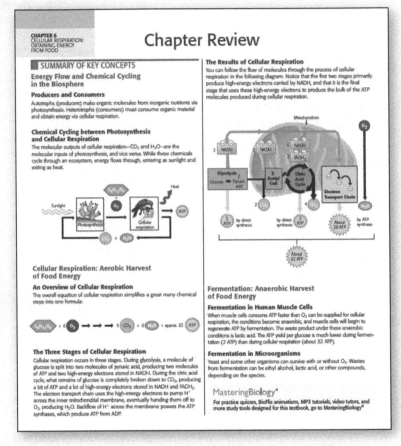

MasteringBiology®

MasteringBiology provides a wide range of activities and study tools to match your learning style, including BioFlix animations, MP3 audio tutorials, interactive practice quizzes, and more. Your instructor can assign activities for extra practice to monitor your progress in the course.

◀ **NEW! Essential Biology videos** introduce you to key concepts and vocabulary, and are narrated by authors Eric Simon and Kelly Hogan. Topics include the **Scientific Method, Molecules of Life, DNA Replication, Mechanisms of Evolution, Ecological Principles,** and more.

Learn Before, During, and After Class

MasteringBiology®

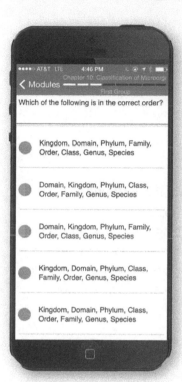

BEFORE CLASS

NEW! **Dynamic Study Modules** help you acquire, retain, and recall information faster and more efficiently than ever before. The convenient practice questions and detailed review explanations can be accessed on the go using a smartphone, tablet, or computer.

Mastering
Continuously Adaptive

Pre-lecture Assignments

Learning Catalytics, Mastering Media

Homework, Quizzing, and Testing

DURING CLASS

NEW! **Learning Catalytics is a "bring your own device" assessment and classroom activity system** that expands the possibilities for student engagement. Using Learning Catalytics, instructors can deliver a wide range of auto-gradable or open-ended questions that test content knowledge and build critical thinking skills using eighteen different answer types.

AFTER CLASS

- **Over 100 Coaching Activities** are created by the textbook author team and help you focus on learning key concepts and building your biology vocabulary.

- **NEW!** **Everyday Biology videos** briefly explore interesting and relevant biology topics that relate to concepts in the course.

Instructors: Extensive Resources for You

Extensive resources save valuable time both in course prep and during class.

- The **Test Bank** provides a variety of test questions, many art- or scenario-based, in both TestGen® and Microsoft® Word.

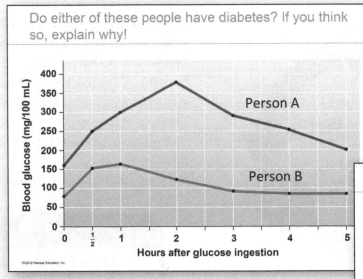

Do either of these people have diabetes? If you think so, explain why!

EXPANDED! Current Topic PowerPoint® **presentations** include new topics such as DNA Profiling, Stem Cells and Cloning, Diabetes, Biodiversity, and more. Each Powerpoint Presentation includes instructor teaching tips and active learning strategies to help you easily create a high-interest, active lecture.

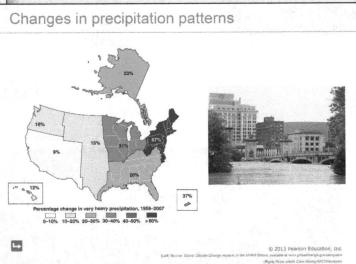

MasteringBiology®

Instructor media resources such as PowerPoint slides, animations, teaching tips videos, and more can be accessed and downloaded from the Instructor Resources area of MasteringBiology.

The **Instructor Exchange** provides successful, class-tested active learning techniques and analogies from biology instructors around the nation, offering a springboard for quick ideas to create more compelling lectures. Co-author Kelly Hogan moderates contributions to the exchange.

Preface

This is a wonderful time to teach and learn biology. Opportunities to marvel at the natural world and the life within it abound. It's difficult to view a news website without finding stories that touch on biology and its intersection with society. In addition, the world of pop culture is rich with books, movies, TV shows, comic strips, and video games that feature biological wonders and challenge us to think about important biological concepts and their implications. Although some people *say* that they don't like biology (or, more often, science in general), nearly everyone will admit to an inborn biophilia. After all, most of us keep pets, tend a garden, enjoy zoos and aquariums, or appreciate time spent outdoors. Furthermore, nearly everyone realizes that the subject of biology has a significant impact on his or her own life through its connections to medicine, biotechnology, agriculture, environmental issues, forensics, and myriad other areas. But despite the inborn affinity that nearly everyone has for biology, it can be a struggle for nonscientists to delve into the subject. Our primary goal in writing *Campbell Essential Biology with Physiology* is to help teachers motivate and educate the next generation of citizens by tapping into the inherent curiosity about life that we all share.

Goals of the Book

Although our world is rich with "teachable moments" and learning opportunities, the explosion of knowledge we have already witnessed in the 21st century threatens to bury a curious person under an avalanche of information. "So much biology, so little time" is the universal lament of biology educators. Neil Campbell conceived of *Campbell Essential Biology with Physiology* as a tool to help teachers and students focus on the most important areas of biology. To that end, the book is organized into six core areas: cells, genes, evolution, ecology, animals, and plants. Dr. Campbell's vision, which we carry on and extend in this edition, has enabled us to keep *Campbell Essential Biology with Physiology* manageable in size and thoughtful in the development of the concepts that are most fundamental to understanding life. We've aligned this new edition with today's "less is more" approach in biology education for nonscience majors—where the emphasis is on fewer topics and more focused explanations—and we never allow the content we do include to be diluted. Toward that end, in this new edition we removed some of the most technical details and terminology, which we hope will help nonscience major students to focus on the key topics in biology.

We formulated our approach after countless conversations with teachers and students in which we noticed some important trends in how biology is taught. In particular, many teachers identify three goals: (1) to engage students by relating the core content to their lives and the greater society; (2) to clarify the process of science by showing how it is applied in the real world and to give students practice in applying scientific and critical thinking skills themselves; and (3) to demonstrate how evolution serves as biology's unifying theme. To help achieve these goals, every chapter of this book includes three important features. First, a chapter-opening essay called Biology and Society highlights a connection between the chapter's core content and students' lives. Second, an essay called The Process of Science (found in the body of the chapter) describes how the scientific process has illuminated the topic at hand, using a classic or modern experiment as an example. Third, a chapter-closing Evolution Connection essay relates the chapter to biology's unifying theme of evolution. To maintain a cohesive narrative throughout each chapter, the content is tied together with a unifying chapter thread, a relevant high-interest topic that is woven throughout the three chapter essays and is touched on several additional times in the chapter. Thus, this unifying chapter thread ties together the three pedagogical goals of the course using a topic that is compelling and relevant to students.

New to This Edition

We hope that this latest edition of *Campbell Essential Biology with Physiology* goes even further in helping students relate the material to their lives, understand the process of science, and appreciate how evolution is the unifying theme of biology. To this end, we've added significant new features and content to this edition:

- **Clarifying the importance of biology to students' lives.** Every student taking an introductory biology course should be made keenly aware of the myriad ways that biology affects his or her own life. To help put such issues front and center, and to "prime the learning pump" before diving into the content, we have included a new feature at the start of each chapter called Why It Matters. Every chapter begins with a series of attention-grabbing facts in conjunction with compelling photographs that illustrate the importance of that chapter's topic to students' lives. These high-interest facts appear again in the chapter narrative, typeset in a design meant to capture students' attention and placed adjacent to the science discussion that explains the fact. Examples include: Why Macromolecules Matter ("A long-distance runner who carbo-loads the night before a race is banking glycogen to be used the next day"), Why Ecology Matters ("Producing the beef for a hamburger requires eight times as much land as producing the soybeans for a soyburger") and Why Hormones Matter ("Strike a pose before a job interview and you just might decrease the hormone that triggers stress").

- **Major themes in biology incorporated throughout the book.** In 2009, the American Association for the

Advancement of Science published a document that served as a call to action in undergraduate biology education. The principles of this document, which is titled "Vision and Change," are becoming widely accepted throughout the biology education community. "Vision and Change" presents five core concepts that serve as the foundation of undergraduate biology. In this edition of *Campbell Essential Biology with Physiology*, we repeatedly and explicitly link book content to each of the five themes. For example, the first theme, the relationship of structure to function, is illustrated in Chapter 2 in the discussion of how the unique chemistry of water accounts for its biological properties. The second theme, information flow, is explored in Chapter 10 in the discussion on how genes control traits. The third theme, interconnections within systems, is illustrated in Chapter 18 in the discussion on the global water cycle. The fourth theme, evolution, is called out in Chapter 17 in the discussion on the phylogeny of animals. The fifth theme, energy transformations, is explored in Chapter 6 in the discussion on the flow of energy through ecosystems. Readers will find at least one major theme called out per chapter, which will help students see the connections between these major themes and the course content, and instructors will have myriad easy-to-reference examples to help underscore these five themes.

- **New unifying chapter threads.** As discussed earlier, every chapter in *Campbell Essential Biology with Physiology* has a unique unifying chapter thread—a high-interest topic that helps to demonstrate the relevance of the chapter content. The chapter thread is incorporated into the three main essays of each chapter (Biology and Society, The Process of Science, and Evolution Connection) and appears throughout the chapter text. This fifth edition features many new chapter threads and essays, each of which highlights a current topic that applies biology to students' lives and to the greater society. For example, Chapter 2 presents a new thread on radioactivity, including discussions of its use in health care and as a tool to test evolutionary hypotheses. Chapter 15 features a new thread on human microbiota, including a recent investigation into the possible role of microbiota in obesity and an exploration of how the change from a hunter-gatherer lifestyle to a diet heavy in processed starch and sugar selected for oral bacteria that cause tooth decay. Chapter 24 offers a new thread on vaccines, introducing the importance of vaccinating an entire community and the reason why a new influenza vaccine is required each year.

- **Developing data literacy.** Many nonscience-major students express anxiety when faced with numerical data, yet the ability to interpret data can help with many

important decisions we all face. To help foster critical thinking skills, we have incorporated a new feature called Interpreting Data into the end-of-chapter assessments. These questions, one per chapter, offer students the opportunity to practice their science literacy skills. For example, in Chapter 10, students are asked to examine historical data of flu mortality; in Chapter 15, students are tasked with calculating how quickly bacteria can multiply on unrefrigerated food; and in Chapter 24, students are presented with a graph illustrating the prevalence of food allergies in children and asked to determine what conclusions can be drawn from the data. We hope that practice examining these simple yet relevant data sets will help students be more comfortable when they must confront numerical data in their own lives.

- **Updated content and figures.** As we do in every edition, we have made many significant updates to the content presented in the book. Examples of new or updated material include new discussions on epigenetics, metagenomics, and RNA interference; an examination of new genomic information on Neanderthals; updated climate change statistics; a discussion of advances in fetal genetic testing; and an updated discussion of new threats to biodiversity. We have also included nearly a dozen new examples of DNA profiling and a cutting-edge exploration of genetically modified foods. We also strive with each new edition to update our photos and illustrations. New figures include examples that show how a prion protein can cause brain damage (Figure 3.20), how a breast cancer drug inhibits cancer cells (Figure 25.15), how angioplasty can repair diseased arteries (Figure 23.15), and how real data from DNA profiling can exonerate wrongly accused individuals (Figure 12.16).

- **New analogies.** As part of our continuing effort to help students visualize and relate to biology concepts, we have included numerous new analogies in this edition. For example, in Chapter 4, we compare the significant differences between prokaryotic and eukaryotic cells to the differences between a bicycle and an SUV. In Chapter 8, we compare the process of DNA winding into chromosomes with the act of winding yarn into a skein. In Chapter 27, we analogize humans seeing an array of colors with only three types of photoreceptor cones to how a printer can print an array of colors from only three colors included in a toner cartridge. We also have included new analogies in visual format, such as how dominoes relate to an action potential moving along an axon (Figure 27.4). Additional examples, both narrative and visual, bring biological scale into focus, such as a 4,600-mile road trip that is used to help students imagine the scale of biological evolution on Earth (Figure 15.1).

- **MasteringBiology updates.** New whiteboard-style animated videos provide students with an introduction to key biological concepts so students can arrive to class better prepared to explore applications or dive into any topic more deeply. New Everyday Biology videos, produced by the BBC, promote connections between concepts and biology in everyday life, and Evaluating Science in the Media activities teach students how to be wise consumers of scientific information and coach them through critically evaluating the validity of scientific information on the Internet. New Scientific Thinking activities encourage students to develop scientific reasoning skills as they explore a current area of research and allows instructors to easily assess student mastery of these skills.
- **Teaching the Issues.** Because many instructors, including the authors, prefer to use current topics to demonstrate the relevance of biology to students' lives, we've expanded our series of Current Topic Instructor PowerPoints© with this edition. New topics include DNA Profiling, Stem Cells and Cloning, Diabetes, Biodiversity, and more. Each PowerPoint© Presentation includes instructor teaching tips and active learning strategies to easily create a high-interest, active lecture.

Attitudes about science and scientists are often shaped by a single required science class—*this* class. We hope to tap into the innate appreciation of nature we all share and nurture this affection into a genuine love of biology. In this spirit, we hope that this textbook and its supplements will encourage all readers to make biological perspectives a part of their personal worldviews. Please let us know how we are doing and how we can improve the next edition of *Campbell Essential Biology with Physiology*.

ERIC SIMON
Department of Biology and Health Science
New England College
Henniker, NH 03242
SimonBiology@gmail.com

JEAN DICKEY
Department of Biology
Clemson University
Clemson, SC 29634
dickeyj@clemson.edu

KELLY HOGAN
Department of Biology
University of North Carolina
Chapel Hill, NC 27599
leek@email.unc.edu

JANE REECE
C/O Pearson Education
1301 Sansome Street
San Francisco, CA 94111
JaneReece@cal.berkeley.edu

Acknowledgments

Throughout the process of planning and writing *Campbell Essential Biology with Physiology*, the author team has had the great fortune of collaborating with an extremely talented group of publishing professionals and educators. Although the responsibility for any shortcomings lies solely with us, the merits of the book and its supplements reflect the contributions of a great many dedicated colleagues.

First and foremost, we must acknowledge our huge debt to Neil Campbell, the original author of this book and a source of ongoing inspiration for each of us. Although this edition has been carefully and thoroughly revised—to update its science, its connections to students' lives, its pedagogy, and its currency—it remains infused with Neil's founding vision and his commitment to share biology with introductory students.

This book could not have been completed without the efforts of the *Campbell Essential Biology with Physiology* team at Pearson Education. Leading the team is acquisitions editor Alison Rodal, who is tireless in her pursuit of educational excellence and who inspires all of us to constantly seek better ways to help teachers and students. We also thank the Pearson Science executive team for their supportive leadership, in particular managing director of Arts, Science, Business and Engineering Paul Corey, vice president of science editorial Adam Jaworski, editor-in-chief Beth Wilbur, director of development Barbara Yien, executive editorial manager Ginnie Simione Jutson, and director of media development Lauren Fogel.

It is no exaggeration to say that the talents of the best editorial team in the industry are evident on every page of this book. The authors were continuously guided with great patience and skill by senior development editors Debbie Hardin, Julia Osborne, and Susan Teahan. We owe this editorial team—which include the wonderfully capable and friendly editorial assistant Alison Cagle—a deep debt of gratitude for their talents and hard work.

Once we formulated our words and images, the production and manufacturing teams transformed them into the final book. Project manager Lori Newman and program manager Leata Holloway oversaw the production process and kept everyone and everything on track. We also thank program manager team lead Mike Early and project manager team lead David Zielonka for their careful oversight. We hope you will agree that every edition of *Campbell Essential Biology with Physiology* is distinguished by continuously updated and beautiful photography. For that we thank photo researcher Kristin Piljay, who constantly dazzles us with her keen ability to locate memorable images.

For the production and composition of the book, we thank senior project editor Norine Strang of S4Carlisle Publishing Services, whose professionalism and commitment to the quality of the finished product is visible throughout. The authors owe much to copyeditor Joanna Dinsmore and proofreader Pete Shanks for their keen eyes and attention to detail. We thank design manager Derek Bacchus (who is also responsible for the stunning cover design) and Gary Hespenheide of Hespenheide Design for the beautiful interior design, and we are grateful to Kristina Seymour and the artists at Precision Graphics for rendering clear and compelling illustrations. We also thank rights and permissions project manager Donna Kalal, manager of rights and permissions Rachel Youdelman, and text permissions project manager William Opaluch for keeping us within bounds. In the final stages of production, the talents of manufacturing buyer Stacy Weinberger shone.

Most instructors view the textbook as just one piece of the learning puzzle, with the book's supplements and media completing the picture. We are lucky to have a *Campbell Essential Biology with Physiology* supplements team that is fully committed to the core goals of accuracy and readability. Project Manager Libby Reiser expertly coordinated the supplements, a difficult task given their number and variety. We thank media project manager Eddie Lee for his work on the excellent Instructor Resources DVD that accompanies the text. We owe particular gratitude to the supplements authors, especially the indefatigable and eagle-eyed Ed Zalisko of Blackburn College, who wrote the Instructor Guide and the PowerPoint© Lectures; the highly skilled and multitalented Hilary Engebretson, of Whatcom Community College, who revised the Quiz Shows and Clicker questions; and Jean DeSaix (University of North Carolina at Chapel Hill), Justin Shaffer (University of California, Irvine), Kristen Miller (University of Georgia), and Suann Yang (Presbyterian College), our collaborative team of test bank authors for ensuring excellence in our assessment program. The authors also thank Justin Shaffer (University of California, Irvine), Suzanne Wakim (Butte Community College), and Eden Effert (Eastern Illinois University) for their fine work on the issues-based presentation Campbell Current Topics PowerPoint© Presentations. In addition, the authors thank Reading Quiz authors Amaya Garcia Costas, Montana State University, and Cindy Klevickis, James Madison University; Reading Quiz accuracy reviewer Veronica Menendez; Practice Test author Chris Romero, Front Range Community College; and Practice Test accuracy reviewer Justin Walgaurnery, University of Hawaii.

We wish to thank the talented group of publishing professionals who worked on the comprehensive media program that accompanies *Campbell Essential Biology with Physiology*. The team members dedicated to MasteringBiology™ are true "game changers" in the field of biology education. We thank content producer for media Daniel Ross for coordinating our multimedia plan. Vital contributions were also made by associate Mastering media producer Taylor Merck, senior content producer Lee Ann Doctor, and web developer Leslie Sumrall. We also thank Tania Mlawer and Sarah Jensen for their efforts to make our media products the best in the industry.

As educators and writers, we are very lucky to have a crack marketing team. Executive marketing manager Lauren Harp, director of marketing Christy Lesko, and field marketing manager Amee Mosely seemed to be everywhere at once as they helped us achieve our authorial goals by keeping us constantly focused on the needs of students and instructors. For their amazing efforts with our marketing materials, we also thank copywriter supervisor Jane Campbell and designer Howie Severson.

We also thank the Pearson Science sales representatives, district and regional managers, and learning technology specialists for representing *Campbell Essential Biology with Physiology* on campuses. These representatives are our lifeline to the greater educational community, telling us what you like (and don't like) about this book and the accompanying supplements and media. Their enthusiasm for helping students makes them not only ideal ambassadors but also our partners in education. We urge all educators to take full advantage of the wonderful resource offered by the Pearson sales team.

Eric Simon would like to thank his colleagues at New England College for their support and for providing a model of excellence in education, in particular, Lori Bergeron, Deb Dunlop, Mark Mitch, Maria Colby, Sachie Howard, and Mark Watman. Eric would also like to acknowledge the contributions of Jim Newcomb for lending his keen eye for accuracy; Jay Withgott for sharing his expertise; Elyse Carter Vosen for providing much-needed social context; Jamey Barone for her sage sensitivity; and Amanda Marsh for her expert eye, sharp attention to detail, tireless commitment, constant support, compassion, and wisdom.

At the end of these acknowledgments, you'll find a list of the many instructors who provided valuable information about their courses, reviewed chapters, and/or conducted class tests of *Campbell Essential Biology with Physiology* with their students. All of our best ideas spring from the classroom, so we thank them for their efforts and support.

Most of all, we thank our families, friends, and colleagues, who continue to tolerate our obsession with doing our best for science education.

ERIC SIMON, JEAN DICKEY, KELLY HOGAN, JANE REECE

REVIEWERS OF THIS EDITION

Shazia Ahmed
Texas Woman's University

Tami Asplin
North Dakota State

TJ Boyle
Blinn College, Bryan Campus

Miriam Chavez
University of New Mexico, Valencia

Joe W. Conner
Pasadena City College

Michael Cullen
University of Evansville

Terry Derting
Murray State University

Danielle Dodenhoff
California State University, Bakersfield

Hilary Engebretson
Whatcom Community College

Holly Swain Ewald
University of Louisville

J. Yvette Gardner
Clayton State University

Sig Harden
Troy University

Jay Hodgson
Armstrong Atlantic State University

Sue Hum-Musser
Western Illinois University

Corey Johnson
University of North Carolina

Gregory Jones
Santa Fe College, Gainesville, Florida

Arnold J. Karpoff
University of Louisville

Tom Kennedy
Central New Mexico Community College

Erica Lannan
Prairie State College

Grace Lasker
Lake Washington Institute of Technology

Bill Mackay
Edinboro University

Mark Manteuffel
St. Louis Community College

Diane Melroy
University of North Carolina Wilmington

Kiran Misra
Edinboro University

Susan Mounce
Eastern Illinois University

Zia Nisani
Antelope Valley College

Michelle Rogers
Austin Peay State University

Bassam M. Salameh
Antelope Valley College

Carsten Sanders
Kuztown University

Justin Shaffer
University of California, Irvine

Jennifer Smith
Triton College

Ashley Spring
Eastern Florida State College

Michael Stevens
Utah Valley University

Chad Thompson
Westchester Community College

Melinda Verdone
Rock Valley College

Eileen Walsh
Westchester Community College

Kathy Watkins
Central Piedmont Community College

Wayne Whaley
Utah Valley University

Holly Woodruff (Kupfer)
Central Piedmont Community College

REVIEWERS OF PREVIOUS EDITIONS

Marilyn Abbott
Lindenwood College

Tammy Adair
Baylor University

Felix O. Akojie
Paducah Community College

Shireen Alemadi
Minnesota State University, Moorhead

William Sylvester Allred, Jr.
Northern Arizona University

Megan E. Anduri
California State University, Fullerton

Estrella Z. Ang
University of Pittsburgh

David Arieti
Oakton Community College

C. Warren Arnold
Allan Hancock Community College

Mohammad Ashraf
Olive-Harvey College

Heather Ashworth
Utah Valley University

Bert Atsma
Union County College

Yael Avissar
Rhode Island College

Barbara J. Backley
Elgin Community College

Gail F. Baker
LaGuardia Community College

Neil Baker
Ohio State University

Kristel K. Bakker
Dakota State University

Andrew Baldwin
Mesa Community College

Linda Barham
Meridian Community College

Charlotte Barker
Angelina College

Verona Barr
Heartland Community College

S. Rose Bast
Mount Mary College

Sam Beattie
California State University, Chico

Rudi Berkelhamer
University of California, Irvine

Penny Bernstein
Kent State University, Stark Campus

Suchi Bhardwaj
Winthrop University

Donna H. Bivans
East Carolina University

Andrea Bixler
Clarke College

Brian Black
Bay de Noc Community College

Allan Blake
Seton Hall University

Karyn Bledsoe
Western Oregon University

Judy Bluemer
Morton College

Sonal Blumenthal
University of Texas at Austin

Lisa Boggs
Southwestern Oklahoma State University

Dennis Bogyo
Valdosta State University

David Boose
Gonzaga University

Virginia M. Borden
University of Minnesota, Duluth

James Botsford
New Mexico State University

Cynthia Bottrell
Scott Community College

Richard Bounds
Mount Olive College

Cynthia Boyd

Hawkeye Community College

Robert Boyd
Auburn University

B. J. Boyer
Suffolk County Community College

Mimi Bres
Prince George's Community College

Patricia Brewer
University of Texas at San Antonio

Jerald S. Bricker
Cameron University

Carol A. Britson
University of Mississippi

George M. Brooks
Ohio University, Zanesville

Janie Sue Brooks
Brevard College

Steve Browder
Franklin College

Evert Brown
Casper College

Mary H. Brown
Lansing Community College

Richard D. Brown
Brunswick Community College

Steven Brumbaugh
Green River Community College

Joseph C. Bundy
University of North Carolina at Greensboro

Carol T. Burton
Bellevue Community College

Rebecca Burton
Alverno College

Warren R. Buss
University of Northern Colorado

Wilbert Butler
Tallahassee Community College

Miguel Cervantes-Cervantes
Lehman College, City University of New York

Maitreyee Chandra
Diablo Valley College

Bane Cheek
Polk Community College

Thomas F. Chubb
Villanova University

Reggie Cobb
Nash Community College

Pamela Cole
Shelton State Community College

William H. Coleman
University of Hartford

Jay L. Comeaux
McNeese State University

James Conkey
Truckee Meadows Community College

Karen A. Conzelman
Glendale Community College

Ann Coopersmith

Maui Community College

Erica Corbett
Southeastern Oklahoma State University

James T. Costa
Western Carolina University

Pat Cox
University of Tennessee, Knoxville

Laurie-Ann Crawford
Hawkeye Community College

Pradeep M. Dass
Appalachian State University

Paul Decelles
Johnson County Community College

Galen DeHay
Tri County Technical College

Cynthia L. Delaney
University of South Alabama

Jean DeSaix
University of North Carolina at Chapel Hill

Elizabeth Desy
Southwest State University

Edward Devine
Moraine Valley Community College

Dwight Dimaculangan
Winthrop University

Deborah Dodson
Vincennes Community College

Diane Doidge
Grand View College

Don Dorfman
Monmouth University

Richard Driskill
Delaware State University

Lianne Drysdale
Ozarks Technical Community College

Terese Dudek
Kishawaukee College

Shannon Dullea
North Dakota State College of Science

David A. Eakin
Eastern Kentucky University

Brian Earle
Cedar Valley College

Ade Ejire
Johnston Community College

Dennis G. Emery
Iowa State University

Renee L. Engle-Goodner
Merritt College

Virginia Erickson
Highline Community College

Carl Estrella
Merced College

Marirose T. Ethington
Genesee Community College

Paul R. Evans
Brigham Young University

Zenephia E. Evans

Purdue University

Jean Everett
College of Charleston

Dianne M. Fair
*Florida Community College
at Jacksonville*

Joseph Faryniarz
Naugatuck Valley Community College

Phillip Fawley
Westminster College

Lynn Fireston
Ricks College

Jennifer Floyd
Leeward Community College

Dennis M. Forsythe
The Citadel

Angela M. Foster
Wake Technical Community College

Brandon Lee Foster
Wake Technical Community College

Carl F. Friese
University of Dayton

Suzanne S. Frucht
Northwest Missouri State University

Edward G. Gabriel
Lycoming College

Anne M. Galbraith
University of Wisconsin, La Crosse

Kathleen Gallucci
Elon University

Gregory R. Garman
Centralia College

Wendy Jean Garrison
University of Mississippi

Gail Gasparich
Towson University

Kathy Gifford
Butler County Community College

Sharon L. Gilman
Coastal Carolina University

Mac Given
Neumann College

Patricia Glas
The Citadel

Ralph C. Goff
Mansfield University

Marian R. Goldsmith
University of Rhode Island

Andrew Goliszek
*North Carolina Agricultural and Technical
State University*

Tamar Liberman Goulet
University of Mississippi

Curt Gravis
Western State College of Colorado

Larry Gray
Utah Valley State College

Tom Green

West Valley College

Robert S. Greene
Niagara University

Ken Griffin
Tarrant County Junior College

Denise Guerin
Santa Fe Community College

Paul Gurn
Naugatuck Valley Community College

Peggy J. Guthrie
University of Central Oklahoma

Henry H. Hagedorn
University of Arizona

Blanche C. Haning
Vance-Granville Community College

Laszlo Hanzely
Northern Illinois University

Sherry Harrel
Eastern Kentucky University

Reba Harrell
Hinds Community College

Frankie Harris
Independence Community College

Lysa Marie Hartley
Methodist College

Janet Haynes
Long Island University

Michael Held
St. Peter's College

Consetta Helmick
University of Idaho

J. L. Henriksen
Bellevue University

Michael Henry
Contra Costa College

Linda Hensel
Mercer University

Jana Henson
Georgetown College

James Hewlett
Finger Lakes Community College

Richard Hilton
Towson University

Juliana Hinton
McNeese State University

Phyllis C. Hirsch
East Los Angeles College

W. Wyatt Hoback
University of Nebraska at Kearney

Elizabeth Hodgson
York College of Pennsylvania

A. Scott Holaday
Texas Tech University

Robert A. Holmes
Hutchinson Community College

R. Dwain Horrocks
Brigham Young University

Howard L. Hosick

Washington State University

Carl Huether
University of Cincinnati

Celene Jackson
Western Michigan University

John Jahoda
Bridgewater State College

Dianne Jennings
Virginia Commonwealth University

Richard J. Jensen
Saint Mary's College

Scott Johnson
Wake Technical Community College

Tari Johnson
Normandale Community College

Tia Johnson
Mitchell Community College

Greg Jones
Santa Fe Community College

John Jorstad
Kirkwood Community College

Tracy L. Kahn
University of California, Riverside

Robert Kalbach
Finger Lakes Community College

Mary K. Kananen
Pennsylvania State University, Altoona

Thomas C. Kane
University of Cincinnati

Arnold J. Karpoff
University of Louisville

John M. Kasmer
Northeastern Illinois University

Valentine Kefeli
Slippery Rock University

Dawn Keller
Hawkeye College

John Kelly
Northeastern University

Cheryl Kerfeld
University of California, Los Angeles

Henrik Kibak
*California State University,
Monterey Bay*

Kerry Kilburn
Old Dominion University

Joyce Kille-Marino
College of Charleston

Peter King
Francis Marion University

Peter Kish
*Oklahoma School of Science and
Mathematics*

Robert Kitchin
University of Wyoming

Cindy Klevickis
James Madison University

Richard Koblin

Oakland Community College

H. Roberta Koepfer
Queens College

Michael E. Kovach
Baldwin-Wallace College

Jocelyn E. Krebs
University of Alaska, Anchorage

Ruhul H. Kuddus
Utah Valley State College

Nuran Kumbaraci
Stevens Institute of Technology

Holly Kupfer
Central Piedmont Community College

Gary Kwiecinski
The University of Scranton

Roya Lahijani
Palomar College

James V. Landrum
Washburn University

Lynn Larsen
Portland Community College

Brenda Leady
University of Toledo

Siu-Lam Lee
University of Massachusetts, Lowell

Thomas P. Lehman
Morgan Community College

William Leonard
Central Alabama Community College

Shawn Lester
Montgomery College

Leslie Lichtenstein
Massasoit Community College

Barbara Liedl
Central College

Harvey Liftin
Broward Community College

David Loring
Johnson County Community College

Eric Lovely
Arkansas Tech University

Lewis M. Lutton
Mercyhurst College

Maria P. MacWilliams
Seton Hall University

Mark Manteuffel
St. Louis Community College

Lisa Maranto
Prince George's Community College

Michael Howard Marcovitz
Midland Lutheran College

Angela M. Mason
*Beaufort County Community
College*

Roy B. Mason
Mt. San Jacinto College

John Mathwig
College of Lake County

Lance D. McBrayer
Georgia Southern University

Bonnie McCormick
University of the Incarnate Word

Katrina McCrae
Abraham Baldwin Agricultural College

Tonya McKinley
Concord College

Mary Anne McMurray
Henderson Community College

Maryanne Menvielle
California State University, Fullerton

Ed Mercurio
Hartnell College

Timothy D. Metz
Campbell University

Andrew Miller
Thomas University

David Mirman
Mt. San Antonio College

Nancy Garnett Morris
Volunteer State Community College

Angela C. Morrow
University of Northern Colorado

Patricia S. Muir
Oregon State University

James Newcomb
New England College

Jon R. Nickles
University of Alaska, Anchorage

Jane Noble-Harvey
University of Delaware

Michael Nosek
Fitchburg State College

Jeanette C. Oliver
Flathead Valley Community College

David O'Neill
Community College of Baltimore County

Sandra M. Pace
Rappahannock Community College

Lois H. Peck
University of the Sciences, Philadelphia

Kathleen E. Pelkki
Saginaw Valley State University

Jennifer Penrod
Lincoln University

Rhoda E. Perozzi
Virginia Commonwealth University

John S. Peters
College of Charleston

Pamela Petrequin
Mount Mary College

Paula A. Piehl
Potomac State College of West Virginia University

Bill Pietraface
State University of New York Oneonta

Gregory Podgorski

Utah State University

Rosamond V. Potter
University of Chicago

Karen Powell
Western Kentucky University

Martha Powell
University of Alabama

Elena Pravosudova
Sierra College

Hallie Ray
Rappahannock Community College

Jill Raymond
Rock Valley College

Dorothy Read
University of Massachusetts, Dartmouth

Nathan S. Reyna
Howard Payne University

Philip Ricker
South Plains College

Todd Rimkus
Marymount University

Lynn Rivers
Henry Ford Community College

Jennifer Roberts
Lewis University

Laurel Roberts
University of Pittsburgh

April Rottman
Rock Valley College

Maxine Losoff Rusche
Northern Arizona University

Michael L. Rutledge
Middle Tennessee State University

Mike Runyan
Lander University

Travis Ryan
Furman University

Tyson Sacco
Cornell University

Sarmad Saman
Quinsigamond Community College

Pamela Sandstrom
University of Nevada, Reno

Leba Sarkis
Aims Community College

Walter Saviuk
Daytona Beach Community College

Neil Schanker
College of the Siskiyous

Robert Schoch
Boston University

John Richard Schrock
Emporia State University

Julie Schroer
Bismarck State College

Karen Schuster
Florida Community College at Jacksonville

Brian W. Schwartz

Columbus State University

Michael Scott
Lincoln University

Eric Scully
Towson State University

Lois Sealy
Valencia Community College

Sandra S. Seidel
Elon University

Wayne Seifert
Brookhaven College

Susmita Sengupta
City College of San Francisco

Patty Shields
George Mason University

Cara Shillington
Eastern Michigan University

Brian Shmaefsky
Kingwood College

Rainy Inman Shorey
Ferris State University

Cahleen Shrier
Azusa Pacific University

Jed Shumsky
Drexel University

Greg Sievert
Emporia State University

Jeffrey Simmons
West Virginia Wesleyan College

Frederick D. Singer
Radford University

Anu Singh-Cundy
Western Washington University

Kerri Skinner
University of Nebraska at Kearney

Sandra Slivka
Miramar College

Margaret W. Smith
Butler University

Thomas Smith
Armstrong Atlantic State University

Deena K. Spielman
Rock Valley College

Minou D. Spradley
San Diego City College

Robert Stamatis
Daytona Beach Community College

Joyce Stamm
University of Evansville

Eric Stavney
Highline Community College

Bethany Stone
University of Missouri, Columbia

Mark T. Sugalski
New England College

Marshall D. Sundberg
Emporia State University

Adelaide Svoboda
Nazareth College

Sharon Thoma
Edgewood College

Kenneth Thomas
Hillsborough Community College

Sumesh Thomas
Baltimore City Community College

Betty Thompson
Baptist University

Paula Thompson
Florida Community College

Michael Anthony Thornton
Florida Agriculture and Mechanical University

Linda Tichenor
University of Arkansas, Fort Smith

John Tjepkema
University of Maine, Orono

Bruce L. Tomlinson
State University of New York, Fredonia

Leslie R. Towill
Arizona State University

Bert Tribbey
California State University, Fresno

Nathan Trueblood
California State University, Sacramento

Robert Turner
Western Oregon University

Michael Twaddle
University of Toledo

Virginia Vandergon
California State University, Northridge

William A. Velhagen, Jr.
Longwood College

Leonard Vincent
Fullerton College

Jonathan Visick
North Central College

Michael Vitale

Daytona Beach Community College
Lisa Volk
Fayetteville Technical Community College

Daryle Waechter-Brulla
University of Wisconsin, Whitewater

Stephen M. Wagener
Western Connecticut State University

Sean E. Walker
California State University, Fullerton

James A. Wallis
St. Petersburg Community College

Helen Walter
Diablo Valley College

Kristen Walton
Missouri Western State University

Jennifer Warner
University of North Carolina at Charlotte

Arthur C. Washington
Florida Agriculture and Mechanical University

Dave Webb
St. Clair County Community College

Harold Webster
Pennsylvania State University, DuBois

Ted Weinheimer
California State University, Bakersfield

Lisa A. Werner
Pima Community College

Joanne Westin
Case Western Reserve University

Wayne Whaley
Utah Valley State College

Joseph D. White
Baylor University

Quinton White
Jacksonville University

Leslie Y. Whiteman
Virginia Union University

Rick Wiedenmann
New Mexico State University at Carlsbad

Peter J. Wilkin
Purdue University North Central

Bethany Williams
California State University, Fullerton

Daniel Williams
Winston-Salem University

Judy A. Williams
Southeastern Oklahoma State University

Dwina Willis
Freed Hardeman University

David Wilson
University of Miami

Mala S. Wingerd
San Diego State University

E. William Wischusen
Louisiana State University

Darla J. Wise
Concord College

Michael Womack
Macon State College

Bonnie Wood
University of Maine at Presque Isle

Jo Wen Wu
Fullerton College

Mark L. Wygoda
McNeese State University

Calvin Young
Fullerton College

Shirley Zajdel
Housatonic Community College

Samuel J. Zeakes
Radford University

Uko Zylstra
Calvin College

The publishers would like to thank the following for their contribution to the Global Edition:

Contributor
Clemens Kiecker, King's College London

Reviewers
Shefali Sabharanjak, Ph.D.
Quek Choon Lau, Ngee Ann Polytechnic
Adriaan Engelbrecht, University of the Western Cape

1 Introduction: Biology Today

If you've ever wondered what an unusual or especially beautiful animal is called, you're curious about taxonomy.

Although you may not realize it, you use the scientific method every day.

One of the primary missions of the Mars rover is to search for signs of life.

Biology All Around Us BIOLOGY AND SOCIETY

An Innate Passion for Life

Do you like biology? Wait, let's put this question another way: Do you have a pet? Are you concerned with fitness or healthy eating? Have you ever visited a zoo or an aquarium for fun, taken a nature hike, or gathered shells on the beach? Do you like watching TV shows about sharks or dinosaurs? If you answered yes to any of these questions—well, then, it turns out that you do like biology!

Most of us have an inherent interest in life, an inborn curiosity about the natural world that leads us to study animals and plants and their habitats. We wrote *Essential Biology* to help you—a student with little or no college-level science experience—harness your innate enthusiasm for life. We'll use this passion to help you develop an understanding of the discipline of biology, one that you can apply to your own life and to the society in which you live. We believe that such a biological perspective is essential for any educated person, which is why we named our book *Essential Biology*. So, whatever your reasons for taking this course—even if only to fulfill your school's science requirement—you'll soon discover that exploring life is relevant and important to you, no matter your background or goals.

An inborn curiosity about nature. This student is interacting with a woolly monkey (*Lagothrix lagotricha*) during a school trip to the Amazon River in Peru.

To reinforce the fact that biology affects your everyday life in many ways, every chapter of *Essential Biology* opens with an essay—called Biology and Society—that will help you see the relevance of that chapter's material. Topics as varied as medical uses of radiation (Chapter 2), the importance of a flu shot (Chapter 10), and the community of microscopic organisms that live in and on your body (Chapter 15) help to illustrate biology's scope and show how the subject of biology is woven into the fabric of society. Throughout *Essential Biology*, we'll continuously emphasize these connections, pointing out many examples of how each topic can be applied to your life and the lives of those you care about.

The Scientific Study of Life

Now that we've established our goal—to examine how biology affects your life—a good place to start is with a basic definition: **Biology** is the scientific study of life. But have you ever looked up a word in the dictionary only to find that you need to look up some of the words within that definition to make sense of the original word? The definition of biology, although seemingly simple, raises more questions: What is a scientific study? And what does it mean to be alive? To help you get started with your investigation of biology, this first chapter of *Essential Biology* expands on important concepts within the definition of biology. First, we'll place the study of life in the broader context of science. Next, we'll investigate the nature of life by surveying the properties and scope of life. Finally, we'll introduce a series of broad themes you will encounter throughout your investigation of life, themes that serve as organizing principles for the information you will learn. Most important, throughout this chapter (and, indeed, throughout all of *Essential Biology*), we'll continue to provide examples of how biology affects *your* life, highlighting the relevance of this subject to society and everyone in it. ☑

are many nonscientific ways that life can be studied. For example, extended meditation is a valid way of studying the nature of life—this approach might be well suited to a philosophy class, for example—but it does not qualify as biology because it is not a *scientific* means of studying life. How, then, do we tell the difference between science and other ways of trying to make sense of nature?

Science is an approach to understanding the natural world that is based on inquiry—a search for information, explanations, and answers to specific questions. This basic human drive to understand our natural world is manifest in two main scientific approaches: discovery science, which is mostly about *describing* nature, and hypothesis-driven science, which is mostly about *explaining* nature. Most scientists practice a combination of these two forms of inquiry.

Discovery Science

Scientists seek natural causes for natural phenomena. This limits the scope of science to the study of structures and processes that we can verifiably observe and measure, either directly or indirectly with the help of tools and technology, such as microscopes **(Figure 1.1)**. Recorded observations are called **data**, and data are the items of information on which scientific inquiry is based. This dependence on verifiable data demystifies nature and distinguishes science from supernatural beliefs. Science can neither prove nor disprove that ghosts, deities, or spirits cause storms, eclipses, or illnesses,

☑ **CHECKPOINT**
Define biology.

Answer: Biology is the scientific study of life.

The Process of Science

Recall the definition at the heart of this chapter: Biology is the scientific study of life. This leads to an obvious first question: What is a scientific study? Notice that biology is not defined as the "study of life" because there

▼ **Figure 1.1 The protist *Paramecium* viewed with three different types of microscopes.**
Photographs taken with microscopes are called micrographs. Throughout this textbook, micrographs will have size notations along the side. For example, "LM 300×" indicates that the micrograph was taken with a light microscope and the objects are magnified to 300 times their original size.

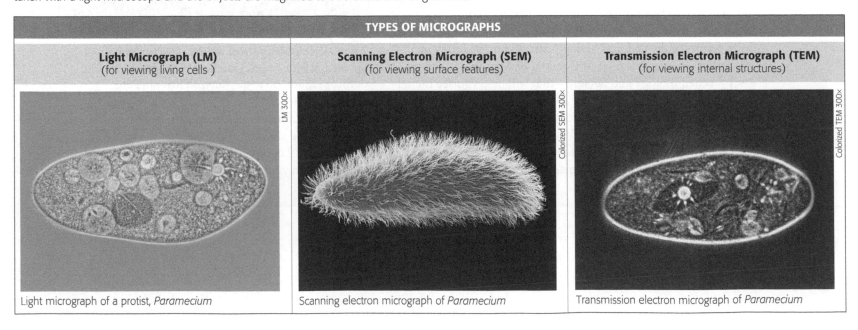

TYPES OF MICROGRAPHS		
Light Micrograph (LM) (for viewing living cells)	**Scanning Electron Micrograph (SEM)** (for viewing surface features)	**Transmission Electron Micrograph (TEM)** (for viewing internal structures)
Light micrograph of a protist, *Paramecium*	Scanning electron micrograph of *Paramecium*	Transmission electron micrograph of *Paramecium*

◄ Figure 1.2 **Careful observation
and measurement: the raw data for
discovery science.** Dr. Jane Goodall
spent decades recording her observations
of chimpanzee behavior during field
research in the jungles of Tanzania.

because such explanations are not measurable and are
therefore outside the bounds of science.

Verifiable observations and measurements are the
data of **discovery science**. In our quest to describe
nature accurately, we discover its structure. Charles
Darwin's careful description of the diverse plants and
animals he observed in South America is an example of
discovery science (as you'll learn in Chapter 13). More
recently, Jane Goodall spent decades observing and
recording the behavior of chimpanzees living in the jungles of Tanzania (Figure 1.2). And even more recently,
molecular biologists have sequenced and analyzed huge
amounts of DNA (an effort discussed in Chapter 12),
gathering data that shed light on the genetic basis of life.

Hypothesis-Driven Science

The observations of discovery science motivate us
to ask questions and seek explanations. Ideally,
such investigations make use of the scientific
method. As a formal process of inquiry, the

scientific method consists of a series of steps (Figure 1.3).
These steps provide a loose guideline for scientific investigations. There is no single formula for successfully discovering something new; instead, the scientific method
suggests a broad outline for how discovery might proceed.
The scientific method is a bit like an incomplete recipe:
A basic outline of steps to be followed is presented, but
the details of the dish are left to the cook. Similarly, working scientists do not typically follow the steps of the scientific method rigidly; different scientists proceed through
the scientific method in different ways.

Most modern scientific investigations can be described
as hypothesis-driven science. A **hypothesis** is a tentative
answer to a question—a proposed explanation for a set
of observations. A good hypothesis immediately leads to
predictions that can be tested by experiments. Although

Experiment *does
not support*
hypothesis; revise
hypothesis or
pose new one.

Revise

Observation	**Question**	**Hypothesis**	**Prediction**	**Experiment**
The TV remote doesn't work.	What's wrong with the remote?	The remote's batteries are dead.	If I replace the batteries, the remote will work.	I replace the batteries with new ones.

Experiment *supports*
hypothesis;
make additional
predictions
and test them.

► Figure 1.3 **Applying the scientific
method to a common problem.**

Although you may not realize it, you use the scientific method every day.

we don't think of it in those terms, we all use hypotheses in solving everyday problems. Imagine that you've completed your homework for the day and are going to reward yourself with some time in front of the TV. You press the power button on your TV remote, but the TV fails to turn on. That the TV does not turn on is an observation. The question that arises is obvious: Why didn't the remote turn on the TV? You could imagine a dozen possible explanations, but you can't possibly investigate them all simultaneously. Instead, you would focus on just one explanation (perhaps the most likely one based on past experience) and test it. That initial explanation is your hypothesis. For example, in this case, a reasonable hypothesis is that the batteries in the remote are dead.

Once a hypothesis is formed, an investigator can make predictions about what results are expected if that hypothesis is correct. We then test the hypothesis by performing an experiment to see whether or not the results are as predicted. This logical testing takes the form of "If . . . then" logic:

Observation: The TV remote doesn't work.

Question: What's wrong with the remote?

Hypothesis: The TV remote doesn't work because its batteries are dead.

Prediction: If I replace the batteries, then the remote will work.

Experiment: I replace the batteries with new ones.

Let's say that after you replace the batteries the remote still doesn't work. You would then formulate a second hypothesis and test it. Perhaps, for example, the TV is unplugged. Or you put in the new batteries incorrectly. You could continue to conduct additional experiments and formulate additional hypotheses until you reach a satisfactory conclusion to your initial question. As you do this, you are following the scientific method, and you are acting as a scientist.

Let's back up and examine what you would probably *not* do in this scenario: You most likely would not blame the malfunctioning remote on supernatural spirits, nor are you likely to meditate on the cause of the observed phenomenon. Your natural instinct is to formulate a hypothesis and then test it; the scientific method is probably your "go-to" method for solving problems. In fact, the scientific method is so deeply embedded in our society and in the way we think that most of us use it automatically (although we don't use the terminology presented here). The scientific method is therefore just a formalization of how you already think and act.

In every chapter of *Essential Biology*, we include examples of how the scientific method was used to learn about the material under discussion. In each of these sections (titled The Process of Science), we will, as a reminder, highlight the steps in the scientific method. The questions we will address include: Does lactose intolerance have a genetic basis (Chapter 3)? Why do dog coats come in so many varieties (Chapter 9)? Do the organisms living in your intestine affect your weight (Chapter 15)? As you become increasingly scientifically literate, you will arm yourself with the tools you need to evaluate claims that you hear. We are all bombarded by information every day—via commercials, websites, magazine articles, and so on—and it can be hard to filter out the bogus from the truly worthwhile. Having a firm grasp of science as a process of inquiry can therefore help you in many ways outside the classroom.

It is important to note that scientific investigations are not the only way of knowing nature. A comparative religion course would be a good way to learn about the diverse stories that focus on a supernatural creation of Earth and its life. Science and religion are two very different ways of trying to make sense of nature. Art is yet another way to make sense of the world around us. A broad education should include exposure to all these different ways of viewing the world. Each of us synthesizes our worldview by integrating our life experiences and multidisciplinary education. As a science textbook and part of that multidisciplinary education, *Essential Biology* showcases life in a purely scientific context. ✓

✓ CHECKPOINT

1. If you observe the squirrels on your campus and collect data on their dietary habits, what kind of science are you performing? If you come up with a tentative explanation for their dietary behavior and then test your idea, what kind of science are you performing?
2. Place these steps of the scientific method in their proper order: experiment, hypothesis, observation, prediction, results, question, revise/repeat.

Answers: **1.** *discovery science; hypothesis-driven science* **2.** *observation, question, hypothesis, prediction, experiment, results, revise/repeat*

▼ **Figure 1.4 Some properties of life.** An object is considered alive if and only if it displays all of these properties simultaneously.

(a) Order

(b) Regulation

(c) Growth and development

(d) Energy processing

Theories in Science

Many people associate facts with science, but accumulating facts is not the primary goal of science. A telephone book is an impressive catalog of factual information, but it has little to do with science. It is true that facts, in the form of verifiable observations and repeatable experimental results, are the prerequisites of science. What really advances science, however, are new theories that tie together a number of observations that previously seemed unrelated. The cornerstones of science are the explanations that apply to the greatest variety of phenomena. People like Isaac Newton, Charles Darwin, and Albert Einstein stand out in the history of science not because they discovered a great many facts but because their theories had such broad explanatory power.

What is a scientific theory, and how is it different from a hypothesis? A scientific **theory** is much broader in scope than a hypothesis. A theory is a comprehensive explanation supported by abundant evidence, and it is general enough to spin off many new testable hypotheses. This is a hypothesis: "White fur is an adaptation that helps polar bears survive in an Arctic habitat." And this is another, seemingly unrelated hypothesis: "The unusual bone structure in a hummingbird's wings is an evolutionary adaptation that provides an advantage in gathering nectar from flowers." In contrast, the following theory ties together those seemingly unrelated hypotheses: "Adaptations to the local environment evolve by natural selection." This particular theory is one that we will describe later in this chapter.

Theories only become widely accepted by scientists if they are supported by an accumulation of extensive and varied evidence and if they have not been contradicted by any scientific data. The use of the term *theory* by scientists contrasts with our everyday usage, which implies untested speculation ("It's just a theory!"). In fact, we use the word "theory" in our everyday speech the way that a scientist uses the word "hypothesis." As you will soon learn, natural selection qualifies as a scientific theory because of its broad application and because it has been validated by a large number of observations and experiments. It is therefore not proper to say that natural selection is "just" a theory to imply that it is untested or lacking in evidence. In fact, any scientific theory is backed up by a wealth of supporting evidence, or else it wouldn't be referred to as a theory. ☑

☑ **CHECKPOINT**

You arrange to meet a friend for dinner at 6 P.M., but when the appointed hour comes, she is not there. You wonder why. Another friend says, "My theory is that she forgot." If your friend were speaking like a scientist, what would she have said?

Answer: "My hypothesis is that she forgot."

The Nature of Life

Recall once again our basic definition: Biology is the scientific study of life. Now that we have an understanding of what constitutes a scientific study, we can turn to the next question raised by this definition: What is life? Or, to put it another way, what distinguishes living things from nonliving things? The phenomenon of **life** seems to defy a simple, one-sentence definition. Yet even a small child instinctively knows that a dog or a bug or a plant is alive but a rock is not.

If I placed an object in front of you and asked you whether it was alive, what would you do? Would you poke it to see if it reacts? Would you watch it closely to see if it moves or breathes? Would you dissect it to look at its parts? Each of these ideas is closely related to how biologists actually define life: We recognize life mainly by what living things do. To start our investigation of biology, let's look at some properties that are shared by all living things.

The Properties of Life

Figure 1.4 highlights seven of the properties and processes associated with life. An object is generally considered to be alive if it displays all of these properties simultaneously. (a) *Order*. All living things exhibit

(e) Response to the environment

(f) Reproduction

(g) Evolution

complex but ordered organization, as seen in the structure of a pinecone. (b) *Regulation*. The environment outside an organism may change drastically, but the organism can adjust its internal environment, keeping it within appropriate limits. When it senses its body temperature dropping, a lizard can bask on a rock to absorb heat. (c) *Growth and development*. Information carried by DNA controls the pattern of growth and development in all organisms, including the crocodile. (d) *Energy processing*. Organisms take in energy and use it to perform all of life's activities; they emit energy as heat. A cheetah obtains energy by eating its kill, uses this energy to power running and other work, and continuously emits body heat into the environment. (e) *Response to the environment*. All organisms respond to environmental stimuli. A carnivorous Venus flytrap closes its leaves rapidly in response to the environmental stimulus of an insect touching the plant's sensory hairs. (f) *Reproduction*. Organisms reproduce their own kind. Thus, monkeys reproduce only monkeys—never lizards or cheetahs. (g) *Evolution*. Reproduction underlies the capacity of populations to change (evolve) over time. For example, the giant leaf insect (*Phyllium giganteum*) has evolved in a way that provides camouflage in its environment. Evolutionary change is a central, unifying phenomenon of all life.

Although we have no proof that life has ever existed anywhere other than Earth, biologists speculate that extraterrestrial life, if it exists, could be recognized by the same properties listed in Figure 1.4. The Mars rover *Curiosity* (**Figure 1.5**), which has

▲ Figure 1.5 **A view from the Mars rover *Curiosity* searching for signs of life.**

been exploring the surface of the red planet since 2012, contains several instruments designed to identify biosignatures, substances that provide evidence of past or present life. For example, *Curiosity* is using a suite of onboard instruments to detect chemicals that could provide evidence of energy processing by microscopic organisms. As of yet, no definitive signs of the properties of life have been detected, and the search continues. ☑

One of the primary missions of the Mars rover is to search for signs of life.

Life in Its Diverse Forms

The tarsier shown in **Figure 1.6** is just one of about 1.8 million identified species on Earth that displays all of the properties outlined in Figure 1.4. The diversity of known life—all the species that have been identified and named—includes at least 290,000 plants, 52,000 vertebrates (animals with backbones), and 1 million insects (more than half of all known forms of life). Biologists add thousands of newly identified species to the list each year. Estimates of the total number of species range from 10 million to more than 100 million. Whatever the actual number turns out to

◄ Figure 1.6 **A small sample of biological diversity.** A primate called a tarsier sits in a tree in a rainforest within the Philippines. The scientific name for this species is *Tarsius syrichta*.

DOMAIN BACTERIA

Colorized TEM 10,000X

DOMAIN ARCHAEA

TEM 18,500X

DOMAIN EUKARYA

Kingdom Plantae

Kingdom Fungi

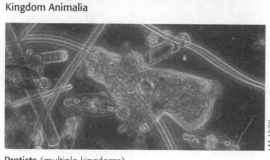

Kingdom Animalia

Protists (multiple kingdoms)

LM 150X

▲ Figure 1.7 **The three domains of life.**

be, the enormous diversity of life presents organizational challenges to biologists who study it.

Grouping Species: The Basic Concept

To make sense of nature, people tend to group diverse items according to similarities. We may speak of "squirrels" and "butterflies," even though we recognize that each group actually includes many different species. A **species** is generally defined as a group of organisms that live in the same place and time and have the potential to interbreed with one another in nature to produce healthy offspring (more on this in Chapter 14). We may even sort groups into broader categories, such as rodents (which include squirrels) and insects (which include butterflies). **Taxonomy**, the branch of biology that names and classifies species, is the arrangement of species into a hierarchy of broader and broader groups. Have you ever seen a fish, or found a mushroom, or watched a bird and wondered what kind it was? If so, you were asking a question of taxonomy. Before we dive into biodiversity in greater detail in later chapters, let's summarize the broadest units of classification of life.

The Three Domains of Life

On the broadest level, biologists divide the diversity of life into three domains: Bacteria, Archaea, and Eukarya **(Figure 1.7)**. Every organism on Earth belongs to one of these three domains. The first two domains, Bacteria and Archaea, identify two very different groups of organisms that have prokaryotic cells—that is, relatively small and simple cells that lack a nucleus or other compartments bounded by internal membranes. All the eukaryotes (organisms with eukaryotic cells—that is, relatively large and complex cells that contain a nucleus and other membrane-enclosed compartments) are grouped into the domain Eukarya.

The domain Eukarya in turn includes three smaller divisions called kingdoms—Plantae, Fungi, and Animalia. Most members of the three kingdoms are multicellular. The kingdoms are distinguished partly by how the organisms obtain food. Plants produce their own sugars and other foods by photosynthesis. Fungi are mostly decomposers, obtaining food by digesting dead organisms and organic wastes. Animals—the kingdom to which we belong—obtain food by ingesting (eating) and digesting other organisms. Those eukaryotes that do not fit into any of the three kingdoms fall into a catch-all group called the protists. Most protists are single-celled; they include microscopic organisms such as amoebas. But protists also include certain multicellular forms, such as seaweeds. Scientists are in the process of organizing protists into multiple kingdoms, although they do not yet agree on exactly how to do this. ☑

If you've ever wondered what an unusual or especially beautiful animal is called, you're curious about taxonomy.

☑ CHECKPOINT

1. Name the three domains of life. To which do you belong?
2. Name three kingdoms found within the domain Eukarya. Name a fourth group within this domain.

Answers: 1. Bacteria, Archaea, Eukarya; 2. Plantae, Fungi, Animalia; the protists

Major Themes in Biology

Biology is a big subject that, as new discoveries unfold every day, grows continuously in breadth and depth. Although we often focus on the details, it is important to recognize that there are broad themes running throughout the subject. These overarching principles unify all aspects of biology, from the microscopic world of cells to the global environment. Focusing on a few big picture ideas that cut across many topics within biology can help organize and make sense of all the information you will learn.

In this section, we'll describe five unifying themes that recur throughout our investigation of biology (**Figure 1.8**). You'll encounter these themes throughout subsequent chapters; we'll highlight each theme as it appears by using the icons from Figure 1.8 and repeating the section headings used below.

Evolution Evolution

Just as you have a family history, each species on Earth today represents one twig on a branching tree of life that extends back in time through ancestral species more and more remote. Species that are very similar, such as the brown bear and the polar bear, share a common ancestor at a relatively recent branch point on the tree of life (**Figure 1.9**). In addition, all bears can be traced back much farther in time to an ancestor common also to squirrels, humans, and all other mammals. All mammals have hair and milk-producing mammary glands, and such similarities are what we would expect if all mammals descended from a common ancestor, a first mammal. And mammals, reptiles, and all other vertebrates share a common ancestor even more ancient than the common ancestor to mammals. Going farther back still, at the cellular level, all life displays striking similarities. For example, all living cells are surrounded by an outer plasma membrane of similar makeup and use structures called ribosomes to produce proteins.

The scientific explanation for the common characteristics found throughout such diverse species is evolution, the process that has transformed life on Earth from its earliest beginnings to the extensive variety we see today. Evolution is the fundamental principle of life and the core theme that unifies all of biology. The theory of evolution by natural selection, first described by Charles Darwin more than 150 years ago, is the one principle that makes sense of everything we know about living organisms. Any student of biology should begin by first understanding evolution. Evolution can help us investigate and understand every aspect of life, from the tiny organisms that occupy the most remote habitats, to the diversity of species in our local environment, to the stability of the global environment. In this section, we will present a basic introduction to this important topic. To emphasize evolution as the core theme of biology, we end each chapter of *Essential Biology* with an Evolution Connection section. You will learn, for example, how evolution can shed light on such topics as our quest to develop better biofuels (Chapter 7), how cancer cells grow and spread through the body (Chapter 10), and the development of antibiotic-resistant bacteria (Chapter 13).

The Darwinian View of Life

The evolutionary view of life came into focus in 1859 when English naturalist Charles Darwin published one

▼ Figure 1.8 **Five unifying themes that run throughout the discipline of biology.**

MAJOR THEMES IN BIOLOGY				
Evolution	**Structure/Function**	**Information Flow**	**Energy Transformations**	**Interconnections within Systems**
Evolution by natural selection is biology's core unifying theme and can be seen at every level in the hierarchy of life.	The structure of an object, such as a molecule or a body part, provides insight into its function, and vice versa.	Within biological systems, information stored in DNA is transmitted and expressed.	All biological systems depend on obtaining, converting, and releasing energy and matter.	All biological systems, from molecules to ecosystems, depend on interactions between components.

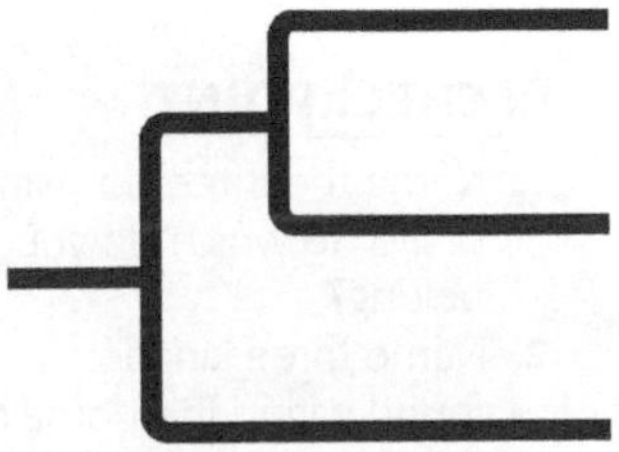

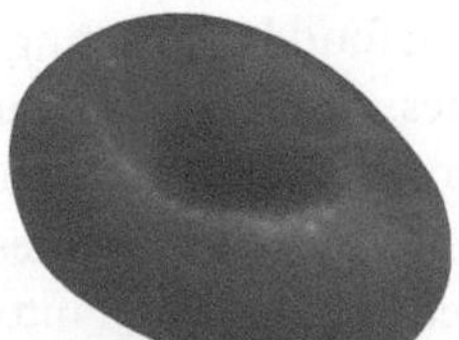

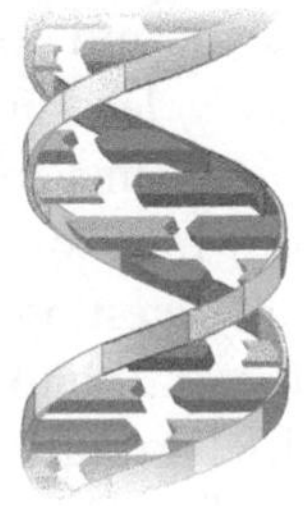

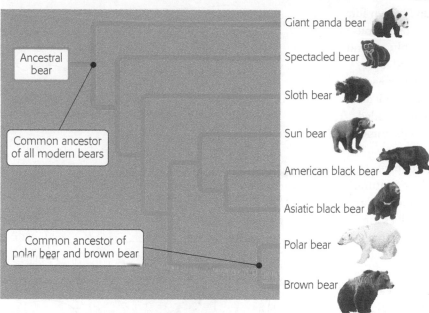

Ancestral bear

Common ancestor of all modern bears

Common ancestor of polar bear and brown bear

Giant panda bear

Spectacled bear

Sloth bear

Sun bear

American black bear

Asiatic black bear

Polar bear

Brown bear

▲ **Figure 1.9 An evolutionary tree of bears.** This tree represents a hypothesis (a tentative model) based on both the fossil record and a comparison of DNA sequences among modern bears. As new evidence about the evolutionary history of bears emerges, this tree will inevitably change.

reproductive success that Darwin called **natural selection** because the environment "selects" only certain heritable traits from those already existing. Natural selection does not promote or somehow encourage changes, but rather it serves to "edit" those changes that have already occurred. And the product of natural selection is adaptation, the accumulation of variations in a population over time. Returning to the example of bear evolution in Figure 1.9, one familiar adaptation in bears is fur color. Polar bears and brown (grizzly) bears each

▼ **Figure 1.10 Charles Darwin (1809–1882), *The Origin of Species*, and blue-footed boobies he observed on the Galápagos Islands.**

of the most important and influential books ever written: *On the Origin of Species by Means of Natural Selection* (Figure 1.10). First, Darwin presented a large amount of evidence in support of the evolutionary view that species living today descended from a succession of ancestral species. Darwin called this process "descent with modification." Darwin's phrase was an insightful one, because it captures the duality of life's unity (which is possible because of a shared descent) and diversity (which is possible because of gradual modification). In the Darwinian view, for example, the diversity of bears is based on different modifications of a common ancestor from which all bears descended.

Second, Darwin proposed a mechanism for descent with modification: the process of natural selection. In the struggle for existence, those individuals with heritable traits best suited to the local environment are more likely to survive and leave the greatest number of healthy offspring. Therefore, the traits that enhance survival and reproductive success will be represented in greater numbers in the next generation. It is this unequal

▲ Figure 1.11 **Finches of the Galápagos Islands.** Charles Darwin personally collected these finches during his time in the Galapagos.

display an evolutionary adaptation (white and brown fur, respectively) that resulted from natural selection operating in their respective environments. Presumably, natural selection tended to favor the fur color that provided each bear lineage with an appearance that provided an advantage in its home territory.

We now recognize many examples of natural selection in action. A classic example involves the finches (a kind of bird) of the Galápagos Islands **(Figure 1.11)**. Over a span of two decades, researchers working on these isolated islands measured changes in beak size in a population of ground finch that prefers to eat small seeds. In dry years, when the preferred small seeds are in short supply, the birds instead have mostly large seeds to eat. Birds with larger, stronger beaks have a feeding advantage in such an environment, and greater reproductive success, and thus the average beak depth for the population increases during dry years. During wet years, small seeds become more abundant. Because smaller beaks are more efficient for eating the plentiful small seeds, the average beak depth decreases over generations. Such changes in structure are measurable evidence of natural selection in action.

The world is rich with examples of natural selection. Consider the development of antibiotic-resistant bacteria **(Figure 1.12)**. Dairy and cattle farmers often add antibiotics to feed because doing so results in larger, more profitable animals. **1** The members of the bacteria population will, through random chance, vary in their susceptibility to the antibiotic. **2** Once the environment has been changed with the addition of antibiotics, some bacteria will succumb quickly and die, whereas others will survive. **3** Those that do survive will have the potential to multiply, producing offspring that will likely inherit the traits that enhance survival. **4** Over many generations, bacteria that are resistant to antibiotics will thrive in greater and greater numbers. Thus, feeding antibiotics to cows may promote the evolution of antibiotic-resistant bacterial populations that are not susceptible to standard drug treatments.

Observing Artificial Selection

Darwin found convincing evidence of the power of natural selection in examples of *artificial* selection, the purposeful breeding of domesticated plants and animals by humans. People have been modifying other species

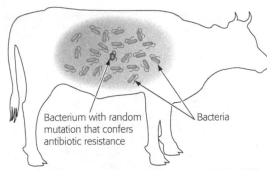

1 **Population with varied inherited traits.** Initially, the population of bacteria varies in its ability to resist the antibiotic. By random chance, a few bacteria are somewhat resistant.

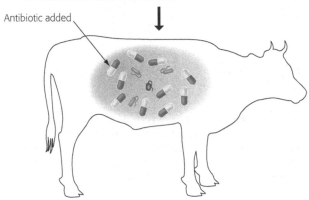

2 **Elimination of individuals with certain traits.** The majority of bacteria are susceptible to the effects of the antibiotic and so will die off. The few resistant bacteria will tend to survive.

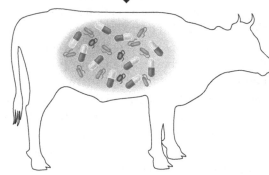

3 **Reproduction of survivors.** The selective pressure of the antibiotic favors survival and reproductive success of the few resistant bacteria. Thus, genes for anitbiotic resistance are passed along to the next generation in greater frequency.

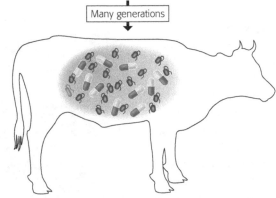

4 **Increasing frequency of traits that enhance survival and reproductive success.** Generation after generation, the bacteria population adapts to its environment through natural selection.

▲ Figure 1.12 **Natural selection in action.**

for millennia by selecting breeding stock with certain traits. For example, the plants we grow for food today bear little resemblance to their wild ancestors. Have you ever had wild blueberries or wild strawberries? They are much different (and, in many ways, much less desirable) than their modern kin. This is because we have customized crop plants through many generations of artificial selection by selecting different parts of the plant to accentuate. All the vegetables shown in **Figure 1.13** (and more) have a common ancestor in one species of wild mustard (shown in the center of the figure). The power of selective breeding is also apparent in our pets, which

have been bred for looks and for useful characteristics. For example, all domesticated dogs are descended from wolves, but people in different cultures have customized hundreds of dog breeds as different as basset hounds and Saint Bernards **(Figure 1.14)**. The tremendous variety of modern dogs reflects thousands of years of artificial selection. You may not realize it, but you are exposed to the products of artificial selection every day!

Darwin's publication of *On the Origin of Species* fueled an explosion in biological research that continues today. Over the past century and a half, a tremendous amount of evidence has accumulated in support of Darwin's theory

► Figure 1.13 **The artificial selection of food crops.**

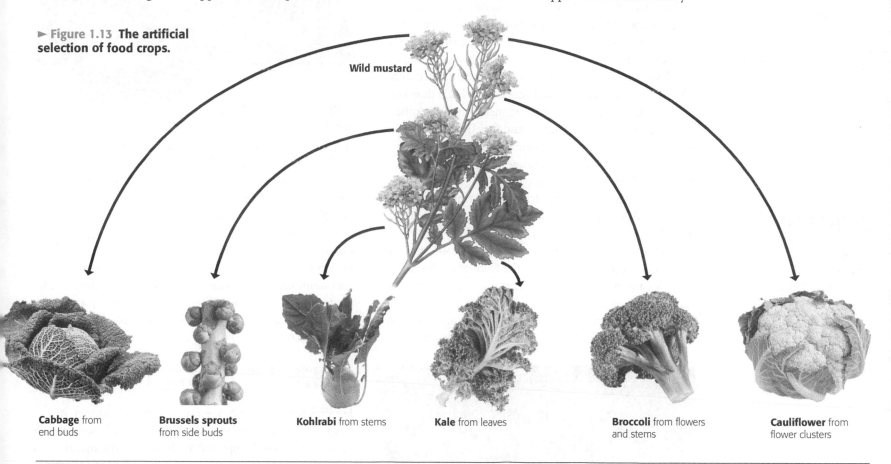

Wild mustard

Cabbage from end buds

Brussels sprouts from side buds

Kohlrabi from stems

Kale from leaves

Broccoli from flowers and stems

Cauliflower from flower clusters

▼ Figure 1.14 **The artificial selection of pets.**

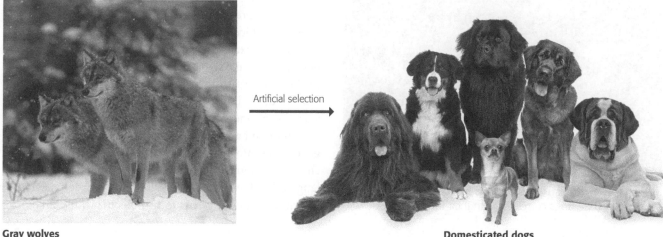

Artificial selection

Gray wolves

Domesticated dogs

of evolution by natural selection, making it one of biology's best-demonstrated, most comprehensive, and longest-held theories. Throughout *Essential Biology*, you will learn more about how natural selection works and see other examples of how natural selection affects your life. ☑

 **Structure/ Function** # The Relationship of Structure to Function

When considering useful objects in your home, you may realize that form and function are related. A chair, for example, cannot have just any shape; it must have a stable base to stand on and a flat area for holding your weight. The function of the chair constrains the possible shapes it may have. Similarly, within biological systems, structure (the shape of something) and function (what it does) are often related, with each providing insight into the other.

The correlation of structure and function can be seen at every level of biological organization. Consider your lungs, which function to exchange gases with the environment: Your lungs bring in oxygen (O_2) and take out carbon dioxide (CO_2). The structure of your lungs correlates with this function (**Figure 1.15**). Increasingly smaller branches of your lungs end in millions of tiny sacs in which the gases cross from the air to your blood, and vice versa. This branched structure (the form of the lungs) provides a tremendous surface area over which a very high volume of air may pass (the function of the lungs). Cells, too, display a correlation of structure and function. As oxygen enters the blood in the lungs, for example,

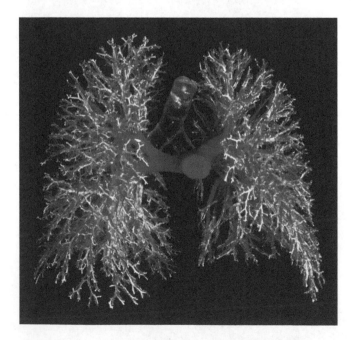

▲ **Figure 1.15 Structure/function: human lungs.** The structure of your lungs correlates with their function.

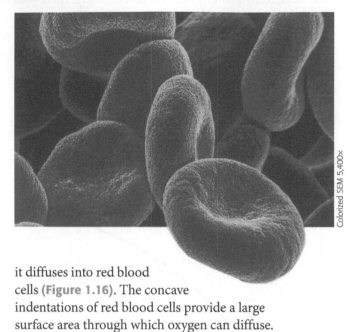

▼ **Figure 1.16 Structure/function: red blood cells.** As oxygen enters the blood in the lungs, it diffuses into red blood cells.

Colorized SEM 5,400x

it diffuses into red blood cells (**Figure 1.16**). The concave indentations of red blood cells provide a large surface area through which oxygen can diffuse.

Throughout the textbook, we will see the structure/function principle apply to all levels of biological organization, from the structure of cells and their components, to DNA replication, to the internal organization of plants and animals. Some specific examples of the correlation of structure and function that you will encounter include explaining why ice floats (Chapter 2), the importance of a protein's shape (Chapter 3), and structural adaptations within the bodies of plants (Chapter 16). ☑

Information Flow # Information Flow

For life's functions to proceed in an orderly manner, information must be received, transmitted, and used. Such information flow is apparent at all levels of biological organization. At the microscopic scale, every cell contains information in the form of **genes**, hereditary units of information consisting of specific sequences of DNA passed on from the previous generation. At the organismal level, as every multicellular organism develops from an embryo, information exchanged among cells enables the overall body plan to take shape in an organized fashion (as you'll see in Chapter 8). Once the organism is mature, information about the internal conditions of its body is used to keep those conditions within a range that allows for life.

Although bacteria and humans inherit different genes, that information is encoded in an identical chemical language common to all organisms. In fact, the language of life has an alphabet of just four letters. The chemical names of DNA's four molecular building blocks are abbreviated as A, G, C, and T (Figure 1.17). A typical gene is hundreds to thousands of chemical "letters" in length. A gene's meaning to a cell is encoded in its specific sequence of these letters, just as the message of this sentence is encoded in its arrangement of the 26 letters of the English alphabet.

The entire set of genetic information that an organism inherits is called its **genome**. The nucleus of each human cell contains a genome that is about 3 billion chemical letters long. In recent years, scientists have tabulated virtually the entire sequence of the genome from humans and hundreds of other organisms. As this work progresses, biologists will continue to learn the functions of the genes and how their activities are coordinated in the development and functioning of an organism. The emerging field of genomics—a branch of biology that studies whole genomes—is a striking example of how the flow of information can inform our study of life at its many levels.

How is all this information used within your body? At any given moment, your genes are producing thousands of different proteins that control your body's processes. (You'll learn the details of how proteins are

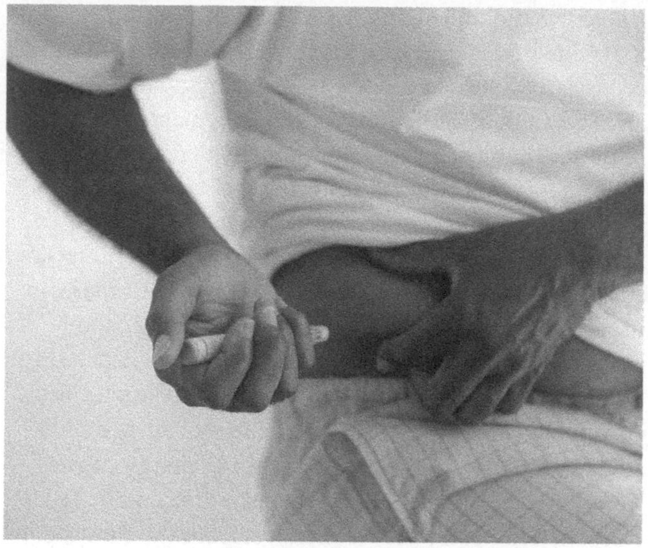

▲ Figure 1.18 **Biotechnology.** Since the 1970s, applications of biology have revolutionized medicine.

produced from DNA in Chapter 10.) Food is broken down, new body tissues are built, cells divide, signals are sent—all under the control of proteins, all of which are built from information stored in your DNA. For example, structures within your body assess the amount of glucose in your blood and, in response, secrete varying amounts of hormones that keep blood glucose levels within acceptable values. The information in one of your genes translates to "Make insulin." Insulin, produced by cells within the pancreas, is a chemical that helps regulate your body's use of glucose as a fuel.

People with type 1 diabetes often have a mutation (error) in a different gene that causes the body's immune cells to attack and destroy the insulin-producing pancreas cells. The breakdown of the normal flow of information within the body leads to disease. Some people with diabetes regulate their sugar levels by injecting themselves with insulin produced by genetically engineered bacteria. These bacteria can make insulin because the human gene has been transplanted into them. This example of genetic engineering was one of the earliest successes of biotechnology, a field that has transformed the pharmaceutical industry and extended millions of lives (Figure 1.18). And biotechnology is only possible because biological information, written in the universal chemical language of DNA, is used in a similar manner by all life on Earth. ☑

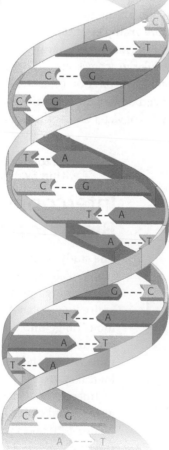

► Figure 1.17 **The language of DNA.** Every molecule of DNA is constructed from four kinds of chemical building blocks that are chained together, shown here as simple shapes and letters.

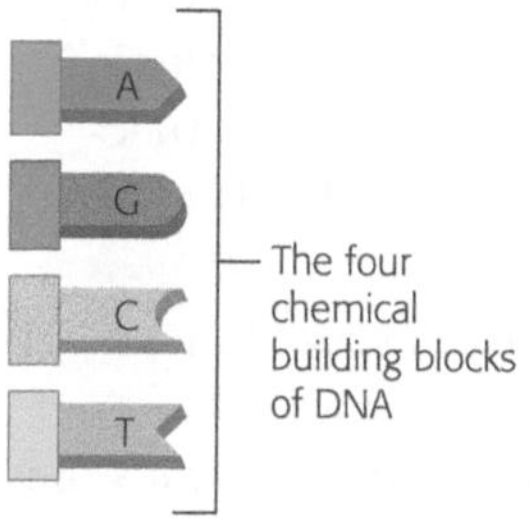

— The four chemical building blocks of DNA

A DNA molecule

☑ CHECKPOINT

Which is bigger: gene or genome?

Answer: genome (because your genome consists of all your genes)

![Energy Transformations] **Pathways that Transform Energy and Matter**

Movement, growth, reproduction, and the various cellular activities of life are work, and work requires energy. The input of energy, primarily from the sun, and the transformation of energy from one form to another make life possible (Figure 1.19). At their source, most ecosystems are solar powered. The energy that enters an ecosystem as sunlight is captured by plants and other photosynthetic organisms (producers) that absorb the sun's energy and convert it, storing it as chemical bonds within sugars and other complex molecules. These molecules then become food for a series of consumers, such as animals, that feed on producers. Consumers can use the food as a source of energy by breaking chemical bonds or as building blocks for making molecules needed by the organism. In other words, the molecules consumed can be used as both a source of energy and a source of matter. In the process of these energy conversions between and within organisms, some energy is converted to heat, which is then lost from the ecosystem. Thus, energy flows through an ecosystem, entering as light and exiting as heat. This is represented by wavy lines in Figure 1.19.

Every object in the universe, both living and nonliving, is composed of matter. In contrast to energy flowing through an ecosystem, matter is recycled within an ecosystem. This is represented by the blue circle in Figure 1.19. For example, minerals that plants absorb from the soil can eventually be recycled back into to the soil when plants are decomposed by microorganisms. Decomposers, such as fungi and many bacteria, break down waste products and the remains of dead organisms, changing complex molecules into simple nutrients. The action of decomposers ensures that nutrients are

available to be taken up from the soil by plants again, thereby completing the cycle.

Within all living cells, a vast network of interconnected chemical reactions (collectively referred to as metabolism) continually converts energy from one form to another as matter is recycled. For example, as food molecules are broken down into simpler molecules, energy stored in the chemical bonds is released. This energy can be captured and used by the body (to power muscle contractions, for example). The atoms that made up the food can then be recycled (to build new muscle tissue, for example). Within all living organisms—there is a never-ending "chemical square dance" in which molecules swap chemical partners as they receive, convert, and release matter and energy. The importance of energy and matter transformations is made clear by examining what happens when they are disrupted. Cyanide is one of the deadliest known poisons. Ingesting just 200 milligrams (about half the size of a tablet of aspirin) causes death in humans. Cyanide is so toxic because it blocks an essential step within the metabolic pathway that harvests energy from glucose. When even a single protein within this pathway becomes inhibited, cells lose the ability to extract the energy stored in the chemical bonds of glucose. The rapid death that follows is a macabre illustration of the importance of energy and matter transformations to life. Throughout your study of biology, you will see more examples of the how living organisms regulate the transformation of energy and matter, from microscopic cellular processes such as photosynthesis (Chapter 7) and cellular respiration (Chapter 8), to ecosystem-wide cycles of carbon and other nutrients (Chapter 20), to global cycles of water across the planet (Chapter 18). ☑

☑ CHECKPOINT

What is the key difference between how energy and matter move in ecosystems?

Answer: Energy moves through an ecosystem (entering and exiting), whereas matter is recycled within an ecosystem.

▼ Figure 1.19 **Nutrient and energy flow in an ecosystem.** Nutrients are recycled within an ecosystem, whereas energy flows into and then out of an ecosystem.

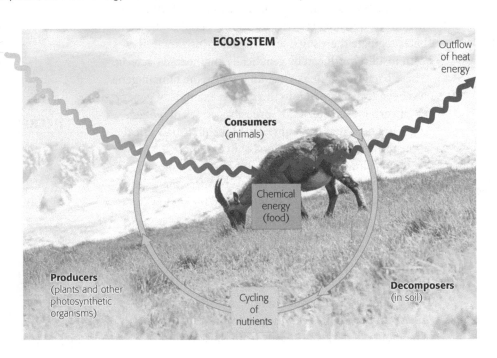

Inflow of light energy

ECOSYSTEM

Outflow of heat energy

Consumers (animals)

Chemical energy (food)

Producers (plants and other photosynthetic organisms)

Cycling of nutrients

Decomposers (in soil)

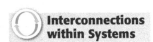

Interconnections within Systems

Interconnections within Biological Systems

The study of life extends from the microscopic scale of the molecules and cells that make up organisms to the global scale of the entire living planet. We can divide this enormous range into different levels of biological organization. There are many interconnections within and between these levels of biological systems.

Imagine zooming in from space to take a closer and closer look at life on Earth. **Figure 1.20** takes you on a tour that spans the full scope of life. The top of the figure shows the global scale of the entire **biosphere**, which

▼ Figure 1.20 **Zooming in on life.**

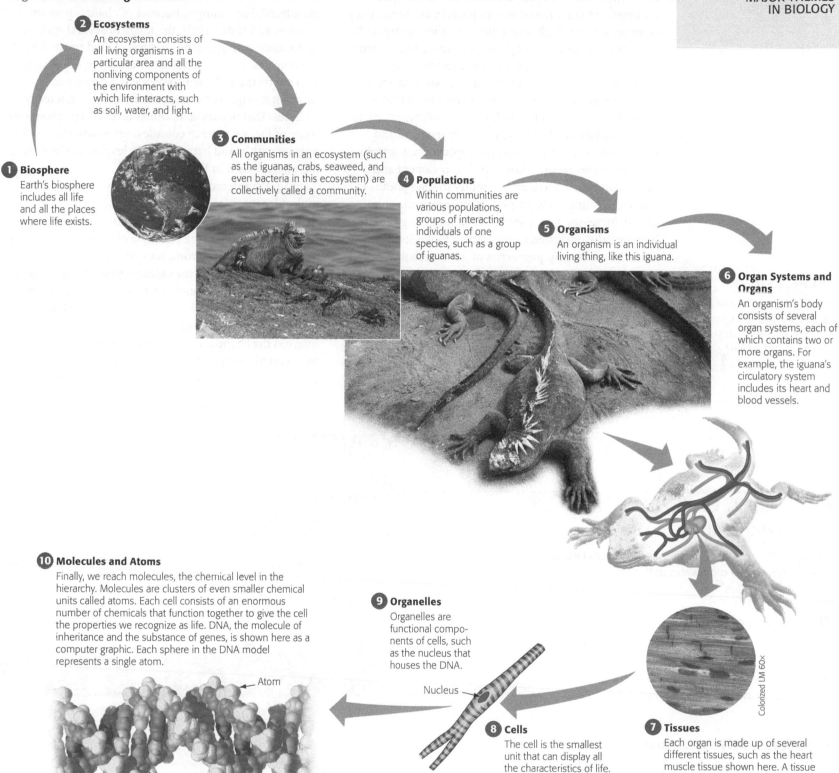

2 Ecosystems

An ecosystem consists of all living organisms in a particular area and all the nonliving components of the environment with which life interacts, such as soil, water, and light.

1 Biosphere

Earth's biosphere includes all life and all the places where life exists.

3 Communities

All organisms in an ecosystem (such as the iguanas, crabs, seaweed, and even bacteria in this ecosystem) are collectively called a community.

4 Populations

Within communities are various populations, groups of interacting individuals of one species, such as a group of iguanas.

5 Organisms

An organism is an individual living thing, like this iguana.

6 Organ Systems and Organs

An organism's body consists of several organ systems, each of which contains two or more organs. For example, the iguana's circulatory system includes its heart and blood vessels.

10 Molecules and Atoms

Finally, we reach molecules, the chemical level in the hierarchy. Molecules are clusters of even smaller chemical units called atoms. Each cell consists of an enormous number of chemicals that function together to give the cell the properties we recognize as life. DNA, the molecule of inheritance and the substance of genes, is shown here as a computer graphic. Each sphere in the DNA model represents a single atom.

Atom

9 Organelles

Organelles are functional components of cells, such as the nucleus that houses the DNA.

Nucleus

Colorized LM 60x

8 Cells

The cell is the smallest unit that can display all the characteristics of life.

7 Tissues

Each organ is made up of several different tissues, such as the heart muscle tissue shown here. A tissue consists of a group of similar cells performing a specific function.

consists of all the environments on Earth that support life—including soil; oceans, lakes, and other bodies of water; and the lower atmosphere. At the other extreme of biological size and complexity are microscopic molecules such as DNA, the chemical responsible for

inheritance. Zooming outward from bottom to top in the figure, you can see that it takes many molecules to build a cell, many cells to make a tissue, multiple tissues to make an organ, and so on. At each new level, novel properties emerge that are absent from the preceding

51

one. These emergent properties are due to the specific arrangement and interactions of parts in an increasingly complex system. Such properties are called emergent because they emerge as complexity increases. For example, life emerges at the level of the cell; a test tube full of molecules is not alive. The saying "the whole is greater than the sum of its parts" captures this idea. Emergent properties are not unique to life. A box of camera parts won't do anything, but if the parts are arranged and interact in a certain way, you can capture photographs. Add structures from a phone, and your camera and phone can interact to gain the ability to send photos to friends instantly. New properties emerge as the complexity increases. Compared to such nonliving examples, however, the unrivaled complexity of biological systems makes the emergent properties of life especially fascinating to study.

Consider another example of interconnectedness within biological systems, one that operates on a much larger scale: global climate. As the composition of the atmosphere changes, patterns of temperature are changing across the planet. This affects the makeup of ecosystems by changing patterns of weather and availability of water. In turn, biological communities and populations are altered. For example, because of changing weather patterns and water levels, the range of several species of disease-causing mosquitoes has drifted north, bringing illnesses (such as malaria) to areas not previously affected by them. When someone is bit by a malaria-carrying mosquito, his or her body will be affected by an illness that occurs at a cellular level. Throughout our study of life, we will see countless interconnections that operate within and between every level of the biological hierarchy shown in Figure 1.20.

From the interactions within the biosphere to the molecular machinery within cells, biologists are investigating life at its many levels. Zooming in at ever-finer resolution illustrates the principle of reductionism—the approach of reducing complex systems to simpler components that are more manageable to study. Reductionism is a powerful strategy in biology. For example, by studying the molecular structure of DNA that had been extracted from cells, James Watson and Francis Crick inferred the chemical basis of biological inheritance. In this reductionist spirit, we will begin our investigation of biology by studying the chemistry of life (Chapter 2). ☑

☑ CHECKPOINT

What is the smallest level of biological organization that can display all the characteristics of life?

Answer: a cell

Chapter Review

▉ SUMMARY OF KEY CONCEPTS

The Scientific Study of Life

Biology is the scientific study of life. It is important to distinguish scientific investigations from other ways of thinking, and to distinguish living objects from nonliving ones.

The Process of Science

Only scientific means of investigating life qualify as biology.

Discovery Science

Describing the natural world with verifiable data is the hallmark of discovery science.

Hypothesis-Driven Science

A scientist formulates a hypothesis (tentative explanation) to account for observations of the natural world. The hypothesis may then be tested via the steps of the scientific method:

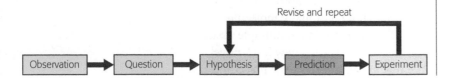

Revise and repeat

Observation → Question → Hypothesis → Prediction → Experiment

Theories in Science

A theory is a broad and comprehensive statement about the world that is supported by the accumulation of a great deal of verifiable evidence.

The Nature of Life

The Properties of Life

All life displays a common set of characteristics:

Order   Regulation   Growth and development   Energy processing

Response to the environment   Reproduction   Evolution

Life in Its Diverse Forms

Biologists organize living organisms into three domains. The domain Eukarya is further divided into three kingdoms (distinguished partly by their means of obtaining food) and one catch-all group:

Life			
Prokaryotes		**Eukaryotes**	
		Plantae Fungi Animalia / Protists (all other eukaryotes)	
		Three kingdoms	
Domain Bacteria	**Domain Archaea**	**Domain Eukarya**	

Major Themes in Biology

Throughout your study of biology, you will frequently come upon examples of five unifying themes: evolution, the relationship of structure to function, the flow of information through biological systems, energy transformations, and interconnections with biological systems.

MAJOR THEMES IN BIOLOGY				
Evolution	**Structure/ Function**	**Information Flow**	**Energy Transformations**	**Interconnections within Systems**

Evolution

Charles Darwin established the ideas of evolution ("descent with modification") via natural selection (unequal reproductive success) in his 1859 publication *The Origin of Species*. Natural selection leads to adaptations to the environment, which—when passed from generation to generation—is the mechanism of evolution.

The Relationship of Structure to Function

At all levels of biology, structure and function are related. Changing structure often results in an altered function, and learning about a component's function will often give insight into its structure.

Information Flow

Throughout living systems, information is stored, transmitted, and used. Within your body, genes provide instructions for building proteins, which perform many of life's tasks.

Pathways That Transform Energy and Matter

Within ecosystems, nutrients are recycled, but energy flows through.

Interconnections within Biological Systems

Life can be studied on many scales, from molecules to the entire biosphere. As complexity increases, novel properties emerge. For example, the cell is the smallest unit that can possibly display all of the characteristics of life.

MasteringBiology®

For practice quizzes, BioFlix animations, MP3 tutorials, video tutors, and more study tools designed for this textbook, go to MasteringBiology®

SELF-QUIZ

1. Which of the following is *not* a property of life?
   a. Populations of organisms rarely change over time.
   b. Living things exhibit complex but ordered organization.
   c. Organisms take in energy and use it to perform all of life's activities.
   d. Organisms reproduce their own kind.

2. Place the following levels of biological organization in order from smallest to largest: atom, biosphere, cell, ecosystem, molecule, organ, organism, population, tissue. Which is the smallest level capable of demonstrating all of the characteristics of life?

3. Plants use the process of photosynthesis to convert the energy in sunlight to chemical energy in the form of sugar. While doing so, they consume carbon dioxide and water and release oxygen. Explain how this process functions in both the cycling of chemical nutrients and the flow of energy through an ecosystem.

4. For each of the following organisms, match its description to its most likely domain and/or kingdom:
   a. A foot-tall organism capable of producing its own food from sunlight
   b. A microscopic, simple, nucleus-free organism found living in a riverbed
   c. An inch-tall organism growing on the forest floor that consumes material from dead leaves
   d. A thimble-sized organism that feeds on algae growing in a pond

   1. Bacteria
   2. Eukarya/Animalia
   3. Eukarya/Fungi
   4. Eukarya/Plantae

5. How does natural selection cause a population to become adapted to its environment over time?

6. Which of the following are the proper components of the scientific method?

   a. experiment, conclusion, application

   b. question, observation, experiment, analysis, prediction

   c. observation, question, hypothesis, prediction, experiment, results, conclusion

   d. observation, question, opinion, conclusion, hypothesis

7. Which of the following statements best distinguishes hypotheses from theories in science?

   a. Theories are hypotheses that have been proved.

   b. Hypotheses are tentative guesses; theories are correct answers to questions about nature.

   c. Hypotheses usually are narrow in scope; theories have broad explanatory power.

   d. Hypotheses and theories mean essentially the same thing in science.

8. _____ selection accounts for the different breeds of domesticated dogs.

9. Match each of the following terms to the phrase that best describes it.

   a. Natural selection   1. A testable idea

   b. Evolution   2. Descent with modification

   c. Hypothesis   3. Unequal reproductive success

   d. Biosphere   4. All life-supporting environments on Earth

Answers to these questions can be found in Appendix: Self-Quiz Answers.

THE PROCESS OF SCIENCE

10. Figure 1.12 depicts the selection of a trait—the resistance to an antibiotic—in a population of bacteria. Let us assume that this resistance is conferred by a gene that helps to neutralize or inactivate the antibiotic and is only carried by the resistant bacteria. What would happen to the bacterial population if this antibiotic is banned? Develop a hypothesis that is consistent with Darwin's theory of natural selection.

11. **Interpreting Data** Trans fats are a form of dietary fat that can have significant health consequences. The graph below represents data from a 2004 study that compared the amount of trans fats found in adipose tissue (body fat) of 79 heart attack patients and 167 people who did not have a heart attack. Write a one-sentence summary of the results presented in the graph.

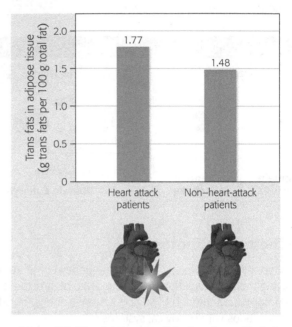

Data from: P. M. Clifton et al., *Trans* fatty acids in adipose tissue and the food supply are associated with myocardial infarction. *Journal of Nutrition* 134: 874–879 (2004).

BIOLOGY AND SOCIETY

12. The development of both drugs and household items typically involves biomedical and/or biochemical research. Select three drugs and three household items, and devise hypothetical flowcharts, like the one depicted in Figure 1.3, which delineate the process of development of each of these.

13. Check the presence of the word "biological" (or "bio") on the containers of food and cleaning products. How does its use on commercial products relate to the scientific definition of biology?

Unit 1
Cells

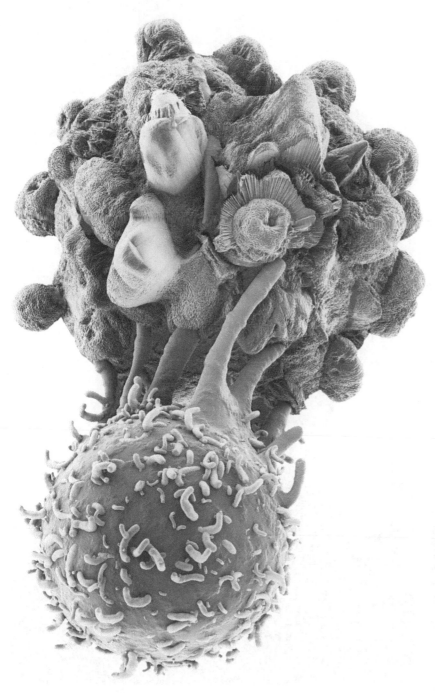

2 Essential Chemistry for Biology

Chapter Thread: **Radioactivity**

3 The Molecules of Life

Chapter Thread: **Lactose Intolerance**

4 A Tour of the Cell

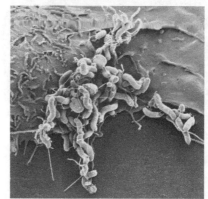

Chapter Thread: **Humans Versus Bacteria**

5 The Working Cell

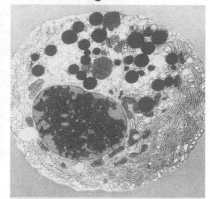

Chapter Thread: **Nanotechnology**

6 Cellular Respiration: Obtaining Energy from Food

Chapter Thread: **Exercise Science**

7 Photosynthesis: Using Light to Make Food

Chapter Thread: **Biofuels**

2 Essential Chemistry for Biology

Too little of the essential element copper in your diet causes anemia, but too much causes kidney and liver damage. ▶

Sodium is an explosive solid and chlorine is a poisonous gas, but when combined they form a common ingredient in your diet: table salt. ▶

▲ Lemon juice has roughly the same acidity as the food-digesting chemicals in your stomach.

Radioactivity BIOLOGY AND SOCIETY

Radiation and Health

The word "radioactive" probably sets off alarm bells in your mind: "Danger! Hazardous!" It is true that radiation, high-energy particles emitted by radioactive substances, can penetrate living tissues and kill cells by damaging DNA. But you probably know that radiation can also be medically beneficial by, for example, helping to treat cancer. So what determines whether radiation is harmful or helpful to an organism's health?

Radiation is most dangerous when exposure is uncontrolled and covers most or all of the body, as happens when a person is exposed to radioactive fallout from a nuclear detonation or accident. In contrast, controlled medical radiation therapy exposes only a small part of the body to a precise dosage of radiation. For example, when treating cancer, carefully calibrated radiation beams are aimed from several angles, intersecting only at the tumor. This provides a deadly dose to cancerous cells but mostly spares surrounding healthy tissues. Radiation therapy is also used to treat Graves' disease, a condition in which an overactive thyroid gland (located in the neck) causes a variety of physical symptoms, including shaking, swelling behind the eyeballs, and heart irregularities. Individuals with Graves' disease may be treated with a "cocktail" containing radioactive iodine. Because the thyroid produces a variety of hormones that use iodine, the radioactive iodine accumulates in this gland, where it then provides a steady low dose of radiation that can, over time, destroy enough thyroid tissue to reduce symptoms.

Pros and cons of radiation. Radiation can be harmful if it is released into the environment, but physicians use controlled doses of radiation to diagnose and treat several diseases.

What makes something radioactive? To understand this question, we have to look to the most basic level of all living things: the atoms that make up all matter. Many questions about life—for example, why is radiation harmful to cells?—can be reduced to questions about chemicals and their interactions—for example, how does radiation affect the atoms in living issues? Knowledge of chemistry is therefore essential to understanding biology. In this chapter, we'll review some basic chemistry that you can apply throughout your study of life. We'll start with an examination of molecules, atoms, and their components. Next, we'll discuss water, one of life's most important molecules, and its crucial role in sustaining life on Earth.

Some Basic Chemistry

Why would a biology textbook include a chapter on chemistry? Well, take any biological system apart and you eventually end up at the chemical level. In fact, you can think of your body as a big watery container of chemicals undergoing a continuous series of chemical reactions. Viewed this way, your metabolism—the sum total of all the chemical reactions that occur in your body—is like a giant square dance, with chemical partners constantly swapping atoms as they move to and fro. Beginning at this basic biological level, let's explore the chemistry of life.

Matter: Elements and Compounds

You and everything that surrounds you is made of matter, the physical "stuff" of the universe. Matter is found on Earth in three physical states: solid, liquid, and gas. Defined more formally, **matter** is anything that occupies space and has mass. **Mass** is a measure of the amount of material in an object. All matter is composed of chemical elements. An **element** is a substance that cannot be broken down into other substances by chemical reactions. Think of it this way: When you burn wood, you are left with ash. But when you burn ash, you only get more ash. That is because wood is a complex mixture of elements, while ash is a pure element (carbon) that cannot be further broken down. There are 92 naturally occurring elements; examples are carbon, oxygen, and gold. Each element has a symbol derived from its English, Latin, or German name. For instance, the symbol for gold, Au, is from the Latin word *aurum*. All the elements—the 92 that occur naturally and

Too little of the essential element copper in your diet causes anemia, but too much causes kidney and liver damage.

several dozen that are human-made—are listed in the **periodic table of the elements**, a familiar fixture in any chemistry or biology lab (**Figure 2.1**; see Appendix B for a full version).

Of the naturally occurring elements, 25 are essential to people. (Other organisms need fewer; plants, for example, typically need 17.) Four of these elements—oxygen (O), carbon (C), hydrogen (H), and nitrogen (N)—make up about 96% of the weight of the body (**Figure 2.2**). Much of the remaining 4% is accounted for by 7 elements, most of which are probably familiar to you, such as calcium (Ca). Calcium, important for building strong bones and teeth, is found abundantly in milk and dairy products as well as sardines and green, leafy vegetables (kale and broccoli, for example).

Less than 0.01% of your weight is made up of 14 trace elements. **Trace elements** are required in only very small amounts, but you cannot live without them. The average person, for example, needs only a tiny speck of iodine each day. Iodine is an essential ingredient of hormones produced by the thyroid gland, located in the neck. An iodine deficiency causes the thyroid gland to enlarge, a condition called goiter. Therefore, consuming foods that are naturally rich in iodine—such as green vegetables, eggs, kelp, and dairy products—prevents goiter. The addition of iodine to table salt ("iodized salt") has nearly eliminated goiter in industrialized nations, but many thousands of people in developing countries are still affected (**Figure 2.3**). Another trace element is fluorine,

Mercury (Hg)

Copper (Cu)

Lead (Pb)

► **Figure 2.1 Abbreviated periodic table of the elements.** In the full periodic table (see Appendix B), each entry contains the element symbol in the center, with the atomic number above and the atomic mass below. The element highlighted here is carbon (C).

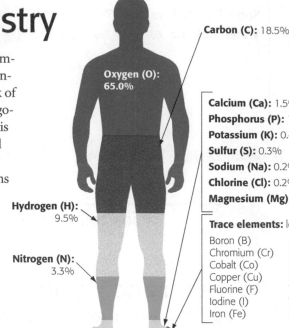

Carbon (C): 18.5%

Oxygen (O): 65.0%

Calcium (Ca): 1.5%
Phosphorus (P): 1.0%
Potassium (K): 0.4%
Sulfur (S): 0.3%
Sodium (Na): 0.2%
Chlorine (Cl): 0.2%
Magnesium (Mg): 0.1%

Hydrogen (H): 9.5%

Nitrogen (N): 3.3%

Trace elements: less than 0.01%
Boron (B)   Manganese (Mn)
Chromium (Cr)   Molybdenum (Mo)
Cobalt (Co)   Selenium (Se)
Copper (Cu)   Silicon (Si)
Fluorine (F)   Tin (Sn)
Iodine (I)   Vanadium (V)
Iron (Fe)   Zinc (Zn)

▲ **Figure 2.2 Chemical composition of the human body.** Notice that just 4 elements make up 96% of your weight.

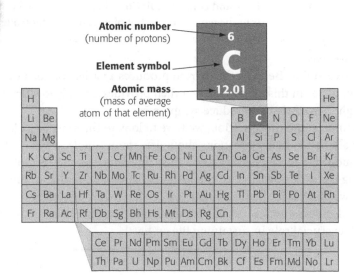

Atomic number (number of protons)

Element symbol

Atomic mass (mass of average atom of that element)

6

C

12.01

▼ Figure 2.3 Diet and goiter.

Goiter, an enlargement of the thyroid gland, shown here in a Malaysian woman, can occur when a person's diet does not include enough iodine, a trace element.

Eating iodine-rich foods can prevent goiter.

which (in the form of fluoride) is added to dental products and drinking water and helps maintain healthy bones and teeth. Many prepared foods are fortified with trace mineral elements. Look at the side of a cereal box and you'll probably see iron listed; you can actually see the iron yourself if you crush the cereal and stir a magnet through it. Be grateful for this additive: It helps prevent anemia due to iron deficiency, one of the most common nutritional deficiencies among Americans.

Elements can combine to form **compounds**, substances that contain two or more elements in a fixed ratio. In everyday life, compounds are much more common than pure elements. Familiar examples of compounds include table salt and water. Table salt is sodium chloride, NaCl, consisting of equal parts of the elements sodium (Na) and chlorine (Cl). A molecule of water, H_2O, has two atoms of hydrogen and one atom of oxygen. Most of the compounds in living organisms contain several different elements. DNA, for example, contains carbon, nitrogen, oxygen, hydrogen, and phosphorus. ☑

> Sodium is an explosive solid and chlorine is a poisonous gas, but when combined they form a common ingredient in your diet: table salt.

Atoms

Each element is made up of one kind of atom, and the atoms in an element are different from the atoms of other elements. An **atom** is the smallest unit of matter that still retains the properties of an element. In other words, the smallest amount of the element carbon is one carbon atom. Just how small is this "piece" of carbon? It would take about a million carbon atoms to stretch across the period at the end of this sentence.

The Structure of Atoms

Atoms are composed of subatomic particles, of which the three most important are protons, electrons, and neutrons. A **proton** is a subatomic particle with a single unit of positive electrical charge (+). An **electron** is a subatomic particle with a single negative charge (−). A **neutron** is electrically neutral (has no charge).

Figure 2.4 shows a simplified model of an atom of the element helium (He), the lighter-than-air gas used to inflate party balloons. Each atom of helium has 2 neutrons (◉) and 2 protons (⊕) tightly packed into the **nucleus**, the atom's central core. Two electrons (⊖) move around the nucleus in a spherical cloud at nearly the speed of light. The electron cloud is much bigger than the nucleus. If the atom were the size of a baseball stadium, the nucleus would be the size of a pea on the pitcher's mound and the electrons would be two gnats buzzing around the bleachers. When an atom has an equal number of protons and electrons, its net electrical charge is zero and so the atom is neutral.

All atoms of a particular element have the same unique number of protons. This number is the element's **atomic number**. Thus, an atom of helium, with 2 protons, has an atomic number of 2, and no other element has 2 protons. The periodic table of elements (Appendix B) lists elements in order of atomic number. Note that in these atoms, the atomic number is also the number of electrons. A standard atom of any element has an equal number of protons and electrons, and thus its net electrical charge is 0 (zero). An atom's **mass number** is the sum of the number of protons and neutrons. For helium, the mass number is 4. The mass of a proton and the mass of a neutron are almost identical and are expressed in a unit of measurement called the dalton. Protons and neutrons each have masses close to 1 dalton. An electron has only about 1/2,000 the mass of a proton, so its mass is approximated as zero. An atom's **atomic mass**, which is listed in the periodic table as the bottom number (under the element symbol), is close to its mass number—the sum of its protons and neutrons—but may differ slightly because it represents an average of all the naturally occurring forms of that element.

Isotopes

Some elements can exist in different forms called **isotopes**, which have the same numbers of protons and electrons as a standard atom of that element but different numbers

▼ Figure 2.4 A simplified model of a helium atom. This model shows the subatomic particles in an atom of helium. The electrons move very fast, creating a spherical cloud of negative charge surrounding the positively charged nucleus.

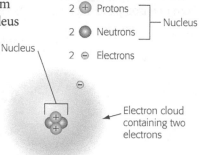

2 ⊕ Protons	Nucleus
2 ◉ Neutrons	
2 ⊖ Electrons	

Nucleus

Electron cloud containing two electrons

☑ **CHECKPOINT**

How many of the naturally occurring elements are used by your body? Which four are the most abundant in living cells?

Answer: 25; oxygen, carbon, hydrogen, and nitrogen

Table 2.1	Isotopes of Carbon		
	Carbon-12	Carbon-13	Carbon-14
Protons	6 ⌐ mass number	6 ⌐ mass number	6 ⌐ mass number
Neutrons	6 ⌐ 12	7 ⌐ 13	8 ⌐ 14
Electrons	6	6	6

☑ **CHECKPOINT**

By definition, all atoms of carbon have exactly 6 _____, but the number of _____ varies from one isotope to another.

Answer: protons; neutrons

of neutrons. In other words, isotopes are forms of an element that differ in mass. As shown in **Table 2.1**, the isotope carbon-12 (named for its mass number), which has 6 neutrons and 6 protons, makes up about 99% of all naturally occurring carbon. Most of the other 1% of carbon on Earth is the isotope carbon-13, which has 7 neutrons and 6 protons. A third isotope, carbon-14, which has 8 neutrons and 6 protons, occurs in minute quantities. All three isotopes have 6 protons—otherwise, they would not be carbon. Both carbon-12 and carbon-13 are stable isotopes, meaning that their nuclei remain intact more or less forever. The isotope carbon-14, on the other hand, is radioactive. A **radioactive isotope** is one in which the nucleus decays spontaneously, shedding particles and energy.

Radiation from decaying isotopes can damage cellular molecules and thus can pose serious health risks. In 1986, the explosion of a nuclear reactor at Chernobyl, Ukraine, released large amounts of radioactive isotopes, killing 30 people within a few weeks. Millions of people in the surrounding areas were exposed, causing an estimated 6,000 cases of thyroid cancer. The 2011 post-tsunami Fukushima nuclear disaster in Japan, did not result in any immediate deaths due to radiation exposure, but scientists are carefully monitoring the people who live in the area to watch for any long-term health consequences.

Natural sources of radiation can also pose a threat. Radon, a radioactive gas, can cause lung cancer. Radon may contaminate buildings where underlying rocks naturally contain the radioactive element uranium. Homeowners can install a radon detector or test their home to ensure that radon levels are safe.

Although radioactive isotopes can cause harm when uncontrolled, they have many uses in biological research and medicine. In the Biology and Society section, we discussed how radioactivity can be used to treat diseases such as cancer and Graves' disease. Let's take a look at another beneficial use of radioactivity: the diagnosis of disease. ☑

Radioactivity THE PROCESS OF SCIENCE

Can Radioactive Tracers Identify Brain Diseases?

Cells use radioactive isotopes the same way they use nonradioactive isotopes of the same element. Once the cell takes up a radioactive isotope, the location and concentration of the isotope can be detected because of the radiation it emits. This makes radioactive isotopes useful as tracers—biological spies, in effect—for monitoring living organisms. For example, a medical diagnostic tool called a PET scan works by detecting small amounts of radiation emitted by radioactive materials that were purposefully introduced into the body (**Figure 2.5**).

In 2012, researchers published a study that used PET scans to investigate Alzheimer's disease. In Alzheimer's disease, a patient gradually loses his or her memory and can become confused, forgetful, and unable to perform normal daily tasks. The disease inevitably leads to a loss of bodily functions and death. A definitive diagnosis of Alzheimer's is difficult because it is hard to distinguish from other age-related disorders. Early detection and treatment of Alzheimer's could benefit many patients and their families.

The **observation** that the brains of people with Alzheimer's are often filled with clumps of a protein called amyloid led the researchers to **question** whether these clumps could be detected by a PET scan. The researchers formed the **hypothesis** that a molecule called florbetapir, which contains the radioactive isotope fluorine-18, could be detected by PET scans after it binds to amyloid deposits in living patients. The researchers' **prediction** was that using florbetapir during PET scans could help with diagnosis.

Their **experiment** involved 229 patients who had been diagnosed with mental decline. Of these, 113 patients showed amyloid deposits in their PET scans. This information led doctors to change the diagnosis in 55% of the patients, sometimes changing the diagnosis to Alzheimer's and sometimes changing it to a different disease. Furthermore, the PET scan data led to changes in treatment (such as different drugs) in 87% of cases. These **results** indicate that radioisotope scans can indeed alter diagnoses and affect treatment. Researchers hope that this will lead to improved outcomes for patients suffering from this debilitating condition.

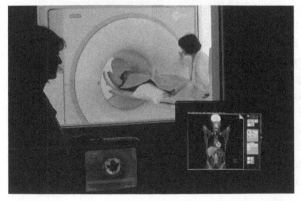

▲ **Figure 2.5 A PET scan.** This monitor shows the images produced by a PET scanner. PET scans can be used to diagnose several diseases, including epilepsy, cancer, and Alzheimer's disease.

Chemical Bonding and Molecules

Of the three subatomic particles we've discussed—protons, neutrons, and electrons—only electrons are directly involved in chemical reactions. The number of electrons in an atom determines the chemical properties of that atom. Chemical reactions enable atoms to transfer or share electrons. These interactions usually result in atoms staying close together, held by attractions called **chemical bonds**. In this section, we will discuss three types of chemical bonds: ionic, covalent, and hydrogen bonds.

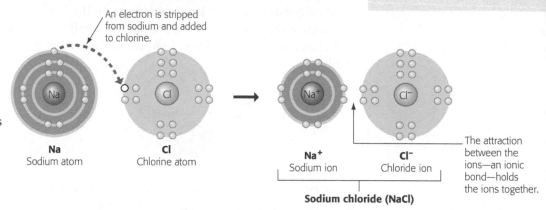

▼ Figure 2.6 **Electron transfer and ionic bonding.** When a sodium atom and a chlorine atom meet, the electron transfer between the two atoms results in two ions with opposite charges.

An electron is stripped from sodium and added to chlorine.

Na
Sodium atom

Cl
Chlorine atom

Na⁺
Sodium ion

Cl⁻
Chloride ion

The attraction between the ions—an ionic bond—holds the ions together.

Sodium chloride (NaCl)

Ionic Bonds

Table salt is an example of how the transfer of electrons can bond atoms together. As discussed earlier, the two ingredients of table salt are the elements sodium (Na) and chlorine (Cl). When in close proximity to each other, a chlorine atom strips an electron from a sodium atom (**Figure 2.6**). Before this electron transfer, both the sodium and chlorine atoms are electrically neutral. Because electrons are negatively charged, the electron transfer moves one unit of negative charge from sodium to chlorine. This action makes both atoms **ions**, atoms or molecules that are electrically charged as a result of gaining or losing electrons. In this case, the loss of an electron results in the sodium ion having a charge of $+1$, whereas chlorine's gain of an electron results in it having a charge of -1. The sodium ion (Na^+) and chloride ion (Cl^-) are then held together by an **ionic bond**, the attraction between oppositely charged ions. Compounds, such as table salt, that are held together by ionic bonds are called ionic compounds. (Note that negatively charged ions often have names ending in "-ide," like "chloride" or "fluoride.") ☑

Covalent Bonds

In contrast to the complete *transfer* of electrons in ionic bonds, a **covalent bond** forms when two atoms *share* one or more pairs of electrons. Of the bonds we've discussed, covalent bonds are the strongest; these are the bonds that hold atoms together in a **molecule**. For example, in **Figure 2.7**, you can see that each of the two hydrogen atoms in a molecule of formaldehyde (CH_2O, a common disinfectant and preservative) shares one pair of electrons with the carbon atom. The oxygen atom shares two pairs of electrons with the carbon, forming a double bond. Notice that each atom of hydrogen (H) can form one covalent bond, oxygen (O) can form two, and carbon (C) can form four.

☑ **CHECKPOINT**

When a lithium ion (Li^+) joins a bromide ion (Br^-) to form lithium bromide, the resulting bond is a(n) _____ bond.

Answer: ionic

▼ Figure 2.7 **Alternative ways to represent a molecule.** A molecular formula, such as CH_2O, tells you the number of each kind of atom in a molecule but not how they are attached together. This figure shows four common ways of representing the arrangement of atoms in molecules.

Name (molecular formula)	Electron configuration	Structural formula	Space-filling model	Ball-and-stick model
	Shows how each atom completes its outer shell by sharing electrons	Represents each covalent bond (a pair of shared electrons) with a line	Shows the shape of a molecule by symbolizing atoms with color-coded balls	Represents atoms with "balls" and bonds with "sticks"
Formaldehyde (CH_2O)				

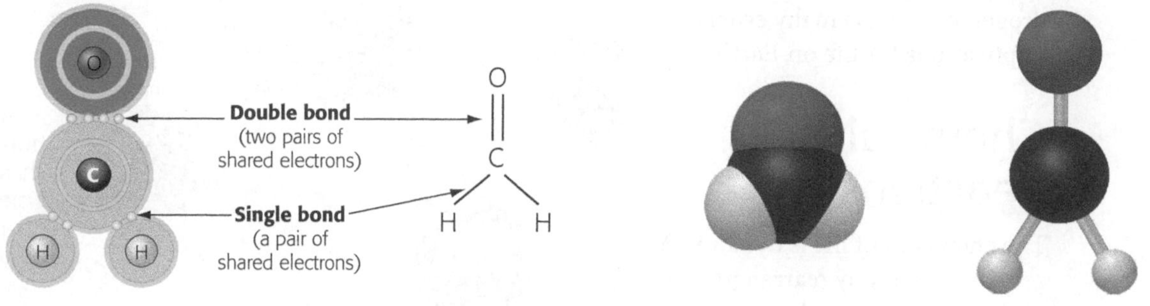

Double bond (two pairs of shared electrons)

Single bond (a pair of shared electrons)

Hydrogen Bonds

A molecule of water (H_2O) consists of two hydrogen atoms joined to one oxygen atom by single covalent bonds. (The covalent bonds between the atoms are represented as the "sticks" here in a ball-and-stick illustration, and the atoms are shown as the "balls"):

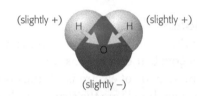

However, the electrons are not shared equally between the oxygen and hydrogen atoms. The two yellow arrows shown in the space-filling model here indicate the stronger pull on the shared electrons that oxygen has compared with its hydrogen partners:

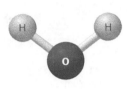

(slightly +)   (slightly +)

(slightly −)

The unequal sharing of negatively charged electrons, combined with its V shape, makes a water molecule polar. A **polar molecule** is one with an uneven distribution of charge that creates two poles, one positive pole and one negative pole. In the case of water, the oxygen end of the molecule has a slight negative charge, and the region around the two hydrogen atoms is slightly positive.

The polarity of water results in weak electrical attractions between neighboring water molecules. Because opposite charges attract, water molecules tend to orient such that a hydrogen atom from one water molecule is near the oxygen atom of an adjacent water molecule. These weak attractions are called **hydrogen bonds (Figure 2.8)**. As you will see later in this chapter, the ability of water to form hydrogen bonds has many crucial implications for life on Earth.

Chemical Reactions

The chemistry of life is dynamic. Your cells are constantly rearranging molecules by breaking existing chemical

bonds and forming new ones in a "chemical square dance." Such changes in the chemical composition of matter are called **chemical reactions**. An example of a chemical reaction is the breakdown of hydrogen peroxide (a common disinfectant that you may have poured on a cut):

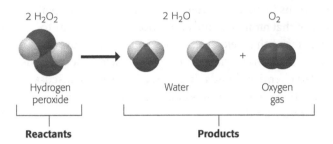

$2\ H_2O_2$ $2\ H_2O$ O_2

Hydrogen   Water   Oxygen
peroxide   gas

Reactants   **Products**

Let's translate the chemical shorthand: Two molecules of hydrogen peroxide ($2\ H_2O_2$) react to form two molecules of water ($2\ H_2O$) and one molecule of oxygen (O_2, which is responsible for the fizzing that happens when hydrogen peroxide interacts with blood). The arrow in this equation indicates the conversion of the starting materials, the **reactants** ($2\ H_2O_2$), to the **products** ($2\ H_2O$ and O_2).

Notice that the same total numbers of hydrogen and oxygen atoms are present in reactants (to the left of the arrow) and products (to the right), although they are grouped differently. Chemical reactions cannot create or destroy matter; they can only rearrange it. These rearrangements usually involve the breaking of chemical bonds in reactants and the forming of new bonds in products.

This discussion of water molecules as the product of a chemical reaction is a good conclusion to this section on basic chemistry. Water is a substance so important in biology that we'll take a closer look at its life-supporting properties in the next section. ☑

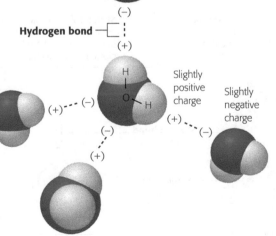

Hydrogen bond

(−)

(+)

(+) --- (−)

Slightly
positive
charge

Slightly
negative
charge

(+) --- (−)

(−)

(+)

◀ **Figure 2.8 Hydrogen bonding in water.** The charged regions of the polar water molecules are attracted to oppositely charged areas of neighboring molecules. Each molecule can hydrogen-bond to a maximum of four partners.

☑ **CHECKPOINT**

Predict the formula for the compound that results when a molecule of sulfur trioxide (SO_3) combines with a molecule of water to produce a single molecule of product. (*Hint*: In chemical reactions, no atoms are gained or lost.)

Answer: H_2SO_4 (sulfuric acid, which is a component of acid rain)

Water and Life

Life on Earth began in water and evolved there for 3 billion years before spreading onto land. Modern life, even land-dwelling life, is still tied to water. You've had personal experience with this dependence on water every time you seek liquids to quench your thirst. Inside your body, your cells are surrounded by a fluid that's composed mostly of water, and your cells themselves range from 70% to 95% in water content.

The abundance of water is a major reason that Earth is habitable. Water is so common that it is easy to overlook the fact that it is an exceptional substance with many extraordinary properties (**Figure 2.9**). We can trace water's unique life-supporting properties to the structure and interactions of its molecules.

 **Structure/Function** Water

The unique properties of water on which all life on Earth depends are a prime example of one of biology's overarching themes: the relationship of structure and function. The structure of water molecules—the polarity and the hydrogen bonding that results (see Figure 2.8)—explains most of water's life-supporting functions. We'll explore four of those properties here: the cohesive nature of water, the ability of water to moderate temperature, the biological significance of ice floating, and the versatility of water as a solvent.

The Cohesion of Water

Water molecules stick together as a result of hydrogen bonding. In a drop of water, any particular set of hydrogen bonds lasts for only a few trillionths of a second, yet at any instant, huge numbers of hydrogen bonds exist between molecules of liquid water. This tendency of molecules of the same kind to stick together, called **cohesion**, is much stronger for water than for most other liquids. The cohesion of water is important in the living world. Trees, for example, depend on cohesion to help transport water from their roots to their leaves (**Figure 2.10**).

Related to cohesion is surface tension, a measure of how difficult it is to stretch or break the surface of a liquid. Hydrogen bonds give water unusually high surface tension, making it behave as though it were coated with an invisible film (**Figure 2.11**). Other

▲ **Figure 2.9 A watery world.** In this photograph, you can see water as liquid (which covers three-quarters of Earth's surface), ice (in the form of snow), and vapor (as steam).

Evaporation from the leaves

Flow of water

Colorized SEM 150×

▲ **Figure 2.10 Cohesion and water transport in plants.** The evaporation of water from leaves pulls water upward from the roots through microscopic tubes in the trunk of the tree. Because of cohesion, the pulling force is relayed through the tubes all the way down to the roots. As a result, water rises against the force of gravity.

Microscopic water-conducting tubes

Cohesion due to hydrogen bonds between water molecules

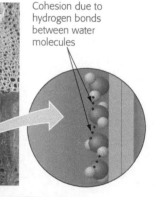

▼ **Figure 2.11 A raft spider walking on water.** The cumulative strength of hydrogen bonds among water molecules allows this spider to walk on pond water without breaking the surface.

liquids have much weaker surface tension; an insect, for example, could not walk on the surface of a cup of gasoline (which is why gardeners sometimes use gas to drown bugs removed from flower bushes).

▲ Figure 2.12 **Sweating as a mechanism of evaporative cooling.**

How Water Moderates Temperature

If you've ever burned your finger on a metal pot while waiting for the water in it to boil, you know that water heats up much more slowly than metal. In fact, because of hydrogen bonding, water has a stronger resistance to temperature change than most other substances.

When water is heated, the heat energy first disrupts hydrogen bonds and then makes water molecules jostle around faster. The temperature of the water doesn't go up until the water molecules start to speed up. Because heat is first used to break hydrogen bonds rather than raise the temperature, water absorbs and stores a large amount of heat while warming up only a few degrees. Conversely, when water cools, hydrogen bonds form, a process that releases heat. Thus, water can release a relatively large amount of heat to the surroundings while the water temperature drops only slightly.

Earth's giant water supply—the oceans, seas, lakes, and rivers—enables temperatures on the planet to stay within limits that permit life by storing a huge amount of heat from the sun during warm periods and giving off heat that warms the air during cold periods. That's why coastal areas generally have milder climates than inland regions. Water's resistance to temperature change also stabilizes ocean temperatures, creating a favorable environment for marine life. You may have noticed that the water temperature at the beach fluctuates much less than the air temperature.

Another way that water moderates temperature is by **evaporative cooling**. When a substance evaporates (changes from a liquid to a gas), the surface of the liquid that remains cools down. This occurs because the molecules with the greatest energy (the "hottest" ones) tend to vaporize first. Think of it like this: If the five fastest runners on your track team quit school, it would lower the average speed of the remaining team. Evaporative cooling helps prevent some land-dwelling creatures from overheating; it's why sweating helps you dissipate excess body heat **(Figure 2.12)**. And the expression "It's not the heat, it's the humidity" has its basis in the difficulty of sweating when the air is already saturated with water vapor.

The Biological Significance of Ice Floating

When most liquids get cold, their molecules move closer together. If the temperature is cold enough, the liquid freezes and becomes a solid. Water, however, behaves differently. When water molecules get cold enough, they move apart, with each molecule staying at "arm's length" from its neighbors, forming ice. A chunk of ice floats because it is less dense than the liquid water in which it is floating. Floating ice is a consequence of hydrogen bonding. In contrast to the short-lived and constantly changing hydrogen bonds in liquid water, those in solid ice last longer, with each molecule bonded to four neighbors. As a result, ice is a spacious crystal **(Figure 2.13)**.

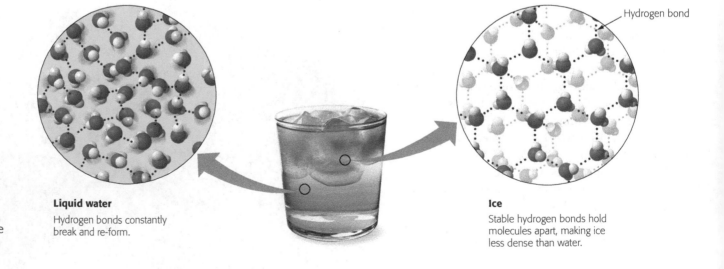

► Figure 2.13 **Why ice floats.** Compare the tightly packed molecules in liquid water with the spaciously arranged molecules in the ice crystal. The less dense ice floats atop the denser water.

Liquid water
Hydrogen bonds constantly break and re-form.

Hydrogen bond

Ice
Stable hydrogen bonds hold molecules apart, making ice less dense than water.

How does floating ice help support life on Earth? When a deep body of water cools and a layer of ice forms on top, the floating ice acts as an insulating "blanket" over the liquid water, allowing life to persist under the frozen surface. But imagine what would happen if ice were denser than water: Ice would sink during winter. All ponds, lakes, and even the oceans would eventually freeze solid without the insulating protection of the top layer of ice. Then, during summer, only the upper few inches of the oceans would thaw. It's hard to imagine life persisting under such conditions.

Water as the Solvent of Life

If you've ever stirred sugar into coffee or added salt to soup, you know that you can dissolve sugar or salt in water. This results in a mixture known as a **solution**, a liquid consisting of a homogeneous mixture of two or more substances. The dissolving agent is called the **solvent**, and any substance that is dissolved is called a **solute**. When water is the solvent, the resulting solution is called an **aqueous solution**. The fluids of organisms are aqueous solutions. For example, tree sap is an aqueous solution consisting of sugar and minerals dissolved in water.

Water can dissolve an enormous variety of solutes necessary for life, providing a medium for chemical reactions. For example, water can dissolve salt ions, as shown in **Figure 2.14**. Each ion becomes surrounded by oppositely charged regions of water molecules. Solutes that are polar molecules, such as sugars, dissolve by orienting locally charged regions of their molecules toward water molecules in a similar way.

We have discussed four special properties of water, each a consequence of water's unique chemical structure. Next, we'll look at aqueous solutions in more detail. ☑

Acids, Bases, and pH

In aqueous solutions, most of the water molecules are intact. However, some of the water molecules break apart into hydrogen ions (H^+) and hydroxide ions (OH^-). A balance of these two highly reactive ions is critical for the proper functioning of chemical processes within organisms.

A chemical compound that releases H^+ to a solution is called an **acid**. One example of a strong acid is hydrochloric acid (HCl), the acid in your stomach that aids in digestion of food. In solution, HCl breaks apart into the ions H^+ and Cl^-. A **base** (or alkali) is a compound that accepts H^+ and removes them from solution. Some bases, such as sodium hydroxide (NaOH), do this by releasing OH^-, which combines with H^+ to form H_2O.

To describe the acidity of a solution, chemists use the **pH scale**, a measure of the hydrogen ion (H^+) concentration in a solution. The scale ranges from 0 (most acidic) to 14 (most basic).

Lemon juice has roughly the same acidity as the food-digesting chemicals in your stomach.

☑ **CHECKPOINT**

1. Explain why, if you pour very carefully, you can actually "stack" water slightly above the rim of a cup.
2. Explain why ice floats.

Answers: 1. Surface tension due to water's cohesion will keep the water from spilling over. 2. Ice is less dense than liquid water because the more stable hydrogen bonds lock the molecules into a spacious crystal.

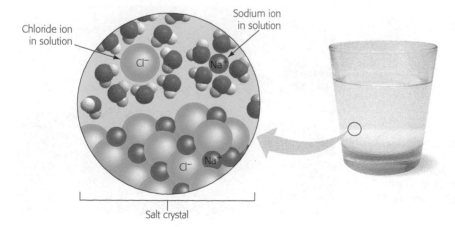

Chloride ion in solution

Sodium ion in solution

Cl^- Na^+

Cl^- Na^+

Salt crystal

◀ **Figure 2.14 A crystal of table salt (NaCl) dissolving in water.** As a result of electrical charge attractions, H_2O molecules surround the sodium and chloride ions, dissolving the crystal in the process.

Each pH unit represents a tenfold change in the concentration of H^+ **(Figure 2.15)**. For example, lemon juice at pH 2 has 100 times more H^+ than an equal amount of tomato juice at pH 4. Aqueous solutions that are neither acidic nor basic (such as pure water) are said to be neutral; they have a pH of 7. They do contain some H^+ and OH^-, but the concentrations of the two ions are equal. The pH of the solution inside most living cells is close to 7.

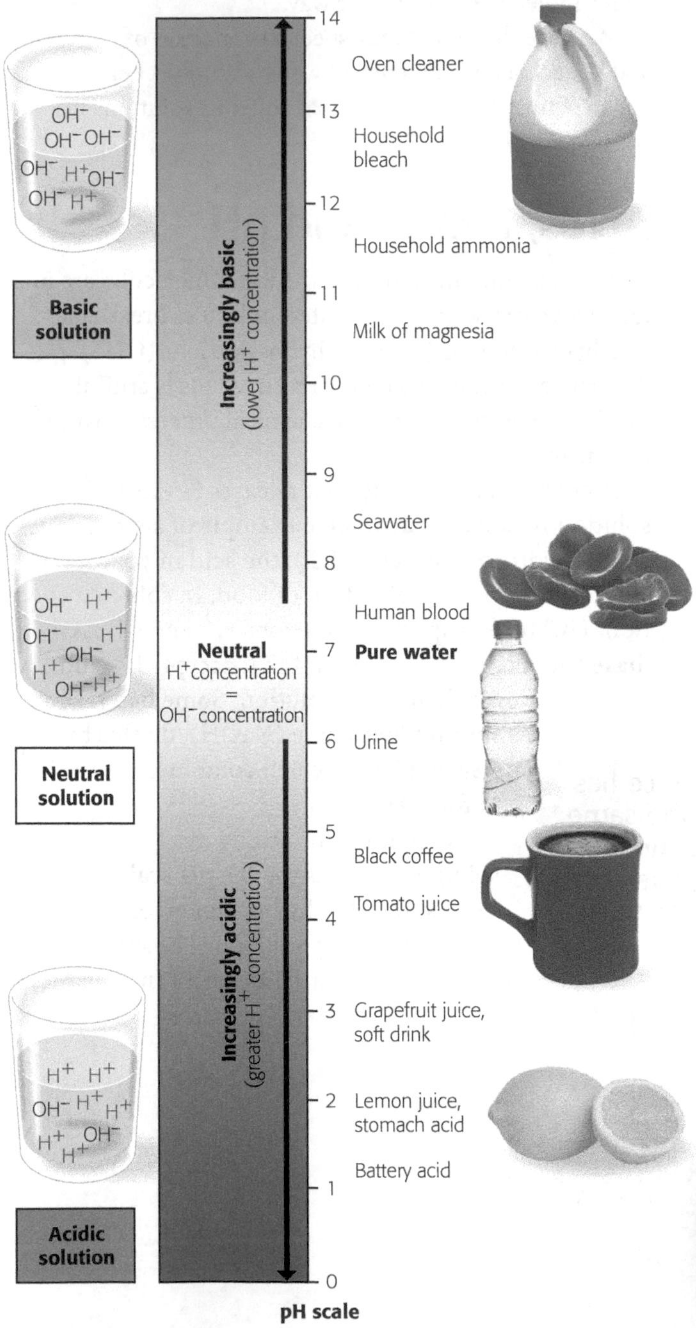

Even a slight change in pH can be harmful to an organism because the molecules in cells are extremely sensitive to H^+ and OH^- concentrations. Biological fluids contain **buffers**, substances that minimize changes in pH by accepting H^+ when that ion is in excess and donating H^+ when it is depleted. For example, buffer in contact lens solution helps protect the surface of the eye from potentially painful changes in pH. This buffering process, however, is not foolproof, and changes in environmental pH can profoundly affect ecosystems. For example, about 25% of the carbon dioxide (CO_2) generated by people (primarily by burning fossil fuels) is absorbed by the oceans. When CO_2 dissolves in seawater, it reacts with water to form carbonic acid **(Figure 2.16)**, which lowers ocean pH. The resulting ocean acidification can greatly change marine environments. Oceanographers have calculated that the pH of the ocean is lower now than at any time in the past 420,000 years, and it is continuing to drop.

The effects of ocean acidification—including coral bleaching and changes in metabolism among a wide variety of sea creatures—are daunting reminders that the chemistry of life is linked to the chemistry of the environment. It reminds us, too, that chemistry happens on a global scale, because industrial processes in one region of the world often cause ecosystem changes in another part of the world. ☑

☑ CHECKPOINT

Compared with a solution of pH 8, the same volume of a solution at pH 5 has _____ times more hydrogen ions (H^+). This second solution is considered a(n) _____.

Answer: 1,000; acid

Basic solution

OH⁻
OH⁻ OH⁻
OH⁻ H⁺OH⁻
OH⁻ H⁺

Increasingly basic (lower H^+ concentration)

Neutral solution

OH⁻ H⁺
OH⁻   H⁺
H⁺ OH⁻
OH⁻H⁺

Neutral
H^+ concentration
=
OH^- concentration

Acidic solution

H⁺ H⁺
OH⁻ H⁺ H⁺
H⁺ OH⁻
H⁺

Increasingly acidic (greater H^+ concentration)

- 14
- 13 — Oven cleaner
- — Household bleach
- 12 — Household ammonia
- 11
- — Milk of magnesia
- 10
- 9
- — Seawater
- 8 — Human blood
- 7 — **Pure water**
- 6 — Urine
- 5
- — Black coffee
- 4 — Tomato juice
- 3 — Grapefruit juice, soft drink
- 2 — Lemon juice, stomach acid
- — Battery acid
- 1
- 0

pH scale

▲ **Figure 2.15 The pH scale.** A solution having a pH of 7 is neutral, meaning that its H^+ and OH^- concentrations are equal. The lower the pH below 7, the more acidic the solution, or the greater its excess of H^+ compared with OH^-. The higher the pH above 7, the more basic the solution, or the greater the deficiency of H^+ relative to OH^-.

▼ **Figure 2.16 Ocean acidification by atmospheric CO_2.** After dissolving in seawater, CO_2 reacts to form carbonic acid. This acid then undergoes further chemical reactions that disrupt coral growth. Such acidification can cause drastic changes in important marine ecosystems.

CO_2
Carbon dioxide

CO_2 + H_2O → H_2CO_3
Carbon dioxide   Water   Carbonic acid

Radioactivity EVOLUTION CONNECTION

Radioactivity as an Evolutionary Clock

Throughout this chapter, we have highlighted ways—both helpful and harmful—that radioactivity can affect the health of living organisms. In addition to detecting and treating diseases, another helpful application of radioactivity involves the natural process of radioactive decay, which can be used to obtain important data about the evolutionary history of life on Earth.

Fossils—both preserved imprints and remains of dead organisms—are reliable chronological records of life because we can determine their ages through radiometric dating **(Figure 2.17)**, which is based on the decay of radioactive isotopes. For example, carbon-14 is a radioactive isotope with a half-life of 5,700 years. It is present in trace amounts in the environment. ❶ A living organism assimilates the different isotopes of an element in proportions that reflect their relative abundances in the environment. In this example, carbon-14 is taken up in trace quantities, along with much larger quantities of the more common carbon-12. ❷ At the time of death, the organism ceases to take in carbon from the environment. From this moment forward, the amount of carbon-14 relative to carbon-12 in the fossil declines: The carbon-14 in the body decays into carbon-12, but no new carbon-14 is added. Because the half-life of carbon-12 is known, the ratio of the two isotopes (carbon-14 to carbon-12) is a reliable indicator of the age of the fossil. In this case, it takes 5,700 years for half of the radioactive carbon-14 to decay; half of the remainder is present after another 5,700 years; and so on. ❸ A fossil's age can be estimated by measuring the ratio of the two isotopes to learn how many half-life reductions have occurred since it died. For example, if the ratio of carbon-14 to carbon-12 in this fossil was found to be $\frac{1}{8}$th that of the environment, this fossil would be about 17,100 (5,700 × 3 half-lives) years old.

Using such techniques, scientists can estimate the ages of fossils from around the world and place them in an ordered sequence called the fossil record. The fossil record is one of the most important and convincing sets of evidence that led Charles Darwin to formulate the theory of natural selection (as you'll see later, in Chapter 14). Every time a new fossil is dated, it can be placed within the vast history of life on Earth.

▼ **Figure 2.17 Radiometric dating.** Living organisms incorporate the carbon-14 isotope (represented here with blue dots). But no new carbon-14 is introduced once an organism dies, and the carbon-14 that remains slowly decays into carbon-12. By measuring the amount of carbon-14 in a fossil, scientists can estimate its age.

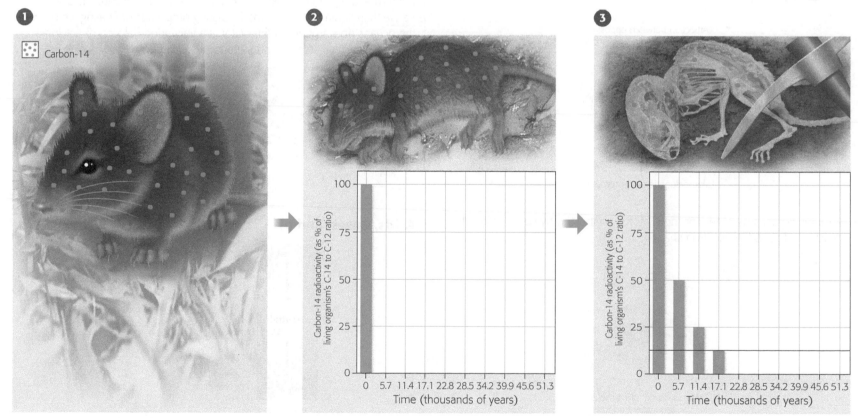

Chapter Review

■ SUMMARY OF KEY CONCEPTS

Some Basic Chemistry

Matter: Elements and Compounds

Matter consists of elements and compounds, which are combinations of two or more elements. Of the 25 elements essential for life, oxygen, carbon, hydrogen, and nitrogen are the most abundant in living matter.

Atoms

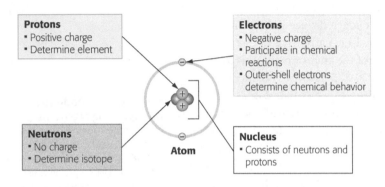

Protons
- Positive charge
- Determine element

Electrons
- Negative charge
- Participate in chemical reactions
- Outer-shell electrons determine chemical behavior

Neutrons
- No charge
- Determine isotope

Atom

Nucleus
- Consists of neutrons and protons

Chemical Bonding and Molecules

Transfer of one or more electrons produces attractions between oppositely charged ions:

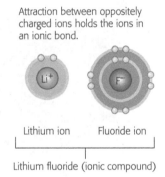

Attraction between oppositely charged ions holds the ions in an ionic bond.

Lithium ion   Fluoride ion

Lithium fluoride (ionic compound)

A molecule consists of two or more atoms connected by covalent bonds, which are formed by electron sharing:

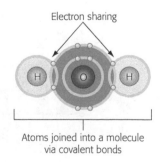

Electron sharing

Atoms joined into a molecule via covalent bonds

Water is a polar molecule; the slightly positively charged H atoms in one water molecule may be attracted to the partial negative charge of O atoms in neighboring water molecules, forming weak but important hydrogen bonds:

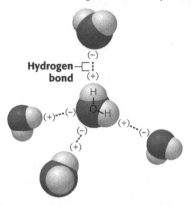

Hydrogen bond

Chemical Reactions

By breaking bonds in reactants and forming new bonds in products, chemical reactions rearrange matter.

Water and Life

Structure/Function: Water

The cohesion (sticking together) of water molecules is essential to life. Water moderates temperature by absorbing heat in warm environments and releasing heat in cold environments. Evaporative cooling also helps stabilize the temperatures of oceans and organisms. Ice floats because it is less dense than liquid water, and the insulating properties of floating ice prevent the oceans from freezing solid. Water is a very good solvent, dissolving a great variety of solutes to produce aqueous solutions.

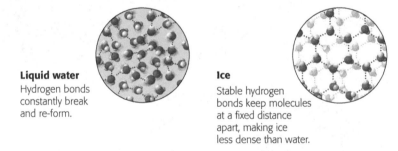

Liquid water
Hydrogen bonds constantly break and re-form.

Ice
Stable hydrogen bonds keep molecules at a fixed distance apart, making ice less dense than water.

Acids, Bases, and pH

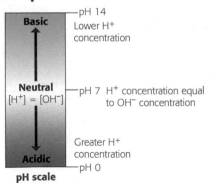

Basic — pH 14 — Lower H^+ concentration

Neutral — $[H^+] = [OH^-]$ — pH 7 — H^+ concentration equal to OH^- concentration

Acidic — pH 0 — Greater H^+ concentration

pH scale

MasteringBiology®

SELF-QUIZ

1. An atom can be changed into an ion by adding or removing _____. An atom can be changed into a different isotope by adding or removing _____.

2. Isotopes of an element have the same number of _____ and different numbers of _____.

3. A sodium atom has 11 protons, and the most common isotope of sodium has 12 neutrons. A radioactive isotope of sodium has 11 neutrons. What are the atomic numbers and mass numbers of the stable and radioactive forms of sodium?

4. Why are radioactive isotopes useful as tracers in research on the chemistry of life?

5. What is chemically nonsensical about this structure?

H—C≡C—H

6. Why is it unlikely that two neighboring water molecules would be arranged like this?

7. Which of the following is not a chemical reaction?
   a. Sugar ($C_6H_{12}O_6$) and oxygen gas (O_2) combine to form carbon dioxide (CO_2) and water (H_2O).
   b. Sodium metal and chlorine gas unite to form sodium chloride.
   c. Hydrogen gas combines with oxygen gas to form water.
   d. Ice melts to form liquid water.

8. Some people in your study group say they don't understand why water is considered a polar molecule. You explain that water is a polar molecule because
   a. the oxygen atom is found between the two hydrogen atoms.
   b. the oxygen atom attracts the hydrogen atoms.
   c. the oxygen end of the molecule has a slight negative charge, and the hydrogen end has a slight positive charge.
   d. both hydrogen atoms are at one end of the molecule, and the oxygen atom is at the other end.

9. Explain how the unique properties of water result from the fact that water is a polar molecule.

10. Explain why it is unlikely that life could exist on a planet where the main liquid is methane, a nonpolar molecule that does not form hydrogen bonds.

11. A can of cola consists mostly of sugar dissolved in water, with some carbon dioxide gas that makes it fizzy and makes the pH less than 7. Describe the cola using the following terms: solute, solvent, acidic, aqueous solution.

Answers to these questions can be found in Appendix: Self-Quiz Answers.

THE PROCESS OF SCIENCE

12. *Helicobacter pylori*, a bacterium that causes gastric inflammation, produces urease, an enzyme that splits urea into carbon dioxide (CO_2) and ammonia. Humans do not produce urease. The "urea breath test" uses carbon isotope labeling to rapidly detect the presence of *Helicobacter pylori* in the intestinal tract of a patient. How might this test work?

13. This diagram shows the arrangement of electrons around the nucleus of a fluorine atom (left) and a potassium atom (right). Predict what kind of bond forms when a fluorine atom comes into contact with a potassium atom.

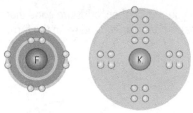

Fluorine atom   Potassium atom

14. **Interpreting Data** As discussed in the text and shown in Figure 2.17, radiometric dating can be used to determine the age of biological materials. The calculation of age depends on the half-life of the radioisotope under consideration. For example, carbon-14 makes up about 1 part per trillion of naturally occurring carbon. When an organism dies, its body contains 1 carbon-14 atom per trillion total carbon atoms. After 5,700 years (the half-life of carbon-14), its body contains one-half as much carbon-14; the other half has decayed and is no longer present. French scientists used carbon-14 dating to determine the age of prehistoric wall paintings in the Niaux caves. They determined that the paintings were made using natural dyes approximately 13,000 years ago. Using your knowledge of radiometric dating, how much carbon-14 must the scientists have found in the cave paintings to support this result? Express your answer in both percentage of carbon-14 remaining and in parts of carbon-14 per trillion.

BIOLOGY AND SOCIETY

15. Critically evaluate this statement: "It's paranoid to worry about contaminating the environment with chemical wastes; this stuff is just made of the same atoms that were already present in our environment."

16. A major source of the CO_2 that causes ocean acidification is emissions from coal-burning power plants. One way to reduce these emissions is to use nuclear power to produce electricity. The proponents of nuclear power contend that it is the only way that the United States can increase its energy production while reducing air pollution, because nuclear power plants emit little or no acid-precipitation-causing pollutants. What are some of the benefits of nuclear power? What are the possible costs and dangers? Do you think we ought to increase our use of nuclear power to generate electricity? If a new power plant were to be built near your home, would you prefer it to be a coal-burning plant or a nuclear plant?

3 The Molecules of Life

If you drink coffee, odds are good that you spooned beets into your cup today—in the form of table sugar.

A long-distance runner who carbo-loads the night before a race is banking glycogen to be used the next day.

The structure of your DNA is nearly indistinguishable from the DNA of a mosquito or an elephant: The differences in animal species result from the way nucleotides are arranged.

Chocolate melts in your mouth (and in your hands) because of the structure of cocoa butter, which is high in saturated fats.

Lactose Intolerance BIOLOGY AND SOCIETY

Got Lactose?

You've probably seen ads that associate a milk mustache with good health. Indeed, milk is a very healthy food: It's rich in protein, minerals, and vitamins, and can be low in fat. But for the majority of adults in the world, a glass of milk delivers a heavy dose of digestive discomfort that can include bloating, gas, and abdominal pain. These are symptoms of lactose intolerance, the inability to digest lactose, the main sugar found in milk.

For people with lactose intolerance, the problem starts when lactose enters the small intestine. To absorb this sugar, digestive cells in the small intestine must produce a molecule called lactase. Lactase is an enzyme, a protein that helps drive chemical reactions—in this case, the breakdown of lactose into smaller sugars. Most people are born with the ability to digest lactose. But after about age 2, lactase levels in most people decline significantly. Lactose that is not broken down in the small intestine passes into the large intestine, where bacteria feed on it and belch out gaseous by-products. An accumulation of gas produces uncomfortable symptoms. Sufficient lactase can thus mean the difference between delight and discomfort when someone drinks a tall glass of chocolate milk.

There is no treatment for the underlying cause of lactose intolerance: the decreased production of lactase. So what are the options for people who do not produce enough of this enzyme? The first option is

Lactose-rich chocolate milk. Lactose digestion results from interactions between several classes of the body's molecules.

avoiding lactose-containing foods; abstinence is always the best form of protection! Alternatively, substitutes are available, such as milk made from soy or almonds, or cow's milk that has been pretreated with lactase. If you are lactose intolerant and pizza is your staple food, you can buy lactose-free cheese (made from soy) or you can make your own mozzarella from lactose-free milk. Also, lactase in pill form can be taken along with food to ease digestion by artificially providing the enzyme that the body naturally lacks.

Lactose intolerance illustrates one way the interplay of biological molecules can affect your health. Such molecular interactions, repeated in countless variations, drive all biological processes. In this chapter, we'll explore the structure and function of large molecules that are essential to life. We'll start with an overview of carbon and then examine four classes of molecules: carbohydrates, lipids, proteins, and nucleic acids. Along the way, we'll look at where these molecules occur in your diet and the important roles they play in your body.

Organic Compounds

A cell is mostly made up of water, but the rest is mainly carbon-based molecules. When it comes to the chemistry of life, carbon plays the leading role. This is because carbon is unparalleled in its ability to form the skeletons of large, complex, diverse molecules that are necessary for life's functions. The study of carbon-based molecules, which are called **organic compounds**, lies at the heart of any study of life.

Carbon Chemistry

Carbon is a versatile molecule ingredient because an atom of carbon can share electrons with other atoms in four covalent bonds that can branch off in four directions. Because carbon can use one or more of its bonds to attach to other carbon atoms, it is possible to construct an endless diversity of carbon skeletons varying in size and branching pattern **(Figure 3.1)**. Thus, molecules with multiple carbon "intersections" can form very elaborate shapes. The carbon atoms of organic compounds can also bond with other elements, most commonly hydrogen, oxygen, and nitrogen.

One of the simplest organic compounds is methane, CH_4, with a single carbon atom bonded to four hydrogen atoms **(Figure 3.2)**. Methane is abundant in natural gas and is also produced by prokaryotes that live in swamps (in the form of swamp gas) and in the digestive

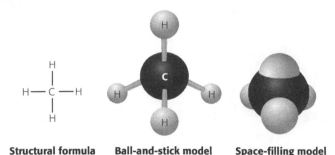

Structural formula Ball-and-stick model Space-filling model

▲ **Figure 3.2 Methane, a simple organic compound.**

tracts of grazing animals, such as cows. Larger organic compounds (such as octane, with eight carbons) are the main molecules in the gasoline we burn in cars and other machines. Organic compounds are also important fuels in your body; the energy-rich parts of fat molecules have a structure similar to gasoline **(Figure 3.3)**.

The unique properties of an organic compound depend not only on its carbon skeleton but also on the atoms attached to the skeleton. In an organic compound, the groups of atoms directly involved in chemical reactions are called **functional groups**. Each functional group plays a particular role during chemical reactions. Two examples of functional groups are the hydroxyl group (—OH, found in alcohols such as isopropyl rubbing alcohol) and the carboxyl group (—COOH, found in all proteins). Many biological molecules have two or more functional groups. Keeping in mind this basic scheme—carbon skeletons with attached functional groups—we are now ready to explore how our cells make large molecules out of smaller ones.

▼ **Figure 3.1 Variations in carbon skeletons.** All of these examples are organic compounds that consist only of carbon and hydrogen. Notice that each carbon atom forms four bonds and each hydrogen atom forms one bond. One line represents a single bond (sharing of one pair of electrons) and two lines represent a double bond (sharing of two pairs of electrons).

Carbon skeletons vary in length

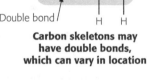

Double bond

Carbon skeletons may have double bonds, which can vary in location

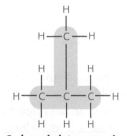

Carbon skeletons may be unbranched or branched

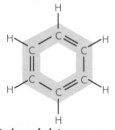

Carbon skeletons may be arranged in rings

▼ **Figure 3.3 Hydrocarbons as fuel.** Energy-rich organic compounds in gasoline provide fuel for machines, and energy-rich molecules in fats provide fuel for cells.

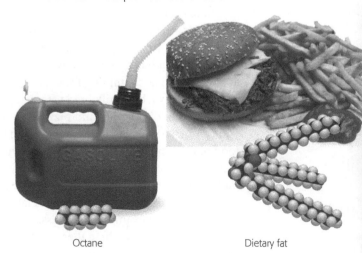

Octane

Dietary fat

Giant Molecules from Smaller Building Blocks

On a molecular scale, the members of three categories of biological molecules—carbohydrates (such as those found in starchy foods like French fries and bagels), proteins (such as enzymes and the components of your hair), and nucleic acids (such as DNA)—are gigantic; in fact, biologists call them **macromolecules** (*macro* means "big"). Despite the size of macromolecules, their structures can be easily understood because they are **polymers**, large molecules made by stringing together many smaller molecules called **monomers**. A polymer is like a necklace made by joining together many monomer "beads" or a train made from a chain of box cars. Although, at first, the structure of a long train may seem very complex, it can be easily understood by asking two questions: What kind of cars make up the train? How are they joined together? Similarly, even large and complex biological macromolecules can be understood by considering which monomers make them up and how those monomers are attached together.

Cells link monomers together to form a polymer through a **dehydration reaction**. As the name implies, this chemical reaction involves removing a molecule of water (Figure 3.4a). For each monomer added to a chain, a water molecule (H_2O) is formed by the release of two hydrogen atoms and one oxygen atom from the reactants. This same type of dehydration reaction occurs regardless of the specific monomers involved and the type of polymer the cell is producing. Therefore, you'll see this type of reaction repeated many times throughout this chapter.

Organisms not only make macromolecules but also break them down. For example, you must digest macromolecules in food to make their monomers available to your cells, which can then rebuild the monomers into the macromolecules that make up your body. Converting macromolecules is like taking apart a car (food) made of interlocking toy blocks and then using these blocks to assemble a new car of your own design (your own body's molecules). The breakdown of polymers occurs by a process called **hydrolysis** (Figure 3.4b). Hydrolysis means to break (*lyse*) with water (*hydro*). Cells break bonds between monomers by adding water to them, a process that is essentially the reverse of a dehydration reaction. Imagine disassembling a long train by throwing water at the junctions between cars. Each time you add water, one more car or set of cars is broken free. You learned in the Biology and Society section about one real-world example of a hydrolysis reaction: the breakdown of lactose into its monomers by the enzyme lactase. ☑

▼ Figure 3.4 **Synthesis and breakdown of polymers.** For simplicity, the only atoms shown in these diagrams are hydrogens and hydroxyl groups (—OH) in strategic locations.

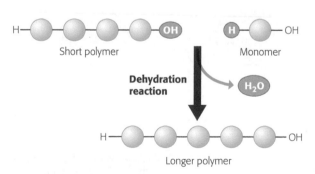

(a) Building a polymer chain. A polymer grows in length when an incoming monomer and the monomer at the end of the polymer each contribute atoms to form a water molecule. The monomers replace the lost covalent bonds with a bond between each other.

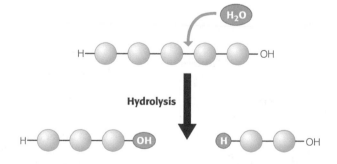

(b) Breaking a polymer chain. Hydrolysis reverses the process by adding a water molecule, which breaks the bond between two monomers, creating two smaller molecules from one bigger one.

Large Biological Molecules

There are four categories of large, important biological molecules found in all living creatures: carbohydrates, lipids, proteins, and nucleic acids. For each category, we'll explore the structure and function of these large molecules by first learning about the smaller molecules used to build them.

Carbohydrates

Carbohydrates, or "carbs," are a class of molecules that includes sugars and polymers of sugars. Some examples are the small sugar molecules dissolved in soft drinks and the long starch molecules in spaghetti and bread. In animals, carbohydrates are a primary source of dietary energy and raw material for manufacturing other kinds of organic compounds; in plants, they serve as a building material for much of the plant body.

Monosaccharides

Simple sugars, or **monosaccharides** (from the Greek *mono*, single, and *sacchar*, sugar), are the monomers of carbohydrates; they cannot be broken down into smaller sugars. Common examples are glucose, found in soft drinks, and fructose, found in fruit. Both of these simple sugars are also in honey (**Figure 3.5**). The molecular formula for glucose is $C_6H_{12}O_6$. Fructose has the same

formula, but its atoms are arranged differently. Glucose and fructose are examples of **isomers**, molecules that have the same molecular formula but different structures. Isomers are like anagrams—words that contain the same letters in a different order, such as *heart* and *earth*. Because molecular shape is so important, seemingly minor differences in the arrangement of atoms give isomers different properties, such as how they react with other molecules. In this case, the rearrangement of functional groups makes fructose taste much sweeter than glucose.

It is convenient to draw sugars as if their carbon skeletons were linear. When dissolved in water, however, many monosaccharides form rings when one end of the molecule forms a bond with another part of the molecule (**Figure 3.6**). You'll notice this ring shape in the depiction of the structure of many carbohydrates in this chapter.

Monosaccharides, particularly glucose, are the main fuel molecules for cellular work. Like an automobile engine consuming gasoline, your cells break down glucose molecules and extract their stored energy, giving off carbon dioxide as "exhaust." The rapid conversion of glucose to cellular energy is why a solution of glucose dissolved in water (often called dextrose) is given as an IV to sick or injured patients; the glucose provides an immediate energy source to tissues in need of repair. ☑

► **Figure 3.5**
Monosaccharides (simple sugars). Glucose and fructose, both of which are found in honey, are isomers, molecules with the same atoms arranged differently.

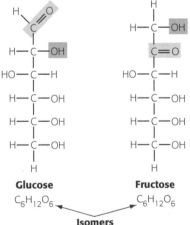

Glucose
$C_6H_{12}O_6$

Fructose
$C_6H_{12}O_6$

Isomers
(same formula, different arrangements)

▼ **Figure 3.6 The ring structure of glucose.**

(a) Linear and ring structures. The carbon atoms are numbered so you can relate the linear and ring versions of the molecule. As the double arrows indicate, ring formation is a reversible process, but at any instant in an aqueous solution, most glucose molecules are rings.

(b) Abbreviated ring structure. In this book, we'll use this abbreviated ring symbol for glucose. Each unmarked corner represents a carbon and its attached atoms.

Disaccharides

A **disaccharide**, or double sugar, is constructed from two monosaccharides by a dehydration reaction. The disaccharide lactose, sometimes called "milk sugar," is made from the monosaccharides glucose and galactose (Figure 3.7). Another common disaccharide is maltose, naturally found in germinating seeds. It is used in making beer, malt whiskey and liquor, malted milk shakes, and malted milk ball candy. A molecule of maltose consists of two glucose monomers joined together.

The most common disaccharide is sucrose (table sugar), which consists of a glucose monomer linked to a fructose monomer. Sucrose is the main carbohydrate in plant sap, and it nourishes all the parts of the plant. Sugar manufacturers extract sucrose from the stems of sugarcane or (much more often in the United States) the roots of sugar beets. Another common sweetener is high-fructose corn syrup (HFCS), made through a commercial process that uses an enzyme to convert natural glucose in corn syrup to the much sweeter fructose. HFCS is a clear, goopy liquid containing about 55% fructose; it is much cheaper than sucrose and easier to mix into drinks and processed foods. If you read the label on a soft drink, you're likely to find that high-fructose corn syrup is one of the first ingredients listed (Figure 3.8).

The average American consumes about 45 kilograms (kg)—a whopping 100 pounds—of sweeteners per year, mainly in the form of sucrose and HFCS. This national "sweet tooth" persists in spite of our growing awareness about how sugar can negatively affect our health. Sugar is a major cause of tooth decay, and overconsumption increases the risk of developing diabetes and heart disease. Moreover, high sugar consumption tends to replace eating more varied and nutritious foods. The description of sugars as "empty calories" is accurate in the sense that most sweeteners contain only negligible

> If you drink coffee, odds are good that you spooned beets into your cup today—in the form of table sugar.

▼ Figure 3.7 Disaccharide (double sugar) formation. To form a disaccharide, two simple sugars are joined by a dehydration reaction, in this case forming a bond between monomers of glucose and galactose to make the double sugar lactose.

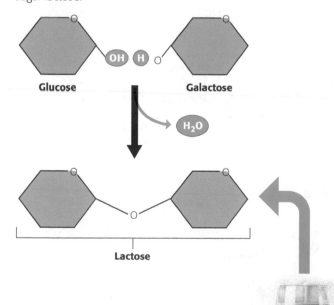

◄ Figure 3.8 High-fructose corn syrup. Many processed foods include high-fructose corn syrup, an artificial sweetener made by chemically treating sugars extracted from corn.

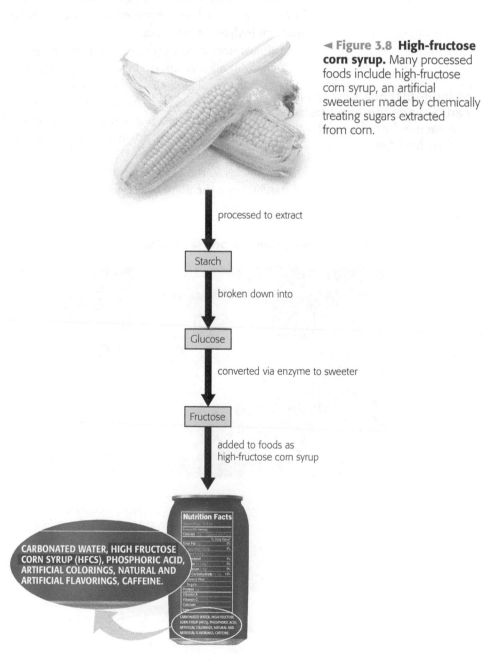

processed to extract

Starch

broken down into

Glucose

converted via enzyme to sweeter

Fructose

added to foods as high-fructose corn syrup

CARBONATED WATER, HIGH FRUCTOSE CORN SYRUP (HFCS), PHOSPHORIC ACID, ARTIFICIAL COLORINGS, NATURAL AND ARTIFICIAL FLAVORINGS, CAFFEINE.

☑ **CHECKPOINT**

How and why do manufacturers produce HFCS?

Answer: They convert glucose to the sweeter fructose. HFCS is cheaper and more easily blended with processed foods.

amounts of nutrients other than carbohydrates. For good health, we also require proteins, fats, vitamins, and minerals. And we need to include substantial amounts of complex carbohydrates—that is, polysaccharides—in our diet. Let's examine these macromolecules next. ☑

Polysaccharides

Complex carbohydrates, or **polysaccharides**, are long chains of sugars—polymers of monosaccharides. One familiar example is starch, a storage polysaccharide found in plants. **Starch** consists of long strings of glucose monomers **(Figure 3.9a)**. Plant cells store starch, providing a sugar stockpile that can be tapped when needed. Potatoes and grains, such as wheat, corn, and rice, are the major sources of starch in our diet. Animals can digest starch because enzymes within their digestive systems break the bonds between glucose monomers through hydrolysis reactions.

Animals store excess glucose in the form of a polysaccharide called **glycogen**. Like starch, glycogen is a polymer of glucose monomers, but glycogen is more extensively branched **(Figure 3.9b)**. Most of your glycogen is stored in liver and muscle cells, which break down the glycogen to release glucose when you need energy. This is why some athletes "carbo-load," consuming large amounts of starchy foods the night before an athletic event. The

> A long-distance runner who carbo-loads the night before a race is banking glycogen to be used the next day.

starch is converted to glycogen, which is then available for rapid use during physical activity the next day.

Cellulose, the most abundant organic compound on Earth, forms cable-like fibrils in the tough walls that enclose plant cells and is a major component of wood and other structural components of plants **(Figure 3.9c)**. We take advantage of that structural strength when we use lumber as a building material. Cellulose is also a polymer of glucose, but its glucose monomers are linked together in a unique way. Unlike the glucose linkages in starch and glycogen, those in cellulose cannot be broken by any enzyme produced by animals. Grazing animals and wood-eating insects such as termites are able to derive nutrition from cellulose because microorganisms inhabiting their digestive tracts break it down. The cellulose in plant foods that you eat, commonly known as dietary fiber (your grandma might call it "roughage") passes through your digestive tract unchanged. Because it remains undigested, fiber does not provide nutrients, but it does help keep your digestive system healthy. The passage of cellulose stimulates cells lining the digestive tract to secrete mucus, which allows food to pass smoothly. The health benefits of dietary fiber include lowering the risk of heart disease, diabetes, and gastrointestinal disease. However, most Americans do not get the recommended levels of fiber in their diet. Foods rich in fiber include fruits and vegetables, whole grains, bran, and beans.

▼ Figure 3.9 **Three common polysaccharides.**

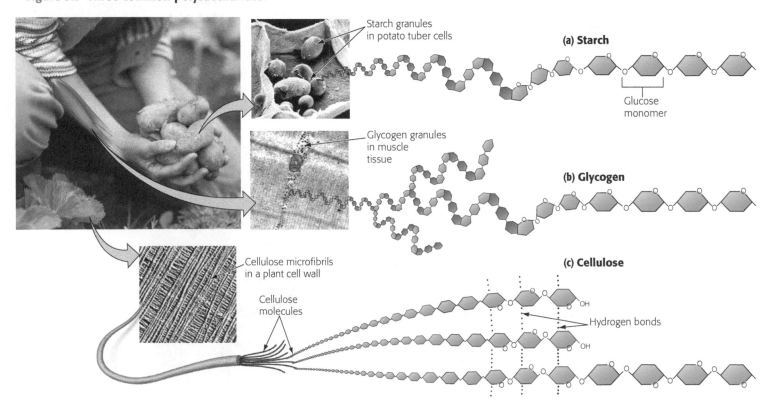

Starch granules in potato tuber cells

Glycogen granules in muscle tissue

Cellulose microfibrils in a plant cell wall

Cellulose molecules

(a) Starch

Glucose monomer

(b) Glycogen

(c) Cellulose

Hydrogen bonds

Lipids

Almost all carbohydrates are **hydrophilic** ("water-loving") molecules that dissolve readily in water. In contrast, **lipids** are **hydrophobic** ("water-fearing"); they do not mix with water. You've probably seen this chemical behavior when you combine oil and vinegar: The oil, which is a type of lipid, separates from the vinegar, which is mostly water **(Figure 3.10)**. If you shake vigorously, you can force a temporary mixture long enough to douse your salad, but what remains in the bottle will quickly separate. Lipids also differ from carbohydrates, proteins, and nucleic acids in that they are neither huge macromolecules nor are they necessarily polymers built from repeating monomers. Lipids are a diverse group of molecules made from different molecular building blocks. In this section, we'll look at two types of lipids: fats and steroids.

Fats

A typical **fat** consists of a glycerol molecule joined with three fatty acid molecules by dehydration reactions **(Figure 3.11a)**. The resulting fat molecule is called a **triglyceride (Figure 3.11b)**, a term you may hear in the results of a blood test. A fatty acid is a long molecule that stores a lot of energy. A pound of fat packs more than twice as much energy as a pound of carbohydrate. The downside to this energy efficiency in fats is that it is very difficult for a person trying to lose weight to "burn off" excess fat. We stock these long-term food stores in specialized reservoirs called adipose cells, which swell and shrink when we deposit and withdraw fat from them. This adipose tissue, or body fat, not only stores energy but also cushions vital organs and insulates us, helping maintain a constant, warm body temperature.

Notice in Figure 3.11b that the bottom fatty acid bends where there is a double bond in the carbon skeleton. That fatty acid is **unsaturated** because it has fewer than the maximum number of hydrogens at the double bond. The other two fatty acids in the fat molecule lack double bonds in their tails. Those fatty acids are **saturated**, meaning that they contain the maximum number of hydrogen atoms, giving them a straight shape. A saturated fat is one with all three of its fatty acid tails saturated. If one or more of the fatty acids is unsaturated, then it's an unsaturated fat, like the one in Figure 3.11b. A polyunsaturated fat has several double bonds within its fatty acids.

Most animal fats, such as lard and butter, have a relatively high proportion of saturated fatty acids. The linear shape of saturated fatty acids allows these molecules to stack easily (like bricks in a wall), so saturated

◄ **Figure 3.10 The separation of hydrophobic (oil) and hydrophilic (vinegar) components in salad dressing.**

Oil (hydrophobic)

Vinegar (hydrophilic)

▼ **Figure 3.11 The synthesis and structure of a triglyceride molecule.**

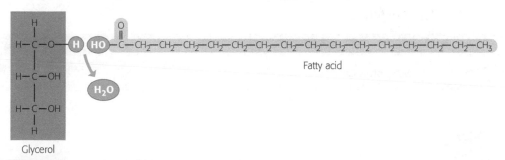

Fatty acid

Glycerol

(a) A dehydration reaction linking a fatty acid to glycerol

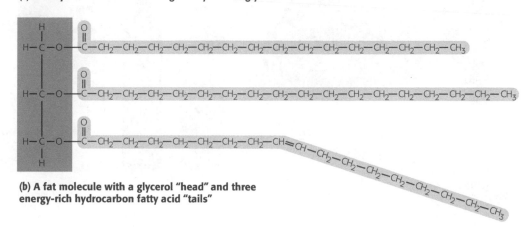

(b) A fat molecule with a glycerol "head" and three energy-rich hydrocarbon fatty acid "tails"

fats tend to be solid at room temperature (Figure 3.12). Diets rich in saturated fats may contribute to cardiovascular disease by promoting atherosclerosis. In this condition, lipid-containing deposits called plaque build up along the inside walls of blood vessels, reducing blood flow and increasing risk of heart attacks and strokes.

Plant and fish fats are relatively high in unsaturated fatty acids. The bent shape of unsaturated fatty acids makes them less likely to form solids (imagine trying to build a wall using bent bricks!). Most unsaturated fats are liquid at room temperature. Fats that are primarily unsaturated include vegetable oils (such as corn and canola oil) and fish oils (such as cod liver oil).

Although plant oils tend to be low in saturated fat, tropical plant fats are an exception. Cocoa butter, a main ingredient in chocolate, contains a mix of saturated and unsaturated fat that gives chocolate a melting point near body temperature. Thus, chocolate stays solid at room temperature but melts in your mouth, creating a pleasing "mouth feel" that is one of the reasons chocolate is so appealing.

Sometimes a food manufacturer wants to use a vegetable oil but needs the food product to be solid, as with margarine or peanut butter. To achieve the desired texture, the manufacturer can convert unsaturated fats to saturated fats by adding hydrogen, a process called **hydrogenation**. Unfortunately, hydrogenation can also create **trans fats**, a type of unsaturated fat that is particularly bad for your health. Since 2006, the U.S. Food and Drug Administration has required that trans fats be listed on nutrition labels. But even if a label on a hydrogenated product says 0 grams (g) of trans fats per serving, the product might actually have up to 0.5 g per serving (which legally can be rounded down to 0). Furthermore, trans fats are often found in fast foods, such as French fries and battered-and-fried foods, that may not have labels. Due to the increased awareness of their unhealthy nature, trans fats are becoming less common as food manufacturers substitute other forms of fat. Indeed, trans fats may soon be a dietary thing of the past. New York City banned restaurants from serving trans fats in 2006, and California followed suit in 2010. The countries of Iceland, Switzerland, and Denmark have effectively removed trans fats from the food supply. In the United States, the Food and Drug Administration has ruled that trans fats are not "generally regarded as safe," a designation that will likely result in a phasing out of trans fats from the American food supply.

Although trans fats should generally be avoided and saturated fats limited, it is not true that *all* fats are unhealthy. In fact, some fats perform important functions within the body and are beneficial and even essential to a healthy diet. For example, fats containing omega-3 fatty acids have been shown to reduce the risk of heart disease and relieve the symptoms of arthritis and inflammatory bowel disease. Some sources of these beneficial fats are nuts and oily fish such as salmon. ☑

Chocolate melts in your mouth (and in your hands) because of the structure of cocoa butter, which is high in saturated fats.

☑ **CHECKPOINT**

What are unsaturated fats? What kinds of unsaturated fat are particularly unhealthy? What kinds of unsaturated fat are most healthful?

Answer: fats with less than the maximum number of hydrogens because of double bonds between some carbons; trans fats; fats containing omega-3 fatty acids

▼ Figure 3.12 **Types of fats.**

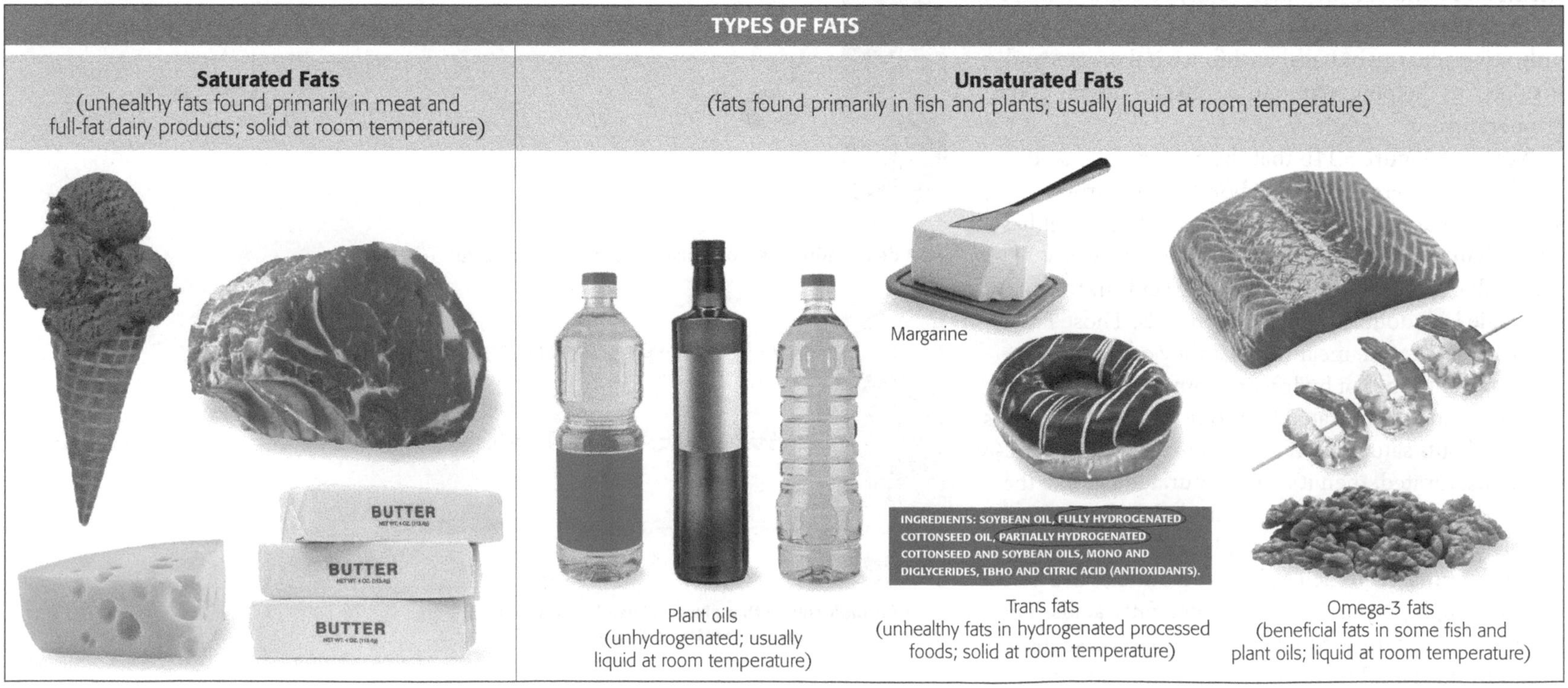

TYPES OF FATS	
Saturated Fats (unhealthy fats found primarily in meat and full-fat dairy products; solid at room temperature)	**Unsaturated Fats** (fats found primarily in fish and plants; usually liquid at room temperature)

Margarine

INGREDIENTS: SOYBEAN OIL, FULLY HYDROGENATED COTTONSEED OIL, PARTIALLY HYDROGENATED COTTONSEED AND SOYBEAN OILS, MONO AND DIGLYCERIDES, TBHQ AND CITRIC ACID (ANTIOXIDANTS).

Plant oils (unhydrogenated; usually liquid at room temperature)

Trans fats (unhealthy fats in hydrogenated processed foods; solid at room temperature)

Omega-3 fats (beneficial fats in some fish and plant oils; liquid at room temperature)

Steroids

Steroids are lipids that are very different from fats in structure and function. All steroids have a carbon skeleton with four fused rings. Different steroids vary in the functional groups attached to this set of rings, and these chemical variations affect their function. One common steroid is cholesterol, which has a bad reputation because of its association with cardiovascular disease. However, cholesterol is a key component of the membranes that surround your cells. It is also the "base steroid" from which your body produces other steroids, such as the hormones estrogen and testosterone (**Figure 3.13**), which are responsible for the development of female and male sex characteristics, respectively.

Anabolic steroids are synthetic variants of testosterone. In human males, testosterone causes buildup of muscle and bone mass during puberty and maintains masculine traits throughout life. Because anabolic steroids resemble testosterone, they mimic some of its effects. Anabolic steroids are prescribed to treat diseases that cause muscle wasting, such as cancer and AIDS. However, individuals sometimes abuse anabolic steroids to build up their muscles quickly. In recent years, many famous athletes have admitted using chemically modified ("designer") performance-enhancing anabolic steroids (**Figure 3.14**). Such revelations have raised questions about the validity of home run records and other athletic accomplishments.

Using anabolic steroids is indeed a fast way to increase body size beyond what hard work can produce. But at what cost? Steroid abuse may cause violent mood swings ("roid rage"), depression, liver damage, high cholesterol, shrunken testicles, reduced sex drive, and infertility. Symptoms related to sexuality occur because artificial anabolic steroids often cause the body to reduce its output of natural sex hormones. Most athletic organizations ban the use of anabolic steroids because of their many potential health hazards coupled with the unfairness of an artificial advantage. ☑

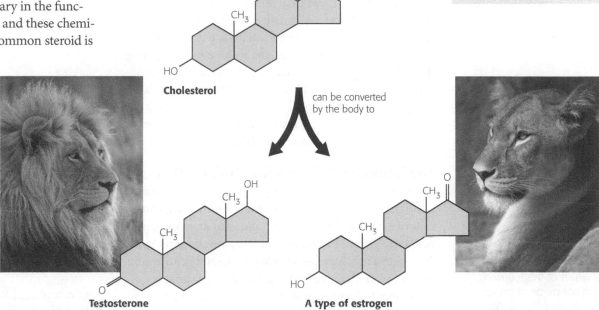

Cholesterol

can be converted by the body to

Testosterone

A type of estrogen

▲ **Figure 3.13 Examples of steroids.** The molecular structures of the steroids shown here are abbreviated by omitting all the atoms that make up the rings. The subtle difference between testosterone and estrogen influences the development of the anatomical and physiological differences between male and female mammals, including lions and humans. This example illustrates the importance of molecular structure to function.

Alex Rodriguez

Mark McGwire

Floyd Landis

Ben Johnson

► **Figure 3.14 Steroids and the modern athlete.** Each of the athletes shown here—baseball players Alex Rodriguez and Mark McGwire, Tour de France cyclist Floyd Landis, and Olympic sprinter Ben Johnson—has admitted to steroid use.

☑ **CHECKPOINT**

What steroid acts as the molecular building block of the human steroid hormones?

Answer: cholesterol

Proteins

A **protein** is a polymer of amino acid monomers. Proteins account for more than 50% of the dry weight of most cells, and they are instrumental in almost everything cells do (**Figure 3.15**). Proteins are the "worker bees" of your body: Chances are, if something is getting done, there is a protein doing it. Your body has tens of thousands of different kinds of proteins, each with a unique three-dimensional shape corresponding to a specific function. In fact, proteins are the most structurally sophisticated molecules in your body.

The Monomers of Proteins: Amino Acids

All proteins are made by stringing together a common set of 20 kinds of amino acids. Every **amino acid** consists of a central carbon atom bonded to four covalent partners. Three of those attachments are common to all 20 amino acids: a carboxyl group (—COOH), an amino group (—NH₂), and a hydrogen atom. The variable component of amino acids is called the side chain (or R group, for radical group); it is attached to the fourth bond of the central carbon (**Figure 3.16a**). Each type of amino acid has a unique side chain which gives that amino acid its special chemical properties (**Figure 3.16b**). Some amino acids have very simple side chains; the amino acid glycine, for example, has a single hydrogen as its side chain. Other amino acids have more complex side chains, some with branches or rings within them. ☑

☑ **CHECKPOINT**

1. Which of the following is *not* made of protein: hair, muscle, cellulose, or enzymes?
2. What are the monomers of all proteins? What is the one part of an amino acid that varies?

Answers: 1. Cellulose is a carbohydrate. 2. amino acids; the side chain

▼ **Figure 3.16 Amino acids.** All amino acids share common functional groups but vary in their side chains.

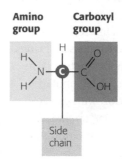

(a) The general structure of an amino acid.

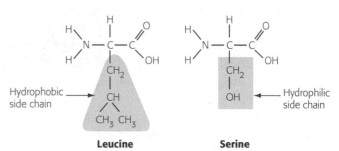

Leucine   Serine

(b) Examples of amino acids with hydrophobic and hydrophilic side chains. The side chain of the amino acid leucine is hydrophobic. In contrast, the side chain of the amino acid serine has a hydroxyl group (—OH), which is hydrophilic.

▼ Figure 3.15 **Some of the varied roles played by proteins.**

MAJOR TYPES OF PROTEINS				
Structural Proteins (provide support)	**Storage Proteins** (provide amino acids for growth)	**Contractile Proteins** (help movement)	**Transport Proteins** (help transport substances)	**Enzymes** (help chemical reactions)
Structural proteins give hair ligaments and horns	Seeds and eggs are rich in storage proteins.	Contractile proteins enable muscles to contract.	The protein hemoglobin within red blood cells transports oxygen.	Some cleaning products use enzymes to help break down molecules.

Structure/Function

Protein Shape

Cells link amino acid monomers together by—can you guess?—dehydration reactions. The bond that joins adjacent amino acids is called a **peptide bond** (**Figure 3.17**). The resulting long chain of amino acids is called a **polypeptide**. A functional protein is one or more polypeptide chains precisely twisted, folded, and coiled into a molecule of unique shape. The difference between a polypeptide and a protein can be likened to the relationship between a long strand of yarn and a sweater. To be functional, the long fiber (the yarn) must be precisely knit into a specific shape (the sweater).

How is it possible to make the huge variety of proteins found in your body from just 20 kinds of amino acids? The answer is arrangement. You know that you can make many different English words by varying the sequence of just 26 letters. Though the protein alphabet is slightly smaller (just 20 "letters"), the "words" are much longer, with a typical polypeptide being hundreds or thousands of amino acids in length. Just as each word is constructed from a unique succession of letters, each protein has a unique linear sequence of amino acids.

The amino acid sequence of each polypeptide determines the three-dimensional structure of the protein. And it is a protein's three-dimensional structure that enables the molecule to carry out its specific function. Nearly all proteins work by recognizing and binding to some other molecule. For example, the specific shape of lactase enables it to recognize and attach to lactose, its molecular target. For all proteins, structure and function are interrelated: What a protein does is a consequence of its shape. The twists and turns of the protein in **Figure 3.18** may appear haphazard, but they represent this protein's specific three-dimensional shape, and without that exact shape, the protein could not do its job.

▼ **Figure 3.18 The structure of a protein.** The chain below, drawn in serpentine fashion so that it fits on the page, shows the amino acid sequence of a polypeptide found in lysozyme, an enzyme in your tears and sweat that helps prevent bacterial infections. The names of the amino acids are given as their three-letter abbreviations; for example, the amino acid alanine is abbreviated "ala." This amino acid sequence folds into a protein of specific shape, shown at bottom in two computer-generated representations. Without this specific shape, the protein could not perform its function.

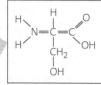

One amino acid (alanine)

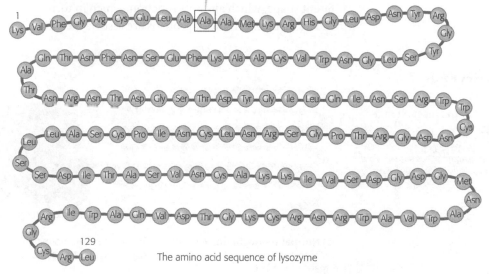

The amino acid sequence of lysozyme

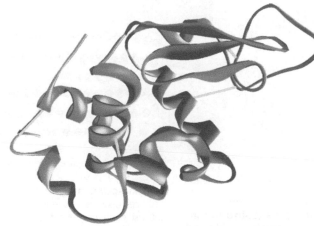

Here you can see how the polypeptide folds into a compact shape.

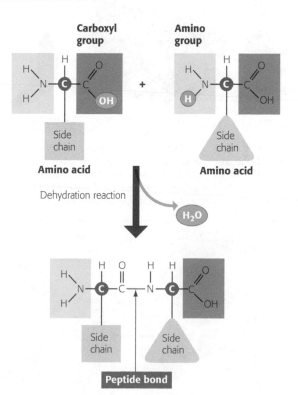

Dehydration reaction

▲ **Figure 3.17 Joining amino acids.** A dehydration reaction links adjacent amino acids by a peptide bond.

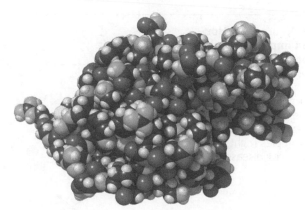

This model allows you to see the details of the protein's structure.

Changing a single letter can drastically affect the meaning of a word—"tasty" versus "nasty," for instance. Similarly, even a slight change in the amino acid sequence can affect a protein's ability to function. For example, the substitution of one amino acid for another at a particular position in hemoglobin, the blood protein that carries oxygen, causes sickle-cell disease, an inherited blood disorder **(Figure 3.19)**. Even though 145 out of the 146 amino acids in hemoglobin are correct, that one change is enough to cause the protein to fold into a different shape, which alters its function, which in turn causes disease. Misfolded proteins are associated with several severe brain disorders. For example, the diseases shown in **Figure 3.20** are all caused by prions, misfolded versions of normal brain proteins. Prions can infiltrate the brain, converting normally folded proteins into the abnormal shape. Clustering of the misfolded proteins eventually disrupts brain function.

In addition to dependence on the amino acid sequence, a protein's shape is sensitive to the environment. An unfavorable change in temperature, pH, or some other factor can cause a protein to unravel. If you cook an egg, the transformation of the egg white from clear to opaque is caused by proteins in the egg white coming apart. One of the reasons why extremely high fevers are so dangerous is that some proteins in the body lose their shape above 104°F.

What determines a protein's amino acid sequence? The amino acid sequence of each polypeptide chain is specified by a gene. And this relationship between genes and proteins brings us to this chapter's last category of large biological molecules: nucleic acids. ☑

☑ CHECKPOINT

How can changing an amino acid alter the function of a protein?

Answer: Changing an amino acid may alter the shape of the protein, which changes its function.

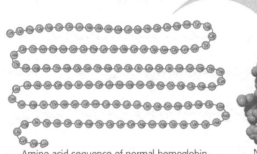

Amino acid sequence of normal hemoglobin

Normal hemoglobin peptide

SEM 500×

Normal red blood cell

(a) Normal hemoglobin. Red blood cells of humans are normally disk-shaped. Each cell contains millions of molecules of the protein hemoglobin, which transports oxygen from the lungs to other organs of the body.

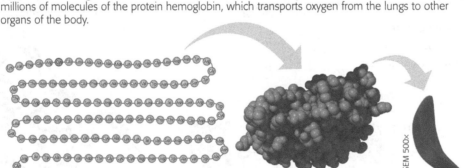

Amino acid sequence of sickle-cell hemoglobin

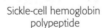

Sickle-cell hemoglobin polypeptide

SEM 500×

Sickled red blood cell

(b) Sickle-cell hemoglobin. A slight change in the amino acid sequence of hemoglobin causes sickle-cell disease. The inherited substitution of one amino acid—valine in place of the amino acid glutamic acid—occurs at the 6th position of the polymer. Such abnormal hemoglobin molecules tend to crystallize, deforming some of the cells into a sickle shape. Someone with the disease suffers from dangerous episodes when the angular cells clog tiny blood vessels, impeding blood flow.

Prions in cattle cause mad cow disease, formally called bovine spongiform encephalopathy (BSE).

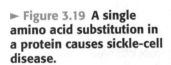

► **Figure 3.19 A single amino acid substitution in a protein causes sickle-cell disease.**

► **Figure 3.20 How an improperly folded protein can lead to brain disease.** Here, you can see how prion proteins bring about the destruction of brain tissue, as well as several examples (right) of the resulting disease.

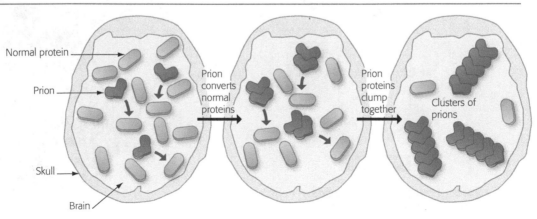

Normal protein

Prion

Skull

Brain

Prion converts normal proteins

Prion proteins clump together

Clusters of prions

First described among one tribe in Papua New Guinea in the 1950s, kuru is a human brain disease transmitted by cannibalistic ingestion of an infected person's brain, which transfers the prion to a new host.

Prions cause fatal weight loss in several animals, including deer, elk, and moose.

Nucleic Acids

Nucleic acids are macromolecules that store information and provide the instructions for building proteins. The name *nucleic* comes from the fact that DNA is found in the nuclei of eukaryotic cells. There are actually two types of nucleic acids: **DNA** (which stands for deoxyribonucleic acid) and **RNA** (for ribonucleic acid). The genetic material that humans and all other organisms inherit from their parents consists of giant molecules of DNA. The DNA resides in the cell as one or more very long fibers called chromosomes. A **gene** is a unit of inheritance encoded in a specific stretch of DNA that programs the amino acid sequence of a polypeptide. Those programmed instructions, however, are written in a chemical code that must be translated from "nucleic acid language" to "protein language" **(Figure 3.21)**. A cell's RNA molecules help make this translation (see Chapter 10).

Nucleic acids are polymers made from monomers called **nucleotides (Figure 3.22)**. Each nucleotide contains three parts. At the center of each nucleotide is a five-carbon sugar (blue in the figure), deoxyribose in DNA and ribose in RNA. Attached to the sugar is a negatively charged phosphate group (yellow) containing a phosphorus atom bonded to oxygen atoms (PO_4^-). Also attached to the sugar is a nitrogen-containing base (green) made of one or two rings. The sugar and phosphate are the same in all nucleotides; only the base varies. Each DNA nucleotide has one of four possible nitrogenous bases: adenine (abbreviated A), guanine (G), cytosine (C), or thymine (T) **(Figure 3.23)**. Thus, all genetic information is written in a four-letter alphabet.

▼ **Figure 3.21 Building a protein.** Within the cell, a gene (a segment of DNA) provides the directions to build a molecule of RNA, which can then be translated into a protein.

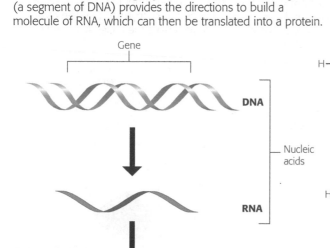

▼ **Figure 3.22 A DNA nucleotide.** A DNA nucleotide monomer consists of three parts: a sugar (deoxyribose), a phosphate, and a nitrogenous (nitrogen-containing) base.

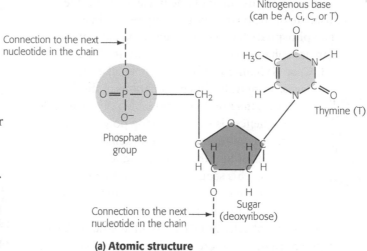

(a) Atomic structure

(b) Symbol used in this book

▼ **Figure 3.23 The nitrogenous bases of DNA.** Notice that adenine and guanine have double-ring structures. Thymine and cytosine have single-ring structures.

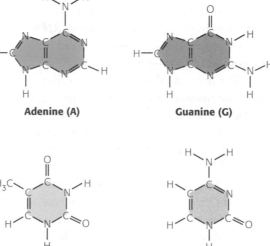

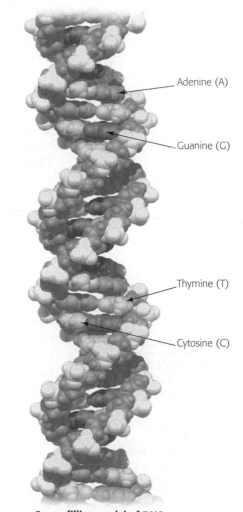

Space-filling model of DNA (showing the four bases in four different colors)

Dehydration reactions link nucleotide monomers into long chains called polynucleotides (Figure 3.24a). In a polynucleotide, nucleotides are joined by covalent bonds between the sugar of one nucleotide and the phosphate of the next. This bonding results in a **sugar-phosphate backbone**, a repeating pattern of sugar-phosphate-sugar-phosphate, with the bases (A, T, C, or G) hanging off the backbone like appendages. With different combinations of the four bases, the number of possible polynucleotide sequences is vast. One long polynucleotide may contain many genes, each a specific series of hundreds or thousands of nucleotides. This sequence is a code that provides instructions for building a specific polypeptide from amino acids.

> The structure of your DNA is nearly indistinguishable from the DNA of a mosquito or an elephant: The differences in animal species result from the way nucleotides are arranged.

A molecule of cellular DNA is double-stranded, with two polynucleotide strands coiled around each other to form a **double helix (Figure 3.24b)**. Think of a candy cane or a barber pole that has two intertwined spirals, one red and one white. In the central core of the helix (corresponding to the interior of the candy cane), the bases along one DNA strand hydrogen-bond to bases along the other strand. The bonds are individually weak, but collectively they zip the two strands together into a very stable double helix formation. To understand how DNA strands are bonded, think of Velcro, in which two strips are held together by hook-and-loop bonds, each of which is weak but which collectively form a tight grip. Because of the way the functional groups hang off the bases, the base pairing in a DNA double helix is specific: The base A can pair only with T, and G can pair only with C. Thus, if you know the sequence of bases along one DNA strand, you also know the sequence along the complementary strand in the double helix. This unique base pairing is the basis of DNA's ability to act as the molecule of inheritance (as discussed in Chapter 10).

There are many similarities between DNA and RNA. Both are polymers of nucleotides, for example, and both are made of nucleotides consisting of a sugar, a phosphate, and a base. But there are three important differences. (1) As its name *ribonucleic acid* denotes, its sugar is ribose rather than deoxyribose. (2) Instead of the base thymine, RNA has a similar but distinct base called uracil (U) (Figure 3.25). Except for the presence of ribose and uracil, an RNA polynucleotide chain is identical to a DNA polynucleotide chain. (3) RNA is usually found in living cells in single-stranded form, whereas DNA usually exists as a double helix.

Now that we've examined the structure of nucleic acids, we'll look at how a change in nucleotide sequence can affect protein production. To illustrate this point, we'll return to a familiar condition. ☑

☑ CHECKPOINT

1. DNA contains _____ polynucleotide strands, each composed of _____ kinds of nucleotides. (Provide two numbers.)

2. If one DNA strand has the sequence GAATGC, what is the sequence of the other strand?

Answers: 1. two; four 2. CTTACG

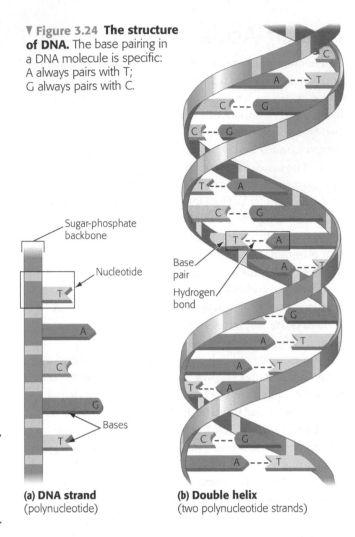

▼ Figure 3.24 **The structure of DNA.** The base pairing in a DNA molecule is specific: A always pairs with T; G always pairs with C.

Sugar-phosphate backbone

Nucleotide

Bases

Base pair

Hydrogen bond

(a) **DNA strand** (polynucleotide)

(b) **Double helix** (two polynucleotide strands)

▼ Figure 3.25 **An RNA nucleotide.** Notice that this RNA nucleotide differs from the DNA nucleotide in Figure 3.22 in two ways: The RNA sugar is ribose rather than deoxyribose, and the base is uracil (U) instead of thymine (T). The other three kinds of RNA nucleotides have the bases A, C, and G, as in DNA.

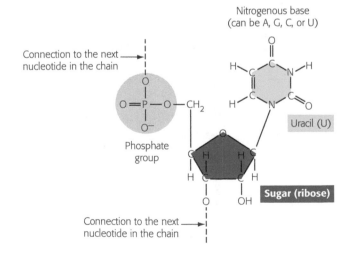

Nitrogenous base (can be A, G, C, or U)

Connection to the next nucleotide in the chain

Phosphate group

Uracil (U)

Sugar (ribose)

Connection to the next nucleotide in the chain

Does Lactose Intolerance Have a Genetic Basis?

The enzyme lactase, like all proteins, is encoded by a DNA gene. A reasonable hypothesis is that lactose-intolerant people have a defect in their lactase gene. However, this hypothesis is not supported by **observation**. Even though lactose intolerance runs in families, most lactose-intolerant people have a normal version of the lactase gene. This raises the following **question**: What is the genetic basis for lactose intolerance?

A group of Finnish and American scientists proposed the **hypothesis** that lactose intolerance can be correlated with a single nucleotide at a particular site within one chromosome. They made the **prediction** that this site would be near, though not within, the lactase gene. In their **experiment**, they examined the genes of 196 lactose-intolerant people from nine Finnish families. Their **results** showed a 100% correlation between lactose intolerance and a nucleotide at a site approximately

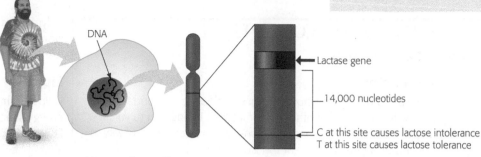

Human cell (DNA in 46 chromosomes) Chromosome 2 (one DNA molecule) Section of chromosome 2

DNA

Lactase gene

14,000 nucleotides

C at this site causes lactose intolerance
T at this site causes lactose tolerance

14,000 nucleotides away from the lactase gene—a relatively short distance in terms of the whole chromosome (Figure 3.26). Other experiments showed that depending on the nucleotide sequence within this region of the DNA molecule, the action of the lactase gene is ramped up or down (in a way that likely involves producing a regulatory protein that interacts with the nucleotides near the lactase gene). This study shows how a small change in a DNA nucleotide sequence can have a major effect on the production of a protein and the well-being of an organism.

◄ **Figure 3.26 A genetic cause of lactose intolerance.** A research study showed a correlation between lactose intolerance and a nucleotide at a specific location on one chromosome.

The Evolution of Lactose Intolerance in Humans

As you'll recall from the Biology and Society section, most people in the world are lactose intolerant as adults and thus do not easily digest the milk sugar lactose. In fact, lactose intolerance is found in 80% of African Americans and Native Americans and 90% of Asian Americans but only in about 10% of Americans of northern European descent. And as discussed in the Process of Science section, lactose intolerance appears to have a genetic basis.

From an evolutionary perspective, it is reasonable to infer that lactose intolerance is rare among northern Europeans because the ability to tolerate lactose offered a survival advantage to their ancestors. In northern Europe's relatively cold climate, only one crop harvest a year is possible. Therefore, herd animals were a main source of food for early humans in that region. Cattle were first domesticated in northern Europe about 9,000 years ago (Figure 3.27). With milk and other dairy products at hand year-round, natural selection would have favored

anyone with a mutation that kept the lactase gene switched on beyond infanthood. In cultures where dairy products were not a staple in the diet, natural selection would not favor such a mutation.

Researchers wondered whether the genetic basis for lactose tolerance in northern Europeans might be present in other cultures that kept dairy herds. To find out, they compared the genetic makeup and lactose tolerance of 43 ethnic groups in East Africa. The researchers identified three other genetic changes that keep the lactase gene permanently active. These mutations appear to have occurred beginning around 7,000 years ago, about the time that archaeological evidence shows domestication of cattle in these African regions.

Genetic changes that confer a selective advantage, such as surviving cold winters or withstanding drought by drinking milk, spread rapidly in these early peoples. Whether or not you can digest milk is therefore an evolutionary record of the cultural history of your ancestors.

▲ **Figure 3.27 A prehistoric cave painting in Lascaux, France, of wild cattle.** The cow-like animal on the right is an auroch, the first species of domesticated cattle in Europe. Aurochs migrated from Asia about 250,000 years ago but became extinct in 1627.

Chapter Review

SUMMARY OF KEY CONCEPTS

Organic Compounds

Carbon Chemistry

Carbon atoms can form large, complex, diverse molecules by bonding to four potential partners, including other carbon atoms. In addition to variations in the size and shape of carbon skeletons, organic compounds vary in the presence and locations of different functional groups.

Giant Molecules from Smaller Building Blocks

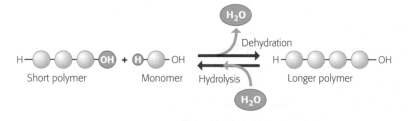

Large Biological Molecules

Large Biological Molecules	Functions	Components	Examples
Carbohydrates	Dietary energy; storage; plant structure	Monosaccharide	Monosaccharides: glucose, fructose; disaccharides: lactose, sucrose; polysaccharides: starch, cellulose
Lipids	Long-term energy storage (fats); hormones (steroids)	Components of a triglyceride	Fats (triglycerides); steroids (testosterone, estrogen)
Proteins	Enzymes, structure, storage, contraction, transport, etc.	Amino acid	Lactase (an enzyme); hemoglobin (a transport protein)
Nucleic acids	Information storage	Nucleotide	DNA, RNA

Carbohydrates

Simple sugars (monosaccharides) provide cells with energy and building materials. Double sugars (disaccharides), such as sucrose, consist of two monosaccharides joined by a dehydration reaction. Polysaccharides are long polymers of sugar monomers. Starch in plants and glycogen in animals are storage polysaccharides. The cellulose of plant cell walls, which is indigestible by animals, is an example of a structural polysaccharide.

Lipids

Lipids are hydrophobic. Fats, a type of lipid, are the major form of long-term energy storage in animals. A molecule of fat, or triglyceride, consists of three fatty acids joined by dehydration reactions to a molecule of glycerol. Most animal fats are saturated, meaning that their fatty acids have the maximum number of hydrogens. Plant oils contain mostly unsaturated fats, having fewer hydrogens in the fatty acids because of double bonding in the carbon skeletons. Steroids, including cholesterol and the sex hormones, are also lipids.

Proteins

There are 20 types of amino acids, the monomers of proteins. They are linked by dehydration reactions to form polymers called polypeptides. A protein consists of one or more polypeptides folded into a specific three-dimensional shape. The shape of a protein determines its function. Changing the amino acid sequence of a polypeptide may alter the shape and therefore the function of the protein. Shape is sensitive to environment, and if a protein loses its shape because of an unfavorable environment, its function may also be lost.

Nucleic Acids

Nucleic acids include RNA and DNA. DNA takes the form of a double helix, two DNA strands (polymers of nucleotides) held together by hydrogen bonds between nucleotide components called bases. There are four kinds of DNA bases: adenine (A), guanine (G), thymine (T), and cytosine (C). A always pairs with T, and G always pairs with C. These base-pairing rules enable DNA to act as the molecule of inheritance. RNA has U (uracil) instead of T.

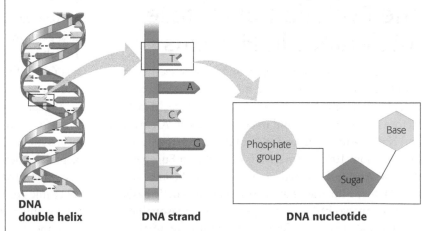

DNA double helix   **DNA strand**   **DNA nucleotide**

MasteringBiology®

For practice quizzes, BioFlix animations, MP3 tutorials, video tutors, and more study tools designed for this textbook, go to MasteringBiology®

SELF-QUIZ

1. One isomer of methamphetamine is the addictive illegal drug known as "crank." Another isomer is a medicine for sinus congestion. How can you explain the differing effects of the two isomers?

2. Monomers are joined together to form larger polymers through _____. Such a reaction releases a molecule of _____.

3. Polymers are broken down into the monomers that make them up through the chemical reaction called _____.

4. Table sugar is _____.
   a. glucose, a monosaccharide
   b. glucose, a disaccharide
   c. sucrose, a monosaccharide
   d. sucrose, a disaccharide

5. When two molecules of glucose ($C_6H_{12}O_6$) are joined together by a dehydration reaction, what are the formulas of the two products? (*Hint*: No atoms are gained or lost.)

6. One molecule of dietary fat is made by joining three molecules of _____ to one molecule of _____. What is the formal name of the resulting molecule?

7. By definition, what type of fatty acid has double bonds?
   a. steroid
   b. triglyceride
   c. unsaturated
   d. saturated

8. Humans and other animals cannot digest wood because they
   a. cannot digest any carbohydrates.
   b. cannot chew it fine enough.
   c. lack the enzyme needed to break down cellulose.
   d. get no nutrients from it.

9. Explain how it could be possible to change an amino acid within a protein but not affect that protein's function.

10. Most proteins can easily dissolve in water. Knowing that, where within the overall three-dimensional shape of a protein would you most likely find hydrophobic amino acids?

11. A shortage of phosphorus in the soil would make it especially difficult for a plant to manufacture
    a. DNA.
    b. proteins.
    c. cellulose.
    d. fatty acids.

12. Nucleic acids are polymers of _____ monomers.

13. Name three similarities between DNA and RNA. Name three differences.

14. What is the structure of a gene? What is the function of a gene?

Answers to these questions can be found in Appendix: Self-Quiz Answers.

THE PROCESS OF SCIENCE

15. A food manufacturer is advertising a new cake mix as fat-free. Scientists at the Food and Drug Administration (FDA) are testing the product to see if it truly lacks fat. Hydrolysis of the cake mix yields glucose, fructose, glycerol, a number of amino acids, and several kinds of molecules with long chains. Further analysis shows that most of the chains have a carboxyl group at one end. What would you tell the food manufacturer if you were a spokesperson for the FDA?

16. Based on your knowledge of the different types of chemical bonds (see Chapter 2) and of the molecular structures involved, predict which of the families of macromolecules discussed in this chapter are soluble in water and which are not, and explain why.

17. **Interpreting Data** Below is a typical food label for one cookie. One gram of fat packs 9 Calories, and 1 gram of carbohydrates or protein packs 4 Calories. The top of the label shows that each cookie contains 140 Calories in total. What percentage of the Calories in this cookie are from fat, carbohydrates, and protein?

Nutrition Facts
Serving Size 1 Cookie (20 g /1 oz)
Servings Per Container 8

Amount Per Serving

Calories 140	Calories from Fat 60
	% Daily Value*
Total Fat 7g	11%
Saturated Fat 3g	15%
Trans Fat 0g	
Cholesterol 10mg	3%
Sodium 80mg	3%
Total Carbohydrate 18g	6%
Dietary Fiber 1g	4%
Sugars 10g	
Protein 2g	

BIOLOGY AND SOCIETY

18. Some amateur and professional athletes take anabolic steroids to help them build strength ("bulk up"). The health risks of this practice are extensively documented. Apart from these health issues, what is your opinion about the ethics of athletes using chemicals to enhance performance? Is this a form of cheating, or is it just part of the preparation required to stay competitive in a sport where anabolic steroids are commonly used? Defend your opinion.

19. Heart disease is the leading cause of death among people in the United States and other industrialized nations. Fast food is a major source of unhealthy fats that contribute significantly to heart disease. Imagine you're a juror sitting on a trial where a fast-food manufacturer is being sued for producing a harmful product. To what extent do you think manufacturers of unhealthy foods should be held responsible for the health consequences of their products? As a jury member, how would you vote?

20. The production of genetically modified (GM) organisms is a controversial topic, and many consumers avoid products that contain GM food. Based on what you have learnt about the chemical nature of DNA and proteins, do you think there is a health risk associated with the consumption of GM food? What may be the wider implications of the production and consumption of such food?

4 A Tour of the Cell

Why Cells Matter

Mushrooms, ▶ amoebas, and you are all made up of the same kind of cells.

The caffeine that gives your cup of tea a kick also protects tea plants ▼ from herbivores.

▲ Without the cytoskeleton, your cells would collapse in on themselves, much like a house collapses when the infrastructure fails.

Humans Versus Bacteria BIOLOGY AND SOCIETY

Antibiotics: Drugs That Target Bacterial Cells

Antibiotics—drugs that disable or kill infectious bacteria—are a marvel of modern medicine. The first antibiotic to be discovered was penicillin in 1920. A revolution in human health rapidly followed. Fatality rates of many diseases (such as bacterial pneumonia and surgical infections) plummeted, saving millions of lives. In fact, human health care improved so quickly and so profoundly that some doctors in the early 1900s predicted the end of infectious diseases altogether. Alas, this did not come to pass (see the Evolution Connection section in Chapter 13 for a discussion of why infectious diseases were not so easily defeated).

The goal of antibiotic treatment is to knock out invading bacteria while doing no damage to the human host. So how does an antibiotic zero in on its target among trillions of human cells? Most antibiotics are so precise because they bind to structures found only in bacterial cells. For example, the common antibiotics erythromycin and streptomycin bind to the bacterial ribosome, a vital cellular structure responsible for the production of proteins. The ribosomes of humans are different enough from those of bacteria that the antibiotics bind only to bacterial ribosomes, leaving human ribosomes unaffected. Ciprofloxacin (commonly referred to as Cipro) is the antibiotic of choice to combat anthrax-causing bacteria. This drug targets an enzyme that bacteria need to maintain their chromosome structure. Your cells can survive just fine in the presence of Cipro because human chromosomes have a sufficiently different makeup than bacterial chromosomes. Other drugs, such as penicillin, ampicillin, and bacitracin, disrupt the synthesis of cell walls, a structure found in most bacteria that is absent from the cells of humans and other animals.

This discussion of how various antibiotics target bacteria underscores the main point of this chapter: To understand how life works—whether in bacteria or in your own body—you first need to learn about cells. On the scale of biological organization, cells occupy a special place: They are the simplest objects that can be alive. Nothing smaller than a cell is capable of displaying all of life's properties. In this chapter, we'll explore the microscopic structure and function of cells. Along the way, we'll further consider how the ongoing battle between humans and infectious bacteria is affected by the cellular structures present on both sides.

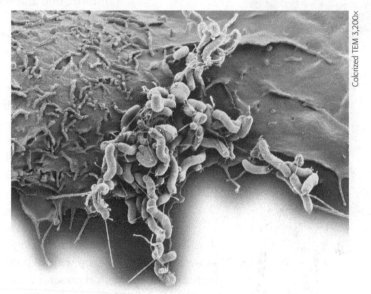

Colorized TEM 3,200x

Two kinds of cells. In this micrograph, *Helicobacter pylori* bacteria (green) can be seen mingling with cells within the human stomach. This species of bacteria causes stomach ulcers.

The Microscopic World of Cells

Each cell in your body is a miniature marvel. If the world's most sophisticated jumbo jet was reduced to microscopic size, its intricacy would pale next to a living cell.

Organisms are either single-celled, such as most prokaryotes and protists, or multicelled, such as plants, animals, and most fungi. Your own body is a cooperative society of trillions of cells of many specialized types. As you read this page, muscle cells allow you to scan your eye across the words, while sensory cells in your eye gather information and send it to brain cells, which interpret the words. Everything you do—every action and every thought—is possible because of processes that occur at the cellular level.

Figure 4.1 shows the size range of cells compared with objects both larger and smaller. Notice that the scale along the left side of the figure increases by powers of 10 to accommodate the range of sizes shown. Starting at the top with 10 meters (m), each subsequent mark represents a tenfold decrease in length. Most cells are between 1 and 100 μm in diameter (yellow region of the figure) and are therefore visible only with a microscope. There are some interesting exceptions: an ostrich egg is a single cell about 6 inches across and weighing about 3 pounds; nerve cells in your body can stretch several feet long; and nerve cells in giant squid can be more than 30 feet long!

How do new living cells arise? First formulated in the 1800s, the **cell theory** states that all living things are composed of cells and that all cells come from earlier cells. So every cell in your body (and in every other living organism on Earth) was formed by division of a previously living cell. (That raises an obvious question: how did the first cell evolve? This fascinating topic will be addressed in Chapter 15.) With that introduction, let's begin to explore the variety of cells found among life on Earth.

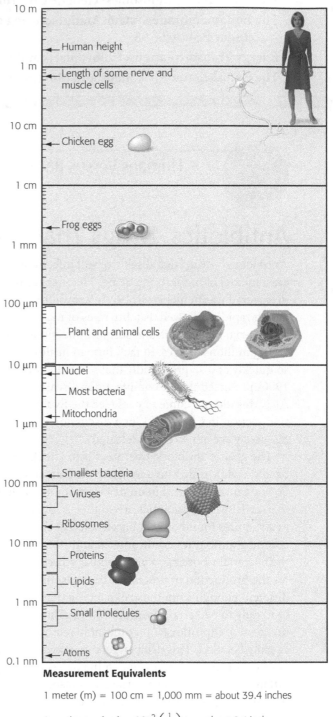

Measurement Equivalents

1 meter (m) = 100 cm = 1,000 mm = about 39.4 inches

1 centimeter (cm) = $10^{-2} \left(\frac{1}{100}\right)$ m = about 0.4 inch

1 millimeter (mm) = $10^{-3} \left(\frac{1}{1,000}\right)$ m = $\frac{1}{10}$ cm

1 micrometer (μm) = 10^{-6} m = 10^{-3} mm

1 nanometer (nm) = 10^{-9} m = 10^{-3} μm

▲ **Figure 4.1 The size range of cells.** Starting at the top of this scale with 10 m (10 meters) and going down, each measurement along the left side marks a tenfold decrease in size.

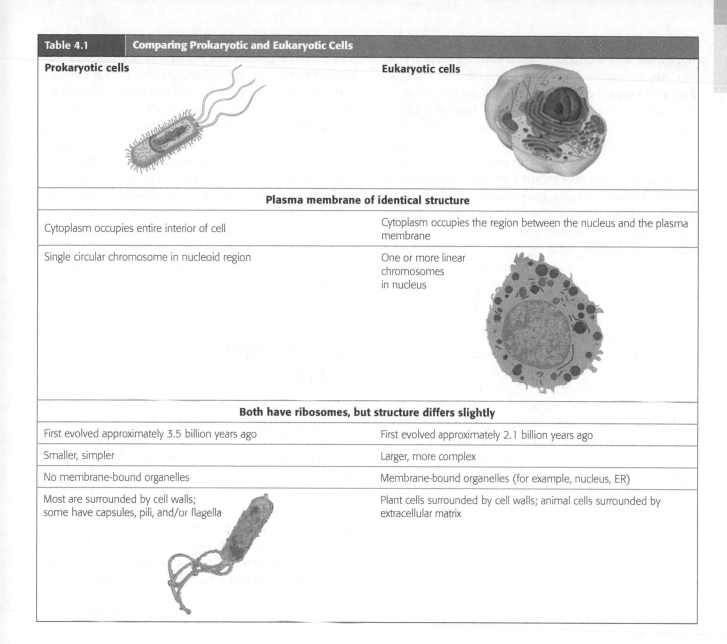

Table 4.1	Comparing Prokaryotic and Eukaryotic Cells
Prokaryotic cells	**Eukaryotic cells**
Plasma membrane of identical structure	
Cytoplasm occupies entire interior of cell	Cytoplasm occupies the region between the nucleus and the plasma membrane
Single circular chromosome in nucleoid region	One or more linear chromosomes in nucleus
Both have ribosomes, but structure differs slightly	
First evolved approximately 3.5 billion years ago	First evolved approximately 2.1 billion years ago
Smaller, simpler	Larger, more complex
No membrane-bound organelles	Membrane-bound organelles (for example, nucleus, ER)
Most are surrounded by cell walls; some have capsules, pili, and/or flagella	Plant cells surrounded by cell walls; animal cells surrounded by extracellular matrix

The Two Major Categories of Cells

The countless cells that exist on Earth can be placed into two basic categories: prokaryotic cells and eukaryotic cells (Table 4.1). **Prokaryotic cells** are found in organisms of the domains Bacteria and Archaea, known as prokaryotes (see Figure 1.7). Organisms of the domain Eukarya—including protists, plants, fungi, and animals—are composed of **eukaryotic cells** and are called eukaryotes.

All cells, whether prokaryotic or eukaryotic, have several features in common. They are all bounded by a barrier called a **plasma membrane**, which regulates the traffic of molecules between the cell and its surroundings. Inside all cells is a thick, jellylike fluid called the **cytosol**, in which cellular components are suspended. All cells have one or more **chromosomes** carrying genes made of DNA. And all cells have **ribosomes** that build proteins according to instructions from the genes. Because of structural differences between bacteria and eukaryotes, as mentioned in the Biology and Society section at the beginning of the chapter, some antibiotics—such as streptomycin—target prokaryotic ribosomes, crippling protein synthesis in the bacterial invaders but not in the eukaryotic host (you).

Although they have many similarities, prokaryotic and eukaryotic cells differ in several important ways. Fossil evidence shows that prokaryotes were the first life on Earth, appearing more than

Mushrooms, amoebas, and you are all made up of the same kind of cells.

3.5 billion years ago. In contrast, the first eukaryotes did not appear until around 2.1 billion years ago. Prokaryotic cells are usually much smaller—about one-tenth the length of a typical eukaryotic cell—and are simpler in structure. Think of a prokaryotic cell as being like a bicycle, whereas a eukaryotic cell is like a sports utility vehicle. Both a bike and an SUV get you from place to place, but one is much smaller and contains many fewer parts than the other. Similarly, prokaryotic cells and eukaryotic cells perform similar functions, but prokaryotes cells are much smaller and less complex. The most significant structural difference between the two types of cells is that eukaryotic cells have **organelles** ("little organs"), membrane-enclosed structures that perform specific functions, and prokaryotic cells do not. The most important organelle is the **nucleus**, which houses most of a eukaryotic cell's DNA. The nucleus is surrounded by a double membrane. A prokaryotic cell lacks a nucleus; its DNA is coiled into a "nucleus-like" region called the **nucleoid**, which is not partitioned from the rest of the cell by membranes.

Consider this analogy: A eukaryotic cell is like an office building that is separated into cubicles. Within each cubicle, a specific function is performed, thus dividing the labor among many internal compartments. One cubicle may hold the accounting department, for example, while another is home to the sales force. The "cubicle walls" within eukaryotic cells are made from membranes that help maintain a unique chemical environment inside each cubicle. In contrast, the interior of a prokaryotic cell is like an open warehouse. The spaces for specific tasks within a "prokaryotic warehouse" are distinct but they are not separated by physical barriers.

Figure 4.2 depicts an idealized prokaryotic cell and a micrograph of an actual bacterium. Surrounding the plasma membrane of most prokaryotic cells is a rigid cell wall, which protects the cell and helps maintain its shape. Recall from the Biology and Society section at the beginning of the chapter that bacterial cell walls are the targets of some antibiotics. In some prokaryotes, a sticky outer coat called a capsule surrounds the cell wall. Capsules provide protection and help prokaryotes stick to surfaces and to other cells in a colony. For example, capsules help bacteria in your mouth stick together to form harmful dental plaque. Prokaryotes can have short projections called pili, which can also attach to surfaces. Many prokaryotic cells have flagella, long projections that propel them through their liquid environment. ☑

An Overview of Eukaryotic Cells

All eukaryotic cells—whether from animals, plants, protists, or fungi—are fundamentally similar to one another and quite different from prokaryotic cells. **Figure 4.3** provides overviews of an idealized animal cell and plant cell. No real cell looks quite like these because living cells have many more copies of most of the structures shown; each of your cells have hundreds of mitochondria and millions of ribosomes, for example. To keep from getting lost on our tour of the cell, throughout this chapter we'll use miniature versions of the diagrams in Figure 4.3 as road maps, highlighting the structure we're

▼ **Figure 4.2 A prokaryotic cell.** A drawing of an idealized prokaryotic cell (right) is shown alongside a micrograph of *Helicobacter pylori* (left), a bacterium that causes stomach ulcers.

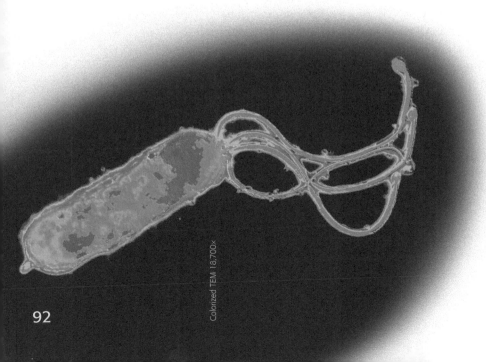

Colorized TEM 18,700×

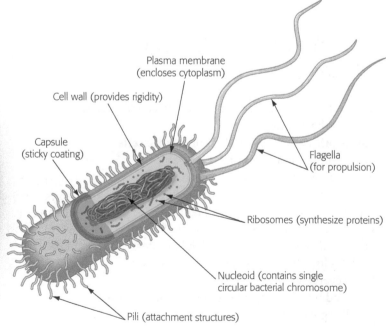

Plasma membrane (encloses cytoplasm)

Cell wall (provides rigidity)

Capsule (sticky coating)

Flagella (for propulsion)

Ribosomes (synthesize proteins)

Nucleoid (contains single circular bacterial chromosome)

Pili (attachment structures)

discussing. Notice that the structures are color-coded; we'll use this color scheme throughout this book.

The region of the cell outside the nucleus and within the plasma membrane is called the **cytoplasm**. (This term is also used to refer to the interior of a prokaryotic cell.) The cytoplasm of a eukaryotic cell consists of various organelles suspended in the liquid cytosol. As you can see in Figure 4.3, most organelles are found in both animal and plant cells. But you'll notice some important differences—for example, only plant cells have chloroplasts (where photosynthesis occurs) and a cell wall (which provides stiffness to plant structures), and only animal cells have lysosomes (bubbles of digestive enzymes surrounded by membranes). In the rest of this section, we'll take a closer look at the architecture of eukaryotic cells, beginning with the plasma membrane. ☑

▼ **Figure 4.3 An idealized animal cell and plant cell.** For now, the labels on the drawings are just words, but these organelles will come to life as we take a closer look at how each part of the cell functions.

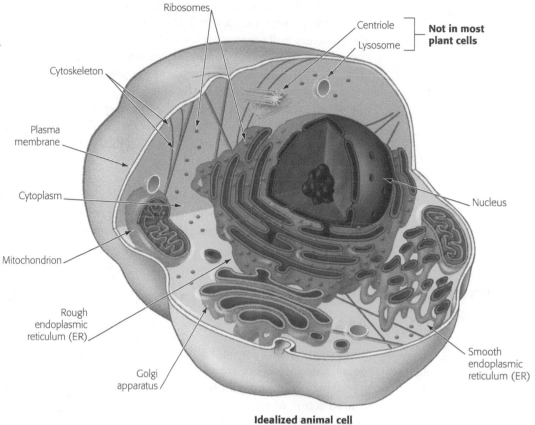

Ribosomes · Centriole · Lysosome — **Not in most plant cells** · Cytoskeleton · Plasma membrane · Cytoplasm · Nucleus · Mitochondrion · Rough endoplasmic reticulum (ER) · Golgi apparatus · Smooth endoplasmic reticulum (ER)

Idealized animal cell

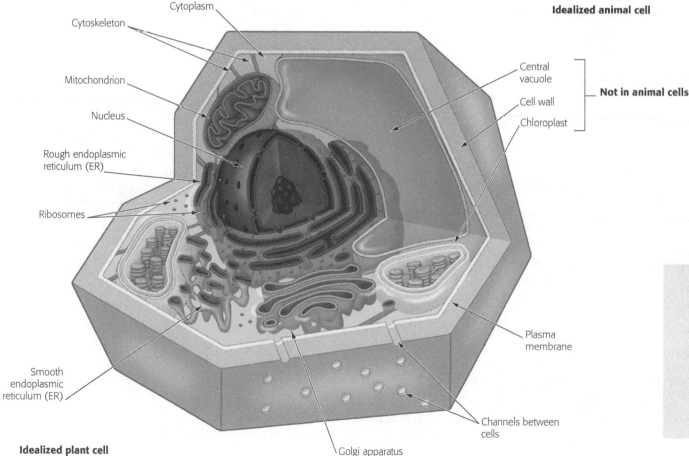

Cytoplasm · Cytoskeleton · Mitochondrion · Nucleus · Rough endoplasmic reticulum (ER) · Ribosomes · Smooth endoplasmic reticulum (ER) · Central vacuole · Cell wall · Chloroplast — **Not in animal cells** · Plasma membrane · Channels between cells · Golgi apparatus

Idealized plant cell

☑ **CHECKPOINT**

1. Name three structures in plant cells that animal cells lack.
2. Name two structures that may be found in animal cells but not in plant cells.

Answers: 1. chloroplasts, a central vacuole, and a cell wall 2. centrioles and lysosomes

Membrane Structure

Before we enter the cell to explore the organelles, let's make a quick stop at the surface of this microscopic world: the plasma membrane. To help you understand the structure and function of the cell membrane, imagine you want to create a new homestead in the wilderness. You will probably want to start by fencing your property to protect it from the outside world. Similarly, the plasma membrane is the boundary that separates the living cell from its nonliving surroundings. The plasma membrane is a remarkable film, so thin that you would have to stack 8,000 of these membranes to equal the thickness of one piece of paper. Yet the plasma membrane can regulate the traffic of chemicals into and out of the cell. As with all things biological, the structure of the plasma membrane correlates with its function.

a major constituent of biological membranes. Each phospholipid is composed of two distinct regions—a "head" with a negatively charged phosphate group and two nonpolar fatty acid "tails." Phospholipids group together to form a two-layer sheet called a **phospholipid bilayer**. As you can see in **Figure 4.4a**, the phospholipids' hydrophilic ("water-loving") heads are arranged to face outward, exposed to the aqueous solutions on both sides of a membrane. Their hydrophobic ("water-fearing") tails are arranged inward, mingling with each other and shielded from water. Suspended in the phospholipid bilayer of most membranes are proteins that help regulate traffic across the membrane; these proteins also perform other functions (**Figure 4.4b**). (You'll learn more about membrane proteins in Chapter 5.)

Membranes are not static sheets of molecules locked rigidly in place, however. In fact, the texture of a cellular membrane is similar to salad oil. The phospholipids and most of the proteins can therefore drift about within the membrane. Thus, a membrane is a **fluid mosaic**—fluid because the molecules can move freely past one another and mosaic because of the diversity of proteins that float like icebergs in the phospholipid sea. Next, we'll see how some bacteria can cause illness by piercing the plasma membrane. ☑

☑ CHECKPOINT

What function does the organization of phospholipids into a bilayer in water serve?

Answer: The bilayer structure shields the hydrophobic tails of the phospholipids from water while exposing the hydrophilic heads to water.

⬤ Structure/ Function — The Plasma Membrane

The plasma membrane and other membranes of the cell are composed mostly of **phospholipids**. The structure of phospholipid molecules is well suited to their function as

▼ Figure 4.4 **The plasma membrane structure.**

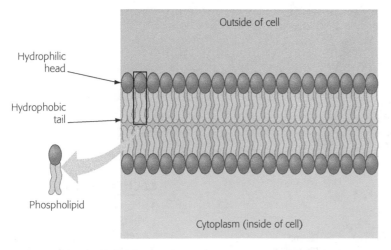

(a) Phospholipid bilayer of membrane. In water, phospholipids arrange themselves into a bilayer. The symbol for a phospholipid that we'll use in this book looks like a lollipop with two wavy sticks. The "head" is the end with the phosphate group, and the two "tails" are chains of carbon and hydrogen. The bilayer arrangement keeps the heads exposed to water while keeping the tails in the oily interior of the membrane.

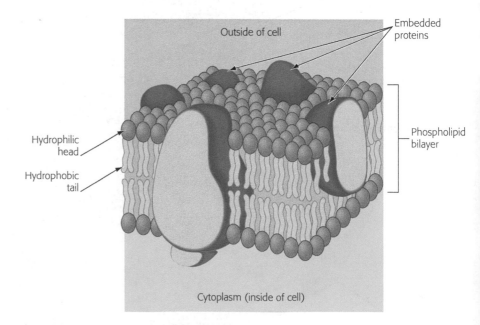

(b) Fluid mosaic model of membrane. Membrane proteins, like the phospholipids, have both hydrophilic and hydrophobic regions.

Humans Versus Bacteria THE PROCESS OF SCIENCE

What Makes a Superbug?

Some bacteria cause disease by rupturing the plasma membrane of human immune cells. One example is a common bacteria called *Staphylococcus aureus* (commonly referred to as "staph" or SA). These bacteria, found on your skin, are usually harmless but may multiply and spread, causing a "staph infection." Staph infections typically occur in hospitals and can cause serious, even life-threatening conditions, such as pneumonia or necrotizing fasciitis ("flesh-eating disease").

Most staph infections can be treated with antibiotics. But particularly dangerous strains of *S. aureus*—known as MRSA (for multidrug-resistant SA)—are unaffected by all of the commonly used antibiotics. In recent years, MRSA infections have become more common in hospitals, gyms, and schools. In one study, scientists from the National Institutes of Health (NIH) studied a particular deadly MRSA strain. They began with the **observation** that other bacteria use a protein called PSM to disable human immune cells by forming holes that rip apart the plasma membrane. This observation led them to **question** whether PSM plays a role in MRSA infection **(Figure 4.5)**. Their **hypothesis** was that MRSA bacteria lacking the ability to produce PSM would be less deadly than normal MRSA strains that produced PSM.

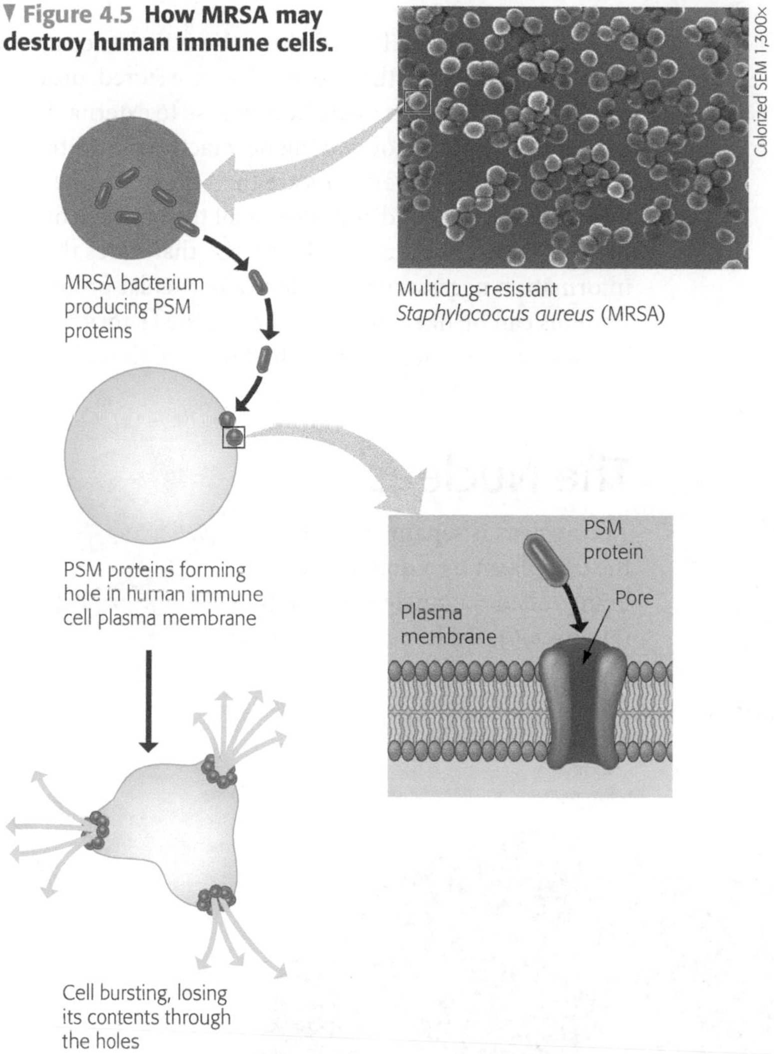

In their **experiment**, scientists infected seven mice with a normal MRSA strain and eight mice with a MRSA strain genetically engineered to not produce PSM. The **results** were striking: All seven mice infected with the normal MRSA strain died, while five of the eight mice infected with the strain that did not produce PSM survived. Immune cells from all of the dead mice had holes within the plasma membrane. The researchers concluded that normal MRSA strains use the

▼ **Figure 4.5 How MRSA may destroy human immune cells.**

MRSA bacterium producing PSM proteins

Multidrug-resistant *Staphylococcus aureus* (MRSA)

Colorized SEM 1,300×

PSM proteins forming hole in human immune cell plasma membrane

Plasma membrane

PSM protein

Pore

Cell bursting, losing its contents through the holes

membrane-destroying PSM protein, but other factors must come into play because three mice died even in the absence of PSM. The deadly effects of MRSA are therefore a reminder of the critical role played by the plasma membrane and another example of the ongoing battle between humans and disease-causing bacteria.

Cell Surfaces

Surrounding their plasma membranes, plant cells have a cell wall made from cellulose fibers, which are long chains of polysaccharide (see Figure 3.9c). The walls protect the cells, maintain cell shape, and keep cells from absorbing so much water that they burst. Plant cells are connected to each other via channels that pass through the cell walls, joining the cytoplasm of each cell to that of its neighbors. These channels allow water and other small molecules to move between cells, integrating the activities of a tissue.

Animal cells lack a cell wall, but most animal cells secrete a sticky coat called the **extracellular matrix**. Fibers made of the protein collagen (also found in your skin, cartilage, bones, and tendons) hold cells together in tissues and can also have protective and supportive functions. In addition, the surfaces of most animal cells contain cell junctions, structures that connect cells together into tissues, allowing the cells to function in a coordinated way. ☑

☑ **CHECKPOINT**

What polysaccharide is the primary component of plant cell walls?

Answer: cellulose

95

The Nucleus and Ribosomes: Genetic Control of the Cell

If you think of the cell as a factory, then the nucleus is its control center. Here, the master plans are stored, orders are given, changes are made in response to external factors, and the process of making new factories is initiated. The factory supervisors are the genes, the inherited DNA molecules that direct almost all the business of the cell. Each gene is a stretch of DNA that stores the information necessary to produce a particular protein. Proteins can be likened to workers on the factory floor because they do most of the actual work of the cell.

☑ **CHECKPOINT**

What is the relationship between chromosomes, chromatin, and DNA?

Answer: *Chromosomes are made of chromatin, which is a combination of DNA and proteins.*

The Nucleus

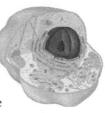

The nucleus is separated from the cytoplasm by a double membrane called the **nuclear envelope** (Figure 4.6). Each membrane of the

nuclear envelope is similar in structure to the plasma membrane: a phospholipid bilayer with associated proteins. Pores in the envelope allow certain materials to pass between the nucleus and the surrounding cytoplasm. (As you'll soon see, among the most important materials that passes between the nucleus and cytoplasm through the nuclear pores are molecules of RNA that carry instructions for building proteins.) Within the nucleus, long DNA molecules and associated proteins form fibers called **chromatin**. Each long chromatin fiber constitutes one chromosome (Figure 4.7). The number of chromosomes in a cell depends on the species; for example, each human body cell has 46 chromosomes, whereas rice cells have 24 and dog cells have 78 (see Figure 8.2 for more examples). The **nucleolus** (shown in Figure 4.6), a prominent structure within the nucleus, is the site where the components of ribosomes are made. We'll examine ribosomes next. ☑

▼ Figure 4.6 **The nucleus.**

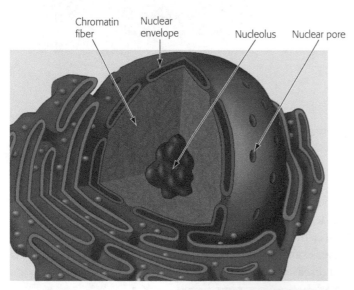

Chromatin fiber — Nuclear envelope — Nucleolus — Nuclear pore

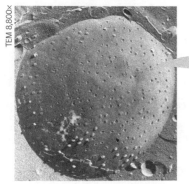

Surface of nuclear envelope

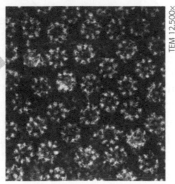

Nuclear pores

▼ Figure 4.7 **The relationship between DNA, chromatin, and a chromosome.**

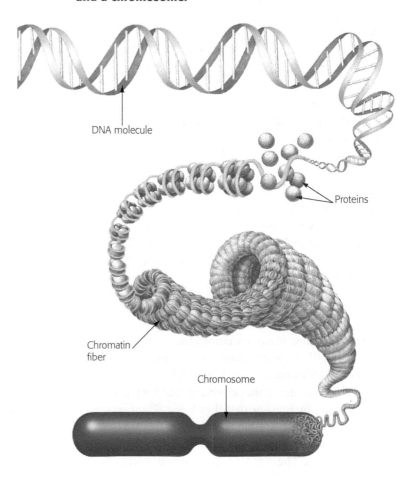

DNA molecule

Proteins

Chromatin fiber

Chromosome

Ribosomes

The small blue dots in the cells in Figure 4.3 and outside the nucleus in Figure 4.6 represent the ribosomes. Ribosomes are responsible for protein synthesis (Figure 4.8). In eukaryotic cells, the components of ribosomes are made in the nucleus and then transported through the pores of the nuclear envelope into the cytoplasm. It is in the cytoplasm that the ribosomes begin their work. Some ribosomes are suspended in the cytosol, making proteins that remain within the fluid of the cell. Other ribosomes are attached to the outside of the nucleus or an organelle called the endoplasmic reticulum (Figure 4.9), making proteins that are incorporated into membranes or secreted by the cell. Free and bound ribosomes are structurally identical, and ribosomes can switch locations, moving between the endoplasmic reticulum and the cytosol. Cells that make a lot of proteins have a large number of ribosomes. For example, each cell in your pancreas that produces digestive enzymes may contain a few million ribosomes.

How DNA Directs Protein Production

Like a company executive, the DNA doesn't actually do any of the work of the cell. Instead, the DNA "executive" issues orders that result in work being done by the protein "workers." Figure 4.10 shows the sequence of events during protein production in a eukaryotic cell (with the DNA and other structures being shown disproportionately large in relation to the nucleus). ❶ DNA transfers its coded information to a molecule called messenger RNA (mRNA). Like a middle manager, the mRNA molecule carries the order to "build this type of protein." ❷ The mRNA exits the nucleus through pores in the nuclear envelope and travels to the cytoplasm, where it binds to a ribosome. ❸ The ribosome moves along the mRNA, translating the genetic message into a protein with a specific amino acid sequence. (You'll learn how the message is translated in Chapter 10.) In this way, information carried by the DNA can direct the work of the entire cell without the DNA ever leaving the protective confines of the nucleus. ☑

☑ CHECKPOINT

1. What is the function of ribosomes?
2. What is the role of mRNA in making a protein?

Answers: 1. protein synthesis 2. A molecule of mRNA carries the genetic message from a gene (DNA) to ribosomes that translate it into protein.

► Figure 4.8 **A computer model of a ribosome in the process of synthesizing a protein.**

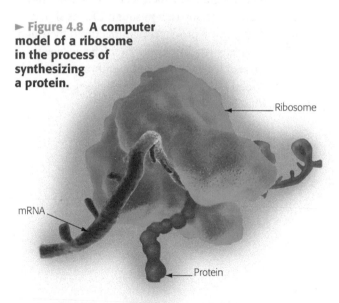

Ribosome

mRNA

Protein

▼ Figure 4.9 **ER-bound ribosomes.**

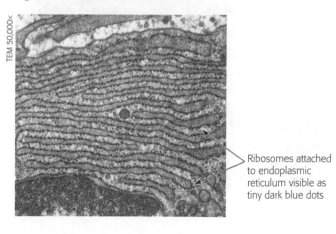

TEM 50,000×

Ribosomes attached to endoplasmic reticulum visible as tiny dark blue dots

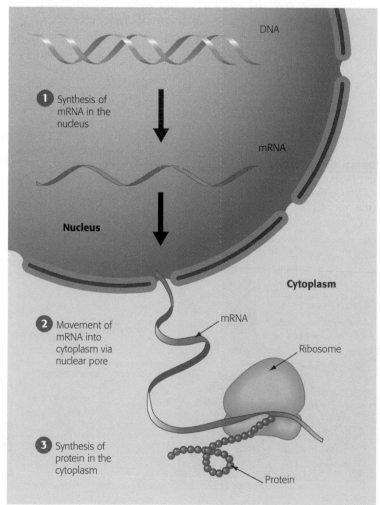

◄ Figure 4.10 **DNA → RNA → protein.** Inherited genes in the nucleus control protein production and hence the activities of the cell.

DNA

❶ Synthesis of mRNA in the nucleus

mRNA

Nucleus

Cytoplasm

❷ Movement of mRNA into cytoplasm via nuclear pore

mRNA

Ribosome

❸ Synthesis of protein in the cytoplasm

Protein

The Endomembrane System: Manufacturing and Distributing Cellular Products

Like an office partitioned into cubicles, the cytoplasm of a eukaryotic cell is partitioned by organelle membranes (see Figure 4.3). Some of the membranes are physically connected, and others are linked by **vesicles** (sacs made of membrane) that transfer membrane segments between organelles. Together, these organelles form the **endomembrane system**. This system includes the nuclear envelope, the endoplasmic reticulum, the Golgi apparatus, lysosomes, and vacuoles.

The Endoplasmic Reticulum

The **endoplasmic reticulum (ER)** is one of the main manufacturing facilities within a cell. It produces an enormous variety of molecules. Connected to the nuclear envelope, the ER forms an extensive labyrinth of tubes and sacs running throughout the cytoplasm **(Figure 4.11)**. A membrane separates the internal ER compartment from the cytosol. There are two components that make up the ER: rough ER and smooth ER. These two types of ER are physically connected but differ in structure and function.

Rough ER

The "rough" in **rough ER** refers to ribosomes that stud the outside of its membrane. One of the functions of rough ER is to make more membrane. Phospholipids made by enzymes of the rough ER are inserted into the ER membrane. In this way, the ER membrane grows, and portions of it can bubble off and be transferred to other parts of the cell. The ribosomes attached to the rough ER produce proteins that will be inserted into the growing ER membrane, transported to other organelles, and eventually exported. Cells that secrete a lot of protein—such as the cells of your salivary glands, which secrete enzymes into your mouth—are especially rich in rough ER. As shown in **Figure 4.12,** ❶ some products manufactured by rough ER are ❷ chemically modified and then ❸ packaged into **transport vesicles**, sacs made of membrane that bud off from the rough ER. The transport vesicles may then be ❹ dispatched to other locations in the cell.

▼ Figure 4.11 **The endoplasmic reticulum (ER).** In this drawing, the flattened sacs of rough ER and the tubes of smooth ER are connected. Notice that the ER is also connected to the nuclear envelope (the nucleus has been omitted from the illustration for clarity).

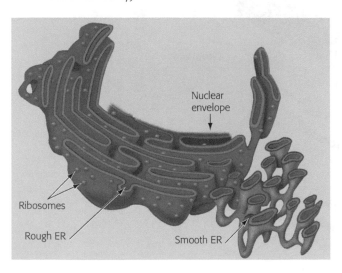

▼ Figure 4.12 **How rough ER manufactures and packages secretory proteins.**

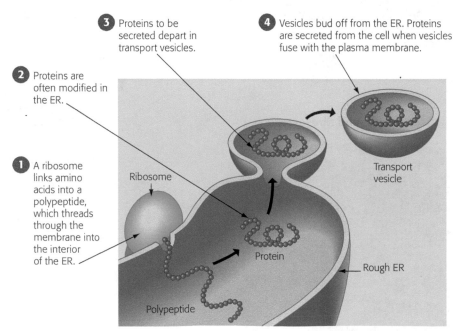

❸ Proteins to be secreted depart in transport vesicles.

❹ Vesicles bud off from the ER. Proteins are secreted from the cell when vesicles fuse with the plasma membrane.

❷ Proteins are often modified in the ER.

❶ A ribosome links amino acids into a polypeptide, which threads through the membrane into the interior of the ER.

Ribosome

Transport vesicle

Protein

Rough ER

Polypeptide

Smooth ER

The "smooth" in **smooth ER** refers to the fact that this organelle lacks the ribosomes that populate the surface of rough ER (see Figure 4.11). A diversity of enzymes built into the smooth ER membrane enables this organelle to perform many functions. One is the synthesis of lipids, including steroids (see Figure 3.13). For example, the cells in ovaries or testes that produce the steroid sex hormones are enriched with smooth ER. In liver cells, enzymes of the smooth ER detoxify circulating drugs such as barbiturates, amphetamines, and some antibiotics (which is why antibiotics don't persist in the bloodstream after combating an infection). As liver cells are exposed to a drug, the amounts of smooth ER and its detoxifying enzymes increase. This can strengthen the body's tolerance of the drug, meaning that higher doses will be required in the future to achieve the desired effect. The growth of smooth ER in response to one drug can also increase tolerance of other drugs. For example, barbiturate use, such as frequently taking sleeping pills, may make certain antibiotics less effective by accelerating their breakdown in the liver.

The Golgi Apparatus

Working in close partnership with the ER, the **Golgi apparatus**, an  organelle named for its discoverer (Italian scientist Camillo Golgi), who first described this structure in 1898, receives, refines, stores, and distributes chemical products of the cell (**Figure 4.13**). You can think of the Golgi apparatus as a detailing facility that receives shipments of newly manufactured cars (proteins), puts on the finishing touches, stores the completed cars, and then ships them out when needed.

Products made in the ER reach the Golgi apparatus in transport vesicles. The Golgi apparatus consists of a stack of membrane plates, looking much like a pile of pita bread. ❶ One side of a Golgi stack serves as a receiving dock for vesicles from the ER. ❷ Proteins within a vesicle are usually modified by enzymes during their transit from the receiving to the shipping side of the Golgi apparatus. For example, molecular identification tags may be added that serve to mark and sort protein molecules into different batches for different destinations. ❸ The shipping side of a Golgi stack is a depot from which finished products can be carried in transport vesicles to other organelles or to the plasma membrane. Vesicles that bind with the plasma membrane transfer proteins to it or secrete finished products to the outside of the cell. ☑

☑ CHECKPOINT

1. What makes rough ER rough?
2. What is the relationship between the Golgi apparatus and the ER in a protein-secreting cell?

Answers: 1. ribosomes attached to the membrane 2. The Golgi apparatus receives proteins from the ER via vesicles, processes the proteins, and then dispatches them in vesicles.

▼ **Figure 4.13 The Golgi apparatus.** The Golgi consists of flattened sacs arranged something like a stack of pita bread. The number of stacks in a cell (from a few to hundreds) correlates with how active the cell is in secreting proteins.

"Receiving" side of the Golgi apparatus

Colorized SEM 130,000×

New vesicle forming

Transport vesicle from rough ER

1

"Receiving" side of the Golgi apparatus

2

New vesicle forming

3

Transport vesicle from the Golgi apparatus

"Shipping" side of the Golgi apparatus

Plasma membrane

Lysosomes

A **lysosome** is a membrane-enclosed sac of digestive enzymes found in animal cells. Most plant cells don't contain lysosomes. Lysosomes develop from vesicles that bud off from the Golgi apparatus. Enzymes within a lysosome can break down large molecules such as proteins, polysaccharides, fats, and nucleic acids. The lysosome provides a compartment where the cell can digest these molecules safely, without unleashing the digestive enzymes on the cell itself.

Lysosomes have several types of digestive functions. Many single-celled protists engulf nutrients into tiny cytoplasmic sacs called food vacuoles. Lysosomes fuse with the food vacuoles, exposing the food to digestive enzymes **(Figure 4.14a)**. Small molecules that result from this digestion, such as amino acids, leave the lysosome and nourish the cell. Lysosomes also help destroy harmful bacteria. For example, your white blood cells ingest bacteria into vacuoles, and lysosomal enzymes that are emptied into these vacuoles rupture the bacterial cell walls. In addition, without harming the cell, a lysosome can engulf and digest parts of another organelle, essentially recycling it by making its molecules available for the construction of new organelles **(Figure 4.14b)**. With the help of lysosomes, a cell can thereby continually renew itself. Lysosomes also have sculpting functions in embryonic development. In an early human embryo, lysosomes release enzymes that digest webbing between fingers of the developing hand.

The importance of lysosomes to cell function and human health is made clear by hereditary disorders called lysosomal storage diseases. A person with such a disease is missing one or more of the digestive enzymes normally found within lysosomes. The abnormal lysosomes become engorged with indigestible substances, and this eventually interferes with other cellular functions. Most of these diseases are fatal in early childhood. In Tay-Sachs disease, for example, lysosomes lack a lipid-digesting enzyme. As a result, nerve cells die as they accumulate excess lipids, ravaging the nervous system. Fortunately, lysosomal storage diseases are rare. ✓

✓ CHECKPOINT

How can defective lysosomes result in excess accumulation of a particular chemical compound in a cell?

Answer: If the lysosomes lack an enzyme needed to break down the compound, the cell will accumulate an excess of that compound.

▼ Figure 4.14 **Two functions of lysosomes.**

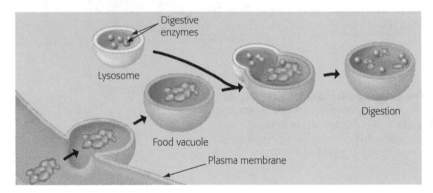

Digestive enzymes

Lysosome

Food vacuole

Plasma membrane

Digestion

(a) A lysosome digesting food

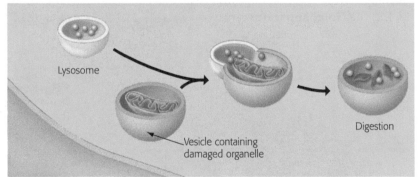

Lysosome

Vesicle containing damaged organelle

Digestion

(b) A lysosome breaking down the molecules of damaged organelles

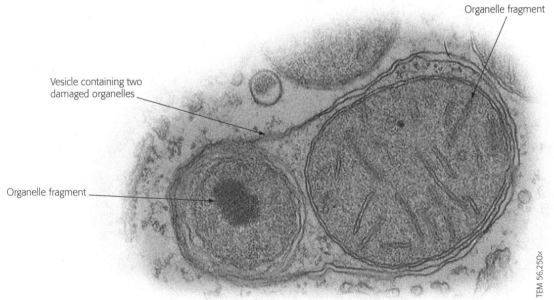

Organelle fragment

Vesicle containing two damaged organelles

Organelle fragment

TEM 56,250×

Vacuoles

Vacuoles are large sacs made of membrane that bud off from the ER or Golgi apparatus. Vacuoles have a variety of functions. For example, Figure 4.14a shows a food vacuole budding from the plasma membrane. Certain freshwater protists have contractile vacuoles that pump out excess water that flows into the cell (**Figure 4.15a**).

Another type of vacuole is a **central vacuole**, a versatile compartment that can account for more than half the volume of a mature plant cell (**Figure 4.15b**). A central vacuole stores organic nutrients, such as proteins stockpiled in the vacuoles of seed cells. It also contributes to plant growth by absorbing water and causing cells to expand. In the cells of flower petals, central

The caffeine that gives your cup of tea a kick also protects tea plants from herbivores.

vacuoles may contain pigments that attract pollinating insects. Central vacuoles may also contain poisons that protect against plant-eating animals. Some important crop plants produce and store large amounts of toxic chemicals—harmful to animals that might graze on the plant but useful to us—such as tobacco plants (which store nicotine) and coffee and tea plants (which store caffeine).

Figure 4.16 will help you review how organelles of the endomembrane system are related. Note that a product made in one part of the endomembrane system may exit the cell or become part of another organelle without crossing a membrane. Also note that membrane made by the ER can become part of the plasma membrane through the fusion of a transport vesicle. In this way, even the plasma membrane is related to the endomembrane system. ☑

☑ **CHECKPOINT**

Place the following cellular structures in the order they would be used in the production and secretion of a protein: Golgi apparatus, nucleus, plasma membrane, ribosome, transport vesicle.

Answer: nucleus, ribosome, transport vesicle, Golgi apparatus, plasma membrane

▼ **Figure 4.15 Two types of vacuoles.**

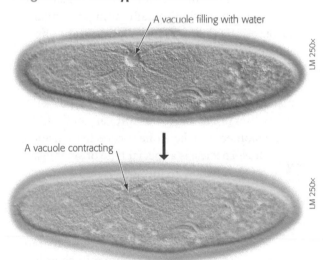

A vacuole filling with water

A vacuole contracting

LM 250x

LM 250x

(a) Contractile vacuole in *Paramecium*. A contractile vacuole fills with water and then contracts to pump the water out of the cell.

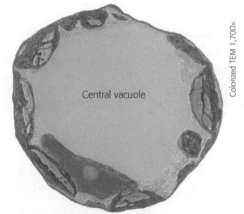

Central vacuole

Colorized TEM 1,700x

(b) Central vacuole in a plant cell. The central vacuole (colorized blue in this micrograph) is often the largest organelle in a mature plant cell.

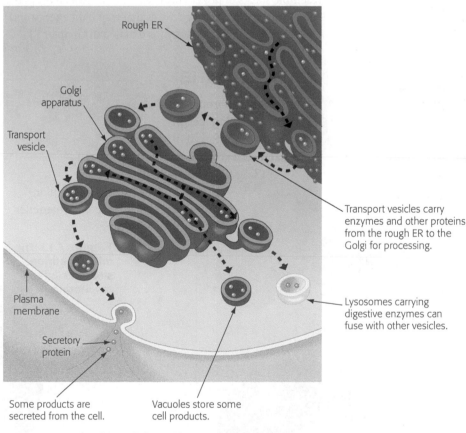

Rough ER

Golgi apparatus

Transport vesicle

Plasma membrane

Secretory protein

Transport vesicles carry enzymes and other proteins from the rough ER to the Golgi for processing.

Lysosomes carrying digestive enzymes can fuse with other vesicles.

Some products are secreted from the cell.

Vacuoles store some cell products.

▲ **Figure 4.16 Review of the endomembrane system.** The dashed arrows show some of the pathways of cell product distribution and membrane migration via transport vesicles.

⚙ Energy Transformations Chloroplasts and Mitochondria

One of the central themes of biology is the transformation of energy: how it enters living systems, is converted from one form to another, and is eventually given off as heat. To follow the energy through living systems, we must consider the two organelles that act as cellular power stations: chloroplasts and mitochondria.

Chloroplasts

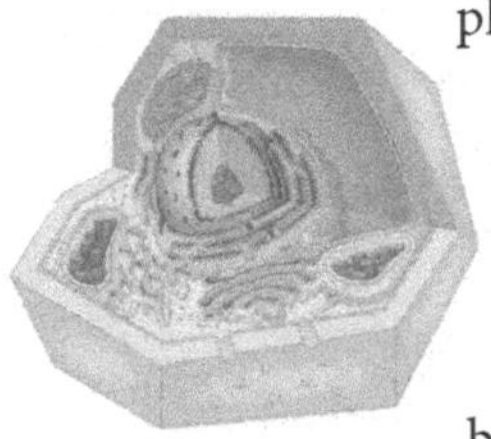

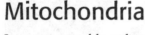

Most of the living world runs on the energy provided by photosynthesis, the conversion of light energy from the sun to the chemical energy of sugar and other organic molecules. **Chloroplasts**, which are unique to the photosynthetic cells of plants and algae, are the organelles that perform photosynthesis.

A chloroplast is partitioned into compartments by two membranes, one inside the other (Figure 4.17). The stroma is a thick fluid found inside the innermost membrane. Suspended in that fluid, the interior of a network of membrane-enclosed disks and tubes forms another compartment. Notice in Figure 4.17 that the disks occur in interconnected stacks called grana (singular, *granum*) that resemble stacks of poker chips. The grana are a chloroplast's solar power packs, the structures that trap light energy and convert it to chemical energy (as detailed in Chapter 7).

Mitochondria

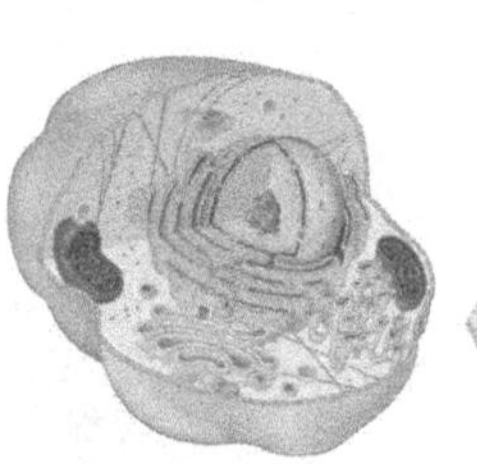

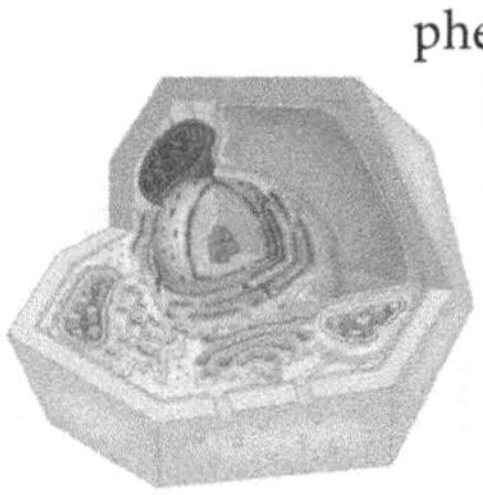

In contrast to chloroplasts (found only in plant cells), mitochondria are found in almost all eukaryotic cells, including those of plants and animals. **Mitochondria** (singular, *mitochondrion*) are the organelles

in which cellular respiration takes place; during cellular respiration, energy is harvested from sugars and transformed into another form of chemical energy called ATP (adenosine triphosphate). Cells use molecules of ATP as a direct energy source.

An envelope of two membranes encloses the mitochondrion, and the inner membrane encloses a thick fluid called the mitochondrial matrix (Figure 4.18). The inner membrane of the envelope has numerous infoldings called cristae. The cristae create a large surface area in which many of the enzymes and other molecules that function in cellular respiration are embedded, thereby maximizing ATP output. (You'll learn more about how mitochondria convert food energy to ATP energy in Chapter 6.)

Besides their ability to provide cellular energy, mitochondria and chloroplasts share another feature: They contain their own DNA that encodes some of their own proteins made by their own ribosomes. Each chloroplast and mitochondrion contains a single circular DNA chromosome that resembles a prokaryotic chromosome. In fact, mitochondria and chloroplasts can grow and pinch in two, reproducing themselves. This is evidence that mitochondria and chloroplasts evolved from ancient free-living prokaryotes that established residence within other, larger host prokaryotes. This phenomenon, where one species lives inside a host species, is a special type of symbiosis (see Chapter 16 for further information). Over time, mitochondria and chloroplasts likely became increasingly interdependent with the host prokaryote, eventually evolving into a single organism with inseparable parts. The DNA found within mitochondria and chloroplasts is therefore likely remnants of this ancient evolutionary event. ✓

☑ CHECKPOINT

1. What does photosynthesis accomplish?
2. What is cellular respiration?

Answers: 1. the conversion of light energy to chemical energy stored in food molecules 2. a process that converts the chemical energy of sugars and other food molecules to chemical energy in the form of ATP

▼ **Figure 4.17 The chloroplast: site of photosynthesis.**

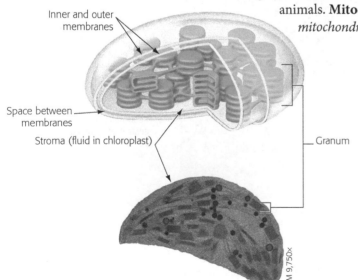

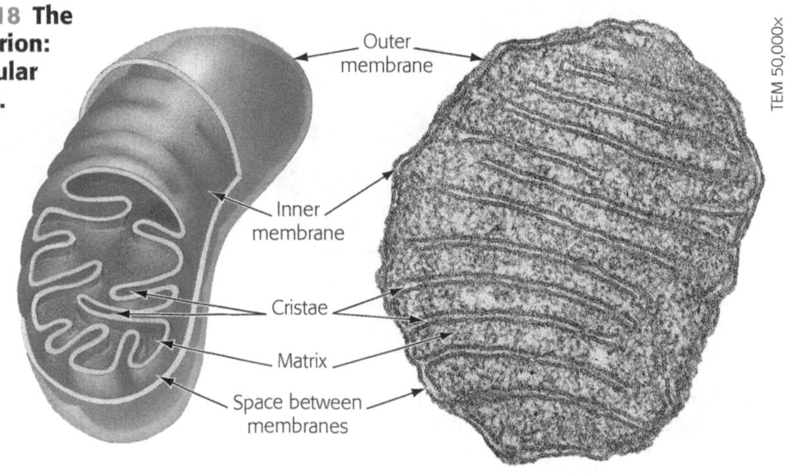

Inner and outer membranes

Space between membranes

Stroma (fluid in chloroplast)

Granum

TEM 9,750x

▼ **Figure 4.18 The mitochondrion: site of cellular respiration.**

Outer membrane

Inner membrane

Cristae

Matrix

Space between membranes

TEM 50,000x

The Cytoskeleton: Cell Shape and Movement

If someone asked you to describe a house, you would most likely mention the various rooms and their locations. You probably would not think to mention the foundation and beams that support the house. Yet these structures perform an extremely important function. Similarly, cells have an infrastructure called the **cytoskeleton**, a network of protein fibers extending throughout the cytoplasm. The cytoskeleton serves as both skeleton and "muscles" for the cell, functioning in support and movement.

Without the cytoskeleton, your cells would collapse in on themselves, much like a house collapses when the infrastructure fails.

Maintaining Cell Shape

One function of the cytoskeleton is to give mechanical support to the cell and maintain its shape. This is especially important for animal cells, which lack rigid cell walls. The cytoskeleton contains several types of fibers made from different types of protein. One important type of fiber forms **microtubules**, hollow tubes of protein (**Figure 4.19a**). The other kinds of cytoskeletal fibers, called intermediate filaments and microfilaments, are thinner and solid.

Just as the bony skeleton of your body helps fix the positions of your organs, the cytoskeleton provides anchorage and reinforcement for many organelles in a cell. For instance, the nucleus is held in place by a "cage" of cytoskeletal filaments. Other organelles use the cytoskeleton for movement. For example, a lysosome might reach a food vacuole by gliding along a microtubule track. Microtubules also guide the movement of chromosomes when cells divide (via the mitotic spindle—see Chapter 8).

A cell's cytoskeleton is dynamic: It can be quickly dismantled in one part of the cell by removing protein subunits and re-formed in a new location by reattaching the subunits. Such rearrangement can provide rigidity in a new location, change the shape of the cell, or even cause the whole cell or some of its parts to move. This process contributes to the amoeboid (crawling) movements of the protist *Amoeba* (**Figure 4.19b**) and movement of some of our white blood cells. ☑

☑ **CHECKPOINT**

From which important class of biological molecules are the microtubules of the cytoskeleton made?

Answer: protein

(b) Microtubules and movement. The crawling movement of an *Amoeba* is due to the rapid degradation and rebuilding of microtubules.

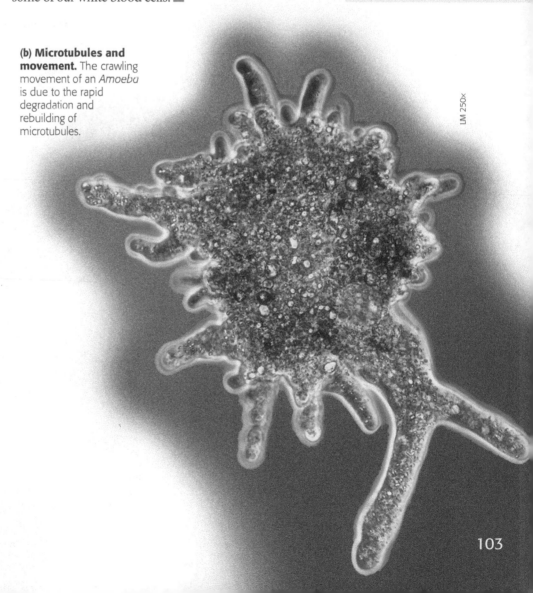

LM 250x

▼ **Figure 4.19 The cytoskeleton.**

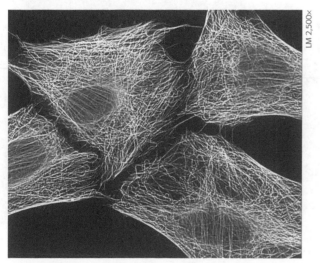

LM 2,500x

(a) Microtubules in the cytoskeleton. In this micrograph of animal cells, the cytoskeleton microtubules are labeled with a fluorescent yellow dye.

Cilia and Flagella

In some eukaryotic cells, microtubules are arranged into structures called flagella and cilia, extensions from a cell that aid in movement. Eukaryotic **flagella** (singular, *flagellum*) propel cells with an undulating, whiplike motion. They often occur singly, such as in human sperm cells **(Figure 4.20a)**, but may also appear in groups on the outer surface of protists. **Cilia** (singular, *cilium*) are generally shorter and more numerous than flagella and move in a coordinated back-and-forth motion, like the rhythmic oars of a crew team. Both cilia and flagella propel various protists through water **(Figure 4.20b)**. Though different in length, number per cell, and beating pattern, cilia and flagella have the same basic architecture. Not all animals have cilia or flagella—many do not—and they are almost never found on plant cells.

Some cilia extend from nonmoving cells that are part of a tissue layer. There, they move fluid over the tissue's surface. For example, cilia lining your windpipe clean your respiratory system by sweeping mucus with trapped debris out of your lungs **(Figure 4.20c)**. Tobacco smoke can inhibit or destroy these cilia, interfering with the normal cleansing mechanisms and allowing more toxin-laden smoke particles to reach the lungs. Frequent coughing—common in heavy smokers—then becomes the body's attempt to cleanse the respiratory system.

Because human sperm rely on flagella for movement, it's easy to understand why problems with flagella can lead to male infertility. Interestingly, some men with a type of hereditary sterility also suffer from respiratory problems. Because of a defect in the structure of their flagella and cilia, the sperm of men afflicted with this disorder cannot swim normally within the female reproductive tract to fertilize an egg (causing sterility), and their cilia do not sweep mucus out of their lungs (causing recurrent respiratory infections). ☑

☑ CHECKPOINT

Compare and contrast cilia and flagella.

Answer: Cilia and flagella have the same basic structure, are made from microtubules, and help move cells or move fluid over cells. Cilia are short and numerous and move back and forth. Flagella are longer, often occurring singly, and they undulate.

▼ Figure 4.20 **Examples of flagella and cilia.**

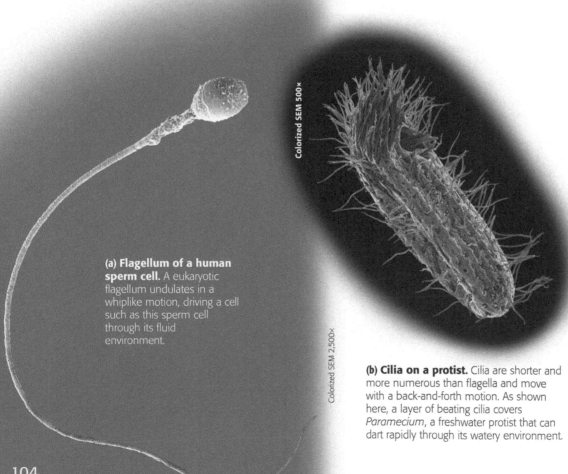

Colorized SEM 2,500×

(a) Flagellum of a human sperm cell. A eukaryotic flagellum undulates in a whiplike motion, driving a cell such as this sperm cell through its fluid environment.

Colorized SEM 500×

(b) Cilia on a protist. Cilia are shorter and more numerous than flagella and move with a back-and-forth motion. As shown here, a layer of beating cilia covers *Paramecium*, a freshwater protist that can dart rapidly through its watery environment.

Colorized SEM 3,000×

(c) Cilia lining the respiratory tract. The cilia lining your respiratory tract sweep mucus with trapped debris out of your lungs. This helps keep your airway clear and prevents infections.

The Evolution of Bacterial Resistance in Humans

Individuals with variations that make them better suited for the local environment will survive and reproduce more often (on average) than those who lack such variations. When the advantageous variations have a genetic basis, the offspring of individuals with the variations will more often also have the favorable adaptions, giving them a survival and reproductive advantage. In this way, repeated over many generations, natural selection promotes evolution of the population.

Within a human population, the persistent presence of a disease can provide a new basis for measuring those individuals who are best suited for survival in the local environment. For example, a recent evolutionary study examined people living in Bangladesh. This population has been exposed to the disease cholera—caused by an infectious bacterium—for millennia (**Figure 4.21**). After cholera bacteria enter a victim's digestive tract (usually through contaminated drinking water), the bacteria produce a toxin that binds to intestinal cells.

There, the toxin alters proteins in the plasma membrane, causing the cells to excrete fluid. The resulting diarrhea, which spreads the bacteria by shedding it back into the environment, can cause severe dehydration and death if untreated.

Because Bangladeshis have lived for so long in an environment that teems with cholera bacteria, one might expect that natural selection would favor those individuals who have some resistance to the bacteria. Indeed, recent studies of people from Bangladesh revealed mutations in several genes that appear to confer an increased resistance to cholera. Researchers discovered one mutation in a gene that encodes for the plasma membrane proteins that are the targets of the cholera bacteria. Although the mechanism is not yet understood, these genes appear to offer a survival advantage by making the proteins somewhat more resistant to attack by the cholera toxin. Because such genes offer a survival advantage within this population, they have slowly spread through the Bangladeshi population over the past 30,000 years. In other words, the Bangladeshi population is evolving increased resistance to cholera.

In addition to providing insight into the recent evolutionary past, data from this study reveal potential ways that humans might thwart the cholera bacterium. Perhaps pharmaceutical companies can exploit the proteins produced by the identified mutations to create a new generation of antibiotics. If so, this will represent another way that biologists have applied lessons learned from our understanding of evolution to improve human health. It also reminds us that we humans, like all life on Earth, are shaped by evolution due to changes in our environment, including the presence of infectious microorganisms that live around us.

▼ **Figure 4.21** People living in Bangladesh appear to have evolved resistance to *Vibrio cholerae*, the bacterium that causes the deadly disease cholera.

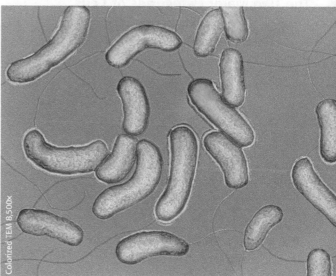

Colorized TEM 8,500×

Chapter Review

SUMMARY OF KEY CONCEPTS

The Microscopic World of Cells

The Two Major Categories of Cells

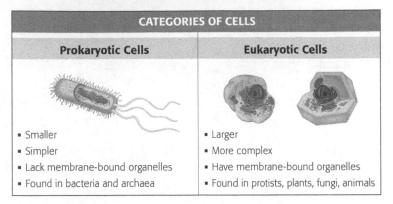

CATEGORIES OF CELLS	
Prokaryotic Cells	**Eukaryotic Cells**
• Smaller	• Larger
• Simpler	• More complex
• Lack membrane-bound organelles	• Have membrane-bound organelles
• Found in bacteria and archaea	• Found in protists, plants, fungi, animals

An Overview of Eukaryotic Cells

Membranes partition eukaryotic cells into a number of functional compartments. The largest organelle is usually the nucleus. Other organelles are located in the cytoplasm, the region outside the nucleus and within the plasma membrane.

Membrane Structure

The Plasma Membrane

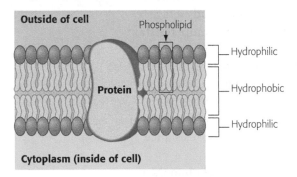

Cell Surfaces

The walls that encase plant cells support plants against the pull of gravity and also prevent cells from absorbing too much water. Animal cells are coated by a sticky extracellular matrix.

The Nucleus and Ribosomes: Genetic Control of the Cell

The Nucleus

An envelope consisting of two membranes encloses the nucleus. Within the nucleus, DNA and proteins make up chromatin fibers; each very long fiber is a single chromosome. The nucleus also contains the nucleolus, which produces components of ribosomes.

Ribosomes

Ribosomes produce proteins in the cytoplasm using messages produced by the DNA.

How DNA Directs Protein Production

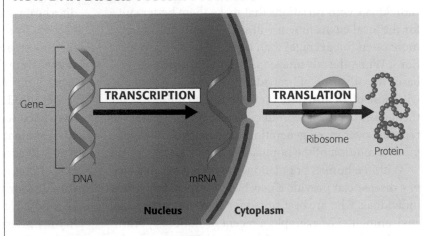

The Endomembrane System: Manufacturing and Distributing Cellular Products

The Endoplasmic Reticulum

The ER consists of membrane-enclosed tubes and sacs within the cytoplasm. Rough ER, named because of the ribosomes attached to its surface, makes membrane and secretory proteins. The functions of smooth ER include lipid synthesis and detoxification.

The Golgi Apparatus

The Golgi apparatus refines certain ER products and packages them in transport vesicles targeted for other organelles or export from the cell.

Lysosomes

Lysosomes, sacs containing digestive enzymes, aid digestion and recycling within the cell.

Vacuoles

Vacuoles include the contractile vacuoles that expel water from certain freshwater protists and the large, multifunctional central vacuoles of plant cells.

Energy Transformations: Chloroplasts and Mitochondria

Chloroplasts and Mitochondria

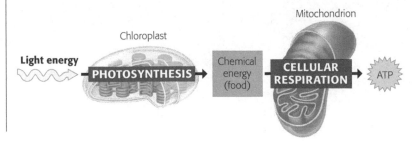

The Cytoskeleton: Cell Shape and Movement

Maintaining Cell Shape

Microtubules are an important component of the cytoskeleton, an organelle that gives support to, and maintains the shape of, cells.

Cilia and Flagella

Cilia and eukaryotic flagella are appendages that aid in movement, and they are made primarily of microtubules. Cilia are short and numerous and move the cell via coordinated beating. Flagella are long, often occur singly, and propel a cell with whiplike movements.

MasteringBiology®

For practice quizzes, BioFlix animations, MP3 tutorials, video tutors, and more study tools designed for this textbook, go to MasteringBiology®

SELF-QUIZ

1. You look into a microscope and view an unknown cell. What might you see that would tell you that the cell is eukaryotic?
   a. DNA
   b. a nucleoid region
   c. a plasma membrane
   d. membrane-enclosed structures called organelles

2. Explain how each word in the term *fluid mosaic* describes the structure of a membrane.

3. Identify which of the following structures includes all the others in the list: rough ER, smooth ER, endomembrane system, the Golgi apparatus.

4. Based on its function in detoxifying drugs, a large amount of _____ ER can be found in liver cells.

5. A type of cell called a lymphocyte makes proteins that are exported from the cell. You can track the path of these proteins within the cell from production through export by labeling them with radioactive isotopes. Identify which of the following structures would be radioactively labeled in your experiment, listing them in the order in which they would be labeled: chloroplasts, Golgi apparatus, plasma membrane, smooth ER, rough ER, nucleus, mitochondria.

6. Name two similarities in the structure or function of chloroplasts and mitochondria. Name two differences.

7. Match the following organelles with their functions:
   a. nucleus   1. locomotion
   b. flagella   2. protein export
   c. mitochondria   3. gene control
   d. Golgi apparatus   4. digestion
   e. lysosomes   5. cellular respiration

8. DNA controls the cell by transmitting genetic messages that result in protein production. Place the following organelles in the order that represents the flow of genetic information from the DNA through the cell: nuclear pores, ribosomes, nucleus, rough ER, Golgi apparatus.

9. Compare and contrast cilia and flagella.

Answers to these questions can be found in Appendix: Self-Quiz Answers.

THE PROCESS OF SCIENCE

10. The cells of plant seeds store oils in the form of droplets enclosed by membranes. Unlike the membranes you learned about in this chapter, the oil droplet membrane consists of a single layer of phospholipids rather than a bilayer. Draw a model for a membrane around an oil droplet. Explain why this arrangement is more stable than a bilayer.

11. Both chloroplasts and mitochondria contain rudimentary genomes, are able to divide independently from their host cells, and are bounded by double membranes. These similarities led to the idea that these organelles could have been derived from ancient prokaryotes that took up residence in eukaryotic ancestor cells. Which of the two organelles is more likely to have invaded the eukaryotic ancestor first and why?

12. **Interpreting Data** A population of bacteria may evolve resistance to a drug over time. Draw a graph that represents such a change. Label the *x*-axis as "time" and the *y*-axis as "population size." Draw a line on the graph that represents how the size of the population might change over time after the introduction of a new antibiotic. Label the point on your line where the new drug was introduced, and then indicate how the population size might change over time after that introduction.

BIOLOGY AND SOCIETY

13. In 2010, Craig Venter's group announced that they have assembled a completely synthetic genome based on the DNA sequence of the bacterium *Mycoplasma mycoides*. They inserted it into a living *Mycoplasma capricolum* cell, whose own genetic material was removed. This experiment resulted in a living and replicating cell that Venter referred to as "the first synthetic cell." In your opinion and based on what you have learnt in the first four chapters, is this description correct? Do you think it is feasible to create an entirely synthetic cell? What are the potential benefits of this type of synthetic biology? What are the dangers associated?

14. Scientists are learning to manipulate living cells in various ways, changing their genetic composition and the way they function. Some biotechnology companies have sought to patent their unique engineered cell lines. Do you think it is in society's best interest to allow cells to be patented? Why or why not? Do you think the same patent rules should apply whether the cells originate from a human or from a bacterium?

5 The Working Cell

Both nerve gas and insecticides work by crippling a vital enzyme. ►

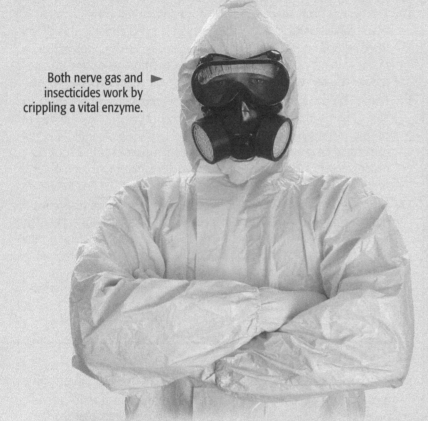

For thousands of years, people have used osmosis to preserve food through salt and sugar curing. ▼

▲ You'd have to walk more than 2 hours to burn the calories in half a pepperoni pizza.

Nanotechnology BIOLOGY AND SOCIETY

Harnessing Cellular Structures

Imagine a tiny movable "car" with balls of carbon atoms for wheels, or a three-dimensional relief map of the world carved onto an object 1,000 times smaller than a grain of sand. These are real-world examples of nanotechnology, the manipulation of materials at the molecular scale. When designing devices of such small size, researchers often turn to living cells for inspiration. After all, you can think of a cell as a machine that continuously and efficiently performs a variety of functions, such as movement, energy processing, and production of various products. Let's consider one example of cell-based nanotechnology and see how it relates to working cells.

Researchers at Cornell University are attempting to harvest the energy-producing capability of human sperm cells. Like other cells, a sperm cell generates energy by breaking down sugars and other molecules that pass through its plasma membrane. Enzymes within the cell carry out a process called glycolysis. During glycolysis, the energy released from the breakdown of glucose is used to produce molecules of ATP. Within a living sperm, the ATP produced during glycolysis and other processes provides the energy that propels the sperm through the female reproductive tract. In an attempt to harness this energy-producing system, the Cornell researchers attached three glycolysis enzymes to a computer chip. The enzymes continued to function in this artificial system, producing energy from sugar. The hope is that a larger set of enzymes can eventually be used to power microscopic robots. Such nanorobots could use glucose from the bloodstream to power the delivery of drugs to body tissues, among many other possible tasks. This example is only a glimpse into the incredible potential of new technologies inspired by working cells.

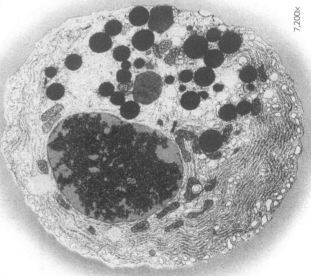

Cellular structures. Even the smallest cell, such as this one from a human pancreas, is a miniature machine of startling complexity.

In this chapter, we'll explore three processes common to all living cells: energy metabolism, the use of enzymes to speed chemical reactions, and transport regulation by the plasma membrane. Along the way, we'll further consider nanotechnologies that mimic the natural activities of living cells.

Some Basic Energy Concepts

Energy makes the world go round—both on a planetary scale and on a cellular scale. But what exactly is energy? Our first step in understanding the working cell is to learn a few basic concepts about energy.

Conservation of Energy

Energy is defined as the capacity to cause change. Some forms of energy are used to perform work, such as moving an object against an opposing force—for example, lifting a barbell against the force of gravity. Imagine a diver climbing to the top of a platform and then diving off **(Figure 5.1)**. To get to the top of the platform, the diver must perform work to overcome the opposing force of gravity. Specifically, chemical energy from food is converted to **kinetic energy**, the energy of motion. In this case, the kinetic energy takes the form of muscle movement propelling the diver to the top of the platform.

What happens to the kinetic energy when the diver reaches the top of the platform? Does it disappear at that point? In fact, it does not. A physical principle known as **conservation of energy** explains that it is not possible to destroy or create energy. Energy can only be converted from one form to another. A power plant, for example, does not make energy; it merely converts it from one form (such as energy stored in coal) to a more convenient form (such as electricity). That's what happens in the diver's climb up the steps. The kinetic energy of muscle movement is stored as **potential energy**, the energy an object has because of its location or structure. The energy contained by water behind a dam or by a compressed spring are examples of potential energy. In our example, the diver at the top of the platform has potential energy because of his elevated location. Then the act of diving off the platform into the water converts the potential energy back to kinetic energy. Life depends on countless similar conversions of energy from one form to another. ☑

☑ CHECKPOINT

Can an object at rest have energy?

Answer: Yes; it can have potential energy because of its location or structure.

► Figure 5.1 **Energy conversions during a dive.**

On the platform, the diver has more potential energy.

Climbing the steps converts kinetic energy of muscle movement to potential energy.

Diving converts potential energy to kinetic energy.

In the water, the diver has less potential energy.

Heat

If energy cannot be destroyed, where has the energy gone in our example when the diver hits the water? The energy has been converted to **heat**, a type of kinetic energy contained in the random motion of atoms and molecules. The friction between the body and its surroundings generated heat in the air and then in the water.

All energy conversions generate some heat. Although releasing heat does not destroy energy, it does make it more difficult to harness for useful work. Heat is energy in its most disordered, chaotic form, the energy of aimless molecular movement.

Entropy is a measure of the amount of disorder, or randomness, in a system. Consider an analogy from your own room. It's easy to increase the chaos—in fact, it seems to happen spontaneously! But it requires the expenditure of significant energy to restore order once again.

Every time energy is converted from one form to another, entropy increases. The energy conversions during the climb up the ladder and the dive from the platform increased entropy as the diver emitted heat to the surroundings. To climb up the steps again for another dive, the diver must use additional stored food energy. This conversion will also create heat and therefore increase entropy. ☑

Chemical Energy

How can molecules derived from the food we eat provide energy for our working cells? The molecules of food, gasoline, and other fuels have a form of potential energy called **chemical energy**, which arises from the arrangement of atoms and can be released by a chemical reaction. Carbohydrates, fats, and gasoline have structures that make them especially rich in chemical energy.

Living cells and automobile engines use the same basic process to make the chemical energy stored in their fuels available for work **(Figure 5.2)**. In both cases, this process breaks organic fuel into smaller waste molecules that have much less chemical energy than the fuel molecules did, thereby releasing energy that can be used to perform work.

For example, the engine of an automobile mixes oxygen with gasoline (which is why all cars require an air intake system) in an explosive chemical reaction that breaks down the fuel molecules and pushes the pistons that eventually move the wheels. The waste products emitted from the car's exhaust pipe are mostly carbon dioxide and water. Only about 25% of the energy that an automobile engine extracts from its fuel is converted to the kinetic energy of the car's movement. Most of

☑ **CHECKPOINT**

Which form of energy is most randomized and difficult to put to work?

Answer: heat

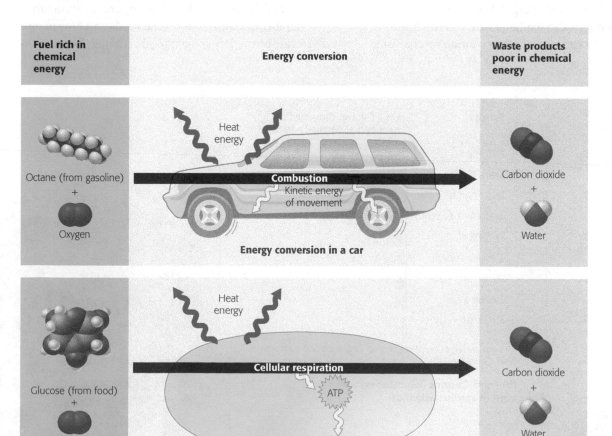

◄ Figure 5.2 **Energy conversions in a car and a cell.** In both a car and a cell, the chemical energy of organic fuel molecules is harvested using oxygen. This chemical breakdown releases energy stored in the fuel molecules and produces carbon dioxide and water. The released energy can be used to perform work.

Fuel rich in chemical energy	Energy conversion	Waste products poor in chemical energy

Octane (from gasoline) + Oxygen

Heat energy

Combustion
Kinetic energy of movement

Carbon dioxide + Water

Energy conversion in a car

Glucose (from food) + Oxygen

Heat energy

Cellular respiration
ATP
Energy for cellular work

Carbon dioxide + Water

Energy conversion in a cell

the rest is converted to heat—so much that the engine would melt if the car's radiator did not disperse heat into the atmosphere. That is why high-end performance cars need sophisticated air flow systems to avoid overheating.

Cells also use oxygen in reactions that release energy from fuel molecules. As in a car engine, the "exhaust" from such reactions in cells is mostly carbon dioxide and water. The combustion of fuel in cells is called cellular respiration, which is a more gradual and efficient "burning" of fuel compared with the explosive combustion in an automobile engine. Cellular respiration is the energy-releasing chemical breakdown of fuel molecules and the storage of that energy in a form the cell can use to perform work. (We will discuss the details of cellular respiration in Chapter 6.) You convert about 34% of your food energy to useful work, such as movement of your muscles. The rest of the energy released by the breakdown of fuel molecules generates body heat. Humans and many other animals can use this heat to keep the body at an almost constant temperature (37°C, or 98.6°F, in the case of humans), even when the surrounding air is much colder. You've probably noticed how quickly a crowded room warms up—it's all that released metabolic heat energy! The liberation of heat energy also explains why you feel hot after exercise. Sweating and other cooling mechanisms enable your body to lose the excess heat, much as a car's radiator keeps the engine from overheating.

☑ CHECKPOINT

According to Figure 5.3, how long would you have to ride your bicycle to burn off the energy in a cheeseburger?

Answer: a little more than a half hour (1 cheeseburger = 295 Calories; 1 hour of cycling consumes 490 Calories, so 295/490 = 0.6 hours, or about 36 minutes)

You'd have to walk more than 2 hours to burn the calories in half a pepperoni pizza.

Food Calories

Read any packaged food label and you'll find the number of calories in each serving of that food. Calories are units of energy. A **calorie** (cal) is the amount of energy that can raise the temperature of 1 gram (g) of water by 1°C. You could actually measure the caloric content of a peanut by burning it under a container of water to convert all of the stored chemical energy to heat and then measuring the temperature increase of the water.

Calories are tiny units of energy, so using them to describe the fuel content of foods is not practical. Instead, it's conventional to use kilocalories (kcal), units of 1,000 calories. In fact, the Calories (capital C) on a food package are actually kilocalories. For example, one peanut has about 5 Calories. That's a lot of energy, enough to increase the temperature of 1 kg (a little more than a quart) of water by 5°C. And just a handful of peanuts contains enough Calories, if converted to heat, to boil 1 kg of water. In living organisms, of course, food isn't used to boil water but instead used to fuel the activities of life. **Figure 5.3** shows the number of Calories in several foods and how many Calories are burned by some typical activities. ☑

▼ **Figure 5.3 Some caloric accounting.**

Food	Food Calories
Cheeseburger	295
Spaghetti with sauce (1 cup)	241
Baked potato (plain, with skin)	220
Fried chicken (drumstick)	193
Bean burrito	189
Pizza with pepperoni (1 slice)	181
Peanuts (1 ounce)	166
Apple	81
Garden salad (2 cups)	56
Popcorn (plain, 1 cup)	31
Broccoli (1 cup)	25

(a) Food Calories (kilocalories) in various foods

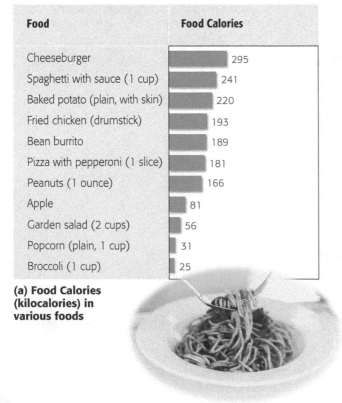

Activity	Food Calories consumed per hour by a 150-pound person*
Running (7 min/mi)	979
Dancing (fast)	510
Bicycling (10 mph)	490
Swimming (2 mph)	408
Walking (3 mph)	245
Dancing (slow)	204
Playing the piano	73
Driving a car	61
Sitting (writing)	28

*Not including energy necessary for basic functions, such as breathing and heartbeat

(b) Food Calories (kilocalories) we burn in various activities

 **Energy Transformations** # ATP and Cellular Work

The carbohydrates, fats, and other fuel molecules we obtain from food cannot be used directly as fuel for our cells. Instead, the chemical energy released by the breakdown of organic molecules during cellular respiration is used to generate molecules of ATP. These molecules of ATP then power cellular work. ATP acts like an energy shuttle, storing energy obtained from food and then releasing it as needed at a later time. Such energy transformations are essential for all life on Earth.

The Structure of ATP

The abbreviation ATP stands for adenosine triphosphate. **ATP** consists of an organic molecule called adenosine plus a tail of three phosphate groups **(Figure 5.4)**. The triphosphate tail is the "business" end of ATP, the part that provides energy for cellular work. Each phosphate group is negatively charged. Negative charges repel each other. The crowding of negative charges in the triphosphate tail contributes to the potential energy of ATP. It's analogous to storing energy by compressing a spring; if you release the spring, it will relax, and you can use that springiness to do some useful work. For ATP power, it is release of the phosphate at the tip of the triphosphate tail that makes energy available to working cells. What remains is **ADP**, adenosine diphosphate (two phosphate groups instead of three, shown on the right side of Figure 5.4).

Phosphate Transfer

When ATP drives work in cells by being converted to ADP, the released phosphate groups don't just fly off into space. ATP energizes other molecules in cells by transferring phosphate groups to those molecules. When a target molecule accepts the third phosphate group, it becomes energized and can then perform work in the cell. Imagine a bicyclist pedaling up a hill. In the muscle cells of the rider's legs, ATP transfers phosphate groups to motor proteins. The proteins then change shape, causing the muscle cells to contract **(Figure 5.5a)**. This contraction provides the mechanical energy needed to propel the rider. ATP also enables the transport of ions and other dissolved substances across the membranes of the rider's nerve cells **(Figure 5.5b)**, helping them send signals to her legs. And ATP drives the production of a cell's large molecules from smaller molecular building blocks **(Figure 5.5c)**.

▼ **Figure 5.4 ATP power.** Each ⓟ in the triphosphate tail of ATP represents a phosphate group, a phosphorus atom bonded to oxygen atoms. The transfer of a phosphate from the triphosphate tail to other molecules provides energy for cellular work.

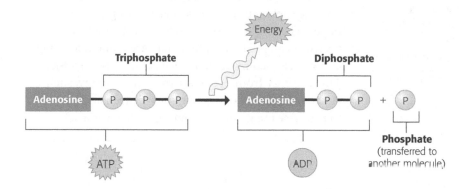

▼ **Figure 5.5 How ATP drives cellular work.** Each type of work shown here is powered when an enzyme transfers phosphate from ATP to a recipient molecule.

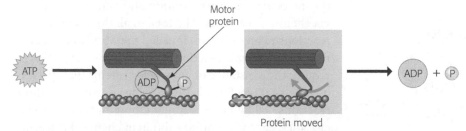

(a) Motor protein performing mechanical work (moving a muscle fiber)

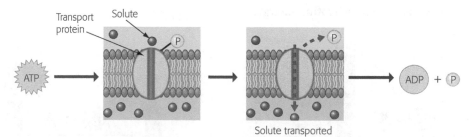

(b) Transport protein performing transport work (importing a solute)

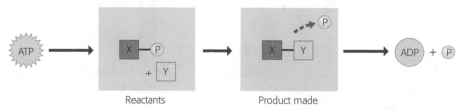

(c) Chemical reactants performing chemical work (promoting a chemical reaction)

The ATP Cycle

Your cells spend ATP continuously. Fortunately, it is a renewable resource. ATP can be restored by adding a phosphate group back to ADP. That takes energy, like recompressing a spring. And that's where food enters the picture. The chemical energy that cellular respiration harvests from sugars and other organic fuels is put to work regenerating a cell's supply of ATP. Cellular work spends ATP, which is recycled when ADP and phosphate are combined using energy released by cellular respiration **(Figure 5.6)**. Thus, energy from processes that yield energy, such as the breakdown of organic fuels, is transferred to processes that consume energy, such as muscle

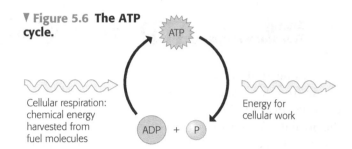

▼ Figure 5.6 **The ATP cycle.**

Cellular respiration: chemical energy harvested from fuel molecules

Energy for cellular work

contraction and other cellular work. The ATP cycle can run at an astonishing pace: Up to 10 million ATPs are consumed and recycled each second in a working muscle cell. ☑

Enzymes

A living organism contains a vast collection of chemicals, and countless chemical reactions constantly change the organism's molecular makeup. In a sense, a living organism is a complex "chemical square dance," with the molecular "dancers" continually changing partners via chemical reactions. The total of all the chemical reactions in an organism is called **metabolism**. But almost no metabolic reactions occur without help. Most require the assistance of **enzymes**, proteins that speed up chemical reactions without being consumed by those reactions. All living cells contain thousands of different enzymes, each promoting a different chemical reaction.

Activation Energy

For a chemical reaction to begin, chemical bonds in the reactant molecules must be broken. (The first step in swapping partners during a square dance is to let go of your current partner's hand.) This process requires that the molecules absorb energy from their surroundings. In other words, for most chemical reactions, a cell has to spend a little energy to make more. You can easily relate this concept to your own life: it takes effort to clean your room, but this will save you more energy in the long run because you won't have to hunt for your belongings. The energy that must be invested to start a reaction is called **activation energy** because it activates the reactants and triggers the chemical reaction.

Enzymes enable metabolism to occur by reducing the amount of activation energy required to break the bonds of reactant molecules. If you think of the activation energy as a barrier to a chemical reaction, an enzyme's function is to lower that barrier **(Figure 5.7)**. It does so by binding to reactant molecules and putting them under physical or chemical stress, making it easier to break their bonds and start a reaction. In our analogy of cleaning your room, this is like a friend offering to help you. You start and end in the same place whether solo or assisted, but your friend's help lowers your activation energy, making it more likely that you'll proceed. Next, we'll return to our theme of nanotechnology to see how enzymes can be engineered to be even more efficient. ☑

▼ Figure 5.7 **Enzymes and activation energy.**

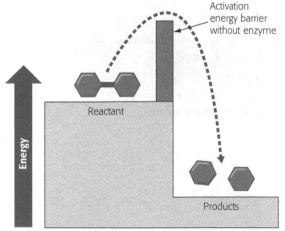

(a) Without enzyme. A reactant molecule must overcome the activation energy barrier before a chemical reaction can break the molecule into products.

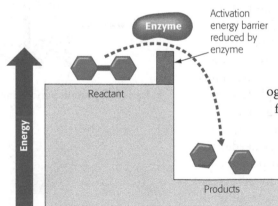

(b) With enzyme. An enzyme speeds the chemical reaction by lowering the activation energy barrier.

Nanotechnology THE PROCESS OF SCIENCE

Can Enzymes Be Engineered?

Like all other proteins, enzymes are encoded by genes. **Observations** of genetic sequences suggest that many of our genes were formed through a type of molecular evolution: One ancestral gene duplicated, and the two copies diverged over time via random genetic changes, eventually becoming distinct genes for enzymes with different functions.

The natural evolution of enzymes raises a **question**: Can laboratory methods mimic this process through artificial selection? A group of researchers at two California biotechnology companies formed the **hypothesis** that an artificial process could be used to modify the gene that codes for the enzyme lactase (which breaks down the sugar lactose) into a new gene coding for a new enzyme with a new function. Their **experiment** used a procedure called directed evolution. In this process, many copies of the gene for the starting lactase enzyme were mutated at random **(Figure 5.8)**. The researchers tested the enzymes resulting from these mutated genes to determine which enzymes best displayed a new activity (in this case, breaking down a different sugar). The genes for the enzymes that did show the new activity were then subjected to several more rounds of duplication, mutation, and screening.

After seven rounds, the **results** indicated that directed evolution had produced a new enzyme with a novel function. Researchers have used similar methods to produce many artificial enzymes with desired properties, such as one that produces an antibiotic with tenfold greater efficiency, ones that remain stable and productive under high-heat industrial conditions, and one that greatly improves the production of cholesterol-lowering drugs. These results show that directed evolution is another example of how scientists can mimic the natural processes of cells for their own purposes.

▼ **Figure 5.8 Directed evolution of an enzyme.** During seven rounds of directed evolution, the lactase enzyme gradually gained a new function.

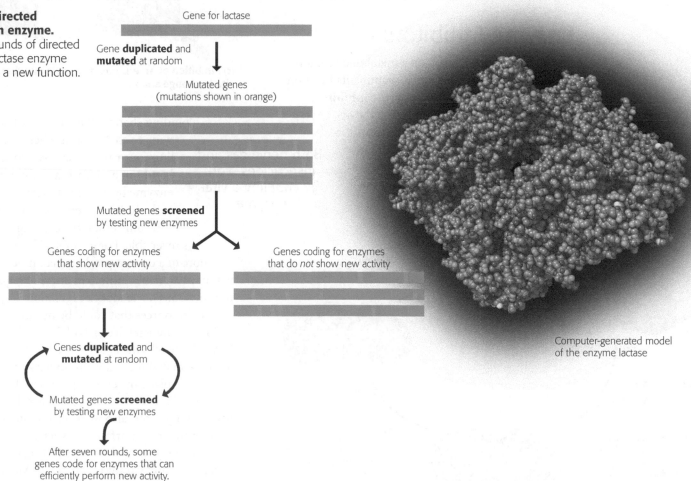

Gene for lactase

Gene **duplicated** and **mutated** at random

Mutated genes (mutations shown in orange)

Mutated genes **screened** by testing new enzymes

Genes coding for enzymes that show new activity

Genes coding for enzymes that do *not* show new activity

Genes **duplicated** and **mutated** at random

Mutated genes **screened** by testing new enzymes

After seven rounds, some genes code for enzymes that can efficiently perform new activity.

Computer-generated model of the enzyme lactase

Structure/Function Enzyme Activity

An enzyme is very selective in the reaction it catalyzes. This selectivity is based on the enzyme's ability to recognize a certain reactant molecule, which is called the enzyme's **substrate**. A region of the enzyme called the **active site** has a shape and chemistry that fits the substrate molecule. The active site is typically a pocket or groove on the surface of the enzyme. When a substrate slips into this docking station, the active site changes shape slightly to embrace the substrate and catalyze the reaction. This interaction is called **induced fit** because the entry of the substrate induces the enzyme to change shape slightly, making the fit between substrate and active site snugger. Think of a handshake: As your hand makes contact with another hand, it changes shape slightly to make a better fit.

After the products are released from the active site, the enzyme can accept another molecule of substrate. In fact, the ability to function repeatedly is a key characteristic of enzymes. **Figure 5.9** follows the action of the enzyme lactase, which breaks down the disaccharide lactose (the substrate). This enzyme is underproduced or defective in lactose-intolerant people. Like lactase, many enzymes are named for their substrates, with an -*ase* ending. ☑

☑ CHECKPOINT

How does an enzyme recognize its substrate?

Answer: The substrate and the enzyme's active site are complementary in shape and chemistry.

Enzyme Inhibitors

Certain molecules can inhibit a metabolic reaction by binding to an enzyme and disrupting its function (**Figure 5.10**). Some of these **enzyme inhibitors** are substrate imposters that plug up the active site. (You can't shake a person's hand if someone else puts a banana in it first!) Other inhibitors bind to the enzyme at a site remote from the active site, but the

Both nerve gas and insecticides work by crippling a vital enzyme.

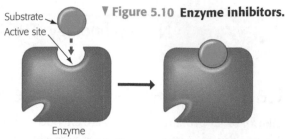

▼ Figure 5.10 **Enzyme inhibitors.**

Substrate
Active site
Enzyme

(a) Enzyme and substrate binding normally

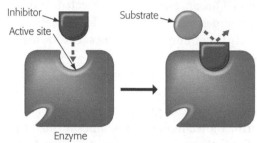

Inhibitor
Active site
Substrate
Enzyme

(b) Enzyme inhibition by a substrate imposter

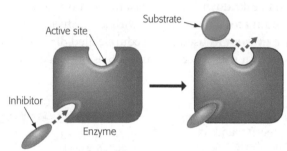

Active site
Substrate
Inhibitor
Enzyme

(c) Inhibition of an enzyme by a molecule that causes the active site to change shape

binding changes the enzyme's shape. (Imagine trying to shake hands when someone is tickling your ribs, causing you to clench your hand.) In each case, an inhibitor disrupts the enzyme by altering its shape—a clear example of the link between structure and function.

In some cases, the binding of an inhibitor is reversible. For example, if a cell is producing more of a certain product than it needs, that product may reversibly inhibit an enzyme required for its production. This feedback regulation keeps the cell from wasting resources that could be put to better use.

Many beneficial drugs work by inhibiting enzymes. Penicillin blocks the active site of an enzyme that bacteria use in making cell walls. Ibuprofen inhibits an enzyme involved in sending pain signals. Many cancer drugs inhibit enzymes that promote cell division. Many toxins and poisons also work as inhibitors. Nerve gases (a form of chemical warfare) irreversibly bind to the active site of an enzyme vital to transmitting nerve impulses, leading to rapid paralysis and death. Many pesticides are toxic to insects because they inhibit this same enzyme.

▼ Figure 5.9 **How an enzyme works.** Our example is the enzyme lactase, named for its substrate, lactose.

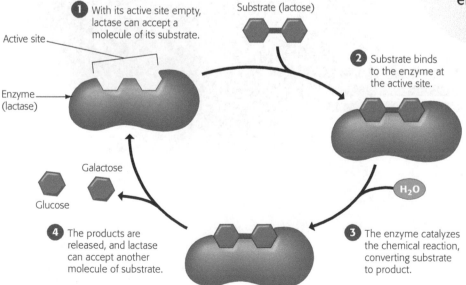

1 With its active site empty, lactase can accept a molecule of its substrate.

Active site

Enzyme (lactase)

Substrate (lactose)

2 Substrate binds to the enzyme at the active site.

H₂O

3 The enzyme catalyzes the chemical reaction, converting substrate to product.

4 The products are released, and lactase can accept another molecule of substrate.

Galactose

Glucose

Membrane Function

So far, we have discussed how cells control the flow of energy and how enzymes affect the pace of chemical reactions. In addition to these vital processes, cells must also regulate the flow of materials to and from the environment. The plasma membrane consists of a double layer of fat (a phospholipid bilayer) with embedded proteins (see Figure 4.4). **Figure 5.11** describes the major functions of these membrane proteins. Of all the functions shown in the figure, one of the most important is the regulation of transport in and out of the cell. A steady traffic of small molecules moves across a cell's plasma membrane in both directions. But this traffic flow is never willy-nilly. Instead, all biological membranes are selectively permeable—that is, they only allow certain molecules to pass. Let's explore this in more detail.

Passive Transport: Diffusion across Membranes

Molecules are restless. They constantly vibrate and wander randomly. One result of this motion is **diffusion**, the movement of molecules spreading out evenly into the available space. Each molecule moves randomly, but the overall diffusion of a population of molecules is usually directional, from a region where the molecules are more concentrated to a region where they are less concentrated. For example, imagine many molecules of perfume inside a bottle. If you remove the bottle top, every molecule of perfume will move randomly about, but the overall movement will be out of the bottle, and the room will eventually smell of the perfume. You could, with great effort, return the perfume molecules to its bottle, but the molecules would never all return spontaneously.

▼ **Figure 5.11 Primary functions of membrane proteins.** An actual cell may have just a few of the types of proteins shown here, and many copies of each particular protein may be present.

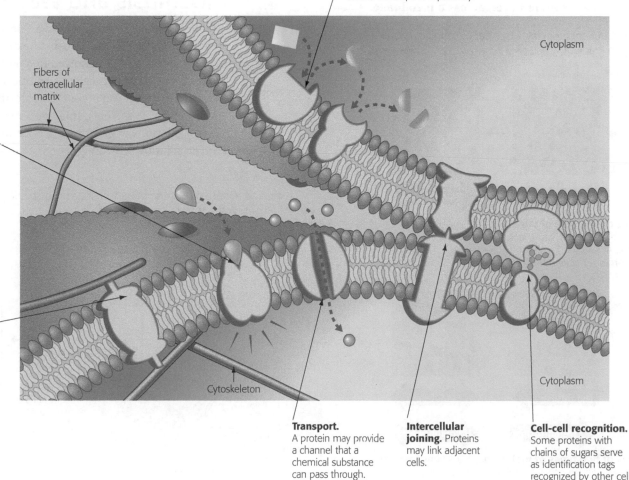

Enzymatic activity. This protein and the one next to it are enzymes, having an active site that fits a substrate. Enzymes may form an assembly line that carries out steps of a pathway.

Cytoplasm

Fibers of extracellular matrix

Cell signaling. A binding site fits the shape of a chemical messenger. The messenger may cause a change in the protein that relays the message to the inside of the cell.

Attachment to the cytoskeleton and extracellular matrix. Such proteins help maintain cell shape and coordinate changes.

Cytoskeleton

Cytoplasm

Transport. A protein may provide a channel that a chemical substance can pass through.

Intercellular joining. Proteins may link adjacent cells.

Cell-cell recognition. Some proteins with chains of sugars serve as identification tags recognized by other cells.

For an example closer to a living cell, imagine a membrane separating pure water from a mixture of dye dissolved in water **(Figure 5.12)**. Assume that this membrane has tiny holes that allow dye molecules to pass. Although each dye molecule moves randomly, there will be a net migration across the membrane to the side that began as pure water. Movement of the dye will continue until both solutions have equal concentrations. After that, there will be a dynamic equilibrium: Molecules will still be moving, but at that point as many dye molecules move in one direction as in the other.

Diffusion of dye across a membrane is an example of **passive transport**—*passive* because the cell does not expend any energy for the diffusion to happen. But remember that the cell membrane is selectively permeable. For example, small molecules such as oxygen (O_2) generally pass through more readily than larger molecules such as amino acids. But the membrane is relatively impermeable to even some very small substances, such as most ions, which are too hydrophilic to pass through the phospholipid bilayer. In passive transport, a substance diffuses down its **concentration gradient**, from where the substance is more concentrated to where it is less concentrated.

In our lungs, for example, there is more oxygen gas (O_2) in the air than in the blood. Therefore, oxygen moves by passive transport from the air into the bloodstream.

Substances that do not cross membranes spontaneously—or otherwise cross very slowly—can be transported via proteins that act as corridors for specific molecules (see Figure 5.11). This assisted transport is called **facilitated diffusion**. For example, water molecules can move through the plasma membrane of some cells via transport proteins—each of which can help 3 billion water molecules per second pass through! People with a rare mutation in the gene that encodes these water-transport proteins have defective kidneys that cannot reabsorb water; such people must drink 20 liters of water every day to prevent dehydration. On the flip side, a common complication of pregnancy is fluid retention, the culprit responsible for swollen ankles and feet, often caused by increased synthesis of water channel proteins. Other specific transport proteins move glucose across cell membranes 50,000 times faster than diffusion. Even at this rate, facilitated diffusion is a type of passive transport because it does not require the cell to expend energy. As in all passive transport, the driving force is the concentration gradient. ☑

☑ **CHECKPOINT**

Why is facilitated diffusion a form of passive transport?

Answer: It uses proteins to transport materials down a concentration gradient without expending energy.

▼ **Figure 5.12 Passive transport: diffusion across a membrane.** A substance will diffuse from where it is more concentrated to where it is less concentrated. Put another way, a substance tends to diffuse down its concentration gradient.

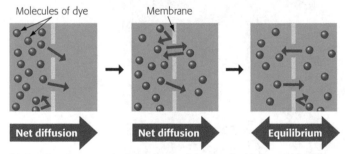

(a) Passive transport of one type of molecule. The membrane is permeable to these dye molecules, which diffuse down the concentration gradient. At equilibrium, the molecules are still restless, but the rate of transport is equal in both directions.

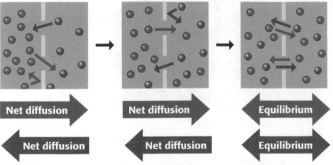

(b) Passive transport of two types of molecules. If solutions have two or more solutes, each will diffuse down its own concentration gradient.

Osmosis and Water Balance

The diffusion of water across a selectively permeable membrane is called **osmosis (Figure 5.13)**. A **solute** is a substance that is dissolved in a liquid solvent, and the resulting mixture is called a solution. For example, a solution of salt water contains salt (the solute) dissolved in water (the solvent). Imagine a membrane separating two solutions with different concentrations of a solute. The solution with a higher concentration of solute is

▼ **Figure 5.13 Osmosis.** A membrane separates two solutions with different sugar concentrations. Water molecules can pass through the membrane, but sugar molecules cannot.

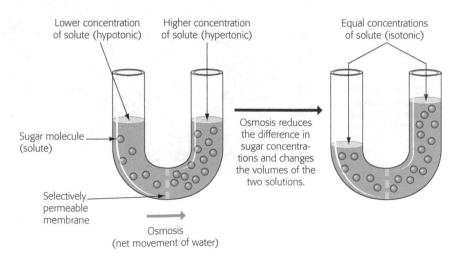

said to be **hypertonic** to the other solution. The solution with the lower solute concentration is said to be **hypotonic** to the other. Note that the hypotonic solution, by having the lower solute concentration, has the higher water concentration (less solute = more water). Therefore, water will diffuse across the membrane along its concentration gradient from an area of higher water concentration (hypotonic solution) to one of lower water concentration (hypertonic solution). This reduces the difference in solute concentrations and changes the volumes of the two solutions.

People can take advantage of osmosis to preserve foods. Salt is often applied to meats—like pork and cod—to cure them; the salt causes water to move out of food-spoiling bacteria and fungi. Food can also be preserved in honey because a high sugar concentration draws water out of food.

When the solute concentrations are the same on both sides of a membrane, water molecules will move at the same rate in both directions, so there will be no net change in solute concentration. Solutions of equal solute concentration are said to be **isotonic**. For example, many marine animals, such as sea stars and crabs, are isotonic to seawater, so that overall they neither gain nor lose water from the environment. In hospitals, intravenous (IV) fluids administered to patients must be isotonic to blood cells to avoid harm.

Water Balance in Animal Cells

The survival of a cell depends on its ability to balance water uptake and loss. When an animal cell is immersed in an isotonic solution, the cell's volume remains constant because the cell gains water at the same rate that it loses water (**Figure 5.14a**, top). But what happens if an animal cell is in contact with a hypotonic solution, which has a lower solute concentration than the cell? Due to osmosis, the cell would gain water, swell, and possibly burst (lyse) like an overfilled water balloon (**Figure 5.14b**, top). A hypertonic environment is also harsh on an animal cell; the cell shrivels from water loss (**Figure 5.14c**, top).

For an animal to survive a hypotonic or hypertonic environment, the animal must have a way to balance the uptake and loss of water. The control of water balance is called **osmoregulation**. For example, a freshwater fish has kidneys and gills that work constantly to prevent an excessive buildup of water in the body. Humans can suffer consequences of osmoregulation failure. Dehydration (consumption of too little water) can cause fatigue and even death. Drinking too much water—called hyponatremia or "water intoxication"—can also cause death by overdiluting necessary ions.

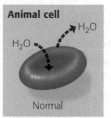
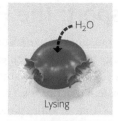
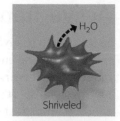
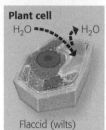
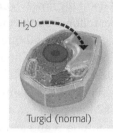
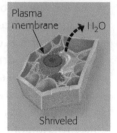

Animal cell
H₂O H₂O
Normal Lysing Shriveled

Plant cell
H₂O H₂O Plasma membrane H₂O
Flaccid (wilts) Turgid (normal) Shriveled

(a) Isotonic solution **(b) Hypotonic solution** **(c) Hypertonic solution**

◄ **Figure 5.14 Osmotic environments.** Animal cells (such as a red blood cell) and plant cells behave differently in different osmotic environments.

For thousands of years, people have used osmosis to preserve food through salt and sugar curing.

Water Balance in Plant Cells

Problems of water balance are somewhat different for cells that have rigid cell walls, such as those from plants, fungi, many prokaryotes, and some protists. A plant cell immersed in an isotonic solution is flaccid (floppy), and the plant wilts (Figure 5.14a, bottom). In contrast, a plant cell is turgid (very firm) and healthiest in a hypotonic environment, with a net inflow of water (Figure 5.14b, bottom). Although the elastic cell wall expands a bit, the back pressure it exerts prevents the cell from taking in too much water and bursting. Turgor is necessary for plants to retain their upright posture and the extended state of their leaves (**Figure 5.15**). However, in a hypertonic environment, a plant cell is no better off than an animal cell. As a plant cell loses water, it shrivels, and its plasma membrane pulls away from the cell wall (Figure 5.14c, bottom). This usually kills the cell. Thus, plant cells thrive in a hypotonic environment, whereas animal cells thrive in an isotonic one. ☑

▼ **Figure 5.15 Plant turgor.** Watering a wilted plant will make it regain its turgor.

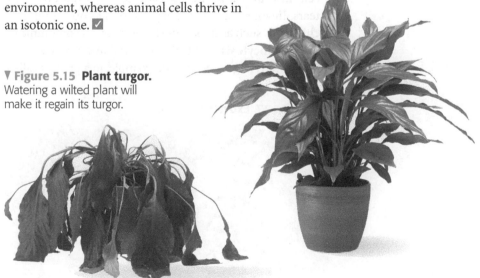

☑ **CHECKPOINT**

1. An animal cell shrivels when it is _____ compared with its environment.
2. The cells of a wilted plant are _____ compared with their environment.

Answers: 1. hypertonic 2. isotonic

Active Transport: The Pumping of Molecules across Membranes

In contrast to passive transport, **active transport** requires that a cell expend energy to move molecules across a membrane. Cellular energy (usually provided by ATP) is used to drive a transport protein that pumps a solute *against* the concentration gradient—that is, in the direction that is opposite the way it would naturally flow **(Figure 5.16)**. Movement against a force, like rolling a boulder uphill against gravity, requires a considerable expenditure of energy.

Active transport allows cells to maintain internal concentrations of small solutes that differ from environmental concentrations. For example, compared with its surroundings, an animal nerve cell has a much higher concentration of potassium ions and a much lower concentration of sodium ions. The plasma membrane helps maintain these differences by pumping sodium out of the cell and potassium into the cell. This particular case of active transport (called the sodium-potassium pump) is vital to the nervous system of most animals. ☑

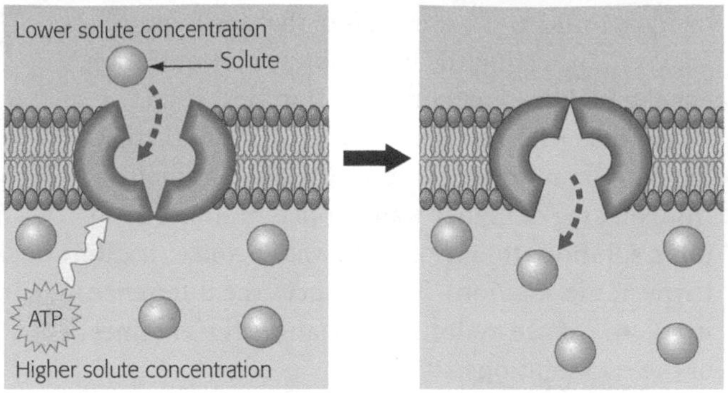

☑ CHECKPOINT

What molecule is the usual energy source for active transport?

Answer: ATP

▲ **Figure 5.16 Active transport.** Transport proteins are specific in their recognition of atoms or molecules. This transport protein (purple) has a binding site that accepts only a certain solute. Using energy from ATP, the protein pumps the solute against its concentration gradient.

Exocytosis and Endocytosis: Traffic of Large Molecules

So far, we've focused on how water and small solutes enter and leave cells by moving through the plasma membrane. The story is different for large molecules such as proteins, which are much too big to fit through the membrane. Their traffic into and out of the cell depends on the ability of the cell to package large molecules inside sacs called vesicles. You have already seen an example of this: During protein production by the cell, secretory proteins exit the cell from transport vesicles that fuse with the plasma membrane, spilling the contents outside the cell (see Figures 4.12 and 4.16). That process is called **exocytosis (Figure 5.17)**. When you cry, for example, cells in your tear glands use exocytosis to export the salty tears. In your brain, the exocytosis of neurotransmitter chemicals such as dopamine helps neurons communicate.

In **endocytosis**, a cell takes material in via vesicles that bud inward **(Figure 5.18)**. For example, in a process called

phagocytosis ("cellular eating"), a cell engulfs a particle and packages it within a food vacuole. Other times, a cell "gulps" droplets of fluid into vesicles. Endocytosis can also be triggered by the binding of certain external molecules to specific receptor proteins built into the plasma membrane. This binding causes the local region of the membrane to form a vesicle that transports the specific substance into the cell. In human liver cells, this process is used to take up cholesterol from the blood. An inherited defect in the receptors on liver cells can lead to an inability to process cholesterol, which can lead to heart attacks at ages as young as 5. Cells of your immune system use endocytosis to engulf and destroy invading bacteria and viruses.

Because all cells have a plasma membrane, it is logical to infer that membranes first formed early in the evolution of life on Earth. In the final section of this chapter, we'll consider the evolution of membranes.

▼ **Figure 5.17 Exocytosis.**

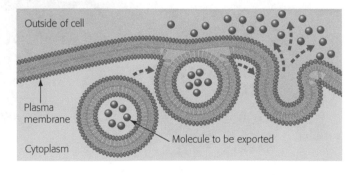

▼ **Figure 5.18 Endocytosis.**

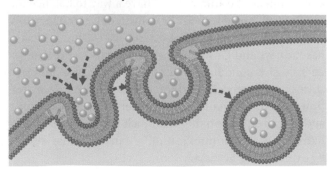

Nanotechnology EVOLUTION CONNECTION

The Origin of Membranes

By simulating conditions found on the early Earth, scientists have been able to demonstrate that many of the molecules important to life can form spontaneously. (See Figure 15.3 and the accompanying text for a description of one such experiment.) Such results suggest that phospholipids, the key ingredients in all membranes, were probably among the first organic compounds that formed from chemical reactions on the early Earth. Once formed, they could self-assemble into simple membranes. When a mixture of phospholipids and water is shaken, for example, the phospholipids organize into bilayers, forming water-filled bubbles of membrane (Figure 5.19). This assembly requires neither genes nor other information beyond the properties of the phospholipids themselves.

The tendency of lipids in water to spontaneously form membranes has led biomedical engineers to produce liposomes (a type of artificial vesicle) that can encase particular chemicals. In the future, these engineered liposomes may be used to deliver nutrients or medications to specific sites within the body. In fact, as of 2012, 12 drugs have been approved for delivery via liposomes, including ones that target fungal infections, influenza, and hepatitis. Thus, membranes—like the other cellular components discussed in the Biology and Society and the Process of Science sections—have inspired novel nanotechnologies.

The formation of membrane-enclosed collections of molecules would have been a critical step in the evolution of the first cells. A membrane can enclose a solution that is different in composition from its surroundings. A plasma membrane that allows cells to regulate their chemical exchanges with the environment is a basic requirement for life. Indeed, all cells are enclosed by a plasma membrane that is similar in structure and function—illustrating the evolutionary unity of life.

▼ Figure 5.19 **The spontaneous formation of membranes: a key step in the origin of life.**

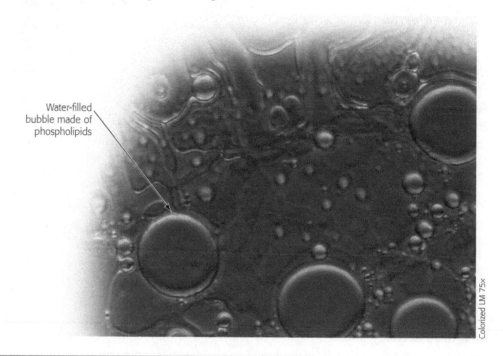

Water-filled bubble made of phospholipids

Colorized LM 75x

Chapter Review

SUMMARY OF KEY CONCEPTS

Some Basic Energy Concepts

Conservation of Energy

Machines and organisms can transform kinetic energy (energy of motion) to potential energy (stored energy) and vice versa. In all such energy transformations, total energy is conserved. Energy cannot be created or destroyed.

Heat

Every energy conversion releases some randomized energy in the form of heat. Entropy is a measure of disorder, or randomness.

Chemical Energy

Molecules store varying amounts of potential energy in the arrangement of their atoms. Organic compounds are relatively rich in such chemical energy. The combustion of gasoline within a car's engine and the breakdown of glucose via cellular respiration within living cells are both examples of how the chemical energy stored in molecules can be converted to useful work.

Food Calories

Food Calories, actually kilocalories, are units used to measure the amount of energy in our foods and the amount of energy we expend in various activities.

ATP and Cellular Work

Your cells recycle ATP: as ATP is broken down to ADP to drive cellular work, new molecules of ATP are built from ADP using energy obtained from food.

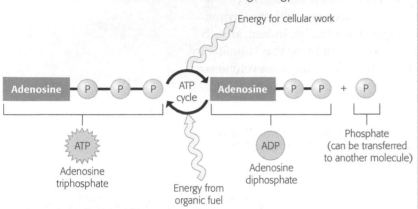

Energy for cellular work

Adenosine | P | P | P — ATP cycle — Adenosine | P | P + P

ATP
Adenosine triphosphate

Energy from organic fuel

ADP
Adenosine diphosphate

Phosphate (can be transferred to another molecule)

Enzymes

Activation Energy

Enzymes are biological catalysts that speed up metabolic reactions by lowering the activation energy required to break the bonds of reactant molecules.

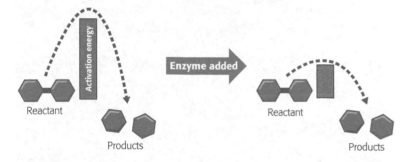

Activation energy

Reactant

Products

Enzyme added

Reactant

Products

Enzyme Activity

The entry of a substrate into the active site of an enzyme causes the enzyme to change shape slightly, allowing for a better fit and thereby promoting the interaction of enzyme with substrate.

Enzyme Inhibitors

Enzyme inhibitors are molecules that can disrupt metabolic reactions by binding to enzymes, either at the active site or elsewhere.

Membrane Function

Proteins embedded in the plasma membrane perform a wide variety of functions, including regulating transport, anchoring to other cells or substances, promoting enzymatic reactions, and recognizing other cells.

Passive Transport, Osmosis, and Active Transport

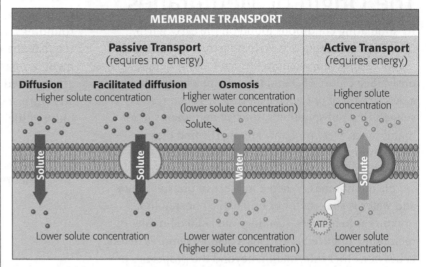

MEMBRANE TRANSPORT

Passive Transport (requires no energy) | **Active Transport** (requires energy)

Diffusion — Higher solute concentration
Facilitated diffusion
Osmosis — Higher water concentration (lower solute concentration)
Solute

Higher solute concentration

Solute | Solute | Water | Solute

Lower solute concentration | Lower water concentration (higher solute concentration) | Lower solute concentration

ATP

Most animal cells require an isotonic environment, with equal concentrations of water within and outside the cell. Plant cells need a hypotonic environment, which causes water to flow inward, keeping walled cells turgid.

Exocytosis and Endocytosis: Traffic of Large Molecules

Exocytosis is the secretion of large molecules within vesicles. Endocytosis is the import of large substances via vesicles into the cell.

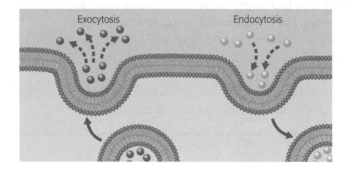

Exocytosis

Endocytosis

MasteringBiology®

For practice quizzes, BioFlix animations, MP3 tutorials, video tutors, and more study tools designed for this textbook, go to MasteringBiology®

SELF-QUIZ

1. Describe the energy transformations that occur when you climb to the top of a stairway.

2. An object at rest has no _____ energy, but it may have _____ energy resulting from its location or structure.

3. The label on a tin of tomato soup says that it contains 56 Calories. If you could convert all of that energy to heat, you could raise the temperature of how much water by 15°C?

4. Why does removing a phosphate group from the triphosphate tail in a molecule of ATP release energy?

5. Your digestive system uses a variety of enzymes to break down large food molecules into smaller ones that your cells can assimilate. A generic name for a digestive enzyme is hydrolase. What is the chemical basis for that name? (*Hint*: Review Figure 3.4.)

6. Explain how an inhibitor can disrupt an enzyme's action without binding to the active site.

7. If someone at the other end of a room smokes a cigarette, you may breathe in some smoke. The movement of smoke is similar to what type of transport?
   a. osmosis
   b. diffusion
   c. facilitated diffusion
   d. active transport

8. Explain why it is not enough just to say that a solution is "hypertonic."

9. What is the primary difference between passive and active transport in terms of concentration gradients?

10. When a person cries, tears are exported from cells through the process of
    a. facilitated diffusion.
    b. active transport.
    c. endocytosis.
    d. exocytosis.

Answers to these questions can be found in Appendix: Self-Quiz Answers.

THE PROCESS OF SCIENCE

11. Cyclic guanosine monophosphate (cGMP) is a signaling molecule that relaxes smooth muscle tissues, thereby allowing blood flow into the corpus cavernosum and triggering penile erection. It is degraded by the enzyme cGMP-specific phosphodiesterase type 5 (PDE5). Sildenafil (Viagra), used to treat erectile dysfunction, has a molecular structure similar to that of cGMP. Propose a model for how sildenafil prevents erectile dysfunction.

12. Gaining and losing weight are matters of caloric accounting: Calories in the food you eat minus Calories that you spend in activity. One pound of human body fat contains approximately 3,500 Calories. Using Figure 5.3, compare ways you could burn off those Calories. How far would you have to run, swim, or walk to burn the equivalent of 1 pound of fat, and how long would it take? Which method of burning Calories

appeals the most to you? The least? How much of each food would you have to consume in order to gain a single pound? How does one pound's worth of food compare with one pound's worth of exercise? Does it seem like an even trade-off?

13. **Interpreting Data** The graph below illustrates the course of a chemical reaction with and without an enzyme. Which curve represents the reaction in the presence of the enzyme? What energy changes are represented by the lines labeled a, b, and c?

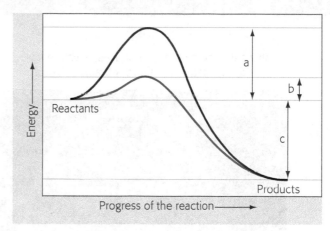

BIOLOGY AND SOCIETY

14. Obesity is a serious health problem for many Americans. Several popular diet plans advocate low-carbohydrate diets. Most low-carb dieters compensate by eating more protein and fat. What are the advantages and disadvantages of such a diet? Should the government regulate the claims of diet books? How should the claims be tested? Should diet proponents be required to obtain and publish data before making claims?

15. Many detox and diet plans recommend drinking large amounts of water —sometimes even up to 5 liters per day. Which properties of water may have led to this recommendation? Is a massively high intake of water always recommendable? How can drinking large amounts of water have a negative physiological effect? Compare recommendations for water intake found on different detoxing/dieting websites, and determine whether they are scientifically sound.

16. Lead acts as an enzyme inhibitor, and it can interfere with the development of the nervous system. One manufacturer of lead-acid batteries instituted a "fetal protection policy" that banned female employees of childbearing age from working in areas where they might be exposed to high levels of lead. These women were transferred to lower-paying jobs in lower-risk areas. Some employees challenged the policy in court, claiming that it deprived women of job opportunities available to men. The U.S. Supreme Court ruled the policy illegal. But many people are uncomfortable about the "right" to work in an unsafe environment. What rights and responsibilities of employers, employees, and government agencies are in conflict? What criteria should be used to decide who can work in a particular environment?

6 Cellular Respiration: Obtaining Energy from Food

About 20% of the energy produced by your body each day is used to sustain your brain. ▼

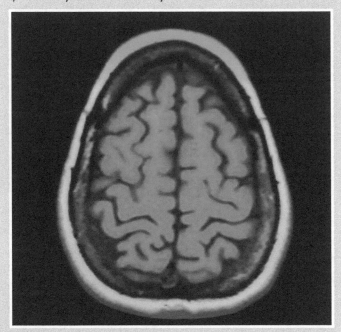

You have something in common with a sports car: You both require an air intake system to burn fuel efficiently. ▼

▲ Similar metabolic processes produce alcohol, pepperoni, soy sauce, rising bread, and acid in your muscles after a hard workout.

Exercise Science BIOLOGY AND SOCIETY

Getting the Most Out of Your Muscles

Serious athletes train extensively to reach the peak of their physical potential. A key aspect of athletic conditioning involves increasing aerobic capacity, the ability of the heart and lungs to deliver oxygen to body cells. For many endurance athletes, such as long-distance runners or cyclists, the rate at which oxygen is provided to working muscles is the limiting factor in their performance.

Why is oxygen so important? Whether you are exercising or just going about your daily tasks, your muscles need a continuous supply of energy to perform work. Muscle cells obtain this energy from the sugar glucose through a series of chemical reactions that depend upon a constant input of oxygen (O_2). Therefore, to keep moving, your body needs a steady supply of O_2.

When there is enough oxygen reaching your cells to support their energy needs, metabolism is said to be aerobic. As your muscles work harder, you breathe faster and deeper to inhale more O_2. If you continue to pick up the pace, you will approach your aerobic capacity, the maximum rate at which O_2 can be taken in and used by your muscle cells and therefore the most strenuous exercise that your body can maintain aerobically. Exercise physiologists (scientists who study how the body works during physical activity) therefore use oxygen-monitoring equipment to precisely determine the maximum possible aerobic output for any given person. Such data allow a well-trained athlete to stay within aerobic limits, ensuring the maximum possible output—in other words, his or her best effort.

If you work even harder and exceed your aerobic capacity, the demand for oxygen in your muscles will outstrip your

The science of exercise. Exercise physiologists can help athletes perform their best by carefully monitoring the consumption of oxygen and the production of carbon dioxide.

body's ability to deliver it; metabolism then becomes anaerobic. With insufficient O_2, your muscle cells switch to an "emergency mode" in which they break down glucose very inefficiently and produce lactic acid as a by-product. As lactic acid and other by-products accumulate, muscle activity is impaired. Your muscles can work under these conditions for only a few minutes before they give out.

Every living organism depends on processes that provide energy. In fact, we need energy to walk, talk, and think—in short, to stay alive. The human body has trillions of cells, all hard at work, all demanding fuel continuously. In this chapter, you'll learn how cells harvest food energy and put it to work with the help of oxygen. Along the way, we'll consider the implications of how the body responds to exercise.

Energy Flow and Chemical Cycling in the Biosphere

All life requires energy. In almost all ecosystems on Earth, this energy originates with the sun. During **photosynthesis**, plants convert the energy of sunlight to the chemical energy of sugars and other organic molecules (as we'll discuss in Chapter 7). Humans and other animals depend on this conversion for our food and more. You're probably wearing clothing made of a product of photosynthesis—cotton. Most of our homes are framed with lumber, which is wood produced by photosynthetic trees. Even textbooks are printed on a material (paper) that can be traced to photosynthesis in plants. But from an animal's point of view, photosynthesis is primarily about providing food.

✓ **CHECKPOINT**

What chemical ingredients do plants require from the environment to synthesize their own food?

Answer: CO₂, H₂O, and soil minerals

Producers and Consumers

Plants and other **autotrophs** ("self-feeders") are organisms that make all their own organic matter—including carbohydrates, lipids, proteins, and nucleic acids—from nutrients that are entirely inorganic: carbon dioxide from the air and water and minerals from the soil. In other words, autotrophs make their own food; they don't need to eat to gain energy to power their cellular processes. In contrast, humans and other animals are heterotrophs ("other-feeders"), organisms that cannot make organic molecules from inorganic ones. Therefore, we must eat organic material to get our nutrients and provide energy for life's processes.

Most ecosystems depend entirely on photosynthesis for food. For this reason, biologists refer to plants and other autotrophs as **producers**. Heterotrophs, in contrast, are **consumers**, because they obtain their food by eating plants or by eating animals that have eaten plants (**Figure 6.1**). We animals and other heterotrophs depend on autotrophs for organic fuel and for the raw organic materials we need to build our cells and tissues. ✓

Chemical Cycling between Photosynthesis and Cellular Respiration

The chemical ingredients for photosynthesis are carbon dioxide (CO_2), a gas that passes from the air into a plant via tiny pores, and water (H_2O), which is absorbed from the soil by the plant's roots. Inside leaf cells, organelles called chloroplasts use light energy to rearrange the atoms of these ingredients to produce sugars—most

▶ Figure 6.1 **Producer and consumer.** A giraffe (consumer) eating leaves produced by a photosynthetic plant (producer).

importantly glucose ($C_6H_{12}O_6$)—and other organic molecules **(Figure 6.2)**. You can think of chloroplasts as tiny solar-powered sugar factories. A by-product of photosynthesis is oxygen gas (O_2) that is released through pores into the atmosphere.

Both animals and plants use the organic products of photosynthesis as sources of energy. A chemical process called cellular respiration uses O_2 to convert the energy stored in the chemical bonds of sugars to another source of chemical energy called ATP. Cells expend ATP for almost all their work. In both plants and animals, the production of ATP during cellular respiration occurs mainly in the organelles called mitochondria (see Figure 4.18).

You might notice in Figure 6.2 that energy takes a one-way trip through an ecosystem, entering as sunlight and exiting as heat. Chemicals, in contrast, are recycled. Notice also in Figure 6.2 that the waste products of cellular respiration are CO_2 and H_2O—the very same ingredients used as inputs for photosynthesis. Plants store chemical energy via photosynthesis and then harvest this energy via cellular respiration. (Note that plants perform *both* photosynthesis to produce fuel molecules *and* cellular respiration to burn them, while animals perform *only* cellular respiration.) Plants usually make more organic molecules than they need for fuel. This photosynthetic surplus provides material for the plant to grow or can be stored (as starch in potatoes, for example). Thus, when you consume a carrot, potato, or turnip, you are eating the energy reservoir that plants (if unharvested) would have used to grow the following spring.

People have always taken advantage of plants' photosynthetic abilities by eating them. More recently, engineers have managed to tap into this energy reserve to produce liquid biofuels, primarily ethanol (see Chapter 7 for a discussion of biofuels). But no matter the end product, you can trace the energy and raw materials for growth back to solar-powered photosynthesis. ✓

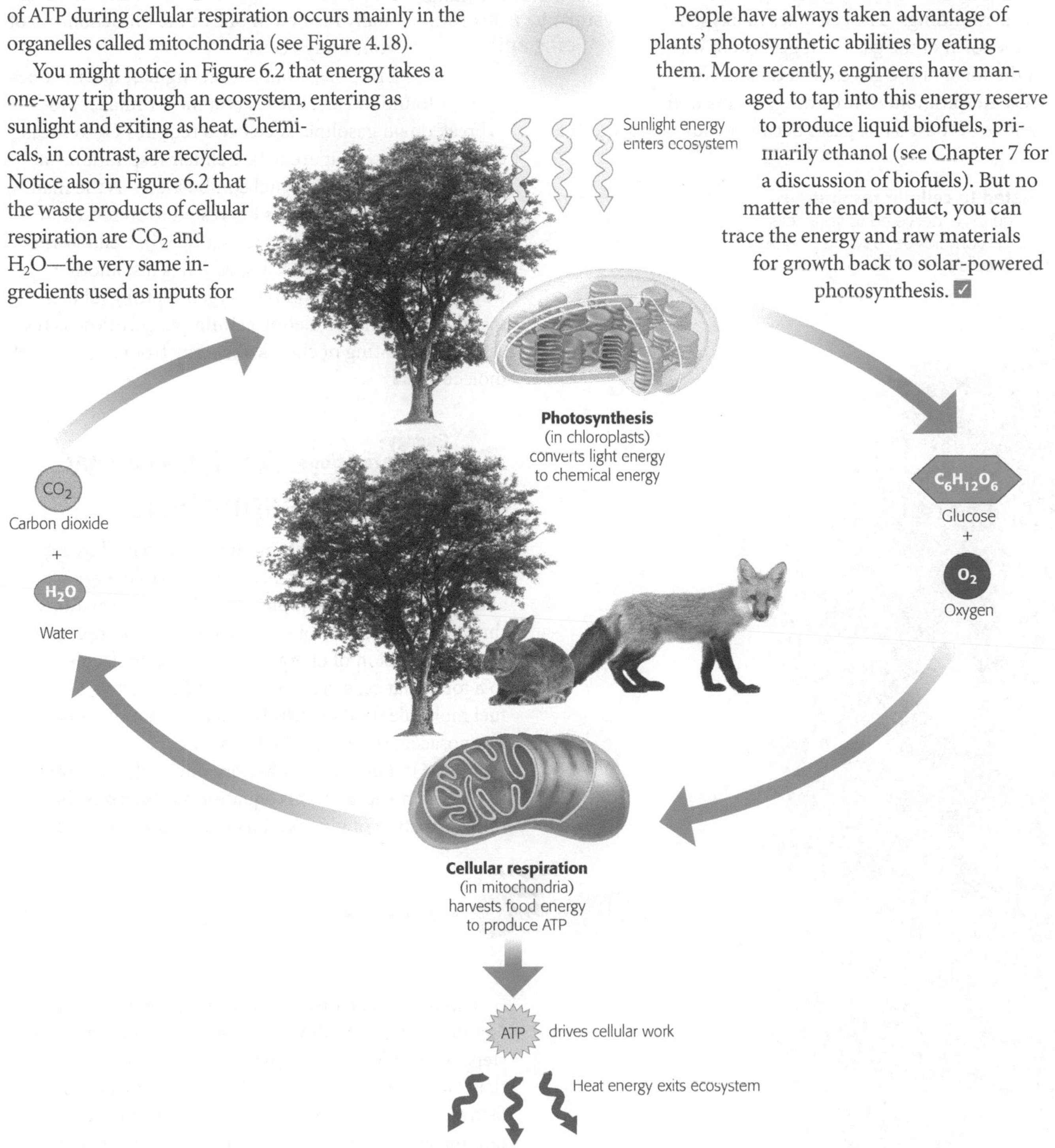

◀ **Figure 6.2 Energy flow and chemical cycling in ecosystems.** Energy flows through an ecosystem, entering as sunlight and exiting as heat. In contrast, chemical elements are recycled within an ecosystem.

Sunlight energy enters ecosystem

Photosynthesis (in chloroplasts) converts light energy to chemical energy

$C_6H_{12}O_6$ Glucose + O_2 Oxygen

CO_2 Carbon dioxide + H_2O Water

Cellular respiration (in mitochondria) harvests food energy to produce ATP

ATP drives cellular work

Heat energy exits ecosystem

Cellular Respiration: Aerobic Harvest of Food Energy

We usually use the word *respiration* to mean breathing. Although respiration on the organismal level should not be confused with cellular respiration, the two processes are closely related **(Figure 6.3)**. Cellular respiration requires a cell to exchange two gases with its surroundings. The cell takes in oxygen in the form of the gas O_2. It gets rid of waste in the form of the gas carbon dioxide, or CO_2. Respiration, or breathing, results in the exchange of these same gases between your blood and the outside air. Oxygen present in the air you inhale diffuses across the lining of your lungs and into your bloodstream. And the CO_2 in your bloodstream diffuses into your lungs and exits your body when you exhale. Every molecule of CO_2 that you exhale was originally formed in one of the mitochondria of your body's cells.

You have something in common with a sports car: You both require an air intake system to burn fuel efficiently.

Internal combustion engines, like the ones found in cars, use O_2 (via the air intakes) to break down gasoline. A cell also requires O_2 to break down its fuel (see Figure 5.2). Cellular respiration—a biological version of internal combustion—is the main way that chemical energy is harvested from food and converted to ATP energy (see Figure 5.6). Cellular respiration is an **aerobic** process, which is just another way of saying that it requires oxygen. Putting all this together, we can now define **cellular respiration** as the aerobic harvesting of chemical energy from organic fuel molecules. ☑

▼ **Figure 6.3 How breathing is related to cellular respiration.** When you inhale, you breathe in O_2. The O_2 is delivered to your cells, where it is used in cellular respiration. Carbon dioxide, a waste product of cellular respiration, diffuses from your cells to your blood and travels to your lungs, where it is exhaled.

Lungs

Cellular respiration

Muscle cells

✳ Energy Transformations An Overview of Cellular Respiration

One of biology's overarching themes is that all living organisms depends on transformations of energy and matter. We see examples of such transformations throughout the study of life, but few are as important as the conversion of energy in fuel (food molecules) to a form that cells can use directly. Most often, the fuel molecule used by cells is glucose, a simple sugar (monosaccharide) with the formula $C_6H_{12}O_6$ (see Figure 3.6). (Less often, other organic molecules are used to gain energy.) This equation summarizes the transformation of glucose during cellular respiration:

$$C_6H_{12}O_6 + 6\ O_2 \rightarrow \rightarrow \rightarrow 6\ CO_2 + 6\ H_2O + \text{approx. } 32\ ATP$$

The series of arrows in this formula represents the fact that cellular respiration consists of many chemical steps. A specific enzyme catalyzes each reaction—more than two dozen reactions in all—in the pathway. In fact, these reactions constitute one of the most important metabolic pathways for nearly every eukaryotic cell—those found in plants, fungi, protists, and animals. This pathway provides the energy these cells need to maintain the functions of life.

The many chemical reactions that make up cellular respiration can be grouped into three main stages: glycolysis, the citric acid cycle, and electron transport. **Figure 6.4** is a road map that will help you follow the three stages of respiration and see where each stage occurs in your cells. During **glycolysis**, a molecule of glucose is split into two molecules of a compound called pyruvic acid. The enzymes for glycolysis are located in the cytoplasm. The **citric acid cycle** (also called the Krebs cycle) completes the breakdown of glucose all the way to CO_2, which is then released as a waste product. The enzymes for the citric acid cycle are dissolved in the fluid within mitochondria. Glycolysis and the citric acid cycle generate a small amount of ATP directly. They generate much more ATP indirectly, via reactions that transfer electrons from fuel molecules to a molecule called NAD$^+$ (nicotinamide adenine dinucleotide) that cells make from niacin, a B vitamin. The electron transfer forms a molecule called **NADH** (the H represents the transfer of hydrogen along with the electrons) that acts as a shuttle carrying high-energy electrons from one area of the cell to another. The third stage of cellular respiration is **electron transport**. Electrons captured from food by the NADH formed in the first two stages are stripped of their energy, a little bit at a time, until they are finally combined with oxygen to form water. The proteins and other molecules that make up electron transport chains are embedded within the inner membrane of the mitochondria. The transport of electrons from NADH to oxygen releases the energy your cells use to make most of their ATP.

The overall equation for cellular respiration shows that the atoms of the reactant molecules glucose and oxygen are rearranged to form the products carbon dioxide and water. But don't lose track of why this process occurs: The main function of cellular respiration is to generate ATP for cellular work. In fact, the process can produce around 32 ATP molecules for each glucose molecule consumed. ☑

☑ **CHECKPOINT**

Which stages of cellular respiration take place in the mitochondria? Which stage takes place outside the mitochondria?

Answer: the citric acid cycle and electron transport; glycolysis

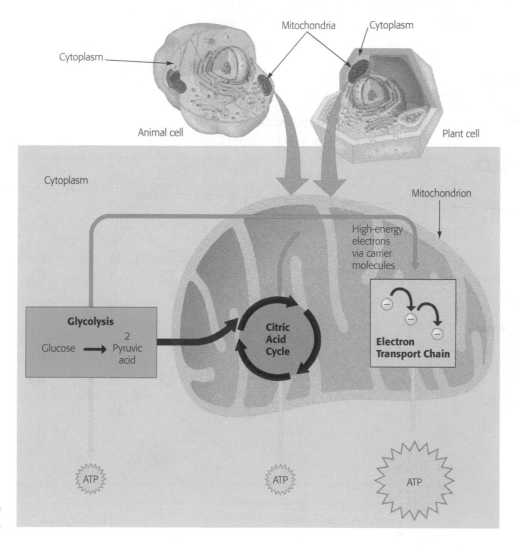

► Figure 6.4 **A road map for cellular respiration.**

The Three Stages of Cellular Respiration

Now that you have a big-picture view of cellular respiration, let's examine the process in more detail. A small version of Figure 6.4 will help you keep the overall process of cellular respiration in plain view as we take a closer look at its three stages.

Stage 1: Glycolysis

The word *glycolysis* means "splitting of sugar" **(Figure 6.5)**, and that's just what happens. ❶ During glycolysis, a six-carbon glucose molecule is broken in half, forming two three-carbon molecules. Notice in Figure 6.5 that the initial split requires an energy investment of two ATP molecules per glucose. ❷ The three-carbon molecules then donate high-energy electrons to NAD$^+$, forming NADH. ❸ In addition to NADH, glycolysis also makes four ATP molecules directly when enzymes transfer phosphate groups

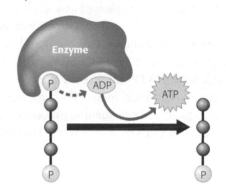

from fuel molecules to ADP **(Figure 6.6)**. Glycolysis thus produces a net of two molecules of ATP per molecule of glucose. (This fact will become important during our later discussion of fermentation.) What remains of the fractured glucose at the end of glycolysis are two molecules of pyruvic acid. The pyruvic acid still holds most of the energy of glucose, and that energy is harvested in the second stage of cellular respiration, the citric acid cycle.

▼ **Figure 6.6 ATP synthesis by direct phosphate transfer.** Glycolysis generates ATP when enzymes transfer phosphate groups directly from fuel molecules to ADP.

▼ **Figure 6.5 Glycolysis.** In glycolysis, a team of enzymes splits glucose, eventually forming two molecules of pyruvic acid. After investing 2 ATP at the start, glycolysis generates 4 ATP directly. More energy will be harvested later from high-energy electrons used to form NADH and from the two molecules of pyruvic acid.

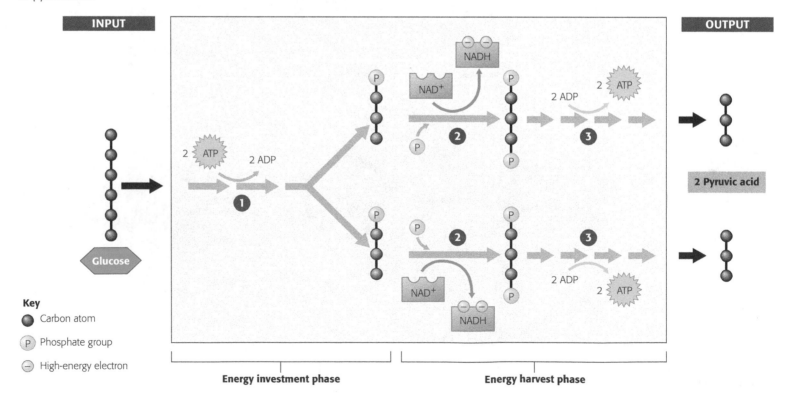

Key
- Carbon atom
- (P) Phosphate group
- ⊖ High-energy electron

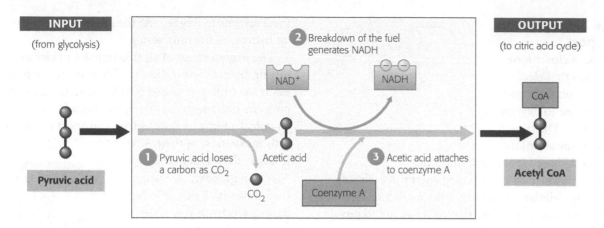

INPUT
(from glycolysis)

② Breakdown of the fuel generates NADH

OUTPUT
(to citric acid cycle)

NAD⁺   NADH

① Pyruvic acid loses a carbon as CO_2

Acetic acid

③ Acetic acid attaches to coenzyme A

CO_2

Coenzyme A

CoA

Pyruvic acid

Acetyl CoA

◄ **Figure 6.7 The link between glycolysis and the citric acid cycle: the conversion of pyruvic acid to acetyl CoA.** Remember that one molecule of glucose is split into two molecules of pyruvic acid. Therefore, the process shown here occurs twice for each starting glucose molecule.

Stage 2: The Citric Acid Cycle

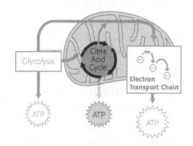

The two molecules of pyruvic acid, the fuel that remains after glycolysis, are not quite ready for the citric acid cycle. The pyruvic acid must be "groomed"—converted to a form the citric acid cycle can use **(Figure 6.7)**. ① First, each pyruvic acid loses a carbon as CO_2. This is the first of this waste product we've seen so far in the breakdown of glucose. The remaining fuel molecules, each with only two carbons left, are called acetic acid (the acid that's in vinegar). ② Electrons are stripped from these molecules and transferred to another molecule of NAD^+, forming more NADH. ③ Finally, each acetic acid is attached to a molecule called coenzyme A (CoA), an enzyme derived from the B vitamin pantothenic acid, to form acetyl CoA. The CoA escorts the acetic acid into

the first reaction of the citric acid cycle. The CoA is then stripped and recycled.

The citric acid cycle finishes extracting the energy of sugar by dismantling the acetic acid molecules all the way down to CO_2 **(Figure 6.8)**. ① Acetic acid joins a four-carbon acceptor molecule to form a six-carbon product called citric acid (for which the cycle is named). For every acetic acid molecule that enters the cycle as fuel, ② two CO_2 molecules eventually exit as a waste product. Along the way, the citric acid cycle harvests energy from the fuel. ③ Some of the energy is used to produce ATP directly. However, the cycle captures much more energy in the form of ④ NADH and ⑤ a second, closely related electron carrier called $FADH_2$. ⑥ All the carbon atoms that entered the cycle as fuel are accounted for as CO_2 exhaust, and the four-carbon acceptor molecule is recycled. We have tracked only one acetic acid molecule through the citric acid cycle here. But because glycolysis splits glucose in two, the citric acid cycle occurs twice for each glucose molecule that fuels a cell. ☑

☑ **CHECKPOINT**

Two molecules of what compound are produced by glycolysis? Does this molecule enter the citric acid cycle?

Answer: Pyruvic acid. No; it is first converted to acetic acid.

► **Figure 6.8 The citric acid cycle.**

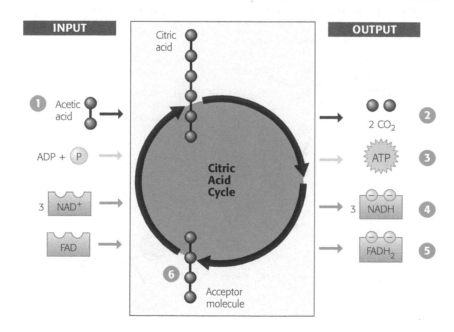

INPUT

Citric acid

① Acetic acid

$ADP + P$

3 NAD^+

FAD

Citric Acid Cycle

OUTPUT

2 CO_2 ②

ATP ③

3 NADH ④

$FADH_2$ ⑤

⑥

Acceptor molecule

Stage 3: Electron Transport

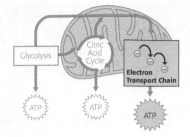

Let's take a closer look at the path that electrons take on their way from glucose to oxygen **(Figure 6.9)**. During cellular respiration, the electrons gathered from food molecules "fall" in a stepwise cascade, losing energy at each step. In this way, cellular respiration unlocks chemical energy in small amounts, bit by bit, that cells can put to productive use. The first stop in the path down the cascade is NAD^+. The transfer of electrons from organic fuel (food) to NAD^+ converts it to NADH. The electrons have now taken one baby step down in their trip from glucose to oxygen. The rest of the cascade consists of an **electron transport chain**.

Each link in an electron transport chain is actually a molecule, usually a protein (shown as purple circles in Figure 6.9). In a series of reactions, each member of the chain transfers electrons. With each transfer, the electrons give up a small amount of energy that can then be used indirectly to generate ATP. The first molecule of the chain accepts electrons from NADH. Thus, NADH carries electrons from glucose and other fuel molecules and deposits them at the top of an electron transport chain. The electrons cascade down the chain, from molecule to molecule, like an electron bucket brigade. The molecule at the bottom of the chain finally "drops"

the electrons to oxygen. At the same time, oxygen picks up hydrogen, forming water.

The overall effect of all this transfer of electrons during cellular respiration is a "downward" trip for electrons from glucose to NADH to an electron transport chain to oxygen. During the stepwise release of chemical energy in the electron transport chain, our cells make most of their ATP. It is actually oxygen, the "electron grabber," at the end, that makes it all possible. By pulling electrons down the transport chain from fuel molecules, oxygen functions somewhat like gravity pulling objects downhill. This role as a final electron acceptor is how the oxygen we breathe functions in our cells and why we cannot survive more than a few minutes without it. Viewed this way, drowning is deadly because it deprives cells of the final "electron grabbers" (oxygen) needed to drive cellular respiration.

The molecules of electron transport chains are built into the inner membranes of mitochondria (see

▼ Figure 6.9 **The role of oxygen in harvesting food energy.** In cellular respiration, electrons "fall" in small steps from food to oxygen, producing water. NADH transfers electrons from food to an electron transport chain. The attraction of oxygen to electrons "pulls" the electrons down the chain.

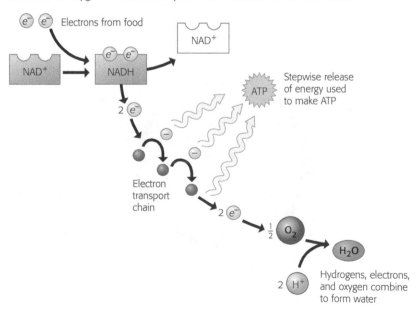

Figure 4.18). Because these membranes are highly folded, their large surface area can accommodate thousands of copies of the electron transport chain—another good example of how biological structure fits function. Each chain acts as a chemical pump that uses the energy released by the "fall" of electrons to move hydrogen ions (H$^+$) across the inner mitochondrial membrane. This pumping causes ions to become more concentrated on one side of the membrane than on the other. Such a difference in concentration stores potential energy, similar to the way water can be stored behind a dam. There

is a tendency for hydrogen ions to gush back to where they are less concentrated, just as there is a tendency for water to flow downhill. The inner membrane temporarily "dams" hydrogen ions.

The energy of dammed water can be harnessed to perform work. Gates in a dam allow the water to rush downhill, turning giant turbines, and this work can be used to generate electricity. Your mitochondria have structures that act like turbines. Each of these miniature machines, called an **ATP synthase**, is constructed from proteins built into the inner mitochondrial membrane, adjacent to the proteins of the electron transport chains. **Figure 6.10** shows a simplified view of how the energy previously stored in NADH and FADH$_2$ can now be used to generate ATP. **1** NADH and **2** FADH$_2$ transfer electrons to an electron transport chain. **3** The electron transport chain uses this energy supply to pump H$^+$ across the inner mitochondrial membrane. **4** Oxygen pulls electrons down the transport chain. **5** The H$^+$ concentrated on one side of the membrane rushes back "downhill" through an ATP synthase. This action spins a component of the ATP synthase, just as water turns the turbines in a dam. **6** The rotation activates parts of the synthase molecule that attach phosphate groups to ADP molecules to generate ATP.

The poison cyanide produces its deadly effect by binding to one of the protein complexes in the electron transport chain (marked with a skull-and-crossbones symbol in Figure 6.10). When bound there, cyanide blocks the passage of electrons to oxygen. This blockage is like clogging a dam. As a result, no H$^+$ gradient is generated, and no ATP is made. Cells stop working, and the organism dies. ☑

☑ **CHECKPOINT**

What is the potential energy source that drives ATP production by ATP synthase?

Answer: a concentration gradient of H$^+$ across the inner membrane of a mitochondrion

▼ Figure 6.10 **How electron transport drives ATP synthase machines.**

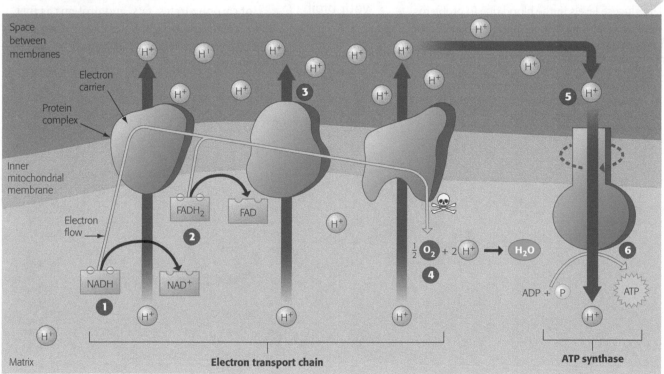

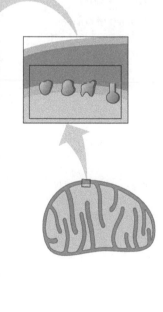

The Results of Cellular Respiration

When taking cellular respiration apart to see how all the molecular nuts and bolts of its metabolic machinery work, it's easy to lose sight of its overall function: to generate about 32 molecules of ATP per molecule of glucose (the actual number can vary by a few, depending on the organism and molecules involved). **Figure 6.11** will help you keep track of the ATP molecules generated. As we discussed, glycolysis and the citric acid cycle each contribute 2 ATP by directly making it. The rest of the ATP molecules are produced by ATP synthase, powered by the "fall" of electrons from food to oxygen. The electrons are carried from the organic fuel to electron transport chains by NADH and FADH$_2$. Each electron pair "dropped" down a transport chain from NADH or FADH$_2$ can power the synthesis of a few ATP. You can visualize the process like this: Energy flows from glucose to carrier molecules and ultimately to ATP.

We have seen that glucose can provide the energy to make the ATP our cells use for all their work. All of the energy-consuming activities of your body—moving your muscles, maintaining your heartbeat and temperature, and even the thinking that goes on within your brain—can be traced back to ATP and, before that, the glucose that was used to make it. The importance of glucose is underscored by the severity of diseases in which glucose balance is disturbed. Diabetes, which affects more than 20 million Americans, is caused by an inability to properly regulate glucose levels in the blood due to problems with the hormone insulin. If left untreated, a glucose imbalance can lead to a variety of problems, including cardiovascular disease, coma, and even death.

> **About 20% of the energy produced by your body each day is used to sustain your brain.**

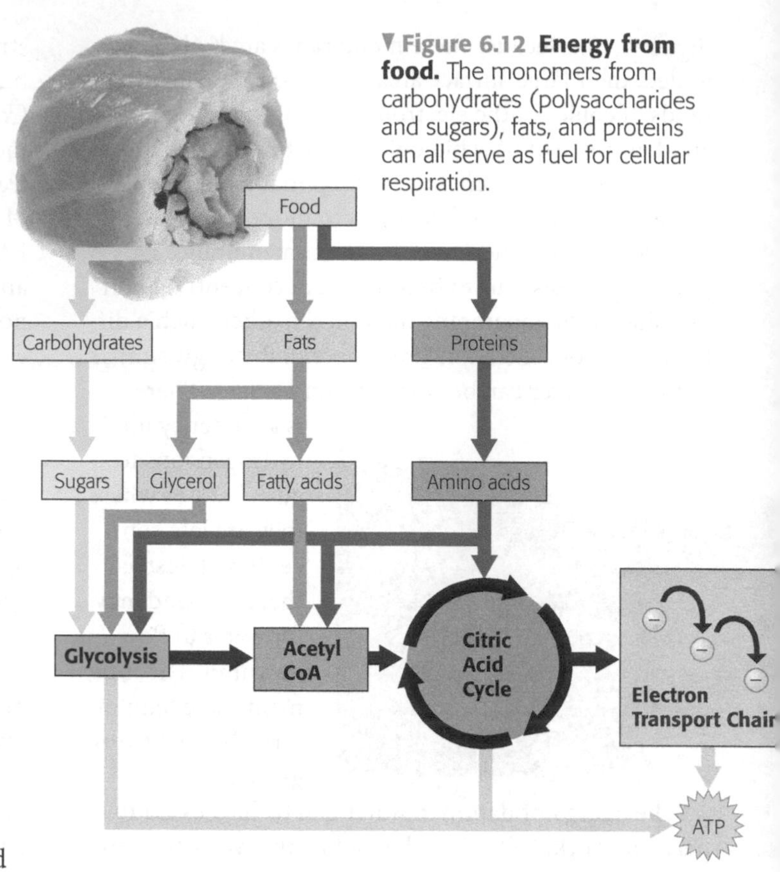

But even though we have concentrated on glucose as the fuel that is broken down during cellular respiration, respiration is a versatile metabolic furnace that can "burn" many other kinds of food molecules. **Figure 6.12** diagrams some metabolic routes for the use of carbohydrates, fats, and proteins as fuel for cellular respiration. Taken together, all of these food molecules make up your calorie-burning metabolism. ☑

☑ CHECKPOINT

Which stage of cellular respiration produces the majority of ATP?

Answer: electron transport

▼ **Figure 6.12 Energy from food.** The monomers from carbohydrates (polysaccharides and sugars), fats, and proteins can all serve as fuel for cellular respiration.

▶ **Figure 6.11 A summary of ATP yield during cellular respiration.**

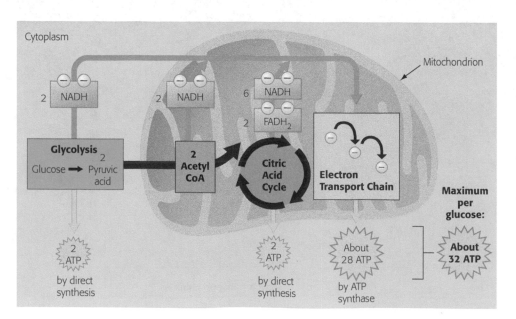

Fermentation: Anaerobic Harvest of Food Energy

Although you must breathe to stay alive, some of your cells can work for short periods without oxygen. This **anaerobic** ("without oxygen") harvest of food energy is called fermentation.

Fermentation in Human Muscle Cells

You know by now that as your muscles work, they require a constant supply of ATP, which is generated by cellular respiration. As long as your blood provides your muscle cells with enough O_2 to keep electrons "falling" down transport chains in mitochondria, your muscles will work aerobically.

But under strenuous conditions, your muscles can spend ATP faster than your bloodstream can deliver O_2; when this happens, your muscle cells begin to work anaerobically. After functioning anaerobically for about 15 seconds, muscle cells will begin to generate ATP by the process of fermentation. **Fermentation** relies on glycolysis, the first stage of cellular respiration. Glycolysis does not require O_2 but does produce two ATP molecules for each glucose molecule broken down to pyruvic acid. That isn't very efficient compared with the 32 or so ATP molecules each glucose molecule generates during cellular respiration, but it can energize muscles for a short burst of activity. However, in such situations your cells will have to consume more glucose fuel per second because so much less ATP per glucose molecule is generated under anaerobic conditions.

To harvest food energy during glycolysis, NAD^+ must be present to receive electrons (see Figure 6.9). This is no problem under aerobic conditions, because the cell regenerates NAD^+ when NADH drops its electron cargo down electron transport chains to O_2. However, this recycling of NAD^+ cannot occur under anaerobic conditions because there is no O_2 to accept the electrons. Instead, NADH disposes of electrons by adding them to the pyruvic acid produced by glycolysis **(Figure 6.13)**. This restores NAD^+ and keeps glycolysis working.

The addition of electrons to pyruvic acid produces a waste product called lactic acid. The lactic acid by-product is eventually transported to the liver, where liver cells convert it back to pyruvic acid. Exercise physiologists have long speculated about the role that lactic acid plays in muscle fatigue, as you'll see next. ☑

☑ CHECKPOINT

How many molecules of ATP can be produced from one molecule of glucose during fermentation?

Answer: two

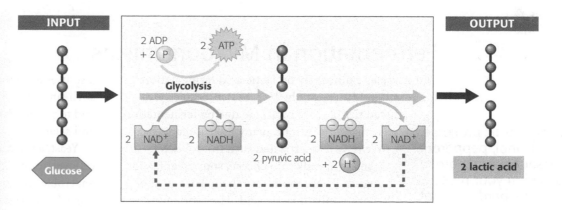

▼ Figure 6.13 **Fermentation: producing lactic acid.** Glycolysis produces ATP even in the absence of O_2. This process requires a continuous supply of NAD^+ to accept electrons from glucose. The NAD^+ is regenerated when NADH transfers the electrons it removed from food to pyruvic acid, thereby producing lactic acid (or other waste products, depending on the species of organism).

Exercise Science THE PROCESS OF SCIENCE

What Causes Muscle Burn?

You may have heard that the burn you experience after hard exercise ("Feel the burn!") is due to the buildup of lactic acid in your muscles. This idea originated with the work of a British biologist named A.V. Hill. Considered one of the founders of the field of exercise physiology, Hill won a 1922 Nobel Prize for his investigations of muscle contraction.

In 1929, Hill performed a classic experiment that began with the **observation** that muscles produce lactic acid under anaerobic conditions. Hill asked the **question**, Does the buildup of lactic acid cause muscle fatigue? To find out, Hill developed a technique for electrically stimulating dissected frog muscles in a laboratory solution. He formed the **hypothesis** that a buildup of lactic acid would cause muscle activity to stop.

Hill's **experiment** tested frog muscles under two different sets of conditions (**Figure 6.14**). First, he showed that muscle performance declined when lactic acid could not diffuse away from the muscle tissue. Next, he showed that when lactic acid was allowed to diffuse away, performance improved significantly. These **results** led Hill to the conclusion that the buildup of lactic acid is the primary cause of muscle failure under anaerobic conditions.

Given his scientific stature (he was considered the world's leading authority on muscle activity), Hill's conclusion went unchallenged for many decades. Gradually, however, evidence that contradicted Hill's results began to accumulate. For example, the effect that Hill demonstrated did not appear to occur at human body temperature. And certain individuals who are unable to accumulate lactic acid have muscles that fatigue

more rapidly, which is the opposite of what you would expect. Recent experiments have directly refuted Hill's conclusions. Research indicates that increased levels of other ions may be to blame, and the role of lactic acid in muscle fatigue remains a hotly debated topic.

The changing view of lactic acid's role in muscle fatigue illustrates an important point about the process of science: It is dynamic and subject to constant adjustment as new evidence is uncovered. This would not have surprised Hill, who himself observed that all scientific hypotheses may become obsolete, and that changing conclusions in light of new evidence is necessary for the advancement of science.

▼ Figure 6.14 **A. V. Hill's 1929 apparatus for measuring muscle fatigue.**

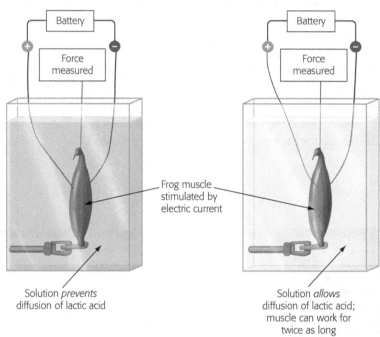

Fermentation in Microorganisms

Our muscles cannot rely on lactic acid fermentation for very long. However, the two ATP molecules produced per glucose molecule during fermentation is enough to sustain many microorganisms. We have domesticated such microbes to transform milk into cheese, sour cream, and yogurt. These foods owe their sharp or sour flavor mainly to lactic acid. The food industry

> Similar metabolic processes produce alcohol, pepperoni, soy sauce, rising bread, and acid in your muscles after a hard workout.

also uses fermentation to produce soy sauce from soybeans, to pickle cucumbers, olives, and cabbage, and to produce meat products like sausage, pepperoni, and salami.

Yeast, a microscopic fungus, is capable of both cellular respiration and fermentation. When kept in an anaerobic environment, yeast cells ferment sugars and other foods to stay alive. As they do, the yeast produce ethyl alcohol as a waste product instead of lactic acid (Figure 6.15). This alcoholic fermentation also releases CO_2. For thousands of years, people have put yeast to work producing alcoholic beverages such as beer and wine. And as every baker knows, the CO_2 bubbles from fermenting yeast also cause bread dough to rise. (The alcohol produced in fermenting bread is burned off during baking.) ☑

☑ CHECKPOINT

What kind of acid builds up in human muscle during strenuous activity?

Answer: lactic acid

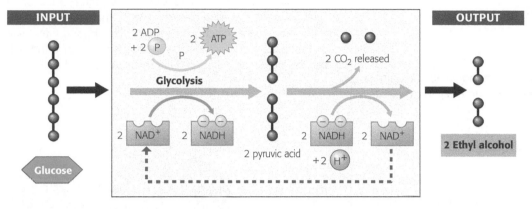

▲ Figure 6.15 **Fermentation: producing ethyl alcohol.** The alcohol produced by yeast as bread rises is burned off during baking.

Exercise Science EVOLUTION CONNECTION

The Importance of Oxygen

In the Biology and Society and the Process of Science sections, we have been reminded of the important role that oxygen plays during aerobic exercise. But in the last section on fermentation, we learned that exercise can continue on a limited basis even under anaerobic (oxygen-free) conditions. Both aerobic and anaerobic respiration start with glycolysis, the splitting of glucose to form pyruvic acid. Glycolysis is thus the universal energy-harvesting process of life.

The role of glycolysis in both respiration and fermentation has an evolutionary basis. Ancient prokaryotes probably used glycolysis to make ATP long before oxygen was present in Earth's atmosphere. The oldest known fossils of bacteria date back more than 3.5 billion years, but significant levels of O_2 did not accumulate in the atmosphere until about 2.7 billion years ago (Figure 6.16). For almost one billion years, prokaryotes must have generated ATP exclusively from glycolysis.

The fact that glycolysis occurs in almost all organisms suggests that it evolved very early in ancestors common to all the domains of life. The location of glycolysis within the cell also implies great antiquity; the pathway does not require any of the membrane-enclosed organelles of the eukaryotic cell, which evolved more than a billion years after the prokaryotic cell. Glycolysis is a metabolic heirloom from early cells that continues to function in fermentation and as the first stage in the breakdown of organic molecules by cellular respiration. The ability of our muscles to function anaerobically can therefore be viewed as a vestige of our ancient ancestors who relied exclusively on this metabolic pathway.

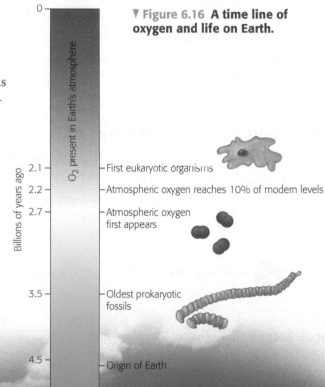

▼ Figure 6.16 **A time line of oxygen and life on Earth.**

O_2 present in Earth's atmosphere

Billions of years ago

0

2.1 — First eukaryotic organisms

2.2 — Atmospheric oxygen reaches 10% of modern levels

2.7 — Atmospheric oxygen first appears

3.5 — Oldest prokaryotic fossils

4.5 — Origin of Earth

Chapter Review

SUMMARY OF KEY CONCEPTS

Energy Flow and Chemical Cycling in the Biosphere

Producers and Consumers

Autotrophs (producers) make organic molecules from inorganic nutrients via photosynthesis. Heterotrophs (consumers) must consume organic material and obtain energy via cellular respiration.

Chemical Cycling between Photosynthesis and Cellular Respiration

The molecular outputs of cellular respiration—CO_2 and H_2O—are the molecular inputs of photosynthesis, and vice versa. While these chemicals cycle through an ecosystem, energy flows through, entering as sunlight and exiting as heat.

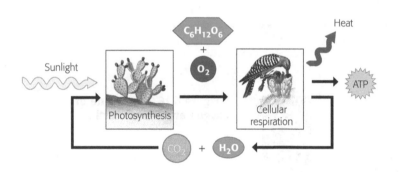

Cellular Respiration: Aerobic Harvest of Food Energy

An Overview of Cellular Respiration

The overall equation of cellular respiration simplifies a great many chemical steps into one formula:

The Three Stages of Cellular Respiration

Cellular respiration occurs in three stages. During glycolysis, a molecule of glucose is split into two molecules of pyruvic acid, producing two molecules of ATP and two high-energy electrons stored in NADH. During the citric acid cycle, what remains of glucose is completely broken down to CO_2, producing a bit of ATP and a lot of high-energy electrons stored in NADH and $FADH_2$. The electron transport chain uses the high-energy electrons to pump H^+ across the inner mitochondrial membrane, eventually handing them off to O_2, producing H_2O. Backflow of H^+ across the membrane powers the ATP synthases, which produce ATP from ADP.

The Results of Cellular Respiration

You can follow the flow of molecules through the process of cellular respiration in the following diagram. Notice that the first two stages primarily produce high-energy electrons carried by NADH, and that it is the final stage that uses these high-energy electrons to produce the bulk of the ATP molecules produced during cellular respiration.

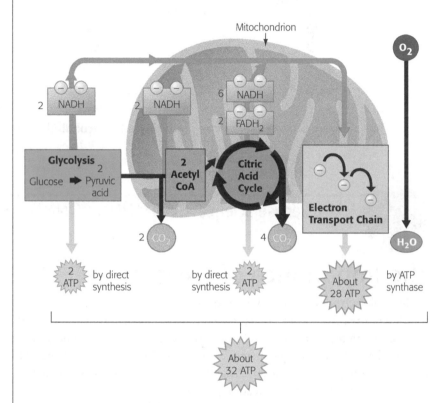

Fermentation: Anaerobic Harvest of Food Energy

Fermentation in Human Muscle Cells

When muscle cells consume ATP faster than O_2 can be supplied for cellular respiration, the conditions become anaerobic, and muscle cells will begin to regenerate ATP by fermentation. The waste product under these anaerobic conditions is lactic acid. The ATP yield per glucose is much lower during fermentation (2 ATP) than during cellular respiration (about 32 ATP).

Fermentation in Microorganisms

Yeast and some other organisms can survive with or without O_2. Wastes from fermentation can be ethyl alcohol, lactic acid, or other compounds, depending on the species.

MasteringBiology®

For practice quizzes, BioFlix animations, MP3 tutorials, video tutors, and more study tools designed for this textbook, go to MasteringBiology®

SELF-QUIZ

1. Which of the following statements is a correct distinction between autotrophs and heterotrophs?
   a. Only heterotrophs require chemical compounds from the environment.
   b. Cellular respiration is unique to heterotrophs.
   c. Only heterotrophs have mitochondria.
   d. Only autotrophs can live on nutrients that are entirely inorganic.

2. Why are plants called producers? Why are animals called consumers?

3. How is your breathing related to your cellular respiration?

4. Which stage of cellular respiration produces the most NADH?

5. The first electron acceptor of cellular respiration is _____.

6. The poison cyanide acts by blocking a key step in the electron transport chain. Knowing this, explain why cyanide kills so quickly.

7. Cells can harvest the most chemical energy from which of the following?
   a. an NADH molecule
   b. a glucose molecule
   c. six CO_2 molecules
   d. two pyruvic acid molecules

8. _____ is a metabolic pathway common to both fermentation and cellular respiration.

9. Physicians find that a child is born with a rare disease in which mitochondria are missing from certain skeletal muscle cells but her muscle cells function. Not surprisingly, they also find that
   a. the muscles contain large amounts of lactic acid following even mild physical exercise.
   b. the muscles contain large amounts of carbon dioxide following even mild physical exercise.
   c. the muscles require extremely high levels of oxygen to function.
   d. the muscle cells cannot split glucose to pyruvic acid.

10. A glucose-fed yeast cell is moved from an aerobic environment to an anaerobic one. For the cell to continue to generate ATP at the same rate, approximately how much glucose must it consume in the anaerobic environment compared with the aerobic environment?

Answers to these questions can be found in Appendix: Self-Quiz Answers.

THE PROCESS OF SCIENCE

11. The process of cellular respiration, which breaks down glucose into CO_2 and H_2O and generates approximately 32 molecules of ATP, takes place in different compartments of the cell—glycolysis occurs in the cytoplasm, the citric acid cycle in the mitochondrial matrix, and the electron transport chain in the inner mitochondrial membrane. In order for this process to get completed, several molecules need to cross the mitochondrial membranes. Which are these molecules, and in which directions do they move? What is the need for this elaborate arrangement?

12. **Interpreting Data** The basal metabolic rate (BMR) is the amount of energy that must be consumed by a person at rest to maintain his or her body weight. BMR depends on several factors, including sex, age, height, and weight. The following graph shows the BMR for a 6'0" 45-year-old male. For this person, how does BMR correlate with weight? How many more calories must a 250-pound man consume to maintain his weight than a 200 pound man? Why would BMR depend on weight?

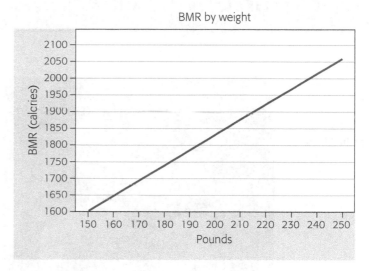

BMR by weight

13. As discussed in the Biology and Society section, the delivery of oxygen to muscles is the limiting factor for many athletes. Some athletes seek to improve their athletic performance through blood doping, which can artificially increase athletic capacity. Other athletes achieve the same result by training at high altitude (which promotes the formation of more red blood cells by the bone marrow). If two athletes achieve exactly the same result—one due to injecting her own blood and one due to training at altitude—why do you think the former is considered cheating but the latter is not? What would you do to enforce antidoping rules in sports at all levels (high school, college, Olympic, professional)?

14. Various cultures have been using fermentation for a long time to produce and conserve foods. Compile a list of the different types of fermentation and to which end they are used.

15. The consumption of alcohol by a pregnant woman can cause a series of birth defects called fetal alcohol syndrome (FAS). Symptoms of FAS include head and facial irregularities, heart defects, mental retardation, and behavioral problems. The U.S. Surgeon General's Office recommends that pregnant women abstain from drinking alcohol, and the government has mandated that a warning label be placed on liquor bottles. Imagine you are a server in a restaurant. An obviously pregnant woman orders a strawberry daiquiri. How would you respond? Is it the woman's right to make those decisions about her unborn child's health? Do you bear any responsibility in the matter? Is a restaurant responsible for monitoring the dietary habits of its customers?

BIOLOGY AND SOCIETY

7 Photosynthesis: Using Light to Make Food

Why Photosynthesis Matters

If you want to reduce the rate of global climate change, plant a tree. ▶

◀ Nearly all life on Earth—including you—can trace its source of energy back to the sun.

Protecting yourself from short wavelengths of light can be lifesaving. ▶

140

Biofuels BIOLOGY AND SOCIETY

A Greasy Crime Wave

In September 2013, police in Ocala, Florida, arrested two men and charged them with organized fraud and grand theft. Their crime? The men were caught red-handed with more than 700 gallons of stolen used cooking oil pilfered from a variety of local eateries. Why would anyone steal that nasty stuff? The reason is simple: Remnants of restaurant deep fryers, sometimes called "liquid gold," fetch about $2 per pound when sold to recyclers. That makes the burglars' haul worth more than $5,000. Why is grease so valuable?

As fossil fuel supplies dwindle and prices rise, the need for reliable, renewable sources of energy increases. In response, scientists are researching better ways to harness biofuels, energy obtained from living material. Some researchers focus on burning plant matter directly (wood pellet boilers, for example), and others focus on using plant material to produce biofuels that can be burned.

There are several types of biofuels. Bioethanol is a type of alcohol (the same kind found in alcoholic drinks) that is made from wheat, corn, sugar beets, and other food crops. Starch made naturally by plants is converted to glucose and then fermented to ethanol by microorganisms such as single-celled algae. Bioethanol can be used directly as a fuel source in specially designed vehicles, but it is more commonly used as a gasoline additive that can increase fuel efficiency while decreasing vehicle emissions. You may have noticed a sticker on a gas pump that declares the percentage of ethanol in that gasoline; most cars today run on a blend of 85% gasoline and 15% ethanol. Many car manufacturers are producing "flexible-fuel" vehicles that can run on any combination of gasoline and bioethanol. Although bioethanol does reduce carbon emissions and is a renewable resource, its production raises the prices of food crops (which become more expensive as acreage is diverted to biofuel production).

Cellulosic ethanol is a form of bioethanol made from cellulose found in nonedible plant material such as wood, grass, or scraps from crops. Biodiesel, the most common biofuel in Europe, is made from plant oils such as recycled frying oil. Like bioethanol, it can be used on its own or as an emissions-reducing additive to standard diesel. In a strange twist, rising values for diesel have sparked a greasy crime wave as thieves tap into this new and largely unguarded source of raw material. Today, only about 2.7% of the world's fuel used for driving is provided by biofuels, but the International Energy Agency has set a goal of 25% by 2050.

When we derive energy from biofuels, we are actually tapping into the energy of the sun, which drives photosynthesis in plants. Photosynthesis is the process by which plants use light to make sugars from carbon dioxide—sugars that are food for the plant and the starting point for most of our own food. In this chapter, we'll first examine some basic concepts of photosynthesis; then we'll look at the specific mechanisms involved in this process.

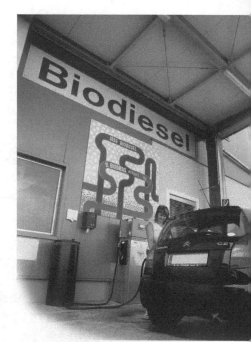

Using biofuels. Biofuels are added to most gasoline in the United States.

The Basics of Photosynthesis

The process of photosynthesis is the ultimate source of energy for nearly every ecosystem on Earth. **Photosynthesis** is a process whereby plants, algae (which are protists), and certain bacteria transform light energy into chemical energy, using carbon dioxide and water as starting materials and releasing oxygen gas as a by-product. The chemical energy produced via photosynthesis is stored in the bonds of sugar molecules.

> Nearly all life on Earth—including you—can trace its source of energy back to the sun.

Organisms that generate their own organic matter from inorganic ingredients are called autotrophs (see Chapter 6). Plants and other organisms that do this by photosynthesis—photoautotrophs—are the producers for most ecosystems **(Figure 7.1)**. Not only do photoautotrophs feed us, they also clothe us (as the source of cotton fibers), house us (wood), and provide energy for warmth, light, and transportation (biofuels).

Chloroplasts: Sites of Photosynthesis

Photosynthesis in plants and algae occurs within light-absorbing organelles called **chloroplasts** (see Chapter 4, especially Figure 4.17). All green parts of a plant have chloroplasts and thus can carry out photosynthesis. In most plants, however, the leaves have the most chloroplasts (about 500,000 per square millimeter of leaf surface—that's equivalent to about 300 million chloroplasts in a leaf the size of a standard postage stamp). Their green color is from **chlorophyll**, a pigment (light-absorbing molecule) in the chloroplasts that plays a central role in converting solar energy to chemical energy.

Chloroplasts are concentrated in the interior cells of leaves **(Figure 7.2)**, with a typical cell containing 30–40 chloroplasts. Carbon dioxide (CO_2) enters a leaf, and oxygen (O_2) exits, by way of tiny pores called **stomata** (singular, *stoma*, meaning "mouth"). The carbon dioxide that enters the leaf is the source of carbon for much of the body of the plant, including the sugars and starches that we eat. So the bulk of the body of a plant derives from the air, not the soil. As proof of this idea, consider hydroponics, a means of growing plants using only air and water; no soil whatsoever is involved. In addition to carbon dioxide, photosynthesis requires water, which is absorbed by the plant's roots and transported to the leaves, where veins carry it to the photosynthetic cells.

Membranes within the chloroplast form the framework where many of the reactions of photosynthesis occur. Like a mitochondrion, a chloroplast has a double-membrane envelope. The chloroplast's inner membrane encloses a compartment filled with **stroma**, a thick fluid. (It's easy to confuse two terms associated with photosynthesis: *stomata* are pores through which

▼ Figure 7.1 **A diversity of photoautotrophs.**

PHOTOSYNTHETIC AUTOTROPHS		
Plants (mostly on land)	**Photosynthetic Protists** (aquatic)	**Photosynthetic Bacteria** (aquatic)
Forest plants	Kelp, a large, multicellular alga	Micrograph of cyanobacteria

gases are exchanged, and *stroma* is the fluid within the chloroplast.) Suspended in the stroma are interconnected membranous sacs called **thylakoids**. The thylakoids are concentrated in stacks called **grana** (singular, *granum*). The chlorophyll molecules that capture light energy are built into the thylakoid membranes. The structure of a chloroplast—with its stacks of disks—aids its function by providing a large surface area for the reactions of photosynthesis. ☑

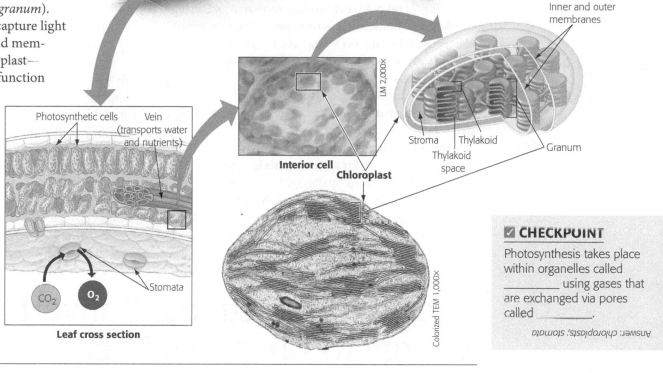

Inner and outer membranes

LM 2,000×

Interior cell

Chloroplast

Stroma
Thylakoid
Thylakoid
space
Granum

▶ **Figure 7.2 Journey into a leaf.** This series of blowups takes you into a leaf's interior, then into a plant cell, and finally into a chloroplast, the site of photosynthesis.

Photosynthetic cells   Vein (transports water and nutrients)

CO_2 O_2 Stomata

Colonized TEM 1,000×

Leaf cross section

☑ **CHECKPOINT**

Photosynthesis takes place within organelles called _____ using gases that are exchanged via pores called _____.

Answer: chloroplasts; stomata

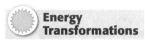 **Energy Transformations** An Overview of Photosynthesis

The following chemical equation, simplified to highlight the relationship between photosynthesis and cellular respiration, provides a summary of the reactants and products of photosynthesis:

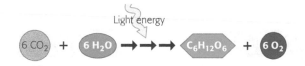

Light energy

$$6\ CO_2 + 6\ H_2O \longrightarrow\longrightarrow\longrightarrow C_6H_{12}O_6 + 6\ O_2$$

Notice that the reactants of photosynthesis—carbon dioxide (CO_2) and water (H_2O)—are the same as the waste products of cellular respiration (see Figure 6.2). Also notice that photosynthesis produces what respiration uses—glucose ($C_6H_{12}O_6$) and oxygen (O_2). In other words, photosynthesis recycles the "exhaust" of cellular respiration and rearranges its atoms to produce food and oxygen. Photosynthesis is a chemical transformation that requires a lot of energy, and sunlight absorbed by chlorophyll provides that energy.

Recall that cellular respiration is a process of electron transfer (see Chapter 6). A "fall" of electrons from food molecules to oxygen to form water releases the energy that mitochondria can use to make ATP (see Figure 6.9). The opposite occurs in photosynthesis: Electrons are boosted "uphill" and added to carbon dioxide to produce

sugar. Hydrogen is moved along with the electrons being transferred from water to carbon dioxide. This transfer of hydrogen requires the chloroplast to split water molecules into hydrogen and oxygen. The hydrogen is transferred along with electrons to carbon dioxide to form sugar. The oxygen escapes through stomata in leaves into the atmosphere as O_2, a waste product of photosynthesis.

The overall equation for photosynthesis is a simple summary of a complex process. Like many energy-producing processes within cells, photosynthesis is a multistep chemical pathway, with each step in the path producing products that are used as reactants in the next step. This is a clear example of one of biology's major themes: the use of metabolic pathways to obtain, process, and store energy. To help get a better overview, let's take a look at the two stages of photosynthesis: the light reactions and the Calvin cycle **(Figure 7.3)**.

▼ **Figure 7.3 A road map for photosynthesis.** We'll use a smaller version of this road map for orientation as we take a closer look at the light reactions and the Calvin cycle.

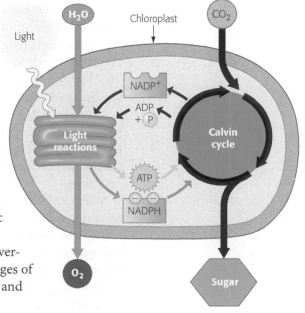

H_2O Chloroplast CO_2

Light

$NADP^+$
$ADP + P$

Light reactions

Calvin cycle

ATP

NADPH

O_2

Sugar

In the **light reactions**, chlorophyll in the thylakoid membranes absorbs solar energy (the "photo" part of photosynthesis), which is then converted to the chemical energy of ATP (the molecule that drives most cellular work) and **NADPH** (an electron carrier). During the light reactions, water is split, providing a source of electrons and giving off O_2 gas as a by-product.

The **Calvin cycle** uses the products of the light reactions to power the production of sugar from carbon dioxide (the "synthesis" part of photosynthesis). The enzymes that drive the Calvin cycle are dissolved in the stroma. ATP generated by the light reactions provides the energy for sugar synthesis. And the NADPH produced by the light reactions provides the high-energy electrons that drive the synthesis of glucose from carbon dioxide. Thus, the

If you want to reduce the rate of global climate change, plant a tree.

Calvin cycle indirectly depends on light to produce sugar because it requires the supply of ATP and NADPH produced by the light reactions.

The initial incorporation of carbon from CO_2 into organic compounds is called **carbon fixation**. This process has important implications for global climate, because the removal of carbon from the air and its incorporation into plant material can help reduce the concentration of carbon dioxide in the atmosphere. Deforestation, which removes a lot of photosynthetic plant life, thereby reduces the ability of the biosphere to absorb carbon. Planting new forests can have the opposite effect of fixing carbon from the atmosphere, potentially reducing the effect of the gases that contribute to global climate change. ☑

The Light Reactions: Converting Solar Energy to Chemical Energy

Chloroplasts are solar-powered sugar factories. Let's look at how they convert sunlight to chemical energy.

The Nature of Sunlight

Sunlight is a type of energy called radiation or electromagnetic energy. Electromagnetic energy travels through space as rhythmic waves, like the ripples made by a pebble dropped into a pond. The distance between the crests of two adjacent waves is called a **wavelength**. The full range of radiation, from the very short wavelengths of gamma rays to the very long wavelengths of radio signals, is called the **electromagnetic spectrum (Figure 7.4)**. Visible light is the fraction of the spectrum that our eyes see as different colors.

When sunlight shines on a pigmented material, certain wavelengths (colors) of the visible light are absorbed and disappear from the light that is reflected by the material. For example, we see a pair of jeans as blue because pigments in the fabric absorb the other colors, leaving only light in the blue part of the spectrum to be reflected from the fabric to our eyes. In the 1800s,

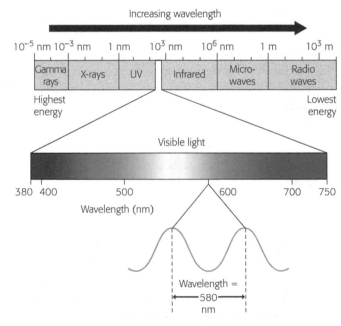

▲ Figure 7.4 **The electromagnetic spectrum.** The middle of the figure expands the thin slice of the spectrum that is visible to us as different colors of light, from about 380 nanometers (nm) to about 750 nm in wavelength. The bottom of the figure shows electromagnetic waves of one particular wavelength of visible light.

botanists (biologists who study plants) discovered that only certain wavelengths of light are used by plants, as we'll see next.

What Colors of Light Drive Photosynthesis?

In 1883, German biologist Theodor Engelmann made the **observation** that certain bacteria living in water tend to cluster in areas with higher oxygen concentrations. He already knew that light passed through a prism would separate into the different wavelengths (colors). Engelmann soon began to **question** whether he could use this information to determine which wavelengths of light work best for photosynthesis.

Engelmann's **hypothesis** was that oxygen-seeking bacteria would congregate near regions of algae performing the most photosynthesis (and hence producing the most oxygen). Engelmann began his **experiment** by laying a string of freshwater algal cells within a drop of water on a microscope slide. He then added oxygen-sensitive bacteria to the drop. Next, using a prism, he created a spectrum of light and shined it on the slide. His **results**, summarized in **Figure 7.5**, showed that most bacteria congregated around algae exposed to red-orange and blue-violet light, with very few bacteria moving to the area of green light. Other experiments have since verified that chloroplasts absorb light mainly in the blue-violet and red-orange part of the spectrum and that those wavelengths of light are the ones mainly responsible for photosynthesis.

Variations of this classic experiment are still performed today. For example, biofuel researchers test different species of algae to determine which wavelengths of light result in optimal fuel production. Biofuel facilities of the future may use a variety of species that take advantage of the full spectrum of light that shines down on them.

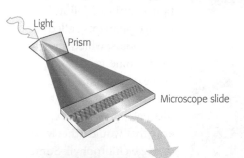

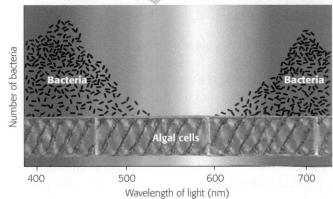

◄ **Figure 7.5 Investigating how light wavelength affects photosynthesis.** When algal cells are placed on a microscope slide, oxygen-seeking bacteria migrate toward algae exposed to certain colors of light. These results suggest that blue-violet and orange-red wavelengths best drive photosynthesis, while green wavelengths do so only a little bit.

Chloroplast Pigments

The selective absorption of light by leaves explains why they appear green to us; light of that color is poorly absorbed by chloroplasts and is thus reflected or transmitted toward the observer (**Figure 7.6**). Energy cannot be destroyed, so the absorbed energy must be converted to other forms. Chloroplasts contain several different pigments that absorb light of different wavelengths.

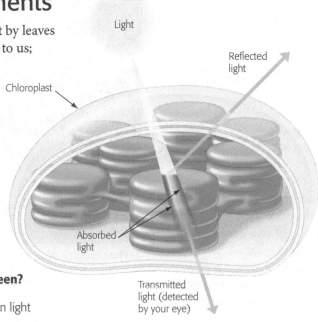

► **Figure 7.6 Why are leaves green?** Chlorophyll and other pigments in chloroplasts reflect or transmit green light while absorbing other colors.

Chlorophyll *a*, the pigment that participates directly in the light reactions, absorbs mainly blue-violet and red light. A very similar molecule, chlorophyll *b*, absorbs mainly blue and orange light. Chlorophyll *b* does not participate directly in the light reactions, but it conveys absorbed energy to chlorophyll *a*, which then puts the energy to work in the light reactions.

Chloroplasts also contain a family of yellow-orange pigments called carotenoids, which absorb mainly blue-green light. Some carotenoids have a protective function: They dissipate excess light energy that would otherwise damage chlorophyll. Some carotenoids are human nutrients: beta-carotene (a bright orange/red pigment found in pumpkins, sweet potatoes, and carrots) is converted to vitamin A in the body, and lycopene (a bright red pigment found in tomatoes, watermelon, and red peppers) is an antioxidant that is being studied for potential anti-cancer properties. Additionally, the spectacular colors of fall foliage in some parts of the world are due partly to the yellow-orange light reflected from carotenoids **(Figure 7.7)**. The decreasing temperatures in autumn cause a decrease in the levels of chlorophyll, allowing the colors of the longer-lasting carotenoids to be seen in all their fall glory.

All of these chloroplast pigments are built into the thylakoid membranes (see Figure 7.2). There the pigments are organized into light-harvesting complexes called photosystems, our next topic. ☑

▲ **Figure 7.7 Photosynthetic pigments.** Falling autumn temperatures cause a decrease in the levels of green chlorophyll within the foliage of leaf-bearing trees. This decrease allows the colors of the carotenoids to be seen.

☑ CHECKPOINT

What is the specific name of the pigment that absorbs energy during the light reactions?

Answer: chlorophyll a

How Photosystems Harvest Light Energy

Thinking about light as waves explains most of light's properties. However, light also behaves as discrete packets of energy called photons. A **photon** is a fixed quantity of light energy. The shorter the wavelength of light, the greater the energy of a photon. A photon of violet light, for example, packs nearly twice as much energy as a photon of red light. This is why short-wavelength light—such as ultraviolet light and X-rays—can be damaging; photons at these wavelengths carry enough energy to damage proteins and DNA, potentially leading to cancerous mutations.

Protecting yourself from short wavelengths of light can be lifesaving.

When a pigment molecule absorbs a photon, one of the pigment's electrons gains energy. This electron is now said to be "excited"; that is, the electron has been raised from its starting state (called the ground state) to an excited state. The excited state is highly unstable, so an excited electron usually loses its excess energy and falls back to its ground state almost immediately **(Figure 7.8a)**. Most pigments release heat energy as their light-excited electrons fall back to their ground state. (That's why a surface with a lot of pigment, such as a black driveway, gets so hot on a sunny day.) But some pigments emit light as well as heat after absorbing

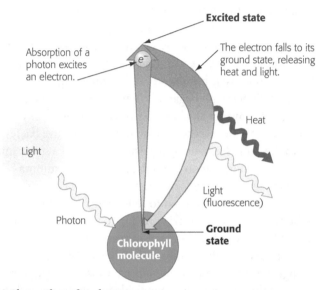

Excited state

Absorption of a photon excites an electron.

The electron falls to its ground state, releasing heat and light.

e^-

Heat

Light

Light (fluorescence)

Photon

Chlorophyll molecule

Ground state

(a) Absorption of a photon

(b) Fluorescence of a glow stick. Breaking a vial within a glow stick starts a chemical reaction that excites electrons within a fluorescent dye. As the electrons fall from their excited state to the ground state, the excess energy is emitted as light.

▲ **Figure 7.8 Excited electrons in pigments.**

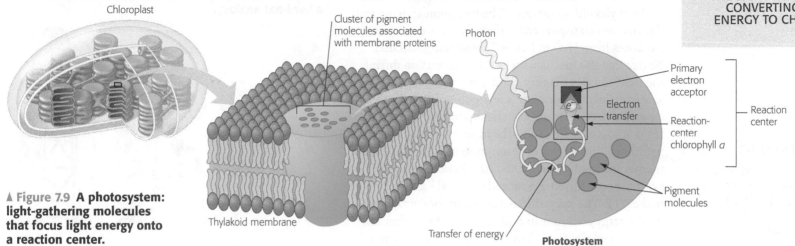

▲ **Figure 7.9 A photosystem: light-gathering molecules that focus light energy onto a reaction center.**

photons. The fluorescent light emitted by a glow stick is caused by a chemical reaction that excites electrons of a fluorescent dye **(Figure 7.8b)**. The excited electrons quickly fall back down to their ground state, releasing energy in the form of fluorescent light.

In the thylakoid membrane, chlorophyll molecules are organized with other molecules into photosystems. Each **photosystem** has a cluster of a few hundred pigment molecules, including chlorophylls *a* and *b* and some carotenoids **(Figure 7.9)**. This cluster of pigment molecules functions as a light-gathering antenna. When a photon strikes one of the pigment molecules, the energy jumps from molecule to molecule until it arrives at the reaction center of the photosystem. The reaction center consists of chlorophyll *a* molecules that sit next to another molecule called a primary electron acceptor. This primary electron acceptor traps the light-excited electron (e^-) from the chlorophyll *a* in the reaction center. Another team of molecules built into the thylakoid membrane then uses that trapped energy to make ATP and NADPH. ☑

☑ **CHECKPOINT**

What is the role of a reaction center during photosynthesis?

Answer: A reaction center transfers a light-excited photon from pigment molecules to molecules that can use this trapped energy to drive chemical reactions.

How the Light Reactions Generate ATP and NADPH

Two photosystems cooperate in the light reactions **(Figure 7.10)**. **①** Photons excite electrons in the chlorophyll of the first photosystem. These photons are then trapped by the primary electron acceptor. This photosystem then replaces the lost electrons by extracting new ones from water. This is the step that releases O_2 during photosynthesis. **②** Energized electrons from the first photosystem pass down an electron transport chain to the second photosystem. The chloroplast uses the energy released by this electron "fall" to make ATP. **③** The second photosystem transfers its light-excited electrons to $NADP^+$, reducing it to NADPH.

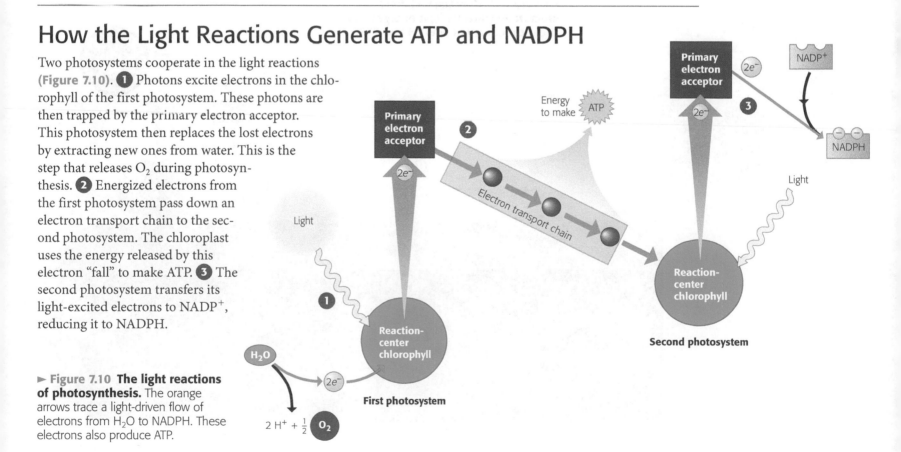

▶ **Figure 7.10 The light reactions of photosynthesis.** The orange arrows trace a light-driven flow of electrons from H_2O to NADPH. These electrons also produce ATP.

147

Figure 7.11 shows the location of the light reactions in the thylakoid membrane. The two photosystems and the electron transport chain that connects them transfer electrons from H_2O to $NADP^+$, producing NADPH. Notice that the mechanism of ATP production during the light reactions is very similar to the mechanism we saw in cellular respiration (see Figure 6.10). In both cases, an electron transport chain pumps hydrogen ions (H^+) across a membrane—the inner mitochondrial membrane in the case of respiration and the thylakoid membrane in photosynthesis. And in both cases, ATP synthases use the energy stored by the H^+ gradient to make ATP. The main difference is that food provides the high-energy electrons in cellular respiration, whereas light-excited electrons flow down the transport chain during photosynthesis. The traffic of electrons shown in Figures 7.10 and 7.11 is analogous to the cartoon in **Figure 7.12**.

We have seen how the light reactions absorb solar energy and convert it to the chemical energy of ATP and NADPH. Notice again, however, that the light reactions did not produce any sugar. That's the job of the Calvin cycle, as we'll see next. ☑

▼ **Figure 7.12 The light reactions illustrated using a hard-hat analogy.**

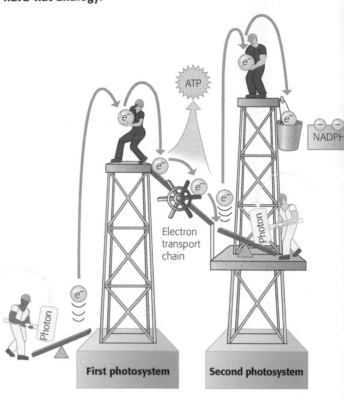

▼ **Figure 7.11 How the thylakoid membrane converts light energy to the chemical energy of NADPH and ATP.**

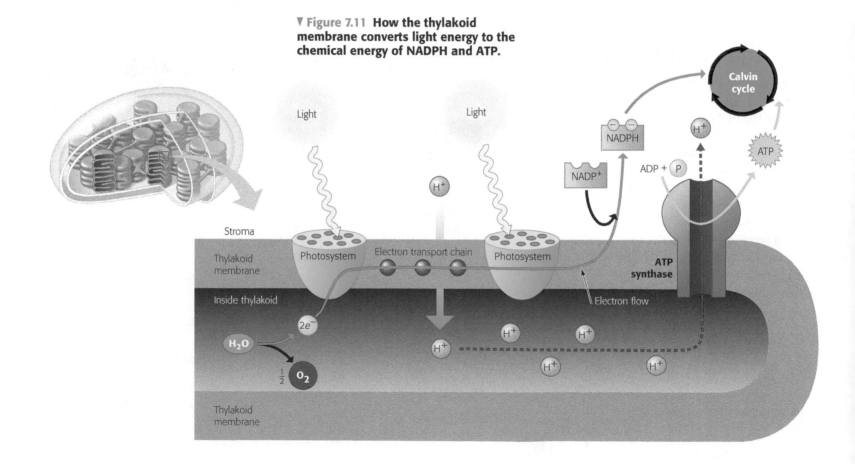

The Calvin Cycle: Making Sugar from Carbon Dioxide

If chloroplasts are solar-powered sugar factories, then the Calvin cycle is the actual sugar-manufacturing machinery. This process is called a cycle because its starting material is regenerated. With each turn of the cycle, there are chemical inputs and outputs. The inputs are CO_2 from the air as well as ATP and NADPH produced by the light reactions. Using carbon from CO_2, energy from ATP, and high-energy electrons from NADPH, the Calvin cycle constructs an energy-rich sugar molecule called glyceraldehyde 3-phosphate (G3P). The plant cell can then use G3P as the raw material to make the glucose and other organic compounds (such as cellulose and starch) that it needs. **Figure 7.13** presents the basics of the Calvin cycle, emphasizing inputs and outputs. Each ● symbol represents a carbon atom, and each ⓟ symbol represents a phosphate group. ☑

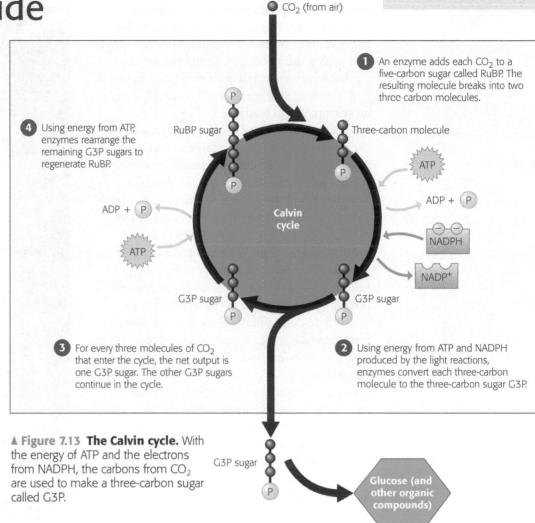

1 An enzyme adds each CO_2 to a five-carbon sugar called RuBP. The resulting molecule breaks into two three-carbon molecules.

2 Using energy from ATP and NADPH produced by the light reactions, enzymes convert each three-carbon molecule to the three-carbon sugar G3P.

3 For every three molecules of CO_2 that enter the cycle, the net output is one G3P sugar. The other G3P sugars continue in the cycle.

4 Using energy from ATP, enzymes rearrange the remaining G3P sugars to regenerate RuBP.

▲ Figure 7.13 **The Calvin cycle.** With the energy of ATP and the electrons from NADPH, the carbons from CO_2 are used to make a three-carbon sugar called G3P.

Biofuels EVOLUTION CONNECTION

Creating a Better Biofuel Factory

Throughout this chapter, we've studied how plants convert solar energy to chemical energy via photosynthesis. Such transformations are vital to our welfare and to Earth's ecosystems. As discussed in the Biology and Society section, scientists are attempting to tap into the "green energy" of photosynthesis to produce biofuels. But the production of biofuels is highly inefficient. In fact, it is usually far more costly to produce biofuels than to extract the equivalent amount of fossil fuels.

Biomechanical engineers are working to solve this dilemma by turning to an obvious example: evolution by natural selection. In nature, organisms with genes that make them better suited to their local environment will, on average, more often survive and pass those genes on to the next generation. Repeated over many generations, genes that enhance survival within that environment will become more common, and the species evolves.

When trying to solve an engineering problem, scientists can impose their own desired outcomes using a process called directed evolution (see the Process of Science section in Chapter 5 for another example).

☑ CHECKPOINT

What is the function of NADPH in the Calvin cycle?

Answer: It provides the high-energy electrons that are added to CO_2 to form G3P (a sugar).

During this process, scientists in the laboratory (instead of the natural environment) determine which organisms are the fittest. Directed evolution of biofuel production often involves microscopic algae (Figure 7.14) rather than plants because algae are easier to manipulate and maintain within the laboratory. Furthermore, some algae produce nearly half their own body weight in hydrocarbons that are only a few chemical steps away from useful biofuels.

In a typical directed evolution experiment, the researcher starts with a large collection of individual alga—sometimes naturally occurring species and sometimes transgenic algae that have been engineered to carry useful genes, such as fungal genes for enzymes that break down cellulose. The algae are exposed to mutation-promoting chemicals. This produces a highly varied collection of algae that can be screened for the desired outcome: the ability to produce the most useful biofuel in the largest quantity. The tiny fraction of total algae that can best perform this task is grown and subjected to another round of mutation and selection. After many repetitions, the algae may slowly improve their ability to efficiently produce biofuels. Many research laboratories—some within major petroleum companies—are using such methods and may someday produce an alga that can provide the ultimate source of green energy, an achievement that would highlight how lessons from natural evolution can be applied to improve our lives.

▼ **Figure 7.14 Microscopic biofuel factories.** This researcher is monitoring a reaction chamber in which microscopic algae are using light to produce biofuels.

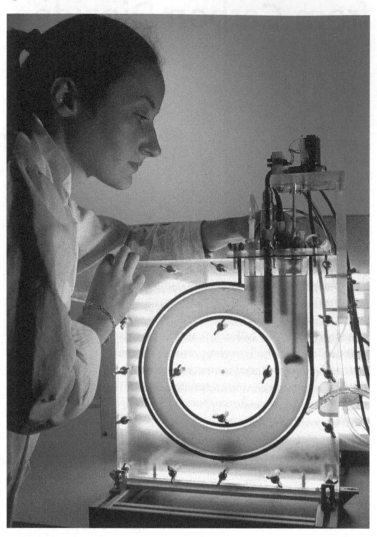

Chapter Review

■ SUMMARY OF KEY CONCEPTS

The Basics of Photosynthesis

Photosynthesis is a process whereby light energy is transformed into chemical energy stored as bonds in sugars made from carbon dioxide and water.

Chloroplasts: Sites of Photosynthesis

Chloroplasts contain a thick fluid called stroma surrounding a network of membranes called thylakoids.

Energy Transformation: An Overview of Photosynthesis

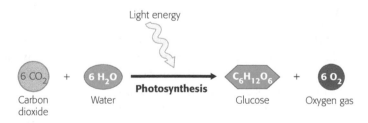

Light energy

$6\ CO_2$ + $6\ H_2O$ → **Photosynthesis** → $C_6H_{12}O_6$ + $6\ O_2$

Carbon dioxide   Water   Glucose   Oxygen gas

The overall process of photosynthesis can be divided into two stages connected by energy- and electron-carrying molecules:

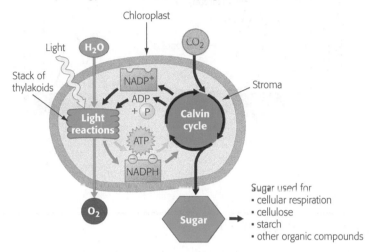

The Light Reactions: Converting Solar Energy to Chemical Energy

The Nature of Sunlight

Visible light is part of the spectrum of electromagnetic energy. It travels through space as waves. Different wavelengths of light appear as different colors; shorter wavelengths carry more energy.

Chloroplast Pigments

Pigment molecules absorb light energy of certain wavelengths and reflect other wavelengths. We see the reflected wavelengths as the color of the pigment. Several chloroplast pigments absorb light of various wavelengths and convey it to other pigments, but it is the green pigment chlorophyll *a* that participates directly in the light reactions.

How Photosystems Harvest Light Energy and How the Light Reactions Generate ATP and NADPH

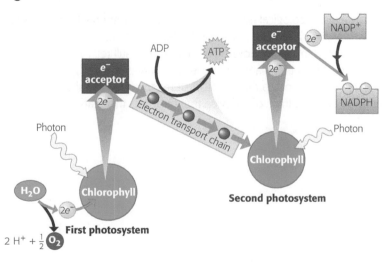

The Calvin Cycle: Making Sugar from Carbon Dioxide

Within the stroma (fluid) of the chloroplast, carbon dioxide from the air and ATP and NADPH produced during the light reactions are used to produce G3P, an energy-rich sugar molecule that can be used to make glucose and other organic molecules.

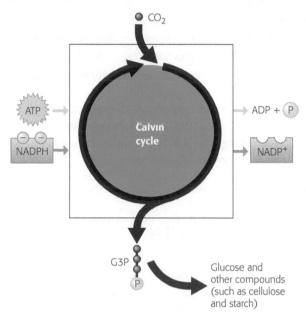

MasteringBiology®

For practice quizzes, BioFlix animations, MP3 tutorials, video tutors, and more study tools designed for this textbook, go to MasteringBiology®

SELF-QUIZ

1. The light reactions take place in the structures of the chloroplast called the _____, while the Calvin cycle takes place in the _____.

2. In terms of the spatial organization of photosynthesis within the chloroplast, what is the advantage of the light reactions producing NADPH and ATP on the stroma side of the thylakoid membrane?

3. Which of the following equations best summarizes photosynthesis?
   a. $6\ CO_2 + 6\ H_2O + 6\ O_2 \rightarrow C_6H_{12}O_6$
   b. $6\ CO_2 + 6\ H_2O \rightarrow C_6H_{12}O_6 + 6\ O_2$
   c. $6\ CO_2 + 6\ O_2 \rightarrow C_6H_{12}O_6 + 6\ H_2O$
   d. $C_6H_{12}O_6 + 6\ O_2 \rightarrow 6\ CO_2 + 6\ H_2O$

4. Explain how the name "photosynthesis" describes what this process accomplishes.

5. Chlorophyll *b* primarily absorbs blue light. What color has chlorophyll *b* got?

6. The carbon atoms that enter the Calvin cycle as CO_2 eventually end up in _____.

7. Which of the following are produced by reactions that take place in the thylakoids and are consumed by reactions in the stroma?
   a. CO_2 and H_2O
   b. $NADP^+$ and ADP
   c. ATP and NADPH
   d. glucose and O_2

8. The reactions of the Calvin cycle are not directly dependent on light, and yet they usually do not occur at night. Why?

9. Of the following metabolic processes, which one is common to photosynthesis and cellular respiration?
   a. reactions that convert light energy to chemical energy
   b. reactions that split H_2O molecules and release O_2
   c. reactions that store energy by pumping H^+ across membranes
   d. reactions that convert CO_2 to sugar

Answers to these questions can be found in Appendix: Self-Quiz Answers.

THE PROCESS OF SCIENCE

10. Oxygen is a highly reactive element that easily combines with (oxidizes) other elements and compounds. For example, it forms five different oxides with iron, which is by mass the most abundant element on Earth. Given its highly reactive nature, how is oxygen maintained at a level of 16% in the Earth's atmosphere? Based on the information in this chapter, draw a cycle of reactions that keeps the concentration of oxygen steady and describes its flow between different interconnected systems of the biosphere. Was the atmospheric concentration of oxygen higher or lower before the emergence of life?

11. Suppose you wanted to discover whether the oxygen atoms in the glucose produced by photosynthesis come from H_2O or CO_2. Explain how you could use a radioactive isotope to find out.

12. **Interpreting Data** The graph above right is called an absorption spectrum. Each line on the graph is made by shining light of varying wavelengths through a sample. For each wavelength, the amount of that light absorbed by the sample is recorded. This graph combines three such measurements, one each for the pigments chlorophyll *a*, chlorophyll *b*, and the carotenoids. Notice that the graphs for the chlorophyll pigments match the data presented in Figure 7.5. Imagine a plant that lacks chlorophyll and relies only on carotenoids for photosynthesis. What colors of light would work best for this plant? How would this plant appear to your eye?

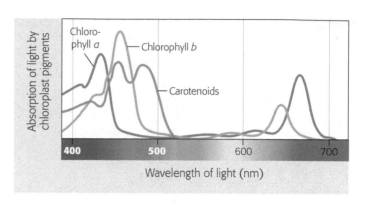

BIOLOGY AND SOCIETY

13. There is strong evidence that Earth is getting warmer because of an intensified greenhouse effect resulting from increased CO_2 emissions from industry, vehicles, and the burning of forests. Global climate change could influence agriculture, melt polar ice, and flood coastal regions. In response to these threats, 192 parties have accepted the Kyoto Protocol, which calls for mandatory reductions of greenhouse gas emissions in 30 industrialized nations by 2012. The United States has signed but not ratified (put into effect) the agreement, instead proposing a more modest set of voluntary goals allowing businesses to decide whether they wish to participate and providing tax incentives to encourage them to do so. The reasons given for rejecting the agreement are that it might hurt the American economy and that some less industrialized countries (such as India) are exempted from it, even though they produce a lot of pollution. Do you agree with this decision? In what ways might efforts to reduce greenhouse gases hurt the economy? How can those costs be weighed against the costs of global climate change? Should poorer nations carry an equal burden to reduce their emissions?

14. In this age of ecological awareness, generation of energy from sunlight is gaining popularity. Drawbacks of this technology are the use of toxic heavy metals in some photovoltaic cells and its still relatively low cost efficiency. Given that plants are experts in harvesting sunlight, would it not make sense to use their photosystems to convert sunlight into energy? Can you think of a device that performs this feat?

Unit 2 Genetics

8 Cellular Reproduction: Cells from Cells

Chapter Thread: **Life with and without Sex**

9 Patterns of Inheritance

Chapter Thread: **Dog Breeding**

10 The Structure and Function of DNA

Chapter Thread: **The Deadliest Virus**

11 How Genes Are Controlled

Chapter Thread: **Cancer**

12 DNA Technology

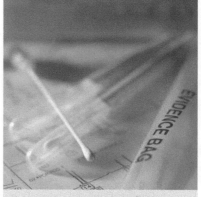

Chapter Thread: **DNA Profiling**

8 Cellular Reproduction: Cells from Cells

Why Cellular Reproduction Matters

▲ In certain species of sea star, a severed arm may be able to regrow a whole new body.

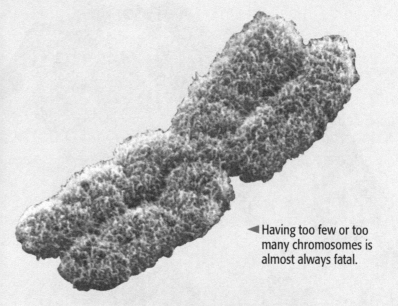

◄ Having too few or too many chromosomes is almost always fatal.

If stretched out, the DNA ► in any one of your cells would be taller than you.

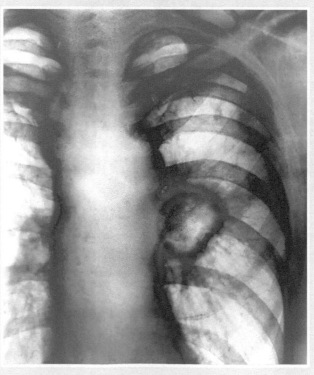

▲ Every tumor is the result of a malfunction in cell division.

Life with and without Sex BIOLOGY AND SOCIETY

Virgin Birth of a Dragon

Zookeepers at the Chester Zoo in England were startled to discover that Flora, a female Komodo dragon—*Varanus komodoensis*, the largest living species of lizard, capable of growing up to 10 feet long—had laid a clutch of 25 eggs. It wasn't surprising that a captive Komodo dragon would breed. In fact, Flora was at the

zoo for that very reason: She was one of two female Komodo dragons that were part of a captive breeding program intended to help repopulate the species. What made Flora's clutch of eggs so remarkable is that she had not yet been in the company of a male, let alone mated with one. As far as anyone knew, Komodo dragons, like the vast majority of animal species, create offspring only through sexual reproduction involving the union of a male's sperm and a female's egg. But despite Flora's virginity, eight of her eggs developed normally and hatched into live, healthy Komodo dragons.

DNA analysis confirmed that Flora's offspring derived their genes solely from her. The new dragons must have resulted from parthenogenesis, the production of offspring by a female without involvement of a male. Parthenogenesis is one form of asexual reproduction, the creation of a new generation without participation of sperm and egg. Parthenogenesis is rare among vertebrates (animals with backbones), although it has been documented in species as diverse as sharks (including the hammerhead), domesticated birds, and now Komodo dragons. Soon, zoologists identified a second Komodo at a different zoo who had also borne young by parthenogenesis. This same Komodo dragon later had additional offspring via sexual reproduction—indicating that this species is capable of switch-

The Komodo dragon. The Komodo is the world's largest lizard and is found in the wild only on three islands in Indonesia.

ing between two reproductive modes. Biologists are investigating the evolutionary basis of this phenomenon and considering what implications it may have on efforts to repopulate this rare species.

The ability of organisms to procreate is the one characteristic that best distinguishes living things from nonliving matter. All organisms—from bacteria to lizards to you—are the result of repeated cell divisions. The perpetuation of life therefore depends on cell division, the production of new cells. In this chapter, we'll look at how individual cells are copied and then see how cell reproduction underlies the process of sexual reproduction. Throughout our discussion, we'll consider examples of asexual and sexual reproduction among both plants and animals.

What Cell Reproduction Accomplishes

When you hear the word *reproduction*, you probably think of the birth of new organisms. But reproduction actually occurs much more often at the cellular level. Consider the skin on your arm. Skin cells are constantly reproducing themselves and moving outward toward the surface, replacing dead cells that have rubbed off. This renewal of your skin goes on throughout your life. And when your skin is injured, additional cell reproduction helps heal the wound.

When a cell undergoes reproduction, or **cell division**, the two "daughter" cells that result are genetically identical to each other and to the original "parent" cell. (Biologists traditionally use the word *daughter* in this context to refer to offspring cells, but of course cells lack gender.) Before the parent cell splits into two, it duplicates its **chromosomes**, the structures that contain most of the cell's DNA. Then, during cell division, each daughter cell receives one identical set of chromosomes from the original parent cell.

As summarized in **Figure 8.1**, cell division plays several important roles in the lives of organisms. For example, within your body, millions of cells must divide every second to replace damaged or lost cells. Another function of cell division is growth. All of the trillions of cells in your body are the result of repeated cell divisions that began in your mother's body with a single fertilized egg cell.

Another vital function of cell division is reproduction. Many single-celled organisms, such as amoebas,

In certain species of sea star, a severed arm may be able to regrow a whole new body.

reproduce by dividing in half, and the offspring are genetic replicas of the parent. Because it does not involve fertilization of an egg by a sperm, this type of reproduction is called **asexual reproduction**. Offspring produced by asexual reproduction inherit all of their chromosomes from a single parent and are thus genetic duplicates.

Many multicellular organisms can reproduce asexually as well. For example, some sea star species have the ability to grow new individuals from fragmented pieces. And if you've ever grown a houseplant from a clipping, you've observed asexual reproduction in plants. In asexual reproduction, there is one simple principle of inheritance: The lone parent and each of its offspring have identical genes. The type of cell division responsible for asexual reproduction and for the growth and maintenance of multicellular organisms is called mitosis.

Sexual reproduction is different; it requires fertilization of an egg by a sperm. The production of **gametes**—egg and sperm—involves a special type of cell division called meiosis, which occurs only in reproductive organs. As we'll discuss later, a gamete has only half as many chromosomes as the parent cell that gave rise to it.

In summary, two kinds of cell division are involved in the lives of sexually reproducing organisms: mitosis for growth and maintenance and meiosis for reproduction. The remainder of the chapter is divided into two main sections, one devoted to each type of cell division. ☑

☑ CHECKPOINT

Ordinary cell division produces two daughter cells that are genetically identical. Name three functions of this type of cell division. Which of these functions occur in your body?

Answer: cell replacement, growth of an organism, asexual reproduction of an organism; only the first two occur in your body

► **Figure 8.1 Three functions of cell division by mitosis.**

FUNCTIONS OF CELL DIVISION BY MITOSIS

Cell Replacement

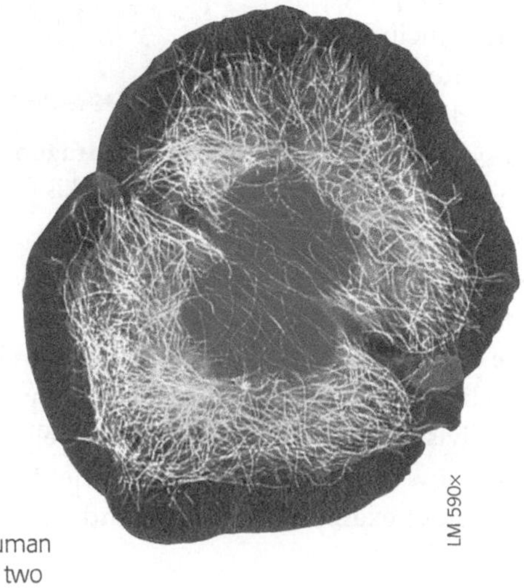

LM 590×

Division of a human kidney cell into two

Growth via Cell Division

Colorized SEM 810×

The cells of an early human embryo

The Cell Cycle and Mitosis

Almost all of the genes of a eukaryotic cell—around 21,000 in humans—are located on chromosomes in the cell nucleus. (The main exceptions are genes on small DNA molecules found in mitochondria and chloroplasts.) Because chromosomes are the lead players in cell division, we'll focus on them before turning our attention to the cell as a whole.

If stretched out, the DNA in any one of your cells would be taller than you.

Most of the time, the chromosomes exist as thin fibers that are much longer than the nucleus they are stored in. In fact, if fully extended, the DNA in just one of your cells would be more than 6 feet long! Chromatin in this state is too thin to be seen using a light microscope. As a cell

Eukaryotic Chromosomes

Each eukaryotic chromosome contains one very long DNA molecule, typically bearing thousands of genes. The number of chromosomes in a eukaryotic cell depends on the species **(Figure 8.2)**. For example, human body cells have 46 chromosomes, while the body cells of a dog have 78 and those of a koala bear have 16. Chromosomes are made up of a material called **chromatin**, fibers composed of roughly equal amounts of DNA and protein molecules. The protein molecules help organize the chromatin and help control the activity of its genes.

▶ Figure 8.2 **The number of chromosomes in the cells of selected mammals.** Notice that humans have 46 chromosomes and that the number of chromosomes does not correspond to the size or complexity of an organism.

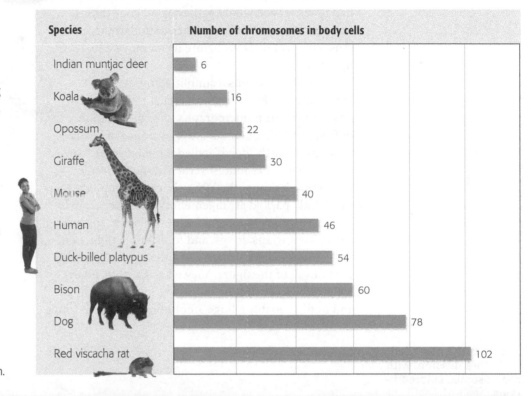

Species	Number of chromosomes in body cells
Indian muntjac deer	6
Koala	16
Opossum	22
Giraffe	30
Mouse	40
Human	46
Duck-billed platypus	54
Bison	60
Dog	78
Red viscacha rat	102

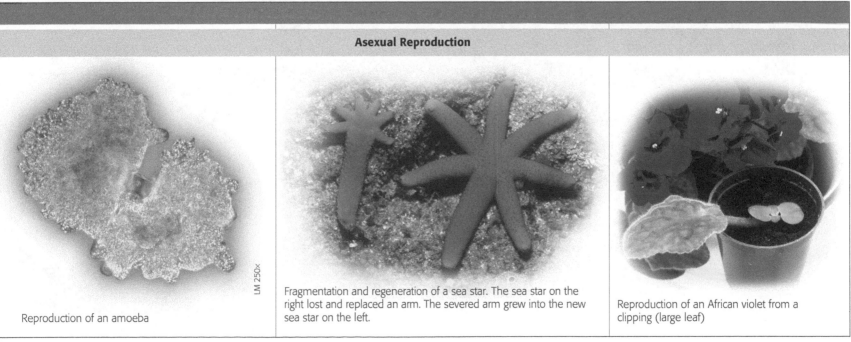

Asexual Reproduction

LM 250x

Reproduction of an amoeba

Fragmentation and regeneration of a sea star. The sea star on the right lost and replaced an arm. The severed arm grew into the new sea star on the left.

Reproduction of an African violet from a clipping (large leaf)

prepares to divide, its chromatin fibers coil up, forming compact chromosomes that become visible under a light microscope **(Figure 8.3)**.

Such long molecules of DNA can fit into the tiny nucleus because within each chromosome the DNA is packed into an elaborate, multilevel system of coiling and folding. A crucial aspect of DNA packing is the association of the DNA with small proteins called **histones**. Why is it necessary for a cell's chromosomes to be compacted in this way? Imagine that your belongings are spread out around your room. If you had to move, you would gather up all your things and put them in small containers. Similarly, a cell must compact its DNA before it can move it to a new cell.

Figure 8.4 presents a simplified model of DNA packing. First, histones attach to the DNA. In electron micrographs, the combination of DNA and histones has the appearance of beads on a string. Each "bead," called a **nucleosome**, consists of DNA wound around several histone molecules. When not dividing, the DNA of active genes takes on this lightly packed arrangement. When preparing to divide, chromosomes condense even more: the beaded string itself wraps, loops, and folds into a tight, compact structure, as you can see in the chromosome at the bottom of the figure. Viewed as a whole, Figure 8.4 gives a sense of how successive levels of coiling and folding enable a huge amount of DNA to fit into a

cell's tiny nucleus. Think of a DNA as a length of yarn; a chromosome is then like a skein of yarn, one very long piece that is folded into a tight package for easier handling.

▼ **Figure 8.4 DNA packing in a eukaryotic chromosome.** Successive levels of coiling of DNA and associated proteins ultimately result in highly compacted chromosomes. The fuzzy appearance of the final chromosome at the bottom arises from the intricate twists and folds of the chromatin fibers.

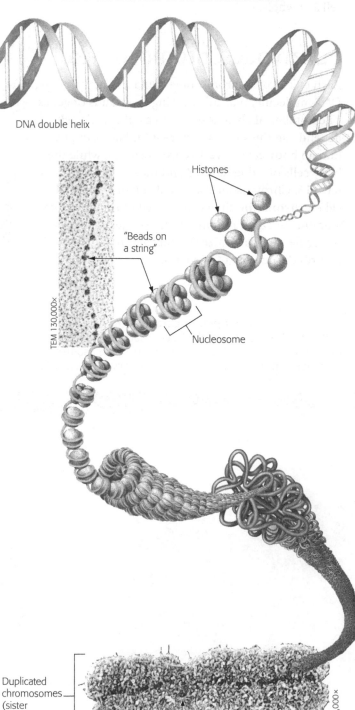

DNA double helix

Histones

"Beads on a string"

Nucleosome

TEM 130,000×

Duplicated chromosomes (sister chromatids)

Centromere

TEM 9,000×

▼ **Figure 8.3 A plant cell just before division, with chromosomes colored by stains.**

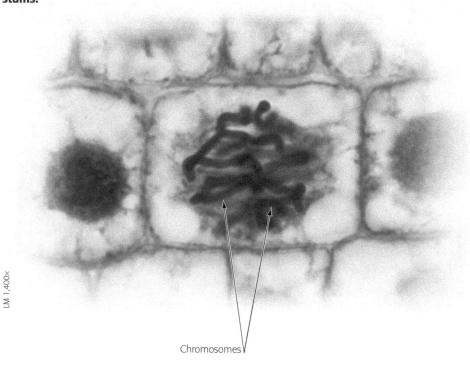

LM 1,400×

Chromosomes

Information Flow Duplicating Chromosomes

Think of the chromosomes as being like a detailed instruction manual on how to run a cell; during division, the original cell must pass on a copy of the manual to the new cell while also retaining a copy for itself. Before a cell begins the division process, it must therefore duplicate all of its chromosomes. The DNA molecule of each chromosome is copied through the process of DNA replication (discussed in detail in Chapter 10), and new histone protein molecules attach as needed. The result is that—at this point—each chromosome consists of two copies called **sister chromatids**, which contain identical genes. At the bottom of Figure 8.4, the two sister chromatids are joined together most tightly at a narrow "waist" called the **centromere**.

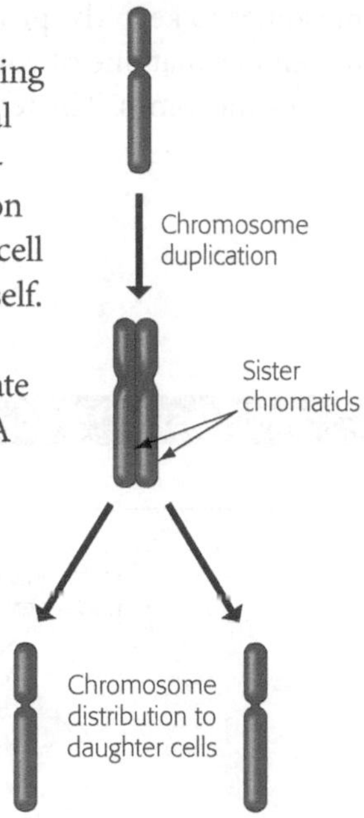

When the cell divides, the sister chromatids of a duplicated chromosome separate from each other (**Figure 8.5**). Once separated from its sister, each chromatid is considered a full-fledged chromosome, and it is identical to the original chromosome. One of the new chromosomes goes to one daughter cell, and the other goes to the other daughter cell. In this way, each daughter cell receives a complete and identical set of chromosomes. A dividing human skin cell, for example, has 46 duplicated chromosomes, and each of the two daughter cells that result from it has 46 single chromosomes.

The Cell Cycle

The rate at which a cell divides depends on its role within the organism's body. Some cells divide once a day, others less often, and some highly specialized cells, such as mature muscle cells, do not divide at all.

The **cell cycle** is the ordered sequence of events that extends from the time a cell is first formed from a dividing parent cell until its own division into two cells. Think of the cell cycle as the "lifetime" of a cell, from its "birth" to its own reproduction. As **Figure 8.6** shows, most of the cell cycle is spent in **interphase**. Interphase is a time when a cell goes about its usual business, performing

Chromosome duplication

Sister chromatids

Chromosome distribution to daughter cells

▲ **Figure 8.5 Duplication and distribution of a single chromosome.** During cell reproduction, the cell duplicates each chromosome and distributes the two copies to the daughter cells.

its normal functions within the organism. For example, during interphase a cell in your stomach lining might make and release enzyme molecules that aid in digestion. While in interphase, a cell roughly doubles everything in its cytoplasm. It increases its supply of proteins, increases the number of many of its organelles (such as mitochondria and ribosomes), and grows in size. Typically, interphase lasts for at least 90% of the cell cycle.

From the standpoint of cell reproduction, the most important event of interphase is chromosome duplication, when the DNA in the nucleus is precisely doubled. The period when this occurs is called the S phase (for DNA *synthesis*). The interphase periods before and after the S phase are called the G_1 and G_2 phases, respectively (G stands for *gap*). During G_1, each chromosome is single, and the cell performs its normal functions. During G_2 (after DNA duplication during the S phase), each chromosome in the cell consists of two identical sister chromatids, and the cell prepares to divide.

The part of the cell cycle when the cell is actually dividing is called the **mitotic (M) phase**. It includes two overlapping stages, mitosis and cytokinesis. In **mitosis**, the nucleus and its contents, most importantly the duplicated chromosomes, divide and are evenly distributed, forming two daughter nuclei. During **cytokinesis**, the cytoplasm (along with all the organelles) is divided in two. The combination of mitosis and cytokinesis produces two genetically identical daughter cells, each fully equipped with a nucleus, cytoplasm, organelles, and plasma membrane. ☑

☑ CHECKPOINT

1. A duplicated chromosome consists of two sister _____ joined together at the _____.
2. What are the two broadest divisions of the cell cycle? What two processes are involved in the actual duplication of the cell?

Answers: **1.** *chromatids; centromere* **2.** *interphase and the mitotic phase; mitosis and cytokinesis*

▼ **Figure 8.6 The eukaryotic cell cycle.** The cell cycle extends from the "birth" of a cell (just after the point indicated by the dark blue arrow at the bottom of the cycle), resulting from cell reproduction, to the time the cell itself divides in two. (During interphase, the chromosomes are diffuse masses of thin fibers; they do not actually appear in the rodlike form you see here.)

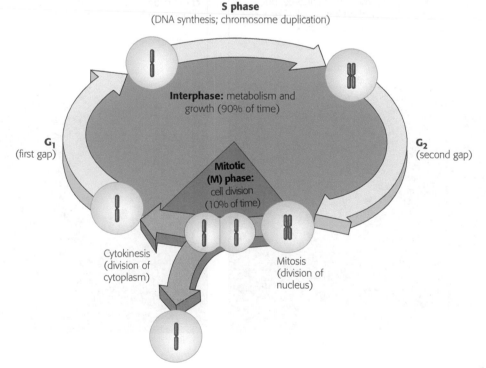

S phase
(DNA synthesis; chromosome duplication)

Interphase: metabolism and growth (90% of time)

G_1
(first gap)

G_2
(second gap)

Mitotic (M) phase: cell division (10% of time)

Cytokinesis (division of cytoplasm)

Mitosis (division of nucleus)

Mitosis and Cytokinesis

Figure 8.7 illustrates the cell cycle for an animal cell using drawings, descriptions, and photomicrographs. The micrographs running along the bottom row of the page show dividing cells from a salamander, with chromosomes depicted in blue. The drawings in the top row include details that are not visible in the micrographs. In these cells, we illustrate just four chromosomes to keep the process a bit simpler to follow; remember that one of your cells actually contains 46 chromosomes. The text within the figure describes

▼ **Figure 8.7 Cell reproduction: A dance of the chromosomes.** After the chromosomes duplicate during interphase, the elaborately choreographed stages of mitosis—prophase, metaphase, anaphase, and telophase—distribute the duplicate sets of chromosomes to two separate nuclei. Cytokinesis then divides the cytoplasm, yielding two genetically identical daughter cells.

INTERPHASE	PROPHASE

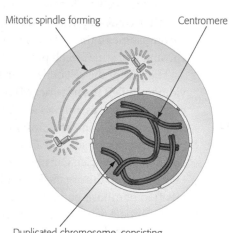

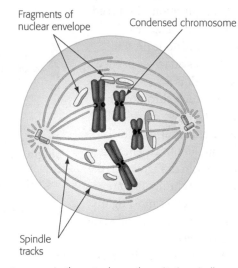

Interphase is the period of cell growth when the cell makes new molecules and organelles. At the point shown here, late interphase (G_2), the cytoplasm contains two centrosomes. Within the nucleus, the chromosomes are duplicated, but they cannot be distinguished individually because they are still in the form of loosely packed chromatin fibers.

During prophase, changes occur in both nucleus and cytoplasm. In the nucleus, the chromatin fibers coil, so that the chromosomes become thick enough to be seen individually with a light microscope. Each chromosome exists as two identical sister chromatids joined together at the narrow "waist" of the centromere. In the cytoplasm, the mitotic spindle begins to form. Late in prophase, the nuclear envelope breaks into pieces. The spindle tracks attach to the centromeres of the chromosomes and move the chromosomes toward the center of the cell.

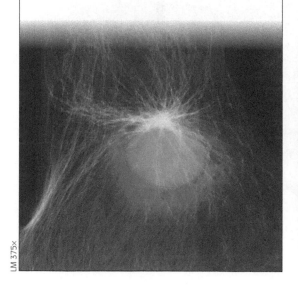

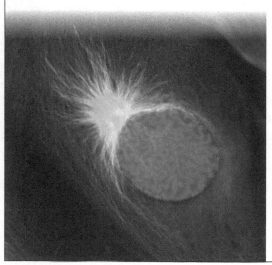

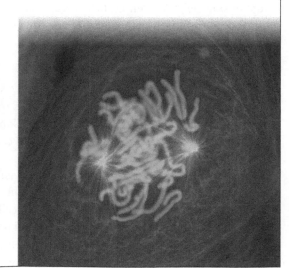

LM 375x

the events occurring at each stage. Study this figure carefully (it has a lot of information and it's important!) and notice the striking changes in the nucleus and other cellular structures.

Biologists distinguish four main stages of mitosis: **prophase**, **metaphase**, **anaphase**, and **telophase**. The timing of these stages is not precise, and they overlap a bit. Think of stages in your own life—infancy, childhood, adulthood, old age—and you'll realize that the stages run into each other and vary from individual to individual; so it is with the stages of mitosis.

The chromosomes are the stars of the mitotic drama, and their movements depend on the **mitotic spindle**, a football-shaped structure of microtubule tracks (colored green in the figure) that guides the separation of the two sets of daughter chromosomes. The tracks of spindle microtubules grow from structures within the cytoplasm called centrosomes.

METAPHASE	ANAPHASE	TELOPHASE

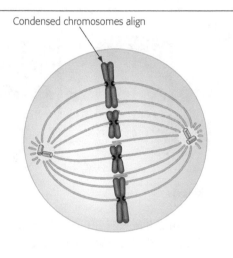

Condensed chromosomes align

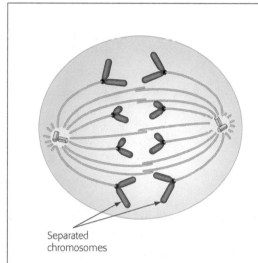

Separated chromosomes

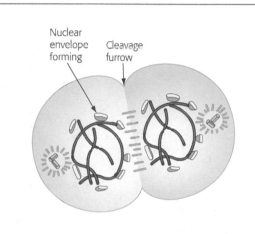

Nuclear envelope forming   Cleavage furrow

The mitotic spindle is now fully formed. The centromeres of all the chromosomes line up between the two poles of the spindle. For each chromosome, the tracks of the mitotic spindle attached to the two sister chromatids pull toward opposite poles. This tug of war keeps the chromosomes in the middle of the cell.

Anaphase begins suddenly when the sister chromatids of each chromosome separate. Each is now considered a full-fledged (daughter) chromosome. The chromosomes move toward opposite poles of the cell as the spindle tracks shorten. Simultaneously, the tracks not attached to chromosomes lengthen, pushing the poles farther apart and elongating the cell.

Telophase begins when the two groups of chromosomes have reached opposite ends of the cell. Telophase is the reverse of prophase: Nuclear envelopes form, the chromosomes uncoil, and the spindle disappears. Mitosis, the division of one nucleus into two genetically identical daughter nuclei, is now finished. Cytokinesis, the division of the cytoplasm, usually occurs with telophase. In animals, a cleavage furrow pinches the cell in two, producing two daughter cells.

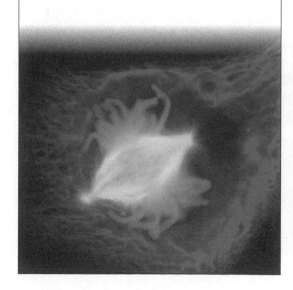

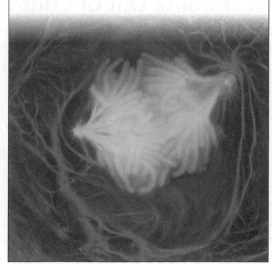

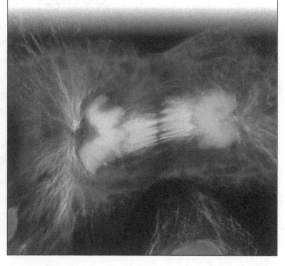

▼ Figure 8.8 **Cytokinesis in animal and plant cells.**

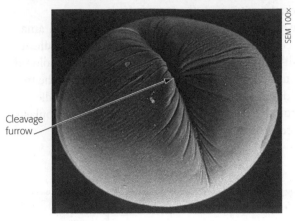

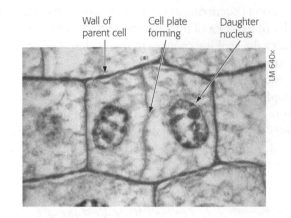

Wall of parent cell

Cell plate forming

Daughter nucleus

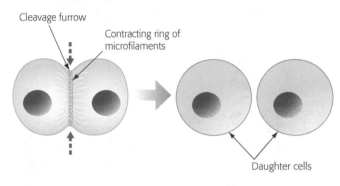

Cleavage furrow

Contracting ring of microfilaments

Daughter cells

(a) Animal cell cytokinesis

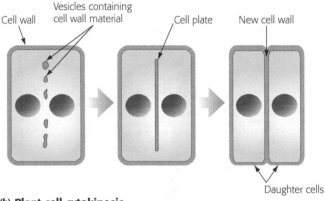

Cell wall

Vesicles containing cell wall material

Cell plate

New cell wall

Daughter cells

(b) Plant cell cytokinesis

Cytokinesis, the division of the cytoplasm into two cells, usually begins during telophase, overlapping the end of mitosis. In animal cells, the cytokinesis process is known as **cleavage**. The first sign of cleavage is the appearance of a cleavage furrow, an indentation at the equator of the cell. A ring of microfilaments in the cytoplasm just under the plasma membrane contracts, like the pulling of a drawstring on a hooded sweatshirt, deepening the furrow and pinching the parent cell in two **(Figure 8.8a)**.

Cytokinesis in a plant cell occurs differently. Vesicles containing cell wall material collect at the middle of the cell. The vesicles fuse, forming a membranous disk called the **cell plate**. The cell plate grows outward, accumulating more cell wall material as more vesicles join it. Eventually, the membrane of the cell plate fuses with the plasma membrane, and the cell plate's contents join the parental cell wall. The result is two daughter cells **(Figure 8.8b)**. ☑

Cancer Cells: Dividing Out of Control

For a plant or animal to grow and maintain its tissues normally, it must be able to control the timing of cell division—speeding up, slowing down, or turning the process off/on as needed. The sequential events of the cell cycle are directed by a **cell cycle control system** that consists of specialized proteins within the cell. These proteins integrate information from the environment and from other body cells and send "stop" and "go-ahead" signals at certain key points during the cell cycle. For example, the cell cycle normally halts within the G_1 phase of interphase unless the cell receives a go-ahead signal via certain control proteins. If that signal never arrives,

Every tumor is the result of a malfunction in cell division.

the cell will switch into a permanently nondividing state. Some of your nerve and muscle cells, for example, are arrested this way. If the go-ahead signal is received and the G_1 checkpoint is passed, the cell will usually complete the rest of the cycle.

What Is Cancer?

Cancer, which currently claims the lives of one out of every five people in the United States and other industrialized nations, is a disease of the cell cycle. Cancer cells do not respond normally to the cell cycle control system; they divide excessively and may invade other tissues of the body. If unchecked, cancer cells may continue to divide until they kill the host. Cancer cells are thus referred to as "immortal" since, unlike other human cells, they will never cease dividing. In fact, thousands of laboratories around the world today use HeLa cells, a laboratory strain of human cells that were originally obtained from a woman named Henrietta Lacks, who died of cervical cancer in 1951.

The abnormal behavior of cancer cells begins when a single cell undergoes genetic changes (mutations) in one or more genes that encode for proteins in the cell cycle control system. These changes cause the cell to grow abnormally. The immune system generally recognizes and destroys such cells. However, if the cell evades destruction, it may proliferate to form a **tumor**, an abnormally growing mass of body cells. If the abnormal cells remain at the original site, the lump is called a **benign tumor**. Benign tumors can cause problems if they grow large and disrupt certain organs, such as the brain, but often they can be completely removed by surgery and are rarely deadly.

In contrast, a **malignant tumor** is one that has the potential to spread into neighboring tissues and other parts of the body, forming new tumors **(Figure 8.9)**. A malignant tumor may or may not have actually begun to spread, but once it does, it will soon displace normal tissue and interrupt organ function. An individual with a malignant tumor is said to have **cancer**. The spread of cancer cells beyond their original site is called **metastasis**. Cancers are named according to where they originate. Liver cancer, for example, always begins in liver tissue and may spread from there.

Cancer Treatment

Once a tumor starts growing in the body, how can it be treated? There are three main types of cancer treatment. Surgery to remove a tumor is usually the first step. For many benign tumors, surgery may be sufficient. If it is not, doctors turn to treatments that attempt to stop cancer cells from dividing. In **radiation therapy**, parts of the body that have cancerous tumors are exposed to concentrated beams of high-energy radiation, which often harm cancer cells more than normal cells. Radiation therapy is often effective against malignant tumors that have not yet spread. However, there is sometimes enough damage to normal body cells to produce side effects, such as nausea and hair loss.

Chemotherapy, the use of drugs to disrupt cell division, is used to treat widespread or metastatic tumors. Chemotherapy drugs work in a variety of ways. Some prevent cell division by interfering with the mitotic spindle. For example, paclitaxel (trade name Taxol) freezes the spindle after it forms, keeping it from functioning. Paclitaxel is made from a chemical discovered in the bark of the Pacific yew, a tree found mainly in the northwestern United States. It has fewer side effects than many other anticancer drugs and seems to be effective against some hard-to-treat cancers of the ovary and breast. Another drug, vinblastine, prevents the mitotic spindle from forming in the first place. Vinblastine was first obtained from the periwinkle plant, which is native to the tropical rain forests of Madagascar. Given these examples, preserving biodiversity may be the key to discovering the next generation of lifesaving anticancer drugs.

Cancer Prevention and Survival

Although cancer can strike anyone, there are certain lifestyle changes you can make to reduce your chances of developing cancer or increase your chances of surviving it. Not smoking, exercising adequately (usually defined as at least 150 minutes of moderate exercise each week), avoiding overexposure to the sun, and eating a high-fiber, low-fat diet can all help reduce the likelihood of getting cancer. Seven types of cancer can be easily detected: skin and oral (via physical exam), breast (via self-exams or mammograms for higher-risk women and women 50 and older), prostate (via rectal exam), cervical (via Pap smear), testicular (via self-exam), and colon (via colonoscopy). Regular visits to the doctor can help identify tumors early, which is the best way to increase the chance of successful treatment. ☑

☑ **CHECKPOINT**

What differentiates a benign tumor from a malignant tumor?

Answer: A benign tumor remains at its point of origin, whereas a malignant tumor can spread. ☑

▼ Figure 8.9 **Growth and metastasis of a malignant tumor of the breast.**

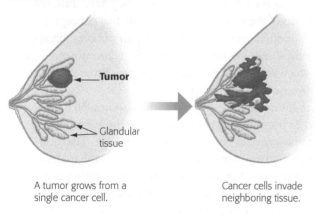

A tumor grows from a single cancer cell.

Cancer cells invade neighboring tissue.

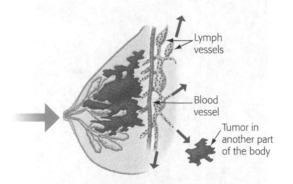

Metastasis: Cancer cells spread through lymph and blood vessels to other parts of the body.

Meiosis, the Basis of Sexual Reproduction

Only maple trees produce more maple trees, only goldfish make more goldfish, and only people make more people. These simple facts of life have been recognized for thousands of years and are reflected in the age-old saying, "Like begets like." But in a strict sense, "Like begets like" applies only to asexual reproduction, where offspring inherit all their DNA from a single parent. Asexual offspring are exact genetic replicas of that one parent and of each other, and their appearances are very similar.

The family photo in **Figure 8.10** makes the point that in a sexually reproducing species, like does not exactly beget like. You probably resemble your biological parents more closely than you resemble strangers, but you do not look exactly like your parents or your siblings—unless you are an identical twin. Each offspring of sexual reproduction inherits a unique combination of genes from its two parents, and this combined set of genes programs a unique combination of traits. As a result, sexual reproduction can produce tremendous variety among offspring.

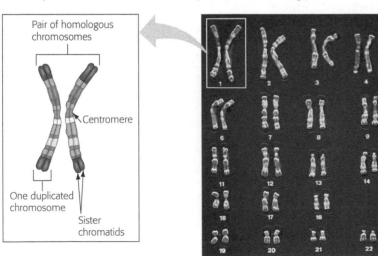

▲ Figure 8.10 **The varied products of sexual reproduction.** Every child inherits a unique combination of genes from his or her parents and displays a unique combination of traits.

Sexual reproduction depends on the cellular processes of meiosis and fertilization. But before discussing these processes, we need to return to chromosomes and the role they play in the life cycle of sexually reproducing organisms.

Homologous Chromosomes

If we examine cells from different individuals of a single species—sticking to one sex, for now—we find that they have the same number and types of chromosomes. Viewed with a microscope, your chromosomes would look exactly like those of Angelina Jolie (if you're a woman) or Brad Pitt (if you're a man).

A typical body cell, called a **somatic cell**, has 46 chromosomes in humans. A technician can break open a human cell in metaphase of mitosis, stain the chromosomes with dyes, take a picture with the aid of a microscope, and arrange the chromosomes in matching pairs by size. The resulting display is called a **karyotype (Figure 8.11)**. Notice in the figure that each chromosome is duplicated, with two sister chromatids joined along their length; within the white box, for example, the left "stick" is actually a pair of sister chromatids stuck together (as shown in the drawing to the left). Notice also that almost every chromosome has a twin that resembles it in length and centromere position; in the figure, the white box surrounds one set of twin chromosomes. The two chromosomes of such a matching pair, called **homologous chromosomes**, carry genes controlling the same inherited characteristics. For example, if a gene influencing freckles is located at a particular place on one chromosome—within the yellow band in the drawing in Figure 8.11, for instance—then the homologous chromosome has that same gene in the same location. However, the two homologous chromosomes may have different versions of the same gene. Let's restate this concept because it often confuses students: A pair of homologous chromosomes has two nearly identical chromosomes, each of which consists of two identical sister chromatids after chromosome duplication.

In human females, the 46 chromosomes fall neatly into 23 homologous pairs. For a male, however, the chromosomes in one pair do not look alike. This non-matching pair, only partly homologous, is the male's sex chromosomes. **Sex chromosomes** determine a person's sex (male versus female). In mammals, males have one X chromosome and one Y chromosome. Females have

▼ Figure 8.11 **Pairs of homologous chromosomes in a human male karyotype.** This karyotype shows 22 completely homologous pairs (autosomes) and a 23rd pair that consists of an X chromosome and a Y chromosome (sex chromosomes). With the exception of X and Y, the homologous chromosomes of each pair match in size, centromere position, and staining pattern.

Pair of homologous chromosomes

Centromere

One duplicated chromosome

Sister chromatids

LM 3,600×

two X chromosomes. (Other organisms have different systems; in this chapter, we focus on humans.) The remaining chromosomes (44 in humans), found in both males and females, are called **autosomes**. For both autosomes and sex chromosomes, you inherited one chromosome of each pair from your mother and the other from your father.

Gametes and the Life Cycle of a Sexual Organism

The **life cycle** of a multicellular organism is the sequence of stages leading from the adults of one generation to the adults of the next. Having two sets of chromosomes, one inherited from each parent, is a key factor in the life cycle of humans and all other species that reproduce sexually. **Figure 8.12** shows the human life cycle, emphasizing the number of chromosomes.

Humans (as well as most other animals and many plants) are **diploid** organisms because all body cells contain pairs of homologous chromosomes. In other words, all your chromosomes come in matching sets. This is similar to shoes in your closet: You may have 46 shoes, but they are organized as 23 pairs, with the members of each pair being nearly identical to each other. The total number of chromosomes, 46 in humans, is the diploid number (abbreviated $2n$). The gametes, egg and sperm cells, are not diploid. Made by meiosis in an ovary or testis, each gamete has a single set of chromosomes: 22 autosomes plus a sex chromosome, either X or Y. A cell with a single chromosome set is called a **haploid** cell; it has only one member of each pair of homologous chromosomes. To visualize the haploid state, imagine your closet containing only one shoe from each pair. For humans, the haploid number, n, is 23.

In the human life cycle, a haploid sperm fuses with a haploid egg in a process called **fertilization**. The resulting fertilized egg, called a **zygote**, is diploid. It has two sets of chromosomes, one set from each parent. The life cycle is completed as a sexually mature adult develops from the zygote. Mitotic cell division ensures that all somatic cells of the human body receive a copy of all of the zygote's 46 chromosomes. Thus, every one of the trillions of cells in your body can trace its ancestry back through mitotic divisions to the single zygote produced when your father's sperm and your mother's egg fused about nine months before you were born (although you probably don't want to dwell on those details!).

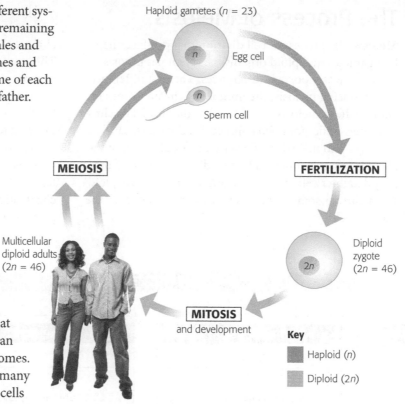

Figure 8.12 The human life cycle. In each generation, the doubling of chromosome number that results from fertilization is offset by the halving of chromosome number during meiosis.

Producing haploid gametes by meiosis keeps the chromosome number from doubling in every generation. To illustrate, **Figure 8.13** tracks one pair of homologous chromosomes. ❶ Each of the chromosomes is duplicated during interphase (before mitosis). ❷ The first division, meiosis I, segregates the two chromosomes of the homologous pair, packaging them in separate (haploid) daughter cells. But each chromosome is still doubled. ❸ Meiosis II separates the sister chromatids. Each of the four daughter cells is haploid and contains only a single chromosome from the pair of homologous chromosomes.

▼ Figure 8.13 **How meiosis halves chromosome number.**

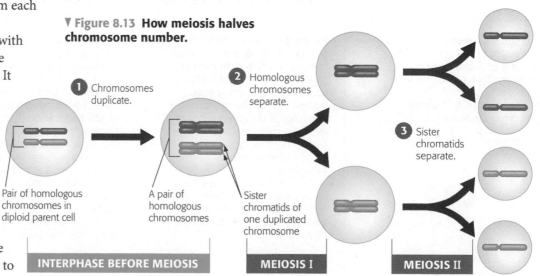

The Process of Meiosis

Meiosis, the process of cell division that produces haploid gametes in diploid organisms, resembles mitosis, but with two important differences. The first difference is that, during meiosis, the number of chromosomes is cut in half. In meiosis, a cell that has duplicated its chromosomes undergoes two consecutive divisions, called meiosis I and meiosis II. Because one duplication of the chromosomes is followed by two divisions, each of the four daughter cells resulting from meiosis has a haploid set of chromosomes—half as many chromosomes as the starting cell.

The second difference of meiosis compared with mitosis is an exchange of genetic material—pieces of chromosomes—between homologous chromosomes. This exchange, called crossing over, occurs during the first prophase of meiosis. We'll look more closely at crossing over later. For now, study **Figure 8.14**, including the text below it, which describes the stages of meiosis in detail for a hypothetical animal cell containing four chromosomes.

As you go through Figure 8.14, keep in mind the difference between homologous chromosomes and sister chromatids: The two chromosomes of a homologous pair are individual chromosomes that were inherited

▼ Figure 8.14 **The stages of meiosis.**

	MEIOSIS I: HOMOLOGOUS CHROMOSOMES SEPARATE

INTERPHASE	PROPHASE I	METAPHASE I	ANAPHASE I
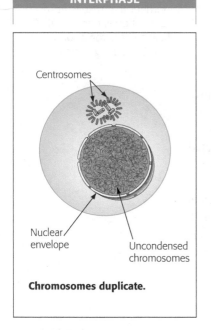	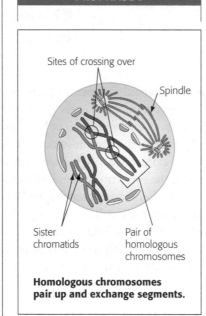	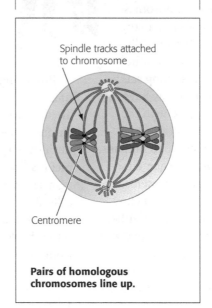	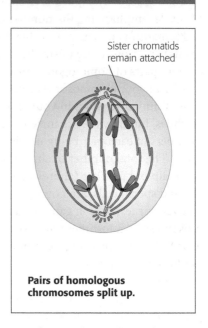
Chromosomes duplicate.	**Homologous chromosomes pair up and exchange segments.**	**Pairs of homologous chromosomes line up.**	**Pairs of homologous chromosomes split up.**

As with mitosis, meiosis is preceded by an interphase during which the chromosomes duplicate. Each chromosome then consists of two identical sister chromatids. The chromosomes consist of uncondensed chromatin fibers.

Prophase I As the chromosomes coil up, special proteins cause the homologous chromosomes to stick together in pairs. The resulting structure has four chromatids. Within each set, chromatids of the homologous chromosomes exchange corresponding segments—they "cross over." Crossing over rearranges genetic information.

As prophase I continues, the chromosomes coil up further, a spindle forms, and the homologous pairs are moved toward the center of the cell.

Metaphase I At metaphase I, the homologous pairs are aligned in the middle of the cell. The sister chromatids of each chromosome are still attached at their centromeres, where they are anchored to spindle tracks. Notice that for each chromosome pair, the spindle tracks attached to one homologous chromosome come from one pole of the cell, and the tracks attached to the other chromosome come from the opposite pole. With this arrangement, the homologous chromosomes are poised to move toward opposite poles of the cell.

Anaphase I The attachment between the homologous chromosomes of each pair breaks, and the chromosomes now migrate toward the poles of the cell. *In contrast to mitosis, the sister chromatids migrate as a pair instead of splitting up.* They are separated not from each other but from their homologous partners.

from different parents, one from the mother and one from the father. The members of a pair of homologous chromosomes in Figure 8.14 (and later figures) are identical in size and shape but colored in the illustrations differently (red versus blue) to remind you that they differ in this way. In the interphase just before meiosis, each chromosome duplicates to form sister chromatids that remain together until anaphase of meiosis II. Before crossing over occurs, sister chromatids are identical and carry the same versions of all their genes. ✓

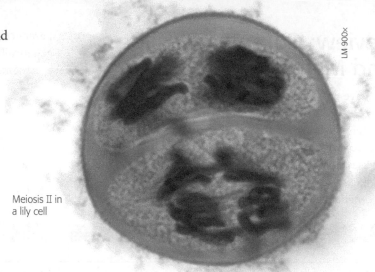

LM 900×

Meiosis II in a lily cell

☑ CHECKPOINT

If a single diploid somatic cell with 18 chromosomes undergoes meiosis and produces sperm, the result will be _____ sperm, each with _____ chromosomes. (Provide two numbers.)

Answer: four; nine

MEIOSIS II: SISTER CHROMATIDS SEPARATE

TELOPHASE I AND CYTOKINESIS	PROPHASE II	METAPHASE II	ANAPHASE II	TELOPHASE II AND CYTOKINESIS

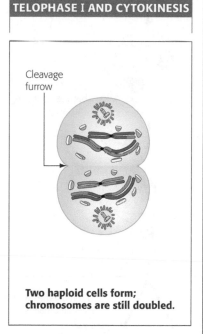

Cleavage furrow

Two haploid cells form; chromosomes are still doubled.

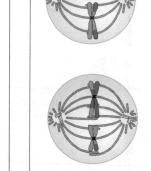

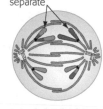
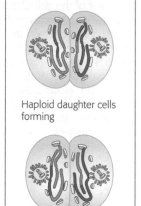

Sister chromatids separate

Haploid daughter cells forming

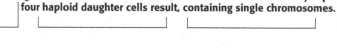

During another round of cell division, the sister chromatids finally separate; four haploid daughter cells result, containing single chromosomes.

Telophase I and Cytokinesis In telophase I, the chromosomes arrive at the poles of the cell. When they finish their journey, each pole has a haploid chromosome set, although each chromosome is still in duplicate form. Usually, cytokinesis occurs along with telophase I, and two haploid daughter cells are formed.

The Process of Meiosis II Meiosis II is essentially the same as mitosis. The important difference is that meiosis II starts with a haploid cell that has *not* undergone chromosome duplication during the preceding interphase.

During prophase II, a spindle forms and moves the chromosomes toward the middle of the cell. During metaphase II, the chromosomes are aligned as they are in mitosis, with the tracks attached to the sister chromatids of each chromosome coming from opposite poles.

In anaphase II, the centromeres of sister chromatids separate, and the sister chromatids of each pair move toward opposite poles of the cell. In telophase II, nuclei form at the cell poles, and cytokinesis occurs at the same time. There are now four haploid daughter cells, each with single chromosomes.

167

Review: Comparing Mitosis and Meiosis

You have now learned the two ways that cells of eukaryotic organisms divide (Figure 8.15). Mitosis—which provides for growth, tissue repair, and asexual reproduction— produces daughter cells that are genetically identical to the parent cell. Meiosis, needed for sexual reproduction, yields genetically unique haploid daughter cells—cells with only one member of each homologous chromosome pair.

For both mitosis and meiosis, the chromosomes duplicate only once, in the preceding interphase. Mitosis involves one division of the nucleus and cytoplasm (duplication, then division in half), producing two

▶ **Figure 8.15 Comparing mitosis and meiosis.** The events unique to meiosis occur during meiosis I: In prophase I, duplicated homologous chromosomes pair along their lengths, and crossing over occurs between homologous (nonsister) chromatids. In metaphase I, pairs of homologous chromosomes (rather than individual chromosomes) are aligned at the center of the cell. During anaphase I, sister chromatids of each chromosome stay together and go to the same pole of the cell as homologous chromosomes separate. At the end of meiosis I, there are two haploid cells, but each chromosome still has two sister chromatids.

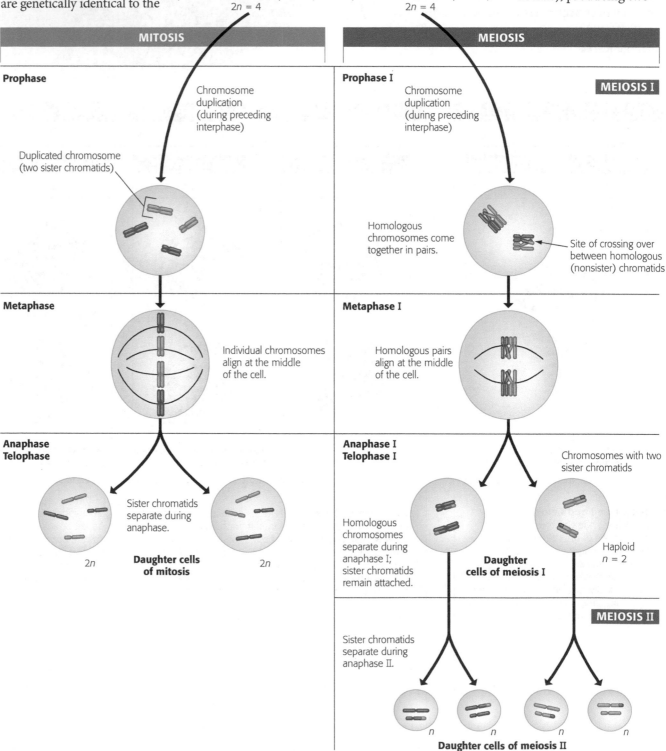

diploid cells. Meiosis entails two nuclear and cytoplasmic divisions (duplication, division in half, then division in half again), yielding four haploid cells.

Figure 8.15 compares mitosis and meiosis, tracing these two processes for a diploid parent cell with four chromosomes. As before, homologous chromosomes are those matching in size. (Imagine that the red chromosomes were inherited from the mother and the blue chromosomes from the father.) Notice that all the events unique to meiosis occur during meiosis I. Meiosis II is virtually identical to mitosis in that it separates sister chromatids. But unlike mitosis, meiosis II yields daughter cells with a haploid set of chromosomes. ☑

The Origins of Genetic Variation

As we discussed earlier, offspring that result from sexual reproduction are genetically different from their parents and from one another. How does meiosis produce such genetic variation?

Independent Assortment of Chromosomes

Figure 8.16 illustrates one way in which meiosis contributes to genetic variety. The figure shows how the arrangement of homologous chromosomes at metaphase of meiosis I affects the resulting gametes. Once again, our example is from a hypothetical diploid organism with four chromosomes (two pairs of homologous chromosomes), with colors used to differentiate homologous chromosomes (red for chromosomes inherited from the mother and blue for chromosomes from the father).

When aligned during metaphase I, the side-by-side orientation of each homologous pair of chromosomes is a matter of chance—either the red or blue chromosome may be on the left or right. Thus, in this example, there are two possible ways that the chromosome pairs can align during metaphase I. In possibility 1, the chromosome pairs are oriented with both red chromosomes on the same side (blue/red and blue/red). In this case, each of the gametes produced at the end of meiosis II has only red or only blue chromosomes (combinations a and b). In possibility 2, the chromosome pairs are oriented differently (blue/red and red/blue). This arrangement produces gametes with one red and one blue chromosome (combinations c and d). Thus, with the two possible arrangements shown in this example, the organism will produce gametes with four different combinations of chromosomes. For a species with more than two pairs of chromosomes, such as humans, every chromosome pair orients independently of all the others at metaphase I. (Chromosomes X and Y behave as a homologous pair in meiosis.)

☑ **CHECKPOINT**

True or false: Both mitosis and meiosis are preceded by chromosome duplication.

Answer: *true*

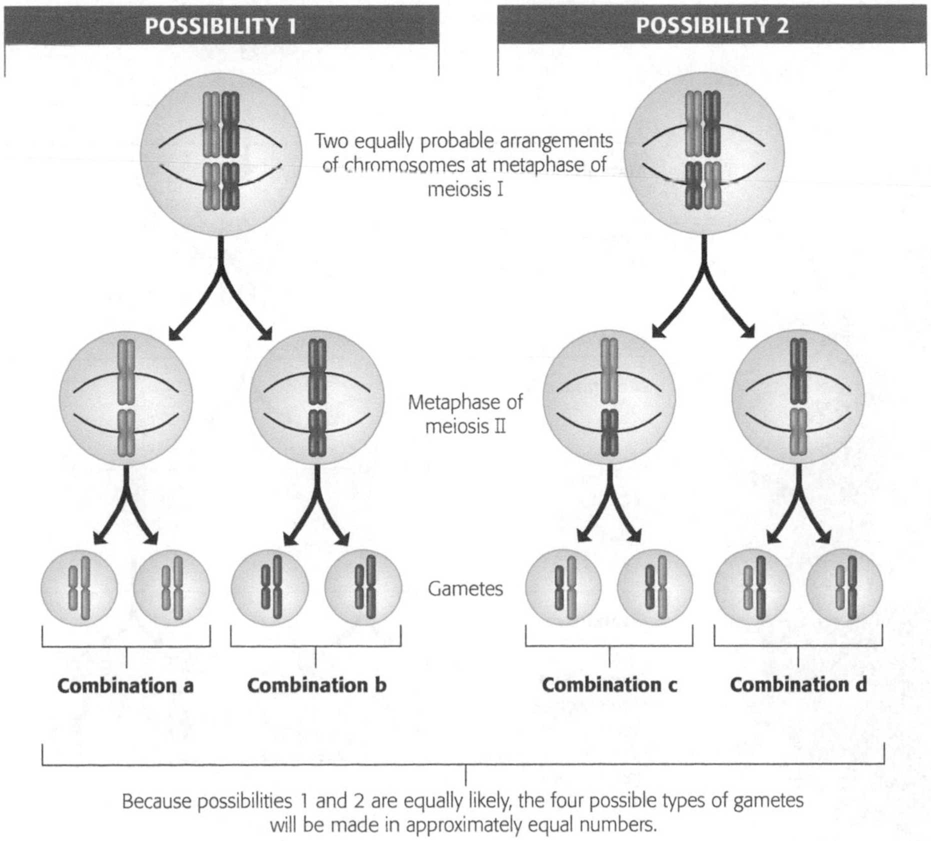

POSSIBILITY 1 POSSIBILITY 2

Two equally probable arrangements of chromosomes at metaphase of meiosis I

Metaphase of meiosis II

Gametes

Combination a   **Combination b**   **Combination c**   **Combination d**

Because possibilities 1 and 2 are equally likely, the four possible types of gametes will be made in approximately equal numbers.

◀ Figure 8.16 **Results of alternative arrangements of chromosomes at metaphase of meiosis I.** The arrangement of chromosomes at metaphase I determines which chromosomes will be packaged together in the haploid gametes.

For any species, the total number of chromosome combinations that can appear in gametes is 2^n, where n represents the haploid number. For the hypothetical organism in Figure 8.16, $n = 2$, so the number of chromosome combinations is 2^2, or 4. For a human ($n = 23$), there are 2^{23}, or about 8 million, possible chromosome combinations! This means that every gamete a person produces contains one of about 8 million possible combinations of maternal and paternal chromosomes. When you consider that a human egg cell with about 8 million possibilities is fertilized at random by a human sperm cell with about 8 million possibilities (**Figure 8.17**), you can see that a single man and a single woman can produce zygotes with 64 trillion combinations of chromosomes!

Crossing Over

So far, we have focused on genetic variety in gametes and zygotes at the whole-chromosome level. We'll now take a closer look at **crossing over**, the exchange of corresponding segments between nonsister chromatids of homologous chromosomes, which occurs during prophase I of meiosis. **Figure 8.18** shows crossing over between two homologous chromosomes and the resulting gametes. At the time that crossing over begins very early in prophase I, homologous chromosomes

are closely paired all along their lengths, with a precise gene-by-gene alignment.

The exchange of segments between nonsister chromatids—one maternal chromatid and one paternal chromatid of a homologous pair—adds to the genetic variety resulting from sexual reproduction. In Figure 8.18, if there were no crossing over, meiosis could produce only two types of gametes: the ones ending up with chromosomes that exactly match the parents' chromosomes, either all blue or all red (as in Figure 8.16). With crossing over, gametes arise

▼ **Figure 8.17 The process of fertilization: a close-up view.** Here you see many human sperm contacting an egg. Only one sperm can add its chromosomes to produce a zygote.

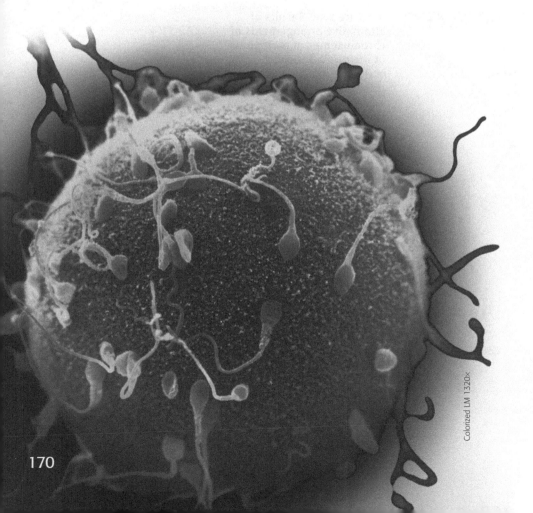

Colorized LM 1320x

▼ **Figure 8.18 The results of crossing over during meiosis for a single pair of homologous chromosomes.** A real cell has multiple pairs of homologous chromosomes that produce a huge variety of recombinant chromosomes in the gametes.

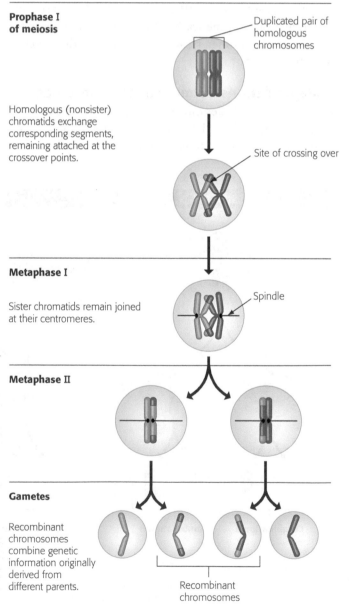

Prophase I of meiosis

Duplicated pair of homologous chromosomes

Homologous (nonsister) chromatids exchange corresponding segments, remaining attached at the crossover points.

Site of crossing over

Metaphase I

Sister chromatids remain joined at their centromeres.

Spindle

Metaphase II

Gametes

Recombinant chromosomes combine genetic information originally derived from different parents.

Recombinant chromosomes

with chromosomes that are partly from the mother and partly from the father. These chromosomes are called "recombinant" because they result from genetic recombination, the production of gene combinations different from those carried by the parental chromosomes.

Because most chromosomes contain thousands of genes, a single crossover event can affect many genes. When we also consider that multiple crossovers can occur in each pair of homologous chromosomes, it's not surprising that gametes and the offspring that result from them are so incredibly varied. ✓

✓ CHECKPOINT

Name two events during meiosis that contribute to genetic variety among gametes. During what stages of meiosis does each occur?

Answer: crossing over between homologous chromosomes during prophase 1 and independent orientation/assortment of the pairs of homologous chromosomes at metaphase 1

Life with and without Sex THE PROCESS OF SCIENCE

Do All Animals Have Sex?

As discussed in the Biology and Society section, some species such as Komodo dragons can reproduce via both sexual and asexual routes. Although some animal species can reproduce asexually, very few animals reproduce *only* asexually. In fact, evolutionary biologists have traditionally considered asexual reproduction an evolutionary dead end (for reasons we'll discuss in the Evolution Connection section at the end of the chapter).

To investigate a case in which asexual reproduction seemed to be the norm, researchers from Harvard University studied a group of animals called bdelloid rotifers **(Figure 8.19)**. This class of nearly microscopic freshwater invertebrates includes more than 300 known species. Despite hundreds of years of **observations**, no one had ever found bdelloid rotifer males or evidence of sexual reproduction. But the possibility remained that bdelloids had sex very infrequently or that the males were impossible to recognize by appearance. Thus, the Harvard research team posed the following **question**: Does this entire class of animals reproduce solely by asexual means?

The researchers formed the **hypothesis** that bdelloid rotifers have indeed thrived for millions of years without sexually reproducing. But how could this hypothesis be tested? In most species, the two versions of a gene in a pair of homologous chromosomes are very similar due to the constant trading of genes during sexual reproduction. If a species has survived without sex for millions of years, the researchers reasoned, then changes in the DNA sequences of homologous genes should accumulate independently, and the two versions of the genes should have significantly diverged from each other over time. This led to the **prediction** that bdelloid rotifers would display much more variation in their pairs of homologous genes than most organisms.

In a simple but elegant **experiment**, the researchers compared the sequences of a particular gene in bdelloid and non-bdelloid rotifers. Their **results** were striking. Among non-bdelloid rotifers that reproduce sexually, the two homologous versions of the gene were nearly identical, differing by only 0.5% on average. In contrast, the two versions of the same gene in bdelloid rotifers differed by 3.5–54%. These data provided strong evidence that bdelloid rotifers have evolved for millions of years without any sexual reproduction.

► Figure 8.19
A bdelloid rotifer.

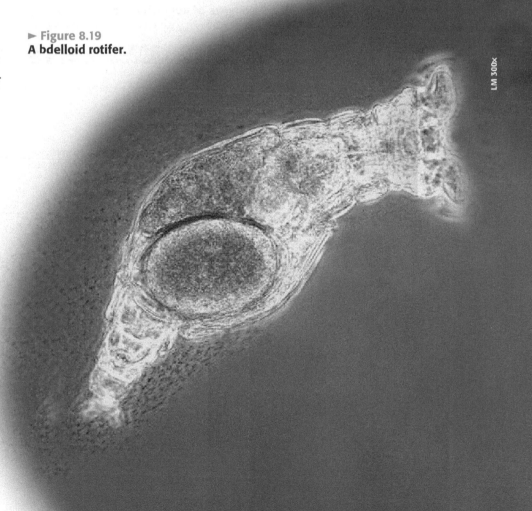

LM 300x

When Meiosis Goes Awry

So far, our discussion of meiosis has focused on the process as it normally and correctly occurs. But what happens when there is an error in the process? Such a mistake can result in genetic abnormalities that range from mild to severe to fatal.

How Accidents during Meiosis Can Alter Chromosome Number

Within the human body, meiosis occurs repeatedly as the testes or ovaries produce gametes. Almost always, chromosomes are distributed to daughter cells without any errors. But occasionally there is a mishap, called a **nondisjunction**, in which the members of a chromosome pair fail to separate at anaphase. Nondisjunction can occur during meiosis I or II (Figure 8.20). In either case, gametes with abnormal numbers of chromosomes are the result.

Figure 8.21 shows what can happen when an abnormal gamete produced by nondisjunction unites with a normal gamete during fertilization. When a normal sperm fertilizes an egg cell with an extra chromosome,

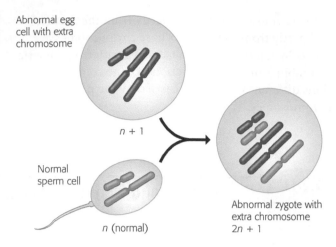

▲ Figure 8.21 **Fertilization after nondisjunction in the mother.**

the result is a zygote with a total of $2n + 1$ chromosomes. Because mitosis duplicates the chromosomes as they are, the abnormality will be passed to all embryonic cells. If the organism survives, it will have an abnormal karyotype and probably a medical disorder caused by the abnormal number of genes. ☑

▼ Figure 8.20 **Two types of nondisjunction.** In both examples in the figure, the cell at the top is diploid (2n), with two pairs of homologous chromosomes.

NONDISJUNCTION IN MEIOSIS I	NONDISJUNCTION IN MEIOSIS II

Meiosis I

Nondisjunction: Pair of homologous chromosomes fails to separate.

Meiosis II

Nondisjunction: Sister chromatids fail to separate.

Gametes

Number of chromosomes

$n + 1$ $n + 1$ $n - 1$ $n - 1$ $n + 1$ $n - 1$ n n

Abnormal gametes   **Abnormal gametes**   **Normal gametes**

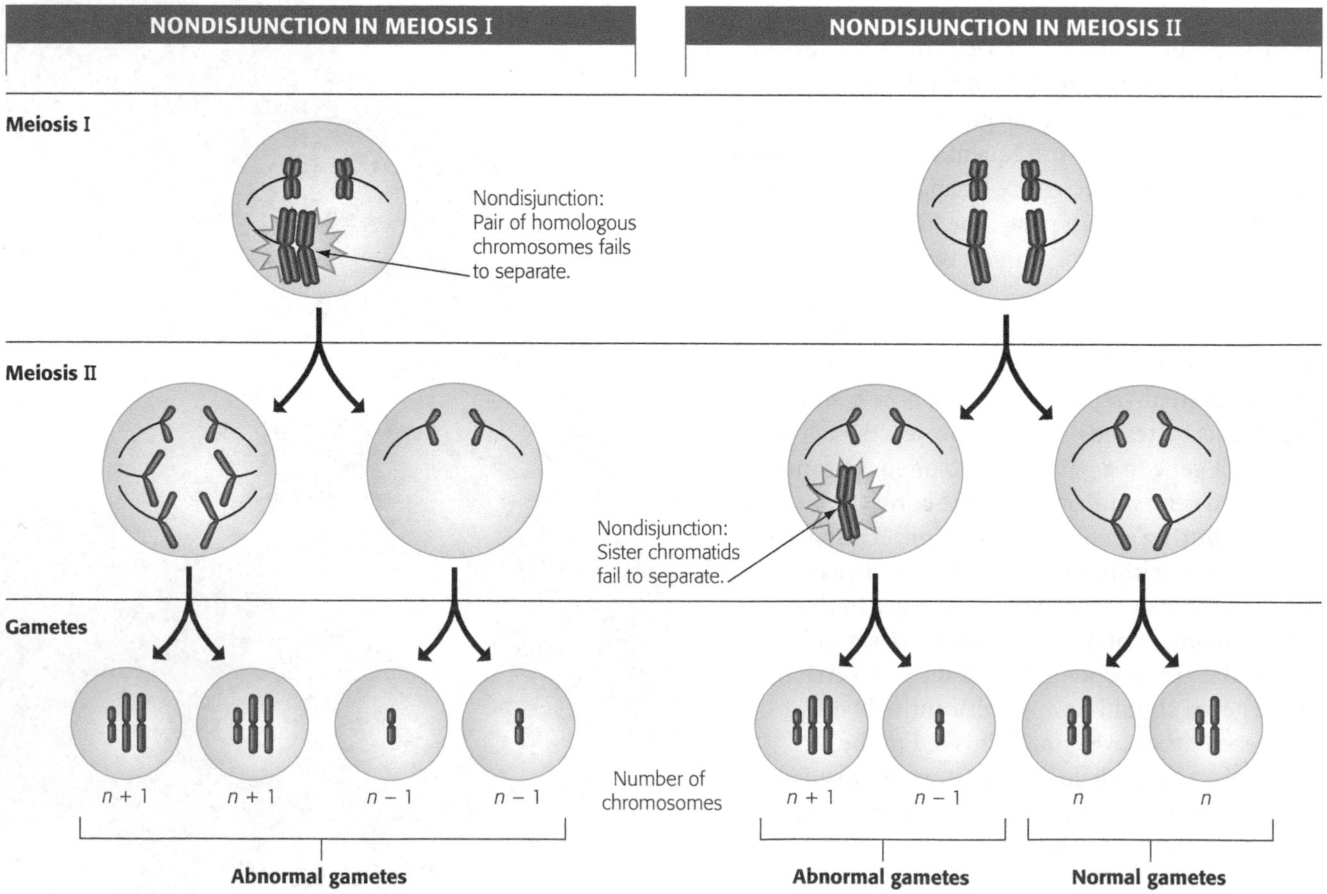

Down Syndrome: An Extra Chromosome 21

Figure 8.11 showed a normal human complement of 23 pairs of chromosomes. Compare it with the karyotype in **Figure 8.22**; besides having two X chromosomes (because it's from a female), the karyotype in Figure 8.22 has three number 21 chromosomes. Because this person is triploid for chromosome 21 (instead of the usual diploid condition),

Having too few or too many chromosomes is almost always fatal.

▼ **Figure 8.22 Trisomy 21 and Down syndrome.** This child displays the characteristic facial features of Down syndrome. The karyotype (bottom) shows trisomy 21; notice the three copies of chromosome 21.

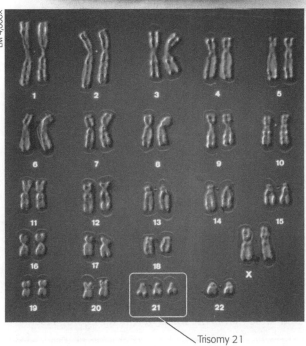

LM 4,000×

Trisomy 21

they have 47 chromosomes in total. This condition is called **trisomy 21**.

In most cases, a human embryo with an atypical number of chromosomes is spontaneously aborted (miscarried) long before birth, often before the woman is even aware that she is pregnant. In fact, some doctors speculate that miscarriages due to genetic defects occur in nearly one-quarter of all pregnancies, although this number is difficult to verify. However, some aberrations in chromosome number seem to upset the genetic balance less drastically, and individuals with such abnormalities can survive. These people usually have a characteristic set of symptoms, called a syndrome. A person with trisomy 21 has a condition called **Down syndrome** (named after John Langdon Down, an English physician who first described this condition in 1866).

Affecting about 1 out of every 700 children, trisomy 21 is the most common chromosome number abnormality and the most common serious birth defect in the United States. Down syndrome includes characteristic facial features—frequently a fold of skin at the inner corner of the eye, a round face, and a flattened nose—as well as short stature, heart defects, and susceptibility to leukemia and Alzheimer's disease. People with Down syndrome usually have a life span shorter than normal. They also exhibit varying degrees of developmental delays. However, some individuals with the syndrome may live to middle age or beyond, and many are socially adept and can function well within society. Although no one is sure why, the risk of Down syndrome increases with the age of the mother, climbing to about 1% risk for mothers at age 40. The fetuses of pregnant women age 35 and older are therefore candidates for chromosomal prenatal screenings (see Chapter 9). ☑

Abnormal Numbers of Sex Chromosomes

Trisomy 21 is an autosomal nondisjunction. Nondisjunction in meiosis can also lead to abnormal numbers of sex chromosomes, X and Y. Unusual numbers of sex chromosomes seem to upset the genetic balance less than unusual numbers of autosomes. This may be because the Y chromosome is very small and carries relatively few genes. Furthermore, mammalian cells normally operate with only one functioning X chromosome because other copies of the chromosome become inactivated in each cell (see Chapter 11).

☑ **CHECKPOINT**

What does it mean to refer to a disease as a "syndrome," as with AIDS?

Answer: A syndrome displays a set of multiple symptoms rather than just a single symptom.

Table 8.1	Abnormalities of Sex Chromosome Number in Humans		
Sex Chromosomes	**Syndrome**	**Origins of Nondisjunction**	**Frequency in Population**
XXY	Klinefelter syndrome (male)	Meiosis in egg or sperm formation	$\frac{1}{2,000}$
XYY	None (normal male)	Meiosis in sperm formation	$\frac{1}{2,000}$
XXX	None (normal female)	Meiosis in egg or sperm formation	$\frac{1}{1,000}$
XO	Turner syndrome (female)	Meiosis in egg or sperm formation	$\frac{1}{5,000}$

Table 8.1 lists the most common human sex chromosome abnormalities. An extra X chromosome in a male, making him XXY, produces a condition called Klinefelter syndrome. If untreated, men with this disorder have male sex organs, but the testes are abnormally small, the individual is sterile, and he often has breast enlargement and other feminine body contours. These symptoms can be reduced through administration of the sex hormone testosterone. Klinefelter syndrome is also found in individuals with more than three sex chromosomes, such as XXYY, XXXY, or XXXXY. These abnormal numbers of sex chromosomes result from multiple nondisjunctions.

Human males with a single extra Y chromosome (XYY) do not have any well-defined syndrome, although they tend to be taller than average. Females with an extra X chromosome (XXX) cannot be distinguished from XX females except by karyotype.

Females who are lacking an X chromosome are designated XO; the O indicates the absence of a second sex chromosome. These women have Turner syndrome. They have a characteristic appearance, including short stature and often a web of skin extending between the neck and shoulders. Women with Turner syndrome are of normal intelligence, but are sterile. If left untreated, they have poor development of breasts and other secondary sex characters. Administration of estrogen can alleviate those symptoms. The XO condition is the sole known case where having only 45 chromosomes is not fatal in humans.

Notice the crucial role of the Y chromosome in determining a person's sex. In general, a single Y chromosome is enough to produce biological maleness, regardless of the number of X chromosomes. The absence of a Y chromosome results in biological femaleness. ☑

☑ **CHECKPOINT**

Why is an individual more likely to survive with an abnormal number of sex chromosomes than an abnormal number of autosomes?

Answer: because the Y chromosome is very small and extra X chromosomes are inactivated

Life with and without Sex EVOLUTION CONNECTION

The Advantages of Sex

Throughout this chapter, we've examined cell division within the context of reproduction. Like the Komodo dragon discussed in the Biology and Society section, many species (including a few dozen animal species, but many more among the plant kingdom) can reproduce both sexually and asexually (Figure 8.23). An important advantage of asexual reproduction is that there is no need for a partner. Asexual reproduction may thus confer an evolutionary advantage when organisms are sparsely distributed (on an isolated island, for example) and unlikely to meet a mate. Furthermore, if an organism is superbly suited to a stable environment, asexual reproduction has the advantage of passing on its entire genetic legacy intact. Asexual reproduction also eliminates the need to expend energy forming gametes and copulating with a partner.

In contrast to plants, the vast majority of animals reproduce by sexual means. There are exceptions, such as the few species that can reproduce via parthenogenesis and the bdelloid rotifers discussed in the Process of Science section. But most animals reproduce only through sex. Therefore, sex must enhance evolutionary fitness. But how? The answer remains elusive. Most hypotheses focus on the unique combinations of genes formed during meiosis and fertilization. By producing offspring of varied genetic makeup, sexual reproduction may enhance survival by speeding adaptation to a changing environment. Another idea is that shuffling genes during sexual reproduction might reduce the incidence of harmful genes more rapidly. But for now, one of biology's most basic questions—Why have sex?—remains a hotly debated topic that is the focus of much ongoing research.

▼ Figure 8.23 **Sexual and asexual reproduction.** Many plants, such as this strawberry, have the ability to reproduce both sexually (via flowers that produce fruit) and asexually (via runners).

Runner ⟶

Chapter Review

■ SUMMARY OF KEY CONCEPTS

What Cell Reproduction Accomplishes

Cell reproduction, also called cell division, produces genetically identical daughter cells:

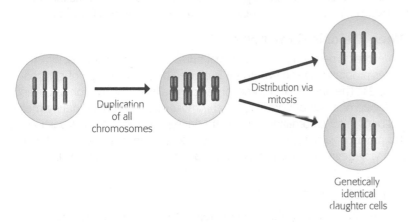

Some organisms use mitosis (ordinary cell division) to reproduce. This is called asexual reproduction, and it results in offspring that are genetically identical to the lone parent and to each other. Mitosis also enables multicellular organisms to grow and develop and to replace damaged or lost cells. Organisms that reproduce sexually, by the union of a sperm with an egg cell, carry out meiosis, a type of cell division that yields gametes with only half as many chromosomes as body (somatic) cells.

The Cell Cycle and Mitosis

Eukaryotic Chromosomes

The genes of a eukaryotic genome are grouped into multiple chromosomes in the nucleus. Each chromosome contains one very long DNA molecule, with many genes, that is wrapped around histone proteins. Individual chromosomes are coiled up and therefore visible with a light microscope only when the cell is in the process of dividing; otherwise, they are in the form of thin, loosely packed chromatin fibers.

Information Flow: Duplicating Chromosomes

Because chromosomes contain the information needed to control cellular processes, they must be copied and distributed to daughter cells. Before a cell starts dividing, the chromosomes duplicate, producing sister chromatids (containing identical DNA) joined together at the centromere.

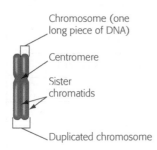

The Cell Cycle

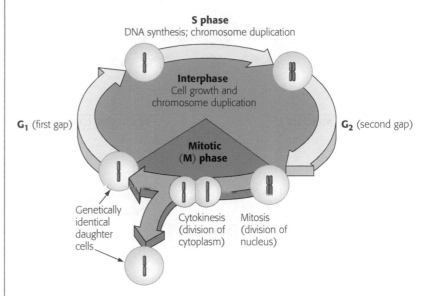

Mitosis and Cytokinesis

Mitosis is divided into four phases: prophase, metaphase, anaphase, and telophase. At the start of mitosis, the chromosomes coil up and the nuclear envelope breaks down (prophase). Next, a mitotic spindle made of microtubule tracks moves the chromosomes to the middle of the cell (metaphase). The sister chromatids then separate and are moved to opposite poles of the cell (anaphase), where two new nuclei form (telophase). Cytokinesis overlaps the end of mitosis. In animals, cytokinesis occurs by cleavage, which pinches the cell in two. In plants, a membranous cell plate divides the cell in two. Mitosis and cytokinesis produce genetically identical cells.

Cancer Cells: Dividing Out of Control

When the cell cycle control system malfunctions, a cell may divide excessively and form a tumor. Cancer cells may grow to form malignant tumors, invade other tissues (metastasize), and even kill the host. Surgery can remove tumors, and radiation and chemotherapy are effective as treatments because they interfere with cell division. You can increase the likelihood of surviving some forms of cancer through lifestyle changes and regular screenings.

Meiosis, the Basis of Sexual Reproduction

Homologous Chromosomes

The somatic cells (body cells) of each species contain a specific number of chromosomes; human cells have 46, made up of 23 pairs of homologous chromosomes. The chromosomes of a homologous pair carry genes for the same characteristics at the same places. Mammalian males have X and Y sex chromosomes (only partly homologous), and females have two X chromosomes.

Gametes and the Life Cycle of a Sexual Organism

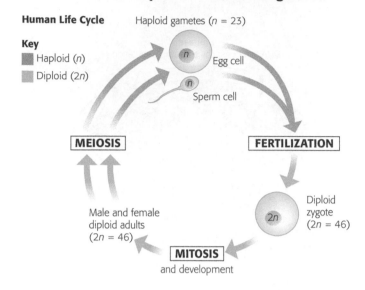

Human Life Cycle

Haploid gametes (*n* = 23)

Key

■ Haploid (*n*)
■ Diploid (2*n*)

n Egg cell

n Sperm cell

MEIOSIS

FERTILIZATION

Male and female diploid adults
(2*n* = 46)

Diploid zygote
(2*n* = 46)
2*n*

MITOSIS
and development

The Process of Meiosis

Meiosis, like mitosis, is preceded by chromosome duplication. But in meiosis, the cell divides twice to form four daughter cells. The first division, meiosis I, starts with the pairing of homologous chromosomes. In crossing over, homologous chromosomes exchange corresponding segments. Meiosis I separates the members of the homologous pairs and produces two daughter cells, each with one set of (duplicated) chromosomes. Meiosis II is essentially the same as mitosis; in each of the cells, the sister chromatids of each chromosome separate.

Review: Comparing Mitosis and Meiosis

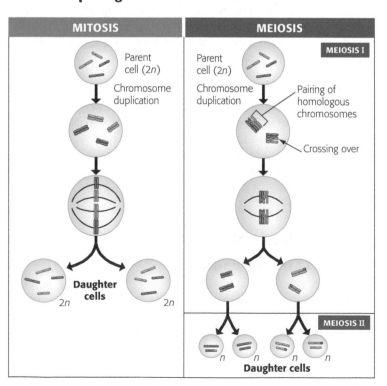

| MITOSIS | MEIOSIS |

Parent cell (2*n*)

Chromosome duplication

Daughter cells
2*n*   2*n*

MEIOSIS I

Parent cell (2*n*)

Chromosome duplication

Pairing of homologous chromosomes

Crossing over

MEIOSIS II

n   *n*   *n*   *n*

Daughter cells

The Origins of Genetic Variation

Because the chromosomes of a homologous pair come from different parents, they carry different versions of many of their genes. The large number of possible arrangements of chromosome pairs at metaphase of meiosis I leads to many different combinations of chromosomes in eggs and sperm. Random fertilization of eggs by sperm greatly increases the variation. Crossing over during prophase of meiosis I increases variation still further.

When Meiosis Goes Awry

Sometimes a person has an abnormal number of chromosomes, which causes problems. Down syndrome is caused by an extra copy of chromosome 21. The abnormal chromosome count is a product of nondisjunction, the failure of a pair of homologous chromosomes to separate during meiosis I or of sister chromatids to separate during meiosis II. Nondisjunction can also produce gametes with extra or missing sex chromosomes, which lead to varying degrees of malfunction but do not usually affect survival.

MasteringBiology®

For practice quizzes, BioFlix animations, MP3 tutorials, video tutors, and more study tools designed for this textbook, go to MasteringBiology®

SELF-QUIZ

1. Which of the following is not a function of mitosis in humans?
   a. repair of wounds
   b. growth
   c. production of gametes from diploid cells
   d. replacement of lost or damaged cells

2. In what sense are the two daughter cells produced by mitosis identical?

3. Why is it difficult to observe individual chromosomes during interphase?

4. A biochemist observes cells growing in the laboratory. A cell that completes the cell cycle without undergoing cytokinesis will
   a. have less genetic material than it started with.
   b. not have completed anaphase.
   c. have its chromosomes lined up in the middle of the cell.
   d. have two nuclei.

5. Which two phases of mitosis are essentially opposites in terms of changes in the nucleus?

6. Complete the following table to compare mitosis and meiosis:

	Mitosis	Meiosis
a. Number of chromosomal duplications		
b. Number of cell divisions		
c. Number of daughter cells produced		
d. Number of chromosomes in daughter cells		
e. How chromosomes line up during metaphase		
f. Genetic relationship of daughter cells to parent cells		
g. Functions performed in the human body		

7. If an intestinal cell in a dog contains 78 chromosomes, a dog sperm cell would contain _____ chromosomes.

8. A micrograph of a dividing cell from a mouse shows 19 chromosomes, each consisting of two sister chromatids. During which stage of meiosis could this micrograph have been taken? (Explain your answer.)

9. Tumors that remain at their site of origin are called _____, and tumors that have cells that can migrate to other body tissues are called _____.

10. For a species with four pairs of chromosomes, _____ chromosome combinations are possible.

11. Although nondisjunction is a random event, there are many more individuals with an extra chromosome 21, which causes Down syndrome, than individuals with an extra chromosome 3 or chromosome 16. Propose an explanation for this.

Answers to these questions can be found in Appendix: Self-Quiz Answers.

THE PROCESS OF SCIENCE

12. In adult humans, some cells remain mitotically active, whereas others have stopped dividing (they remain in the G_0 phase of the cell cycle). For example, skin and blood cells are continuously renewed by the division of stem cells, whereas the nervous system and heart muscle cells are incapable of further mitosis. Some cells stop dividing, but can re-enter the cell cycle in order to replace tissues, for example, following an injury. Cancer chemotherapy targets malignant cells by blocking their division. Which normal tissues will be more affected by chemotherapy and which ones less? In the light of this, how can you explain the typical side effects of chemotherapy?

13. You prepare a slide with a thin slice of an onion root tip. You see the following view in a light microscope. Identify the stage of mitosis for each of the outlined cells, a–d.

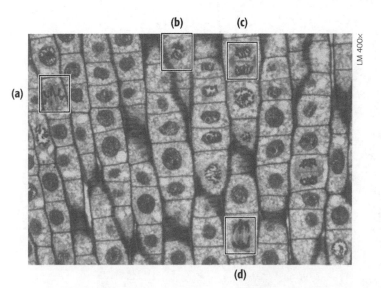

14. Interpreting Data The graph in the right hand column shows the incidence of Down syndrome in the offspring of normal parents as the age of the mother increases. For women under the age of 30, how many infants with Down syndrome are born per 1,000 births? How many for women at age 40? At age 50? How many times more likely is a 50-year-old woman to give birth to a baby with Down syndrome than a 30-year-old woman?

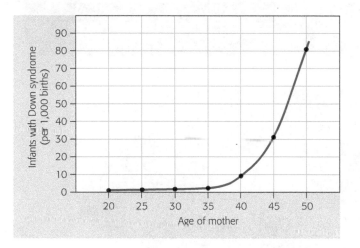

BIOLOGY AND SOCIETY

15. As described in the Biology and Society section, some species (including some endangered species) have been shown to be capable of reproduction via parthenogenesis, without participation of a mate. What implications might this have on efforts to repopulate endangered species? What drawbacks might a parthenogenesis program have on the well-being of the species it targets?

16. Every year, about a million Americans are diagnosed with cancer. This means that about 75 million Americans now living will eventually have cancer, and one in five will die of the disease. There are many kinds of cancers and many causes of the disease. For example, smoking causes most lung cancers. Overexposure to ultraviolet rays in sunlight causes most skin cancers. There is evidence that a high-fat, low-fiber diet is a factor in breast, colon, and prostate cancers. And agents in the workplace, such as asbestos and vinyl chloride, are also implicated as causes of cancer. Hundreds of millions of dollars are spent each year in the search for effective treatments for cancer, yet far less money is spent on preventing cancer. Why might this be true? What kinds of lifestyle changes can you make to help reduce your risk of cancer? What kinds of prevention programs could be initiated or strengthened to encourage these changes? What factors might impede such changes and programs? Should we devote more of our resources to treating cancer or to preventing it? Defend your position.

17. The risk of trisomy 21, the cause of Down syndrome, and other chromosomal abnormalities grows with the age of the pregnant woman. Are there other considerations that make pregnancy at an older age risky? Should modern medical strategies to achieve pregnancy, such as hormone fertility treatment and artificial insemination, be limited to women below a certain age?

Patterns of Inheritance

Why Genetics Matters

▲ Your mother's genes had no role in determining your sex.

Hemophilia is more prevalent among European royalty than
▼ commoners because of past intermarriage.

Because the ▶
environment
helps shape your
appearance, identical
twins aren't identical in
every way.

Dog Breeding BIOLOGY AND SOCIETY

Our Longest-Running Genetic Experiment

The adorable canine in the photo to the right is a purebred Cavalier King Charles Spaniel. If you were to mate two purebred "Cavs," you'd expect the offspring to display the traits that distinguish this breed, such as a silky coat, long elegant ears, and gentle eyes. This is a reasonable expectation because each purebred dog has a well-documented pedigree that includes several generations of ancestors with similar genetic makeup and appearance. But similarities among purebred Cavaliers extend beyond mere appearance, Cavs are generally energetic, obedient, sweet, and gentle—traits that make these dogs particularly well suited to act as companion or therapy dogs. Such behavioral similarities suggest that dog breeders can select for personality as well as physical traits. Cavaliers have four standard coat types—Blenheim (chestnut brown on white, seen at right, named after Blenheim Palace in England, where such dogs were first bred), black and tan, ruby, and tricolor (black, white, and tan)—reminding us that even within a purebred line, there remains significant variation of certain traits.

Purebred pooches are living proof that dogs are more than man's best friend: They are also one of our longest-running genetic experiments. Evidence (which we'll explore at the end of this chapter) suggests that people have selected and mated dogs with preferred traits for more than 15,000 years. Nearly every modern Cavalier, for example, can trace its ancestry back to a single pair that were brought to America by a breeder in 1952. These original dogs were chosen due to their desired shape and temperament. Similar choices were made for every modern dog breed. Over thousands of years, such genetic tinkering has led to the incredible variety of body types and behaviors we have today, from huge, docile Great Danes to tiny, spunky Chihuahuas.

Breeding a best friend. Dogs, such as this Cavalier King Charles Spaniel, are one of humankind's longest running genetic experiments.

Although people have been applying genetics for thousands of years—by breeding food crops (such as wheat, rice, and corn) as well as domesticated animals (such as cows, sheep, and goats)—the biological principles underlying genetics have only recently been understood. In this chapter, you will learn the basic rules of how genetic traits are passed from generation to generation and how the behavior of chromosomes (the topic of Chapter 8) accounts for these rules. In the process, you will learn how to predict the ratios of offspring with particular traits. At several points in the chapter, we'll return to the subject of dog breeding to help illustrate genetic principles.

Genetics and Heredity

Heredity is the transmission of traits from one generation to the next. **Genetics**, the scientific study of heredity, began in the 1860s, when an Augustinian monk named Gregor Mendel (Figure 9.1) deduced its fundamental principles by breeding garden peas. Mendel lived and worked in an abbey in Brunn, Austria (now Brno, in the Czech Republic). Strongly influenced by his study of physics, mathematics, and chemistry at the University of Vienna, his research was both experimentally and mathematically rigorous, and these qualities were largely responsible for his success.

In a paper published in 1866, Mendel correctly argued that parents pass on to their offspring discrete genes (which he termed "heritable factors") that are responsible for inherited traits, such as purple flowers or round seeds in pea plants. (It is interesting to note that Mendel's publication came just seven years after Darwin's 1859 publication of *On the Origin of Species*, making the 1860s a banner decade in the advent of modern biology.) In his paper, Mendel stressed that genes retain their individual identities generation after generation, no matter how they are mixed up or temporarily masked.

reach the carpel. When he wanted to fertilize one plant with pollen from a different plant, he pollinated the plants by hand, as shown in Figure 9.3. Thus, Mendel was always sure of the parentage of his new plants.

Each of the characters Mendel chose to study, such as flower color, occurred in two distinct traits. Mendel worked with his plants until he was sure he had purebred varieties—that is, varieties for which self-fertilization produced offspring all identical to the parent. For instance, he identified a purple-flowered variety that, when self-fertilized, always produced offspring plants that had all purple flowers.

Next Mendel was ready to ask what would happen when he crossed different purebred varieties with each other. For example, what offspring would result if plants with purple flowers and plants with white flowers were

▲ Figure 9.1 Gregor Mendel.

In an Abbey Garden

Mendel probably chose to study garden peas because they were easy to grow and they came in many readily distinguishable varieties. For example, one variety has purple flowers and another variety has white flowers. A heritable feature that varies among individuals, such as flower color, is called a **character**. Each variant of a character, such as purple or white flowers, is called a **trait**.

Perhaps the most important advantage of pea plants as an experimental model was that Mendel could strictly control their reproduction. The petals of a pea flower (Figure 9.2) almost completely enclose the egg-producing organ (the carpel) and the sperm-producing organs (the stamens). Consequently, in nature, pea plants usually self-fertilize because sperm-carrying pollen grains released from the stamens land on the tip of the egg-containing carpel of the same flower. Mendel could ensure self-fertilization by covering a flower with a small bag so that no pollen from another plant could

▼ Figure 9.2 The structure of a pea flower. To reveal the reproductive organs—the stamens and carpel—one of the petals has been removed in this drawing.

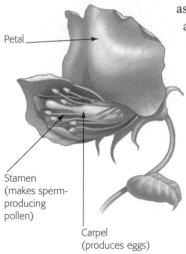

Petal

Stamen (makes sperm-producing pollen)

Carpel (produces eggs)

▼ Figure 9.3 Mendel's technique for cross-fertilizing pea plants.

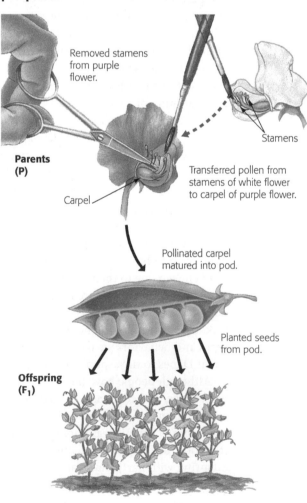

Removed stamens from purple flower.

Stamens

Parents (P)

Carpel

Transferred pollen from stamens of white flower to carpel of purple flower.

Pollinated carpel matured into pod.

Planted seeds from pod.

Offspring (F₁)

cross-fertilized as shown in Figure 9.3? The offspring of two different purebred varieties are called **hybrids**, and the cross-fertilization itself is referred to as a genetic **cross**. The parental plants are called the **P generation**, and their hybrid offspring are the **F_1 generation** (F for *filial*, from the Latin for "son" or "daughter"). When F_1 plants self-fertilize or fertilize each other, their offspring are the **F_2 generation**. ☑

Mendel's Law of Segregation

Mendel performed many experiments in which he tracked the inheritance of characters, such as flower color, that occur as two alternative traits (**Figure 9.4**). The results led him to formulate several hypotheses about inheritance. Let's look at some of his experiments and follow the reasoning that led to his hypotheses.

▼ **Figure 9.4 The seven characters of pea plants studied by Mendel.** Each character comes in the two alternative traits shown here.

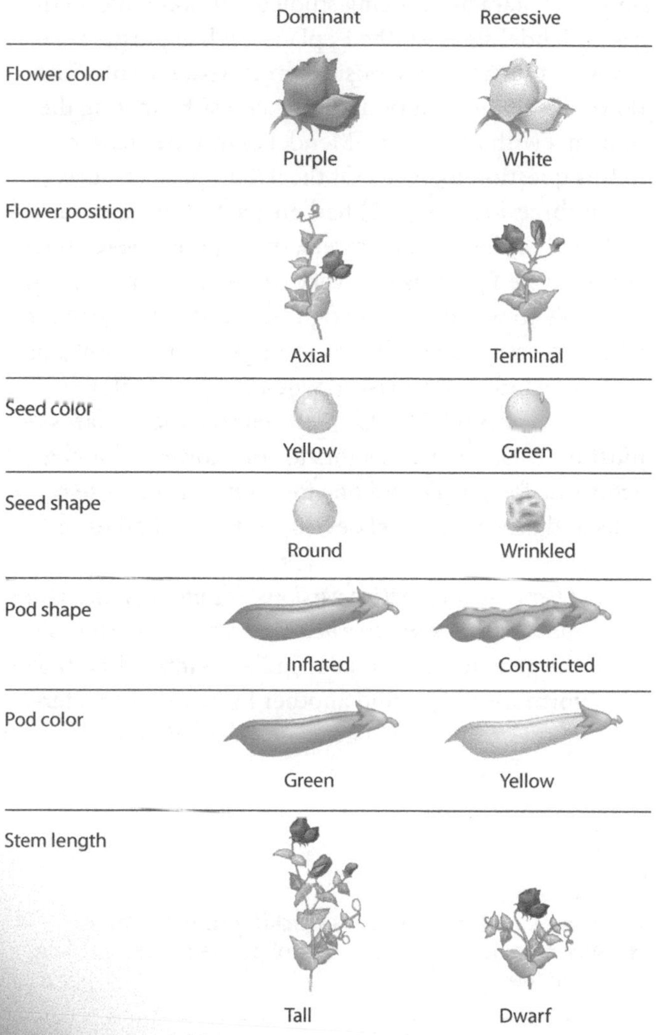

	Dominant	Recessive
Flower color	Purple	White
Flower position	Axial	Terminal
Seed color	Yellow	Green
Seed shape	Round	Wrinkled
Pod shape	Inflated	Constricted
Pod color	Green	Yellow
Stem length	Tall	Dwarf

☑ **CHECKPOINT**

Why was the development of purebred pea plant varieties critical to Mendel's work?

Answer: Purebred varieties allowed Mendel to predict the outcome of specific crosses and therefore to run controlled experiments.

Monohybrid Crosses

Figure 9.5 shows a cross between a purebred pea plant with purple flowers and a purebred pea plant with white flowers. This is called a **monohybrid cross** because only one character is being studied—flower color in this case. Mendel saw that the F₁ plants all had purple flowers. Was the factor responsible for inheritance of white flowers now lost as a result of the cross? By mating the F₁ plants with each other, Mendel found the answer to this question to be no. Of the 929 F₂ plants he bred, about three-fourths (705) had purple flowers and one-fourth (224) had white flowers; that is, there were about three purple F₂ plants for every white plant, or a 3:1 ratio of purple to white. Mendel figured out that the gene for white flowers did not disappear in the F₁ plants but was somehow hidden or masked when the purple-flower factor was present. He also deduced that the F₁ plants must have carried two factors for the flower-color character, one for purple and one for white. From these results and others, Mendel developed four hypotheses:

1. *There are alternative versions of genes that account for variations in inherited characters.* For example, the gene for flower color in pea plants exists in one form for purple and another for white. The alternative versions of a gene are called **alleles**.

2. *For each inherited character, an organism inherits two alleles, one from each parent.* These alleles may be the same or different. An organism that has two identical alleles for a gene is said to be **homozygous** for that gene (and is called a homozygote). An organism that has two different alleles for a gene is said to be **heterozygous** for that gene (and is called a heterozygote).

3. *If the two alleles of an inherited pair differ, then one determines the organism's appearance and is called the* **dominant allele**; *the other has no noticeable effect on the organism's appearance and is called the* **recessive allele**. Geneticists use uppercase italic letters (such as *P*) to represent dominant alleles and lowercase italic letters (such as *p*) to represent recessive alleles.

4. *A sperm or egg carries only one allele for each inherited character because the two alleles for a character segregate (separate) from each other during the production of gametes.* This statement is called the **law of segregation**. When sperm and egg unite at fertilization, each contributes its alleles, restoring the paired condition in the offspring.

Figure 9.6 illustrates Mendel's law of segregation, which explains the inheritance pattern shown in Figure 9.5. Mendel's hypotheses predict that when alleles segregate during gamete formation in the F₁ plants, half the gametes will receive a purple-flower allele (*P*) and the other half a white-flower allele (*p*). During pollination among the F₁ plants, the gametes unite randomly. An egg with a purple-flower allele has an equal chance of being fertilized by a sperm with a purple-flower allele or one with a white-flower allele (that is, a *P* egg may fuse with a *P* sperm or a *p* sperm). Because the same is true for an egg with a white-flower allele (a *p* egg with a *P* sperm or *p* sperm), there are a total of four equally likely combinations of sperm and egg.

The diagram at the bottom of Figure 9.6, called a **Punnett square**, repeats the cross shown in Figure 9.5 in a way that highlights the four possible combinations of gametes and the resulting four possible offspring in the F₂ generation. Each square represents an equally probable product of fertilization. For example, the box in the upper right corner of the Punnett square shows the genetic combination resulting from a *p* sperm fertilizing a *P* egg.

According to the Punnett square, what will be the physical appearance of these F₂ offspring? One-fourth of the plants have two alleles specifying purple flowers (*PP*); clearly, these plants will have purple flowers. One-half (two-fourths) of the F₂ offspring have inherited one

▼ **Figure 9.5 Mendel's cross tracking one character (flower color).** Note the 3:1 ratio of purple flowers to white flowers in the F₂ generation.

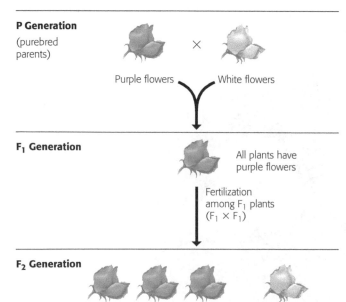

P Generation
(purebred parents)

Purple flowers × White flowers

F₁ Generation

All plants have purple flowers

Fertilization among F₁ plants (F₁ × F₁)

F₂ Generation

3/4 of plants have purple flowers

1/4 of plants have white flowers

▼ Figure 9.6 The law of segregation.

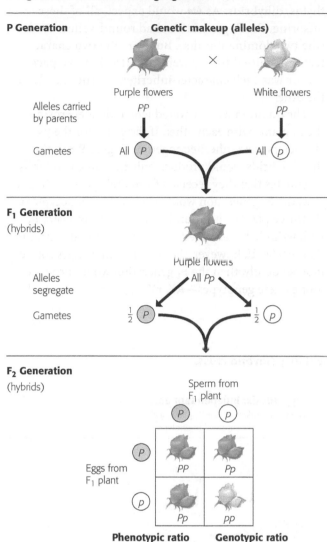

P Generation

Genetic makeup (alleles)

Purple flowers
PP

White flowers
pp

Alleles carried by parents

Gametes

All P

All p

F₁ Generation (hybrids)

Purple flowers
All *Pp*

Alleles segregate

Gametes

$\frac{1}{2}$ P

$\frac{1}{2}$ p

F₂ Generation (hybrids)

Sperm from F₁ plant

P p

Eggs from F₁ plant

P *PP*   *Pp*

p *Pp*   *pp*

Phenotypic ratio
3 purple:1 white

Genotypic ratio
1 *PP*:2 *Pp*:1 *pp*

Mendel found that each of the seven characters he studied had the same inheritance pattern: A parental trait disappeared in the F₁ generation, only to reappear in one-fourth of the F₂ offspring. The underlying mechanism is explained by Mendel's law of segregation: Pairs of alleles segregate (separate) during gamete formation; the fusion of gametes at fertilization creates allele pairs again. Research since Mendel's day has established that the law of segregation applies to all sexually reproducing organisms, including people.

Genetic Alleles and Homologous Chromosomes

Before continuing with Mendel's experiments, let's consider how an understanding of chromosomes (see Chapter 8) fits with what we've said about genetics so far. The diagram in **Figure 9.7** shows a pair of homologous chromosomes—chromosomes that carry alleles of the same genes. Recall that every diploid cell, whether from a pea plant or a person, has pairs of homologous chromosomes. One member of each pair comes from the organism's female parent and the other member of each pair comes from the male parent. Each labeled band on the chromosomes in the figure represents a gene **locus** (plural, *loci*), a specific location of a gene along the chromosome. You can see the connection between Mendel's law of segregation and homologous chromosomes: Alleles (alternative versions) of a gene reside at the same locus on homologous chromosomes. However, the two chromosomes may bear either identical alleles or different ones at any one locus. In other words, the organisms may be homozygous or heterozygous for the gene at that locus. We will return to the chromosomal basis of Mendel's law later in the chapter. ✓

✓ CHECKPOINT

1. Genes come in different versions called _____. What term describes the condition where the two copies are identical? What term describes the condition where the two copies are different?
2. If two plants have the same genotype, must they have the same phenotype? If two plants have the same phenotype, must they have the same genotype?
3. You carry two alleles for every trait. Where did these alleles come from?

Answers: 1. alleles; homozygous; heterozygous. 2. Yes; No: One could be homozygous for the dominant allele, whereas the other is heterozygous. 3. One is from your father via his sperm and one is from your mother via her egg.

allele for purple flowers and one allele for white flowers (*Pp*); like the F₁ plants, these plants will also have purple flowers, the dominant trait. (Note that *Pp* and *pP* are equivalent and usually written as *Pp*.) Finally, one-fourth of the F₂ plants have inherited two alleles specifying white flowers (*pp*) and will express this recessive trait. Thus, Mendel's model accounts for the 3:1 ratio that he observed in the F₂ generation.

Geneticists distinguish between an organism's physical appearance, called its **phenotype** (such as purple or white flowers), and its genetic makeup, called its **genotype** (such as *PP*, *Pp*, or *pp*). Now we can see that Figure 9.5 shows only the phenotypes and Figure 9.6 both the genotypes and phenotypes in our sample cross. For the F₂ plants, the ratio of plants with purple flowers to those with white flowers (3:1) is called the phenotypic ratio. The genotypic ratio is 1(*PP*):2(*Pp*):1(*pp*).

▼ Figure 9.7 The relationship between alleles and homologous chromosomes.
The matching colors of corresponding loci highlight the fact that homologous chromosomes carry alleles for the same genes at the same positions along their lengths.

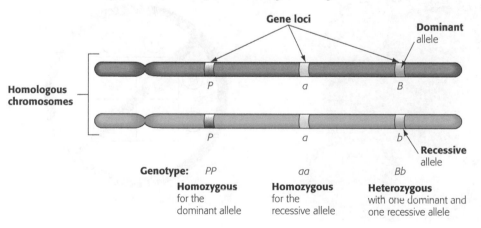

Gene loci

Dominant allele

Homologous chromosomes

P a B

P a b

Recessive allele

Genotype: *PP*   *aa*   *Bb*

Homozygous for the dominant allele

Homozygous for the recessive allele

Heterozygous with one dominant and one recessive allele

Mendel's Law of Independent Assortment

In addition to flower color, Mendel studied two other pea plant characters: seed shape (round versus wrinkled) and seed color (yellow versus green). From tracking these characters one at a time in monohybrid crosses, Mendel knew that the allele for round shape (designated *R*) was dominant to the allele for wrinkled shape (*r*) and that the allele for yellow seed color (*Y*) was dominant to the allele for green seed color (*y*). What would result from a **dihybrid cross**, the mating of parental varieties differing in two characters? Mendel crossed homozygous plants having round-yellow seeds (genotype *RRYY*) with plants having wrinkled-green seeds (*rryy*). As shown in **Figure 9.8**, the union of *RY* and *ry* gametes from the P generation yielded

hybrids heterozygous for both characters (*RrYy*)—that is, dihybrids. As we would expect, all of these offspring, the F_1 generation, had round-yellow seeds (the two dominant traits). But were the two characters transmitted from parents to offspring as a package, or was each character inherited independently of the other?

The question was answered when Mendel crossed the F_1 plants with each other. If the genes for the two characters were inherited together (Figure 9.8a), then the F_1 hybrids would produce only the same two kinds of gametes that they received from their parents. In that case, the F_2 generation would show a 3:1 phenotypic ratio (three plants with round-yellow seeds for every one with wrinkled-green seeds), as in the Punnett square in Figure 9.8a. If, however, the two seed characters sorted independently, then the F_1 generation would produce four gamete genotypes—*RY*, *rY*, *Ry*, and *ry*—in equal

▼ **Figure 9.8 Testing alternative hypotheses for gene assortment in a dihybrid cross.**
Only hypothesis B was supported by Mendel's data.

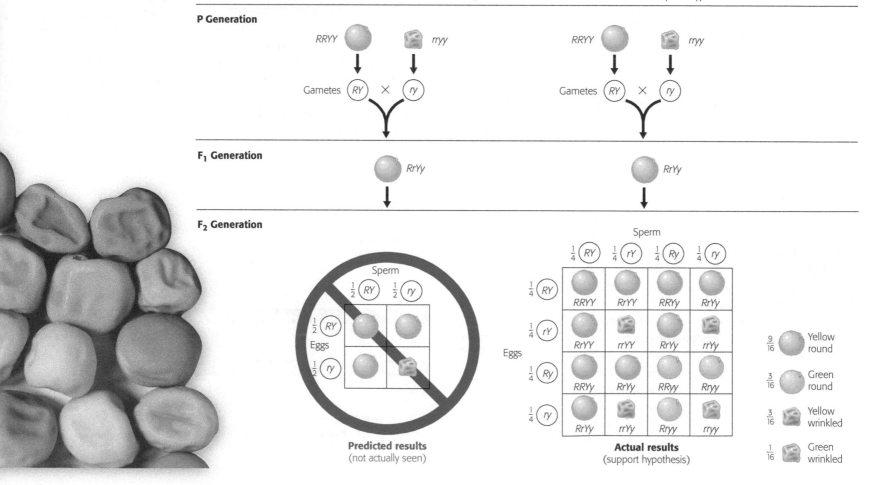

(a) Hypothesis: Dependent assortment
Leads to prediction that F_2 plants will have seeds that match the parents, either round-yellow or wrinkled-green

(b) Hypothesis: Independent assortment
Leads to prediction that F_2 plants will have four different seed phenotypes

P Generation

RRYY   *rryy*

Gametes   *RY* × *ry*

F_1 Generation

RrYy

F_2 Generation

Predicted results
(not actually seen)

Actual results
(support hypothesis)

$\frac{9}{16}$ Yellow round

$\frac{3}{16}$ Green round

$\frac{3}{16}$ Yellow wrinkled

$\frac{1}{16}$ Green wrinkled

quantities. The Punnett square in Figure 9.8b shows all possible combinations of alleles that can result in the F₂ generation from the union of four kinds of sperm with four kinds of eggs. If you study the Punnett square, you'll see that it predicts nine different genotypes in the F₂ generation. These nine genotypes will produce four different phenotypes in a ratio of 9:3:3:1.

The Punnett square in Figure 9.8b also reveals that a dihybrid cross is equivalent to two monohybrid crosses occurring simultaneously. From the 9:3:3:1 ratio, we can see that there are 12 plants in the F₂ generation with round seeds compared to 4 with wrinkled seeds, and 12 yellow-seeded plants compared to four green-seeded ones. These 12:4 ratios each reduce to 3:1, which is the F₂ ratio for a monohybrid cross. Mendel tried his seven pea characters in various dihybrid combinations and always observed a 9:3:3:1 ratio (or two simultaneous 3:1 ratios) of phenotypes in the F₂ generation. These results supported the hypothesis that *each pair of alleles segregates independently of the other pairs of alleles during* *gamete formation*. In other words, the inheritance of one character has no effect on the inheritance of another. This is called Mendel's **law of independent assortment**.

For another application of the law of independent assortment, examine the dog breeding experiment described in **Figure 9.9**. The inheritance of two characters in Labrador retrievers is controlled by separate genes: black versus chocolate coat color, and normal vision versus the eye disorder progressive retinal atrophy (PRA). Black Labs have at least one copy of an allele called *B*. The *B* allele is dominant to *b*, so only the coats of dogs with genotype *bb* are chocolate in color. The allele that causes PRA, called *n*, is recessive to allele *N*, which is necessary for normal vision. Thus, only dogs of genotype *nn* become blind from PRA. If you mate two doubly heterozygous (*BbNn*) Labs (bottom of Figure 9.9), the phenotypic ratio of the offspring (F₂) is 9:3:3:1. These results resemble the F₂ results in Figure 9.8, demonstrating that the coat color and PRA genes are inherited independently. ☑

☑ CHECKPOINT

Looking at Figure 9.9, what is the ratio of black coat to chocolate coat in Labs? Of normal vision to blindness?

Answer: 3:1; 3:1

▼ Figure 9.9 **Independent assortment of genes in Labrador retrievers.** Blanks in the genotypes indicate alleles that can be either dominant or recessive.

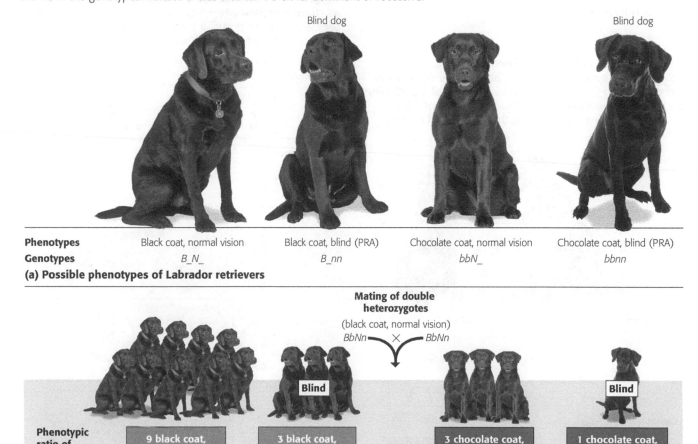

(a) Possible phenotypes of Labrador retrievers

(b) A Labrador dihybrid cross

Using a Testcross to Determine an Unknown Genotype

Suppose you have a Labrador retriever with a chocolate coat. Consulting Figure 9.9, you can tell that its genotype must be *bb*, the only combination of alleles that produces the chocolate-coat phenotype. But what if you have a black Lab? It could have one of two possible genotypes—*BB* or *Bb*—and there is no way to tell which genotype is the correct one by looking at the dog. To determine your dog's genotype, you could perform a **testcross**, a mating between an individual of dominant phenotype but unknown genotype (your black Lab) and a homozygous recessive individual—in this case, a *bb* chocolate Lab.

Figure 9.10 shows the offspring that could result from such a mating. If, as shown on the left, the black Lab parent's genotype is *BB*, we would expect all the offspring to be black, because a cross between genotypes *BB* and *bb* can produce only *Bb* offspring. On the other hand, if the black Lab parent is *Bb*, we would expect both black (*Bb*) and chocolate (*bb*) offspring. Thus, the appearance of the offspring may reveal the original black dog's genotype. ☑

▼ **Figure 9.10 A Labrador retriever testcross.** To determine the genotype of a black Lab, it can be crossed with a chocolate Lab (homozygous recessive, *bb*). If all the offspring are black, the black parent most likely had genotype *BB*. If any of the offspring are chocolate, the black parent must be heterozygous (*Bb*).

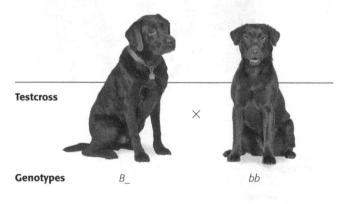

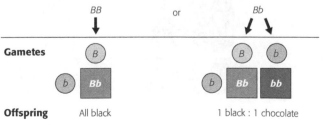

The Rules of Probability

Mendel's strong background in mathematics served him well in his studies of inheritance. For instance, he understood that genetic crosses obey the rules of probability—the same rules that apply when tossing coins, rolling dice, or drawing cards. An important lesson we can learn from tossing coins is that for each and every toss, the probability of heads is $\frac{1}{2}$. Even if heads has landed five times in a row, the probability of the next toss coming up heads is still $\frac{1}{2}$. In other words, the outcome of any particular toss is unaffected by what has happened on previous attempts. Each toss is an independent event.

If two coins are tossed simultaneously, the outcome for each coin is an independent event, unaffected by the other coin. What is the chance that both coins will land heads-up? The probability of such a dual event is the product of the separate probabilities of the independent events—for the coins, $\frac{1}{2} \times \frac{1}{2} = \frac{1}{4}$. This is called the **rule of multiplication**, and it holds true for independent events that occur in genetics as well as coin tosses, as shown in **Figure 9.11**. In our dihybrid cross of Labradors (see Figure 9.9), the genotype of the F_1 dogs

▼ **Figure 9.11 Segregation of alleles and fertilization as chance events.** When heterozygotes (*Bb*) form gametes, segregation of alleles during sperm and egg formation is like two separately tossed coins (that is, two independent events).

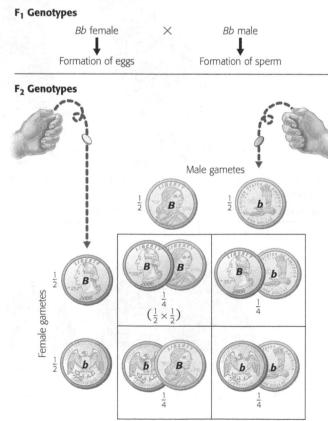

for coat color was *Bb*. What is the probability that a particular F₂ dog will have the *bb* genotype? To produce a *bb* offspring, both egg and sperm must carry the *b* allele. The probability that an egg from a *Bb* dog will have the *b* allele is $\frac{1}{2}$, and the probability that a sperm will have the *b* allele is also $\frac{1}{2}$. By the rule of multiplication, the probability that two *b* alleles will come together at fertilization is $\frac{1}{2} \times \frac{1}{2} = \frac{1}{4}$. This is exactly the answer given by the Punnett square in Figure 9.11. If we know the genotypes of the parents, we can predict the probability for any genotype among the offspring. By applying the rules of probability to segregation and independent assortment, we can solve some rather complex genetics problems. ☑

Family Pedigrees

Mendel's laws apply to the inheritance of many human traits. **Figure 9.12** illustrates alternative forms of three human characters that are each thought to be determined by simple dominant-recessive inheritance of one gene. (The genetic basis of many other human characters—such as eye and hair color—is much more complex and poorly understood.) If we call the dominant allele of any such gene *A*, the dominant phenotype results from either the homozygous genotype *AA* or the heterozygous genotype *Aa*. Recessive phenotypes result only from the homozygous genotype *aa*. In genetics, the word *dominant* does not imply that a phenotype is either normal or more common than a recessive phenotype; **wild-type traits** (those seen most often in nature) are not necessarily specified by dominant alleles. In genetics, dominance means that a heterozygote (*Aa*), carrying only a single copy of a dominant allele, displays the dominant phenotype. By contrast, the phenotype of the corresponding recessive allele is seen only in a homozygote (*aa*). Recessive traits may in fact be more common in the population than dominant ones. For example, the absence of freckles (*ff*) is more common than their presence (*FF* or *Ff*).

How can human genetics be studied? Researchers working with pea plants or Labrador retrievers can perform testcrosses. But geneticists who study people obviously cannot control the mating of their research participants. Instead, they must analyze the results of matings that have already occurred. First, the geneticist collects as much information as possible about a family's history for a trait. Then the researcher assembles this information into a family tree, called a **pedigree**. (You may associate pedigrees with purebred animals such as racehorses and champion dogs, but they can be used to represent human matings just as well.) To analyze a pedigree, the geneticist applies logic and Mendel's laws.

☑ **CHECKPOINT**

Using a standard 52-card deck, what is the probability of being dealt an ace? What about being dealt an ace or a king? What about being dealt an ace and then another ace?

Answer: $\frac{1}{13}$ (4 aces/52 cards); $\frac{2}{13}$ ($\frac{4}{52}$); $\frac{4}{52} \times \frac{3}{51}$ (because there are 3 aces left in a deck with 51 cards remaining) = 0.0045, or $\frac{1}{221}$

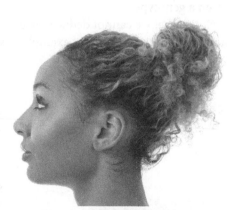
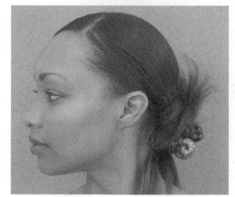

◄ **Figure 9.12 Examples of inherited human traits thought to be controlled by a single gene.**

DOMINANT TRAITS

Freckles

Widow's peak

Free earlobe

RECESSIVE TRAITS

No freckles

Straight hairline

Attached earlobe

187

Let's apply this approach to the example in **Figure 9.13**, a pedigree tracing the incidence of free versus attached earlobes. The letter *F* stands for the dominant allele for free earlobes, and *f* symbolizes the recessive allele for attached earlobes. In the pedigree, ☐ represents a male, ◯ represents a female, colored symbols (■ and ●) indicate that the person has the trait being investigated (in this case, attached earlobes), and an unshaded symbol represents a person who does not have the trait (that person has free earlobes). The earliest (oldest) generation is at the top of the pedigree, and the most recent generation is at the bottom.

By applying Mendel's laws, we can deduce that the attached allele is recessive because that is the only way that Kevin can have attached earlobes when neither of his parents (Hal and Ina) do. We can therefore label all the individuals with attached earlobes in the pedigree (that is, all those with colored circles or squares) as homozygous recessive (*ff*).

Mendel's laws enable us to deduce the genotypes for most of the people in the pedigree. For example, Hal and Ina must have carried the *f* allele (which they passed on to Kevin) along with the *F* allele that gave them free earlobes. The same must be true of Aaron and Betty because they both have free earlobes but Fred and Gabe have attached earlobes. For each of these cases, Mendel's laws and simple logic allow us to definitively assign a genotype.

Notice that we cannot deduce the genotype of every member of the pedigree. For example, Lisa must have at least one *F* allele, but she could be *FF* or *Ff*. We cannot distinguish between these two possibilities using the available data. Perhaps future generations will provide the data that solve this mystery. ☑

Human Disorders Controlled by a Single Gene

The human genetic disorders listed in **Table 9.1** are known to be inherited as dominant or recessive traits controlled by a single gene. These disorders therefore show simple inheritance patterns like the ones Mendel studied in pea plants. The genes involved are all located on autosomes, chromosomes other than the sex chromosomes X and Y.

Recessive Disorders

Most human genetic disorders are recessive. They range in severity from harmless to life-threatening. Most people who have recessive disorders are born to normal parents who are both heterozygotes—that is, parents who are **carriers** of the recessive allele for the disorder but appear normal themselves.

Using Mendel's laws, we can predict the fraction of affected offspring that is likely to result from a marriage between two carriers. Consider a form of inherited deafness caused by a recessive allele. Suppose two heterozygous carriers (*Dd*) have a child. What is the probability that the child will be deaf? As the Punnett square

▼ Figure 9.13 **A family pedigree showing inheritance of free versus attached earlobes.**

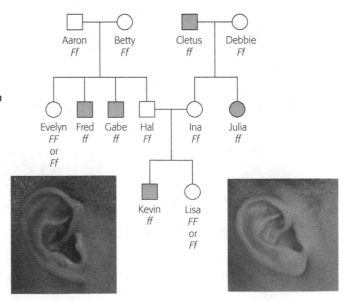

First generation (grandparents)

Aaron *Ff* — Betty *Ff*   Cletus *ff* — Debbie *Ff*

Second generation (parents, aunts, and uncles)

Evelyn *FF* or *Ff*   Fred *ff*   Gabe *ff*   Hal *Ff*   Ina *Ff*   Julia *ff*

Third generation (brother and sister)

Kevin *ff*   Lisa *FF* or *Ff*

Female   Male

● ☐ Attached

◯ ☐ Free

Table 9.1	Some Autosomal Disorders in People
Disorder	**Major Symptoms**
Recessive Disorders	
Albinism	Lack of pigment in skin, hair, and eyes 
Cystic fibrosis	Excess mucus in lungs, digestive tract, liver; increased susceptibility to infections; death in early childhood unless treated
Phenylketonuria (PKU)	Accumulation of phenylalanine in blood; lack of normal skin pigment; mental retardation unless treated 
Sickle-cell disease	Sickled red blood cells; damage to many tissues 
Tay-Sachs disease	Lipid accumulation in brain cells; mental deficiency; blindness; death in childhood
Dominant Disorders	
Achondroplasia	Dwarfism 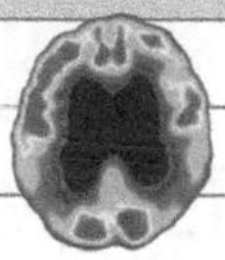
Alzheimer's disease (one type)	Mental deterioration; usually strikes late in life
Huntington's disease	Mental deterioration and uncontrollable movements; strikes in middle age
Hypercholesterolemia	Excess cholesterol in blood; heart disease

in **Figure 9.14** shows, each child of two carriers has a $\frac{1}{4}$ chance of inheriting two recessive alleles. Thus, we can say that about one-fourth of the children of this couple are likely to be deaf. We can also say that a hearing child from such a family has a $\frac{2}{3}$ chance of being a carrier (that is, on average, two out of three of the offspring with the hearing phenotype will be *Dd*). We can apply

▼ Figure 9.14 **Predicted offspring when both parents are carriers for a recessive disorder.**

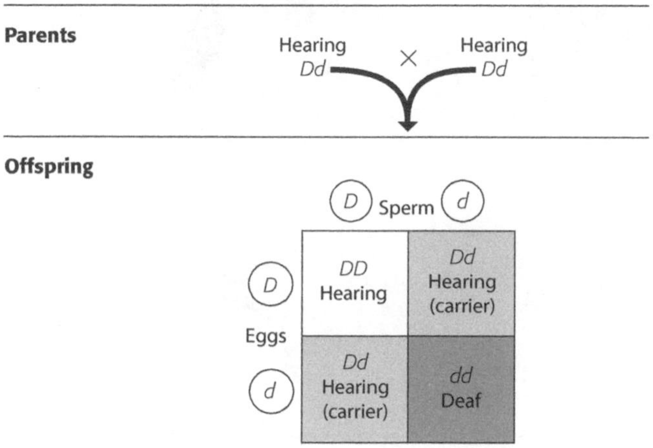

this same method of pedigree analysis and prediction to any genetic trait controlled by a single gene.

The most common lethal genetic disease in the United States is **cystic fibrosis (CF)**. Affecting about 30,000 Americans, the recessive CF allele is carried by about 1 in 31 Americans. A person with two copies of this allele has cystic fibrosis, which is characterized by an excessive secretion of very thick mucus from the lungs, pancreas, and other organs. This mucus can interfere with breathing, digestion, and liver function and makes the person vulnerable to recurrent bacterial infections. Although there is no cure for this fatal disease, a special diet, antibiotics to prevent infection, frequent pounding of the chest and back to clear the lungs, and other treatments can greatly extend life. Once invariably fatal in childhood, advances in CF treatment have raised the median survival age of Americans with CF to 37. ☑

Dominant Disorders

A number of human disorders are caused by dominant alleles. Some are harmless, such as extra fingers and toes or digits that are webbed. A serious but nonlethal disorder caused by a dominant allele is achondroplasia, a form of dwarfism in which the head and torso develop

☑ **CHECKPOINT**
A man and a woman who are both carriers of cystic fibrosis have had three children without cystic fibrosis. If the couple has a fourth child, what is the probability that the child will have the disorder?

Answer: $\frac{1}{4}$ (The genotypes and phenotypes of their other children are irrelevant.)

189

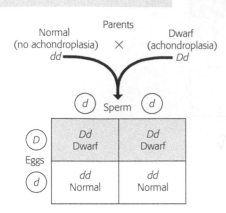

Parents
Normal
(no achondroplasia) × Dwarf (achondroplasia)
dd — *Dd*

▲ **Figure 9.15 A Punnett square illustrating a family with and without achondroplasia.**

normally but the arms and legs are short. The homozygous dominant genotype (*AA*) causes death of the embryo, and therefore only heterozygotes (*Aa*), individuals with a single copy of the defective allele, have this disorder. This also means that a person with achondroplasia has a 50% chance of passing the condition on to any children (Figure 9.15). Therefore, all those who do not have achondroplasia, more than 99.99% of the population, are homozygous for the recessive allele (*aa*). This example makes it clear that a dominant allele is not necessarily more common in a population than the corresponding recessive allele.

Dominant alleles that cause lethal disorders are much less common than lethal recessive alleles. One example is the allele that causes Huntington's disease, a degeneration of the nervous system that usually does not begin until middle age. Once the deterioration of the nervous system begins, it is irreversible and inevitably fatal. Because the allele for Huntington's disease is dominant, any child born to a parent with the allele has a 50% chance of inheriting the allele and the disorder. This example makes it clear that a dominant allele is not necessarily "better" than the corresponding recessive allele.

One way that geneticists study human traits is to find similar genes in animals, which can then be studied in greater detail through controlled matings and other experiments. Let's return to the chapter thread—dog breeding—to investigate the furry coats of man's best friend.

 Dog Breeding THE PROCESS OF SCIENCE

What Is the Genetic Basis of Coat Variation in Dogs?

You've probably made the **observation** that compared with most mammals, dogs come in a wide variety of physical types. For instance, dog coats can be short or long, straight or curly or wire-haired (having a "moustache" and "eyebrows"). Sometimes a single breed, such as the fox terrier, may display two or more of these variations (Figure 9.16).

In 2005, the complete genome of a dog—the sequence of all the DNA of a female boxer named Tasha—was published. Since that time, canine geneticists have added a wealth of data from other breeds. In 2009, an international group of researchers set out to investigate the **question** of the genetic basis for canine coats. They proposed the **hypothesis** that a comparison of genes from a wide variety of dogs with different coats would identify the genes responsible. They made the **prediction** that mutations in just a few genes could account for the coat appearance. Their **experiment** compared DNA sequences in 622 dogs from dozens of breeds. Their **results** identified three genes that in different combinations produced seven different coat appearances, from very short hair to full, thick, wired hair. The difference between the two dogs in the photo, for example, is due to a change in a single gene that regulates keratin, a protein that is one of the primary structural components of hair.

This experiment shows how the extreme range of phenotypes in dogs combined with the availability of genome sequences can be used to provide insight into interesting genetic questions. Indeed, similar research has uncovered the genetic basis of other dog traits, such as body size, hairlessness, and coat color.

▼ **Figure 9.16 Smooth versus wired fox terrier.** The fox terrier, like several other breeds, comes in smooth (left) and wired (right) coats.

Genetic Testing

Until relatively recently, the onset of symptoms was the only way to know if a person had inherited an allele that might lead to disease. Today, there are many tests that can detect the presence of disease-causing alleles in an individual's genome.

Most genetic tests are performed during pregnancy if the prospective parents are aware that they have an increased risk of having a baby with a genetic disease. Genetic testing before birth usually requires the collection of fetal cells. In amniocentesis, a physician uses a needle to extract about 2 teaspoons of the fluid that bathes the developing fetus **(Figure 9.17)**. In chorionic villus sampling, a physician inserts a narrow, flexible tube through the mother's vagina and into her uterus, removing some placental tissue. Once cells are obtained, they can be screened for genetic diseases.

Because amniocentesis and chorionic villus sampling have risks of complications, these techniques are usually reserved for situations in which the possibility of a genetic disease is high. Alternatively, blood tests on the mother at 15 to 20 weeks of pregnancy can help identify fetuses at risk for certain birth defects. The most widely used blood test measures levels of a protein called AFP in the mother's blood; abnormally high levels may indicate developmental defects in the fetus, while abnormally low levels may indicate Down syndrome. For a more complete risk profile, a woman's doctor may order a "triple screen test," which measures AFP as well as two other hormones produced by the placenta. Abnormal levels of these substances in the maternal blood may also point to a risk of Down syndrome. Newer genetic screening procedures involve isolating tiny amounts of fetal cells or DNA released into the mother's bloodstream. Because they are more accurate and can be performed earlier and more safely than other tests, these newer technologies are gradually replacing more invasive screening methods.

As genetic testing becomes more routine, geneticists are working to make sure that the tests do not cause more problems than they solve. Geneticists stress that patients seeking genetic testing should receive counseling both before and after to explain the test and to help them cope with the results. Identifying a genetic disease early can give families time to prepare—emotionally, medically, and financially. Advances in biotechnology offer possibilities for reducing human suffering, but not before key ethical issues are resolved. The dilemmas posed by human genetics reinforce one of this book's themes: the immense social implications of biology. ☑

☑ CHECKPOINT

Peter is a 28-year-old man whose father died of Huntington's disease. Peter's mother shows no signs of the disease. What is the probability that Peter has inherited Huntington's disease?

Answers: ½

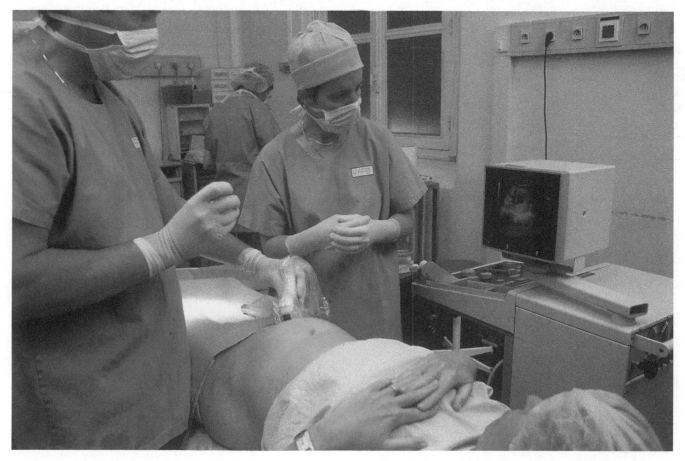

◄ **Figure 9.17**
Amniocentesis. A doctor uses a sonogram to help guide the removal of fetal cells for genetic testing.

Variations on Mendel's Laws

Mendel's two laws explain inheritance in terms of genes that are passed along from generation to generation according to simple rules of probability. These laws are valid for all sexually reproducing organisms, including garden peas, Labrador retrievers, and people. But just as the basic rules of musical harmony cannot account for all the rich sounds of a symphony, Mendel's laws stop short of explaining some patterns of genetic inheritance. In fact, for most sexually reproducing organisms, cases in which Mendel's rules can strictly account for the patterns of inheritance are relatively rare. More often, the observed inheritance patterns are more complex. Next, we'll look at several extensions to Mendel's laws that help account for this complexity.

Incomplete Dominance in Plants and People

The F_1 offspring of Mendel's pea crosses always looked like one of the two parent plants. In such situations, the dominant allele has the same effect on the phenotype whether present in one or two copies. But for some characters, the appearance of F_1 hybrids falls between the phenotypes of the two parents, an effect called **incomplete dominance**. For instance, when red snapdragons are crossed with white snapdragons, all the F_1 hybrids have pink flowers **(Figure 9.18)**. And in the F_2 generation, the genotypic ratio and the phenotypic ratio are the same: 1:2:1.

We also see examples of incomplete dominance in people. One case involves a recessive allele (h) that causes hypercholesterolemia, dangerously high levels of cholesterol in the blood. Normal individuals are homozygous dominant, HH. Heterozygotes (Hh) have blood cholesterol levels about twice what is normal. Such heterozygotes are very prone to cholesterol buildup in artery walls and may have heart attacks from blocked heart arteries by their mid-30s. Hypercholesterolemia is even more serious in homozygous individuals (hh). Homozygotes have about five times the normal amount of blood cholesterol and may have heart attacks as early as age 2. If we look at the molecular basis for hypercholesterolemia, we can understand the intermediate phenotype of heterozygotes **(Figure 9.19)**. The H allele specifies a cell-surface receptor protein that liver cells use to mop up excess low-density lipoprotein (LDL, or "bad cholesterol") from the blood. With only half as many receptors as HH individuals, heterozygotes can remove much less excess cholesterol.

▼ **Figure 9.18 Incomplete dominance in snapdragons.**
Compare this diagram with Figure 9.6, where one of the alleles displays complete dominance.

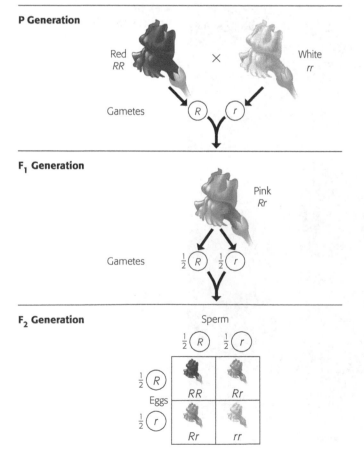

▼ **Figure 9.19 Incomplete dominance in human hypercholesterolemia.**
LDL receptors on liver cells promote the breakdown of cholesterol carried in the bloodstream by LDL (low-density lipoprotein). This process helps prevent the accumulation of cholesterol in the arteries. Having too few receptors allows dangerous levels of LDL to build up in the blood.

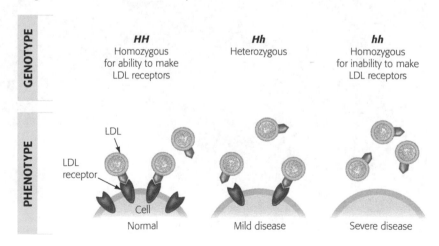

ABO Blood Groups: An Example of Multiple Alleles and Codominance

So far, we have discussed inheritance patterns involving only two alleles per gene (H versus h, for example). But most genes can be found in populations in more than two forms, known as multiple alleles. Although each individual carries, at most, two different alleles for a particular gene, in cases of multiple alleles, more than two possible alleles exist in the population.

The **ABO blood groups** in humans involve three alleles of a single gene. Various combinations of these three alleles produce four phenotypes: A person's blood type may be A, B, AB, or O. These letters refer to two carbohydrates, designated A and B, that may be found on the surface of red blood cells **(Figure 9.20)**. A person's red blood cells may be coated with carbohydrate A (giving them type A blood), carbohydrate B (type B), both (type AB), or neither (type O). (In case you are wondering, the "positive" and "negative" notations associated with blood types—referred to as the Rh blood group system—are due to inheritance of a separate, unrelated gene.)

Matching compatible blood groups is critical for safe blood transfusions. If a donor's blood cells have a carbohydrate (A or B) that is foreign to the recipient, then the recipient's immune system produces blood proteins called antibodies that bind to the foreign carbohydrates and cause the donor blood cells to clump together, potentially killing the recipient.

The four blood groups result from various combinations of the three different alleles: I^A (for the ability to make substance A), I^B (for B), and i (for neither A nor B). Each person inherits one of these alleles from each parent. Because there are three alleles, there are six possible genotypes, as listed in Figure 9.21. Both the I^A and I^B alleles are dominant to the i allele. Thus, I^AI^A and I^Ai people have type A blood, and I^BI^B and I^Bi people have type B. Recessive homozygotes (ii) have type O blood with neither carbohydrate. Finally, people of genotype I^AI^B make *both* carbohydrates. In other words, the I^A and I^B alleles are **codominant**, meaning that both alleles are expressed in heterozygous individuals (I^AI^B) who have type AB blood. Notice that type O blood reacts with no others, making such a person a universal donor. A person with type AB blood is a universal receiver. Be careful to distinguish codominance (the expression of both alleles) from incomplete dominance (the expression of one intermediate trait). ☑

✓ **CHECKPOINT**

1. Why is a testcross unnecessary to determine whether a snapdragon with red flowers is homozygous or heterozygous?
2. Maria has type O blood, and her sister has type AB blood. What are the genotypes of the girls' parents?

Answers: **1.** Only plants homozygous for the dominant allele have red flowers; heterozygotes have pink flowers. **2.** One parent is I^Ai, and the other parent is I^Bi.

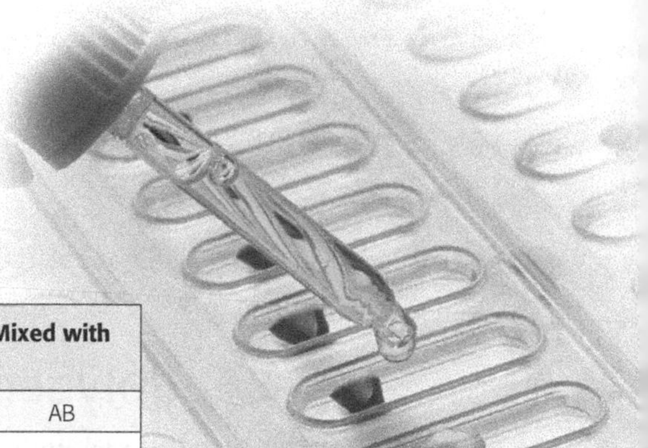

▼ **Figure 9.20 Multiple alleles for the ABO blood groups.** The three versions of the gene responsible for blood type may produce carbohydrate A (allele I^A), carbohydrate B (allele I^B), or neither carbohydrate (allele i). Because each person carries two alleles, six genotypes are possible that result in four different phenotypes. The clumping reaction that occurs between antibodies and foreign blood cells is the basis of blood-typing (shown in the photograph at right) and of the adverse reaction that occurs when someone receives a transfusion of incompatible blood.

Blood Group (Phenotype)	Genotypes	Red Blood Cells	Antibodies Present in Blood	Reactions When Blood from Groups Below Is Mixed with Antibodies from Groups at Left			
				O	A	B	AB
A	I^AI^A or I^Ai	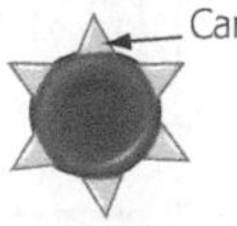 Carbohydrate A	Anti-B	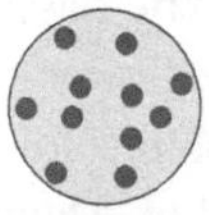	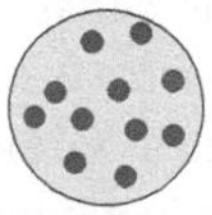	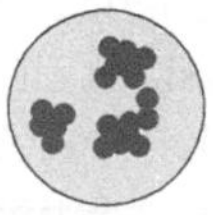	
B	I^BI^B or I^Bi	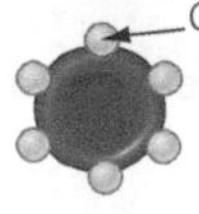 Carbohydrate B	Anti-A	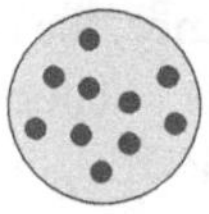	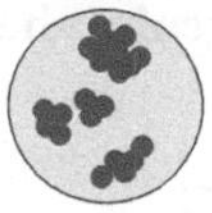	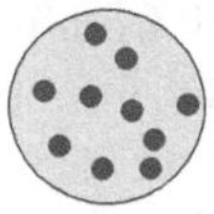	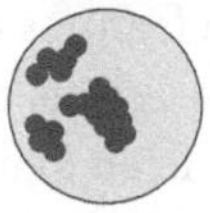
AB	I^AI^B		—	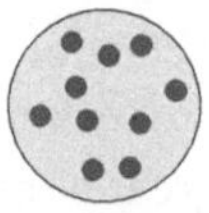	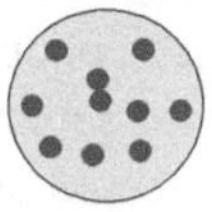	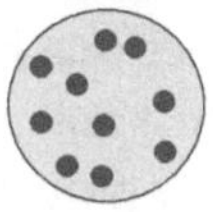	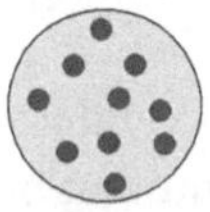
O	ii		Anti-A Anti-B	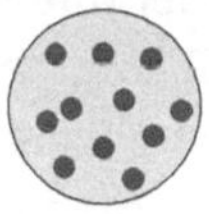	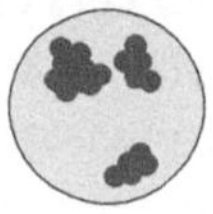	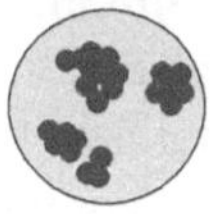	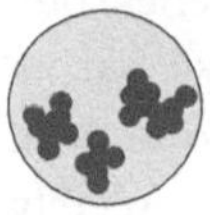

▼ **Figure 9.21 Sickle-cell disease: multiple effects of a single human gene.**

Individual homozygous for sickle-cell allele

Sickle-cell (abnormal) hemoglobin

Abnormal hemoglobin crystallizes into long, flexible chains, causing red blood cells to become sickle-shaped.

Colorized SEM 4,000×

Sickled cells can lead to a cascade of symptoms, such as weakness, pain, organ damage, and paralysis.

 Structure/Function

Pleiotropy and Sickle-Cell Disease

Our genetic examples to this point have been cases in which each gene specifies only one hereditary character. But in many cases, one gene influences several characters, a property called **pleiotropy**.

An example of pleiotropy in humans is sickle-cell disease (sometimes called sickle-cell anemia), a disorder characterized by a diverse set of symptoms. The direct effect of the sickle-cell allele is to make red blood cells produce abnormal hemoglobin proteins (see Figure 3.19). These abnormal molecules tend to link together and crystallize, especially when the oxygen content of the blood is lower than usual because of high altitude, overexertion, or respiratory ailments. As the hemoglobin crystallizes, the normally disk-shaped red blood cells deform to a sickle shape with jagged edges (**Figure 9.21**). As is so often the case in biology, altering structure affects function. Because of their shape, sickled cells do not flow smoothly in the blood and tend to accumulate and clog tiny blood vessels. Blood flow to body parts is reduced, resulting in periodic fever, severe pain, and damage to the heart, brain, and kidneys. The abnormal sickled cells are destroyed by the body, causing anemia and general weakness. Blood transfusions and drugs may relieve some of the symptoms, but there is no cure, and sickle-cell disease kills about 100,000 people in the world annually. ☑

Polygenic Inheritance

Mendel studied genetic characters that could be classified on an either-or basis, such as purple or white flower color. However, many characters, such as human skin color and height, vary along a continuum in a population. Many such features result from **polygenic inheritance**, the additive effects of two or more genes on a single phenotypic character. (This is the logical opposite of pleiotropy, in which one gene affects several characters.)

There is evidence that height in people is controlled by several genes that are inherited separately. (In actuality, human height is probably affected by a great number of genes, but we'll simplify here.) Let's consider three genes, with a tall allele for each (*A*, *B*, and *C*) contributing one "unit" of tallness to the phenotype and being incompletely dominant to the other alleles (*a*, *b*, and *c*).

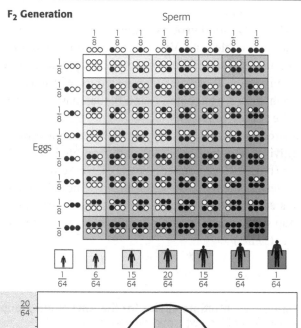

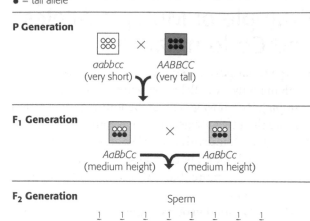

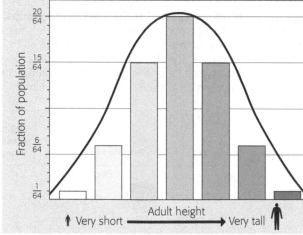

▲ **Figure 9.22 A model for polygenic inheritance of height.**

A person who is *AABBCC* would be very tall, while an *aabbcc* individual would be very short. An *AaBbCc* person would be of intermediate height. Because the alleles have an additive effect, the genotype *AaBbCc* would produce the same height as any other genotype with just three tallness alleles, such as *AABbcc*. The Punnett square in **Figure 9.22** shows all possible genotypes from a mating of two triple heterozygotes (*AaBbCd*). The row of figures below the Punnett square shows the seven height phenotypes that would theoretically result. This hypothetical example shows how inheritance of three genes could lead to seven different versions of a trait at the frequencies indicated by the bars in the graph at the bottom of the figure.

Epigenetics and the Role of Environment

If we examine a real human population for height, we would see more height phenotypes than just seven. The true range might be similar to the entire distribution of height under the bell-shaped curve in Figure 9.22. In fact, no matter how carefully we characterize the genes for height, a purely genetic description will always be incomplete. This is because height is also influenced by environmental factors, such as nutrition and exercise.

Many phenotypic characters result from a combination of heredity and environment. For example, the leaves of a tree all have the same genotype, but they vary in size, shape, and color, depending on exposure to wind and sun and the tree's nutritional state. For people, exercise alters build; experience improves performance on intelligence tests; and social and cultural factors can greatly affect appearance. As geneticists learn more and more about our genes, it's becoming clear that many human characters—such as risk of heart disease, cancer, alcoholism, and schizophrenia—are influenced by both genes *and* environment.

Whether human characters are more influenced by genes or by the environment—nature or nurture—is a very old and hotly contested issue. For some characters, such as the ABO blood group, a given genotype mandates an exact phenotype, and the environment plays no role whatsoever. In contrast, the number of blood cells in a drop of your blood varies quite a bit, depending on such factors as the altitude, your physical activity, and whether or not you have a cold.

Spending time with identical twins will convince anyone that environment, and not just genes, affects a person's traits (**Figure 9.23**). In general, only genetic influences are inherited, and any effects of the environment are not usually passed to the next generation. In recent years, however, biologists have begun to recognize the importance of **epigenetic inheritance**, the transmission of traits by mechanisms not directly involving DNA sequence. For example, DNA and/or protein components of chromosomes can be chemically modified by adding or removing chemical groups. Over a lifetime, the environment plays a role in these changes, which may explain how one identical twin can suffer from a genetically based disease whereas the other twin does not, despite their identical genomes. Recent research has shown that young identical twins are essentially indistinguishable in terms of their epigenetic markers, but epigenetic differences accumulate as they age, resulting in substantial differences in gene expression between older twins. Epigenetic modifications—and the changes in gene activity that result—may even be carried on to the next generation. For example, some research suggests that epigenetic changes may underlie certain instincts in animals, allowing behaviors learned by one generation (such as avoiding certain stimuli) to be passed on to the next generation via chromosomal modifications.

Unlike alterations to the DNA sequence, chemical changes to the chromosomes can be reversed (by processes that are not yet fully understood). Research into the importance of epigenetic inheritance is a very active area of biology, and new discoveries that alter our understanding of genetics are sure to follow. ✓

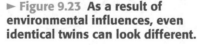

> ▶ **Figure 9.23 As a result of environmental influences, even identical twins can look different.**

Because the environment helps shape your appearance, identical twins aren't identical in every way.

✓ CHECKPOINT

If you produced cloned mice, how would you predict the number of epigenetic differences among the clones to change as they age?

Answer: You would expect the number of epigenetic differences to be low at first and then gradually increase as the clones age.

The Chromosomal Basis of Inheritance

It was not until many years after Mendel's death that biologists understood the significance of his work. Cell biologists worked out the processes of mitosis and meiosis in the late 1800s (see Chapter 8). Then, around 1900, researchers began to notice parallels between the behavior of chromosomes and the behavior of genes. One of biology's most important concepts began to emerge.

The **chromosome theory of inheritance** states that genes are located at specific positions (loci) on chromosomes and that the behavior of chromosomes during meiosis and fertilization accounts for inheritance patterns. Indeed, it is chromosomes that undergo segregation and independent assortment during meiosis and thus account for Mendel's laws. **Figure 9.24** correlates the results of the dihybrid cross in Figure 9.8b with the movement of chromosomes through meiosis. Starting with two purebred parental plants, the diagram follows two genes on different chromosomes—one for seed shape (alleles R and r) and one for seed color (alleles Y and y)—through the F_1 and F_2 generations. ☑

Linked Genes

Realizing that genes are on chromosomes when they segregate leads to some important conclusions about the ways genes can be inherited. The number of genes in a cell is far

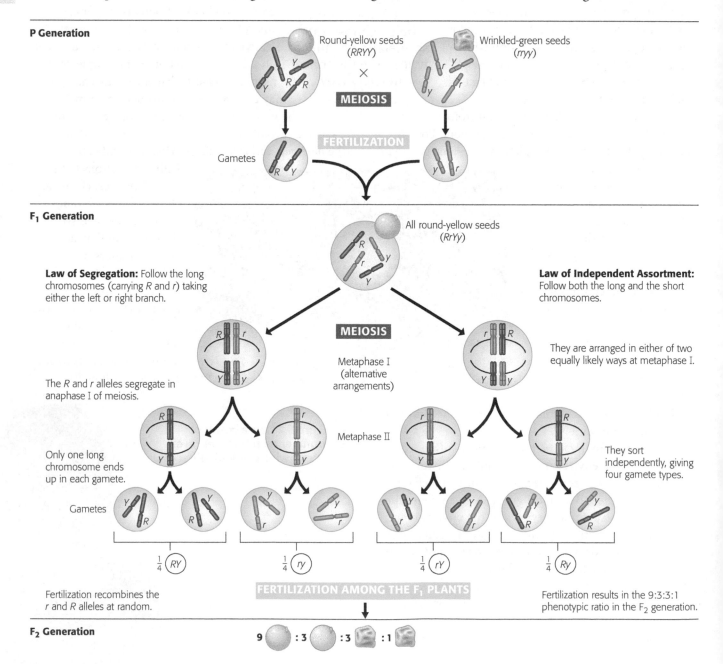

P Generation

Round-yellow seeds (*RRYY*) × Wrinkled-green seeds (*rryy*)

MEIOSIS

FERTILIZATION

Gametes

F₁ Generation

All round-yellow seeds (*RrYy*)

Law of Segregation: Follow the long chromosomes (carrying *R* and *r*) taking either the left or right branch.

Law of Independent Assortment: Follow both the long and the short chromosomes.

MEIOSIS

Metaphase I (alternative arrangements)

They are arranged in either of two equally likely ways at metaphase I.

The *R* and *r* alleles segregate in anaphase I of meiosis.

Metaphase II

Only one long chromosome ends up in each gamete.

They sort independently, giving four gamete types.

Gametes

$\frac{1}{4}$ (*RY*) $\frac{1}{4}$ (*ry*) $\frac{1}{4}$ (*rY*) $\frac{1}{4}$ (*Ry*)

Fertilization recombines the *r* and *R* alleles at random.

FERTILIZATION AMONG THE F₁ PLANTS

Fertilization results in the 9:3:3:1 phenotypic ratio in the F₂ generation.

F₂ Generation

9 : 3 : 3 : 1

► **Figure 9.24 The chromosomal basis of Mendel's laws.**

greater than the number of chromosomes; each chromosome therefore carries hundreds or thousands of genes. Genes located near each other on the same chromosome, called **linked genes**, tend to travel together during meiosis and fertilization. Such genes are often inherited as a set and therefore often do not follow Mendel's law of independent assortment. For closely linked genes, the inheritance of one does in fact correlate with the inheritance of the other, producing results that do not follow the standard ratios predicted by Punnett squares. In contrast, genes located far apart on the same chromosome usually do sort independently because of crossing over (see Figure 8.18).

The patterns of genetic inheritance we've discussed so far have always involved genes located on autosomes, not on the sex chromosomes. We're now ready to look at the role of sex chromosomes and the inheritance patterns exhibited by the characters they control. As you'll see, genes located on sex chromosomes produce some unusual patterns of inheritance. ☑

Sex Determination in Humans

Many animals, including all mammals, have a pair of sex chromosomes—designated X and Y—that determine an individual's sex (**Figure 9.25**). Individuals with one X chromosome and one Y chromosome are males; XX individuals are females. Human males and females both have 44 autosomes (chromosomes other than sex chromosomes). As a result of chromosome segregation during

meiosis, each gamete contains one sex chromosome and a haploid set of autosomes (22 in humans). All eggs contain a single X chromosome. Of the sperm cells, half contain an X chromosome and half contain a Y chromosome. An offspring's sex depends on whether the sperm cell that fertilizes the egg bears an X or a Y.

Sex-Linked Genes

Besides bearing genes that determine sex, the sex chromosomes also contain genes for characters unrelated to maleness or femaleness. A gene located on a sex chromosome is called a **sex-linked gene**. The human X chromosome contains approximately 1,100 genes, whereas the Y chromosome contains genes that encode for only about 25 proteins (most of which affect only the testes); therefore, most sex-linked genes are found on the X chromosome.

A number of human conditions, including red-green colorblindness, hemophilia, and a type of muscular dystrophy, result from sex-linked recessive alleles. Red-green colorblindness is a common sex-linked disorder that is caused by a malfunction of light-sensitive cells in the eyes. (Color blindness is actually a large class of disorders involving several sex-linked genes, but we will focus on just one specific type of colorblindness here.) A person with normal color vision can see more than 150 colors. In contrast, someone with red-green colorblindness can see fewer than 25. **Figure 9.26** shows a simple test for red-green

Your mother's genes had no role in determining your sex.

☑ **CHECKPOINT**

What are linked genes? Why do they often disobey Mendel's law of independent assortment?

Answer: Genes located near each other on the same chromosome. Because they often travel together during meiosis and fertilization, linked genes tend to be inherited together rather than independently.

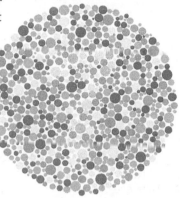

▲ **Figure 9.26 A test for red-green colorblindness.** Can you see a green numeral 7 against the reddish background? If not, you probably have some form of red-green colorblindness, a sex-linked trait.

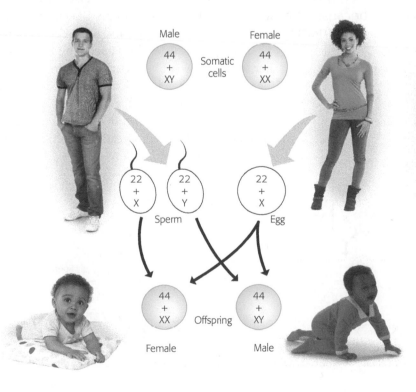

Male — Somatic cells — Female

44 + XY

44 + XX

22 + X | 22 + Y — Sperm

22 + X — Egg

44 + XX — Female

Offspring

44 + XY — Male

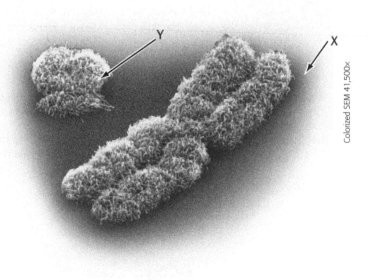

Y   X

Colorized SEM 41,500x

▲ **Figure 9.25 The chromosomal basis of sex determination in humans.** The micrograph above shows human X and Y chromosomes in a duplicated state.

colorblindness. Mostly males are affected, but heterozygous females have some defects, too.

Because they are located on the sex chromosomes, sex-linked genes exhibit unusual inheritance patterns. **Figure 9.27a** (below) illustrates what happens when a colorblind male has offspring with a homozygous female with normal color vision. All the children have normal color vision, suggesting that the allele for wild-type (normal) color vision is dominant. If a female carrier mates with a male who has normal color vision, the classic 3:1 phenotypic ratio of normal color vision to colorblindness appears among the children (**Figure 9.27b**). However, there is a surprising twist: The colorblind trait shows up only in males. All the females have normal vision, while half the males are colorblind and half are normal. This is because the gene involved in this inheritance pattern is located exclusively on the X chromosome; there is no corresponding locus on the Y. Thus, females (XX) carry two copies of the gene for this character, and males (XY) carry only one. Because the colorblindness allele is recessive, a female will be colorblind only if she receives that allele on both X chromosomes (**Figure 9.27c**). For a male, however, a single copy of the recessive allele confers colorblindness. For this reason, recessive sex-linked traits are expressed much more frequently in men than in women. For example, colorblindness is about 20-fold more common among males than among females. This inheritance pattern is why people often say that certain genes "skip a generation"—because they are passed from a male (in generation 1) to a female carrier (who does not express it in generation 2) back to a male (in generation 3).

CHECKPOINT

1. What is meant by a sex-linked gene?
2. White eye color is a recessive sex-linked trait in fruit flies. If a white-eyed *Drosophila* female is mated with a red-eyed (wild-type) male, what do you predict for the numerous offspring?

Answers: 1. a gene that is located on a sex chromosome, usually the X chromosome 2. All female offspring will be heterozygous ($X^R X^r$), with red eyes; all male offspring will be white-eyed ($X^r Y$).

▲ Figure 9.28 **Hemophilia in the royal family of Russia.** The photograph shows Queen Victoria's granddaughter Alexandra, her husband Nicholas, who was the last czar of Russia, their son Alexis, and their daughters. In the pedigree, half-colored symbols represent heterozygous carriers of the hemophilia allele and fully colored symbols represent a person with hemophilia.

Hemophilia is a sex-linked recessive trait with a long, well-documented history. Hemophiliacs bleed excessively when injured because they have inherited an abnormal allele for a factor involved in blood clotting. The most seriously affected individuals may bleed to death after relatively minor bruises or cuts. A high incidence of hemophilia plagued the royal families of Europe. Queen Victoria (1819–1901) of England was a carrier of the hemophilia allele. She passed it on to one of her sons and two of her daughters. Through marriage, her daughters then introduced the disease into the royal families of Prussia, Russia, and Spain. In this way, the former practice of strengthening international alliances by marriage effectively spread hemophilia through the royal families of several nations (**Figure 9.28**). ✓

Hemophilia is more prevalent among European royalty than commoners because of past intermarriage.

▼ Figure 9.27 **Inheritance of colorblindness, a sex-linked recessive trait.** We use an uppercase *N* for the dominant, normal color-vision allele and *n* for the recessive, colorblind allele. To indicate that these alleles are on the X chromosome, we show them as superscripts to the letter X. The Y chromosome does not have a gene locus for vision; therefore, the male's phenotype results entirely from the sex-linked gene on his single X chromosome.

Key
- ☐ Unaffected individual
- ▨ Carrier
- ▨ Colorblind individual

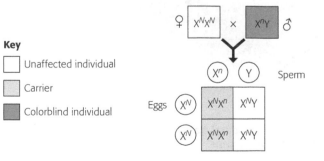

(a) **Normal female × colorblind male**

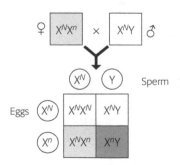

(b) **Carrier female × normal male**

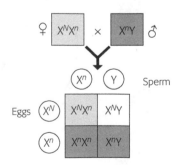

(c) **Carrier female × colorblind male**

Dog Breeding EVOLUTION CONNECTION

Barking Up the Evolutionary Tree

As we've seen throughout this chapter, dogs are more than man's best friend: They are also one of our longest-running genetic experiments. About 15,000 years ago, in East Asia, people began to cohabit with ancestral canines that were predecessors of both modern wolves and dogs. As people moved into permanent, geographically isolated settlements, populations of canines were separated from one another and eventually became inbred.

Different groups of people chose dogs with different traits. A 2010 study indicated that small dogs were first bred within early agricultural settlements of the Middle East around 12,000 years ago. Elsewhere, herders selected dogs for controlling flocks while hunters chose dogs for retrieving prey. Continued over millennia, such genetic tinkering has resulted in a diverse array of dog body types and behaviors. In each breed, a distinct genetic makeup results in a distinct set of physical and behavioral traits.

As discussed in the Process of Science section, our understanding of canine evolution took a big leap forward when researchers sequenced the complete genome of a dog. Using the genome sequence and other data, canine geneticists produced an evolutionary tree based on a genetic analysis of 85 breeds **(Figure 9.29)**. The analysis shows that the canine family tree includes a series of well-defined branch points. Each fork represents a purposeful selection by people that produced a genetically distinct subpopulation with specific desired traits.

The genetic tree shows that the most ancient breeds, those most closely related to the wolf, are Asian species such as the shar-pei and Akita. Subsequent genetic splits created distinct breeds in Africa (basenji), the Arctic (Alaskan malamute and Siberian husky), and the Middle East (Afghan hound and saluki). The remaining breeds, primarily of European ancestry, were developed most recently and can be grouped by genetic makeup into those bred for guarding (for example, the rottweiler), herding (such as sheepdogs), and hunting (including the Labrador retriever and beagle). The formulation of an evolutionary tree for the domestic dog shows that new technologies can provide important insights into genetic and evolutionary questions about life on Earth.

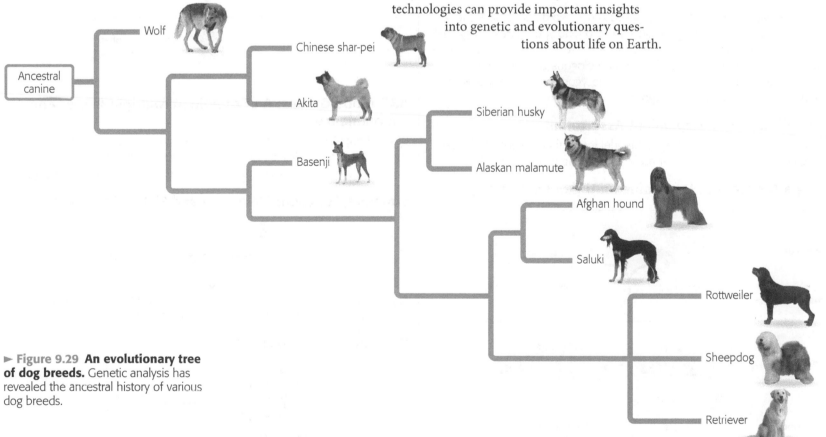

► **Figure 9.29 An evolutionary tree of dog breeds.** Genetic analysis has revealed the ancestral history of various dog breeds.

Chapter Review

■ SUMMARY OF KEY CONCEPTS

Genetics and Heredity

Gregor Mendel, the first to study the science of heredity by analyzing patterns of inheritance, emphasized that genes retain permanent identities.

In an Abbey Garden

Mendel started with purebred varieties of pea plants representing two alternative variants of a hereditary character. He then crossed the different varieties and traced the inheritance of traits from generation to generation.

Mendel's Law of Segregation

Pairs of alleles separate during gamete formation; fertilization restores the pairs.

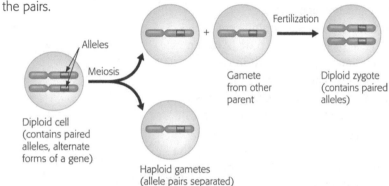

If an individual's genotype (genetic makeup) has two different alleles for a gene and only one influences the organism's phenotype (appearance), that allele is said to be dominant and the other allele recessive. Alleles of a gene reside at the same locus, or position, on homologous chromosomes. When the allele pair matches, the organism is homozygous; when the alleles are different, the organism is heterozygous.

Mendel's Law of Independent Assortment

By following two characters at once, Mendel found that the alleles of a pair segregate independently of other allele pairs during gamete formation.

Using a Testcross to Determine an Unknown Genotype

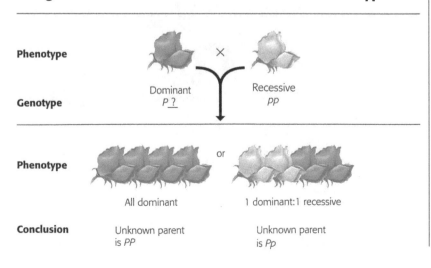

The Rules of Probability

Inheritance follows the rules of probability. The chance of inheriting a recessive allele from a heterozygous parent is $\frac{1}{2}$. The chance of inheriting it from both of two heterozygous parents is $\frac{1}{2} \times \frac{1}{2} = \frac{1}{4}$, illustrating the rule of multiplication for calculating the probability of two independent events.

Family Pedigrees

The inheritance of many human traits, from freckles to genetic diseases, follows Mendel's laws and the rules of probability. Geneticists can use family pedigrees to determine patterns of inheritance and individual genotypes.

Human Disorders Controlled by a Single Gene

For traits that vary within a population, the one most commonly found in nature is called the wild type. Many inherited disorders in humans are controlled by a single gene with two alleles. Most of these disorders, such as cystic fibrosis, are caused by autosomal recessive alleles. A few, such as Huntington's disease, are caused by dominant alleles.

Variations on Mendel's Laws

Incomplete Dominance in Plants and People

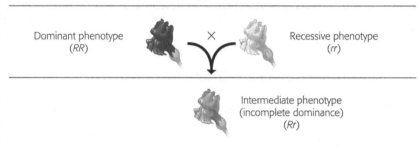

ABO Blood Groups: An Example of Multiple Alleles and Codominance

Within a population, there are often multiple kinds of alleles for a character, such as the three alleles for the ABO blood groups. The alleles determining blood groups are codominant; that is, both are expressed in a heterozygote.

Structure/Function: Pleiotropy and Sickle-Cell Disease

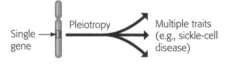

In pleiotropy, one gene (such as the sickle-cell disease gene) can affect many characters (such as the multiple symptoms of the disease).

Polygenic Inheritance

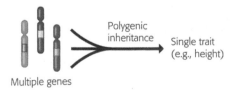

Epigenetics and the Role of Environment

Many phenotypic characters result from a combination of genetic and environmental effects, but only genetic influences are biologically heritable. Epigenetic inheritance, the transmission of traits from one generation to the next by means of chemical modifications to DNA and proteins, may explain how environmental factors can affect genetic traits.

The Chromosomal Basis of Inheritance

Genes are located on chromosomes. The behavior of chromosomes during meiosis and fertilization accounts for inheritance patterns.

Linked Genes

Certain genes are linked: They tend to be inherited as a set because they lie close together on the same chromosome.

Sex Determination in Humans

In humans, sex is determined by whether a Y chromosome is present. A person who inherits two X chromosomes develops as a female. A person who inherits one X and one Y chromosome develops as a male.

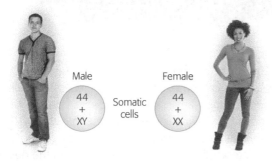

Male
44
+
XY

Somatic cells

Female
44
+
XX

Sex-Linked Genes

Their inheritance of genes on the X chromosome reflects the fact that females have two homologous X chromosomes, but males have only one. Most sex-linked human disorders, such as red-green colorblindness and hemophilia, are due to recessive alleles and are seen mostly in males. A male receiving a single sex-linked recessive allele from his mother will have the disorder; a female has to receive the allele from both parents to be affected.

Sex-Linked Traits			
Female: Two alleles	Genotype	$X^N X^N$	$X^N X^n$	$X^n X^n$
	Phenotype	Normal female	Carrier female	Affected female (rare)
Male: One allele	Genotype	$X^N Y$		$X^n Y$
	Phenotype	Normal male		Affected male

MasteringBiology®

For practice quizzes, BioFlix animations, MP3 tutorials, video tutors, and more study tools designed for this textbook, go to MasteringBiology®

SELF-QUIZ

1. The genetic makeup of an organism is called its _____, and the physical traits of an organism are called its _____.

2. Which of Mendel's laws is represented by each statement?
   a. Alleles of each pair of homologous chromosomes separate independently during gamete formation.
   b. Alleles segregate during gamete formation; fertilization creates pairs of alleles once again.

3. Edward was found to be heterozygous (Ss) for the sickle-cell trait. The alleles represented by the letters S and s are
   a. on the X and Y chromosomes.
   b. linked.
   c. on homologous chromosomes.
   d. both present in each of Edward's sperm cells.

4. A purebred plant that produces yellow seeds is crossed with a purebred plant that produces green seeds. The seeds of all of the offspring are yellow. Why?
   a. The yellow allele is recessive to the green allele.
   b. The yellow allele is dominant to the green allele.
   c. All of the offspring are homozygous yellow.
   d. The alleles are codominant.

5. Mendel crossed purebred purple-flowered plants with purebred white-flowered plants, and all of the resulting offspring produced purple flowers. The offspring are all _____, and the allele for purple flowers is _____.
   a. heterozygotes; recessive
   b. heterozygotes; dominant
   c. homozygotes; recessive
   d. homozygotes; dominant

6. All the offspring of a white hen and a black rooster are gray. The simplest explanation for this pattern of inheritance is
   a. pleiotropy.
   b. sex linkage.
   c. codominance.
   d. incomplete dominance.

7. A man who has type B blood and a woman who has type A blood could have children of which of the following phenotypes? (Hint: Review Figure 9.20.)
   a. A, B, or O
   b. AB only
   c. AB or O
   d. A, B, AB, or O

8. Duchenne muscular dystrophy is a sex-linked recessive disorder characterized by a progressive loss of muscle tissue. Neither Rudy nor Carla has Duchenne muscular dystrophy, but their first son does have it. If the couple has a second child, what is the probability that he or she will also have the disease?

9. Adult height in people is at least partially hereditary; tall parents tend to have tall children. But people come in a range of sizes, not just tall or short. What extension of Mendel's model could produce this variation in height?

10. A purebred brown mouse is repeatedly mated with a purebred white mouse, and all their offspring are brown. If two of these brown offspring are mated, what fraction of the F_2 mice will be brown?

11. How could you determine the genotype of one of the brown F_2 mice in problem 10? How would you know whether a brown mouse is homozygous? Heterozygous?

12. Tim and Jan both have freckles (a dominant trait), but their son Michael does not. Show with a Punnett square how this is possible. If Tim and Jan have two more children, what is the probability that *both* of them will have freckles?

13. Achondroplasia is a form of dwarfism caused by a dominant allele. The homozygous dominant genotype causes death, so individuals who have this condition are all heterozygotes. If a person with achondroplasia mates with a person who does not have achondroplasia, what percentage of their children would be expected to have achondroplasia?

14. Why was Henry VIII wrong to blame his wives for always producing daughters?

15. Both parents of a boy are phenotypically normal, but their son suffers from hemophilia, a sex-linked recessive disorder. Draw a pedigree that shows the genotypes of the three individuals. What fraction of the couple's children are likely to suffer from hemophilia? What fraction are likely to be carriers?

16. Heather was surprised to discover that she suffered from red-green colorblindness. She told her biology professor, who said, "Your father is colorblind, too, right?" How did her professor know this? Why did her professor not say the same thing to the colorblind males in the class?

17. In rabbits, black hair depends on a dominant allele, *B*, and brown hair on a recessive allele, *b*. Short hair is due to a dominant allele, *S*, and long hair to a recessive allele, *s*. If a true-breeding black short-haired male is mated with a brown long-haired female, describe their offspring. What will be the genotypes of the offspring? If two of these F_1 rabbits are mated, what phenotypes would you expect among their offspring? In what proportions?

Answers to these questions can be found in Appendix: Self-Quiz Answers.

THE PROCESS OF SCIENCE

18. In 1981, a stray cat with unusual curled-back ears was adopted by a family in Lakewood, California. Hundreds of descendants of this cat have since been born, and cat fanciers hope to develop the "curl" cat into a show breed. The curl allele is apparently dominant and carried on an autosome. Suppose you owned the first curl cat and wanted to develop a purebred variety. Describe tests that would determine whether the curl gene is dominant or recessive and whether it is autosomal or sex-linked.

19. **Interpreting Data** As shown in the Punnett square below, one variety of deafness is caused by an autosomal recessive allele. Two parents who do not show any signs of the disease but are carriers could therefore have a child who is deaf, because that child could inherit one recessive deafness-causing gene from each parent. Imagine that a deaf male mates with a hearing female. We know the deaf male must have the genotype *dd*, but the female could be either *Dd* or *DD*. Such a mating is essentially a testcross, like the one shown in Figure 9.10. If the parents' first child has hearing, can you say with certainty what the mother's genotype must be? What if the couple has four children (none twins), all with hearing—can you say with certainty the mother's genotype? What would it take for a definitive genotype to be assigned?

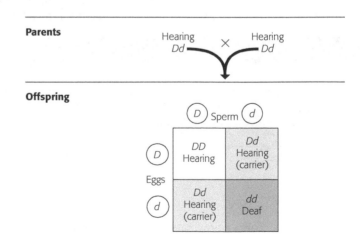

BIOLOGY AND SOCIETY

20. There are now nearly 200 recognized breeds of dog, from the Affenpinscher to the Yorkshire Terrier. But several of these suffer from medical problems due to the inbreeding required to establish the breed. For example, nearly every Cavalier King Charles (discussed in the Biology and Society essay) suffers from heart murmurs caused  by a genetically defective heart valve. Such problems are likely to remain as long as the organizations that oversee dog breeding maintain strict pedigree requirements. Some people are suggesting that every breed be allowed to mix with others to help introduce new gene lines free of the congenital defects. Why do you think the governing societies are resistant to such cross-breed mixing? What would you do if you were in charge of addressing the genetic defects that currently plague some breeds?

21. In the section on prenatal genetic testing, we have briefly assessed the possible impact of this diagnostic technology on the parents of the unborn baby and the need to provide counseling before and after the test. However, genetic testing is also likely to have an impact on the society, for example, by changing the way we perceive and deal with congenital disorders and genetic defects. In your opinion, what are the social risks associated with this technology?

22. Imagine that scientists discover a single gene that is responsible for a "negative" or "unwanted" personality trait, such as having a very high level of aggression. Do you think that this discovery would lead to an increased acceptance of this trait because it is an inescapable consequence of an individual's genetic makeup? Would the affected individuals be considered "disabled"? What are the other possible consequences of such a discovery?

10 The Structure and Function of DNA

Why Molecular Biology Matters

A single molecular "typo" in DNA can result in a life-threatening disease.

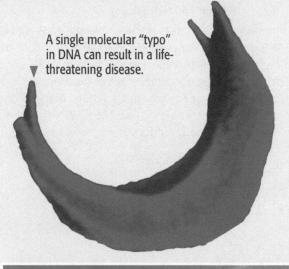

Because all life on Earth shares a universal genetic code, your DNA could be used to genetically modify a monkey.

Mad cow disease is caused by an abnormal molecule of protein.

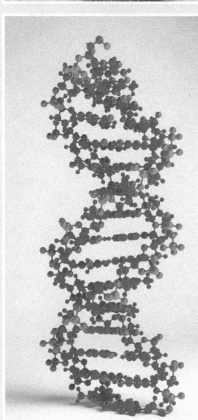

Enzymes help maintain the integrity of your DNA to greater than 99.999% accuracy.

The Deadliest Virus BIOLOGY AND SOCIETY

The First 21st-Century Pandemic

In 2009, a cluster of unusual flu cases was reported in and around Mexico City. Despite a near total shutdown of the city, the new flu virus, called 2009 H1N1, quickly spread to California and Texas. This strain was originally misnamed the "swine flu"; in fact, pigs had little impact on the spread of this virus, which, like most flu viruses, is passed from person to person through respiratory fluids. Nonetheless, the new strain received heavy media attention and caused considerable worry among the general public. In June 2009, the World Health Organization (WHO) declared H1N1 to be the first influenza pandemic (global epidemic) of the 21st century; the last flu pandemic was in 1968. In response, the WHO unveiled a comprehensive effort to contain H1N1, including increased surveillance, development of rapid testing procedures, the issuance of travel warnings, recommendations for increased sanitation (hand washing, etc.), and the production, stockpiling, and distribution of antiviral drugs. By 2010, cases of H1N1 had been confirmed in 214 countries.

The H1N1 influenza virus. Citizens of Russia try to protect themselves from H1N1 infection in late fall 2009.

Scientists soon determined that H1N1 was a hybrid flu strain, created when a previously known flu virus (which itself originated through a combination of viruses from birds, swine, and people) mixed with an Asian swine flu virus. This novel combination of genes produced some unusual features in the H1N1 strain. Most significantly, it infected young, healthy people, whereas the flu usually affects the elderly or people who are already sick. Many countries participated in the WHO-coordinated response, including widespread distribution of a new vaccine. As a result, the virus was contained, and WHO declared the pandemic over in August 2010. The WHO confirmed that this virus killed about 18,000 people, and estimates of the total number of unreported deaths topped 250,000.

You may wonder, What's the big deal about the flu? The flu is not merely a seasonal inconvenience. In fact, the influenza virus may be the deadliest pathogen known to science. In a typical year in the United States, more than 20,000 people die from influenza infection. And that is considered a good year. Once every few decades, a new flu strain explodes on the scene, causing pandemics and widespread death. H1N1 pales in comparison to the deadliest outbreak: the pandemic of 1918–1919, which in just 18 months killed 40 million people worldwide—that's more people than have died of AIDS since it was discovered more than 30 years ago. Given the nature of the threat, it's no surprise that health workers remain vigilant, always aware that a new and deadly flu virus may appear at any time.

The flu virus, like all viruses, consists of a relatively simple structure of nucleic acid (RNA in this case) and protein. Combating any virus requires a detailed understanding of life at the molecular level. In this chapter, we will explore the structure of DNA, how it replicates and mutates, and how it controls the cell by directing the synthesis of RNA and protein.

DNA: Structure and Replication

DNA was known to be a chemical component of cells by the late 1800s, but Gregor Mendel and other early geneticists did their work without any knowledge of DNA's role in heredity. By the late 1930s, experimental studies had convinced most biologists that one specific kind of molecule, rather than a complex chemical mixture, is the basis of inheritance. Attention focused on chromosomes, which were already known to carry genes. By the 1940s, scientists knew that chromosomes consist of two types of chemicals: DNA and protein. And by the early 1950s, a series of discoveries had convinced the scientific world that DNA was the molecule that acts as the hereditary material. This breakthrough ushered in the field of **molecular biology**, the study of heredity at the molecular level.

What came next was one of the most celebrated quests in the history of science: the effort to figure out the structure of DNA. A good deal was already known about DNA. Scientists had identified all its atoms and knew how they were bonded to one another. What was not understood was the specific three-dimensional arrangement of atoms that gives DNA its unique properties—the capacity to store genetic information, copy it, and pass it from

generation to generation. The race was on to discover the link between the structure and function of this important molecule. We will describe that momentous discovery shortly. First, let's review the underlying chemical structure of DNA and its chemical cousin RNA.

DNA and RNA Structure

Both DNA and RNA are nucleic acids, which consist of long chains (polymers) of chemical units (monomers) called **nucleotides**. (For an in-depth refresher, see Figures 3.21–3.25.) A diagram of a nucleotide polymer, or **polynucleotide**, is shown in **Figure 10.1**. Polynucleotides can be very long and may have any sequence of the four different types of nucleotides (abbreviated A, C, T, and G), so a tremendous variety of polynucleotide chains is possible.

Nucleotides are joined together by covalent bonds between the sugar of one nucleotide and the phosphate of the next. This results in a repeating pattern of sugar-phosphate-sugar-phosphate, which is known as a **sugar-phosphate backbone**. The nitrogenous bases are arranged like ribs that project from this backbone. You can think of a polynucleotide as a long ladder split in

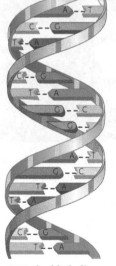

DNA double helix

► Figure 10.1 **The chemical structure of a DNA polynucleotide.** A molecule of DNA contains two polynucleotides, each a chain of nucleotides. Each nucleotide consists of a nitrogenous base, a sugar (blue), and a phosphate group (gold).

Polynucleotide

Phosphate group

Nitrogenous base

A

Sugar

C

DNA nucleotide

T

G

G

Sugar-phosphate backbone

Phosphate group

O=P—O—CH₂

O⁻

H₃C—C

C

N

H

H

C

C

N

O

Thymine (T)

Sugar
(deoxyribose)

DNA nucleotide

half longwise, with rungs that come in four colors. The sugars and phosphates make up the side of the ladder, with the sugars acting as the half-rungs.

Moving from left to right across Figure 10.1, we can zoom in to see that each nucleotide consists of three components: a nitrogenous base, a sugar (blue), and a phosphate group (gold). Examining a single nucleotide even more closely, we see the chemical structure of its three components. The phosphate group, with a phosphorus atom (P) at its center, is the source of the *acid* in nucleic acid. Each phosphate has a negative charge on one of its oxygen atoms. The sugar has five carbon atoms (shown in red): four in its ring and one extending above the ring. The ring also includes an oxygen atom. The sugar is called *deoxyribose* because, compared with the sugar ribose, it is missing an oxygen atom. The full name for **DNA** is *deoxyribonucleic acid*, with *nucleic* referring to DNA's location in the nuclei of eukaryotic cells. The nitrogenous base (thymine, in our example) has a ring of nitrogen and carbon atoms with various chemical groups attached. Nitrogenous bases are basic (having a high pH, the opposite of acidic), hence their name.

The four nucleotides found in DNA differ only in their nitrogenous bases (see Figure 3.23 for a review).

The bases can be divided into two types. **Thymine (T)** and **cytosine (C)** are single-ring structures. **Adenine (A)** and **guanine (G)** are larger, double-ring structures. Instead of thymine, RNA has a similar base called **uracil (U)**. And RNA contains a slightly different sugar than DNA (ribose instead of deoxyribose, accounting for the names <u>R</u>NA vs. <u>D</u>NA). Other than that, RNA and DNA polynucleotides have the same chemical structure. **Figure 10.2** is a computer graphic of a piece of RNA polynucleotide about 20 nucleotides long. ☑

Cytosine
Uracil
Adenine
Guanine
Phosphate
Sugar (ribose)

▶ **Figure 10.2 An RNA polynucleotide.** The yellow used for the phosphorus atoms and the blue of the sugar atoms make it easy to spot the sugar-phosphate backbone.

☑ **CHECKPOINT**

Compare and contrast the chemical components of DNA and RNA.

Answer: Both are polymers of nucleotides (a sugar + a nitrogenous base + a phosphate group). In RNA, the sugar is ribose; in DNA, it is deoxyribose. Both RNA and DNA have the bases A, C, and G, but DNA has T and RNA has U.

Watson and Crick's Discovery of the Double Helix

The celebrated partnership that solved the puzzle of DNA structure began soon after a 23-year-old newly minted American Ph.D. named James D. Watson journeyed to Cambridge University in England. There, a more senior scientist, Francis Crick, was studying protein structure with a technique called X-ray crystallography. While visiting the laboratory of Maurice Wilkins at King's College in London, Watson saw an X-ray image of DNA produced by Wilkins's colleague Rosalind Franklin. The data produced by Franklin turned out to be the key to the puzzle. A careful study of the image enabled Watson to figure out that the basic shape of DNA is a helix (spiral) with a uniform diameter. The thickness of the helix suggested that it was made up of two polynucleotide strands—in other words, a **double helix**. But how were the nucleotides arranged in the double helix?

Using wire models, Watson and Crick began trying to construct a double helix that conformed to all known data about DNA (**Figure 10.3**). Watson placed the backbones on the outside of the model, forcing the nitrogenous bases to swivel to the interior of the molecule. As he did this, it occurred to him that the four kinds of bases must pair in a specific way. This idea of specific base pairing was a flash of inspiration that enabled Watson and Crick to solve the DNA puzzle.

▼ **Figure 10.3 Discoverers of the double helix.**

James Watson (left) and **Francis Crick.** The discoverers of the structure of DNA are shown in 1953 with their model of the double helix.

Rosalind Franklin Using X-rays, Franklin generated some of the key data that provided insight into the structure of DNA.

At first, Watson imagined that the bases paired like with like—for example, A with A, C with C. But that kind of pairing did not fit with the fact that the DNA molecule has a uniform diameter. An AA pair (made of two double-ringed bases) would be almost twice as wide as a CC pair (made of two single-ringed bases), resulting in an uneven molecule. It soon became apparent that a double-ringed base on one strand must always be paired with a single-ringed base on the opposite strand. Moreover, Watson and Crick realized that the chemical structure of each kind of base dictated the pairings even more specifically.

Each base has protruding chemical groups that can best form hydrogen bonds with just one appropriate partner. This would be like having four colors of snap-together puzzle pieces and realizing that only certain colors can snap together (e.g., red can only snap together with blue). Similarly, adenine can best form hydrogen bonds with thymine, and guanine with cytosine. In the biologist's shorthand: A pairs with T, and G pairs with C. A is also said to be "complementary" to T, and G to C.

If you imagine a polynucleotide strand as a half ladder, then you can picture the model of the DNA double helix proposed by Watson and Crick as a full ladder twisted into a spiral (**Figure 10.4**). Figure 10.5 shows three more detailed representations of the double helix. The ribbonlike diagram in **Figure 10.5a** symbolizes the bases with shapes that emphasize their complementarity. **Figure 10.5b** is a more chemically precise version showing only four base pairs, with the helix untwisted and the individual hydrogen bonds specified by dashed lines. **Figure 10.5c** is a computer model showing part of a double helix in detail.

Although the base-pairing rules dictate the side-by-side combinations of bases that form the rungs of the double helix, they place no restrictions on the sequence of nucleotides along the length of a DNA strand. In fact, the sequence of bases can vary in countless ways.

In 1953, Watson and Crick rocked the scientific world with a succinct paper proposing their molecular model for

► **Figure 10.4 A rope-ladder model of a double helix.** The ropes at the sides represent the sugar-phosphate backbones. Each rung stands for a pair of bases connected by hydrogen bonds.

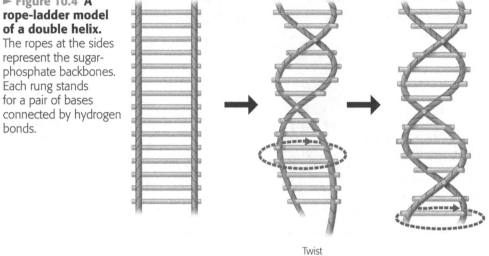

Twist

▼ **Figure 10.5 Three representations of DNA.**

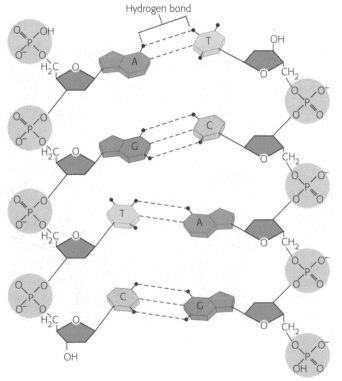

Hydrogen bond

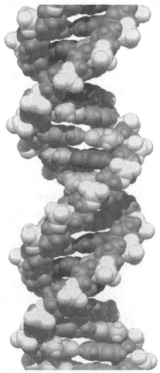

(a) Ribbon model. The sugar-phosphate backbones are blue ribbons, and the bases are complementary shapes in shades of green and orange.

(b) Atomic model. In this more chemically detailed structure, you can see the individual hydrogen bonds (dashed lines). You can also see that the strands run in opposite directions: Notice that the sugars on the two strands are upside down with respect to each other.

(c) Computer model. Each atom is shown as a sphere, creating a space-filling model.

DNA. Few milestones in the history of biology have had as broad an impact as their double helix, with its A-T and C-G base pairing. In 1962, Watson, Crick, and Wilkins received the Nobel Prize for their work. (Franklin deserved a share of the prize, but she had died from cancer in 1958, and Nobel prizes are never granted posthumously.)

In their 1953 paper, Watson and Crick wrote that the structure they proposed "immediately suggests a possible copying mechanism for the genetic material." In other words, the structure of DNA points toward a molecular explanation for life's unique properties of reproduction and inheritance. Understanding how the arrangement of parts within DNA affects it actions within a cell is an excellent example of biology's important theme of the relationship of structure to function, as we see next.

Enzymes help maintain the integrity of your DNA to greater than 99.999% accuracy.

Structure/ Function DNA Replication

Every cell contains a DNA "cookbook" that provides complete information on how to make and maintain that cell. When a cell reproduces, it must duplicate this information, providing one copy to the new offspring cell while keeping one copy for itself. Thus, each cell must have a means of copying the DNA instructions. In a clear demonstration of how the structure of a biological system can provide insight into its function, Watson and Crick's model of DNA suggests that each DNA strand serves as a mold, or template, to guide reproduction of the other strand. If you know the sequence of bases in one strand of the double helix, you can very easily determine the sequence of bases in the other strand by applying the base-pairing rules: A pairs with T (and T with A), and G pairs with C (and C with G). For example, if one polynucleotide has the sequence AGTC, then the complementary polynucleotide in that DNA molecule must have the sequence TCAG.

Figure 10.6 shows how this model can account for the direct copying of a piece of DNA. The two strands of parental DNA separate, and each becomes a template for the assembly of a complementary strand from a supply of free nucleotides. The nucleotides are lined up one at a time along the template strand in accordance with the base-pairing rules. Enzymes link the nucleotides to form new DNA strands. The completed new molecules, identical to the parental molecule, are known as daughter DNA molecules (no gender should be inferred from this name).

The process of DNA replication requires the cooperation of more than a dozen enzymes and other proteins. The enzymes that make the covalent bonds between the nucleotides of a new DNA strand are called **DNA polymerases**. As an incoming nucleotide base-pairs with its complement on the template strand, a DNA polymerase adds it to the end of the growing daughter strand. The process is both fast (a rate of 50 nucleotides per second is typical) and amazingly accurate, with fewer than one in a billion bases incorrectly paired. In addition to their roles in DNA replication, DNA polymerases and some of the associated proteins can repair DNA that has been damaged by toxic chemicals or high-energy radiation, such as X-rays and ultraviolet light.

DNA replication begins on a double helix at specific sites, called origins of replication. Replication then proceeds in both directions, creating what are called replication "bubbles" (**Figure 10.7**). The parental DNA strands open up as daughter strands elongate on both sides of each bubble. The DNA molecule of a typical eukaryotic chromosome has many origins where replication can start simultaneously, shortening the total time needed for the process. Eventually, all the bubbles merge, yielding two completed double-stranded daughter DNA molecules.

DNA replication ensures that all the body cells in a multicellular organism carry the same genetic information. It is also the means by which genetic information is passed along to offspring. ☑

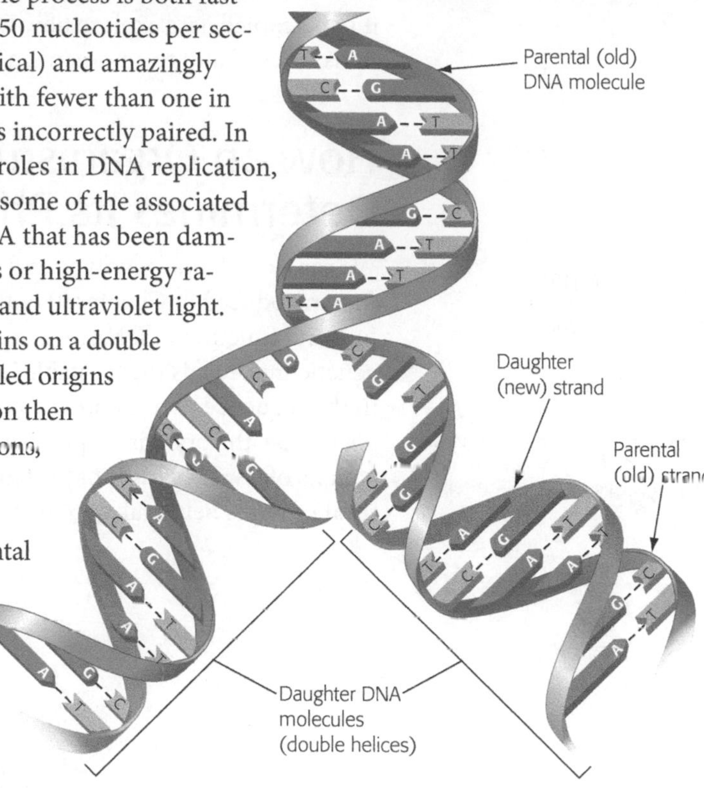

Parental (old) DNA molecule

Daughter (new) strand

Parental (old) strand

Daughter DNA molecules (double helices)

▲ **Figure 10.6 DNA replication.** Replication results in two daughter DNA molecules, each consisting of one old strand and one new strand. The parental DNA untwists as its strands separate, and the daughter DNA rewinds as it forms.

▼ **Figure 10.7 Multiple "bubbles" in replicating DNA.**

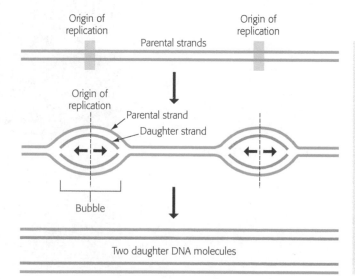

Origin of replication

Origin of replication

Parental strands

Origin of replication

Parental strand

Daughter strand

Bubble

Two daughter DNA molecules

☑ **CHECKPOINT**

1. How does complementary base pairing make DNA replication possible?
2. What enzymes connect nucleotides together during DNA replication?

Answers: **1.** When the two strands of the double helix separate, each serves as a template on which nucleotides can be arranged by specific base pairing into new complementary strands. **2.** DNA polymerases

 **Information Flow**

From DNA to RNA to Protein

Now that we've seen how the structure of DNA allows it to be copied, let's explore how DNA provides instructions to a cell and to an organism as a whole.

How an Organism's Genotype Determines Its Phenotype

We can now define genotype and phenotype (terms first introduced in Chapter 9) with regard to the structure and function of DNA. An organism's *genotype*, its genetic makeup, is the heritable information contained in the sequence of nucleotide bases in its DNA. The *phenotype*, the organism's physical traits, arises from the actions of a wide variety of proteins. For example, structural proteins help make up the body of an organism,

and enzymes catalyze the chemical reactions that are necessary for life.

DNA specifies the synthesis of proteins. However, a gene does not build a protein directly. Instead, DNA dispatches instructions in the form of RNA, which in turn programs protein synthesis. This fundamental concept in biology is summarized in **Figure 10.8**. The molecular "chain of command" is from DNA in the nucleus (purple area in the figure) to RNA to protein synthesis in the cytoplasm (blue area). The two stages are **transcription**, the transfer of genetic information from DNA into an RNA molecule, and **translation**, the transfer of the information from RNA into a polypeptide (protein strand). The relationship between genes and proteins is thus one of information flow: The function of a DNA gene is to dictate the production of a polypeptide. ☑

☑ **CHECKPOINT**

What are transcription and translation?

Answer: Transcription is the transfer of genetic information from DNA to RNA. Translation is the use of the information in an RNA molecule for the synthesis of a polypeptide.

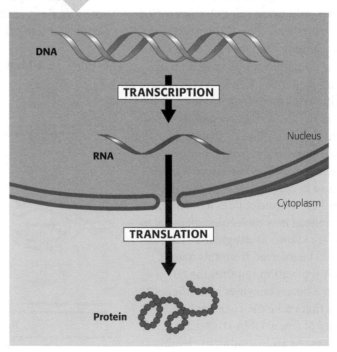

▼ **Figure 10.8 The flow of genetic information in a eukaryotic cell.** A sequence of nucleotides in the DNA is transcribed into a molecule of RNA in the cell's nucleus. The RNA travels to the cytoplasm, where it is translated into the specific amino acid sequence of a protein.

From Nucleotides to Amino Acids: An Overview

Genetic information in DNA is transcribed into RNA and then translated into polypeptides, which then fold into proteins. But how do these processes occur? Transcription and translation are linguistic terms, and it is useful to think of nucleic acids and proteins as having languages. To understand how genetic information passes from genotype to phenotype, we need to see how the chemical language of DNA is translated into the different chemical language of proteins.

What exactly is the language of nucleic acids? Both DNA and RNA are polymers made of nucleotide monomers strung together in specific sequences that convey information, much as specific sequences of letters convey information in English. In DNA, the monomers are the four types of nucleotides, which differ in their nitrogenous bases (A, T, C, and G). The same is true for RNA, although it has the base U instead of T.

The language of DNA is written as a linear sequence of nucleotide bases, such as the blue sequence you see on the enlarged DNA strand in **Figure 10.9**. Every gene is made up of a specific sequence of bases, with special sequences marking the beginning and the end. A typical gene is a few thousand nucleotides in length.

When a segment of DNA is transcribed, the result is an RNA molecule. The process is called transcription because the nucleic acid language of DNA has simply been rewritten (transcribed) as a sequence of bases of RNA; the language is still that of nucleic acids. The nucleotide bases of the RNA molecule are complementary to those on the DNA strand. As you will soon see, this is because the RNA was synthesized using the DNA as a template.

Translation is the conversion of the nucleic acid language to the polypeptide language. Like nucleic acids, polypeptides are straight polymers, but the monomers that make them up—the letters of the polypeptide alphabet—are the 20 amino acids common to all organisms (represented as purple shapes in Figure 10.9). The sequence of nucleotides of the RNA molecule dictates the sequence of amino acids of the polypeptide. But remember, RNA is only a messenger; the genetic information that dictates the amino acid sequence originates in DNA.

What are the rules for translating the RNA message into a polypeptide? In other words, what is the correspondence between the nucleotides of an RNA molecule and the amino acids of a polypeptide? Keep in mind that there are only four different kinds of nucleotides in DNA (A, G, C, T) and RNA (A, G, C, U). During translation, these four must somehow specify 20 amino acids. If each nucleotide base coded for one amino acid, only 4 of the 20 amino acids could be accounted for. In fact, triplets of bases are the smallest "words" of uniform length that can specify all the amino acids. There can be 64 (that is, 4^3) possible code words of this type—more than enough to specify the 20 amino acids. Indeed, there are enough triplets to allow more than one coding for each amino acid. For example, the base triplets AAA and AAG both code for the same amino acid.

Experiments have verified that the flow of information from gene to protein is based on a triplet code. The genetic instructions for the amino acid sequence of a polypeptide chain are written in DNA and RNA as a series of three-base words called **codons**. Three-base codons in the DNA are transcribed into complementary three-base codons in the RNA, and then the RNA codons are translated into amino acids that form a polypeptide. As summarized in Figure 10.9, one DNA codon (three nucleotides) → one RNA codon (three nucleotides) → one amino acid. Next we turn to the codons themselves. ☑

☑ **CHECKPOINT**

How many nucleotides are necessary to code for a polypeptide that is 100 amino acids long?

Answer: 300

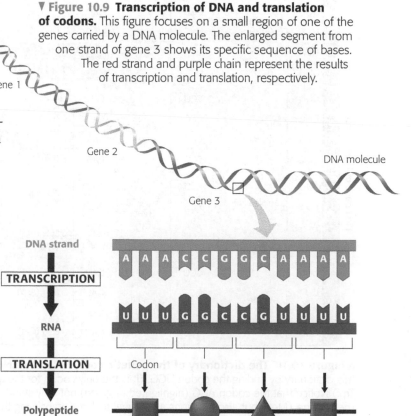

▼ **Figure 10.9 Transcription of DNA and translation of codons.** This figure focuses on a small region of one of the genes carried by a DNA molecule. The enlarged segment from one strand of gene 3 shows its specific sequence of bases. The red strand and purple chain represent the results of transcription and translation, respectively.

Gene 1

Gene 2

Gene 3

DNA molecule

DNA strand

TRANSCRIPTION

RNA

TRANSLATION

Polypeptide

A A A C C G G C A A A A

U U U G G C C G U U U U

Codon

Amino acid

The Genetic Code

The **genetic code** is the set of rules that convert a nucleotide sequence in RNA to an amino acid sequence. As

> Because all life on Earth shares a universal genetic code, your DNA could be used to genetically modify a monkey.

Figure 10.10 shows, 61 of the 64 triplets code for amino acids. The triplet AUG has a dual function: It codes for the amino acid methionine (abbreviated Met) and can also provide a signal for the start of a polypeptide chain. Three codons (UAA, UAG, and UGA) do not designate amino acids. They are the stop codons that instruct the ribosomes to end the polypeptide.

Notice in Figure 10.10 that a given RNA triplet always specifies a given amino acid. For example, although codons UUU and UUC both specify phenylalanine (Phe), neither of them ever represents any other amino acid. The codons in the figure are the triplets found in RNA. They have a straightforward, complementary relationship to the codons in DNA. The nucleotides making up the codons occur in a linear order along the DNA and RNA, with no gaps separating the codons.

The genetic code is nearly universal, shared by organisms from the simplest bacteria to the most complex plants and animals. The universality of the genetic vocabulary suggests that it arose very early in evolution and was passed on over the eons to all the organisms living on Earth today. In fact, such universality is the key to modern DNA technologies. Because diverse organisms share a common genetic code, it is possible to program one species to produce a protein from another species by transplanting DNA (**Figure 10.11**). This allows scientists to mix and match genes from various species—a procedure with many useful genetic engineering applications in agriculture, medicine, and research (see Chapter 12 for further discussion of genetic engineering). Besides having practical purposes, a shared genetic vocabulary also reminds us of the evolutionary kinship that connects all life on Earth. ☑

▼ **Figure 10.11** **A pig expressing a foreign gene.** The glowing porker in the middle was created when researchers incorporated a jelly (jellyfish) gene for a protein called green fluorescent protein (GFP) into the DNA of a standard pig.

Second base of RNA codon

First base of RNA codon	U	C	A	G	Third base of RNA codon
U	UUU ⎤ Phenylalanine (Phe) UUC ⎦ UUA ⎤ Leucine (Leu) UUG ⎦	UCU ⎤ UCC ⎥ Serine (Ser) UCA ⎥ UCG ⎦	UAU ⎤ Tyrosine (Tyr) UAC ⎦ UAA Stop UAG Stop	UGU ⎤ Cysteine (Cys) UGC ⎦ UGA Stop UGG Tryptophan (Trp)	U C A G
C	CUU ⎤ CUC ⎥ Leucine (Leu) CUA ⎥ CUG ⎦	CCU ⎤ CCC ⎥ Proline (Pro) CCA ⎥ CCG ⎦	CAU ⎤ Histidine (His) CAC ⎦ CAA ⎤ Glutamine (Gln) CAG ⎦	CGU ⎤ CGC ⎥ Arginine (Arg) CGA ⎥ CGG ⎦	U C A G
A	AUU ⎤ AUC ⎥ Isoleucine (Ile) AUA ⎦ AUG Met or start	ACU ⎤ ACC ⎥ Threonine (Thr) ACA ⎥ ACG ⎦	AAU ⎤ Asparagine (Asn) AAC ⎦ AAA ⎤ Lysine (Lys) AAG ⎦	AGU ⎤ Serine (Ser) AGC ⎦ AGA ⎤ Arginine (Arg) AGG ⎦	U C A G
G	GUU ⎤ GUC ⎥ Valine (Val) GUA ⎥ GUG ⎦	GCU ⎤ GCC ⎥ Alanine (Ala) GCA ⎥ GCG ⎦	GAU ⎤ Aspartic acid (Asp) GAC ⎦ GAA ⎤ Glutamic acid (Glu) GAG ⎦	GGU ⎤ GGC ⎥ Glycine (Gly) GGA ⎥ GGG ⎦	U C A G

▲ **Figure 10.10** **The dictionary of the genetic code, listed by RNA codons.** Practice using this dictionary by finding the codon UGG. (It is the only codon for the amino acid tryptophan, Trp.) Notice that the codon AUG (highlighted in green) not only stands for the amino acid methionine (Met), but also functions as a signal to "start" translating the RNA at that place. Three of the 64 codons (highlighted in red) function as "stop" signals that mark the end of a genetic message, but do not encode any amino acids.

☑ **CHECKPOINT**

An RNA molecule contains the nucleotide sequence CCAUUUACG. Using Figure 10.10, translate this sequence into the corresponding amino acid sequence.

Answer: Pro-Phe-Thr

Transcription: From DNA to RNA

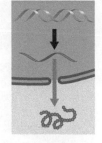

Let's look more closely at transcription, the transfer of genetic information from DNA to RNA. If you think of your DNA as a cookbook, then transcription is the process of copying one recipe onto an index card (a molecule of RNA) for immediate use. **Figure 10.12a** is a close-up view of this process. As with DNA replication, the two DNA strands must first separate at the place where the process will start. In transcription, however, only one of the DNA strands serves as a template for the newly forming RNA molecule; the other strand is unused. The nucleotides that make up the new RNA molecule take their place one at a time along the DNA template strand by forming hydrogen bonds with the nucleotide bases there. Notice that the RNA nucleotides follow the usual base-pairing rules, except that U, rather than T, pairs with A. The RNA nucleotides are linked by the transcription enzyme **RNA polymerase**.

Figure 10.12b is an overview of the transcription of an entire gene. Special sequences of DNA nucleotides tell the RNA polymerase where to start and where to stop the transcribing process.

❶ Initiation of Transcription

The "start transcribing" signal is a nucleotide sequence called a **promoter**, which is located in the DNA at the beginning of the gene. A promoter is a specific place where RNA polymerase attaches. The first phase of transcription, called initiation, is the attachment of RNA polymerase to the promoter and the start of RNA synthesis. For any gene, the promoter dictates which of the two DNA strands is to be transcribed (the particular strand varies from gene to gene).

❷ RNA Elongation

During the second phase of transcription, elongation, the RNA grows longer. As RNA synthesis continues, the RNA strand peels away from its DNA template, allowing the two separated DNA strands to come back together in the region already transcribed.

❸ Termination of Transcription

In the third phase, termination, the RNA polymerase reaches a special sequence of bases in the DNA template called a **terminator**. This sequence signals the end of the gene. At this point, the polymerase molecule detaches from the RNA molecule and the gene, and the DNA strands rejoin.

In addition to producing RNA that encodes amino acid sequences, transcription makes two other kinds of RNA that are involved in building polypeptides. We discuss these kinds of RNA a little later. ☑

▼ **Figure 10.12 Transcription.**

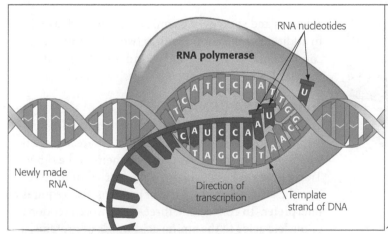

(a) A close-up view of transcription. As RNA nucleotides base-pair one by one with DNA bases on one DNA strand (called the template strand), the enzyme RNA polymerase (orange) links the RNA nucleotides into an RNA chain.

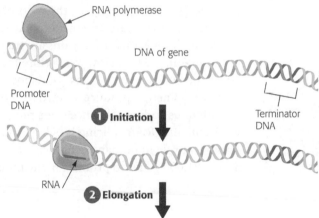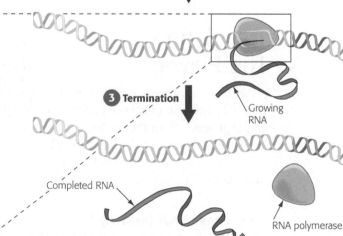

(b) Transcription of a gene. The transcription of an entire gene occurs in three phases: initiation, elongation, and termination of the RNA. The section of DNA where the RNA polymerase starts is called the promoter; the place where it stops is called the terminator.

The Processing of Eukaryotic RNA

In the cells of prokaryotes, which lack nuclei, the RNA transcribed from a gene immediately functions as **messenger RNA (mRNA)**, the molecule that is translated into protein. But this is not the case in eukaryotic cells. The eukaryotic cell not only localizes transcription in the nucleus but also modifies, or processes, the RNA transcripts there before they move to the cytoplasm for translation by the ribosomes.

One kind of RNA processing is the addition of extra nucleotides to the ends of the RNA transcript. These additions, called the **cap** and **tail**, protect the RNA from attack by cellular enzymes and help ribosomes recognize the RNA as mRNA.

Another type of RNA processing is made necessary in eukaryotes by noncoding stretches of nucleotides that interrupt the nucleotides that actually code for amino acids. It is as if nonsense words were randomly interspersed within a recipe that you copied. Most genes of plants and animals, it turns out, include such internal noncoding regions, which are called **introns**. The coding regions—the parts of a gene that are expressed—are called **exons**. As **Figure 10.13** illustrates, both exons and introns are transcribed from DNA into RNA. However, before the RNA leaves the nucleus, the introns are removed, and the exons are joined to produce an mRNA molecule with a continuous coding sequence. This process is called **RNA splicing**. RNA splicing is believed to play a significant role in humans in allowing our approximately 21,000 genes to produce many thousands more

☑ CHECKPOINT

Why is a final mRNA often shorter than the DNA gene that coded for it?

Answer: because introns are removed from the RNA

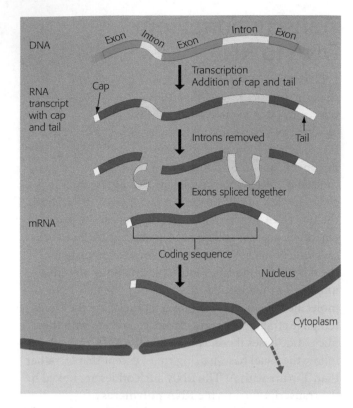

▲ Figure 10.13 **The production of messenger RNA (mRNA) in a eukaryotic cell.** Note that the molecule of mRNA that leaves the nucleus is substantially different from the molecule of RNA that was first transcribed from the gene. In the cytoplasm, the coding sequence of the final mRNA will be translated.

polypeptides. This is accomplished by varying the exons that are included in the final mRNA.

With capping, tailing, and splicing completed, the "final draft" of eukaryotic mRNA is ready for translation. ☑

Translation: The Players

As we have already discussed, translation is a conversion between different languages—from the nucleic acid language to the protein language—and it involves more elaborate machinery than transcription.

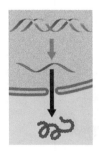

Messenger RNA (mRNA)

The first important ingredient required for translation is the mRNA produced by transcription. Once it is present, the machinery used to translate mRNA requires

enzymes and sources of chemical energy, such as ATP. In addition, translation requires two other important components: ribosomes and a kind of RNA called transfer RNA.

Transfer RNA (tRNA)

Translation of any language into another requires an interpreter, someone or something that can recognize the words of one language and convert them to the other. Translation of the genetic message carried in mRNA into the amino acid language of proteins also requires an interpreter. To convert the three-letter words (codons) of nucleic acids to the amino acid words of proteins, a cell uses a molecular interpreter, a type of RNA called **transfer RNA (tRNA)**, depicted in **Figure 10.14**.

▼ **Figure 10.14 The structure of tRNA.** At one end of the tRNA is the site where an amino acid will attach (purple), and at the other end is the three-nucleotide anticodon where the mRNA will attach (light green).

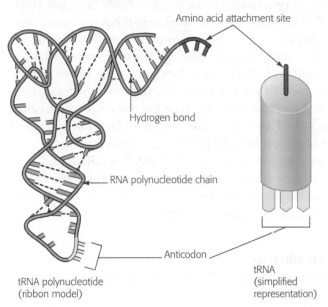

Amino acid attachment site

Hydrogen bond

RNA polynucleotide chain

Anticodon

tRNA polynucleotide (ribbon model)

tRNA (simplified representation)

A cell that is producing proteins has in its cytoplasm a supply of amino acids. But amino acids themselves cannot recognize the codons arranged in sequence along messenger RNA. It is up to the cell's molecular interpreters, tRNA molecules, to match amino acids to the appropriate codons to form the new polypeptide. To perform this task, tRNA molecules must carry out two distinct functions: (1) pick up the appropriate amino acids and (2) recognize the appropriate codons in the mRNA. The unique structure of tRNA molecules enables them to perform both tasks.

As shown on the left in Figure 10.14, a tRNA molecule is made of a single strand of RNA—one polynucleotide chain—consisting of about 80 nucleotides. The chain twists and folds upon itself, forming several double-stranded regions in which short stretches of RNA base-pair with other stretches. At one end of the folded molecule is a special triplet of bases called an **anticodon**. The anticodon triplet is complementary to a codon triplet on mRNA. During translation, the anticodon on the tRNA recognizes a particular codon on the mRNA by using base-pairing rules. At the other end of the tRNA molecule is a site where one specific kind of amino acid attaches. Although all tRNA molecules are similar, there are slightly different versions of tRNA for each amino acid.

Ribosomes

Ribosomes are the organelles in the cytoplasm that coordinate the functioning of mRNA and tRNA and actually make polypeptides. As you can see in **Figure 10.15a**, a ribosome consists of two subunits. Each subunit is made up of proteins and a considerable amount of yet another kind of RNA, **ribosomal RNA (rRNA)**. A fully assembled ribosome has a binding site for mRNA on its small subunit and binding sites for tRNA on its large subunit. **Figure 10.15b** shows how two tRNA molecules get together with an mRNA molecule on a ribosome. One of the tRNA binding sites, the P site, holds the tRNA carrying the growing polypeptide chain, while another, the A site, holds a tRNA carrying the next amino acid to be added to the chain. The anticodon on each tRNA base-pairs with a codon on the mRNA. The subunits of the ribosome act like a vise, holding the tRNA and mRNA molecules close together. The ribosome can then connect the amino acid from the tRNA in the A site to the growing polypeptide. ☑

☑ **CHECKPOINT**

What is an anticodon?

Answer: An anticodon is the base triplet of a tRNA molecule that couples the tRNA to a complementary codon in the mRNA. The base pairing of anticodon to codon is a key step in translating mRNA to a polypeptide.

▼ **Figure 10.15 The ribosome.**

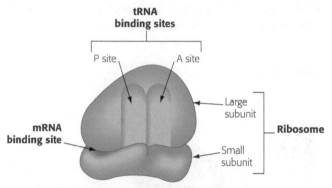

tRNA binding sites

P site   A site

Large subunit

mRNA binding site

Ribosome

Small subunit

(a) A simplified diagram of a ribosome. Notice the two subunits and sites where mRNA and tRNA molecules bind.

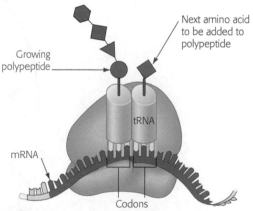

Growing polypeptide

Next amino acid to be added to polypeptide

tRNA

mRNA

Codons

(b) The "players" of translation. When functioning in polypeptide synthesis, a ribosome holds one molecule of mRNA and two molecules of tRNA. The growing polypeptide is attached to one of the tRNAs.

Translation: The Process

Translation is divided into the same three phases as transcription: initiation, elongation, and termination.

Initiation

▼ Figure 10.16 **A molecule of mRNA.**

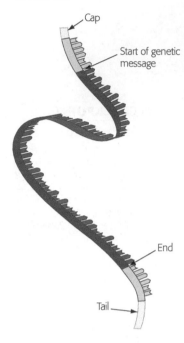

Cap

Start of genetic message

End

Tail

This first phase brings together the mRNA, the first amino acid with its attached tRNA, and the two sub-units of a ribosome. An mRNA molecule, even after splicing, is longer than the genetic message it carries (Figure 10.16). Nucleotide sequences at either end of the molecule (pink) are not part of the message, but along with the cap and tail in eukaryotes, they help the mRNA bind to the ribosome. The initiation process determines exactly where translation will begin so that the mRNA codons will be translated into the correct sequence of amino acids. Initiation occurs in two steps, as shown in Figure 10.17 ❶. An mRNA molecule binds to a small ribosomal subunit. A special initiator tRNA then binds to the **start codon**, where translation is to begin on the mRNA. The initiator tRNA carries the amino acid methionine (Met); its anticodon, UAC, binds to the start codon, AUG ❷. A large ribosomal subunit binds to the small one, creating a functional ribosome. The initiator tRNA fits into the P site on the ribosome.

Elongation

Once initiation is complete, amino acids are added one by one to the first amino acid. Each addition occurs in the three-step elongation process shown in **Figure 10.18**. ❶ The anticodon of an incoming tRNA molecule, carrying its amino acid, pairs with the mRNA codon in the A site of the ribosome. ❷ The polypeptide leaves the tRNA in the P site and attaches to the amino acid on the tRNA in the A site. The ribosome creates a new peptide bond. Now the chain has one more amino acid. ❸ The P site tRNA now leaves the ribosome, and the ribosome moves the remaining tRNA, carrying the growing polypeptide, to the P site. The mRNA and tRNA move as a unit. This movement brings into the A site the next mRNA codon to be translated, and the process can start again with step 1.

Termination

Elongation continues until a **stop codon** reaches the ribosome's A site. Stop codons—UAA, UAG, and UGA—do not code for amino acids but instead tell translation to stop. The completed polypeptide, typically several hundred amino acids long, is freed, and the ribosome splits back into its subunits. ☑

▼ Figure 10.17 **The initiation of translation.**

☑ **CHECKPOINT**

Which of the following does not participate directly in translation: ribosomes, transfer RNA, messenger RNA, DNA?

Answer: DNA

► Figure 10.18
The elongation of a polypeptide. The dashed red arrows indicate movement.

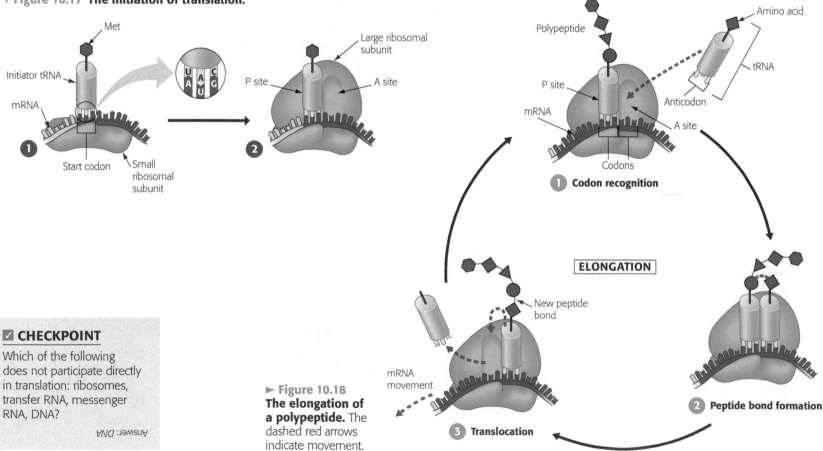

Review: DNA → RNA → Protein

Figure 10.19 reviews the flow of genetic information in the cell, from DNA to RNA to protein. In eukaryotic cells, transcription (DNA → RNA) occurs in the nucleus, and the RNA is processed before it enters the cytoplasm. Translation (RNA → protein) is rapid; a single ribosome can make an average-sized polypeptide in less than a minute. As it is made, a polypeptide coils and folds, assuming its final three-dimensional shape.

What is the overall significance of transcription and translation? These are the processes whereby genes control the structures and activities of cells—or, more broadly, the way the genotype produces the phenotype. The flow of information originates with the specific sequence of nucleotides in a DNA gene. The gene dictates the transcription of a complementary sequence of nucleotides in mRNA. In turn, the information within the mRNA specifies the sequence of amino acids in a polypeptide. Finally, the proteins that form from the polypeptides determine the appearance and capabilities of the cell and organism.

For decades, the DNA → RNA → protein pathway was believed to be the sole means by which genetic information controls traits. In recent years, however, this notion has been challenged by discoveries that point to more complex roles for RNA. (We will explore some of these special properties of RNA in Chapter 11.) ☑

☑ **CHECKPOINT**

1. Transcription is the synthesis of _____, using _____ as a template.
2. Translation is the synthesis of _____, with one _____ determining each amino acid in the sequence.
3. Which organelle coordinates translation?

Answers: 1. mRNA; DNA 2. protein (polypeptides); codon 3. ribosomes

▼ **Figure 10.19 A summary of transcription and translation.** This figure summarizes the main stages in the flow of genetic information from DNA to protein in a eukaryotic cell.

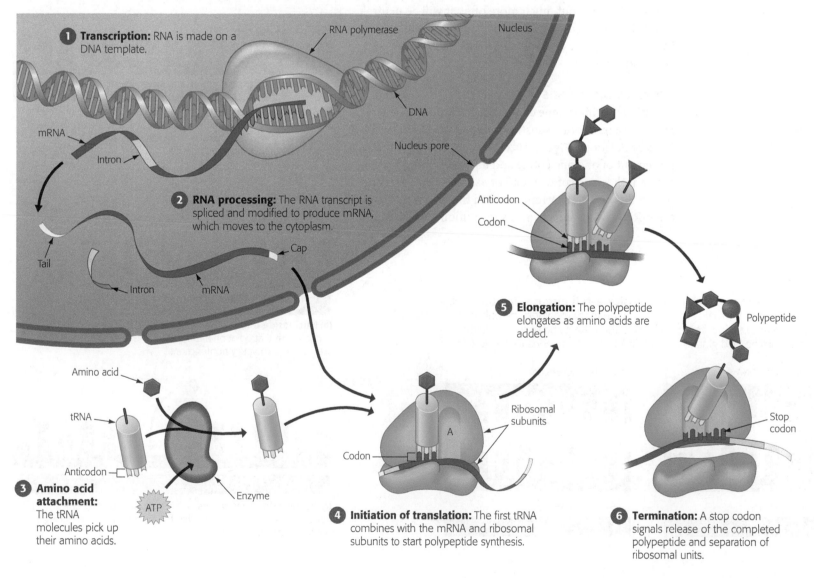

217

Mutations

Since discovering how genes are translated into proteins, scientists have been able to describe many heritable differences in molecular terms. For instance, sickle-cell disease can be traced to a change in a single amino acid in one of the polypeptides in the hemoglobin protein (see Figure 3.19). This difference is caused by a single nucleotide difference in the DNA coding for that polypeptide **(Figure 10.20)**.

Any change in the nucleotide sequence of a cell's DNA is called a **mutation**. Mutations can involve large regions of a chromosome or just a single nucleotide pair, as in sickle-cell disease. Occasionally, a base substitution leads to an improved protein or one with new capabilities that enhance the success of the mutant organism and its descendants. Much more often, though, mutations are harmful. Think of a mutation as a typo in a recipe; occasionally, such a typo might lead to an improved recipe, but much more often it will be neutral, mildly bad, or disastrous. Let's consider how mutations involving only one or a few nucleotide pairs can affect gene translation.

Types of Mutations

Mutations within a gene can be divided into two general categories: nucleotide substitutions and nucleotide insertions or deletions **(Figure 10.21)**. A substitution is the replacement of one nucleotide and its base-pairing partner with another nucleotide pair. For example, in the second row in Figure 10.21, A replaces G in the fourth codon of the mRNA. What effect can a substitution have? Because

A single molecular "typo" in DNA can result in a life-threatening disease.

the genetic code is redundant, some substitution mutations have no effect at all. For example, if a mutation causes an mRNA codon to change from GAA to GAG, no change in the protein product would result because GAA and GAG both code for the same amino acid (Glu). Such a change is called a silent mutation. In our recipe example, changing "1¼ cup sugar" to "1¼ cup sugor"

▼ **Figure 10.21 Three types of mutations and their effects.** Mutations are changes in DNA, but they are shown here in mRNA and the polypeptide product.

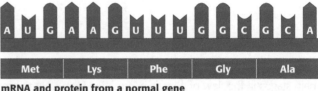

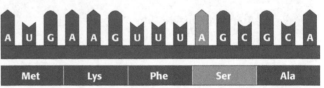

(a) Base substitution. Here, an A replaces a G in the fourth codon of the mRNA. The result in the polypeptide is a serine (Ser) instead of a glycine (Gly). This amino acid substitution may or may not affect the protein's function.

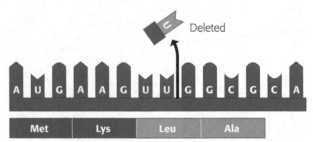

(b) Nucleotide deletion. When a nucleotide is deleted, all the codons from that point on are misread. The resulting polypeptide is likely to be completely nonfunctional.

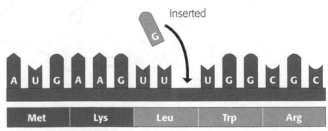

(c) Nucleotide insertion. As with a deletion, inserting one nucleotide disrupts all codons that follow, most likely producing a nonfunctional polypeptide.

▼ **Figure 10.20 The molecular basis of sickle-cell disease.** The sickle-cell allele differs from its normal counterpart, a gene for hemoglobin, by only one nucleotide (orange). This difference changes the mRNA codon from one that codes for the amino acid glutamic acid (Glu) to one that codes for valine (Val).

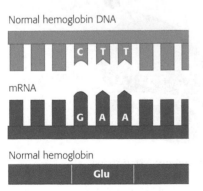

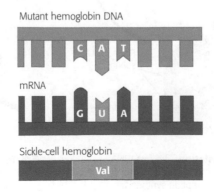

would probably be translated the same way, just like the translation of a silent mutation does not change the meaning of the message.

Other substitutions involving a single nucleotide do change the amino acid coding. Such mutations are called missense mutations. For example, if a mutation causes an mRNA codon to change from GGC to AGC, the resulting protein will have a serine (Ser) instead of a glycine (Gly) at this position. Some missense mutations have little or no effect on the shape or function of the resulting protein; imagine changing a recipe from "1¼ cups sugar" to "1⅓ cups sugar"—this will probably have a negligible effect on your final product. However, other substitutions, as we saw in the sickle-cell case, cause changes in the protein that prevent it from performing normally. This would be like changing "1¼ cups sugar" to "6¼ cups sugar"—this one change is enough to ruin the recipe.

Some substitutions, called nonsense mutations, change an amino acid codon into a stop codon. For example, if an AGA (Arg) codon is mutated to a UGA (stop) codon, the result will be a prematurely terminated protein, which probably will not function properly. In our recipe analogy, this would be like stopping food preparation before the end of the recipe, which is almost certainly going to ruin the dish.

Mutations involving the deletion or insertion of one or more nucleotides in a gene, called frameshift mutations, often have disastrous effects (see Figure 10.21b and c). Because mRNA is read as a series of nucleotide triplets during translation, adding or subtracting nucleotides may alter the triplet grouping of the genetic message. All the nucleotides after the insertion or deletion will be regrouped into different codons. Consider this recipe example: Add one cup egg nog. Deleting the second letter produces an entirely nonsensical message— ado nec upe ggn og—which will not produce a useful product. Similarly, a frameshift mutation most often produces a nonfunctioning polypeptide.

Mutagens

Mutations can occur in a number of ways. Spontaneous mutations result from random errors during DNA replication or recombination. Other sources of mutation are physical and chemical agents called **mutagens**. The most common physical mutagen is high-energy radiation, such as X-rays and ultraviolet (UV) light. Chemical mutagens are of various types. One type, for example, consists of chemicals that are similar to normal DNA bases but that base-pair incorrectly when incorporated into DNA.

Because many mutagens can act as carcinogens, agents that cause cancer, you would do well to avoid them as much as possible. What can you do to avoid exposure to mutagens? Several lifestyle practices can help, including not smoking and wearing protective clothing and sunscreen to minimize direct exposure to the sun's UV rays. But such precautions are not foolproof, and it is not possible to avoid mutagens (such as UV radiation and secondhand smoke) entirely.

Although mutations are often harmful, they can also be beneficial, both in nature and in the laboratory. Mutations are one source of the rich diversity of genes in the living world, a diversity that makes evolution by natural selection possible **(Figure 10.22)**. Mutations are also essential tools for geneticists. Whether naturally occurring or created in the laboratory, mutations are responsible for the different alleles needed for genetic research. ☑

☑ **CHECKPOINT**

1. What would happen if a mutation changed a start codon to some other codon?
2. What happens when one nucleotide is lost from the middle of a gene?

Answers: **1.** mRNA transcribed from the mutated gene would be nonfunctional because ribosomes would not initiate translation. **2.** In the mRNA, the reading of the triplets downstream from the deletion is shifted, leading to a long string of incorrect amino acids in the polypeptide.

▼ **Figure 10.22 Mutations and diversity.** Mutations are one source of the diversity of life visible in this scene from the Isle of Staffa, in the North Atlantic.

Viruses and Other Noncellular Infectious Agents

Viruses share some of the characteristics of living organisms, such as having genetic material in the form of nucleic acid packaged within a highly organized structure. A virus is generally not considered alive, however, because it is not cellular and cannot reproduce on its own. (See Figure 1.4 to review the properties of life.) A **virus** is an infectious particle consisting of little more than "genes in a box": a bit of nucleic acid wrapped in a protein coat and, in some cases, an envelope of membrane **(Figure 10.23)**. A virus cannot reproduce on its own, and thus it can multiply only by infecting a living cell and directing the cell's molecular machinery to make more

viruses. In this section, we'll look at viruses that infect different types of host organisms, starting with bacteria.

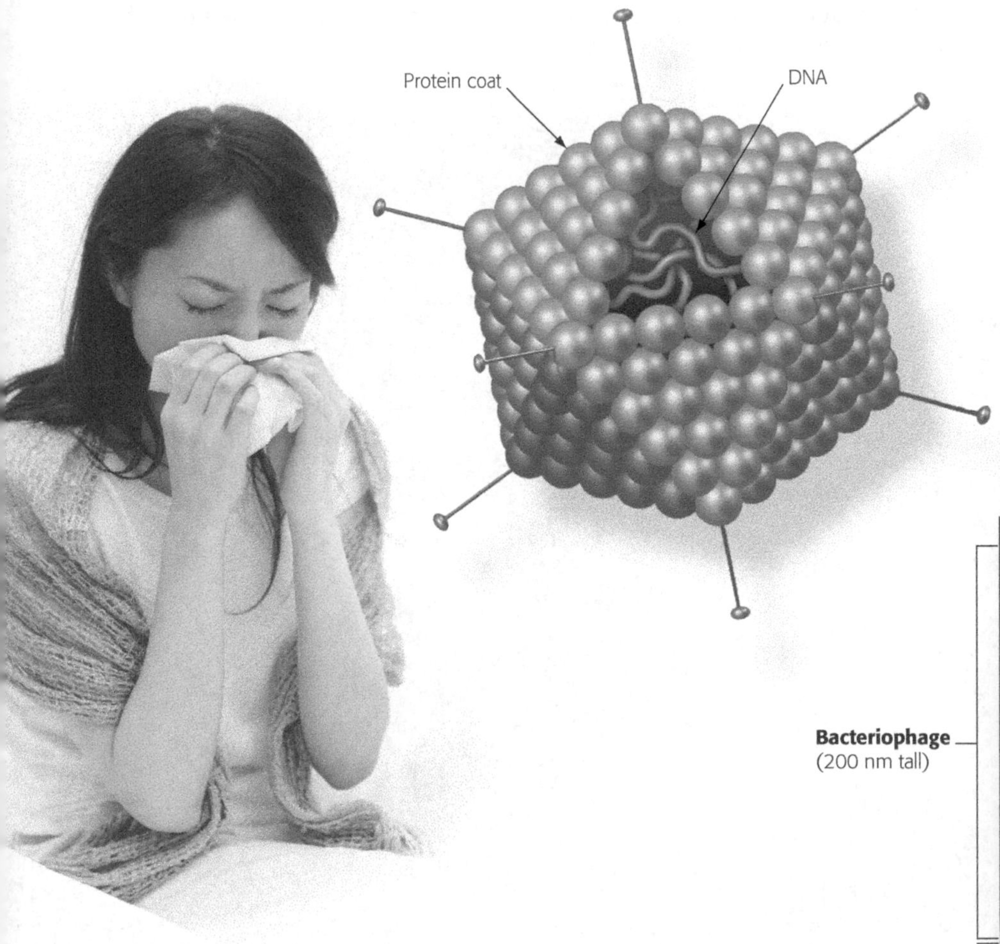

Bacteriophages

Viruses that attack bacteria are called **bacteriophages** ("bacteria-eaters"), or **phages** for short. **Figure 10.24** shows a micrograph of a bacteriophage called T4 infecting an *Escherichia coli* bacterium. The phage consists of a molecule of DNA enclosed within an elaborate structure made of proteins. The "legs" of the phage bend when they touch the cell surface. The tail is a hollow rod enclosed in a springlike sheath. As the legs bend, the spring compresses, the bottom of the rod punctures the cell membrane, and the viral DNA passes from inside the head of the virus into the cell.

Once they infect a bacterium, most phages enter a reproductive cycle called the **lytic cycle**. The lytic cycle gets its name from the fact that after many copies of the phage are produced within the bacterial cell, the bacterium lyses (breaks open). Some viruses can also reproduce by an alternative route—the **lysogenic cycle**. During a lysogenic cycle, viral DNA replication occurs without phage production or the death of the cell.

▼ Figure 10.24 **Bacteriophages (viruses) infecting a bacterial cell.**

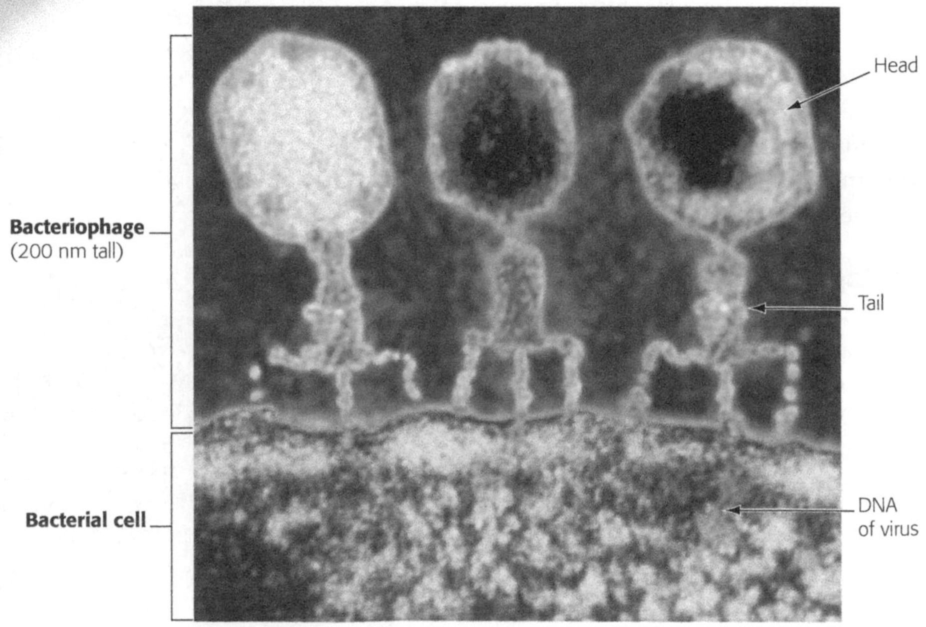

Protein coat — DNA

Head

Bacteriophage (200 nm tall)

Tail

Bacterial cell

DNA of virus

Colorized TEM 225,000×

▲ Figure 10.23 **Adenovirus.** A virus that infects the human respiratory system, an adenovirus consists of DNA enclosed in a protein coat shaped like a 20-sided polyhedron, shown here in a computer-generated model that is magnified approximately 500,000 times the actual size. At each corner of the polyhedron is a protein spike, which helps the virus attach to a susceptible cell.

Figure 10.25 illustrates the two kinds of cycles for a phage named lambda that can infect *E. coli* bacteria. At the start of infection, ❶ lambda binds to the outside of a bacterium and injects its DNA inside. ❷ The injected lambda DNA forms a circle. In the lytic cycle, this DNA immediately turns the cell into a virus-producing factory. ❸ The cell's own machinery for DNA replication, transcription, and translation is hijacked by the virus and used to produce copies of the virus. ❹ The cell lyses, releasing the new phages.

In the lysogenic cycle, ❺ the viral DNA is inserted into the bacterial chromosome. Once there, the phage DNA is referred to as a **prophage**, and most of its genes are inactive. Survival of the prophage depends on the reproduction of the cell where it resides. ❻ The host cell replicates the prophage DNA along with its cellular DNA and then, upon dividing, passes on both the prophage and the cellular DNA to its two daughter cells. A single infected bacterium can quickly give rise to a large population of bacteria that all carry prophages. The prophages may remain in the bacterial cells indefinitely. ❼ Occasionally, however, a prophage leaves its chromosome; this event may be triggered by environmental conditions such as exposure to a mutagen. Once separate, the lambda DNA usually switches to the lytic cycle, which results in the production of many copies of the virus and lysing of the host cell.

Sometimes the few prophage genes active in a lysogenic bacterial cell can cause medical problems. For example, the bacteria that cause diphtheria, botulism, and scarlet fever would be harmless to people if it were not for the prophage genes they carry. Certain of these genes direct the bacteria to produce toxins that make people ill. ☑

✔ **CHECKPOINT**

Describe one way some viruses can perpetuate their genes without immediately destroying the cells they infect.

Answer: Some viruses can insert their DNA into the DNA of the cell they infect (the lysogenic cycle). The viral DNA is replicated along with the cell's DNA every time the cell divides.

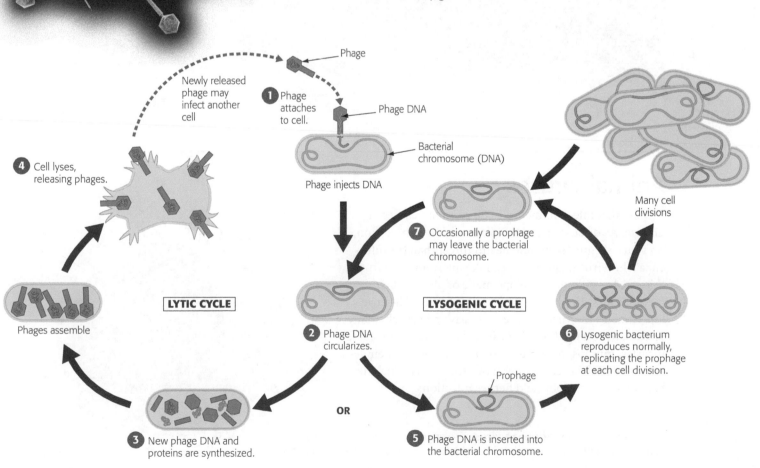

Phage lambda

E. coli

▼ Figure 10.25 **Alternative phage reproductive cycles.** Certain phages can undergo alternative reproductive cycles. After entering the bacterial cell, the phage DNA can either integrate into the bacterial chromosome (lysogenic cycle) or immediately start the production of progeny phages (lytic cycle), destroying the cell. Once it enters a lysogenic cycle, the phage's DNA may be carried in the host cell's chromosome for many generations.

Phage

Newly released phage may infect another cell

❶ Phage attaches to cell.

Phage DNA

Bacterial chromosome (DNA)

Phage injects DNA

❹ Cell lyses, releasing phages.

Many cell divisions

LYTIC CYCLE

LYSOGENIC CYCLE

❼ Occasionally a prophage may leave the bacterial chromosome.

Phages assemble

❷ Phage DNA circularizes.

❻ Lysogenic bacterium reproduces normally, replicating the prophage at each cell division.

Prophage

OR

❸ New phage DNA and proteins are synthesized.

❺ Phage DNA is inserted into the bacterial chromosome.

221

Plant Viruses

Viruses that infect plant cells can stunt plant growth and diminish crop yields. Most known plant viruses have RNA rather than DNA as their genetic material. Many of them, like the tobacco mosaic virus (TMV) shown in **Figure 10.26**, are rod-shaped with a spiral arrangement of proteins surrounding the nucleic acid. TMV, which infects tobacco and related plants, causing discolored spots on the leaves, was the first virus ever discovered (in 1930).

To infect a plant, a virus must first get past the plant's epidermis, an outer protective layer of cells. For this reason, a plant damaged by wind, chilling, injury, or insects is more susceptible to infection than a healthy plant. Some insects carry and transmit plant viruses, and farmers and gardeners may unwittingly spread plant viruses through the use of pruning shears and other tools.

There is no cure for most viral plant diseases, and agricultural scientists focus on preventing infection and on breeding or genetically engineering varieties of crop plants that resist viral infection. In Hawaii, for example, the spread of papaya ringspot potyvirus (PRSV) by aphids wiped out the native papaya (Hawaii's second largest crop) in certain island regions. But since 1998, farmers have been able to plant a genetically engineered PRSV-resistant strain of papaya, and papayas have been reintroduced into their old habitats. ☑

▼ **Figure 10.26 Tobacco mosaic virus.** The photo shows the mottling of leaves in tobacco mosaic disease. The rod-shaped virus causing the disease has RNA as its genetic material.

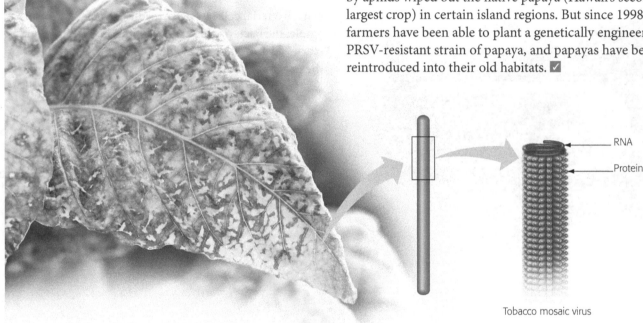

RNA

Protein

Tobacco mosaic virus

Animal Viruses

Viruses that infect animal cells are common causes of disease. As discussed in the Biology and Society section, no virus is a greater human health threat than the influenza (flu) virus (**Figure 10.27**). Like many animal viruses, this one has an outer envelope made of phospholipid membrane, with projecting spikes of protein. The envelope enables the virus to enter and leave a host cell. Many viruses, including those that cause the flu, common cold, measles, mumps, AIDS, and polio, have RNA as their genetic material. Diseases caused by DNA viruses include hepatitis, chicken pox, and herpes infections.

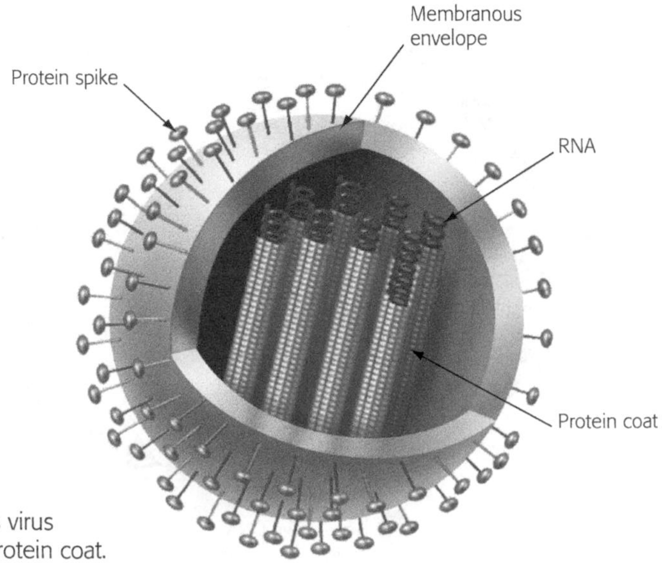

Membranous envelope

Protein spike

RNA

Protein coat

▶ **Figure 10.27 An influenza virus.** The genetic material of this virus consists of eight separate molecules of RNA, each wrapped in a protein coat.

Figure 10.28 shows the reproductive cycle of the mumps virus, a typical RNA virus. Once a common childhood disease characterized by fever and swelling of the salivary glands, mumps has become quite rare in industrialized nations due to widespread vaccination. When the virus contacts a susceptible cell, protein spikes on its outer surface attach to receptor proteins on the cell's plasma membrane. ❶ The viral envelope fuses with the cell's membrane, allowing the protein-coated RNA to enter the cytoplasm. ❷ Enzymes then remove the protein coat. ❸ An enzyme that entered the cell as part of the virus uses the virus's RNA genome as a template for making complementary strands of RNA. The new strands have two functions: ❹ They serve as mRNA for the synthesis of new viral proteins, and ❺ they serve as templates for synthesizing new viral genome RNA. ❻ The new coat proteins assemble around the new viral RNA. ❼ Finally, the viruses leave the cell by cloaking themselves in plasma membrane. In other words, the virus obtains its envelope from the cell, budding off the cell without necessarily rupturing it.

Not all animal viruses reproduce in the cytoplasm. For example, herpesviruses—which cause chicken pox, shingles, cold sores, and genital herpes—are enveloped DNA viruses that reproduce in a host cell's nucleus, and they get their envelopes from the cell's nuclear membrane. Copies of the herpesvirus DNA usually remain behind in the nuclei of certain nerve cells. There they remain dormant until some sort of stress, such as a cold, sunburn, or emotional stress, triggers virus production, resulting in unpleasant symptoms. Once acquired, herpes infections may flare up repeatedly throughout a person's life. More than 75% of American adults carry herpes simplex 1 (which causes cold sores), and more than 20% carry herpes simplex 2 (which causes genital herpes).

The amount of damage a virus causes the body depends partly on how quickly the immune system responds to fight the infection and partly on the ability of the infected tissue to repair itself. We usually recover completely from colds because our respiratory tract tissue can efficiently replace damaged cells. In contrast, the poliovirus attacks nerve cells, which are not usually replaceable. The damage to such cells by polio is permanent. In such cases, the only medical option is to prevent the disease with vaccines.

How effective are vaccines? We'll examine this question next using the example of the flu vaccine. ☑

☑ CHECKPOINT

Why is infection by herpesvirus permanent?

Answer: because herpesvirus leaves viral DNA in the nuclei of nerve cells

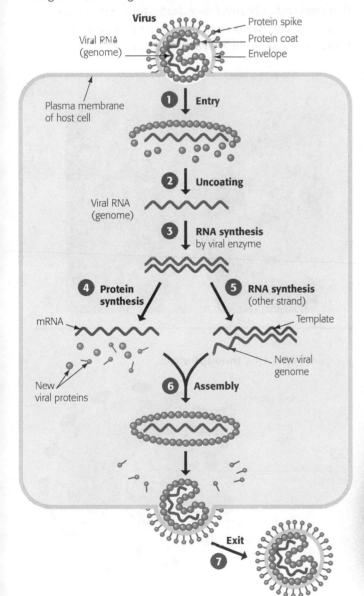

▼ **Figure 10.28 The reproductive cycle of an enveloped virus.** This virus is the one that causes mumps. Like the flu virus, it has a membranous envelope with protein spikes, but its genome is a single molecule of RNA.

Virus
- Protein spike
- Protein coat
- Envelope
- Viral RNA (genome)

Plasma membrane of host cell

❶ **Entry**

❷ **Uncoating**

Viral RNA (genome)

❸ **RNA synthesis** by viral enzyme

❹ **Protein synthesis**

❺ **RNA synthesis** (other strand)

mRNA

Template

New viral genome

New viral proteins

❻ **Assembly**

Exit
❼

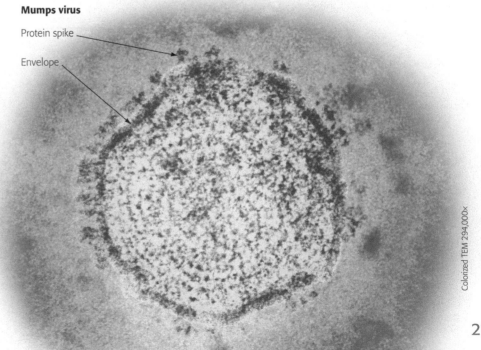

Mumps virus

Protein spike

Envelope

Colorized TEM 294,000x

Do Flu Vaccines Protect the Elderly?

Yearly flu vaccinations are recommended for nearly all people over the age of six months. But how can we be sure they are effective? Because elderly people often have weaker immune systems than younger people and because the elderly account for a significant slice of total health-care spending, they are an important population for vaccination efforts. Epidemiologists (scientists who study the distribution, causes, and control of diseases in populations) have made the **observation** that vaccination rates among the elderly rose from 15% in 1980 to 65% in 1996. This observation has led them to ask an important and basic **question:** Do flu vaccines decrease the mortality rate as a result of the flu among those elderly people who receive them? To find out, researchers investigated data from the general population. Their **hypothesis** was that elderly people who were immunized would have fewer hospital stays and deaths during the winter after vaccination. Their **experiment** followed tens of thousands of people over the age of 65 during the ten flu seasons of the 1990s. The **results** are summarized in **Figure 10.29**. People who were vaccinated had a 27% less chance of being hospitalized during the next flu season and a 48% less chance of dying. But could some factor other than flu shots be at play? For example, maybe people who choose to be vaccinated are healthier for other reasons. As a control, the researchers examined health data for the summer (when flu is not a factor). During these months, there was no difference in the hospitalization rates and only 16% fewer deaths for the immunized, suggesting that flu vaccines provide a significant health benefit among the elderly during the flu season.

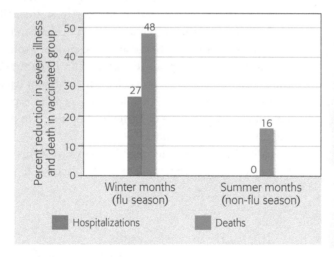

► Figure 10.29 **The effect of flu vaccines on the elderly.** Receiving a flu vaccine greatly reduced the risk of hospitalization and death in the flu season following the shot. The reduction was much smaller or nonexistent in later summer months.

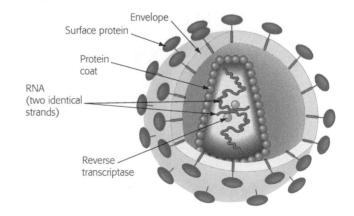

HIV, the AIDS Virus

The devastating disease **AIDS** (<u>a</u>cquired <u>i</u>mmuno<u>d</u>eficiency <u>s</u>yndrome) is caused by **HIV** (<u>h</u>uman <u>i</u>mmunodeficiency <u>v</u>irus), an RNA virus with some nasty twists. In outward appearance, HIV **(Figure 10.30)** resembles the mumps virus. Its envelope enables HIV to enter and leave a cell much the way the mumps virus does. But HIV has a different mode of reproduction. It is a **retrovirus**, an RNA virus that reproduces by means of a DNA molecule, the reverse of the usual DNA → RNA flow of genetic information. These viruses carry molecules of an enzyme called **reverse transcriptase**, which catalyzes reverse transcription: the synthesis of DNA on an RNA template.

▼ Figure 10.30 **HIV, the AIDS virus.**

Figure 10.31 illustrates what happens after HIV RNA is uncoated in the cytoplasm of a cell. The reverse transcriptase (green) **1** uses the RNA as a template to make a DNA strand and then **2** adds a second, complementary DNA strand. **3** The resulting double-stranded viral DNA then enters the cell nucleus and inserts itself into the chromosomal DNA, becoming a **provirus**. Occasionally, the provirus is **4** transcribed into RNA **5** and translated into viral proteins. **6** New viruses assembled from these components eventually leave the cell and can then infect other cells. This is the standard reproductive cycle for retroviruses.

HIV infects and eventually kills several kinds of white blood cells that are important in the body's immune system. The loss of such cells causes the body to become susceptible to other infections that it would normally be able to fight off. Such secondary infections cause the syndrome (a collection of symptoms) that eventually

▼ Figure 10.31 **The behavior of HIV nucleic acid in an infected cell.**

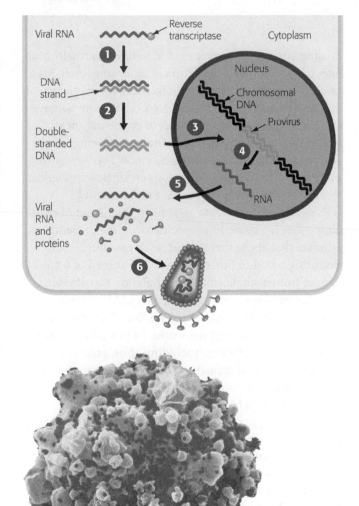

HIV (red dots) infecting a white blood cell

SEM 5,500×

kills AIDS patients. Since it was first recognized in 1981, HIV has infected tens of millions of people worldwide, resulting in millions of deaths.

Although there is as yet no cure for AIDS, its progression can be slowed by two categories of anti-HIV drugs. Both types of medicine interfere with the reproduction of the virus. The first type inhibits the action of enzymes called proteases, which help produce the final versions of HIV proteins. The second type, which includes the drug AZT, inhibits the action of the HIV enzyme reverse transcriptase. The key to AZT's effectiveness is its shape. The shape of a molecule of AZT is very similar to the shape of part of the T (thymine) nucleotide (Figure 10.32). In fact, AZT's shape is so similar to the T nucleotide that AZT can bind to reverse transcriptase, essentially taking the place of T. But unlike thymine, AZT cannot be incorporated into a growing DNA chain. Thus, AZT "gums up the works," interfering with the synthesis of HIV DNA. Because this synthesis is an essential step in the reproductive cycle of HIV, AZT may block the spread of the virus within the body.

Many HIV-infected people in the United States and other industrialized countries take a "drug cocktail" that contains both reverse transcriptase inhibitors and protease inhibitors, and the combination seems to be much more effective than the individual drugs in keeping the virus at bay and extending patients' lives. In fact, the death rate from HIV infection can be lowered by 80% with proper treatment. However, even in combination, the drugs do not completely rid the body of the virus. Typically, HIV reproduction and the symptoms of AIDS return if a patient discontinues the medications. Because AIDS has no cure yet, prevention (namely, avoiding unprotected sex, and staying away from needle sharing) is the only healthy option. ☑

☑ **CHECKPOINT**

Why is HIV called a retrovirus?

Answer: Because it synthesizes DNA from its RNA genome. This is the reverse ("retro") of the usual DNA → RNA information flow.

▼ Figure 10.32 **AZT and the T nucleotide.** The anti-HIV drug AZT (right) has a chemical shape very similar to part of the T (thymine) nucleotide of DNA.

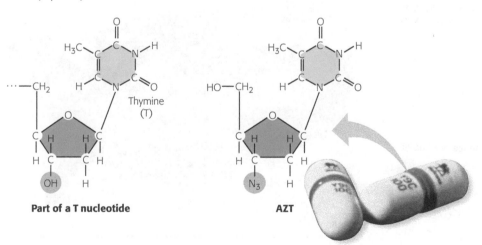

Part of a T nucleotide   AZT

Viroids and Prions

Viruses may be small and simple, but they are giants compared to two other classes of pathogens: viroids and prions. Viroids are small, circular RNA molecules that infect plants. Viroids do not encode proteins but can nonetheless replicate in host plant cells using cellular enzymes. These small RNA molecules may cause disease by interfering with the regulatory systems that control plant growth.

Even stranger are infectious proteins called **prions**. Prions cause a number of brain diseases in various animal species, including scrapie in sheep and goats, chronic wasting disease in deer and elk, and mad cow disease (formally called bovine spongiform encephalopathy, or BSE), which infected more than

Mad cow disease is caused by an abnormal molecule of protein.

2 million cattle in the United Kingdom in the 1980s. In humans, prions cause Creutzfeldt-Jakob disease, an extremely rare, incurable, and inevitably fatal deterioration of the brain.

How can a protein cause disease? A prion is thought to be a misfolded form of a protein normally present in brain cells. When a prion enters a cell containing the normal form of protein, the prion somehow converts the normal protein molecules to the misfolded prion version. The abnormal proteins clump together, which may lead to loss of brain tissue (although how this occurs is the subject of much debate and ongoing research). To date, there is no known cure for prion diseases, so hope rests on understanding and preventing the process of infection. ☑

☑ CHECKPOINT

What makes prions so unusual as pathogens?

Answer: Prions, unlike any other infectious agent, have no nucleic acid (DNA or RNA).

The Deadliest Virus EVOLUTION CONNECTION

Emerging Viruses

Viruses that suddenly come to the attention of medical scientists are called **emerging viruses**. H1N1 (discussed in the Biology and Society section) is one example; another is West Nile virus, which appeared in North America in 1999 and has since spread to all 48 contiguous U.S. states. West Nile virus is spread primarily by mosquitoes, which carry the virus in blood sucked from one victim and can transfer it to another victim. West Nile virus cases surged in 2012, especially in Texas, claiming nearly 300 lives (Figure 10.33).

How do such viruses burst on the human scene, giving rise to new diseases? One way is by the mutation of

existing viruses. RNA viruses tend to have unusually high rates of mutation because errors in replicating their RNA genomes are not subject to proofreading mechanisms that help reduce errors during DNA replication. Some mutations enable existing viruses to evolve into new strains that can cause disease in individuals who have developed resistance to the ancestral virus. This is why we need yearly flu vaccines: Mutations create new influenza virus strains to which people have no immunity.

New viral diseases also arise from the spread of existing viruses from one host species to another. Scientists estimate that about three-quarters of new human diseases originated in other animals. The spread of a viral disease from a small, isolated population can also lead to widespread epidemics. For instance, AIDS went unnamed and virtually unnoticed for decades before it began to spread around the world. In this case, technological and social factors, including affordable international travel, blood transfusions, sexual activity, and the abuse of intravenous drugs, allowed a previously rare human disease to become a global scourge.

Acknowledging the persistent threat that viruses pose to human health, Nobel Prize winner Joshua Lederberg once warned: "We live in evolutionary competition with microbes. There is no guarantee that we will be the survivors." If we someday manage to control HIV, influenza, and other emerging viruses, this success will likely arise from our understanding of molecular biology.

► **Figure 10.33 Mapping the West Nile virus outbreak of 2012.**

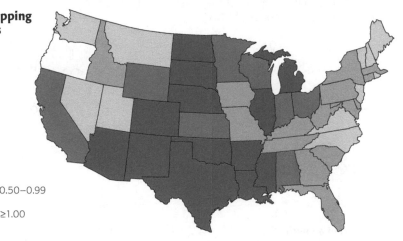

Incidence per 100,000

0.00	0.50–0.99
0.01–0.024	≥1.00
0.25–0.49	

Chapter Review

SUMMARY OF KEY CONCEPTS

DNA: Structure and Replication

DNA and RNA Structure

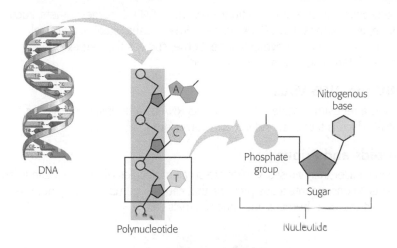

DNA
Polynucleotide
Nucleotide

Nitrogenous base
Phosphate group
Sugar

A
C
T

	DNA	RNA
Nitrogenous base	C G A T	C G A U
Sugar	Deoxy-ribose	Ribose
Number of strands	2	1

Structure/Function: Watson and Crick's Discovery of the Double Helix

Watson and Crick worked out the three-dimensional structure of DNA: two polynucleotide strands wrapped around each other in a double helix. Hydrogen bonds between bases hold the strands together. Each base pairs with a complementary partner: A with T, and G with C.

DNA Replication

The structure of DNA, with its complementary base pairing, allows it to function as the molecule of heredity through DNA replication.

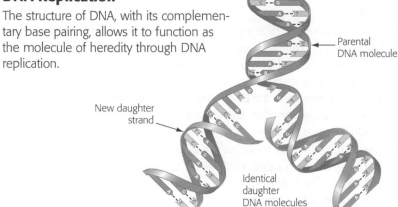

Parental DNA molecule

New daughter strand

Identical daughter DNA molecules

Information Flow: From DNA to RNA to Protein

How an Organism's Genotype Determines Its Phenotype

The information constituting an organism's genotype is carried in the sequence of its DNA bases. The genotype controls phenotype through the expression of proteins.

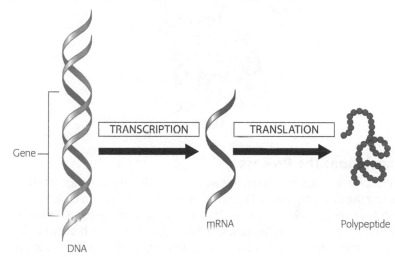

Gene

TRANSCRIPTION

TRANSLATION

DNA

mRNA

Polypeptide

From Nucleotides to Amino Acids: An Overview

The DNA of a gene is transcribed into RNA using the usual base-pairing rules, except that an A in DNA pairs with U in RNA. In the translation of a genetic message, each triplet of nucleotide bases in the RNA, called a codon, specifies one amino acid in the polypeptide.

The Genetic Code

In addition to codons that specify amino acids, the genetic code has one codon that is a start signal and three that are stop signals for translation. The genetic code is redundant: There is more than one codon for most amino acids.

Transcription: From DNA to RNA

In transcription, RNA polymerase binds to the promoter of a gene, opens the DNA double helix there, and catalyzes the synthesis of an RNA molecule using one DNA strand as a template. As the single-stranded RNA transcript peels away from the gene, the DNA strands rejoin.

The Processing of Eukaryotic RNA

The RNA transcribed from a eukaryotic gene is processed before leaving the nucleus to serve as messenger RNA (mRNA). Introns are spliced out, and a cap and tail are added.

Translation: The Players

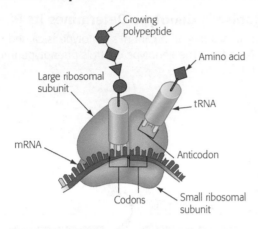

Translation: The Process

In initiation, a ribosome assembles with the mRNA and the initiator tRNA bearing the first amino acid. Beginning at the start codon, the codons of the mRNA are recognized one by one by tRNAs bearing succeeding amino acids. The ribosome bonds the amino acids together. With each addition, the mRNA moves by one codon through the ribosome. When a stop codon is reached, the completed polypeptide is released.

Review: DNA → RNA → Protein

The sequence of codons in DNA, via the sequence of codons in mRNA, spells out the primary structure of a polypeptide.

Mutations

Mutations are changes in the DNA base sequence, caused by errors in DNA replication or recombination or by mutagens. Substituting, deleting, or inserting nucleotides in a gene has varying effects on the polypeptide and organism.

Type of Mutation	Effect
Substitution of one DNA base for another	**Silent** mutations result in no change to amino acids.
	Missense mutations swap one amino acid for another.
	Nonsense mutations change an amino acid codon to a stop codon.
Insertions or **deletions** of DNA nucleotides	Frameshift mutations can alter the triplet grouping of codons and greatly change the amino acid sequence.

Viruses and Other Noncellular Infectious Agents

Viruses are infectious particles consisting of genes packaged in protein.

Bacteriophages

When phage DNA enters a lytic cycle inside a bacterium, it is replicated, transcribed, and translated. The new viral DNA and protein molecules then assemble into new phages, which burst from the cell. In the lysogenic cycle, phage DNA inserts into the cell's chromosome and is passed on to generations of daughter cells. Much later, it may initiate phage production.

Plant Viruses

Viruses that infect plants can be a serious agricultural problem. Most have RNA genomes. Viruses enter plants via breaks in the plant's outer layers.

Animal Viruses

Many animal viruses, such as flu viruses, have RNA genomes; others, such as hepatitis viruses, have DNA. Some animal viruses "steal" a bit of cell membrane as a protective envelope. Some, such as the herpesvirus, can remain latent inside cells for long periods.

HIV, the AIDS Virus

HIV is a retrovirus. Inside a cell it uses its RNA as a template for making DNA, which is then inserted into a chromosome.

Viroids and Prions

Even smaller than viruses, viroids are small molecules of RNA that can infect plants. Prions are infectious proteins that cause a number of degenerative brain diseases in humans and other animals.

MasteringBiology®

For practice quizzes, BioFlix animations, MP3 tutorials, video tutors, and more study tools designed for this textbook, go to MasteringBiology®

SELF-QUIZ

1. A molecule of DNA contains two polymer strands called _____, made by bonding together many monomers called _____.

2. Name the three parts of every nucleotide.

3. The backbone of DNA consists of
   a. nitrogenous bases.
   b. a repeating sugar-nucleotide-sugar-nucleotide pattern.
   c. a repeating sugar-phosphate-sugar-phosphate pattern.
   d. paired nucleotides.

4. A scientist inserts a radioactively labeled DNA molecule into a bacterium. The bacterium replicates this DNA molecule and distributes one daughter molecule (double helix) to each of two daughter cells. How much radioactivity will the DNA in each of the two daughter cells contain? Why?

5. Which mRNA nucleotide triplet encodes the amino acid tryptophan (see Figure 10.10)? During translation, an amino-acid-conjugated tRNA binds to an mRNA nucleotide triplet via its anticodon. What is the nucleotide sequence of the tryptophan's tRNA anticodon? What is the corresponding original codon on the DNA molecule that the mRNA is transcribed from? Are there any other amino acids that are encoded by a single codon?

6. Describe the process by which the information in a gene is transcribed and translated into a protein. Correctly use these terms in your description: tRNA, amino acid, start codon, transcription, mRNA, gene, codon, RNA polymerase, ribosome, translation, anticodon, peptide bond, stop codon.

7. Match the following molecules with the cellular process or processes in which they are primarily involved.

a. ribosomes
b. tRNA
c. DNA polymerases
d. RNA polymerase
e. mRNA

1. DNA replication
2. transcription
3. translation

8. A mutation within a gene that will insert a premature stop codon in mRNA would

a. result in a polypeptide that is one amino acid shorter than the one produced prior to the mutation.
b. result in a shortened polypeptide chain.
c. change the location at which transcription of the next gene begins.
d. have the same effect as deleting a single nucleotide in the gene.

9. Scientists have discovered how to put together a bacteriophage with the protein coat of phage A and the DNA of phage B. If this composite phage were allowed to infect a bacterium, the phages produced in the cell would have

a. the protein of A and the DNA of B.
b. the protein of B and the DNA of A.
c. the protein and DNA of A.
d. the protein and DNA of B.

10. How do some viruses reproduce without ever having DNA?

11. HIV requires an enzyme called _____ to convert its RNA genome to a DNA version. Why is this enzyme a particularly good target for anti-AIDS drugs? (*Hint*: Would you expect such a drug to harm the human host?)

Answers to these questions can be found in Appendix: Self-Quiz Answers.

THE PROCESS OF SCIENCE

12. In 1958, Matthew Meselson and Franklin Stahl grew bacteria in a medium enriched with the rare, heavy nitrogen isotope ^{15}N. The DNA extracted from these bacteria can be separated from the DNA extracted from the bacteria grown in normal, mostly ^{14}N-containing medium by density gradient ultracentrifugation—the two DNA fractions separate into distinct layers within the gradient. What did Meselson and Stahl observe when they transferred bacteria that were initially cultured in ^{15}N medium to ^{14}N medium just long enough for one more round of cell division? What does this experiment demonstrate?

13. In a classic 1952 experiment, biologists Alfred Hershey and Martha Chase labeled two batches of bacteriophages, one with radioactive sulfur (which only tags protein) and the other with radioactive phosphorus (which only tags DNA). In separate test tubes, they allowed each batch of phages to bind to nonradioactive bacteria and inject its DNA. After a few minutes, they separated the bacterial cells from the viral parts that remained outside the bacterial cells and measured the radioactivity of both portions. What results do you think they obtained? How would these results help them to determine which viral component—DNA or protein—was the infectious portion?

14. Interpreting Data The graph below summarizes the number of children who died of all strains of flu from 2007 until 2013. Each bar represents the number of child deaths occurring in one week. Why does the graph have the shape it does, with a series of peaks and valleys? Looking over the Biology and Society section at the start of the chapter, why does the graph reach its highest points near the middle? Based on these data, when does flu season begin and end in a typical year?

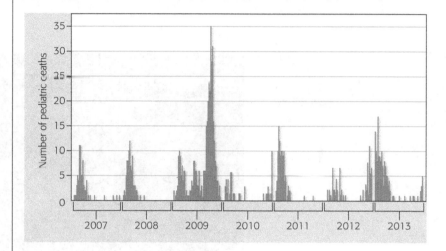

BIOLOGY AND SOCIETY

15. Scientists at the National Institutes of Health (NIH) have worked out thousands of sequences of genes and the proteins they encode, and similar analyses are being carried out at universities and private companies. Knowledge of the nucleotide sequences of genes might be used to treat genetic defects or produce lifesaving medicines. NIH and some U.S. biotechnology companies have applied for patents on their discoveries. In Britain, the courts have ruled that a naturally occurring gene cannot be patented. Do you think individuals and companies should be able to patent genes and gene products? Before answering, consider the following: What are the purposes of a patent? How might the discoverer of a gene benefit from a patent? How might the public benefit? What negative effects might result from patenting genes?

16. Your college roommate seeks to improve her appearance by visiting a tanning salon. How would you explain the dangers of this to her?

17. Flu vaccines have been shown to be safe, are very reliable at reducing the risk of hospitalization or death from influenza, and are inexpensive. Should children be required to obtain a flu vaccine before going to school? What about hospital workers before reporting to work? Defend your answers to these questions.

11 How Genes Are Controlled

Why Gene Regulation Matters

Someday a DNA chip might report on the activities of all your genes. ▶

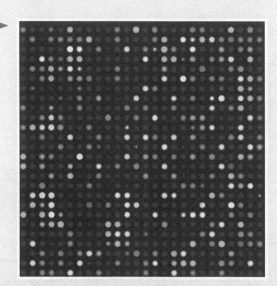

▼ Cloning may help save the giant panda from extinction.

▲ Lifestyle choices you make can dramatically affect your risk of cancer.

Cancer BIOLOGY AND SOCIETY

Tobacco's Smoking Gun

When European explorers returned from their first voyages to the Americas, they brought back tobacco, a common trade item among Native Americans. The novelty of smoking tobacco quickly spread through Europe. To keep up with demand, the southern United States soon became a major tobacco producer. Over time, smoking increased in popularity around the world, and by the 1950s, about half of all Americans smoked more than a pack of cigarettes each day. In these early days, little attention was paid to purported health risks; in fact, cigarette advertising often touted the "health benefits" of tobacco, claiming that use of the product had a soothing effect on the throat and helped the smoker remain calm and lose weight.

By the 1960s, however, doctors began to notice a disturbing trend: The rate of lung cancer had increased dramatically. Although the disease was rare in 1930, by 1955 it had become the deadliest form of cancer among American men. In fact, by 1990, lung cancer was killing more than twice as many men each year as any other type of cancer. But a few vocal skeptics, mostly tied to groups with an economic interest in the tobacco industry, doubted the link between smoking and cancer. They pointed out that the evidence was purely statistical or based on animal studies; no direct proof, they said, had been found that tobacco smoke causes cancer in humans.

The "smoking gun" of proof was found in 1996 when researchers added one component of tobacco smoke, called BPDE, to human lung cells growing in the lab. The researchers showed that BPDE binds to a gene within these cells called *p53*. That gene codes for a protein that helps suppress the formation of tumors. The researchers showed that BPDE causes mutations in the *p53* gene that deactivate the protein; with this important tumor-suppressor protein deactivated, tumors grow. This work directly linked a chemical in tobacco smoke to the formation of human lung tumors. Since that time, a mountain of experimental data and statistical studies has removed any scientific doubt of the link between smoking and cancer.

How can a mutation in a gene lead to cancer? It turns out that many cancer-associated genes encode proteins that turn other genes on or off. When these proteins malfunction, the cell may become cancerous. In fact, the ability to properly control which genes are active at any given time is crucial to normal cell function. How genes are controlled and how the regulation of genes affects cells and organisms—including ways this topic affects your own life—are the subjects of this chapter.

Colorized SEM 450×

Human cancer cells. These tumor cells have lost the ability to control their growth.

How and Why Genes Are Regulated

Every cell in your body—and, indeed, all the cells in the body of every sexually reproducing organism—was produced through successive rounds of mitosis starting from the zygote, the original cell that formed after fusion of sperm and egg. Mitosis exactly duplicates the chromosomes. Therefore, every cell in your body has the same DNA as the zygote. To put it another way: Every somatic (body) cell contains every gene. However, the cells in your body are specialized in structure and function; a neuron, for example, looks and acts nothing like a red blood cell. But if every cell contains identical genetic instructions, how do cells develop differently from one another? To help you understand this idea, imagine that every restaurant in your hometown uses the same cookbook. If that were the case, how could each restaurant develop a unique menu? The answer is obvious: Even though each restaurant has the same cookbook, different restaurants pick and choose different recipes from this book to prepare. Similarly, cells with the same genetic information can develop into different types of cells through **gene regulation**, mechanisms that turn on certain genes while other genes remain turned off. Regulating gene activity allows for specialization of cells within the body, just as regulating which recipes are used allows for varying menus in multiple restaurants.

As an example of gene regulation, consider the development of a unicellular zygote into a multicellular organism. During embryonic growth, groups of cells follow different paths, and each group becomes a particular kind of tissue. In the mature organism, each cell type—neuron or red blood cell, for instance—has a different pattern of turned-on genes.

What does it mean to say that genes are turned on or off? Genes determine the nucleotide sequence of specific mRNA molecules, and mRNA in turn determines the sequence of amino acids in proteins (in summary: DNA → RNA → protein; see Chapter 10). A gene that is turned on is being transcribed into mRNA, and that message is being translated into specific proteins. The overall process by which genetic information flows from genes to proteins is called **gene expression**.

As an illustration of this principle, **Figure 11.1** shows the patterns of gene expression for four genes in three different specialized cells of an adult human. Note that the genes for "housekeeping" enzymes, such as those that provide energy via glycolysis, are "on" in all the cells. In contrast, the genes for some proteins, such as insulin and hemoglobin, are expressed only by particular kinds of cells. One protein, hemoglobin, is not expressed in any of the cell types shown in the figure. ☑

Gene Regulation in Bacteria

To understand how a cell can regulate gene expression, consider the relatively simple case of bacteria. In the course of their lives, bacteria must regulate their genes in response to environmental changes. For example,

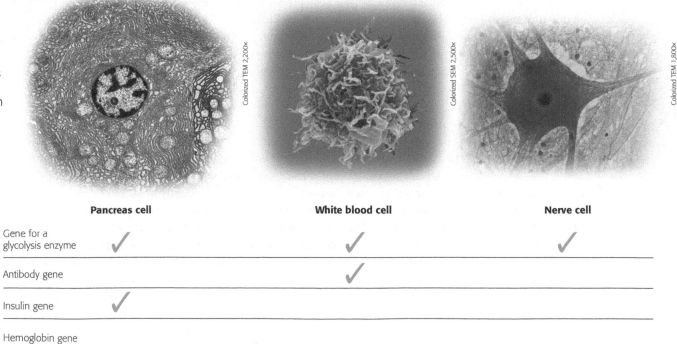

► Figure 11.1 **Patterns of gene expression in three types of human cells.** Different types of cells express different combinations of genes. The specialized proteins whose genes are represented here are an enzyme involved in glucose digestion; an antibody, which aids in fighting infection; insulin, a hormone made in the pancreas; and the oxygen transport protein hemoglobin, which is expressed only in red blood cells.

	Pancreas cell	White blood cell	Nerve cell
Gene for a glycolysis enzyme	✓	✓	✓
Antibody gene		✓	
Insulin gene	✓		
Hemoglobin gene			

Key

 = Active gene

when a nutrient is plentiful, bacteria do not squander valuable resources to make the nutrient from scratch. Bacterial cells that can conserve resources and energy have a survival advantage over cells that are unable to do so. Thus, natural selection has favored bacteria that express only the genes whose products are needed by the cell.

Imagine an *Escherichia coli* bacterium living in your intestines. It will be bathed in various nutrients, depending on what you eat. If you drink a milk shake, for example, there will be a sudden rush of the sugar lactose. In response, *E. coli* will express three genes for enzymes that enable the bacterium to absorb and digest this sugar. After the lactose is gone, these genes are turned off; the bacterium does not waste its energy continuing to produce these enzymes when they are not needed. Thus, a bacterium can adjust its gene expression to changes in the environment.

How does a bacterium "know" if lactose is present or not? In other words, how does the presence or absence of lactose influence the activity of the genes that code for the lactose enzymes? The key is the way the three lactose-digesting genes are organized: They are adjacent in the DNA and turned on and off as a single unit. This regulation is achieved through short stretches of DNA that help turn all three genes on and off at once, coordinating their expression. Such a cluster of related genes and sequences that control them is called an **operon** (Figure 11.2). The operon considered here, the *lac* (short for lactose) operon, illustrates principles of gene regulation that apply to a wide variety of prokaryotic genes.

How do DNA control sequences turn genes on or off? One control sequence, called a **promoter** (green in the figure), is the site where the enzyme RNA polymerase attaches and initiates transcription—in our example, transcription of the genes for lactose-digesting enzymes. Between the promoter and the enzyme genes, a DNA segment called an **operator** (yellow) acts as a switch that is turned on or off, depending on whether a specific protein is bound there. The operator and protein together determine whether RNA polymerase can attach to the promoter and start transcribing the genes (light blue). In the *lac* operon, when the operator switch is turned on, all the enzymes needed to metabolize lactose are made at once.

The top half of Figure 11.2 shows the *lac* operon in "off" mode, its status when there is no lactose available. Transcription is turned off because ❶ a protein called a **repressor** ( ) binds to the operator ( ) and ❷ physically blocks the attachment of RNA polymerase ( ) to the promoter ( ).

The bottom half of Figure 11.2 shows the operon in "on" mode, when lactose is present. The lactose ( )

interferes with attachment of the *lac* repressor to the operator by ❶ binding to the repressor and ❷ changing the repressor's shape. In its new shape ( ), the repressor cannot bind to the operator, and the operator switch remains on. ❸ RNA polymerase is no longer blocked, so it can now bind to the promoter and from there ❹ transcribe the genes for the lactose enzymes into mRNA. ❺ Translation produces all three lactose enzymes (purple).

Many operons have been identified in bacteria. Some are quite similar to the *lac* operon, whereas others have somewhat different mechanisms of control. For example, operons that control amino acid synthesis cause bacteria to stop making these molecules when they are already present in the environment, saving materials and energy for the cells. In these cases, the amino acid *activates* the repressor. Armed with a variety of operons, *E. coli* and other prokaryotes can thrive in frequently changing environments. ☑

☑ **CHECKPOINT**

A mutation in *E. coli* makes the *lac* operator unable to bind the active repressor. How would this mutation affect the cell? Why would this effect be a disadvantage?

Answer: *The cell would wastefully produce the enzymes for lactose metabolism continuously, even in the absence of lactose.*

▼ **Figure 11.2 The *lac* operon of *E. coli*.**

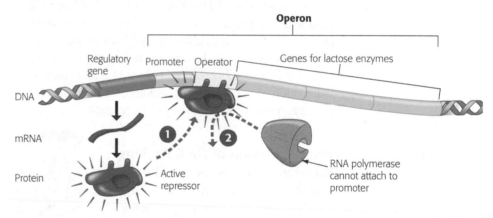

Operon turned off (lactose absent)

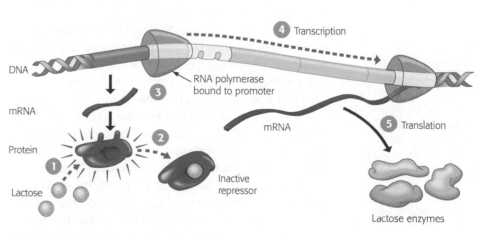

Operon turned on (lactose inactivates repressor)

Gene Regulation in Eukaryotic Cells

Eukaryotes, especially multicellular ones, have more sophisticated mechanisms than bacteria for regulating the expression of their genes. This is not surprising because a prokaryote, being a single cell, does not require the elaborate regulation of gene expression that leads to cell specialization in multicellular eukaryotic organisms. A bacterium does not have neurons that need to be different from blood cells, for example.

The pathway from gene to protein in eukaryotic cells is a long one, providing a number of points where the process can be turned on or off, speeded up or slowed down. Picture the series of pipes that carry water from your local reservoir to a faucet in your home. At various points, valves control the flow of water. We use this analogy in **Figure 11.3** to illustrate the flow of genetic information from a eukaryotic chromosome—a reservoir of genetic information—to an active protein that has been made in the cell's cytoplasm. The multiple mechanisms that control gene expression are analogous to the control valves in your water pipes. In the figure, each control knob indicates a gene expression "valve." All these knobs represent possible control points, although only one or a few control points are likely to be important for a typical protein.

Using a reduced version of Figure 11.3 as a guide, we will explore several ways that eukaryotes can control gene expression, starting within the nucleus.

The Regulation of DNA Packing

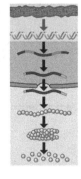

Eukaryotic chromosomes may be in a more or less condensed state, with the DNA and accompanying proteins more or less tightly wrapped together (see Figure 8.4). DNA packing tends to prevent gene expression by preventing RNA polymerase and other transcription proteins from binding to the DNA.

Cells may use DNA packing for the long-term inactivation of genes. One intriguing case is seen in female mammals, where one X chromosome in each somatic cell is highly compacted and almost entirely inactive. This **X chromosome inactivation** first takes place early in embryonic development, when one of the two X chromosomes in each cell is inactivated at random. After one X chromosome is inactivated in each embryonic cell, all of that cell's descendants will have the same X chromosome turned off. Consequently, if a female has different versions of a gene on each of her X chromosomes, about half of her cells will express one version, while the other half will express the alternate version **(Figure 11.4)**. ✓

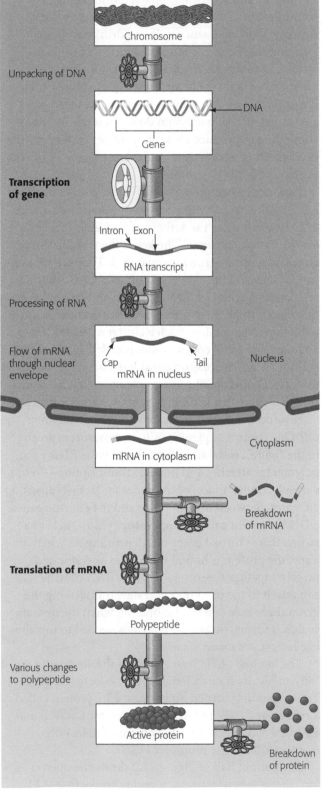

▼ **Figure 11.3 The gene expression "pipeline" in a eukaryotic cell.** Each valve in the pipeline represents a stage at which the pathway from chromosome to functioning protein can be regulated, turned on or off, or speeded up or slowed down. Throughout this discussion we will use a miniature version of this figure to keep track of the stages as they are discussed.

Chromosome

Unpacking of DNA

DNA

Gene

Transcription of gene

Intron | Exon

RNA transcript

Processing of RNA

Flow of mRNA through nuclear envelope

Cap | Tail

mRNA in nucleus

Nucleus

Cytoplasm

mRNA in cytoplasm

Breakdown of mRNA

Translation of mRNA

Polypeptide

Various changes to polypeptide

Active protein

Breakdown of protein

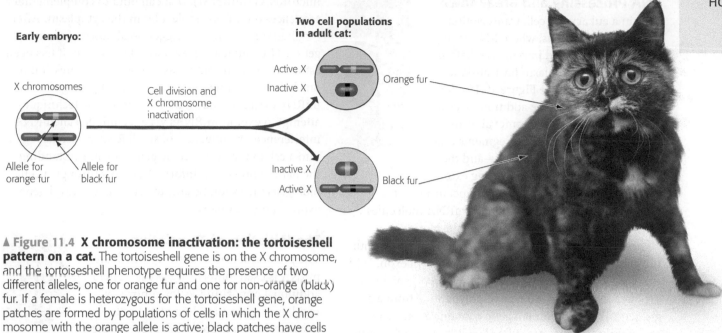

Early embryo:

X chromosomes

Allele for
orange fur

Allele for
black fur

Cell division and
X chromosome
inactivation

**Two cell populations
in adult cat:**

Active X

Inactive X

Orange fur

Inactive X

Active X

Black fur

▲ Figure 11.4 **X chromosome inactivation: the tortoiseshell pattern on a cat.** The tortoiseshell gene is on the X chromosome, and the tortoiseshell phenotype requires the presence of two different alleles, one for orange fur and one for non-orange (black) fur. If a female is heterozygous for the tortoiseshell gene, orange patches are formed by populations of cells in which the X chromosome with the orange allele is active; black patches have cells in which the X chromosome with the non-orange allele is active.

The Initiation of Transcription

The initiation of transcription (whether transcription starts or not) is the most important stage for regulating gene expression. In both prokaryotes and eukaryotes, regulatory proteins bind to DNA and turn the transcription of genes on and off. Unlike prokaryotic genes, however, most eukaryotic genes are not grouped into operons. Instead, each eukaryotic gene usually has its own promoter and other control sequences.

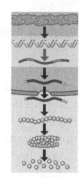

As illustrated in **Figure 11.5**, transcriptional regulation in eukaryotes is complex, typically involving many proteins (collectively called **transcription factors**; shown in purple in the figure) acting in concert to bind to DNA sequences called **enhancers** (yellow) and to the promoter (green). The DNA-protein assembly promotes the binding of RNA polymerase (orange) to the promoter. Genes coding for related enzymes, such as those in a metabolic pathway, may share a specific kind of enhancer (or collection of enhancers), allowing these genes to be activated at the same time. Not shown in the figure are repressor proteins, which may bind to DNA sequences called **silencers**, inhibiting the start of transcription.

In fact, repressor proteins that turn genes off are less common in eukaryotes than **activators**, proteins that turn genes on by binding to DNA. Activators act by making it easier for RNA polymerase to bind to the promoter. The use of activators is efficient because a typical animal or plant cell needs to turn on (transcribe) only a small percentage of its genes, those required for the cell's specialized structure and function. The "default" state for most genes in multicellular eukaryotes seems to be off, with the exception of "housekeeping" genes for routine activities such as the digestion of glucose. ☑

▼ Figure 11.5 **A model for turning on a eukaryotic gene.** A large assembly of transcription factors (proteins shown in purple) and several control sequences in the DNA are involved in initiating the transcription of a eukaryotic gene.

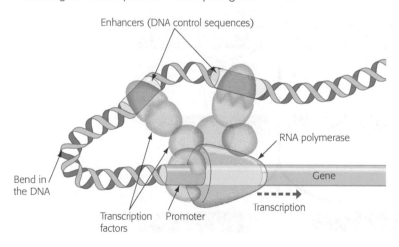

Enhancers (DNA control sequences)

RNA polymerase

Bend in
the DNA

Gene

Transcription
factors

Promoter

Transcription

RNA Processing and Breakdown

Within a eukaryotic cell, transcription occurs in the nucleus, where RNA transcripts are processed into mRNA before moving to the cytoplasm for translation by the ribosomes (see Figure 10.19). RNA processing includes the addition of a cap and a tail, as well as the removal of any introns—noncoding DNA segments that interrupt the genetic message—and the splicing together of the remaining exons.

Within a cell, exon splicing can occur in more than one way, generating different mRNA molecules from the same starting RNA molecule. Notice in **Figure 11.6**, for example, that one mRNA ends up with the green exon and the other with the brown exon. With this sort of **alternative RNA splicing**, an organism can produce more than one type of polypeptide from a single gene. A typical human gene contains about ten exons; nearly all genes are spliced in at least two different ways, and some are spliced hundreds of different ways.

After an mRNA is produced in its final form, its "lifetime" can be highly variable, from hours to weeks to months. Controlling the timing of mRNA breakdown provides another opportunity for regulation. But all mRNAs are eventually broken down and their parts recycled.

microRNAs

Recent research has established an important role for a variety of small single-stranded RNA molecules, called

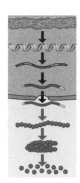

microRNAs (miRNAs), that can bind to complementary sequences on mRNA molecules in the cytoplasm. After binding, some miRNAs trigger breakdown of their target mRNA, whereas others block translation. It has been estimated that miRNAs may regulate the expression of one half of all human genes, a striking figure given that miRNAs were unknown 20 years ago. New techniques attempt to exploit miRNAs via a technique called RNA interference, the injection of small RNA molecules into a cell to turn off specific genes. By understanding the natural process of information flow through a cell, biologists may soon be able to artificially control gene expression in humans.

The Initiation of Translation

The process of translation—in which an mRNA is used to make a protein—offers additional opportunities for control by regulatory molecules. Red blood cells, for instance, have a protein that prevents the translation of hemoglobin mRNA unless the cell has a supply of heme, an iron-containing chemical group essential for hemoglobin function.

Protein Activation and Breakdown

The final opportunities for regulating gene expression occur after translation. For example, the hormone insulin is synthesized as one long, inactive polypeptide that must be chopped into pieces before it comes active **(Figure 11.7)**. Other proteins require chemical modification before they become active.

Another control mechanism operating after translation is the selective breakdown of proteins. Some proteins that trigger metabolic changes in cells are broken down within a few minutes or hours. This regulation allows a cell to adjust the kinds and amounts of its proteins in response to changes in its environment. ☑

☑ CHECKPOINT

After a gene is transcribed in the nucleus, how is the transcript modified to become mRNA? After the mRNA reaches the cytoplasm, what are four control mechanisms that can regulate the amount of active protein in the cell?

Answer: by RNA processing, including the addition of cap and tail and RNA splicing; control by microRNAs, initiation of translation, activation of the protein, and breakdown of the protein

▼ **Figure 11.6 Alternative RNA splicing: producing multiple mRNAs from the same gene.** Two different cells can use the same DNA gene to synthesize different mRNAs and proteins. In this example, one mRNA has ended up with exon 3 (brown) and the other with exon 4 (green). These mRNAs, which are just two of many possible outcomes, can then be translated into different proteins.

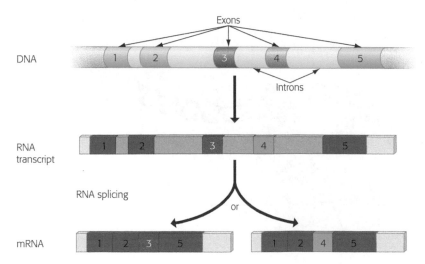

▼ **Figure 11.7 The formation of an active insulin molecule.** Only in its final form, with a central region removed, does insulin act as a hormone.

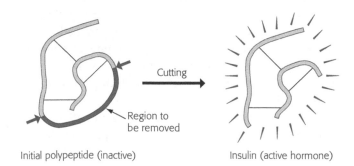

Initial polypeptide (inactive)   Insulin (active hormone)

Information Flow — Cell Signaling

The control of gene expression is a good example of one of biology's important themes: information flow. Through regulation, a cell can alter its activities in response to signals from the environment. So far, we have considered gene regulation only within a single cell. In a multicellular organism, the process can cross cell boundaries, allowing information to be communicated between and among cells. For example, a cell can produce and secrete chemicals, such as hormones, that affect gene regulation in another cell. Consider an analogy from your own experience: In grade school, did you ever station a classmate near the door to signal the teacher's return? Information from outside the room (the teacher's approach) was used to alter behavior within the classroom (stop messing around!). In a similar way, cells use protein "lookouts" to convey information into the cell, resulting in changes to cellular functions.

A signal molecule can act by binding to a receptor protein and initiating a **signal transduction pathway**, a series of molecular changes that converts a signal received outside a cell to a specific response inside the target cell. **Figure 11.8** shows an example of cell-to-cell signaling in which the target cell's response is the transcription (turning on) of a gene. ❶ First, the signaling cell secretes the signal molecule ( ). ❷ This molecule binds to a specific receptor protein ( ) embedded in the target cell's plasma membrane. ❸ The binding activates a signal transduction pathway consisting of a series of relay proteins (green) within the target cell. Each relay molecule activates the next. ❹ The last relay molecule in the series activates a transcription factor ( ) that ❺ triggers the transcription of a specific gene. ❻ Translation of the mRNA produces a protein that can then perform the function originally called for by the signal. ☑

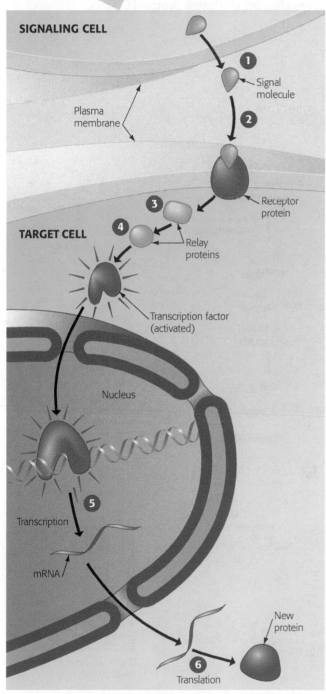

▲ **Figure 11.8 A cell-signaling pathway that turns on a gene.** The coordination of cellular activities in a multicellular organism depends on cell-to-cell signaling that helps regulate genes.

Homeotic Genes

Cell-to-cell signaling and the control of gene expression are particularly important during early embryonic development, when a single-celled zygote develops into a multicellular organism. Master control genes called **homeotic genes** regulate groups of other genes that determine what body parts will develop in which locations. For example, one set of homeotic genes in fruit flies instructs cells in the midbody to form legs. Elsewhere, these homeotic genes remain turned off, while others are turned on. Mutations in homeotic genes can produce bizarre effects. For example, fruit flies with mutations in homeotic genes may have extra sets of legs growing from their head (Figure 11.9).

▶ **Figure 11.9 The effect of homeotic genes.** The strange mutant fruit fly shown at the bottom results from a mutation in a homeotic (master control) gene.

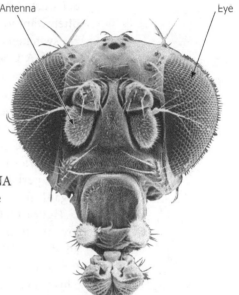

Normal head due to presence of normal homeotic gene

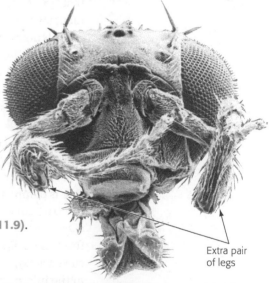

Head with extra legs growing due to presence of mutant homeotic gene

237

☑ **CHECKPOINT**

How can a mutation
in just one homeotic
gene drastically affect
an organism's physical
appearance?

Answer: Because homeotic genes
control many other genes, a single
change can affect the expression
of many of the proteins that control
appearance.

One of the most significant biological discoveries in recent years uncovered the fact that similar homeotic genes help direct embryonic development in nearly every eukaryotic organism examined so far, including yeasts, plants, earthworms, frogs, chickens, mice, and humans. These similarities suggest that these homeotic genes arose very early in the history of life and that the genes have remained remarkably unchanged over eons of animal evolution. ☑

Someday a DNA chip might report on the activities of all your genes.

which genes were being transcribed in the starting cells. Researchers can thus learn which genes are active in different tissues, at different times, or in tissues from individuals in different states of health. Such information may contribute to a better understanding of diseases and suggest new therapies. For example, comparing patterns of gene expression in breast cancer tumors with noncancerous breast tissue has resulted in more effective treatment protocols.

DNA Microarrays: Visualizing Gene Expression

Scientists who study gene regulation often want to determine which genes are switched on or off in a particular cell. A **DNA microarray** is a slide with thousands of different kinds of single-stranded DNA fragments attached in a tightly spaced array (grid). Each DNA fragment is obtained from a particular gene; a single microarray thus carries DNA from thousands of genes, perhaps even all the genes of an organism.

Figure 11.10 outlines how microarrays are used. ❶ A researcher collects all of the mRNA transcribed in a particular type of cell at a given moment. This collection of mRNA is mixed with reverse transcriptase, a viral enzyme that ❷ produces DNA that is complementary to each mRNA sequence. This **complementary DNA (cDNA)** is synthesized using nucleotides that have been modified to fluoresce (glow). The fluorescent cDNA collection thus represents all of the genes being actively transcribed in the cell. ❸ A small amount of the fluorescently labeled cDNA mixture is added to the DNA fragments of the microarray. If a molecule in the cDNA mixture is complementary to a DNA fragment at a particular location on the grid, the cDNA molecule binds to it, becoming fixed there. ❹ After unbound cDNA is rinsed away, the remaining cDNA glows in the microarray. The pattern of glowing spots enables the researcher to determine

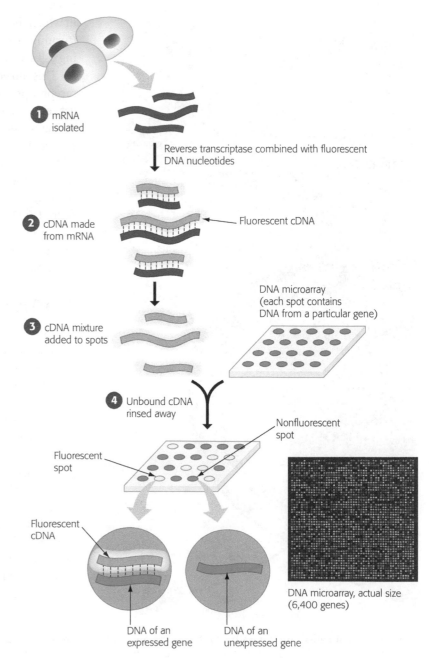

❶ mRNA isolated

Reverse transcriptase combined with fluorescent DNA nucleotides

❷ cDNA made from mRNA — Fluorescent cDNA

DNA microarray (each spot contains DNA from a particular gene)

❸ cDNA mixture added to spots

❹ Unbound cDNA rinsed away

Nonfluorescent spot

Fluorescent spot

Fluorescent cDNA

DNA of an expressed gene

DNA of an unexpressed gene

DNA microarray, actual size (6,400 genes)

▲ Figure 11.10 **Visualizing gene expression using a DNA microarray.**

Cloning Plants and Animals

Now that we have examined how gene expression is regulated, we will devote the rest of this chapter to discussing how gene regulation affects two important processes: cloning and cancer.

The Genetic Potential of Cells

One of the most important take-home lessons from this chapter is that all body cells contain a complete complement of genes, even if they are not expressing all of them. If you've ever grown a plant from a small cutting, you've seen evidence of this yourself: A single differentiated plant cell can undergo cell division and give rise to a complete adult plant. On a larger scale, the technique described in **Figure 11.11** can be used to produce hundreds or thousands of genetically identical organisms—clones—from the cells of a single plant.

Plant cloning is now used extensively in agriculture. For some plants, such as orchids, cloning is the only commercially practical means of reproducing plants. In other cases, cloning has been used to reproduce a plant with specific desirable traits, such as high fruit yield or resistance to disease. Seedless plants (such as seedless grapes, watermelons, and oranges) cannot reproduce sexually, leaving cloning as the sole means of mass-producing these common foods.

Is this sort of cloning possible in animals? A good indication that some animal cells can also tap into their full genetic potential is **regeneration**, the regrowth of lost body parts. When a salamander loses a tail, for example, certain cells in the tail stump reverse their differentiated state, divide, and then differentiate again to give rise to a new tail. Many other animals, especially among the invertebrates (sea stars and sponges, for example), can regenerate lost parts, and isolated pieces of a few relatively simple animals can dedifferentiate and then develop into an entirely new organism (see Figure 8.1).

▶ **Figure 11.11 Test-tube cloning of an orchid.** Tissue removed from the stem of an orchid plant and placed in growth medium may begin dividing and eventually grow into an adult plant. The new plant is a genetic duplicate of the parent plant. This process proves that mature plant cells can reverse their differentiation and develop into all the specialized cells of an adult plant.

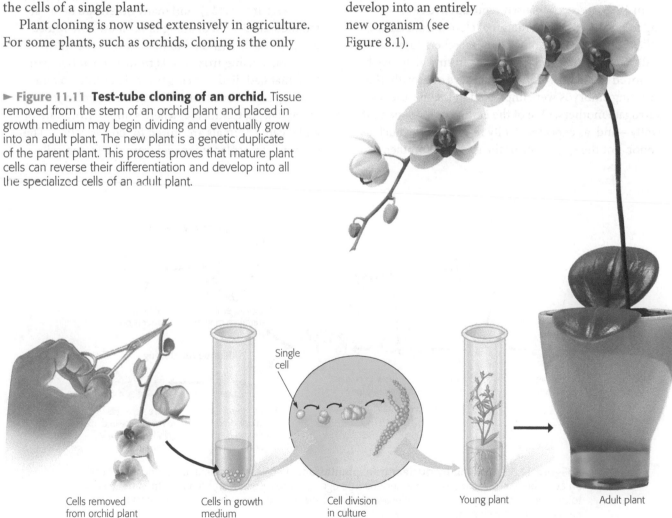

Cells removed from orchid plant

Cells in growth medium

Single cell

Cell division in culture

Young plant

Adult plant

Reproductive Cloning of Animals

Animal cloning is achieved through a procedure called **nuclear transplantation** (Figure 11.12). First performed in the 1950s on frog embryos and in the 1990s on adult mammals, nuclear transplantation involves replacing the nucleus of an egg cell or a zygote with a nucleus removed from an adult body cell. If properly stimulated, the recipient cell may then begin to divide. Repeated cell divisions form a hollow ball of about 100 cells. At this point, the cells may be used for different purposes, as indicated by the two branches in Figure 11.12.

If the animal to be cloned is a mammal, further development requires implanting the early embryo into the uterus of a surrogate mother (Figure 11.12, upper branch). The resulting animal will be a "clone" (genetic copy) of the donor. This type of cloning is called **reproductive cloning** because it results in the birth of a new animal.

In 1996, researchers used reproductive cloning to produce the first mammal cloned from an adult cell, a sheep named Dolly. The researchers fused specially treated sheep cells with eggs from which they had removed the nuclei. After several days of growth, the resulting embryos were implanted in the uteruses of surrogate mothers. One of the embryos developed into Dolly—and, as expected, Dolly resembled the nucleus donor, not the egg donor or the surrogate mother.

Cloning may help save the giant panda from extinction.

Practical Applications of Reproductive Cloning

Since the first success in 1996, researchers have cloned many species of mammals, including mice, horses, dogs, mules, cows, pigs, rabbits, ferrets, camels, goats, and cats (Figure 11.13a). Why would anyone want to do this? In agriculture, farm animals with specific sets of desirable traits might be cloned to produce identical herds. In research, genetically identical animals can provide perfect "control animals" for experiments. The pharmaceutical industry is experimenting with cloning animals for potential medical use (Figure 11.13b). For example, researchers have produced pig clones that lack a gene for a protein that can cause immune system rejection in humans. Organs from such pigs may one day be used in human patients who require life-saving transplants.

Perhaps the most intriguing use of reproductive cloning is to restock populations of endangered animals. Among the rare animals that have been cloned are a wild mouflon (a small European sheep), a gaur (an Asian ox), and gray wolves (Figure 11.13c), and many others are being attempted. In 2003, a banteng (a Javanese cow whose numbers have dwindled to just a few in the wild) was cloned using frozen cells from a zoo-raised banteng that had died 23 years prior. Scientists obtained banteng skin tissue from "The Frozen Zoo," a facility in San Diego, California, where samples from rare or endangered animals are stored for conservation. The scientists transplanted nuclei from the frozen cells

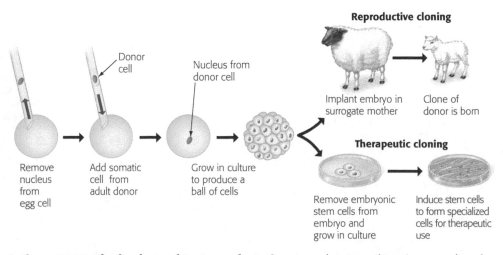

▲ **Figure 11.12 Cloning by nuclear transplantation.** In nuclear transplantation, a nucleus from an adult body cell is injected into a nucleus-free egg cell. The resulting embryo may then be used to produce a new organism (reproductive cloning, shown in the upper branch) or to provide stem cells (therapeutic cloning, lower branch).

into nucleus-free eggs from dairy cows. The resulting embryos were implanted into surrogate cows, leading to the birth of a healthy baby banteng. This success shows that it is possible to produce a baby even when a female of the donor species is unavailable. Scientists may someday be able to use similar cross-species methods to clone an animal from a recently extinct species.

The use of cloning to repopulate endangered species holds tremendous promise. However, cloning may also create new problems. Conservationists argue that cloning may detract from efforts to preserve natural habitats. They correctly point out that cloning does not increase genetic diversity and is therefore not as beneficial to endangered species as natural reproduction. In addition, an increasing body of evidence suggests that cloned animals are less healthy than animals produced via fertilization: Many cloned animals exhibit defects such as susceptibility to obesity, pneumonia, liver failure, and premature death. Dolly the cloned

sheep, for example, was euthanized in 2003 after suffering complications from a lung disease usually seen only in much older sheep. She was 6 years old, while her breed has a life expectancy of 12 years. Some evidence suggests that chromosomal changes in the cloned animals are the cause, but the effects of cloning on animal health are still being investigated.

Human Cloning

The cloning of various mammals has heightened speculation that humans could be cloned. Critics point out the many practical and ethical objections to human cloning. Practically, cloning of mammals is extremely difficult and inefficient. Only a small percentage of cloned embryos (usually less than 10%) develop normally, and they appear less healthy than naturally born kin. Ethically, the discussion about whether or not people should be cloned—and if so, under what circumstances—is far from settled. Meanwhile, the research and the debate continue. ☑

☑ **CHECKPOINT**

Imagine that mouse coat color is always passed down from parent to offspring. Suppose a nucleus from an adult body cell of a black mouse is injected into an egg removed from a white mouse, and then the embryo is implanted into a brown mouse. What would be the color of the resulting cloned mice?

Answer: black, the color of the nucleus donor

▼ Figure 11.13 **Reproductive cloning of mammals.**

(a) **The first clone.** Dolly the sheep, shown in 1996 with her lone parent, was the first mammal cloned from an adult cell.

(b) **Cloning for medical use.** These piglets are clones of a pig that was genetically modified to lack a protein that causes transplant rejection in humans.

(c) **Clones of endangered animals**

Mouflon lamb with mother

Banteng

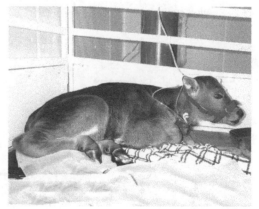

Gaur

Therapeutic Cloning and Stem Cells

The lower branch of Figure 11.12 shows the process of **therapeutic cloning**. The purpose of this procedure is not to produce a living organism but rather to produce embryonic stem cells.

Embryonic Stem Cells

In mammals, **embryonic stem cells (ES cells)** are obtained by removing cells from an early embryo and growing them in laboratory culture. Embryonic stem cells can divide indefinitely, and under the right conditions—such as the presence of certain growth-stimulating proteins—can (hypothetically) develop into a wide variety of different specialized cells (**Figure 11.14**). If scientists can discover the right conditions, they may be able to grow cells for the repair of injured or diseased organs. Some people speculate, for example, that ES cells may one day be used to replace cells damaged by spinal cord injuries or heart attacks. The use of embryonic stem cells in therapeutic cloning is controversial, however, because the removal of ES cells destroys the embryo.

Umbilical Cord Blood Banking

Another source of stem cells is blood collected from the umbilical cord and placenta at birth (**Figure 11.15**). Such stem cells appear to be partially differentiated. In 2005, doctors reported that an infusion of umbilical cord blood stem cells appeared to cure some babies of Krabbe disease, a fatal inherited disorder of the nervous system. Other people have received cord blood as a treatment for leukemia. To date, however, most attempts at umbilical cord blood therapy have not been successful. At present, the American Academy of Pediatrics recommends cord blood banking only for babies born into families with a known genetic risk.

Adult Stem Cells

Embryonic stem cells are not the only stem cells available to researchers. **Adult stem cells** can also generate replacements for some of the body's cells. Adult stem cells are further along the road to differentiation than ES cells and can therefore give rise to only a few related types of specialized cells. For example, stem cells in bone marrow generate different kinds of blood cells. Adult stem cells from donor bone marrow have long been used as a source of immune system cells in patients whose own immune systems have been destroyed by disease or cancer treatments.

Because no embryonic tissue is involved in their harvest, adult stem cells are less ethically problematic than ES cells. However, many researchers hypothesize that only the more versatile ES cells are likely to lead to groundbreaking advances in human health. Recent research has shown that some adult cells, such as human skin cells, may be reprogrammed to act like ES cells. In the near future, such cells may prove to be both therapeutically useful and ethically clear. ☑

☑ **CHECKPOINT**

How do the results of reproductive cloning and therapeutic cloning differ?

Answer: Reproductive cloning results in the production of a live individual; therapeutic cloning produces stem cells.

▼ Figure 11.14 **Differentiation of embryonic stem cells in culture.** Scientists hope to someday discover growth conditions that will stimulate cultured stem cells to differentiate into specialized cells.

▼ Figure 11.15 **Umbilical cord blood banking.** Just after birth, a doctor inserts a needle into the umbilical cord and extracts ¼ to ½ cup of blood. The umbilical cord blood (inset), rich in stem cells, is frozen and kept in a blood bank, where it is available if needed for medical treatment.

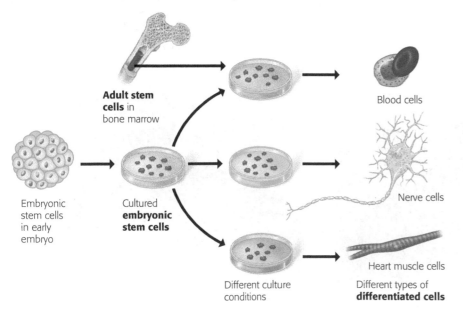

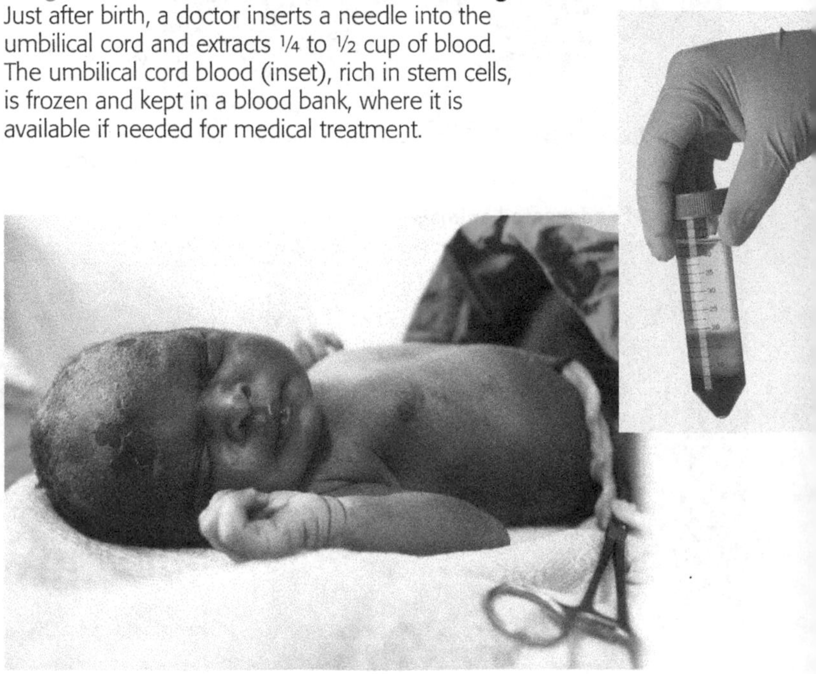

Adult stem cells in bone marrow

Blood cells

Nerve cells

Heart muscle cells

Embryonic stem cells in early embryo

Cultured **embryonic stem cells**

Different culture conditions

Different types of **differentiated cells**

The Genetic Basis of Cancer

Cancer includes a variety of diseases in which cells escape from the control mechanisms that normally limit their growth and division (as introduced in Chapter 8). This escape involves changes in gene expression.

Genes That Cause Cancer

One of the earliest clues to the role of genes in cancer was the discovery in 1911 of a virus that causes cancer in chickens. Viruses that cause cancer can become permanent residents in host cells by inserting their nucleic acid into the DNA of host chromosomes. Over the last century, researchers have identified a number of viruses that harbor cancer-causing genes. One example is the human papillomavirus (HPV), which can be transmitted through sexual contact and is associated with several types of cancer, including cervical cancer.

Oncogenes and Tumor-Suppressor Genes

In 1976, American molecular biologists J. Michael Bishop, Harold Varmus, and their colleagues made a startling discovery. They found that a cancer-causing chicken virus contains a cancer-causing gene that is an altered version of a normal chicken gene. A gene that causes cancer is called an **oncogene** ("tumor gene"). Subsequent research has shown that the chromosomes of many animals, including humans, contain genes that can be converted to oncogenes. A normal gene with the potential to become an oncogene is called a **proto-oncogene**. (These terms can be confusing, so let's repeat them: A *proto-oncogene* is a normal, healthy gene that, if changed, can become a cancer-causing *oncogene*.) A cell can acquire an oncogene from a virus or from the mutation of one of its own proto-oncogenes.

How can a change in a gene cause cancer? Searching for the normal roles of proto-oncogenes in the cell, researchers found that many of these genes code for **growth factors**—proteins that stimulate cell division—or for other proteins that affect the cell cycle. When all these proteins are functioning normally, in the right amounts at the right times, they help keep the rate of cell division at an appropriate level. When they malfunction—if a growth factor becomes hyperactive, for example—cancer (uncontrolled cell growth) may result.

For a proto-oncogene to become an oncogene, a mutation must occur in the cell's DNA. **Figure 11.16** illustrates three kinds of changes in DNA that can produce active oncogenes. In all three cases, abnormal gene expression stimulates the cell to divide excessively.

Changes in genes whose products inhibit cell division are also involved in cancer. These genes are called **tumor-suppressor genes** because the proteins they encode normally help prevent uncontrolled cell growth **(Figure 11.17)**. Any mutation that keeps a growth-inhibiting protein from being made or from functioning may contribute to the development of cancer. Researchers have identified many mutations in both tumor-suppressor and growth factor genes that are associated with cancer, as we'll discuss next.

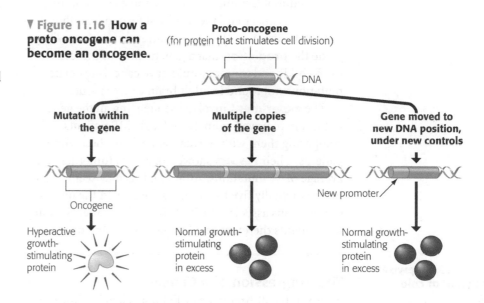

▼ **Figure 11.16 How a proto-oncogene can become an oncogene.**

Proto-oncogene
(for protein that stimulates cell division)

DNA

Mutation within the gene

Oncogene

Hyperactive growth-stimulating protein

Multiple copies of the gene

Normal growth-stimulating protein in excess

Gene moved to new DNA position, under new controls

New promoter

Normal growth-stimulating protein in excess

▼ **Figure 11.17 Tumor-suppressor genes.**

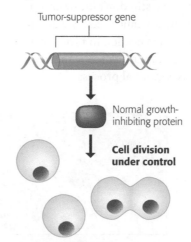

Tumor-suppressor gene

Normal growth-inhibiting protein

Cell division under control

(a) Normal cell growth. A tumor-suppressor gene normally codes for a protein that inhibits cell growth and division. Such genes help prevent cancerous tumors from arising or spreading.

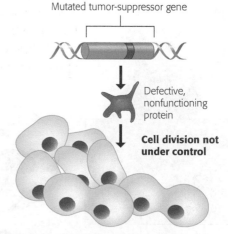

Mutated tumor-suppressor gene

Defective, nonfunctioning protein

Cell division not under control

(b) Uncontrolled cell growth (cancer). When a mutation in a tumor-suppressor gene makes its protein defective, cells that are usually under the control of the normal protein may divide excessively, forming a tumor.

Cancer THE PROCESS OF SCIENCE

Are Childhood Tumors Different?

Medical researchers have made many **observations** of specific mutations that can lead to cancer. They therefore **question** whether different kinds of cancers are associated with specific mutations. A large research team led by the Johns Hopkins Kimmel Cancer Center in Baltimore formed the **hypothesis** that young patients with medulloblastoma (MB)—the most common pediatric brain cancer and the deadliest form of childhood cancer (Figure 11.18)—harbor unique mutations. They made the **prediction** that a genetic map of MB cells from childhood tumors would have cancer-associated mutations not found in adult brain cancer tissue.

The **experiment** involved sequencing all the genes in tumors removed from 22 pediatric MB patients and comparing them with normal tissue from these same patients. Their **results** showed that each tumor had an average of 11 mutations. Although this may seem like a lot, it is actually five to ten times fewer than the number of mutations associated with MB in adult patients. Young MB patients therefore seem to have fewer, but deadlier,

mutations. When they investigated the role that the mutated genes play, the research team found that some help control DNA packing, whereas others play a role in the development of organs. The researchers hope that this new knowledge about the genetic basis of MB may be used to develop new therapies for this often fatal disease.

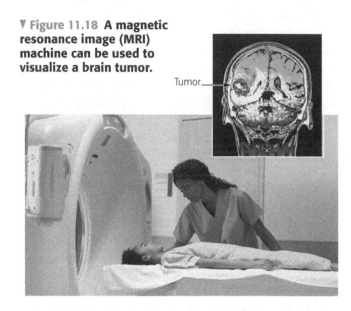

▼ Figure 11.18 **A magnetic resonance image (MRI) machine can be used to visualize a brain tumor.**

Tumor

The Progression of a Cancer

Nearly 150,000 Americans will be stricken by cancer of the colon (the main part of the large intestine) this year. One of the best-understood types of human cancer, colon cancer illustrates an important principle about how cancer develops: Although we still don't know exactly how a particular cell becomes a cancer cell, we do know that more than one mutation is needed to produce a full-fledged cancer cell. As in many cancers, the development of colon cancer is a gradual process.

As shown in **Figure 11.19, ❶** colon cancer begins when an oncogene arises through mutation, causing unusually frequent division of normal-looking cells in the colon lining. ❷ Later, additional DNA mutations (such as the inactivation of a tumor-suppressor gene) cause the growth of a small benign tumor (called a polyp) in the colon wall. The cells of the polyp look normal, although they divide unusually frequently. If detected during a colonoscopy, suspicious polyps can usually be removed before they become a serious risk. ❸ Further mutations

▼ Figure 11.19 **Stepwise development of colon cancer.**

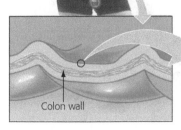

Colon wall

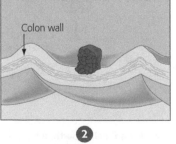

Colon wall

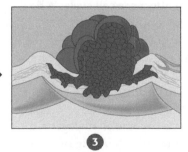

	❶	❷	❸
Cellular changes:	Increased cell division	Growth of benign tumor	Growth of malignant tumor
DNA changes:	Oncogene activated	Tumor-suppressor gene inactivated	Second tumor-suppressor gene inactivated

eventually lead to formation of a malignant tumor—a tumor that has the potential to metastasize (spread). It typically takes at least six DNA mutations (usually creating at least one active oncogene and disabling at least one tumor-suppressor gene) before a cell becomes fully cancerous.

The development of a malignant tumor is accompanied by a gradual accumulation of mutations that convert proto-oncogenes to oncogenes and knock out tumor-suppressor genes (Figure 11.20). The requirement for several DNA mutations—usually four or more—explains why cancers can take a long time to develop. This requirement may also help explain why the incidence of cancer increases with age; the longer we live, the more likely we are to accumulate mutations that cause cancer.

Inherited Cancer

The fact that multiple genetic changes are required to produce a cancer cell helps explain the observation that cancers can run in families. An individual inheriting an oncogene or a mutant version of a tumor-suppressor gene is one step closer to accumulating the necessary mutations for cancer to develop than is an individual without any such mutations. Geneticists are therefore devoting much effort to identifying inherited cancer mutations so that predisposition to certain cancers can be detected early in life.

About 15% of colorectal cancers, for example, involve inherited mutations. There is also evidence that inheritance plays a role in 5–10% of patients with breast cancer, a disease that strikes one out of every ten American women (Figure 11.21). Mutations in either or both of two genes—called *BRCA1* (pronounced "braca-1") and *BRCA2*—are found in at least half of inherited breast cancers. Both *BRCA1* and *BRCA2* are considered tumor-suppressor genes because the normal versions protect against breast cancer. A woman who inherits one mutant

▲ Figure 11.21 **Breast cancer.** In 2013, at age 37, the actress Angelina Jolie underwent a preventive double mastectomy after learning she had a mutant *BRCA1* gene. Jolie's mother, grandmother, and aunt all died young from breast or ovarian cancer.

BRCA1 allele has a 60% probability of developing breast cancer before the age of 50, compared with only a 2% probability for an individual lacking the mutations. Tests using DNA sequencing can now detect these mutations. Unfortunately, these tests are of limited use because surgical removal of the breasts and/or ovaries is the only preventive option currently available to women who carry the mutant genes. ☑

☑ CHECKPOINT

How can a mutation in a tumor-suppressor gene contribute to the development of cancer?

Answer: A mutated tumor-suppressor gene may produce a defective protein that normally inhibits cell division and therefore normally suppresses tumors.

▼ Figure 11.20 **Accumulation of mutations in the development of a cancer cell.**
Mutations leading to cancer accumulate in a lineage of cells. In this figure, colors distinguish the normal cells from cells with one or more mutations, leading to increased cell division and cancer. Once a cancer-promoting mutation occurs (orange band on chromosome), it is passed to all the descendants of the cell carrying it.

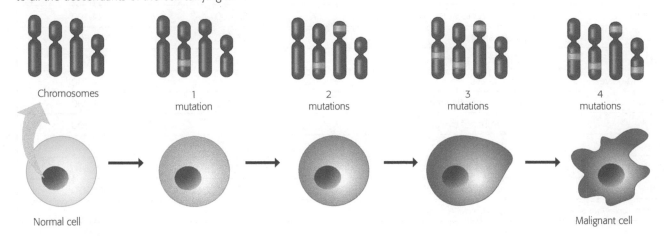

Chromosomes   1 mutation   2 mutations   3 mutations   4 mutations

Normal cell   Malignant cell

Cancer Risk and Prevention

Cancer is the second-leading cause of death (after heart disease) in most industrialized countries. Death rates due to certain forms of cancer have decreased in recent years, but the overall cancer death rate is still on the rise, currently increasing at about 1% per decade.

Although some cancers occur spontaneously, most cancers arise from mutations that are caused by **carcinogens**, cancer-causing agents found in the environment. Mutations often result from decades of exposure to carcinogens. One of the most potent carcinogens is ultraviolet (UV) radiation. Excessive exposure to UV radiation from the sun can cause skin cancer, including a deadly type called melanoma. You can decrease your risk by using sun protection (clothing, lotion, hats, etc.).

The one substance known to cause more cases and types of cancer than any other is tobacco. By a wide margin, more people die from lung cancer (nearly 160,000 Americans in 2014) than from any other form of cancer. Most tobacco-related cancers are due to smoking cigarettes, but smoking cigars, inhaling secondhand smoke, and smokeless tobacco also pose risks. As **Table 11.1** indicates, tobacco use, sometimes in combination with alcohol consumption, causes several types of cancer. Exposure to some of the most lethal carcinogens is often a matter of individual choice: Tobacco use, the consumption of alcohol, and excessive time spent in the sun are all avoidable behaviors that affect cancer risk.

Some food choices significantly reduce a person's odds of developing cancer. For instance, eating 20–30 g of plant fiber daily (about the amount found in seven apples), while eating less animal fat may help prevent colon cancer. There is also evidence that certain substances in fruits and vegetables, including vitamins C and E and certain compounds related to vitamin A, may help protect against a variety of cancers. Cabbage and its relatives, such as broccoli and cauliflower, are thought to be especially rich in substances that help prevent cancer, although some of the specific substances have not yet been identified. Determining how diet influences cancer has become an important focus of nutrition research.

The battle against cancer is being waged on many fronts, and there is reason for optimism in the progress being made. It is especially encouraging that we can help reduce our risk of acquiring some of the most common forms of cancer by the choices we make in our daily lives. ☑

> Lifestyle choices you make can dramatically affect your risk of cancer.

☑ CHECKPOINT

Of all known behavioral factors, which one causes the most cancer cases and deaths?

Answer: tobacco use

Table 11.1	Cancer in the United States (Ranked by Number of Cases)		
Cancer	**Known or Likely Carcinogens or Factors**	**Estimated Cases (2014)**	**Estimated Deaths (2014)**
Breast	Estrogen; possibly dietary fat	235,000	40,400
Prostate	Testosterone; possibly dietary fat	233,000	29,500
Lung	Cigarette smoke	224,000	159,000
Colon and rectum	High dietary fat; low dietary fiber	136,800	50,310
Skin	Ultraviolet light	81,200	13,000
Lymphomas	Viruses (for some types)	80,000	20,100
Bladder	Cigarette smoke	74,700	15,600
Uterus	Estrogen	65,000	12,600
Kidney	Cigarette smoke	63,900	13,900
Leukemias	X-rays; benzene; viruses (for some types)	52,400	24,100
Pancreas	Cigarette smoke	46,400	39,600
Liver	Alcohol; hepatitis viruses	33,200	23,000
Brain and nerve	Trauma; X-rays	23,400	14,300
Stomach	Table salt; cigarette smoke	22,200	11,000
Ovary	Large number of ovulation cycles	22,000	14,300
Cervix	Viruses; cigarette smoke	12,400	4,000
All other types		259,940	101,010
Total		1,665,540	585,720

Data from: Cancer Facts and Figures 2014 (American Cancer Society Inc.).

The Evolution of Cancer in the Body

The theory of evolution describes natural selection acting on populations. Recently, medical researchers have been using an evolutionary perspective to gain insight into the development of tumors, such as the bone tumor shown in **Figure 11.22**. Evolution drives the growth of a tumor—which can be thought of as a population of cancer cells—and also affects how those cells respond to cancer treatments.

Recall that there are several assumptions behind Darwin's theory of natural selection (see Chapter 1). Let's consider how each one can be applied to cancer. First, all evolving populations have the potential to produce more offspring than can be supported by the environment. Cancer cells, with their uncontrolled growth, clearly demonstrate overproduction. Second, there must be variation among individuals of

the population. Studies of tumor cell DNA, like the one described in the Process of Science section, show genetic variability within tumors. Third, variations in the population must affect survival and reproductive success. Indeed, the accumulation of mutations in cancer cells renders them less susceptible to normal mechanisms of reproductive control. Mutations that enhance survival of malignant cancer cells are passed on to that cell's descendants. In short, a tumor evolves.

Viewing the progression of cancer through the lens of evolution helps explain why there is no easy "cure" for cancer but may also pave the way for novel therapies. For example, some researchers are attempting to "prime" tumors for treatment by increasing the reproductive success of only those cells that will be susceptible to a chemotherapy drug. Our understanding of cancer, like all other aspects of biology, benefits from an evolutionary perspective.

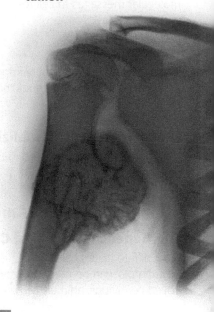

▼ **Figure 11.22 X-ray of shoulder and upper arm, revealing a large bone tumor.**

Chapter Review

■ SUMMARY OF KEY CONCEPTS

How and Why Genes Are Regulated

The various types of cells in a multicellular organism owe their distinctiveness to different combinations of genes being turned on and off via gene regulation in each cell type.

Gene Regulation in Bacteria

An operon is a cluster of genes with related functions together with their promoter and other DNA sequences that control their transcription. For example, the *lac* operon allows *E. coli* to produce enzymes for lactose use only when the sugar is present.

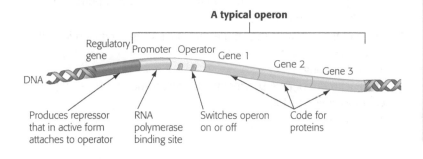

A typical operon

DNA

Regulatory gene — Produces repressor that in active form attaches to operator

Promoter — RNA polymerase binding site

Operator — Switches operon on or off

Gene 1 / Gene 2 / Gene 3 — Code for proteins

Gene Regulation in Eukaryotic Cells

In the nucleus of eukaryotic cells, there are several possible control points in the pathway of gene expression.

- DNA packing tends to block gene expression by preventing access of transcription proteins to the DNA. An extreme example is X chromosome inactivation in the cells of female mammals.

- The most important control point in both eukaryotes and prokaryotes is at gene transcription. Various regulatory proteins interact with DNA and with each other to turn the transcription of eukaryotic genes on or off.

- There are also opportunities for the control of eukaryotic gene expression after transcription, when introns are cut out of the RNA and a cap and tail are added to process RNA transcripts into mRNA.

- In the cytoplasm, presence of microRNAs may block the translation of an mRNA, and various proteins may regulate the start of translation.

- Finally, the cell may activate the finished protein in various ways (for instance, by cutting out portions or chemical modification). Eventually, the protein may be selectively broken down.

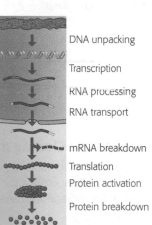

DNA unpacking
Transcription
RNA processing
RNA transport
mRNA breakdown
Translation
Protein activation
Protein breakdown

Information Flow: Cell Signaling

Cell-to-cell signaling is key to the development and function-ing of multicellular organisms. Signal transduction pathways convert molecular messages to cell responses, such as the transcription of particular genes.

Homeotic Genes

Evidence for the evolutionary importance of gene regulation is apparent in homeotic genes, master genes that regulate other genes that in turn control embryonic development.

DNA Microarrays: Visualizing Gene Expression

DNA microarrays can be used to determine which genes are turned on in a particular cell type.

Cloning Plants and Animals

The Genetic Potential of Cells

Most differentiated cells retain a complete set of genes, so an orchid plant, for example, can be made to grow from a single orchid cell. Under con-trolled conditions, animals can also be cloned.

Reproductive Cloning of Animals

Nuclear transplantation is a procedure whereby a donor cell nucleus is inserted into an egg from which the nucleus has been removed. First demonstrated in frogs in the 1950s, reproductive cloning was used in 1996 to clone a sheep from an adult cell and has since been used to create many other cloned animals.

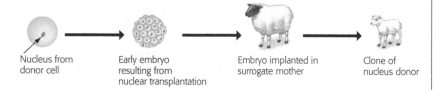

Nucleus from donor cell → Early embryo resulting from nuclear transplantation → Embryo implanted in surrogate mother → Clone of nucleus donor

Therapeutic Cloning and Stem Cells

The purpose of therapeutic cloning is to produce embryonic stem cells for medical uses. Embryonic, umbilical cord, and adult stem cells all show promise for therapeutic uses.

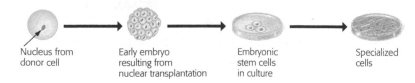

Nucleus from donor cell → Early embryo resulting from nuclear transplantation → Embryonic stem cells in culture → Specialized cells

The Genetic Basis of Cancer

Genes That Cause Cancer

Cancer cells, which divide uncontrollably, can result from mutations in genes whose protein products regulate the cell cycle.

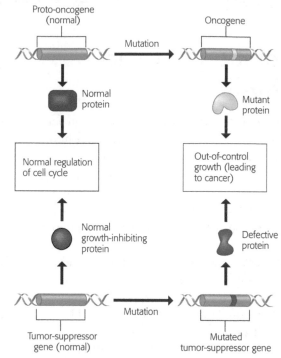

Proto-oncogene (normal) — Mutation → Oncogene

Normal protein / Mutant protein

Normal regulation of cell cycle / Out-of-control growth (leading to cancer)

Normal growth-inhibiting protein / Defective protein

Tumor-suppressor gene (normal) — Mutation → Mutated tumor-suppressor gene

Many proto-oncogenes and tumor-suppressor genes code for proteins active in signal transduction pathways regulating cell division. Mutations of these genes cause malfunction of the pathways. Cancer results from a series of genetic changes in a cell lineage. Researchers have identified many genes that, when mutated, promote the development of cancer.

Cancer Risk and Prevention

Reducing exposure to carcinogens (which induce cancer-causing mutations) and making other healthful lifestyle choices can help reduce cancer risk.

MasteringBiology®

For practice quizzes, BioFlix animations, MP3 tutorials, video tutors, and more study tools designed for this textbook, go to MasteringBiology®

SELF-QUIZ

1. Your bone cells, muscle cells, and skin cells look different because
   a. different kinds of genes are present in each kind of cell.
   b. they are present in different organs.
   c. different genes are active in each kind of cell.
   d. different mutations have occurred in each kind of cell.

2. A group of prokaryotic genes with related functions that are regulated as a single unit, along with the control sequences that perform this regulation, is called a(n) _____.

3. Which of these plays a role in the regulation of transcription in both prokaryotic and eukaryotic cells?
   a. RNA splicing
   b. attachment of RNA polymerase to the promoter
   c. transcription factors
   d. gene operons

4. A eukaryotic gene was inserted into the DNA of a bacterium. The bacterium then transcribed this gene into mRNA and translated the mRNA into protein. The protein produced was useless and contained many more amino acids than the protein made by the eukaryotic cell. Why?
   a. The mRNA was not spliced as it is in eukaryotes.
   b. Eukaryotes and prokaryotes use different genetic codes.
   c. Repressor proteins interfered with transcription and translation.
   d. Ribosomes were not able to bind to tRNA.

5. How is it that the cells in different body tissues are able to perform different functions?

6. Is the genetic information in an animal cloned by nuclear transplantation identical to that of the nuclear donor?

7. The most common procedure for cloning an animal is _____.

8. What are the possible uses of reproductive cloning?

9. Which of the following is a substantial difference between embryonic stem cells and the stem cells found in adult tissues?
   a. In laboratory culture, only adult stem cells are immortal.
   b. In nature, only embryonic stem cells give rise to all the different types of cells in the organism.
   c. Only adult stem cells can be made to differentiate in the laboratory.
   d. Only embryonic stem cells are in every tissue of the adult body.

10. Name three potential sources of stem cells.

11. What is the difference between oncogenes and proto-oncogenes? How can one turn into the other? What function do proto-oncogenes serve?

12. A mutation in a single gene may cause a major change in the body of a fruit fly, such as an extra pair of legs or wings. Yet it takes many genes to produce a wing or leg. How can a change in just one gene cause such a big change in the body? What are such genes called?

Answers to these questions can be found in Appendix: Self-Quiz Answers.

THE PROCESS OF SCIENCE

13. Study the depiction of the *lac* operon in Figure 11.2. Normally, the genes are turned off when lactose is not present. Lactose activates the genes, which code for enzymes that enable the cell to use lactose. Mutations can alter the function of this operon. Predict how the following mutations would affect the function of the operon in the presence and absence of lactose:
   a. mutation of regulatory gene; repressor will not bind to lactose
   b. mutation of operator; repressor will not bind to operator
   c. mutation of regulatory gene; repressor will not bind to operator
   d. mutation of promoter; RNA polymerase will not attach to promoter

14. The human body has a far greater variety of proteins than genes, a fact that seems to highlight the importance of alternative RNA splicing, which allows several different mRNAs to be made from a single gene.

Suppose you have samples of two types of adult cells from one person. Design an experiment using microarrays to determine whether or not the different gene expression is due to alternative RNA splicing.

15. Because a cat must have both orange and non-orange alleles to be tortoiseshell (see Figure 11.4), we would expect only female cats, which have two X chromosomes, to be tortoiseshell. Normal male cats (XY) can carry only one of the two alleles. Male tortoiseshell cats are rare and usually sterile. What might you guess their genotype to be?

16. Design two complementary experimental approaches that test whether the *Antennapedia* gene is a homeotic (master control) gene of leg development in the fruit fly.

17. **Interpreting Data** Review Table 11.1, which lists the number of diagnoses and the number of deaths for many types of cancer. We can estimate the deadliness of each type of cancer by dividing the number of deaths by the number of cases. (This is not accurate, however, because someone diagnosed during a particular year will not necessarily die during that same year, but it is a useful approximation.) If nearly everyone who received a diagnosis of one particular type of cancer dies, that ratio will be near 1 (or 100% deadly). If many more people receive diagnoses than die, the ratio will be near 0 (or near 0% deadly). Using this criteria, which two forms of cancer listed are the deadliest? The least deadly? What is the overall ratio for all types of cancer? What does this tell you about cancer survivability in general?

BIOLOGY AND SOCIETY

18. The possibility of cloning humans is a highly controversial topic. Experts warn of the low success rates of cloning mammals, potential health problems, and a potentially shortened life span. However, there are also advocates of human cloning who point out the possibility of eliminating defective genes and of generating specific cells from a patient for therapeutic purposes. What are your opinions on human cloning?

19. There are genetic tests available for several types of "inherited cancer." The results from these tests cannot usually predict that someone will get cancer within a particular amount of time. Rather, they indicate only that a person has an increased risk of developing cancer. For many of these cancers, lifestyle changes cannot decrease a person's risk. Therefore, some people consider the tests useless. If your close family had a history of cancer and a test were available, would you want to get screened? Why or why not? What would you do with this information? If a sibling decided to get screened, explain whether you would want to know the results.

12 DNA Technology

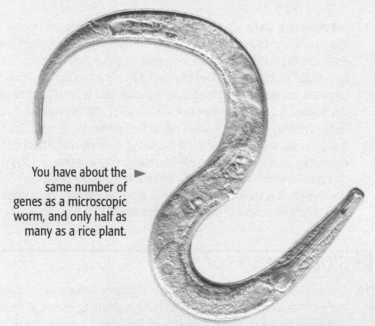

You have about the same number of genes as a microscopic worm, and only half as many as a rice plant.

Millions of individuals with diabetes live healthier lives thanks to insulin made by bacteria.

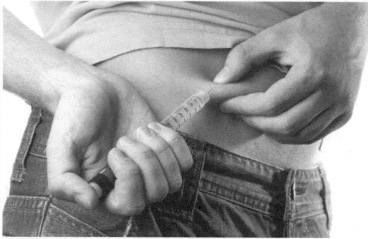

Someday, genetically modified potatoes could prevent tens of thousands of children from dying of cholera.

DNA Profiling BIOLOGY AND SOCIETY

Using DNA to Establish Guilt and Innocence

It was a horrific crime in the heart of the nation's capital: On February 24, 1981, a man broke into an apartment, just miles from the White House, and attacked a 27-year-old woman. After binding and raping her, the perpetrator stole traveler's checks and ran off. The victim had only glimpses of her attacker. Several weeks later, a police officer thought that 18-year-old Kirk Odom resembled a sketch of the perpetrator. A month later, the victim picked Odom out of a police lineup.

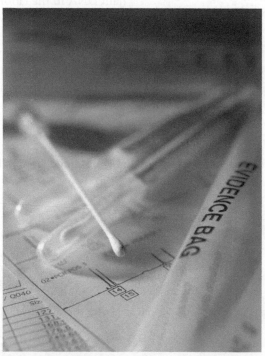

A DNA profile. Even minuscule bits of evidence can provide a DNA profile.

At trial, a Special Agent of the FBI testified that hair found in the victim's clothing, when viewed under a microscope, was "indistinguishable" from Odom's hair. Despite Odom having an alibi for his whereabouts on the night of the crime, after just a few hours of deliberation the jury convicted Odom of several charges, including rape while armed. He was sentenced to 20 to 66 years in prison. Odom served more than 20 years before being released on lifelong parole as a registered sex offender. However, in February 2011, prompted by recent discoveries that indicated that the FBI's microscopic hair analysis technique was flawed—and subsequent overturned convictions in other cases—a motion was filed to reopen the Odom case for DNA testing.

Modern forensic DNA analysis hinges on a simple fact: The cells of every person (except identical twins) contain unique DNA. DNA profiling is the analysis of DNA samples to determine whether they come from the same individual. In the Odom case, the government located sheets, clothing, and hair evidence from the 1981 crime scene. In the more recent analysis, the results were conclusive and irrefutable: The DNA left at the crime scene did *not* match Odom's. In fact, the DNA matched a different man, a convicted sex offender (who was never charged because the statute of limitations on the crime had expired). On July 13, 2012—Odom's 50th birthday—the court acknowledged that the original hair analysis data was erroneous and that Kirk Odom had "suffered a terrible injustice." After 30 years, Odom was officially declared innocent by the court.

A steady stream of stories such as this one demonstrates that DNA technology can provide evidence of either guilt or innocence. Beyond the courtroom, DNA technology has led to some of the most remarkable scientific advances in recent years: Crop plants have been genetically modified to produce their own insecticides; human genes are being compared with those of other animals to help shed light on what makes us distinctly human; and significant advances have been made toward detecting and curing fatal genetic diseases. This chapter will describe these and other uses of DNA technology and explain how various DNA techniques are performed. We'll also examine some of the social, legal, and ethical issues that lie at the intersection of biology and society.

Genetic Engineering

You may think of **biotechnology**, the manipulation of organisms or their components to make useful products, as a modern phenomenon, but it actually dates back to the dawn of civilization. Consider such ancient practices as using yeast to make bread and beer and the selective breeding of livestock. But when people use the term *biotechnology* today, they are usually referring to DNA technology, modern laboratory techniques for studying and manipulating genetic material. Using the methods of DNA technology, scientists can modify specific genes and move them between organisms as different as bacteria, plants, and animals. Organisms that have acquired one or more genes by artificial means are called **genetically modified (GM) organisms**. If the newly acquired gene is from another organism, typically of another species, the recombinant organism is called a **transgenic organism**.

In the 1970s, the field of biotechnology exploded with the invention of methods for making recombinant DNA in the laboratory. Scientists can construct **recombinant DNA** by combining pieces of DNA from two different sources—often from different species—to form a single DNA molecule. Recombinant DNA technology is widely used in **genetic engineering**, the direct manipulation of genes for practical purposes. Scientists have genetically engineered bacteria to mass-produce a variety of useful chemicals, from cancer drugs to pesticides. Scientists have also transferred genes from bacteria to plants and from one animal species to another (**Figure 12.1**). Such engineering can serve a variety of purposes, from basic research (What does this gene do?) to medical applications (Can we create animal models for this human disease?).

Recombinant DNA Techniques

Although genetic engineering can be performed on a variety of organisms, bacteria (*Escherichia coli*, in particular) are the workhorses of modern biotechnology. To manipulate genes in the laboratory, biologists often use bacterial **plasmids**, which are small, circular DNA molecules that duplicate separately from the larger bacterial chromosome (**Figure 12.2**). Because plasmids can carry virtually any gene and are passed from one generation of bacteria to the next, they are key tools for **gene cloning**, the production of multiple identical copies of a gene-carrying piece of DNA. Gene cloning methods are central to the production of useful products from genetically engineered organisms. ☑

☑ CHECKPOINT

What is biotechnology? What is recombinant DNA?

Answer: *the manipulation of organisms or their parts to produce a useful product; a molecule containing DNA from two different sources, often different species*

▼ Figure 12.1 **Genetic engineers produced glowing fish by transferring a gene for a fluorescent protein originally obtained from jellies ("jellyfish").**

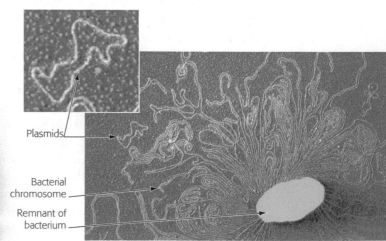

Plasmids

Bacterial chromosome

Remnant of bacterium

▲ Figure 12.2 **Bacterial plasmids.** The micrograph shows a bacterial cell that has been ruptured, revealing one long chromosome and several smaller plasmids. The inset is an enlarged view of a single plasmid.

How to Clone a Gene

Consider a typical genetic engineering challenge: A genetic engineer at a pharmaceutical company identifies a gene of interest that codes for a valuable protein, such as a potential new drug. The biologist wants to manufacture the protein on a large scale. **Figure 12.3** illustrates a way to accomplish this by using recombinant DNA techniques.

To start, the biologist isolates two kinds of DNA: bacterial plasmids that will serve as **vectors** (gene carriers, shown in blue in the figure) and DNA from another organism that includes the gene of interest (shown in yellow). This foreign DNA may be from any type of organism, even a human. ❶ The DNA from the two sources is joined together, resulting in recombinant DNA plasmids. ❷ The recombinant plasmids are then mixed with bacteria. Under the right conditions, the bacteria take up the recombinant plasmids. ❸ Each bacterium, carrying its recombinant plasmid, is allowed to reproduce via cell division to form a **clone**, a group of identical cells descended from a single original cell. As the bacteria multiply, the foreign gene carried by the recombinant plasmid is also copied. ❹ The transgenic bacteria with the gene of interest can then be grown in large tanks, producing the protein in marketable quantities. The end products of gene cloning may be copies of the gene itself, to be used in additional genetic engineering projects, or the protein product of the cloned gene, to be harvested and used. ☑

☑ **CHECKPOINT**

Why are plasmids valuable tools for the production of recombinant DNA?

Answer: Plasmids can carry virtually any foreign gene and are replicated by their bacterial host cells.

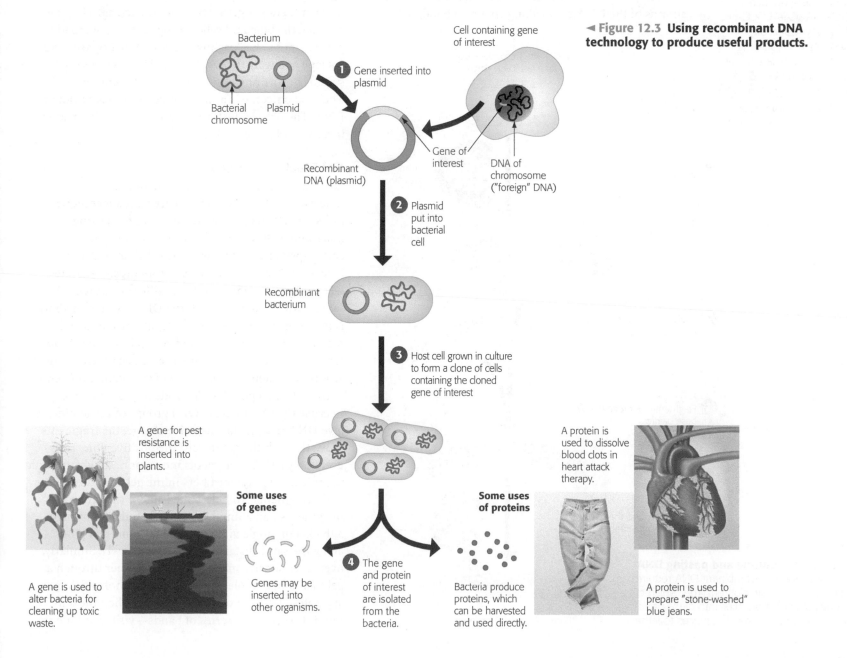

◄ Figure 12.3 **Using recombinant DNA technology to produce useful products.**

Bacterium

Bacterial chromosome   Plasmid

❶ Gene inserted into plasmid

Cell containing gene of interest

Recombinant DNA (plasmid)

Gene of interest

DNA of chromosome ("foreign" DNA)

❷ Plasmid put into bacterial cell

Recombinant bacterium

❸ Host cell grown in culture to form a clone of cells containing the cloned gene of interest

A gene for pest resistance is inserted into plants.

A gene is used to alter bacteria for cleaning up toxic waste.

Some uses of genes

Genes may be inserted into other organisms.

❹ The gene and protein of interest are isolated from the bacteria.

Some uses of proteins

Bacteria produce proteins, which can be harvested and used directly.

A protein is used to dissolve blood clots in heart attack therapy.

A protein is used to prepare "stone-washed" blue jeans.

Cutting and Pasting DNA with Restriction Enzymes

As shown in Figure 12.3, recombinant DNA is created by combining two ingredients: a bacterial plasmid and the gene of interest. To understand how these DNA molecules are spliced together, you need to learn how enzymes cut and paste DNA.

The cutting tools used for making recombinant DNA are bacterial enzymes called **restriction enzymes**. Biologists have identified hundreds of restriction enzymes, each recognizing a particular short DNA sequence, usually four to eight nucleotides long. For example, one restriction enzyme only recognizes the DNA sequence GAATTC, whereas another recognizes GGATCC. The DNA sequence recognized by a particular restriction enzyme is called a **restriction site**. After a restriction enzyme binds to its restriction site, it cuts the two strands of the DNA by breaking chemical bonds at

☑ **CHECKPOINT**

If you mix a restriction enzyme that cuts within the sequence AATC with DNA of the sequence CGAATCTAGCAATCGCGA, how many restriction fragments will result? (For simplicity, the sequence of only one of the two DNA strands is listed.)

Answer: Cuts at two restriction sites will yield three restriction fragments.

specific points within the sequence, like a pair of highly specific molecular scissors.

The top of **Figure 12.4** shows a piece of DNA (blue) that contains one restriction site for a particular restriction enzyme. ① The restriction enzyme cuts the DNA strands between the bases A and G within the recognition sequence, producing pieces of DNA called **restriction fragments**. The staggered cuts yield two double-stranded DNA fragments with single-stranded ends, called "sticky ends." Sticky ends are the key to joining DNA restriction fragments originating from different sources. ② Next, a piece of DNA from another source (yellow) is added. Notice that the yellow DNA has single-stranded ends identical in base sequence to the sticky ends on the blue DNA because the same restriction enzyme was used to cut both types of DNA. ③ The complementary ends on the blue and yellow fragments stick together by base pairing. ④ The union between the blue and yellow fragments is then made permanent by the "pasting" enzyme **DNA ligase**. This enzyme connects the DNA pieces into continuous strands by forming bonds between adjacent nucleotides. The final outcome is a single molecule of recombinant DNA. The process just described explains what happens in step 1 of Figure 12.3. ☑

Gel Electrophoresis

To separate and visualize DNA fragments of different lengths, researchers carry out a technique called **gel electrophoresis**, a method for sorting macromolecules—usually proteins or nucleic acids—primarily by their electrical charge and size. **Figure 12.5** shows how gel electrophoresis separates DNA fragments obtained from different sources. A sample with many copies of the DNA from each source is placed in a separate well (hole) at one end of a flat, rectangular gel. The gel is a thin slab of jellylike material that acts as a molecular sieve. A negatively charged electrode is then attached to the DNA-containing end of the gel and a positive electrode to the other end. Because the phosphate (PO_4^-) groups of nucleotides give DNA fragments a negative charge, the fragments move through the gel toward the positive pole. However, longer DNA fragments move more slowly through the thicket of polymer fibers in the gel than do shorter DNA fragments. To help you visualize the process, imagine a small animal scampering quickly through a thicket of jungle vines, while a large animal plods along the same distance much more slowly. Similarly, over time, shorter molecules move farther through a gel than longer molecules. Gel electrophoresis thus separates DNA fragments by length. When the current is turned off, a series of bands—visible as blue

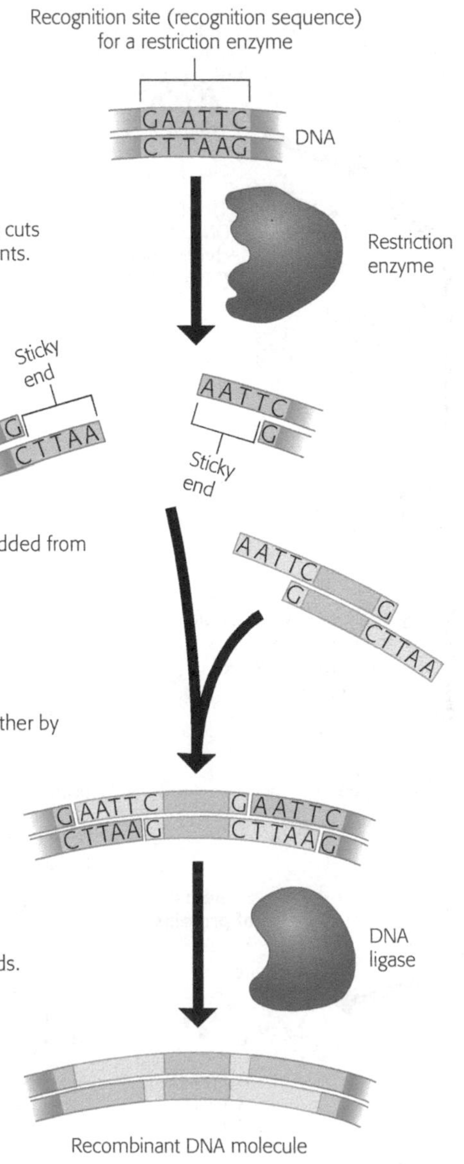

► **Figure 12.4 Cutting and pasting DNA.** The production of recombinant DNA requires two enzymes: a restriction enzyme, which cuts the original DNA molecules into pieces, and DNA ligase, which pastes the pieces together.

▼ **Figure 12.5 Gel electrophoresis of DNA molecules.** The photo shows DNA fragments of various sizes visibly stained on the gel.

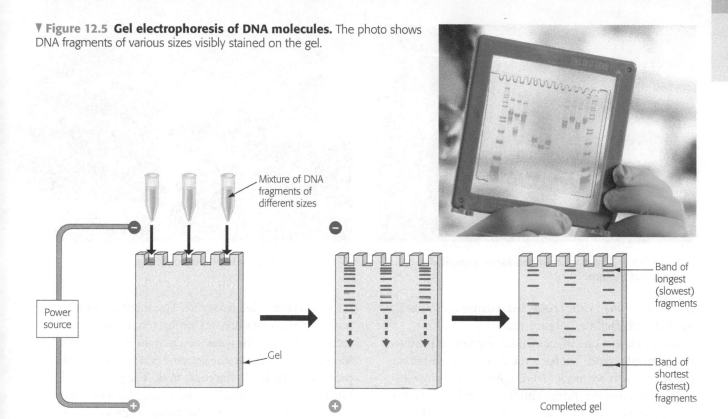

Mixture of DNA fragments of different sizes

Power source

Gel

Band of longest (slowest) fragments

Band of shortest (fastest) fragments

Completed gel

smudges in the photograph of the gel—is left in each column of the gel. Each band is a collection of DNA fragments of the same length. The bands can be made visible by staining, by exposure onto photographic film (if the DNA is radioactively labeled), or by measuring fluorescence (if the DNA is labeled with a fluorescent dye). ✓

Pharmaceutical Applications

By transferring the gene for a desired protein into a bacterium, yeast, or other kind of cell that is easy to grow in culture, scientists can produce large quantities of useful proteins that are present naturally only in small amounts. In this section, you'll learn about some applications of recombinant DNA technology.

Humulin is human insulin produced by genetically modified bacteria **(Figure 12.6)**. In humans, insulin is a protein normally made by the

Millions of individuals with diabetes live healthier lives thanks to insulin made by bacteria.

pancreas. Insulin functions as a hormone and helps regulate the level of glucose in the blood. If the body fails to produce enough insulin, the result is type 1 diabetes. There is no cure, so people with this disease must inject themselves daily with doses of insulin for the rest of their lives.

Because human insulin is not readily available, diabetes was historically treated using insulin from cows and pigs. This treatment was problematic, however. Pig and cow insulins can cause allergic reactions in people because their chemical structures differ slightly from that of human insulin. In addition, by the 1970s, the supply of beef and pork pancreas available for insulin extraction could not keep up with the demand.

In 1978, scientists working at a biotechnology company chemically synthesized DNA fragments and linked them to form the two genes that code for the two polypeptides that make up human insulin (see Figure 11.7). They then inserted these artificial genes into *E. coli* host cells. Under proper growing conditions, the transgenic bacteria cranked out large quantities of the human protein. In 1982, Humulin hit the market as the world's first genetically engineered pharmaceutical product. Today, it is produced around the clock in gigantic fermentation vats filled with a liquid culture of bacteria. Each day, more than 4 million people with

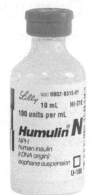

► **Figure 12.6 Humulin, human insulin produced by genetically modified bacteria.**

☑ **CHECKPOINT**

You use a restriction enzyme to cut a long DNA molecule that has three copies of the enzyme's recognition sequence clustered near one end. When you separate the restriction fragments by gel electrophoresis, how do you expect the bands to appear?

Answer: three bands near the positive pole at the bottom of the gel (small fragments) and one band near the negative pole at the top of the gel (large fragment)

▲ Figure 12.7 **A factory that produces genetically engineered insulin.**

▲ Figure 12.8 **A genetically modified goat.**

diabetes use the insulin collected, purified, and packaged at such facilities (Figure 12.7).

Insulin is just one of many human proteins produced by genetically modified bacteria. Another example is human growth hormone (HGH). Abnormally low levels of this hormone during childhood and adolescence can cause dwarfism. Because growth hormones from other animals are not effective in people, HGH was an early target of genetic engineers. Before genetically engineered HGH became available in 1985, children with an HGH deficiency could only be treated with scarce and expensive supplies of HGH obtained from human cadavers. Another important pharmaceutical product produced by genetic engineering is tissue plasminogen activator (abbreviated as tPA), a natural human protein that helps dissolve blood clots. If administered shortly after a stroke, tPA reduces the risk of additional strokes and heart attacks.

Besides bacteria, yeast and mammalian cells can also be used to produce medically valuable human proteins. For example, genetically modified mammalian cells growing in laboratory cultures are currently used to produce a hormone called erythropoietin (EPO) that stimulates production of red blood cells. EPO is used to treat anemia; unfortunately, some athletes abuse the drug to seek the advantage of artificially high levels of oxygen-carrying red blood cells (a practice called "blood doping"). In recent decades, genetic engineers have even developed transgenic plant cells that can produce human drugs. Because they are easily grown in culture and are unlikely to be contaminated by human pathogens (such as viruses), some scientists believe that carrots may be the drug factories of the future!

Genetically modified whole animals are also used to produce drugs. **Figure 12.8** shows a transgenic goat that carries a gene

for an enzyme called lysozyme. This enzyme, found naturally in breast milk, has antibacterial properties. In another example, the gene for a human blood protein has been inserted into the genome of a goat so that the protein is secreted in the goat's milk. The protein is then purified from the milk. Because transgenic animals are difficult to produce, researchers may create a single transgenic animal and then breed or clone it. The resulting herd of transgenic animals could then serve as a grazing pharmaceutical factory—"pharm" animals.

DNA technology is also helping medical researchers develop vaccines. A vaccine is a harmless variant or derivative of a disease-causing microbe—such as a bacterium or virus—that is used to prevent an infectious disease. When a person is inoculated, the vaccine stimulates the immune system to develop lasting defenses against the microbe. For many viral diseases, the only way to prevent serious harm from the illness is to use vaccination to prevent the illness in the first place. For example, the vaccine against hepatitis B, a disabling and sometimes fatal liver disease, is produced by genetically engineered yeast cells that secrete a protein found on the microbe's outer surface.

Genetically Modified Organisms in Agriculture

Since ancient times, people have selectively bred agricultural crops to make them more useful (see Figure 1.13). Today, DNA technology is quickly replacing traditional breeding programs as scientists work to improve the productivity of agriculturally important plants and animals.

In the United States today, more than 80% of the corn crop, more than 90% of the soybean crop, and about 75% of the cotton crop

Someday, genetically modified potatoes could prevent tens of thousands of children from dying of cholera.

are genetically modified. **Figure 12.9** shows corn that has been genetically engineered to resist attack by an insect called the European corn borer. Growing insect-resistant plants reduces the need for chemical insecticides. In another example, modified strawberry plants produce bacterial proteins that act as a natural antifreeze, protecting the delicate plants from the damages of cold weather. Potatoes and rice have been engineered to produce harmless proteins derived from the cholera bacterium; researchers hope that these modified foods will one day serve as an edible vaccine against cholera, a disease that kills thousands of children in developing nations every year. In India, the insertion of a natural but rare saltwater-resistance gene has enabled new varieties of rice to thrive in water three times as salty as seawater, allowing food to be grown in drought-stricken or flooded regions.

Scientists are also using genetic engineering to improve the nutritional value of crop plants **(Figure 12.10)**. One example is "golden rice 2," a transgenic variety of rice that carries genes from daffodils and corn. This rice could help prevent vitamin A deficiency and resulting blindness, especially in developing nations that depend on rice as a staple crop. Cassava, a starchy root crop that is a staple for nearly 1 billion people in developing nations, has similarly been modified to produce increased levels of iron and beta-carotene (which is converted to vitamin A in the body). However, controversy surrounds the use of GM foods, as we'll discuss at the end of the chapter.

Genetic engineers are targeting agricultural animals as well as plant crops. Although no transgenic animals are

▲ Figure 12.10 **Genetically modified staple crops.** "Golden rice 2," the yellow grains shown here (top) alongside ordinary rice, has been genetically modified to produce high levels of beta-carotene, a molecule that the body converts to vitamin A. Transgenic cassava (bottom), a starchy root crop that serves as the main food source for nearly a billion people, has been modified to produce extra nutrients.

▼ **Figure 12.9 Genetically modified corn.** The corn plants in this field carry a bacterial gene that helps prevent infestation by the European corn borer (inset).

currently being sold as food, the Food and Drug Administration (FDA) has issued regulatory guidelines for their eventual introduction. Scientists might, for example, identify in one variety of cattle a gene that causes the development of larger muscles (which make up most of the meat we eat) and transfer it to other cattle or even to chickens. Researchers have genetically modified pigs to carry a roundworm gene whose protein converts less healthy fatty acids to omega-3 fatty acids. Meat from the modified pigs contains four to five times as much healthy omega-3 fat as regular pork. AquAdvantage is a trade name for Atlantic salmon that have been genetically modified to reach market size in half the normal time (18 months vs. 3 years). The FDA is currently reviewing AquAdvantage salmon; it may become the first GM animal approved for consumption in the United States. As of late 2014, the review continues. ☑

☑ CHECKPOINT

What is a genetically modified organism?

Answer: one that carries DNA introduced through artificial means

257

Human Gene Therapy

We've seen that bacteria, plants, and nonhuman animals can be genetically modified—so what about humans? **Human gene therapy** is intended to treat disease by introducing new genes into an afflicted person. In cases where a single defective gene causes a disorder, the mutant version of a gene may be replaced or supplemented with the normal allele. This could potentially correct a genetic disorder, perhaps permanently. In other cases, genes are inserted and expressed only long enough to treat a medical problem.

Figure 12.11 summarizes one approach to human gene therapy. The procedure closely resembles the gene cloning process shown in steps 1 through 3 of Figure 12.3, but in this instance human cells, rather than bacteria, are the targets. ❶ A gene from a normal individual is cloned, converted to an RNA version, and then inserted into the RNA genome of a harmless virus. ❷ Bone marrow cells are taken from the patient and infected with the recombinant virus. ❸ The virus inserts a DNA copy of its genome, including the normal human gene, into the DNA of the patient's cells. ❹ The engineered cells are then injected back into the patient. The normal gene is transcribed and translated within the patient's body, producing the desired protein. Ideally, the nonmutant version of the gene would be inserted into cells that multiply throughout a person's life. Bone marrow cells, which include the stem cells that give rise to all the types of blood cells, are prime candidates. If the procedure succeeds, the cells will multiply permanently and produce a steady supply of the missing protein, curing the patient.

The promise of gene therapy thus far exceeds actual results, but there have been some successes. In 2009, an international research team conducted a trial that focused on a form of progressive blindness linked to a defect in a gene responsible for producing light-detecting pigments in the eye. The researchers found that a single injection of a virus carrying the normal gene into one eye of affected children improved vision in that eye, sometimes enough to allow normal functioning, without significant side effects. The other eye was left untreated as a control.

From 2000 to 2011, gene therapy was used to cure 22 children with severe combined immunodeficiency (SCID), a fatal inherited disease caused by a defective gene that prevents development of the immune system, requiring patients to remain isolated within protective "bubbles." Unless treated with a bone marrow transplant, which is effective only 60% of the time, SCID patients quickly die from infections by microbes that most of us easily fend off. In these cases, researchers

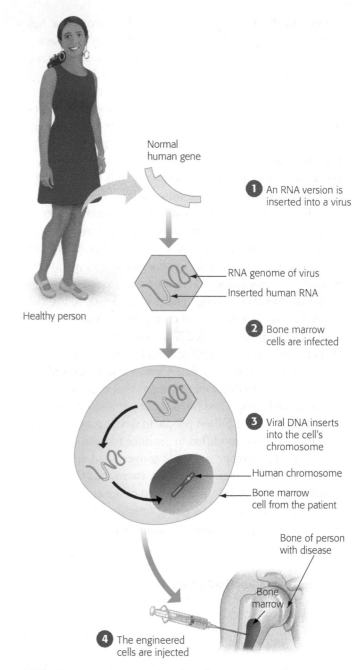

Normal human gene

Healthy person

❶ An RNA version is inserted into a virus

RNA genome of virus

Inserted human RNA

❷ Bone marrow cells are infected

❸ Viral DNA inserts into the cell's chromosome

Human chromosome

Bone marrow cell from the patient

Bone of person with disease

Bone marrow

❹ The engineered cells are injected

▲ Figure 12.11 **One approach to human gene therapy.**

periodically removed immune system cells from the patients' blood, infected them with a virus engineered to carry the normal allele of the defective gene, then reinjected the blood into the patient. The treatment cured the patients of SCID, but there have been some serious side effects: Four of the treated patients developed leukemia, and one died after the inserted gene activated an oncogene (see Chapter 11), creating cancerous blood cells. Gene therapy remains promising, but there is very little evidence to date of safe and effective application. Active research continues, with new, tougher safety guidelines in place that are meant to minimize dangers. ☑

☑ **CHECKPOINT**

Why are bone marrow stem cells ideally suited as targets for gene therapy?

Answer: because bone marrow stem cells multiply throughout a person's life

DNA Profiling and Forensic Science

When a crime is committed, body fluids (such as blood or semen) or small pieces of tissue (such as skin beneath a victim's fingernails) may be left at the scene or on the victim or assailant. As discussed in the Biology and Society section at the start of the chapter, such evidence can be examined by **DNA profiling**, the analysis of DNA samples to determine whether they come from the same individual. Indeed, DNA profiling has rapidly transformed the field of **forensics**, the scientific analysis of evidence for crime scene investigations and other legal proceedings. To produce a DNA profile, scientists compare DNA sequences that vary from person to person.

Figure 12.12 presents an overview of a typical investigation using DNA profiling. ❶ First, DNA samples are isolated from the crime scene, suspects, victims, or other evidence. ❷ Next, selected sequences from each DNA sample are amplified (copied many times) to produce a large sample of DNA fragments. ❸ Finally, the amplified DNA fragments are compared. All together, these steps provide data about which samples are from the same individual and which samples are unique.

▼ **Figure 12.12 Overview of DNA profiling.** In this example, DNA from suspect 1 does not match DNA found at the crime scene, but DNA from suspect 2 does match.

DNA Profiling Techniques

In this section, you'll learn about techniques for making a DNA profile.

The Polymerase Chain Reaction (PCR)

The **polymerase chain reaction (PCR)** is a technique by which a specific segment of DNA can be amplified: targeted and copied quickly and precisely. Through PCR, a scientist can obtain enough DNA from even minute amounts of blood or other tissue to allow a DNA profile to be constructed. In fact, a microscopic sample with as few as 20 cells can be sufficient for PCR amplification.

In principle, PCR is simple. A DNA sample is mixed with nucleotides, the DNA replication enzyme DNA polymerase, and a few other ingredients. The solution is then exposed to cycles of heating (to separate the DNA strands) and cooling (to allow double-stranded DNA to re-form). During these cycles, specific regions of each molecule of DNA are replicated, doubling the amount of that DNA (**Figure 12.13**). The result of this

▼ **Figure 12.13 DNA amplification by PCR.** The polymerase chain reaction (PCR) is a method for making many copies of a specific segment of DNA. Each round of PCR, performed on a tabletop thermal cycler (shown at top), doubles the total quantity of DNA.

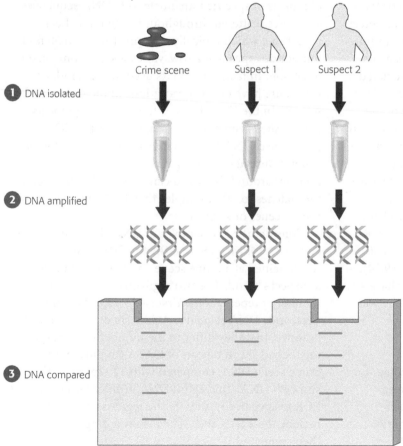

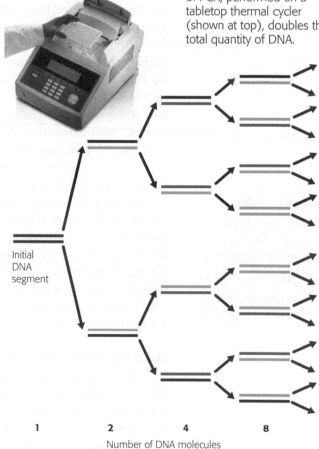

1 2 4 8
Number of DNA molecules

259

chain reaction is an exponentially growing population of identical DNA molecules. The key to automated PCR is an unusually heat-stable DNA polymerase, first isolated from prokaryotes living in hot springs (such as those shown in Figure 13.15B). Unlike most proteins, this enzyme can withstand the heat at the start of each cycle.

A DNA molecule within a starting sample is likely to be very long. But, most often, only a very small target region of that large DNA molecule needs to be amplified. The key to amplifying one particular segment of DNA and no others is the use of **primers**, short (usually 15–20 nucleotides long), chemically synthesized single-stranded DNA molecules. For each experiment, specific primers are chosen that are complementary to sequences found only at each end of the target sequence. The primers thus bind to sequences that flank the target sequence, marking the start and end points for the segment of DNA to be amplified. Beginning with a single DNA molecule and the appropriate primers, automated PCR can generate hundreds of billions of copies of the desired sequence in a few hours.

In addition to forensic applications, PCR can be used in the treatment and diagnosis of disease. For example, because the sequence of the genome of HIV (the virus that causes AIDS) is known, PCR can be used to amplify, and thus detect, HIV in blood or tissue samples. In fact, PCR is often the best way to detect this otherwise elusive virus. Medical scientists can now diagnose hundreds of human genetic disorders by using PCR with primers that target the genes associated with these disorders. The amplified DNA product is then studied to reveal the presence or absence of the disease-causing mutation. Among the genes for human diseases that have been identified are those for sickle-cell disease, hemophilia, cystic fibrosis, Huntington's disease, and Duchenne muscular dystrophy. Individuals afflicted with such diseases can often be identified before the onset of symptoms, even before birth, allowing for preventative medical care to begin. PCR can also be used to identify symptomless carriers of potentially harmful recessive alleles (see Figure 9.14). Parents may thus be informed of whether they have a risk of bearing a child with a rare disease that they do not themselves display. ✓

Short Tandem Repeat (STR) Analysis

How do you prove that two samples of DNA come from the same person? You could compare the entire genomes found in the two samples. But such an approach is impractical because the DNA of two humans of the same sex is 99.9% identical. Instead, forensic scientists typically compare about a dozen short segments of noncoding repetitive DNA that are known to vary between people. Have you ever seen a puzzle in a magazine that presents two nearly identical photos and asks you to find the few differences between them? In a similar way, scientists can focus on the few areas of difference in the human genome, ignoring the identical majority.

Repetitive DNA, which makes up much of the DNA that lies between genes in humans, consists of nucleotide sequences that are present in multiple copies in the genome. Some of this DNA consists of short sequences repeated many times tandemly (one after another); such a series of repeats in the genome is called a **short tandem repeat (STR)**. For example, one person might have the sequence AGAT repeated 12 times in a row at one place in the genome, the sequence GATA repeated 35 times at a second place, and so on; another person is likely to have the same sequences at the same places but with a different number of repeats. Like genes that cause physical traits, these stretches of repetitive DNA are more likely to be an exact match between relatives than between unrelated individuals.

STR analysis is a method of DNA profiling that compares the lengths of STR sequences at specific sites in the genome. The standard STR analysis procedure used by law enforcement compares the number of repeats of specific four-nucleotide DNA sequences at 13 sites scattered throughout the genome. Each repeat site, which typically contains from 3 to 50 four-nucleotide repeats in a row, varies widely from person to person. In fact, some STRs used in the standard procedure have up to 80 variations in the number of repeats. In the United States, the number of repeats at each site is entered into a database called CODIS (Combined DNA Index System) administered by the Federal Bureau of Investigation. Law enforcement agencies around the world can access CODIS to search for matches to DNA samples they have obtained from crime scenes or suspects.

Consider the two samples of DNA shown in **Figure 12.14**. Imagine that the top DNA segment was obtained at a crime scene and the bottom from a suspect's blood. The two segments have the same number of repeats at the first site: 7 repeats of the four-nucleotide DNA sequence AGAT (in orange). Notice, however, that they differ in the number of repeats at the second site: 8 repeats of GATA (in purple) in the crime scene DNA, compared with 12 repeats in the suspect's DNA. To create a DNA profile, a scientist uses PCR to specifically amplify the regions of DNA that include these STR sites. The resulting fragments are then compared.

✓ CHECKPOINT

Why is only the slightest trace of DNA at a crime scene often sufficient for forensic analysis?

Answer: because PCR can be used to produce enough molecules for analysis

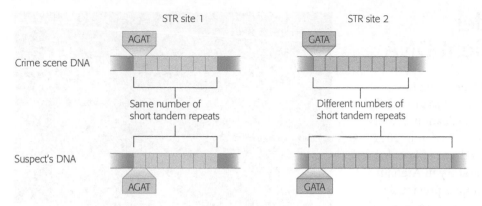

STR site 1

STR site 2

Crime scene DNA

AGAT

GATA

Same number of short tandem repeats

Different numbers of short tandem repeats

Suspect's DNA

AGAT

GATA

◀ **Figure 12.14 Short tandem repeat (STR) sites.** Scattered throughout the genome, STR sites contain tandem repeats of four-nucleotide sequences. The number of repetitions at each site can vary from individual to individual. In this figure, both DNA samples have the same number of repeats (7) at the first STR site, but different numbers (8 versus 12) at the second.

Figure 12.15 shows the gel that would result from using gel electrophoresis to separate the DNA fragments from the example in Figure 12.14. (This figure simplifies the process; an actual STR analysis uses more sites than 2 and uses a different method to visualize the results.) The differences in the locations of the bands reflect the different lengths of the DNA fragments. This gel would provide evidence that the crime scene DNA did not come from the suspect.

As happened in the case discussed in the Biology and Society section, DNA profiling can provide evidence of either guilt or innocence. As of 2014, lawyers at the Innocence Project, a nonprofit legal organization located in New York City, have helped to exonerate more than 310 convicted criminals in 35 states, including 18 who were on death row. The average sentence served by those who were exonerated was 14 years. In nearly half of these cases, DNA profiling has also identified the true perpetrators. **Figure 12.16** presents some data from a real case in which STR analysis

proved a convicted man innocent and also helped identify the true perpetrator.

Just how reliable is a genetic profile? In forensic cases using STR analysis with the 13 standard markers, the probability of two people having identical DNA profiles is somewhere between one chance in 10 billion and one in several trillion. (The exact probability depends on the frequency of the individual's particular markers in the general population.) Thus, despite problems that can still arise from insufficient data, human error, or flawed evidence, genetic profiles are now accepted as compelling evidence by legal experts and scientists alike. ☑

☑ **CHECKPOINT**

What are STRs, and why are they useful for DNA profiling?

Answer: STRs (short tandem repeats) are nucleotide sequences repeated many times in a row within the human genome. STRs are valuable for DNA profiling because different people have different numbers of repeats at the various STR sites.

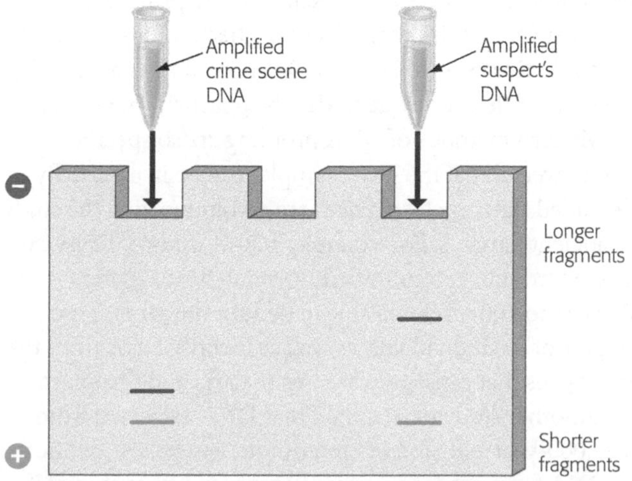

▼ **Figure 12.16 DNA profiling: proof of innocence and guilt.** In 1984, Earl Washington was convicted and sentenced to death for a 1982 rape and murder. In 2000, STR analysis showed conclusively that he was innocent. Because every person has two chromosomes, each STR site is represented by two numbers of repeats. The table shows the number of repeats for three STR markers in three samples: from semen found on the victim, from Washington, and from another man who was in prison after an unrelated conviction. These and other STR data (not shown) exonerated Washington and led the other man to plead guilty to the murder.

Amplified crime scene DNA

Amplified suspect's DNA

−

Longer fragments

Shorter fragments

+

▲ **Figure 12.15 Visualizing STR fragment patterns.** This figure shows the bands that would result from gel electrophoresis of the STR sites illustrated in Figure 12.14. Notice that one of the bands from the crime scene DNA does not match one of the bands from the suspect's DNA.

Source of sample	STR marker 1	STR marker 2	STR marker 3
Semen on victim	17,19	13,16	12,12
Earl Washington	16,18	14,15	11,12
Kenneth Tinsley	17,19	13,16	12,12

Investigating Murder, Paternity, and Ancient DNA

Since its introduction in 1986, DNA profiling has become a standard tool of forensics and has provided crucial evidence in many famous investigations. After the death of terrorist leader Osama bin Laden in 2011, U.S. Special Forces members obtained a sample of his DNA. Within hours, a military laboratory in Afghanistan compared the tissue against samples previously obtained from several of bin Laden's relatives, including his sister who had died of brain cancer in a Boston hospital in 2010. Although facial recognition and an eyewitness identification provided preliminary evidence, it was DNA that provided a conclusive match, officially ending the long hunt for the notorious terrorist.

DNA profiling can also be used to identify murder victims. The largest such effort in history took place after the World Trade Center attack on September 11, 2001. Forensic scientists in New York City worked for years to identify more than 20,000 samples of victims' remains. DNA profiles of tissue samples from the disaster site were matched to DNA profiles from tissue known to be from the victims or their relatives. More than half of the victims identified at the World Trade Center site were recognized solely by DNA evidence, providing closure to many grieving families. Since that time, the victims of other atrocities, such as mass killings during civil wars in Europe and Africa, have been identified using DNA profiling techniques. In 2010, for example, DNA analysis was used to identify the remains of war crime victims who had been buried in mass graves in Bosnia 15 years earlier. DNA profiling can also be used to identify victims of natural disasters. After a tsunami devastated southern Asia the day after Christmas 2004, DNA profiling was used to identify hundreds of victims, mostly foreign tourists.

Comparing the DNA of a mother, her child, and the purported father can settle a question of paternity. Sometimes paternity is of historical interest: DNA profiling proved that Thomas Jefferson or a close male relative fathered a child with an enslaved woman, Sally Hemings. In another historical case, researchers wished to investigate whether any descendants of Marie Antoinette (Figure 12.17), one-time queen of France, survived the French Revolution. DNA extracted from a preserved heart said to belong to her son was compared to DNA extracted from a lock of Marie's hair. A DNA match proved that her last known heir had, in fact, died in jail during the revolution. More recently, a former backup singer for the "Godfather of Soul" James Brown sued his estate after his death, claiming that her child was Brown's son. A DNA paternity test proved her claim, and 25% of Brown's estate was awarded to the mother and child.

▲ **Figure 12.17 Marie Antoinette.** DNA profiling proved that Louis (depicted with his mother in this 1785 painting), the son of the Queen of France, did not survive the French Revolution.

DNA profiling can also help protect endangered species by conclusively proving the origin of contraband animal products. For example, analysis of seized elephant tusks can pinpoint the location of the poaching, allowing enforcement officials to increase surveillance and prosecute those responsible. In 2014, three tiger poachers in India were sentenced to five years in jail after DNA profiling matched the dead tigers' flesh to tissue under the poachers' fingernails.

Modern methods of DNA profiling are so specific and powerful that the DNA samples can be in a partially degraded state. Such advances are revolutionizing the study of ancient remains. For example, a 2014 study of DNA extracted from five mummified Egyptian heads (dating from 800 B.C. to 100 A.D.) was able to deduce the geographic origins of the individuals, as well as identify DNA from the pathogens that cause the diseases malaria and toxoplasmosis. Another study determined that DNA extracted from a 27,000-year-old Siberian mammoth was 98.6% identical to DNA from modern African elephants. Other studies involving a large collection of mammoth samples suggested that the last populations of the huge beasts migrated from North America to Siberia, where separate species interbred and, eventually, died out several thousand years ago.

Bioinformatics

In the past decade, new experimental techniques have generated enormous volumes of data related to DNA sequences. The need to make sense of an ever-increasing flood of information has spawned the field of **bioinformatics**, the application of computational methods to the storage and analysis of biological data. In this section, we'll explore some of the methods by which sequence data are accumulated, as well as many of the practical ways such knowledge can be put to use.

DNA Sequencing

Researchers can exploit the principle of complementary base pairing to determine the complete nucleotide sequence of a DNA molecule. This process is called **DNA sequencing**. In one standard procedure, the DNA is first cut into fragments, and then each fragment is sequenced (**Figure 12.18**). In the past decade, "next-generation sequencing" techniques have been developed that can simultaneously sequence thousands or hundreds of thousands of fragments, each of which can be 400–1,000 nucleotides long. This technology makes it possible to sequence nearly a million nucleotides per hour! This is an example of "high-throughput" DNA technology, which is currently the method of choice for studies where massive numbers of DNA samples—even representing an entire genome—are being sequenced. In "third-generation sequencing," a single, very long DNA molecule is sequenced on its own. Several groups of scientists have been working on the idea of moving a single strand of a DNA molecule through a very small pore in a membrane (a nanopore), detecting the bases one by one by their interruption of an electrical current.

▼ **Figure 12.18 A DNA sequencer.** This high-throughput DNA sequencing machine can process half a billion bases in a single 10-hour run.

The idea is that each type of base interrupts the current for a slightly different length of time. Such techniques, if perfected, may usher in a new era of faster, even more affordable sequencing.

Genomics

Improved DNA sequencing techniques have transformed the way in which we can explore fundamental biological questions about evolution and how life works. A major leap forward occurred in 1995 when a team of scientists announced that it had determined the nucleotide sequence of the entire genome of *Haemophilus influenzae*, a bacterium that can cause several human diseases, including pneumonia and meningitis. **Genomics**, the study of complete sets of genes (genomes), was born.

The first targets of genomics research were bacteria, which have relatively little DNA (as you can see in **Table 12.1**). But soon the attention of genomics researchers

Table 12.1	Some Important Sequenced Genomes*		
Organism	**Year Completed**	**Size of Genome (in base pairs)**	**Approximate Number of Genes**
Haemophilus influenzae (bacterium)	1995	1.8 million	1,700
Saccharomyces cerevisiae (yeast)	1996	12 million	6,300
Escherichia coli (bacterium)	1997	4.6 million	4,400
Caenorhabditis elegans (roundworm)	1998	100 million	20,100
Drosophila melanogaster (fruit fly)	2000	165 million	14,000
Arabidopsis thaliana (mustard plant)	2000	120 million	25,500
Oryza sativa (rice)	2002	430 million	42,000
Homo sapiens (human)	2003	3.0 billion	21,000
Rattus norvegicus (lab rat)	2004	2.8 billion	20,000
Pan troglodytes (chimpanzee)	2005	3.1 billion	20,000
Macaca mulatta (macaque)	2007	2.9 billion	22,000
Ornithorhynchus anatinus (duck-billed platypus)	2008	1.8 billion	18,500
Prunus persica (peach)	2013	227 million	27,900

*Some of the values listed are likely to be revised as genome analysis continues.

263

turned toward more complex organisms with much larger genomes. Baker's yeast (*Saccharomyces cerevisiae*) was the first eukaryote to have its full sequence determined, and the roundworm *Caenorhabditis elegans* was the first multicellular organism. Other sequenced animals include the fruit fly (*Drosophila melanogaster*) and lab rat (*Rattus norvegicus*), both model organisms for genetics research. Among the sequenced plants are *Arabidopsis thaliana*, a type of mustard plant used as a model organism, and rice (*Oryza sativa*), one of the world's most economically important crops.

As of 2014, the genomes of thousands of species have been published, and tens of thousands more are in progress. The majority of organisms sequenced to date are prokaryotes, including more than 4,000 bacterial species and nearly 200 archaea. Hundreds of eukaryotic genomes—including protists, fungi, plants, and animals both invertebrate and vertebrate—have also been completed. Genome sequences have been determined for cells from several cancers, for ancient humans, and for the many bacteria that live in the human intestine. ✓

Genome-Mapping Techniques

Genomes are most often sequenced using a technique called the **whole-genome shotgun method**. The first step is to chop the entire genome into fragments using restriction enzymes. Next, all the fragments are cloned and sequenced. Finally, computers running specialized mapping software reassemble the millions of overlapping short sequences into a single continuous sequence for every chromosome—an entire genome (**Figure 12.19**).

The DNA sequences determined by many research groups in the United States are deposited in GenBank, a database that is available to anyone via the Internet. You can browse it yourself at the National Center for Biotechnology Information: www.ncbi.nlm.nih.gov. GenBank includes the sequences of more than a hundred billion base pairs of DNA! The database is constantly updated, and the amount of data it contains doubles every 18 months.

Any sequence in the database can be retrieved and analyzed. For example, software can compare a collection of sequences from different species and diagram them as an evolutionary tree based on the sequence relationships. Bioinformatics has thereby revolutionized evolutionary biology by opening a vast new reservoir of data that can test evolutionary hypotheses. Next we'll discuss a particularly notable example of a sequenced animal genome—our own.

The Human Genome

The **Human Genome Project** was a massive scientific endeavor to determine the nucleotide sequence of all the DNA in the human genome and to identify the location and sequence of every gene. The project began in 1990 as

▼ **Figure 12.19 Genome sequencing.** In the photo at the bottom, a technician performs a step in the whole-genome shotgun method (depicted in the diagram).

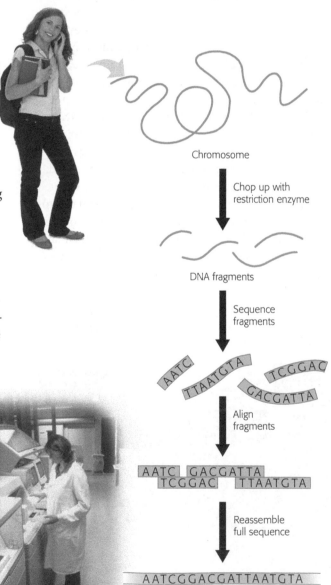

Chromosome

Chop up with restriction enzyme

DNA fragments

Sequence fragments

AATC
TTAATGTA
TCGGAC
GACGATTA

Align fragments

AATC GACGATTA
TCGGAC TTAATGTA

Reassemble full sequence

AATCGGACGATTAATGTA

an effort by government-funded researchers from six countries. Several years into the project, private companies joined the effort. At the completion of the project, more than 99% of the genome had been determined to 99.999% accuracy. (There remain a few hundred gaps of unknown sequence that require special methods to figure out.) This ambitious project has provided a wealth of data that may illuminate the genetic basis of what it means to be human.

The chromosomes in the human genome (22 autosomes plus the X and Y sex chromosomes) contain approximately 3 billion nucleotide pairs of DNA. If you imagine this sequence printed in letters (A, T, C, and G) the same size you see on this page, the sequence would fill a stack of books 18 stories high! However, the biggest surprise from the Human Genome Project is the relatively small number of human genes—currently estimated to be about 21,000—very close to the number found in a roundworm!

The human genome was a major challenge to sequence because, like the genomes of most complex eukaryotes, only a small amount of our total DNA consists of genes that code for proteins, tRNAs, or rRNAs. Most complex eukaryotes have a huge amount of noncoding DNA—about 98% of human DNA is of this type. Some of this noncoding DNA is made up of gene control sequences such as promoters, enhancers, and microRNAs (see Chapter 11). Other noncoding regions include introns and repetitive DNA (some of which is used in DNA profiling). Some noncoding DNA is important to our health, with certain regions known to carry disease-causing mutations. But the function of most noncoding DNA remains unknown.

The human genome sequenced by government-funded scientists was actually a reference genome compiled from a group of individuals. As of today, the complete genomes of many individuals have been completed. Whereas sequencing the first human genome took 13 years and cost $100 million, we are rapidly approaching the day when an individual's genome can be sequenced in a matter of hours for less than $1,000.

Scientists have even begun to gather sequence data from our extinct relatives. In 2013, scientists sequenced the entire genome of a 130,000-year-old female Neanderthal (*Homo neanderthalensis*). Using DNA extracted from a toe bone found in a Siberian cave, the resulting genome was nearly as complete as that from a modern human. Analysis of the Neanderthal genome revealed evidence of interbreeding with *Homo sapiens*. A 2014 study provided evidence that many present-day humans of European and Asian descent (but not African descent) carry Neanderthal-derived genes that influence the production of keratin, a protein that is a key structural component of hair, nails, and skin. Modern humans appear to have

You have about the same number of genes as a microscopic worm, and only half as many as a rice plant.

inherited the gene from Neanderthals around 70,000 years ago and then passed it on to their descendants. Such studies provide valuable insight into our own evolutionary tree.

Bioinformatics can also provide insights into our evolutionary relationships with nonhuman animals. In 2005, researchers completed the genome sequence for our closest living relative on the evolutionary tree of life, the chimpanzee (*Pan troglodytes*). Comparisons with human DNA revealed that we share 96% of our genome. Genomic scientists are currently finding and studying the important differences, shedding scientific light on the age-old question of what makes us human.

The potential benefits of knowing many human genomes are enormous. Thus far, more than 2,000 disease-associated genes have been identified. A recent example involved Behcet's disease, a painful and life threatening illness that involves swelling of blood vessels throughout the body. Researchers have long known that this disease is found most commonly among people living along the ancient trade route in Asia called the Silk Road (Figure 12.20). In 2013, researchers conducted a genome-wide search for genetic differences among Turkish people with and without the disease. They discovered four regions of the genome that are associated with the disease. The nearby genes are implicated in the immune system's ability to destroy invading microbes, to recognize infection sites, and to regulate autoimmune diseases. Interestingly, the function of the fourth gene has never been identified, but its close association with Behcet's disease may help researchers pinpoint its role. Next, we'll examine a much more common disease that may benefit from genomic analysis.

▼ **Figure 12.20 The Silk Road.** Behcet's disease is most commonly found along the Silk Road, part of which is shown in this map using modern place names.

Key

— = Silk Road

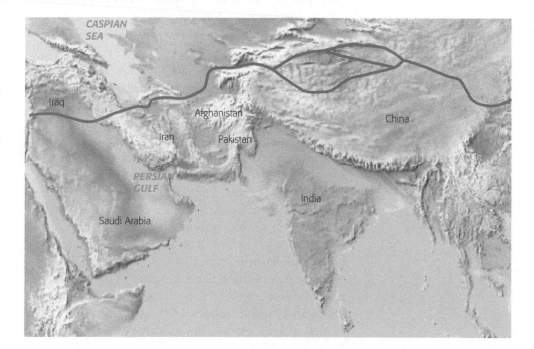

DNA Profiling THE PROCESS OF SCIENCE

Can Genomics Cure Cancer?

Lung cancer, which kills more Americans every year than any other type of cancer, has long been the target of searches for effective chemotherapy drugs. One drug used to treat lung cancer, called gefitinib, targets the protein encoded by the *EGFR* gene. This protein is found on the surface of cells that line the lungs and is also found in lung cancer tumors.

Unfortunately, treatment with gefitinib is ineffective for many patients. While studying the effectiveness of gefitinib, researchers at the Dana-Farber Cancer Institute in Boston made the **observation** that a few patients responded quite well to the drug. This posed a **question:** Are genetic differences among lung cancer patients responsible for the differences in gefitinib's effectiveness? The researchers' **hypothesis** was that mutations in the *EGFR* gene were causing the different responses to gefitinib. The team made the **prediction** that DNA profiles focusing on the *EGFR* gene would reveal different DNA sequences in the tumors of responsive patients compared with the tumors of unresponsive patients. The researchers' **experiment** involved sequencing the *EGFR* gene in cells extracted from the tumors of five patients who responded to the drug and four who did not.

The **results** were quite striking: All five tumors from gefitinib-responsive patients had mutations in *EGFR*, whereas none of the other four tumors did (**Figure 12.21**). Although the small sample size means that additional tests are needed, these results suggest that doctors can use DNA profiling techniques to screen lung cancer patients

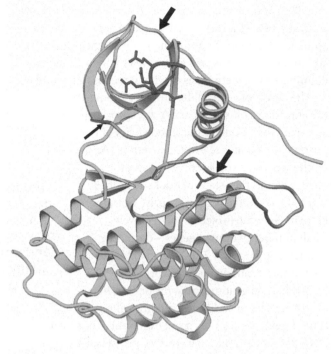

▲ Figure 12.21 **The *EGFR* protein: Fighting cancer with genomics.** Mutations (located at sites indicated by black arrows) in the *EGFR* protein can affect the ability of a cancer-fighting drug to destroy lung tumors. Here, the amino acid backbone of the protein is shown in green, with some important regions highlighted in orange, blue, and red.

for those who are most likely to benefit from treatment with this drug. In broader terms, this work suggests that fast and cheap sequencing may usher in an age of "personal genomics," where individual genetic differences among people will be put to routine medical use.

Applied Genomics

In 2001, a 63-year-old Florida man died from inhalation anthrax, a disease caused by breathing spores of the bacterium *Bacillus anthracis*. Because he was the first victim of this disease in the United States since 1976 (and coming so soon after the 9/11 terrorist attacks the month before), his death was immediately suspicious. By the end of the year, four more people had died from inhaling anthrax. Law enforcement officials realized that someone was sending anthrax spores through the mail (**Figure 12.22**). The United States was facing an unprecedented bioterrorist attack.

In the investigation that followed, one of the most helpful clues turned out to be the anthrax spores themselves. Investigators sequenced the genomes of the mailed anthrax spores. They quickly established that all

of the mailed spores were genetically identical to a laboratory subtype stored in a single flask at the U.S. Army Medical Research Institute of Infectious Diseases in Fort Detrick, Maryland. Based in part on this evidence, the FBI named an army research scientist as a suspect in the case. Although never charged, that suspect committed suicide in 2008; the case officially remains unsolved.

The anthrax case is just one example of the investigative powers of genomics. Sequence data also provided strong evidence that a Florida dentist transmitted HIV to several patients and that a single natural strain of West Nile virus can infect both birds and people. A 2013 study used DNA sequencing to prove that cancerous skin cells that had spread to the brain had done so after fusing with red blood cells provided by a bone marrow donor. These results provided researchers with new insight into how cancer spreads throughout the body.

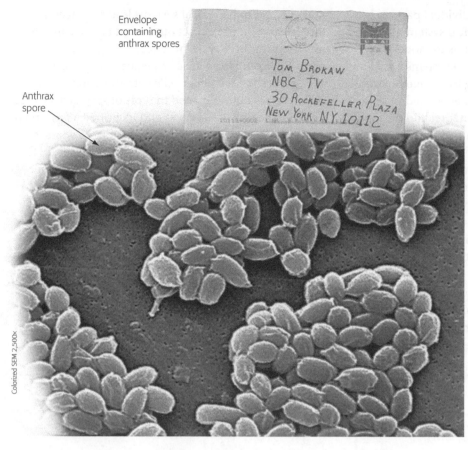

Envelope containing anthrax spores

Anthrax spore

Colorized SEM 2,500×

▲ **Figure 12.22 The 2001 anthrax attacks.** In 2001, envelopes containing anthrax spores caused five deaths.

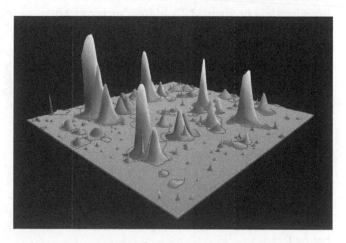

contribute to our survival. The ability to sequence the DNA of mixed populations eliminates the need to culture each species separately in the lab, making it more efficient to study microbial species.

The successes in genomics have encouraged scientists to begin similar systematic studies of the full protein sets that genomes encode (proteomes), an approach called **proteomics** (**Figure 12.23**). The number of different proteins in humans far exceeds the number of different genes (about 100,000 proteins versus about 21,000 genes). And because proteins, not genes, actually carry out the activities of the cell, scientists must study when and where proteins are produced and how they interact to understand the functioning of cells and organisms.

Genomics and proteomics enable biologists to approach the study of life from an increasingly global perspective. Biologists are now compiling catalogs of genes and proteins—that is, listings of all the "parts" that contribute to the operation of cells, tissues, and organisms. As such catalogs become complete, researchers are

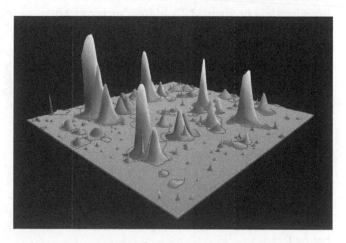

▲ **Figure 12.23 Proteomics.** Each peak on this three-dimensional graph represents one protein separated by gel electrophoresis. The height of the peak correlates with the amount of that protein. By identifying every protein in a sample, researchers can gain a fuller understanding of the complete biological system.

Interconnections within Systems Systems Biology

The computational power provided by the tools of bioinformatics allows the study of whole sets of genes and their interactions, as well as the comparison of genomes from different species. Genomics is a rich source of new insights into fundamental questions about genome organization, regulation of gene expression, embryonic development, and evolution.

Technological advances have also led to metagenomics, the study of DNA from an environmental sample. When obtained directly from the environment, a sample will likely contain genomes from many species. After the whole sample is sequenced, computer software sorts out the partial sequences from different species and assembles them into the individual specific genomes. So far, this approach has been applied to microbial communities found in environments as diverse as the Sargasso Sea and the human intestine. A 2012 study cataloged the astounding diversity of the human "microbiome"—the many species of bacteria that coexist within and upon our bodies and that

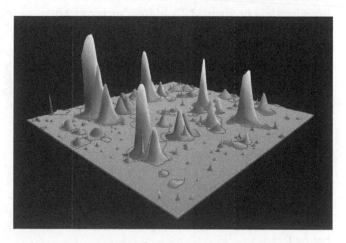

shifting their attention from the individual parts to how these parts work together in biological systems. This approach, called systems biology, aims to model the dynamic behavior of whole biological systems based on the study of the interactions among the system's parts. Because of the vast amounts of data generated in these types of studies, advances in computer technology and bioinformatics have been crucial in making systems biology possible.

Such analyses may have many practical applications. For example, proteins associated with specific diseases may be used to aid diagnosis (by developing tests that search for a particular combination of proteins) and treatment (by designing drugs that interact with the proteins involved). As high-throughput techniques become more rapid and less expensive, they are increasingly being applied to the problem of cancer. The Cancer Genome Atlas project is a simultaneous investigation by multiple research teams of a large group of interacting genes and gene products. This project aims to determine how changes in biological systems lead to cancer. A three-year pilot project set out to find all the common mutations in three types of cancer—lung, ovarian, and brain—by comparing gene sequences and patterns of gene expression in cancer cells with those in normal cells. The results confirmed the role of several genes suspected to be linked to cancer and identified a few previously unknown ones, suggesting possible new targets for therapies. The research approach proved so fruitful for these three types of cancer that the project has been extended to ten other types of cancer, chosen because they are common and often lethal in humans.

Systems biology is a very efficient way to study emergent properties, novel properties that arise at each successive level of biological complexity as a result of the arrangement of building blocks at the underlying level. The more we can learn about the arrangement and interactions of the components of genetic systems, the deeper will be our understanding of whole organisms. ☑

CHECKPOINT

☑ **CHECKPOINT**

What is the difference between genomics and proteomics?

Answer: Genomics concerns the complete set of an organism's genes, whereas proteomics concerns the complete set of an organism's proteins.

Safety and Ethical Issues

As soon as scientists realized the power of DNA technology, they began to worry about potential dangers. Early concerns focused on the possibility of creating hazardous new disease-causing organisms. What might happen, for instance, if cancer-causing genes were transferred into infectious bacteria or viruses? To address such concerns, scientists developed a set of guidelines that have become formal government regulations in the United States and some other countries.

One safety measure in place is a set of strict laboratory procedures to protect researchers from infection by engineered microbes and to prevent microbes from accidentally leaving the laboratory (Figure 12.24). In addition, strains of microbes to be used in recombinant DNA experiments are genetically crippled to ensure that they cannot survive outside the laboratory. As a further precaution, certain obviously dangerous experiments have been banned.

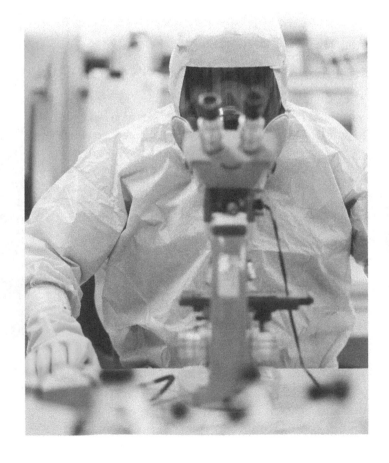

► Figure 12.24 **Maximum-security laboratory.** A scientist in a high-containment laboratory wears a biohazard suit, used for working with dangerous microorganisms.

The Controversy over Genetically Modified Foods

Today, most public concern about possible hazards centers on genetically modified (GM) foods. GM strains account for a significant percentage of several staple crops in the United States, Argentina, and Brazil; together these countries account for more than 80% of the world's supply of GM crops. Controversy about the safety of these foods is an important political issue (Figure 12.25). For example, the European Union has suspended the introduction into the market of new GM crops and considered banning the import of all GM foodstuffs. In the United States and other countries where the GM revolution has proceeded relatively unnoticed (until recently), mandatory labeling of GM foods is now being debated.

Advocates of a cautious approach are concerned that crops carrying genes from other species might harm the environment or be hazardous to human health (by, for example, introducing new allergens, molecules that can cause allergic reactions, into foods). A major worry is that transgenic plants might pass their new genes to close relatives in nearby wild areas. We know that lawn and crop grasses, for example, commonly exchange genes with wild relatives via pollen transfer. If domestic plants carrying genes for resistance to herbicides, diseases, or insect pests pollinated wild plants, the offspring might become "superweeds" that would be very difficult to control. However, researchers may be able to prevent the escape of such plant genes in various ways—for example, by engineering plants so that they cannot breed. Concern has also been raised that the widespread use of GM seeds may reduce natural genetic diversity, leaving crops susceptible to catastrophic die-offs in the event of a sudden change to the environment or introduction of a new pest. Although the U.S. National Academy of Sciences released a study finding no scientific evidence that transgenic crops pose any special health or environmental risks, the authors of the study also recommended more stringent long-term monitoring to watch for unanticipated environmental impacts.

Negotiators from 130 countries (including the United States) agreed on a Biosafety Protocol that requires exporters to identify GM organisms present in bulk food

▼ Figure 12.25 **Opposition to genetically modified organisms (GMOs).** Protesters in Oregon voice their displeasure against GMOs.

☑ **CHECKPOINT**

What is the main concern about adding genes for herbicide resistance to crop plants?

Answer: the possibility that the genes could escape, via cross-pollination, to wild plants that are closely related to the crop species

shipments and allows importing countries to decide whether the shipments pose environmental or health risks. The United States declined to sign the agreement, but it went into effect anyway because the majority of countries were in favor of it. Since then, European countries have, on occasion, refused crops from the United States and other countries for fear that they contain GM crops, leading to trade disputes.

Governments and regulatory agencies throughout the world are grappling with how to facilitate the use of biotechnology in agriculture, industry, and medicine while ensuring that new products and procedures are safe. In the United States, all genetic engineering projects are evaluated for potential risks by a number of regulatory agencies, including the Food and Drug Administration, the Environmental Protection Agency, the National Institutes of Health, and the Department of Agriculture. ☑

Ethical Questions Raised by Human DNA Technologies

Human DNA technology raises legal and ethical questions—few of which have clear answers. Consider, for example, how the treatment of dwarfism with injections of human growth hormone (HGH) produced by genetically engineered cells might be extended beyond its current use. Should parents of short but hormonally normal children be able to seek HGH treatment to make their kids taller? If not, who decides which children are "tall enough" to be excluded from treatment? In addition to technical challenges, human gene therapy also provokes ethical questions. Some critics believe that tampering with human genes in any way is immoral or unethical. Other observers see no fundamental difference between the transplantation of genes into somatic cells and the transplantation of organs.

Genetic engineering of gametes (sperm or ova) and zygotes has been accomplished in lab animals. It has not been attempted in humans because such a procedure raises very difficult ethical questions. Should we try to eliminate genetic defects in our children and their descendants? Should we interfere with evolution in this way? From a long-term perspective, the elimination of unwanted versions of genes from the gene pool could backfire. Genetic variety is a necessary ingredient for the adaptation of a species as environmental conditions change with time. Genes that are damaging under some conditions may be advantageous under others (one example is the sickle-cell allele—see the Evolution Connection in Chapter 17). Are we willing to risk making genetic changes that could be detrimental to our species in the future?

Similarly, advances in genetic profiling raise privacy issues (**Figure 12.26**). If we were to create a DNA profile of every person at birth, then theoretically we could match nearly every violent crime to a perpetrator because it is virtually impossible for someone to commit a violent crime without leaving behind DNA evidence. But are we, as a society, prepared to sacrifice our genetic privacy, even for such worthwhile goals? In 2014, the U.S. Supreme Court, by a 5–4 vote, upheld the practice of collecting DNA samples from suspects at the time of their arrest (before they had been convicted). Ruling that obtaining DNA is "like fingerprinting and photographing, a legitimate police booking procedure that is reasonable under the Fourth Amendment," the Supreme Court decision will likely usher in an era of expanded use of DNA profiling in many aspects of police work.

As more information becomes available about our personal genetic makeup, some people question whether greater access to this information is always beneficial. For example, mail-in kits (**Figure 12.27**) have become available that can tell healthy people their relative risk of developing various diseases (such as Parkinson's and Crohn's) later in life. Some argue that such information helps families to prepare. Others worry that the tests prey on our fears without offering any real benefit because certain diseases, such as Parkinson's, are not currently preventable or treatable. Other tests, however, such as for breast cancer risk, may help a person make changes that can prevent disease. How can we identify truly useful tests?

There is also a danger that information about disease-associated genes

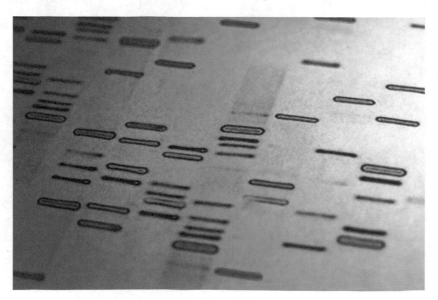

◄ Figure 12.26 **Access to genetic information raises privacy concerns.**

could be abused. One issue is the possibility of discrimination and stigmatization. In response, the U.S. Congress passed the Genetic Information Nondiscrimination Act of 2008. Title I of the act prohibits insurance companies from requesting or requiring genetic information during an application for health insurance. Title II provides similar protections in employment.

A much broader ethical question is how do we really feel about wielding one of nature's powers—the evolution of new organisms? Some might ask if we have any right to alter an organism's genes—or to create new organisms. DNA technologies raise many complex issues that have no easy answers. It is up to you, as a participating citizen, to make informed choices. ☑

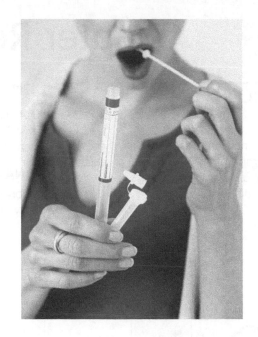

☑ **CHECKPOINT**

Why does genetically modifying a human gamete raise different ethical questions than genetically modifying a human somatic (body) cell?

Answer: A genetically modified somatic cell will affect only the patient. Modifying a gamete will affect an unborn individual as well as all of his or her descendants.

▶ Figure 12.27 **Personalized genetic testing.** This kit can be used to send saliva for genetic analysis. The results can indicate a person's risk of developing certain diseases.

 DNA Profiling EVOLUTION CONNECTION

The Y Chromosome as a Window on History

The human Y chromosome passes essentially intact from father to son. Therefore, by comparing Y DNA, researchers can learn about the ancestry of human males. DNA profiling can thus provide data about recent human evolution.

Geneticists have discovered that about 8% of males currently living in central Asia have Y chromosomes of striking genetic similarity. Further analysis traced their common genetic heritage to a single man living about 1,000 years ago. In combination with historical records, the data led to the speculation that the Mongolian ruler Genghis Khan (Figure 12.28) may have been responsible for the spread of the chromosome to nearly 16 million men living today. A similar study of Irish men suggested that nearly 10% of them were descendants of Niall of the Nine Hostages, a warlord who lived during the 1400s. Another study of Y DNA seemed to confirm the claim by the Lemba people of southern Africa that they are descended from ancient Jews (Figure 12.29). Sequences of Y DNA distinctive of the Jewish priestly caste called Kohanim are found at high frequencies among the Lemba.

Comparison of Y chromosome DNA profiles is part of a larger effort to learn more about the human genome. Other research efforts are extending genomic studies to many more species. These studies will advance our understanding of all aspects of biology, including health and ecology, as well as evolution. In fact, comparisons of the completed genome sequences of bacteria, archaea, and eukaryotes first supported the theory that these are the three fundamental domains of life—a topic we discuss further in the next unit, "Evolution and Diversity."

▲ Figure 12.28 **Genghis Khan.**

▶ Figure 12.29 **Lemba people of southern Africa.**

Chapter Review

■ SUMMARY OF KEY CONCEPTS

Genetic Engineering

DNA technology, the manipulation of genetic material, is a relatively new branch of biotechnology, the use of organisms to make helpful products. DNA technology often involves the use of recombinant DNA, the combination of nucleotide sequences from two different sources.

Recombinant DNA Techniques

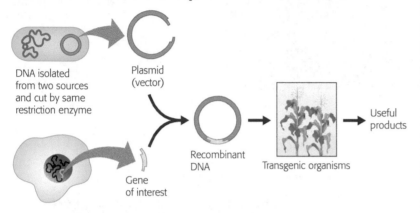

DNA isolated from two sources and cut by same restriction enzyme

Plasmid (vector)

Gene of interest

Recombinant DNA

Transgenic organisms

Useful products

Pharmaceutical Applications

By transferring a human gene into a bacterium or other easy-to-grow cell, scientists can mass-produce valuable human proteins to be used as drugs or vaccines.

Genetically Modified Organisms in Agriculture

Recombinant DNA techniques have been used to create genetically modified organisms, organisms that carry artificially introduced genes. Nonhuman cells have been engineered to produce human proteins, genetically modified food crops, and transgenic farm animals. A transgenic organism is one that carries artificially introduced genes, typically from a different species.

Human Gene Therapy

A virus can be modified to include a normal human gene. If this virus is injected into the bone marrow of a person suffering from a genetic disease, the normal human gene may be transcribed and translated, producing a normal human protein that may cure the genetic disease. This technique has been used in gene therapy trials involving a number of inherited diseases. There have been both successes and failures to date, and research continues.

DNA Profiling and Forensic Science

Forensics, the scientific analysis of legal evidence, has been revolutionized by DNA technology. DNA profiling is used to determine whether two DNA samples come from the same individual.

DNA Profiling Techniques

Short tandem repeat (STR) analysis compares DNA fragments using the polymerase chain reaction (PCR) and gel electrophoresis.

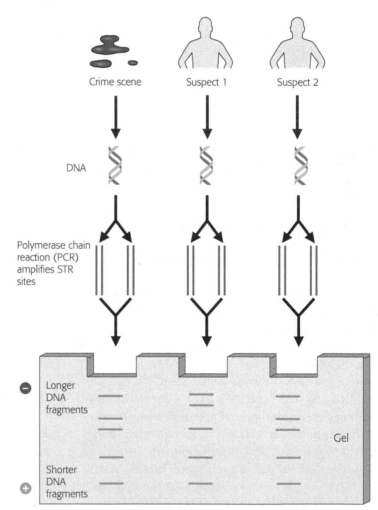

DNA fragments compared by gel electrophoresis
(Bands of shorter fragments move faster toward the positive pole.)

Investigating Murder, Paternity, and Ancient DNA

DNA profiling can be used to establish innocence or guilt of a criminal suspect, identify victims, determine paternity, and contribute to basic research.

Bioinformatics

DNA Sequencing

Automated machines can now sequence many thousands of DNA nucleotides per hour.

Genomics

Advances in DNA sequencing have ushered in the era of genomics, the study of complete genome sets.

Genome-Mapping Techniques

The whole-genome shotgun method involves sequencing DNA fragments from an entire genome and then assembling the sequences.

The Human Genome

The nucleotide sequence of the human genome is providing a wealth of useful data. The 24 different chromosomes of the human genome contain about 3 billion nucleotide pairs and 21,000 genes. The majority of the genome consists of noncoding DNA.

Applied Genomics

Comparing genomes can aid criminal investigations and basic biological research.

Interconnections within Systems: Systems Biology

Success in genomics has given rise to proteomics, the systematic study of the full set of proteins found in organisms. Genomics and proteomics both contribute to systems biology, the study of how many parts work together within complex biological systems.

Safety and Ethical Issues

The Controversy over Genetically Modified Foods

The debate about genetically modified crops centers on whether they might harm humans or damage the environment by transferring genes through cross-pollination with other species.

Ethical Questions Raised by Human DNA Technologies

As members of society we must become educated about DNA technologies so that we can intelligently address the ethical questions raised by their use.

MasteringBiology®

For practice quizzes, BioFlix animations, MP3 tutorials, video tutors, and more study tools designed for this textbook, go to MasteringBiology®

SELF-QUIZ

1. Which of the following best describes recombinant DNA?
   a. DNA that results from bacterial conjugation
   b. DNA that includes pieces from two different sources
   c. An alternate form of DNA that is the product of a mutation
   d. DNA that carries a translocation

2. The enzyme used to bind DNA fragments together is _____.

3. In making recombinant DNA, what is the benefit of using a restriction enzyme that cuts DNA in a staggered fashion?

4. A paleontologist has recovered a bit of organic material from the 400-year-old preserved skin of an extinct dodo. She would like to compare DNA from the sample with DNA from living birds. The most useful method for initially increasing the amount of dodo DNA available for testing is _____.

5. Why do DNA fragments containing STR sites from different people tend to migrate to different locations during gel electrophoresis?

6. Gel electrophoresis separates DNA fragments on the basis of differences in their _____.
   a. nucleotide sequences
   b. lengths
   c. hydrogen bonds between base pairs
   d. pH

7. After a gel electrophoresis procedure is run, the pattern of bars in the gel shows
   a. the order of bases in a particular gene.
   b. the presence of various-sized fragments of DNA.
   c. the order of genes along particular chromosomes.
   d. the exact location of a specific gene in the genome.

8. Name the steps of the whole-genome shotgun method.

9. Put the following steps of human gene therapy in the correct order.
   a. Virus is injected into patient.
   b. Human gene is inserted into a virus.
   c. Normal human gene is isolated and cloned.
   d. Normal human gene is transcribed and translated in the patient.

Answers to these questions can be found in Appendix: Self-Quiz Answers.

THE PROCESS OF SCIENCE

10. Large-scale sequencing of the mitochondrial DNA of humans across the globe revealed significantly higher sequence diversity among Africans than among non-Africans. What does this tell us about the origin of humans?

11. **Interpreting Data** When comparing genomes from different species, biologists often calculate the genome density, the number of genes per number of nucleotides in the genome. Refer to Table 12.1. You can calculate the gene density of each species by dividing the number of genes by the size of the genome (usually expressed in Mb, which is mega base pairs, or 1 million base pairs). Using a spreadsheet, calculate the gene density for every species in the table. (Don't forget that 1 billion = 1,000 million; for example, humans have 3,000 Mb.) How does the gene density of bacteria compare to humans? Humans and roundworms have nearly the same number of genes, but how do the gene densities of these two species compare? Can you identify any general correlation between gene density and the size or complexity of an organism?

12. Listed below are 4 of the 13 genome sites used to create a standard DNA profile. Each site consists of a number of short tandem repeats: sets of 4 nucleotides repeated in a row within the genome. For each site, the number of repeats found at that site for this individual are listed:

Chromosome number	Genetic site	# of repeats
3	D3S1358	4
5	D5S818	10
7	D7S820	5
8	D8S1179	22

Imagine you perform a PCR procedure to create a DNA profile for this individual. Which of the following four gels correctly represents the DNA profile of this person?

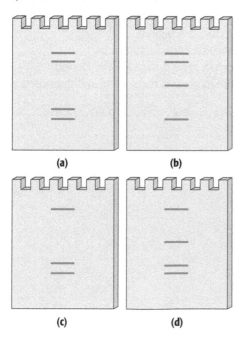

(a)   (b)

(c)   (d)

BIOLOGY AND SOCIETY

13. In the not-too-distant future, gene therapy may be used to treat many inherited disorders. What do you think are the most serious ethical issues to face before human gene therapy is used on a large scale? Explain.

14. Today, it is fairly easy to make transgenic plants and animals. What are some safety and ethical issues raised by this use of recombinant DNA technology? What are some dangers of introducing genetically engineered organisms into the environment? What are some reasons for and against leaving such decisions to scientists? Who should make these decisions?

15. In October 2002, the government of the African nation of Zambia announced that it was refusing to distribute 15,000 tons of corn donated by the United States, enough corn to feed 2.5 million Zambians for three weeks. The government rejected the corn because it was likely to contain genetically modified kernels. The government made the decision after its scientific advisers concluded that the studies of the health risks posed by GM crops "are inconclusive." Do you agree with Zambia's decision to turn away the corn? Why or why not? In your answer, consider that at the time, Zambia was facing food shortages, and 35,000 Zambians were expected to starve to death over the next six months. How do the risks posed by GM crops compare to the risk of starvation?

16. From 1977 to 2000, 12 convicts were executed in Illinois. During that same period, 13 death row inmates were exonerated based on DNA evidence. In 2000, the governor of Illinois declared a moratorium on all executions in his state because the death penalty system was "fraught with errors." Do you support the Illinois governor's decision? What rights should death penalty inmates have with regard to DNA testing of old evidence? Who should pay for this additional testing?

Unit 3
Evolution and Diversity

13 **How Populations Evolve**

Chapter Thread: **Evolution Past, Present, and Future**

14 **How Biological Diversity Evolves**

Chapter Thread: **Mass Extinctions**

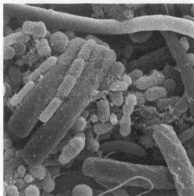

15 **The Evolution of Microbial Life**

Chapter Thread: **Human Microbiota**

16 **The Evolution of Plants and Fungi**

Chapter Thread: **Plant-Fungus Interactions**

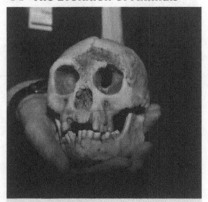

17 **The Evolution of Animals**

Chapter Thread: **Human Evolution**

How Populations Evolve

Thanks to natural selection, populations of more than 500 insect species are invulnerable to
▼ the most widely used pesticides.

▲ If it weren't for artificial selection, you'd be eating tomatoes the size of blueberries.

In some ► species, males engage in nonlethal combat in the hope of mating with females.

▲ Populations of endangered species may be doomed by their lack of genetic diversity.

Evolution Past, Present, and Future BIOLOGY AND SOCIETY

Evolution Today

Life on Earth is gloriously diverse. More than 1.3 million species have been identified, and scientists estimate that several million more await discovery. How did Earth come to be populated by this abundance of diversity? The answer is evolution. In studying evolution, we reach hundreds of millions of years into the past to trace the history of life up until the present time. It isn't all about stones and bones, though. Evolution is happening right now, somewhere near you.

Consider the biological diversity in your surroundings. The organisms that are our closest neighbors are those that are best adapted to a human-dominated environment—the birds that feed on seeds and insects in suburban yards or city parks; the plants that thrive in vacant lots or sidewalk cracks; the insects and other pests that feast on our discarded food, our crops, and even our blood. Then there's the vast variety of microscopic life teeming everywhere, including both beneficial and disease-causing species that have adapted to the human body. All of these organisms are products of evolution, and they continue to evolve generation by generation in the present day. As you will learn in this chapter, the environment plays a powerful role in evolution. Human activities—agriculture, mining, logging, development, the burning of fossil fuels, and the use of medicine, to name just a few—may alter the environments of organisms in ways that result in rapid, observable evolution. Evolutionary changes can also be studied in settings affected by natural environmental changes. Within a hundred thousand years, a blink of the eye in geologic time, lakes may appear or disappear; and newly minted volcanic islands can be clothed in plants and animals over a span of decades, centuries, or millennia, depending on the nearness of land.

White-tailed deer. Suburban landscapes often include patches of forest that provide food and shelter for deer, such as this one nibbling on garden plants.

An understanding of evolution informs all of biology, from exploring life's molecules to analyzing ecosystems. And applications of evolutionary biology are transforming medicine, agriculture, biotechnology, and conservation biology. In this chapter, you'll learn how the process of evolution works and read about verifiable, measurable examples of evolution that affect our world.

The Diversity of Life

For all of human history, people have named, described, and classified the inhabitants of the natural world. As trade and exploration connected all regions of the planet, these tasks became increasingly complex. For example, a scholar who sought to describe all the types of plants known to the Greeks in 300 B.C. had about 500 species to distinguish. Today, scientists recognize roughly 400,000 plant species.

By the 1700s, it was clear that a unified system of naming and classifying was needed. Agreement eventually coalesced around a scheme introduced by Carolus Linnaeus, a Swedish scientist. His system, which is still in use today, is the basis for **taxonomy**, the branch of biology concerned with identifying, naming, and classifying species. The Linnaean system includes a method of naming species and a hierarchical classification of species into broader groups of organisms.

Naming and Classifying the Diversity of Life

In the Linnaean system, each species is given a two-part Latinized name, or **binomial**. The first part of a binomial is the **genus** (plural, *genera*), a group of closely related species. For example, the genus of large cats is *Panthera*. The second part of a binomial is used to distinguish species within a genus. The two parts must be used together to name a species. Thus, the scientific name for the leopard is *Panthera pardus*. Notice that the first letter of the genus is capitalized and that the whole binomial is italicized. For instance, a newly discovered spider in the genus *Aptostichus* was named *Aptostichus stephencolberti*, after the television personality.

The Linnaean binomial solved the problem of the ambiguity of common names. Referring to an animal as a squirrel or a plant as a daisy is not specific—there are many species of squirrels and daisies. In addition, people in different regions may use the same common name for different species. For example, the flowers called bluebells in Scotland, England, Texas, and the eastern United States are actually four unrelated species.

Linnaeus also introduced a system for grouping species into a hierarchy of categories. The first step of this classification is built into the binomial. For example, the genus *Panthera* contains three other species: the lion (*Panthera leo*), the tiger (*Panthera tigris*), and the jaguar (*Panthera onca*). Beyond the grouping of species within genera, taxonomy extends to progressively broader categories of classification. It places similar genera in the same **family**, puts families into **orders**, orders into **classes**, classes into **phyla** (singular, *phylum*), phyla into **kingdoms**, and kingdoms into **domains**. Figure 13.1 places the leopard in this taxonomic scheme of groups within groups. The resulting classification of a particular organism is somewhat like a postal address identifying a person in a particular apartment, in a building with many apartments, on a street with many apartment buildings, in a city with many streets, and so on.

Grouping organisms into broader categories is a way to structure our understanding of the world. However, the criteria used to define more inclusive groups such as families, orders, and classes are ultimately arbitrary. After you learn about the processes by which the diversity of life evolved, we will introduce a classification system based on an understanding of evolutionary relationships (see Chapter 14). ☑

◄ **Figure 13.1 Hierarchical classification.** Taxonomy classifies species—the least inclusive groups—into increasingly broad categories. *Panthera pardus* is one of four species (indicated here with yellow boxes) in the genus *Panthera*; *Panthera* is a genus (orange box) in the family Felidae, and so on.

Species: *Panthera pardus*

Genus: *Panthera*

Family: Felidae

Order: Carnivora

Class: Mammalia

Phylum: Chordata

Kingdom: Animalia

Bacteria — **Domain:** Eukarya — Archaea

Explaining the Diversity of Life

While early naturalists and philosophers sought to describe and organize the diversity of life, they also sought to explain its origin. The explanation accepted by present-day biologists is the evolutionary theory proposed by Charles Darwin in his best-known book, *On the Origin of Species by Means of Natural Selection*, published in 1859. Before we introduce Darwin's theory, however, let's take a brief look at the scientific and cultural context that made the theory of evolution such a radical idea in Darwin's time.

The Idea of Fixed Species

The Greek philosopher Aristotle, whose ideas had an enormous impact on Western culture, generally held the view that species are fixed, permanent forms that do not change over time. Judeo-Christian culture reinforced this idea with a literal interpretation of the biblical book of Genesis, which tells the story of each form of life being individually created in its present-day form. In the 1600s, religious scholars used biblical accounts to estimate the age of Earth at 6,000 years. Thus, the idea that all living species came into being relatively recently and are unchanging in form dominated the intellectual climate of the Western world for centuries.

At the same time, however, naturalists were also grappling with the interpretation of **fossils**—imprints or remains of organisms that lived in the past. Although fossils were thought to be the remains of living creatures, many were puzzling. For example, if "snakestones" (Figure 13.2a) were the coiled bodies of snakes, then why were none ever found with an intact head? Could some fossils represent species that had become extinct? Stunning discoveries in the early 1800s, including fossilized skeletons of a gigantic sea creature dubbed an ichthyosaur, or fish-lizard (Figure 13.2b), convinced many naturalists that extinctions had indeed occurred.

Lamarck and Evolutionary Adaptations

Fossils told of other changes in the history of life, too. Naturalists compared fossil forms with living species and noted patterns of similarities and differences. In the early 1800s, French naturalist Jean-Baptiste de Lamarck suggested that the best explanation for these observations is that life evolves. Lamarck explained evolution as the refinement of traits that equip organisms to perform successfully in their environments. He proposed that by using or not using its body parts, an individual may develop certain traits that it passes on to its offspring. For example, some birds have powerful beaks that enable them to crack tough seeds. Lamarck suggested that these strong beaks are the cumulative result of ancestors exercising their beaks during feeding and passing that acquired beak power on to offspring. However, simple observations provide evidence against the inheritance of acquired traits: A carpenter who builds up strength and stamina through a lifetime of pounding nails with a heavy hammer will not pass enhanced biceps on to children. Although Lamarck's idea of how species evolve was mistaken, his proposal that species evolve as a result of interactions between organisms and their environments helped set the stage for Darwin. ☑

☑ **CHECKPOINT**

How do fossils contradict the idea of fixed species?

Answer: They don't match any living creature.

▼ Figure 13.2 **Fossils that perplexed naturalists in the 1800s.**

(a) "Snakestone." This fossil is actually a mollusc called an ammonite, an extinct relative of the present-day nautilus (see Figure 17.13). Ammonites of this type ranged in size from about several inches to more than 7 feet in diameter.

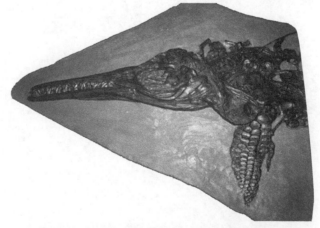

(b) Icthyosaur skull and paddle-like forelimb. These marine reptiles—some more than 50 feet long—ruled the oceans for 155 million years before becoming extinct about 90 million years ago. The enormous eye is thought to be an adaptation to the dim light of the deep sea.

Charles Darwin and *The Origin of Species*

Although Charles Darwin was born more than 200 years ago—on the very same day as Abraham Lincoln—his work had such an extraordinary impact that many scientists mark his birthday with a celebration of his contributions to biology. How did Darwin become a rock star of science?

As a boy, Darwin was fascinated with nature. He loved collecting insects and fossils, as well as reading books about nature. His father, an eminent physician, could see no future for his son as a naturalist and sent him to medical school. But young Darwin, finding medicine boring and surgery before the days of anesthesia horrifying, quit medical school. His father then enrolled him at Cambridge University with the intention that he should become a clergyman. After college, however, Darwin returned to his childhood interests rather than following the career path mapped out by his father. At the age of 22, he began a sea voyage on the HMS *Beagle* that helped frame his theory of evolution.

Darwin's Journey

The *Beagle* was a survey ship. Although it stopped at many locations around the world, its main task was charting poorly known stretches of the South American coast (Figure 13.3). Darwin, a skilled naturalist, spent most of his time on shore doing what he enjoyed most— exploring the natural world. He collected thousands of specimens of fossils and living plants and animals. He also kept detailed journals of his observations. For a naturalist from a small, temperate country, seeing the glorious diversity of unfamiliar life-forms on other continents was a revelation. He carefully noted the characteristics of plants and animals that made them well suited to such diverse environments as the jungles of Brazil, the grasslands of Argentina, the towering peaks of the Andes, and the desolate and frigid lands at the southern tip of South America.

Observations

Many of Darwin's observations indicated that geographic proximity is a better predictor of relationships among organisms than similarity of environment. For example, the plants and animals living in temperate regions of South America more closely resembled species living in tropical regions of that continent than

▼ **Figure 13.3 The voyage of the *Beagle*.**

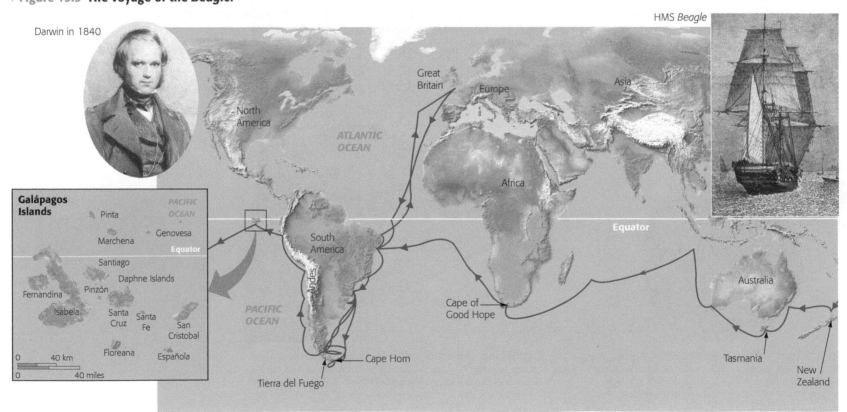

species living in similarly temperate regions of Europe. And the South American fossils Darwin found, though clearly examples of species different from living ones, were distinctly South American in their resemblance to the contemporary plants and animals of that continent. For instance, he collected fossilized armor plates resembling those of living armadillo species. Paleontologists later reconstructed the creature to which the armor belonged—which turned out to be an extinct armadillo the size of a Volkswagen Beetle.

Darwin was particularly intrigued by the geographic distribution of organisms on the Galápagos Islands. The Galápagos are relatively young volcanic islands about 900 kilometers (540 miles) off the Pacific coast of South America. Most of the animals that inhabit these remote islands are found nowhere else in the world, but they resemble South American species.

Darwin noticed that Galápagos marine iguanas—with a flattened tail that aids in swimming—are similar to, but distinct from, land-dwelling iguanas on the islands and on the South American mainland. Furthermore, each island had its own distinct variety of giant tortoise **(Figure 13.4)**, the strikingly unique inhabitants for which the islands were named (*galápago* means "tortoise" in Spanish).

New Insights

While on his voyage, Darwin was strongly influenced by the newly published *Principles of Geology* by Scottish geologist Charles Lyell. The book presented the case for an ancient Earth sculpted over millions of years by gradual geological processes that continue today. Darwin witnessed the power of natural forces to change Earth's surface firsthand when he experienced an earthquake that raised part of the coastline of Chile almost a meter.

By the time Darwin returned to Great Britain five years after the *Beagle* first set sail, he had begun to seriously doubt that Earth and all its living organisms had been specially created only a few thousand years earlier. As he reflected on his observations, analyzed his specimen collections, and discussed his work with colleagues, he concluded that the evidence was better explained by the hypothesis that present-day species are the descendants of ancient ancestors that they still resemble in some ways. Over time, differences gradually accumulated by a process that Darwin called "descent with modification," his phrase to describe evolution. Unlike others who had explored the idea that organisms had changed over time, however, Darwin also proposed a scientific mechanism for *how* life evolves, a process he called natural selection. In **natural selection**, individuals with certain inherited traits are more likely to survive and reproduce than are individuals with other traits. He hypothesized that as the descendants of a remote ancestor spread into various habitats over millions and millions of years, natural selection resulted in diverse modifications, or **evolutionary adaptations**, that fit them to specific ways of life in their environment.

▼ Figure 13.4 **Two varieties of Galápagos tortoise.**

(a) The thick, domed shell and short neck and legs are characteristic of tortoises found on wetter islands that have more abundant, dense vegetation.

(b) Saddleback shells have an arch at the front, which allows the long neck to emerge. This, along with longer legs, enables the tortoise to stretch higher to reach the scarce vegetation on dry islands.

✓ **CHECKPOINT**

What was the most
significant difference
between Darwin's theory
of evolution and the ideas
about evolution that had
been proposed previously?

*Answer: Darwin also proposed a
mechanism (natural selection) for
how evolution occurs.*

 **Evolution** Darwin's Theory

Darwin spent the next two decades compiling and writing about evidence for evolution. He realized that his ideas would cause an uproar, however, and he delayed publishing. Meanwhile, Darwin learned that Alfred Russel Wallace, a British naturalist doing fieldwork in Indonesia, had conceived a hypothesis almost identical to Darwin's. Not wanting to have his life's work eclipsed by Wallace's, Darwin finally published *The Origin of Species*, a book that supported his hypothesis with immaculate logic and hundreds of pages of evidence drawn from observations and experiments in biology, geology, and paleontology. The hypothesis of evolution set forth in *The Origin of Species* has since generated predictions that have been tested and verified by more than 150 years of research. Consequently, scientists regard Darwin's concept of evolution by means of natural selection as a **theory**—a widely accepted explanatory idea that is broader in scope than a hypothesis, generates new hypotheses, and is supported by a large body of evidence.

In the next several pages, we examine lines of evidence for Darwin's theory of **evolution**, the idea that living species are descendants of ancestral species that were different from present-day ones. Then we will return to natural selection as the mechanism for evolutionary change. With our current understanding of how this mechanism works, we extend Darwin's definition of evolution to include genetic changes in a population from generation to generation. ✓

Evidence of Evolution

Evolution leaves observable signs. Such clues to the past are essential to any historical science. Historians of human civilization can study written records from earlier times. But they can also piece together the evolution of societies by recognizing vestiges of the past in modern cultures. Even if we did not know from written documents that Spaniards colonized the Americas, we would deduce this from the Hispanic stamp on Latin American culture. Similarly, biological evolution has left evidence in fossils, as well as in today's organisms.

Evidence from Fossils

Fossils—imprints or remains of organisms that lived in the past—document differences between past and present organisms and show that many species have become extinct. The soft parts of a dead organism usually decay rapidly, but the hard parts of an animal that are rich in minerals, such as the bones and teeth of vertebrates and the shells of clams and snails, may remain as fossils. **Figure 13.5** (facing page) illustrates some of the ways that organisms can fossilize.

Not all fossils are the actual remnants of organisms. Some, such as the ammonite in Figure 13.2a, are casts. A cast forms when a dead organism that was buried in sediment decomposes and leaves an empty "mold" that is later filled by minerals dissolved in water. The minerals harden within the mold, making a replica of the organism. You may have seen crime scene investigators on TV shows use fast-acting plaster in the same way to make casts of footprints or tire tracks. Fossils may also be imprints, such as footprints or burrows, that remain after the organism decays. **Paleontologists** (scientists who study fossils) also eagerly examine coprolites—fossilized feces—for clues about the diets and digestive systems of extinct animals.

In rare instances, an entire organism, including its soft parts, is encased in a medium that prevents decomposition. Examples include insects trapped in amber (fossilized tree resin) and mammoths, bison, and even prehistoric humans frozen in ice or preserved in bogs.

Many fossils are found in fine-grained sedimentary rocks formed from the sand or mud that settles to the bottom of seas, lakes, swamps, and other aquatic habitats, covering dead organisms. Over millions of years, new layers of sediment are deposited atop older ones and compress them into layers of rock called strata (singular, *stratum*). Thus, the fossils in a particular stratum provide a glimpse of some of the organisms that lived in the area at the time the layer formed. Because younger strata are on top of older layers, the relative ages of fossils can be determined by the layer in which they are found. (Radiometric dating—see Figure 2.17—can be used to determine the approximate ages of fossils.) As a result, the sequence in which fossils appear within layers of sedimentary rocks is a historical record of life on

Earth. The **fossil record** is this ordered sequence of fossils as they appear in the rock layers, marking the passage of geologic time (see Figure 14.12).

Of course, as Darwin acknowledged, the fossil record is incomplete. Many of Earth's organisms did not live in areas that favor fossilization. Many fossils that did form were in rocks later distorted or destroyed by geologic processes. Furthermore, not all fossils that have been preserved are accessible to paleontologists. Even with its limitations, however, the fossil record is remarkably detailed. And although the incompleteness of the fossil record may seem frustrating (wouldn't it be nice to have all our questions answered!), it makes paleontology an unexpectedly thrilling occupation. Like a mystery series in which new clues are uncovered in each episode, the thousands of fossils newly discovered each year give paleontologists new opportunities to test hypotheses about how the diversity of life evolved. In the next section, you'll learn how fossils are revealing answers to a very old puzzle. ☑

☑ **CHECKPOINT**

Why are older fossils generally in deeper rock layers than younger fossils?

Answer: Sedimentation places younger rock layers on top of older ones.

▼ Figure 13.5 **A fossil gallery.**

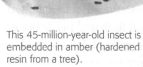
Sedimentary fossils are formed when minerals seep into and replace organic matter. This petrified (stone) tree in Arizona's Petrified Forest National Park is about 190 million years old.

Sedimentary rocks are the richest hunting grounds for paleontologists, scientists who study fossils. This researcher is excavating a fossilized dinosaur skeleton from sandstone in Dinosaur National Monument, located in Utah and Colorado.

This 45-million-year-old insect is embedded in amber (hardened resin from a tree).

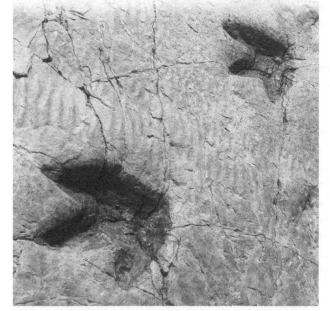

A dinosaur left these footprints 120 million years ago in what is now northern Spain. Biologists often study footprints to learn how extinct animals moved.

These tusks belong to a whole 23,000-year-old mammoth, which scientists discovered in Siberian ice in 1999.

Did Whales Evolve from Land-Dwelling Mammals?

In *The Origin of Species*, Darwin predicted the existence of fossils of transitional forms linking very different groups of organisms. The first such fossil was discovered shortly after Darwin's book was published—the 150-million-year-old *Archaeopteryx*, a fossil that displayed a combination of reptilian and birdlike features. Thousands of fossil discoveries have since shed light on the evolutionary origin of birds from a lineage of dinosaurs, as well as the origins of many other groups of plants and animals, including the transition of fish to amphibian and the evolution of mammals from a reptilian ancestor.

The origin of whales is one of the most intriguing evolutionary transitions. Whales are cetaceans, a group that also includes dolphins and porpoises. As you probably know, cetaceans are mammals that are thoroughly adapted to their aquatic environment. For example, their ears are internal and highly specialized for underwater hearing. They have forelimbs in the form of flippers but lack hind limbs. Overall, cetaceans are so very different from other mammals that scientists have long puzzled over their origins. In the 1960s, **observations** of fossil teeth led paleontologists to form the **hypothesis** that whales were the descendants of primitive hoofed, wolflike carnivores. They made the **prediction** that transitional fossils would show reduced hind limb and pelvic bones as whales evolved from these terrestrial, four-limbed

ancestors. Rather than a controlled **experiment**, paleontologists tested their hypothesis by making detailed measurements and other observations of fossils accumulated over 20 years of extraordinary discoveries in Pakistan and Egypt. The **results** supported their hypothesis. **Figure 13.6** illustrates the diminishing size of the hind limb and pelvic bones in a few of the specimens studied. As you will learn shortly, however, a different kind of evidence later called this conclusion into question.

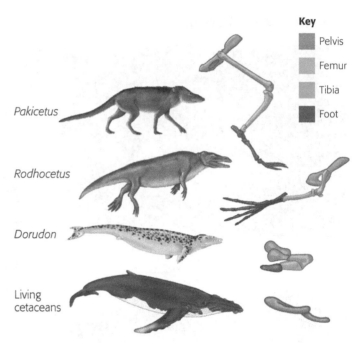

Key
- Pelvis
- Femur
- Tibia
- Foot

Pakicetus

Rodhocetus

Dorudon

Living cetaceans

▲ Figure 13.6 **Transitional forms in the evolution of whales.**

Evidence from Homologies

A second type of evidence for evolution comes from analyzing similarities among different organisms. Evolution is a process of descent with modification—characteristics present in an ancestral organism are altered over time by natural selection as its descendants face different environmental conditions. In other words, evolution is a remodeling process. As a result, related species can have characteristics that have an underlying similarity yet function differently. Similarity resulting from common ancestry is known as **homology**.

Darwin cited the anatomical similarities among vertebrate forelimbs as evidence of common ancestry. As **Figure 13.7** shows, the same skeletal elements make up the forelimbs of humans, cats, whales, and bats. The functions of these forelimbs differ. A whale's flipper does

not do the same job as a bat's wing, so if these structures had been uniquely engineered, then we would expect that their basic designs would be very different. The logical explanation instead is that the arms, forelegs, flippers, and wings of these different mammals are variations on an anatomical structure of an ancestral organism that over millions of years has become adapted to different functions. Biologists call such anatomical similarities in different organisms homologous structures—features that often have different functions but are structurally similar because of common ancestry.

As a result of advances in **molecular biology**, the study of the molecular basis of genes and gene expression, present-day scientists have a much deeper understanding of homologies than Darwin did. Just as your genetic background is recorded in the DNA you inherit from your parents, the evolutionary history of each

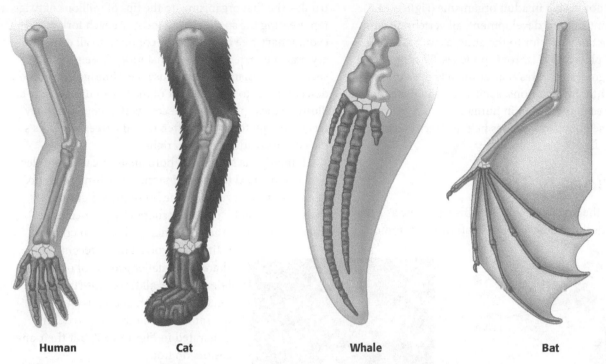

| Human | Cat | Whale | Bat |

▲ Figure 13.7 **Homologous structures: anatomical signs of descent with modification.** The forelimbs of all mammals are constructed from the same skeletal elements. (Homologous bones in each of these four mammals are colored the same.) The hypothesis that all mammals descended from a common ancestor predicts that their forelimbs, though diversely adapted, would be variations on a common anatomical theme.

species is documented in the DNA inherited from its ancestral species. If two species have homologous genes with sequences that match closely, biologists conclude that these sequences must have been inherited from a relatively recent common ancestor. Conversely, the greater the number of sequence differences between species, the more distant is their last common ancestor. Molecular comparisons between diverse organisms have allowed biologists to develop and test hypotheses about the evolutionary divergence of major branches on the tree of life.

Darwin's boldest hypothesis was that all life-forms are related. Molecular biology provides strong evidence for this claim: All forms of life use the same genetic language of DNA and RNA, and the genetic code—how RNA triplets are translated into amino acids—is essentially universal (see Figure 10.10). Thus, it is likely that all species descended from common ancestors that used this code. Because of these molecular homologies, bacteria engineered with human genes can produce human proteins such as insulin and human growth hormone. But molecular homologies go beyond a shared genetic code. For example, organisms as dissimilar as humans and bacteria share homologous genes inherited from a very distant common ancestor.

Geneticists have also uncovered hidden molecular homologies. Organisms may retain genes that have lost their function through mutations, even though homologous genes in related species are fully functional. Many

of these inactive genes have been identified in humans. For example, one such gene encodes an enzyme known as GLO that is used in making vitamin C. Almost all mammals have a metabolic pathway to make this essential vitamin from glucose. Although humans and other primates have functional genes for the first three steps in the pathway, the inactive GLO gene prevents us from making vitamin C—we must get sufficient amounts in our diet to maintain health.

Some of the most interesting homologies are "leftover" structures that are of marginal or perhaps no importance to the organism. These **vestigial structures** are remnants of features that served important functions in the organism's ancestors. For example, the small pelvis and hind-leg bones of ancient whales are vestiges (traces) of their walking ancestors. The eye remnants that are buried under scales in blind species of cave fishes—a vestige of their sighted ancestors—are another example. Humans have vestigial structures, too. When we're cold or agitated, we often get goose bumps caused by small muscles under the skin that make the body hair stand on end. The same response is more visible (and more functional) in a bird that fluffs up its feathery insulation (see Figure 18.9) or a cat that bristles when threatened.

An understanding of homology can also explain observations about embryonic development that are otherwise puzzling. For example, comparing early stages of development in different animal species reveals

285

✅ **CHECKPOINT**

How does the need for
dietary vitamin C show that
humans are more closely
related to other primates
than to other mammals?

Answer: Humans and primates need vitamin C because a gene needed to make it in the body is nonfunctional. A homologous gene in most other mammals is functional. Thus, the gene's function must have been lost by mutation in an early primate or primate ancestor and inherited in that form by both humans and other primates.

similarities not visible in adult organisms (**Figure 13.8**). At some point in their development, all vertebrate embryos have a tail posterior to the anus, as well as structures called pharyngeal (throat) pouches. These pouches are homologous structures that ultimately develop to have very different functions, such as gills in fishes and parts of the ears and throat in humans.

Next we see how homologies help us trace evolutionary descent. ✅

Evolutionary Trees

Darwin was the first to visualize the history of life as a tree in which patterns of descent branch off from a common

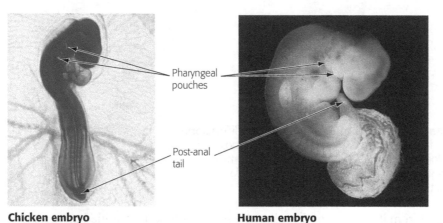

► **Figure 13.8 Evolutionary signs from comparative embryology.** At the early stage of development shown here, the kinship of vertebrates is unmistakable. Notice, for example, the pharyngeal pouches and tails in both the chicken embryo and the human embryo.

Pharyngeal pouches

Post-anal tail

Chicken embryo

Human embryo

trunk—the first organism—to the tips of millions of twigs representing the species living today. At each fork of the evolutionary tree is an ancestor common to all evolutionary branches extending from that fork. Closely related species share many traits because their lineage of common descent traces to a recent fork of the tree of life. Biologists illustrate these patterns of descent with an **evolutionary tree**, although today they often turn the trees sideways so they can be read from left to right.

Homologous structures, both anatomical and molecular, can be used to determine the branching sequence of an evolutionary tree. Some homologous characters, such as the genetic code, are shared by all species because they date to the deep ancestral past. In contrast, traits that evolved more recently are shared by smaller groups of organisms. For example, all tetrapods (from the Greek *tetra*, four, and *pod*, foot) have the same basic limb bone structure illustrated in Figure 13.7, but their ancestors do not.

Figure 13.9 is an evolutionary tree of tetrapods (amphibians, mammals, and reptiles, including birds) and their closest living relatives, the lungfishes. In this diagram, each branch point represents the common ancestor of all species that

▼ Figure 13.9 **An evolutionary tree of tetrapods (four-limbed animals).**

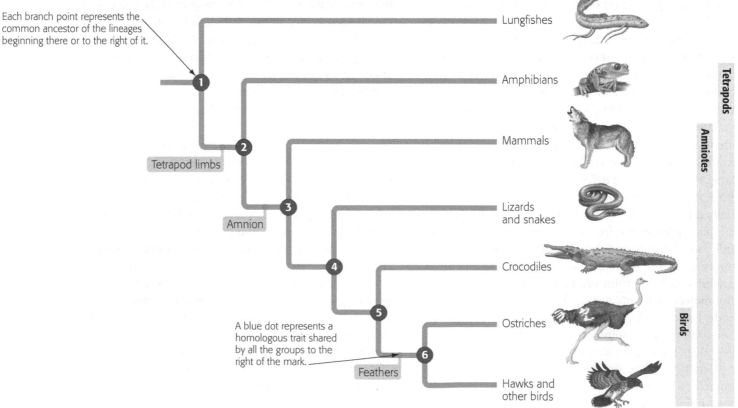

Each branch point represents the common ancestor of the lineages beginning there or to the right of it.

Tetrapod limbs

Amnion

A blue dot represents a homologous trait shared by all the groups to the right of the mark.

Feathers

Lungfishes

Amphibians

Mammals

Lizards and snakes

Crocodiles

Ostriches

Hawks and other birds

Tetrapods

Amniotes

Birds

descended from it. For example, lungfishes and all tetrapods descended from ancestor **1**. Three homologies are shown by the blue dots on the tree—tetrapod limbs, the amnion (a protective embryonic membrane), and feathers. Tetrapod limbs were present in common ancestor **2** and hence are found in its descendants (the tetrapods). The amnion was present in ancestor **3** and thus is shared only by mammals and reptiles, which are known as amniotes. Feathers were present only in **6** ancestor and hence are found only in birds.

Evolutionary trees are hypotheses reflecting our current understanding of patterns of evolutionary descent. Some trees, such as the one in Figure 13.9, are based on a convincing combination of fossil, anatomical, and molecular data. Others are more speculative because sufficient data are not yet available. ☑

Whale Evolution Revisited

Now let's resume the story of whale evolution. As we discussed, beginning in the 1970s, paleontologists unearthed a remarkable series of transitional fossils that supported the hypothesis that whales evolved from hoofed, wolflike carnivores. However, molecular biologists using DNA analysis to infer relationships among living animals found a close relationship between whales and hippopotamuses, which are members of a group of mostly herbivorous, cloven-hoofed mammals that includes pigs, deer, and camels (**Figure 13.10**). Consequently, they proposed an alternative hypothesis of whale evolution: that whales and hippos are both descendants of a cloven-hoofed ancestor.

Paleontologists were taken aback by the contradictory results. Nevertheless, openness to new evidence is a hallmark of science, and the paleontologists had an idea

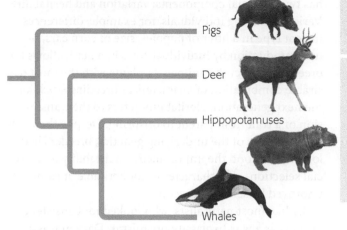

Pigs

Deer

Hippopotamuses

Whales

▲ **Figure 13.10 An evolutionary tree of present-day whales and herbivorous cloven-hoofed mammals.**
Data from: M. Nikaido et al., Phylogenetic relationships among Cetartiodactyls based on insertions of short and long interspersed elements: hippopotamuses are the closest extant relatives of whales. *Proceedings of the National Academy of Sciences USA* 96: 10261–10266 (1999).

☑ **CHECKPOINT**

In Figure 13.9, which number represents the most recent common ancestor of humans and canaries?

Answer: The most recent ancestor of humans (mammals) and canaries (birds) is **3**

for resolving the issue. Cloven-hoofed mammals have a unique ankle bone. Like most fossilized skeletons, the specimens of early cetaceans that had been found were incomplete—none included an ankle bone. If the ancestor of whales was a wolflike carnivore, then the shape of its ankle bone would be similar to most present-day mammals. Two fossils discovered in 2001 provided the answer. Both *Pakicetus* and *Rodhocetus* (see Figure 13.6) had the distinctive ankle bone of a cloven-hoofed mammal, a result that supported the hypothesis based on the DNA analysis. Thus, as is often the case in science, scientists are becoming more certain about the evolutionary origin of whales as mounting evidence from different lines of inquiry converge.

Natural Selection as the Mechanism for Evolution

Now that you have learned about the lines of evidence supporting Darwin's theory of descent with modification, let's look at Darwin's explanation of *how* life evolves. Because he hypothesized that species formed gradually over long periods of time, Darwin knew that he would not be able to study the evolution of new species by direct observation. But he did have a way to gain insight into the process of incremental change: the practices used by plant and animal breeders.

All domesticated plants and animals are the products of selective breeding from wild ancestors. For example,

the baseball-sized tomatoes grown today are very different from their Peruvian ancestors, which were not much larger than blueberries. Having conceived the notion that **artificial selection**—the selective breeding of domesticated plants and animals to promote the occurrence of desirable traits in the offspring—was the key to understanding evolutionary change, Darwin bred pigeons to gain firsthand experience. He acquired further insight through conversations with farmers about livestock breeding. He learned that artificial selection

If it weren't for artificial selection, you'd be eating tomatoes the size of blueberries.

287

has two essential components: variation and heritability. Variation among individuals, for example, differences in coat type in a litter of puppies, size of corn ears, or milk production by individual cows in a herd, allows the breeder to select the animals or plants with the most desirable combination of characters as breeding stock for the next generation. Heritability refers to the transmission of a trait from parent to offspring. Despite their lack of knowledge of the underlying genetics, breeders had long understood the importance of heritability in artificial selection—if a character is not heritable, it cannot be improved by selective breeding.

Unlike most naturalists, who looked for consistency of traits as a way to classify organisms, Darwin was a careful observer of variations between individuals. He knew that individuals in natural populations have small but measurable differences, for example, variations in color and markings (Figure 13.11). But what forces in nature determined which individuals became the breeding stock for the next generation?

Darwin found inspiration in an essay written by economist Thomas Malthus, who contended that much of human suffering—disease, famine, and war—was the consequence of human populations increasing faster than food supplies and other resources. Darwin applied Malthus's idea to populations of plants and animals, reasoning that the resources of any given environment are limited. The production of more individuals than the environment can support leads to a struggle for existence, with only some offspring surviving in each generation (Figure 13.12). Of the many eggs laid, young born, and seeds spread, only a tiny fraction complete development and leave offspring themselves. The rest are eaten, starved, diseased, unmated, or unable to reproduce for other reasons. The essence of natural selection

▲ Figure 13.12 **Overproduction of offspring.** This sea slug, a mollusc related to snails (see Figure 17.13), is laying thousands of eggs embedded in the yellow ribbon around its body. Only a tiny fraction of the eggs will actually give rise to offspring that survive and reproduce.

is this unequal reproduction. In the process of natural selection, individuals whose traits better enable them to obtain food, escape predators, or tolerate physical conditions will survive and reproduce more successfully, passing these adaptive traits to their offspring.

Darwin reasoned that if artificial selection can bring about significant change in a relatively short period of time, then natural selection could modify species considerably over hundreds or thousands of generations. Over vast spans of time, many traits that adapt a population to its environment will accumulate. If the environment changes, however, or if individuals move to a new environment, natural selection will select for adaptations to these new conditions, sometimes producing changes that result in the origin of a completely new species.

Next, let's look at an example of how natural selection operates.

Natural Selection in Action

Look at any natural environment and you will see the products of natural selection—adaptations that suit organisms to their environment. But can we see natural selection in action? Yes, indeed! Biologists have documented evolutionary change in thousands of scientific studies.

An unsettling example of natural selection in action is the evolution of pesticide resistance in hundreds of insect species. Pesticides control insects and prevent them from eating crops or transmitting diseases. But whenever a new type of pesticide is used to control pests, the outcome

▼ Figure 13.11 **Color variation within a population of Asian lady beetles.**

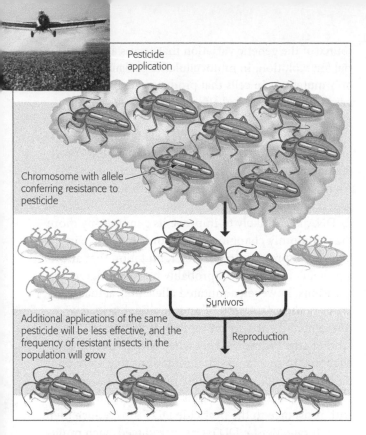

▲ **Figure 13.13 Evolution of pesticide resistance in insect populations.** By spraying crops with poisons to kill insect pests, people have unwittingly favored the reproductive success of insects with inherent resistance to the poisons.

Pesticide application

Chromosome with allele conferring resistance to pesticide

Survivors

Reproduction

Additional applications of the same pesticide will be less effective, and the frequency of resistant insects in the population will grow

is similar **(Figure 13.13)**: A relatively small amount of poison initially kills most of the insects, but subsequent applications are less and less effective. The few survivors of the first pesticide wave are individuals that are genetically resistant, carrying an allele (alternative form of a gene) that somehow enables them to survive the chemical attack. So the poison kills most members of the population, leaving the resistant survivors to reproduce and pass the alleles for pesticide resistance to their

offspring. Thus, the proportion of pesticide-resistant individuals increases in each generation.

Key Points about Natural Selection

Before we move on, let's summarize how natural selection works in bringing about evolutionary change.

Natural selection affects individual organisms—in Figure 13.13, each insect either survived the pesticide or was killed by it. However, individuals do not evolve. Rather, it is the population—the group of organisms—that evolves over time as adaptive traits become more common in the group and other traits change or disappear. Thus, evolution refers to generation-to-generation changes in populations.

Natural selection can amplify or diminish only heritable traits. Although an organism may, during its lifetime, acquire characters that help it survive, such acquired characters cannot be passed on to offspring.

Natural selection is more an editing process than a creative mechanism. A pesticide does not create new alleles that allow insects to survive. Rather, the presence of the pesticide leads to natural selection for insects in the population that already have those alleles.

Natural selection is not goal-directed; it does not lead to perfectly adapted organisms. Whereas artificial selection is a deliberate attempt by humans to produce individuals with specific traits, natural selection is the result of environmental factors that vary from place to place and over time. A trait that is favorable in one situation may be useless—or even detrimental—in different circumstances. And, as you will see, some adaptations are compromises. ☑

Thanks to natural selection, populations of more than 500 insect species are invulnerable to the most widely used pesticides.

☑ **CHECKPOINT**

Explain why the following statement is incorrect: "Pesticides cause pesticide resistance in insects."

Answer: An environmental factor does not create new traits such as pesticide resistance, but instead favors traits that are already represented in the population.

The Evolution of Populations

In *The Origin of Species*, Darwin provided evidence that life on Earth has evolved over time, and he proposed that natural selection, in favoring some heritable traits over others, was the primary mechanism for that change. But how do the variations that are the raw material for natural selection arise in a population? And how are these variations passed along from parents to offspring? Darwin did not know that Gregor Mendel (see Figure 9.1) had already answered these questions. Although both men lived and worked at around the same time, Mendel's work was largely ignored by the scientific community. Its rediscovery in 1900 set the stage for understanding the genetic differences on which evolution is based.

Sources of Genetic Variation

You have no trouble recognizing friends in a crowd. Each person has a unique genome, reflected in individual phenotypic variations such as appearance and other traits. Indeed, individual variation occurs in all species, as illustrated by the garter snakes in **Figure 13.14**. In addition to obvious physical differences, such as the snakes' colors and patterns, most populations have a great deal of phenotypic variation that can be observed only at the molecular level, such as an enzyme that detoxifies a pesticide. Of course, not all variation in a population is heritable. Phenotype—the expressed traits of an organism—results from a combination of the genotype, which is inherited, and many environmental influences. For instance, if you have dental work to straighten and whiten your teeth, you will not pass your environmentally produced smile to your offspring. Only the genetic component of variation is relevant to natural selection. Many of the characters that vary in a population result from the combined effect of several genes. Other features, such as Mendel's purple and white pea flowers or human blood types, are determined by a single gene locus, with different alleles producing distinct phenotypes. But where do these alleles come from?

Mutation

New alleles originate by mutation, a change in the nucleotide sequence of DNA. Thus, mutation is the ultimate source of the genetic variation that serves as raw material for evolution. In multicellular organisms, however, only mutations in cells that produce gametes can be passed to offspring and affect a population's genetic variability.

A change as small as a single nucleotide in a protein-coding gene can have a significant effect on phenotype, as in sickle-cell disease (see Figure 9.21). An organism is a refined product of thousands of generations of past selection, and a random change in its DNA is not likely to improve its genome any more than randomly changing some words on a page is likely to improve a story. In fact, mutation that affects a protein's function will probably be harmful. On rare occasions, however, a mutated allele may actually improve the adaptation of an individual to its environment and enhance its reproductive success. This kind of effect is more likely when the environment is changing in such a way that mutations that were once disadvantageous are favorable under the new conditions. For instance, mutations that endow houseflies with resistance to the pesticide DDT also reduce their growth rate. Before DDT was introduced, such mutations were a handicap to the flies that had them. But once DDT was part of the environment, the mutant alleles were advantageous, and natural selection increased their frequency in fly populations.

Chromosomal mutations that delete, disrupt, or rearrange many gene loci at once are almost certain to be harmful. But duplication of a gene or small pieces of DNA through errors in meiosis can provide an important source of genetic variation. If a repeated segment of DNA can persist over the generations, mutations may accumulate in the duplicate copies without affecting the function of the original gene, eventually leading to new genes with novel functions. This process may have played a major role in evolution. For example, the remote ancestors of mammals carried a single gene for detecting odors that has since been duplicated repeatedly. As a result, mice have about 1,300 different genes that encode smell receptors. It is likely that such dramatic increases helped early mammals by enabling them to distinguish among many different smells. And repeated duplications of genes that control development are linked to the origin of vertebrate animals from an invertebrate ancestor.

▼ Figure 13.14 **Variation in a garter snake population.** These four garter snakes, which belong to the same species, were all captured in one Oregon field. The behavior of each physical type is correlated with its coloration. When approached, spotted snakes, which blend in with their background, generally freeze. In contrast, snakes with stripes, which make it difficult to judge the speed of motion, usually flee rapidly when approached.

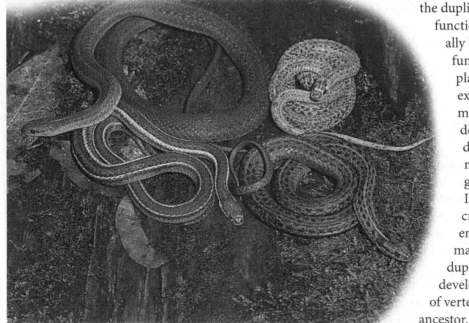

In prokaryotes, mutations can quickly generate genetic variation in a population. Because bacteria multiply so rapidly, a beneficial mutation can increase in frequency in a matter of hours or days. And because bacteria are haploid, with a single allele for each gene, a new allele can have an effect immediately. Mutation rates in animals and plants average about 1 in every 100,000 genes per generation. For these organisms, low mutation rates, long time spans between generations, and diploid genomes prevent most mutations from significantly affecting genetic variation from one generation to the next.

Sexual Reproduction

In organisms that reproduce sexually, most of the genetic variation in a population results from the unique combination of alleles that each individual inherits. (Of course, the origin of those allele variations is past mutations.)

Fresh assortments of existing alleles arise every generation from three random components of sexual reproduction: independent orientation of homologous chromosomes at metaphase I of meiosis (see Figure 8.16), crossing over (see Figure 8.18), and random fertilization. During meiosis, pairs of homologous chromosomes, one set inherited from each parent, trade some of their genes by crossing over. These homologous chromosomes separate into gametes independently of other chromosome pairs. Thus, gametes from any individual vary extensively in their genetic makeup. Finally, each zygote made by a mating pair has a unique assortment of alleles resulting from the random union of sperm and egg.

Populations as the Units of Evolution

One common misconception about evolution is that individual organisms evolve during their lifetimes. It is true that natural selection acts on individuals: Each individual's combination of traits affects its survival and reproductive success. But the evolutionary impact of natural selection is only apparent in the changes in a population of organisms over time.

A **population** is a group of individuals of the same species that live in the same area and interbreed. We can measure evolution as a change in the prevalence of certain heritable traits in a population over a span of generations. The increasing proportion of resistant insects in areas sprayed with pesticide is one example. Natural selection favored insects with alleles for pesticide resistance. As a result, these insects left more offspring than nonresistant individuals, changing the genetic makeup of the next generation's population.

Different populations of the same species may be geographically isolated from each other to such an extent that an exchange of genetic material never or only rarely occurs. Such isolation is common in populations confined to different lakes **(Figure 13.15)** or islands. For example, each population of Galápagos tortoise is restricted to its own island. Not all populations have such sharp boundaries, however; members of a population typically breed with one another and are therefore more closely related to each other than they are to members of a different population.

In studying evolution at the population level, biologists focus on the **gene pool**, which consists of all copies of every type of allele at every locus in all members of the population. For many loci, there are two or more alleles in the gene pool. For example, in a housefly population, there may be two alleles relating to DDT breakdown, one that codes for an enzyme that breaks down DDT and one for a version of the enzyme that does not. In populations living in fields sprayed with DDT, the allele for the enzyme conferring resistance will

▼ Figure 13.15 **Isolated lakes in Denali National Park and Preserve, Alaska.**

☑ **CHECKPOINT**

Which process, mutation or sexual reproduction, results in most of the generation-to-generation variability in human populations? Why?

Answer: sexual reproduction, because humans have a relatively long generation span and mutations have relatively little effect in a single generation

increase in frequency and the other allele will decrease in frequency. When the relative frequencies of alleles in a population change like this over a number of generations, evolution is taking place.

Next, we'll explore how to test whether evolution is occurring in a population. ☑

Analyzing Gene Pools

Imagine a wildflower population with two varieties of blooms that are different colors (**Figure 13.16**). An allele for red flowers, which we will symbolize by *R*, is dominant to an allele for white flowers, symbolized by *r*. These are the only two alleles for flower color in the gene pool of this hypothetical plant population. Now, let's say that 80%, or 0.8, of all flower-color loci in the gene pool have the *R* allele. We'll use the letter *p* to represent the relative frequency of the *R* allele in the population. Thus, $p = 0.8$. Because there are only two alleles in this example, the *r* allele must be present at the other 20% (0.2) of the gene pool's flower-color loci. (This accounts for 100% of the flower-color loci in the gene pool, or a relative frequency of 1.) Let's use the letter *q* for the frequency of the *r* allele in the population. For the wildflower population, $q = 0.2$. And since there are only two alleles for flower color, we can express their frequencies as follows:

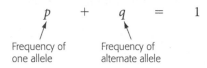

Notice that if we know the frequency of either allele in the gene pool, we can subtract it from 1 to calculate the frequency of the other allele.

From the frequencies of alleles, we can also calculate the frequencies of different genotypes in the population if the gene pool is completely stable (not evolving). In the wildflower population, what is the probability of producing an *RR* individual by "drawing" two *R* alleles from the pool of gametes? (Here we apply the rule of multiplication that you learned in Chapter 9; review Figure 9.11.) The probability of drawing an *R* sperm multiplied by the probability of drawing an *R* egg is $p \times p = p^2$, or $0.8 \times 0.8 = 0.64$. In other words, 64% of the plants in the population will have the *RR* genotype. Applying the same math, we also know the frequency of *rr* individuals in the population: $q^2 = 0.2 \times 0.2 = 0.04$. Thus, 4% of the plants are *rr*, giving them white flowers. Calculating the frequency of heterozygous individuals, *Rr*, is trickier. That's because the heterozygous genotype can form in two ways, depending on whether the sperm or egg supplies the dominant allele. So the frequency of the *Rr* genotype is $2pq$, which is $2 \times 0.8 \times 0.2 = 0.32$. In our imaginary wildflower population, 32% of the plants are *Rr*, with red flowers. **Figure 13.17** reviews these calculations graphically.

Now we can write a general formula for calculating the frequencies of genotypes in a gene pool from the frequencies of alleles, and vice versa:

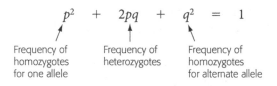

▼ **Figure 13.17 A mathematical swim in the gene pool.** Each of the four boxes in the Punnett square corresponds to a probable "draw" of alleles from the gene pool.

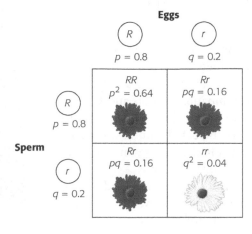

▼ **Figure 13.16 A population of wildflowers with two varieties of color.**

Notice that the frequencies of all genotypes in the gene pool must add up to 1. This formula is called the Hardy-Weinberg equation, named for the two scientists who derived it.

Population Genetics and Health Science

Public health scientists use the Hardy-Weinberg equation to calculate the percentage of a human population that carries the allele for certain inherited diseases. Consider phenylketonuria (PKU), which is an inherited inability to break down the amino acid phenylalanine. If untreated, the disorder has serious effects on brain development. PKU occurs in about 1 out of 10,000 babies born in the United States. Newborn babies are now routinely tested for PKU, and symptoms can be prevented if individuals living with the disease follow a strict diet that limits phenylalanine. In addition to occurring naturally, phenylalanine is found in the widely used artificial sweetener aspartame (**Figure 13.18**).

PKU is caused by a recessive allele (that is, one that must be present in two copies to produce the phenotype). Thus, we can represent the frequency of individuals in the U.S. population born with PKU with the q^2 term in the Hardy-Weinberg formula. For one PKU occurrence per 10,000 births, $q^2 = 0.0001$. Therefore, q, the frequency of the recessive allele in the population,

equals the square root of 0.0001, or 0.01. And p, the frequency of the dominant allele, equals $1 - q$, or 0.99.

Now let's calculate the frequency of carriers, who are heterozygous individuals who carry the PKU allele in a single copy and may pass it on to offspring. Carriers are represented in the formula by $2pq$: $2 \times 0.99 \times 0.01$, or 0.0198. Thus, the Hardy-Weinberg formula tells us that about 2% of the U.S. population are carriers for the PKU allele. Estimating the frequency of a harmful allele is essential for any public health program dealing with genetic diseases.

Microevolution as Change in a Gene Pool

As stated earlier, evolution can be measured as changes in the genetic composition of a population over time. It helps, as a basis of comparison, to know what to expect if a population is not evolving. A nonevolving population is in genetic equilibrium, which is also known as **Hardy-Weinberg equilibrium**. The population's gene pool remains constant. From generation to generation, the frequencies of alleles (p and q) and genotypes (p^2, $2pq$, and q^2) are unchanged. Sexual shuffling of genes cannot by itself change a large gene pool. Because a generation-to-generation change in allele frequencies of a population is evolution viewed on the smallest scale, it is sometimes referred to as **microevolution**. ☑

▼ Figure 13.18
A warning to individuals with PKU.

INGREDIENTS: SORBITOL, MAGNESIUM STEARATE, ARTIFICIAL FLAVOR, **ASPARTAME†** (SWEETENER), ARTIFICIAL COLOR (YELLOW 5 LAKE, BLUE 1 LAKE), ZINC GLUCONATE. **†PHENYLKETONURICS: CONTAINS PHENYLALANINE**

☑ **CHECKPOINT**

1. Which term in the Hardy-Weinberg formula ($p^2 + 2pq + q^2 = 1$) corresponds to the frequency of individuals with *no* alleles for the recessive disease PKU?
2. Define microevolution.

Answers: 1. p^2 2. Microevolution is a change in a population's frequencies of alleles.

Mechanisms That Alter Allele Frequencies in a Population

Now that we've defined microevolution as changes in a population's genetic makeup from generation to generation, we come to an obvious question: What mechanisms can change a gene pool? Natural selection is the most important, because it is the only process that promotes adaptation. We'll examine natural selection in more detail shortly. But first, we look at two other mechanisms of evolutionary change: genetic drift, which is due to chance, and gene flow, the exchange of alleles between neighboring populations.

Genetic Drift

Flip a coin 1,000 times, and a result of 700 heads and 300 tails would make you very suspicious about that coin.

But flip a coin 10 times, and an outcome of 7 heads and 3 tails would seem within reason. With a smaller sample, there is a greater chance of deviation from an idealized result—in this case, an equal number of heads and tails.

Let's apply this coin toss logic to a population's gene pool. If a new generation draws its alleles at random from the previous generation, then the larger the population (the sample size), the better the new generation will represent the gene pool of the previous generation. Thus, one requirement for a gene pool to maintain the status quo is a large population size. The gene pool of a small population may not be accurately represented in the next generation because of sampling error. The changed gene pool is analogous to the erratic outcome from a small sample of coin tosses.

Figure 13.19 applies this concept of sampling error to a small population of wildflowers. Chance causes the frequencies of the alleles for red (*R*) and white (*r*) flowers to change over the generations. And that fits our definition of microevolution. This evolutionary mechanism, a change in the gene pool of a population due to chance, is called **genetic drift**. But what would cause a population to shrink down to a size where there is genetic drift? Two ways this can occur are the bottleneck effect and the founder effect, both of which we explore next.

The Bottleneck Effect

Disasters such as earthquakes, floods, and fires may kill large numbers of individuals, producing a small surviving population that is unlikely to have the same genetic makeup as the original population. Again, the gene pool of the surviving population is a small sample of the genetic diversity originally present. By chance, certain alleles may be overrepresented among the survivors. Other alleles may be underrepresented. And some alleles may be eliminated. Chance may continue to change the gene pool for many generations until the population is again large enough for sampling errors to be insignificant.

The analogy illustrated in **Figure 13.20** shows why genetic drift due to a drastic reduction in population size is called the **bottleneck effect**.

▼ **Figure 13.20 The bottleneck effect.** The colored marbles in this analogy represent three alleles in an imaginary population. Shaking just a few of the marbles through the bottleneck is like an environmental disaster that drastically reduces the size of a population. Compared with the predisaster population, purple marbles are overrepresented in the new population, green marbles are underrepresented, and orange marbles are absent—all by chance. Similarly, a population that passes through a "bottleneck" event emerges with reduced variability.

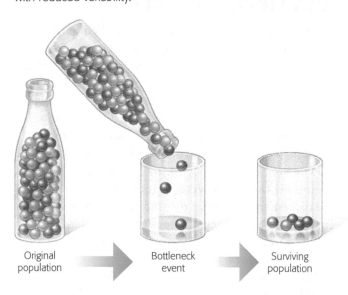

Original population   Bottleneck event   Surviving population

▼ **Figure 13.19 Genetic drift.** This hypothetical wildflower population consists of only ten plants. Due to random change over the generations, genetic drift can eliminate some alleles, as is the case for the *r* allele in generation 3 of this imaginary population.

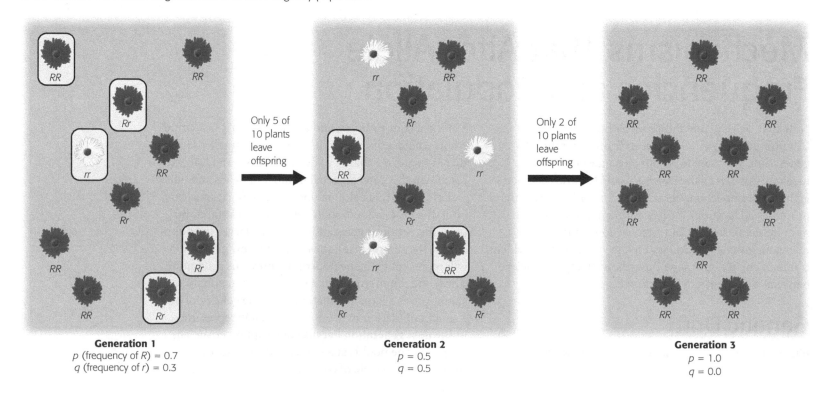

Generation 1
p (frequency of *R*) = 0.7
q (frequency of *r*) = 0.3

Only 5 of 10 plants leave offspring

Generation 2
p = 0.5
q = 0.5

Only 2 of 10 plants leave offspring

Generation 3
p = 1.0
q = 0.0

Passing through a "bottleneck"—a severe reduction in population size—decreases the overall genetic variability in a population because at least some alleles are likely to be lost from the gene pool. We can see this concept at work in the potential loss of individual variation, and hence adaptability, in drastically reduced populations of endangered species.

One such endangered species is the cheetah (Figure 13.21). The fastest of all running animals, cheetahs are magnificent cats that were once widespread in Africa and Asia. Like many African mammals, the number of cheetahs fell dramatically during the last ice age (around 10,000 years ago). At that time, the species suffered a severe bottleneck, possibly as a result of disease, human hunting, and periodic droughts. The South African cheetah population may have suffered a second bottleneck during the 1800s, when farmers hunted the animals to near extinction. Today, only a few small populations of cheetahs exist in the wild. Genetic variability in these populations is very low. In addition, the cheetahs remaining in Africa are being crowded into nature preserves and parks as human demands on the land increase. Along with crowding comes an increased potential for the spread of disease. With so little variability, the cheetah has a reduced capacity to adapt to such environmental challenges. Although captive breeding programs can boost cheetah population sizes, the species' pre-bottleneck genetic diversity can never be restored. ☑

Populations of endangered species may be doomed by their lack of genetic diversity.

The Founder Effect

When a few individuals colonize an isolated island, lake, or other new habitat, the genetic makeup of the colony is only a sample of the gene pool in the larger population. The smaller the colony (in other words, the smaller the sample size), the less likely it is to be representative of all the genetic diversity present in the population from which the colonists emigrated. If the colony succeeds, genetic drift will continue to change the frequency of alleles randomly until the population is large enough for genetic drift to be minimal. The type of genetic drift resulting from the establishment of a small, new population whose gene pool differs from that of the parent population is called the **founder effect**.

Numerous examples of the founder effect have been identified in geographically or socially isolated human populations. In such situations, disease-causing alleles that are rare in the larger population may became common in a small colony. For example, Amish and Mennonite communities in North America were founded by small numbers of European immigrants in the 1700s, and individuals within the community have since intermarried, remaining genetically separate from the larger population (Figure 13.22). Dozens of genetic diseases that are extremely rare elsewhere are relatively common in these communities. On the other hand, the high frequency of genetic diseases in populations has enabled genetic researchers to identify the mutation responsible for certain genetic disorders. In some cases, a disorder is treatable if detected early.

▲ Figure 13.22 **The founder effect.** Small, isolated populations often have high frequencies of alleles that are rare in large populations.

☑ CHECKPOINT

Would you expect modern cheetahs to have more genetic variation or less than cheetahs did 1,000 years ago?

Answer: less, because the bottleneck effect reduces genetic variability

▼ Figure 13.21 **Implications of the bottleneck effect in conservation biology.** Some endangered species, such as the cheetah, have low genetic variability. As a result, they are less adaptable to environmental changes, such as new diseases, than are species with a greater resource of genetic variation.

☑ **CHECKPOINT**

Which mechanism of
microevolution has been
most affected by the
increased ease of people
traveling throughout the
world?

Answer: gene flow

▲ **Figure 13.23 Gene
flow.** The pollen of some
plants can be carried by the
wind for hundreds of miles,
allowing gene flow to occur
between distant populations.

☑ **CHECKPOINT**

What is the best measure
of relative fitness?

*Answer: the number of fertile
offspring an individual leaves*

Gene Flow

Another source of evolutionary change is **gene flow**,
which is genetic exchange with another population. A
population may gain or lose alleles when fertile individuals move into or out of the population or when gametes
(such as plant pollen) are transferred between populations (Figure 13.23). For example, consider our hypothetical wildflower population in Figure 13.16. Suppose
a neighboring population consists entirely of white-
flowered individuals. A wind storm may blow pollen
to our wildflowers from the neighboring population,
resulting in a higher frequency of the white-flower allele
in the next generation—a microevolutionary change.

Gene flow tends to reduce differences between
populations. If it is extensive enough, gene flow can
eventually join neighboring populations into a single
population with a common gene pool. As people began
to move about the world more freely, gene flow became
an important agent of microevolutionary change in
populations that were previously isolated. ☑

Natural Selection:
A Closer Look

Genetic drift, gene flow, and even mutation can cause
microevolution. But only by rare chance would these
events result in improving a population's fit to its environment. In natural selection, on the other hand, only
the events that produce genetic variation (mutation and
sexual reproduction) are random. The process of natural selection, in which individuals better adapted to the
environment are more likely to survive and reproduce,
is *not* random. Consequently, only natural selection
consistently leads to adaptive evolution—evolution
that results in a better fit between organisms and their
environment.

The adaptations of organisms include many striking examples. Consider the blue-footed booby
(Figure 13.24), one of the most memorable creatures
that Darwin encountered on the Galápagos Islands. The
bird's body and bill are streamlined like a torpedo, mini-
mizing friction as it dives from heights up to 24 m (over
75 feet) into the shallow water below. To pull out of this
high-speed dive once it hits the water, the booby uses its
large tail as a brake. Those remarkable blue feet are an
essential requirement for the male's reproductive success. Female boobies prefer males with the brightest blue
feet. Thus, the male's courtship display is a dance that
features frequent flashes of his colorful assets.

Such adaptations are the result of natural selection.
By consistently favoring some alleles over others, natural

▲ **Figure 13.24 A blue-footed booby.** Although the booby's
large, webbed feet are highly advantageous in the water, they
are clumsy for walking on land.

selection improves the match between organisms and
their environment. However, the environment may
change over time. As a result, what constitutes a "good
match" between an organism and its environment is a
moving target, making adaptive evolution a continuous,
dynamic process.

Evolutionary Fitness

The commonly used phrase "survival of the fittest" is
misleading if we take it to mean head-to-head competition between individuals. Reproductive success, the key
to evolutionary success, is generally more subtle and
passive. In a varying population of moths, certain individuals may produce more offspring than others because
their wing colors hide them from predators better. Plants
in a wildflower population may differ in reproductive
success because some attract more pollinators, owing
to slight variations in flower color, shape, or fragrance.
In a given environment, such traits can lead to greater
relative fitness, the contribution an individual makes to
the gene pool of the next generation *relative to* the contributions of other individuals. The fittest individuals in
the context of evolution are those that produce the largest number of viable, fertile offspring and thus pass on
the most genes to the next generation. ☑

Three General Outcomes of Natural Selection

Imagine a population of mice with individuals ranging in fur color from very light to very dark gray. If we graph the number of mice in each color category, we get a bell-shaped curve like the one shown at the top of **Figure 13.25**. If natural selection favors certain fur-color phenotypes over others, the population of mice will change over the generations. Three general outcomes are possible, depending on which phenotypes are favored. These three modes of natural selection are called directional selection, disruptive selection, and stabilizing selection.

Directional selection shifts the overall makeup of a population by selecting in favor of one extreme phenotype—the darkest mice, for example **(Figure 13.25a)**. Directional selection is most common when the local environment changes or when organisms migrate to a new environment. An actual example is the shift of insect populations toward a greater frequency of pesticide-resistant individuals.

Disruptive selection can lead to a balance between two or more contrasting phenotypes in a population **(Figure 13.25b)**. A patchy environment, which favors different phenotypes in different patches, is one situation associated with disruptive selection. The variations seen in the snake population in Figure 13.14 result from disruptive selection.

Stabilizing selection favors intermediate phenotypes **(Figure 13.25c)**. Such selection typically occurs in relatively stable environments, where conditions tend to reduce physical variation. This evolutionary conservatism works by selecting against the more extreme phenotypes. For example, stabilizing selection keeps the majority of human birth weights between 3 and 4 kg (approximately 6.5 to 9 pounds). For babies much lighter or heavier than this, infant mortality is greater.

Of the three selection modes, stabilizing selection occurs in most situations, resisting change in well-adapted populations. Evolutionary spurts occur when a population is stressed by a change in the environment or by migration to a new place. When challenged with a new set of environmental problems, a population either adapts through natural selection or dies off in that locale. The fossil record tells us that the population's extinction is the most common result. Those populations that do survive crises may change

enough to be designated new species. (You'll learn more about this in Chapter 14.) ☑

Sexual Selection

Darwin was the first to explore the implications of **sexual selection**, a form of natural selection in which individuals with certain traits are more likely than other individuals to obtain mates. The males and females of an animal species obviously have different reproductive organs. But they may also have secondary sexual traits, noticeable differences not directly associated with reproduction or survival. This distinction in appearance, called **sexual dimorphism**, is often manifested in a size difference. Among male vertebrates, sexual dimorphism may also be evident in adornment, such as manes on

☑ CHECKPOINT

The thickness of fur in a bear population increases over several generations as the climate in the region becomes colder. This is an example of which type of selection: directional, disruptive, or stabilizing?

Answer: directional

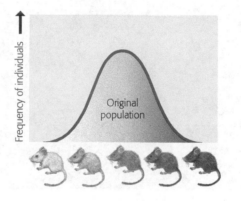

◄ **Figure 13.25 Three possible outcomes for selection working on fur color in imaginary populations of mice.** The large downward arrows symbolize the pressure of natural selection working against certain phenotypes.

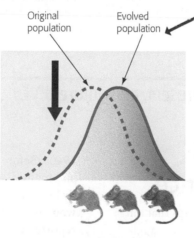

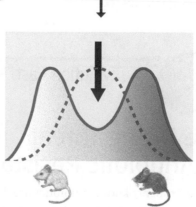

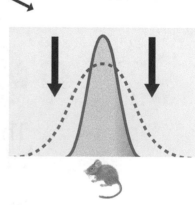

(a) Directional selection shifts the overall makeup of the population by favoring variants at one extreme. In this case, the trend is toward darker color, perhaps because the landscape has been shaded by the growth of trees, making darker mice less noticeable to predators.

(b) Disruptive selection favors variants at opposite extremes over intermediate individuals. Here, the relative frequencies of very light and very dark mice have increased. Perhaps the mice have colonized a patchy habitat where a background of light soil is studded with dark rocks.

(c) Stabilizing selection culls extreme variants from the population, in this case eliminating individuals that are unusually light or dark. The trend is toward reduced phenotypic variation and increased frequency of an intermediate phenotype.

297

lions, antlers on deer, or colorful plumage on peacocks and other birds (Figure 13.26a).

In some species, secondary sex structures may be used to compete with members of the same sex (usually males) for mates. Contests may involve physical combat, but are more often ritualized displays (Figure 13.26b). Such selection is common in species in which the winner acquires a harem of mates—an obvious boost to that male's evolutionary fitness.

> In some species, males engage in nonlethal combat in the hope of mating with females.

In a more common type of sexual selection, individuals of one sex (usually females) are choosy in selecting their mates. Males with the largest or most colorful adornments are often the most attractive to females. The extraordinary feathers of a peacock's tail are an example of this sort of "choose me!" statement. Every time a female chooses a mate based on a certain appearance or behavior, she perpetuates the alleles that caused her to make that choice and allows a male with that particular phenotype to perpetuate his alleles.

What is the advantage to females of being choosy? One hypothesis is that females prefer male traits that are correlated with "good" alleles. In several bird species, research has shown that traits preferred by females, such as bright beaks or long tails, are related to overall male health.

▼ Figure 13.26 **Sexual dimorphism.**

(a) **Sexual dimorphism in a finch species.** Among vertebrates, including this pair of green-winged Pytilia (native to Africa), males (right) are usually the showier sex.

(b) **Competing for mates.** Male Spanish ibex engage in nonlethal combat for the right to mate with females.

Evolution Past, Present, and Future EVOLUTION CONNECTION

The Rising Threat of Antibiotic Resistance

As you probably know, antibiotics are drugs that kill infectious microorganisms. But antibiotics are a relatively recent development in medicine. Before antibiotics, people often died from diseases such as whooping cough, and a minor wound—a razor nick or a scratch from a rose thorn—could result in a fatal infection. A revolution in human health followed the introduction of penicillin, the first widely used antibiotic, in the 1940s. Before long, more antibiotics were developed, and many diseases that had once been fatal could easily be cured. But even amid the enthusiasm for the new wonder drugs, signs of trouble began to appear. Doctors reported cases of bacterial infections that did not respond to antibiotics. By 1952, a researcher had identified the reason: Some bacteria had genetic traits that enabled them to resist the killing power of the drugs. In the same way that pesticides select for resistant insects, antibiotics select for resistant bacteria. A gene that codes for an enzyme that breaks down an antibiotic or a mutation that alters the site where an antibiotic binds can make a bacterium and its offspring resistant to that antibiotic. Again we see both the random and nonrandom aspects of natural selection—the random genetic mutations in bacteria and the nonrandom selective effects as the environment favors the antibiotic-resistant phenotype.

Ironically, our enthusiasm for the curative powers of antibiotics has encouraged the evolution of antibiotic-resistant bacteria. Livestock producers add antibiotics to animal feed, a practice that may select for bacteria resistant to standard antibiotics. Doctors may overprescribe antibiotics—for example, to patients with viral infections, which do not respond to antibiotic treatment. You may also be part of the problem. If you stop taking a prescribed antibiotic as soon as you feel better, instead of taking the full course of medication your doctor has prescribed, you are allowing mutant bacteria that may be killed more slowly by the drug to survive and multiply. Subsequent mutations in such bacteria may lead to full-blown antibiotic resistance.

Natural selection for antibiotic resistance is particularly strong in hospitals, where antibiotic use is extensive. A formidable "superbug" known as MRSA (methicillin-resistant *Staphylococcus aureus*) can cause "flesh-eating disease" and potentially fatal systemic (whole-body) infections. Alarmingly, a growing number of incidents of MRSA infection now begin in community settings, such as athletic facilities, schools, and military barracks (Figure 13.27).

MRSA is not the only antibiotic-resistant microorganism. According to a report issued by the Centers for Disease Control (CDC), 17 bacterial infections are no longer treatable with standard antibiotics; most of these are considered urgent or serious threats to public health. The most recent "superbug" to emerge is a strain of the bacteria that cause gonorrhea, a sexually transmitted disease. Public health officials fear that as this strain spreads, gonorrhea will become an incurable disease.

▲ Figure 13.27 **Cautioning athletes about MRSA.**

Drug resistance has also evolved in disease-causing viruses (for example, HIV and influenza) and parasites (for example, the organism that causes malaria).

Medical and pharmaceutical researchers are racing to develop new antibiotics and other drugs. However, experience suggests that our battle against the evolution of drug-resistant bacteria will continue for the foreseeable future.

Chapter Review

SUMMARY OF KEY CONCEPTS

The Diversity of Life

Naming and Classifying the Diversity of Life

In the Linnaean system of classification, each species is assigned a two-part name. The first part is the genus, and the second part is unique for each species within the genus. In the taxonomic hierarchy, domain > kingdom > phylum > class > order > family > genus > species.

Explaining the Diversity of Life

Present-day biologists accept Darwin's theory of evolution by means of natural selection as the best explanation for the diversity of life. However, when Darwin published his theory in 1859, it was a radical departure from the prevailing views.

Charles Darwin and *The Origin of Species*

Darwin's Journey

During his around-the-world voyage on the *Beagle*, Darwin observed adaptations of organisms that inhabited diverse environments. In particular, Darwin was struck by the geographic distribution of organisms on the Galápagos Islands, off the South American coast. When Darwin considered his observations in light of new evidence for a very old Earth that changed slowly, he arrived at ideas that were at odds with the long-held notion of a young Earth populated by unchanging species.

Evolution: Darwin's Theory

In his book *On the Origin of Species by Means of Natural Selection*, Darwin made two proposals: (1) Existing species descended from ancestral species, and (2) natural selection is the mechanism of evolution.

Evidence of Evolution

Evidence from Fossils

The fossil record shows that organisms have appeared in a historical sequence, and many fossils link ancestral species with those living today. For example, the evolution of whales from ancestral land-dwelling animals is documented by fossils of transitional forms.

Evidence from Homologies

Structural and molecular homologies reveal evolutionary relationships. Closely related species often have similar stages in their embryonic development. All species share a common genetic code, suggesting that all forms of life are related through branching evolution from the earliest organisms.

Evolutionary Trees

An evolutionary tree represents a succession of related species, with the most recent at the tips of the branches. Each branch point represents a common ancestor of all species that radiate from it.

Natural Selection as the Mechanism for Evolution

Darwin proposed natural selection as the mechanism that produces adaptive evolutionary change. In a population that varies, individuals best suited for a particular environment are more likely to survive and reproduce than those that are less suited to that environment.

Natural Selection in Action

Natural selection has been observed in many scientific studies, including in the evolution of pesticide-resistant insects.

Key Points about Natural Selection

Individuals do not evolve. Only heritable traits, not those acquired during an individual's lifetime, can be amplified or diminished by natural selection. Natural selection only works on existing variation—new variation does not arise in response to an environmental change. Natural selection does not produce perfect organisms.

The Evolution of Populations

Sources of Genetic Variation

Mutation and sexual reproduction produce genetic variation. Mutation is the ultimate source of genetic variation. Individual mutations have little short-term effect on a large gene pool, but in the long term, mutation is the source of genetic variation.

Populations as the Units of Evolution

A population, members of the same species living in the same time and place, is the smallest biological unit that can evolve.

Analyzing Gene Pools

A gene pool consists of all the alleles in all the individuals making up a population. The Hardy-Weinberg formula can be used to calculate the frequencies of genotypes in a gene pool from the frequencies of alleles, and vice versa:

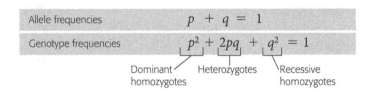

Allele frequencies	$p + q = 1$
Genotype frequencies	$p^2 + 2pq + q^2 = 1$

Dominant homozygotes — Heterozygotes — Recessive homozygotes

Population Genetics and Health Science

The Hardy-Weinberg formula can be used to estimate the frequency of a harmful allele, which is useful information for public health programs dealing with genetic diseases.

Microevolution as Change in a Gene Pool

Microevolution is generation-to-generation change in allele frequencies in a population.

Mechanisms That Alter Allele Frequencies in a Population

Genetic Drift

Genetic drift is a change in the gene pool of a small population due to chance. A bottleneck event (a drastic reduction in population size) and the founder effect (occurring in a new population started by a few individuals) are two situations leading to genetic drift.

Gene Flow

A population may gain or lose alleles by gene flow, which is genetic exchange with another population.

Natural Selection: A Closer Look

Of all causes of evolution, only natural selection promotes evolutionary adaptations. Relative fitness is the contribution an individual makes to the gene pool of the next generation relative to the contributions of other individuals. The outcome of natural selection may be directional, disruptive, or stabilizing. Secondary sexual traits (such as sex-specific plumage or behaviors) can promote sexual selection, a type of natural selection in which mating preferences are determined by inherited traits.

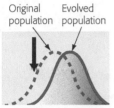 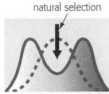

Original population Evolved population Pressure of natural selection

Directional selection   **Disruptive selection**   **Stabilizing selection**

MasteringBiology®

For practice quizzes, BioFlix animations, MP3 tutorials, video tutors, and more study tools designed for this textbook, go to MasteringBiology®

SELF-QUIZ

1. Place these levels of classification in order from least inclusive to most inclusive: class, domain, family, genus, kingdom, order, phylum, species.

2. Which of the following is a true statement about Charles Darwin?
   a. He was the first to discover that living things can change, or evolve.
   b. He based his theory on the inheritance of acquired traits.
   c. He proposed natural selection as the mechanism of evolution.
   d. He was the first to realize that Earth is more than 6,000 years old.

3. How did the insights of Lyell and other geologists influence Darwin's thinking about evolution?

4. In a population with two alleles for a particular genetic locus, B and b, the allele frequency of B is 0.7. If this population is in Hardy-Weinberg equilibrium, what is the frequency of heterozygotes? What is the frequency of homozygous dominants? What is the frequency of homozygous recessives?

5. Define fitness from an evolutionary perspective.

6. Which of the following processes is the ultimate source of the genetic variation that serves as raw material for evolution?
   a. sexual reproduction   c. genetic drift
   b. mutation   d. natural selection

7. Which of the following is *not* a requirement of natural selection?
   a. genetic variation
   b. catastrophic events
   c. differential reproductive success
   d. overproduction of offspring

8. Compare and contrast how the bottleneck effect and the founder effect can lead to genetic drift.

9. Garter snakes with different color patterns behave differently when threatened. Of the three general outcomes of natural selection (directional, disruptive, or stabilizing), this example illustrates _____.

10. Which of the following is a characteristic of a population in Hardy-Weinberg equilibrium?
    a. The population is subject to natural selection.
    b. The population is not evolving.
    c. Genetic drift is occurring.
    d. Gene flow in and out of the population occurs.

Answers to these questions can be found in Appendix: Self-Quiz Answers.

THE PROCESS OF SCIENCE

11. **Interpreting Data** A population of snails has recently become established in a new region. The snails are preyed on by birds that break the snails open on rocks, eat the soft bodies, and leave the shells. The snails occur in both striped and unstriped forms. In one area, researchers counted both live snails and broken shells. Their data are summarized here:

	Striped Shells	Unstriped Shells
Number of live snails	264	296
Number of broken snail shells	486	377
Total	750	673

Based on these data, which snail form is subject to more predation by birds? Predict how the frequencies of striped and unstriped individuals might change over time.

12. Five years after the experiment described in the previous question, the researchers repeat the snail count in order to test their prediction. Surprisingly, they find that the ratio of broken shells to live snails has gone down for the striped and up for the unstriped snails. They notice that a weed with striped leaves has newly spread in the examined region. Formulate a hypothesis that explains this observation.

BIOLOGY AND SOCIETY

13. To what extent are people in a technological society exempt from natural selection? Explain your answer.

14. "Social Darwinism" suggests that the principle of "survival of the fittest" is applicable to the human society. Is this theory a valid approach? Is it based on science?

14 How Biological Diversity Evolves

If you drove off the rim of the Grand Canyon, you would pass by 40 rock layers and several hundred million years of geologic history before hitting the ground.

Alleles from two wild grasses could be used to make a better ear of corn.

Piecing together evolutionary histories shows who's related to whom.

Mass Extinctions BIOLOGY AND SOCIETY

The Sixth Mass Extinction

The fossil record reveals that the evolutionary history of life on Earth has been episodic, with long, relatively stable periods punctuated by brief, cataclysmic ones. During these upheavals, new species formed and others died out in great numbers.

Extinctions are inevitable in a changing world, but the fossil record reveals a few instances of great change, times when the majority of life on Earth—between 50% and 90% of living species—suddenly died out, vanishing forever. Scientists have documented five such mass extinctions during the last 540 million years. Today, human activities are modifying the global environment to such an extent that many species are disappearing at an alarming rate. In the past 400 years—a very short time on a geologic scale—more than 1,000 species are known to have become extinct. Scientists estimate that this is 100 to 1,000 times the extinction rate seen in most of the fossil record.

Are we in the midst of a sixth mass extinction? In an extensive analysis published in *Nature* in 2011, researchers compared data from the fossil record of the "big five" mass extinctions with data from the modern era. The good news is that the current loss of biodiversity does not yet qualify as a mass extinction. The bad news is that we're teetering on the brink. The loss of species that are now at critical risk of extinction—like the sea otter pictured here—would push our planet into a period of mass extinction. When the researchers include species that are endangered or threatened (lower categories of risk) in their calculations, the picture becomes very grim indeed. In contrast to the ancient mass extinctions, which unfolded over hundreds of thousands of years, a human-driven sixth mass extinction could be completed in just a few centuries. And, as with prior mass extinctions, life on Earth may take millions of years to recover.

An endangered species. Oil spills, entanglement in commercial fishing nets, and disease are among the causes of declining sea otter populations.

But the fossil record also shows a creative side to the destruction. Mass extinctions can pave the way for the evolution of many diverse species from a common ancestor, such as the diversification of mammals after the extinction of dinosaurs. Accordingly, we'll begin this chapter by discussing the birth of new species and then examine how biologists trace the evolution of biological diversity. We'll also take a closer look at how scientists classify living organisms.

The Origin of Species

Natural selection, a microevolutionary mechanism, explains the striking ways in which organisms are suited to their environment. But what accounts for the tremendous diversity of life, the millions of species that have existed during Earth's history? This question intrigued Darwin, who referred to it in his diary as "that mystery of mysteries—the first appearance of new beings on this Earth."

When, as a young man, Darwin visited the Galápagos Islands (see Figure 13.3), he realized that he was visiting a place of origins. Though the volcanic islands were geologically young, they were already home to many plants and animals known nowhere else in the world. Among these unique inhabitants were marine iguanas **(Figure 14.1)**, Galápagos tortoises (see Figure 13.4), and numerous species of small birds called finches, which you will learn more about in this chapter (see Figure 14.11). Surely, Darwin thought, not all of these species could have been among the original colonists. Some of them must have evolved later on, the myriad descendants of the original colonists, modified by natural selection from those original ancestors.

In the century and a half since the publication of Darwin's *On the Origin of Species by Means of Natural Selection*, new discoveries and technological advances— especially in molecular biology—have given scientists a wealth of new information about the evolution of life on Earth. For example, researchers have explained the genetic patterns underlying the homology of vertebrate limbs (see Figure 13.7). Hundreds of thousands more fossil discoveries have been cataloged, including many of the transitional (intermediate) forms predicted by Darwin. In a fascinating convergence of old and new techniques, researchers have even been able to investigate the genetic material of certain fossils, including our ancient relatives, the Neanderthals (see Figure 14.23). New dating methods have confirmed that Earth is billions of years old, much older than even the most radical geologists of Darwin's time proposed. As you'll learn in this chapter, these dating methods have also enabled researchers to determine the ages of fossils and rocks, providing valuable insight into evolutionary relationships between groups of organisms. In addition, our enhanced understanding of geologic processes, such as the changing positions of continents, explains some of the geographic distributions of organisms and fossils that puzzled Darwin and his contemporaries.

In this chapter, you'll learn how evolution has woven the rich tapestry of life, beginning with **speciation**, the process in which one species splits into two or more species. Other topics include the origin of evolutionary novelty, such as the wings and feathers of birds and the large brains of humans, and the impact of mass extinctions, which clear the way for new adaptive explosions, such as the diversification of mammals following the disappearance of most of the dinosaurs.

▼ **Figure 14.1 A marine iguana (right), an example of the unique species inhabiting the Galápagos.** Darwin noticed that Galápagos marine iguanas—with a flattened tail that aids in swimming—are similar to, but distinct from, land-dwelling iguanas on the islands and on the South American mainland (left).

What Is a Species?

Species is a Latin word meaning "kind" or "appearance." Indeed, as children we learn to distinguish between the kinds of plants and animals—between dogs and cats, for example, or between roses and dandelions—from differences in their appearance. Although the basic idea of species as distinct life-forms seems intuitive, devising a more formal definition is not so easy.

One way of defining a species (and the main definition used in this book) is the **biological species concept**. It defines a **species** as a group of populations whose members have the potential to interbreed with one another in nature and produce fertile offspring (offspring that can reproduce) **(Figure 14.2)**. Geography and culture may conspire to keep a Manhattan businesswoman and a Mongolian dairyman apart. But if the two did meet and mate, they could have viable babies who develop into fertile adults because all humans belong to the same species. In contrast, humans and chimpanzees, despite having a shared evolutionary history, are distinct species because they can't successfully interbreed.

We cannot apply the biological species concept to all situations. For example, basing the definition of species on reproductive compatibility excludes organisms that only reproduce asexually (producing offspring from a single parent), such as most prokaryotes. And because fossils are obviously not currently reproducing sexually, they cannot be evaluated by the biological species concept. In response to such challenges, biologists have developed other ways to define species. For example, most of the species named so far have been classified based on measurable physical traits such as number and type of teeth or flower structures. Another approach defines a species as the smallest group of individuals sharing a common ancestor and forming one branch on the tree of life. Yet another approach proposes defining a species solely on the basis of molecular data, a sort of bar code that identifies each species.

Each species concept is useful, depending on the situation and the questions being asked. The biological species concept, however, is particularly useful when focusing on how species originate, that is, when we ask: What prevents a member of one group from successfully interbreeding with a member of another group? You'll learn about the variety of answers to that question next. ☑

☑ **CHECKPOINT**

According to the biological species concept, what defines a species?

Answer: the ability of its members to interbreed with one another and produce fertile offspring in a natural setting

▼ Figure 14.2 **The biological species concept is based on reproductive compatibility rather than physical similarity.**

Similarity between different species. The eastern meadowlark (left) and the western meadowlark (right) are very similar in appearance, but they are separate species and do not interbreed.

Diversity within one species. Humans, as diverse in appearance as we are, belong to a single species (*Homo sapiens*) and can interbreed.

Reproductive Barriers between Species

Clearly, a fly will not mate with a frog or a fern. But what prevents closely related species from interbreeding? What, for example, maintains the species boundary between the eastern meadowlark and the western meadowlark (shown in Figure 14.2)? Their geographic ranges overlap in the Great Plains region, and they are so similar that only expert birders can tell them apart. And yet, these two bird species do not interbreed.

A **reproductive barrier** is anything that prevents individuals of closely related species from interbreeding. Let's examine the different kinds of reproductive barriers that isolate the gene pools of species (Figure 14.3). We can classify reproductive barriers as either prezygotic or postzygotic, depending on whether they block interbreeding before or after the formation of zygotes (fertilized eggs).

Prezygotic barriers prevent mating or fertilization between species (Figure 14.4). The barrier may be time-based (temporal isolation). For example, western spotted skunks breed in the fall, but the eastern species breeds in late winter. Temporal isolation keeps the species from mating even where they coexist on the Great Plains. In other cases, species live in the same region but not in the same habitats (habitat isolation). For example, one species of North American garter snake lives mainly in water, while a closely related species lives on land. Traits that enable individuals to recognize potential mates, such as a particular odor, coloration, or courtship ritual, can also function as reproductive barriers (behavioral isolation). In many bird species, for example,

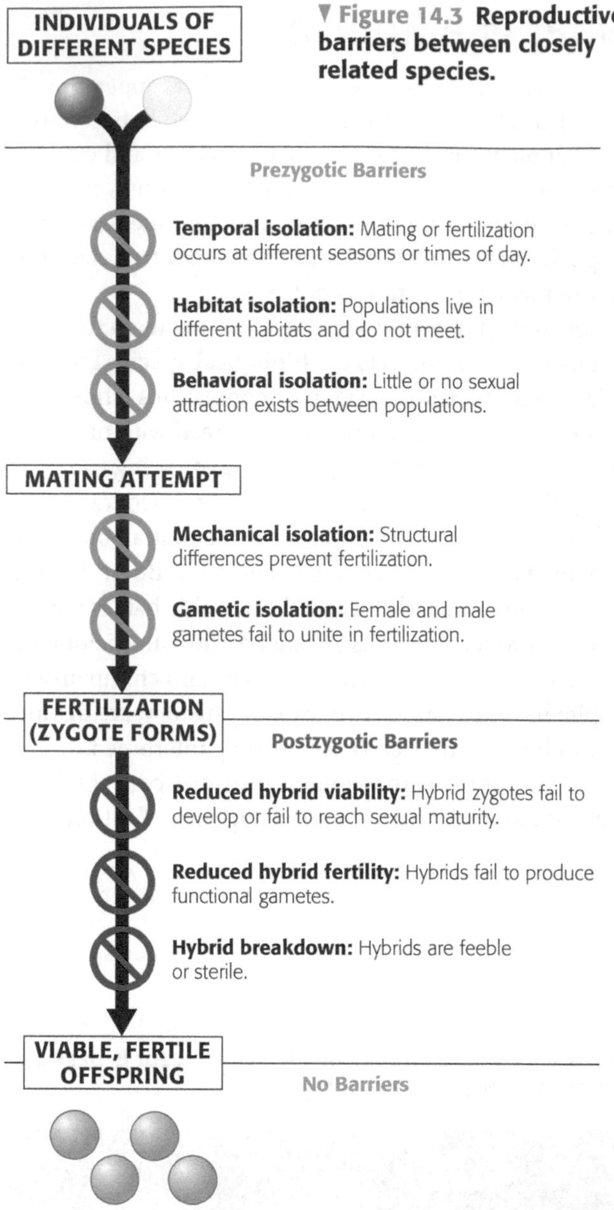

▼ **Figure 14.3** Reproductive barriers between closely related species.

INDIVIDUALS OF DIFFERENT SPECIES

Prezygotic Barriers

Temporal isolation: Mating or fertilization occurs at different seasons or times of day.

Habitat isolation: Populations live in different habitats and do not meet.

Behavioral isolation: Little or no sexual attraction exists between populations.

MATING ATTEMPT

Mechanical isolation: Structural differences prevent fertilization.

Gametic isolation: Female and male gametes fail to unite in fertilization.

FERTILIZATION (ZYGOTE FORMS)

Postzygotic Barriers

Reduced hybrid viability: Hybrid zygotes fail to develop or fail to reach sexual maturity.

Reduced hybrid fertility: Hybrids fail to produce functional gametes.

Hybrid breakdown: Hybrids are feeble or sterile.

VIABLE, FERTILE OFFSPRING

No Barriers

► **Figure 14.4 Prezygotic barriers.** Prezygotic barriers prevent mating or fertilization.

PREZYGOTIC BARRIERS

Temporal Isolation

These two closely related species of skunks mate at different times of the year.

Habitat Isolation

These two closely related species of garter snakes do not mate because one lives in the water while the other lives on land.

courtship behavior is so elaborate that individuals are unlikely to mistake a bird of a different species as one of their kind. In still other cases, the egg-producing and sperm-producing structures of different species are anatomically incompatible (mechanical isolation). For example, the two closely related species of snails in Figure 14.4 cannot join their male and female sex organs because their shells spiral in opposite directions. In still other cases, gametes (eggs and sperm) of different species are incompatible, preventing fertilization (gametic isolation). Gametic isolation is very important when fertilization is external. Male and female sea urchins of many species release eggs and sperm into the sea, but fertilization occurs only if species-specific molecules on the surface of egg and sperm attach to each other.

Postzygotic barriers operate if interspecies mating actually occurs and results in hybrid zygotes (**Figure 14.5**). (In this context, "hybrid" means that the egg comes from one species and the sperm from another species.) In some cases, hybrid offspring die before reaching reproductive maturity (reduced hybrid viability). For example, although certain closely related salamander species will hybridize, the offspring fail to develop normally because of genetic incompatibilities between the two species. In other cases of hybridization, offspring may become vigorous adults, but are infertile (reduced hybrid fertility). A mule, for example, is the hybrid offspring of a female horse and a male donkey. Mules are sterile—they cannot successfully breed with each other. Thus, horses and donkeys remain distinct species. In other cases, the first-generation hybrids are viable and fertile, but when these hybrids mate with one another or with either parent species, the offspring are feeble or sterile (hybrid breakdown). For example,

different species of cotton plants can produce fertile hybrids, but the offspring of the hybrids do not survive.

In summary, reproductive barriers form the boundaries around closely related species. In most cases, it is not a single reproductive barrier but some combination of two or more that keeps species isolated. Next, we examine situations that make reproductive isolation and speciation possible. ✓

▼ **Figure 14.5 Postzygotic barriers.** Postzygotic barriers prevent development of fertile adults.

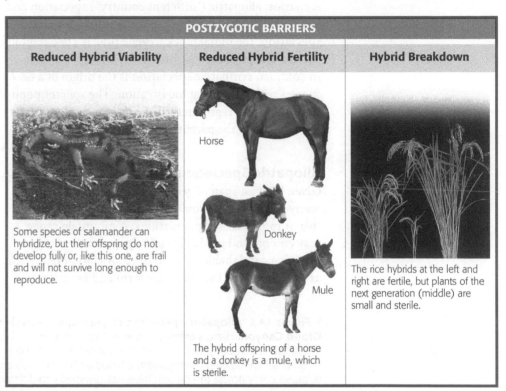

POSTZYGOTIC BARRIERS

Reduced Hybrid Viability

Some species of salamander can hybridize, but their offspring do not develop fully or, like this one, are frail and will not survive long enough to reproduce.

Reduced Hybrid Fertility

Horse

Donkey

Mule

The hybrid offspring of a horse and a donkey is a mule, which is sterile.

Hybrid Breakdown

The rice hybrids at the left and right are fertile, but plants of the next generation (middle) are small and sterile.

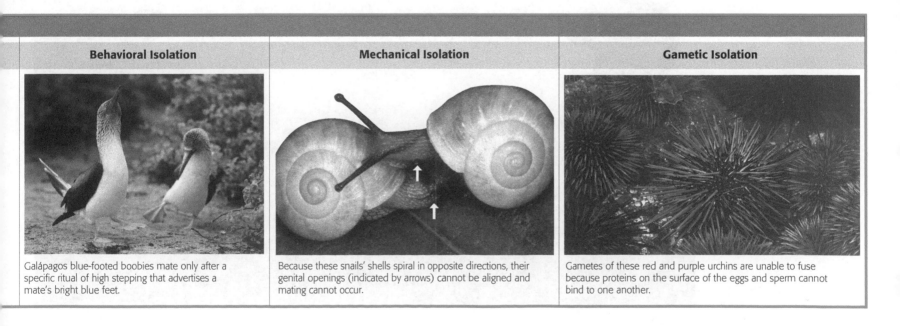

Behavioral Isolation

Galápagos blue-footed boobies mate only after a specific ritual of high stepping that advertises a mate's bright blue feet.

Mechanical Isolation

Because these snails' shells spiral in opposite directions, their genital openings (indicated by arrows) cannot be aligned and mating cannot occur.

Gametic Isolation

Gametes of these red and purple urchins are unable to fuse because proteins on the surface of the eggs and sperm cannot bind to one another.

⌑ Evolution Mechanisms of Speciation

A key event in the origin of many species occurs when a population is somehow cut off from other populations of the parent species. With its gene pool isolated, the splinter population can follow its own evolutionary course. Changes in its allele frequencies caused by genetic drift and natural selection will not be diluted by alleles entering from other populations (gene flow). Such reproductive isolation can result from two general scenarios: allopatric ("different country") speciation and sympatric ("same country") speciation. In **allopatric speciation**, the initial block to gene flow is a geographic barrier that physically isolates the splinter population. In contrast, **sympatric speciation** is the origin of a new species without geographic isolation. The splinter population becomes reproductively isolated even though it is in the midst of the parent population.

Allopatric Speciation

Given the vast span of geologic time, we can imagine many scenarios that could fragment a population into two or more isolated populations. A mountain range may emerge and gradually split a population of organisms that can inhabit only lowlands. A land bridge, such as the Isthmus of Panama, may form and separate the marine life on either side. A large lake may subside and form several smaller lakes, with their populations now isolated. Glaciation could force small populations into ice-free areas that would remain isolated for millennia until the glaciers receded.

How formidable must a geographic barrier be to keep allopatric populations apart? The answer depends partly on the ability of the organisms to move about. Birds, mountain lions, and coyotes can cross mountain ranges, rivers, and canyons. Such barriers also do not hinder the windblown pollen of pine trees or the spread of seeds carried by animals capable of crossing the barrier. In contrast, small rodents may find a deep canyon or a wide river an impassable barrier (**Figure 14.6**).

Speciation is more common for a small, isolated population because it is more likely than a large population to have its gene pool changed substantially by both genetic drift and natural selection. But for each small, isolated population that becomes a new species, many more simply perish in their new environment. Life on the frontier is harsh, and most pioneer populations become extinct.

Even if a small, isolated population survives, it does not necessarily evolve into a new species. The population may adapt to its local environment and begin to look very different from the ancestral population, but that doesn't necessarily make it a new species. Speciation occurs with the evolution of reproductive barriers between the isolated population and its parent

▼ **Figure 14.6 Allopatric speciation of antelope squirrels on opposite rims of the Grand Canyon.** Harris's antelope squirrel (*Ammospermophilus harrisii*) is found on the south rim of the Grand Canyon. Just a few miles away on the north rim is the closely related white-tailed antelope squirrel (*Ammospermophilus leucurus*). Birds and other organisms that can disperse easily across the canyon have not diverged into different species on opposite rims.

Ammospermophilus harrisii

Ammospermophilus leucurus

► **Figure 14.7 Possible outcomes after geographic isolation of populations.** In this diagram, the orange and green arrows track populations over time. The mountain symbolizes a period of geographic isolation during which time genetic changes may occur in both populations. After a period of time, the populations are no longer separated by a geographic barrier (right side of diagram) and come back into contact. If the populations can interbreed freely (top diagram), speciation has not occurred. If the populations cannot interbreed (bottom), then speciation has occurred.

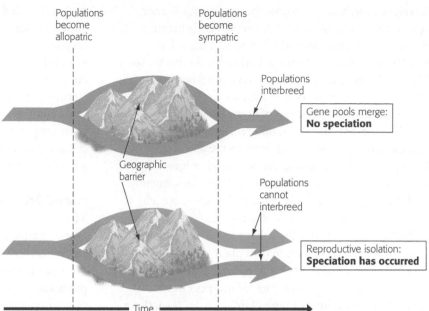

Populations become allopatric

Populations become sympatric

Populations interbreed

Gene pools merge: **No speciation**

Geographic barrier

Populations cannot interbreed

Reproductive isolation: **Speciation has occurred**

Time

population. In other words, if speciation occurs during geographic separation, the new species cannot breed with its ancestral population even if the two populations should come back into contact at some later time (**Figure 14.7**). ☑

Sympatric Speciation

Speciation doesn't necessarily require vast spans of time, nor is geographic isolation a prerequisite. A species may originate from an accident during cell division that results in an extra set of chromosomes, a condition called **polyploidy**. Examples of polyploid speciation have been found in some animal species, especially fish and amphibians (**Figure 14.8**). However, it is most common in

plants—an estimated 80% of present-day plant species are descended from ancestors that arose by polyploid speciation (**Figure 14.9**).

Two distinct forms of polyploid speciation have been observed. In one form, polyploidy arises from a single parent species. For example, a failure of cell division might double the chromosome number from the original diploid number ($2n$) to tetraploid ($4n$). Because the polyploid individual cannot produce fertile hybrids with its parent species, immediate reproductive isolation results.

A second form of polyploid speciation can occur when two different species interbreed and produce hybrid offspring. Most instances of polyploid speciation in plants resulted from such hybridizations. How did these

▼ **Figure 14.8 Gray tree frog.** This amphibian is thought to have originated by polyploid speciation.

▼ **Figure 14.9 Chinese hibiscus.** Many ornamental varieties of this species have been produced through polyploidy, which may make the flowers larger or increase the number of petals.

☑ **CHECKPOINT**

What mechanism accounts for most observed instances of sympatric speciation? Why might this be the case?

Answer: Accidents of cell division that result in polyploidy. Polyploidy produces "instant" reproductive isolation.

interspecies hybrids overcome the postzygotic barrier of sterility (see Figure 14.5)? A mule is sterile because its parents' chromosomes don't match. A horse has 64 chromosomes (32 pairs); a donkey has 62 chromosomes (31 pairs). Therefore, a mule has 63 chromosomes. Recall that gametes are produced by meiosis, a cell division process that involves pairing of homologous chromosomes (see Figure 8.14). Structural differences between the chromosomes of horses and donkeys prevent them from pairing correctly, and the odd number leaves one chromosome without a potential partner. Consequently, mules don't produce viable gametes. Now, let's see why plants don't have the same problem.

In **Figure 14.10**, species A hybridizes with species B. Like a mule, the resulting hybrid plant has an odd number of chromosomes, and its chromosomes are not homologous. However, the hybrid may be able to reproduce asexually, as many plants can do. If so, there may eventually be an error in cell division that results in polyploidy and thus reproductive isolation from the parent. Biologists have identified several plant species that originated via polyploidy within the past 150 years.

Many of the plant species we grow for food are polyploids, including oats, potatoes, bananas, strawberries, peanuts, apples, sugarcane, and wheat. The wheat used for bread is a hybrid of three different parent species and has six sets of chromosomes, two sets from each parent. Plant geneticists use chemicals to induce errors in cell division and generate new polyploids in the laboratory. By harnessing this evolutionary process, they can produce new hybrid species with desirable qualities.

Biologists have also identified cases in which subpopulations appear to be in the process of sympatric speciation. In some cases, subgroups of a population become adapted for exploiting food sources in different habitats,

▼ **Figure 14.10 Sympatric speciation in a plant.** The gametes from two different species result in a sterile hybrid, which may undergo asexual reproduction. Such a hybrid may eventually form a new species by polyploidy.

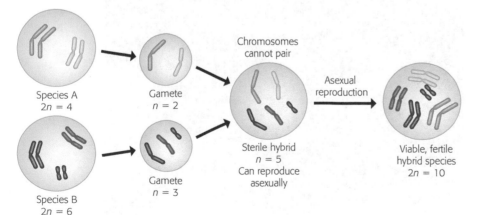

such as the shallow versus deep habitats of a lake. In another example, a type of sexual selection in which female fish choose mates based on color has contributed to rapid reproductive isolation. Because of its direct effect on reproductive success, sexual selection can interrupt gene flow within a population and may therefore be an important factor in sympatric speciation. But the most frequently observed mechanism of sympatric speciation involves large-scale genetic changes that occur in a single generation. ☑

Island Showcases of Speciation

Volcanic islands, such as the Galápagos and Hawaiian island chains, are initially devoid of life. Over time, colonists arrive via ocean currents or winds. Some of these organisms gain a foothold and establish new populations. In their new environment, these populations may diverge significantly from their distant parent populations. In addition, islands that have physically diverse habitats and that are far enough apart to permit populations to evolve in isolation but close enough to allow occasional dispersals to occur are often the sites of multiple speciation events.

The Galápagos Islands, which were formed by underwater volcanoes from 5 million to 1 million years ago, are one of the world's great showcases of speciation. They are home to numerous plants, snails, reptiles, and birds that are found nowhere else on Earth. For example, the islands have 14 species of closely related finches, which are often called Darwin's finches because he collected them during his around-the-world voyage (see Figure 13.3). These birds share many finch-like traits, but they differ in their feeding habits and their beaks, which are specialized for what they eat. Their various foods include insects, large or small seeds, cactus fruits, and even eggs of other species. The woodpecker finch uses cactus spines or twigs as tools to pry insects from trees. The "vampire" finch supplements its diet of seeds and insects by pecking wounds on the backs of seabirds and drinking their blood. **Figure 14.11** (facing page) shows some of these birds, with their distinctive beaks adapted for their specific diets. The finches differ in their habitats as well as their beaks—some live in trees and others spend most of their time on the ground.

How might Darwin's finch species have evolved from a small population of ancestral birds that colonized one of the islands? Completely isolated on the island, the founder population may have changed significantly as natural selection adapted it to the new environment, and thus it became a new species. Later, a few individuals of this new species may have migrated to a neighboring island, where, under different conditions, this new founder population was changed enough through

Cactus-seed-eater (cactus finch)

Tool-using insect-eater (woodpecker finch)

Blood, seed, and insect-eater (vampire finch)

▲ Figure 14.11 **Galápagos finches with beaks adapted for specific diets.**

process of speciation is often extremely slow. So you may be surprised to learn that we *can* see speciation occurring. Consider that life has been evolving over hundreds of millions of years and will continue to evolve. The species living today represent a snapshot, a brief instant in this vast span of time. The environment continues to change—sometimes rapidly due to human impact—and natural selection continues to act on affected populations. It is reasonable to assume that some of these populations are changing in ways that could eventually lead to speciation. Studying populations as they diverge gives biologists a window on the process of speciation.

Researchers have documented at least two dozen cases in which populations are currently diverging as they use different food resources or breed in different habitats. Numerous cases involve insects exploiting different food plants. In one well-studied example, a subpopulation of a fly that feeds on hawthorn fruits found a new resource when American colonists planted apple trees. Although the two fly populations are still regarded as subspecies, researchers have identified mechanisms that severely restrict gene flow between them. In other cases, biologists have identified animal populations that are diverging as a result of differences in male courtship behavior.

Although biologists are continually making observations and devising experiments to study evolution in progress, much of the evidence for evolution comes from the fossil record. So what does the fossil record say about the time frame for speciation—the length of time between when a new species forms and when its populations diverge enough to produce another new species? In one survey of 84 groups of plants and animals, the time for speciation ranged from 4,000 to 40 million years. Such long time frames tell us that it has taken vast spans of time for life on Earth to evolve.

As you've seen, speciation may begin with small differences. However, as speciation occurs again and again, these differences accumulate and may eventually lead to new groups that differ greatly from their ancestors, as in the origin of whales from four-legged land animals (see Figure 13.6). The cumulative effects of multiple speciations, as well as extinctions, have shaped the dramatic changes documented in the fossil record. We begin to examine such changes next.

natural selection to become yet another new species. Some of these birds may then have recolonized the first island and coexisted there with the original ancestral species if reproductive barriers kept the species distinct. Multiple rounds of colonization and speciation on the many separate islands of the Galápagos probably followed. Today, each of the Galápagos Islands has several species of finches, with as many as ten on some islands. Reproductive isolation due to species-specific songs helps keep the species separate.

Observing Speciation in Progress

In contrast to microevolutionary change, which may be apparent in a population within a few generations, the

Earth History and Macroevolution

Having examined how new species arise, we are ready to turn our attention to macroevolution. **Macroevolution** is evolutionary change above the species level, for example, the origin of a new group of organisms through a series of speciation events. Macroevolution also includes the impact of mass extinctions on the diversity of life and its subsequent recovery. An understanding of macroevolution begins with a look at the span of geologic time over which life's diversity has evolved.

The Fossil Record

Fossils are evidence of organisms that lived in the past (see Figure 13.5). The strata (layers) of sedimentary rocks provide a record of life on Earth—each rock layer contains a local sample of the organisms that existed at the time the sediment was deposited. Thus, the fossil record, the sequence in which fossils appear in rock strata, is an archive of macroevolution. For example, scan the wall of the Grand Canyon from rim to floor and you look back through hundreds of millions of years **(Figure 14.12)**. Younger strata formed atop older ones; correspondingly, younger fossils are found in layers closer to the surface, whereas the deepest strata contain the oldest fossils. However, this only gives us the ages of fossils relative to each other. Like peeling off layers of wallpaper in an old house, we can infer the order in which the layers were applied but not the year that each layer was added.

If you drove off the rim of the Grand Canyon, you would pass by 40 rock layers and several hundred million years of geologic history before hitting the ground.

By studying many sites, geologists have established a **geologic time scale** that divides Earth's history into a sequence of geologic periods. The time line presented in **Table 14.1** is separated into four broad divisions: the Precambrian (a general term for the time before about 540 million years ago), followed by the Paleozoic, Mesozoic, and Cenozoic eras. Each of these divisions represents a distinct age in the history of Earth and its life. The boundaries between eras are marked by mass extinctions, when many forms of life disappeared from the fossil record and were replaced by species that diversified from the survivors.

The most common method geologists use to learn the ages of rocks and the fossils they contain is **radiometric dating**, a method based on the decay of radioactive isotopes (see Figure 2.17). For example, a living organism contains both the common isotope carbon-12 and the radioactive isotope carbon-14 in the same ratio as that present in the atmosphere. Once an organism dies, it stops accumulating carbon, and the stable carbon-12 in its tissues does not change. Carbon-14, however, spontaneously decays to another element. Carbon-14 has a half-life of 5,730 years, so half the carbon-14 in a specimen decays in about 5,730 years, half the remaining carbon-14 decays in the next 5,730 years, and so on. Scientists measure the ratio of carbon-14 to carbon-12 in a fossil to calculate its age.

Carbon-14 is useful for dating relatively young fossils—up to about 75,000 years old. There are isotopes with longer half-lives, such as uranium-235 (half-life 71.3 million years) and potassium-40 (half-life 1.3 billion years). However, organisms don't incorporate those elements into their bodies. Therefore, scientists use indirect methods to date older fossils. One commonly used method is to date layers of volcanic rock or ash above and below the sedimentary layer in which fossils are found. By inference, the age of the fossils is between those two dates. Potassium-argon dating is often used for volcanic rock, for example. Isotopes of uranium are useful for other types of ancient rock.

▼ **Figure 14.12 Strata of sedimentary rock at the Grand Canyon.** The Colorado River has cut through more than a mile of rock, exposing sedimentary strata that are like huge pages from the book of life. Each stratum entombs fossils that represent some of the organisms from that period of Earth's history.

Table 14.1	The Geologic Time Scale				

Geologic Time	Period	Epoch	Age (millions of years ago)	Some Important Events in the History of Life	Relative Time Span
Cenozoic era	Quaternary	Recent		Historical time	Cenozoic
			0.01		Mesozoic
		Pleistocene		Ice ages; humans appear	
			1.8		Paleozoic
	Tertiary	Pliocene		Origin of genus *Homo*	
			5		
		Miocene		Continued speciation of mammals and angiosperms	
			23		
		Oligocene		Origins of many primate groups, including apes	
			34		
		Eocene		Angiosperm dominance increases; origins of most living mammalian orders	
			56		
		Paleocene		Major speciation of mammals, birds, and pollinating insects	
			65		
Mesozoic era	Cretaceous			Flowering plants (angiosperms) appear; many groups of organisms, including most dinosaur lineages, become extinct at end of period (Cretaceous extinctions)	
			145		
	Jurassic			Gymnosperms continue as dominant plants; dinosaurs become dominant	
			200		
	Triassic			Cone-bearing plants (gymnosperms) dominate landscape; speciation of dinosaurs, early mammals, and birds	
			251		
Paleozoic era	Permian			Extinction of many marine and terrestrial organisms (Permian extinctions); speciation of reptiles; origins of mammal-like reptiles and most living orders of insects	
			299		
	Carboniferous			Extensive forests of vascular plants; first seed plants; origin of reptiles; amphibians become dominant	Pre-cambrian
			359		
	Devonian			Diversification of bony fishes; first amphibians and insects	
			416		
	Silurian			Early vascular plants dominate land	
			444		
	Ordovician			Marine algae are abundant; colonization of land by diverse fungi, plants, and animals	
			488		
	Cambrian			Origin of most living animal phyla (Cambrian explosion)	
			542		
Precambrian			600	Diverse algae and soft-bodied invertebrate animals appear	
			635	Oldest animal fossils	
			2,100	Oldest eukaryotic fossils	
			2,700	Oxygen begins accumulating in atmosphere	
			3,500	Oldest fossils known (prokaryotes)	
			4,600	Approximate time of origin of Earth	

Plate Tectonics and Biogeography

If photographs of Earth were taken from space every 10,000 years and then spliced together, it would make a remarkable movie. The seemingly "rock solid" continents we live on drift about Earth's surface. According to the theory of **plate tectonics**, the continents and seafloors form a thin outer layer of solid rock, called the crust, which covers a mass of hot, viscous material called the mantle. The crust is not one continuous expanse, however. It is divided into giant, irregularly shaped plates that float atop the mantle **(Figure 14.13)**. In a process called continental drift, movements in the mantle cause the plates to move. The boundaries of some plates are hotspots of geologic activity. In some cases, we feel an immediate, violent symptom of this activity, as when an earthquake signals that two plates are scraping past or colliding with each other **(Figure 14.14)**. Although most movement is extremely slow, not much faster than the speed at which fingernails grow, continents have wandered thousands of miles over the long course of earth's history.

By reshaping the physical features of the planet and altering the environments in which organisms live, continental drift has had a tremendous impact on the evolution of life's diversity. Two chapters in the continuing saga of continental drift had an especially strong influence on life. About 250 million years ago, near the end of the Paleozoic era, plate movements brought all the previously separated landmasses together into a supercontinent called Pangaea, which means "all land" **(Figure 14.15,** facing page). Imagine some of the possible effects on life. Species that had been evolving in isolation came together and competed. As the landmasses joined, the total amount of shoreline was reduced. There is also evidence that the ocean basins increased in depth, lowering sea level and draining the shallow coastal seas. Then, as now, most

▲ Figure 14.14 **A tsunami caused by an earthquake off the coast of Japan in March 2011.** Japan sits atop four different plates. Frequent earthquakes occur as the plates move and bump against each other.

marine species inhabited shallow waters, and the formation of Pangaea destroyed a considerable amount of that habitat. It was probably a long, traumatic period for terrestrial life as well. The continental interior, which was drier and had a more erratic climate than the coastal regions, increased in area substantially when the land came together. Changing ocean currents also undoubtedly affected land life as well as sea life. Thus, the formation of Pangaea had a tremendous environmental impact that reshaped biological diversity by causing extinctions and providing new opportunities for diversification of the survivors.

The second dramatic chapter in the history of continental drift began during the mid-Mesozoic era. Pangaea started to break up, causing geographic isolation of colossal proportions. As the landmasses drifted apart, each continent became a separate evolutionary arena as its climates changed and its organisms diverged.

The history of continental mergers and separations explains many patterns of **biogeography**, the study of the past and present distribution of organisms. For example, almost all the animals and plants that live on Madagascar, a large island located off the southern coast of Africa, are unique—they diversified from ancestral populations after Madagascar was isolated from Africa and India. The more than 50 species of lemurs that currently inhabit Madagascar, for instance, evolved from a common ancestor over the past 40 million years.

▼ Figure 14.13 **Earth's tectonic plates.** The red dots indicate zones where violent geologic activity, such as earthquakes and volcanic eruptions, take place. Arrows indicate direction of continental drift.

Key

● Zones of violent tectonic activity

↑ Direction of movement

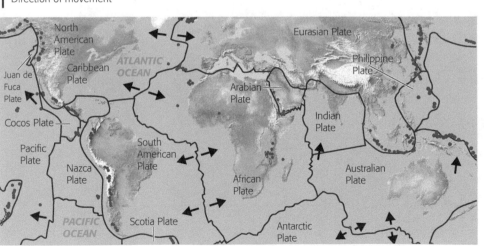

Continental drift also separated Australia from other landmasses. Australia and its neighboring islands are home to more than 200 species of marsupials, most of which are found nowhere else in the world (Figure 14.16). Marsupials are mammals such as kangaroos, koalas, and wombats whose young complete their embryonic development in a pouch outside the mother's body. The rest of the world is dominated by eutherian (placental) mammals, whose young complete their development in the mother's uterus. Looking at a current map of the world, you might hypothesize that marsupials evolved only on the island continent of Australia. But marsupials are not unique to Australia. More than a hundred species live in Central and South America. North America is also home to a few, including the Virginia opossum. The distribution of marsupials only makes sense in the context of continental drift—marsupials must have originated when the continents were joined. Fossil evidence suggests that marsupials originated in what is now Asia and later dispersed to the tip of South America while it was still connected to Antarctica. They made their way to Australia before continental drift separated Antarctica from Australia, setting "afloat" a great raft of marsupials. The few early eutherians that lived in Australia became extinct, whereas on other continents, most marsupials became extinct. Isolated on Australia, marsupials evolved and diversified, filling ecological roles analogous to those filled by eutherians on other continents. ☑

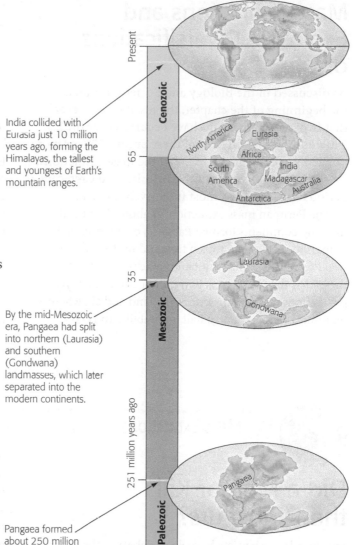

India collided with Eurasia just 10 million years ago, forming the Himalayas, the tallest and youngest of Earth's mountain ranges.

By the mid-Mesozoic era, Pangaea had split into northern (Laurasia) and southern (Gondwana) landmasses, which later separated into the modern continents.

Pangaea formed about 250 million years ago.

Present

65

35

251 million years ago

Cenozoic

Mesozoic

Paleozoic

◄ Figure 14.15 **The history of plate tectonics.** The continents continue to drift, though not at a rate that's likely to cause any motion sickness for their passengers.

☑ **CHECKPOINT**

How many continents did Earth have at the time of Pangaea?

Answer: one

Tasmanian devil, a carnivore

▼ Figure 14.16 **Australian marsupials.** The continent of Australia is home to many unique plants and animals, including diverse marsupials, mammals that have ecological roles filled by eutherians on other continents.

Sugar glider, an omnivore

Koala, an herbivore

Mass Extinctions and Explosive Diversifications of Life

As discussed in the Biology and Society section at the beginning of the chapter, the fossil record reveals that five mass extinctions have occurred over the last 540 million years. In each of these events, 50% or more of Earth's species died out. Of all the mass extinctions, those marking the ends of the Permian and Cretaceous periods have been the most intensively studied.

The Permian mass extinction, at about the time the merging continents formed Pangaea, claimed about 96% of marine species and took a tremendous toll on terrestrial life as well. Equally notable is the mass extinction at the end of the Cretaceous period. For 150 million years prior, dinosaurs dominated Earth's land and air, whereas mammals were few and small, resembling today's rodents. Then, about 65 million years ago, most of the dinosaurs became extinct, leaving behind only the descendants of one lineage, the birds. Remarkably, the massive die-off (which also included half of all other species) occurred in less than 10 million years—a brief period in geologic time.

But there is a flip side to the destruction. Each massive dip in species diversity has been followed by explosive diversification of certain survivors. Extinctions seem to have provided the surviving organisms with new environmental opportunities. For example, mammals existed for at least 75 million years before undergoing an explosive increase in diversity just after the Cretaceous period. Their rise to prominence was undoubtedly associated with the void left by the extinction of the dinosaurs. The world would be a very different place today if many dinosaur lineages had escaped the Cretaceous extinctions or if none of the mammals had survived.

 Mass Extinctions THE PROCESS OF SCIENCE

Did a Meteor Kill the Dinosaurs?

For decades, scientists have been debating the cause of the rapid dinosaur die-off that occurred 65 million years ago. Many **observations** provide clues. The fossil record shows that the climate had cooled and that shallow seas were receding from continental lowlands. It also shows that many plant species died out. Perhaps the most telling evidence was discovered by physicist Luis Alvarez and his geologist son Walter Alvarez, both of the University of California, Berkeley. In 1980, they found that rock deposited around 65 million years ago contains a thin layer of clay rich in iridium, an element very rare on Earth but common in meteors and other extraterrestrial material that occasionally falls to Earth. This discovery led the Alvarez team to ask the following **question**: Is the iridium layer the result of fallout from a huge cloud of dust that billowed into the atmosphere when a large meteor or asteroid hit Earth?

The father and son formed the **hypothesis** that the mass extinction 65 million years ago was caused by the impact of an extraterrestrial object. This hypothesis makes a clear **prediction**: A huge impact crater of the right age should be found somewhere on Earth's surface. (This is a good example of using verifiable observations rather than a direct **experiment** to test a hypothesis; see Chapter 1.) In 1981, two petroleum geologists found the **results** predicted by the Alvarezes' hypothesis: the Chicxulub crater, located near Mexico's Yucatán Peninsula in the Caribbean Sea **(Figure 14.17)**. This impact site, about 180 km wide (about 112 miles) and dating from the predicted time, was created when a meteor or asteroid about 10 km in diameter (about 6 miles, more than the length of 100 football fields) slammed into Earth, releasing thousands of times more energy than is stored in the world's combined stockpile of nuclear weapons. Such a cloud could have blocked sunlight and disturbed climate severely for months, perhaps killing off many plant species and, later, the animals that depended on those plants for food.

Debate continues about whether this impact alone caused the dinosaurs to die out or whether other factors—such as continental movements or volcanic activity—also contributed. Most scientists agree, however, that the collision that created the Chicxulub crater could indeed have been a major factor in global climatic change and mass extinctions.

▼ Figure 14.17 **Trauma for planet Earth and its Cretaceous life.**

Chicxulub crater

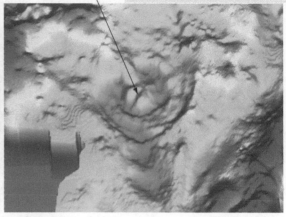

An artist's representation of the impact of an asteroid or comet.

The impact's immediate effect was most likely a cloud of hot vapor and debris that could have killed many of the plants and animals in North America within hours.

The 65-million-year-old Chicxulub impact crater is located in the Caribbean Sea near the Yucatán Peninsula of Mexico. The horseshoe shape of the crater and the pattern of debris in sedimentary rocks indicate that an asteroid or comet struck at a low angle from the southeast.

Mechanisms of Macroevolution

The fossil record can tell us what the great events in the history of life have been and when they occurred. Continental drift and mass extinctions followed by the diversification of survivors provide a big-picture view of how those changes came about. But now scientists are increasingly able to explain the basic biological mechanisms that underlie the macroevolutionary changes seen in the fossil record.

Large Effects from Small Genetic Changes

Scientists working at the interface of evolutionary biology and developmental biology—the research field abbreviated **evo-devo**—are studying how slight genetic changes can become magnified into major structural differences between species. Homeotic genes, the master control genes, program development by controlling the rate, timing, and spatial pattern of changes in an organism's form as it develops from a zygote into an adult. These genes determine such basic developmental events as where a pair of wings or legs will appear on a fruit fly (see Figure 11.9). A subtle change in the developmental program can have profound effects. Accordingly, changes in the number, nucleotide sequence, and regulation of homeotic genes have led to the huge diversity in body forms.

Changes in rate of developmental events explains changes in the homologous limb bones of vertebrates (see Figure 13.7). Increased growth rates produced the extra-long "finger" bones in bat wings. Slower growth rates of leg and pelvic bones led to the eventual loss of hind limbs in whales (see Figure 13.6). On the other hand, the loss of limbs in the evolution of snakes from a four-limbed lizard-like ancestor resulted from different spatial patterns of expression of homeotic genes.

Striking evolutionary transformations may also result from a change in the timing of developmental events. **Figure 14.18** is a photograph of an axolotl, a salamander that illustrates a phenomenon called **paedomorphosis**, the retention in the adult of body structures that were juvenile features in an ancestral species. The axolotl grows to full size and reproduces without losing its external gills, a juvenile feature in most species of salamanders.

Gills

◄ Figure 14.18
Paedomorphosis. The axolotl, a salamander, becomes an adult (shown here) and reproduces while retaining certain tadpole characteristics, including gills.

317

Paedomorphosis has also been important in human evolution. Humans and chimpanzees are even more alike in body form as fetuses than they are as adults. In the fetuses of both species, the skulls are rounded and the jaws are small, making the face rather flat **(Figure 14.19)**. As development proceeds, accelerated growth in the jaw produces the elongated skull, sloping forehead, and massive jaws of an adult chimpanzee. In the human lineage, genetic changes that slowed the growth of the jaw relative to other parts of the skull produced an adult whose head proportions still resembled that of a child—and that of a baby chimpanzee. Our large skull and complex brain are among our most distinctive features. The human brain is proportionately larger than the chimpanzee brain because growth of the organ is switched off much later in human development. Compared with the brain of a chimpanzee, our brain continues to grow for several more years, which can be interpreted as the prolonging of a juvenile process.

Next, we see how the process of evolution can produce new, complex structures.

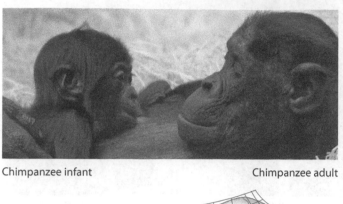

Chimpanzee infant   Chimpanzee adult

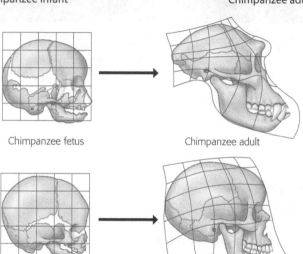

Chimpanzee fetus   Chimpanzee adult

Human fetus   Human adult (paedomorphic features)

▲ **Figure 14.19 Comparison of human and chimpanzee skull development.** Starting with fetal skulls that are very similar (left), the differential growth rates of the bones making up the skulls produce adult heads with very different proportions. The grid lines will help you relate the fetal skulls to the adult skulls.

The Evolution of Biological Novelty

The two squirrels in Figure 14.6 are different species, but they are very similar animals that live very much the same way. How do we account for the dramatic differences between dissimilar groups—squirrels and birds, for example? Let's see how the Darwinian theory of gradual change can explain the evolution of intricate structures such as eyes or of new kinds of structures (novel structures) such as feathers.

Structure/Function   **Adaptation of Old Structures for New Functions**

The feathered flight of birds is a perfect marriage of structure and function. Consider the evolution of feathers, which are clearly essential to avian aeronautics. In a flight feather, separate filaments called barbs emerge from a central shaft that runs from base to tip. Each barb is linked to the next by tiny hooks that act much like the teeth of a zipper, forming a tightly connected sheet of barbs that is strong but flexible. In flight, the shapes and arrangements of various feathers produce lift, smooth airflow, and help with steering and balance. How did such a beautifully intricate structure evolve? Reptilian features apparent in fossils of *Archaeopteryx*, one of the earliest birds, offered clues in Darwin's time **(Figure 14.20)**, but the definitive answer came in 1996. Birds were not the first feathered animals on Earth—dinosaurs were.

The first feathered dinosaur to be discovered, a 130-million-year-old fossil found in northeastern China, was named *Sinosauropteryx* ("Chinese lizard-wing"). About the size of a turkey, it had short arms and ran on its hind legs, using its long tail for balance. Its

▼ **Figure 14.20 An extinct bird.** Called *Archaeopteryx* ("ancient wing"), this animal lived near tropical lagoons in central Europe about 150 million years ago. Despite its feathers, *Archaeopteryx* has many features in common with reptiles. *Archaeopteryx* is not considered an ancestor of today's birds. Instead, it probably represents an extinct side branch of the bird lineage.

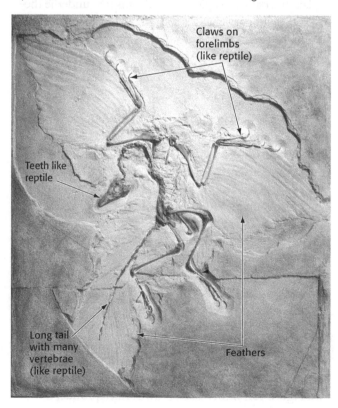

Claws on forelimbs (like reptile)

Teeth like reptile

Long tail with many vertebrae (like reptile)

Feathers

unimpressive plumage consisted of a downy covering of hairlike feathers. Since the discovery of *Sinosauropteryx*, thousands of fossils of feathered dinosaurs have been found and classified into more than 30 different species. Although none was unequivocally capable of flying, many of these species had elaborate feathers that would be the envy of any modern bird. But the feathers seen in these fossils could not have been used for flight, nor would their reptilian anatomy have been suited to flying. So if feathers evolved before flight, what was their function? Their first utility may have been for insulation. It is possible that longer, winglike forelimbs and feathers, which increased the surface area of these forelimbs, were co-opted for flight after functioning in some other capacity, such as mating displays, thermoregulation, or camouflage (all functions that feathers still serve today). The first flights may have been only short glides to the ground or from branch to branch in tree-dwelling species. Once flight itself became an advantage, natural selection would have gradually remodeled feathers and wings to fit their additional function.

Structures such as feathers that evolve in one context but become co-opted for another function are called exaptations. However, exaptation does not mean that a structure evolves in anticipation of future use. Natural selection cannot predict the future; it can only improve an existing structure in the context of its current use. ☑

From Simple to Complex Structures in Gradual Stages

Most complex structures have evolved in small steps from simpler versions having the same basic function—a process of refinement rather than the sudden appearance of complexity. Consider the amazing camera-like eyes of vertebrates and squids. Although these complex eyes evolved independently, the origin of both can be traced from a simple ancestral patch of photoreceptor cells through a series of incremental modifications that benefited their owners at each stage. Indeed, there appears to have been a single evolutionary origin of light-sensitive cells, and all animals with eyes—vertebrates and invertebrates alike—share the same master genes that regulate eye development.

Figure 14.21 illustrates the range of complexity in the structure of eyes among present-day molluscs, a large and diverse phylum of animals. Simple patches of pigmented cells enable limpets, single-shelled molluscs that cling to seaside rocks, to distinguish light from dark. When a shadow falls on them, they hold on more tightly—a behavioral adaptation that reduces the risk of being eaten. Other molluscs have eyecups that have no lenses or other means of focusing images but can indicate light direction. In those molluscs that do have complex eyes, the organs probably evolved in small steps of adaptation. You can see examples of such small steps in Figure 14.21.

▼ Figure 14.21 **A range of eye complexity among molluscs.** The complex eye of the squid evolved in small steps. Even the simplest eye was useful to its owner.

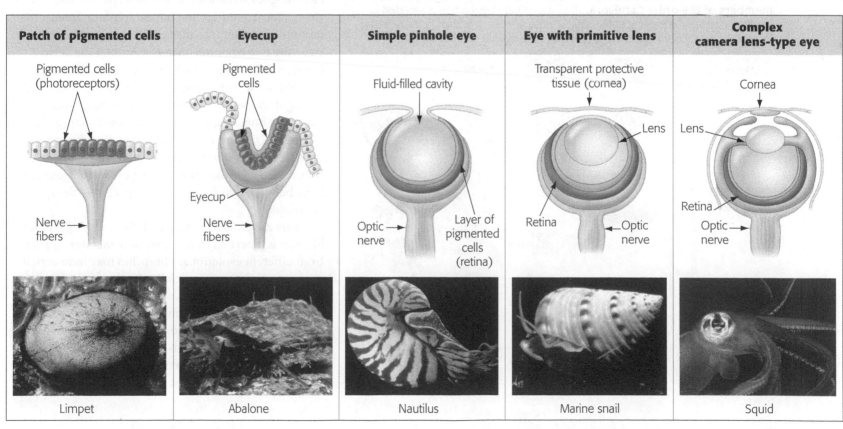

Patch of pigmented cells	Eyecup	Simple pinhole eye	Eye with primitive lens	Complex camera lens-type eye
Pigmented cells (photoreceptors) / Nerve fibers	Pigmented cells / Eyecup / Nerve fibers	Fluid-filled cavity / Optic nerve / Layer of pigmented cells (retina)	Transparent protective tissue (cornea) / Retina / Lens / Optic nerve	Cornea / Lens / Retina / Optic nerve
Limpet	Abalone	Nautilus	Marine snail	Squid

Classifying the Diversity of Life

The Linnaean system of taxonomy (see Figure 13.1) is quite a useful method of organizing life's diversity into groups. Ever since Darwin, however, biologists have had a goal beyond simple organization: to have classification reflect evolutionary relationships. In other words, how an organism is named and classified should reflect its place within the evolutionary tree of life. **Systematics**, which includes taxonomy, is a discipline of biology that focuses on classifying organisms and determining their evolutionary relationships.

Classification and Phylogeny

Biologists use **phylogenetic trees** to depict hypotheses about the evolutionary history, or phylogeny, of species. These branching diagrams reflect the hierarchical classification of groups nested within more inclusive groups. The tree in **Figure 14.22** shows the classification of some carnivores and their probable evolutionary relationships. Note that each branch point represents the divergence of two lineages from a common ancestor. (You may recall Figure 13.9, which is a phylogenetic tree of tetrapods.)

Understanding phylogeny can have practical applications. For example, maize (corn) is an important food crop worldwide; it also provides us with snack favorites such as popcorn, tortilla chips, and corn dog batter. Thousands of years of artificial selection (selective breeding) transformed a scrawny grass with small ears of rock-hard kernels into the maize we know today. In the process, much of the plant's original genetic variation was stripped away. By constructing a phylogeny of maize, researchers have identified two species of wild grasses that may be maize's closest living relatives. The genomes of these plants may harbor alleles that offer disease resistance or other useful traits that could be transferred into cultivated maize by crossbreeding or genetic engineering—insurance against future disease outbreaks or other environmental changes that might threaten corn crops.

Alleles from two wild grasses could be used to make a better ear of corn.

▼ **Figure 14.22 The relationship of classification and phylogeny for some members of the order Carnivora.** The hierarchical classification is reflected in the finer and finer branching of the phylogenetic tree. Each branch point in the tree represents an ancestor common to species to the right of that branch point.

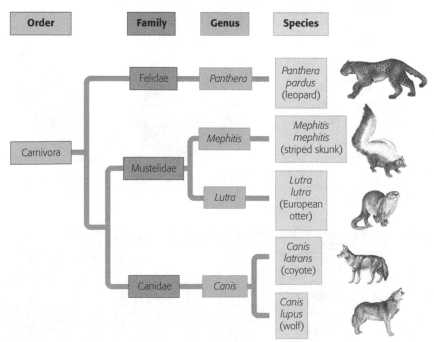

Order	Family	Genus	Species

Carnivora
- Felidae — Panthera — *Panthera pardus* (leopard)
- Mustelidae
  - Mephitis — *Mephitis mephitis* (striped skunk)
  - Lutra — *Lutra lutra* (European otter)
- Canidae — Canis
  - *Canis latrans* (coyote)
  - *Canis lupus* (wolf)

Identifying Homologous Characters

Homologous structures in different species may vary in form and function but exhibit fundamental similarities because they evolved from the same structure in a common ancestor. Among the vertebrates, for instance, the whale forelimb is adapted for steering in the water, whereas the bat wing is adapted for flight. Nonetheless, there are many basic similarities in the bones supporting these two structures (see Figure 13.7). Thus, homologous structures are one of the best sources of information for phylogenetic relationships. The greater the number of homologous structures between two species, the more closely the species are related.

There are pitfalls in the search for homology: Not all likeness is inherited from a common ancestor. Species from different evolutionary branches may have certain structures that are superficially similar if natural selection has shaped analogous adaptations. This is called **convergent evolution**. Similarity due to convergence is called **analogy**, not homology. For example, the wings of insects and those of birds are analogous flight equipment: They evolved independently and are built from entirely different structures.

▲ Figure 14.23 **Artist's reconstruction of Neanderthal.** DNA extracted from Neanderthals, extinct members of the human family, has allowed scientists to study their evolutionary relationship with modern humans.

Comparing the embryonic development of two species can often reveal homology that is not apparent in the mature structures (for example, see Figure 13.8). There is another clue to distinguishing homology from analogy: The more complex two similar structures are, the less likely it is they evolved independently. For example, compare the skulls of a human and a chimpanzee (see Figure 14.19). Although each is a fusion of many bones, they match almost perfectly, bone for bone. It is highly improbable that such complex structures matching in so many details could have separate origins. Most likely, the genes required to build these skulls were inherited from a common ancestor.

If homology reflects common ancestry, then comparing the DNA sequences of organisms gets to the heart of their evolutionary relationships. The more recently two species have branched from a common ancestor, the more similar their DNA sequences should be. Scientists have sequenced well over 150 billion bases of DNA from thousands of species. This enormous database has fueled a boom in the study of phylogeny and clarified many evolutionary relationships. In addition, some fossils are preserved in such a way that DNA fragments can be extracted for comparison with living organisms **(Figure 14.23)**. ☑

Inferring Phylogeny from Homologous Characters

Once homologous characters—characters that reflect an evolutionary relationship—have been identified

for a group of organisms, how are these characters used to construct phylogenies? The most widely used approach is called cladistics. In **cladistics**, organisms are grouped by common ancestry. A **clade** (from the Greek word for "branch") consists of an ancestral species and all its evolutionary descendants—a distinct branch in the tree of life. Thus, identifying clades makes it possible to construct classification schemes that reflect the branching pattern of evolution.

Cladistics is based on the Darwinian concept of "descent with modification from a common ancestor"— species have some characters in common with their ancestors, but they also differ from them. To identify clades, scientists compare an ingroup with an outgroup **(Figure 14.24)**. The ingroup (for example, the three mammals in Figure 14.24) is the group of species that is actually being analyzed. The outgroup (in Figure 14.24, the iguana, representing reptiles) is a species or group of species known to have diverged before the lineage that contains the groups being studied. By comparing members of the ingroup with each other and with the outgroup, we can determine what characters distinguish the ingroup from the outgroup. All the mammals in the ingroup have hair and mammary glands. These characters were present in the ancestral mammal, but not in the outgroup. Next, gestation, the carrying of offspring in the uterus within the female parent, is absent from the duck-billed platypus (which lays eggs with a shell). From this absence we might infer that the duck-billed platypus represents an early branch point in the mammalian clade. Proceeding in this manner, we can construct a phylogenetic tree. Each branch represents the divergence of two groups from a

Piecing together evolutionary histories shows who's related to whom.

▼ Figure 14.24 **A simplified example of cladistics.**

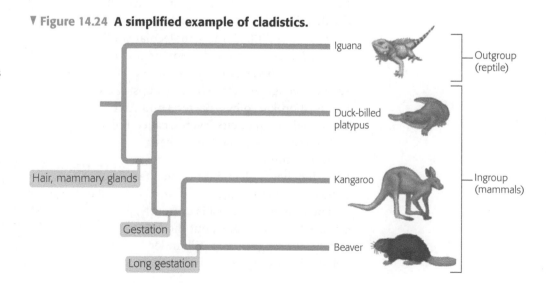

Iguana — Outgroup (reptile)

Duck-billed platypus

Hair, mammary glands

Kangaroo — Ingroup (mammals)

Gestation

Beaver

Long gestation

common ancestor, with the emergence of a lineage possessing one or more new features. The sequence of branching represents the order in which they evolved and when groups last shared a common ancestor. In other words, cladistics focuses on the changes that define the branch points in evolution.

The cladistics approach to phylogeny clarifies evolutionary relationships that were not always apparent in other taxonomic classifications. For instance, biologists traditionally placed birds and reptiles in separate classes of vertebrates (class Aves and class Reptilia, respectively). This classification, however, is inconsistent with cladistics. An inventory of homologies indicates that birds and crocodiles make up one clade, and lizards and snakes form another. If we go back as far as the ancestor that crocodiles share with lizards and snakes to make up a clade, then the class Reptilia must also include birds. The tree in **Figure 14.25** is thus more consistent with cladistics than with traditional classifications. ☑

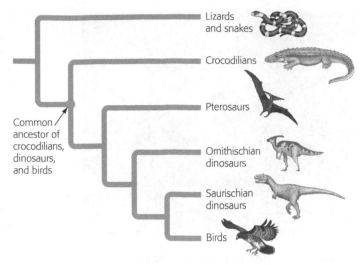

▲ Figure 14.25 **How cladistics is shaking phylogenetic trees.** Strict application of cladistics sometimes produces phylogenetic trees that conflict with classical taxonomy.

Classification: A Work in Progress

Phylogenetic trees are hypotheses about evolutionary history. Like all hypotheses, they are revised (or in some cases rejected) based on new evidence. Molecular systematics and cladistics are combining to remodel phylogenetic trees and challenge traditional classifications.

Linnaeus divided all known forms of life between the plant and animal kingdoms, and the two-kingdom system prevailed in biology for over 200 years. In the mid-1900s, the two-kingdom system was replaced by a five-kingdom system that placed all prokaryotes in one kingdom and divided the eukaryotes among four other kingdoms.

In the late 20th century, molecular studies and cladistics led to the development of a **three-domain system** (**Figure 14.26**). This current scheme recognizes three basic groups: two domains of prokaryotes—Bacteria and Archaea—and one domain of eukaryotes, called Eukarya. The domains Bacteria and Archaea differ in a number of important structural, biochemical, and functional features (see Chapter 15).

The domain Eukarya is currently divided into kingdoms, but the exact number of kingdoms is still under debate. Biologists generally agree on

the kingdoms Plantae, Fungi, and Animalia. These kingdoms consist of multicellular eukaryotes that differ in structure, development, and modes of nutrition. Plants make their own food by photosynthesis. Fungi live by decomposing the remains of other

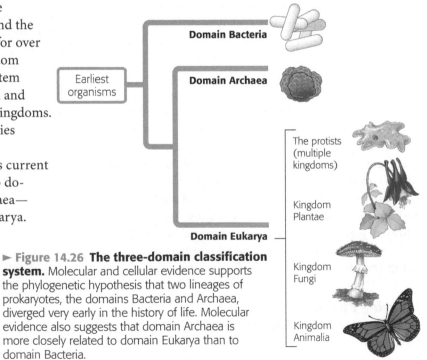

► Figure 14.26 **The three-domain classification system.** Molecular and cellular evidence supports the phylogenetic hypothesis that two lineages of prokaryotes, the domains Bacteria and Archaea, diverged very early in the history of life. Molecular evidence also suggests that domain Archaea is more closely related to domain Eukarya than to domain Bacteria.

organisms and absorbing small organic molecules. Most animals live by ingesting food and digesting it within their bodies.

The remaining eukaryotes, the protists, include all those eukaryotes that do not fit the definition of plant, fungus, or animal—effectively, a taxonomic grab bag. Most protists are unicellular (amoebas, for example). But the protists also include certain large, multicellular organisms that are believed to be direct descendants of unicellular protists. For example, many biologists classify the seaweeds as protists because they are more closely related to some single-celled algae than they are to true plants.

It is important to understand that classifying Earth's diverse species is a work in progress as we learn more about organisms and their evolution. Charles Darwin envisioned the goals of modern systematics when he wrote in *The Origin of Species*, "Our classifications will come to be, as far as they can be so made, genealogies." ☑

☑ **CHECKPOINT**

What lines of evidence caused biologists to develop the three-domain system of classification?

Answer: molecular studies and cladistics

 Mass Extinctions EVOLUTION CONNECTION

Rise of the Mammals

In this chapter, you've read about mass extinctions and their effects on the evolution of life on Earth. In the fossil record, each mass extinction was followed by a period of evolutionary change. Many new species arose as survivors became adapted to occupy new habitats or fill community roles vacated by extinctions.

For example, fossil evidence indicates that the number of mammal species increased dramatically after the extinction of most of the dinosaurs around 65 million years ago **(Figure 14.27)**. Although mammals originated 180 million years ago, fossils older than 65 million years indicate that they were mostly small and not very diverse. Early mammals may have been eaten or outcompeted by the larger and more diverse dinosaurs. With the disappearance of most of the dinosaurs, mammals expanded greatly in both diversity and size, filling the ecological roles once occupied by dinosaurs. Had it not been for the dinosaur extinction, mammals may never have expanded their territories and become the predominant land animals. Therefore, we humans may owe our existence to the demise of older species. Through the process of evolution by natural selection, this pattern of death and renewal is repeated throughout the history of life on Earth.

▼ **Figure 14.27 The increase in mammalian species after the extinction of dinosaurs.** Although mammals originated more than 150 million years ago, they did not begin to widely diverge until after the demise of the dinosaurs.

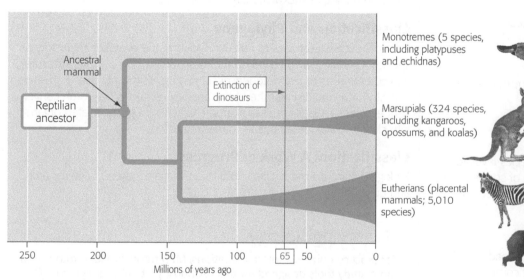

American black bear

Chapter Review

◼ SUMMARY OF KEY CONCEPTS

The Origin of Species

What Is a Species?

The diversity of life evolved through speciation, the process in which one species splits into two or more species. According to the biological species concept, a species is a group of populations whose members have the potential to interbreed in nature to produce fertile offspring. The biological species concept is just one of several possible ways to define species.

Reproductive Barriers between Species

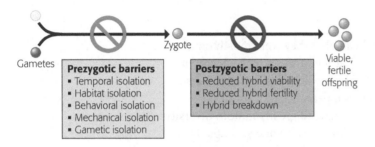

Gametes

Zygote

Viable, fertile offspring

Prezygotic barriers
- Temporal isolation
- Habitat isolation
- Behavioral isolation
- Mechanical isolation
- Gametic isolation

Postzygotic barriers
- Reduced hybrid viability
- Reduced hybrid fertility
- Hybrid breakdown

Evolution: Mechanisms of Speciation

When the gene pool of a population is severed from other gene pools of the parent species, the splinter population can follow its own evolutionary course.

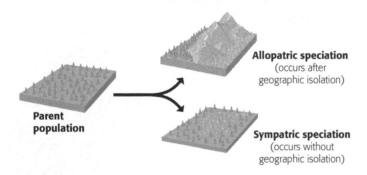

Parent population

Allopatric speciation
(occurs after geographic isolation)

Sympatric speciation
(occurs without geographic isolation)

Hybridization leading to polyploids is a common mechanism of sympatric speciation in plants.

Earth History and Macroevolution

Macroevolution refers to evolutionary change above the species level, for example, the origin of evolutionary novelty and new groups of species and the impact of mass extinctions on the diversity of life and its subsequent recovery.

The Fossil Record

Geologists have established a geologic time scale with four broad divisions: Precambrian, Paleozoic, Mesozoic, and Cenozoic. The most common method for determining the ages of fossils is radiometric dating.

Plate Tectonics and Biogeography

Earth's crust is divided into large, irregularly shaped plates that float atop the viscous mantle. About 250 million years ago, plate movements brought all the landmasses together into the supercontinent Pangaea, causing extinctions and providing new opportunities for the survivors to diversify. About 180 million years ago, Pangaea began to break up, causing geographic isolation. Continental drift accounts for patterns of biogeography, such as the diversity of unique marsupials in Australia.

Mass Extinctions and Explosive Diversifications of Life

The fossil record reveals long, relatively stable periods punctuated by mass extinctions followed in turn by explosive diversification of certain survivors. For example, during the Cretaceous extinctions, about 65 million years ago, the world lost an enormous number of species, including most of the dinosaurs. Mammals greatly increased in diversity after the Cretaceous period.

Mechanisms of Macroevolution

Large Effects from Small Genetic Changes

A subtle change in the genes that control a species' development can have profound effects. In paedomorphosis, for example, the adult retains body features that were strictly juvenile in ancestral species.

The Evolution of Biological Novelty

An exaptation is a structure that evolves in one context and gradually becomes adapted for other functions. Most complex structures have evolved incrementally from simpler versions having the same function.

Classifying the Diversity of Life

Systematics, which includes taxonomy, focuses on classifying organisms and determining their evolutionary relationships.

Classification and Phylogeny

The goal of classification is to reflect phylogeny, the evolutionary history of species. Classification is based on the fossil record, homologous structures, and comparisons of DNA sequences. Homology (similarity based on shared ancestry) must be distinguished from analogy (similarity based on convergent evolution). Cladistics uses shared characters to group related organisms into clades—distinctive branches in the tree of life.

Classification: A Work in Progress

Biologists currently classify life into a three-domain system: Bacteria, Archaea, and Eukarya.

MasteringBiology®

For practice quizzes, BioFlix animations, MP3 tutorials, video tutors, and more study tools designed for this textbook, go to MasteringBiology®

SELF-QUIZ

1. Distinguish between microevolution, speciation, and macroevolution.

2. Bird guides once listed the myrtle warbler and Audubon's warbler as distinct species that lived side by side in parts of their ranges. However, recent books describe them as the eastern and western forms of a single species, the yellow-rumped warbler. Apparently, the two kinds of warblers
   a. live in the same areas.
   b. successfully interbreed.
   c. are almost identical in appearance.
   d. are merging to form a single species.

3. Identify each of the following reproductive barriers as prezygotic or postzygotic.
   a. One lilac species lives on acidic soil, another on basic soil.
   b. Mallard and pintail ducks mate at different times of the year.
   c. Two species of leopard frogs have different mating calls.
   d. Hybrid offspring of two species of jimsonweed always die before reproducing.
   e. Pollen of one kind of pine tree cannot fertilize another kind.

4. Why is a small, isolated population more likely to undergo speciation than a large one?

5. Many species of plants and animals adapted to desert conditions probably did not arise there. Their success in living in deserts could be due to _____, structures that originally had one use but became adapted for different functions.

6. Plate tectonics has been responsible for instances of all of the following *except*
   a. volcanic explosions.
   b. sympatric speciation.
   c. allopatric speciation.
   d. mass extinction.

7. The animals and plants of India are almost completely different from the species in nearby Southeast Asia. Why might this be true?
   a. They have become separated by convergent evolution.
   b. The climates of the two regions are completely different.
   c. India is in the process of separating from the rest of Asia.
   d. India was a separate continent until relatively recently.

8. Uranium-235, with a half-life of 713,000,000 years, decays to lead-207. A paleontologist determines that a rock sample has one-quarter of the uranium-235 content it had when it formed. From this information, you can conclude that the rock sample is approximately _____ billion years old.

9. Why are biologists careful to distinguish similarities due to homology from similarities due to analogy when constructing phylogenetic trees?

10. In the three-domain system, which domain contains eukaryotic organisms?

Answers to these questions can be found in Appendix: Self-Quiz Answers.

THE PROCESS OF SCIENCE

11. Morphs are distinct populations within one species. Three differently sized morphs of the large-eared horseshoe bat (*Rhinolophus philippinensis*) echolocate their preys at different ultrasound harmonics. What type of speciation are these morphs likely to have undergone? How may this have led to reproductive isolation among the morphs?

12. **Interpreting Data** The carbon-14/carbon-12 ratio of a fossilized skull is about 6.25% that of the skull of a present-day animal. Using the graph below, what is the approximate age of the fossil?

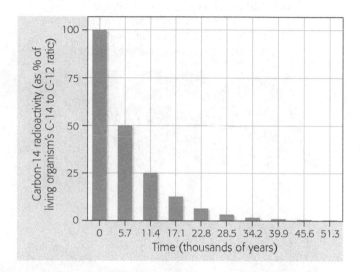

BIOLOGY AND SOCIETY

13. Can cloning be a solution to the alarming extinction rate of certain species? Justify your answer. Are you for or against reviving species that have been extinct for a long time out of scientific curiosity?

14. The red wolf (*Canis rufus*) was once widespread in the southeastern United States but was declared to be extinct in the wild. Biologists bred captive red wolf individuals and reintroduced them into areas of eastern North Carolina, where they are federally protected as endangered species. The current wild population is estimated to be about 100 individuals. However, a new threat to red wolves has arisen: hybridization with coyotes (*Canis latrans*), which have become more numerous in the areas inhabited by red wolves. Although red wolves and coyotes differ in morphology and DNA, they are capable of interbreeding and producing fertile offspring. Social behavior is the main reproductive barrier between the species and is more easily overcome when same-species mates are rare. For this reason, some people think that the endangered status of the red wolf should be withdrawn and resources should not be spent to protect what is not a "pure" species. Do you agree? Why or why not?

15 The Evolution of Microbial Life

Why Microorganisms Matter

If your family took a vacation in which you traveled 1 mile for every million years in the history of life, you'd still be asking, "Are we there yet?" after driving from Miami to Seattle.

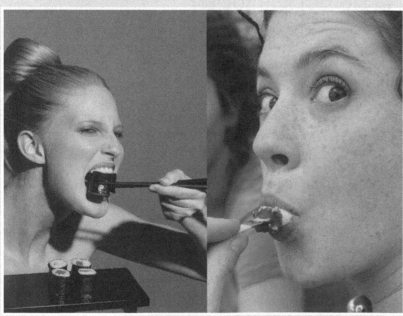

▲ Seaweeds aren't just used for wrapping sushi—they're in your ice cream, too.

According to a ▶ recent study, infection by the parasite *Toxoplasma* makes mice lose their fear of cats.

▲ You have microorganisms to thank for the clean water you drink every day.

Human Microbiota BIOLOGY AND SOCIETY

Our Invisible Inhabitants

You probably know that your body contains trillions of individual cells, but did you know that they aren't all "you"? In fact, microorganisms residing in and on your body outnumber your own cells by 10 to 1. That means 100 trillion bacteria, archaea, and protists call your body home. Your skin, mouth, and nasal passages and your digestive and urogenital tracts are prime real estate for these microorganisms. Although each individual is so tiny that it would have to be magnified hundreds of times for you to see it, the weight of your microbial residents totals two to five pounds.

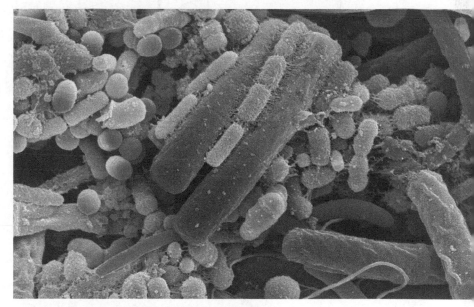

We acquire our microbial communities during the first two years of life, and they remain fairly stable thereafter. However, modern life is taking a toll on that stability. We alter the balance of these communities by taking antibiotics, purifying our water, sterilizing our food, attempting to germproof our surroundings, and scrubbing our skin and teeth. Scientists hypothesize that disrupting our microbial communities may increase our susceptibility to infectious diseases, predispose us to certain cancers, and contribute to conditions such as asthma and other allergies, irritable bowel syndrome, Crohn's disease, and autism. Researchers are even investigating whether having the wrong microbial community could make us fat. In addition, scientists are studying

Colorized scanning electron micrograph of bacteria on a human tongue (14,500×).

how our microbial communities have evolved over the course of human history. As you'll discover in the Evolution Connection section at the end of this chapter, for example, dietary changes invited decay-causing bacteria to make themselves at home on our teeth.

Throughout this chapter, you will learn about the benefits and drawbacks of human-microbe interactions. You will also sample a bit of the remarkable diversity of prokaryotes and protists. This chapter is the first of three that explore the magnificent diversity of life. And so it is fitting that we begin with the prokaryotes, Earth's first life-form, and the protists, the bridge between unicellular eukaryotes and multicellular plants, fungi, and animals.

Major Episodes in the History of Life

To put our survey of diversity into perspective, let's look at a brief overview of major events in the history of life on Earth. Our planet's history began 4.6 billion years ago, a time span that is difficult to grasp. To visualize this immense scale, imagine taking a road trip across North America in which each mile traveled is the equivalent of passing through 1 million years. Our journey will take us from Kamloops, British Columbia, in Canada, to the finish line of the Boston Marathon in Boston, Massachusetts, on a route that covers 4,600 miles (Figure 15.1).

Setting out from Kamloops, we head southwest to Seattle, Washington, and then south toward San Francisco,

California. By the time we reach the California border, nearly 750 million years have passed, and the first rocks have formed on Earth's cooling surface. Our arrival at the Golden Gate Bridge coincides with the appearance of the first cells in the fossil record. After 1,100 million years, life on Earth has begun! Those earliest organisms were all **prokaryotes**, having cells that lack true nuclei. You'll learn more about the origin of these first cells in the next section.

Conditions on the young Earth were very different from conditions today. One difference that was critical to the origin and

▶ Figure 15.1 **Some major episodes in the history of life.** On this 4,600-mile metaphorical road trip, each mile equals 1 million years in Earth's history.

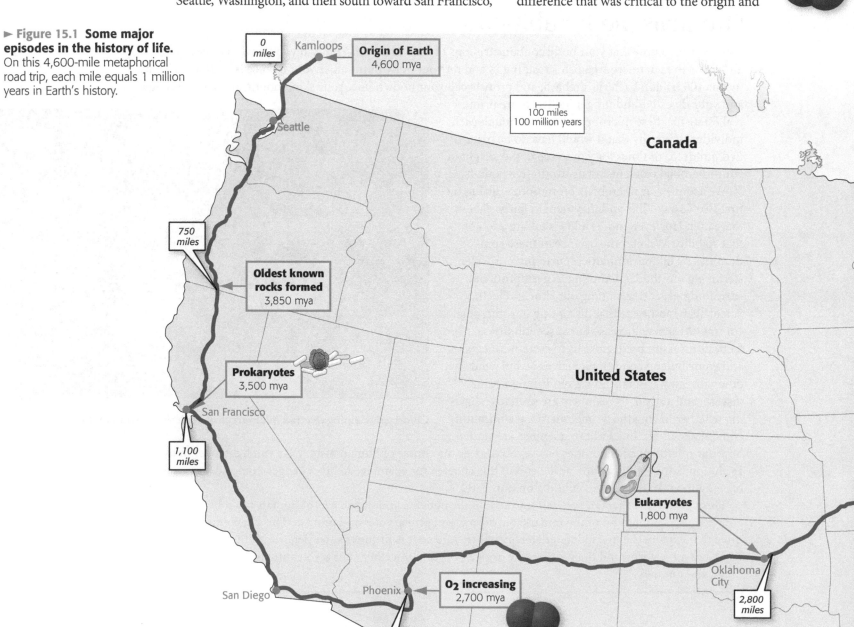

evolution of life was the lack of atmospheric O_2. As we continue driving south on our metaphoric journey, diverse metabolic pathways are evolving among the prokaryotes. However, we won't reach our next milestone for another 800 million years. By that time we have passed through San Diego and turned east across the desert. At Phoenix, Arizona, Earth is 2,700 million years old, and atmospheric O_2 has begun to increase as a result of photosynthesis by autotrophic prokaryotes.

Nine hundred million years later, just past Oklahoma City, we find the first fossils of eukaryotic organisms. **Eukaryotes** are composed of one or more cells that contain nuclei and many other membrane-bound organelles absent in prokaryotic cells. Eukaryotic cells evolved from ancestral host cells that engulfed smaller prokaryotes. The mitochondria of our cells and those of every other eukaryote are descendants of those prokaryotes, as are the chloroplasts of plants and algae.

Prokaryotes had been on Earth for 1.7 billion years before eukaryotes evolved. The appearance of these more complex cells, however, launched a period of tremendous diversification of eukaryotic forms. These new organisms were the protists. Protists are mostly microscopic and unicellular, and as you will learn in this chapter, they are represented today by a great diversity of species.

The next great event in the evolution of life was multicellularity. The oldest fossils that are clearly multicellular are 1.2 billion years old, or about midway between St. Louis, Missouri, and Terre Haute, Indiana, on our road trip. The organisms that left these fossils were tiny and not at all complex.

It isn't until 600 million years later (roughly 600 million years ago) that large, diverse, multicellular organisms appeared in the fossil record. At this point, we have traveled 4,000 miles—4 billion years—and now find ourselves in Erie, near the western edge of Pennsylvania (coincidentally, where one of the authors was born). Less than 15% of our journey remains, and we still have not encountered much diversity. But that's about to change. A great diversification of animals, the so-called Cambrian explosion, marked the beginning of the Paleozoic era, about 542 million years ago (see Table 14.1). By the end of that period, all the major animal body plans, as well as all the major groups, had evolved. The colonization of land by plants, fungi, and insects also occurred during the Paleozoic. This evolutionary transition began about 500 million years ago, a time that, in our road trip, corresponds to reaching Buffalo, New York.

By the time we get to Albany, New York, we're in the middle of the Mesozoic era, sometimes called the age of dinosaurs. At the end of the Mesozoic, 65 million years ago, we find ourselves about halfway across the state of Massachusetts. As we draw closer to Boston, more and more familiar organisms begin to dominate the landscape—flowering plants, birds, and mammals, including primates.

Modern humans, *Homo sapiens*, appeared roughly 195,000 years ago. At that point, we are less than two blocks shy of the finish line of our journey from the origin of Earth to the present day. Thus, spans of time that seem lengthy in the context of our own lives are brief moments in the history of life on Earth.

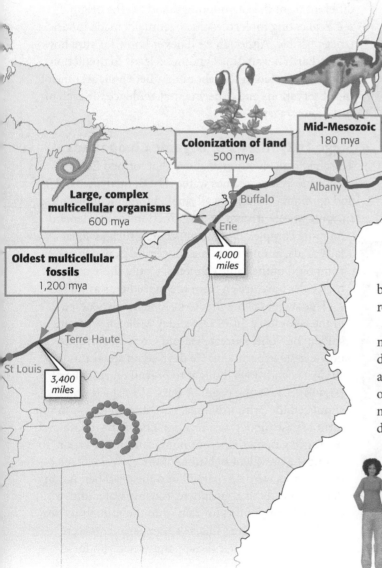

Large, complex multicellular organisms
600 mya

Colonization of land
500 mya

Mid-Mesozoic
180 mya

Homo sapiens
0.195 mya

Boston

Albany

4,600 miles

Buffalo

Erie

4,000 miles

Oldest multicellular fossils
1,200 mya

Terre Haute

St Louis

3,400 miles

If your family took a vacation in which you traveled 1 mile for every million years in the history of life, you'd still be asking, "Are we there yet?" after driving from Miami to Seattle.

The Origin of Life

Looking at the abundant diversity of life around us, it is difficult to imagine that Earth was not always like this. But early in its 4.6-billion-year history, our planet had no oceans, no lakes, very little oxygen, and no life at all. It was bombarded by debris left over from the formation of the solar system—gigantic chunks of rock and ice whose impact with Earth generated heat so fierce that it vaporized all available water. Indeed, for the first several hundred million years of its existence, conditions on the young Earth were so harsh that it's doubtful life could have originated; if it did, it could not have survived.

Although conditions had become considerably calmer, the Earth of 4 billion years ago was still in violent turmoil. Water vapor had condensed into oceans on the planet's cooling surface, but volcanic eruptions belched gases such as carbon dioxide, methane, and ammonia and other nitrogen compounds into its atmosphere (Figure 15.2). Although lack of atmospheric O_2 would be deadly to most of Earth's present-day inhabitants, this environment made the origin of life possible—O_2 is a corrosive agent that tends to disrupt chemical bonds and would have prevented the formation of complex molecules.

Life is an emergent property that arises from the specific arrangement and interactions of its molecular parts (see Figure 1.20). To learn how life originated from nonliving substances, biologists draw on research

☑ **CHECKPOINT**

One reason why the spontaneous generation of life on Earth could not occur today is the abundance of _____ in our modern atmosphere.

Answer: oxygen (O_2)

from the fields of chemistry, geology, and physics. In the next sections, we'll look at the attributes that must have arisen during the origin of life. Scientists agree on these properties, but which of several plausible scenarios led to them is still the subject of vigorous debate. ☑

A Four-Stage Hypothesis for the Origin of Life

According to one hypothesis for the origin of life, the first organisms were products of chemical evolution in four stages: (1) the synthesis of small organic molecules, such as amino acids and nucleotide monomers; (2) the joining of these small molecules into macromolecules, including proteins and nucleic acids; (3) the packaging of all these molecules into pre-cells, droplets with membranes that maintained an internal chemistry different from the surroundings; and (4) the origin of self-replicating molecules that eventually made inheritance possible. Although we'll never know for sure how life on Earth began, this hypothesis leads to predictions that can be tested in the laboratory. Let's look at some of the observations and experimental evidence for each of these four stages.

Stage 1: Synthesis of Organic Compounds

The chemicals mentioned above that formed Earth's early atmosphere, such as water (H_2O), methane (CH_4), and ammonia (NH_3), are all small, inorganic molecules. In contrast, the structures and functions of life depend on more complex organic molecules, such as sugars, fatty acids, amino acids, and nucleotides, which are composed of the same elements. Could these complex molecules have arisen from the ingredients available?

This stage was the first to be extensively studied in the laboratory. In 1953, Stanley Miller, a graduate student of Nobel laureate Harold Urey, performed a now-classic experiment. He devised an apparatus to simulate conditions thought to prevail on early Earth (Figure 15.3, facing page). A flask of warmed water simulated the primordial sea. An "atmosphere"—in the form of gases added to a reaction chamber—contained hydrogen gas, methane, ammonia, and water vapor. To mimic the prevalent lightning of the early Earth, electrical sparks were discharged into the chamber. A condenser cooled the atmosphere, causing water and any dissolved compounds to "rain" into the miniature "sea."

The results of the Miller-Urey experiments were front-page news. After the apparatus had run for a

▼ Figure 15.2 **An artist's rendition of conditions on early Earth.**

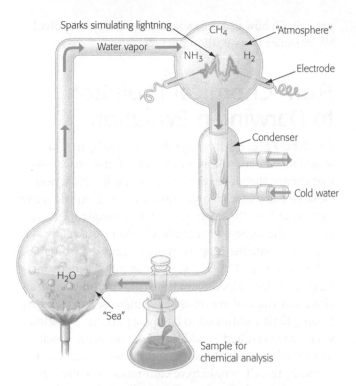

▲ **Figure 15.3 Apparatus used to simulate early-Earth chemistry in Urey and Miller's experiments.**

Labels in figure: Sparks simulating lightning; Water vapor; CH_4; "Atmosphere"; NH_3; H_2; Electrode; Condenser; Cold water; H_2O; "Sea"; Sample for chemical analysis

week, an abundance of organic molecules essential for life, including amino acids, the monomers of proteins, had collected in the "sea." Many laboratories have since repeated Miller's experiment using various atmospheric mixtures and have also produced organic compounds. In 2008, one of Miller's former graduate students discovered some samples from an experiment that Miller had designed with a different atmosphere, one that would mimic volcanic conditions. Reanalyzing these samples using modern equipment, he identified additional organic compounds that had been synthesized. Indeed, 22 amino acids had been produced under Miller's simulated volcanic conditions, compared with the 11 produced with the atmosphere in his original 1953 experiment.

Scientists are also testing other hypotheses for the origin of organic molecules on Earth. Some researchers are exploring the hypothesis that life may have begun in submerged volcanoes or deep-sea hydrothermal vents, gaps in Earth's crust where hot water and minerals gush into deep oceans. These environments, among the most extreme environments in which life exists today, could have provided the initial chemical resources for life.

Another intriguing hypothesis proposes that meteorites were the source of Earth's first organic molecules. Fragments of a 4.5-billion-year-old meteorite that fell to Earth in Australia in 1969 contain more than 80 types of amino acids, some in large amounts. Recent studies have shown that this meteorite also contains other key organic molecules, including lipids, simple sugars, and nitrogenous bases such as uracil.

Stage 2: Abiotic Synthesis of Polymers

Once small organic molecules were present on Earth, how were they linked together to form polymers such as proteins and nucleic acids without the help of enzymes and other cellular equipment? Researchers have brought about such polymerization in the laboratory by dripping solutions of organic monomers onto hot sand, clay, or rock. The heat vaporizes the water in the solutions and concentrates the monomers on the underlying material. Some of the monomers then spontaneously bond together to form polymers. On the early Earth, raindrops or waves may have splashed dilute solutions of organic monomers onto fresh lava or other hot rocks and then washed polypeptides and other polymers back into the sea. There they could accumulate in great quantities (since no life existed to consume them). ☑

Stage 3: Formation of Pre-Cells

The cell membrane forms a boundary that separates a living cell and its functions from the external environment. A key step in the origin of life would have been the isolation of a collection of organic molecules within a membrane. We'll call these molecular aggregates pre-cells—not really cells, but molecular packages with some of the properties of life. Within a confined space, certain combinations of molecules could be concentrated and interact more efficiently.

Researchers have demonstrated that pre-cells could have formed spontaneously from fatty acids (see Figure 3.11). The presence of a specific type of clay thought to be common on early Earth greatly speeds up the rate of spontaneous pre-cell formation. Unlike the plasma membranes of today's cells, these rudimentary membranes were quite porous, allowing organic monomers such as RNA nucleotides and amino acids to cross freely. However, polymers that formed inside the pre-cells were too large to exit. In addition to physically corralling molecules into a fluid-filled space, these pre-cells may have had certain properties of life. They may have had the ability to use chemical energy and grow. Most intriguingly, pre-cells made in laboratory experiments can divide to produce new pre-cells—a simple form of reproduction.

Stage 4: Origin of Self-Replicating Molecules

Life is defined partly by the process of inheritance, which is based on self-replicating molecules. Today's cells store their genetic information as DNA. They

☑ **CHECKPOINT**

What is the name of the chemical reaction whereby monomers are linked together into polymers? (*Hint*: Review Figure 3.4.)

Answer: *dehydration reaction*

transcribe the information into RNA and then translate RNA messages into specific enzymes and other proteins (see Figure 10.9). This mechanism of information flow probably emerged gradually through a series of small changes to much simpler processes.

What were the first genes like? One hypothesis is that they were short strands of RNA that replicated without the assistance of proteins. In laboratory experiments, short RNA molecules can assemble spontaneously from nucleotide monomers in the absence of enzymes **(Figure 15.4)**. The result is a population of RNA molecules, each with a random sequence of monomers. Some of the molecules self-replicate, but their success at this reproduction varies. What happens can be described as molecular evolution: The RNA varieties that replicate fastest increase their frequency in the population.

In addition to the experimental evidence, there is another reason the idea of RNA genes in the primordial world is plausible. Cells actually have RNAs that can act as enzymes; they are called ribozymes. Perhaps early ribozymes catalyzed their own replication. That would help with the "chicken and egg" paradox of which came first, enzymes or genes. Maybe the "chicken and egg" came together in the same RNA molecules. The molecular biology of today may have been preceded by an ancient "RNA world."

The pre-cells described above could assemble spontaneously, reproduce, and grow. These abilities, however, would only result in endless copies of the original pre-cell. So how did cells acquire the ability to evolve? We consider this question next. ☑

From Chemical Evolution to Darwinian Evolution

Recall how present-day organisms evolve by natural selection. Genetic variation arises by mutations, errors that change the nucleotide sequence of the DNA. Some variations increase an organism's chance of reproductive success and are thus perpetuated in subsequent generations. In the same way, natural selection would have begun to shape the properties of pre-cells that contained self-replicating RNA. Those that contained genetic information that helped them grow and reproduce more efficiently than others would have increased in number, passing their abilities on to later generations. Mutations would have resulted in additional variation on which natural selection could work, and the most successful of these pre-cells would have continued to evolve. Of course, the gap between such pre-cells and even the simplest of modern cells is enormous. But with millions of years of incremental changes through natural selection, these molecular cooperatives could have become more and more cell-like. At some point during this time, pre-cells passed a fuzzy border to become true cells. The stage was then set for the evolution of diverse life-forms, changes that we see documented in the fossil record.

▼ Figure 15.4 **Self-replication of RNA "genes."**

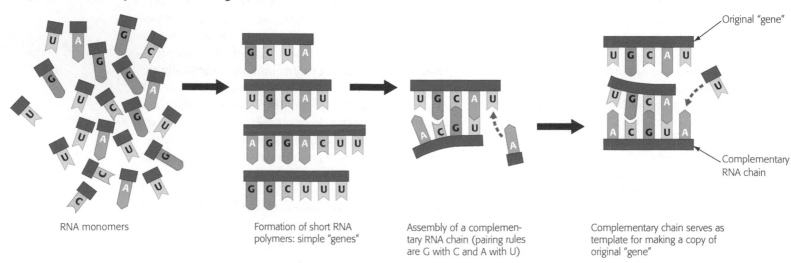

RNA monomers

Formation of short RNA polymers: simple "genes"

Assembly of a complementary RNA chain (pairing rules are G with C and A with U)

Complementary chain serves as template for making a copy of original "gene"

Original "gene"

Complementary RNA chain

Prokaryotes

The history of prokaryotic life is a success story spanning billions of years. Prokaryotes lived and evolved all alone on Earth for about 2 billion years. They have continued to adapt and flourish on a changing Earth and in turn have helped to modify the planet. In this section, you will learn about prokaryotic structure and function, diversity, relationships, and ecological significance.

They're Everywhere!

Today, prokaryotes are found wherever there is life, including in and on the bodies of multicellular organisms. The collective biological mass (biomass) of prokaryotes is at least ten times that of all eukaryotes. Prokaryotes also thrive in habitats too cold, too hot, too salty, too acidic, or too alkaline for any eukaryote (Figure 15.5). Scientists are just beginning to investigate the extensive prokaryotic diversity in the oceans. Biologists have even discovered prokaryotes living on the walls of a gold mine 3.3 km (2 miles) below Earth's surface.

Though individual prokaryotes are small organisms (Figure 15.6), they are giants in their collective impact on Earth and its life. We hear most about the relatively few species that cause illness. Bacterial infections are responsible for about half of all human diseases, including tuberculosis, cholera, many sexually transmitted infections, and certain types of food poisoning. However, prokaryotic life is much more than just a rogues' gallery. The Biology and Society section introduced our **microbiota**, the community of microorganisms that live in and on our bodies. Each of us harbors several hundred different species and genetic strains of prokaryotes, including a few whose positive effects are well studied. For example, some of our intestinal inhabitants supply essential vitamins and enable us to extract nutrition from food molecules that we can't otherwise digest. Many of the bacteria that live on our skin perform helpful housekeeping functions such as decomposing dead skin cells. Prokaryotes also guard the body against disease-causing intruders.

It would be hard to overstate the importance of prokaryotes to the health of the environment. Prokaryotes living in the soil and at the bottom of lakes, rivers, and oceans help to decompose dead organisms and other

◄ **Figure 15.5 A window to early life?** An instrument on a research submarine samples the water around a hydrothermal vent more than 1.5 km (about a mile) below the ocean's surface. Prokaryotes that live near the vent use the emitted gases as an energy source. This environment, which is very dark, hot, and under high pressure, is among the most extreme in which life is known to exist.

► **Figure 15.6 Bacteria on the point of a pin.** The orange rods are individual bacteria, each about 5 μm long, on the point of a pin. Besides highlighting their tiny size, this micrograph will help you understand why a pin prick can cause infection.

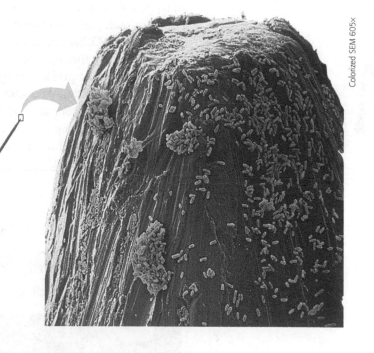

Colorized SEM 605x

waste materials, returning vital chemical elements such as nitrogen to the environment. If prokaryotes were to disappear, the chemical cycles that sustain life would come to a halt, and all forms of eukaryotic life would also be doomed. In contrast, prokaryotic life would undoubtedly persist in the absence of eukaryotes, as it once did for 2 billion years.

 **Structure/Function**

Prokaryotes

Prokaryotes have a cellular organization fundamentally different from that of eukaryotes. Whereas eukaryotic cells have a membrane-enclosed nucleus and numerous other membrane-enclosed organelles, prokaryotic cells lack these structural features (see Figures 4.2 and 4.3). And nearly all species of prokaryotes have cell walls exterior to their plasma membranes. Despite the simplicity of the cell structures, though, prokaryotes display a remarkable range of diversity. In this section, you'll learn about aspects of their form, reproduction, and nutrition that help these organisms survive in their environments.

Prokaryotic Forms

Determining cell shape by microscopic examination is an important step in identifying a prokaryote. The micrographs in **Figure 15.7** show the three most common shapes. Spherical prokaryotic cells are called **cocci** (singular, *coccus*). Rod-shaped prokaryotes are called **bacilli** (singular, *bacillus*). Spiral-shaped prokaryotes include spirochetes, different species of which cause syphilis and Lyme disease.

Although all prokaryotes are unicellular, the cells of some species usually exist as groups of two or more cells. For example, cocci that occur in clusters are called staphylococci. Other cocci, including the bacterium that causes strep throat, occur in chains; they are called streptococci. Some prokaryotes grow branching chains of cells **(Figure 15.8a)**. And some species even exhibit a simple division of labor among specialized types of cells **(Figure 15.8b)**. Among unicellular species, moreover, there are some giants that actually dwarf most eukaryotic cells **(Figure 15.8c)**.

▼ **Figure 15.8 A diversity of prokaryotic shapes and sizes.**

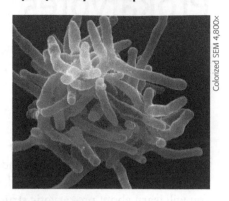

(a) Actinomycete. An actinomycete is a mass of branching chains of rod-shaped cells. These bacteria are common in soil, where they secrete antibiotics that inhibit the growth of other bacteria. Various antibiotic drugs, such as streptomycin, are obtained from actinomycetes.

Colorized SEM 4,800×

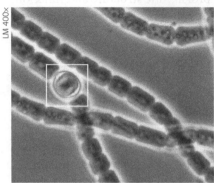

LM 400×

(b) Cyanobacteria. These photosynthetic cyanobacteria exhibit division of labor. The box highlights a cell that converts atmospheric nitrogen to ammonia, which can then be incorporated into amino acids and other organic compounds.

LM 18×

(c) Giant bacterium. The larger white blob in this photo is the marine bacterium *Thiomargarita namibiensis*. This prokaryotic cell is more than 0.5 mm in diameter, about the size of the fruit fly's head below it.

▼ **Figure 15.7 Three common shapes of prokaryotic cells.**

SHAPES OF PROKARYOTIC CELLS		
Spherical (cocci)	**Rod-shaped (bacilli)**	**Spiral**

Colorized SEM 10,000× (Spherical)
Colorized SEM 10,000× (Rod-shaped)
Colorized TEM 30,000× (Spiral)

About half of all prokaryotic species are mobile. Many of those that travel have one or more flagella that propel the cells away from unfavorable places or toward more favorable places, such as nutrient-rich locales.

In many natural environments, prokaryotes attach to surfaces in a highly organized colony called a **biofilm**. A biofilm may consist of one or several species of prokaryotes, and it may include protists and fungi as well. As a biofilm becomes larger and more complex, it develops into a "city" of microbes. Communicating by chemical signals, members of the community engage in division of labor, including defense against invaders, among other activities.

Biofilms can form on almost any type of surface, such as rocks, organic material (including living tissue), metal, and plastic. A biofilm known as dental plaque (**Figure 15.9**) can cause tooth decay, a topic you'll learn more about in the Evolution Connection section at the end of this chapter. Biofilms are common among bacteria that cause disease in humans. For instance, ear infections and urinary tract infections are often the result of biofilm-forming bacteria. Biofilms of harmful bacteria can also form on implanted medical devices such as catheters, replacement joints, and pacemakers. The structure of biofilms makes these infections especially difficult to defeat. Antibiotics may not be able to penetrate beyond the outer layer of cells, leaving much of the community intact.

Prokaryotic Reproduction

Many prokaryotes can reproduce at a phenomenal rate if conditions are favorable. The cells copy their DNA almost continuously and divide again and again by the

▼ Figure 15.9 **Dental plaque, a biofilm that forms on teeth.**

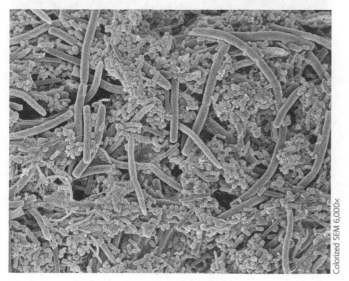

Colorized SEM 6,000×

▼ Figure 15.10 **Household sponge contaminated with bacteria (red, green, yellow, and blue objects in this colorized micrograph).**

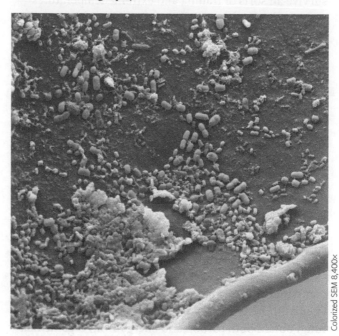

Colorized SEM 8,400×

process of **binary fission**. Dividing by binary fission, a single cell becomes 2 cells, which then become 4, 8, 16, and so on. Some species can produce a new generation in only 20 minutes under optimal conditions. If reproduction continued unchecked at this rate, a single prokaryote could give rise to a colony outweighing Earth in only three days! Also, each time DNA is replicated prior to binary fission, spontaneous mutations occur. As a result, rapid reproduction generates a great deal of genetic variation in a prokaryotic population. If the environment changes, an individual that possesses a beneficial gene can quickly take advantage of the new conditions. For example, exposure to antibiotics may select for antibiotic resistance in a bacterial population (see the Evolution Connection section at the end of Chapter 13).

Fortunately, few prokaryotic populations can sustain exponential growth for long. Environments are usually limiting in resources such as food and space. Prokaryotes also produce metabolic waste products that may eventually pollute the colony's environment. Still, you can understand why kitchen sponges may harbor large numbers of bacteria (**Figure 15.10**) and why food can spoil so rapidly. Refrigeration retards food spoilage not because the cold kills bacteria, but because most microorganisms reproduce very slowly at such low temperatures.

Some prokaryotes can survive during very harsh conditions by forming specialized cells called endospores. An **endospore** is a thick-coated, protective cell produced within the prokaryotic cell when the prokaryote is exposed to unfavorable conditions. The endospore can survive all sorts of trauma and extreme temperatures—not even boiling water kills most of these resistant cells. And when the environment becomes more hospitable, the endospore can absorb water and resume growth. To ensure that all cells, including endospores, are killed when laboratory equipment is sterilized, microbiologists use an appliance called an autoclave, a pressure cooker that applies high-pressure steam at a temperature of 121°C (250°F). The food-canning industry uses similar methods to kill endospores of dangerous bacteria such as *Clostridium botulinum*, the cause of the potentially fatal disease botulism. ✓

Prokaryotic Nutrition

You are probably familiar with the methods by which multicellular organisms obtain energy and carbon, the two main resources needed for synthesizing organic compounds. Plants use carbon dioxide and the sun's energy in the process of photosynthesis; animals and fungi obtain both carbon and energy from organic matter. These modes of nutrition are both common among prokaryotes as well **(Figure 15.11)**. But the metabolic capabilities of prokaryotes are far more diverse than those of eukaryotes. Some species harvest energy from inorganic substances such as ammonia (NH_3) and hydrogen sulfide (H_2S). Soil bacteria that obtain energy

from inorganic nitrogen compounds are essential to the chemical cycle that makes nitrogen available to plants (see Figure 20.34). Because they don't depend on sunlight for energy, these prokaryotes can thrive in conditions that seem totally inhospitable to life—even between layers of rocks buried hundreds of feet below Earth's surface! Near hydrothermal vents, where scalding water and hot gases surge into the sea more than a mile below the surface, bacteria that use sulfur compounds as a source of energy support diverse animal communities.

The metabolic talents of prokaryotes make them excellent symbiotic partners with animals, plants, and fungi. **Symbiosis** ("living together") is a close association between organisms of two or more species. In some cases of symbiosis, both organisms benefit from the partnership. For example, many of the animals that inhabit hydrothermal vent communities, including the giant tube worm shown in **Figure 15.12**, harbor sulfur bacteria within their bodies. The animals absorb sulfur compounds from the water. The bacteria use the compounds as an energy source to convert CO_2 from seawater into organic molecules that, in turn, provide food for their hosts. Similarly, photosynthesis by cyanobacteria

▼ **Figure 15.12 Giant tube worm.** These animals, which can grow up to 2 m (over 6 ft) long, depend on symbiotic prokaryotes to supply them with food.

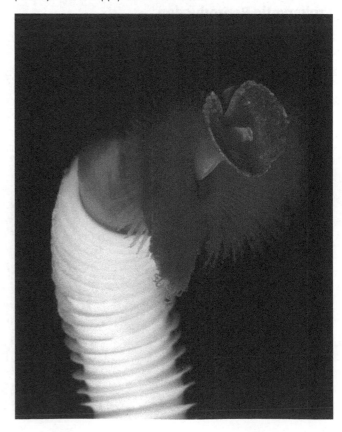

▼ **Figure 15.11 Two methods of obtaining energy and carbon.**

LM 65×

(a) *Oscillatoria.* This is a species of cyanobacteria, a type of prokaryote that undergoes photosynthesis.

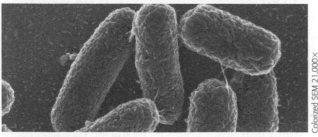

Colorized SEM 21,000×

(b) *Salmonella.* These bacteria, which cause a type of food poisoning, obtain energy and carbon from organic matter—in this case, living human cells.

▲ **Figure 15.13** *Azolla* **(water fern, inset) floating in rice paddy.** These small plants grow rapidly and cover the water's surface, which helps exclude weeds. Decomposition of the short-lived plants releases nitrogen that fertilizes the rice.

provides food for a fungal partner in the symbiosis that forms some lichens (see the Evolution Connection section at the end of Chapter 16).

In addition to photosynthesis, many cyanobacteria are also capable of nitrogen fixation, the process of converting atmospheric nitrogen (N_2) into a form usable by plants. Symbiosis with cyanobacteria gives plants such as the water fern *Azolla* an advantage in nitrogen-poor environments. This tiny, floating plant has been used to boost rice production for more than a thousand years (**Figure 15.13**). Other nitrogen-fixing bacteria live symbiotically in root nodules of plants in the legume family, a large group that includes many economically important species such as beans, soybeans, peas, and peanuts.

The Ecological Impact of Prokaryotes

As a result of their nutritional diversity, prokaryotes perform a variety of ecological services that are essential to our well-being. Let's turn our attention now to the vital role that prokaryotes play in sustaining the biosphere.

Prokaryotes and Chemical Recycling

Not too long ago, the atoms making up the organic molecules in your body were part of inorganic compounds found in soil, air, and water, as they will be again one day. Life depends on the recycling of chemical elements between the biological and physical components of ecosystems. Prokaryotes play essential roles in these chemical cycles. For example, nearly all the nitrogen that plants use to make proteins and nucleic acids comes from prokaryotic metabolism in the soil. In turn, animals get their nitrogen compounds from plants.

Another vital function of prokaryotes, mentioned earlier in the chapter, is the breakdown of organic wastes and dead organisms. Prokaryotes decompose organic matter and, in the process, return elements to the environment in inorganic forms that can be used by other organisms. If it were not for such decomposers, carbon, nitrogen, and other elements essential to life would become locked in the organic molecules of corpses and waste products. (You'll learn more about the role that prokaryotes play in chemical cycling in Chapter 20.)

Putting Prokaryotes to Work

People have put the metabolically diverse prokaryotes to work in cleaning up the environment. **Bioremediation** is the use of organisms to remove pollutants from water, air, or soil. One example of bioremediation is the use of prokaryotic decomposers to treat our sewage. Raw sewage is first passed through a series of screens and shredders, and solid matter settles out from the liquid waste. This solid matter, called sludge, is then gradually added to a culture of anaerobic prokaryotes, including both bacteria and archaea. The microbes decompose the organic matter in the sludge, converting it to material that can be used as landfill or fertilizer. The liquid wastes may then be passed through a trickling filter system consisting of a long horizontal bar that slowly rotates, spraying liquid wastes onto a bed of rocks (**Figure 15.14**). Aerobic prokaryotes and fungi growing on the rocks remove much of the organic matter from the liquid. The outflow from the rock bed is then sterilized and released back into the environment.

You have microorganisms to thank for the clean water you drink every day.

▼ **Figure 15.14** **Putting microbes to work in sewage treatment facilities.** This is a diagram of a trickling filter system, which uses bacteria, archaea, and fungi to treat liquid wastes after sludge is removed.

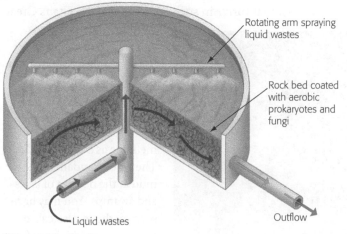

Rotating arm spraying liquid wastes

Rock bed coated with aerobic prokaryotes and fungi

Liquid wastes

Outflow

© Pearson Education, Inc.

☑ **CHECKPOINT**

How do bacteria help
restore the atmospheric
CO_2 required by plants for
photosynthesis?

*Answer: By decomposing the
organic molecules of dead
organisms and organic refuse such
as leaf litter, bacteria release carbon
from the organic matter in the form
of CO_2.*

Bioremediation has also become an important tool for cleaning up toxic chemicals released into the soil and water by industrial processes. Naturally occurring prokaryotes capable of degrading pollutants such as oil, solvents, and pesticides are often present in contaminated soil, but environmental workers may use methods of speeding up their activity. In **Figure 15.15**, an airplane is spraying chemical dispersants on oil from the disastrous 2010 Deepwater Horizon spill in the Gulf of Mexico. Like detergents that help clean greasy dishes, these chemicals break oil into smaller droplets that offer more surface area for microbial attack. Prokaryotes are also helping to decontaminate old mining sites where the soil and water are laced with heavy metals and other poisons. We are just beginning to explore the great potential that prokaryotes offer for bioremediation. In the future, genetically engineered microbes may be put to

▲ Figure 15.15 **Spraying chemical dispersants on an oil spill in the Gulf of Mexico.**

work cleaning up the wide variety of toxic waste products that continue to accumulate in our landscapes and landfills. ☑

The Two Main Branches of Prokaryotic Evolution: Bacteria and Archaea

By comparing diverse prokaryotes at the molecular level, biologists have identified two major branches of prokaryotic evolution: **bacteria** and **archaea**. Thus, life is organized into three domains—**Bacteria**, **Archaea**, and **Eukarya** (review Figure 14.26). Although bacteria and archaea have prokaryotic cell organization in common, they differ in many structural and physiological characteristics. In this section, we'll focus on the special characteristics of archaea before turning our attention to bacteria.

Archaea are abundant in many habitats, including places where few other organisms can survive. One group of archaea, the extreme thermophiles ("heat lovers"), live in very hot water **(Figure 15.16)**; some archaea even populate the deep-ocean vents that gush near-boiling water, such as the one shown in Figure 15.5. Another group is the extreme halophiles ("salt lovers"), archaea that thrive in such environments as Utah's Great Salt Lake, the Dead Sea, and seawater-evaporating ponds used to produce salt.

A third group of archaea are methanogens, which live in anaerobic (oxygen-free) environments and give off methane as a waste product. They are abundant in the mud at the bottom of lakes and swamps. You may have seen methane, also called marsh gas, bubbling up from

a swamp. Methanogens flourish in the anaerobic conditions of solid waste landfills, where the large amount of methane they produce is a significant contributor to global warming (see Figure 18.43). Many municipalities collect this methane and use it as a source of energy **(Figure 15.17)**.

Great numbers of methanogens also inhabit the digestive tracts of animals. In humans, intestinal gas is largely the result of their metabolism. More important, methanogens aid digestion in cattle, deer, and other animals that depend heavily on cellulose for their nutrition. Normally, bloating does not occur in these animals because they regularly expel large volumes of gas produced by the methanogens. (And that may be more than you wanted to know about these gas-producing microbes!)

Archaea are also abundant in more moderate conditions, especially the oceans, where they can be found at all depths. They are a substantial fraction of the prokaryotes in waters 150 m (nearly 500 ft) below the surface and half of the prokaryotes that live below 1,500 m (0.9 mi). Archaea are thus one of the most abundant cell types in Earth's largest habitat.

Bacteria That Cause Disease

Although most bacteria are harmless or even beneficial to us, a tiny minority of species cause a disproportionate amount of misery. Bacteria and other organisms that cause disease are called **pathogens**. We're healthy most of the time because our body's defenses check the growth of pathogen populations. Occasionally, the balance shifts in favor of a pathogen, and we become ill. Even some of

▼ Figure 15.16 **Heat-loving archaea.** In this photo, you can see yellow and orange colonies of heat-loving prokaryotes growing in the Abyss Pool, West Thumb Geyser, in Yellowstone Park, Wyoming.

▲ Figure 15.17 **Pipes for collecting gas generated by methanogenic archaea from a landfill.**

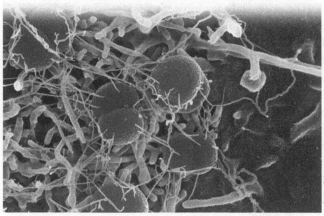

◄ Figure 15.18 **Bacteria that cause meningitis.** *Neisseria meningitidis,* an endotoxin-producing pathogen.

Colonized SEM 18,000x

the bacteria that are normal residents of the human body can make us sick when our defenses have been weakened by poor nutrition or a viral infection.

Most pathogenic bacteria cause disease by producing a poison—either an exotoxin or an endotoxin. **Exotoxins** are proteins that bacterial cells secrete into their environment. For example, *Staphylococcus aureus* (abbreviated *S. aureus*) produces several exotoxins. Although *S. aureus* is commonly found on the skin and in the nasal passages, if it enters the body through a wound, it can cause serious disease. One of its exotoxins causes layers of skin to slough off ("flesh-eating disease"); another can produce a potentially deadly disease called toxic shock syndrome, which has been associated with improper use of tampons. *S. aureus* exotoxins are also a leading cause of food poisoning. If food contaminated with *S. aureus* is not refrigerated, the bacteria reproduce and release exotoxins so potent that ingesting a millionth of a gram causes vomiting and diarrhea. Once the food has been contaminated, even boiling it will not destroy the exotoxins.

Endotoxins are chemical components of the outer membrane of certain bacteria. All endotoxins induce the same general symptoms: fever, aches, and sometimes a dangerous drop in blood pressure (septic shock). Septic shock triggered by an endotoxin of the pathogen that causes bacterial meningitis **(Figure 15.18)** can kill a healthy person in a matter of days, or even hours. Because the bacteria are easily transmitted among people living in close contact, many colleges require students to be vaccinated against this disease. Other examples of endotoxin-producing bacteria include the species of *Salmonella* that cause food poisoning and typhoid fever.

Sanitation is generally the most effective way to prevent bacterial disease. The installation of water treatment and sewage systems continues to be a public health priority throughout the world. Antibiotics have been discovered that can cure most bacterial diseases.

However, resistance to widely used antibiotics has evolved in many of these pathogens.

In addition to sanitation and antibiotics, a third defense against bacterial disease is education. A case in point is Lyme disease, which is caused by a spirochete bacterium carried by ticks **(Figure 15.19)**. Disease-carrying ticks live on deer and field mice but also bite humans. Lyme disease usually starts as a red rash shaped like a bull's-eye around a tick bite. Antibiotics can cure the disease if administered within a month of exposure. If untreated, Lyme disease can cause debilitating arthritis, heart disease, and nervous system disorders. Because there is no vaccine, the best defense against Lyme disease is public education about avoiding tick bites and the importance of seeking treatment if a rash develops.

▼ Figure 15.19 **Lyme disease, a bacterial disease transmitted by ticks.** The bacterium that causes Lyme disease (shown in micrograph at right) is carried from deer to humans by ticks.

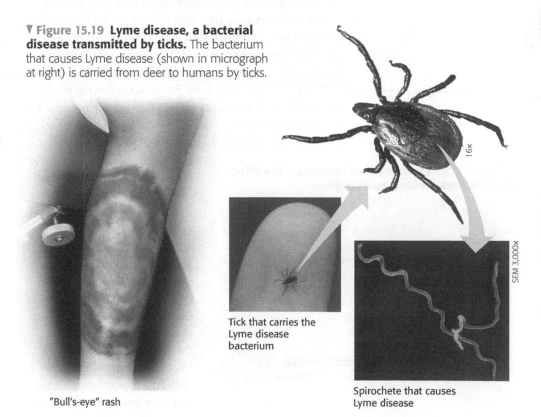

"Bull's-eye" rash

Tick that carries the Lyme disease bacterium

16x

SEM 3,000x

Spirochete that causes Lyme disease

The potential of some pathogens to cause serious harm has led to their use as biological weapons. One of the greatest threats is from endospores of the bacterium that causes anthrax. When anthrax endospores enter the lungs, they germinate, and the bacteria multiply, producing an exotoxin that eventually accumulates to lethal levels in the blood. Although the bacteria can be killed by certain antibiotics, the antibiotics don't eliminate the toxin already in the body. As a result, inhalation anthrax has a very high death rate. In an incident in 2001, five people died when anthrax spores were mailed to members of the news media and the U.S. Senate.

Another bacterium considered to have dangerous potential as a weapon is *Clostridium botulinum*. Unlike other biological agents, the weapon form of *C. botulinum* is the exotoxin it produces, botulinum, rather than the living microbes. Botulinum, the deadliest poison on Earth, blocks transmission of the nerve signals that cause muscle contraction, resulting in paralysis of the muscles required for breathing. Thirty grams of botulinum, a bit more than an ounce, could kill every person in the United States. On the other hand, the minute amount of botulinum in Botox is used for cosmetic purposes. When the toxin is injected under the skin, it relaxes the facial muscles that cause wrinkles. ☑

Human Microbiota THE PROCESS OF SCIENCE

Are Intestinal Microbiota to Blame for Obesity?

As you learned in the Biology and Society section, our bodies are home to trillions of bacteria that cause no harm or are even beneficial to our health. In the past decade, researchers have made enormous strides in characterizing our microbiota and have begun to investigate the specific effects of these residents on our physiological processes. Because our intestinal microbes are known to be involved in some aspects of food processing, researchers speculate that they might be involved in obesity. Let's examine how a team of scientists investigated the impact of microbiota on body composition—the amount of fat versus lean body mass.

Using **observations** from previous studies, the scientists asked the following **question**: Can microbiota from an obese person affect the body composition of another person? Although this is the question that we ultimately want answered, researchers routinely test hypotheses in animal models before using human subjects. Mice that have been raised in germ-free conditions have no microbiota, making them ideal subjects for this type of experiment. Therefore, the scientists formed the **hypothesis** that intestinal microbiota of an obese person would increase the amount of body fat in mice. Their **prediction** was that if the hypothesis was correct, then lean, germ-free mice

that received transplants of microbes from the intestines of obese individuals would show a greater increase of body fat than would germ-free mice that received transplants of microbes from the intestines of lean individuals.

The researchers recruited four pairs of female twins for the **experiment**. In each pair, one twin was obese and the other was lean. Microbiota from the feces of each individual were transplanted into separate groups of germ-free mice (**Figure 15.20**). The **results**, shown in **Figure 15.21**, supported the hypothesis. Mice that received microbiota from an obese donor became more obese; mice that received microbiota from a lean donor remained lean.

Is a microbe-based cure for obesity just around the corner? It's not likely. The experiment described here—and many similar experiments—represent an early stage of scientific investigation. A great deal more research is needed to determine whether our microbial residents are responsible for obesity. If that proves to be the case, the next challenge will be figuring out how to safely manipulate the complex ecosystem within our bodies.

▼ Figure 15.20 **Experiment to investigate the effect of microbiota on body composition.**

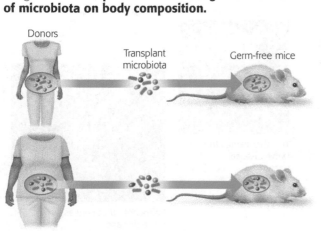

◄ Figure 15.21 **Results of microbiota transplantation experiment.** The graph shows the change in body composition (lean vs. fat mass) of mice that received microbiota from a lean donor (left) or an obese donor (right). Data from: V. K. Ridaura et al., Gut microbiota from twins discordant for obesity modulate metabolism in mice. *Science* 341 (2013). DOI: 10.1126/science.1241214.

Protists

The fossil record indicates that the first eukaryotes evolved from prokaryotes around 2 billion years ago. **Endosymbiosis**, a relationship in which one organism lives inside the cell or cells of a host organism, played a key role in the evolution of eukaryotic cells. Extensive evidence supports the theory that mitochondria and chloroplasts originated as prokaryotes that established residence inside larger host cells. Hosts and endosymbionts became interdependent and eventually became single organisms with inseparable parts. These primal eukaryotes were not only the predecessors of the great variety of modern protists; they were also ancestral to all other eukaryotes—plants, fungi, and animals.

The term protists is not a taxonomic category. At one time, protists were classified in a fourth eukaryotic kingdom (kingdom Protista). However, recent genetic and structural studies shattered the notion of protists as a unified group—some protists are more closely related to fungi, plants, or animals than they are to each other. Hypotheses about protist phylogeny (and thus, classification) are changing rapidly as new information causes scientists to revise their ideas. Although there is general agreement about some relationships, others are hotly disputed. Thus, **protist** is a bit of a catch-all category that includes all eukaryotes that are not fungi, animals, or plants. Most, but not all, protists are unicellular. Because their cells are eukaryotic, even the simplest protists are much more complex than any prokaryote.

One mark of protist diversity is the variety of ways they obtain their nutrition. Some protists are autotrophs, producing their food by photosynthesis. Photosynthetic protists belong to an informal category called **algae** (singular, *alga*), which also includes cyanobacteria. Protistan algae may be unicellular, colonial, or multicellular, such as the one shown in **Figure 15.22a**. Other protists are heterotrophs, acquiring their food from other organisms. Some heterotrophic protists eat bacteria or other protists; some are fungus-like and obtain organic molecules by absorption. Still others are parasitic. A **parasite** derives its nutrition from a living host, which is harmed by the interaction. For example, the parasitic trypanosomes shown among human red blood cells in **Figure 15.22b** cause sleeping sickness, a debilitating disease common in parts of Africa. Yet other protists are mixotrophs, capable of both photosynthesis and heterotrophy. *Euglena* (**Figure 15.22c**), a common inhabitant of pond water, can change its mode of nutrition, depending on availability of light and nutrients.

Protist habitats are also diverse. Most protists are aquatic, living in oceans, lakes, and ponds, but they are found almost anywhere there is moisture, including terrestrial habitats such as damp soil and leaf litter. Others are symbionts that reside in the bodies of various host organisms.

Because the classification of protists remains a work in progress, our brief survey of protists is not organized to correspond with any hypothesis about phylogeny. Instead, we'll look at four informal categories of protists: protozoans, slime molds, unicellular and colonial algae, and seaweeds.

▼ Figure 15.22 **Protist modes of nutrition.**

(a) **An autotroph:** *Caulerpa,* a multicellular alga

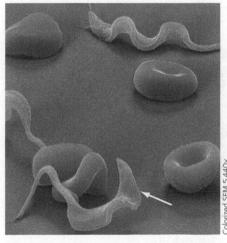

(b) **A heterotroph:** parasitic trypanosome (arrow)

Colorized SEM 5,440x

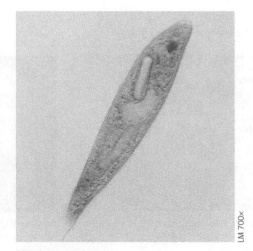

(c) **A mixotroph:** *Euglena*

LM 700x

Protozoans

Protists that live primarily by ingesting food are called **protozoans** (Figure 15.23). Protozoans thrive in all types of aquatic environments. Most species eat bacteria or other protozoans, but some can absorb nutrients dissolved in the water. Protozoans that live as parasites in animals, though in the minority, cause some of the world's most harmful diseases.

Flagellates are protozoans that move by means of one or more flagella. Most species are free-living (nonparasitic). However, this group also includes some nasty parasites that make people sick. An example is *Giardia*, a common waterborne parasite that causes severe diarrhea. People most often pick up *Giardia* by drinking water contaminated with feces containing the parasite. For example, a swimmer in a lake or river might accidentally ingest water, or a hiker might drink contaminated water from a seemingly pristine stream. (Boiling the water first will kill *Giardia*.) Another flagellate is *Trichomonas*, a common sexually transmitted parasite. The parasite travels through the reproductive tract by moving its flagella and undulating part of its membrane. In women, these protozoans feed on white blood cells and bacteria living on the cells lining the vagina. *Trichomonas* also infects the cells lining the male reproductive tract, but limited availability of food results in very small population sizes. Consequently, males typically have no symptoms of infection.

Other flagellates live symbiotically in a relationship that benefits both partners. Termites are notoriously destructive to wooden structures, but they lack enzymes to digest the tough, complex cellulose molecules that make up wood. Flagellates that reside in the termite's digestive tract break down the cellulose into simpler molecules, sharing the bounty with their hosts.

Amoebas are characterized by great flexibility in their body shape and the absence of permanent organelles for locomotion. Most species move and feed by means of **pseudopodia** (singular, *pseudopodium*), temporary extensions of the cell. Amoebas can assume virtually any shape as they creep over rocks, sticks, or mud at the bottom of a pond or ocean. One species of parasitic amoeba causes amoebic dysentery, a disease that is responsible for an estimated 100,000 deaths worldwide every year. Other protozoans with pseudopodia include the **forams**, which have shells. Although they are single-celled, the largest forams grow to a diameter of several centimeters. Ninety percent of forams that have been identified are fossils. The fossilized shells, which are a component of sedimentary rock such as limestone, are excellent markers for relating the ages of rocks in different parts of the world.

Apicomplexans are all parasitic, and some cause serious human diseases. They are named for a structure at their apex (tip) that is specialized for penetrating host cells and tissues. This group of protozoans includes *Plasmodium*, the parasite that causes malaria. Another apicomplexan is *Toxoplasma*, which requires a feline host to complete its complex life cycle. Cats that eat wild birds or rodents may be infected and shed the parasite in their feces. People can become infected by handling cat litter of outdoor cats, but they don't become sick because the immune system keeps the parasite in check. However, a woman who is newly infected with *Toxoplasma* during pregnancy can pass the parasite to her unborn child, who may suffer damage to the nervous system as a result.

Ciliates are protozoans named for their hairlike structures called cilia, which provide movement of the protist and sweep food into the protist's "mouth." Nearly all ciliates are free-living (nonparasitic) and include both heterotrophs and mixotrophs. If you have had the opportunity to explore protist diversity in a droplet of pond water, you may have seen the common freshwater ciliate *Paramecium*. ✓

> According to a recent study, infection by the parasite *Toxoplasma* makes mice lose their fear of cats.

✓ CHECKPOINT

According to a recent study, mice infected with *Toxoplasma* lose their fear of cats. How would altering the behavior of mice in this way be a beneficial adaptation for the *Toxoplasma* parasite?

Answer: Fearless mice are more likely to be caught and eaten by a cat, thus transmitting the parasite to a host in which it can complete its life cycle (that is, successfully reproduce).

▼ **Figure 15.23 A diversity of protozoans.**

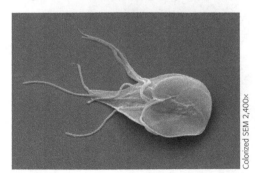

A flagellate: *Giardia*. This flagellated protozoan parasite can colonize and reproduce within the human intestine, causing disease.

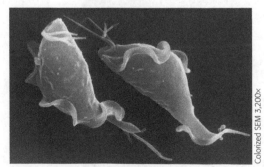

Another flagellate: *Trichomonas*. This common sexually transmitted parasite causes an estimated 5 million new infections each year.

An amoeba. This amoeba is beginning to wrap pseudopodia around an algal cell in preparation for engulfing it.

Slime Molds

Slime molds are multicellular protists related to amoebas. Although slime molds were once classified as fungi, DNA analysis showed that they arose from different evolutionary lineages. Like many fungi, slime molds feed on dead plant material. Two distinct types of slime molds have been identified.

In plasmodial slime molds, the feeding body is an amoeboid mass called a plasmodium (**Figure 15.24**) that extends pseudopodia among the leaf litter and other decaying material on a forest floor. A plasmodium can measure several centimeters across, with its network of fine filaments taking in bacteria and bits of dead organic matter amoeboid-style. Large as it is, though, the plasmodium is actually a single cell with many nuclei. Its mass of cytoplasm is not divided by plasma membranes. When the plasmodium runs out of food,

or its environment dries up, the slime mold grows reproductive structures. Stalked bodies that arise from the plasmodium bear tough-coated spores that can survive harsh conditions. If you look closely at a decaying log or the mulch in landscaped areas, you may notice these spore-bearing structures. Like the endospores of anthrax bacteria, slime mold spores can absorb water and grow once favorable conditions return.

In cellular slime molds, the feeding stage consists of solitary amoeboid cells that function independently of each other rather than a plasmodium (**Figure 15.25**). But when food is in short supply, the amoeboid cells swarm together to form a slug-like colony that moves and functions as a single unit. After a brief period of mobility, the colony extends a stalk and develops into a multicellular reproductive structure.

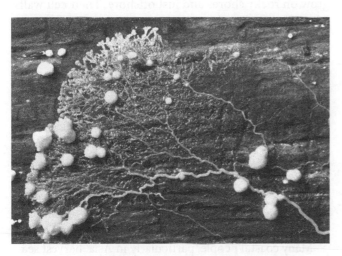

▲ **Figure 15.24 A plasmodial slime mold.** The weblike form of the slime mold's feeding stage (yellow structure) is an adaptation that enlarges the organism's surface area, increasing its contact with food, water, and oxygen.

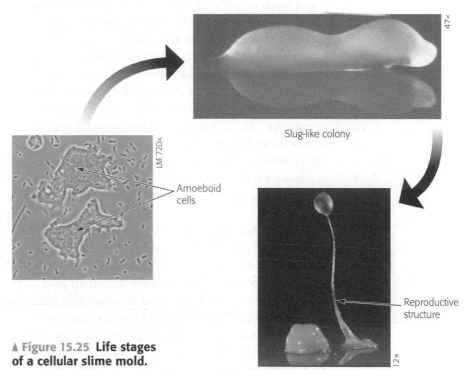

Slug-like colony

Amoeboid cells

Reproductive structure

▲ **Figure 15.25 Life stages of a cellular slime mold.**

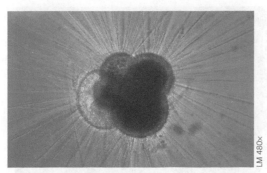

A foram. A foram cell secretes a shell made of organic material hardened with calcium carbonate. The pseudopodia, seen here as thin rays, extend through small pores in the shell.

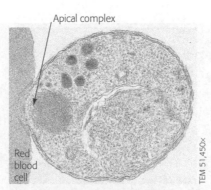

Apical complex

Red blood cell

An apicomplexan. *Plasmodium*, which causes malaria, enters red blood cells of its human host. The parasite feeds on the host cell from within, eventually destroying it.

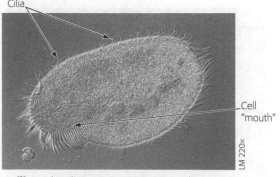

Cilia

Cell "mouth"

A ciliate. The ciliate *Paramecium* uses its cilia to move through pond water. Cilia lining the oral groove keep a current of water containing food moving toward the cell "mouth."

Unicellular and Colonial Algae

Algae are protists and cyanobacteria whose photosynthesis supports food chains in freshwater and marine ecosystems. Researchers are currently trying to harness their ability to convert light energy to chemical energy for another purpose—to produce biofuels. We'll look at four groups of protistan algae. Three of these groups—dinoflagellates, diatoms, and green algae—are unicellular, and one group is colonial.

Many unicellular algae are components of **phytoplankton**, the photosynthetic organisms, mostly microscopic, that drift near the surfaces of ponds, lakes, and oceans. Dinoflagellates are abundant in the vast aquatic pastures of phytoplankton. Each **dinoflagellate** species has a characteristic shape reinforced by external plates made of cellulose **(Figure 15.26a)**. The beating of two flagella in perpendicular grooves produces a spinning movement. Dinoflagellate blooms—population explosions—sometimes cause warm coastal waters to turn pinkish orange, a phenomenon known as a red tide. Toxins produced by some red-tide dinoflagellates have caused massive fish kills and are poisonous to humans as well. One group of dinoflagellates resides within the cells of reef-building corals. Without these algal partners, corals could not build and sustain the massive reefs that provide the food, living space, and shelter that support the splendid diversity of the reef community.

Diatoms have glassy cell walls containing silica, the mineral used to make glass **(Figure 15.26b)**. The cell wall consists of two halves that fit together like the bottom and lid of a shoe box. Diatoms store their food reserves in the form of an oil that provides buoyancy, keeping diatoms floating as plankton near the sunlit surface. The organic remains of diatoms that lived hundreds of millions of years ago are thought to be the main component of oil deposits. But why wait millions of years? Researchers are currently working on methods of growing diatoms and processing the oil into biodiesel.

Green algae are named for their grass-green chloroplasts. Unicellular green algae flourish in most freshwater lakes and ponds, as well as many home pools and aquariums. The green algal group also includes colonial forms, such as *Volvox*, shown in **Figure 15.26c**. Each *Volvox* colony is a hollow ball of flagellated cells (the small green dots in the photo) that are very similar to certain unicellular green algae. The balls within the balls in Figure 15.25c are daughter colonies that will be released when the parent colonies rupture. Of all photosynthetic protists, green algae are the most closely related to plants.

Seaweeds

Defined as large, multicellular marine algae, **seaweeds** grow on rocky shores and just offshore. Their cell walls have slimy and rubbery substances that cushion their bodies against the agitation of the waves. Some seaweeds are as large and complex as many plants. And although the word *seaweed* implies a plantlike appearance, the similarities between these algae and plants are a consequence of convergent evolution. In fact, the closest relatives of seaweeds are certain unicellular algae, which is why many biologists include seaweeds with the protists. Seaweeds are classified into three different groups **(Figure 15.27)**, based partly on the types of pigments present in their chloroplasts: green algae, red algae, and brown algae (some of which are known as kelp).

Many coastal people, particularly in Asia, harvest seaweeds for food. For example, in Japan and Korea, some seaweed species, including brown algae called kombu, are ingredients in soups. Other seaweeds, such as red algae called nori, are used to wrap sushi. Marine algae are rich

▼ Figure 15.26 **Unicellular and colonial algae.**

Colonized SEM 667×

(a) **Dinoflagellate:** Note the wall of protective plates

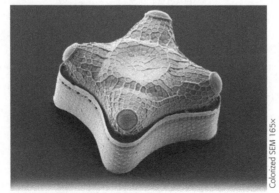

Colonized SEM 165×

(b) **Diatom:** Note the glassy walls

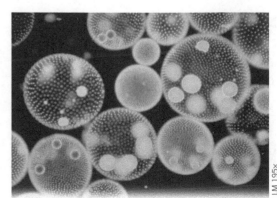

LM 195×

(c) **Volvox:** A colonial green alga

▼ Figure 15.27 **The three major groups of seaweeds.**

Green algae. This sea lettuce is an edible species that inhabits the intertidal zone, where the land meets the ocean.

Red algae. These seaweeds are most abundant in the warm coastal waters of the tropics.

Brown algae. This group includes the largest seaweeds, known as kelp, which grow as marine "forests."

in iodine and other essential minerals, but much of their organic material consists of unusual polysaccharides that humans cannot digest. They are eaten mostly for their rich tastes and unusual textures. The gel-forming substances in the cell walls of seaweeds are widely used as thickeners for such processed foods as puddings, ice cream, and salad dressing. The seaweed extract called agar provides the gel-forming base for the media microbiologists use to culture bacteria in petri dishes. ☑

Seaweeds aren't just used for wrapping sushi—they're in your ice cream, too.

Human Microbiota EVOLUTION CONNECTION

The Sweet Life of *Streptococcus mutans*

Did you ever wonder why candy causes cavities? A biofilm-forming species of bacteria called *Streptococcus mutans* is the culprit. *S. mutans* thrives in the anaerobic environment found in the tiny crevices in tooth enamel. Using sucrose (table sugar) to make a sticky polysaccharide, the bacteria glue themselves in place and build up thick deposits of plaque. Unless you make an effort to remove it, the plaque becomes mineralized, turning into a crusty tartar that must be scraped off your teeth by a dental hygienist (Figure 15.28). Within this fortress, *S. mutans* ferments sugars to obtain energy, releasing lactic acid as a by-product. The acid attacks tooth enamel and eventually eats through it. Other bacteria then use the entrance and infect the soft tissue in the interior of the tooth.

Early humans were hunter-gatherers, living on food foraged in the wild. In a major cultural shift, this lifestyle was replaced by agriculture, which provided a diet rich in carbohydrates from grain. A later dietary shift brought processed flour and sugar to the table. Each change in diet altered the environment inhabited by our oral microbiota. Studies of prehistoric human remains have correlated dental disease with these changes in diet.

Recent research links *S. mutans* directly to these increases in tooth decay. In one study, researchers analyzed DNA from tartar on the teeth of prehistoric humans who lived in Europe from 7,500 to 400 years ago. The tartar of hunter-gatherers included many species of bacteria but few known to cause tooth decay. *S. mutans* first appeared as one member of a diverse bacterial community after the agricultural diet had been established. Diversity of the oral microbiota dropped dramatically about 400 years ago, around the time sugar was introduced into the diet, and *S. mutans* became the overwhelmingly dominant species. Thus, we can infer that natural selection in the high-sugar environment favored *S. mutans*.

What adaptations gave *S. mutans* an advantage over other species? Another research team investigated genetic changes that allowed *S. mutans* to thrive in a high-sugar environment. Their results turned up more than a dozen genes that improved the ability of *S. mutans* to metabolize sugars and survive increased acidity. They also discovered chemical weapons produced by *S. mutans* that kill harmless bacteria—their competitors for space in the limited terrain of the human oral cavity.

It appears that *S. mutans* took full advantage of an evolutionary opportunity afforded by the human sweet tooth. Having adapted to new environmental conditions and ousted its competitors, it is now firmly entrenched as the dominant member of the mouth's microbial community.

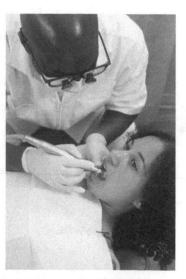

▲ Figure 15.28 **Checking the effects of *Streptococcus mutans*.**

Chapter Review

SUMMARY OF KEY CONCEPTS

Major Episodes in the History of Life

Major Episode	Millions of Years Ago
Plants and fungi colonize land	500
Fossils of large, diverse multicellular organisms	600
Oldest fossils of multicellular organisms	1,200
Oldest eukaryotic fossils	1,800
Beginning of atmospheric accumulation of O_2	2,700
Oldest prokaryotic fossils	3,500
Origin of Earth	4,600

The Origin of Life

A Four-Stage Hypothesis for the Origin of Life

One scenario suggests that the first organisms were products of chemical evolution in four stages:

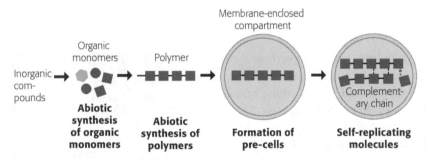

Inorganic compounds → **Abiotic synthesis of organic monomers** → **Abiotic synthesis of polymers** → **Formation of pre-cells** (Membrane-enclosed compartment) → **Self-replicating molecules** (Complementary chain)

From Chemical Evolution to Darwinian Evolution

Over millions of years, natural selection favored the most efficient pre-cells, which evolved into the first prokaryotic cells.

Prokaryotes

They're Everywhere!

Prokaryotes are found wherever there is life and greatly outnumber eukaryotes. Prokaryotes thrive in habitats where eukaryotes cannot live. A few prokaryotic species cause serious disease, but most are either benign or beneficial to other forms of life.

Structure/Function: Prokaryotes

Prokaryotic cells lack nuclei and other membrane-enclosed organelles. Most have cell walls. Prokaryotes exhibit three common shapes.

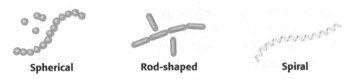

Spherical   **Rod-shaped**   **Spiral**

About half of all prokaryotic species are mobile, most of these using flagella to move. Some prokaryotes can survive extended periods of harsh conditions by forming endospores. Many prokaryotes can reproduce by binary fission at high rates if conditions are favorable, but growth is usually restricted by limited resources.

Prokaryotes include species that obtain energy from the sun (like plants) and from organic material (like animals and fungi). Some species derive energy from inorganic substances such as ammonia (NH_3) or hydrogen sulfide (H_2S). Some prokaryotes live in symbiotic associations with animals, plants, or fungi.

The Ecological Impact of Prokaryotes

Prokaryotes help recycle chemical elements between the biological and physical components of ecosystems. People use prokaryotes to remove pollutants from water, air, and soil in the process called bioremediation.

The Two Main Branches of Prokaryotic Evolution: Bacteria and Archaea

The prokaryotic lineage includes domains Bacteria and Archaea. Many archaea are "extremophiles" capable of surviving under conditions (such as high heat or salt concentrations) that would kill other forms of life; other archaea are found in more moderate environments. Some bacteria cause disease, mostly by producing exotoxins or endotoxins. Sanitation, antibiotics, and education are the best defenses against bacterial disease.

Protists

Protists are unicellular eukaryotes and their closest multicellular relatives.

Protozoans

Protozoans, including flagellates, amoebas, apicomplexans, and ciliates, primarily live in aquatic environments and ingest their food.

Slime Molds

Slime molds (including plasmodial slime molds and cellular slime molds) resemble fungi in appearance and lifestyle as decomposers but are not at all closely related.

Unicellular and Colonial Algae

Unicellular algae, including dinoflagellates, diatoms, and unicellular green algae, are photosynthetic protists that support food chains in freshwater and marine ecosystems.

Seaweeds

Seaweeds—which include green, red, and brown algae—are large, multicellular marine algae that grow on and near rocky shores.

MasteringBiology®

For practice quizzes, BioFlix animations, MP3 tutorials, video tutors, and more study tools designed for this textbook, go to MasteringBiology®

SELF-QUIZ

1. Place these events in the history of life on Earth in the order that they occurred.
   a. accumulation of O_2 in Earth's atmosphere
   b. colonization of land by plants and fungi
   c. diversification of animals (Cambrian explosion)
   d. origin of eukaryotes
   e. origin of humans
   f. origin of multicellular organisms
   g. origin of prokaryotes

2. Place the following steps in the origin of life in the order that they are hypothesized to have occurred.
   a. integration of abiotically produced molecules into membrane-enclosed pre-cells
   b. origin of the first molecules capable of self-replication
   c. abiotic joining of organic monomers into polymers
   d. abiotic synthesis of organic monomers
   e. natural selection among pre-cells

3. DNA replication relies on the enzyme DNA polymerase. Why does this suggest that the earliest genes were made from RNA?

4. Contrast exotoxins with endotoxins.

5. Humans have symbiotic relationships with prokaryotes. *E. coli*, the main constituent of the gut flora, provides its host with vitamin K. What benefit does *E. coli* derive?

6. The bacteria that cause tetanus can be killed only by prolonged heating at temperatures considerably above boiling. What does this suggest about tetanus bacteria?

7. How is the process used in sewage treatment similar to the decomposition of leaf litter in a forest?

8. Are algae the only photosynthetically active nonplants?

9. Which of the following has an organization that may be similar to an evolutionary stage between unicellularity and multicellularity?
   a. *Thiobacillus*
   b. *Volvox*
   c. amoeba
   d. plasmodium

10. Which protozoan group consists solely of parasitic forms?
    a. apicomplexans
    b. flagellates
    c. ciliates
    d. amoebas

Answers to these questions can be found in Appendix: Self-Quiz Answers.

THE PROCESS OF SCIENCE

11. *Acetabularia* are enormously large (2–4 cm long), single-celled green algae that look somewhat like mushrooms. They consist of a cap, a stalk, and a root-like structure called "rhizoid" that contains the large cell nucleus. In 1943, Joachim Hämmerling exchanged the nuclei of an *Acetabularia mediterranea* (which forms a flat cap) with that of an *Acetabularia crenulata* (which forms a castellated cap). What result would you expect from this experiment? Which basic concept does it confirm?

12. **Interpreting Data** Because bacteria divide by binary fission, the population size doubles every generation. Suppose the generation time for bacteria that cause food poisoning (for example, *Staphylococcus aureus* or *Salmonella*) is 30 minutes at room temperature. The number of cells in a population can be calculated using this formula.

Initial number of cells $\times\ 2^{(\text{Number of generations})}$ = Population size

For example, if a dish of potato salad is contaminated with 10 bacteria, the bacterial population after 1 hour (2 generations) is $10 \times 2^2 = 40$. Fill in the table below to show how the bacterial population increases when a dish of potato salad is left on the kitchen counter after dinner instead of being refrigerated overnight.

Time (Hours)	Number of Generations	Number of Bacteria
0	0	10
1	2	40
2	4	
3	6	
4	8	
5	10	
6	12	
8	16	
10	20	
12	24	

Why does the rate of increase change over time? Describe how a graph of the data would look.

BIOLOGY AND SOCIETY

13. Allergies are caused by an excessive reaction of the immune system to normally harmless substances, such as certain foods, dust mites, and pollen. Surveys indicate that allergies are on the rise, with about one-third of the population affected at some point in their lives. The rise of allergies can be noticed in all the countries that are undergoing industrial development. Propose an explanation for this.

14. Probiotics, foods and supplements that contain living microorganisms, are thought to cure problems of the digestive tract by restoring the natural balance of its microbial community. Sales of these products total billions of dollars a year. Explore the topic of probiotics and evaluate the scientific evidence for their beneficial effects. A good starting point is the website of the U.S. Food and Drug Administration, the agency that regulates advertising claims of health benefits of dietary supplements (www.fda.gov/Food/DietarySupplements/default.htm).

16 The Evolution of Plants and Fungi

Why Plants and Fungi Matter

If you were to hug one of the largest giant sequoia trees, you would need more than a dozen friends to help you reach around its circumference.

If you've ever had mushroom pizza, you've snacked on the fungi's reproductive structure.

Fungi may not always be pretty, but if you find the right one, it could pay your way through college.

The most expensive coffee in the world has passed through the digestive tract of a catlike animal called a civet.

Plant-Fungus Interactions BIOLOGY AND SOCIETY

The Diamond of the Kitchen

To the untrained eye, a truffle (the fungus, not the chocolate) is an unappealing lump that you probably wouldn't eat on a dare. But despite their unappetizing appearance, truffles are treasured by gourmets as "diamonds of the kitchen." They command prices up to hundreds of dollars per ounce for the best specimens. A 3.3-pound white truffle—an unusually large individual of the rarest variety—brought a record $330,000 at auction. What's the attraction? Rather than their taste, truffles are prized for their powerful scent, which has been described as earthy or moldy. A little truffle goes a long way. Chefs only need to add a few shavings from a truffle to infuse a dish with its delectable essence.

Truffles are the subterranean reproductive bodies of certain fungi. Their job is to produce spores, single cells that are capable of growing into a new fungus, just as a seed can grow into a new plant. It is generally advantageous for seeds or spores to start their lives in a new place, away from the ground inhabited by their parents. And that's where the potent truffle smell comes in. Attracted by the heady aroma, certain animals excavate and eat the fungi and later deposit the hardy spores in their feces. Human noses are not sensitive enough to locate the buried treasures, so truffle hunters use pigs or trained dogs to sniff out their quarry.

Thinly sliced black truffles top a pasta dish. Black truffles, which can be cultivated, are much less expensive than the white variety, which are found only in a small region of Italy.

Besides being culinary gems, truffles represent the essential role of fungi as the hidden power behind the throne of the plant kingdom. The roots of most plants are surrounded—in some cases even permeated—by a finely woven web of fungal filaments. Truffles, for example, have such a relationship with certain species of trees, which is why skilled truffle hunters look for their treasures beneath oak and hazelnut trees. This association is an example of symbiosis, an interaction in which one species lives in or on another species. The ultrathin fungal filaments reach into pockets between the soil grains that are too small for roots to enter, absorbing water and inorganic nutrients and passing them to the plant. The plant returns the favor by supplying the fungus with sugars and other organic molecules. This relationship between fungus and root is evident in some of the oldest plant fossils, suggesting that the mutually beneficial symbiosis was crucial to the colonization of land.

Colonizing Land

What exactly is a plant? A **plant** is a multicellular eukaryote that carries out photosynthesis and has a set of adaptations for living on land. Photosynthesis distinguishes plants from the animal and fungal kingdoms, which are also made up of eukaryotic, multicellular organisms. Large algae, including seaweeds, are also eukaryotic, multicellular, and photosynthetic. However, they lack terrestrial adaptations and thus are classified as protists rather than plants (see Figure 15.27). It is true that some, such as water lilies, have returned to the water, but they evolved from terrestrial ancestors (just as several species of aquatic mammals, such as whales, evolved from terrestrial mammals).

Terrestrial Adaptations of Plants

Why does life on land require a special set of adaptations? Consider how algae fare after being washed up on the

▼ Figure 16.1 **Structural adaptations of algae and plants.**

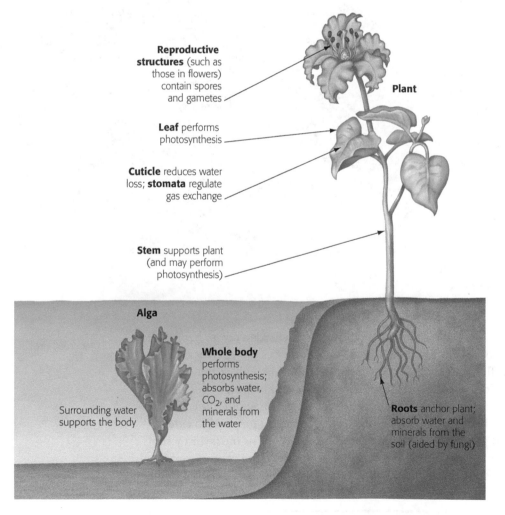

Reproductive **structures** (such as those in flowers) contain spores and gametes

Plant

Leaf performs photosynthesis

Cuticle reduces water loss; **stomata** regulate gas exchange

Stem supports plant (and may perform photosynthesis)

Alga

Whole body performs photosynthesis; absorbs water, CO_2, and minerals from the water

Surrounding water supports the body

Roots anchor plant; absorb water and minerals from the soil (aided by fungi)

beach. Bodies that were upright in the buoyant water go limp on land and soon shrivel in the drying air. In addition, algae are not equipped to obtain the carbon dioxide needed for photosynthesis from the air. Clearly, living on land poses different problems than living in water. In this section, we discuss some of the terrestrial adaptations that distinguished plants from algae and allowed plants to colonize land. As you will see in later sections, it took more than 100 million years to complete the transition. The earliest land plants lacked some of the adaptations that made subsequent groups more successful in the terrestrial environment.

 **Structure/ Function** **Adaptations of the Plant Body**

For algae, carbon dioxide and minerals are available by diffusion from the surrounding water **(Figure 16.1)**. Resources on land are found in two very different places: Carbon dioxide is mainly available in the air, whereas mineral nutrients and water are found mainly in the soil. Thus, the complex bodies of plants have organs specialized in different ways to function in these two environments. Subterranean organs called **roots** anchor the plant in soil and absorb minerals and water from the soil. Aboveground, **shoots** are organ systems that consist of photosynthetic leaves supported by stems.

Roots typically have many fine branches that thread among the grains of soil, providing a large surface area that maximizes contact with mineral-bearing water in the soil. In addition, as you learned in the Biology and Society section in the beginning of this chapter, most plants have symbiotic fungi associated with their roots. These root-fungus combinations, called **mycorrhizae** ("fungus roots"), enlarge the root's functional surface area **(Figure 16.2)**. For their part, the fungi absorb water and essential minerals from the soil and provide these materials to the plant. The sugars produced by the plant nourish the fungi. Mycorrhizae are key adaptations that made it possible for plants to live on land.

Shoots also show structural adaptations to the terrestrial environment. Leaves are the main photosynthetic organs of most plants. Exchange of carbon dioxide (CO_2) and oxygen (O_2) between the atmosphere and the photosynthetic interior of a leaf occurs via **stomata** (singular, *stoma*), the microscopic pores found on a leaf's surface (see Figure 7.2). A waxy layer called the **cuticle** coats the leaves and other aerial parts of most plants, helping the plant body retain water (see Figure 18.8b).

To transport vital materials between roots and shoots, most plants have **vascular tissue**, a network

▼ Figure 16.2 Mycorrhizae: symbiotic associations of fungi and roots. The finely branched filaments of the fungus (white in the photo) provide an extensive surface area for absorption of water and minerals from the soil.

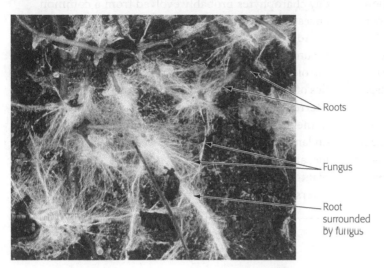

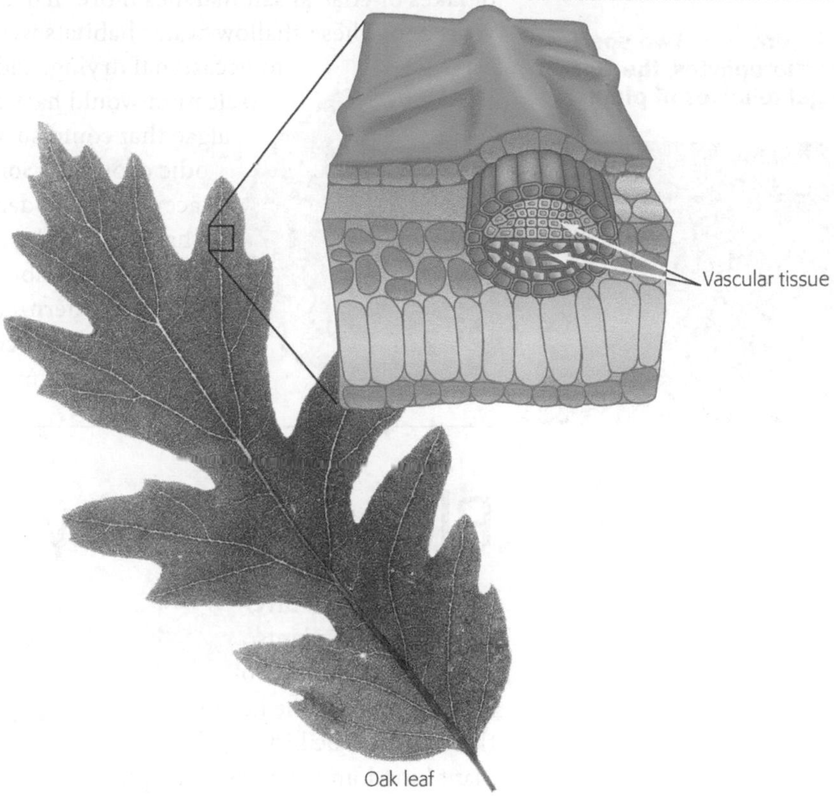

Roots

Fungus

Root surrounded by fungus

▼ Figure 16.3 Network of vascular tissue in a leaf. The vascular tissue is visible as yellow veins on the underside of the leaf in the photograph.

Vascular tissue

Oak leaf

of tube-shaped cells that branch throughout the plant **(Figure 16.3)**. There are two types of vascular tissue. One type is specialized for transporting water and minerals from roots to leaves, and the other distributes sugars from the leaves to the roots and other nonphotosynthetic parts of the plant.

Vascular tissue also solved the problem of structural support on land. The cell walls of many of the cells in vascular tissue are hardened by a chemical called **lignin**. The structural strength of lignified vascular tissue, otherwise known as wood, is amply demonstrated by its use as a building material.

Reproductive Adaptations

Adapting to land also required a new mode of reproduction. For algae, the surrounding water ensures that gametes (sperm and eggs) and developing offspring stay moist. The aquatic environment also provides a means of dispersing the gametes and offspring. Plants, however, must keep their gametes and developing offspring from drying out in the air. Plants produce their gametes in a structure that allows gametes to develop without dehydrating. In addition, the egg remains within tissues of the mother plant and is fertilized there. In plants, but not algae, the zygote (fertilized egg) develops into an embryo while still contained within the female parent, which protects the embryo and keeps it from dehydrating **(Figure 16.4)**. Further adaptations in some plant groups allow sperm to travel via air and improve the survival of offspring during dispersal. ✓

▼ Figure 16.4 The protected embryo of a plant. Internal fertilization, with sperm and egg combining within a moist chamber on the female plant, is an adaptation for living on land. The female plant continues to nourish and protect the plant embryo, which develops from the zygote.

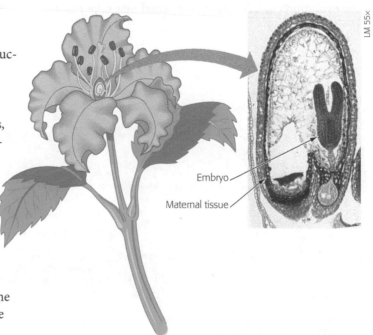

LM 55×

Embryo

Maternal tissue

☑ CHECKPOINT

Name some adaptations of plants for living on land.

Answer: any of the following: cuticle, stomata, vascular tissue, lignin-hardened cell walls, protected gametes and embryos, and differentiation of the body into aerial shoots and subterranean roots

351

The Origin of Plants from Green Algae

The algal ancestors of plants carpeted moist fringes of lakes or coastal salt marshes more than 500 million years ago. These shallow-water habitats were subject to occasional drying, and natural selection would have favored algae that could survive periodic droughts. Some species accumulated adaptations that enabled them to live permanently above the water line. A modern-day lineage of green algae, the **charophytes** (Figure 16.5), may resemble one of these early plant ancestors. Plants and present-day charophytes probably evolved from a common ancestor.

Adaptations making life on dry land possible had accumulated by about 470 million years ago, the age of the oldest known plant fossils. The evolutionary novelties of these first land plants opened the new frontier of a terrestrial habitat. Early plant life would have thrived in the new environment. Bright sunlight was abundant on land, the atmosphere had a wealth of carbon dioxide, and at first there were relatively few pathogens and plant-eating animals. The stage was set for an explosive diversification of plant life.

▼ Figure 16.5 **Two species of charophytes, the closest algal relatives of plants.**

LM 265×

Plant Diversity

As we survey the diversity of modern plants, remember that the evolutionary past is the key to the present. The history of the plant kingdom is a story of adaptation to diverse terrestrial habitats.

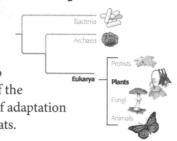

▼ Figure 16.6 **Highlights of plant evolution.** This phylogenetic tree highlights the evolution of structures that allowed plants to move onto land; these structures still exist in modern plants. As we survey the diversity of plants, miniature versions of this tree will help you place each plant group in its evolutionary context.

Highlights of Plant Evolution

The fossil record chronicles four major periods of plant evolution, which are also evident in the diversity of modern plants (Figure 16.6). Each stage is marked by the evolution of structures that opened new opportunities on land.

❶ After plants originated from an algal ancestor approximately 470 million years ago, early diversification gave rise to nonvascular plants, including mosses, liverworts, and hornworts. These plants, called **bryophytes**, lack true roots and leaves. Bryophytes also lack lignin, the wall-hardening material that enables other plants to stand tall. Without lignified cell walls, bryophytes have weak upright support. The most familiar bryophytes are **mosses**. A mat of moss actually consists of many plants growing in a tight pack, holding one another up. Structures that protect the gametes and embryos are a terrestrial adaptation that originated in bryophytes.

❷ The second period of plant evolution, begun about 425 million years ago, was the diversification of

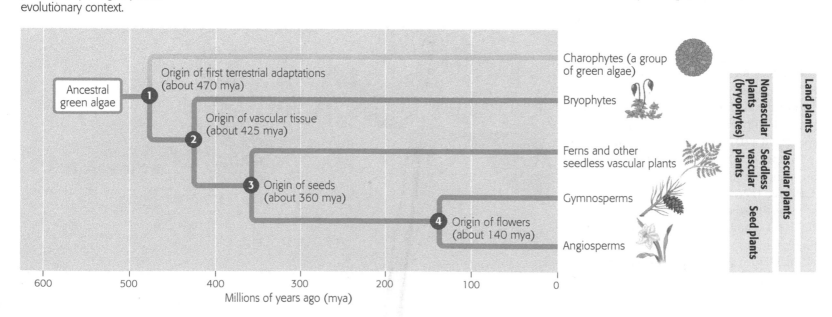

plants with vascular tissue. The presence of conducting tissues hardened with lignin allowed vascular plants to grow much taller, rising above the ground to achieve significant height. The earliest vascular plants lacked seeds. Today, this seedless condition is retained by **ferns** and a few other groups of vascular plants.

❸ The third major period of plant evolution began with the origin of the seed about 360 million years ago. Seeds advanced the colonization of land by further protecting plant embryos from drying and other hazards. A **seed** consists of an embryo packaged along with a store of food within a protective covering. The seeds of early seed plants were not enclosed in any specialized chambers. These plants gave rise to the **gymnosperms** ("naked seeds"). Today, the most widespread and diverse gymnosperms are the **conifers**, consisting mainly of cone-bearing trees, such as pines.

❹ The fourth major episode in the evolutionary history of plants was the emergence of flowering plants, or **angiosperms** ("contained seeds"), at least 140 million years ago. The **flower** is a complex reproductive structure that bears seeds within protective chambers called ovaries. This contrasts with the naked seeds of gymnosperms. The great majority of living plants—some 250,000 species—are angiosperms, including all our fruit and vegetable crops, grains and other grasses, and most trees.

With these highlights as our framework, we are now ready to survey the four major groups of modern plants: bryophytes, ferns, gymnosperms, and angiosperms **(Figure 16.7).** ☑

▼ Figure 16.7 **The major groups of plants.**

PLANT DIVERSITY			
Bryophytes (nonvascular plants) 	**Ferns** (seedless vascular plants) 	**Gymnosperms** (naked-seed plants) 	**Angiosperms** (flowering plants) 

Bryophytes

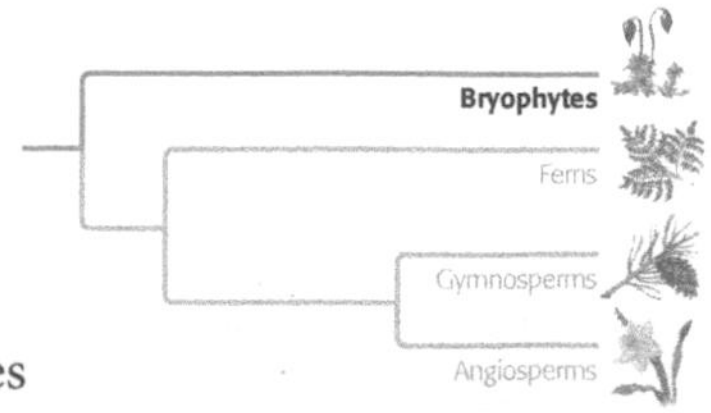

Mosses, which are bryophytes, may sprawl as low mats over acres of land **(Figure 16.8).** Mosses display two of the key terrestrial adaptations that made the move onto land possible: (1) a waxy cuticle that helps prevent dehydration and (2) the retention of developing embryos within the female plant. However, mosses are not totally liberated from their ancestral aquatic habitat. Mosses need water to reproduce because their sperm need to swim to reach eggs located within the female plant. (A film of rainwater or dew is enough moisture for the sperm to travel.) In addition, because most mosses have no vascular

Bryophytes
Ferns
Gymnosperms
Angiosperms

▼ Figure 16.8 **A peat moss bog in Scotland.** Mosses are bryophytes, which are nonvascular plants. Sphagnum mosses, collectively called peat moss, carpet at least 3% of Earth's land surface. They are most commonly found in high northern latitudes. The ability of peat moss to absorb and retain water makes it an excellent addition to garden soil.

353

tissue to carry water from soil to aerial parts of the plant, they need to live in damp places.

If you examine a mat of moss closely, you can see two distinct forms of the plant. The green, spongelike plant that is the more obvious is called the **gametophyte**. Careful examination will reveal the other form of the moss, called a **sporophyte**, growing out of a gameto-phyte as a stalk with a capsule at its tip (**Figure 16.9**). The cells of the gametophyte are haploid—they have one set of chromosomes (see Figure 8.12). In contrast, the sporophyte is made up of diploid cells (with two chro-mosome sets). These two different stages of the plant life cycle are named for the types of reproductive cells they produce. Gametophytes produce gametes (sperm and eggs), while sporophytes produce spores. A **spore** is a haploid cell that can develop into a new individual without

fusing with another cell (two gametes must fuse to form a zygote). Spores usually have tough coats that enable them to survive in harsh environments. Seedless plants, including mosses and ferns, disperse their offspring as spores rather than as multicellular seeds.

The gametophyte and sporophyte are alternating generations that take turns producing each other. Gametophytes produce gametes that unite to form zygotes, which develop into new sporophytes. And sporophytes produce spores that give rise to new ga-metophytes. This type of life cycle, called **alternation of generations**, occurs only in plants and multicellular green algae (**Figure 16.10**). Among plants, mosses and other bryophytes are unique in having the gameto-phyte as the larger, more obvious plant. As we continue our survey of plants, we'll see an increasing dominance of the sporophyte as the more highly developed generation. ☑

▼ **Figure 16.9 The two forms of a moss.** The feathery plant we generally know as a moss is the gametophyte. The stalk with the capsule at its tip is the sporophyte.

Spore capsule

Sporophytes

Gametophytes

▶ **Figure 16.10 Alternation of generations.** Plants have life cycles very different from ours. Each of us is a diploid individual; the only haploid stages in the human life cycle, as for nearly all animals, are sperm and eggs. By contrast, plants have alternating generations: Diploid (2*n*) individuals (sporophytes) and haploid (*n*) individuals (gametophytes) generate each other in the life cycle.

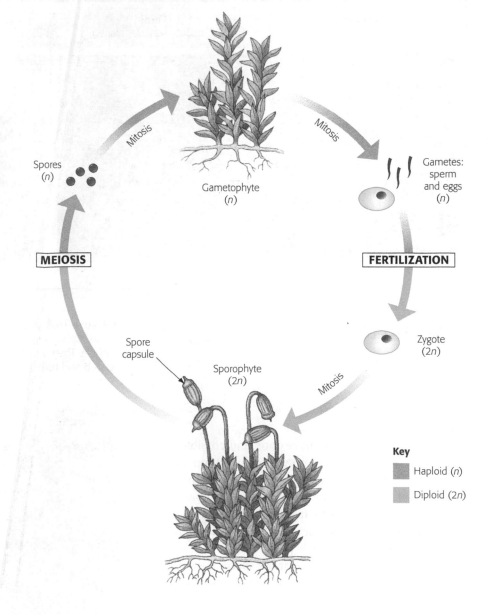

Spores
(*n*)

Gametophyte
(*n*)

Mitosis

Mitosis

Gametes:
sperm
and eggs
(*n*)

MEIOSIS

FERTILIZATION

Zygote
(2*n*)

Spore
capsule

Sporophyte
(2*n*)

Mitosis

Key

Haploid (*n*)

Diploid (2*n*)

Ferns

The evolution of vascular tissue allowed ferns to colonize a greater variety of habitats than mosses. Ferns are by far the most diverse seedless vascular plants, with more than 12,000 known species. However, the sperm of ferns, like those of mosses, have flagella and must swim through a film of water to fertilize eggs. Most ferns inhabit the tropics, although many species are found in temperate forests, such as many woodlands in the United States (Figure 16.11).

During the Carboniferous period, from about 360 to 300 million years ago, ancient ferns were part of a much greater diversity of seedless plants that formed vast, swampy tropical forests over much of what is now Eurasia and North America (Figure 16.12). As the plants died, they fell into stagnant wetlands and did not decay completely. Their remains formed thick organic deposits. Later, seawater flooded the swamps, marine sediments covered the organic deposits, and pressure and heat gradually converted them to coal. Coal is black sedimentary rock made up of fossilized plant material. Like coal, oil and natural gas also formed from the remains of long-dead organisms; thus, all three are known as **fossil fuels**. Since the Industrial Revolution, coal has been a crucial source of energy for people. However, burning these fossil fuels releases CO_2 and other gases that contribute to global climate change (see Figure 18.46). ☑

▼ Figure 16.11 **Ferns (seedless vascular plants).** The ferns in the foreground are growing on the forest floor in Redwood National Park, California. The fern generation familiar to us is the sporophyte generation. You would have to crawl on the forest floor and explore with careful hands and sharp eyes to find fern gametophytes (upper right), tiny plants growing on or just below the soil surface.

New sporophyte

Gametophyte

Cluster of spore capsules

"Fiddlehead" (young leaf ready to unfurl)

◀ Figure 16.12 **A "coal forest" of the Carboniferous period.** This painting, based on fossil evidence, reconstructs one of the great seedless forests. Most of the large trees belong to ancient groups of seedless vascular plants that are represented by just a few present-day species. The plants near the base of the trees are ferns.

☑ **CHECKPOINT**

Why are ferns able to grow taller than mosses?

Answer: Vascular tissue hardened with lignin allows ferns to stand taller and transport nutrients farther.

Gymnosperms

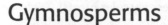

"Coal forests" dominated the North American and Eurasian landscapes until near the end of the Carboniferous period. At that time, the global climate turned drier and colder, and the vast swamps began to disappear. This climatic change provided an opportunity for seed plants, which can complete their life cycles on dry land and withstand long, harsh winters. Of the earliest seed plants, the most successful were the gymnosperms, and several kinds grew along with the seedless plants in the Carboniferous swamps. Their descendants include the conifers, or cone-bearing plants.

If you were to hug one of the largest giant sequoia trees, you would need more than a dozen friends to reach around its circumference.

Conifers

Perhaps you have had the fun of hiking or skiing through a forest of conifers, the most common gymnosperms. Pines, firs, spruces, junipers, cedars, and redwoods are all conifers. A broad band of coniferous forests covers much of northern Eurasia and North America and extends southward in mountainous regions **(Figure 16.13)**. Today, about 190 million acres of coniferous forests in the United States are designated national forests.

Conifers are among the tallest, largest, and oldest organisms on Earth. Coastal redwoods, native to the northern California coast, are the world's tallest trees—up to 110 m, the height of a 33-story building. Giant sequoias, relatives of redwoods that grow in the Sierra Nevada mountains of California, are massive. One, known as the General Sherman tree, is about 84 m (275 feet) high and outweighs the combined weight of a dozen space shuttles. Bristlecone pines, another species of California conifer, are among the oldest organisms alive. A specimen discovered in 2012 is more than 5,000 years old; it was a seedling when people invented writing.

Nearly all conifers are evergreens, meaning they retain leaves throughout the year. Even during winter, they perform a limited amount of photosynthesis on sunny days. And when spring comes, conifers already have fully developed leaves that can take advantage of the sunnier days. The needle-shaped leaves of pines and firs are also adapted to survive dry seasons. A thick cuticle covers the leaf, and the stomata are located in pits, further reducing water loss.

Coniferous forests are highly productive; you probably use products harvested from them every day. For example, conifers provide much of our lumber for building and wood pulp for paper production. What we call wood is actually an accumulation of vascular tissue with lignin, which gives the tree structural support.

Evolution Terrestrial Adaptations of Seed Plants

Compared with ferns, most gymnosperms have three additional adaptations that make survival in diverse terrestrial habitats possible: (1) further reduction of the gametophyte, (2) pollen, and (3) seeds.

The first adaptation is an even greater development of the diploid sporophyte compared with the haploid gametophyte generation **(Figure 16.14)**. A pine tree or

▼ **Figure 16.13 A coniferous forest in Tetlin National Wildlife Refuge, Alaska.** Coniferous forests are widespread in northern North America and Eurasia; conifers also grow in the Southern Hemisphere, though they are less numerous there.

► **Figure 16.14 Three variations on alternation of generations in plants.**

Key

⬛ Haploid (*n*)

⬜ Diploid (2*n*)

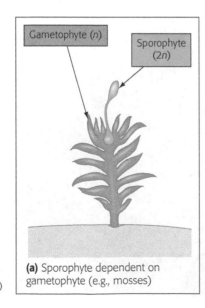

(a) Sporophyte dependent on gametophyte (e.g., mosses)

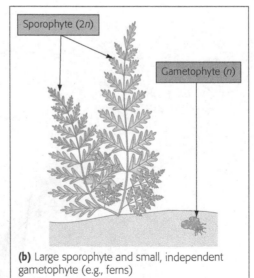

(b) Large sporophyte and small, independent gametophyte (e.g., ferns)

Sporophyte (2*n*)

Gametophyte (*n*)

(c) Reduced gametophyte dependent on sporophyte (seed plants)

other conifer is a sporophyte with tiny gametophytes living in its cones (**Figure 16.15**). In contrast to what is seen in bryophytes and ferns, gymnosperm gametophytes are totally dependent on and protected by the tissues of the parent sporophyte.

A second adaptation of seed plants to dry land came with the evolution of pollen. A **pollen grain** is actually the much-reduced male gametophyte; it houses cells that will develop into sperm. In the case of conifers, **pollination**, the delivery of pollen from the male parts of a plant to the female parts of a plant, occurs via wind. This mechanism for sperm transfer contrasts with the swimming sperm of mosses and ferns. In seed plants, the use of tough, airborne pollen that carries sperm to egg is a terrestrial adaptation that led to even greater success and diversity of plants on land.

The third important terrestrial adaptation of seed plants is the seed itself. A seed consists of a plant embryo packaged along with a food supply within a protective coat. Seeds develop from **ovules**, structures that contain the female gametophytes (**Figure 16.16**). In conifers, the ovules are located on the scales of

female cones. Once released from the parent plant, the seed can remain dormant for days, months, or even years. Under favorable conditions, the seed can then **germinate**, or sprout: Its embryo emerges through the seed coat as a seedling. Some seeds drop close to their parents, while others are carried far by the wind or animals. ☑

PLANT DIVERSITY

☑ **CHECKPOINT**

Contrast the modes of sperm delivery in ferns and conifers.

Answer: The flagellated sperm of ferns must swim through water to reach eggs. In contrast, the airborne pollen of conifers delivers sperm to eggs in ovules without the need to go through water.

▼ Figure 16.15 **A pine tree, the sporophyte, bearing two types of cones containing gametophytes.** Each scale of the female cone is actually a modified leaf that bears a structure called an ovule containing a female gametophyte. Male cones release clouds of millions of pollen grains, the male gametophytes. Some of these pollen grains land on female cones on trees of the same species. The sperm can fertilize eggs in the ovules of the female cones. The ovules eventually develop into seeds.

Ovule-producing cones; the scales contain female gametophytes

Scale

Pollen-producing cones; they produce male gametophytes

Ponderosa pine

▼ Figure 16.16 **From ovule to seed.**

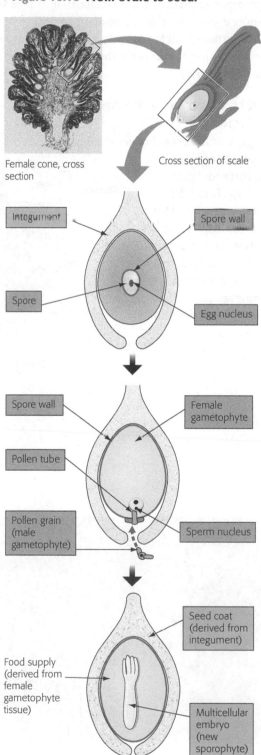

Female cone, cross section

Cross section of scale

Key

Haploid (*n*)

Diploid (2*n*)

Integument

Spore wall

Spore

Egg nucleus

(a) Ovule. The sporophyte produces spores within a tissue surrounded by a protective integument, which may be multilayered. The spore develops into a female gametophyte, which produces an egg nucleus.

Spore wall

Female gametophyte

Pollen tube

Pollen grain (male gametophyte)

Sperm nucleus

(b) Fertilized ovule. After pollination, the pollen grain grows a tiny tube that enters the ovule, where it releases a sperm nucleus that fertilizes the egg.

Seed coat (derived from integument)

Food supply (derived from female gametophyte tissue)

Multicellular embryo (new sporophyte)

(c) Seed. Fertilization triggers the transformation of ovule to seed. The fertilized egg (zygote) develops into a multicellular embryo; the rest of the gametophyte forms a tissue that stockpiles food; and the integument of the ovule hardens to become the seed coat.

Angiosperms

Angiosperms dominate
the modern landscape.
About 250,000 angio-
sperm species have been
identified, compared to about
700 species of gymnosperms. Several unique
adaptations account for the success of angiosperms.
For example, refinements in vascular tissue make water
transport even more efficient in angiosperms than in
gymnosperms. Of all terrestrial adaptations, however, it
is the flower that accounts for the unparalleled success
of the angiosperms.

Bryophytes

Ferns

Gymnosperms

Angiosperms

Flowers, Fruits, and the Angiosperm Life Cycle

No organism makes a showier display of its sex
life than the angiosperm.
From roses to dande-
lions, flowers are the
site of procreation. This
showiness helps to attract
go-betweens—insects and other
animals—that transfer pollen
from one flower to another of
the same species. Angiosperms
that rely on wind pollination,
including grasses and many trees, have much smaller,
less flamboyant flowers. In those species, the plant's
reproductive energy is allocated to making massive
amounts of pollen for release into the wind.

A flower is a short stem bearing modified leaves
that are attached in concentric circles at its base
(**Figure 16.17**). The outer layer consists of the **sepals**,
which are usually green. They enclose the flower before
it opens (think of the green "wrapping" on a rosebud).
When the sepals are peeled away, the next layer is the
petals, which are often colorful—these are the showy
structures that attract pollinators. Plucking off the petals
reveals the **stamens**, the male reproductive structures.
Pollen grains develop in the **anther**, a sac at the top of
each stamen. At the center of the flower
is the **carpel**, the female reproductive
structure. It includes the **ovary**, a protective chamber containing one or more
ovules, in which eggs develop.
The sticky tip of the carpel,
called the **stigma**, traps
pollen. As you can see in
Figure 16.18, the basic structure of a flower can exist in
many beautiful variations.

Petal

Stamen — Anther, Filament

Stigma, Style, Ovary — **Carpel**

Ovule

Sepal

◄ **Figure 16.17 Structure of a flower.**

▼ **Figure 16.18 A diversity of flowers.**

Prickly pear cactus

Bleeding heart

California poppy

Summer snapdragon

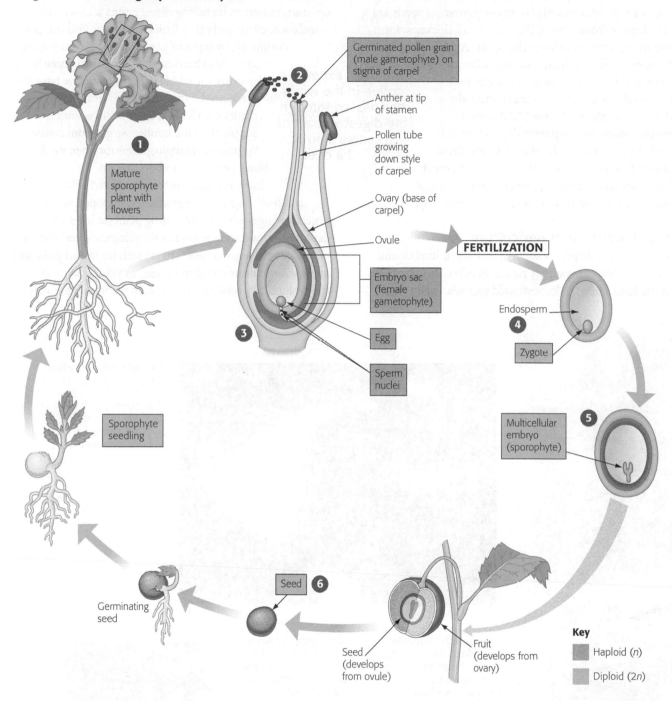

▼ Figure 16.19 **The angiosperm life cycle.**

In angiosperms, as in gymnosperms, the sporophyte generation is dominant and produces the gametophyte generation within its body. **Figure 16.19** highlights key stages in the angiosperm life cycle. ❶ The flower is part of the sporophyte plant. As in gymnosperms, the pollen grain is the male gametophyte of angiosperms. The female gametophyte is located within an ovule, which in turn resides within a chamber of the ovary. ❷ After a pollen grain lands on the stigma, a pollen tube grows down to the ovule and ❸ releases a sperm nucleus that fertilizes an egg within the embryo sac. ❹ This produces a zygote, which ❺ develops into an embryo. The tissue surrounding the embryo develops into nutrient-rich **endosperm**, which will provide a food supply for the growing plant. ❻ The whole ovule develops into a seed, which can germinate and develop into a new sporophyte to begin the cycle anew. The seed's enclosure within an ovary is what distinguishes angiosperms (from the Greek *angion*, container, and *sperm*, seed) from gymnosperms (from the Greek *gymnos*, naked), which have a naked seed.

A **fruit** is the ripened ovary of a flower. Thus, fruits are produced only by angiosperms. As seeds are developing from ovules, the ovary wall thickens, forming the fruit that encloses the seeds. A pea pod is an example of a fruit with seeds (mature ovules, the peas) encased in the ripened ovary (the pod). Fruits protect and help disperse seeds. As **Figure 16.20** demonstrates, many angiosperms depend on animals to disperse seeds, often passing them through the digestive tract. Conversely, most land animals, including humans, rely on angiosperms as a food source, directly or indirectly. ☑

The most expensive coffee in the world has passed through the digestive tract of a catlike animal called a civet.

and chickens. More than 90% of the plant kingdom is made up of angiosperms, including cereal grains such as wheat and corn, citrus and other fruit trees, coffee and tea, and cotton. Many types of garden produce—tomatoes, squash, strawberries, and oranges, to name just a few—are the edible fruits of plants we have domesticated. Fine hardwoods from flowering plants such as oak, cherry, and walnut trees supplement the lumber we get from conifers. We also grow angiosperms for fiber, medications, perfumes, and decoration.

Early humans probably collected wild seeds and fruits. Agriculture gradually developed as people began sowing seeds and cultivating plants to have a more dependable food source. And as they domesticated certain plants, people began to select those with improved yield and quality. Agriculture can thus be seen as yet another facet of the evolutionary relationship between plants and animals.

Angiosperms and Agriculture

Whereas gymnosperms supply most of our lumber and paper, angiosperms supply nearly all of our food—as well as the food eaten by domesticated animals, such as cows

▼ **Figure 16.20 Fruits and seed dispersal.** Different types of fruits are adapted for different methods of dispersal.

Wind dispersal. Some angiosperms depend on wind for seed dispersal. Here, milkweed pods open to release masses of seeds (brown) carried by silken parachutes (part of the seed coat).

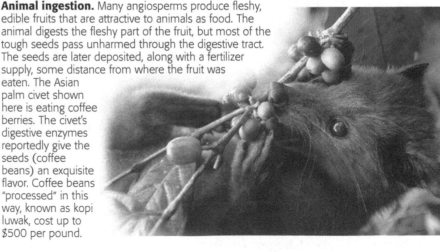

Animal transportation. Some fruits are adapted to hitch free rides on animals. The cockleburs attached to the fur of this dog may be carried miles before opening and releasing seeds.

Animal ingestion. Many angiosperms produce fleshy, edible fruits that are attractive to animals as food. The animal digests the fleshy part of the fruit, but most of the tough seeds pass unharmed through the digestive tract. The seeds are later deposited, along with a fertilizer supply, some distance from where the fruit was eaten. The Asian palm civet shown here is eating coffee berries. The civet's digestive enzymes reportedly give the seeds (coffee beans) an exquisite flavor. Coffee beans "processed" in this way, known as kopi luwak, cost up to $500 per pound.

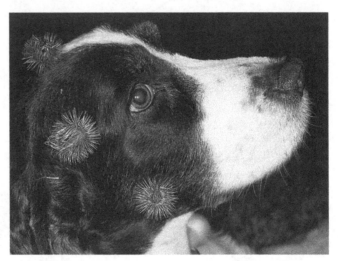

Plant Diversity as a Nonrenewable Resource

The ever-increasing human population, with its demand for space and natural resources, is extinguishing plant species at an unprecedented rate. The problem is especially critical for forest ecosystems, which are home to as many as 80% of the world's plant and animal species. Deforestation has been an ongoing human activity for centuries. Worldwide, less than 25% of the original forests remain; the figure in the contiguous United States is around 10%. Typically, forests are cut down to harvest timber or to clear land for housing or large-scale agriculture (Figure 16.21). Much of the forest that remains is tropical, and it is disappearing at an estimated 150,000 km^2 per year (about 58,000 mi^2), which is roughly equal to the area of Illinois. Government policies in Brazil have produced a marked slowdown in Amazonian deforestation. Nevertheless, the global rate of tropical deforestation has increased by 2,101 km^2 (more than 800 mi^2) per year over the past decade.

Why does the loss of tropical forests matter? In addition to forests being centers of biodiversity, millions of people worldwide depend on these forests for their livelihood. There are other practical reasons to be concerned about the loss of plant diversity represented in tropical forests. More than 120 prescription drugs are made from substances derived from plants (Table 16.1). Pharmaceutical companies were led to most of these species by local peoples who use the plants in preparing their traditional medicines. Today, researchers are seeking to combine their scientific skills with the stores of local knowledge in partnerships that will develop new drugs and also benefit the local economies.

Scientists are working to slow the loss of plant diversity, in part by researching sustainable ways for people to benefit from forests. The goal of such efforts is to encourage management practices that use forests as resources without damaging them. The solutions we propose must be economically realistic; people who live where there are tropical rain forests must be able to make a living. But if the only goal is profit for the short term, then the destruction will continue until the forests are gone. We need to appreciate the rain forests and other ecosystems as living treasures that can regenerate only slowly. Only then will we learn to work with them in ways that preserve their biological diversity for the future.

Throughout our survey of plants in this chapter, we have seen how entangled the botanical world is with other terrestrial life. We switch our attention now to that other group of organisms that moved onto land with plants: the kingdom Fungi. ☑

☑ CHECKPOINT

In what way are forests renewable resources? In what way are they not?

Answer: Forests are renewable in the sense that new trees can grow where old growth has been removed by logging. Habitats that are permanently destroyed cannot be replaced, however, so forests must be harvested in a sustainable manner.

▼ Figure 16.21 **Cultivated land bordering a tropical forest in Uganda.** Bwindi Impenetrable National Park (on the right) is renowned for its biodiversity, which includes half the world's remaining mountain gorillas.

Table 16.1	A Sampling of Medicines Derived from Plants		
Compound	**Source**		**Example of Use**
Atropine	Belladonna plant		Pupil dilator in eye exams
Digitalin	Foxglove		Heart medication
Menthol	Wild mint		Ingredient in cough medicines, decongestants
Morphine	Opium poppy		Pain reliever
Quinine	Quinine tree		Malaria preventive
Paclitaxel (Taxol)	Pacific yew		Ovarian cancer drug
Tubocurarine	Curare tree		Muscle relaxant during surgery
Vinblastine	Periwinkle		Leukemia drug

Source: Adapted from Randy Moore et al., Botany, 2nd ed. Dubuque, IA: Brown, 1998, Table 2.2, p. 37.

Fungi

The word *fungus* often evokes unpleasant images. Fungi rot timbers, spoil food, and afflict people with athlete's foot and worse.

However, ecosystems would collapse without fungi to decompose dead organisms, fallen leaves, feces, and other organic materials. Fungi recycle vital chemical elements back to the environment in forms that other organisms can assimilate. And you have already learned that nearly all plants have mycorrhizae, fungus-root associations that help plants absorb minerals and water from the soil. In addition to these ecological roles, fungi have been used by people in various ways for centuries. We eat fungi (mushrooms and extremely expensive truffles, for instance), culture fungi to produce antibiotics and other drugs, add them to dough to make bread rise, culture them in milk to produce a variety of cheeses, and use them to ferment beer and wine.

> Fungi may not always be pretty, but if you find the right one, it could pay your way through college.

Fungi are eukaryotes, and most are multicellular, but many have body structures and modes of reproduction unlike those of any other organism **(Figure 16.22)**. Molecular studies indicate that fungi and animals arose from a common ancestor around 1.5 billion years ago. The oldest undisputed fossils of fungi, however, are only about 460 million years old, perhaps because the ancestors of terrestrial fungi were microscopic and fossilized poorly. Despite appearances, a mushroom is more closely related to you than it is to any plant!

Biologists who study fungi have described more than 100,000 species, but there may be as many as 1.5 million. Classifying fungi is an ongoing area of research. (One widely accepted phylogenetic tree divides the kingdom Fungi into five groups.) You are probably familiar with several kinds of fungi, including mushrooms, mold, and yeast. In this section, we'll discuss the characteristics common to all fungi and then survey their wide-ranging ecological impact. ☑

☑ CHECKPOINT

Name three ways that we benefit from fungi in our environment.

Answer: Fungi help recycle nutrients by decomposing dead organisms; mycorrhizae help plants absorb water and nutrients; some fungi serve us as food.

▼ Figure 16.22 **A gallery of diverse fungi.**

A "fairy ring." Some mushroom-producing fungi poke up "fairy rings," which can appear on a lawn overnight. A ring develops at the edge of the main body of the fungus, which consists of an underground mass of tiny filaments (hyphae) within the ring. As the underground fungal mass grows outward from its center, the diameter of the fairy ring produced at its expanding perimeter increases annually.

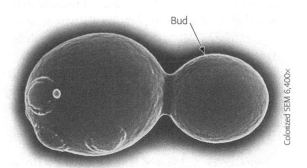

Colonized SEM 6,400×

Budding yeast. Yeasts are unicellular fungi. This yeast cell is reproducing asexually by a process called budding.

Bracket fungi. These are the reproductive structures of a fungus that absorbs nutrients as it decomposes a fallen tree on a forest floor. The green "leafy" material is lichens, which are described in the Evolution Connection section in this chapter.

Colonized SEM 2,800×

Mold. Molds grow rapidly on their food sources, which are often our food sources as well. The mold on this orange reproduces asexually by producing chains of microscopic spores (inset) that are dispersed via air currents.

Corn smut. This parasitic fungus plagues corn growers. To gourmets, however, it is a delicacy known as *huitlacoche*.

Structure/Function Characteristics of Fungi

We'll begin our look at the structure and function of fungi with an overview of how fungi obtain nutrients.

Fungal Nutrition

Fungi are heterotrophs that acquire their nutrients by **absorption**. In this mode of nutrition, small organic molecules are absorbed from the surrounding medium. A fungus digests food outside its body by secreting powerful digestive enzymes into the food. The enzymes decompose complex molecules to simpler compounds that the fungus can absorb. Fungi absorb nutrients from such nonliving organic material as fallen logs, animal corpses, and the wastes of live organisms.

Fungal Structure

The bodies of most fungi are constructed of threadlike filaments called **hyphae** (singular, *hypha*). Fungal hyphae are minute threads of cytoplasm surrounded by a plasma membrane and cell wall. The cell walls of fungi differ from the cellulose walls of plants. Fungal cell walls are usually built mainly of chitin, a strong but flexible polysaccharide that is also found in the external skeletons of insects. Most fungi have multicellular hyphae, which consist of chains of cells separated by cross-walls with pores. In many fungi, cell-to-cell channels allow ribosomes, mitochondria, and even nuclei to flow between cells.

Fungal hyphae branch repeatedly, forming an interwoven network called a **mycelium** (plural, *mycelia*), the feeding structure of the fungus **(Figure 16.23)**. Fungal mycelia usually escape our notice because they are often subterranean, but they can be huge. In fact, scientists have discovered that the mycelium of one humongous fungus in Oregon is 5.5 km—that's 3.4 miles!—in diameter and spreads through 2,200 acres of forest. This fungus is at least 2,600 years old and weighs hundreds of tons, qualifying it as one of Earth's oldest and largest organisms.

A mycelium maximizes contact with its food source by mingling with the organic matter it is decomposing and absorbing. A bucketful of rich organic soil may contain as much as a kilometer of hyphae. A fungal mycelium grows rapidly, adding hyphae as it branches within its food. The great majority of fungi are nonmotile; they cannot run, swim, or fly in search of food. But the mycelium makes up for the lack of mobility by swiftly extending the tips of its hyphae into new territory. ☑

Fungal Reproduction

The mushroom in Figure 16.23 is actually made up of tightly packed hyphae. Mushrooms arise from an underground mycelium. While the mycelium obtains food from organic material via absorption, the function of the mushroom is reproduction. Unlike a truffle, which relies on animals to disperse its spores, a mushroom pops up above ground to disperse its spores on air currents.

Fungi typically reproduce by releasing haploid spores that are produced either sexually or asexually. The output of spores is mind-boggling. For example, puffballs, which are the reproductive structures of certain fungi, can spew clouds containing trillions of spores. Easily carried by wind or water, spores germinate to produce mycelia if they land in a moist place where there is food. Spores thus function in dispersal and account for the wide geographic distribution of many species of fungi. The airborne spores of fungi have been found more than 160 km (100 miles) above Earth. Closer to home, try leaving a slice of bread out for a week and you will observe the furry mycelia that grow from the invisible spores raining down from the surrounding air.

If you've ever had mushroom pizza, you've snacked on the fungi's reproductive structure.

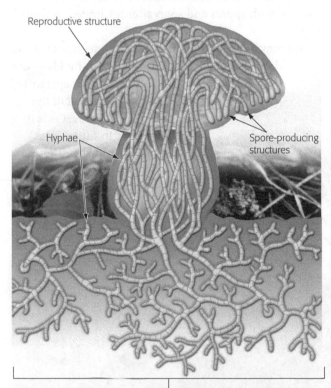

Reproductive structure

Hyphae

Spore-producing structures

Mycelium

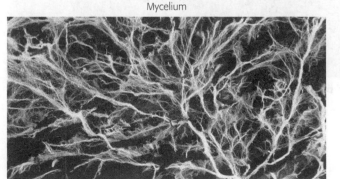

◄ **Figure 16.23 The fungal mycelium.** A mushroom consists of tightly packed hyphae that extend upward from a much more massive mycelium of hyphae growing underground. The photo at the bottom shows a mycelium made up of the cottony threads that decompose organic litter.

☑ CHECKPOINT

Describe how the structure of a fungal mycelium reflects its function.

Answer: The extensive network of hyphae puts a large surface area in contact with the food source.

The Ecological Impact of Fungi

Fungi have been major players in terrestrial communities ever since plants and fungi together moved onto land. Let's examine a few examples of how fungi continue to have an enormous ecological impact, including numerous interactions with people.

Fungi as Decomposers

Fungi and bacteria are the principal decomposers that keep ecosystems stocked with the inorganic nutrients essential for plant growth. Without decomposers, carbon, nitrogen, and other elements would accumulate in nonliving organic matter. Plants and the animals they feed would starve because elements taken from the soil would not be returned.

Fungi are well adapted as decomposers of organic refuse. Their invasive hyphae enter the tissues and cells of dead organisms and digest polymers, including the cellulose of plant cell walls. A succession of fungi, in concert with bacteria and, in some environments, invertebrate animals, is responsible for the complete breakdown of organic litter. The air is so loaded with fungal spores that as soon as a leaf falls or an insect dies, it is covered with spores and soon after infiltrated by fungal hyphae.

We may applaud fungi that decompose forest litter or dung, but it's a different story when molds attack our food or our shower curtains. A significant amount of the world's fruit harvest is lost each year to fungal attack. And a wood-digesting fungus does not distinguish between a fallen oak limb and the oak planks of a boat. During the Revolutionary War, the British lost more ships to fungal rot than to enemy attack. What's more, soldiers stationed in the tropics during World War II watched as their tents, clothing, and boots were destroyed by molds.

Parasitic Fungi

Parasitism is a relationship in which two species live in contact and one organism benefits while the other is harmed. Parasitic fungi absorb nutrients from the cells or body fluids of living hosts. Of the 100,000 known species of fungi, about 30% make their living as parasites.

About 500 species of fungi are known to be parasitic in humans and other animals. A baffling disease known as coccidioidomycosis, or valley fever, causes devastating illness in some people; others suffer only mild flu-like symptoms. People contract the disease when they inhale the spores of a fungus that lives in the soil of the southwestern United States. In recent years, researchers have noticed an uptick in reported cases of valley fever, perhaps due to changes in climate patterns or development of once-rural areas inhabited by the fungus. Less serious fungal diseases include vaginal yeast infections and ringworm, so named because it appears as circular red areas on the skin. The ringworm fungus can infect almost any skin surface, where it produces intense itching and sometimes blisters. One species attacks the feet, causing athlete's foot. Another species is responsible for the misery known as jock itch.

The great majority of fungal parasites infect plants. American chestnut and American elm trees, once common in forests, fields, and city streets, were devastated by fungal epidemics in the 20th century (**Figure 16.24a**). Fungi are also serious agricultural pests, and some of the fungi that attack food crops are toxic to humans. The seed heads of many kinds of grain and grasses, including rye, wheat, and oats, are sometimes infected with fungal growths called ergots (**Figure 16.24b**). Consumption of flour made from ergot-infested grain can cause hallucinations, temporary insanity, and death. In fact, lysergic acid, the raw material from which the hallucinogenic drug LSD is made, has been isolated from ergots. This fact may help explain a centuries-old mystery, as we'll see next.

(a) **An American elm tree killed by Dutch elm disease fungus**

Ergots

(b) **Ergots on rye**

◀ **Figure 16.24 Parasitic fungi that cause plant disease.** (a) The parasitic fungus that causes Dutch elm disease evolved with European species of elm trees, and it is relatively harmless to them. But it has been deadly to American elms since it was accidentally introduced in 1926. (b) Ergots, parasitic fungi, are the dark structures on these rye seed heads.

Did a Fungus Lead to the Salem Witch Hunt?

In January 1692, eight young girls in the town of Salem, Massachusetts, began to act bizarrely. The girls suffered from incomprehensible speech, odd skin sensations, convulsions, and hallucinations. The worried community blamed the girls' symptoms on witchcraft and began to accuse one another. By the time the hysteria ended that autumn, more than 150 villagers had been accused of witchcraft, and 20 of them had been hanged as a result. Finding the cause behind the "Salem witch hunt" has long intrigued historians.

In 1976, a University of California psychology graduate student offered a new explanation. She began with the **observation** that the symptoms reported by the girls were consistent with ergot poisoning (Figure 16.25). This led her to **question** whether an ergot outbreak could have been behind the witch hunt. The researcher tested her **hypothesis** by examining the historical records, making the **prediction** that facts consistent with ergot poisoning would be uncovered.

Her **results** were suggestive, though not conclusive. Agricultural records confirm that rye—the principal host for ergot—grew abundantly around Salem at that time and that the growing season of 1691 had been particularly warm and wet, conditions under which ergot thrives. This suggests that the rye crop consumed during the winter of 1691–1692 could easily have been contaminated. The summer of 1692, when the accusations began to die down, was dry, consistent with an ergot die-off. Most important, the reported symptoms appear consistent with those of ergot poisoning. These clues suggest (but do not prove) that the girls, and perhaps others in Salem, were in the grips of ergot-induced illness. Some historians dispute this idea, and other hypotheses have been proposed. Conclusive evidence may never be found, but this story reinforces the unifying thread of this chapter—the importance of the interaction of plants and fungi—and illustrates how the scientific method can be applied in a wide variety of academic disciplines.

▼ **Figure 16.25 Ergot and the Salem witch hunt.** Ergot poisoning may have been the catalyst for the Salem witch hunt of 1692.

Commercial Uses of Fungi

It would not be fair to fungi to end our discussion with an account of diseases. In addition to their positive global impact as decomposers, fungi also have a number of practical uses for people.

Most of us have eaten mushrooms, although we may not have realized that we were ingesting the reproductive extensions of subterranean fungi. Your grocery store probably stocks Portobello, shiitake, and oyster mushrooms along with common button mushrooms. If you like to cook with mushrooms, you can buy a mushroom "garden"—mycelium embedded in a rich food source—that makes it easy to grow your own. Some enthusiasts gather edible fungi from fields and forests (Figure 16.26), but only experts should dare to eat wild fungi. Some poisonous species resemble edible ones, and there are no simple rules to help the novice distinguish one from the other.

Other fungi are used in food production. The distinctive flavors of certain kinds of cheeses, including Roquefort and blue cheese, come from the fungi used to ripen them. And people have used yeasts (unicellular fungi) for thousands of years to produce alcoholic beverages and cause bread to rise (see Figure 6.15).

▼ **Figure 16.26 Fungi eaten by people.**

Chicken of the woods. This fleshy shelf fungus is said to taste like chicken.

Chanterelle mushrooms. Gourmet mushrooms are highly prized by chefs for their earthy flavors and interesting textures.

Giant puffball. These humongous fungi can grow to more than 2 feet in diameter. They are only edible when immature, before the spores form. Beware: Deadly fungi often resemble small puffballs.

☑ **CHECKPOINT**

1. What is athlete's foot?
2. What do you think is the natural function of the antibiotics that fungi produce in their native environments?

Answers: 1. Athlete's foot is infection of the foot's skin with ringworm fungus. 2. The antibiotics block the growth of microorganisms, especially bacteria, that compete with the fungi for nutrients and other resources.

Fungi are medically valuable as well. Some fungi produce antibiotics that are used to treat bacterial diseases. In fact, the first antibiotic discovered was penicillin, made by the common mold *Penicillium* (Figure 16.27). Researchers are also investigating fungal products that show potential as anticancer drugs.

As sources of medicines and food, as decomposers, and as partners with plants in mycorrhizae, fungi play vital roles in life on Earth. ☑

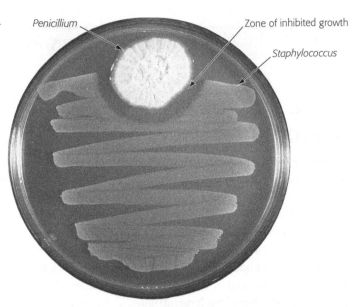

Penicillium — Zone of inhibited growth — Staphylococcus

► Figure 16.27 **Fungal production of an antibiotic.** Penicillin is made by the common mold *Penicillium*. In this petri dish, the clear area between the mold and the growing *Staphylococcus* bacteria is where the antibiotic produced by the *Penicillium* inhibits the growth of the bacteria.

Plant-Fungus Interactions EVOLUTION CONNECTION

A Mutually Beneficial Symbiosis

▼ Figure 16.28 **Lichens: symbiotic associations of fungi and algae.** Lichens generally grow very slowly, sometimes less than a millimeter per year. Some lichens are thousands of years old, rivaling the oldest plants as Earth's elders. The close relationship between the fungal and algal partners is evident in the microscopic blowup of a lichen.

Discussing fungi in the same chapter as plants may seem to indicate that these two kingdoms are close relatives, but as we've discussed, fungi are more closely related to animals than to plants. However, the success of plants on land and the great diversity of fungi are interconnected; neither could have populated the land without the other.

Evolution is not just about the origin and adaptation of individual species. Relationships between species are also an evolutionary product. For example, prokaryotes and protists partner with a variety of organisms in mutually beneficial symbioses (see Figures 15.12 and 15.13). Bacteria living in the roots of certain plants provide nitrogen compounds to their host and receive food in exchange. We even have our own symbiotic bacteria that help keep our skin healthy and produce certain vitamins in our intestines. Particularly relevant to this chapter is the symbiotic association of fungi and plant roots—mycorrhizae—that made life's move onto land possible.

Lichens, symbiotic associations of unicellular algae or photosynthetic bacteria held in a mass of fungal hyphae, are striking examples of how intimate these relationships can be. At a distance, it is easy to mistake lichens for mosses or other simple plants growing on rocks, rotting logs, or trees (Figure 16.28). The partners are so closely entwined that they appear to be a single organism. The fungus receives food from its photosynthetic partner. The fungal mycelium, in turn, provides a suitable habitat for the algae, helping the algae absorb and retain water and minerals. The merger is so complete that lichens are actually named as species, as though they were individual organisms.

After protists and plants, fungi is the third group of eukaryotes that we have surveyed so far. Strong evidence suggests that they evolved from protist ancestors that also gave rise to the fourth and most diverse group of eukaryotes: the animals, the topic of the next chapter.

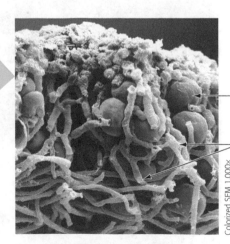

Algal cell

Fungal hyphae

Colorized SEM 1,000×

Chapter Review

SUMMARY OF KEY CONCEPTS

Colonizing Land

Terrestrial Adaptations of Plants

Plants are multicellular photosynthetic eukaryotes with adaptations for living on land.

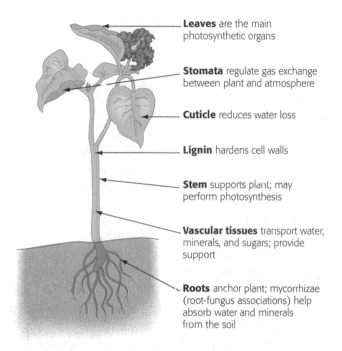

Leaves are the main photosynthetic organs

Stomata regulate gas exchange between plant and atmosphere

Cuticle reduces water loss

Lignin hardens cell walls

Stem supports plant; may perform photosynthesis

Vascular tissues transport water, minerals, and sugars; provide support

Roots anchor plant; mycorrhizae (root-fungus associations) help absorb water and minerals from the soil

The Origin of Plants from Green Algae

Plants evolved from a group of multicellular green algae called charophytes.

Plant Diversity

Highlights of Plant Evolution

Four periods of plant evolution are marked by terrestrial adaptations in major plant groups.

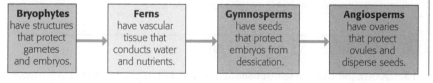

| **Bryophytes** have structures that protect gametes and embryos. | **Ferns** have vascular tissue that conducts water and nutrients. | **Gymnosperms** have seeds that protect embryos from dessication. | **Angiosperms** have ovaries that protect ovules and disperse seeds. |

Bryophytes

 The most familiar bryophytes are mosses. Mosses display two key terrestrial adaptations: a waxy cuticle that prevents dehydration and the retention of developing embryos within the female plant's body. Mosses are most common in moist environments because their sperm must swim to the eggs and because they lack lignin in their cell walls and thus cannot stand tall. Bryophytes are unique among plants in having the gametophyte as the dominant generation in the life cycle.

Ferns

Ferns are seedless plants that have vascular tissues but still use flagellated sperm to fertilize eggs. During the Carboniferous period, giant ferns were among the plants that decayed to thick deposits of organic matter, which were gradually converted to coal.

Gymnosperms

 A drier and colder global climate near the end of the Carboniferous period favored the evolution of the first seed plants. The most successful were the gymnosperms, represented by conifers. Needle-shaped leaves with thick cuticles and sunken stomata are adaptations to dry conditions. Conifers and most other gymnosperms have three additional terrestrial adaptations: (1) further reduction of the haploid gametophyte and greater development of the diploid sporophyte; (2) sperm-bearing pollen, which doesn't require water for transport; and (3) seeds, which consist of a plant embryo packaged along with a food supply inside a protective coat.

Angiosperms

 Angiosperms supply nearly all our food and much of our fiber for textiles. The evolution of the flower and more efficient water transport help account for the success of the angiosperms. The dominant stage is a sporophyte with gametophytes in its flowers. The female gametophyte is located within an ovule, which in turn resides within a chamber of the ovary. Fertilization of an egg in the female gametophyte produces a zygote, which develops into an embryo. The whole ovule develops into a seed. The seed's enclosure within an ovary is what distinguishes angiosperms from gymnosperms, which have naked seeds. A fruit is the ripened ovary of a flower. Fruits protect and help disperse seeds. Angiosperms are a major food source for animals, while animals aid plants in pollination and seed dispersal. Agriculture constitutes a unique kind of evolutionary relationship among plants, people, and other animals.

Plant Diversity as a Nonrenewable Resource

Deforestation to meet the demand of human activities for space and natural resources is causing the extinction of plant species at an unprecedented rate. The problem is especially critical for tropical forests.

Fungi

Characteristics of Fungi

Fungi are unicellular or multicellular eukaryotes; they are heterotrophs that digest their food externally and absorb the nutrients from the environment. They are more closely related to animals than to plants. A fungus usually consists of a mass of threadlike hyphae, forming a mycelium. The cell walls of fungi are mainly composed of chitin. Although most fungi are nonmotile, a mycelium can grow very quickly, extending the tips of its hyphae into new territory. Mushrooms are reproductive structures that extend from the underground mycelium. Fungi reproduce and disperse by releasing spores that are produced either sexually or asexually.

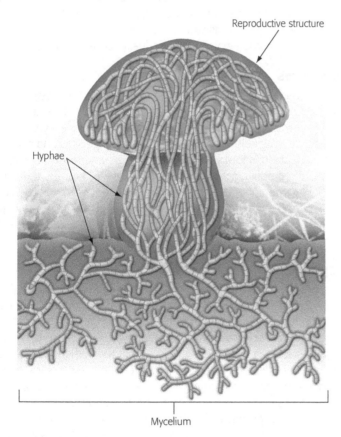

Reproductive structure

Hyphae

Mycelium

The Ecological Impact of Fungi

Fungi and bacteria are the principal decomposers of ecosystems. Many molds destroy fruit, wood, and human-made materials. About 500 species of fungi are known to be parasites of people and other animals. Fungi are also commercially important as food and in baking, beer and wine production, and the manufacture of antibiotics.

MasteringBiology®

For practice quizzes, BioFlix animations, MP3 tutorials, video tutors, and more study tools designed for this textbook, go to MasteringBiology®

SELF-QUIZ

1. Which of the following structures is common to all four major plant groups? vascular tissue, flowers, seeds, cuticle, pollen

2. Angiosperms are distinguished from all other plants because only angiosperms have reproductive structures called _____.

3. Complete the following analogies:
   a. Gametophyte is to haploid as _____ is to diploid.
   b. _____ are to conifers as flowers are to _____.
   c. Ovule is to seed as ovary is to _____.

4. Nutritionally, all fungi are
   a. chemoheterotrophs.
   b. photoautotrophs.
   c. chemoautotrophs.
   d. parasites.

5. Of the following, which is the earliest step in the formation of fossil fuels?
   a. subjecting organic matter to extreme pressure.
   b. fungi and bacteria converting the remains of organisms to inorganic nutrients.
   c. incomplete decomposition of organic matter.
   d. subjecting organic matter to extreme heat.

6. You discover a new species of plant. Under the microscope, you find that it produces flagellated sperm. A genetic analysis shows that its dominant generation has diploid cells. What kind of plant do you have?

7. How does the evergreen nature of pines and other conifers adapt the plants for living where the growing season is very short?

8. Which of the following terms includes all others in the list? angiosperm, fern, vascular plant, gymnosperm, seed plant

9. Plant diversity is greatest in
   a. tropical forests.
   b. the temperate forests of Europe.
   c. deserts.
   d. the oceans.

10. Why are many fruits green when their seeds are immature?

11. Like plants, fungi have _____; however, in plants they are composed of _____, whereas in fungi they are composed of _____.

12. Contrast the heterotrophic nutrition of a fungus with your own heterotrophic nutrition.

Answers to these questions can be found in Appendix: Self-Quiz Answers.

THE PROCESS OF SCIENCE

13. Quorn, the major meat substitute product in the UK, is based on a mycoprotein extracted from the fungus *Fusarium venenatum*. What characteristics of this fungus do you think make it ideal for food production? From an ecological point of view, what may be the advantages of eating a fungal protein over meat? When Quorn was released in the US in 2002,

the Center for Science in the Public Interest (CSPI) had a few concerns about the product. One of those was that Quorn was marketed as a "mushroom product." Why would the CSPI take offense at this labeling? Speculate why the company might have chosen this labeling.

14. **Interpreting Data** Airborne pollen of wind-pollinated plants such as pines, oaks, weeds, and grasses causes seasonal allergy symptoms in many people. As global warming lengthens the growing season for plants, scientists predict longer periods of misery for allergy sufferers. However, global warming does not affect all regions equally (see Figure 18.44). The table below shows the length of the average season in nine locations for ragweed pollen, an allergen that affects millions of people. Calculate the change in length of the pollen season from 1995 to 2009 for each location and graph this information against latitude. Is there a latitudinal trend in the length of ragweed season? You may want to record the data on a map to help you visualize the geographic locations at which samples were taken.

Location	Latitude (°N)	Length of Pollen Season in 1995 (days)	Length of Pollen Season in 2009 (days)	Change in Length of Pollen Season (days)
Average Length of Ragweed Pollen Season in Nine Locations (averages obtained from at least 15 years of data)				
Oklahoma City, OK	35.47	88	89	
Rogers, AR	36.33	64	69	
Papillion, WI	41.15	69	80	
Madison, WI	43.00	64	76	
La Crosse, WI	43.80	58	71	
Minneapolis, MN	45.00	62	78	
Fargo, ND	46.88	36	52	
Winnipeg, MB, Canada	50.07	57	82	
Saskatoon, SK, Canada	52.07	44	71	

Data from: L. Ziska et al., Recent warming by latitude associated with increased length of ragweed pollen season in central North America. *Proceedings of the National Academy of Sciences* 108: 4248–4251 (2011).

BIOLOGY AND SOCIETY

15. Antibiotics are widely used to prevent infections in livestock, and studies have revealed that many physicians readily prescribe antibiotics to patients suffering from the common cold or may even prescribe them in a preventative fashion. What are the possible negative consequences of the excessive usage of antibiotics?

16. As you learned in this chapter, many prescription drugs are derived from natural plant products. Numerous other plant substances, including caffeine and nicotine, have effects in the human body, as well. There is also a wide array of plant products, in the form of pills, powders, or teas, marketed as herbal medicines. Some people prefer taking these "natural" products to pharmaceuticals. Others use herbal supplements to boost energy, promote weight loss, strengthen the immune system, relieve stress, and more. The U.S. Federal Drug Administration, which approves pharmaceuticals, is also responsible for regulating herbal remedies. What does the label "FDA-approved" on an herbal remedy mean? How does that compare to FDA approval of a drug? The FDA website http://www.fda.gov/ForConsumers/default.htm is a good place to start your research. (Note that the FDA classifies herbal remedies as dietary supplements.)

The Evolution of Animals

Why Animal Diversity Matters

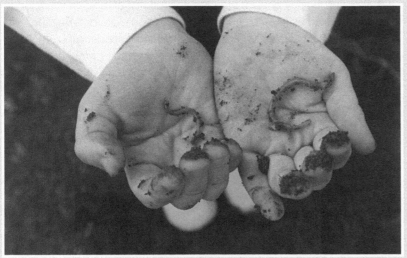

If you want to find nightcrawlers for fishing bait, look in a field of dairy cows.

Evolutionary biologists have discovered the answer to the riddle "Which came first, the chicken or the egg?"

According to recent DNA analysis, many of us have a bit of Neanderthal in our genes.

If all the arthropods on Earth were divided equally among the human population, we would each get about 140 million.

Human Evolution BIOLOGY AND SOCIETY

The Discovery of the Hobbit People

In 2003, Australian anthropologists digging on the Indonesian island of Flores stumbled upon bones of some highly unusual people, including the nearly complete skeleton of an adult female. Next to a modern woman, this fully grown individual would stand about waist-high. Features of her skull, such as the shape and thickness of the bones, were human-like, but its size was proportional to her tiny body, and her brain was the size of a chimpanzee's. Surprisingly, the bones were accompanied by tools for hunting and butchering animals, along with evidence of cooking fires. Most startling of all, the remains dated to roughly 18,000 years ago, a time when scientists had thought that *Homo sapiens* was the only surviving human species. Since the initial discovery, researchers have unearthed the bones of a dozen or so more of these miniature humans.

The discoverers attributed their astonishing find to a previously unknown species, which they named *Homo floresiensis* and nicknamed "hobbits." They speculated that a band of ancestral humans arrived on Flores millions of years ago from Africa and, within the isolated environment of the island, evolved into the diminutive *Homo floresiensis*. There is precedent for animal evolution of this type: Biologists have discovered island-bound dwarf populations of deer, elephants, and hippos. One hypothesis is that a lack of predators favors the evolution of smaller, more energy-efficient forms. Controversy erupted as soon as the research team announced their findings. Skeptical scientists suggested that the bones were from *Homo sapiens* with diseases that cause skeletal malformations. As you'll find out in the Process of Science section, the more scientists learn about the "hobbit" people, the more puzzling their findings become.

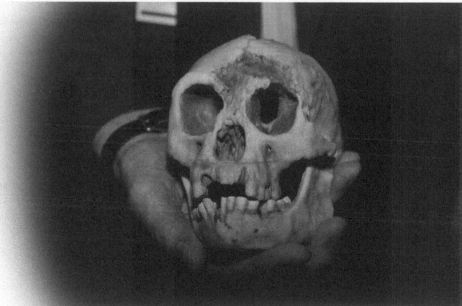

A skull from one of Indonesia's "hobbit people." Scientists are debating whether the skull is from an ancient human-like species.

Homo sapiens and *Homo floresiensis* are just two of the 1.3 million species of animals that have been named and described by biologists. This amazing diversity arose through hundreds of millions of years of evolution as natural selection shaped animal adaptations to Earth's many environments. In this chapter, we'll look at the 9 most abundant and widespread of the roughly 35 phyla (major groups) in the kingdom Animalia. We'll give special attention to the major milestones in animal evolution and conclude by reconnecting with the fascinating subject of human evolution.

The Origins of Animal Diversity

Animal life began in Precambrian seas with the evolution of multicellular creatures that ate other organisms. We are among their descendants.

What Is an Animal?

Animals are eukaryotic, multicellular, heterotrophic organisms that obtain nutrients by eating. This mode of nutrition contrasts animals with plants and other organisms that construct organic molecules through photosynthesis. It also contrasts with fungi, which obtain nutrients by absorption after digesting the food outside their bodies (see Figure 16.23). Most animals digest food within their bodies after ingesting other organisms, dead or alive, whole or by the piece **(Figure 17.1)**.

Animal cells lack the cell walls that provide strong support in the bodies of plants and fungi. And most animals have muscle cells for movement and nerve cells that control the muscles. The most complex animals can use their muscular and nervous systems for many functions other than eating. Some species even use massive networks of nerve cells called brains to think.

Most animals are diploid and reproduce sexually; eggs and sperm are the only haploid cells. The life cycle of a sea star **(Figure 17.2)** includes basic stages found in most animal life cycles. **1** Male and female adult animals make haploid gametes by meiosis, and **2** an egg and a sperm fuse, producing a zygote. **3** The zygote divides by mitosis, forming **4** an early embryonic stage called a **blastula**, which is usually a hollow ball of cells. **5** In most animals, one side of the blastula folds inward, forming a stage called a **gastrula**. **6** The gastrula develops into a saclike embryo with inner, outer, and middle cell layers and an opening at one end. After the gastrula stage, many animals develop directly into adults. Others, such as the sea star, develop into a **larva** **7**, an immature individual that looks different from the adult animal. (A tadpole, for another example, is a larval frog.) **8** The larva undergoes a major change of body form, called **metamorphosis**, in becoming an adult capable of reproducing sexually. ☑

▼ Figure 17.1 **Nutrition by ingestion, the animal way of life.** Few animals ingest a piece of food as large as the gazelle being eaten by this rock python. The snake will spend two weeks or more digesting its meal.

▼ Figure 17.2 **The life cycle of a sea star as an example of animal development.**

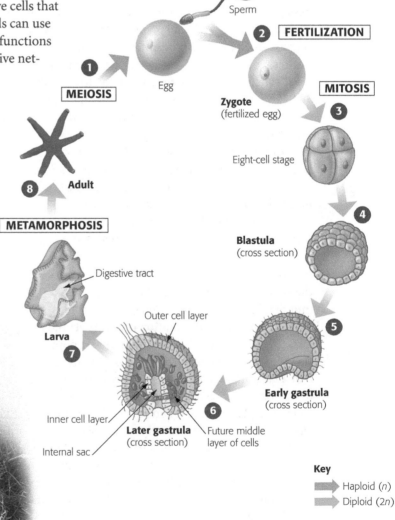

Sperm

2 FERTILIZATION

1

MEIOSIS

Egg

Zygote
(fertilized egg)

MITOSIS

3

Eight-cell stage

4

Blastula
(cross section)

5

Early gastrula
(cross section)

6

Outer cell layer

Future middle layer of cells

Later gastrula
(cross section)

Inner cell layer

Internal sac

Larva

7

Digestive tract

METAMORPHOSIS

8 Adult

Key

Haploid (*n*)

Diploid (*2n*)

Early Animals and the Cambrian Explosion

Scientists hypothesize that animals evolved from a colonial flagellated protist (**Figure 17.3**). Although molecular data point to a much earlier origin, the oldest animal fossils that have been found are 550–575 million years old. Animal evolution must have already been under way for some time prior to that—the fossils reveal a variety of shapes, and sizes range from 1 cm to 1 m in length (**Figure 17.4**).

Animal diversification appears to have accelerated rapidly from 535 to 525 million years ago, during the Cambrian period. Because so many animal body plans and new phyla appear in the fossils from such an evolutionarily short time span, biologists call this episode the Cambrian explosion. The most celebrated source of Cambrian fossils is located in the mountains of British Columbia, Canada. The Burgess Shale, as it is known, provided a cornucopia of perfectly preserved animal fossils. In contrast to the Precambrian animals, many Cambrian animals had hard parts such as shells, and many are clearly related to existing animal groups. For example, scientists have classified more than a third of the species found in the Burgess Shale as arthropods, the group that includes present-day crabs, shrimps, and insects (**Figure 17.5**). Other fossils are more difficult to place. Some are downright weird, like the spiky creature near the center of the drawing, known as *Hallucigenia*, and *Opabinia*, the five-eyed predator grasping a worm with the long, flexible appendage that protrudes in front of its mouth.

What ignited the Cambrian explosion? Scientists have proposed several hypotheses, including increasingly complex predator-prey relationships and an increase in atmospheric oxygen. But whatever the cause of the rapid diversification, it is likely that the set of "master control" genes—the genetic framework for complex bodies—was already in place. Much of the diversity in body form among the animal phyla is associated with variations in where and when these genes are expressed within developing embryos.

In the last half billion years, animal evolution has to a large degree merely generated variations of the animal forms that originated in the Cambrian seas. Continuing research will help test hypotheses about the Cambrian explosion. But even as the explosion becomes less mysterious, it will seem no less wondrous. ☑

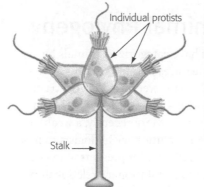

◀ **Figure 17.3 Hypothetical common ancestor of animals.** As you'll learn shortly, the individual cells of this colonial flagellated protist resemble the feeding cells of sponges.

Individual protists

Stalk

☑ **CHECKPOINT**

Why is animal evolution during the early Cambrian referred to as an "explosion"?

Answer: because a great diversity of animals evolved in a relatively short time span

▶ **Figure 17.4 Fossils of Precambrian animals.** All of the oldest animal fossils are impressions of soft-bodied animals. Most of them do not appear to be related to any living group of animals.

Sea pen, possibly related to present-day colonial Cnidarians

Impression of upper surface of *Tribrachidium heraldicum*, which had a hemispheric shape and three-part symmetry unlike any living animal (up to 5 cm across)

▼ **Figure 17.5 A Cambrian seascape.** This drawing is based on fossils from the Burgess Shale. The flat-bodied animals are extinct arthropods called trilobites. A photo of a fossil trilobite is shown at the right.

Evolution Animal Phylogeny

Historically, biologists have categorized animals by "body plan"—general features of body structure. Distinctions between body plans were used to construct phylogenetic trees showing the evolutionary relationships among animal groups. More recently, a wealth of genetic data has allowed evolutionary biologists to modify and refine groups. **Figure 17.6** represents a revised set of hypotheses about the evolutionary relationships among nine major animal phyla based on both structural and genetic similarities.

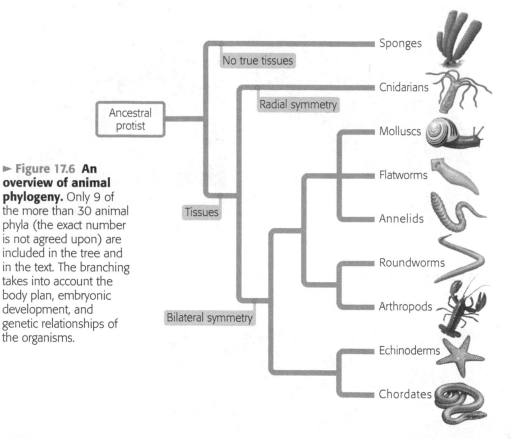

► **Figure 17.6 An overview of animal phylogeny.** Only 9 of the more than 30 animal phyla (the exact number is not agreed upon) are included in the tree and in the text. The branching takes into account the body plan, embryonic development, and genetic relationships of the organisms.

▼ **Figure 17.7 Body symmetry.**

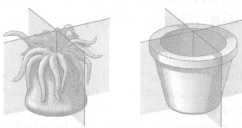

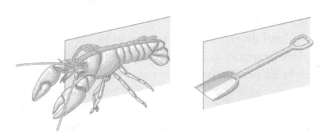

Radial symmetry. Parts radiate from the center, so any slice through the central axis divides into mirror images.

Bilateral symmetry. Only one slice can divide left and right sides into mirror-image halves.

that first encounters food, danger, and other stimuli when traveling. In most bilateral animals, a nerve center in the form of a brain is at the head end, near a concentration of sense organs such as eyes. Thus, bilateral symmetry is an adaptation that aids movement, such as crawling, burrowing, or swimming. Indeed, many radial animals are stationary, whereas most bilateral animals are mobile.

Among bilateral animals, analysis of embryonic development sets the echinoderms and chordates apart from the evolutionary branch that includes molluscs, flatworms, annelids, roundworms, and arthropods. Molluscs, flatworms, and annelids have genetic similarities that are not shared by roundworms and arthropods.

The evolution of body cavities also helped lead to more complex animals. A **body cavity** (**Figure 17.8**, facing page) is a fluid-filled space separating the digestive tract from the outer body wall. The cavity enables the internal organs to grow and move independently of the outer body wall, and the fluid cushions them from injury. In soft-bodied animals such as earthworms, the fluid is under pressure and functions as a hydrostatic skeleton. Of the phyla shown in Figure 17.6, only sponges, cnidarians, and flatworms lack a body cavity.

With the overview of animal evolution in Figure 17.6 as our guide, we're ready to take a closer look at the nine most numerous animal phyla. ✔

A major branch point in animal evolution distinguishes sponges from all other animals based on structural complexity. Unlike more complex animals, sponges lack true tissues, groups of similar cells that perform a function (such as nervous tissue). A second major evolutionary split is based on body symmetry: radial versus bilateral (**Figure 17.7**). Like the flowerpot, the sea anemone has **radial symmetry**, identical all around a central axis. The shovel has **bilateral symmetry**, which means there's only one way to split it into two equal halves—right down the midline. A bilateral animal, such as the lobster in Figure 17.7, has a definite "head end"

✔ CHECKPOINT

1. In the phylogeny shown in Figure 17.6, chordates (the phylum that includes humans) are most closely related to which other animal phylum?
2. A round pizza displays _____ symmetry, whereas a slice of pizza displays _____ symmetry.

Answers: 1. echinoderms 2. radial; bilateral

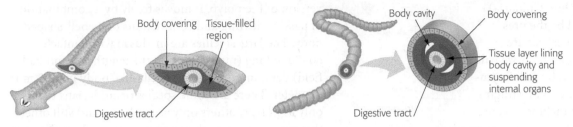

▼ **Figure 17.8 Body plans of bilateral animals.** The various organ systems of these animals develop from the three tissue layers that form in the embryo.

Body covering Tissue-filled region

Digestive tract

Body cavity Body covering

Tissue layer lining body cavity and suspending internal organs

Digestive tract

(a) No body cavity: for example, flatworm

(b) Body cavity: for example, earthworm

Major Invertebrate Phyla

Living on land as we do, our sense of animal diversity is biased in favor of vertebrates, animals with a backbone, such as amphibians, reptiles, and mammals. However, vertebrates make up less than 5% of all animal species. If we were to sample the animals in an aquatic habitat, such as a pond, tide pool, or coral reef, or if we were to consider the millions of insects that share our terrestrial world, we would find ourselves in the realm of **invertebrates**, animals without backbones. We give special attention to the vertebrates simply because we humans are among the backboned ones. However, by exploring the other 95% of the animal kingdom—the invertebrates—we'll discover an astonishing diversity of beautiful creatures that too often escape our notice.

Is drawn through the pores into a central cavity and then flows out of the sponge through a larger opening **(Figure 17.9)**. Flagella on specialized cells called choanocytes sweep water through the sponge's porous body. Tiny nets encircling the flagella trap bacteria and other food particles, which the choanocytes then engulf. Specialized cells called amoebocytes pick up food from the choanocytes, digest it, and carry the nutrients to other cells. Amoebocytes also manufacture the fibers that make up a sponge's skeleton. In some sponges, these fibers are sharp and spur-like, like the ones shown in Figure 17.9. Other sponges have softer, more flexible skeletons; these pliable, honeycombed skeletons are often used as natural sponges in the bath or to wash cars. ☑

Sponges

Sponges (phylum Porifera) are stationary animals that appear so immobile that you might mistake them for plants. The simplest of all animals, sponges probably evolved very early from colonial protists. They have no nerves or muscles, but their individual cells can sense and react to changes in the environment. The cell layers of sponges are loose associations that are not considered true tissues. Sponges range in height from 1 cm (about half an inch) to 2 m (over 6 ft). Although some live in fresh water, the majority of the 5,500 or so species of sponges are marine.

Sponges are examples of suspension feeders, animals that collect food particles from water passed through some type of food-trapping equipment. The body of a sponge resembles a sac perforated with holes. Water

Sponges

Cnidarians

Molluscs

Flatworms

Annelids

Roundworms

Arthropods

Echinoderms

Chordates

▼ **Figure 17.9 Anatomy of a sponge.** To obtain enough food to grow by 3 ounces, a sponge must filter roughly 275 gallons of water through its body, enough to fill 3.5 typical-size bathtubs.

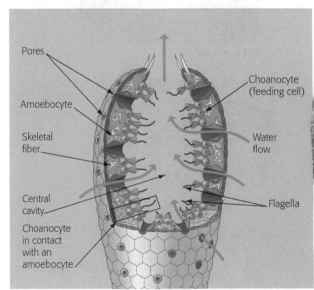

Pores

Amoebocyte

Skeletal fiber

Central cavity

Choanocyte in contact with an amoebocyte

Choanocyte (feeding cell)

Water flow

Flagella

Cnidarians

Cnidarians (phylum Cnidaria) are characterized by the presence of body tissues—as are all the remaining animals we will discuss—as well as by radial symmetry and tentacles with stinging cells. Cnidarians include sea anemones, hydras, corals, and jellies (sometimes called jellyfish, though they are not fish). Most of the 10,000 cnidarian species are marine.

The basic body plan of a cnidarian is a sac with a central digestive compartment, the **gastrovascular cavity**. A single opening to this cavity functions as both mouth and anus. This basic body plan has two variations: the stationary **polyp** and the floating **medusa** (Figure 17.10). Polyps adhere to larger objects and extend their tentacles, waiting for prey. Examples of the polyp body plan are corals, sea anemones, and hydras. A medusa (plural, *medusae*) is a flattened, mouth-down version of the polyp. It moves freely by a combination of passive drifting and contractions of its bell-shaped body. The largest jellies are medusae with tentacles 60–70 m long (more than half the length of a football field) dangling from an umbrella-like body up to 2 m in diameter. There are some species of cnidarians that live only as polyps, others only as medusae, and still others that pass through both a medusa stage and a polyp stage in their life cycle.

Cnidarians are carnivores that use tentacles arranged in a ring around the mouth to capture prey and push the food into the gastrovascular cavity, where digestion begins. The undigested remains are eliminated through the mouth/anus. The tentacles are armed with batteries of cnidocytes ("stinging cells") that function in defense and in the capture of prey (Figure 17.11). The phylum Cnidaria is named for these stinging cells. ☑

▼ **Figure 17.10 Polyp and medusa forms of cnidarians.** Note that cnidarians have two tissue layers, distinguished in the diagrams by blue and yellow. The gastrovascular cavity has only one opening, which functions as both mouth and anus.

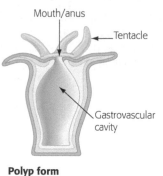
Polyp form

Coral

Sea anemone

Hydra

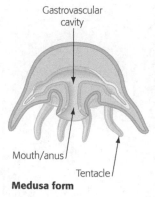
Medusa form

Jelly

▼ **Figure 17.11 Cnidocyte action.** When a trigger on a tentacle is stimulated by touch, a fine thread shoots out from a capsule. Some cnidocyte threads entangle prey, while others puncture the prey and inject a poison.

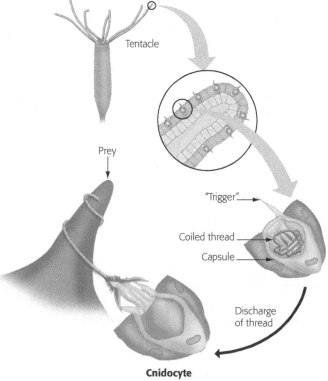
Cnidocyte

Molluscs

Snails and slugs, oysters and clams, and octopuses and squids are all **molluscs** (phylum Mollusca). Molluscs are soft-bodied animals, but most are protected by a hard shell. Many molluscs feed by extending a file-like organ called a **radula** to scrape up food. For example, the radula of some aquatic snails slides back and forth like a backhoe, scraping and scooping algae off rocks. You can observe a radula in action by watching a snail graze on the glass wall of an aquarium. In cone snails, a group of predatory marine molluscs, the radula is modified to inject venom into prey. The sting of some cone snails is painful, or even fatal, to people.

There are 100,000 known species of molluscs, with most being marine animals. All molluscs have a similar body plan (Figure 17.12). The body has three main parts: a muscular foot, usually used for movement; a visceral mass containing most of the internal organs; and a fold of tissue called the mantle. The **mantle** drapes over the visceral mass and secretes the shell if one is present. The three major groups of molluscs are gastropods, bivalves, and cephalopods (Figure 17.13).

Most **gastropods**, including snails, are protected by a single spiraled shell into which the animal can retreat when threatened. Slugs and sea slugs lack shells. Many gastropods have a distinct head with eyes at the tips of tentacles (think of a garden snail). Marine, freshwater, and terrestrial gastropods make up about three-quarters of the living mollusc species.

The **bivalves**, including clams, oysters, mussels, and scallops, have shells divided into two halves hinged together. None of the bivalves have a radula. There are both marine and freshwater species, with most being sedentary, using their muscular foot for digging and anchoring in sand or mud.

Cephalopods are all marine animals and generally differ from gastropods and sedentary bivalves in that their bodies are fast and agile. A few have large, heavy shells, but in most the shell is small and internal (as in squids) or missing (as in octopuses). Cephalopods have large brains and sophisticated sense organs, which contribute to their success as mobile predators. They use beak-like jaws and a radula to crush or rip prey apart. The mouth is at the base of the foot, which is drawn out into several long tentacles for catching and holding prey. The colossal squid, discovered in the ocean depths near Antarctica, is the largest living invertebrate. Scientists estimate that this massive cephalopod grows to an average length of 13 m—as long as a school bus. ☑

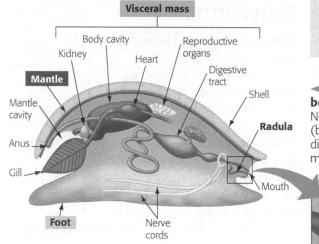

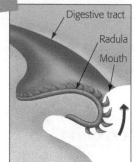

◀ **Figure 17.12 The general body plan of a mollusc.** Note the small body cavity (brown) and the complete digestive tract, with both mouth and anus (pink).

☑ **CHECKPOINT**

Classify these molluscs: A garden snail is an example of a _____; a clam is an example of a _____; a squid is an example of a _____.

Answer: gastropod; bivalve; cephalopod

▼ **Figure 17.13 Mollusc diversity.**

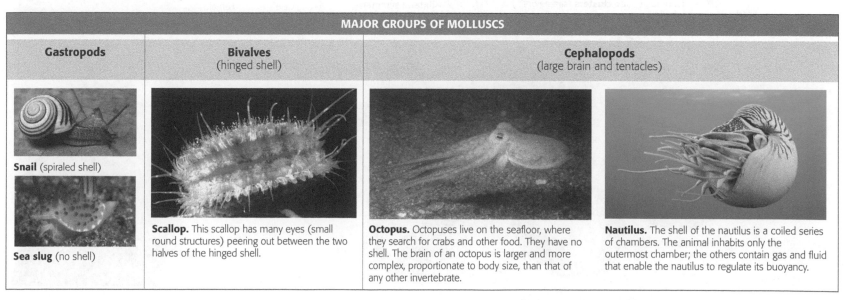

MAJOR GROUPS OF MOLLUSCS

Gastropods	Bivalves (hinged shell)	Cephalopods (large brain and tentacles)	

Snail (spiraled shell)

Sea slug (no shell)

Scallop. This scallop has many eyes (small round structures) peering out between the two halves of the hinged shell.

Octopus. Octopuses live on the seafloor, where they search for crabs and other food. They have no shell. The brain of an octopus is larger and more complex, proportionate to body size, than that of any other invertebrate.

Nautilus. The shell of the nautilus is a coiled series of chambers. The animal inhabits only the outermost chamber; the others contain gas and fluid that enable the nautilus to regulate its buoyancy.

Flatworms

Flatworms (phylum Platyhelminthes) are the simplest animals with bilateral symmetry. True to their name, these worms are ribbonlike and range from about 1 mm to about 20 m (about 65 ft) in length. Most flatworms have a gastrovascular cavity with a single opening. There are about 20,000 species of flatworms living in marine, freshwater, and damp terrestrial habitats **(Figure 17.14)**.

The gastrovascular cavity of free-living flatworms called planarians is highly branched, providing an extensive surface area for the absorption of nutrients. When the animal feeds, a muscular tube projects through the mouth and sucks food in. Planarians live on the undersurfaces of rocks in freshwater ponds and streams.

Parasitic flatworms include blood flukes called schistosomes, which are a major health problem in the tropics. These worms have suckers that attach to the inside of the blood vessels near the human host's intestines. Infection by these flatworms causes a long-lasting disease called schistosomiasis, with such symptoms as severe abdominal pain, anemia, and dysentery. Although schistosomes are not found in the United States, more than 200 million people around the world are infected by these parasites each year.

Tapeworms parasitize many vertebrates, including people. Most tapeworms have a very long, ribbonlike body with repeated parts. There is no mouth and no gastrovascular cavity. The head of a tapeworm is equipped with suckers and hooks that lock the worm to the intestinal lining of the host. Bathed in partially digested food in the intestines of its host, the tapeworm simply absorbs nutrients across its body surface. Behind the head is a long ribbon of units that are little more than sacs of sex organs. At the back of the worm, mature units containing thousands of eggs break off and leave the host's body with the feces. People can become infected with tapeworms by eating undercooked beef, pork, or fish containing tapeworm larvae. The larvae are microscopic, but the adults can reach lengths of 2 m (6.5 ft) in the intestine. Such large tapeworms can cause intestinal blockage and rob enough nutrients from the host to cause nutritional deficiencies. Fortunately, an orally administered drug can kill the adult worms. ☑

☑ **CHECKPOINT**

Flatworms are the simplest animals to display a body plan that is _____.

Answer: bilaterally symmetric

▼ **Figure 17.14 Flatworm diversity.**

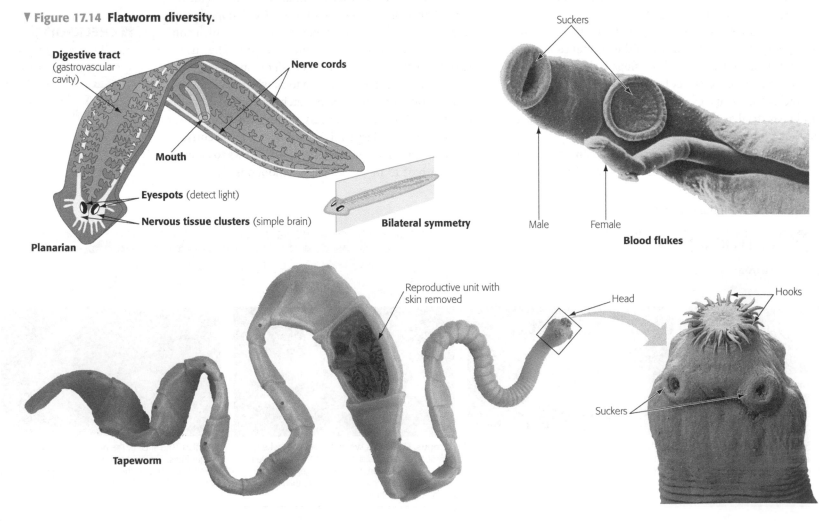

Digestive tract (gastrovascular cavity)

Nerve cords

Mouth

Eyespots (detect light)

Nervous tissue clusters (simple brain)

Planarian

Bilateral symmetry

Suckers

Male   Female

Blood flukes

Reproductive unit with skin removed

Head

Hooks

Suckers

Tapeworm

Annelids

Annelids (phylum Annelida) are worms that have **body segmentation**, which is the subdivision of the body along its length into a series of repeated parts called segments. In annelids, the segments look like a set of fused rings. There are about 16,500 annelid species, ranging in length from less than 1 mm to the giant Australian earthworm, which can grow up to 3 m (nearly 10 ft) long. Annelids live in damp soil, the sea, and most freshwater habitats. There are three main groups: earthworms, polychaetes, and leeches **(Figure 17.15)**.

Annelids exhibit two characteristics shared by all other bilateral animals except flatworms. One is a **complete digestive tract**, which is a digestive tube with two openings: a mouth and an anus. A complete digestive tract can process food and absorb nutrients as a meal moves in one direction from one specialized digestive organ to the next. In people, for example, the mouth, stomach, and intestines act as digestive organs. A second characteristic is a body cavity (see Figure 17.8b).

Earthworms, like all annelids, are segmented both externally and internally **(Figure 17.16)**. The body cavity is partitioned by walls (only two segment walls are fully shown here). Many of the internal structures, such as the nervous system (yellow in the figure) and organs that dispose of fluid wastes (green) are repeated in each segment. Segmental blood vessels include one main heart and five pairs of accessory hearts. The digestive tract, however, is not segmented; it passes through the segment walls from the mouth to the anus.

Earthworms eat their way through the soil, extracting nutrients from organic matter as the soil passes through the digestive tract. Undigested material is eliminated as castings through the anus. Farmers and gardeners value earthworms because the animals aerate the soil and increase the amount of mineral nutrients available to plants. Because crops provide the earthworms' food, land use and farming practices have a huge effect on earthworm populations. In one study, for example, corn fields that were plowed yearly averaged 39,000 earthworms per acre, while populations in unplowed dairy pastures numbered at least 1,333,000 million per acre.

If you want to find nightcrawlers for fishing bait, look in a field of dairy cows.

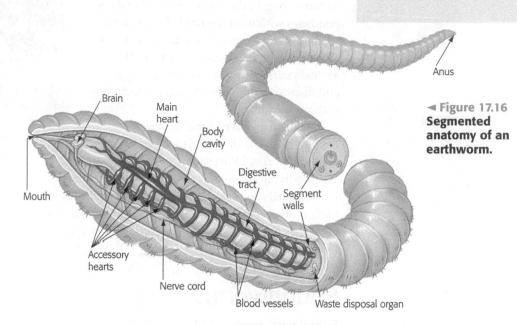

◄ **Figure 17.16** **Segmented anatomy of an earthworm.**

Brain · Main heart · Body cavity · Anus · Digestive tract · Segment walls · Mouth · Accessory hearts · Nerve cord · Blood vessels · Waste disposal organ

▼ **Figure 17.15** **Annelid diversity.**

MAJOR GROUPS OF ANNELIDS

Earthworms

This giant Australian earthworm is bigger than many snakes. Perhaps you've slipped on slimy worms, but imagine actually tripping over one!

Polychaetes

Polychaetes have segmental appendages that function in movement and as gills. This species is called a Christmas tree worm. The feathery, spiral structures are a pair of gills from a single worm.

Leeches

The parasitic fish leech has a sucker on each end of its body. It attaches one sucker to a rock or plant. It then extends its body to wave freely in the water, using the other sucker to latch on to a passing host.

☑ **CHECKPOINT**

The body plan of an
annelid displays _____,
meaning that the body
is divided into a series of
repeated regions.

Answer: segmentation

In contrast to earthworms, most **polychaetes** are marine, mainly crawling on or burrowing in the seafloor. Segmental appendages with hard bristles help the worm wriggle about in search of small invertebrates to eat. The appendages also increase the animal's surface area for taking up oxygen and disposing of metabolic wastes, including carbon dioxide.

The third group of annelids, **leeches**, are notorious for the bloodsucking habits of some species. However, most species are free-living carnivores that eat small invertebrates such as snails and insects. A few terrestrial species inhabit moist vegetation in the tropics, but the majority of leeches live in fresh water. A European freshwater

Medicinal leech

species called *Hirudo medicinalis* is used for the treatment of circulatory complications. Most commonly, leeches are applied after reconstructive microsurgery in which limbs or digits are reattached. Because arteries (which transport blood into a reattached area) are easier to reconnect than veins (which transport blood out), blood can pool in the reattached area and stagnate, starving the healing tissue of oxygen. Medicinal leeches have razor-like jaws with hundreds of tiny teeth that cut through the skin. They secrete saliva containing an anesthetic and an anticoagulant into the wound. The anesthetic makes the bite virtually painless, and the anticoagulant prevents clotting as the leech drains excess blood from the wound. ☑

Roundworms

Roundworms (also called **nematodes**, members of the phylum Nematoda) get their common name from their cylindrical body, which is usually tapered at both ends (**Figure 17.17**). Roundworms are among the most numerous and widespread of all animals. About 25,000 species of roundworms are known, and perhaps ten times that number actually exist. Roundworms range in length from about 1 mm to 1 m. They

Sponges
Cnidarians
Molluscs
Flatworms
Annelids
Roundworms
Arthropods
Echinoderms
Chordates

are found in most aquatic habitats, in wet soil, and as parasites in the body fluids and tissues of plants and animals.

Free-living roundworms are important decomposers. They live virtually everywhere there is decaying organic matter, and their numbers are huge. Ninety thousand nematodes have been found in a single rotting apple. Recently, researchers even found nematodes living 2 miles underground, where they survive by grazing on microbes. Other species of nematodes thrive as parasites in plants and animals. Some are major agricultural pests that attack the roots of plants. At least 50 parasitic roundworm species infect people; they include pinworms, hookworms, and the parasite that causes trichinosis. ☑

☑ **CHECKPOINT**

Which phylum is most
closely related to the
roundworms? (*Hint*: Refer to
the phylogenetic tree.)

Answer: arthropods

▼ **Figure 17.17 Roundworm diversity.**

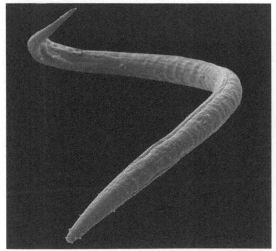

(a) A free-living roundworm. This species has the classic roundworm shape: cylindrical with tapered ends. The ridges indicate muscles that run the length of the body.

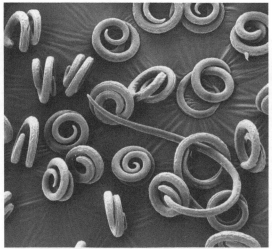

(b) Parasitic roundworms in pork. The potentially fatal disease trichinosis is caused by eating undercooked pork infected with *Trichinella* roundworms. The worms (shown here in pork tissue) burrow into a person's intestine and then invade muscle tissue.

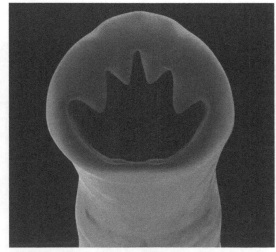

(c) Head of hookworm. Hookworms sink their hooks into the wall of the host's small intestine and feed on blood. Although the worms are small (less than 1 cm), a severe infestation can cause serious anemia.

Arthropods

Arthropods (phylum Arthropoda) are named for their jointed appendages. Crustaceans (such as crabs and lobsters), arachnids (such as spiders and scorpions), and insects (such as grasshoppers and moths) are examples of arthropods (**Figure 17.18**). Zoologists estimate that the total arthropod population numbers about a billion billion (10^{18}) individuals. Researchers have identified more than a million arthropod species, mostly insects. In fact, two out of every three species of life that have been scientifically described are arthropods. And arthropods are represented in nearly all habitats of the biosphere. In species diversity, distribution, and sheer numbers, arthropods must be regarded as the most successful animal phylum.

Sponges
Cnidarians
Molluscs
Flatworms
Annelids
Roundworms
Arthropods
Echinoderms
Chordates

If all the arthropods on Earth were divided equally among the human population, we would each get about 140 million.

◀ **Figure 17.18 Arthropod diversity.**

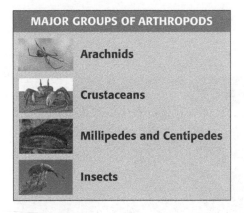

MAJOR GROUPS OF ARTHROPODS

Arachnids

Crustaceans

Millipedes and Centipedes

Insects

General Characteristics of Arthropods

Arthropods are segmented animals. In contrast with the repeating similar segments of annelids, however, arthropod segments and their appendages have become specialized for a great variety of functions. This evolutionary flexibility contributed to the great diversification of arthropods. Specialization of segments (or of fused groups of segments) provides for an efficient division of labor among body regions. For example, the appendages of different segments may be adapted for walking, feeding, sensory reception, swimming, or defense (**Figure 17.19**).

The body of an arthropod is completely covered by an **exoskeleton**, an external skeleton. This coat is constructed from layers of protein and a tough polysaccharide called chitin. The exoskeleton can be a thick, hard armor over some parts of the body (such as the head), yet be paper-thin and flexible in other locations (such as the joints). The exoskeleton protects the animal and provides points of attachment for the muscles that move the appendages. There are, of course, advantages to wearing hard parts on the outside. Our own skeleton is interior to most of our soft tissues, an arrangement that doesn't provide much protection from injury. But our skeleton does offer the advantage of being able to grow along with the rest of our body. In contrast, a growing arthropod must occasionally shed its old exoskeleton and secrete a larger one. This process, called molting, leaves the animal temporarily vulnerable to predators and other dangers. The next five pages explore the major groups of arthropods. ☑

☑ **CHECKPOINT**

Seafood lovers look forward to soft-shell crab season, when almost the entire animal can be eaten without cracking a hard shell to get at the meat. An individual crab is only a "soft-shell" for a matter of hours; harvesters therefore capture crabs and keep them in holding tanks until they are ready. What process makes soft-shell crabs possible?

Answer: Crabs have a thick exoskeleton that must be shed (molted) to allow the crab's body to grow. Soft-shell crabs have molted but have not yet secreted a new hard shell.

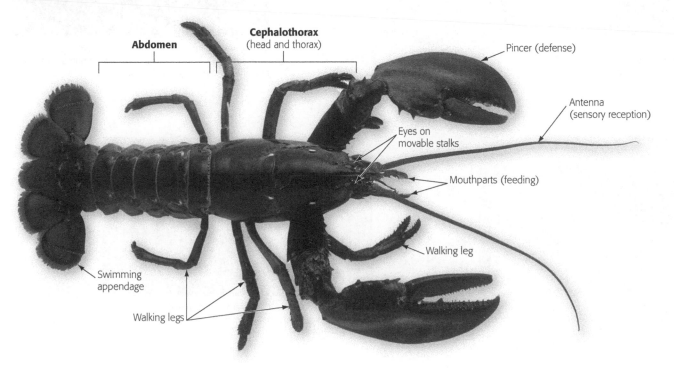

Abdomen

Cephalothorax
(head and thorax)

Pincer (defense)

Antenna
(sensory reception)

Eyes on movable stalks

Mouthparts (feeding)

Walking leg

Swimming appendage

Walking legs

◀ **Figure 17.19 Anatomy of a lobster, a crustacean.** The whole body, including the appendages, is covered by an exoskeleton. The body is segmented, but this characteristic is obvious only in the abdomen.

Arachnids

Arachnids include scorpions, spiders, ticks, and mites (Figure 17.20). Most arachnids live on land. Members of this arthropod group usually have four pairs of walking legs and a specialized pair of feeding appendages. In spiders, these feeding appendages are fang-like and equipped with poison glands. As a spider uses these appendages to immobilize and dismantle its prey, it spills digestive juices onto the torn tissues and sucks up its liquid meal.

▼ Figure 17.20 **Arachnid characteristics and diversity.**

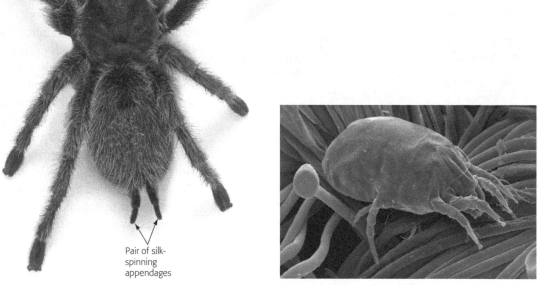

Pair of feeding appendages

Leg (four pairs)

Pair of silk-spinning appendages

Scorpion. Scorpions have a pair of large pincers that function in defense and food capture. The tip of the tail bears a poisonous stinger. Scorpions sting people only when prodded or stepped on.

Dust mite. This microscopic house dust mite is a ubiquitous scavenger in our homes. Dust mites are harmless except to people who are allergic to the mites' feces.

Spider. Like most spiders, including the tarantula in the large photo above, this black widow spins a web of liquid silk, which solidifies as it comes out of specialized glands. A black widow's venom can kill small prey but is rarely fatal to humans.

Wood tick. Wood ticks and other species carry bacteria that cause Rocky Mountain spotted fever. Lyme disease is carried by several different species of ticks.

Crustaceans

Crustaceans, which are nearly all aquatic, include delectable species of crabs, lobsters, crayfish, and shrimps (Figure 17.21). Barnacles, which anchor themselves to rocks, boat hulls, and even whales, are also crustaceans. One group of crustaceans, the isopods, is represented on land by pill bugs. All of these animals exhibit the arthropod characteristic of multiple pairs of specialized appendages.

▼ **Figure 17.21 Crustacean characteristics and diversity.**

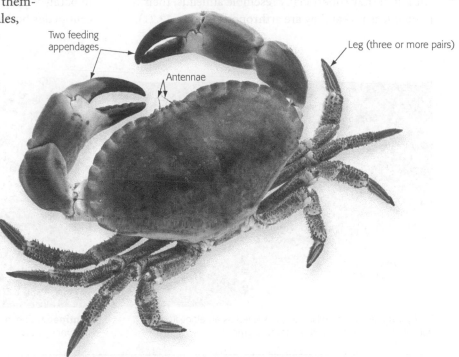

Two feeding appendages

Antennae

Leg (three or more pairs)

Crab. Ghost crabs are common along shorelines throughout the world. They scurry along the surf's edge and then quickly bury themselves in sand.

Shrimp. Naturally found in Pacific waters from Africa to Asia, giant prawns are widely cultivated as food.

Pill bug. Commonly found in moist locations with decaying leaves, such as under logs, pill bugs get their name from their tendency to roll up into a tight ball when they sense danger.

Crayfish. The red swamp crayfish, cultivated worldwide for food, is native to the southeastern United States. When released outside its native range, it competes aggressively with other species, endangering the local ecosystem.

Barnacles. Barnacles are stationary crustaceans with exoskeletons hardened into shells by calcium carbonate (lime). The jointed appendages projecting from the shell capture small plankton.

Millipedes and Centipedes

Millipedes and **centipedes** are terrestrial arthropods that have similar segments over most of the body. Although they superficially resemble annelids, their jointed legs reveal they are arthropods (**Figure 17.22**).

Millipedes eat decaying plant matter. They have two pairs of short legs per body segment. Centipedes are carnivores, with a pair of poison claws used in defense and to paralyze prey, such as cockroaches and flies. Each of a centipede's body segments bears a single pair of legs.

► Figure 17.22 **Millipedes and centipedes.**

Two pairs of legs per segment

One pair of legs per segment

Millipede. Like most millipedes, this one has an elongated body with two pairs of legs per trunk segment.

Centipede. Centipedes can be found in dirt and leaf litter. Their venomous claws can harm cockroaches and spiders but not people.

Insect Anatomy

Like the grasshopper in **Figure 17.23**, most **insects** have a three-part body: head, thorax, and abdomen. The head usually bears a pair of sensory antennae and a pair of eyes. The mouthparts of insects are adapted for particular kinds of eating—for example, for biting and chewing plant material in grasshoppers, for lapping up fluids in houseflies, and for piercing skin and sucking blood in mosquitoes. Most adult insects have three pairs of legs and one or two pairs of wings, all extending from the thorax.

Flight is obviously one key to the great success of insects. An animal that can fly can escape many predators, more readily find food and mates, and disperse to new habitats much faster than an animal that must crawl on the ground. Because their wings are extensions of the exoskeleton and not true appendages, insects can fly without sacrificing legs. By contrast, the flying vertebrates—birds and bats—have one of their two pairs of legs modified for wings, which explains why these vertebrates are generally not very swift on the ground.

▼ Figure 17.23 **Anatomy of a grasshopper.**

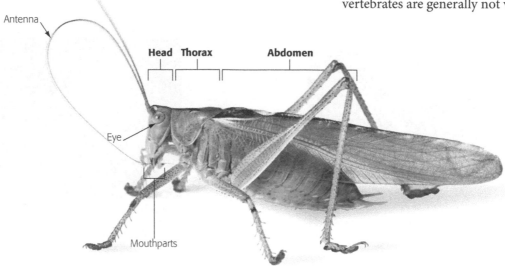

Antenna

Head Thorax Abdomen

Eye

Mouthparts

Insect Diversity

In species diversity, insects outnumber all other forms of life combined (Figure 17.24). They live in almost every terrestrial habitat and in fresh water, and flying insects fill the air. Insects are rare in the seas, where crustaceans are the dominant arthropods. The oldest insect fossils date back to about 400 million years ago. Later, the evolution of flight sparked an explosion in insect variety.

▼ Figure 17.24 **Insect diversity.**

Rhinoceros beetle. Only males have a "horn," which is used for fighting other males and for digging.

Robber fly. This predatory insect injects prey with enzymes that liquefy the tissues, and then suck in the resulting fluid.

Candy-striped leafhopper. Leafhoppers are tiny insects—about the width of the nail on your little finger—that can jump 40 times their body length.

Praying mantis. There are more then 2,000 species of praying mantises throughout the world.

Greater arid-land katydid. Often called the red-eyed devil, this scary-looking predator displays its wings and spiny legs when threatened.

Weevil. Light reflected from varying thicknesses of chitin produces the brilliant colors of this weevil.

Buckeye butterfly. The eyespots on the wings of this butterfly may startle predators.

Blue dasher dragonfly. The huge multifaceted eyes of dragonflies wrap around the head for nearly 360° vision. Each of the four wings works independently, giving dragonflies excellent maneuverability as they pursue prey.

Many insects undergo metamorphosis in their development. In the case of grasshoppers and some other insect groups, the young resemble adults but are smaller and have different body proportions. The animal goes through a series of molts, each time looking more like an adult, until it reaches full size. In other cases, insects have distinctive larval stages specialized for eating and growing that are known by such names as maggots (fly larvae), grubs (beetle larvae), or caterpillars (larvae of moths and butterflies). The larval stage looks entirely different from the adult stage, which is specialized for dispersal and reproduction. Metamorphosis from the larva to the adult occurs during a pupal stage **(Figure 17.25)**.

Forensic entomologists (forensic scientists who study insects) use their knowledge of insect life cycles to help solve criminal cases. For example, blowfly maggots feed on decaying flesh. Female blowflies, which can smell a dead body from up to a mile away, typically arrive and lay their eggs in a fresh corpse within minutes. By knowing the length of each stage in the life cycle of a blowfly, an entomologist can determine how much time has passed since death occurred.

Animals so numerous, diverse, and widespread as insects are bound to affect the lives of all other terrestrial organisms, including people, in many ways. On the one hand, we depend on bees, flies, and other insects to pollinate our crops and orchards. On the other hand, insects are carriers of the microbes that cause many human diseases, such as malaria and West Nile disease. Insects also compete with people for food by eating our field crops. In an effort to minimize their losses, farmers in the United States spend billions of dollars each year on pesticides, spraying crops with massive doses of insecticide poisons. But try as they might, not even humans have significantly challenged the preeminence of insects and their arthropod kin. Rather, the evolution of pesticide resistance has caused humans to change their pest-control methods (see Figure 13.13). ☑

▼ **Figure 17.25 Metamorphosis of a monarch butterfly.**

The **larva (caterpillar)** spends its time eating and growing, molting as it grows.

After several molts, the larva becomes a **pupa** encased in a cocoon.

Within the pupa, the larval organs break down and adult organs develop from cells that were dormant in the larva.

Finally, the **adult** emerges from the cocoon.

The butterfly flies off and reproduces, nourished mainly by calories stored when it was a caterpillar.

Echinoderms

The **echinoderms** (phylum Echinodermata) are named for their spiny surfaces (*echin* is Greek for "spiny"). Among the echinoderms are sea stars, sea urchins, sea cucumbers, and sand dollars (**Figure 17.26**).

Echinoderms include about 7,000 species, all of them marine. Most move slowly, if at all. Echinoderms lack body segments, and most have radial symmetry as adults. Both the external and the internal parts of a sea star, for instance, radiate from the center like the spokes of a wheel. In contrast to the adult, the larval stage of echinoderms is bilaterally symmetrical. This supports other evidence that echinoderms are not closely related to other radial animals, such as cnidarians, that never show bilateral symmetry. Most echinoderms have an **endoskeleton** (interior skeleton) constructed from hard plates just beneath the skin. Bumps and spines of this endoskeleton account for the animal's rough or prickly surface. Unique to echinoderms is the **water vascular system**, a network of water-filled canals that circulate water throughout the echinoderm's body, facilitating gas exchange (the entry of O_2 and the removal of CO_2) and waste disposal. The water vascular system also branches into extensions called tube feet. A sea star or sea urchin pulls itself slowly over the seafloor using its suction-cup-like tube feet. Sea stars also use their tube feet to grip prey during feeding.

Looking at sea stars and other adult echinoderms, you may think they have little in common with humans and other vertebrates. But as shown by the phylogenetic tree at the left, echinoderms share an evolutionary branch with chordates, the phylum that includes vertebrates. Analysis of embryonic development can differentiate the echinoderms and chordates from the evolutionary branch that includes molluscs, flatworms, annelids, roundworms, and arthropods. With this context in mind, we're now ready to make the transition in our discussion from invertebrates to vertebrates. ☑

*Phylogenetic tree showing: Sponges, Cnidarians, Molluscs, Flatworms, Annelids, Roundworms, Arthropods, **Echinoderms**, Chordates*

☑ **CHECKPOINT**

Contrast the skeleton of an echinoderm with that of an arthropod.

Answer: An echinoderm has an endoskeleton; an arthropod has an exoskeleton.

▼ **Figure 17.26 Echinoderm diversity.**

Sea star. When a sea star encounters an oyster or clam, it grips the mollusc's shell with its tube feet (see inset) and positions its mouth next to the narrow opening between the two halves of the prey's shell. The sea star then pushes its stomach out through its mouth and the crack in the mollusc's shell.

Sea urchin. In contrast to sea stars, sea urchins are spherical and have no arms. If you look closely, you can see the long tube feet projecting among the spines. Unlike sea stars, which are mostly carnivorous, sea urchins mainly graze on seaweed and other algae.

Tube feet

Sea cucumber. On casual inspection, this sea cucumber does not look much like other echinoderms. However, a closer look would reveal many echinoderm traits, including five rows of tube feet.

Sand dollar. Live sand dollars have a skin of short, movable spines covering a rigid skeleton. A set of five pores (arranged in a star pattern) allows seawater to be drawn into the sand dollar's body.

Vertebrate Evolution and Diversity

Most of us are curious about our family ancestry. Biologists are also interested in the larger question of tracing human ancestry within the animal kingdom. In this section, we trace the evolution of the vertebrates, the group that includes humans and their closest relatives. All vertebrates have endoskeletons, a characteristic shared with most echinoderms. However, vertebrate endoskeletons are unique in having a skull and a backbone, a series of bones called vertebrae (singular, *vertebra*), for which the group is named **(Figure 17.27)**. Our first step in tracing the vertebrate lineage is to determine where vertebrates fit in the animal kingdom.

▼ **Figure 17.27 A vertebrate endoskeleton.** This snake skeleton, like those of all vertebrates, has a skull and a backbone consisting of vertebrae.

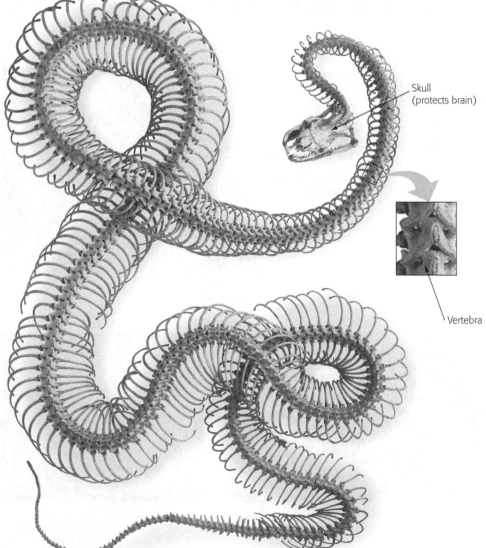

Skull
(protects brain)

Vertebra

Characteristics of Chordates

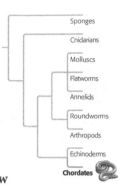

Sponges
Cnidarians
Molluscs
Flatworms
Annelids
Roundworms
Arthropods
Echinoderms
Chordates

The last phylum in our survey of the animal kingdom is the phylum Chordata. **Chordates** share four key features that appear in the embryo and sometimes in the adult **(Figure 17.28)**. These four chordate characteristics are (1) a **dorsal, hollow nerve cord**, (2) a **notochord**, which is a flexible, longitudinal rod located between the digestive tract and the nerve cord, (3) **pharyngeal slits**, which are grooves in the pharynx, the region of the digestive tube just behind the mouth, and (4) a **post-anal tail**, which is a tail to the rear of the anus. Though these chordate characteristics are often difficult to recognize in the adult animal, they are always present in chordate embryos. For example, the notochord, for which our phylum is named, persists in adult humans only in the form of the cartilage disks that function as cushions between the vertebrae. Back injuries described as "ruptured disks" or "slipped disks" refer to these notochord remnants.

Body segmentation is another chordate characteristic. Chordate segmentation is apparent in the backbone of vertebrates (see Figure 17.27) and is also evident in the segmental muscles of all chordates (see the chevron-shaped—<<<<—muscles in the lancelet in Figure 17.29). Segmental musculature is not so obvious in adult humans unless one is motivated to sculpt "washboard abs."

Two groups of chordates, **tunicates** and **lancelets** **(Figure 17.29)**, are invertebrates. All other chordates

▼ **Figure 17.28 Chordate characteristics.**

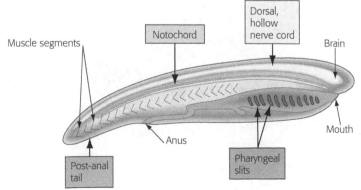

Muscle segments

Notochord

Dorsal, hollow nerve cord

Brain

Mouth

Pharyngeal slits

Anus

Post-anal tail

are **vertebrates**, which retain the basic chordate characteristics but have additional features that are unique—including, of course, the backbone. **Figure 17.30** is an overview of chordate and vertebrate evolution that will provide a context for our survey. ☑

▼ **Figure 17.29 Invertebrate chordates.**

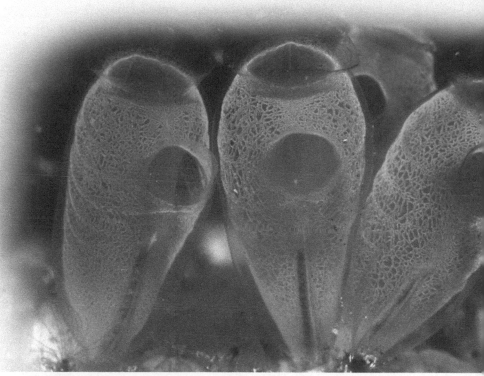

Lancelet. This marine invertebrate owes its name to its bladelike shape. Only a few centimeters long, lancelets wiggle backward into the gravel, leaving their mouth exposed, and filter tiny food particles from the seawater.

Tunicates. The tunicate, or sea squirt, is a stationary animal that filters food from the water. These pastel sea squirts get their nickname from their coloration and the fact that they can quickly expel water to startle intruders.

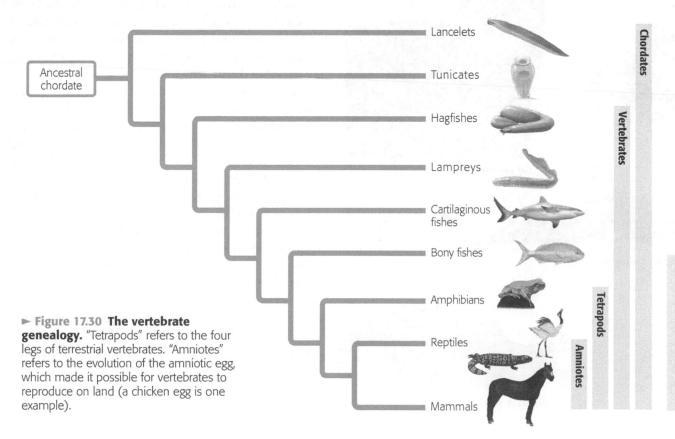

► **Figure 17.30 The vertebrate genealogy.** "Tetrapods" refers to the four legs of terrestrial vertebrates. "Amniotes" refers to the evolution of the amniotic egg, which made it possible for vertebrates to reproduce on land (a chicken egg is one example).

☑ **CHECKPOINT**

During our early embryonic development, what four features do we share with invertebrate chordates such as lancelets?

Answer: (1) dorsal, hollow nerve cord, (2) notochord, (3) pharyngeal slits, (4) post-anal tail

Fishes

The first vertebrates were aquatic and probably evolved during the early Cambrian period about 542 million years ago. In contrast with most living vertebrates, they lacked jaws, hinged bone structures that work the mouth.

Two types of jawless fishes survive today: hagfishes and lampreys. Present-day hagfishes scavenge dead or dying animals on the cold, dark seafloor (**Figure 17.31a**). When threatened, a hagfish exudes an enormous amount of slime from special glands on the sides of its body. Recently, some hagfishes have become endangered because their skin is used to make "eel-skin" belts, purses, and boots. Most species of lampreys are parasites that use their jawless mouths as suckers to attach to the sides of large fish (**Figure 17.31b**). The rasping tongue then penetrates the skin, allowing the lamprey to feed on its victim's blood and tissues.

We know from the fossil record that the first jawed vertebrates were fishes that evolved about 440 million years ago. They had two pairs of fins, making them agile swimmers. Some early fishes were active predators up to 10 m (33 ft) in length that could chase prey and bite off chunks of flesh. Even today, most fishes are carnivores.

Cartilaginous fishes, such as sharks and rays, have a flexible skeleton made of cartilage (**Figure 17.31c**). Most sharks are adept predators because they are fast swimmers with streamlined bodies, acute senses, and powerful jaws. A shark does not have keen eyesight, but its sense of smell is very sharp. In addition, special electrosensors on the head can detect minute electrical fields produced by muscle contractions in nearby animals. Sharks also have a **lateral line system**, a row of sensory organs running along each side of the body. Sensitive to changes in water pressure, the lateral line system enables a shark to detect minor vibrations caused by animals swimming in its neighborhood. There are about 1,000 living species of cartilaginous fishes, nearly all of them marine.

The skeletons of **bony fishes** are reinforced by calcium (**Figure 17.31d**). Bony fishes have a lateral line system, a keen sense of smell, and excellent eyesight. On each side of the head, a protective flap called the **operculum** (plural, *opercula*) covers a chamber housing the gills, feathery external organs that extract oxygen from water. Movement of the operculum allows the fish to breathe without swimming. By contrast, sharks lack opercula and must swim to pass water over their gills. The need to move water over the gills is why a shark must keep moving to stay alive. Also unlike sharks, bony fishes have a **swim bladder**, a gas-filled sac that enables the fish to control its buoyancy. Thus, many bony fishes can conserve energy by remaining almost motionless, in contrast to sharks, which sink if they stop swimming.

Most bony fishes, including familiar species such as tuna, trout, and goldfish, are **ray-finned fishes**. Their fins are webs of skin supported by thin, flexible skeletal rays, the feature for which the group was named. There are approximately 27,000 species of ray-finned fishes, the greatest number of species of any vertebrate group.

A second evolutionary branch includes the **lobe-finned fishes**. In contrast to the ray-finned fishes, their fins are muscular and are supported by stout bones that are homologous to amphibian limb bones. Early lobe-fins lived in coastal wetlands and may have used their fins to "walk" underwater. Today, three lineages of lobe-fins survive. The coelacanth is a deep-sea dweller once thought to be extinct. The lungfishes are represented by several Southern Hemisphere species that inhabit stagnant waters and gulp air into lungs connected to the pharynx. The third lineage of lobe-fins adapted to life on land and gave rise to amphibians, the first terrestrial vertebrates. ☑

☑ **CHECKPOINT**

A shark has a _____ skeleton, whereas a tuna has a _____ skeleton.

Answer: cartilaginous; bony

Lancelets
Tunicates
Hagfishes
Lampreys
Cartilaginous fishes
Bony fishes
Amphibians
Reptiles
Mammals

▼ **Figure 17.31 Fish diversity.**

(a) Hagfish

(b) Lamprey (inset: mouth)

(c) Shark, a cartilaginous fish

Lateral line

Operculum

(d) Bony fish

Amphibians

In Greek, the word *amphibios* means "living a double life." Most **amphibians** exhibit a mixture of aquatic and terrestrial adaptations. Most species are tied to water because their eggs, lacking shells, dry out quickly in the air. A frog may spend much of its time on land, but it lays its eggs in water **(Figure 17.32a)**. An egg develops into a larva called a tadpole, a legless, aquatic algae-eater with gills, a lateral line system resembling that of fishes, and a long-finned tail. The tadpole undergoes a radical metamorphosis when changing into

a frog **(Figure 17.32b)**. When a young frog crawls onto shore and begins life as a terrestrial insect-eater, it has four legs, air-breathing lungs instead of gills, external eardrums, and no lateral line system. But even as adults, amphibians are most abundant in damp habitats, such as swamps and rain forests. This is partly because amphibians depend on their moist skin to supplement lung function in exchanging gases with the environment. Thus, even those frogs that are adapted to relatively dry habitats spend much of their time in humid burrows or under piles of moist leaves. The amphibians of today, including frogs and salamanders, account for about 12% of all living vertebrates, or about 6,000 species **(Figure 17.32c)**.

Amphibians were the first vertebrates to colonize land. They descended from fishes that had lungs and fins with muscles and skeletal supports strong enough to enable some movement, however clumsy, on land **(Figure 17.33)**. The fossil record chronicles the evolution of four-limbed amphibians from fishlike ancestors. Terrestrial vertebrates—amphibians, reptiles, and mammals—are collectively called **tetrapods**, which means "four feet." ☑

Lancelets
Tunicates
Hagfishes
Lampreys
Cartilaginous fishes
Bony fishes
Amphibians
Reptiles
Mammals

☑ CHECKPOINT

Amphibians were the first _____, four-footed terrestrial vertebrates.

Answer: tetrapods

▼ **Figure 17.32 Amphibian diversity.**

(a) Frog eggs

(b) Tadpole and adult tree frog (right)

Malayan horned frog

Texas barred tiger salamander
(c) Frogs and salamanders: the two major groups of amphibians

▼ **Figure 17.33 The origin of tetrapods.**

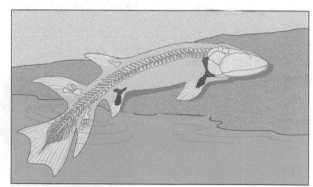

Lobe-finned fish. Fossils of some lobe-finned fishes have skeletal supports extending into their fins.

Early amphibian. Fossils of early amphibians have limb skeletons that probably functioned in helping them move on land.

391

Reptiles

Reptiles (including birds) and mammals are **amniotes**. The evolution of amniotes from an amphibian ancestor included many adaptations for living on land. The adaptation that gives the group its name is the **amniotic egg**, a fluid-filled egg with a waterproof shell that encloses the developing embryo **(Figure 17.34)**.

Evolutionary biologists have discovered the answer to the riddle "Which came first, the chicken or the egg?"

The amniotic egg functions as a self-contained "pond" that enables amniotes to complete their life cycle on land.

The **reptiles** include snakes, lizards, turtles, crocodiles, alligators, and birds, along with a number of extinct groups, including most of the dinosaurs. The European grass snake in Figure 17.34 displays two reptilian adaptations to living on land: scaled waterproof skin, preventing dehydration in dry air, and amniotic eggs with shells, providing a watery, nutritious internal environment where the embryo can develop. These adaptations allowed reptiles to break their ancestral ties to aquatic habitats. Reptiles cannot breathe through their dry skin and so obtain most of their oxygen through their lungs.

Nonbird Reptiles

Nonbird reptiles are sometimes referred to as "cold-blooded" animals because they do not use their metabolism extensively to control body temperature. Reptiles do regulate body temperature, but largely through behavioral adaptations. For example, many lizards regulate their internal temperature by basking in the sun when the air is cool and seeking shade when the air is too warm. Because lizards and other nonbird reptiles absorb external heat rather than generating much of their own, they are said to be **ectotherms**, a term more accurate than "cold-blooded." By heating directly with solar energy rather than through the metabolic breakdown of food, a nonbird reptile can survive on less than 10% of the calories required by a mammal of equivalent size.

As successful as reptiles are today, they were far more widespread, numerous, and diverse during the Mesozoic era, which is sometimes known as the "age of reptiles." Reptiles diversified extensively during that era, producing a dynasty that lasted until about 65 million years ago. Dinosaurs, the most diverse reptile group, included the largest animals ever to inhabit land. Some were gentle giants that lumbered about while browsing vegetation. Others were voracious carnivores that chased their larger prey on two legs.

The age of reptiles began to fade about 70 million years ago. Around that time, the global climate became cooler and more variable. This was a period of mass extinctions that claimed all the dinosaurs by about 65 million years ago, except for one lineage (see Table 14.1). That lone surviving lineage is represented today by the reptilian group we know as birds. ☑

☑ **CHECKPOINT**

What is an amniotic egg?

Answer: a shelled egg surrounding fluid that contains an embryo

▶ Figure 17.34 **Reptile diversity.**

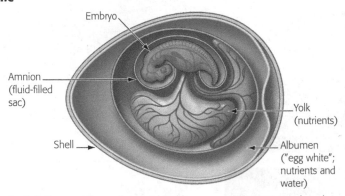

Aminiotic egg. The embryo and its life-support system are enclosed in a waterproof shell.

Embryo

Amnion (fluid-filled sac)

Shell

Yolk (nutrients)

Albumen ("egg white"; nutrients and water)

Snake. The shells of nonbird reptile eggs are leathery and flexible. This European grass snake is nonvenomous. When threatened, it may hiss and strike. If the bluff fails, it goes limp, pretending to be dead.

Lizard. The Gila monster, a desert-dweller of the Southwest, is the only venomous lizard native to the United States. Although large (up to 2 feet long), it moves too slowly to pose any danger to people.

Lancelets
Tunicates
Hagfishes
Lampreys
Cartilaginous fishes
Bony fishes
Amphibians
Reptiles
Mammals

Birds

You may have noticed that the reptilian egg resembles the more familiar chicken egg. Birds have scaly skin on their legs and feet; even feathers—their signature feature—are modified scales. Genetic and fossil evidence shows that **birds** are indeed reptiles, having evolved from a lineage of small, two-legged dinosaurs called theropods. Today, birds look quite different from reptiles because of their distinctive flight equipment—almost all of the 10,000 living bird species are airborne. The few flightless species, including the ostrich and the penguin, evolved from flying ancestors.

Almost every element of bird anatomy is adapted in some way that enhances flight. The bones have a honeycombed structure that makes them strong but light. (The wings of airplanes have the same basic construction.) For example, a huge seagoing species called the frigate bird has a wingspan of more than 2 m (6.6 ft), but its whole skeleton weighs only about 113 g (a mere 4 ounces, about as much as an iPhone). Another adaptation that reduces the weight of birds is the absence of some internal organs found in other vertebrates. Female birds, for instance, have only one ovary instead of a pair. Also, today's birds are toothless, an adaptation that trims the weight of the head, preventing uncontrolled nosedives. Birds do not chew food in the mouth but grind it in the gizzard, a muscular chamber of the digestive tract near the stomach.

Flying requires a great expenditure of energy and an active metabolism. Unlike other reptiles, birds are **endotherms**, meaning they use their own metabolic heat to maintain a warm, constant body temperature.

A bird's most obvious flight equipment is its wings. Bird wings are airfoils that illustrate the same principles of aerodynamics as the wings of an airplane (Figure 17.35). A bird's flight motors are its powerful breast muscles, which are anchored to a keel-like breastbone. It is mainly these flight muscles that we call "white meat" on chicken and turkey breasts. Some birds, such as eagles and hawks, have wings adapted for soaring on air currents and flap their wings only occasionally. Other birds, including hummingbirds, excel at maneuvering but must flap continuously to stay aloft. Feathers are made of the same protein that forms the scales of reptiles. Feathers may have functioned first as insulation, helping birds retain body heat, or for courtship displays. Only later were they adapted as flight gear ☑

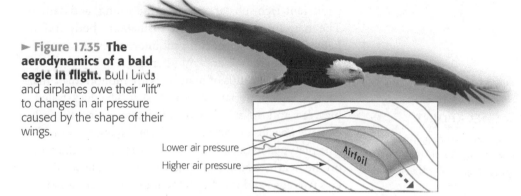

▶ Figure 17.35 **The aerodynamics of a bald eagle in flight.** Both birds and airplanes owe their "lift" to changes in air pressure caused by the shape of their wings.

Lower air pressure

Higher air pressure

Airfoil

Crocodile. Found throughout central and southern Africa, the Nile crocodile can grow up to 20 feet long and weigh nearly 2,000 pounds.

Birds. These red-crowned cranes, native to China, are performing an elaborate courtship dance.

Dinosaur. This *Herrerasaurus* skeleton is from a carnivorous theropod discovered in Argentina.

Mammals

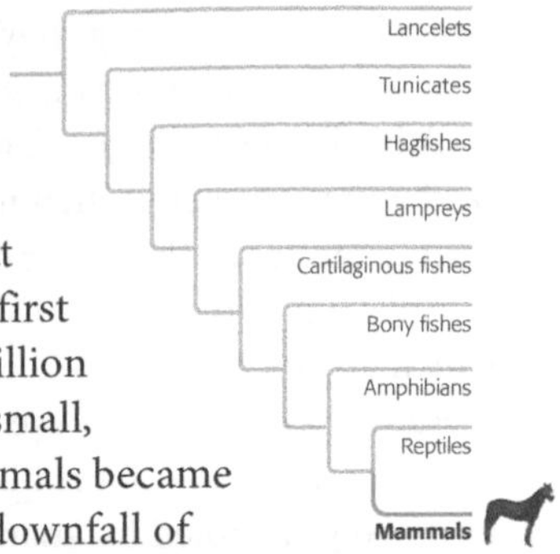

There are two major lineages of amniotes: one that led to the reptiles and one that produced the mammals. The first **mammals** arose about 200 million years ago and were probably small, nocturnal insect-eaters. Mammals became much more diverse after the downfall of the dinosaurs. Most mammals are terrestrial. These include nearly 1,000 species of winged mammals, the bats. And the roughly 80 species of dolphins, porpoises, and whales are totally aquatic. The blue whale—an endangered mammal that grows to lengths of nearly 30 m (about as long as a basketball court)—is the largest animal that has ever lived. Mammals have two unique characteristics: mammary glands (which produce milk, a nutrient-rich substance to feed the young) and hair. The main function of hair is to insulate the body and help maintain a warm, constant internal temperature; like birds, mammals are endotherms.

The major groups of mammals are monotremes, marsupials, and eutherians **(Figure 17.36)**. The duck-billed platypus and the echidna, or spiny anteater, are the only existing species of **monotremes**, egg-laying mammals. The platypus lives along rivers in eastern Australia and on the nearby island of Tasmania. The female usually lays two eggs and incubates them in a leaf nest. After hatching, the young nurse by licking up milk secreted by rudimentary mammary glands onto the mother's fur.

Most mammals are born rather than hatched. During pregnancy in marsupials and eutherians, the embryos are nurtured inside the mother by an organ called the **placenta**. Consisting of both embryonic and maternal tissues, the placenta joins the embryo to the mother within the uterus. The embryo receives oxygen and nutrients from maternal blood that flows close to the embryonic blood system in the placenta.

Marsupials, the so-called pouched mammals, include kangaroos, koalas, and opossums. These mammals have a brief pregnancy and give birth to embryonic offspring. After birth, the tiny offspring make their way to an external pouch on the mother's abdomen, where they attach to mammary glands. Nearly all marsupials live in Australia, New Zealand, and North and South America. Australian marsupials have diversified extensively, filling terrestrial habitats that on the other continents are occupied by eutherian mammals (see Figure 14.16).

Eutherians are also called **placental mammals** because their placentas provide a more intimate and longer-lasting association between the mother and her developing young than do marsupial placentas. Eutherians make up almost 95% of the 5,300 species of living mammals. Dogs, cats, cows, rodents, rabbits, bats, and whales are all examples of eutherian mammals. One of the eutherian groups is the primates, which include monkeys, apes, and humans. ✓

Lancelets
Tunicates
Hagfishes
Lampreys
Cartilaginous fishes
Bony fishes
Amphibians
Reptiles
Mammals

☑ **CHECKPOINT**

What are two defining characteristics of mammals?

Answer: mammary glands and hair

▼ **Figure 17.36 Mammalian diversity.**

MAJOR GROUPS OF MAMMALS		
Monotremes (hatched from eggs)	**Marsupials** (embryonic at birth)	**Eutherians** (fully developed at birth)

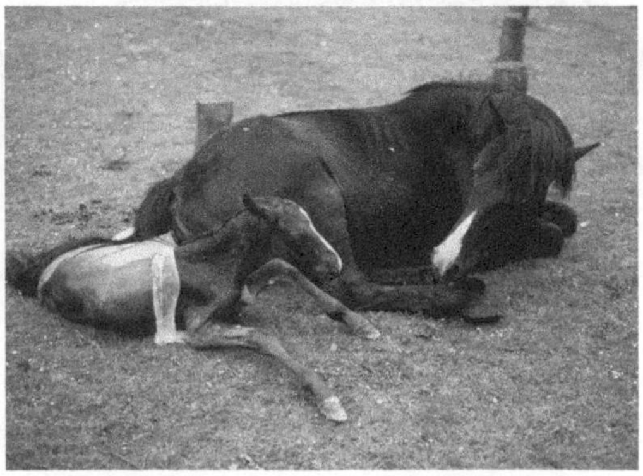

Monotremes, such as this duck-billed platypus, are the only mammals that lay eggs. Like other mammals, platypus mothers nourish their young with milk.

The young of marsupials are born very early in their development. The newborn kangaroo will finish its growth while nursing from a nipple in its mother's pouch.

In eutherians (placental mammals), young develop within the uterus of the mother. There they are nurtured by the flow of blood through the dense network of vessels in the placenta. This newborn foal is coated by remnants of the placenta.

The Human Ancestry

We have now traced animal phylogeny to the **primates**, the mammalian group that includes us—*Homo sapiens*—and our closest kin. To understand what that means, we must follow our ancestry back to the trees, where some of our most treasured traits originated.

The Evolution of Primates

Primate evolution provides a context for understanding human origins. The fossil record supports the hypothesis that primates evolved from insect-eating mammals during the late Cretaceous period, about 65 million years ago. Those early primates were small, arboreal (tree-dwelling) mammals. Thus, primates were first distinguished by characteristics that were shaped, through natural selection, by the demands of living in the trees. For example, primates have limber shoulder joints, which make it possible to swing from branch to branch. The agile hands of primates can hang on to branches and manipulate food. Nails have replaced claws in many primate species, and the fingertips are very sensitive. The eyes of primates are close together on the front of the face. The overlapping fields of vision of the two eyes enhance depth perception, an obvious advantage when swinging in trees. Excellent eye-hand coordination is also important for arboreal maneuvering. Parental care is essential for young animals in the trees. Mammals devote more energy to caring for their young than most other vertebrates, and primates are among the most attentive parents of all mammals. Most primates have single births and nurture their offspring for a long time. Although humans never lived in trees, we retain in modified form many traits that originated there.

Taxonomists divide the primates into three main groups **(Figure 17.37)**. The first includes lemurs, lorises, and bush babies. These primates live in Madagascar, southern Asia, and Africa. Tarsiers, small nocturnal tree-dwellers found only in Southeast Asia, form the second group of primates. The third group of primates, **anthropoids**, includes monkeys and apes. All monkeys in the New World (the Americas) are arboreal and are distinguished by prehensile (grasping) tails that function as an extra appendage for swinging. If you see a monkey in a zoo swinging by its tail, you know it's from the New World. Although some Old World (African and Asian) monkeys are also arboreal, their tails are not prehensile. And many Old World monkeys, including baboons, macaques, and mandrills, are mainly ground-dwellers.

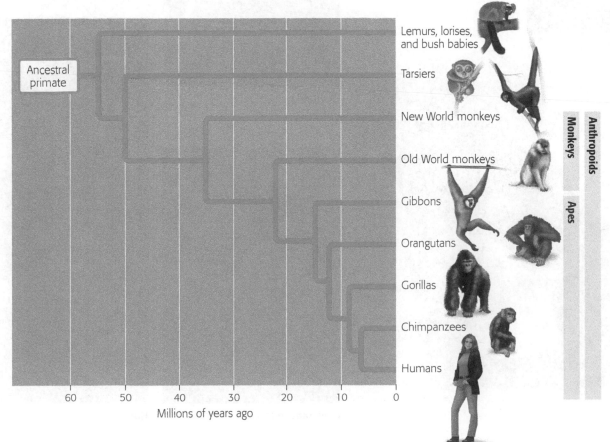

◄ **Figure 17.37 Primate phylogeny.**

395

Anthropoids also have a fully opposable thumb; that is, they can touch the tips of all four fingers with their thumb.

Our closest anthropoid relatives are the nonhuman apes: gibbons, orangutans, gorillas, and chimpanzees. They live only in tropical regions of the Old World. Except for some gibbons, apes are larger than monkeys, with relatively long arms, short legs, and no tail.

Although all apes are capable of living in trees, only gibbons and orangutans are primarily arboreal. Gorillas and chimpanzees are highly social. Apes have larger brains proportionate to body size than monkeys, and their behavior is more adaptable. And, of course, the apes include humans. **Figure 17.38** shows examples of primates.

▼ **Figure 17.38 Primate diversity.**

Red ruffed lemur

Tarsier

Black spider monkey
(New World monkey)

Orangutan (ape)

Patas monkey (Old World monkey)

Gibbon (ape)

Gorilla (ape)

Chimpanzee (ape)

Human

The Emergence of Humankind

Humanity is one very young twig on the tree of life. In the continuum of life spanning 3.5 billion years, the fossil record and molecular systematics indicate that humans and chimpanzees have shared a common African ancestry for all but the last 6–7 million years (see Figure 17.37). Put another way, if we compressed the history of life to a year, the human branch has existed for only 18 hours.

Some Common Misconceptions

Certain misconceptions about human evolution persist in the minds of many, long after these myths have been debunked by the fossil evidence. One of these myths is expressed in the question "If chimpanzees were our ancestors, then why do they still exist?" In fact, scientists do not think that humans evolved from chimpanzees. Rather, as illustrated in Figure 17.37, the lineages that led to present-day humans and chimpanzees diverged from a common ancestor several million years ago. Each branch then evolved separately. As an analogy, consider a large family reunion attended by the descendants of a man who was born in 1830. Although the attendees share this common ancestor who lived several generations ago, they may be only distantly related to each other, as sixth or seventh cousins. Similarly, chimpanzees are not our parent species, but more like our very distant phylogenetic cousins, related through a common ancestor that lived hundreds of thousands of generations in the past.

Another myth envisions human evolution as a ladder with a series of steps leading directly from an ancestral anthropoid to *Homo sapiens*. This is often illustrated as a parade of fossil **hominins** (members of the human family) becoming progressively more modern as they march across the page. In fact, as **Figure 17.39** shows, there were times in hominin history when several human species coexisted. Scientists have identified about 20 species of fossil hominins, demonstrating that human phylogeny is more like a multibranched bush than a ladder, with our species being the tip of the only twig that still lives.

Although evidence from hundreds of thousands of hominin fossils debunks these and other myths about human evolution, many fascinating questions about our ancestry remain. The discovery of each new hominin fossil brings scientists a little closer to solving the puzzle of how we became human. In the following pages, you'll learn about some of the important clues that have been discovered so far.

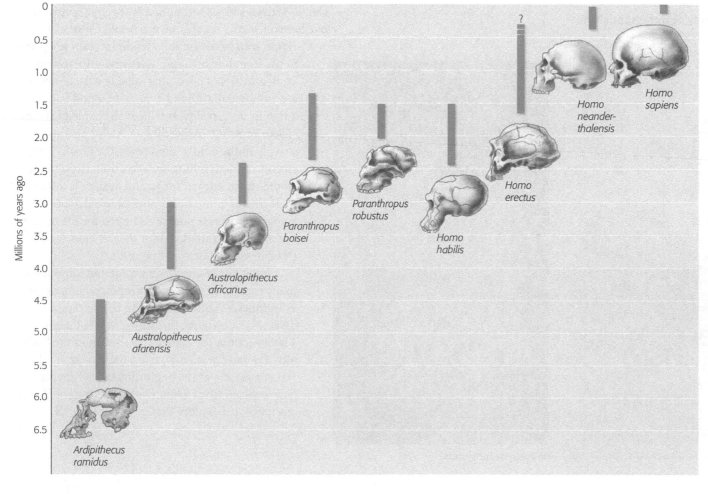

◀ **Figure 17.39 A time line of human evolution.** The orange bars indicate the time span during which each species lived. Notice that there have been times when two or more hominin species coexisted. The skulls are all drawn to the same scale so you can compare brain sizes.

397

Australopithecus and the Antiquity of Bipedalism

Present-day humans and chimpanzees clearly differ in two major physical features: Humans are bipedal (walk upright) and have much larger brains. When did these features emerge? In the early 1900s, scientists hypothesized that increased brain size was the initial change that separated hominins from other apes. That hypothesis was overturned when a team of researchers in Ethiopia (Figure 17.40a) unearthed a stunning 3.24-million-year-old female hominin that had a small brain and walked on two legs. Officially named *Australopithecus afarensis*, but nicknamed Lucy by her discoverers, the individual was only about 3 feet tall and with a head about the size of a softball. Corroborating evidence of early bipedalism was found soon after—the footprints of two upright-walking hominins preserved in a 3.6-million-year-old layer of volcanic ash (Figure 17.40b). Since these initial discoveries, many more *Australopithecus afarensis* fossils have been found, including most of the skeleton of a 3-year-old member of the species. Other species of *Australopithecus* have also been discovered. The first analysis of the most recent addition to the genus, *Australopithecus sediba*, was published in 2010.

☑ CHECKPOINT

Which hominin species was the first to walk upright? Which was the first to spread beyond Africa?

Answer: Australopithecus afarensis; Homo erectus

▼ **Figure 17.40 The antiquity of upright posture.**

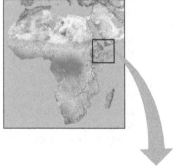

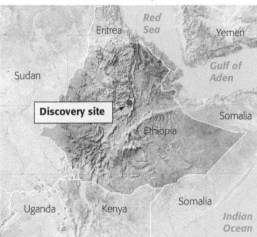

(a) Afar region of Ethiopia, where Lucy was discovered

(b) Ancient footprints

Scientists are now certain that bipedalism is a very old trait. Another lineage known as "robust" australopiths—including *Paranthropus boisei* and *Paranthropus robustus* in Figure 17.39—were also small-brained bipeds. So was *Ardipithecus ramidus*, the oldest hominin shown in Figure 17.39.

Homo habilis and the Evolution of Inventive Minds

Enlargement of the human brain is first evident in fossils from East Africa dating to about 2.4 million years ago. Thus, the fundamental human trait of an enlarged brain evolved a few million years after bipedalism.

Anthropologists have found skulls with brain capacities intermediate in size between those of the latest *Australopithecus* species and those of *Homo sapiens*. Simple handmade stone tools are sometimes found with the larger-brained fossils, which have been dubbed *Homo habilis* ("handy man"). After walking upright for about 2 million years, humans were finally beginning to use their manual dexterity and big brains to invent tools that enhanced their hunting, gathering, and scavenging on the African savanna.

Homo erectus and the Global Dispersal of Humanity

The first species to extend humanity's range from Africa to other continents was *Homo erectus*. Skeletons of *Homo erectus* dating to 1.8 million years ago found in the Republic of Georgia represent the oldest known fossils of hominins outside Africa. *Homo erectus* was taller than *Homo habilis* and had a larger brain capacity. Intelligence enabled this species to continue succeeding in Africa and also to survive in the colder climates of the north. *Homo erectus* resided in huts or caves, built fires, made clothes from animal skins, and designed stone tools. In anatomical and physiological adaptations, *Homo erectus* was poorly equipped for life outside the tropics, but made up for the deficiencies with cleverness and social cooperation.

Eventually, *Homo erectus* migrated to populate many regions of Asia and Europe, moving as far as Indonesia. This species gave rise to regionally diverse descendants, including, as you'll learn in a later section, the Neanderthals. Could the "hobbits" (*Homo floresiensis*) from the Biology and Society section at the beginning of this chapter have evolved from a long-isolated group of *Homo erectus*? We'll explore that question next. ☑

Human Evolution THE PROCESS OF SCIENCE

Who Were the Hobbit People?

How do scientists test hypotheses about events that occurred in the distant past? One approach involves using fossils, the historical record of life on Earth. If you have watched TV shows like *Bones*, in which forensic scientists use skeletal remains to solve crimes, you know that even a small piece of a skeleton can provide a wealth of information to an expert. Researchers have applied this method of hypothesis testing to the scientific debate over the "hobbit" people.

The first **observations** made by the researchers who discovered hominin fossils on the island of Flores revealed that the new fossils did not belong to any known species. This led researchers to ask the **question**, Where does this hominin fit in our evolutionary history? The scientists formed the **hypothesis** that the hobbits evolved from an isolated population of *Homo erectus,* the ancient hominin whose far-flung travels predated those of *Homo sapiens.* They made the **prediction** that key traits of the new species, such as its skull characteristics and body proportions, would resemble those of a miniature *Homo erectus* individual. Rather than a controlled **experiment**, the investigation consisted of making detailed measurements and other observations of the new fossils and comparing them with data from *Homo erectus* fossils. The initial **results** supported their hypothesis.

As you have learned, however, preliminary conclusions are often overturned by new evidence. In the past several years, further analyses have been carried out, and additional specimens have been examined. Many scientists now think that the evidence supports an alternative hypothesis: *Homo floresiensis* is most closely related to *Homo habilis,*

a smaller and more ancient species than *Homo erectus.* Other researchers continue to test the hypothesis that the hobbits are not a species at all but a *Homo sapiens* population with a disorder that caused bone malformations.

How can scientists determine which hypothesis is correct? By accumulating further evidence. While some researchers continue to excavate the site where *Homo floresiensis* was discovered (Figure 17.41), others are widening the search to additional locations. The most helpful information would come from finding a second skull or unearthing bones or teeth from which DNA could be extracted and analyzed. Meanwhile, the mystery of the hobbit people continues.

▼ **Figure 17.41 Searching for hobbits.** Researchers continue to excavate Liang Bua Cave, on the Indonesian island of Flores, where *Homo floresiensis* was discovered.

Homo neanderthalensis

Homo floresiensis is not the only mystery on our family tree. The first discovery of fossilized remains in the Neander Valley in Germany 150 years ago, *Homo neanderthalensis*—commonly called Neanderthals—stimulated the public's imagination. This intriguing species had a large brain and hunted big game with tools made from stone and wood. Neanderthals were living in Europe as far back as 350,000 years ago and spread to the Near East, central Asia, and southern Siberia. By 28,000 years ago, the species was extinct. Who were the Neanderthals?

Analysis of DNA extracted from Neanderthal fossils proved that humans are not the descendants of Neanderthals, as was once thought. Rather, humans and Neanderthals shared a common ancestor, and their lineages diverged about 400,000 years ago. You may

be surprised to learn that sequencing of Neanderthal genomes suggests that interbreeding between Neanderthals and some populations of *Homo sapiens* left a genetic legacy in our species. Roughly 2% of the genomes of most present-day humans came from Neanderthals. Africans are the exception—their DNA carries no detectable trace of Neanderthal ancestry. Scientists also learned that at least some Neanderthals had pale skin and red hair (see Figure 14.23). In the most recent analysis of a Neanderthal genome, published in 2014, researchers identified specific genes and gene regulation sequences (see Figure 11.3) that distinguish modern humans from Neanderthals. Clues such as these will help scientists understand the genetic differences that were important in the evolution of *Homo sapiens.*

According to recent DNA analysis, many of us have a bit of Neanderthal in our genes.

The Origin and Dispersal of *Homo sapiens*

Evidence from fossils and DNA studies is coming together to support a compelling hypothesis about how our own species, *Homo sapiens*, emerged and spread around the world. The oldest known fossils of *Homo sapiens* were discovered in Ethiopia and date from 160,000 to 195,000 years ago. These early humans lacked the heavy browridges of *Homo erectus* and *Homo neanderthalensis* and were more slender, suggesting that they belong to a distinct lineage. The Ethiopian fossils support molecular evidence about the origin of humans: DNA studies strongly suggest that all living humans can trace their ancestry back to a single African *Homo sapiens* lineage that began 160,000 to 200,000 years ago.

The oldest fossils of *Homo sapiens* outside Africa are from the Middle East and date back about 115,000 years.

Evidence suggests that our species emerged from Africa in one or more waves, spreading first into Asia and then to Europe, Southeast Asia, Australia, and finally to the New World (North and South America) **(Figure 17.42)**. The date of the first arrival of humans in the New World is uncertain, although the generally accepted evidence suggests a minimum of 15,000 years ago.

Certain uniquely human traits have allowed for the development of human societies. The primate brain continues to grow after birth, and the period of growth is longer for a human than for any other primate. The extended period of human development also lengthens the time parents care for their offspring, which contributes to the child's ability to benefit from the experiences of earlier generations. This is the basis of human culture—social transmission of accumulated knowledge, customs, beliefs, and art over generations **(Figure 17.43)**. The major means of this transmission is language, spoken and written. Humans have evolved culturally as well as biologically.

Nothing has had a greater impact on life on Earth than *Homo sapiens*. The global consequences of human evolution have been enormous. Cultural evolution made modern *Homo sapiens* a new force in the history of life—a species that could defy its physical limitations. We do not have to wait to adapt to an environment through natural selection; we simply change the environment to meet our needs.

We are the most numerous and widespread of all large animals, and wherever we go, we bring environmental change faster than many species can adapt. In the next unit, on ecology, we'll examine the interactions of humans—as well as other species—with the environment. ☑

▼ Figure 17.42 **The spread of *Homo sapiens* (dates given as years before present, BP)**

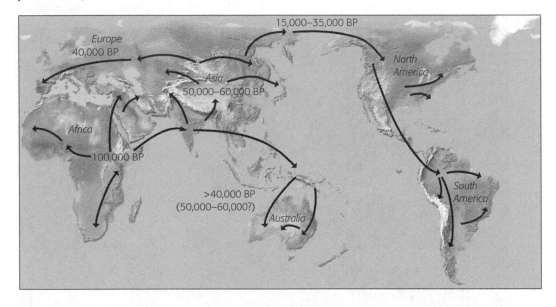

Europe
40,000 BP

15,000–35,000 BP

North America

Asia
50,000–60,000 BP

Africa

100,000 BP

>40,000 BP
(50,000–60,000?)

Australia

South America

☑ CHECKPOINT

1. Humans first evolved on which continent?
2. When would a *Homo sapiens* individual have had an opportunity to meet a Neanderthal?

Answers: 1. Africa 2. between the time Homo sapiens reached the regions inhabited by Neanderthals (no sooner than 115,000 years ago) and the time when the Neanderthals died out (28,000 years ago)

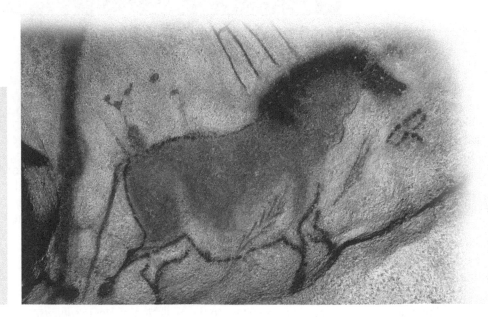

◄ Figure 17.43 **Art history goes back a long way.** Beautiful ancient art, such as this 30,000-year-old painting from Lascaux Cave in France, is just one example of our cultural roots in early societies.

Human Evolution EVOLUTION CONNECTION

Are We Still Evolving?

Imagine that you could take a time machine 100,000 years into the past and bring back a *Homo sapiens* man. If you dressed him in jeans and a T-shirt and took him for a stroll around campus, chances are that no one would look twice. Did we stop evolving after becoming *Homo sapiens*? In some ways, yes. The human body has not changed much in the past 100,000 years. And by the time *Homo sapiens* began to travel out of Africa, all of the complex characteristics that define our humanity, including our big-brained intelligence and our capacity for language and symbolic thought, had already evolved.

But as humans wandered far from their site of origin and settled in diverse environments, populations encountered different selective forces. Some traits of people today reflect evolutionary responses of ancient ancestors to their physical and cultural environment. For example, the high frequency of sickle hemoglobin (see Figure 9.21) in certain populations is an adaptation that protects against the deadly disease malaria. In other malarial regions, a group of inherited blood disorders called thalassemia serve the same adaptive function.

Diet has had a significant effect on human evolution, too. For example, the ability to digest lactose as adults (see the Evolution Connection section of Chapter 3) evolved in populations that kept dairy herds. Reliance by early farmers on starchy crops such as rice or tubers also left a genetic trace: extra copies of the gene that encodes the starch-digesting enzyme amylase.

One of the most striking differences among people is skin color (Figure 17.44). The loss of skin pigmentation in humans who migrated north from Africa is thought to be an adaptation to low levels of ultraviolet (UV) radiation in northern latitudes. Dark pigment blocks the UV radiation necessary for synthesizing vitamin D—essential for proper bone development—in the skin. Recent research has turned up numerous other examples of adaptations that enabled us to colonize Earth's varied environments. For instance, Tibetans live at altitudes up to 14,000 feet (2.6 miles), where the air has 40% less oxygen than at sea level. Researchers have identified genes that have undergone evolutionary changes in response to this challenging environment (Figure 17.45). High-altitude adaptations have also been discovered in populations that inhabit the Andes Mountains in South America. Despite evolutionary tweaks such as these, however, we remain a single species.

▼ **Figure 17.45 Tibetans, a population adapted to living at high altitude.**

▼ **Figure 17.44 People with different adaptations to UV radiation.**

Chapter Review

SUMMARY OF KEY CONCEPTS

The Origins of Animal Diversity

What Is an Animal?

Animals are eukaryotic, multicellular, heterotrophic organisms that obtain nutrients by ingestion. Most animals reproduce sexually and develop from a zygote to a blastula and then to a gastrula. After the gastrula stage, some develop directly into adults, whereas others pass through a larval stage.

Early Animals and the Cambrian Explosion

Animals probably evolved from a colonial flagellated protist. Precambrian animals were soft-bodied. Animals with hard parts appeared during the Cambrian period. Between 535 and 525 million years ago, animal diversity increased rapidly.

Evolution: Animal Phylogeny

Major branches of animal evolution are defined by two key evolutionary differences: the presence or absence of tissues and radial versus bilateral body symmetry. A tissue-lined body cavity evolved in a number of later branches.

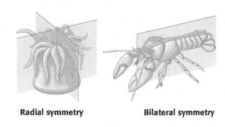

Radial symmetry   Bilateral symmetry

Major Invertebrate Phyla

This tree shows the eight major invertebrate phyla, as well as chordates, which include a few invertebrates.

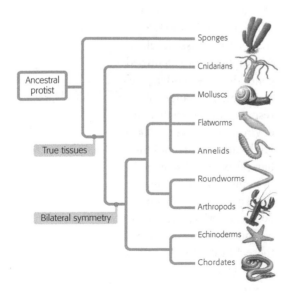

Sponges

Sponges (phylum Porifera) are stationary animals with porous bodies but no true tissues. Specialized cells draw water through pores in the sides of the body and trap food particles.

Cnidarians

Cnidarians (phylum Cnidaria) have radial symmetry, a gastrovascular cavity with a single opening, and tentacles with stinging cnidocytes. The body is either a stationary polyp or a floating medusa.

Molluscs

Molluscs (phylum Mollusca) are soft-bodied animals often protected by a hard shell. The body has three main parts: a muscular foot, a visceral mass, and a fold of tissue called the mantle.

MOLLUSCS		
Gastropods	**Bivalves**	**Cephalopods**

Flatworms

Flatworms (phylum Platyhelminthes) are the simplest bilateral animals. They may be free-living (such as planarians) or parasitic (such as tapeworms).

Annelids

Annelids (phylum Annelida) are segmented worms with complete digestive tracts. They may be free-living or parasitic.

Roundworms

Roundworms, also called nematodes (phylum Nematoda), are unsegmented and cylindrical with tapered ends. They may be free-living or parasitic.

Arthropods

Arthropods (phylum Arthropoda) are segmented animals with an exoskeleton and specialized, jointed appendages.

ARTHROPODS			
Arachnids	**Crustaceans**	**Millipedes and Centipedes**	**Insects**

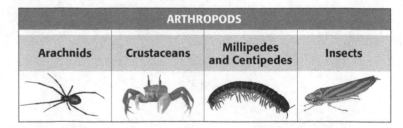

Echinoderms

Echinoderms (phylum Echinodermata) are stationary or slow-moving marine animals that lack body segments and possess a unique water vascular system. Bilaterally symmetric larvae usually change to radially symmetric adults. Echinoderms have a bumpy endoskeleton.

Vertebrate Evolution and Diversity

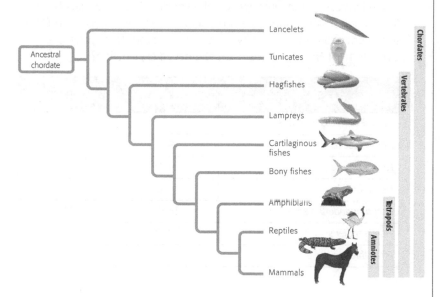

Characteristics of Chordates

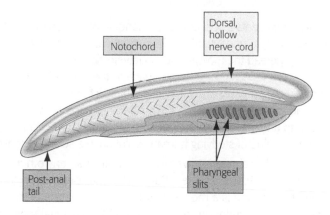

Tunicates and lancelets are invertebrate chordates. The vast majority of chordates are vertebrates, possessing a skull and backbone.

Fishes

Hagfishes and lampreys are jawless vertebrates. Cartilaginous fishes, such as sharks, are mostly predators with powerful jaws and a flexible skeleton made of cartilage. Bony fishes have a stiff skeleton reinforced by calcium. Bony fishes are further classified into ray-finned fishes and lobe-finned fishes (including lungfishes).

Amphibians

Amphibians are tetrapod vertebrates that usually deposit their eggs (lacking shells) in water. Aquatic larvae typically undergo a radical metamorphosis into the adult stage. Their moist skin requires that amphibians spend much of their adult life in humid environments.

Reptiles

Reptiles are amniotes, vertebrates that develop in a fluid-filled egg enclosed by a shell. Reptiles include terrestrial ectotherms with lungs and waterproof skin covered by scales. Scales and amniotic eggs enhanced reproduction on land. Birds are endothermic reptiles with wings, feathers, and other adaptations for flight.

Mammals

Mammals are endothermic vertebrates with mammary glands and hair. There are three major groups of mammals: Monotremes lay eggs; marsupials use a placenta but give birth to tiny embryonic offspring that usually complete development while attached to nipples inside the mother's pouch; and eutherians, or placental mammals, use their placenta in a longer-lasting association between the mother and her developing young.

MAMMALS		
Monotremes	**Marsupials**	**Eutherians**

The Human Ancestry

The Evolution of Primates

The first primates were small, arboreal mammals that evolved from insect-eating mammals about 65 million years ago. Anthropoids consist of New World monkeys (with prehensile tails), Old World monkeys (without prehensile tails), apes, and humans.

The Emergence of Humankind

Chimpanzees and humans evolved from a common ancestor about 6–7 million years ago. Species of the genus *Australopithecus*, which lived at least 4 million years ago, walked upright and had a small brain. Enlargement of the human brain in *Homo habilis* came later, about 2.4 million years ago. *Homo erectus* was the first species to extend humanity's range from its birthplace in Africa to other continents. *Homo erectus* gave rise to regionally diverse descendants, including the Neanderthals (*Homo neanderthalensis*). Current data indicate a relatively recent dispersal of modern Africans that gave rise to today's human diversity.

MasteringBiology®

For practice quizzes, BioFlix animations, MP3 tutorials, video tutors, and more study tools designed for this textbook, go to MasteringBiology®

SELF-QUIZ

1. Bilateral symmetry in the animal kingdom is best correlated with
   a. an ability to sense equally in all directions.
   b. the presence of a skeleton.
   c. motility and active predation and escape.
   d. development of a body cavity.

2. Identify which of the following categories includes all others in the list: arthropod, arachnid, insect, butterfly, crustacean, millipede.

3. _____ are the most diverse group of arthropods.

4. Features unique to mammals include
   a. the presence of hair.
   b. extended parental care of the young.
   c. being endotherms.
   d. having no egg-laying members.

5. What is the name of the phylum to which humans belong? For what anatomical structure is the phylum named? Where in your body is a derivative of this anatomical structure found?

6. Fossils suggest that the first major trait distinguishing human primates from other primates was _____.

7. Which of the following types of animals is not included in the human ancestry? (*Hint*: See Figure 17.30.)
   a. a bird
   b. a bony fish
   c. an amphibian
   d. a primate

8. Put the following list of species in order, from the oldest to the most recent: *Homo erectus*, *Australopithecus* species, *Homo habilis*, *Homo sapiens*.

9. Match each of the following animals to its phylum:
   a. lancelet   1. Platyhelminthes
   b. clam   2. Arthropoda
   c. tapeworm   3. Porifera
   d. dust mite   4. Chordata
   e. sponge   5. Mollusca

Answers to these questions can be found in Appendix: Self-Quiz Answers.

THE PROCESS OF SCIENCE

10. Imagine that you are a marine biologist. As part of your exploration, you dredge up an unknown animal from the seafloor. Describe some of the characteristics you should look at to determine the phylum to which the creature should be assigned.

11. Vegan and vegetarian diets are increasingly popular. While vegans consume no animals or animal products, many vegetarians are less strict. Talk to acquaintances who describe themselves as vegetarians, or who follow a meat-free diet, and determine which taxonomic groups they avoid eating (see Figures 17.6 and 17.30). Try to generalize about their diet. What do they consider "meat"? For example, do they avoid eating vertebrates but eat some invertebrates? Do they avoid only birds and mammals? What about fish? Do they eat dairy products or eggs?

12. Which evolutionary changes might have taken place to allow humans to develop a complex language?

13. **Interpreting Data** Average brain size, relative to body mass, gives a rough indication of the intelligence of a species. Graph the data below for hominins. Which species is most similar to *Homo floresiensis*? Does this information support the hypothesis that *Homo floresiensis* is a dwarf form of *Homo erectus* or an alternative explanation?

Hominin Species	Mean Brain Volume (cm³)	Mean Body Mass (kg)
Australopithecus afarensis	440	37
Homo erectus	940	58
Homo floresiensis	420	32
Homo habilis	610	34
Homo neanderthalensis	1480	65
Homo sapiens	1330	64
Paranthropus boisei	490	41

BIOLOGY AND SOCIETY

14. This chapter presents a scientific understanding of human origins. Science is one approach to understanding the natural world (as you learned in Chapter 1). Reread the Biology and Society section and underline words or phrases that reflect a scientific approach to the study of *Homo floresiensis*. You are probably also familiar with at least one way of understanding human origins that is outside the scope of science. How does that approach to understanding life differ from a scientific one? What might be the potential value of studying human evolution in a scientific context?

15. Coral reefs harbor a greater diversity of animals than any other environment in the sea. Australia's Great Barrier Reef has been protected as a marine reserve and is a mecca for scientists and nature enthusiasts. Elsewhere, such as in Indonesia and the Philippines, coral reefs are in danger. Many reefs have been depleted of fish, and runoff from the shore has covered coral with sediment. Nearly all the changes in the reefs can be traced back to human activities. What kinds of activities do you think might be contributing to the decline of the reefs? What are some reasons to be concerned about this decline? Do you think the situation is likely to improve or worsen in the future? Why? What might the local people do to halt the decline? Should the more industrialized countries help? Why or why not?

16. Many of us, including many anthropologists and zoologists, think that humans are at a special position among all animals due to a range of characteristics, such as an upright posture, high intelligence, handling of complex tools, development of a complex language system, a prolonged childhood, and an extended lifespan. However, many of these characteristics can be observed in other species as well. Are we really unique among animals? Justify your answer.

Unit 4
Ecology

18 **An Introduction to Ecology and the Biosphere**

Chapter Thread: **Global Climate Change**

19 **Population Ecology**

Chapter Thread: **Biological Invasions**

20 **Communities and Ecosystems**

Chapter Thread: **Biodiversity in Decline**

An Introduction to Ecology and the Biosphere

Airborne pollutants from car exhaust can combine with water and return to Earth as acid precipitation far away—evidence that the water cycle operates on a global scale.

Your "goose bumps" are caused by the same muscles that fluff up geese on a cold day.

Some forms of life on Earth would survive if the sun stopped shining—but we would not be among them.

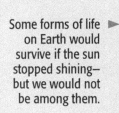

Your home may be infested with "vampire" devices that consume electricity even while you sleep.

Global Climate Change BIOLOGY AND SOCIETY

Penguins, Polar Bears, and People in Peril

Ninety-seven percent of climate scientists agree: The global climate is changing. The change is driven by a rapid rise in temperatures—on average, 0.8°C (about 1.4°F) over the past century, mostly over the last 30 years. The current rate of warming is *ten times* faster than the average rate during the warm-up that followed the last ice age. What do we know about climate change now, and what can we expect for the future?

The northernmost regions of the Northern Hemisphere and the Antarctic Peninsula have heated up the most. In parts of Alaska, for example, winter temperatures have risen by 5–6°F. The permanent Arctic sea ice is shrinking; each summer brings thinner ice and more open water. Polar bears, which stalk their prey on ice and need to store up body fat for the warmer months when there is no ice, are showing signs of starvation as their winter hunting grounds melt away. At the other end of the planet, diminishing sea ice near the Antarctic Peninsula limits the access of Adélie penguins to their food supply, and spring blizzards of unprecedented frequency and severity are taking a heavy toll on their eggs and chicks. But these charismatic animals are just the canaries in the coal mine whose distress alerts us to our own danger. We are already feeling the effects of climate change in more frequent and larger wildfires, deadly heat waves, and altered precipitation patterns that bring drought to some regions and torrential downpours to others.

An Adélie penguin. Climate change is bad news for some populations of Adélie penguins.

What does climate change mean for the future of life on Earth? Any predictions that scientists make now about future impacts of global climate change are based on incomplete information. Much remains to be discovered about species diversity and about the complex interactions of organisms (living things) with each other and with their environments. There is overwhelming evidence that human enterprises are responsible for the changes that are occurring. How we respond to this crisis will determine whether circumstances improve or worsen. And the process begins with understanding the basic concepts of ecology, which we start to explore in this chapter.

An Overview of Ecology

In your study of biology so far, you have learned about the diversity of life on Earth and about the molecular and cellular structures and processes that make life tick. **Ecology**, the scientific study of the interactions between organisms and their environments, offers a different perspective on life—biology from the skin out, so to speak.

Humans have always had an interest in other organisms and their environments. As hunters and gatherers, prehistoric people had to learn where and when game and edible plants could be found in greatest abundance. Naturalists, from Aristotle to Darwin and beyond, made the process of observing and describing organisms in their natural habitats an end in itself rather than simply a means of survival. We can still gain valuable insight from this discovery-based approach of watching nature and recording its structure and processes (**Figure 18.1**). As you might expect, hypothesis-driven science performed in natural environments is fundamental to ecology. But ecologists also test hypotheses using laboratory experiments, where conditions can be simplified and controlled. And some ecologists take a theoretical approach, devising mathematical and computer models, which enable them to simulate large-scale experiments that are impossible to conduct in the field.

▼ **Figure 18.1 Discovery science in a rain forest canopy.** A biologist collects insects in a rain forest on the eastern slope of the Andes mountain range in Argentina.

Ecology and Environmentalism

Technological innovations have enabled people to colonize just about every environment on Earth. Even so, our survival depends on Earth's resources, which have been profoundly altered by human activities (**Figure 18.2**). Global climate change is just one of the many environmental issues that have stirred public concern in recent decades. Some of our industrial and agricultural practices have contaminated the air, soil, and water. Our relentless quest for land and other resources has endangered a lengthy list of plant and animal species and has even driven some to extinction.

The science of ecology can provide the understanding needed to solve environmental problems. But these problems cannot be solved by ecologists alone, because they require making decisions based on values and ethics. On a personal level, each of us makes daily choices that affect our ecological impact. And legislators and corporations, motivated by environmentally aware voters and consumers, must address questions that have wider implications: How should the use of land and water be regulated? Should we try to save all species or just certain ones? What alternatives to environmentally destructive practices can be developed? How can we balance environmental impacts with economic needs?

▼ **Figure 18.2 Human impact on the environment.** A man paddles a canoe through a trash-clogged waterway in Manila, capital of the Philippines.

Interconnections within Systems
A Hierarchy of Interactions

Many different factors can potentially affect an organism's interaction with its environment. **Biotic factors**—all of the organisms in the area—make up the living component of the environment. Other organisms may compete with an individual for food and other resources, prey upon it, or change its physical and chemical environment. **Abiotic factors** make up the environment's nonliving component and include chemical and physical factors, such as temperature, light, water, minerals, and air. An organism's **habitat**, the specific environment it lives in, includes the biotic and abiotic factors of its surroundings.

When we study the interactions between organisms and their environments, it is convenient to divide ecology into four increasingly comprehensive levels: organismal ecology, population ecology, community ecology, and ecosystem ecology.

An **organism** is an individual living thing. **Organismal ecology** is concerned with the evolutionary adaptations that enable organisms to meet the challenges posed by their abiotic environments. The distribution of organisms is limited by the abiotic conditions they can tolerate. For example, amphibians such as the salamander in **Figure 18.3a** cannot live in cold climates because they gain most of their body warmth by absorbing heat from their surroundings. Temperature and precipitation shifts due to global climate change have already affected the distributions of some salamander species, and many more will feel the impact in the coming decades.

The next level of organization in ecology is the **population**, a group of individuals of the same species living in a particular geographic area. **Population ecology** concentrates mainly on factors that affect population density and growth (**Figure 18.3b**). Biologists who study endangered species are especially interested in this level of ecology.

A **community** consists of all the organisms that inhabit a particular area; it is an assemblage of populations of different species. Questions in **community ecology** focus on how interactions between species, such as predation and competition, affect community structure and organization (**Figure 18.3c**).

An **ecosystem** includes all the abiotic factors in addition to the community of species in a certain area. For example, a savanna ecosystem includes not only the organisms, such as diverse plants and animals, but also the soil, water sources, sunlight, and other abiotic factors of the environment. In **ecosystem ecology**, questions concern energy flow and the cycling of chemicals among the various biotic and abiotic factors (**Figure 18.3d**).

▼ **Figure 18.3 Examples of questions at different levels of ecology.**

(a) Organismal ecology. What range of temperatures can a red salamander tolerate?

(b) Population ecology. What factors affect the survival of emperor penguin chicks?

(c) Community ecology. How do predators such as this beech marten affect the diversity of rodents in a community?

(d) Ecosystem ecology. What processes recycle vital chemical elements such as nitrogen within a savanna ecosystem in Africa?

The **biosphere** is the global ecosystem—the sum of all the planet's ecosystems, or all of life and where it lives. The most complex level in ecology, the biosphere includes the atmosphere to an altitude of several kilometers, the land down to water-bearing rocks about 1,500 m (almost a mile) deep, lakes and streams, caves, and the oceans to a depth of several kilometers. But despite its grand scale, organisms within the biosphere are linked; events in one part may have far-reaching effects. ☑

☑ CHECKPOINT

What does the ecosystem level of classification have in common with the community level of classification? What does the ecosystem level include that the community level does not?

Answer: all the biotic factors of the area; the abiotic factors of the area

Living in Earth's Diverse Environments

Whether you have seen the world by traveling or through television and movies, you have probably noticed that there are striking regional patterns in the distribution of life. For example, some terrestrial areas, such as the tropical forests of South America and Africa, are home to plentiful plant life, whereas other areas, such as deserts, are relatively barren. Coral reefs are alive with vibrantly colored organisms; other parts of the ocean appear empty by comparison.

The distribution of life varies on a local scale, too. In the aerial view of a New Zealand wilderness in **Figure 18.4**, we can see a mixture of forest, a large lake, a meandering river, and mountains. Within these different environments, variation occurs on an even smaller scale. For example, we would find that the lake has several different habitats, and each habitat has a characteristic community of organisms.

▼ Figure 18.4 **Local variation of the environment in a New Zealand wilderness.**

Abiotic Factors of the Biosphere

Patterns in the distribution of life mainly reflect differences in the abiotic factors of the environment. Let's look at several major abiotic factors that influence where organisms live.

Energy Source

All organisms require a usable source of energy to live. Solar energy from sunlight, captured by chlorophyll during the process of photosynthesis, powers most ecosystems. In the image shown in **Figure 18.5**, colors are keyed to the relative abundance of chlorophyll. Green areas on land indicate high densities of plant life. Orange areas on land, including the Sahara region of Africa and much of the western United States, are much less productive. Green regions of the ocean contain an abundance of algae and photosynthetic bacteria compared to darker regions.

Lack of sunlight is seldom the most important factor limiting plant growth for terrestrial ecosystems, although shading by trees does create intense competition for light among plants growing on forest floors. In many aquatic environments, however, light cannot penetrate beyond certain depths. As a result, most photosynthesis in a body of water occurs near the surface. Surprisingly, life also thrives in environments that are completely dark. A mile or more below

> Some forms of life on Earth would survive if the sun stopped shining—but we would not be among them.

the ocean's surface lies the remarkable world of hydrothermal vents, sites near the adjoining edges of giant plates of Earth's crust where molten rock and hot gases surge upward from Earth's interior. Towering chimneys, some as tall as a nine-story building, emit scalding water and hot gases (**Figure 18.6**). These ecosystems are powered by bacteria that derive energy from the oxidation of inorganic chemicals such as hydrogen sulfide. Bacteria with similar metabolic talents support communities of cave-dwelling organisms.

▼ Figure 18.5 **Distribution of life in the biosphere.** In this image of Earth, colors are keyed to the relative abundance of chlorophyll, which correlates with the regional densities of photosynthetic organisms.

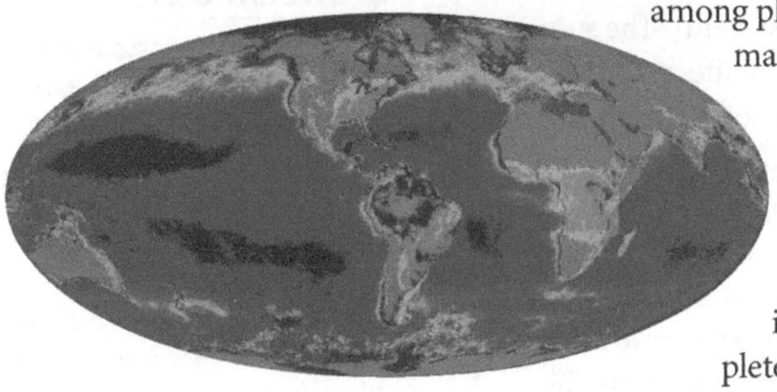

▼ Figure 18.6 **A deep-sea hydrothermal vent.** "Black smokers" west of Vancouver Island spew plumes of hot gases from Earth's interior. Giant tube worms (inset), annelids that may grow to 2 m long, are members of the vent community.

Temperature

Temperature is an important abiotic factor because of its effect on metabolism. Few organisms can maintain a sufficiently active metabolism at temperatures close to 0°C (32°F), and temperatures above 45°C (113°F) destroy the enzymes of most organisms. Most organisms function best within a specific range of environmental temperatures. For example, the American pika (Figure 18.7) has a high body temperature well suited to the chilly climate of its mountain habitat. On warm days, however, pikas must take refuge in crevices where pockets of cold air prevent fatal overheating. In winter, pikas depend on a blanket of snow to insulate their shelters from the perilous cold.

▲ Figure 18.7 **An American pika.** This diminutive relative of rabbits lives at high elevations in the western United States and Canada.

Water

Water is essential to all life. Aquatic organisms are surrounded by water, but they face problems of water balance if their own solute concentration does not match that of their surroundings (see Figure 5.14). For terrestrial organisms, the primary threat is drying out in the air. Many land animals have watertight coverings that reduce water loss, such as reptilian scales (Figure 18.8a). Most plants have waxy coatings on their leaves and other aerial parts (Figure 18.8b). Wax from carnauba palm leaves is valued for the glossy, waterproof coat it imparts to polishing products for cars, surfboards, furniture, and shoes. Carnauba wax is an ingredient of many other products as well, including cosmetics such as lipstick and mascara.

Inorganic Nutrients

The distribution and abundance of photosynthetic organisms, including plants, algae, and photosynthetic bacteria, depend on the availability of inorganic nutrients such as compounds of nitrogen and phosphorus. Plants obtain these nutrients from the soil. Soil structure, pH, and nutrient content often play major roles in determining the distribution of plants. In many aquatic ecosystems, low levels of nitrogen and phosphorus limit the growth of algae and photosynthetic bacteria.

Other Aquatic Factors

Several abiotic factors are important in aquatic, but not terrestrial, ecosystems. While terrestrial organisms have a plentiful supply of oxygen from the air, aquatic organisms must depend on oxygen dissolved in water. This is a critical factor for many species of fish. Cold, fast-moving water has a higher oxygen content than warm or stagnant water. Salinity (saltiness), currents, and tides also play a role in many aquatic ecosystems.

Other Terrestrial Factors

Some abiotic factors affect terrestrial, but not aquatic, ecosystems. For example, wind is often an important factor on land. Wind increases an organism's rate of water loss by evaporation. The resulting increase in evaporative cooling can be advantageous on a hot summer day, but it can cause dangerous wind chill in the winter. In some ecosystems, frequent occurrences of natural disturbances such as storms or fire play a role in the distribution of organisms. ☑

☑ CHECKPOINT

Why is solar energy such an important factor for most ecosystems?

Answer: Solar energy captured by the process of photosynthesis provides most of the organic fuel and building material for the organisms in those ecosystems.

▼ Figure 18.8 **Watertight coverings.**

(a) Scales on a Carolina anole lizard.

(b) Beaded water droplets showing the water repellency of the leaf's waxy coating.

The Evolutionary Adaptations of Organisms

The ability of organisms to live in Earth's diverse environments demonstrates the close relationship between the fields of ecology and evolutionary biology. Charles Darwin was an ecologist, although he predated the word *ecology*. It was the geographic distribution of organisms and their exquisite adaptations to specific environments that provided Darwin with evidence for evolution. Evolutionary adaptation via natural selection results from the interactions of organisms with their environments, which brings us back to our definition of ecology. Thus, events that occur in the short term, during the course of an individual's lifetime, may translate into effects over the longer scale of evolutionary time. For example, because the availability of water affects a plant's growth and ultimately its reproductive success, precipitation has an impact on the gene pool of a plant population. After a period of lower-than-average rainfall, drought-resistant individuals may be more prevalent in a plant population. Organisms also evolve in response to biotic interactions, such as predation and competition. ☑

✓ CHECKPOINT

How are the fields of ecology and evolution linked?

Answer: The process of evolutionary adaptation via natural selection results from the interactions of organisms with their environments (ecology).

▼ **Figure 18.9 A mourning dove demonstrating its physiological response to cold weather.**

Your "goose bumps" are caused by the same muscles that fluff up geese on a cold day.

Adjusting to Environmental Variability

The abiotic factors in a habitat may vary from year to year, seasonally, or over the course of a day. An individual's abilities to adjust to environmental changes that occur during its lifetime are themselves adaptations refined by natural selection. For instance, if you see a bird on a cold day, it may look unusually fluffy **(Figure 18.9)**. Small muscles in the skin raise the bird's feathers, a physiological response that traps insulating pockets of air. Some species of birds adjust to seasonal cold by growing heavier feathers. And some bird species respond to the onset of cold weather by migrating to warmer regions—a behavioral response. Note that these responses occur during the lifetime of an individual, so they are not examples of evolution, which is change in a population over time.

Physiological Responses

Like birds, mammals can adjust to a cold day by contracting skin muscles—in this case attached to hairs—to create a temporary layer of insulation. (Our own muscles do this, too, but we just get "goose bumps" instead of a furry insulation.) The blood vessels in the skin also constrict, which slows the loss of body heat. In both cases, the adjustment occurs in just seconds.

A gradual, though still reversible, physiological adjustment that occurs in response to an environmental change is called **acclimation**. For example, suppose you moved from Boston, which is essentially at sea level, to the mile-high city of Denver, where there is less oxygen. One physiological response to your new environment would be a gradual increase in the number of your red blood cells, which transport O_2 from your lungs to other parts of your body. Acclimation can take days or weeks. This is why high-altitude climbers, such as those attempting to scale Mount Everest, need extended stays at a high-elevation base camp before proceeding to the summit.

The ability to acclimate is generally related to the range of environmental conditions a species naturally experiences. Species that live in very warm climates, for example, usually cannot acclimate to extreme cold. Among vertebrates, birds and mammals can generally tolerate the greatest temperature extremes because, as endotherms, they use their metabolism to regulate internal temperature. In contrast, ectothermic reptiles can tolerate only a more limited range of temperatures **(Figure 18.10)**.

▼ **Figure 18.10 The number of lizard species in different regions of the contiguous United States.** Notice that there are fewer and fewer lizard species in more northern regions. This reflects lizards' ectothermic physiology, which depends on environmental heat for keeping the body warm enough for the animal to be active.

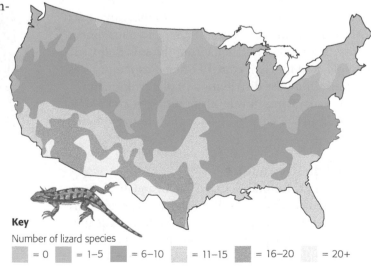

Key

Number of lizard species

■ = 0   ■ = 1–5   ■ = 6–10   ■ = 11–15   ■ = 16–20   ■ = 20+

Anatomical Responses

Many organisms respond to environmental challenge with some type of change in body shape or structure. When the change is reversible, the response is an example of acclimation. Many mammals, for example, grow a heavier coat of fur before the winter cold sets in and shed it when summer comes. In some animals, fur or feather color changes seasonally as well, camouflaging the animal against winter snow and summer vegetation (**Figure 18.11**).

Other anatomical changes are irreversible over the lifetime of an individual. Environmental variation can affect growth and development so much that there may be remarkable differences in body shape within a population. You can see an example in **Figure 18.12**, which shows the "flagging" that wind causes in certain trees. In general, plants are more anatomically changeable than animals. Rooted and unable to move to a better location, plants rely entirely on their anatomical and physiological responses to survive environmental fluctuations.

▲ Figure 18.11 **The arctic fox in winter and summer coats.**

▼ Figure 18.12 **Wind as an abiotic factor that shapes trees.** The mechanical disturbance of the prevailing wind hinders limb growth on the windward side of this fir tree near the timberline in the Rocky Mountains, while limbs on the other side grow normally. This anatomical response is an evolutionary adaptation that reduces the number of limbs that are broken during strong winds.

Behavioral Responses

In contrast to plants, most animals can respond to an unfavorable change in the environment by moving to a new location. Such movement may be fairly localized. For example, many desert ectotherms, including reptiles, maintain a reasonably constant body temperature by shuttling between sun and shade. Some animals are capable of migrating great distances in response to such environmental cues as the changing seasons. Many migratory birds overwinter in Central and South America, returning to northern latitudes to breed during summer. And we humans, with our large brains and available technology, have an especially rich range of behavioral responses available to us (**Figure 18.13**). ☑

☑ CHECKPOINT

What is acclimation?

Answer: a gradual, reversible change in anatomy or physiology in response to an environmental change

▼ Figure 18.13 **Behavioral responses have expanded the geographic range of humans.** Dressing for the weather is a thermoregulatory behavior unique to people.

Biomes

The abiotic factors you learned about in the previous section are largely responsible for the distribution of life on Earth. (You'll learn about the role of biotic factors in species distribution in Chapter 20.) Using various combinations of these factors, ecologists have categorized Earth's environments into biomes. A **biome** is a major terrestrial or aquatic life zone, characterized by vegetation type in terrestrial biomes and the physical environment in aquatic biomes. In this section, we'll briefly survey the aquatic biomes, followed by the terrestrial biomes.

Aquatic biomes, which occupy roughly 75% of Earth's surface, are determined by their salinity and other physical factors. Freshwater biomes (lakes, streams and rivers, and wetlands) typically have salt concentrations of less than 1%. The salt concentrations of marine biomes (oceans, intertidal zones, and coral reefs) are generally around 3%.

CHECKPOINT

Why does sewage cause heavy algal growth in lakes?

Answer: Sewage adds mineral nutrients that stimulate growth of the algae.

Freshwater Biomes

Freshwater biomes cover less than 1% of Earth, and they contain a mere 0.01% of its water. But they harbor a disproportionate share of biodiversity—an estimated 6% of all described species. Moreover, we depend on freshwater biomes for drinking water, crop irrigation, sanitation, and industry.

Freshwater biomes fall into two broad groups: standing water, which includes lakes and ponds, and flowing water, such as rivers and streams. The difference in water movement results in profound differences in ecosystem structure.

Lakes and Ponds

Standing bodies of water range from small ponds only a few square meters in area to large lakes, such as North America's Great Lakes, that are thousands of square kilometers (**Figure 18.14**).

◄ Figure 18.14 **A satellite view of the Great Lakes.**

In lakes and large ponds, communities of plants, algae, and animals are distributed according to the depth of the water and the distance from shore (**Figure 18.15**). Shallow water near shore and the upper layer of water away from shore make up the **photic zone**, so named because light is available for photosynthesis. Microscopic algae and cyanobacteria grow in the photic zone, joined by rooted plants and floating plants such as water lilies in the photic area near shore. If a lake or pond is deep enough or murky enough, it has an **aphotic zone**, where light levels are too low to support photosynthesis.

The **benthic realm** is at the bottom of all aquatic biomes. Made up of sand and organic and inorganic sediments, the benthic realm is occupied by communities of organisms that may include algae, aquatic plants, worms, insect larvae, molluscs, and microorganisms. Dead material that "rains" down from the productive surface waters of the photic zone is a major source of food for animals of the benthic realm.

The mineral nutrients nitrogen and phosphorus typically regulate the growth of **phytoplankton**, the collective name for microscopic algae and cyanobacteria that drift near the surfaces of aquatic biomes. Many lakes and ponds are affected by large inputs of nitrogen and phosphorus from sewage and runoff from fertilized lawns and farms. These nutrients often produce heavy growth of algae, which reduces light penetration. When the algae die and decompose, a pond or lake can suffer serious oxygen depletion, killing fish that are adapted to high-oxygen conditions. ☑

▼ Figure 18.15 **Zones in a lake.**

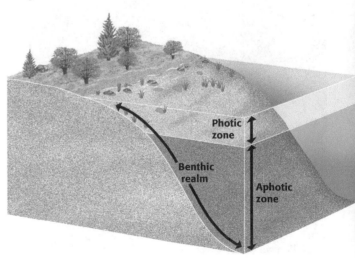

Photic zone

Benthic realm

Aphotic zone

▲ Figure 18.16 **A stream in the Appalachian Mountains.**

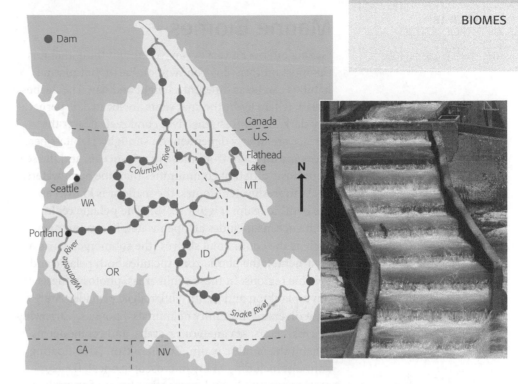

▲ Figure 18.17 **Damming the Columbia River basin.** This map shows only the largest of the 250 dams that have altered freshwater ecosystems throughout the Pacific Northwest. These great concrete obstacles make it difficult for salmon to swim upriver to their breeding streams, though many dams now have "fish ladders" that provide detours (inset).

Rivers and Streams

Rivers and streams, which are bodies of flowing water, generally support communities of organisms quite different from those of lakes and ponds **(Figure 18.16)**. A river or stream changes greatly between its source (perhaps a spring or snowmelt in the mountains) and the point at which it empties into a lake or the ocean. Near a source, the water is usually cold, low in nutrients, and clear. The channel is often narrow, with a swift current that does not allow much silt to accumulate on the bottom. The current also inhibits the growth of phytoplankton; most of the organisms found here are supported by the photosynthesis of algae attached to rocks or by organic material (such as leaves) carried into the stream from the surrounding land. The most abundant benthic animals are usually insects that eat algae, leaves, or one another. Trout are often the predominant fishes, locating their food, including insects, mainly by sight in the clear water.

Downstream, a river or stream typically widens and slows. There the water is usually warmer and may be murkier because of sediments and phytoplankton suspended in it. Worms and insects that burrow into mud are often abundant, as are waterfowl, frogs, and catfish and other fishes that find food more by scent and taste than by sight.

People have altered rivers by constructing dams to control flooding, to provide reservoirs of drinking water, or to generate hydroelectric power. In many cases, dams have completely changed the downstream ecosystems, altering the rate and volume of water flow and affecting fish and invertebrate populations **(Figure 18.17)**. Many streams and rivers have also been affected by pollution from human activities.

Wetlands

A **wetland** is a transitional biome between an aquatic ecosystem and a terrestrial one. Freshwater wetlands include swamps, bogs, and marshes **(Figure 18.18)**. Covered with water either permanently or periodically, wetlands support the growth of aquatic plants and are rich in species diversity. Migrating waterfowl and many other birds depend on wetland "pit stops" for food and shelter during their journeys. In addition, wetlands provide water storage areas that reduce flooding. Wetlands also improve water quality by trapping pollutants such as metals and organic compounds in their sediments.

► Figure 18.18 **A wetland near Kent, Ohio.**

Marine Biomes

Gazing out over a vast ocean, you might think that it is the most uniform environment on Earth. But marine habitats can be as different as night and day. The deepest ocean, where hydrothermal vents are located, is perpetually dark. In contrast, the vivid coral reefs nearer the surface are utterly dependent on sunlight. Habitats near shore are different from those in mid-ocean, and the seafloor hosts different communities than the open waters.

As in freshwater biomes, the seafloor is known as the benthic realm (**Figure 18.19**). The **pelagic realm** of the oceans includes all open water. In shallow areas, such as the continental shelves (the submerged parts of continents), the photic zone includes both pelagic and benthic regions. In these sunlit areas, photosynthesis by phytoplankton and multicellular algae provides energy for a diverse community of animals. Sponges, burrowing worms, clams, sea anemones, crabs, and echinoderms inhabit the benthic realm. **Zooplankton** (free-floating animals, including many microscopic ones), fishes, marine mammals, and many other types of animals are abundant in the pelagic photic zone.

The **coral reef** biome occurs in the photic zone of warm tropical waters in scattered locations around the globe (**Figure 18.20**). A coral reef is built up slowly by successive generations of coral animals—a diverse group of cnidarians that secrete a hard external skeleton—and by multicellular algae encrusted with limestone. Unicellular algae live within the coral's cells, providing the coral with food. The physical structure and productivity of coral reefs support a huge variety of invertebrates and fishes.

The photic zone extends down a maximum of 200 m (about 656 feet) in the ocean. Although there is not enough light for photosynthesis between 200 and 1,000 m (down a little more than half a mile), some light does reach these depths of the aphotic zone. This dimly lit world, sometimes called the twilight zone, is dominated by a fascinating variety of small fishes and crustaceans. Food sinking from the photic zone provides some sustenance for these animals. In addition, many of them migrate to the surface at night to feed. Some fishes in the twilight zone have enlarged eyes, enabling them to see in the very dim light, and light-emitting organs that attract mates and prey.

Below 1,000 m—a depth greater than the height of two Empire State Buildings—the ocean is completely and permanently dark. Adaptation to this environment has produced many bizarre-looking creatures. Most of the benthic organisms here are deposit feeders, animals that consume dead organic material in the sediments on the seafloor. Crustaceans, polychaete worms, sea anemones, and echinoderms such as sea cucumbers, sea stars,

▼ **Figure 18.19 Ocean life.** (Zone depths and organisms not drawn to scale.)

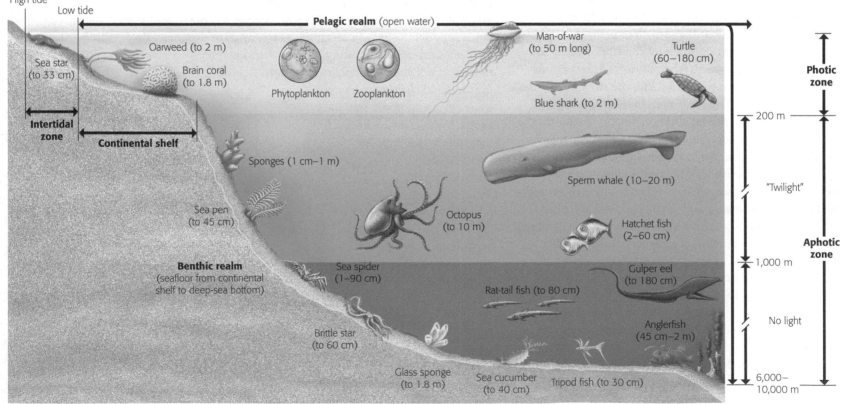

▲ Figure 18.20 **A coral reef in the Red Sea off the coast of Egypt.**

▼ Figure 18.21 **Organisms clinging to the rocks of an intertidal zone on the Pacific coast of Washington State.**

and sea urchins are common. Food is scarce, however. The density of animals is low except at hydrothermal vents, the prokaryote-powered ecosystems mentioned earlier (see Figure 18.6).

The marine environment also includes distinctive biomes, such as the intertidal zone and estuaries, where the ocean interfaces with land or with fresh water. In the **intertidal zone**, where the ocean meets land, the shore is pounded by waves during high tide and exposed to the sun and drying winds during low tide. The rocky intertidal zone is home to many sedentary organisms, such as algae, barnacles, and mussels, which attach to rocks and are thus prevented from being washed away (Figure 18.21). On sandy beaches, suspension-feeding worms, clams, and predatory crustaceans bury themselves in the ground.

Figure 18.22 shows an **estuary**, a transition area between a river and the ocean. The saltiness of estuaries ranges from nearly that of fresh water to that of the ocean. With their waters enriched by nutrients from rivers, estuaries, like freshwater wetlands, are among the most productive areas on Earth. Oysters, crabs, and many fishes live or reproduce in estuaries. Estuaries are also crucial nesting and feeding areas for waterfowl. Mudflats and salt marshes are extensive coastal wetlands that often border estuaries.

For centuries, people viewed the ocean as a limitless resource, harvesting its bounty with increasingly effective and indiscriminate technologies and using it as a dumping ground for wastes. The negative effects of these practices are now becoming clear. Populations of commercial fish species are declining. Small bits of plastic debris float beneath the surface of vast swaths of the Pacific Ocean, concentrated by converging currents in a region dubbed the "Great Pacific Garbage Patch."

Many marine habitats are polluted by nutrients or toxic chemicals; it will be years before the full extent of damage from the massive Deepwater Horizon oil spill in the Gulf of Mexico in 2010 is known. Because of their proximity to land, estuaries are especially vulnerable. Many have been completely replaced by development on landfill; other threats include pollution and alteration of freshwater inflow. Coral reefs are imperiled by ocean acidification and rising sea surface temperatures due to global warming.

Meanwhile, our knowledge of marine biomes is woefully incomplete. A massive international census of marine life, completed in 2010, announced the discovery of more than 6,000 new species. ☑

☑ CHECKPOINT

What are phytoplankton? Why are they essential to other oceanic life?

Answer: Phytoplankton are photosynthetic algae and bacteria. They are food for animals in the photic zone; those animals in turn may become food for animals in the aphotic zone.

▼ Figure 18.22 **Waterfowl in an estuary on the southeast coast of England.**

417

How Climate Affects Terrestrial Biome Distribution

Terrestrial biomes are determined primarily by climate, especially temperature and rainfall. Before we survey these biomes, let's look at the broad patterns of global climate that help explain their locations.

Earth's global climate patterns are largely the result of the input of solar energy, which warms the atmosphere, land, and water, and of the planet's movement in space. Because of Earth's curvature, the intensity of sunlight varies according to latitude (**Figure 18.23**). The equator receives the greatest intensity of solar radiation and thus has the highest temperatures, which in turn evaporate water from Earth's surface. As this warm, moist air rises, it cools, diminishing its ability to hold moisture. The water vapor condenses into clouds and rain falls (**Figure 18.24**). This process largely explains why rain forests are concentrated in the **tropics**—the region from the Tropic of Cancer to the Tropic of Capricorn.

After losing moisture over equatorial zones, dry high-altitude air masses spread away from the equator until they cool and descend at latitudes of about 30° north and south. Many of the world's great deserts—the Sahara in North Africa and the Arabian on the Arabian Peninsula, for example—are centered at these latitudes because of the dry air they receive.

Latitudes between the tropics and the Arctic Circle and the Antarctic Circle are called **temperate zones**. Generally, these regions have milder climates than the tropics or the polar regions. Notice in Figure 18.24 that some of the descending, dry air moves into the latitudes above 30°. At first these air masses pick up moisture, but they tend to drop it as they cool at higher latitudes. This is why the north and south temperate zones tend to be relatively wet. Coniferous forests dominate the landscape at the wet but cool latitudes around 60° north.

Proximity to large bodies of water and the presence of landforms such as mountain ranges also affect climate. Oceans and large lakes moderate climate by absorbing heat when the air is warm and releasing heat to cold air. Mountains affect climate in two major ways. First, air temperature drops as elevation increases. As a result, driving up a tall mountain offers a quick tour of several biomes. **Figure 18.25** shows the scenery you might encounter on a journey from the scorching lowlands of the Sonoran Desert to a cool coniferous forest at an elevation of 11,000 feet above sea level.

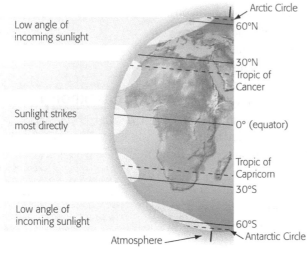

▼ Figure 18.23 **Uneven heating of Earth.**

▼ Figure 18.24 **How uneven heating of Earth produces various climates.**

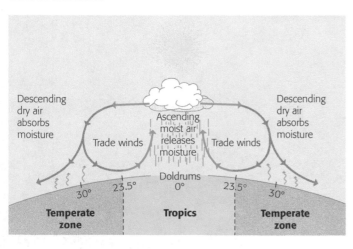

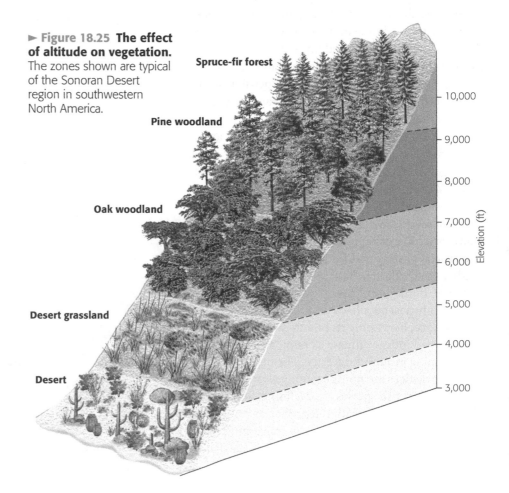

► Figure 18.25 **The effect of altitude on vegetation.** The zones shown are typical of the Sonoran Desert region in southwestern North America.

Second, mountains can block the flow of moist air from a coast, causing radically different climates on opposite sides of a mountain range. In the example shown in **Figure 18.26**, moist air moves in off the Pacific Ocean and encounters the Coast Range in California. Air flows upward, cools at higher altitudes, and drops a large amount of rainfall. The world's tallest trees, the coastal redwoods, thrive here. Precipitation increases again farther inland as the air moves up and over higher mountains (the Sierra Nevada). By the time it reaches the eastern side of the Sierra, the air contains little moisture; as this dry air descends, it absorbs moisture. As a result, there is little precipitation on the eastern side of the mountains. This effect, called a rain shadow, is responsible for the desert that covers much of central Nevada. ☑

▼ Figure 18.26 **How mountains affect rainfall.**

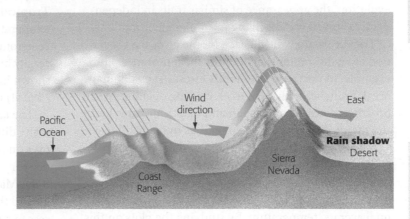

☑ **CHECKPOINT**
Why is there so much rainfall in the tropics?

Answer: Air at the equator rises as it is warmed by direct sunlight. As the air rises, it cools. This causes cloud formation and rainfall because cool air holds less moisture than warm air.

Terrestrial Biomes

Terrestrial ecosystems are grouped into biomes primarily on the basis of their vegetation type (**Figure 18.27**). By providing food, shelter, and nesting sites for animals, as well as much of the organic material for the decomposers that recycle mineral nutrients, plants build the foundation for the communities of organisms typical of each biome. The geographic distribution of plants, and thus of biomes, largely depends on climate, with temperature and rainfall often the key factors determining the kind of biome that exists in a particular region. If the climate in two geographically separate areas is similar, the same type of biome may occur in both. Coniferous forests, for instance, extend in a broad band across North America, Europe, and Asia.

Each biome is characterized by a type of biological community rather than an assemblage of particular species. For example, the groups of species living in the deserts of southwestern North America and in the Sahara Desert of Africa are different, but both groups are adapted to desert conditions. Organisms in widely

► Figure 18.27 **A map of the major terrestrial biomes.** Although this map has sharp boundaries, biomes actually grade into one another. We'll use smaller versions of this map, highlighted by color coding, during our closer look at the terrestrial biomes in the next several pages.

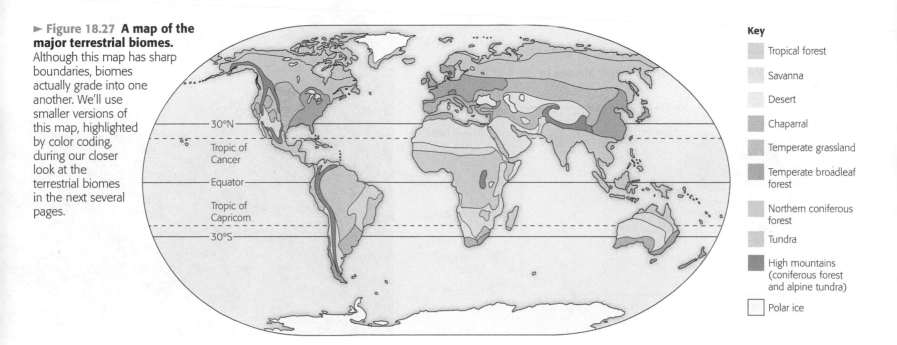

Key
Tropical forest
Savanna
Desert
Chaparral
Temperate grassland
Temperate broadleaf forest
Northern coniferous forest
Tundra
High mountains (coniferous forest and alpine tundra)
Polar ice

separated biomes may look alike because of convergent evolution, the appearance of similar traits in independently evolved species living in similar environments.

Local variation within each biome gives the vegetation a patchy, rather than a uniform, appearance. For example, in northern coniferous forests, snowfall may break branches and small trees, causing openings where broadleaf trees such as aspen and birch can grow. Local storms and fires also create openings in many biomes.

The graph in **Figure 18.28** shows the range of precipitation and temperature that characterize terrestrial biomes. The *x*-axis shows the range of annual average precipitation, and the *y*-axis displays the range of annual average temperature. By studying the plots on this graph, we can compare these abiotic factors in different biomes. For example, although the range of precipitation in temperate broadleaf forests is similar to that of northern coniferous forests, the lower range of temperatures in northern coniferous forests reveals a significant difference in the abiotic environments of these two biomes. Grasslands are typically drier than forests, and deserts are drier still.

Today, concern about global warming is generating intense interest in the effect of climate on vegetation patterns. Using powerful new tools such as satellite imagery, scientists are documenting shifts in latitude of biome borders, decreases in snow and ice coverage, and changes in the length of the growing season. At the same time, many natural biomes have been fragmented and altered by human activity. We'll discuss both of these issues after we survey the major terrestrial biomes, beginning near the equator and generally approaching the poles.

☑ CHECKPOINT

Why are climbing plants common in tropical rain forests?

Answer: Climbing is a plant adaptation for reaching sunlight in a closed canopy, where little sunlight reaches the forest floor.

Tropical Forest

Tropical forests occur in equatorial areas where the temperature is warm and days are 11–12 hours long year-round. The type of vegetation is determined primarily by rainfall. Tropical rain forests, like the one shown in **Figure 18.29**, receive 200–400 cm (6.6 to 13 *feet*!) of rain per year.

The layered structure of tropical rain forests provides many different habitats. Treetops form a closed canopy over one or two layers of smaller trees and a shrub understory. Few plants grow in the deep shade of the forest floor. Many trees are covered by woody vines growing toward the light. Other plants, such as orchids, gain access to sunlight by growing on the branches or trunks of tall trees. Scattered trees reach full sunlight by towering above the canopy. Many of the animals also dwell in trees, where food is abundant. Monkeys, birds, insects, snakes, bats, and frogs find food and shelter many meters above the ground.

Rainfall is less plentiful in other tropical forests. Tropical dry forests predominate in lowland areas that have a prolonged dry season or scarce rainfall at any time. The plants found there are a mixture of thorny shrubs and trees and succulents. In regions with distinct wet and dry seasons, tropical deciduous trees are common. ☑

▼ **Figure 18.29 Tropical rain forest in Borneo.**

▼ **Figure 18.28 A climate graph for some major biomes in North America.**

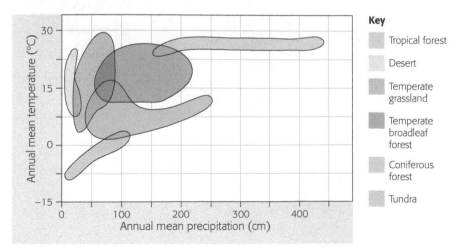

Key

Tropical forest

Desert

Temperate grassland

Temperate broadleaf forest

Coniferous forest

Tundra

Savanna

Savannas, such as the one shown in **Figure 18.30**, are dominated by grasses and scattered trees. The temperature is warm year-round. Rainfall averages 30–50 cm (roughly 12–20 inches) per year, with dramatic seasonal variation.

Fire, caused by lightning or human activity, is an important abiotic factor in the savanna. The grasses survive burning because the growing points of their shoots are below ground. Other plants have seeds that sprout rapidly after a fire. Poor soil and lack of moisture, along with fire and grazing animals, prevent the establishment of most trees. The luxuriant growth of grasses and small broadleaf plants during the rainy season provides a rich food source for plant-eating animals.

Many of the world's large grazing mammals and their predators inhabit savannas. African savannas are home to zebras and many species of antelope, as well as to lions and cheetahs. Several species of kangaroo are the dominant grazers of Australian savannas. Oddly, though, the large grazers are not the dominant plant-eaters in savannas. That distinction belongs to insects, especially ants and termites. Other animals include burrowers such as mice, moles, gophers, and ground squirrels.

Desert

Deserts are the driest of all biomes, characterized by low and unpredictable rainfall—less than 30 cm (about 12 inches) per year. Some deserts are very hot, with daytime soil surface temperatures above 60°C (140°F) and large daily temperature fluctuations. Other deserts, such as those west of the Rocky Mountains and the Gobi Desert, spanning northern China and southern Mongolia, are relatively cold. Air temperatures in cold deserts may fall below −30°C (−22°F).

Desert vegetation typically includes water-storing plants, such as cacti, and deeply rooted shrubs. Various snakes, lizards, and seed-eating rodents are common inhabitants. Arthropods such as scorpions and insects also thrive in the desert. Evolutionary adaptations of desert plants and animals include a remarkable array of mechanisms that conserve water. For example, the "pleated" stem of saguaro cacti **(Figure 18.31)** enables the plants to expand when they absorb water during wet periods. Some desert mice *never* drink, deriving all their water from the metabolic breakdown of the seeds they eat. Protective adaptations that deter feeding by mammals and insects, such as spines on cacti and poisons in the leaves of shrubs, are common in desert plants. ☑

☑ **CHECKPOINT**

1. How does the savanna climate vary seasonally?
2. What abiotic factor characterizes deserts?

Answers: 1. *Temperature stays about the same year-round, but rainfall varies dramatically.* 2. *Rainfall is low and unpredictable.*

▼ Figure 18.30 **Savanna in the Serengeti Plain in Tanzania.**

▼ Figure 18.31 **Sonoran Desert.**

Chaparral

The climate that supports **chaparral** vegetation results mainly from cool ocean currents circulating offshore, producing mild, rainy winters. Summers are hot and dry. This biome is limited to small coastal areas, some in California (Figure 18.32). The largest region of chaparral surrounds the Mediterranean Sea; in fact, Mediterranean is another name for this biome. Dense, spiny, evergreen shrubs dominate chaparral. Annual plants are also common during the wet winter and spring months. Animals characteristic of the chaparral are deer, fruit-eating birds, seed-eating rodents, and lizards and snakes.

Chaparral vegetation is adapted to periodic fires caused by lightning. Many plants contain flammable chemicals and burn fiercely, especially where dead brush has accumulated. After a fire, shrubs use food reserves stored in the surviving roots to support rapid shoot regeneration. Some chaparral plants produce seeds that will germinate only after a hot fire. The ashes of burned vegetation fertilize the soil with mineral nutrients, promoting regrowth of the plant community. Houses do not fare as well. The firestorms that race through the densely populated canyons of Southern California can be devastating to the human inhabitants.

☑ **CHECKPOINT**

1. What is one way that homeowners in chaparral areas can protect their neighborhoods from fire?
2. How do people now use most of the North American land that was once temperate grassland?

Answers: **1.** by keeping the area clear of dead brush, which is flammable **2.** for farming

Temperate Grassland

Temperate grasslands have some of the characteristics of tropical savannas, but they are mostly treeless, except along rivers or streams, and are found in regions of relatively cold winter temperatures. Rainfall, averaging between 25 and 75 cm per year (approximately 10–30 inches), with frequent severe droughts, is too low to support forest growth. Periodic fires and grazing by large mammals also prevent invasion by woody plants. These grazers include the bison and pronghorn in North America, the wild horses and sheep of the Asian steppes, and kangaroos in Australia. As in the savanna, however, the dominant plant-eaters are invertebrates, especially grasshoppers and soil-dwelling nematodes.

Without trees, many birds nest on the ground. Many small mammals, such as rabbits, voles, ground squirrels, prairie dogs, and pocket gophers, dig burrows to escape predators. Temperate grasslands like the one shown in Figure 18.33 once covered much of central North America.

Because grassland soil is both deep and rich in nutrients, these habitats provide fertile land for agriculture. Most grassland in the United States has been converted to cropland or pasture, and very little natural prairie exists today. ☑

▼ **Figure 18.32 Chaparral in California.**

▼ **Figure 18.33 Temperate grassland in Saskatchewan, Canada.**

Temperate Broadleaf Forest

Temperate broadleaf forests occur throughout mid-latitudes where there is sufficient moisture to support the growth of large trees. Annual precipitation is relatively high at 75–150 cm (30–60 inches) and typically distributed evenly around the year. Temperature varies seasonally over a wide range, with hot summers and cold winters. In the Northern Hemisphere, dense stands of deciduous trees are trademarks of temperate forests, such as the one pictured in **Figure 18.34**. Deciduous trees drop their leaves before winter, when temperatures are too low for effective photosynthesis and water lost by evaporation is not easily replaced from frozen soil.

Numerous invertebrates live in the soil and the thick layer of leaf litter that accumulates on the forest floor. Some vertebrates, such as mice, shrews, and ground squirrels, burrow for shelter and food, while others, including many species of birds, live in the trees. Predators include bobcats, foxes, black bears, and mountain lions. Many mammals that inhabit these forests enter a dormant winter state called hibernation, and some bird species migrate to warmer climates.

Virtually all the original temperate broadleaf forests in North America were cut for timber or cleared for agriculture or development. These forests tend to recover after disturbance, however, and today we see deciduous trees growing in undeveloped areas over much of their former range.

Coniferous Forest

Cone-bearing evergreen trees such as pine, spruce, fir, and hemlock dominate **coniferous forests** in the Northern Hemisphere. (Other kinds of conifers grow in parts of South America, Africa, and Australia.) The northern coniferous forest, or **taiga** (Figure 18.35), is the largest terrestrial biome on Earth, stretching in a broad band across North America and Asia south of the Arctic Circle. Taiga is also found at cool, high elevations in more temperate latitudes—for example, in much of the mountainous region of western North America. The taiga is characterized by long, snowy winters and short, wet summers that are sometimes warm. The slow decomposition of conifer needles in the thin, acidic soil makes few nutrients available for plant growth. The conical shape of many conifers prevents too much snow from accumulating on their branches and breaking them. Animals of the taiga include moose, elk, hares, bears, wolves, grouse, and migratory birds. The Asian taiga is home to the dwindling number of Siberian tigers that remain in the wild.

The **temperate rain forests** of coastal North America (from Alaska to Oregon) are also coniferous forests. Warm, moist air from the Pacific Ocean supports this unique biome, which, like most coniferous forests, is dominated by a few tree species, typically hemlock, Douglas fir, and redwood. These forests are heavily logged, and the old-growth stands of trees are rapidly disappearing. ☑

☑ CHECKPOINT

1. How does the loss of leaves function as an adaptation of deciduous trees to cold winters?
2. What type of trees are characteristic of the taiga?

Answers: **1.** by reducing loss of water from the trees when that water cannot be replaced because of frozen soil **2.** conifers such as pine, spruce, fir, and hemlock

▼ **Figure 18.34**
Temperate broadleaf forest in Vermont in autumn.

Temperature range | Precipitation

▼ **Figure 18.35**
Northern coniferous forest in Finland, with the sky lit by the northern lights.

Temperature range | Precipitation

Tundra

Tundra covers expansive areas of the Arctic between the taiga and polar ice. **Permafrost** (permanently frozen subsoil), bitterly cold temperatures, and high winds are responsible for the absence of trees and other tall plants in the arctic tundra shown in **Figure 18.36**. The arctic tundra receives very little annual precipitation. However, water cannot penetrate the underlying permafrost, so melted snow and ice accumulate in pools on the shallow topsoil during the short summer.

Tundra vegetation includes small shrubs, grasses, mosses, and lichens. When summer arrives, flowering plants grow quickly and bloom in a rapid burst. Caribou, musk oxen, wolves, and small rodents called lemmings are among the mammals found in the arctic tundra. Many migratory birds use the tundra as a summer breeding ground. During the brief but productive warm season, the marshy ground supports the aquatic larvae of insects, providing food for migratory waterfowl, and clouds of mosquitoes often fill the tundra air.

On very high mountaintops at all latitudes, including the tropics, high winds and cold temperatures create plant communities called alpine tundra. Although these communities are similar to arctic tundra, there is no permafrost beneath alpine tundra.

Polar Ice

Polar ice covers the land at high latitudes north of the arctic tundra in the Northern Hemisphere and in Antarctica in the Southern Hemisphere (**Figure 18.37**). The temperature in these regions is extremely cold year-round, and precipitation is very low. Only a small portion of these landmasses is free of ice or snow, even during the summer. Nevertheless, small plants, such as mosses and lichens, eke out a living, and invertebrates such as nematodes, mites, and wingless insects called springtails inhabit the frigid soil. Nearby sea ice provides feeding platforms for large animals such as polar bears (in the Northern Hemisphere), penguins (in the Southern Hemisphere), and seals. Seals, penguins, and other marine birds visit the land to rest and breed. The polar marine biome provides the food that sustains these birds and mammals. In the Antarctic, penguins feed at sea, eating a variety of fish, squids, and small shrimplike crustaceans known as krill. Antarctic krill, an important food source for many species of fish, seals, squids, seabirds, and filter-feeding whales as well as penguins, depend on sea ice for breeding and as a refuge from predators. As the amount and duration of sea ice decline as a consequence of global climate change, krill habitat is shrinking. ☑

▼ Figure 18.36 **Arctic tundra in Yukon Territory, Canada.**

▼ Figure 18.37 **Polar ice in Antarctica.**

Interconnections within Systems

The Water Cycle

Biomes are not self-contained units. Rather, all parts of the biosphere are linked by the global water cycle, illustrated in **Figure 18.38**, and by nutrient cycles (see Chapter 20). Consequently, events in one biome may reverberate throughout the biosphere.

As you learned earlier in this chapter, water and air move in global patterns driven by solar energy. Precipitation and evaporation continuously move water between the land, oceans, and the atmosphere. Water also evaporates from plants, which pull it from the soil in a process called transpiration.

Over the oceans, evaporation exceeds precipitation. The result is a net movement of water vapor to clouds that are carried by winds from the oceans across the land. On land, precipitation exceeds evaporation and transpiration. The excess precipitation may stay on the surface, or it may trickle through the soil to become groundwater. Both surface water and groundwater eventually flow back to the sea, completing the water cycle.

Just as the water draining from your shower carries dead skin cells from your body along with the day's grime, the water washing over and through the ground carries traces of the land and its history. For example, water flowing from land to the sea carries with it silt (fine soil particles) and chemicals such as fertilizers and pesticides.

Erosion from coastal development has caused silt to muddy the waters of some coral reefs, dimming the light available to the photosynthetic algae that power the reef community. Chemicals in surface water may travel hundreds of miles by stream and river to the ocean, where currents then carry them even farther from their point of origin. For instance, traces of pesticides and chemicals from industrial wastes have been found in marine mammals in the Arctic and in deep-sea octopuses and squids. Airborne pollutants such as nitrogen oxides and sulfur oxides, which combine with water to form acid precipitation, are distributed by the water cycle, too.

Human activity also affects the global water cycle itself in a number of important ways. One of the main sources of atmospheric water is transpiration from the dense vegetation making up tropical rain forests. The destruction of these forests changes the amount of water vapor in the air. Pumping large amounts of groundwater to the surface for irrigation increases the rate of evaporation over land and may deplete groundwater supplies. In addition, global warming affects the water cycle in complex ways that will have far-reaching effects on precipitation patterns. We'll consider some of these environmental impacts in the following sections. ☑

Airborne pollutants from car exhaust can combine with water and return to Earth as acid precipitation far away—evidence that the water cycle operates on a global scale.

☑ CHECKPOINT

What is the main way that living organisms contribute to the water cycle?

Answer: Plants move water from the ground to the air via transpiration.

▼ Figure 18.38 **The global water cycle.**

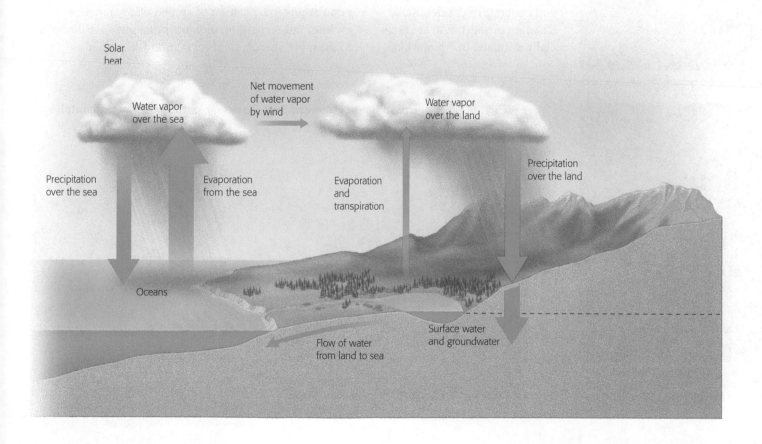

Solar heat

Water vapor over the sea

Net movement of water vapor by wind

Water vapor over the land

Precipitation over the sea

Evaporation from the sea

Evaporation and transpiration

Precipitation over the land

Oceans

Flow of water from land to sea

Surface water and groundwater

Human Impact on Biomes

For hundreds of years, people have been using increasingly effective technologies to capture or produce food, to extract resources from the environment, and to build cities. It is now clear that the environmental costs of these enterprises are staggering. In this section, you'll see some examples of how human activities are affecting forest and freshwater resources. Throughout the remainder of this unit, you'll learn about the role of ecological knowledge in achieving **sustainability**, the goal of developing, managing, and conserving Earth's resources in ways that meet the needs of people today without compromising the ability of future generations to meet their needs.

▼ Figure 18.39 **Satellite photos of the Rondonia area of the Brazilian rain forest.**

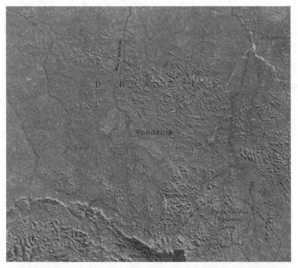

1975. In 1975, the forest in this remote region was virtually intact.

2001. Same area in 2001, after a paved highway through the region brought loggers and farmers. The "fishbone" pattern marks the new network of roads carved through the forest.

Forests

The map in Figure 18.27 shows the terrestrial biomes that would be expected to flourish under the prevailing climatic conditions. However, about three-quarters of Earth's land surface has been altered by thousands of years of human occupation. Most of the land that we've appropriated is used for agriculture; another hefty chunk is covered by the asphalt and concrete of development. Changes in vegetation are especially dramatic in regions like tropical forests that escaped large-scale human intervention until recently. Satellite photos of a small area in Brazil show how thoroughly a landscape can be altered in a short amount of time (Figure 18.39).

Every year, more and more forested land is cleared for agriculture. You might think that this land is needed to feed new mouths as the human population continues to grow, but that's not entirely the case. Unsustainable agricultural practices have degraded much of the world's cropland so severely that it is unusable. Researchers estimate that replacing worn-out farmland accounts for up to 80% of the deforestation occurring today. Tropical forests, such as the one in Figure 18.40, are also being cleared to grow palm oil for products such as cosmetics and a long list of packaged foods, including cookies, crackers, potato chips, chocolate products, and soups. Other forests are being lost to logging, mining, and air pollution, problems that are hitting coniferous forests especially hard. (As we mentioned in the previous section, most temperate broadleaf forests were replaced by human enterprises long ago.) Land that hasn't been directly converted to food production and living space also bears the imprint of our presence. Roads penetrate regions that are otherwise unaltered, bringing pollution to the wilderness, providing avenues for new diseases to emerge, and slicing vast tracts of biome into segments that are too small to support a full array of species.

Land uses that provide resources such as food, fuel, and shelter are clearly beneficial to us. But natural ecosystems also provide services that support the human population—purification of air and water, nutrient cycling, and recreation, to name just a few. (We'll return to the topic of ecosystem services in Chapter 20.)

▼ Figure 18.40 **Tropical forest in Indonesia that was clear-cut for a palm oil plantation.**

Fresh Water

The impact of human activities on freshwater ecosystems may pose an even greater threat to life on Earth—including ourselves—than the damage to terrestrial ecosystems. Freshwater ecosystems are being polluted by large amounts of nitrogen and phosphorus compounds that run off from heavily fertilized farms or from livestock feedlots. A wide variety of other pollutants, such as industrial wastes, also contaminate freshwater habitats, drinking water, and groundwater. Some regions of the world face dire shortages of water as a result of the overuse of groundwater for irrigation, extended droughts (partially caused by global climate change), or poor water management practices.

Las Vegas, the population center of Clark County, Nevada, is one example of a city whose water resources are increasingly stressed by drought and overuse. **Figure 18.41a** is a satellite photo of Las Vegas in 1973, when the population of Clark County was 319,400. **Figure 18.41b** shows the same area 40 years later, when the population had swelled to more than 2 million. In contrast to the disappearance of greenery in the photos of Brazilian rain forest, the mark of human activities in Figure 18.41b is the notable expansion of greenery—the result of watering lawns and golf courses. Las Vegas is situated in a high valley in the Mojave Desert. Where does it get the water to turn barren desert into green fields?

Las Vegas taps underground aquifers for some water, but its main water supply is Lake Mead. Lake Mead is an enormous reservoir formed by the Hoover Dam on the Colorado River, which in turn receives almost all of its water from snowmelt in the Rocky Mountains. With decreased annual snowfall, attributable largely to global warming, the flow of the Colorado has greatly diminished. The water level in Lake Mead has dropped drastically **(Figure 18.42)**, and parched cities and farms farther downstream are pleading for more water.

To ensure an adequate water supply for the future, Las Vegas is looking for new sources of water. Among other options, Las Vegas is eyeing the abundant supply of groundwater in the northern end of the valley where it lies. Although sparsely populated, that area is home to many ranchers whose livelihoods depend on the groundwater. It is also home to numerous endangered species. Not surprisingly, environmentalists and residents of the north valley are resisting efforts to pipe its groundwater to Las Vegas.

Nevada is just one of many places where the hard realities of climate change are beginning to affect daily life. Battles over water resources are shaping up throughout the arid West and Southwest of the United States, where changing precipitation patterns due to global warming are projected to continue the drought for many years to come. In other regions of the world, including China, India, and North Africa, the increasing demands of economic, agricultural, and population growth are straining water resources that are already scarce.

While policymakers are dealing with current crises and planning how to manage resources in the future, researchers are seeking methods of sustainable agriculture and water use. Basic ecological research is an essential component of ensuring that sufficient food and water will be available for people now—and for the generations to come. Next, we take a closer look at a major threat to sustainability: global climate change. ☑

▼ **Figure 18.42 Low water level in Lake Mead.** The white "bathtub ring" is caused by mineral deposits on rocks that were once submerged.

▼ **Figure 18.41 Satellite photos of Las Vegas, Nevada.**

(a) May 1973

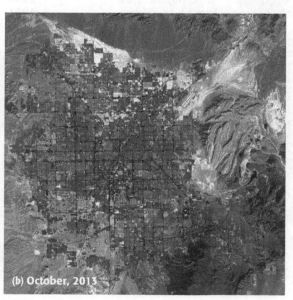

(b) October, 2013

Global Climate Change

Rising concentrations of carbon dioxide (CO_2) and certain other gases in the atmosphere are changing global climate patterns. This was the overarching conclusion of the assessment report released by the Intergovernmental Panel on Climate Change (IPCC) in 2014. Thousands of scientists and policymakers from more than 100 countries participated in producing the report, which is based on data published in thousands of scientific papers. Thus, there is no debate among scientists about whether climate change is occurring. In this section, you'll learn why it is occurring, how it is affecting the biosphere, and what you can do about it.

The Greenhouse Effect and Global Warming

Why is Earth's atmosphere becoming warmer? A useful analogy is a greenhouse, which is used to grow plants when the weather outside is too cold. Its transparent glass or plastic walls allow solar radiation to pass through, but they also trap some of the heat that accumulates inside the building. On a smaller scale, consider how hot it gets in a closed car on a sunny day. Similarly, certain gases in Earth's atmosphere are transparent to solar radiation but absorb or reflect heat. Some of these so-called **greenhouse gases** are natural, including CO_2, water vapor, and methane. Others, such as chlorofluorocarbons (CFCs, found in some aerosol sprays and refrigerants), are synthetic. As **Figure 18.43** shows, greenhouse gases act as a blanket that traps heat in the atmosphere. This heating effect, often called the **greenhouse effect**, is highly beneficial. Without it, the average air temperature on Earth would be a frigid −18°C (−2.4°F), far too cold for most life as we know it. However, increasing the insulation that the blanket provides is making Earth overly warm.

The signature effect of rapidly increasing greenhouse gases is the steady increase in the average global temperature, which has risen 0.8°C (1.4°F) over the last 100 years, with 75% of that increase occurring over the last three decades. Further increases of 2–4.5°C are likely by the end of this century, according to the 2014 IPPC report. Ocean temperatures are also rising, in deeper layers as well as at the surface. But the temperature increases are not distributed evenly around the globe. The largest increases are in the northernmost regions of the Northern Hemisphere and parts of Antarctica **(Figure 18.44)**. ☑

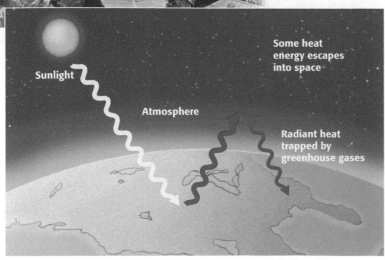

▼ Figure 18.43 **The greenhouse effect.** The atmosphere traps heat in the same way that glass keeps heat inside a greenhouse.

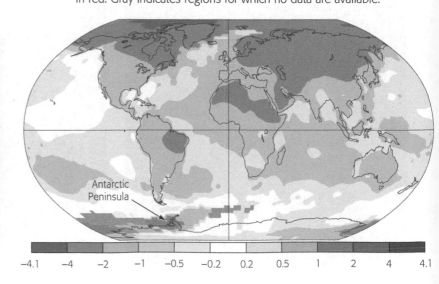

▼ Figure 18.44 **Differences in average temperatures during 2004–2013 compared with long-term averages during 1951–1980, in °C.** The largest temperature increases are shown in red. Gray indicates regions for which no data are available.

The Accumulation of Greenhouse Gases

After many years of data collection and debate, the vast majority of scientists are confident that human activities have caused the rising concentrations of greenhouse gases. Major sources of emissions include agriculture, landfills, and the burning of wood and fossil fuels (oil, coal, and natural gas).

Let's take a closer look at CO_2, the dominant greenhouse gas. For 650,000 years, the atmospheric concentration of CO_2 did not exceed 300 parts per million (ppm); the concentration before the Industrial Revolution was 280 ppm. In 2013, the average atmospheric CO_2 was 396 ppm, and continuing to rise (Figure 18.45). The levels of other greenhouse gases have increased dramatically, too. Remember that CO_2 is removed from the atmosphere by the process of photosynthesis and stored in organic molecules such as carbohydrates. (See Figure 6.2.) These molecules are eventually broken down by cellular respiration, releasing CO_2. Overall, uptake of CO_2 by photosynthesis roughly equals the release of CO_2 by cellular respiration (Figure 18.46). However, extensive deforestation has significantly decreased the incorporation of CO_2 into organic material. At the same time, CO_2 is flooding into the atmosphere from the burning of fossil fuels and wood, a process that releases CO_2 from organic material much more rapidly than cellular respiration.

CO_2 is also exchanged between the atmosphere and the surface waters of the oceans. For decades, the oceans have acted as massive sponges, soaking up considerably more CO_2 than they have released. But now, the excess CO_2 has made the oceans more acidic, a change that could have a profound effect on marine communities. As ocean acidification worsens, many species of plankton and marine animals such as corals and molluscs will be unable to build their shells or exoskeletons. Their demise will remove critical links from marine food webs and ultimately damage marine ecosystems around the world. ✓

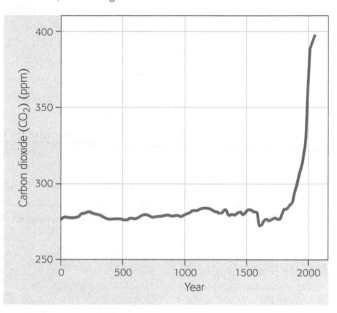

▼ Figure 18.45 **Atmospheric concentration of CO₂.** Notice that the concentration was relatively stable until the Industrial Revolution, which began in the late 1700s.

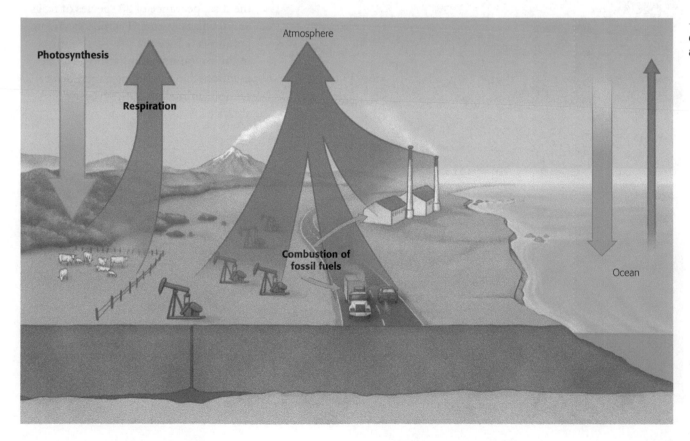

◀ Figure 18.46 **How CO₂ enters and leaves the atmosphere.**

Photosynthesis

Respiration

Atmosphere

Combustion of fossil fuels

Ocean

☑ CHECKPOINT

What is the major source of CO_2 released by human activities?

Answer: burning fossil fuels

Global Climate Change THE PROCESS OF SCIENCE

How Does Climate Change Affect Species Distribution?

As you've learned in this chapter, abiotic factors of the environment are fundamental determinants of where organisms live. It stands to reason that changes in temperature and precipitation patterns will have a significant effect on the distribution of life. With rising temperatures, the ranges of many species have already shifted toward the poles or to higher elevations. For example, shifts in the ranges of many bird species have been reported; the Inuit peoples living north of the Arctic Circle have sighted robins in the region for the first time.

Let's examine how a team of ecologists investigated the impact of climate change on European butterflies. Using the **observation** that the average temperature in Europe has risen 0.8°C (about 1.4°F) and that butterflies are sensitive to temperature change, the researchers asked the **question**: Have the ranges of butterflies changed in response to the temperature changes? This question led to the **hypothesis** that butterfly range boundaries are shifting in line with the warming trend. The researchers **predicted** that butterfly species will be establishing new populations to the north of their former ranges, and populations at the southern edges of their ranges will become extinct. The **experiment** involved analyzing historical data on the ranges of 35 species of butterflies in Europe. The **results** showed that more than 60% of the species have pushed their northern range boundaries poleward over the last century, some by as much as 150 miles. The southern boundaries have simultaneously contracted for some species, but not for others. **Figure 18.47** shows the range shift for the species *Argynnis paphia*.

While some organisms have the dispersal ability and the room for northward population shifts, species that live on mountaintops or in polar regions have nowhere to go. For example, researchers in Costa Rica have reported the disappearance of 20 species of frogs and toads as warmer Pacific Ocean temperatures reduce the dry-season mists in their mountain habitats. And as we mentioned in the Biology and Society section, animals at both poles are also at risk.

► **Figure 18.47 Northward shift of *Argynnis paphia*.** On the map, orange represents the butterfly's range in 1970; its 1997 range is shown in light green.

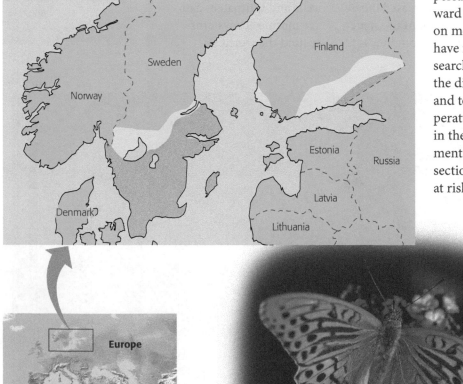

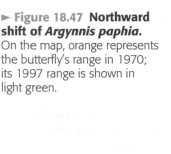

Argynnis paphia (silver-washed fritillary butterfly)

Effects of Climate Change on Ecosystems

The sentiment expressed by the poet John Donne that "no man is an island" is equally true for every other species in nature: Every species needs others to survive. Climate change is knocking some of these interactions out of sync. In temperate and polar climates, life cycle events of many plants and animals are triggered by warming temperatures. Across the Northern Hemisphere, the warm temperatures of spring are arriving earlier. Satellite images show earlier greening of the landscape, and flowering occurs sooner. A variety of species, including birds and frogs, have begun their breeding seasons earlier. But for other species, the environmental cue that spring has arrived is day length, which is not affected by climate change. Consequently, the winter white fur of snowshoe hares may be conspicuous against a greening landscape, or plants may bloom before pollinators have emerged.

The combined effects of climate change on forest ecosystems in western North America have spawned catastrophic wildfire seasons (Figure 18.48). In these regions, spring snowmelt in the mountains releases water into streams that sustain forest moisture levels over the summer dry season. With the earlier arrival of spring, snowmelt begins earlier and dwindles away before the dry season ends. As a result, the fire season is lasting longer. Meanwhile, bark beetles, which bore into conifers to lay their eggs, have benefited from global warming. Healthy trees can fight off the pests, but drought-stressed trees are too weak to resist (Figure 18.49). As a bonus for the beetles, the warmer weather allows them to reproduce twice a year rather than once. In turn, vast numbers of dead trees add fuel to a fire. Wildfires burn longer, and the number of acres burned has increased dramatically.

The map of terrestrial biomes (see Figure 18.27), which is primarily determined by temperature and rainfall, is also changing. Melting permafrost is shifting the boundary of the tundra northward as shrubs and conifers are able to stretch their ranges into the previously frozen ground. Prolonged droughts are extending the boundaries of deserts. Scientists also predict that great expanses of the Amazonian rain forest will gradually become savanna as increased temperatures dry out the soil.

Global climate change has significant consequences for people, too, as changing temperature and precipitation patterns affect food production, the availability of fresh water, and the structural integrity of buildings and roads. All projections point to an even greater impact in the future. Unlike other species, however, humans can take action to reduce greenhouse gas emissions and maybe even reverse the warming trend. ☑

▼ Figure 18.49 **Pines in Colorado infested by bark beetles.** Red or light-colored foliage indicates dead or dying trees. Green trees are still healthy.

▼ Figure 18.48 **A wildfire in Yosemite National Park, California, August 2013.**

Looking to Our Future

From 1990 to 2013, emissions of greenhouse gases increased 61%, and they continue to rise. At this rate, further climate change is inevitable. However, with effort, ingenuity, and international cooperation, we may be able to begin reducing emissions.

Given the vast scope and complexity of the problem, you might think that your own actions would have little impact on greenhouse gas emissions. But it was the collective activities of individuals that caused—and are still causing—emissions to rise. The amount of greenhouse gas emitted as a result of the actions of a single individual is that person's **carbon footprint** (from the fact that the most important greenhouse gas is CO_2). A carbon footprint can be estimated using a set of rough calculations; several different calculators are available online.

Home energy use is one major contributor to the carbon footprint. It's easy to reduce your energy consumption by turning off the lights, TV, and other electrical appliances when they aren't in use. Unplug "vampire" devices—electronics such as cell phone chargers, videogame consoles, and other computer, video, and audio equipment that draw electricity even when they aren't being used. You can also switch to energy-efficient lightbulbs.

Transportation is another significant part of the carbon footprint. If you have a car, keep it well maintained, consolidate trips, share rides with friends, and use alternative means of transportation whenever possible.

Manufactured goods are a third category in the carbon footprint. Every item you purchase generated its own carbon footprint in the process of going from raw materials to store shelves. You can reduce your carbon emissions by not buying unnecessary goods and by recycling—or better yet, reusing—items instead of putting them in the trash **(Figure 18.50)**. Landfills are the largest human-related source of methane, a greenhouse gas more potent than CO_2. The methane is released by prokaryotes that decompose (break down) landfill waste.

Changes in your eating habits can also shrink your carbon footprint. Like landfills, the digestive system of cattle depends on methane-producing bacteria. The methane released by cattle and by the bacteria that decompose their manure accounts for about 20% of methane emissions in the United States. Thus, replacing beef and dairy products in your diet with fish, chicken, eggs, and vegetables reduces your carbon footprint. In addition, eating locally grown fresh foods may lower the greenhouse gas emissions that result from food processing and transportation **(Figure 18.51)**. Many websites offer additional suggestions for reducing your carbon footprint. ☑

Your home may be infested with "vampire" devices that consume electricity even while you sleep.

▼ **Figure 18.50 Turning trash into treasure.** If you have things that you no longer need, don't trash them when you move out of your dorm or apartment. Environmentally conscious students at the University of North Carolina collect and sell discarded items, donating the proceeds to charity.

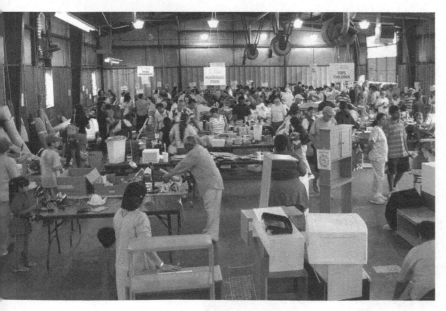

▼ **Figure 18.51 Eating locally grown foods may reduce your carbon footprint—and they taste good, too!**

Global Climate Change EVOLUTION CONNECTION

Climate Change as an Agent of Natural Selection

Environmental change has always been a part of life; in fact, it is a key ingredient of evolutionary change. Will evolutionary adaptation counteract the negative effects of climate change on organisms? Researchers have documented microevolutionary shifts in a few populations, including red squirrels, a few bird species, and a tiny mosquito (Figure 18.52a). It appears that some populations, especially those with high genetic variability and short life spans, may avoid extinction by means of evolutionary adaptation. However, evolutionary adaptation is unlikely to save long-lived species, such as polar bears and penguins, that are experiencing rapid habitat loss (Figure 18.52b). The rate of climate change is incredibly fast compared with major climate shifts in evolutionary history. If climate change continues on its present course, thousands of species—the IPCC estimates as many as 30% of all plants and animals—will face extinction by midcentury.

▼ Figure 18.52 **Which species will survive climate change?**

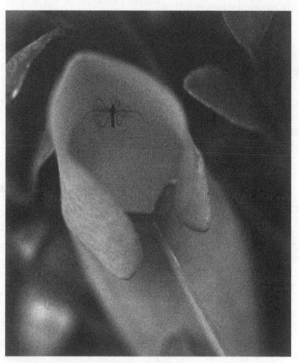

(a) Pitcher plant mosquito. The pitcher plant mosquito, pictured inside the carnivorous plant for which it is named, may be able to evolve quickly enough.

(b) Polar bear. Polar bears, which depend on sea ice for feeding, are not likely to make it.

Chapter Review

SUMMARY OF KEY CONCEPTS

An Overview of Ecology

Ecology is the scientific study of interactions between organisms and their environments. The environment includes abiotic (nonliving) and biotic (living) components. Ecologists use observation, experiments, and computer models to test hypothetical explanations of these interactions.

Ecology and Environmentalism

Human activities have had an impact on all parts of the biosphere. Ecology provides the basis for understanding and addressing these environmental problems.

A Hierarchy of Interactions

Ecologists study interactions at four increasingly complex levels.

Organismal ecology (individual) → Population ecology (group of individuals) → Community ecology (all organisms in a particular area) → Ecosystem ecology (all organisms and abiotic factors)

Living in Earth's Diverse Environments

The biosphere is an environmental patchwork in which abiotic factors affect the distribution and abundance of organisms.

Abiotic Factors of the Biosphere

Abiotic factors include the availability of sunlight, water, nutrients, and temperature. In aquatic habitats, dissolved oxygen, salinity, current, and tides are also important. Additional factors in terrestrial environments include wind and fire.

The Evolutionary Adaptations of Organisms

Adaptation via natural selection results from the interactions of organisms with their environments.

Adjusting to Environmental Variability

Organisms also have adaptations that enable them to cope with environmental variability, including physiological, behavioral, and anatomical responses to changing conditions.

Biomes

A biome is a major terrestrial or aquatic life zone.

Freshwater Biomes

Freshwater biomes include lakes, ponds, rivers, streams, and wetlands. Lakes vary, depending on depth, with regard to light penetration (photic and aphotic zones), temperature, nutrients, oxygen levels, and community structure. Rivers change greatly from their source to the point at which they empty into a lake or ocean. The bottom of an aquatic biome is its benthic realm.

Marine Biomes

Marine life is distributed into distinct realms (benthic and pelagic) and zones (photic, aphotic, and intertidal) according to the depth of the water, degree of light penetration, distance from shore, and open water versus deep-sea bottom. Marine biomes include the pelagic realm and the benthic realm of the oceans, coral reefs, intertidal zones, and estuaries. Coral reefs, which occur in warm tropical waters above the continental shelf, have an abundance of biological diversity. An ecosystem found near hydrothermal vents in the deep ocean is powered by chemical energy from Earth's interior instead of sunlight. Estuaries, located where a freshwater river or stream merges with the ocean, are some of the most biologically productive environments on Earth.

How Climate Affects Terrestrial Biome Distribution

The geographic distribution of terrestrial biomes is based mainly on regional variations in climate. Climate is largely determined by the uneven distribution of solar energy on Earth. Proximity to large bodies of water and the presence of landforms such as mountains also affect climate.

Terrestrial Biomes

Most terrestrial biomes are named for their climate and predominant vegetation. The major terrestrial biomes include tropical forest, savanna, desert, chaparral, temperate grassland, temperate broadleaf forest, coniferous forest, tundra, and polar ice. If the climate in two geographically separate areas is similar, the same type of biome may occur in both.

Interconnections within Systems: The Water Cycle

The global water cycle links aquatic and terrestrial biomes. Human activities are disrupting the water cycle.

Human Impact on Biomes

Land use by humans has altered vast tracts of forest and degraded the services provided by natural ecosystems. Unsustainable agricultural practices have depleted cropland fertility. Human activities have polluted freshwater ecosystems, which are vital for life. Agriculture, population growth, drought, and declining snowfall are all factors in the rapid depletion of freshwater resources in some regions.

Global Climate Change

The Greenhouse Effect and Global Warming

So-called greenhouse gases, including CO_2 and methane, increase the amount of heat retained in Earth's atmosphere. The accumulation of these gases has caused increases in the average global temperature.

The Accumulation of Greenhouse Gases

Human activities, especially the burning of fossil fuels, are responsible for the rise in greenhouse gases over the past century. Release of CO_2 has exceeded the amount that can be absorbed by natural processes.

Effects of Climate Change on Ecosystems

Climate change is disrupting interactions between species. Devastating wildfires are among the effects of climate change in certain ecosystems. Climate change is also shifting biome boundaries.

Looking to Our Future

Each person has a carbon footprint—that person's responsibility for a portion of global greenhouse gas emissions. We can take action to reduce our carbon footprints.

MasteringBiology®

For practice quizzes, BioFlix animations, MP3 tutorials, video tutors, and more study tools designed for this textbook, go to MasteringBiology®

▌ SELF-QUIZ

1. Place these levels of ecological study in order from the least to the most comprehensive: community ecology, ecosystem ecology, organismal ecology, population ecology.

2. Name several abiotic factors that might affect the community of organisms living inside a home fish tank.

3. The physical and physiological changes experienced by astronauts who spend months in space are an example of _____.

4. In what part of the ocean are phytoplankton found?
a. pelagic realm
b. intertidal zone
c. benthic realm
d. aphotic zone

5. Identify the following biomes on the graph below: tundra, coniferous forest, desert, grassland, temperate broadleaf forest, and tropical rain forest.

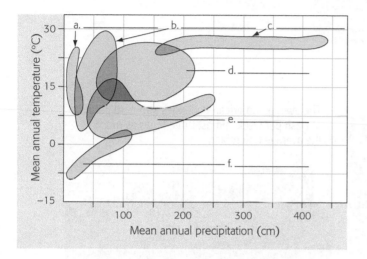

6. We are on a coastal hillside on a hot, dry summer day among evergreen shrubs that are adapted to fire. We are most likely standing in a _____ biome.

7. What three abiotic factors account for the rarity of trees in arctic tundra?

8. What human activity is responsible for the greatest amount of deforestation?

9. What is the greenhouse effect? How is the greenhouse effect related to global warming?

10. Temperature increases due to global warming have been greatest
a. close to the poles.
b. in the sea.
c. in the tropics.
d. in deserts.

11. What populations of organisms are most likely to survive climate change via evolutionary adaptation?

Answers to these questions can be found in Appendix: Self-Quiz Answers.

THE PROCESS OF SCIENCE

12. Predict the regional climate and the prevailing biome of a town located in a hilly region on the Pacific coast, 37° north of the equator.

13. Interpreting Data This graph shows average monthly temperature and precipitation for a city in the Northern Hemisphere. Based on the biome descriptions on pages 385–390, in which biome is this city located? Explain your answer.

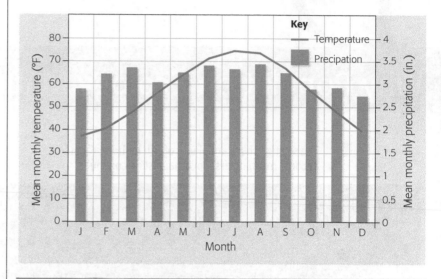

BIOLOGY AND SOCIETY

14. Some people are not convinced that human-induced global climate change is a real phenomenon. Using your knowledge of the scientific process (see Chapter 1) and the information from this chapter, develop arguments you could use to explain the scientific basis for saying that global climate change is truly occurring and that people are responsible for it.

15. The concept of carbon footprint can be applied to a single person, an organization, a product, a city, or a country. What are the practical limitations of this estimate of greenhouse gas emissions?

16. In the summer of 2007, unprecedented drought conditions brought the city of Atlanta, Georgia, within weeks of running out of water. Most of Atlanta's water comes from Lake Lanier, a reservoir that the Army Corps of Engineers created by damming the Chattahoochee River. With Lake Lanier drying up from lack of rainfall, Georgia petitioned the Corps, which manages the dam, to reduce the amount of water released downstream. The Corps refused, citing its obligation under the Endangered Species Act to protect the habitats of a species of sturgeon (a fish) and two species of mussel (a mollusc). Objections were also raised by Alabama and Florida, where hundreds of towns, recreational facilities, and power plants depend on the water released downstream. Some people thought that Atlanta authorities had brought the water shortage on themselves by allowing developers to build without considering whether adequate water was available. Florida also argued that reduction of freshwater inflows would harm its oyster fisheries. How would you prioritize the competing claims on the water from Lake Lanier? Who should allocate scarce water resources? How can cities and states plan more wisely for future shortages?

19 Population Ecology

The saying "There are plenty of fish in the sea" could become meaningless if overharvesting continues.

The average American generates about 40 pounds of trash per week, almost three times as much as the average Colombian.

Counting one person per second, it would take more than 225 years to count all the people alive today.

If environmental resources were unlimited, a single pair of elephants would give rise to a population of 19 million in 750 years.

436

Biological Invasions BIOLOGY AND SOCIETY

Invasion of the Lionfish

Lionfish, with their graceful, flowing fins, bold stripes, and eye-catching array of spines, are striking members of tropical reef communities. They are also favorites of saltwater aquarium enthusiasts—especially the red lionfish, a native to the coral reefs of the South Pacific and Indian Oceans. There are a few drawbacks to owning a red lionfish, however. The spines are venomous and can inflict an intensely painful sting. Lionfish are merciless predators, so any tankmates must be chosen with care. And they are large. A 2-inch juvenile can rapidly become an 18-inch adult that requires, at minimum, a 120-gallon tank. Apparently, some aquarium owners who regretted their purchase released their lionfish into the wild.

The red lionfish, a beautiful but deadly invader, is a threat to coral reef communities. This lionfish was photographed off the coast of North Carolina, halfway around the world from its native habitat.

Freed from the competitors and predators of their native reefs as well as their tanks, red lionfish have multiplied exponentially. Within a few years of the first sightings off the southeastern coast of Florida, lionfish populations had spread up the East Coast. They have since invaded islands and coastlines throughout the Atlantic and Caribbean regions and are now swarming into the Gulf of Mexico. The speed of the onslaught has stunned scientists, who are just beginning to document its devastating effects on native ecosystems. Lionfish consume prodigious numbers of fish, including species that are key to maintaining the legendary diversity of reef communities and juveniles of economically important fishes such as grouper and snapper. Some biologists think our best hope of stopping the lionfish invasion is for *us* to consume *them*. The National Oceanic and Atmospheric Administration (NOAA) has launched an "Eat Lionfish" campaign to encourage human predation on the tasty fish.

For as long as people have traveled the globe, they have carried—intentionally or accidentally—thousands of species to new habitats. Many of these non-native species have established populations that spread far and wide, leaving environmental havoc in their wake. We humans, too, have multiplied and spread far from our point of origin, radically changing our environment in the process. As you explore population ecology in this chapter, you'll also learn about trends in human population growth and other applications of this area of ecological research.

An Overview of Population Ecology

Ecologists usually define a **population** as a group of individuals of a single species that occupy the same general area at the same time. These individuals rely on the same resources, are influenced by the same environmental factors, and are likely to interact and breed with one another. For example, the red lionfish that live in the vicinity of a specific reef are a population.

Population ecology is concerned with changes in population size and the factors that regulate populations over time. A population ecologist might describe a population in terms of its size (number of individuals), age structure (proportion of individuals of different ages), or density (number of individuals per unit area or volume). Population ecologists also study population dynamics, the interactions between biotic and abiotic factors that cause variation in population size. One important aspect of population dynamics—and a major topic for this chapter—is population growth.

Population ecology plays a key role in applied research. For example, it provides critical information for conservation and restoration projects (**Figure 19.1**). Population ecology is used to develop sustainable fisheries and to manage wildlife populations. Studying the population ecology of pests provides insight into controlling how they spread.

Population ecologists also study human population growth, one of the most critical environmental issues of our time.

Let's consider what a snapshot of a population might look like. The first question is, Which individuals are included in this population? A population's geographic boundaries may be natural, as with lionfish inhabiting a particular coral reef. But ecologists often define a population's boundaries in more arbitrary ways that fit their research questions. For example, an ecologist studying the contribution of asexual reproduction to the population growth of sea anemones might define a population as all the anemones of one species in a particular tide pool. Another researcher studying the effects of hunting on deer might define a population as all the deer within a particular state. Yet another researcher, attempting to understand the spread of HIV/AIDS, might study the HIV infection rate of the human population in one nation or throughout the world.

▼ **Figure 19.1 Ecologists getting up close and personal with members of the populations they study.**

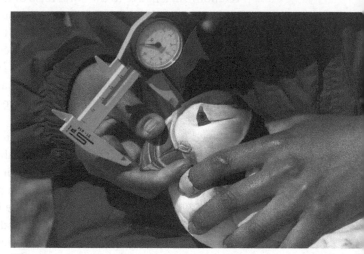

(b) A researcher collects data on a puffin for a project to restore nesting colonies of the small seabird in the Gulf of Maine.

(a) After sedating a black bear in La Mauricie National Park, Quebec, Canada, a biologist attaches a radio collar to track its movements.

(c) A researcher marks meerkats in the Kuruman River Reserve in Northern Cape, South Africa.

Population Density

Our snapshot of a population would include **population density**, the number of individuals of a species per unit area or volume of the habitat: the number of largemouth bass per cubic kilometer (km^3) of a lake, for example, or the number of oak trees per square kilometer (km^2) in a forest, or the number of nematodes per cubic meter (m^3) in the forest's soil. In rare cases, an ecologist can actually count all the individuals within the boundaries of the population. For example, we could count the total number of oak trees (say, 200) in a forest covering 50 km^2 (about 19 square miles). The population density would be the total number of trees divided by the area, or 4 trees per square kilometer ($4/km^2$).

In most cases, however, it is impractical or impossible to count all individuals in a population. Instead, ecologists use a variety of sampling techniques to estimate population density. For example, they might estimate the density of alligators in the Florida Everglades based on a count of individuals in a few sample plots of 1 km^2 each. Generally speaking, the larger the number and size of sample plots, the more accurate the estimates. Population densities may also be estimated by indicators such as number of bird nests or rodent burrows **(Figure 19.2)** rather than by actual counts of organisms.

Keep in mind that population density is not a constant number. It changes when individuals are born or die and when new individuals enter the population (immigration) or leave it (emigration).

Population Age Structure

The **age structure** of a population—the distribution of individuals in different age-groups—reveals information that is not apparent from population density. For instance, age structure can provide insight into the history of a population's survival or reproductive success and how it relates to environmental factors. **Figure 19.3** shows the age structure of males in a population of cactus finches on the Galápagos Islands in 1987. (See Figure 14.11 for other examples of Galápagos finches.) Four-year-old birds, born in 1983, made up almost half the population, while there were not any 2- or 3-year-olds. Why was there such dramatic variation? For their food, cactus finches depend on plants, which in turn depend on rainfall. The 1983 baby boom resulted from unusually wet weather that produced abundant plant growth and food for the finches. Severe droughts in 1984 and 1985 limited the food supply, preventing reproduction and causing many deaths. As we'll see later in this chapter, age structure is also a useful tool for predicting future changes in a population. ☑

☑ **CHECKPOINT**

What does an age structure show?

Answer: the distribution of individuals in different age-groups

▼ Figure 19.2 **An indirect census of a prairie dog population.** We could get a rough estimate of the number of prairie dogs in this colony in Saskatchewan, Canada, by counting the number of burrows constructed by the rodents and then multiplying by the number of animals that use a typical burrow.

▼ Figure 19.3 **Age structure for the males in a population of large cactus finches (inset) on one of the Galápagos Islands in 1987.**

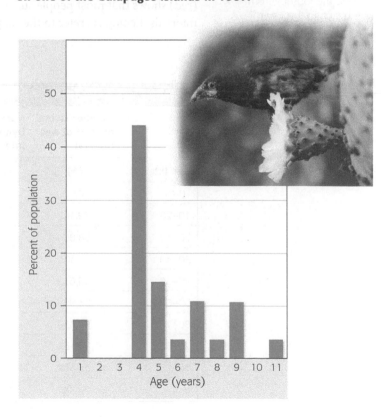

Life Tables and Survivorship Curves

Life tables track survivorship, the chance of an individual in a given population surviving to various ages. The life insurance industry uses life tables to predict how long, on average, a person of a given age will live. Starting with a population of 100,000 people, **Table 19.1** shows the number of people who are expected to be alive at the beginning of each age interval, based on death rates in 2008. For example, 93,999 out of 100,000 people are expected to live to age 50. Their chance of surviving to age 60, shown in the last column of the same row, is 0.94; 94% of 50-year-olds will reach the age of 60. The chance of 80-year-olds surviving to age 90, however, is only 0.402. Population ecologists have adopted this technique and constructed life tables to help them understand the structure and dynamics of various plant and animal species. By identifying the most vulnerable stage of the life cycle, life table data may also help conservationists develop effective measures to protect species whose populations are declining.

Ecologists represent life table data graphically in a **survivorship curve**, a plot of the number of individuals still alive at each age in the maximum life span (**Figure 19.4**). By using a percentage scale instead of actual ages on the *x*-axis, we can compare species with widely varying life spans, such as humans and squirrels, on the same graph. The curve for the human population (red) shows that most people survive to the older age intervals. Ecologists refer to the shape of this curve as

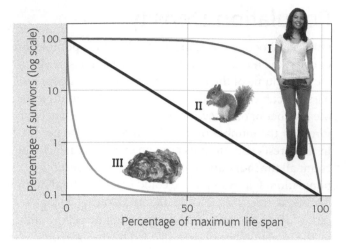

▲ Figure 19.4 **Three idealized types of survivorship curves.**

Type I survivorship. Species that exhibit a Type I curve—humans and many other large mammals—usually produce few offspring but give them good care, increasing the likelihood that they will survive to maturity.

In contrast, a Type III curve (blue) indicates low survivorship for the very young, followed by a period when survivorship is high for those few individuals who live to a certain age. Species with this type of survivorship curve usually produce very large numbers of offspring but provide little or no care for them. Some species of fish, for example, produce millions of eggs at a time, but most of these offspring die as larvae from predation or other causes. Many invertebrates, including oysters, also have Type III survivorship curves.

A Type II curve (black) is intermediate, with survivorship constant over the life span. That is, individuals are no more vulnerable at one stage of the life cycle than another. This type of survivorship has been observed in some invertebrates, lizards, and rodents. ☑

Evolution Life History Traits as Adaptations

A population's pattern of survivorship is an important feature of its **life history**, the set of traits that affect an organism's schedule of reproduction and survival. Some key life history traits are the age at first reproduction, the frequency of reproduction, the number of offspring, and the amount of parental care given. As you might expect from the different types of survivorship curves, life history traits vary among organisms. Let's take a closer look at how natural selection shapes these traits.

As you may recall, reproductive success is key to evolutionary success (see Chapter 13). Accordingly,

Table 19.1	Life Table for the U.S. Population in 2008		
	Number Living at Start of Age Interval	Number Dying During Interval	Chance of Surviving Interval
Age Interval	(N)	(D)	1−(D/N)
0–10	100,000	833	0.992
10–20	99,167	363	0.996
20–30	98,804	941	0.990
30–40	97,863	1,224	0.987
40–50	96,639	2,640	0.973
50–60	93,999	5,643	0.940
60–70	88,356	11,203	0.873
70–80	77,153	21,591	0.720
80–90	55,562	33,215	0.402
90+	22,347	22,347	0.000

you may wonder why all organisms don't simply produce a large number of offspring. One reason is that reproduction is expensive in terms of time, energy, and nutrients—resources that are available in limited amounts. An organism that gives birth to a large number of offspring will not be able to provide a great deal of parental care. Consequently, the combination of life history traits represents trade-offs that balance the demands of reproduction and survival. In other words, life history traits, like anatomical features, are shaped by evolutionary adaptation.

Because selective pressures vary, life histories are very diverse. Nevertheless, ecologists have observed some patterns that are useful for understanding how natural selection influences life history characteristics.

One life history pattern is typified by small-bodied, short-lived species (for example, insects and small rodents) that develop and reach sexual maturity rapidly, have a large number of offspring, and offer little or no parental care. In plants, "parental care" is measured by the amount of nutritional material stocked in each seed. Many small, non-woody plants (dandelions, for example) produce thousands of tiny seeds. Such organisms have an **opportunistic life history**, one that enables the plant or animal to take immediate advantage of favorable conditions. In general, populations with this life history pattern exhibit a Type III survivorship curve.

In contrast, some organisms have an **equilibrial life history**, a pattern of developing and reaching sexual maturity slowly and producing a few well-cared-for offspring. Organisms that have an equilibrial life history are typically larger-bodied, longer-lived species (for example, bears and elephants). Populations with this life history pattern exhibit a Type I survivorship curve. Plants with comparable life history traits include certain trees. For example, coconut palms produce relatively few seeds, but those seeds are well stocked with nutrient-rich material. **Table 19.2** compares key traits of opportunistic and equilibrial life history patterns.

What accounts for the differences in life history patterns? Some ecologists hypothesize that the potential survival rate of the offspring and the likelihood that the adult will live to reproduce again are the critical factors. In a harsh, unpredictable environment, an adult may have just one good shot at reproduction, so it may be an advantage to invest in quantity rather than quality. On the other hand, in an environment where favorable conditions are more dependable, an adult is more likely to survive to reproduce again. Seeds are more likely to fall on fertile ground, and newly emerged animals are more likely to survive to adulthood. In that case, it may be more advantageous for the adult to invest its energy in producing a few well-cared-for offspring at a time.

Of course, there is much more diversity in life history patterns than the two extremes described here. Nevertheless, the contrasting patterns are useful for understanding the interactions between life history traits and our next topic, population growth. ☑

☑ CHECKPOINT

How does the term *opportunistic* capture the key characteristics of that life history pattern?

Answer: An opportunistic life history is characterized by an ability to produce a large number of offspring very rapidly when the environment affords a temporary opportunity for a burst of reproduction.

Dandelions have an opportunistic life history.

Elephants have an equilibrial life history.

Table 19.2	Some Life History Characteristics of Opportunistic and Equilibrial Populations	
Characteristic	**Opportunistic Populations (such as many wildflowers)**	**Equilibrial Populations (such as many large mammals)**
Climate	Relatively unpredictable	Relatively predictable
Maturation time	Short	Long
Life span	Short	Long
Death rate	Often high	Usually low
Number of offspring per reproductive episode	Many	Few
Number of reproductions per lifetime	Usually one	Often several
Timing of first reproduction	Early in life	Later in life
Size of offspring or eggs	Small	Large
Parental care	Little or none	Often extensive

441

Population Growth Models

Population size fluctuates as new individuals are born or move into an area and others die or move out of it. Some populations—for example, trees in a mature forest—are relatively constant over time. Other populations change rapidly, even explosively. Consider a single bacterium that divides every 20 minutes. There would be two bacteria after 20 minutes, four after 40 minutes, eight after 60 minutes, and so on. In just 12 hours, the population would approach 70 billion cells. If reproduction continued at this rate for a day and a half—a mere 36 hours—there would be enough bacteria to form a layer a foot deep over the entire Earth! Population ecologists use idealized models to investigate how the size of a particular population may change over time under different conditions. We'll describe two basic mathematical models that illustrate fundamental concepts of population growth.

If environmental resources were unlimited, a single pair of elephants would give rise to a population of 19 million in 750 years.

The Exponential Population Growth Model: The Ideal of an Unlimited Environment

The first model, known as exponential growth, works like compound interest on a savings account: The principal (population size) grows faster with each interest payment (the individuals added to the population). **Exponential population growth** describes the expansion of a population in an ideal, unlimited environment. In this model, the population size of each new generation is calculated by multiplying the current population size by a constant factor that represents the birth rate minus the death rate. Let's look at how such a population grows. In **Figure 19.5**, we begin with a population of 20 rabbits, indicated on the *y*-axis. Each month, there are more rabbit births than deaths; as a result, the population size increases each month.

Notice in Figure 19.5 that each increase is larger than the previous one. In other words, the larger the population, the faster it grows. The increasing speed of population growth produces a J-shaped curve that is typical of exponential growth. The slope of the curve shows how rapidly the population is growing. At the outset, when the population is small, the curve is almost flat: Over the first 4 months, the population increases by a total of only 37 individuals, an average of 9.25 births per month. By the end of 7 months, the growth rate has increased to an average of 15 births per month. The largest increase is seen in the period from 10 to 12 months, when an average of 85 rabbits are born each month.

Exponential population growth is common in certain situations. For example, a disturbance such as a fire, flood, hurricane, drought, or cold snap may suddenly reduce the size of a population. Organisms that have opportunistic life history patterns can rapidly take advantage of the lack of competition and quickly recolonize the habitat by exponential population growth. Human activity can also be a major cause of disturbance, and plants and animals with opportunistic life history traits commonly occupy road cuts, freshly cleared fields and woodlots, and poorly maintained lawns. However, no natural environment can sustain exponential growth indefinitely. ☑

▼ Figure 19.5 **Exponential growth of a rabbit population.**

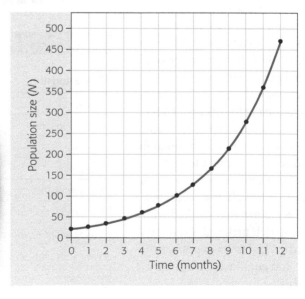

The Logistic Population Growth Model: The Reality of a Limited Environment

Most natural environments do not have an unlimited supply of the resources needed to sustain population growth. Environmental factors that restrict population growth are called **limiting factors**. Limiting factors ultimately control the number of individuals that can occupy a habitat. Ecologists define **carrying capacity** as the maximum population size that a particular environment can sustain. In **logistic population growth**, the growth rate decreases as the population size approaches carrying capacity. When the population is at carrying capacity, the growth rate is zero.

You can see the effect of limiting factors in the graph in **Figure 19.6**, which shows the growth of a population of fur seals on St. Paul Island, off the coast of Alaska. (For simplicity, only the mated bulls were counted. Each has a harem of females, as shown in the photograph.) Before 1925, the seal population on the island remained low—between 1,000 and 4,500 mated bulls—because of uncontrolled hunting. After hunting was controlled,

the population increased rapidly until about 1935, when it began to level off and started fluctuating around a population size of about 10,000 bull seals—the carrying capacity for St. Paul Island. In this instance, the main limiting factor was the amount of space suitable for breeding territories.

The carrying capacity for a population varies, depending on the species and the resources available in the habitat. For example, carrying capacity might be considerably less than 10,000 for a fur seal population on a smaller island with fewer breeding sites. Even in one location, it is not a fixed number. Organisms interact with other organisms in their communities, including predators, pathogens, and food sources, and these interactions may affect carrying capacity. Changes in abiotic factors may also increase or decrease carrying capacity. In any case, the concept of carrying capacity expresses an essential fact of nature: Resources are finite.

Ecologists hypothesize that selection for organisms exhibiting equilibrial life history patterns occurs in environments where the population size is at or near carrying capacity. Because competition for resources is keen under these circumstances, organisms gain an advantage by allocating energy to their own survival and to the survival of their descendants.

Figure 19.7 compares logistic growth (blue) with exponential growth (red). As you can see, the logistic curve is J-shaped at first, but gradually levels off to resemble an S shape as carrying capacity is reached. Both the logistic model and the exponential model of population growth are theoretical ideals. No natural population fits either one perfectly. However, these models are useful starting points for studying population growth. Ecologists use them to predict how populations will grow in certain environments and as a basis for constructing more complex models. ✔

▼ Figure 19.6 **Logistic growth of a seal population.**

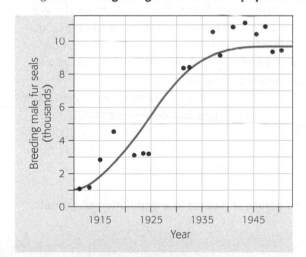

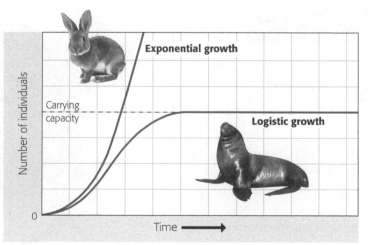

◄ Figure 19.7
Comparison of exponential and logistic growth.

☑ **CHECKPOINT**

What happens when a population reaches its carrying capacity?

Answer: Enough resources are available to sustain that population size, but the population does not continue to increase.

Regulation of Population Growth

Now let's take a closer look at how population growth is regulated in nature. What stops a population from continuing to increase after reaching carrying capacity?

Density-Dependent Factors

Several **density-dependent factors**—limiting factors whose intensity is related to population density—can limit growth in natural populations. The most obvious is **intraspecific competition**, the competition between individuals of the same species for the same limited resources. As a limited food supply is divided among more and more individuals, birth rates may decline as individuals have less energy available for reproduction. Density-dependent factors may also depress a population's growth by increasing the death rate. For example, in a population of song sparrows, both factors reduced the number of offspring that survived and left the nest (**Figure 19.8a**). As the number of competitors for food increased, female song sparrows laid fewer eggs. In addition, the death rate of eggs and nestlings increased with increasing population density.

Plants that grow close together may experience an increased death rate as intraspecific competition for resources increases. And those that do survive will produce fewer flowers, fruits, and seeds than uncrowded individuals. After seeds sprout, gardeners often pull out some of the seedlings to allow sufficient resources for the remaining plants. Intraspecific competition is also the reason that plants purchased from a nursery come with instructions to space the plants a certain distance apart.

A limited resource may be something other than food or nutrients. Like a game of musical chairs, the number of safe hiding places may limit a prey population by exposing some individuals to a greater risk of predation. For example, young kelp perch hide from predators in "forests" of the large seaweed known as kelp (see Figure 15.27). In the experiment shown in **Figure 19.8b**, the proportion of perch eaten by a predator increased with increasing perch density. In many animals that defend a territory, the availability of space may limit reproduction. For instance, the number of nesting sites on rocky islands may limit the population size of oceanic birds such as gannets, which maintain breeding territories (**Figure 19.9**).

In addition to competition for resources, other factors may cause density-dependent deaths in a population. For example, the death rate may climb as a result of increased disease transmission under crowded conditions or the accumulation of toxic waste products.

▼ Figure 19.8 **Density-dependent regulation of population growth.**

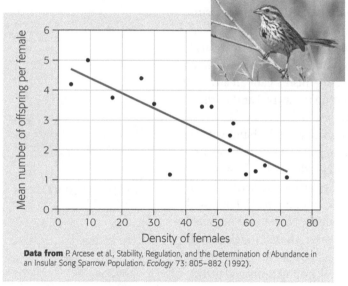

Data from P. Arcese et al., Stability, Regulation, and the Determination of Abundance in an Insular Song Sparrow Population. *Ecology* 73: 805–882 (1992).

(a) Declining reproductive success of song sparrows (inset) with increasing population density.

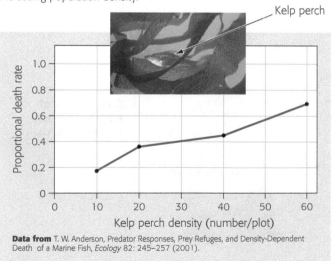

Data from T. W. Anderson, Predator Responses, Prey Refuges, and Density-Dependent Death of a Marine Fish, *Ecology* 82: 245–257 (2001).

(b) Increasing death rate of kelp perch (inset) with increasing population density.

▼ Figure 19.9 **Space as a limiting resource in a population of gannets.**

Density-Independent Factors

In many natural populations, abiotic factors such as weather may limit or reduce population size well before other limiting factors become important. A population-limiting factor whose intensity is unrelated to population density is called a **density-independent factor**. If we look at the growth curve of such a population, we see something like exponential growth followed by a rapid decline rather than a leveling off. **Figure 19.10** shows this effect for a population of aphids, insects that feed on the sugary sap of plants. These and many other insects undergo virtually exponential growth in the spring and then rapidly die off when the weather turns hot and dry in the summer. A few individuals may remain, allowing population growth to resume if favorable conditions return. In some populations of insects—many mosquitoes and grasshoppers, for instance—the adults die off entirely, leaving behind eggs that will initiate population growth the following year. In addition to seasonal changes in the weather, environmental disturbances, such as fire, floods, and storms, can affect a population's size regardless of its density.

Over the long term, most populations are probably regulated by a complex interaction of density-dependent and density-independent factors. Although some populations remain fairly stable in size and are presumably close to a carrying capacity that is determined by biotic factors such as competition or predation, most populations for which we have long-term data do fluctuate.

Population Cycles

Some populations of insects, birds, and mammals undergo dramatic fluctuations in density with remarkable regularity. "Booms" characterized by rapid exponential growth are followed by "busts," during which the population falls back to a minimal level. Lemmings, small rodents that live in the tundra, are a striking example. In lemming populations, boom-and-bust growth cycles occur every three to four years. Some researchers hypothesize that natural changes in the lemmings' food supply may be the underlying cause. Another hypothesis is that stress from crowding during the "boom" triggers hormonal changes that may cause the "bust" by reducing birth rates.

Figure 19.11 illustrates another example—the cycles of the snowshoe hare and the lynx. The lynx is one of the main predators of the snowshoe hare in the far northern forests of Canada and Alaska. About every ten years, both hare and lynx populations show a rapid increase followed by a sharp decline. What causes these boom-and-bust cycles? Since ups and downs in the two populations seem to almost match each other on the graph, does this mean that changes in one directly affect the other? For the hare cycles, there are three main hypotheses. First, cycles may be caused by winter food shortages that result from overgrazing. Second, cycles may be due to predator-prey interactions. Many predators other than lynx, such as coyotes, foxes, and great-horned owls, eat hares, and together these predators might overexploit their prey. Third, cycles may be affected by a combination of food resource limitation and excessive predation. Recent field studies support the hypothesis that the ten-year cycles of the snowshoe hare are largely driven by excessive predation, but are also influenced by fluctuations in the hare's food supplies. Long-term studies are the key to unraveling the complex causes of such population cycles. ☑

☑ **CHECKPOINT**

List some density-dependent factors that limit population growth.

Answer: food and nutrient limitations, insufficient space for territories or nests, increase in disease and predation, accumulation of toxins

▼ Figure 19.10 **Weather change as a density-independent factor limiting growth of an aphid population.**

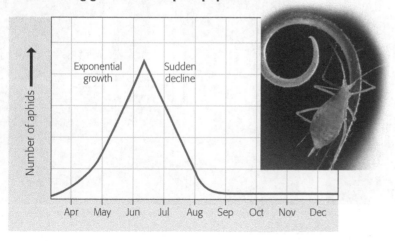

▼ Figure 19.11 **Population cycles of the snowshoe hare and the lynx.**

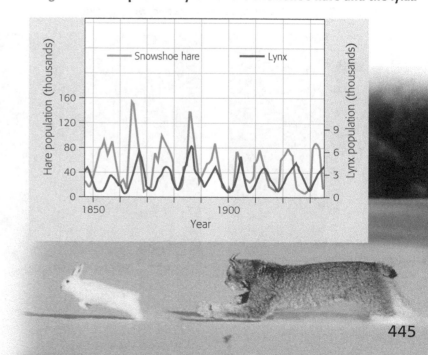

Applications of Population Ecology

To a great extent, we humans have converted Earth's natural ecosystems to ecosystems that produce goods and services for our own benefit. In some cases, we try to increase populations of organisms that we wish to harvest and decrease populations of organisms that we consider pests. Other efforts are aimed at saving populations that are perilously close to extinction. Principles of population ecology help guide us toward these various resource management goals.

Conservation of Endangered Species

The U.S. Endangered Species Act defines an **endangered species** as one that is "in danger of extinction throughout all or a significant portion of its range." **Threatened species** are defined as those that are likely to become endangered in the foreseeable future. Endangered and threatened species are characterized by population sizes that are highly reduced or steadily declining. The challenge for conservationists is to determine the circumstances that threaten a species with extinction and try to remedy the situation.

The red-cockaded woodpecker was one of the first species to be listed as endangered (**Figure 19.12**). The bird requires longleaf pine forests, where it drills its nest holes in mature, living pine trees. Originally found throughout the southeastern United States, the woodpeckers declined in number as suitable habitats were lost to logging and agriculture. Moreover, we have altered the composition of many of the remaining forests by suppressing the fires that are a natural occurrence in these ecosystems. Research revealed that breeding birds tend to abandon nests when vegetation among the pines is thick and higher than about 4.5 m (15 feet). Apparently, the birds require a clear flight path between their home trees and the neighboring feeding grounds. Armed with an understanding of the factors that regulate population growth of this species, wildlife managers protected critical habitat and began a maintenance program that included controlled fires to reduce forest undergrowth. As a result of such measures, populations of red-cockaded woodpeckers are beginning to recover. ☑

Sustainable Resource Management

A goal of wildlife managers, fishery biologists, and foresters is to gather the largest possible harvest while sustaining the productivity of the resource for future harvests. This means maintaining a high population growth rate to replenish the population. According to the logistic growth model, the fastest growth rate occurs when the population size is at roughly half the carrying capacity of the habitat. Theoretically, a resource manager should achieve the best results by harvesting the population down to this level. However, the logistic model assumes that growth rate and carrying capacity

▼ Figure 19.12 **The habitat of the red-cockaded woodpecker.**

A red-cockaded woodpecker perches at the entrance to its nest in a longleaf pine tree.

High, dense undergrowth impedes the woodpeckers' access to feeding grounds.

Low undergrowth offers birds a clear flight path between nest sites and feeding grounds.

are stable over time. Calculations based on these assumptions, which are not realistic for some populations, may lead to unsustainably high harvest levels that ultimately deplete the resource. In addition, human economic and political pressures often outweigh ecological concerns, and scientific information is frequently insufficient.

Fish, the only wild animals still hunted on a large scale, are particularly vulnerable to overharvesting. For example, in the northern Atlantic cod fishery, estimates of cod stocks were too high, and the practice of discarding young cod (not of legal size) at sea caused a higher death rate than was predicted. The fishery collapsed in 1992 and has not recovered (Figure 19.13).

Until the 1970s, marine fisheries concentrated on species such as cod that inhabit the continental shelves (see Figure 18.19). As these resources dwindled, attention turned to deeper waters, most commonly the continental slopes below 600 m. In many of these new locations, however, catches are initially high but then quickly fall off as stocks are depleted. Deeper waters are colder, and food is relatively scarce. Fishes that are adapted to this environment, such as Chilean sea bass and orange roughy, typically grow more slowly, take

The saying "There are plenty of fish in the sea" could become meaningless if overharvesting continues.

longer to reach maturity, and have a lower reproductive rate than continental shelf species. Sustainable catch rates can't be estimated without knowing these essential life history traits for the target species. In addition, knowledge of population ecology alone is not sufficient; sustainable fisheries also require knowledge of community and ecosystem characteristics. ☑

☑ **CHECKPOINT**

Why do managers try to maintain populations of fish and game species at about half their carrying capacity?

Answer: to prevent overharvesting yet maintain lower population levels so that growth rate is high

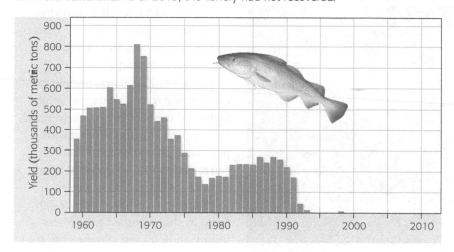

▼ Figure 19.13 **The collapse of the northern cod fishery off Newfoundland.** As of 2013, the fishery had not recovered.

Invasive Species

Like the lionfish featured in the Biology and Society section, organisms that are introduced into non-native habitats can have a devastating effect on the ecosystem. An **invasive species** is a non-native species that has spread far beyond its original point of introduction and causes environmental or economic damage by colonizing and dominating suitable habitats. In the United States alone, there are hundreds of invasive species, including plants, mammals, birds, fishes, arthropods, and molluscs. Worldwide, there are thousands more. Regardless of where you live, an invasive plant or animal is probably living nearby. Invasive species are a leading cause of local extinctions. And the economic costs of invasive species are enormous—an estimated $137 billion a year in the United States.

Not every organism that is introduced to a new habitat is successful, and not every species that survives in its new habitat becomes invasive. There is no single explanation for why any non-native species turns into a damaging pest, but invasive species typically exhibit an opportunistic life history pattern. A female lionfish, for

example, is sexually mature at a year old and can produce 2 million eggs per year.

The life history traits of cheatgrass (*Bromus tectorum*, Figure 19.14), an invasive plant of the arid western United States, have enabled its spectacular success. Its seeds were accidentally carried into the United States with grain from Asia and then spread by livestock. Cheatgrass covers millions of acres of rangeland that was formerly dominated by native grasses and sagebrush, and it claims an additional area the size of Rhode Island every year. Cheatgrass seeds sprout during fall rains, and the roots continue to grow underground through the winter. Already established when the warm spring weather arrives, dense clumps of cheatgrass deprive native species and crops of soil moisture and mineral nutrients. It also produces seeds earlier and in greater abundance than its competitors.

▼ Figure 19.14 **Cheatgrass, an invasive plant.**

After cheatgrass seeds mature in early summer, the plants become extremely dry and flammable, creating abundant fuel that is easily ignited by lightning or a stray spark. Cheatgrass fires are more intense and occur much more frequently than the fires that native plants have evolved to tolerate. After a few fire cycles, the native plants are gone, robbing more than 150 species of birds and mammals of the food and shelter they derive from sagebrush. Global climate change is also hastening the transition of rangeland into fields of cheatgrass. Studies have shown that cheatgrass responds to increased CO_2 levels by growing faster and accumulating more tissue, which in turn becomes more fuel for the fires that extend its domain.

For a non-native organism to become invasive, the biotic and abiotic factors of the new environment must be compatible with the organism's needs and tolerances. For example, Burmese pythons set loose in South Florida—either accidently released by damaging storms or deliberately freed by disenchanted pet owners—found a hot, humid climate similar to their native area. Prey such as birds, mammals, reptiles, and amphibians are readily available, especially in the Everglades. As a result, South Florida is now home to a burgeoning population of the giant reptiles (Figure 19.15). Burmese pythons released in a less favorable environment might survive for a short time, but would not be able to establish a population. ☑

▼ **Figure 19.15 A Burmese python.** This 9-foot-long snake was captured in Florida after devouring a pet cat.

☑ CHECKPOINT

What distinguishes invasive species from organisms that are introduced to non-native habitats but do not become invasive?

Answer: Invasive species spread far from where they are introduced, and they cause environmental or economic damage.

Biological Control of Pests

The absence of biotic factors that limit population growth, such as pathogens, predators, or herbivores, may contribute to the success of invasive species. Accordingly, efforts to eliminate or control these troublesome organisms often focus on **biological control**, the intentional release of a natural enemy to attack a pest population. Agricultural researchers have long been interested in identifying potential biological agents to control insects, weeds, and other organisms that reduce crop yield.

Biological control has been effective in numerous instances, especially with invasive insects and plants. In one classic success story, beetles were brought in to combat St. John's wort, a perennial (long-lived) European weed that invaded the western United States. By the 1940s, St. John's wort (also known as Klamath weed) had overgrown millions of acres of rangeland and pasture, leaving few edible plants for grazing livestock. Researchers imported leaf beetles from the plant's native region that feed exclusively on St. John's wort. The shiny, pea-sized insects reduced the weed to less than 5% of its former abundance, restoring the land's value to ranchers.

One potential pitfall of biological control is the danger that an imported control agent may be as invasive as its target. One cautionary tale comes from introducing the mongoose (Figure 19.16) to control rats. Rats that originated in India and northern Asia were accidentally transported around the world and became invasive in many places. For sugarcane growers, rat infestation meant massive crop damage. Cane planters imported the small Indian mongoose, a fierce little carnivore, to deal with the problem. In time, mongooses were introduced to dozens of natural habitats, including all of the largest Caribbean and Hawaiian islands—and became invasive themselves. Mongooses are not picky eaters, and they have voracious appetites. On island after island, populations of reptiles, amphibians, and ground-nesting birds have declined or vanished as mongoose populations have grown and spread. They also prey on domestic poultry and ruin crops, costing millions of dollars a year. Clearly, rigorous research is needed to assess the safety and efficacy of potential biological control agents.

◄ **Figure 19.16 A small Indian mongoose.**

Can Biological Control Defeat Kudzu?

Let's take a brief look at the search for a biological agent to control kudzu, an invasive vine known as "the plant that ate the South" **(Figure 19.17)**. In the 1930s, the U.S. Department of Agriculture distributed plantings of this Asian import to help control erosion along road cuts and irrigation canals. Today, kudzu covers an estimated 31,000 square kilometers (roughly the combined area of Maryland and Delaware). With its formidable growth rate of up to a foot per day, kudzu climbs over forest trees and blankets the ground in dense greenery. The shoots die back in winter but quickly regenerate from the roots in the spring. Cold winters have limited its acquisition of new territory—the roots don't survive being frozen. However, as global climate change brings warmer winters, kudzu is advancing farther north.

Unlike many invasive species, kudzu does have natural enemies in the United States, but it easily outgrows the damage they inflict. Researchers are investigating the possibility that one of these native pathogens or insects could be manipulated to provide effective control. Several possibilities have already been tested and discarded. For example, experiments showed that larvae of a species of moth have a prodigious appetite for kudzu. Further investigation, however, proved that the larvae actually preferred soybeans—an important crop species closely related to kudzu. At present, a fungal pathogen called *Myrothecium verrucaria* appears to be a promising candidate.

Researchers chose to test *M. verrucaria* because of **observations** that it causes severe disease in other weeds belonging to the same family as kudzu. Preliminary tests in a greenhouse, in controlled environment chambers, and in small outdoor plantings established that *M. verrucaria* kills kudzu when a high enough concentration of *M. verrucaria* spores is sprayed on the plants along with a "wetting agent" (a soap-like substance that reduces the surface tension of water). These findings led researchers to ask the **question**, Will *M. verrucaria* treatment work on an established stand of kudzu in a natural setting? Their **hypothesis** was that the *M. verrucaria* treatment that was most effective in the small outdoor plantings would also be most effective

▲ **Figure 19.17 Kudzu (*Pueraria lobata*).**

in a natural setting. Their **prediction** was that the treatment that sprayed the highest concentration of spores in combination with a wetting agent would produce the highest death rate. The **results** of this field experiment, which are shown in **Figure 19.18**, support the hypothesis. However, these experiments are only the initial steps toward biological control of kudzu. A great deal of research is needed to ensure that the method is safe, effective, and practical.

▼ **Figure 19.18 Biological control of a natural infestation of kudzu with the fungus *Myrothecium verrucaria*.**

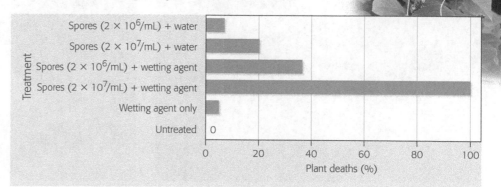

Integrated Pest Management

In contrast to enterprises such as fisheries, which harvest resources from natural ecosystems, agricultural operations create their own highly managed ecosystems. A typical crop population consists of genetically similar individuals (a monoculture) planted in close proximity to each other—a banquet laid out for the many plant-eating animals and pathogenic bacteria, viruses, and fungi in the community. The tilled, fertile ground nurtures weeds as well as crops. Thus, farmers wage an eternal war against pests that compete with their crop for soil minerals, water, and light; that siphon nutrients from the growing plants; or that consume their leaves, roots, fruits, or seeds. At home, you may be engaged in combat against pests on a smaller scale as you attempt to eradicate the weeds, insects, fungi, and bacteria that attack your lawn and garden or the mosquitoes that make your summer evenings miserable.

Like invasive species, most crop pests have an opportunistic life history pattern that enables them to rapidly take advantage of a favorable habitat. The history of agriculture abounds with examples of devastating pest outbreaks. For example, the boll weevil **(Figure 19.19)** is an insect that feeds on cotton plants both as larvae and as adults. Its unstoppable spread across the southern United States in the early 1900s severely damaged local economies and had a lasting impact on the region. The folklore of the invasion includes the song "The Boll Weevil Blues," which has been recorded by many artists, including the White Stripes. Viruses, fungi, bacteria, nematodes, and other plant-eating insects can cause massive damage as well.

When synthetic herbicides and insecticides such as DDT were developed in the 1940s, they quickly became the method of choice in agriculture. However, chemical solutions to pest problems bring numerous problems of their own. These chemicals are pollutants that can be carried great distances by air or water currents. In addition, natural selection may result in populations that are not affected by a pesticide (see Figure 13.13). Furthermore, most insecticides kill both the pest and its natural predators. Because prey species often have a higher reproductive rate than predators, pest populations rapidly rebound before their predators can reproduce. There may be other unintended damage as well, such as killing pollinators that are essential for both agricultural and natural ecosystems.

Integrated pest management (IPM) uses a combination of biological, chemical, and cultural methods for sustainable control of agricultural pests. Researchers are also investigating IPM approaches to invasive species. IPM relies on knowledge of the population ecology of the pest and its associated predators and parasites, as well as plant growth dynamics. In contrast to traditional methods of pest control, IPM advocates tolerating a low level of pests rather than attempting total eradication. Thus, many pest control measures are aimed at lowering the habitat's carrying capacity for the pest population by using pest-resistant varieties of crops, mixed-species plantings, and crop rotation to deprive the pest of a dependable food source. Biological control is also used when possible. For example, many gardeners release ladybird beetles to control aphid infestations **(Figure 19.20)**. Pesticides are applied when necessary, but adherence to the principles of IPM prevents the overuse of chemicals. ☑

☑ **CHECKPOINT**

Why is integrated pest management considered a more sustainable method of pest control than the use of chemicals alone?

Answer: Use of pesticides may result in resistant pest populations, which makes the pesticides less effective. Also, pesticide use is unsustainable in the long run because of problems with pollution.

▲ Figure 19.19 **A boll weevil on a damaged boll (seed pod) of a cotton plant.**

▲ Figure 19.20 **Ladybird beetles feeding on aphids.** Each of these voracious predators can eat as many as 50 aphids per day.

Human Population Growth

Now that we have examined the regulation of population growth in other organisms, what about our own species? Let's begin by looking at the history of the human population and then consider some current and future trends in population growth.

The History of Human Population Growth

In the few seconds it takes you to read this sentence, approximately 21 babies will be born somewhere in the world and 9 people will die. An imbalance between births and deaths is the cause of population growth (or decline), and as the line graph in **Figure 19.21** shows, the human population is expected to continue increasing for at least the next several decades. The bar graph in Figure 19.21 tells a different part of the story. The number of people added to the population each year has been declining since the 1980s. How do we explain these patterns of human population growth?

Let's begin with the rise in world population from approximately 480 million people in 1500 to the current population of more than 7 billion. In the exponential population growth model introduced earlier in this chapter, we assumed that the net rate of increase (birth rate minus death rate) was constant—births and deaths were roughly equal. As a result, population growth depended only on the size of the existing population. Throughout most of human history, this assumption held true. Although parents had many children, the death rate was also high, resulting in a rate of increase

only slightly higher than 0. Consequently, human population growth was initially very slow. (If we extended the x-axis of Figure 19.21 back in time to year 1, when the population was roughly 300 million, the line would be almost flat for 1,500 years.) The 1 billion mark was not reached until the early 1800s. As economic development in Europe and the United States led to advances in nutrition and sanitation, and later, medical care, people took control of their population's growth rate. At first, the death rate decreased while the birth rate remained the same. The net rate of increase rose, and population growth began to pick up steam by the beginning of the 1900s. By midcentury, improvements in nutrition, sanitation, and health care had spread to the developing world, spurring growth at a breakneck pace as birth rates far outstripped death rates.

As the world population skyrocketed from 2 billion in 1927 to 3 billion just 33 years later, some scientists became alarmed. They feared that Earth's carrying capacity would be reached and that density-dependent factors would maintain that population size through human suffering and death. But the overall growth rate peaked in 1962. In the more developed nations, advanced medical care continued to improve survivorship, but effective contraceptives held down the birth rate. As a result, the overall growth rate of the world's population began a downward trend as the difference between birth rate and death rate decreased. In the most developed nations, the overall rate of increase is near zero **(Table 19.3)**. In the developing world, on the other hand, death rates have dropped, but high birth rates persist. As a result, these populations are growing rapidly—of the 77.7 million people added to the world in 2012, nearly 74 million were in developing nations. Thus, the world population continues to increase. ☑

Counting one person per second, it would take more than 225 years to count all the people alive today.

☑ **CHECKPOINT**

Why was the growth rate of the world's population so high during most of the 1900s? What accounts for the recent decrease in the growth rate of the world's population?

Answer: In the 1900s, the death rate decreased dramatically due to improved nutrition, sanitation, and health care, but the birth rate remained high. As a result, the overall population growth rate was high. The recent decrease in growth rate is the result of lower birth rates in some regions.

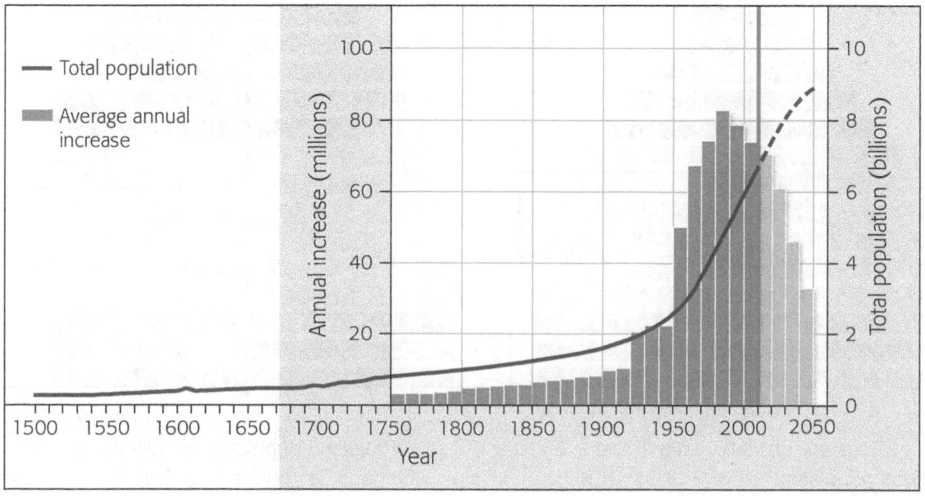

▲ Figure 19.21 **Five centuries of human population growth, projected to 2050.**

Table 19.3	Population Trends in 2012		
Population	Birth Rate per 1,000	Death Rate per 1,000	Growth Rate (%)
World	19.1	7.9	1.1
More developed countries	11.2	10.1	0.3
Less developed countries	20.8	7.4	1.3

Age Structures

Age structures, which were introduced at the beginning of this chapter, are helpful for predicting a population's future growth. **Figure 19.22** shows the estimated and projected age structures of Mexico's population in 1989, 2012, and 2035. In these diagrams, the area to the left of each vertical line represents the number of males in each age-group; females are represented on the right side of the line. The three different colors represent the portion of the population in their prereproductive years (0–14), prime reproductive years (15–44), and postreproductive years (45 and older). Within each of these broader groups, each horizontal bar represents the population in a 5-year age-group.

In 1989, each age-group was larger than the one above it, indicating a high birth rate. The pyramidal shape of this age structure is typical of a population that is growing rapidly. In 2012, the population's growth rate was lower; notice that the three youngest (bottommost) age-groups are roughly the same size. However, the population continues to be affected by its earlier expansion. This situation, which results from the increased proportion of women of childbearing age in the population, is known as **population momentum**. Girls who were 0–14 years old in the 1989 age structure (outlined in pink) were in their reproductive prime in 2012, and girls who were 0–14 years old in 2012 (outlined in blue) will carry the legacy of rapid growth forward to 2035. Putting the brakes on a rapidly expanding population is like stopping a freight train; the actual event takes place long after the decision to do it was made. Even when fertility—the number of live births over a woman's lifetime—is reduced to replacement rate (an average of two children per female), the total population size will continue to increase for several decades. Thus, the percentage of individuals under the age of 15 gives a rough idea of future growth. In less developed countries, about 28% of the population is in this age-group. In contrast, 16.5% of the population of more developed nations is under the age of 15. Population momentum also explains why the total population size worldwide (line graph in Figure 19.21) continues to increase even though fewer people are added to the population each year, as shown by the bar graph in Figure 19.21.

Age structure diagrams may also indicate social conditions. For instance, an expanding population has an increasing need for schools, employment, and infrastructure. A large elderly population requires that extensive resources be allotted to health care. Let's look at trends in the age structure of the United States from 1989 to 2035 **(Figure 19.23)**. The noticeable bulge in the 1989 population (yellow screen) corresponds to the "baby boom" that lasted for about two decades after World War II ended in 1945. The large number of children swelled school enrollments, prompting construction of new schools and creating a demand for teachers. On the other hand, graduates who were born near the end of the boom faced stiff competition for jobs. Because they make up such a large segment of the population, boomers have had an enormous influence on social,

▼ Figure 19.22 **Population momentum in Mexico.**

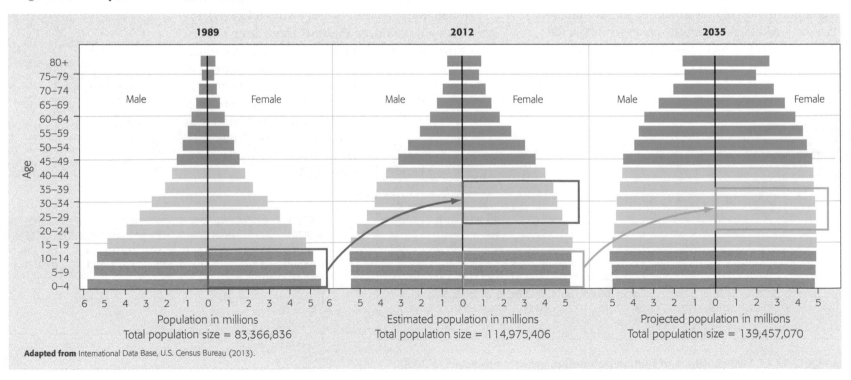

Adapted from International Data Base, U.S. Census Bureau (2013).

▼ **Figure 19.23** **Age structures for the United States in 1989, 2012 (estimated), and 2035 (projected).**

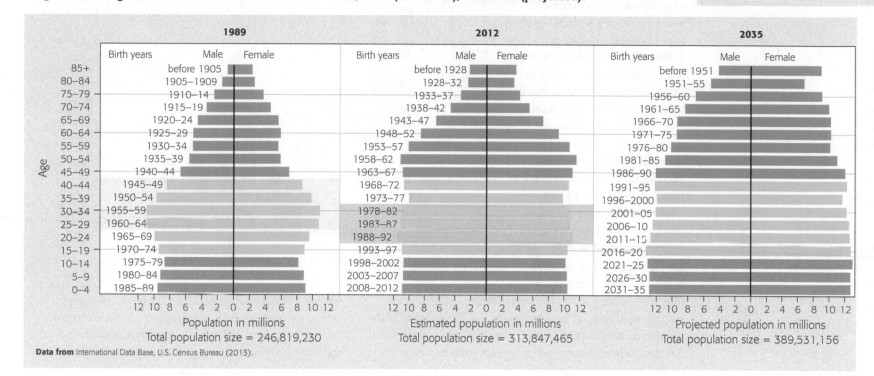

Data from International Data Base, U.S. Census Bureau (2013).

economic, and political trends. They also produced a boomlet of their own, seen in the 0–4 age-group in 1989 and the bump (pink screen) in the 2012 age structure.

Where are the baby boomers now? The leading edge has reached retirement age, which will place pressure on programs such as Medicare and Social Security. In 2012, 60% of the U.S. population was between 20 and 64, the ages most likely to be in the workforce, and 13.5% was over 65. In 2035, the percentages are projected to be 54 and 20, respectively. In part, the increase in the elderly population results from people living longer. The percentage of the population over 80, which was 2.7% in 1985, is projected to rise to nearly 6%—more than 23 million people—in 2035. ☑

Our Ecological Footprint

How large a human population can Earth support? Figure 19.21 shows that the world's population is increasing rapidly, though at a slower rate than it did in the last century. The rate of increase, as well as population momentum, indicates that the populations of most developing nations will continue to increase for the foreseeable future. The U.S. Census Bureau projects a global population of 8 billion within the next 10 years and 9.6 billion by 2050. But these numbers are only part of the story. Trillions of bacteria can live in a petri dish *if* they have

sufficient resources. Do we have sufficient resources to sustain 8 or 9 billion people?

To accommodate all the people expected to live on our planet in the coming decades and improve the diets of those who are currently malnourished or undernourished, world food production must increase dramatically. But agricultural lands are already under pressure. Overgrazing by the world's growing herds of livestock is turning vast areas of grassland into desert. Water use has risen sixfold over the past 70 years, causing rivers to run dry, water for irrigation to be depleted, and levels of groundwater to drop. Changes in precipitation patterns due to global warming are already causing food shortages in some regions of the world. And because so much open space will be needed to support the expanding human population, many other species are expected to become extinct.

The concept of an ecological footprint is one approach to understanding resource availability and usage. An **ecological footprint** is an estimate of the area of land and water required to provide the resources an individual or a nation consumes—for example, food, fuel, and housing—and to absorb the waste it generates, of which carbon emissions are a major component. Comparing our demand for resources with Earth's capacity to renew these resources, or **biocapacity**, gives us a broad view of the sustainability of human activities.

☑ **CHECKPOINT**

Why is the percentage of individuals under the age of 15 a good indication of future population growth?

Answer: These individuals have not yet entered their reproductive years. If they make up a large percentage of the population (a bottom-heavy age structure), future population growth will be high.

When the total area of ecologically productive land on Earth is divided by the global population, we each have a share of about 1.8 global hectares (1 hectare = 2.47 acres; a global hectare is a hectare with world-average ability to produce resources and absorb wastes). When used sustainably, resources such as crops, pastureland, forests, and fishing grounds can regenerate. But according to the World Wildlife Fund, in 2008 (the most recent year for which data are available), the average ecological footprint for the world's population was 2.7 global hectares—roughly 1.5 times the planet's biocapacity per person. By overshooting Earth's biocapacity, we are depleting our resources. The collapse of the northern cod fisheries (see Figure 19.13) illustrates what happens when usage exceeds regenerative capacity.

Figure 19.24 compares the ecological footprints of several countries to the world average footprint (pink line) and Earth's biocapacity (green line). Affluent nations such as the United States and Australia consume a disproportionate amount of resources (Figure 19.25). By this measure, the ecological impact of affluent nations is potentially as damaging as unrestrained population growth in the developing world. So the problem is not just overpopulation, but overconsumption. The world's richest countries, with 15% of the global population, account for 36% of humanity's total footprint. Some researchers estimate that providing everyone with the same standard of living as the United States would require the resources of more than four planet Earths.

To stay within the planet's regenerative capacity, all of humanity would have to live like an average citizen of Colombia or Uzbekistan.

If you would like to learn about your personal resource consumption, a number of online quizzes can provide a rough estimate of your ecological footprint. Like carbon footprint calculators (described in Chapter 18), these tools are useful for learning how to reduce your environmental impact. ☑

The average American generates about 40 pounds of trash per week, almost three times as much as the average Colombian.

☑ **CHECKPOINT**
How does an individual's large ecological footprint affect Earth's carrying capacity?

Answer: *The more resources required to sustain an individual, the lower Earth's carrying capacity will be. (That is, Earth can sustain fewer people if each of those people consumes a large share of available resources.)*

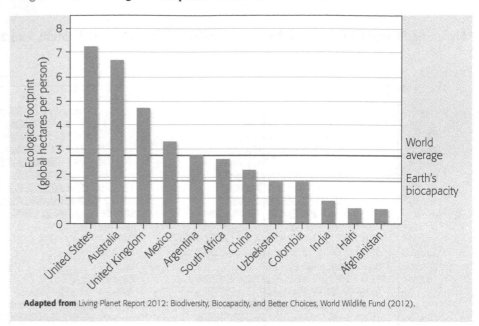

▼ Figure 19.24 **Ecological footprints of several countries.**

Ecological footprint (global hectares per person)

World average

Earth's biocapacity

United States, Australia, United Kingdom, Mexico, Argentina, South Africa, China, Uzbekistan, Colombia, India, Haiti, Afghanistan

Adapted from Living Planet Report 2012: Biodiversity, Biocapacity, and Better Choices, World Wildlife Fund (2012).

▲ Figure 19.25 **Mealtime in an Afghan (left) and an American household (right).** Notice the differences in consumer goods and in resources used for food preparation.

Biological Invasions EVOLUTION CONNECTION

Humans as an Invasive Species

The magnificent pronghorn antelope (*Antilocapra americana*) is the descendant of ancestors that roamed the open plains and shrub deserts of North America millions of years ago (Figure 19.26). With strides that cover 6 m (20 feet) or more at its top speed of 97 km/h (60 mph), it is easily the fastest mammal on the continent. The pronghorn's speed is more than a match for its major predator, the wolf, which typically takes adults that have been weakened by age or illness. What selection pressure promoted such extravagant speed? Ecologists hypothesize that the pronghorn's ancestors were running from the now-extinct American cheetah, a fleet-footed predator that bore some similarities to the more familiar African cheetah.

Cheetahs were not the only danger in the pronghorn's environment. During the Pleistocene epoch, which lasted from 1.8 million to 10,000 years ago, North America was also home to an intimidating list of other predators: lions, jaguars, saber-toothed cats with canines up to 7 inches long, and towering short-faced bears, which stood 11 feet tall and weighed three-quarters of a ton. There were plenty of potential prey for these fearsome predators, including massive ground sloths, bison with horns that spread 10 feet, elephant-like mammoths, a variety of horses and camels, and several species of pronghorns. Of all these species of large mammals, only *Antilocapra americana* remained at the end of the Pleistocene. The others went extinct during a relatively brief period of time that coincided with the spread of humans throughout North America. Although the cause of the extinctions has been hotly disputed, many scientists think that the human invasion, combined with climate change at the end of the last ice age, was responsible. Taken together, changes in the biotic and abiotic environments happened too rapidly for an evolutionary response by these large mammals.

The role of humans in the Pleistocene extinctions was merely a preview of things to come. The human population continues to increase, colonizing almost every corner of the globe. Like other invasive species, we change the environment of the other organisms that share our habitats. As the scope and speed of human-induced environmental changes increase, extinctions are occurring at an accelerating pace. This rapid loss of biodiversity will be our unifying thread in the next chapter.

▼ Figure 19.26 **A pronghorn antelope racing across the North American plains.**

Chapter Review

◼ SUMMARY OF KEY CONCEPTS

An Overview of Population Ecology

A population consists of all the members of a species living in the same place at the same time. Population ecology focuses on the factors that influence a population's size, density, age structure, and growth rate.

Population Density

Population density, the number of individuals of a species per unit area or volume, can be estimated by a variety of sampling techniques.

Population Age Structure

A graph showing the distribution of individuals in different age-groups often provides useful information about the population.

Life Tables and Survivorship Curves

A life table tracks the chance of an individual in a population surviving to various ages. Survivorship curves can be classified into three general types.

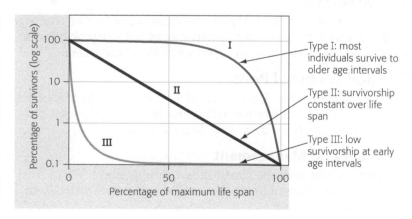

Type I: most individuals survive to older age intervals

Type II: survivorship constant over life span

Type III: low survivorship at early age intervals

Evolution: Life History Traits as Adaptations

Life history traits are shaped by evolutionary adaptation. Most populations probably fall between the extreme opportunistic life histories (reach sexual maturity rapidly; produce many offspring; little or no parental care) of many insects and the equilibrial life histories (develop slowly and produce few, well-cared-for offspring) of many larger-bodied species.

Population Growth Models

The Exponential Population Growth Model: The Ideal of an Unlimited Environment

Exponential population growth is the accelerating increase that occurs when growth is unlimited. The exponential model predicts that the larger a population becomes, the faster it grows.

The Logistic Population Growth Model: The Reality of a Limited Environment

Logistic population growth occurs when growth is slowed by limiting factors. The logistic model predicts that a population's growth rate will be low when the population size is either small or large and highest when the population is at an intermediate level relative to the carrying capacity.

Regulation of Population Growth

Over the long term, most population growth is limited by a mixture of density-independent factors, which affect the same percentage of individuals regardless of population size, and density-dependent factors, which intensify as a population increases in density. Some populations have regular boom-and-bust cycles.

Applications of Population Ecology

Conservation of Endangered Species

Endangered and threatened species are characterized by very small population sizes. One approach to conservation is identifying and attempting to supply the critical combination of habitat factors needed by the population.

Sustainable Resource Management

Resource managers apply principles of population ecology to help determine sustainable harvesting practices.

Invasive Species

Invasive species are non-native organisms that spread far beyond their original point of introduction and cause environmental and economic damage. Typically, invasive species have an opportunistic life history pattern.

Biological Control of Pests

Biological control, the intentional release of a natural enemy to attack a pest population, is sometimes effective against invasive species. However, prospective control agents can themselves become invasive.

Integrated Pest Management

Crop scientists have developed integrated pest management (IPM) strategies—combinations of biological, chemical, and cultural methods—to deal with agricultural pests.

Human Population Growth

The History of Human Population Growth

The human population grew rapidly during the 1900s and is currently more than 7 billion. A shift from high birth and death rates to low birth and death rates has lowered the rate of growth in more developed countries. In developing nations, death rates have dropped, but birth rates are still high.

Age Structures

The age structure of a population affects its future growth. The wide base of the age structure of Mexico in 1989—the 0–14 age-group—predicts continued population growth in the next generation. Population momentum is the continued growth that occurs after a population's high fertility rate has been reduced to replacement rate; it is a result of girls in the 0–14 age-group reaching their childbearing years. Age structures may also indicate social and economic trends, as in the age structure on the right below.

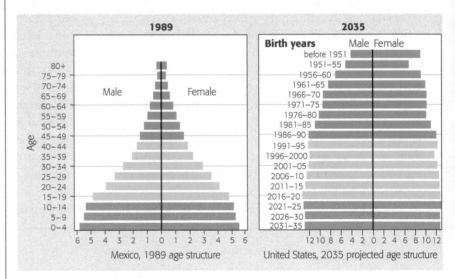

Mexico, 1989 age structure | United States, 2035 projected age structure

Our Ecological Footprint

An ecological footprint represents the amount of land and water needed to produce the resources used by an individual or nation. There is a huge disparity between resource consumption in more developed and less developed nations.

MasteringBiology®

For practice quizzes, BioFlix animations, MP3 tutorials, video tutors, and more study tools designed for this textbook, go to MasteringBiology®

SELF-QUIZ

1. What two values would you need to know to figure out the human population density of your community?

2. If members of a species produce a large number of offspring but provide minimal parental care, then a Type _____ survivorship curve is expected. In contrast, if members of a species produce few offspring and provide them with long-standing care, then a Type _____ survivorship curve is expected.

3. Use this graph of the idealized exponential and logistic growth curves to complete the following.

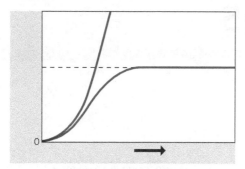

a. Label the axes and curves on the graph.
b. What does the dotted line represent?
c. For each curve, indicate and explain where population growth is the most rapid.
d. Which of these curves better represents global human population growth?

4. Which of these factors operates in a density-dependent manner?
a. blizzard
b. volcanic eruption
c. food supply
d. flood

5. Which life history pattern is typical of invasive species?

6. Throughout most of human history, human population size
a. was at carrying capacity.
b. grew very slowly.
c. showed boom-and-bust cycles.
d. showed skyrocketing growth.

7. A study of the human ecological footprint shows that
a. we have already overshot the planet's capacity to sustain us.
b. Earth can sustain the current population, but not much more.
c. Earth can sustain a population about double the current population.
d. the size of the human population will soon crash.

Answers to these questions can be found in Appendix: Self-Quiz Answers.

THE PROCESS OF SCIENCE

8. What are the limiting factors for a population of tropical fish in an aquarium? How would you test if there is any interspecies competition with another type of fish?

9. Interpreting Data The graph below shows data for population trends in Mexico from 1890 to 2012, with projected trends for 2012–2050. How has Mexico's rate of population growth changed over this time period? How is it expected to change by midcentury? Describe Mexico's projected age structure in 2050.

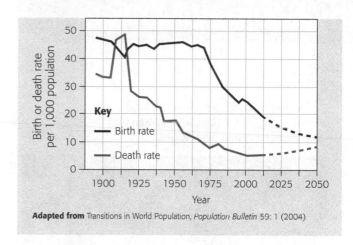

Adapted from Transitions in World Population, *Population Bulletin* 59: 1 (2004).

BIOLOGY AND SOCIETY

10. Experts are enlisting the help of citizen scientists in the battle against invasive species. For example, in 2014, the Florida Fish and Wildlife Commission released a phone app for reporting lionfish sightings, data that will help target fish for removal. A phone app to identify and report the locations of invasive species at more than 100 sites across the United States and Canada is available at www.whatsinvasive.org. Other citizen science projects are listed at www.birds.cornell.edu/citscitoolkit/projects/find/projects-invasive-species. Using these sites or other resources, determine which invasive species are of greatest concern in your area. Learn how to identify them, and find out what control measures are being taken.

11. The tiger, mountain gorilla, spotted owl, giant panda, snow leopard, and grizzly bear are all endangered by human encroachment on their habitats. Why might these animals, which all have equilibrial life history traits, be more easily endangered than animals with opportunistic life history traits? What general type of survivorship curve would you expect these species to exhibit? Explain your answer.

12. Wars and epidemics can limit the growth of the human population. How big has the influence of wars and epidemics been on the human population in the past?

20 Communities and Ecosystems

Tuna fish, which eat high on the food chain, are likely to be contaminated with mercury and other toxins.

◀ Producing the beef for a hamburger (top) requires eight times as much land as producing the soybeans for a soyburger (bottom).

▲ The atoms in your body might have been part of a dinosaur that lived 75 million years ago.

▲ Human encroachment on tropical rain forests has brought new diseases such as Ebola to human populations.

Biodiversity in Decline BIOLOGY AND SOCIETY

Why Biodiversity Matters

As the human population has expanded, hundreds of species have become extinct and thousands more are threatened with extinction. These changes represent a loss in biological diversity, or biodiversity. Biodiversity loss goes hand in hand with the disappearance of natural ecosystems. Only about a quarter of Earth's land surfaces remain untouched by human alterations. We see the evidence of our impact on natural ecosystems every day. We live and work in altered landscapes. And though we may be less aware of it, our impact on the oceans is also extensive.

What is the value of biodiversity? Most people appreciate the direct benefits provided by certain ecosystems. For example, you probably know that we use resources—such as water, wood, and fish—that come from natural or near-natural ecosystems. These resources have economic value, as the massive 2010 oil spill in the Gulf of Mexico dramatically demonstrated. Billions of dollars were lost by fishing, recreation, and other industries as a result of the disaster. But human well-being also depends on less obvious services that healthy ecosystems provide. The coastal wetlands affected by the Gulf oil spill normally act as a buffer against hurricanes, reduce the impact of flooding, and filter pollutants. The wetlands also furnish nesting sites for birds and marine turtles and breeding areas and nurseries for a wide variety of fish and shellfish. Natural ecosystems provide other services as well—such as recycling

A 2014 oil spill in Galveston Bay, Texas, threatened bird sanctuaries on the nearby Bolivar Peninsula, home of this snowy egret.

nutrients, preventing erosion and mudslides, controlling agricultural pests, and pollinating crops. Some scientists have attempted to assign an economic value to these benefits. They arrived at an average annual value of ecosystem services of $33 trillion, almost twice the global gross national product for the year they published their results. Although rough, these estimates make the important point that we cannot afford to take biodiversity for granted.

In this chapter, we'll examine the interactions among organisms and how those relationships determine the characteristics of communities. On a larger scale, we'll explore the dynamics of ecosystems. Finally, we'll consider how scientists are working to save biodiversity. And throughout the chapter, you'll learn how an understanding of ecology can help us manage Earth's resources wisely.

The Loss of Biodiversity

☑ **CHECKPOINT**

How does the loss of genetic diversity endanger a population?

Answer: A population with decreased genetic diversity has less ability to evolve in response to environmental change.

As mentioned previously, **biodiversity** is short for biological diversity, the variety of living things. It includes genetic diversity, species diversity, and ecosystem diversity. Thus, the loss of biodiversity encompasses more than just the fate of individual species.

Genetic Diversity

The genetic diversity within a population is the raw material that makes microevolution and adaptation to the environment possible. If local populations are lost, then the number of individuals in the species declines, and so do the genetic resources for that species. Severe reduction in genetic variation threatens the survival of a species. When an entire species is lost, so are all of its unique genes.

The enormous genetic diversity of all the organisms on Earth has great potential benefit for people. Many researchers and biotechnology leaders are enthusiastic about the potential that genetic "bioprospecting" holds for future development of new medicines, industrial chemicals, and other products. Bioprospecting may also hold the key to the world's food supply. For example, researchers are currently scrambling to stop the spread of a deadly new strain of wheat stem rust, a fungal pathogen that has devastated harvests in eastern Africa and central Asia. At least 75% of the wheat varieties planted worldwide are susceptible to this pathogen, but researchers hope to find a resistance gene in the wild relatives of wheat (**Figure 20.1**). ☑

Species Diversity

In view of the damage we are doing to the biosphere, ecologists believe that we are pushing species toward extinction at an alarming rate. The present rate of species loss may be as much as 100 times higher than at any time in the past 100,000 years. Some researchers estimate that at the current rate of destruction, over half of all currently living plant and animal species will be gone by the end of this century. **Figure 20.2** shows two recent victims. The International Union for the Conservation of Nature (IUCN) compiles scientific assessments of the conservation status of species worldwide. Here are some examples of where things stand:

- Approximately 13% of the 10,004 bird species and almost a quarter of the 4,667 mammalian species assessed are threatened with extinction.

- More than 20% of freshwater fishes in the world either have become extinct during human history or are seriously threatened.

- Roughly 40% of all assessed amphibian species are in danger of extinction.

- Of the approximately 20,000 known plant species in the United States, 200 species have become extinct since dependable records have been kept. More than 10,000 plant species worldwide are in danger of extinction.

▼ **Figure 20.2 Recent additions to the list of human-caused extinctions.**

Clouded leopard. In 2013, scientists gave up hope that the Formosan clouded leopard, a subspecies found only on the island of Taiwan, still exists. This photo shows a similar subspecies living in a zoo.

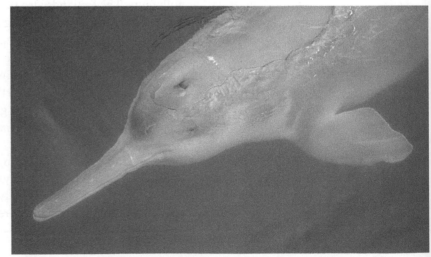

A Chinese river dolphin. This former resident of the Yangtze River, also known as a baiji, fell victim to pollution and habitat loss. The baiji was declared extinct in 2006 after a two-year search failed to find any remaining.

Ecosystem Diversity

Ecosystem diversity is the third component of biological diversity. Recall that an ecosystem includes both the organisms and the abiotic factors in a particular area. Because of the network of interactions among populations of different species within an ecosystem, the loss of one species can have a negative effect on the entire ecosystem. The disappearance of natural ecosystems results in the loss of **ecosystem services**, functions performed by an ecosystem that directly or indirectly benefit people. These vital services include air and water purification, climate regulation, and erosion control. For example, forests absorb and store carbon from the atmosphere, a service that vanishes when forests are destroyed or degraded (see Figure 18.39). Coral reefs not only are rich in species diversity (Figure 20.3) but also provide a wealth of benefits to people, including food, storm protection, and recreation. An estimated 20% of the world's coral reefs have already been destroyed by human activities. A study published in 2011 found that 75% of the remaining reefs are threatened, a percentage expected to top 90% by 2030 if current abuses continue. ☑

▼ **Figure 20.3 A coral reef, a colorful display of biodiversity.**

Causes of Declining Biodiversity

Ecologists have identified four main factors responsible for the loss of biodiversity: habitat destruction and fragmentation, invasive species, overexploitation, and pollution. The ever-expanding size and dominance of the human population are at the root of all four factors. In addition, scientists expect global climate change to become a leading cause of extinctions in the near future (see Chapter 18).

Habitat Destruction

The massive destruction and fragmentation of habitats caused by agriculture, urban development, forestry, and mining pose the single greatest threat to biodiversity (Figure 20.4). According to the IUCN, habitat destruction affects more than 85% of all birds, mammals, and amphibians that are threatened with extinction. The destruction of its forest habitat, along with trade in its gorgeous pelt, doomed the Formosan clouded leopard. The remaining subspecies of clouded leopard, which inhabit the forests of Southeast Asia, are also vulnerable to extinction as a result of deforestation. We'll take a closer look at the consequences of habitat fragmentation later in this chapter.

Invasive Species

Ranking second behind habitat destruction as a cause of biodiversity loss is the introduction of invasive species. Uncontrolled population growth of human-introduced species has caused havoc when the introduced species have competed with, preyed on, or parasitized native species (see Chapter 19). The lack of interactions with other species that could keep the newcomers in check is often a key factor in a non-native species becoming invasive.

☑ **CHECKPOINT**

When ecosystems are destroyed, the services they provide are lost. What are some examples of ecosystem services?

Answer: Possible answers are mentioned in the Biology and Society section as well as on this page. Services include air and water purification, climate regulation, erosion control, recreation, and the provision of resources used by people, such as wood, water, and food.

▼ Figure 20.4 **Habitat destruction.** In a controversial method known as mountaintop removal, mining companies blast the tops off of mountains and then scoop out coal. The earth removed from the mountain is dumped into a neighboring valley.

Overexploitation

Unsustainable marine fisheries (see Figure 19.13) demonstrate how people can overexploit wildlife by harvesting at rates that exceed the ability of populations to rebound. American bison, Galápagos tortoises, and tigers are among the many terrestrial species whose numbers have been drastically reduced by excessive commercial harvesting, poaching, and sport hunting. Overharvesting also threatens some plants, including rare trees such as mahogany and rosewood that produce valuable wood.

Pollution

Air and water pollution **(Figure 20.5)** is a contributing factor in declining populations of hundreds of species worldwide. The global water cycle can transport pollutants from terrestrial to aquatic ecosystems hundreds of miles away. Pollutants that are emitted into the atmosphere may be carried aloft for thousands of miles before falling to earth in the form of acid precipitation. ☑

▲ **Figure 20.5 A pelican stuck in oil from the 2010 Gulf of Mexico disaster.** Wildlife is often the most visible casualty of pollution, but the impact extends throughout the ecosystem.

Community Ecology

On your next walk through a field or woodland, or even across campus or your own backyard, observe the variety of species present. You may see birds in trees, butterflies on flowers, dandelions in the grass of a lawn, or lizards darting for cover as you approach. Each of these organisms interacts with other organisms as it carries out its life activities. An organism's biotic environment includes not just individuals from its own population, but also populations of other species living in the same area. Ecologists call such an assemblage of species living close enough together for potential interaction a **community**. In **Figure 20.6**, the lion, the zebra, the hyena, the vultures, the plants, and the unseen microbes are all members of an ecological community in Kenya.

Interspecific Interactions

Our study of communities begins with **interspecific interactions**—that is, interactions between species. Interspecific interactions can be classified according to the effect on the populations concerned, which may be helpful (**+**) or harmful (**−**). In some cases, two populations in a community vie for a resource such as food or space. The effect of this interaction is generally negative for both species (**−/−**) because neither species has access to the full range of resources offered by the habitat. On the other hand, some interspecific interactions benefit both parties (**+/+**). For example, the interactions between flowers and their pollinators are mutually beneficial. In a third type of interspecific interaction, one species exploits another species as a source of food. The effect of this interaction is clearly beneficial to one population and harmful to the other (**+/−**). In the next several pages, you will learn more about these interspecific interactions and how they affect communities. You will also see that interspecific interactions can be powerful agents of natural selection.

▼ **Figure 20.6 Diverse species interacting in a Kenyan savanna community.**

Interspecific Competition (−/−)

In the logistic model of population growth (see Figure 19.6), increasing population density reduces the amount of resources available for each individual. This intraspecific (within-species) competition for limited resources ultimately limits population growth. In **interspecific competition** (between-species competition), the population growth of a species may be limited by the population densities of competing species as well as by the density of its own population (intraspecific competition).

What determines whether populations in a community compete with each other? Each species has an **ecological niche**, defined as its total use of the biotic and abiotic resources in its environment. For example, the ecological niche of a small bird called the Virginia's warbler **(Figure 20.7a)** includes its nest sites and nest-building materials, the insects it eats, and climatic conditions such as the amount of precipitation and the temperature and humidity that enable it to survive. In other words, the ecological niche encompasses everything the Virginia's warbler needs for its existence. The ecological niche of the orange-crowned warbler **(Figure 20.7b)** includes some of the same resources used by the Virginia's warbler. Consequently, when these two species inhabit the same area, they are competitors.

Ecologists investigated the effects of interspecific competition between populations of these two birds in a community in central Arizona. When they removed either Virginia's warblers or orange-crowned warblers from the study site, members of the remaining species were significantly more successful in raising their offspring. This study showed that interspecific competition can have a direct, negative effect on reproductive fitness.

If the ecological niches of two species are too similar, they cannot coexist in the same place. Ecologists call this the **competitive exclusion principle**, a concept introduced by Russian ecologist G. F. Gause, who demonstrated this effect with an elegant series of experiments. Gause used two closely related species of protists, *Paramecium caudatum* and *P. aurelia*. First, he established the carrying capacity for each species separately under the conditions used to grow them in the laboratory **(Figure 20.8**, top graph). Then he grew the two species in the same habitat. Within two weeks, the *P. caudatum* population had crashed (bottom graph). Gause concluded that the requirements of these two species were so similar that the superior competitor—in this case, *P. aurelia*—deprived *P. caudatum* of essential resources. ☑

▼ Figure 20.8 **Competitive exclusion in laboratory populations of *Paramecium*.**

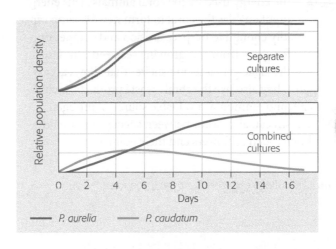

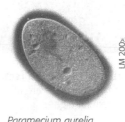

Paramecium aurelia

LM 200×

LM 200×

Paramecium caudatum

▼ Figure 20.7 **Species that use similar resources.**

(a) Virginia's warbler

(b) Orange-crowned warbler

Mutualism (+/+)

In **mutualism**, both species benefit from an interaction. Some mutualisms occur between symbiotic species—those in which the organisms have a close physical association with each other. For example, in the symbiotic root-fungus associations known as mycorrhizae (see the Biology and Society section in Chapter 16), the fungus delivers mineral nutrients to the plant and receives organic nutrients in return. Coral reef ecosystems depend on the mutualism between certain species of coral animals and millions of unicellular algae that live in the cells of each coral polyp **(Figure 20.9)**. Reefs are constructed by successive generations of colonial corals that secrete an external calcium carbonate skeleton. The sugars that the algae produce by photosynthesis provide at least half of the energy used by the coral animals. This energy input enables the coral to form new skeleton rapidly enough to outpace erosion and competition for space from fast-growing seaweeds. In return, the algae gain a secure shelter that allows access to light. They also use the coral's waste products, including CO_2 and ammonia, a valuable source of nitrogen. Mutualism can also occur between species that are not symbiotic, such as flowers and their pollinators.

▲ **Figure 20.9 Mutualism.** The cells of coral polyps are inhabited by unicellular algae.

▼ **Figure 20.10 Cryptic coloration.** Camouflage conceals the pygmy seahorse from predators.

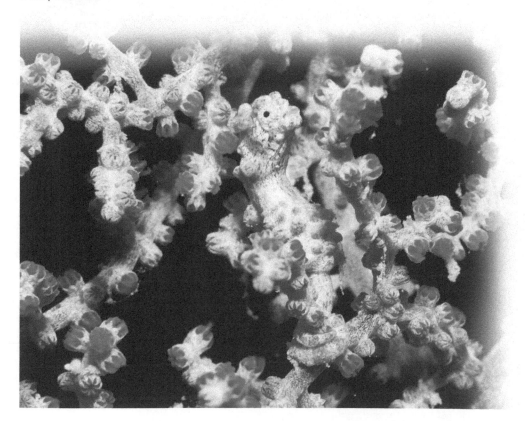

Predation (+/−)

Predation refers to an interaction in which one species (the predator) kills and eats another (the prey). Because predation has such a negative impact on the reproductive success of the prey, numerous adaptations for predator avoidance have evolved in prey populations through natural selection. For example, some prey species, such as the pronghorn antelope, run fast enough to escape their predators (see the Evolution Connection section in Chapter 19). Others, such as rabbits, flee into shelters. Still other prey species rely on mechanical defenses, such as the porcupine's sharp quills or the hard shells of clams and oysters.

Adaptive coloration is a type of defense that has evolved in many species of animals. Camouflage, called **cryptic coloration**, makes potential prey difficult to spot against its background **(Figure 20.10)**. **Warning coloration**, bright patterns of yellow, red, or orange in combination with black, often marks animals with effective chemical defenses. Predators learn to associate these color patterns with undesirable consequences, such as a noxious taste or painful sting, and avoid potential prey with similar markings. The vivid colors of the poison dart frog **(Figure 20.11)**, an inhabitant of Costa Rican rain forests, warn of noxious chemicals in the frog's skin.

A prey species may also gain significant protection through mimicry, a "copycat" adaptation in which one species looks like another. For example, the pattern of alternating red, black, and yellow rings of the harmless scarlet king snake resembles the bold color pattern of the venomous eastern coral snake **(Figure 20.12)**. Some insects have combined protective coloration with adaptations of body structures in elaborate disguises. For instance, there are insects that resemble twigs, leaves, and bird droppings. Some even do a passable imitation of a vertebrate. For example, the colors on the dorsal side

▼ Figure 20.11 **Warning coloration of a poison dart frog.**

▲ **Figure 20.12 Mimicry in snakes.** The color pattern of the nonvenomous scarlet king snake (left) is similar to that of the venomous eastern coral snake (right).

of certain caterpillars are an effective camouflage, but when disturbed, the caterpillars flip over to reveal the snakelike eyespots of their ventral side (**Figure 20.13**). Eyespots that resemble vertebrate eyes are common in several groups of moths and butterflies. A flash of these large "eyes" startles would-be predators. In other species, an eyespot may deflect a predator's attack away from vital body parts.

Herbivory (+/−)

Herbivory is the consumption of plant parts or algae by an animal. Although herbivory is not usually fatal to plants, a plant whose body parts have been partially eaten by an animal must expend energy to replace the loss. Consequently, numerous defenses against herbivores have evolved in plants. Spines and thorns are obvious anti-herbivore devices, as anyone who has plucked a rose from a thorny rosebush or brushed against a spiky cactus knows. Chemical toxins are also very common in plants. Like the chemical defenses of animals, toxins

in plants are distasteful, and herbivores learn to avoid them. Among such chemical weapons are the poison strychnine, produced by a tropical vine called *Strychnos toxifera*; morphine, from the opium poppy; nicotine, produced by the tobacco plant; mescaline, from peyote cactus; and tannins, from a variety of plant species. Other defensive compounds that are not toxic to humans but may be distasteful to herbivores are responsible for the familiar flavors of peppermint, cloves, and cinnamon (**Figure 20.14**). Some plants even produce chemicals that cause abnormal development in insects that eat them. Chemical companies have taken advantage of the poisonous properties of certain plants to produce pesticides. For example, nicotine is used as an insecticide. ☑

☑ CHECKPOINT

People find most bitter-tasting foods objectionable. Why do you suppose we have taste receptors for bitter-tasting chemicals?

Answer: Taste receptors sensitive to bitter chemicals presumably enabled the ancestors of humans to identify potentially toxic plants when they foraged for food, thus allowing them to survive longer.

▼ **Figure 20.13 An insect mimicking a snake.** When disturbed, this sphinx moth larva flips over (left), revealing eyespots that resemble a snake (right).

▼ **Figure 20.14 Flavorful plants.**

Peppermint. Parts of the peppermint plant yield a pungent oil.

Cloves. The cloves used in cooking are the flower buds of this plant.

Cinnamon. Cinnamon comes from the inner bark of this tree.

Parasites and Pathogens (+/−)

Both plants and animals may be victimized by parasites or pathogens. These interactions are beneficial to one species (the parasite or pathogen) and harmful to the other, known as the host. A **parasite** lives on or in a **host** from which it obtains nourishment. Invertebrate parasites include flatworms, such as flukes and tapeworms, and a variety of roundworms, which live inside a host organism's body. External parasites, such as ticks, lice, mites, and mosquitoes, attach to their victims temporarily to feed on blood or other body fluids. Plants are also attacked by parasites, including roundworms and aphids, tiny insects that tap into the phloem to suck plant sap (see Figure 19.10). In any parasite population, reproductive success is greatest for individuals that are best at locating and feeding on their hosts. For example, some aquatic leeches first locate a host by detecting movement in the water and then confirm its identity based on the host's body temperature and chemical cues on its skin.

Pathogens are disease-causing bacteria, viruses, fungi, or protists that can be thought of as microscopic parasites. Non-native pathogens, whose impact can be rapid and dramatic, have provided some opportunities to investigate the effects of pathogens on communities. In one example, ecologists studied the consequences of the epidemic of chestnut blight, a disease caused by a protist. The loss of chestnuts, massive canopy trees that once dominated many forest communities in North America, had a significant impact on community composition and structure. Trees such as oaks and hickories that had formerly competed with chestnuts became more numerous; overall, the diversity of tree species increased. Dead chestnut trees also furnished niches for other organisms, such as insects, cavity-nesting birds, and, eventually, decomposers. ☑

Trophic Structure

Now that we have looked at how populations in a community interact with one another, let's consider the community as a whole. The feeding relationships among the various species in a community are referred to as its **trophic structure**. A community's trophic structure determines the passage of energy and nutrients from plants and other photosynthetic organisms to herbivores and then to predators. The sequence of food transfer between trophic levels is called a **food chain**.

Figure 20.15 shows two food chains, one terrestrial and one aquatic. At the bottom of both chains is the trophic level that supports all others. This level consists of autotrophs, which ecologists call **producers**. Photosynthetic producers transform light energy to chemical energy stored in the bonds of organic compounds. Plants are the main producers on land. In water, the main producers are photosynthetic protists and cyanobacteria, collectively called phytoplankton. Multicellular algae and aquatic plants are also important producers in shallow waters. In a few communities, such as those surrounding hydrothermal vents, the producers are chemosynthetic prokaryotes.

All organisms in trophic levels above the producers are heterotrophs, or consumers. All consumers depend directly or indirectly on the output of producers. **Herbivores**, which eat plants, algae, or phytoplankton, are **primary consumers**. Primary consumers on land include grasshoppers and many other insects, snails, and certain vertebrates, such as grazing mammals and birds that eat seeds and fruits. In aquatic environments, primary consumers include a variety of zooplankton (mainly protists and microscopic animals such as small shrimps) that eat phytoplankton.

▼ Figure 20.15 **Examples of food chains.** The arrows trace the transfer of food from one trophic level to the next in terrestrial and aquatic communities.

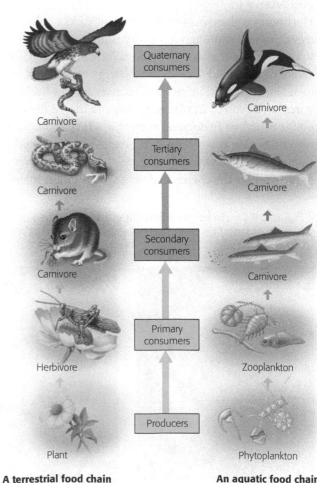

Quaternary consumers

Carnivore

Tertiary consumers

Carnivore

Carnivore

Secondary consumers

Carnivore

Carnivore

Primary consumers

Herbivore

Zooplankton

Producers

Plant

Phytoplankton

A terrestrial food chain   **An aquatic food chain**

Above the primary consumers, the trophic levels are made up of **carnivores**, which eat the consumers from the level below. On land, **secondary consumers** include many small mammals, such as the mouse shown in Figure 20.15 eating an herbivorous insect, and a great variety of birds, frogs, and spiders, as well as lions and other large carnivores that eat grazers. In aquatic communities, secondary consumers are mainly small fishes that eat zooplankton. Higher trophic levels include **tertiary consumers** (third-level consumers), such as snakes that eat mice and other secondary consumers. Most communities have secondary and tertiary consumers. As the figure indicates, some also have a higher level—**quaternary consumers** (fourth-level consumers). These include hawks in terrestrial communities and killer whales in the marine environment. There are no fifth-level consumers—you'll find out why shortly.

Figure 20.15 shows only those consumers that eat living organisms. Some consumers derive their energy from **detritus**, the dead material left by all trophic levels, including animal wastes, plant litter, and dead bodies. Different organisms consume detritus in different stages of decay. **Scavengers**, which are large animals such as crows and vultures, feast on carcasses left behind by predators or speeding cars. The diet of **detritivores** is made up primarily of decaying organic material. Earthworms and millipedes are examples of detritivores. **Decomposers**, mainly prokaryotes and fungi, secrete enzymes that digest molecules in organic material and convert them to inorganic forms. Enormous numbers of microscopic decomposers in the soil and in the mud at the bottom of lakes and oceans break down organic materials to inorganic compounds—the raw materials used by plants and phytoplankton to make new organic materials that may eventually become food for consumers. Many gardeners keep a compost pile, using the services of detritivores and decomposers to break down organic material from kitchen scraps and yard trimmings (Figure 20.16).

Biological Magnification

Organisms can't metabolize many of the toxins produced by industrial wastes or applied as pesticides; after consumption, the chemicals remain in the body. These toxins become concentrated as they pass through a food chain, a process called **biological magnification**. **Figure 20.17** shows the biological magnification of chemicals called PCBs (organic compounds used in electrical equipment until 1977) in a Great Lakes food chain. Zooplankton—the base of the pyramid—feed on phytoplankton contaminated by PCBs in the water. Smelt (small fish) feed on contaminated zooplankton. Because each smelt consumes many zooplankton, the concentration of PCBs is higher in smelt than in zooplankton. For the same reason, the concentration of PCBs in trout is higher than in smelt. The top-level predators—herring gulls in this example—have the highest concentrations of PCBs in the food chain and are the organisms most severely affected by any toxic compounds in the environment. In this case, fewer of the contaminated eggs hatched, resulting in a decline in the reproductive success of herring gulls. Many other synthetic chemicals that cannot be degraded by microorganisms, including DDT and mercury, can also become concentrated through biological magnification.

Tuna fish, which eat high on the food chain, are likely to be contaminated with mercury and other toxins.

▼ Figure 20.16 **Compost bin in a garden.** As it decays, the rich organic material of compost provides a slow-release source of inorganic nutrients for plants.

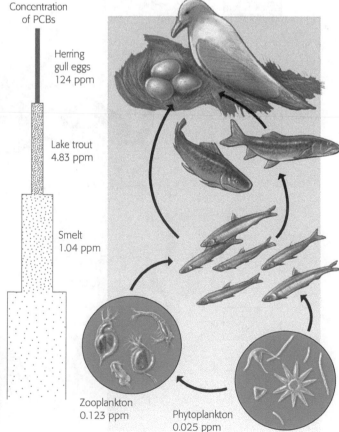

Concentration of PCBs

Herring gull eggs 124 ppm

Lake trout 4.83 ppm

Smelt 1.04 ppm

Zooplankton 0.123 ppm

Phytoplankton 0.025 ppm

◀ Figure 20.17 **Biological magnification of PCBs in a Great Lakes food chain in the early 1960s.** Eating top-level consumers, such as tuna, swordfish, and sharks, from contaminated waters can cause health problems for people, especially during pregnancy

☑ **CHECKPOINT**

You're eating a pizza. At which trophic level(s) are you feeding?

Answer: You're a primary consumer when you eat flour (in the crust) and tomato sauce, and you're a secondary consumer when you eat cheese or meat on the pizza.

Food Webs

Few, if any, communities are so simple that they are characterized by a single, unbranched food chain. Several types of primary consumers usually feed on the same plant species, and one species of primary consumer may eat several different plants. Such branching of food chains occurs at the other trophic levels as well. Thus, the feeding relationships in a community are woven into elaborate **food webs**.

Consider the grasshopper mouse in **Figure 20.18**, the secondary consumer shown crunching on a grasshopper in Figure 20.15. Its diet also includes plants, making it a primary consumer, too. It is an **omnivore**, an animal that eats producers as well as consumers of different levels. The rattlesnake that eats the mouse also feeds on more than one trophic level. The blue arrow leading to the rattlesnake indicates that it eats primary consumers—it is a secondary consumer. The purple arrow shows that it eats secondary consumers, so it is also a tertiary consumer. Sound complicated? Actually, Figure 20.18 shows a simplified food web. An actual food web would involve many more organisms at each trophic level, and most of the animals would have a more diverse diet than shown in the figure. ☑

▶ Figure 20.18 **A simplified food web for a Sonoran desert community.** As in the food chains of Figure 20.15, the arrows in this web indicate "who eats whom," the direction of nutrient transfers. We also continue the color-coding introduced in Figure 20.15 for the trophic levels and food transfers.

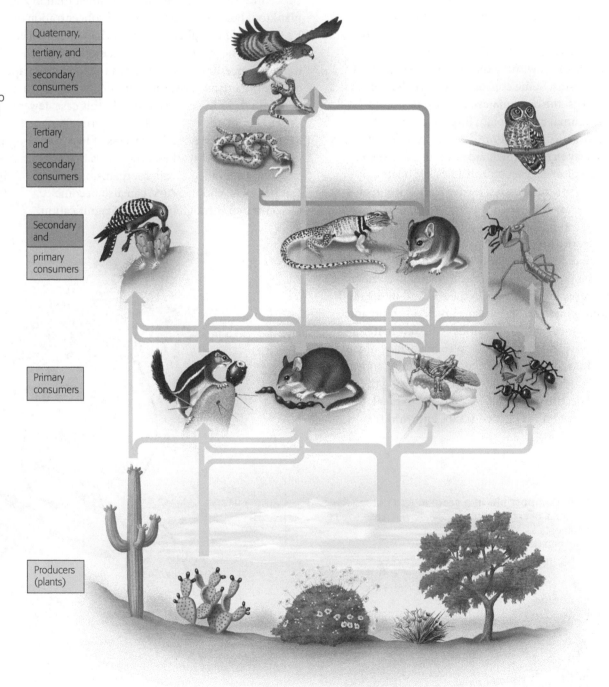

Quaternary, tertiary, and secondary consumers

Tertiary and secondary consumers

Secondary and primary consumers

Primary consumers

Producers (plants)

Species Diversity in Communities

The **species diversity** of a community—the variety of species that make up the community—has two components. The first component is **species richness**, or the number of different species in the community. The other component is the **relative abundance** of the different species, the proportional representation of each species in a community. To understand why both components are important for describing species diversity, imagine walking through the woodlands shown in **Figure 20.19**. On the path through woodland A, you would pass by four different species of trees, but most of the trees you encounter would be the same species. Now imagine walking on a path through woodland B. You would see the same four species of trees that you saw in woodland A—the species richness of the two woodlands is the same. However, woodland B might seem more diverse to you because no single species predominates. As **Figure 20.20** shows, the relative abundance of one species in woodland A is much higher than the relative abundances of the other three species. In woodland B, all four species are equally abundant. As a result, species diversity is greater in woodland B. Because plants provide food and shelter for many animals, a diverse plant community promotes animal diversity.

Although the abundance of a dominant species such as a forest tree can have an impact on the diversity of

▼ **Figure 20.19 Which woodland is more diverse?**

Woodland A

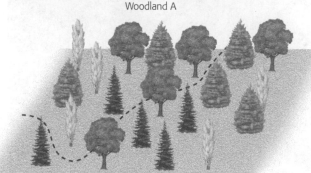

Woodland B

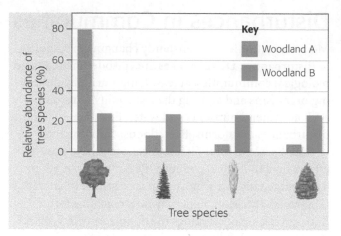

◄ **Figure 20.20 Relative abundance of tree species in woodlands A and B.**

other species in the community, a nondominant species may also exert control over community composition. A **keystone species** is a species whose impact on its community is much larger than its total mass or abundance indicates. The term *keystone species* was derived from the wedge-shaped stone at the top of an arch that locks the other pieces in place. If the keystone is removed, the arch collapses. A keystone species occupies an ecological niche that holds the rest of its community in place.

To investigate the role of a potential keystone species in a community, ecologists compare diversity when the species is present to that when it is absent. Experiments by Robert Paine in the 1960s were among the first to provide evidence of the keystone species effect. Paine manually removed a predator, a sea star of the genus *Pisaster* (**Figure 20.21**), from experimental areas within the intertidal zone of the Washington coast. The result was that *Pisaster*'s main prey, a mussel, outcompeted many of the other shoreline organisms (algae, barnacles, and snails, for instance) for the important resource of space on the rocks. The number of different organisms present in experimental areas dropped from more than 15 species to fewer than 5 species.

Ecologists have identified other species that play a key role in ecosystem structure. For instance, the decline of sea otters off the western coast of Alaska allowed populations of sea urchins, their main prey, to increase. The greater abundance of sea urchins, which consume seaweeds such as kelp, has resulted in the loss of many of the kelp "forests" (see Figure 15.27) and the diversity of marine life that they support. In many ecosystems, however, ecologists are just beginning to understand the complex relationships among species; the value of an individual species may not be apparent until it is gone. ☑

▼ **Figure 20.21 A *Pisaster* sea star.**

☑ CHECKPOINT

How could a community appear to have relatively little diversity even though it is rich in species?

Answer: if one or a few of the diverse species accounted for almost all the organisms in the community, with the other species being rare

Disturbances in Communities

Most communities are constantly changing in response
to disturbances. **Disturbances** are episodes that damage
biological communities, at least temporarily, by destroying organisms and altering the availability of resources
such as mineral nutrients and water. Examples of natural
disturbances are storms, fires, floods, and droughts.

Small-scale natural disturbances often have positive
effects on a biological community. For example, when
a large tree falls in a windstorm, it creates new habitats
(**Figure 20.22**). More light may now reach the forest
floor, giving small seedlings the opportunity to grow;
the depression left by the tree's roots may fill with water
and be used as egg-laying sites by frogs, salamanders,
and numerous insects.

People are by far the most significant agents of
ecological disturbance today. One consequence of
human-caused disturbance is the emergence of previously unknown infectious diseases. Three-quarters of
emerging diseases have jumped to humans from
another vertebrate species. In many cases,
people come into contact with these pathogens through activities such as clearing
land for agriculture, building roads, or
hunting in previously isolated ecosystems.
HIV, which may have been transmitted to
people from the blood of primates butchered
for food, is probably the best-known example.

Others include potentially fatal hemorrhagic fevers, such
as Ebola. Habitat destruction may also cause pathogen-
carrying animals to venture closer to human dwellings in
search of food. ☑

Ecological Succession

Communities change drastically following a severe
disturbance that strips away vegetation and even soil.
The disturbed area may be colonized by a variety
of species, which are gradually replaced by a succession of other species, in a process called **ecological
succession**.

Ecological succession that begins in a virtually
lifeless area with no soil is called **primary succession**
(**Figure 20.23**). Examples of such areas are cooled
lava flows on volcanic islands and the rubble left by
a retreating glacier. Often the only life-forms initially
present are autotrophic bacteria. Lichens and mosses,
which grow from windblown spores, are commonly
the first multicellular producers to colonize the
area. Soil develops gradually as rocks weather
and organic matter accumulates from the
decomposing remains of the early coloniz-
ers. Lichens and mosses are eventually
overgrown by grasses and shrubs that
sprout from seeds blown in from nearby
areas or carried in by animals. Finally, the
area is colonized by plants that become the

Human
encroachment on
tropical rain forests
has brought new
diseases such as
Ebola to human
populations.

▼ Figure 20.22 **A small-scale disturbance.** When this tree fell during a windstorm, its root
system and the surrounding soil uplifted, resulting in a depression that filled with water. The dead
tree, the root mound, and the water-filled depression are new habitats.

▼ Figure 20.23 **Primary succession underway on a
lava flow in Volcanoes National Park, Hawaii.**

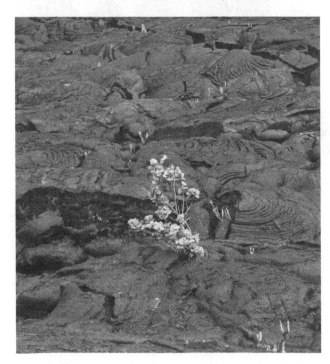

community's prevalent form of vegetation. Primary succession can take hundreds or thousands of years.

Secondary succession occurs where a disturbance has destroyed an existing community but left the soil intact. For example, secondary succession occurs as areas recover from floods or fires (Figure 20.24). Disturbances that lead to secondary succession are also caused by human activities. Even before colonial times, people were clearing the temperate deciduous forests of eastern North America for agriculture and settlements. Some of this land was later abandoned as the soil was depleted of its chemical nutrients or the residents moved west to new territories. Whenever human intervention stops, secondary succession begins. ☑

▼ Figure 20.24 **Secondary succession after a fire.**

☑ **CHECKPOINT**
What is the main abiotic factor that distinguishes primary from secondary succession?

Answer: absence of soil (primary succession) versus presence of soil (secondary succession) at the onset of succession

Ecosystem Ecology

In addition to the community of species in a given area, an **ecosystem** includes all the abiotic factors, such as energy, soil characteristics, and water. Let's look at a small-scale ecosystem—a terrarium—to see how the community interacts with these abiotic factors (Figure 20.25). A terrarium microcosm exhibits the two major processes that sustain all ecosystems: energy flow and chemical cycling. **Energy flow** is the passage of energy through the components of the ecosystem. **Chemical cycling** is the use and reuse of chemical elements such as carbon and nitrogen within the ecosystem.

Energy enters the terrarium in the form of sunlight (yellow arrows). Plants (producers) convert light energy to chemical energy through the process of photosynthesis. Animals (consumers) take in some of this chemical energy in the form of organic compounds when they eat the plants. Detritivores and decomposers in the soil obtain chemical energy when they feed on the dead remains of plants and animals. Every use of chemical energy by organisms involves a loss of some energy to the surroundings in the form of heat (red arrows). Because so much of the energy captured by photosynthesis is lost as heat, this ecosystem would run out of energy if it were not powered by a continuous inflow of energy from the sun.

In contrast to energy flow, chemical cycling (blue arrows in Figure 20.25) involves the transfer of materials within the ecosystem. While most ecosystems have a constant input of energy from sunlight or another source, the supply of the chemical elements used to construct molecules is limited. Chemical elements such as carbon and nitrogen are cycled between the abiotic components of the ecosystem, including air, water, and soil, and the biotic components of the ecosystem (the community). Plants acquire these elements in inorganic form from the air and soil and use them to construct organic molecules. Animals, such as the snail in Figure 20.25, consume some of these organic molecules. When the plants and animals become detritus, decomposers return most of the elements to the soil and air in inorganic form. Some elements are also returned to the air and soil as the by-products of plant and animal metabolism.

In summary, both energy flow and chemical cycling involve the transfer of substances through the trophic levels of the ecosystem. However, energy flows through, and ultimately out of, ecosystems, whereas chemicals are recycled within and between ecosystems.

▼ Figure 20.25 **A terrarium ecosystem.** Though it is small and artificial, this sealed terrarium illustrates the two major ecosystem processes: energy flow and chemical cycling.

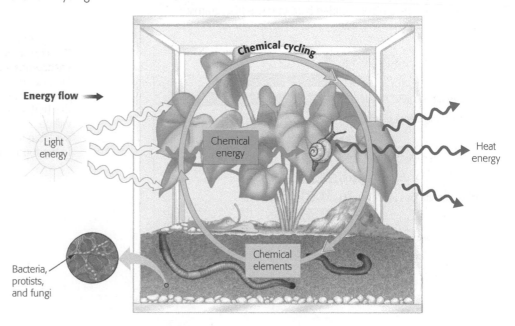

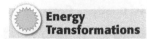
Energy Flow in Ecosystems

All organisms require energy for growth, maintenance, reproduction, and, in many species, locomotion. In this section, we take a closer look at energy flow through ecosystems. Along the way, we'll answer two key questions: What limits the length of food chains? and How do lessons about energy flow apply to people's use of resources?

Primary Production and the Energy Budgets of Ecosystems

Each day, Earth receives about 10^{19} kcal of solar energy, the energy equivalent of about 100 million atomic bombs. Most of this energy is absorbed, scattered, or reflected by the atmosphere or by Earth's surface. Of the visible light that reaches plants, algae, and cyanobacteria, only about 1% is converted to chemical energy by photosynthesis.

Ecologists call the amount, or mass, of living organic material in an ecosystem the **biomass**. The rate at which an ecosystem's producers convert solar energy to the chemical energy stored in organic compounds is called **primary production**. The primary production of the entire biosphere is roughly 165 billion tons of biomass per year.

Different ecosystems vary considerably in their primary production **(Figure 20.26)** as well as in their contribution to the total production of the biosphere. Tropical rain forests are among the most productive terrestrial ecosystems and contribute a large portion of the planet's overall production of biomass. Coral reefs

also have very high production, but their contribution to global production is small because they cover such a small area. Interestingly, even though the open ocean has very low production, it contributes the most to Earth's total primary production because of its huge size—it covers 65% of Earth's surface area. Whatever the ecosystem, primary production sets the spending limit for the energy budget of the entire ecosystem because consumers must acquire their organic fuels from producers. Now let's see how this energy budget is divided among the different trophic levels in an ecosystem's food web. ☑

Ecological Pyramids

When energy flows as organic matter through the trophic levels of an ecosystem, much of it is lost at each link in the food chain. Consider the transfer of organic matter from plants (producers) to herbivores (primary consumers). In most ecosystems, herbivores manage to eat only a fraction of the plant material produced, and they can't digest all of what they do consume. For example, a caterpillar feeding on leaves passes about half the energy in the leaves as feces **(Figure 20.27)**. Another 35% of the energy is expended in cellular respiration. Only about 15% of the energy in the caterpillar's food is transformed into caterpillar biomass. Only this biomass (and the energy it contains) is available to the consumer that eats the caterpillar.

Figure 20.28, called a **pyramid of production**, illustrates the cumulative loss of energy with each transfer in a food chain. Each tier of the pyramid represents all of the organisms in one trophic level, and the width of each tier indicates how much of the chemical energy of the tier below is actually incorporated into the organic matter of that trophic level. Note that producers convert only about 1% of the energy in the sunlight available to them to primary production. In this generalized

▼ **Figure 20.26 Primary production of different ecosystems.**
Primary production is the amount of biomass created by the producers of an ecosystem over a unit of time, in this case a year. Aquatic ecosystems are color-coded blue in these histograms; terrestrial ecosystems are green.

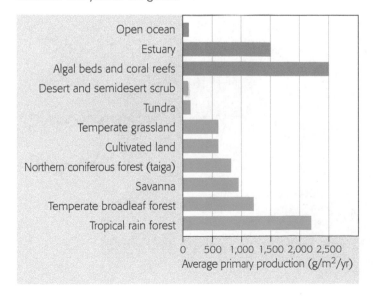

► **Figure 20.27 What becomes of a caterpillar's food?**
Only about 15% of the calories of plant material this herbivore consumes will be stored as biomass available to the next link in the food chain.

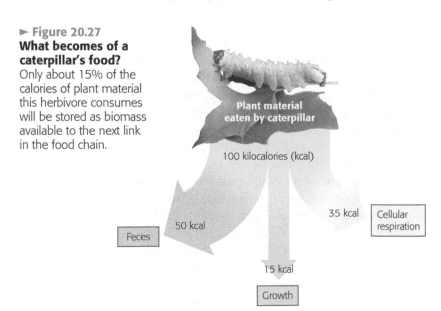

▼ Figure 20.28 **An idealized pyramid of production.**

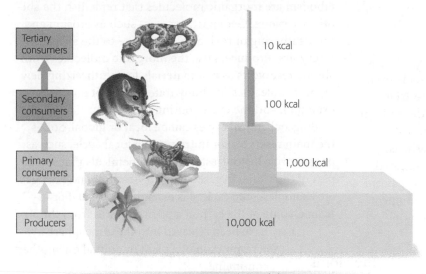

Tertiary consumers — 10 kcal

Secondary consumers — 100 kcal

Primary consumers — 1,000 kcal

Producers — 10,000 kcal

1,000,000 kcal of sunlight

pyramid, 10% of the energy available at each trophic level becomes incorporated into the next higher level. Actual efficiencies of energy transfer are usually in the range of 5–20%. In other words, 80–95% of the energy at one trophic level never reaches the next.

An important implication of this stepwise decline of energy in a trophic structure is that the amount of energy available to top-level consumers is small compared with that available to lower-level consumers. Only a tiny fraction of the energy stored by photosynthesis flows through a food chain to a tertiary consumer, such as a snake feeding on a mouse. This explains why top-level consumers such as lions and hawks require so much geographic territory: It takes a lot of vegetation to support trophic levels so many steps removed from photosynthetic production. You can also understand why most food chains are limited to three to five levels; there is simply not enough energy at the very top of an ecological pyramid to support another

trophic level. There are, for example, no nonhuman predators of lions, eagles, and killer whales; the biomass in populations of these top-level consumers is insufficient to supply yet another trophic level with a reliable source of nutrition.

Ecosystem Energetics and Human Resource Use

The dynamics of energy flow apply to the human population as much as to other organisms. The two production pyramids in **Figure 20.29** are based on the same generalized model used to construct Figure 20.28, with roughly 10% of the energy in each trophic level available for consumption by the next trophic level. The pyramid on the left shows energy flow from producers (represented by corn) to people as primary consumers—vegetarians. The pyramid on the right illustrates energy flow from the same corn crop, with people as secondary consumers, eating beef. Clearly, the human population has less energy available to it when people eat at higher trophic levels than as primary consumers.

Worldwide, only about 20% of agricultural land is used to produce plants for direct human consumption. The rest of the land produces food for livestock. In either case, large-scale agriculture is environmentally expensive. Land is cleared of its native vegetation; fossil fuels are burned; chemical fertilizers and pesticides are applied; and in many regions, water is used for irrigation. Currently, people in many countries cannot afford to buy meat and are vegetarians by necessity. As nations become more affluent, the demand for meat increases—and so do the environmental costs of food production. ☑

Producing the beef for a hamburger requires eight times as much land as producing the soybeans for a soyburger.

▼ Figure 20.29 **Food energy available to the human population at different trophic levels.**

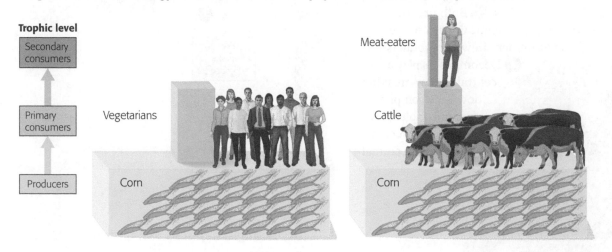

Trophic level

Secondary consumers — Meat-eaters

Primary consumers — Vegetarians / Cattle

Producers — Corn / Corn

Interconnections within Systems

Chemical Cycling in Ecosystems

The sun (or in some cases Earth's interior) supplies ecosystems with a continual input of energy, but aside from an occasional meteorite, there are no extraterrestrial sources of chemical elements. Life, therefore, depends on the recycling of chemicals. While an organism is alive, much of its chemical stock changes continuously, as nutrients are acquired and waste products are released. Atoms present in the complex molecules of an organism at the time of its death are returned to the environment by the action of decomposers, replenishing the pool of inorganic nutrients that plants and other producers use to build new organic matter (**Figure 20.30**). In a sense, each living thing only borrows an ecosystem's chemical elements, returning what is left in its body after it dies. Let's take a closer look at how chemicals cycle between organisms and the abiotic components of ecosystems.

The General Scheme of Chemical Cycling

Because chemical cycles in an ecosystem involve both biotic components (organisms and nonliving organic material) and abiotic (geologic and atmospheric) components, they are called **biogeochemical cycles**. **Figure 20.31** is a general scheme for the cycling of a mineral nutrient within an ecosystem. Note that the cycle has an **abiotic reservoir** (white box) where a chemical accumulates or is stockpiled outside of living organisms. The atmosphere, for example, is an abiotic reservoir for carbon. The water of aquatic ecosystems contains dissolved carbon, nitrogen, and phosphorus compounds.

Let's trace our way around our general biogeochemical cycle. ❶ Producers incorporate chemicals from the abiotic reservoir into organic compounds. ❷ Consumers feed on the producers, incorporating some of the chemicals into their own bodies. ❸ Both producers and consumers release some chemicals back to the environment in waste products. ❹ Decomposers play a central role by breaking down the complex organic molecules in detritus such as plant litter, animal wastes, and dead organisms. The products of this metabolism are inorganic molecules that replenish the abiotic reservoirs. Geologic processes such as erosion and the weathering of rock also contribute to the abiotic reservoirs. Producers use the inorganic molecules from abiotic reservoirs as raw materials for synthesizing new organic molecules (carbohydrates and proteins, for example), and the cycle continues.

Biogeochemical cycles can be local or global. Soil is the main reservoir for nutrients in a local cycle, such as phosphorus. In contrast, for those chemicals that spend part of their time in gaseous form—carbon and nitrogen are examples—the cycling is essentially global. For instance, some of the carbon a plant acquires from the air may have been released into the atmosphere by the respiration of a plant or animal on another continent.

The atoms in your body might have been part of a dinosaur that lived 75 million years ago.

Now let's examine three important biogeochemical cycles more closely: the cycles for carbon, phosphorus, and nitrogen. As you study the cycles, look for the four basic steps we described, as well as the geologic processes that move chemicals around and between ecosystems. In all the diagrams, the main abiotic reservoirs appear in white boxes. ☑

☑ **CHECKPOINT**

What is the role of abiotic reservoirs in biogeochemical cycles? What are the three major abiotic reservoirs?

Answer: Abiotic reservoirs are sources of the nutrients used by producers for synthesizing organic molecules. The atmosphere, soil, and water are the three major abiotic reservoirs.

▼ **Figure 20.31 General scheme for biogeochemical cycles.**

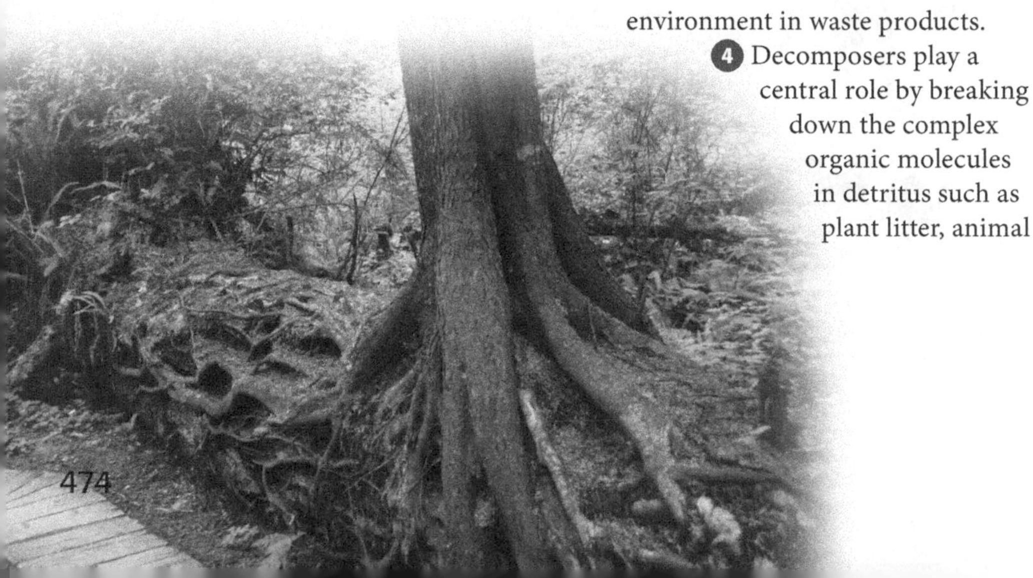

▼ **Figure 20.30 Plant growth on fallen tree.** In the temperate rain forest of Olympic National Park, in Washington State, plants—including other trees—quickly take advantage of the mineral nutrients supplied by decomposing "nurse logs."

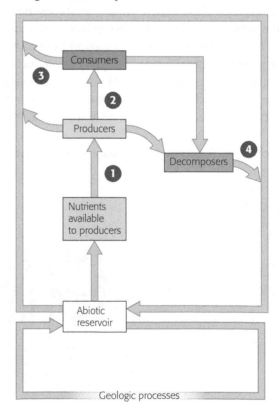

The Carbon Cycle

Carbon, the major ingredient of all organic molecules, has an atmospheric reservoir and cycles globally. Other abiotic reservoirs of carbon include fossil fuels and dissolved carbon compounds in the oceans. The reciprocal metabolic processes of photosynthesis and cellular respiration, which you may recall from previous chapters, are mainly responsible for the cycling of carbon between the biotic and abiotic worlds (Figure 20.32). ① Photosynthesis removes CO_2 from the atmosphere and incorporates it into organic molecules, which are ② passed along the food chain by consumers. ③ Cellular respiration returns CO_2 to the atmosphere. ④ Decomposers break down the carbon compounds in detritus; that carbon, too, is eventually released as CO_2. On a global scale, the return of CO_2 to the atmosphere by respiration closely balances its removal by photosynthesis. However, increasing levels of CO_2 caused by ⑤ the burning of wood and fossil fuels (coal and petroleum) are contributing to global climate change (see Figure 18.46).

The Phosphorus Cycle

Organisms require phosphorus as an ingredient of nucleic acids, phospholipids, and ATP and (in vertebrates) as a mineral component of bones and teeth. In contrast to the carbon cycle and the other major biogeochemical cycles, the phosphorus cycle does not have an atmospheric component. Rocks are the only source of phosphorus for terrestrial ecosystems; in fact, rocks that have high phosphorus content are mined for fertilizer.

At the center of Figure 20.33, ① the weathering (breakdown) of rock gradually adds inorganic phosphate (PO_4^{3-}) to the soil. ② Plants absorb dissolved phosphate from the soil and assimilate it by building the phosphorus atoms into organic compounds. ③ Consumers obtain phosphorus in organic form by eating plants. ④ Phosphates are returned to the soil by the action of decomposers on animal waste and the remains of dead plants and animals. ⑤ Some of the phosphates drain from terrestrial ecosystems into the sea, where they may settle and eventually become part of new rocks. Phosphorus removed from the cycle in this way will not be available to living organisms until ⑥ geologic processes uplift the rocks and expose them to weathering.

Phosphates move from land to aquatic ecosystems much more rapidly than they are replaced, and soil characteristics may also decrease the amount of phosphate available to plants. As a result, phosphate is a limiting factor in many terrestrial ecosystems. Farmers and gardeners often use phosphate fertilizer, such as crushed phosphate rock or bone meal (finely ground bones from slaughtered livestock or fish), to boost plant growth.

▼ Figure 20.32 **The carbon cycle.**

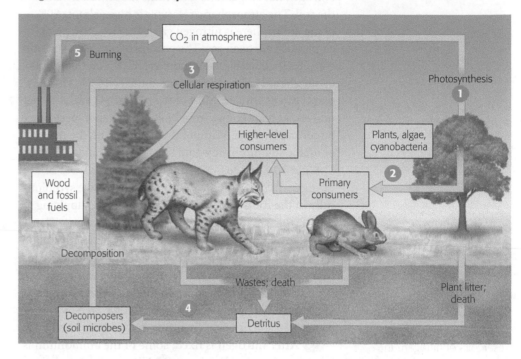

▼ Figure 20.33 **The phosphorus cycle.**

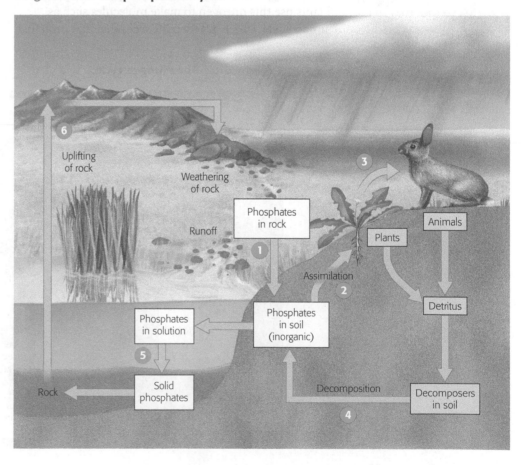

The Nitrogen Cycle

As an ingredient of proteins and nucleic acids, nitrogen is essential to the structure and functioning of all organisms. Nitrogen has two abiotic reservoirs, the atmosphere and the soil. The atmospheric reservoir is huge; almost 80% of the atmosphere is nitrogen gas (N_2). However, plants cannot use nitrogen gas. The process of **nitrogen fixation** converts gaseous N_2 to ammonia (NH_3). NH_3 then picks up another H^+ to become ammonium (NH_4^+), which plants can assimilate. Most of the nitrogen available in natural ecosystems comes from biological fixation performed by certain bacteria. Without these organisms, the reservoir of usable soil nitrogen would be extremely limited.

Figure 20.34 illustrates the actions of two types of nitrogen-fixing bacteria. ❶ Some bacteria live symbiotically in the roots of certain species of plants, supplying their hosts with a direct source of usable nitrogen. The largest group of plants with this mutualistic relationship is the legumes, a family that includes peanuts and soybeans. Many farmers improve soil fertility by alternating crops of legumes, which add nitrogen to the soil, with plants such as corn that require nitrogen fertilizer. ❷ Free-living bacteria in soil or water fix nitrogen, resulting in NH_4^+.

❸ After nitrogen is fixed, some of the ammonium is taken up and used by plants. ❹ Nitrifying bacteria in the soil also convert some of the ammonium to nitrate (NO_3^-), ❺ which is more readily acquired by plants. Plants use this nitrogen to make molecules such as amino acids, which are then incorporated into proteins.

❻ When an herbivore (represented here by a rabbit) eats a plant, it digests the proteins into amino acids and then uses the amino acids to build the proteins it needs. Higher-order consumers get nitrogen from the organic molecules of their prey. Because animals form nitrogen-containing waste products during protein metabolism, consumers excrete some nitrogen into the soil or water. The urine that rabbits and other mammals excrete contains urea, a nitrogen compound that is widely used as fertilizer.

Organisms that are not consumed eventually die and become detritus, which is decomposed by bacteria and fungi. ❼ The decomposition of organic compounds releases ammonium into the soil, replenishing that abiotic reservoir. Under low-oxygen conditions, however, ❽ soil bacteria known as denitrifying bacteria strip the oxygen atoms from nitrates, releasing N_2 back into the atmosphere and depleting the soil of usable nitrogen.

Human activities are disrupting the nitrogen cycle by adding more nitrogen to the biosphere each year than natural processes do. Combustion of fossil fuels and modern agricultural practices are two major sources of nitrogen. For example, many farmers use enormous amounts of synthetic nitrogen fertilizer to supplement natural nitrogen. However, less than half the fertilizer applied is actually used by the crop plants. Some nitrogen escapes to the atmosphere, where it forms nitrous oxide (N_2O), a gas that contributes to global warming. And as you'll learn next, nitrogen fertilizers also pollute aquatic systems. ✓

▼ Figure 20.34 **The nitrogen cycle.**

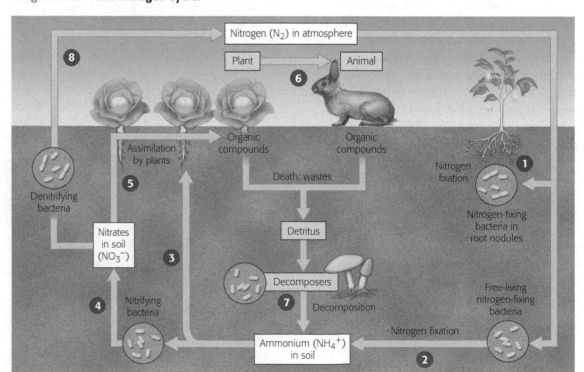

Nutrient Pollution

Low nutrient levels, especially of phosphorus and nitrogen, often limit the growth of algae and cyanobacteria in aquatic ecosystems. Nutrient pollution occurs when human activities add excess amounts of these chemicals to aquatic ecosystems.

In many areas, phosphate pollution comes from agricultural fertilizers and runoff of animal waste from livestock feedlots (where hundreds of animals are penned together). Phosphates are also a common ingredient in dishwasher detergents, making the outflow from sewage-treatment facilities—which also contains phosphorus from human waste—a major source of phosphate pollution. Phosphate pollution of lakes and rivers results in heavy growth of algae and cyanobacteria **(Figure 20.35)**. Microbes consume a great deal of oxygen as they decompose the extra biomass, a process that depletes the water of oxygen. These changes lead to reduced diversity of aquatic species and a much less appealing body of water.

A major source of nitrogen pollution is the large amount of inorganic nitrogen fertilizers routinely applied to crops, lawns, and golf courses. Plants take up some of the nitrogen compounds, and denitrifiers convert some to atmospheric N_2, but nitrate is not bound tightly by soil particles and is easily washed out of the soil by rain or irrigation. As a result, chemical fertilizers often exceed the soil's natural recycling capacity. Runoff from the manure in high-density livestock facilities,

such as cattle feedlots and hog farms, is another significant source of nitrogen pollution. Nitrogen pollution may also come from sewage treatment facilities when extreme conditions (such as unusual storms) or malfunctioning equipment prevent them from meeting water quality standards.

In an example of how far-reaching this problem can be, nitrogen runoff from Midwestern farm fields has been linked to an annual summer "dead zone" in the Gulf of Mexico **(Figure 20.36)**. Vast algal blooms extend outward from where the Mississippi River deposits its nutrient-laden waters. As the algae die, decomposition of the huge quantities of biomass diminishes the supply of dissolved oxygen over an area that ranges from 13,000 km^2 to 22,000 km^2 (from roughly the size of Connecticut to the size of New Jersey). Oxygen depletion disrupts benthic communities, displacing fish and invertebrates that can move and killing organisms that are attached to the substrate. More than 400 recurring and permanent coastal dead zones totaling approximately 245,000 km^2 (about the area of Michigan) have been documented worldwide. ☑

☑ **CHECKPOINT**

How does the excessive addition of mineral nutrients to a pond eventually result in the loss of most fish in the pond?

Answer: Excessive mineral nutrients initially cause population explosions of algae and the organisms that feed on them. The respiration of so much life, especially of the microbes decomposing all the organic refuse, consumes most of the lake's oxygen, which the fish require.

▼ **Figure 20.35 Algal growth resulting from nutrient pollution.** The flat green area in this photo is not a lawn, but rather the surface of a polluted pond.

▼ **Figure 20.36 The Gulf of Mexico dead zone.**

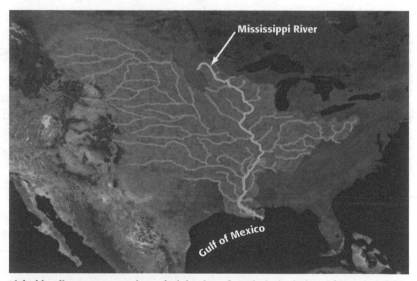

Light blue lines represent rivers draining into the Mississippi River (shown in bright blue). Nitrogen runoff carried by these rivers ends up in the Gulf of Mexico. In the images below, red and orange indicate high concentrations of phytoplankton. Bacteria feeding on dead phytoplankton deplete the water of oxygen, creating a "dead zone."

Summer

Winter

Conservation and Restoration Biology

As we have seen in this unit, many of the environmental problems facing us today have been caused by human enterprises. But the science of ecology is not just useful for telling us how things have gone wrong. Ecological research is also the foundation for finding solutions to these problems and for reversing the negative consequences of ecosystem alteration. Thus, we end the ecology unit by highlighting these beneficial applications of ecological research.

Conservation biology is a goal-oriented science that seeks to understand and counter the loss of biodiversity. Conservation biologists recognize that biodiversity can be sustained only if the evolutionary mechanisms that have given rise to species and communities of organisms continue to operate. Thus, the goal is not simply to preserve individual species but to sustain ecosystems, where natural selection can continue to function, and to maintain the genetic variability on which natural selection acts. The expanding field of **restoration ecology** uses ecological principles to develop methods of returning degraded areas to their natural state.

✓ CHECKPOINT

What is a biodiversity hot spot?

Answer: a relatively small area with a disproportionately large number of species, including endangered species

Biodiversity "Hot Spots"

Conservation biologists are applying their understanding of population, community, and ecosystem dynamics in establishing parks, wilderness areas, and other legally protected nature reserves. Choosing locations for these protected zones often focuses on **biodiversity hot spots**.

These relatively small areas have a large number of endangered and threatened species and an exceptional concentration of **endemic species**, species that are found nowhere else. Together, the "hottest" of Earth's biodiversity hot spots, shown in **Figure 20.37**, total less than 1.5% of Earth's land surface but are home to a third of all species of plants and vertebrates. For example, all of Earth's lemurs—more than 50 species—are endemic to Madagascar, a large island off the eastern coast of Africa. In fact, almost all of the mammals, reptiles, amphibians, and plants that inhabit Madagascar are endemic. There are also hot spots in aquatic ecosystems, such as certain river systems and coral reefs. Because biodiversity hot spots can also be hot spots of extinction, they rank high on the list of areas demanding strong global conservation efforts.

Concentrations of species provide an opportunity to protect many species in very limited areas. However, the "hot spot" designation tends to favor the most noticeable organisms, especially vertebrates and plants. Invertebrates and microorganisms are often overlooked. Furthermore, species endangerment is a global problem, and focusing on hot spots should not detract from efforts to conserve habitats and species diversity in other areas. Finally, even the protection of a nature reserve does not shield organisms from the effects of climate change or other threats, such as invasive species or infectious disease. To stem the tide of biodiversity loss, we will have to address environmental problems globally as well as locally. ✓

▼ Figure 20.37 **Earth's terrestrial biodiversity hot spots (purple).**

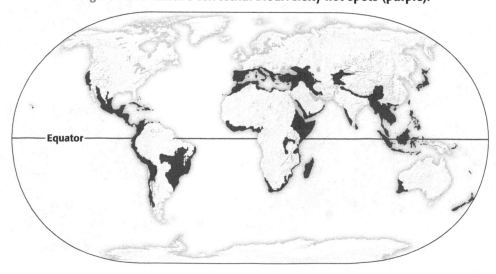

Conservation at the Ecosystem Level

In the past, most conservation efforts focused on saving individual species, and this work continues. (You have already learned about one example, the red-cockaded woodpecker; see Figure 19.12.) More and more, however, conservation biology aims at sustaining the biodiversity of entire communities and ecosystems. On an even broader scale, conservation biology considers the biodiversity of whole landscapes. Ecologically, a **landscape** is a regional assemblage of interacting ecosystems, such as an area with forest, adjacent fields, wetlands, streams, and streamside habitats. **Landscape ecology** is the application of ecological principles to the study of

land-use patterns. Its goal is to make ecosystem conservation a functional part of the planning for land use.

Edges between ecosystems are prominent features of landscapes, whether natural or altered by people (**Figure 20.38**). Edges have their own sets of physical conditions—such as soil type and surface features—that differ from the ecosystems on either side of them. Edges also may have their own type and amount of disturbance. For instance, the edge of a forest often has more blown-down trees than a forest interior because the edge is less protected from strong winds. Because of their specific physical features, edges also have their own communities of organisms. Some organisms thrive in edges because they require resources found only there. For instance, whitetail deer browse on woody shrubs found in edges between woods and fields, so their populations often expand when forests are logged or fragmented by development.

Edges can have both positive and negative effects on biodiversity. A recent study in a tropical rain forest in western Africa indicated that natural edge communities are important sites of speciation. On the other hand, landscapes where human activities have produced edges often have fewer species.

Another important landscape feature, especially where habitats have been severely fragmented, is the **movement corridor**, a narrow strip or series of small clumps of suitable habitat connecting otherwise isolated patches. In places where there is extremely heavy human impact, artificial corridors are sometimes constructed (**Figure 20.39**). Corridors can promote dispersal and help sustain populations, and they are especially important to species that migrate between different habitats seasonally. But a corridor can also be harmful—as, for example, in the spread of disease, especially among small subpopulations in closely situated habitat patches. ☑

▼ Figure 20.38 **Edges between ecosystems within a landscape.**

Natural edges. Forests border grassland ecosystems in Lake Clark National Park, Alaska.

Edges created by human activity. Forest edges surround farmland in the Cotswolds region of south central England.

☑ CHECKPOINT

How is a landscape different from an ecosystem?

Answer: A landscape is more inclusive in that it consists of several interacting ecosystems in the same region.

► Figure 20.39 **An artificial corridor.** This bridge over a road provides an artificial corridor for animals in Banff National Park, Canada.

How Does Tropical Forest Fragmentation Affect Biodiversity?

Long-term studies are essential for learning how we might best conserve biodiversity and other natural resources. One site for such studies is the Biological Dynamics of Forest Fragmentation Project (BDFFP), a 1,000-km^2 ecological "laboratory" located deep in the Amazonian forest of Brazil. When the project was begun in 1979, laws required landowners who cleared forest for ranching or other agricultural operations to leave scattered tracts of untouched forest. Biologists recruited some of these landowners to create reserves in isolated fragments of 1 ha (roughly 2.5 acres), 10 ha, and 100 ha. **Figure 20.40** shows some of these forest "islands." Before each area was isolated from the main forest, a small army of specialists inventoried the organisms present and measured the trees.

Hundreds of researchers have used the BDFFP sites to investigate the effects of forest fragmentation on all

▼ **Figure 20.40 Fragments of forest in the Amazon that were created as part of the Biological Dynamics of Forest Fragmentation Project.** The patch of forest on the right is a 1-ha (2.5-acre) fragment.

levels of ecological study. The initial **observations** for these investigations were gleaned from the results of other ecological studies—for example, the effects of fragmentation in temperate forests or differences in the biodiversity of small islands compared with mainland ecosystems. These observations led many researchers to ask the **question**, How does fragmentation of tropical forests affect species diversity within the fragments? Based on previous research on numerous species, a reasonable **hypothesis** might be that species diversity declines as the size of the forest fragment decreases. An ecologist studying large predators such as jaguars and pumas, which require large hunting territories, might therefore make the **prediction** that predators will be found only in the largest areas. The BDFFP is unique because it allows researchers to test their predictions with new observations that compare species diversity in forest fragments with a comparable control: species diversity in the same area when it was intact. An undisturbed area of 25,400 acres is also available for comparison with fragmented areas. In addition, data can be collected over a period of years or even decades.

Since the BDFFP was established, scientists have studied many different groups of plants and animals. In general, the **results** have shown that fragmentation of forest into smaller pieces leads to a decline in species diversity. Species richness decreases as a result of local extinctions of many species of large mammals, insects, and insectivorous birds. The population density of remaining species often declines. Researchers also documented edge effects, such as those described in the previous section. Changes in abiotic factors along the fragment edges, including increased wind disturbance, higher temperature, and decreased soil moisture, played a role in community alterations. For example, ecologists found that tree mortality was higher than normal at fragment edges. They also observed changes in the composition of communities of invertebrates that inhabit the soil and leaf litter.

Restoring Ecosystems

One of the major strategies in restoration ecology is **bioremediation**, the use of living organisms to detoxify polluted ecosystems. For example, bacteria have been used to clean up old mining sites and oil spills (see Figure 15.15). Researchers are also investigating the potential of using plants to remove toxic substances such as heavy metals and organic pollutants (for example, PCBs) from contaminated soil (**Figure 20.41**).

Some restoration projects have the broader goal of returning ecosystems to their natural state. Such projects may involve replanting vegetation, fencing out non-native animals, or removing dams that restrict water flow. Hundreds of restoration projects are currently under way in the United States. One of the most ambitious endeavors is the Kissimmee River Restoration Project in south-central Florida.

The Kissimmee River was once a shallow, meandering river that wound its way from Lake Kissimmee southward into Lake Okeechobee. During about half of the year, the river flooded into a wide floodplain, creating wetlands that provided habitat for large numbers of birds, fishes, and invertebrates. And as the floods deposited the river's load of nutrient-rich silt on the floodplain, they boosted soil fertility and maintained the water quality of the river.

Between 1962 and 1971, the U.S. Army Corps of Engineers converted the 166-km wandering river to a straight canal 9 m deep, 100 m wide, and 90 km long. This project, designed to allow development on the floodplain, drained approximately 31,000 acres of wetlands, with significant negative impacts on fish and wetland bird populations. Without the marshes to help filter and reduce agricultural runoff, the river transported phosphates and other excess nutrients from Lake Okeechobee to the Everglades ecosystem to the south.

The restoration project involves removing water-control structures such as dams, reservoirs, and channel modifications and filling in about 35 km of the canal (**Figure 20.42**). The first phase of the project was completed in 2004; the rest is scheduled to be completed in 2015. The photo shows a section of the Kissimmee canal that has been plugged, diverting flow into the remnant river channels. Birds and other wildlife have returned in unexpected numbers to the 11,000 acres of wetlands that have been restored. The marshes are filled with native vegetation, and game fishes again swim in the river channels. ☑

☑ CHECKPOINT

The water in the Kissimmee River eventually flows into the Everglades. How will the Kissimmee River Restoration Project affect water quality in the Everglades ecosystem?

Answer: Wetlands filter agricultural runoff, which prevents nutrient pollution from flowing downstream. By restoring this ecosystem service, the project will improve water quality in the Everglades.

▼ Figure 20.41 **Bioremediation using plants.**
A researcher from the U.S. Department of Agriculture investigates the use of canola plants to reduce toxic levels of selenium in contaminated soil.

▼ Figure 20.42 **The Kissimmee River Restoration Project.**

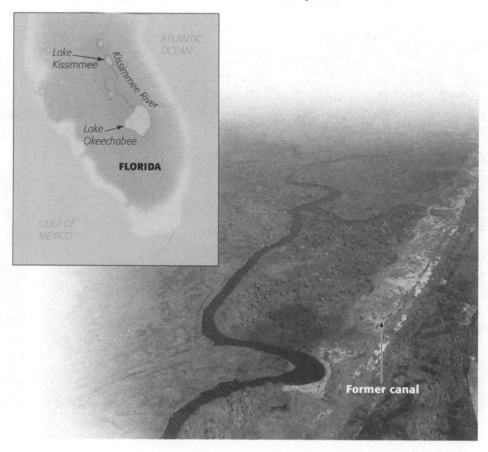

The Goal of Sustainable Development

As the world population grows and becomes more affluent, the demand for the "provisioning" services of ecosystems, such as food, wood, and water, is increasing. Although these demands are currently being met, they are satisfied at the expense of other critical ecosystem services, such as climate regulation and protection against natural disasters. Clearly, we have set ourselves and the rest of the biosphere on a precarious path into the future. How can we achieve **sustainable development**—development that meets the needs of people today without limiting the ability of future generations to meet their needs?

Many nations, scientific associations, corporations, and private foundations have embraced the concept of sustainable development. The Ecological Society of America, the world's largest organization of ecologists, endorses a research agenda called the Sustainable Biosphere Initiative. The goal of this initiative is to acquire the ecological information necessary for the responsible development, management, and conservation of Earth's resources. The research agenda includes the search for ways to sustain the productivity of natural and artificial ecosystems and studies of the relationship between biological diversity, global climate change, and ecological processes.

Sustainable development depends on more than continued research and application of ecological knowledge. It also requires that we connect the life sciences with the social sciences, economics, and humanities. Conservation of biodiversity is only one side of sustainable development; the other side is improving the human condition. Public education and the political commitment and cooperation of nations are essential to the success of this endeavor.

An awareness of our unique ability to alter the biosphere and jeopardize the existence of other species, as well as our own, may help us choose a path toward a sustainable future. The risk of a world without adequate natural resources for all its people is not a vision of the distant future. It is a prospect for your children's lifetime, or perhaps even your own. But although the current state of the biosphere is grim, the situation is far from hopeless. Now is the time to take action by aggressively pursuing greater knowledge about the diversity of life on our planet and by joining with others in working toward long-term sustainability **(Figure 20.43)**. ☑

☑ CHECKPOINT

What is meant by sustainable development?

Answer: development that meets current needs while ensuring an adequate supply of natural and economic resources for future generations

▼ Figure 20.43 **Working toward sustainability.**

Students at the University of Virginia sorted trash from dumpsters to promote recycling. Recycling just one aluminum can saves enough energy to power a laptop for five hours.

A student at California State University, Fresno, pulls weeds in a plot of mustard and kale plants at the university's organic farm.

Can Biophilia Save Biodiversity?

For millions of years, the diversity of life has flourished via evolutionary adaptation in response to environmental change. For many species, however, the pace of evolution can't match the breakneck speed at which humans are changing the environment (see the Evolution Connection sections in Chapters 18 and 19). Perhaps those species are doomed to extinction. Or perhaps they can be saved by one human characteristic that works in their favor: biophilia.

Biophilia, which literally means "love of life," is a term that Edward O. Wilson, one of the world's foremost experts on biodiversity and conservation, uses for the human desire to affiliate with other life in its many forms. People develop close relationships with pets, nurture houseplants, invite avian visitors with backyard feeders, and flock to zoos, gardens, and nature parks (Figure 20.44). Our attraction to pristine landscapes with clean water and lush vegetation is also testimony to our biophilia. Wilson proposes that our biophilia is innate, an evolutionary product of natural selection acting on a brainy species whose survival depended on a close connection to the environment and a practical appreciation of plants and animals. We evolved in natural environments rich in biodiversity, and we still have an affinity for such settings.

It will come as no surprise that many biologists have embraced the concept of biophilia. After all, these are people who have turned their passion for nature into careers. But biophilia strikes a chord with biologists for another reason. If biophilia is evolutionarily embedded in our genome, then there is hope that we can become better custodians of the biosphere. If we all pay more attention to our biophilia, a new environmental ethic could catch on among individuals and societies. And that ethic is a resolve never to knowingly allow a single species to become extinct as a result of our actions or any ecosystem to be destroyed as long as there are reasonable ways to prevent it. Yes, we should be motivated to preserve biodiversity because we depend on it for food, medicine, building materials, fertile soil, flood control, habitable climate, drinkable water, and breathable air. But maybe we can also work harder to prevent the extinction of other forms of life just because it is the ethical thing for us to do.

Biophilia is a fitting capstone for this unit. Modern biology is the scientific extension of our human tendency to feel connected to and curious about all forms of life. People are most likely to save what they appreciate, and most likely to appreciate what they understand. We hope that our discussion of biodiversity has deepened your biophilia and broadened your education.

▼ **Figure 20.44 Biophilia.** Whether we seek other organisms in their own habitats or invite them into ours, we clearly find pleasure in the diversity of life.

Chapter Review

SUMMARY OF KEY CONCEPTS

The Loss of Biodiversity

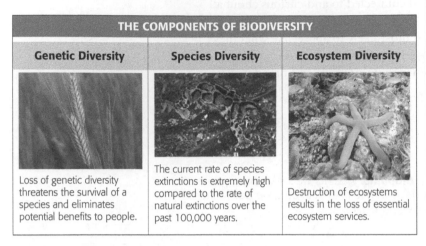

THE COMPONENTS OF BIODIVERSITY		
Genetic Diversity	**Species Diversity**	**Ecosystem Diversity**
Loss of genetic diversity threatens the survival of a species and eliminates potential benefits to people.	The current rate of species extinctions is extremely high compared to the rate of natural extinctions over the past 100,000 years.	Destruction of ecosystems results in the loss of essential ecosystem services.

Causes of Declining Biodiversity

Habitat destruction is the leading cause of extinctions. Invasive species, over-exploitation, and pollution are also significant factors.

Community Ecology

Interspecific Interactions

Populations in a community interact in a variety of ways that can be generally categorized as being beneficial (**+**) or harmful (**−**) to the populations. Because **+/−** interactions (exploitation of one species by another species) may have such a negative impact on the individual that is harmed, defensive evolutionary adaptations are common.

Trophic Structure

The trophic structure of a community defines the feeding relationships among organisms. These relationships are sometimes organized into food chains or food webs. In the process of biological magnification, toxins become more concentrated as they are passed up a food chain to the top predators.

Increasing PCB concentration

Species Diversity in Communities

Diversity within a community includes species richness and relative abundance of different species. A keystone species is a species that has a great impact on the composition of the community despite a relatively low abundance or biomass.

Disturbances in Communities

Disturbances are episodes that damage communities, at least temporarily, by destroying organisms or altering the availability of resources such as mineral nutrients and water. People are the most significant cause of disturbances today.

Ecological Succession

The sequence of changes in a community after a disturbance is called ecological succession. Primary succession occurs where a community arises in a virtually lifeless area with no soil. Secondary succession occurs where a disturbance has destroyed an existing community but left the soil intact.

INTERACTIONS BETWEEN SPECIES IN A COMMUNITY					
Interspecific Interaction	**Effect on Species 1**	**Effect on Species 2**	**Interspecific Interaction**	**Effect on Species 1**	**Effect on Species 2**
Competition	−	−	**Exploitation** Predation	+	−
Mutualism	+	+	Herbivory	+	−
			Parasites and Pathogens	+	−

Ecosystem Ecology

Energy Transformations: Energy Flow in Ecosystems

An ecosystem is a biological community and the abiotic factors with which the community interacts. Energy must flow continuously through an ecosystem, from producers to consumers and decomposers. Chemical elements can be recycled between an ecosystem's living community and the abiotic environment. Trophic relationships determine an ecosystem's routes of energy flow and chemical cycling.

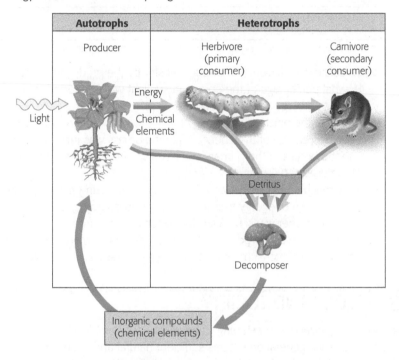

Primary production is the rate at which plants and other producers build biomass. Ecosystems vary considerably in their productivity. Primary production sets the spending limit for the energy budget of the entire ecosystem because consumers must acquire their organic fuels from producers. In a food chain, only about 10% of the biomass at one trophic level is available to the next, resulting in a pyramid of production.

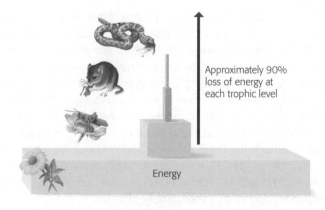

When people eat producers instead of consumers, less photosynthetic production is required, which reduces the impact on the environment.

Interconnections within Systems: Chemical Cycling in Ecosystems

Biogeochemical cycles involve biotic and abiotic components. Each circuit has an abiotic reservoir through which the chemical cycles. Some chemical elements require "processing" by certain microorganisms before they are available to plants as inorganic nutrients. A chemical's specific route through an ecosystem varies with the element and the trophic structure of the ecosystem. Phosphorus is not very mobile and is cycled locally. Carbon and nitrogen spend part of their time in gaseous form and are cycled globally. Runoff of nitrogen and phosphorus, especially from agricultural land, causes algal blooms in aquatic ecosystems, lowering water quality and sometimes depleting the water of oxygen.

Conservation and Restoration Biology

Biodiversity "Hot Spots"

Conservation biology is a goal-oriented science that seeks to counter the loss of biodiversity. The front lines for conservation biology are biodiversity "hot spots," relatively small geographic areas that are especially rich in endangered species.

Conservation at the Ecosystem Level

Increasingly, conservation biology aims at sustaining the biodiversity of entire communities, ecosystems, and landscapes. Edges between ecosystems are prominent features of landscapes, with positive and negative effects on biodiversity. Corridors can promote dispersal and help sustain populations.

Restoring Ecosystems

In some cases, ecologists use microbes or plants to remove toxic substances, such as heavy metals, from ecosystems. Ecologists are working to revitalize some ecosystems by planting native vegetation, removing barriers to wildlife, and other means. The Kissimmee River Restoration Project is an attempt to undo the ecological damage done when the river was engineered into straight channels.

The Goal of Sustainable Development

Balancing the needs of people with the health of the biosphere, sustainable development has the goal of long-term prosperity of human societies and the ecosystems that support them.

MasteringBiology®

For practice quizzes, BioFlix animations, MP3 tutorials, video tutors, and more study tools designed for this textbook, go to MasteringBiology®

SELF-QUIZ

1. Currently, the number one cause of biodiversity loss is _____.
2. According to the concept of competitive exclusion,
   a. two species cannot coexist in the same habitat.
   b. extinction or emigration is the only possible result of competitive interactions.
   c. intraspecific competition results in the success of the best-adapted individuals.
   d. two species cannot share the same niche in a community.

3. The concept of trophic structure emphasizes the
   a. prevalent form of vegetation.
   b. keystone species concept.
   c. feeding relationships within a community.
   d. species richness of the community.

4. Match each organism with its trophic level (you may choose a level more than once).
   a. alga
   b. grasshopper
   c. zooplankton
   d. eagle
   e. fungus

   1. decomposer
   2. producer
   3. tertiary consumer
   4. secondary consumer
   5. primary consumer

5. Why are the top predators in food chains most severely affected by pesticides such as DDT?

6. Over a period of many years, grass grows on a sand dune, then shrubs grow, and then eventually trees grow. This is an example of ecological _____.

7. According to the pyramid of production, why is eating grain-fed beef a relatively inefficient means of obtaining the energy trapped by photosynthesis?

8. Local conditions, such as heavy rainfall or the removal of plants, may limit the amount of nitrogen, phosphorus, or calcium available to a particular terrestrial ecosystem, but the amount of carbon available to the ecosystem is seldom a problem. Why?

9. Species found in only one place on Earth are called _____ species.

10. Movement corridors are
    a. always created by humans.
    b. only found along streamsides.
    c. beneficial because they allow for population dispersal and are not harmful.
    d. harmful because they allow for the spread of diseases and beneficial because they allow for population dispersal.

Answers to these questions can be found in Appendix: Self-Quiz Answers.

THE PROCESS OF SCIENCE

11. An ecologist studying desert plants performed the following experiment. She staked out two identical plots that included a few sagebrush plants and numerous small annual wildflowers. She found the same five wildflower species in similar numbers in both plots. Then she enclosed one of the plots with a fence to keep out kangaroo rats, the most common herbivores in the area. After two years, four species of wildflowers were no longer present in the fenced plot, but one wildflower species had increased dramatically. The unfenced control plot had not changed significantly in species composition. Using the concepts discussed in the chapter, what do you think happened?

12. Beavers are considered to be "ecosystem engineers"—keystone species that actively transform the environment. They cut down old trees and use the wood to build river dams. What are the likely impacts of the beavers' activities on the riparian (river-based) environment? What would be the consequences if beavers disappeared?

13. **Interpreting Data** In a classic study, John Teal measured energy flow in a salt marsh ecosystem. The table below shows some of his results.

Form of energy	Kcal/m²/yr	Efficiency of energy transfer (%)
Sunlight	600,000	n/a
Chemical energy in producers	6,585	
Chemical energy in primary consumers	81	

Data from: J. M. Teal, Energy Flow in the Salt Marsh Ecosystem of Georgia. *Ecology* 43: 614–24 (1962).

   a. Calculate the efficiency of energy transfer by the producers. That is, what percentage of the energy in sunlight was converted into chemical energy and incorporated into plant biomass?
   b. Calculate the efficiency of energy transfer by the primary consumers. What percentage of the energy in plant biomass was incorporated into the bodies of the primary consumers? What became of the rest of the energy (see Figure 20.27)?
   c. How much energy is available for secondary consumers? Based on the efficiency of energy transfer by primary consumers, estimate how much energy will be available to tertiary consumers.
   d. Draw a pyramid of production for the producers, primary consumers, and secondary consumers for this ecosystem (see Figure 20.28).

BIOLOGY AND SOCIETY

14. Some organizations are starting to envision a sustainable society—one in which each generation inherits sufficient natural and economic resources and a relatively stable environment. The Worldwatch Institute, an environmental policy organization, estimates that we must reach sustainability by the year 2030 to avoid economic and environmental collapse. In what ways is our current system not sustainable? What might we do to work toward sustainability, and what are the major roadblocks to achieving it? How would your life be different in a sustainable society?

15. The Biological Dynamics of Forest Fragmentation Project, which has contributed so much to our understanding of threats to biodiversity, is itself endangered. Encouraged by Brazilian government agencies, urban sprawl and intensive forest settlement are closing in on the study site. Activities such as clear-cutting, burning, hunting, and logging threaten the integrity of the surrounding forest. Researchers at the BDFFP, which is jointly operated by a Brazilian research agency and the Smithsonian Tropical Research Institute, hope that they can bring attention to the problem through the Brazilian media and pressure the government into protecting the project. How would you argue the importance of protecting the BDFFP against ecologically destructive activities in a letter to a newspaper editor or to a government official?

16. Mass extinctions have happened several times on Earth. Why then are we worried about the possibility of a loss of biodiversity?

Unit 5
Animal Structure and Function

21 **Unifying Concepts of Animal Structure and Function**

Chapter Thread: **Controlling Body Temperature**

22 **Nutrition and Digestion**

Chapter Thread: **Controlling Your Weight**

23 **Circulation and Respiration**

Chapter Thread: **Athletic Endurance**

24 **The Body's Defenses**

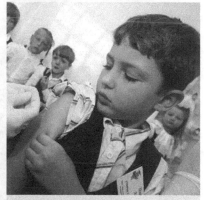

Chapter Thread: **Vaccines**

25 **Hormones**

Chapter Thread: **Steroid Abuse**

26 **Reproduction and Development**

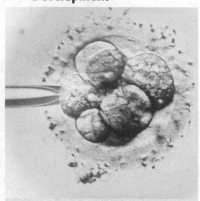

Chapter Thread: **High-Tech Babies**

27 **Nervous, Sensory, and Locomotor Systems**

Chapter Thread: **Extrahuman Senses**

21 Unifying Concepts of Animal Structure and Function

Why Animal Structure and Function Matter

◄ You can't help it when you blush: Involuntary muscle in your blood vessels is to blame.

Diabetes, illicit drug use, sexually transmitted disease—a urine sample can reveal a lot about a person. ▼

Contrary to common thinking, a moderate fever is not always your body's enemy but often is an ally to your immune system in its battle against an infection. ►

Controlling Body Temperature BIOLOGY AND SOCIETY

An Avoidable Tragedy

After a sweltering morning of football practice in northern Florida on August 2, 2011, 16-year-old Don'terio J. Searcy collapsed. Within hours, the junior defensive lineman was dead. The cause of his death was heat stroke, a condition wherein a person experiences such prolonged and extreme heat exposure that the brain's temperature control center fails. Heat stroke is the most severe of related conditions collectively referred to as heat illness. You might think that an athletic teenager suffering from heat illness would be rare, but it isn't. Approximately 10,000 high school athletes are afflicted with the condition each year, and since 1995, more than 40 high school football players have died from heat stroke.

When you play or work out hard, your muscles produce a lot of heat that in turn increases your overall internal temperature. A control center in your brain initiates responses to keep your internal temperature near normal (37°C, or 99°F). You begin to sweat, and as the sweat evaporates, it has a cooling effect. Simultaneously, blood vessels near the surface of your skin expand to release body heat. The padding worn by football players restricts evaporative cooling and traps body heat.

If your body temperature continues to rise, you sweat profusely, and blood vessels near the surface of your skin widen to their maximum. These responses, which could lower your blood pressure enough to cause you to faint, are symptoms of heat exhaustion, your body's final attempt to cool itself. Under extreme heat exhaustion, body temperature rises to the point that it disrupts the brain's temperature control center. The result? Heat stroke. During heat stroke, the body's organs begin to fail, and unless the body is cooled and hydrated, brain damage and then death can follow, as was the case with Searcy. To avoid heat exhaustion, drink plenty of water when exerting yourself and rest if you feel lightheaded or nauseous or your skin flushes excessively.

Heat illness. When athletes practice multiple times a day in hot weather while wearing heavy equipment, they increase their risk for heat illness.

Although heartbreaking, Searcy's death brought about a comforting change: public awareness of heat illness. In his home state of Georgia, policy is now in place that sets limits on practice schedules and the amount of time players wear padding during practice.

In this unit, you will learn how the animal body self-regulates and become aware of evolutionary adaptations that are common to animals. To set the stage for the unit, this chapter describes basic animal structure, from cells to tissues to organs to organ systems, and applies structure to function, using two examples of internal regulation: control of body temperature and balance in bodily fluids.

☑ **CHECKPOINT**

Place these levels of
biological organization
in order from largest to
smallest: cell, organ, organ
system, organism, tissue.

*Answer: organism, organ system,
organ, tissue, cell*

The Structural Organization of Animals

Life is characterized by a hierarchy of organization. In animals, individual and specialized cells group to form tissue, and various tissues combine into functional organs. Organs do not operate alone; they form organ systems. Organ systems, each specialized for certain tasks, function together as an integrated, coordinated unit (**Figure 21.1**).

Levels of the hierarchy interact to perform the functions of life, such as regulating an animal's internal environment. For example, a decrease in your body temperature requires the brain to send signals via the nervous system that trigger the widening of capillaries under the skin and an increase in secretion from sweat-synthesizing cells. As sweat evaporates from the skin, the result is a cooling in temperature throughout the whole body. ☑

▼ **Figure 21.1 Structural hierarchy in a human.**

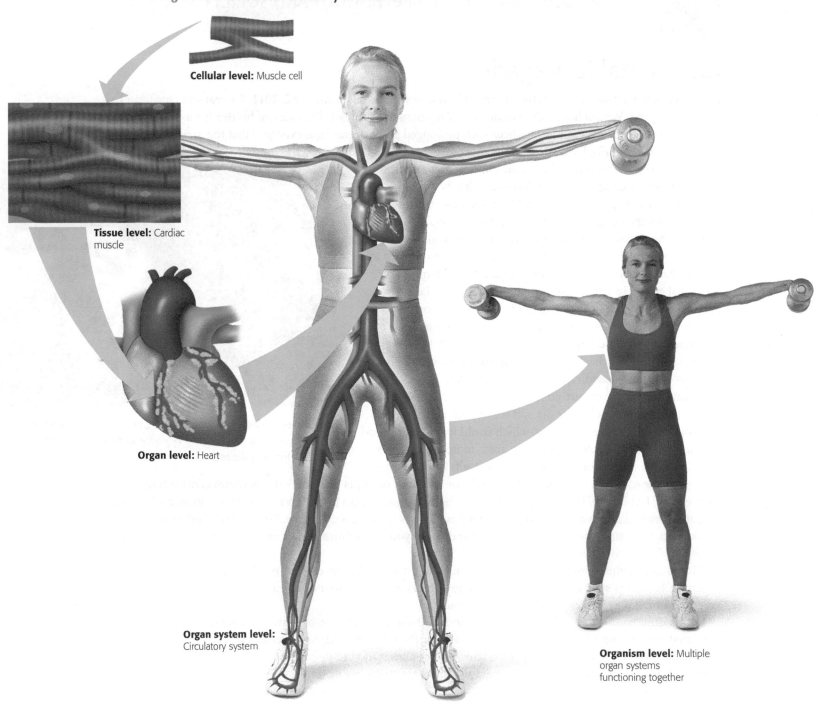

Cellular level: Muscle cell

Tissue level: Cardiac muscle

Organ level: Heart

Organ system level: Circulatory system

Organism level: Multiple organ systems functioning together

Anatomy and Physiology

Structure/Function

Which is the better tool: a hammer or a screwdriver? The answer, of course, depends on what you're trying to do. Given a choice of tools, you would not use a hammer to loosen a screw or a screwdriver to pound a nail. A device's form relates to how it works: Structure correlates with function.

The design of a screwdriver—the cross-shaped tip of a Philips screwdriver, for example—is dictated by the task that the tool must perform—turning a screw with a cross-shaped head. The same principle applies to life at its many levels, from cells to organisms. The structure of the shoulder's ball-and-socket joint fits its function of allowing the upper arm bone to rotate in all directions. Analyzing a biological structure gives us clues about how it works and what it does. Conversely, knowing what a biological structure does provides insight about how it operates and how it is constructed.

When discussing structure and function, biologists distinguish anatomy from physiology. **Anatomy** is the study of the structure of an organism's parts. **Physiology** is the study of the function of those parts. For example, an anatomist might study the arrangement of blood vessels near the surface of the skin, whereas a physiologist might study how those blood vessels widen in response to a sudden rise in temperature. Despite their different approaches, the two disciplines serve the same purpose: to better understand the connections between structure and function.

The correlation of structure and function is a fundamental principle of biology that is evident at all levels of life's hierarchy; it is a principle that will guide us throughout our study of animals **(Figure 21.2)**. But this rule of "design" does not mean that such biological tools as a bird's wings or a mammal's blood vessels are products of purposeful invention. Here the analogy to household tools such as screwdrivers fails, because those tools were designed with specific goals. It is natural selection that refines biological structure. It does so by screening for the most effective variations among individuals of a population—those variations that are most advantageous in the current environment. Generations of selecting for what works best in a particular environment will fit structure to function without any goal-oriented plan. Thus, the structure-function principle is just another facet of one of biology's major themes: evolution. ☑

Tissues

The cell is the basic unit of all living organisms. In almost all animals, including humans, cells rarely act alone but instead are grouped into tissues. A **tissue**, an integrated group of similar cells, performs a specific function. The cells composing a tissue are specialized; they have an overall structure that enables them to perform a specific task. An animal has four main categories of tissue: epithelial tissue, connective tissue, muscle tissue, and nervous tissue. As you'll see, the structure of each type of tissue relates to its specific function.

☑ **CHECKPOINT**

Explain how your complete set of teeth illustrates the "structure-function" principle.

Answers: Your front teeth are sharp and narrow, allowing you to bite or tear off manageable pieces of food, and your back teeth are flat and wide, allowing you to chew and grind the food into small bits.

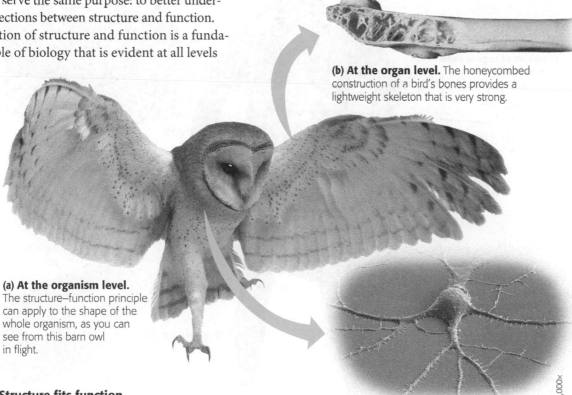

(b) At the organ level. The honeycombed construction of a bird's bones provides a lightweight skeleton that is very strong.

(a) At the organism level. The structure–function principle can apply to the shape of the whole organism, as you can see from this barn owl in flight.

(c) At the cellular level. Nerve cells that control the muscles of a bird or other animal have long extensions that transmit signals.

2,000x

► Figure 21.2 **Structure fits function.**

Epithelial Tissue

Epithelial tissue (also known as **epithelium**) is a tissue that covers the surface of the body and lines organs **(Figure 21.3)**. The architecture of an epithelium illustrates how structure fits function at the tissue level. Your skin contains many layers of tightly bound epithelial cells, forming a protective, waterproof barrier that surrounds your body and keeps it safe from external threats. In contrast to the layered and tight cell structure of the skin, a single thin and leaky layer of epithelial tissue lines capillaries, where it is well suited for the role it plays of exchanging substances with the circulatory system.

The body continuously renews the cells of many epithelial tissues. For example, cells of the digestive tract and skin are renewed every few days or weeks, depending on the specific cell type. Such turnover requires cells to divide rapidly, which increases the risk of an error in cell division, a mistake that can lead to cancer. Compared to other body tissues, epithelium contacts the environment most directly and is therefore exposed to much higher levels of cancer-causing substances, which are known as carcinogens (smoke and UV light, for example). Consequently, about 80% of all cancers arise in epithelial tissue. Cancers of epithelial tissue are collectively called carcinomas; the five most common cancers— breast, prostate, lung, colon, and rectum—are all carcinomas (see Table 11.1).

Connective Tissue

In contrast to epithelium, with its sheets of cells, **connective tissue** contains cells scattered throughout a material called the **extracellular matrix**. The structure of the matrix varies and matches the function of each tissue. Two major functions of the connective tissue are to support and join other tissues.

Figure 21.4 illustrates six of the major types of connective tissue. The most widespread connective tissue in the body of vertebrates (animals with backbones, including humans) is **loose connective tissue**. It binds epithelia to underlying tissues. Think about the skin on a roasted chicken leg; when you pull back the skin, you break loose connective tissue that joins the skin to the underlying muscle. A matrix of protein fibers that provide great strength and elasticity is characteristic of loose connective tissue. You can get a sense of the tissue's elasticity if you consider how quickly your flesh snaps back after someone pinches your cheek. Because production of the matrix decreases with age, older skin doesn't snap back; instead, it tends to sag and wrinkle.

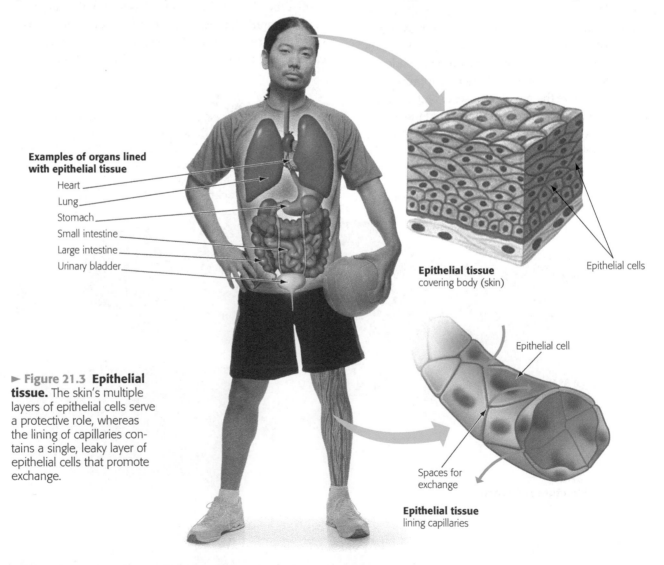

**Examples of organs lined
with epithelial tissue**

Heart
Lung
Stomach
Small intestine
Large intestine
Urinary bladder

Epithelial tissue
covering body (skin)

Epithelial cells

Epithelial cell

Spaces for
exchange

Epithelial tissue
lining capillaries

▶ Figure 21.3 **Epithelial
tissue.** The skin's multiple
layers of epithelial cells serve
a protective role, whereas
the lining of capillaries contains a single, leaky layer of
epithelial cells that promote
exchange.

Three types of connective tissue have cells suspended in dense matrices. **Fibrous connective tissue** has a dense matrix of collagen. It forms tendons, which attach muscles to bones, and ligaments, which securely join bones together at joints. Think of a roasted chicken leg again; pulling a leg bone free from a thigh bone requires force as you tear the ligament between them. The matrix of **cartilage** is strong but flexible. Use your finger to flip your outer ear, and you'll get the idea of how cartilage functions as a flexible, boneless skeleton. Unlike most tissues, cartilage has no blood vessels, so injuries to it (knee injuries, for example) tend to heal very slowly, if at all. **Bone** is a rigid connective tissue with a dense matrix of collagen fibers hardened with deposits of calcium salts. This combination makes bone hard without being brittle.

Adipose tissue stores fat in closely packed cells of a sparse matrix. Functioning as an energy bank, adipose tissue swells when fat from food is deposited into its cells and shrinks when fat is spent from its cells for energy. The fat of adipose tissue insulates and cushions the body as well. The thick layer of adipose tissue in whales called blubber, for example, functions like a blanket (insulates) to keep the animal's internal heat from escaping into the extremely cold ocean waters of its external environment. Press on the bottom of your heel and feel some of the fat that absorbs (cushions) the impact on bones in your feet when you stand, walk, or run.

Unlike the other connective tissues wherein cells compose a solid matrix, **blood** consists of cells suspended in a liquid matrix called plasma. The main function of blood is transporting substances in the plasma from one part of the body to another, but blood also plays major roles in immunity and sealing broken blood vessels (through blood clots).

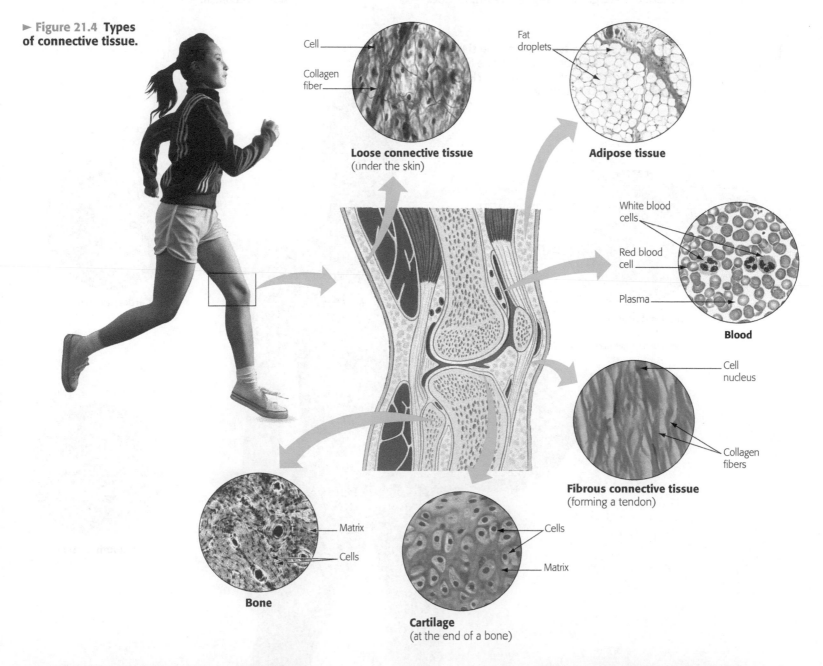

► Figure 21.4 **Types of connective tissue.**

Cell

Collagen fiber

Loose connective tissue (under the skin)

Fat droplets

Adipose tissue

White blood cells

Red blood cell

Plasma

Blood

Cell nucleus

Collagen fibers

Fibrous connective tissue (forming a tendon)

Matrix

Cells

Bone

Cells

Matrix

Cartilage (at the end of a bone)

Muscle Tissue

Muscle is the most abundant tissue in the vast majority of animals. In fact, what we call meat on our dinner plate is mostly animal muscle. **Muscle tissue** consists of bundles of long, thin, cylindrical cells known as muscle fibers. Each muscle fiber has specialized proteins arranged into a structure that contracts (pulls inward) when stimulated by a signal from a nerve. Humans and other vertebrates have three types of muscle tissue, each with unique contractile proteins: skeletal muscle, cardiac (heart) muscle, and smooth muscle **(Figure 21.5)**.

Attached to your bones by tendons, **skeletal muscle** moves your skeleton. Skeletal muscle is responsible for your voluntary movements, such as walking and talking. Humans are born with most of the skeletal muscle fibers they will have for life; new muscle fibers are normally not generated. Weight training, for example, does

not increase the number of skeletal muscle fibers, but it does enlarge the individual skeletal muscle fibers.

The other two types of muscle tissues are involuntary, meaning they contract without any conscious control on your part. **Cardiac muscle** is found only in heart tissue. The contraction of the cardiac muscle produces a coordinated heartbeat. **Smooth muscle** is found in many organs and can contract slowly for a long period of time. The powerful contractions of smooth muscle expel the fetus from the uterus during childbirth. The walls of the intestines are composed of smooth muscle that contracts to move food and waste along. Smooth muscle is found in blood vessels, too. You have undoubtedly felt its involuntary effects when you are embarrassed. Rings of smooth muscle in blood vessels widen, causing blood to quickly flow to your face and neck, and the skin there becomes red and hot—you blush.

You can't help it when you blush: Involuntary muscle in your blood vessels is to blame.

▼ **Figure 21.5 Three types of muscle tissue.** Skeletal muscle is the only muscle tissue type that we can consciously control; cardiac muscle and smooth muscle act involuntarily.

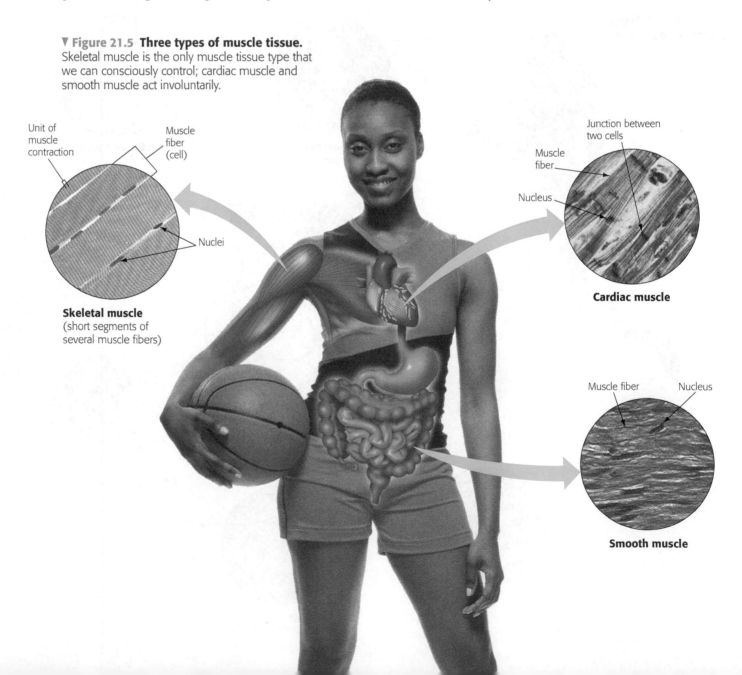

Unit of muscle contraction

Muscle fiber (cell)

Nuclei

Skeletal muscle
(short segments of several muscle fibers)

Junction between two cells

Muscle fiber

Nucleus

Cardiac muscle

Muscle fiber

Nucleus

Smooth muscle

Nervous Tissue

Most animals respond rapidly to stimuli from their environment. For example, if you step onto hot pavement barefoot, you will quickly jump to a cooler surface. This response requires information to be relayed from one part of the body to another. It is **nervous tissue** that makes such communication possible. Nervous tissue is found in your brain and spinal cord, as well as in the nerves that connect these organs to all other parts of your body.

The basic unit of nervous tissue is the **neuron**, or nerve cell (**Figure 21.6**). With their long extensions, neurons can transmit electrical signals very rapidly over long distances. For example, neurons in your leg may be as long as 1 m (3.3 feet), running all the way from the base of your spinal cord to the tips of your toes. Neurons connected to each other form networks; in the brain, these complex interconnections are responsible for what we broadly call "thinking." ☑

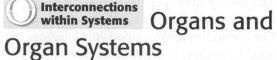 **Interconnections within Systems** **Organs and Organ Systems**

The hierarchical organization of animals is an example of one of the major themes in biology: interconnections within systems. Recall that at each level, novel properties emerge that are absent from the preceding level. After the cellular and tissue levels in the hierarchy is the organ level. An **organ** consists of two or more tissues packaged into one working unit that performs a specific function. An organ performs functions that none of its component tissues can carry out alone. Your heart, brain, and small intestine are examples of organs.

To see how multiple tissues interconnect within a single organ, examine the layered arrangement of tissues in the wall of the small intestine in **Figure 21.7**. The wall is lined with epithelial tissue (the top layer of cells in figure) that secretes mucus and absorbs nutrients.

☑ **CHECKPOINT**

Name the four types of tissue found in most organs.

Answers: epithelial, connective, muscle, and nervous

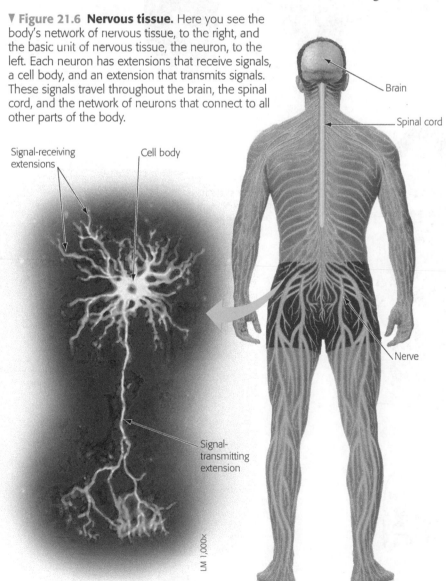

▼ **Figure 21.6 Nervous tissue.** Here you see the body's network of nervous tissue, to the right, and the basic unit of nervous tissue, the neuron, to the left. Each neuron has extensions that receive signals, a cell body, and an extension that transmits signals. These signals travel throughout the brain, the spinal cord, and the network of neurons that connect to all other parts of the body.

Signal-receiving extensions

Cell body

Brain

Spinal cord

Nerve

Signal-transmitting extension

LM 1,000x

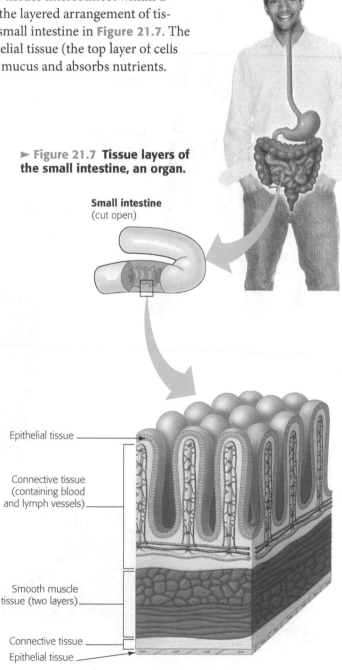

► **Figure 21.7 Tissue layers of the small intestine, an organ.**

Small intestine (cut open)

Epithelial tissue

Connective tissue (containing blood and lymph vessels)

Smooth muscle tissue (two layers)

Connective tissue

Epithelial tissue

Underneath the top layer of epithelial tissue is connective tissue, containing blood capillaries and lymph vessels. The connective tissue supports the functions of the epithelial layer above it as well as the two layers of smooth muscle below it. One more layer of connective tissue binds the epithelial tissue that forms a sheet on the outside surface.

The organs of humans and most other animals are organized into **organ systems**, teams of organs that work together to perform vital body functions. The components of an organ system can be physically connected or they can be dispersed throughout the body. An example of an organ system is your circulatory system. Its main organs are the heart and blood vessels, and its vital

▼ Figure 21.8 **Human organ systems.**

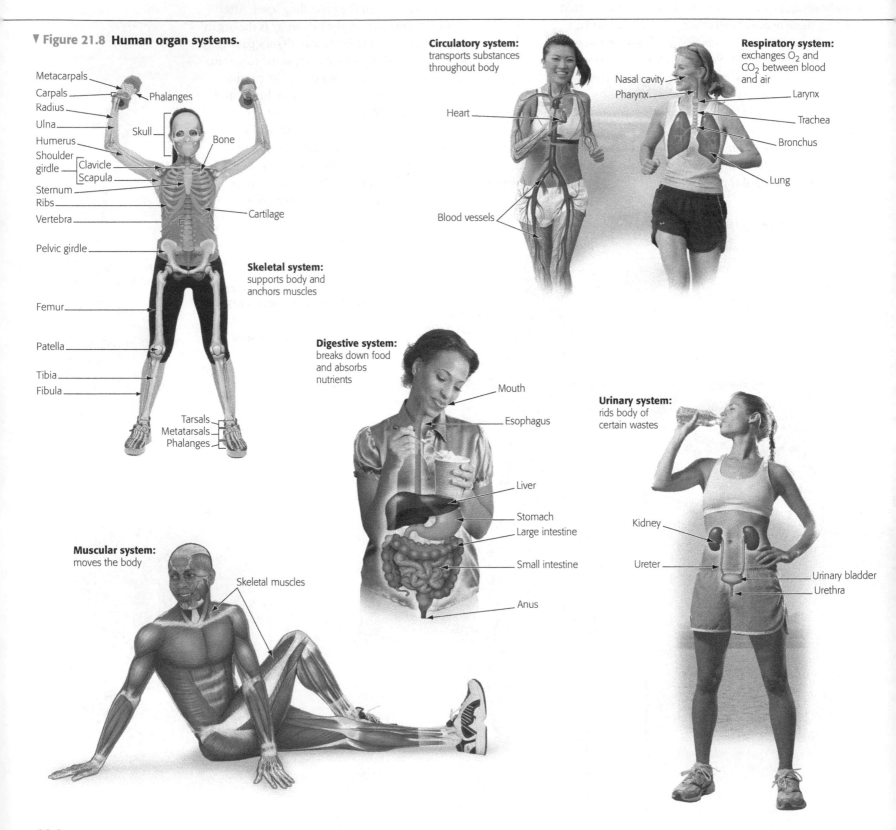

Skeletal system: supports body and anchors muscles

Metacarpals, Carpals, Radius, Ulna, Humerus, Shoulder girdle, Clavicle, Scapula, Sternum, Ribs, Vertebra, Pelvic girdle, Femur, Patella, Tibia, Fibula, Tarsals, Metatarsals, Phalanges, Skull, Bone, Cartilage

Circulatory system: transports substances throughout body

Heart, Blood vessels

Respiratory system: exchanges O_2 and CO_2 between blood and air

Nasal cavity, Pharynx, Larynx, Trachea, Bronchus, Lung

Digestive system: breaks down food and absorbs nutrients

Mouth, Esophagus, Liver, Stomach, Large intestine, Small intestine, Anus

Urinary system: rids body of certain wastes

Kidney, Ureter, Urinary bladder, Urethra

Muscular system: moves the body

Skeletal muscles

function is transport of nutrients, wastes, and other substances to and from cells. **Figure 21.8** presents an overview of 11 major organ systems in vertebrates.

An organism depends on the interconnection of all its organ systems for survival. For instance, skipping a meal lowers your blood glucose level. This change stimulates pancreatic cells of your endocrine system to release a hormone that causes your liver to release glucose into the bloodstream. Your blood glucose level then rises. The failure of any organ system will jeopardize the entire animal because organ systems are dependent on each other. Your body is a whole, living unit that is greater than the sum of its parts. In the next section, we'll look at how the body's organ systems interact with the external environment.

Endocrine system: secretes hormones that regulate body

Hypothalamus
Pituitary gland
Parathyroid gland
Thyroid gland
Adrenal gland
Pancreas

Testis (male)

Ovary (female)

Reproductive system: produces gametes and offspring

Oviduct
Ovary
Uterus
Vagina

Seminal vesicles
Prostate gland
Vas deferens
Penis
Urethra

Testis

Integumentary system: protects body

Hair
Nail
Skin

Lymphatic and immune system: defends against disease

Tonsil
Thymus
Spleen

Appendix

Lymph nodes

Lymphatic vessels

Nervous system: processes sensory information and controls responses

Brain
Sense organ (ear)
Spinal cord
Nerves

Exchanges with the External Environment

Most animals are covered with a protective layer that separates the external world from their internal environment. This does not mean, however, that an animal encloses itself in some sort of protective bubble, isolated from the harsh world outside. Just as a car engine will quickly stall if the air intake is clogged, your body will quickly shut down if deprived of outside oxygen. Every organism is an **open system** that continuously exchanges chemicals and energy with its surroundings. You eat, breathe, defecate, urinate, sweat, and radiate heat—all examples of how you operate as an open system. And the exchange of materials extends down the hierarchy, from every organ system to each individual cell: Nutrients and oxygen must enter every living cell, and carbon dioxide and other wastes must exit.

An animal's size and shape affect its exchanges with its surrounding environment. Every living cell of an animal's body must be bathed in a watery solution, partly because substances must be dissolved in water to cross cell membranes. In a single-celled amoeba, every part of the cell's membrane touches the outside world, where exchange with the watery environment can occur **(Figure 21.9a)**. A hydra

(a multicellular pond-dwelling relative of jellies) has a body wall only two cell layers thick **(Figure 21.9b)**. Both layers of cells are bathed in pond water, which enters the digestive sac through the mouth. Every cell of the hydra can thus exchange materials through direct contact with the aqueous environment.

Exchange with the environment is more complicated for complex, multilayered animals. Each cell in a multicellular organism has a plasma membrane where exchange can occur. But this exchange only works if all the cells of the animal have access to a suitable watery environment. **Figure 21.10** shows a schematic model of an animal body, highlighting the three organ systems—digestive, respiratory, and urinary—that exchange materials with the external environment. Notice also the vital

▼ **Figure 21.9 Contact of simple organisms with the environment.**

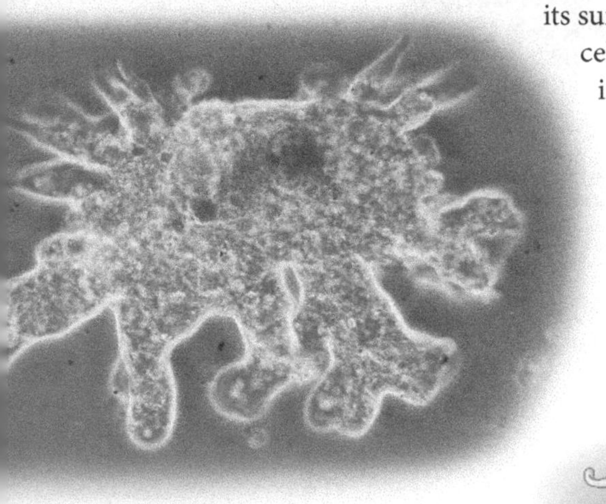

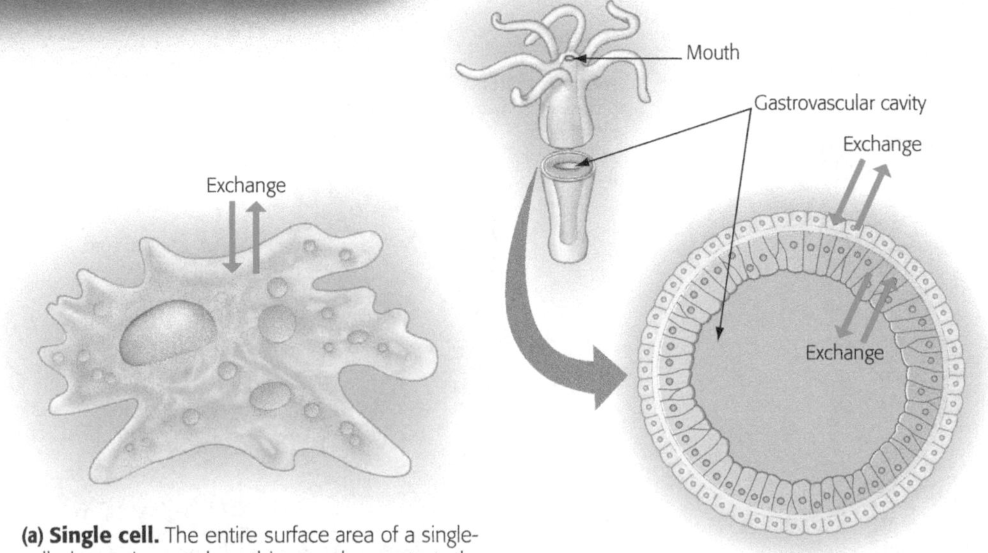

(a) Single cell. The entire surface area of a single-celled organism, such as this amoeba, contacts the environment. Because of its small size, the organism has a large surface area (relative to its volume) through which it exchanges materials with the external world.

Exchange

Mouth

Gastrovascular cavity

Exchange

Exchange

(b) Two cell layers. Although a hydra is multicellular, each one of its cells touches an aqueous envrionment. The body has only two cell layers, both exposed to water.

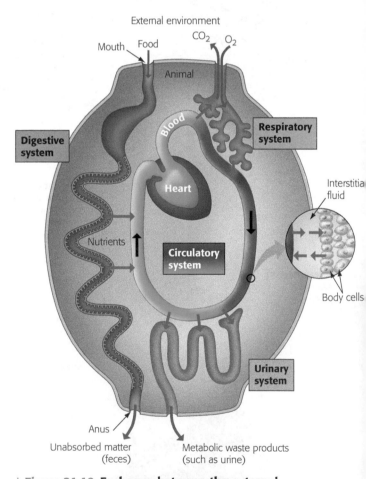

External environment

Mouth Food CO_2 O_2

Animal

Blood

Respiratory system

Digestive system

Heart

Interstitial fluid

Nutrients

Circulatory system

Body cells

Urinary system

Anus

Unabsorbed matter (feces)

Metabolic waste products (such as urine)

▲ **Figure 21.10 Exchange between the external environment and the internal environment of complex animals.** The blue arrows indicate the exchange of materials between the circulatory system and three other systems and between the circulatory system and body cells.

role of the circulatory system: It connects to nearly every organ system as it transports needed materials from the environment to the body's tissues and carries wastes away. Looking at the figure, you can see how nutrients that are absorbed from the digestive tract are distributed throughout the body by the circulatory system. The heart that pumps blood through the circulatory system requires nutrients absorbed from food by the digestive tract and also oxygen (O_2) obtained from the air by the respiratory system. And absorption of nutrients inevitably produces wastes that must be excreted by the urinary system. The circulatory system thereby provides an internal conduit that joins the systems that connect with the outside world.

Complex animals have evolved extensively folded or branched internal surfaces that maximize surface area for exchange with the immediate environment. Take, for example, your lungs and intestines. Your lungs, which exchange oxygen and carbon dioxide with the air you breathe, are not shaped like big balloons but are in fact millions of tiny hollow structures at the tips of finely branched air tubes **(Figure 21.11)**. The epithelium of the lungs has a very large total surface area—about the size of a tennis court. And your intestines are

not smooth on the inside; rather, the internal structure is highly folded and covered with finger-like projections that provide far more surface area for nutrient absorption, as you can see in Figure 21.7. ☑

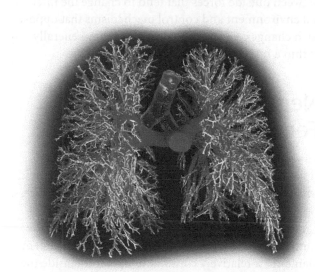

▲ **Figure 21.11 The branched surface area of the human lung.** This plastic model shows the tiny air tubes of the lungs (white) and the thin blood vessels (red) that transport gases between the heart and lungs.

☑ **CHECKPOINT**

Is your body an open or closed system? Why?

Answer: Your body is an open system because the digestive, respiratory, and urinary systems are in direct contact with the external environment.

Regulating the Internal Environment

In this section, you'll learn how animals adjust to a changing environment. Let's begin by looking at the mechanisms that stabilize the internal body despite fluctuations in the external environment.

Homeostasis

One of the body's most important functions is to stay relatively unchanged even when the world around it changes. The internal environment of vertebrates includes the **interstitial fluid** that fills the spaces between cells and exchanges nutrients and wastes with microscopic blood vessels (see Figure 21.10). It is vital for animal survival that the composition of the interstitial fluid in cells remains relatively constant no matter what occurs in the outside world.

Homeostasis, which literally means "steady state," is the tendency to maintain relatively constant conditions in the internal environment even when the external environment changes. In the face of large external changes, the mechanisms of homeostasis normally maintain internal conditions within a range that the animal's metabolism can tolerate **(Figure 21.12)**. For example, a cat's body temperature normally fluctuates by less than

a degree, whether it is hunting on a cold winter night or basking in the sun on a hot summer day. Losing the ability to maintain homeostasis can have significant health consequences. Diabetes, for example, is a loss of glucose homeostasis in the blood.

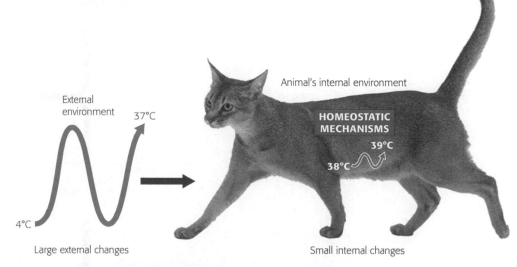

External environment

37°C

4°C

Large external changes

Animal's internal environment

HOMEOSTATIC MECHANISMS

39°C

38°C

Small internal changes

▲ **Figure 21.12 An example of homeostasis.** Despite large fluctuations in outside temperature, the cat's internal temperature is tightly regulated and only fluctuates slightly.

499

The internal environment of an animal always fluctuates slightly in response to internal and external changes. Blood glucose level, for example, rises temporarily after a meal. Homeostasis is a dynamic state, an interplay between outside forces that tend to change the internal environment and control mechanisms that oppose such changes. Changes do occur, but they generally stay within a range that is tolerable for living cells.

Negative and Positive Feedback

Most mechanisms of homeostasis depend on a principle called **negative feedback**, a form of regulation in which the results of a process inhibit that same process. **Figure 21.13** illustrates the concept of negative feedback with an example of household homeostasis, the control of room temperature. A home heating system maintains a relatively constant temperature inside the house despite drastic changes that may occur outside. A key component of this system is a control center—a thermostat—that monitors temperature and switches the heater on and off. Whenever room temperature

drops below the set point (20°C in the figure), the thermostat switches the heater on (bottom pathway in Figure 21.13). When the temperature rises above the set point, the thermostat switches the heater off (top pathway). The basic principle of negative feedback is simple: The result (increased room temperature, for example) of a process (heating of the air) inhibits that very process (by switching the heater off). Negative feedback is the most common mechanism of homeostatic control in animals.

Less commonly, organisms use **positive feedback**, in which the results of a process intensify that same process. For example, when a blood vessel is broken, proteins found in blood are stimulated, resulting in a small netlike clot. The stimulated proteins activate more and more proteins, which cause a larger and larger clot, in a positive-feedback loop that continually enhances the response. The result is a clot large enough to seal the broken blood vessel.

In the rest of this section, we'll examine specific examples of homeostasis. We'll discuss the control of temperature and the control of water gain and loss in a variety of animals. Then we'll focus on the human urinary system. ☑

► Figure 21.13 **An example of negative feedback: control of room temperature.** This is an example of negative feedback because the result of the process (heat) shuts that process down.

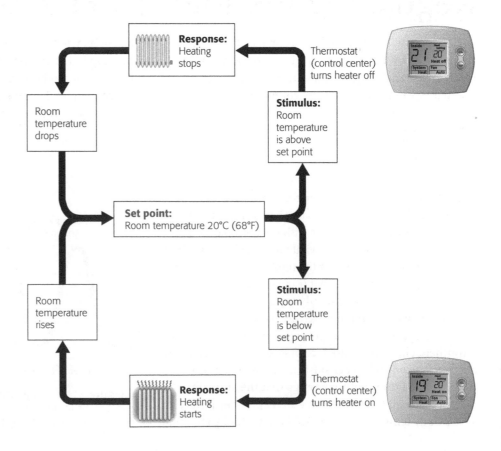

Thermoregulation

As you saw in the Biology and Society section on the dangers of heat illness, internal body temperature must be within a certain range for the animal body to function normally. The homeostatic mechanism that controls temperature is called **thermoregulation**. As you walk from a warm building to the outside on a cold day, your body's temperature will barely fluctuate, despite the drastic change in temperature in the environment around you. The ability to maintain a body temperature substantially warmer than the surrounding environment is characteristic of **endotherms**, animals such as mammals and birds that derive most of their body heat from their own metabolism. In contrast, **ectotherms**, which include most invertebrates, fishes, amphibians, and nonbird reptiles, obtain their body heat primarily by absorbing it from their surroundings.

You have a number of structures and mechanisms that aid in thermoregulation. As shown in **Figure 21.14**, your brain has a control center that maintains your body temperature near 37°C (98.6°F). When your body temperature falls below normal, your brain's control center sends signals that trigger changes that will bring it back to normal: Blood vessels near your body's surface constrict (reducing heat loss from your body surface) and muscles contract, causing you to shiver (increasing heat production). All cells produce heat as a result of cellular respiration, thus working muscles produce a lot of heat (the reason shivering and exercise warm you). When body temperature gets too high, the control center sends signals to dilate the blood vessels near your skin and activate sweat glands, allowing excess heat to escape. Like a home controlled by a thermostat, the temperature of the body does not stay completely constant but fluctuates up and down within an acceptable range.

Fever, an abnormally high internal temperature, is a body-wide response that usually indicates an ongoing fight against infection. When your immune system cells encounter invading microbes, the cells release chemicals that travel through the bloodstream to your brain. These chemicals stimulate your brain's thermostatic control center to raise the body's internal temperature, producing a fever. Many people mistakenly believe that the invading microbes themselves cause a fever. In fact, the cause is usually the body's fight *against* the microbes. A fever of more than 40°C (104°F) may be life-threatening because a temperature that high can damage body proteins. A moderate fever of 38–39°C (100–102°F), however, discourages bacterial growth and speeds the body's internal defenses.

As endotherms, we generate body heat to warm ourselves. Ectotherms generally cannot, but there are exceptions, as we'll see next. ☑

Contrary to common thinking, a moderate fever is not always your body's enemy but often is an ally to your immune system in its battle against an infection.

☑ **CHECKPOINT**

Name two features of the human body that help release heat and two features that help keep the body warm.

Answer: Examples include sweating and dilation of blood vessels to release heat and shivering and constriction of blood vessels to keep the body warm.

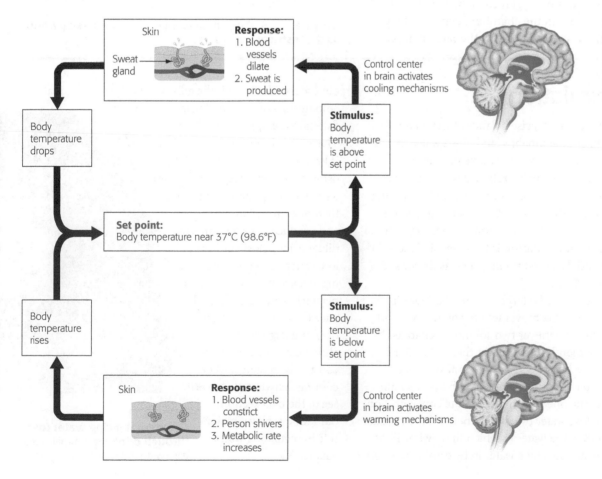

◄ Figure 21.14 **Thermoregulation in the human body.** By comparing this figure with Figure 21.13, you can see how similar negative-feedback controls moderate the temperatures of a room and your body. The control center for body temperature is located in the brain.

Skin

Response:
1. Blood vessels dilate
2. Sweat is produced

Sweat gland

Control center in brain activates cooling mechanisms

Stimulus: Body temperature is above set point

Body temperature drops

Set point: Body temperature near 37°C (98.6°F)

Stimulus: Body temperature is below set point

Body temperature rises

Skin

Response:
1. Blood vessels constrict
2. Person shivers
3. Metabolic rate increases

Control center in brain activates warming mechanisms

How Does a Python Warm Her Eggs?

As just described, ectotherms usually absorb heat from their surroundings. Under certain circumstances, however, some ectotherms can generate their own body heat. An example is the Burmese python (*Python bivittatus*), a very large snake native to the rain forests of southeastern Asia. Like most snakes, the female reproduces by laying eggs. Researchers at the Bronx Zoo, in New York City, made the **observation** that a female Burmese python incubating her eggs wraps her body around them and frequently contracts the muscles in her coils. Researchers formed the **hypothesis** that the snake's muscle contractions elevate its body temperature for transfer of heat to its eggs.

To test their hypothesis, the researchers performed a simple **experiment**. They placed a python and her eggs in a chamber and varied the chamber's temperature. They monitored the rate of the python's muscle contractions and took into account the snake's oxygen uptake, a measure of the rate of cellular respiration. Their **results** showed that the python's oxygen consumption increased when the temperature in the chamber decreased. Oxygen consumption also changed with the rate of muscle

contraction (**Figure 21.15**). Because oxygen consumption generates heat through cellular respiration and increases with the rate of muscle contraction, the researchers concluded that the muscle contractions were the source of the Burmese python's elevated body temperature during egg incubation. As you'll see in the Evolution Connection section, such behavior is just one of many temperature-regulating adaptations that have evolved in animals.

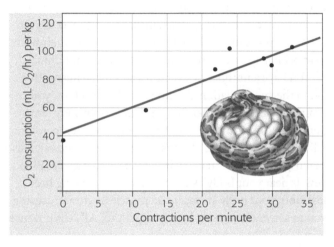

▲ Figure 21.15 **Oxygen consumed by a Burmese python as it constricts its muscles.**

Osmoregulation

Living cells depend on a precise balance of water and solutes. Whether an animal inhabits land, fresh water, or salt water, its cells cannot survive if they take in or lose too much water. That is, a cell may exchange water with the environment as long as the total amount of water entering and exiting the cell is the same. If too much water enters, animal cells will burst; if too much water exits, they will shrivel and die. **Osmoregulation** is the control of the gain or loss of water and dissolved solutes, such as the ions of NaCl and other salts.

Osmoregulation is based largely on regulating solutes because water follows the movement of solutes by osmosis. Osmosis occurs whenever two solutions separated by a cell's permeable membrane differ in their total solute concentrations (see Figure 5.13); there is always a net movement of water from the solution with lower solute concentration to the one with higher solute concentration.

Saltwater fish lose water by osmosis because their tissues contain less salt than the water in which they swim. To osmoregulate, saltwater fish take water in by drinking, pump

out excess salt through the gills, and produce only small amounts of concentrated urine. Freshwater fish have the opposite problem: The external solute concentration is low, so water enters the fish by osmosis. To osmoregulate, freshwater fish take up salt ions through their gills and digestive system and produce large amounts of dilute urine. Most land animals lose water through urinating, defecating, breathing, and perspiring but can counterbalance the loss by eating and drinking (**Figure 21.16**). Still, most land animals are constantly in danger of becoming dehydrated. As we'll see in the next section, our kidneys play a major role in regulating water balance. ☑

▲ Figure 21.16
Counterbalancing water loss through drinking. Animals also gain water by eating.

☑ **CHECKPOINT**

A patient received fluid through an intravenous (IV) line that caused his interstitial fluid to have a very high solute concentration. What will happen to his cells?

Answer: They will shrink as water exits via osmosis.

Interconnections within Systems

Homeostasis in the Urinary System

No doubt you have noticed that when you are dehydrated, your urine is concentrated and yellow, but when you are hydrated, it is dilute and clear. The urinary system plays a central role in osmoregulation, regulating the amount of water and solutes in body fluids by retaining water when we are dehydrated and expelling it when we are hydrated. Besides osmoregulation, the urinary system plays another important role—the excretion of wastes. Chief among the waste products to be excreted in urine is urea, a nitrogen-containing compound produced from the breakdown of proteins and nucleic acids. All cells of the animal body require water balance and waste removal; thus, the urinary system is functionally interconnected to all other systems of the body.

In humans, the main processing centers of the urinary system are the two kidneys. Each is a compact organ, a bit smaller than a fist, located on either side of the abdomen. The kidneys contain nearly 100 miles of thin tubes called **tubules** and an intricate network of capillaries (tiny blood vessels). Every day, all of your blood passes through the capillaries of your kidneys hundreds of times. As the blood circulates, a fraction of it is filtered, and the plasma (the liquid portion of blood) enters the kidney tubules. Once there, the plasma is called **filtrate**. The filtrate contains valuable substances that the body needs to reclaim (such as water and glucose) and other substances such as urea that the body must dispose of. Humans cannot simply excrete all of the filtrate as urine; if we did, we would lose vital nutrients and dehydrate rapidly. Instead, our kidneys refine the filtrate, concentrating the wastes and returning most of the water and useful solutes to the blood. Without at least one functional kidney, a person cannot survive more than a few days unless they undergo a routine medical treatment called dialysis, which removes waste and excess water from the blood.

The anatomy of the human urinary system is shown in Figure 21.17. Starting with the whole system in **Figure 21.17a**, blood to be filtered enters each kidney via a blood vessel called the renal artery. Filtered blood leaves the kidney in the renal vein. **Figure 21.17b** shows a cutaway view of a kidney. Within the kidney, the renal artery branches into millions of thin blood vessels. **Figure 21.17c** shows one branch of the renal artery supplying blood to a nephron via a network of capillaries. A **nephron** consists of a tubule and its associated blood vessels. Each kidney contains about a million nephrons.

The nephrons carry out the functions of the urinary system. Blood pressure forces water and solutes from the blood through a filter at the start of the nephron tubule, creating filtrate. As the filtrate then passes through the tubule, water and needed nutrients are reabsorbed into the bloodstream and wastes are secreted into the filtrate. The filtrate becomes more and more concentrated, resulting in a relatively small quantity of fluid called **urine**. At the end of the tubule, urine leaves the nephron via a collecting duct. Urine collects in the kidney and then leaves via the **ureter**. Urine is stored in the **urinary bladder** (visible in Figure 21.17a). Urine is expelled from the urinary bladder via the **urethra**, a tube that empties near the vagina in females and through the penis in males.

▼ Figure 21.17 **Anatomy of the human urinary system.**

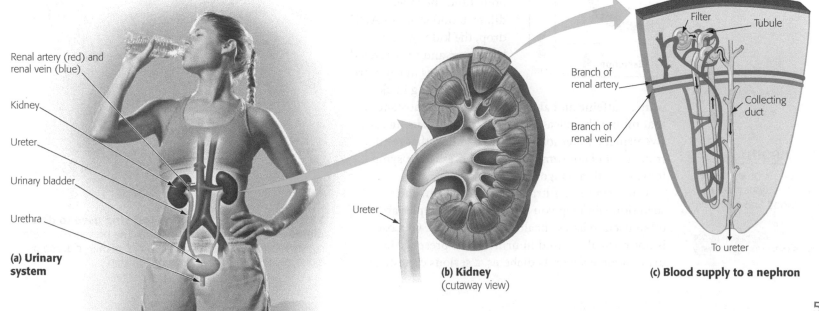

(a) **Urinary system**

(b) **Kidney** (cutaway view)

(c) **Blood supply to a nephron**

Figure 21.18 summarizes the functions performed by each nephron. First, during **filtration**, water and other small molecules are forced out of the blood when it passes through capillary walls into the kidney tubule, forming filtrate. Much like pulling bottles out from a trash bin so they can be recycled, **reabsorption** reclaims water and valuable solutes from the filtrate and returns them to the blood. In **secretion**, certain substances, such as some ions and drugs, are transported into the filtrate. What remains after filtration, reabsorption, and secretion is the urine. Finally, in **excretion**, urine passes from the kidneys to the outside.

Hormones regulate the kidney's nephrons to maintain water balance and are central to the interconnections of the nervous, endocrine, and urinary systems. When the solute concentration of body fluids rises too high (indicating that not enough water is present), the brain signals the endocrine system's pituitary gland to increase the release of a hormone called ADH (antidiuretic hormone) into the blood. When ADH reaches the kidneys, it causes nephrons to reabsorb more water from the filtrate, effectively increasing the body's water content while concentrating the urine. (Urine is dark yellow when concentrated.) Conversely, when body fluids become too dilute, blood levels of ADH drop, the kidneys reabsorb less water, and the excreted urine becomes much more watery, resulting in clear urine. Caffeine and alcohol are diuretics, substances that inhibit the release of ADH, thereby causing excessive urinary water loss. The dehydration that results from alcohol consumption contributes to the symptoms we call "a hangover."

Understanding filtration, reabsorption, and secretion will help you see why urine samples are often used to assess health **(Figure 21.19)**. Glucose is not normally found in urine, so its presence in a urine sample suggests diabetes, a serious condition in which blood glucose level is elevated. The excess glucose in the blood of diabetics enters the filtrate but is not reabsorbed before the urine exits the body.

> *Diabetes, illicit drug use, sexually transmitted disease—a urine sample can reveal a lot about a person.*

Drugs are secreted into urine and can be detected there. For example, a breakdown product of marijuana can be found in urine for days and sometimes weeks after use of the drug. Pregnancy can be confirmed by the presence of a specific hormone excreted only in the urine of pregnant women. If a physician suspects a sexually transmitted disease (STD) is causing a urinary tract infection, a urine sample may be taken to check for bacterial species that don't normally reside in the urinary system and cause STDs. Keep in mind that conditions like pregnancy and diabetes can also be tested for in blood, but urine is more easily accessible.

A seven-year study that attempted to identify all the compounds found in human urine catalogued more than 3,000 compounds. Of course, your urine doesn't contain all of these compounds. Urine differs from person to person. Factors affecting urine composition include the environment, presence of disease, secretions from the various harmless microbes that can live in the lower part of the urethra, and diet.

Regulating the amount of water and solutes in body fluids is just one of many examples of homeostasis we will see in this unit. We end this chapter by exploring how natural selection shapes homeostatic mechanisms in animals. As an example, we'll revisit thermoregulation. ☑

▼ **Figure 21.18 Major functions of a nephron.** Blood enters and exits the nephron via capillaries (shown as red then blue). Most of the fluid (called filtrate) that exits the blood at the beginning of the nephron is eventually reabsorbed.

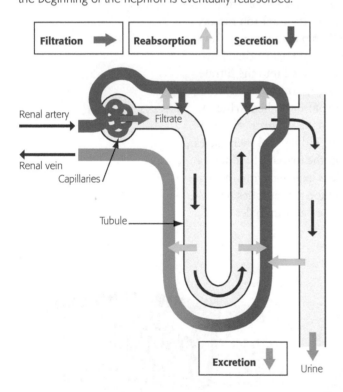

| Filtration ➡ | Reabsorption ⬆ | Secretion ⬇ |

Renal artery
Filtrate
Renal vein
Capillaries
Tubule

| Excretion ⬇ | Urine |

☑ CHECKPOINT

Name the four processes that occur as kidney tubules process blood and create urine.

Answer: filtration, reabsorption, secretion, excretion

▲ **Figure 21.19 A typical urine test strip used to detect diabetes and kidney malfunction.** The presence of certain substances not normally found in urine, such as glucose or blood, results in a color change on the strip.

Controlling Body Temperature EVOLUTION CONNECTION

Adaptations for Thermoregulation

Natural selection has promoted a wide variety of adaptations for thermoregulation—anatomical, physiological, and behavioral—that help animals cope with heat gained from or lost to the environment (Figure 21.20). A major anatomical adaptation in mammals and birds is insulation, consisting of hair (often called fur), feathers, or fat layers. A thin coat of fur in the summer often becomes a thick, insulating coat during the winter. Aquatic mammals (such as seals) and aquatic birds (such as penguins) have a thick insulating layer of fat.

Some adaptations are physiological. In most land mammals and birds, smooth muscles in the skin involuntarily contract in the cold, causing the fur or feathers to raise. This adaptation traps a thicker layer of air next to the warm skin, improving insulation. In humans, muscles raise hair in the cold, causing goose bumps, a vestige from our furry ancestors. In cold weather, hormonal changes tend to boost the metabolic rate of some birds and mammals, increasing their heat production. Shivering produces heat as a metabolic by-product of the contraction of skeletal muscles. Honey bees, for example, survive cold winters by clustering together and shivering

in their hive, and their metabolic activity generates enough heat to keep the colony alive. Panting and sweating are examples of physiological adaptations that greatly increase cooling. Humans produce a watery sweat that evaporates from the skin and cools us. (Covering up the skin with football padding, as discussed in the Biology and Society section, restricts evaporative cooling.)

A variety of behavioral responses can regulate body temperature. Some birds and butterflies migrate seasonally to more suitable climates. Other animals, such as desert lizards, bask in the sun when it is cold and find cool, damp areas or burrows when it is hot. Emperor penguins huddle together to stay warm. Many animals cool themselves by bathing, and a few, such as kangaroos, even spread saliva on parts of their bodies. Dung beetles are unique: They cool off by crawling onto moist balls of feces, which provide cooler surfaces than the surrounding hot sand. All of these mechanisms—anatomical, physiological, and behavioral—demonstrate how a single selection pressure (the need to thermoregulate) can drive the evolution of a multitude of adaptations in a wide variety of environments.

▼ Figure 21.20 **Methods of thermoregulation in animals.**

METHODS OF THERMOREGULATION		
Anatomical Adaptations (such as hair, fat, and feathers)	**Physiological Adaptations** (such as panting, shivering, and sweating)	**Behavioral Adaptations** (such as bathing, basking, hibernating, and migrating)
Seals have a thick layer of fat for insulation.	Sheep pant to cool themselves.	Elephants spray themselves with water to cool off.

Chapter Review

■ SUMMARY OF KEY CONCEPTS

The Structural Organization of Animals

HIERARCHICAL ORGANIZATION OF ANIMALS		
Level	**Description**	**Example**
Cell	The basic unit of all living organisms	Muscle cell
Tissue	A collection of similar cells performing a specific function	Cardiac muscle
Organ	Multiple tissues forming a structure that performs a specific function	Heart
Organ system	A team of organs that work together	Circulatory system
Organism	A living being, which depends on the coordination of all structural levels for homeostasis and survival	Person

Structure/Function: Anatomy and Physiology

At every level, the structure of a body part is correlated with the task it must perform. Anatomy is the study of the structure of organisms, whereas physiology is the study of the function of an organism's structures.

Tissues

Animals have four main kinds of tissue.

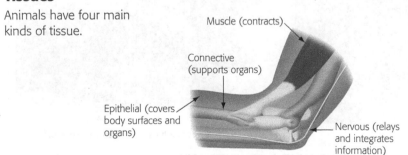

Muscle (contracts)

Connective (supports organs)

Epithelial (covers body surfaces and organs)

Nervous (relays and integrates information)

Interconnections within Systems: Organs and Organ Systems

An organ is a collection of tissues that together perform a specific function; an organ system is a group of organs that together provide a vital function for the organism.

Exchanges with the External Environment

All animals are open systems, exchanging chemicals and energy with the environment. Every cell of a simple organism can exchange materials through direct contact with the environment. Large and complex organisms require indirect exchange between extensively branched internal structures and the environment, usually via a circulatory system.

Regulating the Internal Environment

Homeostasis

Homeostasis is the body's tendency to maintain relatively constant internal conditions despite large fluctuations in the external environment.

Negative and Positive Feedback

In negative feedback, the most common homeostatic mechanism, the results of a process inhibit that process. Less common is positive feedback, in which the results of a process intensify that process.

Thermoregulation

In thermoregulation, homeostatic mechanisms regulate internal body temperature. Endotherms are warmed primarily by heat generated during metabolism. Ectotherms are warmed primarily by heat absorbed from the environment.

Osmoregulation

All organisms balance the gain or loss of water and dissolved solutes. Examples of active regulation include drinking and urinating.

Interconnections within Systems: Homeostasis in the Urinary System

The urinary system expels wastes and regulates solute and water balance. The diagram shown summarizes the structure and function of the nephron, the functional unit of the kidney.

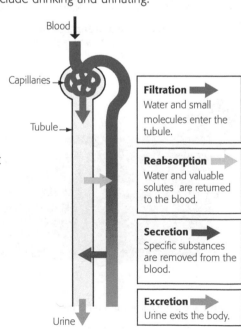

Blood

Capillaries

Tubule

Filtration
Water and small molecules enter the tubule.

Reabsorption
Water and valuable solutes are returned to the blood.

Secretion
Specific substances are removed from the blood.

Excretion
Urine exits the body.

Urine

MasteringBiology®

For practice quizzes, BioFlix animations, MP3 tutorials, video tutors, and more study tools designed for this textbook, go to MasteringBiology®

SELF-QUIZ

1. Which is likely to be a more common cancer in an adult: cancer of the epithelial tissue lining the inside of the throat or cancer of the smooth muscle of the intestines? Why?

2. How might an animal's body shape indicate whether it has a circulatory system?

3. Most homeostatic mechanisms depend on
   a. positive and negative feedback.
   b. negative feedback.
   c. predictable environmental conditions.
   d. predictable internal conditions.

4. The _____, respiratory, and _____ are the three organ systems that directly exchange materials with the external environment.

5. When foraging spiders are too hot, they retreat into their burrows. What type of feedback system does this indicate?
   a. hormonal feedback
   b. positive feedback
   c. negative feedback
   d. positive and negative feedback

6. While visiting a farm on a hot summer day, you see pigs rolling in mud. Based on what you know about thermoregulation, why are the pigs rolling in mud?

7. What is normally found in urine? What are three conditions that can be tested for with a urinalysis?

8. Which of the following is a function of the kidneys?
   a. elimination of urea
   b. reabsorption of valuable solutes
   c. reabsorption of water
   d. all of the above

9. What happens to most of the water that passes from the blood into the kidneys by filtration? What happens to the rest of it?

10. Explain the similarities between how your body temperature is maintained and how a thermostat regulates temperature in a home.

11. Polar bears, like many other hibernating animals, have large amounts of brown fat, a tissue specialized in generating heat. Is this an anatomical, physiological, or behavioral adaptation in response to cold? What about the polar bears' dramatic increase in food intake before the hibernation period?

Answers to these questions can be found in Appendix: Self-Quiz Answers.

THE PROCESS OF SCIENCE

12. **Interpreting Data** Using the data in the graph below, write a conclusion about the thermoregulation of river otters and largemouth bass.

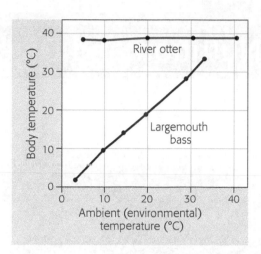

13. Several carrion-feeding species of desert ants, such as the Saharan silver ant (*Cataglyphis bombycina*), can tolerate temperatures up to 70°C. They leave their nests during the hottest period of the day (typically mid-day) to forage for food in a brief "explosive outburst" that usually lasts only a few minutes. What selective pressures may have led to these ants being able to withstand such high temperatures? What is their selective advantage?

BIOLOGY AND SOCIETY

14. On a hot day of summer, you have a training session with your football team. After half an hour of training, you and your teammates start to feel dizzy. Your trainer encourages you to rehydrate yourselves with his homemade iced tea rather than water. Do you think tea is an ideal means of rehydrating on a hot day?

15. In 2007, as part of an on-air radio contest, a woman entered a water-drinking contest to win her three children a gaming system. The goal was to drink more water than anyone else without urinating; she drank 6 liters of water in 3 hours. Afterwards, she vomited and began to suffer a terrible headache. She left the radio station ill and died at home a few hours later. During the contest, a person called the radio station to warn that the competition could be deadly. Pretend you are a journalist doing a story about the woman's death and that you are interviewing different experts, asking questions like the following: Did water kill the woman? Is the radio station at fault for what she did willingly? Write your questions and responses as a mock script or make a podcast.

22 Nutrition and Digestion

▲ Too much vitamin A can turn your skin orange, but too little can jeopardize your vision.

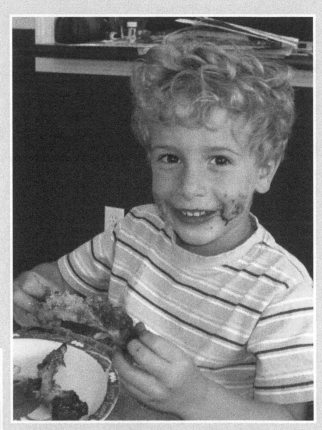

▲ The gallbladder helps digest fats, yet more than 500,000 Americans have theirs removed each year—with no ill effects.

A plate of red beans and rice provides the same essential amino acids as a steak. ►

Controlling Your Weight BIOLOGY AND SOCIETY

The "Secret" to Shedding Pounds

Americans are obsessed with weight. Each year, about one in seven Americans—roughly 50 million—starts a diet. Indeed, weight loss is big business: The leading weight loss programs boast millions of members, and Americans spend billions of dollars each year on diet-related products. But only about 5% of dieters are able to reach their goal weight and maintain it for the long term.

For many of us, the desire to lose weight is appropriate. More than a third of American adults are obese (very overweight), and obesity contributes to heart disease, diabetes, cancer, and other health problems—to the tune of 300,000 preventable deaths per year in the United States. Young adults are often particularly susceptible to sudden weight gain because the switch to independent living can lead to changes in diet and activity levels, often for the worse. And the problem is not limited to the United States: The United Nations World Health Organization recognizes obesity as a major global health problem. For most of us, therefore, any attempt to manage weight is effort well spent.

But with the wealth of fad diets with bogus claims, it can be difficult to determine the best way to shed pounds. Indeed, access to weight loss plans far outpaces our access to reliable data on their effectiveness. The good news is that there really is no trick to managing your weight: Add up the calories from the food you eat and then subtract the calories that your body burns. If you take in more by eating than you burn by activity, you will gain weight. If you burn more than you take in, you will lose weight. Weight control can be summed up in a five-word "secret": Eat less and exercise more! This is easier said than done, of course, but the principle is sound and simple, making weight loss within the reach of everyone's understanding.

Caloric balance alone does not ensure good nutrition. Food must also provide the raw materials for building healthy cells and tissues. You are what you eat: Your health and appearance depend on the quality of your diet and the proper functioning of your digestive system. This chapter focuses on essential concepts of digestion and nutrition, beginning with an overview of how animals process food. Next, we'll focus on the structure and function of the human digestive system. Along the way, we'll come back to the topic of weight loss as a reminder of its importance.

The key to maintaining a healthy weight. Proper intake of calories and adequate exercise are the best steps you can take to maintain a proper weight.

An Overview of Animal Nutrition

Every mealtime reminds us that, as animals, we must eat other organisms to acquire nutrients. Food provides the raw materials we need to build tissue and fuel cellular work. However, food primarily consists of large, complex molecules that are not in a form an animal's cells can use. Thus, the body must break down these nutrients—digest them—to make them useful.

Animal Diets

All animals eat other organisms, dead or alive, whole or by the piece. Beyond that generalization, however, animal diets vary extensively **(Figure 22.1)**. **Herbivores**, such as cattle, gorillas, and sea urchins, feed mainly on plants and/or algae. **Carnivores**, such as lions, snakes, domestic cats, and spiders, mainly eat other animals. **Omnivores**, such as crows, cockroaches, and people, regularly eat animals as well as plants and/or algae. ☑

☑ **CHECKPOINT**

A chicken eats corn and insects. That makes it a(n)
_____.

Answer: omnivore

The Four Stages of Food Processing

Mmmm, pizza. The next time you enjoy a slice, think of its fate as it passes through the four stages of food processing: ingestion, digestion, absorption, and elimination. **Ingestion** is just another word for eating. You ingest pizza when you bite off a piece. **Digestion** is the breakdown of food into molecules small enough for the body to absorb. The tomato sauce on a pizza, for example, is broken down to simple sugars and amino acids. **Absorption** is the uptake of small nutrient molecules by cells lining the digestive tract. For instance, amino acids made available by the breakdown of the cheese protein in pizza are absorbed and transferred to the bloodstream, which distributes them throughout the body. **Elimination** is the disposal of undigested materials left over from food.

▼ **Figure 22.1 Animal diets.** Animals can be classified into three broad groups according to their diet.

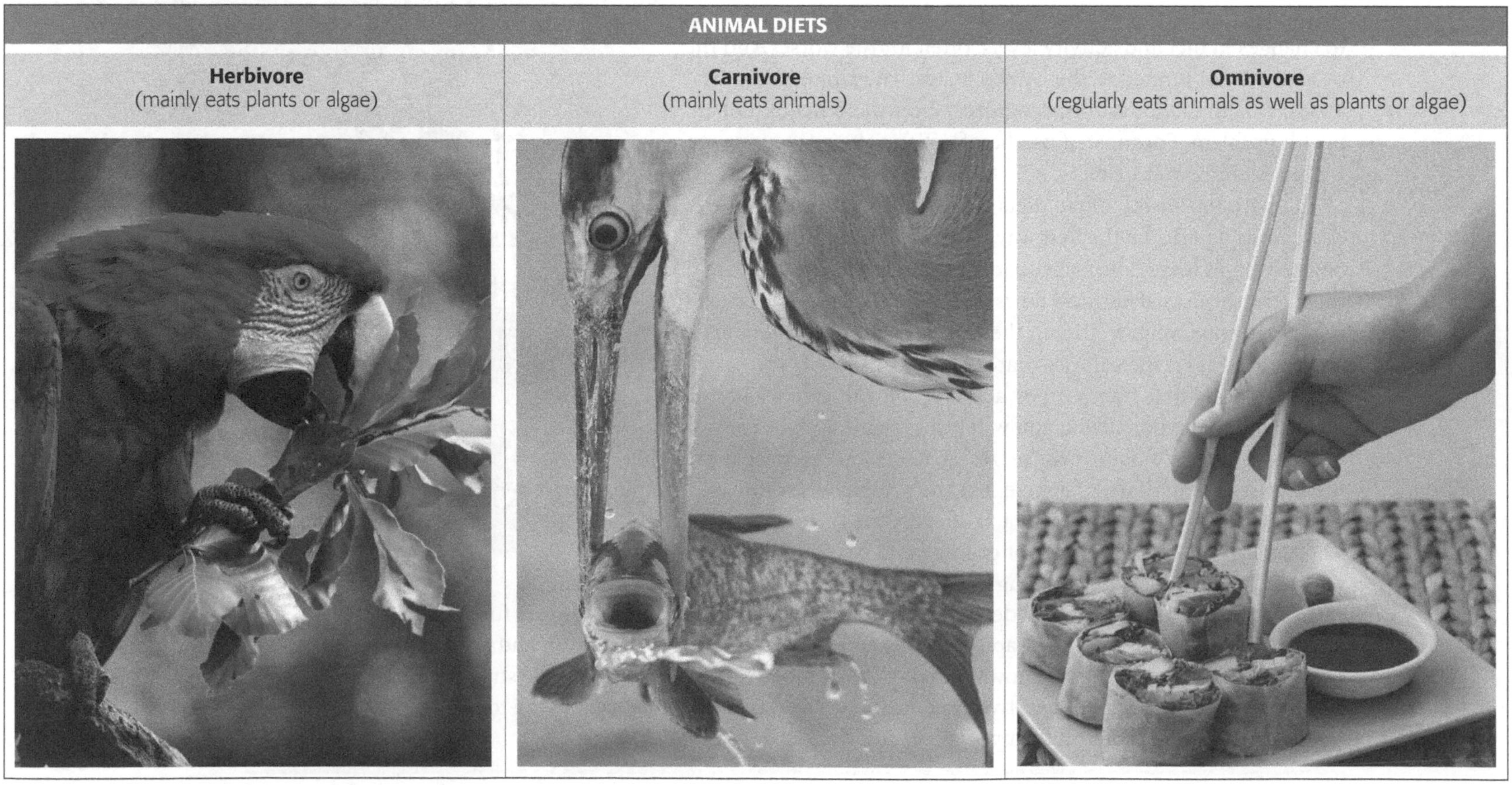

ANIMAL DIETS		
Herbivore (mainly eats plants or algae)	**Carnivore** (mainly eats animals)	**Omnivore** (regularly eats animals as well as plants or algae)

Digestion: A Closer Look

Digestion usually begins with mechanical processes such as chewing. Mechanical digestion breaks chunks of food into small pieces, exposing them to chemical digestion, the breakdown of food by digestive enzymes. Food molecules that are polymers, such as carbohydrates and proteins, are broken down via chemical digestion into monomers (see Chapter 3). For instance, starch is digested to the glucose monomers that make it up.

The dismantling of food molecules is necessary for two reasons. First, these molecules are too large to cross the membranes of animal cells; they must be broken down into molecules that are small enough for cells to absorb. Second, most food molecules—the proteins in cheese, for example—are different from the molecules that make up an animal's body. Your body does not directly use the protein that you eat but instead dismantles it and uses the pieces (amino acids) to build its own new proteins **(Figure 22.2)**. To help envision this process, think of plastic snap-together building blocks: Suppose you have a car made of blocks (food), but you need to build a boat. You must first break down the car (via digestion) into individual blocks (monomers). The individual blocks can then be reassembled into the boat (your own proteins).

Chemical digestion happens via hydrolysis, chemical reactions that break down large biological molecules by the addition of water molecules (see Figure 3.4b). Like most of life's chemical reactions, digestion requires enzymes **(Figure 22.3)**. For example, lipases are enzymes that digest fats, such as those found in pizza's cheese and meat. ☑

☑ CHECKPOINT

1. Place the four stages of food processing in their proper order: absorption, digestion, elimination, ingestion.
2. Food molecules are broken down into smaller molecules by enzymes that catalyze chemical reactions known as _____ reactions.

Answers: 1. ingestion, digestion, absorption, elimination 2. hydrolysis

▼ Figure 22.2 **From cheese protein to human protein.**

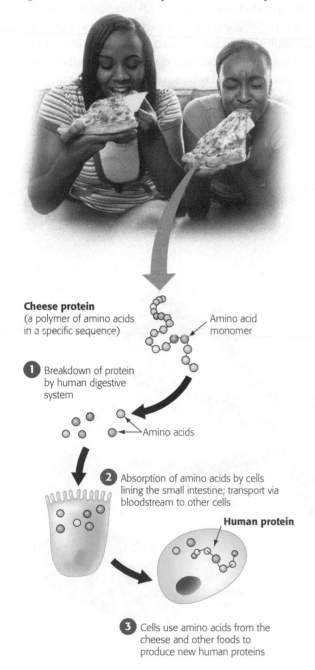

Cheese protein
(a polymer of amino acids in a specific sequence)

Amino acid monomer

1 Breakdown of protein by human digestive system

Amino acids

2 Absorption of amino acids by cells lining the small intestine; transport via bloodstream to other cells

Human protein

3 Cells use amino acids from the cheese and other foods to produce new human proteins

▼ Figure 22.3 **Chemical digestion: hydrolysis of food molecules.** Digestive enzymes catalyze the breakdown of food molecules by the addition of water molecules.

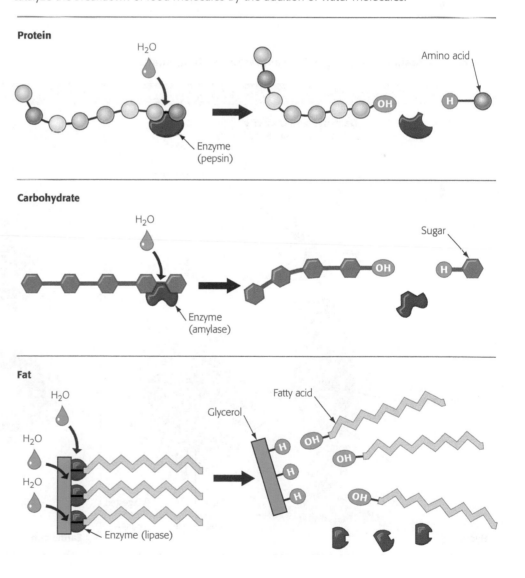

Protein

H_2O

Amino acid

Enzyme (pepsin)

Carbohydrate

H_2O

Sugar

Enzyme (amylase)

Fat

H_2O

H_2O

H_2O

Glycerol

Fatty acid

Enzyme (lipase)

Digestive Compartments

How can an animal digest its food without also digesting its own tissues? After all, digestive enzymes break down the same kinds of biological molecules that animals are made of, and it is obviously important to avoid digesting yourself! A common solution to this problem has evolved in animals: Chemical digestion proceeds safely within some kind of compartment. Such compartments can be within individual cells or they can be formed by cells as part of a digestive organ.

The simplest type of digestion occurs within a cellular organelle. In this process, a cell engulfs food by phagocytosis, forming a vacuole. This food vacuole then fuses with a lysosome containing enzymes, forming a digestive compartment (see Figure 4.14a). As food is digested, small food molecules pass through the vacuole membrane into the cytoplasm and nourish the cell. This type of digestion is common in protists, but sponges are the only animals that digest food solely within their cells.

Most animals have a digestive compartment where food is processed (Figure 22.4). Such compartments allow animals to digest pieces of food that are much larger than a single cell. Simpler animals, including cnidarians (such as hydras and jellies) and flatworms, have a **gastrovascular cavity**, a digestive compartment with a single opening that functions as both the entrance for food (like a mouth) and the exit for undigested wastes (like an anus).

The vast majority of animals, including earthworms and humans, have a digestive tube with two separate openings—a mouth at one end and an anus at the other end. Such a tube is called an **alimentary canal**, or **digestive tract**. Food moves in just one direction through specialized regions that digest and absorb nutrients in a stepwise fashion. This adaptation allows for much more efficient food processing. You know what an assembly line is; an alimentary canal is a *disassembly* line. Undigested wastes are eliminated from the alimentary canal as feces via the anus. ☑

CHECKPOINT

What is the main difference between gastrovascular cavities and alimentary canals?

Answer: Gastrovascular cavities have just one opening; alimentary canals have two (mouth and anus).

▼ Figure 22.4 **Main types of digestive compartments in animals.**

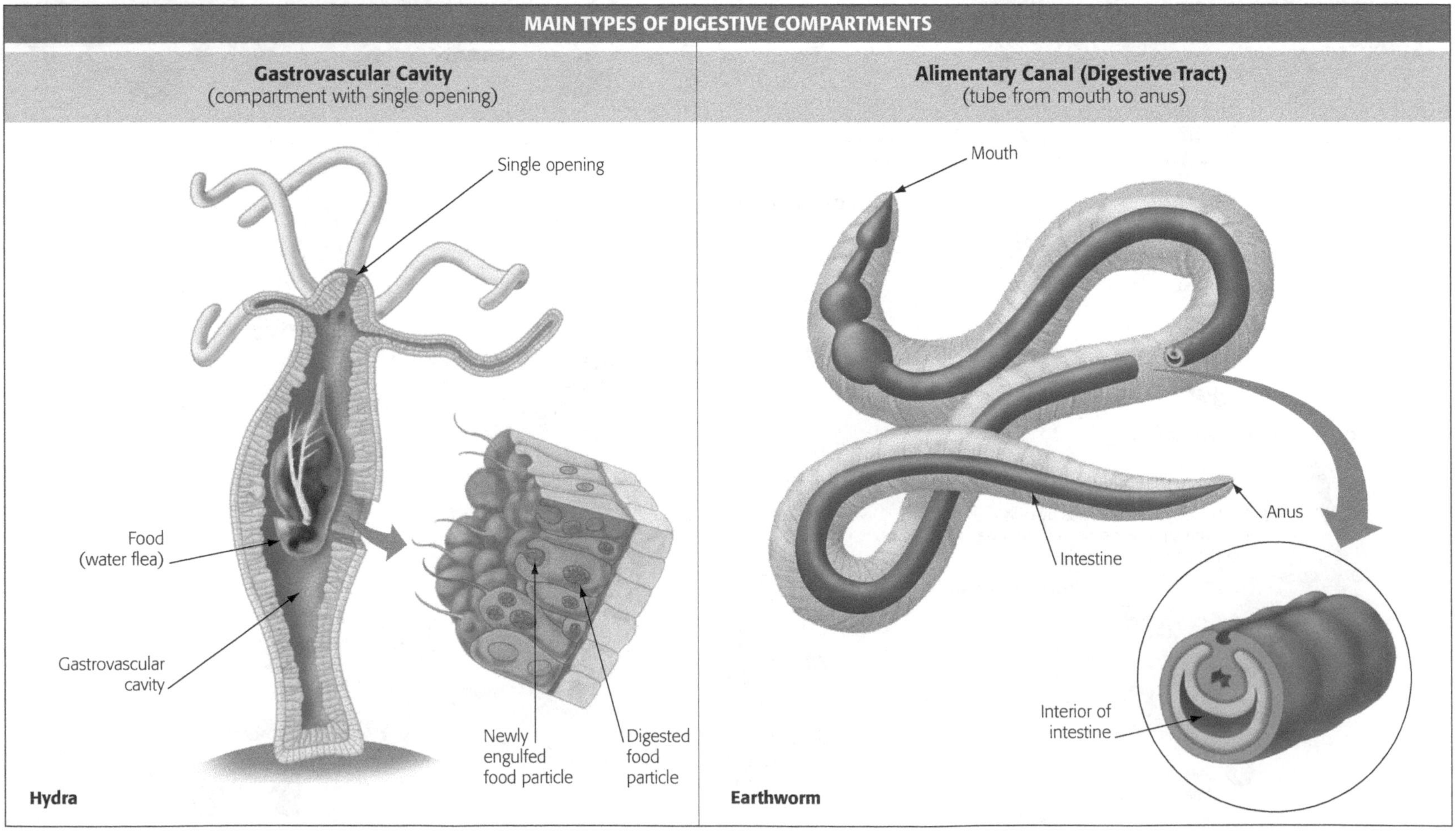

MAIN TYPES OF DIGESTIVE COMPARTMENTS	
Gastrovascular Cavity (compartment with single opening)	**Alimentary Canal (Digestive Tract)** (tube from mouth to anus)

Single opening

Food (water flea)

Gastrovascular cavity

Newly engulfed food particle

Digested food particle

Hydra

Mouth

Anus

Intestine

Interior of intestine

Earthworm

A Tour of the Human Digestive System

We are now ready to follow that slice of pizza through the human alimentary canal, from mouth to anus. It's important to have a good map of the human digestive system so we don't get lost in there.

System Map

The human digestive system (**Figure 22.5**) consists of an alimentary canal and several accessory organs (salivary glands, pancreas, liver, and gallbladder). The accessory organs secrete digestive chemicals into the alimentary canal via ducts (thin tubes). The human alimentary canal totals about 9 m (30 feet) in length. Such a long tube fits within a human body because it folds back and forth over itself with many switchbacks.

The alimentary canal is divided into specialized digestive organs along its length: mouth (oral cavity) → pharynx → esophagus → stomach → small intestine → large intestine (colon and rectum) → anus. You'll get a closer look at the structure and function of the digestive organs on our journey through the alimentary canal.

▶ Figure 22.5 **The human digestive system.** The human digestive system consists of the alimentary canal (black labels) and accessory organs (blue labels).

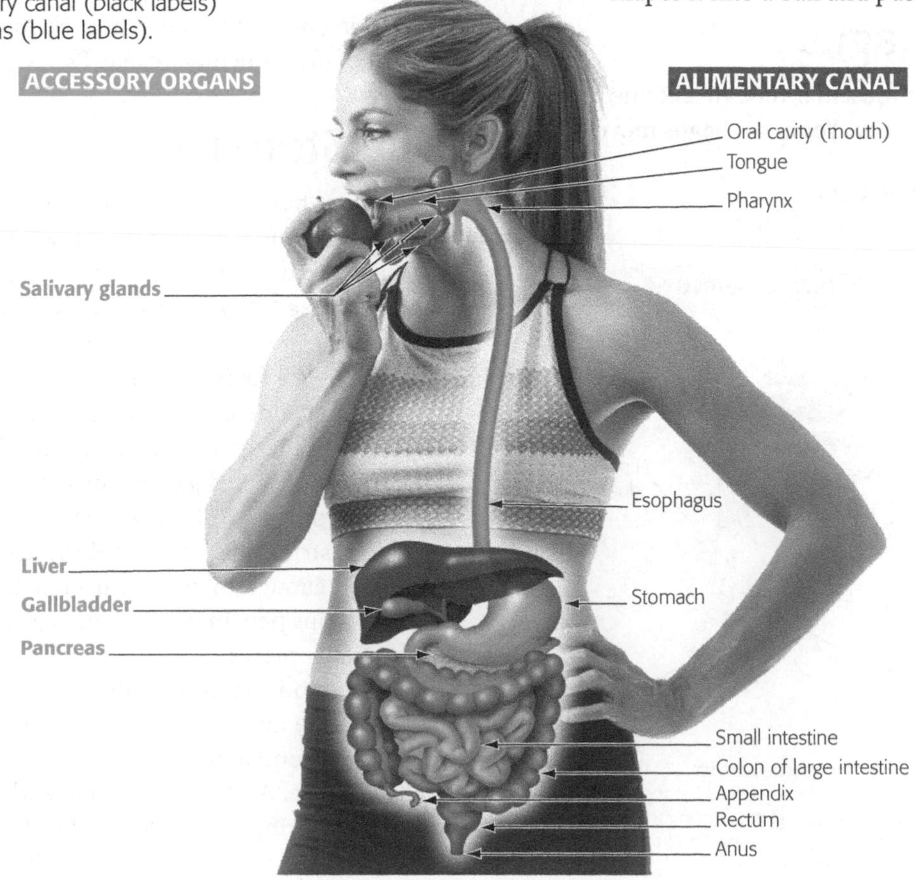

ACCESSORY ORGANS

ALIMENTARY CANAL

- Oral cavity (mouth)
- Tongue
- Pharynx

Salivary glands

Liver
Gallbladder
Pancreas

- Esophagus

- Stomach

- Small intestine
- Colon of large intestine
- Appendix
- Rectum
- Anus

The Mouth

The **mouth**, also known as the **oral cavity**, ingests food and begins to digest it (**Figure 22.6**). Mechanical digestion begins here as the teeth cut, smash, and grind the food. Chewing makes food easier to swallow and exposes more surfaces of the food to digestive juices. Typically, an adult human has 32 teeth. The bladelike incisors are in the middle and help bite off pieces of food. Just to their side are the pointed canines, used for ripping and tearing—think of the fangs of a dog or a wolf. Toward the back are two sets of premolars and three sets of molars, used for crushing and grinding. The third set of molars are called wisdom teeth; some people must have them removed if they put too much pressure on other teeth.

Chemical digestion also begins in the mouth with the secretion of saliva from **salivary glands**. Saliva contains the digestive enzyme salivary amylase. This enzyme breaks down starch, a major ingredient in pizza crust. The muscular **tongue** is also very busy during mealtime. Besides tasting the food, the tongue shapes it into a ball and pushes this food ball to the back of the mouth. Swallowing moves food into the pharynx. ☑

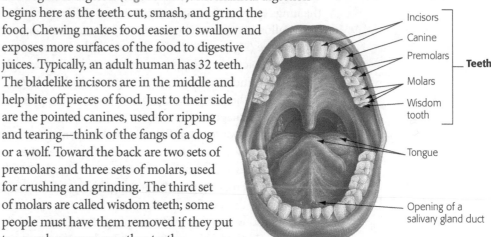

▼ Figure 22.6 **The human mouth and its teeth.**

- Incisors
- Canine
- Premolars — **Teeth**
- Molars
- Wisdom tooth
- Tongue
- Opening of a salivary gland duct

The Pharynx

The chamber called the **pharynx**, located in your throat, is an intersection of the pathways for swallowing and breathing. The pharynx connects the mouth to the esophagus (part of the digestive system). But the pharynx also opens to the trachea, or windpipe, which leads to the lungs (part of the respiratory system). When you're not swallowing, the trachea entrance is open and you can breathe (**Figure 22.7**, left). Air enters the larynx (also called the voice box), flows past the vocal chords, through the trachea, and to your lungs. Men generally have larger larynxes and therefore more prominent Adam's apples formed by cartilage on the outside of the larynx.

When you swallow, a reflex moves the opening of the trachea upward and tips a door-like flap called the epiglottis to close the trachea entrance (see Figure 22.7, right). Like a crossing guard at a dangerous intersection, the epiglottis directs the closing of the trachea, ensuring that the food will go down the esophagus. You can feel this action in the bobbing of your Adam's apple every time you swallow. Occasionally, food begins to "go down the wrong pipe," which irritates the lining of the trachea and triggers a strong coughing reflex that helps keep your airway clear of food. If an object becomes lodged in the trachea, choking can be fatal. The Heimlich maneuver uses air from the lungs to forcibly eject a stuck object.

CHECKPOINT

When you start coughing because food or drink "went down the wrong pipe," the material has entered the _____ instead of the _____.

Answer: trachea (windpipe); esophagus

The Esophagus

The **esophagus** is a muscular tube that connects the pharynx to the stomach. Your esophagus moves food

▼ Figure 22.7 **The epiglottis controls whether the pharynx is open to the lungs (left) or the stomach (right).**

BREATHING	SWALLOWING

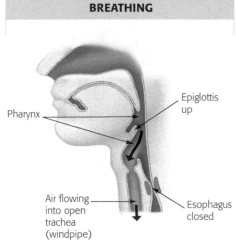

When you're not swallowing food or drink, air travels freely through the trachea (black arrows). The esophagus is closed.

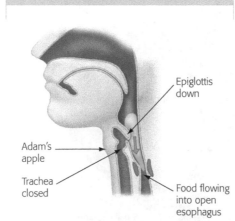

When the food reaches the back of the mouth, a swallowing reflex is triggered. The top of the trachea rises against the epiglottis, closing the air passage, while the esophagus opens. Food travels down the esophagus (green arrow).

▼ Figure 22.8 **The esophagus and peristalsis.** Muscles of the esophageal wall contract just behind the food ball and relax just ahead of it.

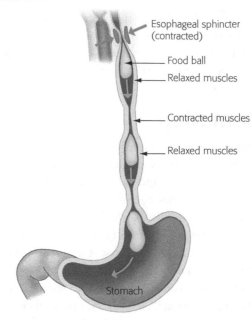

by **peristalsis**, alternating waves of muscular contraction and relaxation that squeeze the food ball along the esophagus (**Figure 22.8**). It is this action that allows a lapping animal to swallow, even when its head is lower than its stomach (like a cat at a water dish or a giraffe at a watering hole). Peristalsis propels food throughout the length of the alimentary canal. ☑

The Stomach

You do not have to eat constantly because the human **stomach** is a large organ that acts as an expandable storage tank, holding enough food to sustain you for several hours (**Figure 22.9**). Like a collapsible canvas water bag that can be stretched out as needed, the elastic wall and accordion-like folds of the stomach allow it to hold more than half a gallon of food and drink.

The cells lining the stomach's interior secrete a digestive fluid called **gastric juice**. Gastric juice is made up of a strong acid, digestive enzymes, and mucus. The acid in gastric juice is hydrochloric acid, and it is concentrated enough to dissolve iron nails. Gastric juice also contains **pepsin**, an enzyme that breaks proteins into smaller pieces.

When food passes from the esophagus into the stomach, the muscular stomach walls begin to churn, mixing the food and gastric juice into a thick soup called **chyme**. At the downstream end of the stomach, a sphincter (a ring of muscle) works like a drawstring to close the stomach off, holding the chyme there for about

▼ Figure 22.9 **The human stomach.**

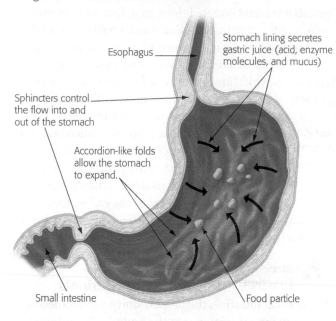

Esophagus

Stomach lining secretes gastric juice (acid, enzyme molecules, and mucus)

Sphincters control the flow into and out of the stomach

Accordion-like folds allow the stomach to expand.

Small intestine

Food particle

2–6 hours. The chyme leaves the stomach for the small intestine one squirt at a time. Continued contractions of stomach muscles after the stomach is empty causes the "stomach growling" that lets everyone know you are hungry.

With all of that acid, what keeps the stomach from digesting itself? Mucus coating the stomach lining helps protect it from gastric juices and from abrasive materials in food. Timing is also a factor: Nerve and hormone signals regulate the secretion of gastric juice so that it is discharged only when there is food in the stomach. Even with these safeguards, gastric juice can still erode the stomach lining, requiring the production of new cells by cell division. In fact, your stomach replaces its lining completely about once every three days!

Stomach Ailments

Gastric juice can be harmful. Occasional backflow of chyme causes pain in the esophagus, commonly but inaccurately called "heartburn" (it should really be called "esophageal burn"). Some people suffer this backflow frequently and severely enough to harm the lining of the esophagus, a condition known as acid reflux or GERD (gastroesophageal reflux disease). GERD can often be treated with lifestyle changes, such as eating small meals and avoiding lying down for a few hours after eating.

If the stomach lining is eroded by gastric juice faster than it can regenerate, painful open sores called gastric ulcers can form in the stomach wall. The cause of most ulcers is not stress, as was once thought, but infection of the stomach lining by an acid-tolerant bacterium called *Helicobacter pylori*. These bacteria damage the coat of mucus, making the lining more accessible to gastric juice. In severe ulcers, the erosion can produce a hole in the stomach wall and cause life-threatening internal bleeding and infection. (The cause of ulcers was established in 1984 when biologist Barry Marshall experimented on himself by drinking beef soup laced with *H. pylori* bacteria; although Marshall won a Nobel Prize for his work, we do not recommend this kind of experimentation!) Most stomach ulcers are treated with antibiotics. Affected people can also get relief by taking medications that contain bismuth, which helps reduce ulcer symptoms and may kill some bacteria.

Weight Loss Surgeries

The most common weight loss surgery in the United States (with about 150,000 operations each year) is gastric bypass. Staples are used to reduce the stomach to about the size of a chicken egg, and the first 18 inches of the small intestine are bypassed by attaching the downstream intestine directly to the reduced stomach pouch (Figure 22.10). As a result, patients quickly feel full, and the body's ability to absorb food is reduced. When accompanied by a healthy lifestyle, weight loss surgeries are successful in 90% of patients. However, all surgeries carry risks (there is about a 1% mortality rate), and patients must carefully monitor their diet to ensure proper nutrition. Weight loss surgeries can be effective, but most health professionals recommend them only as a last resort. ☑

☑ **CHECKPOINT**

Why are antibiotics an effective treatment for most gastric ulcers?

Answer: The antibiotics kill bacteria that damage the stomach lining.

▼ Figure 22.10 **Gastric bypass surgery.** An incision is made in the small intestine about 18 inches from where it joins the stomach. The free end of the intestine is then attached to the smaller stomach pouch, which now has a significantly reduced capacity to store food.

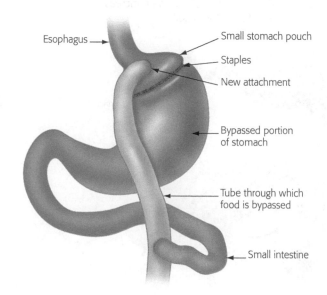

Esophagus

Small stomach pouch

Staples

New attachment

Bypassed portion of stomach

Tube through which food is bypassed

Small intestine

The Small Intestine

The **small intestine**, with a length of about 6 m (about 20 feet), is the longest part of the alimentary canal. Although this intestine isn't small in length, it is small in diameter—only 2.5 cm (about the width of a quarter) across—compared with the large intestine, which is 5 cm across. The small intestine is the major organ for chemical digestion and for absorption of nutrients into the bloodstream. It normally takes about 5–6 hours for food to pass through the small intestine.

The gallbladder helps digest fats, yet more than 500,000 Americans have theirs removed each year—with no ill effects.

Chemical Digestion in the Small Intestine

Do you remember that pizza we've been following through the digestive tract? By this point, mechanical digestion in the mouth and stomach has turned it into a thick, nutrient-rich soup. Chemical digestion by salivary amylase and gastric pepsin has initiated the breakdown of the pizza's starches and proteins. Now the small intestine takes over with an arsenal of enzymes that dismantle the food molecules into smaller molecules. These enzymes are mixed with chyme in the first 25 cm or so (about a foot) of the small intestine, the region called the **duodenum**.

The duodenum receives digestive juices from the pancreas, liver, gallbladder, and the intestinal lining (**Figure 22.11**). The **pancreas** is a large gland that secretes pancreatic juice into the duodenum via a duct. Pancreatic juice neutralizes the stomach acid that enters the duodenum, and it contains enzymes that aid in digestion. As peristalsis propels the mix along the small intestine, these enzymes contribute to the breakdown of food molecules.

Bile is a juice produced by the **liver**, stored in the **gallbladder**, and secreted through a duct into the duodenum. Bile contains salts that break up fats into small droplets that are more susceptible to dismantling by digestive enzymes. Bile contains dark pigments, which are responsible for the dark color of feces. Certain blood or liver disorders can cause bile pigments to accumulate in the skin, producing a yellowing called jaundice. Bile sometimes crystallizes to form gallstones, which can cause pain by obstructing the gallbladder or its ducts. Often the only cure is surgical removal of the gallbladder, which usually has no long-lasting effect on digestion because the liver still produces and secretes bile.

Structure/ Function **Absorption of Nutrients**

In the small intestine, the broken-down food molecules are ready for absorption into the body. Wait a minute! Aren't these nutrients already in the body? Not really. The alimentary canal is a tube running through the body, and its cavity is continuous with the great outdoors. The doughnut analogy shown in **Figure 22.12** should convince you that this is so. Until nutrients actually cross the tissue lining the alimentary canal and enter the bloodstream, they are still *outside* the body. If it were not for nutrient absorption, we could eat and digest huge meals but still starve.

▼ **Figure 22.12 Nutrients within the small intestine are not yet inside the body.** Imagine an elongated doughnut that's analogous to the "tube-within-a-tube" anatomy of the human alimentary canal. Food does not actually enter the body until it is absorbed by cells lining the alimentary canal.

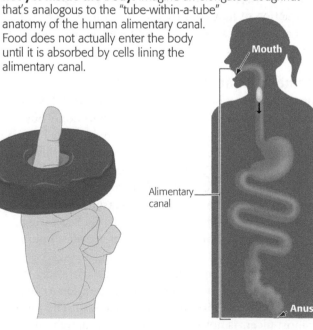

▼ **Figure 22.11 The duodenum.** Chyme squirted from the stomach into the duodenum is mixed with digestive juices from the pancreas, liver, gallbladder, and the lining of the duodenum itself.

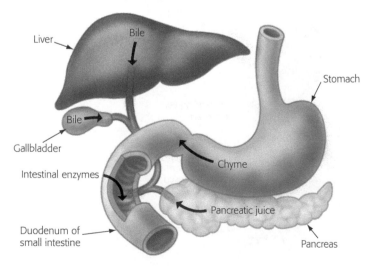

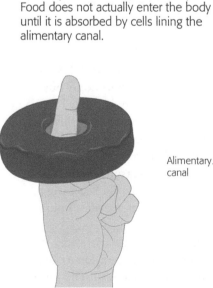

A finger through a hole   **Food through the alimentary canal**

Most digestion is complete by the time what's left of the pizza reaches the end of the duodenum. The next several meters of the small intestine are specialized for nutrient absorption. The structure of the intestinal lining, or epithelium, fits this function **(Figure 22.13)**. The surface area of this epithelium is huge—roughly 300 m², the size of a tennis court. The intestinal lining has large folds, as well as finger-like outgrowths called villi (singular, *villus*), making the epithelium something like the absorptive surface of a fluffy bath towel. Each cell of the epithelium also has microscopic projections called microvilli, which add even more surface area. The structure of the epithelium, with its expansive surface area, is an evolutionary adaptation that correlates with the function of this part of the alimentary canal: the absorption of nutrients. Absorbed small molecules pass from the digestive tract into the network of small blood vessels and lymphatic vessels in the core of each villus.

Normally, the small intestine is free of microorganisms. But improper sanitation can lead to infection by various bacteria. The disease cholera, for example, occurs when the bacterium *Vibrio cholera* multiplies in the small intestine and releases a toxin that leads to profuse watery vomiting and diarrhea, which in turn can lead to dehydration and eventually death. Cholera is primarily found in developing nations or after a natural disaster, where inadequate sanitation can lead to outbreaks.

After nutrients have crossed the cell membranes of the microvilli, they are finally *inside* the body, where the bloodstream and lymph carry them away to distant cells. But our tour of the digestive system is not over yet, because we still have to visit the large intestine. ☑

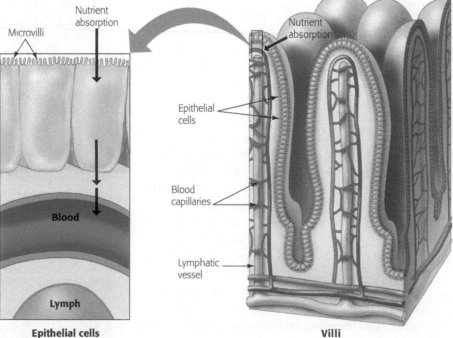

▶ Figure 22.13 **The small intestine and nutrient absorption.** Folds with projections (villi) that have even smaller projections (microvilli) give the small intestine an enormous surface area for nutrient absorption (black arrows). Most nutrients are transported across the microvilli into blood capillaries.

☑ CHECKPOINT

The major sites for absorption of nutrients in the human body are the _____ of the _____.

Answer: *microvilli; small intestine*

The Large Intestine

At only 1.5 m (5 feet) in length, the **large intestine** is shorter than the small intestine but almost twice as wide. Where the two organs join, a sphincter controls the passage of what's left of a meal **(Figure 22.14)**. Nearby is a small finger-like extension called the **appendix**. The appendix contains white blood cells that make minor contributions to the immune system. If the junction between the appendix and the large intestine becomes blocked, appendicitis—a bacterial infection of the appendix—may result. Emergency surgery is usually required to remove the appendix and prevent the spread of infection.

The main portion of the large intestine is the **colon**. The primary function of the colon is to absorb water from the alimentary canal. Every day about 7 L (not quite 2 gallons) of digestive juice spill into your digestive tract. About 90% of the water it contains is absorbed back into your blood and tissue fluids, with the small intestine reclaiming much of the water and the colon finishing the job. As water is absorbed, undigested materials from the meal become more solid as they are conveyed along the colon by peristalsis. It typically takes 12–24 hours for undigested material to move through the large intestine. The end product is **feces**, consisting of undigested material (such as cellulose from peppers on a pizza). About one-third of the dry weight of feces consists of bacteria from the colon. Most colon bacteria are harmless, while some, such as *Escherichia coli*, produce B vitamins and vitamin K that are absorbed through your colon wall and help supplement your diet.

If the lining of the colon is irritated by a viral or bacterial infection (sometimes mistakenly called a "stomach bug"), the colon may be unable to reabsorb water efficiently, resulting in diarrhea. Prolonged diarrhea can cause life-threatening dehydration, particularly among the very young and very old. The opposite problem, constipation, occurs when peristalsis moves feces along too slowly and the colon reabsorbs so much water that the feces become too compacted. Constipation can

result from lack of exercise or from a diet that does not include enough plant fiber.

Several intestinal disorders are characterized by inflammation (painful swelling) of the intestinal wall. Celiac disease results when gluten, a protein found in wheat, triggers an immune reaction that leads to swelling and a lack of nutrient absorption. The only treatment is a lifelong gluten-free diet. An inappropriate immune response also causes Crohn's disease, a chronic inflammation that can periodically flare up along any part of the alimentary canal. Crohn's disease is usually treated with drugs that dampen the immune response.

The **rectum**, the last 15 cm (6 inches) of the large intestine, stores feces until they can be eliminated. Contractions of the colon and the rectum create the urge to defecate. Two rectal sphincters, one voluntary and the other involuntary, regulate the opening of the **anus**. When the voluntary sphincter is relaxed, contractions of the rectum expel feces.

From entrance to exit, we have now followed a pizza slice all the way through the alimentary canal. **Figure 22.15** stretches out the tube to help you review food processing along its length. ☑

▼ **Figure 22.14 The large intestine and its connection to the small intestine.**

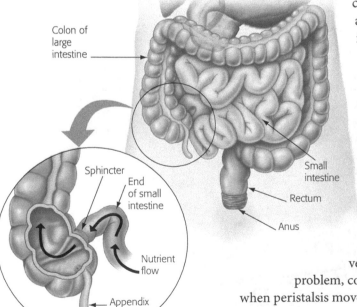

Colon of large intestine

Sphincter

End of small intestine

Nutrient flow

Appendix

Small intestine

Rectum

Anus

▼ **Figure 22.15 Review of food processing in the human alimentary canal.**

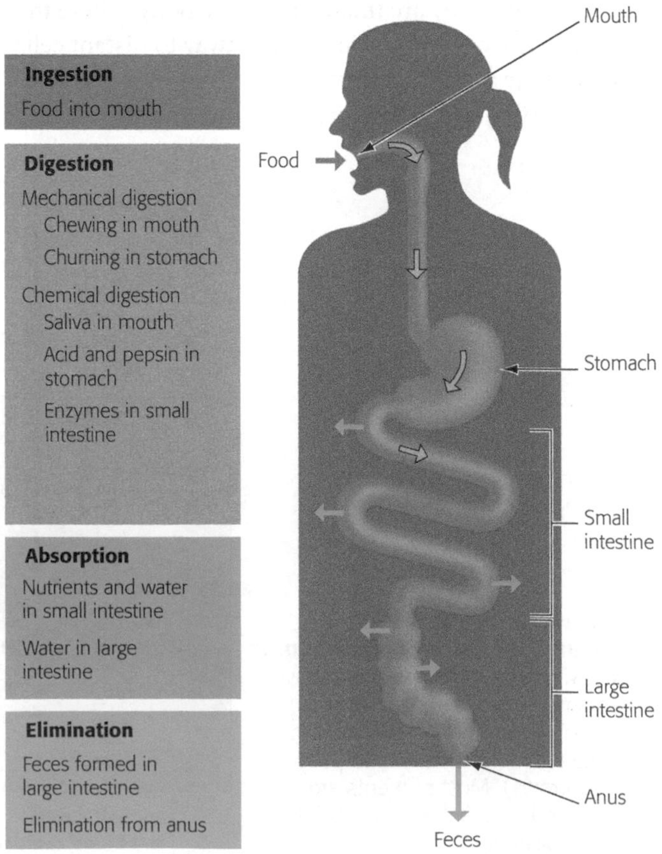

Ingestion	Mouth
Food into mouth	Food
Digestion	
Mechanical digestion	
Chewing in mouth	
Churning in stomach	Stomach
Chemical digestion	
Saliva in mouth	
Acid and pepsin in stomach	
Enzymes in small intestine	Small intestine
Absorption	
Nutrients and water in small intestine	
Water in large intestine	Large intestine
Elimination	
Feces formed in large intestine	
Elimination from anus	Anus
	Feces

Human Nutritional Requirements

For any animal, proper nutrition provides fuel for cellular work, materials for building molecules, and essential nutrients for health. A healthy human diet is rich in whole grains, vegetables, fruits, and calcium, along with moderate quantities of protein from lean meat, eggs, nuts, or beans. Nutritionists recommend limited consumption of fats, sugars, and salt.

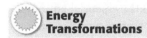 **Energy Transformations Food as Fuel**

One of biology's overarching themes is the transformation of energy and matter. All living cells take in energy, convert it to useful forms, and expel energy. Such transformations are at the heart of your body's use of food as fuel. Cells can extract energy stored in the organic molecules of food through the process of cellular respiration (described in detail in Chapter 6) and expend that energy as cellular work. Using oxygen, cellular respiration breaks down sugar and other food molecules. This process generates many molecules of ATP for cells to use as a direct source of energy and releases carbon dioxide and water as waste "exhaust" (**Figure 22.16**).

Calories

Calories are a measure of the energy stored in your food as well as the energy you expend during daily activities. One **calorie** is the amount of energy required to raise the temperature of a gram of water by 1°C. But such a tiny amount of energy is not very useful from a human perspective. We can scale up with the **kilocalorie** (1 **kcal** = 1,000 calories). Now, for a wrinkle in these definitions: The "Calories" (with an uppercase C) listed on food labels are actually kilocalories. Thus the 280 or so Calories in a slice of thick-crust pepperoni pizza are actually 280 kcal. That's a whole lot of fuel for making

ATP. However, about 60% of our food energy is lost as heat that dissipates to the environment, which is why a crowded room warms up so quickly.

Metabolic Rate

How fast do you "burn" your food? The rate at which your body consumes energy is called your **metabolic rate**. Your overall metabolic rate is equal to your **basal metabolic rate (BMR)**, the amount of energy it takes just to maintain your basic body functions, plus any energy consumption above that base rate. BMRs for people average about 1,300–1,500 kcal per day for adult females and 1,600–1,800 kcal per day for adult males. This is roughly equivalent to the amount of energy used by a ceiling fan. The more active you are, the greater your actual metabolic rate and the more calories your body uses per day. Metabolic rate also depends on other factors, such as body size, age, stress level, and heredity.

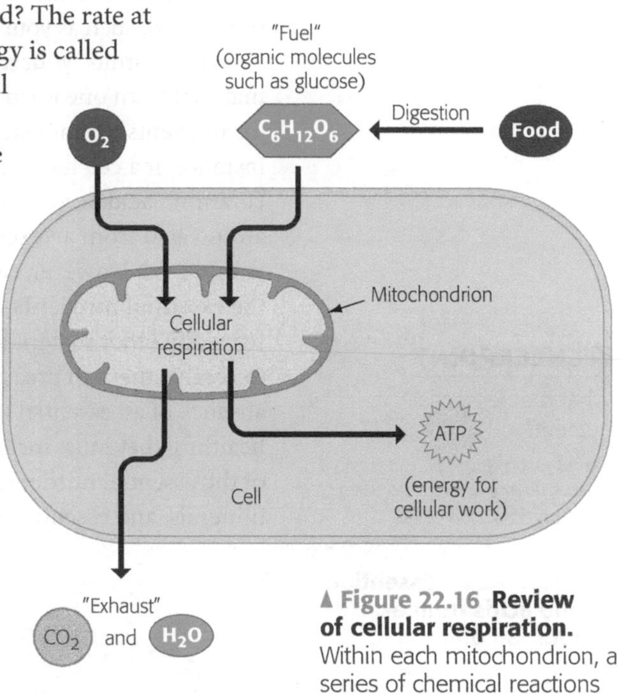

The examples in **Table 22.1** give you an idea of the amount of activity that it takes to use up the kilocalories in several common foods. As discussed in the Biology and Society section, balancing the calories you take in (via your diet) and the calories you expend (via exercise) can help you manage your weight. Caloric balance, however, does not ensure good nutrition, since food is our source of substance as well as our source of energy. ☑

"Fuel" (organic molecules such as glucose)

O_2 $C_6H_{12}O_6$ Digestion   Food

Mitochondrion

Cellular respiration

ATP

Cell   (energy for cellular work)

"Exhaust"

CO_2 and H_2O

▲ **Figure 22.16 Review of cellular respiration.** Within each mitochondrion, a series of chemical reactions extracts energy from fuel molecules and generates ATP, a chemical that provides energy for most cellular work.

Table 22.1	Exercise Required to Burn the Calories (kcal) in Common Foods		
	Jogging (9 min/mi)	**Playing Soccer**	**Walking (20 min/mi)**
kcal "burned" per hour*	775	477	245
Ice cream (1 cup, premium style), 500 kcal	39 min	1 hr, 3 min	2 hr, 2 min
Cheeseburger (quarter-pound), 417 kcal	32 min	52 min	1 hr, 42 min
Pepperoni pizza (1 large slice), 280 kcal	22 min	35 min	1 hr, 8 min
Soft drink (12 oz), 152 kcal	12 min	19 min	37 min
Apple (medium), 100 kcal	8 min	13 min	24 min
Whole wheat bread (1 slice), 65 kcal	5 min	8 min	16 min

*These data are estimated for a person weighing 68 kg (150 pounds).

☑ CHECKPOINT

1. What cellular process ultimately uses the energy stored in food molecules?
2. If you eat 1,000 Calories in a day, would you be likely to gain weight?

Answers: 1. cellular respiration 2. No. Chances are your basal metabolic rate is higher than that, so you would lose a bit of weight.

Food as Building Material

Even if you have stopped growing, your health depends on the continuous repair and maintenance of your tissues. The building materials required for such work are provided by the small organic molecules (monomers) produced during the digestion of food. Your cells can reassemble those smaller molecules into various large biological molecules (polymers), such as your own unique proteins and DNA.

Within limits, your metabolism can change organic material from one form to another and compensate for nutrients that are lacking in your diet. For instance, if a cell has a shortage of a particular amino acid, it may be able to make that amino acid from an excess supply of another amino acid. However, certain substances—the **essential nutrients**—cannot be made from any other materials, so the body needs to receive them in preassembled form. The absence of an essential nutrient makes you ill. A healthful diet must include adequate amounts of all of the essential nutrients: essential amino acids, vitamins, minerals, and essential fatty acids. ✓

A plate of red beans and rice provides the same essential amino acids as a steak.

Essential Amino Acids

All proteins are built from 20 different kinds of amino acids (see Chapter 3). The adult body can manufacture 12 of those amino acids from other compounds. The other 8 are **essential amino acids**: They must be obtained from the diet because human cells cannot make them. (Infants also require a ninth, histidine.)

Different foods contain different proportions of amino acids. Animal proteins, such as those in meat, eggs, and milk, are said to be "complete" because they provide adequate amounts of all the essential amino acids. In contrast, most plant proteins are incomplete, meaning they are deficient in one or more of the essential amino acids. If you are a vegetarian (by choice or, as for much of the world's population, by economic necessity), the key to good nutrition is to eat a variety of plants that together provide all of the essential amino acids. The combination of a grain and a legume (such as beans, peanuts, or peas) often provides the right balance **(Figure 22.17)**. Most societies have a staple meal that includes such a combination.

☑ **CHECKPOINT**

What is an essential nutrient?

Answer: a substance that an animal requires in its diet but cannot make

► **Figure 22.17 Essential amino acids from a vegetarian diet.**

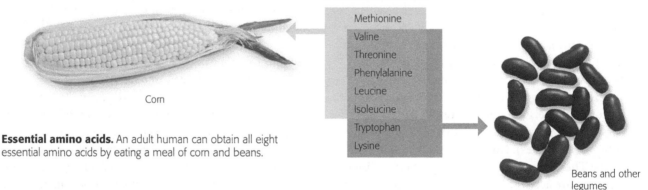

Essential amino acids

Essential amino acids
Methionine
Valine
Threonine
Phenylalanine
Leucine
Isoleucine
Tryptophan
Lysine

Corn

Beans and other legumes

Essential amino acids. An adult human can obtain all eight essential amino acids by eating a meal of corn and beans.

Rice and beans

Bread and peanut butter

Rice and tofu

Complete meals. Many societies have a staple meal that provides a complete set of essential amino acids by combining a grain with a legume.

Vitamins

Organic molecules that are required in the diet in very small amounts are called **vitamins**. For example, humans require vitamin B_{12}, but one tablespoon is enough to provide the daily requirement of nearly a million people. There are 13 vitamins essential to human health (Table 22.2). Most are needed because they assist enzymes.

Too much vitamin A can turn your skin orange, but too little can jeopardize your vision.

Deficiencies in any essential vitamin can cause serious health problems.

People who eat a balanced diet should be able to obtain enough of all needed nutrients in their food. For others, vitamin supplements can fill in the gaps, but supplements should not be used indiscriminately because overdoses of some vitamins (such as A, D, and K) can be harmful. ☑

☑ **CHECKPOINT**

Why do vegetarians have to be particularly careful to get proper amounts of the essential amino acids?

Answer: because few plant products contain all of them

Table 22.2	Vitamins	
Vitamin	**Major Dietary Sources**	**Interesting Facts**
Water-Soluble Vitamins		
Vitamin B_1 (thiamine)	Pork, legumes, peanuts, whole grains	Too little thiamin leads to the disease beriberi; refined grains (like polished white rice) lack thiamin.
Vitamin B_2 (riboflavin)	Dairy products, organ meats, enriched grains, vegetables	Deficiency causes photophobia (aversion to light).
Niacin	Nuts, meats, fish, grains	High doses reduce cholesterol but can cause hot flashes. Too little leads to the disease pellagra, which can be deadly.
Vitamin B_6 (pyridoxine)	Common in most foods: meats, vegetables, whole grains, milk, legumes	Deficiency is rare.
Pantothenic acid	Most foods: meats, fish, dairy, whole grains	Component of coenzyme A. Deficiency is rare but can lead to fatigue.
Folic acid (folate)	Green leafy vegetables, oranges, nuts, legumes, whole grains, fortified foods	Recommended as a supplement for women of childbearing age because it cuts in half the risk of some birth defects involving development of the spinal cord.
Vitamin B_{12}	Animal products: meats, eggs, dairy	It's hard for vegans to get enough B_{12} in their diet; some intestinal disorders (such as Crohn's disease or celiac disease) may cause deficiencies.
Biotin	Legumes, most vegetables, meats, milk, liver, egg yolks	For most people, adequate levels are provided by intestinal bacteria.
Vitamin C (ascorbic acid)	Raw fruits and vegetables, especially citrus fruits, broccoli, cabbage, tomatoes, green peppers, strawberries	Many animals can produce their own vitamin C, but humans cannot. Deficiency causes the disease scurvy, which was a significant health problem during the era of lengthy sea voyages.
Fat-Soluble Vitamins		
Vitamin A	Dark green and orange vegetables and fruits, dairy products	Too little in the diet can lead to vision loss. Too much can cause yellow/orange skin and liver damage.
Vitamin D	Fortified dairy products, egg yolk	Made in human skin in presence of sunlight; aids calcium absorption and bone formation. Too little causes rickets (bone deformities).
Vitamin E (tocopherol)	Green leafy vegetables, oils, nuts, seeds, wheat germ	Antioxidant that helps prevent damage to cell membranes. Deficiency is rare.
Vitamin K	Green vegetables, vegetable oils	Needed for blood clotting. Produced by intestinal bacteria, so newborns and people taking long-term antibiotics may be deficient.

Minerals

The organic molecules in our diet, such as carbohydrates, fats, and proteins, provide the four chemical elements most abundant in our body: carbon, oxygen, hydrogen, and nitrogen. We also require smaller amounts of 21 other chemical elements (see Figure 2.2) that are acquired mainly in the form of inorganic nutrients called **minerals**. Like vitamin deficiencies, mineral deficiencies can cause health problems. For example, calcium is needed as a building material for bones and teeth and for the proper functioning of nerves and muscles. It can be obtained from dairy products, legumes, and dark green vegetables such as spinach. Too little calcium can result in osteoporosis, a degenerative bone disease.

Mineral excesses can also cause problems. For example, we require sodium for our nerves and muscles to function, but the average American consumes about 20 times the required amount of sodium, mainly in the form of salt (sodium chloride) added to processed foods. Excess sodium can contribute to high blood pressure.

Essential Fatty Acids

Our cells make fats and other lipids by combining fatty acids with other molecules, such as glycerol (see Figure 3.11). We can make most of the required fatty acids, in turn, from simpler molecules; those we cannot make, called **essential fatty acids**, we must obtain in our diet. One essential fatty acid, linoleic acid (one of the omega-6 family of fatty acids), is especially important

because it is needed to make some of the phospholipids of cell membranes. Most diets furnish ample amounts of the essential fatty acids, so deficiencies are rare.

Decoding Food Labels

To help consumers assess the nutritional value of packaged foods, the U.S. Food and Drug Administration (FDA) requires two blocks of information on labels **(Figure 22.18)**. One lists the ingredients (by weight) from greatest amount to least. The other lists key nutrients, emphasizing the ones associated with disease and the ones associated with a healthy diet. You'll see a wide variety of data—including Calories, fat, cholesterol, carbohydrates, fiber, protein, vitamins, and minerals—expressed as amounts per serving and as percentages of a daily value. (Daily values are based on a 2,000-Calorie-per-day diet; these values are therefore "one-size-fits-all" numbers that should be used only as rough guidelines.) You should note the serving size and adjust the rest of the nutritional information to reflect your actual serving. For example, if the serving size is 1 cup of cereal but you're eating 2 cups, double all of the other values. If you pay attention, you'll find that serving sizes are often surprising; a single cookie or a snack-sized bag of chips may actually contain multiple servings and therefore many more calories than a quick glance would indicate. Reading food labels carefully can help you make informed choices about what you put into your body. ☑

☑ **CHECKPOINT**

1. What is the difference between vitamins and minerals?
2. How many slices of the bread in Figure 22.18 would an average person have to eat to obtain the daily value of fiber?

Answers: 1. Both are required in the diet for proper health, but vitamins are organic compounds, whereas minerals are elements. 2. nine slices

---

► Figure 22.18 **FDA-required food labels.**

Nutritional Disorders

Considering the central role that nutrients play in a healthy body, it is not surprising that dietary problems can have severe health consequences. In this section, we examine some common nutritional disorders.

Malnutrition

Living in an industrialized country where food is plentiful and most people can afford a decent diet, we may find it hard to relate to starvation. But 800 million people around the world—nearly three times the population of the United States—must cope with hunger. As difficult as it is to imagine, 11,000 children under the age of 5 starve to death *each day*.

The main type of nutritional deficiency is **malnutrition**, health problems caused by an improper or insufficient diet. Malnutrition may be caused by inadequate intake or medical problems such as metabolic or digestive abnormalities. Undernutrition, insufficient caloric intake, may occur when food supplies are disrupted by crises such as drought or war, or when poverty prevents people from obtaining sufficient food.

On a global scale, protein deficiency—insufficient intake of one or more essential amino acids—causes the most suffering. Protein deficiency is most common in less industrialized countries where there is a great gap between food supply and population size. The most reliable sources of essential amino acids are animal products, but these foods are expensive. People forced by economic necessity to get almost all their calories from a single plant staple, such as corn or potatoes, will suffer deficiencies of essential amino acids. Most victims of protein deficiency are children, who are likely to develop poorly both physically and mentally (**Figure 22.19**).

Malnutrition is not always associated with poverty or disorders; it can result from a steady diet of junk food, which offers little nutritional value. A person can therefore be both malnourished and overweight.

Eating Disorders

Millions of Americans, mostly female, are affected by malnutrition not because they lack access to food but because they suffer from an eating disorder. **Anorexia nervosa** is characterized by self-starvation due to an intense and irrational fear of gaining weight, even when the person is underweight. **Bulimia** is a behavioral pattern of binge eating followed by purging through induced vomiting, abuse of laxatives, or excessive exercise. Both disorders are characterized by an obsession with body weight and shape and can result in serious health problems.

The causes of anorexia and bulimia are unknown. Genetics, psychology, and brain chemistry all appear to play a role. Culture also seems to be a factor: Anorexia and bulimia occur almost exclusively in affluent industrialized countries, where food is plentiful but thinness is idealized. Treatment options include counseling and antidepressant medications. Some people suffering from anorexia and bulimia eventually develop healthy eating habits without treatment. But for many, dysfunctional nutrition becomes a long-term problem that can impair health and even lead to death. ☑

▼ **Figure 22.19 Kwashiorkor in a 3-year-old Kenyan boy.** Kwashiorkor, a form of malnutrition due to inadequate protein in the diet, causes fluid to enter the abdominal cavity, producing swelling of the belly.

☑ **CHECKPOINT**

What is the difference between anorexia and bulimia?

Answer: Anorexia is self-starvation; bulimia is a pattern of binge eating and purging.

Obesity

In the United States and many other industrialized countries, overnourishment is the nutritional disorder of greatest concern. **Obesity** is defined as having a too-high **body mass index (BMI)**, a ratio of weight to height. About one-third of all Americans are obese, and another one-third are overweight (a BMI that is between normal and obese). Obesity increases the risk of heart attack, diabetes, cancer, and several other diseases.

It is important to realize that not being as slim or well toned as a magazine model does not mean that you are obese. Researchers continue to debate how heavy we can be before we are considered unhealthy. And keep in mind that BMI is only an approximation that does not, for example, take into account body mass from muscle versus body mass from fat. Reflecting this uncertainty, BMI charts show a range of acceptable values (Figure 22.20). Further complicating matters, a tendency toward obesity is inherited to some extent, as we'll see next. ☑

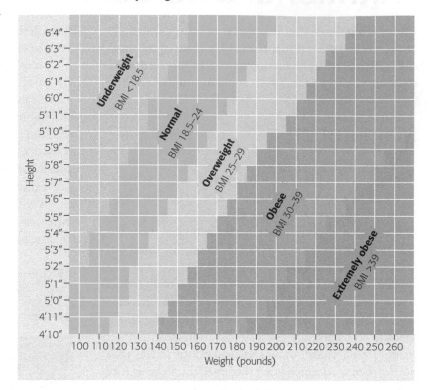

▼ Figure 22.20 **Body mass index (BMI): one measure of healthy weight.**

☑ **CHECKPOINT**

What is the BMI category of a person who is 6 feet 4 inches tall and weighs 230 pounds?

Answer: overweight

Controlling Your Weight THE PROCESS OF SCIENCE

Can a Gene Make You Fat?

Several decades ago, researchers made the **observation** that a mutation in a particular gene in mice leads to a significant increase in body fat (Figure 22.21). An obvious **question** about this so-called *obese* gene is: How does a small change in DNA cause such a large change in the body? Researchers formed the **hypothesis** that the mutant mice become overweight because their defective *obese* gene fails to produce a protein called leptin, which helps control body weight. They made the **prediction** that injecting leptin into mutant mice would overcome the effects of the defective *obese* gene.

▼ Figure 22.21 **A ravenous rodent.** The obese mouse on the left has a defect in a gene that normally produces an appetite-regulating protein.

Their **experiment** is summarized in **Figure 22.22**. The researchers injected some mutant and normal mice with leptin and injected others with saline (as a control). Their **results** were striking: Mutant mice that received leptin ended the study weighing about half as much as those that received saline (compare the top two bars in Figure 22.22). Normal mice receiving leptin lost weight compared with those receiving saline, but the difference was not as large (compare the bottom two bars). The researchers concluded that leptin helps regulate body weight in mice.

How do these results apply to people? Although leptin is present in human blood, the *obese* gene is only one of more than 100 genes that contribute to the complex weight maintenance system seen in humans. So far, this complexity has prevented formulation of effective drugs to treat obesity. More important, genetics only partially explains why certain people must fight hard to control their weight, whereas others can eat all they want without gaining a pound. Genetic defects can't produce obesity without a large intake of fattening foods. The best way to maintain a healthy weight is to eat a balanced diet and get plenty of exercise.

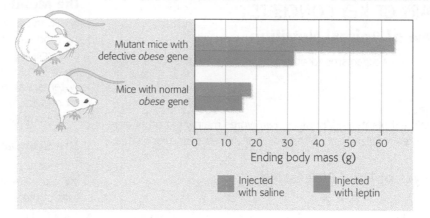

▲ Figure 22.22 **The effect of leptin on mice with the _obese_ gene.** This graph shows the final average weight of normal and mutant mice after receiving injections of saline and leptin.

Controlling Your Weight EVOLUTION CONNECTION

Fat and Sugar Cravings

The majority of Americans consume too many of the high-calorie foods that contribute to obesity. We all know that we need to limit the amount of fats and sugars in our diet, but it sure is hard, isn't it? Most of us crave fatty or sweet foods: cheeseburgers, chips, fries, ice cream, candy. For many of us, such foods are satisfying in a way that other foods are not.

Why do we hunger for unhealthy foods? The seemingly unhelpful trait of craving fat and sugar makes more sense from an evolutionary standpoint. In the natural world, overnourishment is usually an advantage. It is only within the last 100 years that large numbers of people have had access to a reliable supply of food. For most of human history, our ancestors were continually in danger of starvation. On the African savanna, fatty or sweet foods were probably hard to find. In such a feast-or-famine existence, natural selection may have favored individuals who gorged themselves on the rare occasions when rich, fatty foods were available. With their ample reserves, they were more likely than thinner peers to survive the inevitable famines.

Perhaps our modern taste for fats and sugars reflects the advantage it conveyed in our evolutionary history. Of course, today most people typically hunt and gather in grocery stores, restaurants, and cafeterias (**Figure 22.23**). Although we know it is unhealthful, many of us find it difficult to overcome the ancient survival behavior of stockpiling for the next famine.

◄ Figure 22.23 **Modern hunter-gatherers.**

Chapter Review

SUMMARY OF KEY CONCEPTS

An Overview of Animal Nutrition

Animals must eat other organisms to obtain organic molecules for energy and building materials.

Animal Diets

Herbivores mainly eat plants and/or algae, carnivores mainly eat other animals, and omnivores regularly eat animals as well as plants and/or algae.

The Four Stages of Food Processing

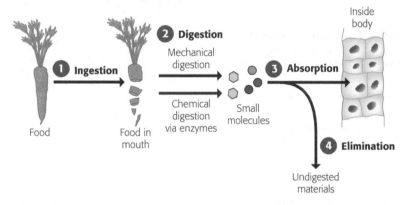

Digestive Compartments

In animals, the breakdown of food occurs within food vacuoles (membrane-enclosed compartments within cells), a gastrovascular cavity (a digestive compartment with a single opening), or an alimentary canal, also called a digestive tract (a tube with two separate openings: mouth and anus).

A Tour of the Human Digestive System

System Map

| Alimentary canal | Accessory organs | Digestion | | Absorption |
		Mechanical	Chemical	
Mouth (oral cavity)	Salivary glands	Chewing	Salivary amylase	
Pharynx and esophagus				
Stomach		Churning	Acid and pepsin (in gastric juice)	
Small intestine	Liver, gallbladder, pancreas		Other enzymes	Nutrients and water
Large intestine				Water
Anus				

The Mouth, the Pharynx, the Esophagus

Within the mouth (oral cavity), teeth function in mechanical digestion, breaking food into smaller pieces. Salivary amylase, an enzyme in saliva, initiates chemical digestion by starting the breakdown of starch into sugars. The pharynx within the throat chamber leads to both the digestive and respiratory systems. When you swallow, the epiglottis closes the trachea and opens the esophagus. The esophagus connects the pharynx to the stomach and moves food along via rhythmic muscular contractions called peristalsis.

The Stomach

The elastic stomach stores food, functions in mechanical digestion by churning, and uses gastric juice for chemical digestion. Mucus secreted by the cells of the stomach lining helps protect the stomach from self-digestion. Sphincters regulate the passage of food into and out of the stomach.

The Small Intestine

Small in diameter but very long, the small intestine is the main organ of chemical digestion and nutrient absorption. Most chemical digestion occurs in the duodenum, the first part of the small intestine, where chyme from the stomach mixes with pancreatic juice, bile, and a digestive juice secreted by the intestinal lining. These digestive juices include enzymes for dismantling large food molecules into small molecules. The rest of the small intestine is specialized for absorbing these small molecules. Across the large surface area of the intestinal lining, with its microvilli upon villi, intestinal cells transport nutrients into capillaries of the circulatory system.

The Large Intestine

The colon, wider and shorter than the small intestine, makes up most of the large intestine. The large intestine finishes reclaiming water that entered the alimentary canal in digestive juices. When undigested wastes reach the rectum, most water has been reabsorbed into the blood. Feces are stored in the rectum until eliminated via the anus.

Human Nutritional Requirements

Energy Transformations: Food as Fuel

Our metabolic rate depends on basal metabolic rate (energy expenditure at complete rest) plus energy burned by additional activity.

Food as Building Material

To repair and maintain tissues, we need building blocks from the breakdown of organic molecules. We also need four types of preassembled essential nutrients.

Essential Nutrients			
Essential Amino Acids	**Vitamins**	**Minerals**	**Essential Fatty Acids**
Required for protein production	Organic molecules required in very small amounts	Essential chemical elements from inorganic compounds	Required to make cell membranes

Decoding Food Labels

The U.S. Food and Drug Administration requires packaged foods to have labels that list ingredients in descending order of abundance and provide information about Calories and specific nutrients.

Nutritional Disorders

Malnutrition

Malnutrition results from inadequate intake of one or more essential nutrients. Protein deficiency is the most common dietary deficiency in less industrialized countries.

Eating Disorders

Anorexia nervosa is an eating disorder characterized by purposefully undereating because of an irrational fear of gaining weight. Bulimia is an eating disorder characterized by cycles of overeating and purging.

Obesity

Defined as an inappropriately high ratio of weight to height, obesity is the most common nutritional disorder in most industrialized countries. About one-third of all Americans are obese, putting them at higher risk for heart disease and diabetes.

MasteringBiology®

For practice quizzes, BioFlix animations, MP3 tutorials, video tutors, and more study tools designed for this textbook, go to MasteringBiology®

SELF-QUIZ

1. Imagine that a new species of large animal has just been discovered. Most animals have an alimentary canal, but this animal *appears* to rely entirely on food vacuoles. Why is this unlikely?

2. A friend says, "It's not ingesting that causes you to gain weight; it's absorption." Is that statement true?

3. _____ is the mechanical and chemical breakdown of food into small molecules, whereas _____ is the uptake of these small molecules by the body's cells.

4. Why don't astronauts in zero gravity have trouble swallowing?

5. Which class of large biological molecules is primarily broken down in the stomach? What enzyme is responsible?

6. What type of organism is responsible for most stomach ulcers? If the ulcers are detected early, how can they be treated?

7. The stomach's gastric glands secrete
   a. digestive enzymes only.
   b. hydrochloric acid only.
   c. mucus, digestive enzymes, and hydrochloric acid.
   d. digestive enzymes and hydrochloric acid.

8. Explain how treatment of a chronic infection with antibiotics for an extended period of time can cause a vitamin K deficiency.

9. Why is the amount of oxygen you consume proportional to your metabolic rate?

10. A friend takes mineral supplements. Which mineral is needed for the proper functioning of nerves, muscles, and bones?
    a. phosphorus   c. sodium
    b. calcium   d. salt

Answers to these questions can be found in Appendix: Self-Quiz Answers.

THE PROCESS OF SCIENCE

11. The FDA recommends that about 12% of our daily calories come from protein. How do we know if our diet is in accordance with those regulations? Nutritional labels typically tell how many Calories are from fat but don't specify how many Calories are from other nutrients. Proteins have approximately 4 Calories per gram. Read this cookie label and calculate the percentage of calories from protein.

Nutrition Facts
Serving Size 1 Cookie (28 g /1 oz)
Servings Per Container 8

Amount Per Serving

Calories 140   Calories from Fat 60

	% Daily Value*
Total Fat 7g	11%
Saturated Fat 3g	15%
Trans Fat 0g	
Cholesterol 10mg	3%
Sodium 80mg	3%
Total Carbohydrate 18g	6%
Dietary Fiber 1g	4%
Sugars 10g	
Protein 2g	

12. **Interpreting Data** The Process of Science section in this chapter describes an experiment involving mice with normal and mutant versions of a gene called *obese*. To determine the role of a specific gene, geneticists often work with two groups of research subjects that differ only in the gene in question (normal version of the gene versus a mutant version of the gene). In this case, researchers compared mice that had two normal copies of the *obese* gene (called "*ob+*") with mice that had two defective copies (called "*ob*"). The researchers found that mice with two copies of the mutant *obese* gene (*ob/ob*) gained significantly more weight than normal (*ob+/ob+*) mice. Next, researchers sought to determine whether this difference was due to the production of a hormone. To find out, researchers surgically linked the circulatory systems of mice, so that any factor circulating in the blood of one mouse would circulate in the blood of the other. The table below summarizes the results. How do these data support the hypothesis that the *obese* gene controls the production of a hormone?

Experiment	Genotypes of mice	Average weight gain per mouse (in grams)
(a)	*ob+/ob+* paired with *ob+/ob+*	8.3
(b)	*ob/ob* paired with *ob/ob*	38.7
(c)	*ob+/ob+* paired with *ob/ob*	8.2

BIOLOGY AND SOCIETY

13. A friend of yours is curious to know if he has a healthy body composition. Would you recommend calculating BMI to him? Why or why not?

14. Do we really need to supplement our diets with vitamins, minerals, and other additives? What could be the consequences of an indiscriminate use of dietary supplements?

23 Circulation and Respiration

A long trip in a confined seat is not just uncomfortable: It could wreak havoc with your circulatory system.

If hemoglobin in your blood could choose between binding to life-sustaining oxygen or poisonous carbon monoxide, it would pick the poison.

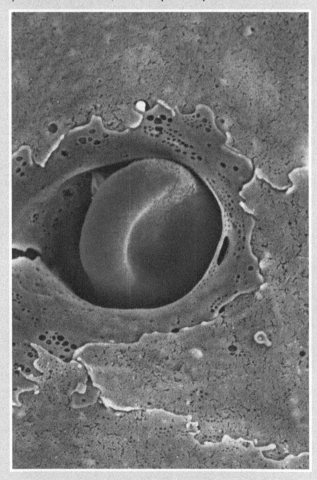

Half a million metal stents, not much wider than spaghetti, save lives every year.

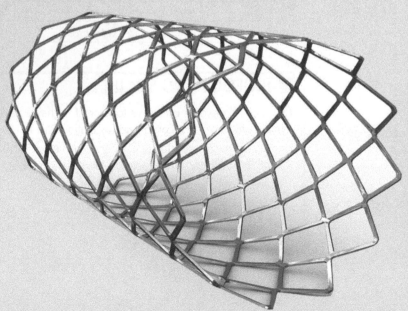

Athletic Endurance BIOLOGY AND SOCIETY

Avoiding "The Wall"

What does it take to be an elite endurance athlete? To be sure, it takes determination and dedication. But is that all? From a biological standpoint, what it really takes is a steady supply of oxygen to muscle cells over an extended period of time. World-class champions have world-class circulatory and respiratory systems.

To understand the endurance of an elite athlete, you need to know what goes on within muscle cells. To perform work over an extended period—over the course of a 26-mile run, for example—muscle cells require a steady supply of oxygen (O_2) and must continuously rid themselves of carbon dioxide (CO_2) waste. Blood, specifically red blood cells, delivers O_2 and removes CO_2. As long as there is enough O_2 reaching the muscle tissue and enough CO_2 being removed, athletes can perform at a steady pace. What limits the stamina of most athletes is the inability of their heart and lungs to deliver the required amount of O_2 to muscle cells. Without enough O_2 in their muscle cells, an athlete will hit "the wall"—that is, they will experience a sudden loss of energy to the extent that they are unable to continue to perform.

It isn't surprising that to avoid hitting the wall athletes put considerable effort into training their bodies to deliver more O_2 to their muscle cells over longer periods of time, a technique known as endurance training. Elite athletes who need an intensive endurance regimen often turn to high-altitude training. The lower atmospheric pressure at high altitude means that fewer molecules of O_2 are taken into the body with each breath. By training under these conditions, an athlete hopes that their body will acclimate by making more red blood cells. If it does, competing at normal altitude will be easier because the increased number of red blood cells will deliver more oxygen to muscles. Some athletes train or sleep in low-O_2 "tents" that simulate high altitudes.

To improve their endurance through means other than training, some athletes increase O_2 supply to their muscles by "blood doping," artificially boosting the number of red blood cells in their circulation. In blood doping, an athlete may inject red blood cells to increase the number of them in circulation or take drugs that temporarily increase their red blood cell count.

Properly functioning circulatory and respiratory systems are essential not just to athletes but to all of us. A failure of either system can result in death. The two systems are so closely interconnected that we will explore both in this chapter. We'll consider challenges in the structure and function of each system and survey the various ways these challenges are overcome. We'll then take a look at the circulatory and respiratory systems in humans, examine what happens when these systems fail, and consider medical options for maintaining the health of the two systems.

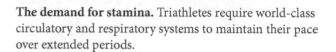

The demand for stamina. Triathletes require world-class circulatory and respiratory systems to maintain their pace over extended periods.

Unifying Concepts of Animal Circulation

Every organism must exchange materials and energy with its environment. In simple animals, such as hydras and jellyfish, nearly all the cells are in direct contact with the outside world. Thus, every body cell can easily exchange materials with the environment by diffusion, the spontaneous movement of molecules from an area of higher concentration to an area of lower concentration (see Figure 5.12). However, most animals are too large or too complex for exchange to take place by diffusion alone. In such animals, a **circulatory system** facilitates the exchange of materials, providing a rapid, long-distance internal transport system that brings resources close enough to cells for diffusion to occur. Once there, the resources that cells need, such as nutrients and O_2, enter the cytoplasm through the plasma membrane. And metabolic wastes, such as CO_2, diffuse from the cells to the circulatory system for disposal.

All but the simplest animals have a circulatory system with three main components: a central pump, a vascular system (a set of tubes), and a circulating fluid. Two main types of circulatory systems have evolved in animals (**Figure 23.1**). Many invertebrates, including most molluscs and all arthropods, have an **open circulatory system**. The system is called "open" because the circulating fluid is pumped through open-ended tubes and flows out among cells. Thus, in open circulatory systems, the circulating fluid is also the interstitial fluid that bathes all cells. Nearly all other animals have a **closed circulatory system**. The system is called "closed" because the circulating fluid, called blood, is pumped within a set of closed tubes and is distinct from the interstitial fluid.

The closed circulatory system of humans and other vertebrates is called a **cardiovascular system**, and it consists of the heart, blood, and blood vessels. Blood circulates to and from the heart through three types of vessels: arteries, capillaries, and veins. **Arteries** carry blood away from the heart, branching into smaller arterioles as they approach organs. The blood then flows from arterioles into networks of tiny vessels called **capillaries** that run through nearly every organ and tissue in the body. The thin walls of capillaries allow

▼ Figure 23.1 **The diversity of circulatory systems.** It is important to note that in double circulation systems, the pulmonary and systemic circuits work simultaneously.

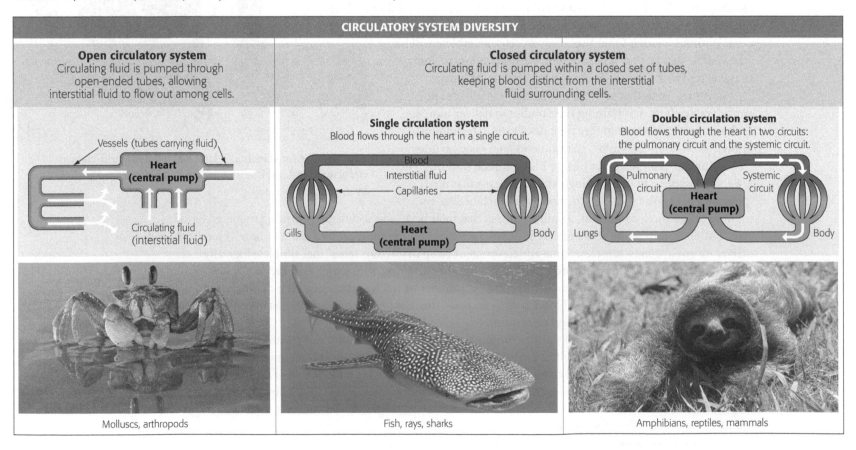

CIRCULATORY SYSTEM DIVERSITY

Open circulatory system
Circulating fluid is pumped through open-ended tubes, allowing interstitial fluid to flow out among cells.

Closed circulatory system
Circulating fluid is pumped within a closed set of tubes, keeping blood distinct from the interstitial fluid surrounding cells.

Vessels (tubes carrying fluid)
Heart (central pump)
Circulating fluid (interstitial fluid)

Single circulation system
Blood flows through the heart in a single circuit.

Blood
Interstitial fluid
Capillaries
Gills
Heart (central pump)
Body

Double circulation system
Blood flows through the heart in two circuits: the pulmonary circuit and the systemic circuit.

Pulmonary circuit
Systemic circuit
Heart (central pump)
Lungs
Body

Molluscs, arthropods

Fish, rays, sharks

Amphibians, reptiles, mammals

exchange between the blood and the interstitial fluid. Capillaries are the functional center of the circulatory system: This is where materials are transferred to and away from surrounding tissues. As blood flows from the capillaries, it enters venules, which in turn converge into larger **veins** that return blood back to the heart, completing the circuit.

As we see in Figure 23.1, two distinct cardiovascular, or closed circulatory, systems exist in vertebrates. In the **single circulation system** found in bony fishes, rays, and sharks, blood flows in one loop, or circuit. In the **double circulation system** found in amphibians, reptiles (including birds), and mammals, blood flows in two loops, the **pulmonary circuit** between the lungs and the heart and the **systemic circuit** between the heart and the rest of the body.

> A long trip in a confined seat is not just uncomfortable: It could wreak havoc with your circulatory system.

Let's examine the two circuits in humans. The pulmonary circuit carries blood between the heart and lungs (**Figure 23.2a**). In the lungs, CO_2 diffuses from the blood into the lungs, while O_2 diffuses from the lungs into the blood. The pulmonary circuit then returns this O_2-rich blood back to the heart. The systemic circuit carries blood between the heart and the rest of the body (**Figure 23.2b**). The blood supplies O_2 to body tissues, while it picks up CO_2. The oxygen-poor blood returns to the heart via the systemic circuit.

Obstruction of the cardiovascular system is dangerous and sometimes deadly. For instance, if a blood clot becomes lodged in a vessel of a lung, it can cause shortness of breath and lung tissue damage. If the clot is large enough, it may completely obstruct blood flow through the pulmonary circuit and cause sudden death as the heart and brain lose access to O_2-rich blood. Clots lodged in the lungs often originate from clots that form in the veins of the legs, a condition known as deep vein thrombosis (DVT). Risk factors for DVT include physical inactivity and dehydration, which sometimes occur when individuals are confined to small seats on long plane flights. To fully grasp how a blood clot formed in the leg can become lodged in the lungs, you need to understand more completely how blood flows through the body, a process we cover next. ☑

☑ **CHECKPOINT**

1. Why can't the human body rely solely on diffusion to provide all needed chemicals to body cells?
2. What kind of blood vessel carries blood away from the heart? What kind of vessel returns blood to the heart?

*Answers: **1.** Diffusion is only effective over short distances. Diffusion alone would take too long to convey materials throughout the large human body. **2.** arteries; veins.*

▼ **Figure 23.2 The pulmonary and systemic circuits in humans.**

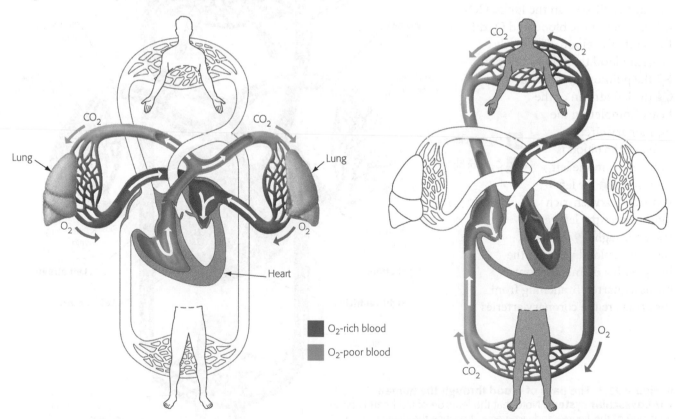

(a) Pulmonary circuit. In organisms with double circulation, the pulmonary circuit transports blood between the heart and lungs.

(b) Systemic circuit. The systemic circuit transports blood between the heart and the rest of the body.

O₂-rich blood

O₂-poor blood

The Human Cardiovascular System

We'll discuss in detail each part of the human cardiovascular system—the heart, the blood vessels, and circulating blood. But first, let's consider the property that emerges from these parts: the ability to transport nutrients to and carry waste away from every cell of the body.

The Path of Blood

Figure 23.3 traces the path of blood as it makes one complete trip around the body. Essential to the system is the four-chambered heart. The heart always receives blood in a chamber called the **atrium** (plural, *atria*) and blood is always pumped away from the heart from a chamber called the **ventricle**. Both sides of the heart have one atrium and one ventricle. The pulmonary and systemic circuits operate simultaneously: The two ventricles pump almost in unison, sending some blood through the pulmonary circuit and the rest through the systemic circuit.

Let's start at ❶ the right atrium, where two large veins empty O_2-poor blood from the body into the heart. The blood is then pumped to ❷ the right ventricle, which pumps the O_2-poor blood to the lungs via ❸ two pulmonary arteries. ❹ As the blood flows through capillaries in the lungs, CO_2 diffuses out of the blood and O_2 diffuses into the blood. The newly O_2-rich blood then flows through ❺ the pulmonary veins to ❻ the left atrium of the heart, completing the pulmonary circuit.

Next, the blood is pumped from the left atrium to ❼ the left ventricle. Oxygen-rich blood leaves the left ventricle through ❽ the aorta, the largest blood vessel in the body, with a diameter about as big as a quarter. Branching from the aorta are the coronary arteries

that supply the heart itself and large arteries that lead to ❾ the head and arms and the abdominal region and legs. Within each organ, arteries lead to arterioles that branch into capillaries. Diffusion of O_2 into the cells and CO_2 out of the cells takes place across the thin walls of the capillaries. Downstream, the capillaries join into venules, which convey the blood back into veins. Oxygen-poor blood from the upper body and head is channeled into ❿ a vein called the superior vena cava (in this case, "superior" means "sitting above"), and oxygen-poor blood from the lower body flows into ⓫ the inferior ("below") vena cava. The superior vena cava and inferior vena cava complete the systemic circuit by returning blood to the heart.

Now, let's discuss the structure and function of each component in the human cardiovascular system. As you read the next section, you may find it useful to refer back to Figure 23.3. ✓

☑ CHECKPOINT

Trace the path of a blood clot from a leg vein to a lung. List each location.

Answer: leg vein, inferior vena cava, right atrium, right ventricle, pulmonary artery, capillary bed of a lung

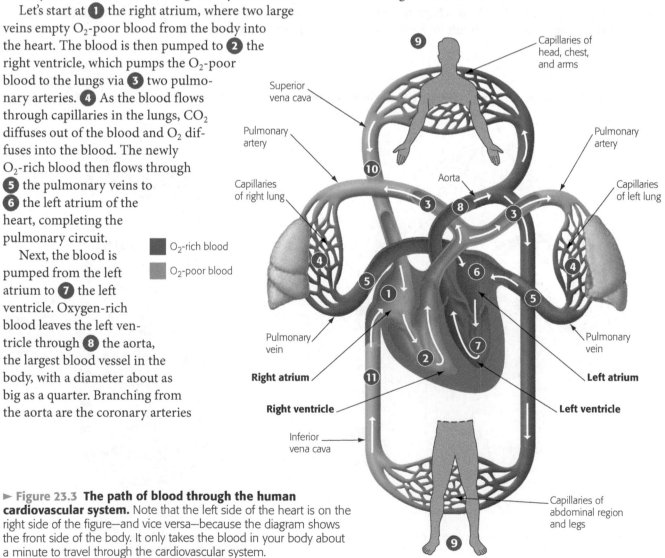

O₂-rich blood

O₂-poor blood

Capillaries of head, chest, and arms

Superior vena cava

Pulmonary artery

Pulmonary artery

Capillaries of right lung

Aorta

Capillaries of left lung

Pulmonary vein

Pulmonary vein

Right atrium

Left atrium

Right ventricle

Left ventricle

Inferior vena cava

Capillaries of abdominal region and legs

▶ **Figure 23.3 The path of blood through the human cardiovascular system.** Note that the left side of the heart is on the right side of the figure—and vice versa—because the diagram shows the front side of the body. It only takes the blood in your body about a minute to travel through the cardiovascular system.

How the Heart Works

The hub of the human cardiovascular system is the **heart**, a muscular organ about the size of a fist located under the breastbone. The heart's four chambers support double circulation and prevent oxygen-rich and oxygen-poor blood of each circuit from mixing. The oxygen-poor blood from body tissues flows into the right atrium, and the right ventricle pumps it to the lungs (**Figure 23.4**). The oxygen-rich blood returning from the lungs doesn't mix with oxygen-poor blood in the heart because it enters a separate chamber, the left atrium. The blood is forcefully pumped out to body tissues from the left ventricle. Amphibians and many reptiles have hearts with only three chambers, and the oxygen-rich blood mixes with oxygen-poor blood in a single ventricle before pumping to the systemic circuit. The evolution of a powerful four-chambered heart was an essential adaptation to support the high metabolic needs of birds and mammals, which are endothermic (Chapter 21). Endotherms use about 10 times as much energy as equal-sized ectotherms. Our four-chambered heart supports our body's need for efficient fuel and oxygen delivery, as well as athletic feats accomplished by elite athletes, as discussed in the Biology and Society section. ☑

▼ **Figure 23.4 Blood flow through the human heart.** The heart contains four chambers, with two atria located above two ventricles. Valves help maintain a one-way flow of blood. Notice that oxygen-poor blood and oxygen-rich blood do not mix.

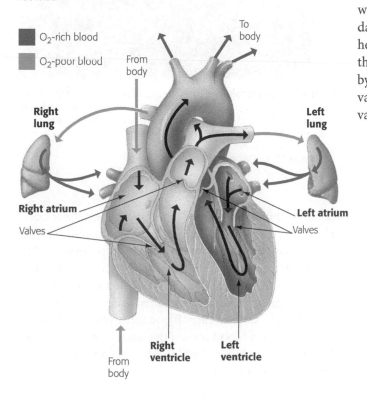

O₂-rich blood

O₂-poor blood

To body

From body

Right lung

Left lung

Right atrium

Left atrium

Valves

Valves

From body

Right ventricle

Left ventricle

The Cardiac Cycle

The cardiac muscles of the heart relax and contract rhythmically in what is called the **cardiac cycle**. One heartbeat makes up a complete circuit of the cardiac cycle. In a healthy adult at rest, the number of beats per minute, or **heart rate**, ranges between 60 and 100. You can measure your heart rate by taking your **pulse**, which is the stretching of arteries with each heartbeat. Athletes have lower resting heart rates because a stronger heart can pump a greater volume of blood with each contraction.

The relaxation phase of the heart cycle is known as **diastole**; the contraction phase is called **systole**. **Figure 23.5** follows the heart through a cycle, which lasts only 0.8 second. ① During diastole, which lasts about 0.4 second, blood returning to the heart flows into all four chambers. ② During the first 0.1 second of systole, the atria contract, forcing all the blood into the ventricles. ③ In the last 0.3 second of systole, the ventricles contract, pumping blood out of the heart and into the aorta and pulmonary arteries.

As it beats, the heart makes a distinctive "lubb-dupp, lubb-dupp" sound. Valves prevent backflow and keep blood moving in the right direction. As the heart valves snap shut, blood recoils against the closed valves and vibrations occur. You can hear these sounds with a stethoscope. A trained ear can detect the sound of a heart murmur, which indicates a defect in one or more of the valves. A serious murmur sounds like a "hisssss" as blood squirts backward through a defective valve. Some people are born with murmurs, while others have valves damaged by infection. Most cases of heart murmur are not serious, and those that are can be corrected by replacing the damaged valve with a mechanical valve or donor tissue.

☑ **CHECKPOINT**

Which chambers pump blood out of the heart? Which chambers receive blood returning to the heart?

Answer: ventricles; atria

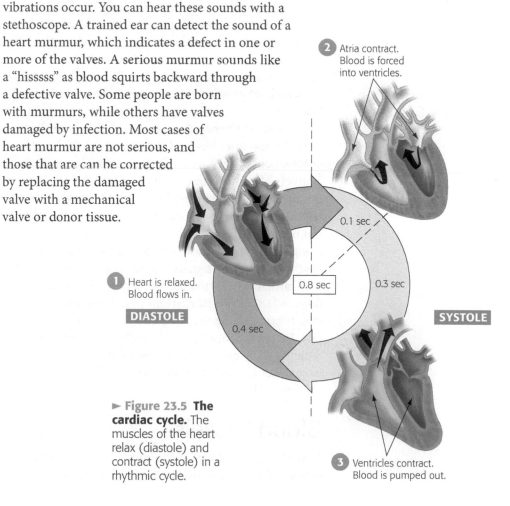

② Atria contract. Blood is forced into ventricles.

① Heart is relaxed. Blood flows in.

DIASTOLE

0.1 sec

0.8 sec

0.3 sec

0.4 sec

SYSTOLE

③ Ventricles contract. Blood is pumped out.

► **Figure 23.5 The cardiac cycle.** The muscles of the heart relax (diastole) and contract (systole) in a rhythmic cycle.

The Pacemaker and the Control of Heart Rate

All the muscle cells in your heart beat in unison under the direction of a conductor called the sinoatrial node, or the **pacemaker**. (Here we are referring to the body's natural pacemaker; we will discuss artificial pacemakers shortly.) The pacemaker is made up of specialized muscle tissue in the wall of the right atrium that generates electrical impulses **(Figure 23.6a)**. ❶ Impulses from the pacemaker spread quickly through the walls of both atria, prompting the atria to contract at the same time. ❷ The impulses then pass to a relay point that delays the signals by about 0.1 second, allowing the atria to empty before the impulses are passed to the ventricles. The ventricles then contract strongly, driving the blood out of the heart. The pacemaker's electrical currents can be detected by electrodes placed on the skin that pick up the heart's "rhythm" and record it as an electrocardiogram (ECG or EKG).

Like a musical conductor leading a band to play a song quickly or slowly, the heart's pacemaker directs muscles of the heart to contract faster or slower under the influence of a variety of signals. Epinephrine (or adrenaline), the "fight-or-flight" hormone released during times of stress, increases heart rate (see Chapter 25). Stimulants, such as caffeine, make the heart beat faster, as does exercise, an adaptation that enables the circulatory system to provide additional oxygen to muscles hard at work. Rest or sleep induces the heart to beat slower.

Sometimes the heart's pacemaker fails to coordinate the electrical impulses, and the muscles of the heart contract out of sync, producing an erratic heart rhythm. Think back to the music analogy: Without a conductor, musicians would play their instruments without regard for other players and the overall sound would not be music. If a heart continually fails to maintain a normal rhythm, an artificial pacemaker that emits rhythmic electrical signals can be surgically implanted into cardiac muscle to maintain a normal heartbeat **(Figure 23.6b)**. Erratic rhythms may cause a heart to stop beating suddenly, a condition termed cardiac arrest. External defibrillators ("paddles") placed on a patient's chest can be used to reset the heart's natural pacemaker by delivering an electrical shock to bring the rhythm back to normal.

Blood Vessels

If we think of the heart as the body's "pump," then the system of arteries, veins, and capillaries connected to

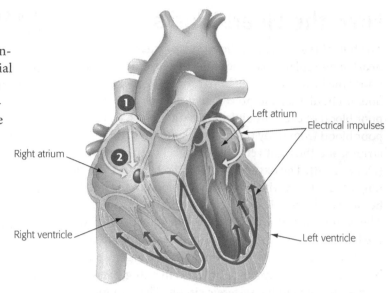

(a) The heart's natural pacemaker. The pacemaker is located in the right atrium. Electrical impulses spread through the heart, first to the atria, then to the ventricles.

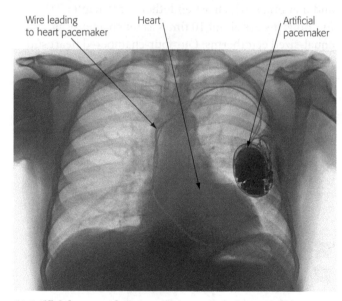

(b) Artificial pacemaker. A small electronic device surgically implanted into cardiac muscle or (as shown here) the chest cavity and connected to the body's pacemaker by a wire can help maintain proper electrical rhythms in a defective heart.

▲ Figure 23.6 **Pacemakers.**

it can be thought of as the "plumbing." The network of blood vessels is extensive. If laid out end to end, the total length of all the blood vessels in an average human adult would be twice Earth's circumference at the equator. If you review Figure 23.3, you'll notice that arteries and veins are distinguished by the direction in which they carry blood: Arteries carry blood *away from* the heart, and veins carry blood *toward* the heart. Capillaries allow for exchange between the bloodstream and the tissue cells (via interstitial fluid).

All blood vessels are lined by a thin layer of tightly packed epithelial cells. Structural differences in the walls of the different kinds of blood vessels correlate with their different functions (Figure 23.7). Capillaries have very thin walls—often just one cell thick—that allow rapid exchange of substances between the blood and the interstitial fluid that bathes tissue cells. The smallest capillaries are so small in width (about eight times smaller than the width of a human hair) that blood cells must pass through them single file. The walls of arteries and veins have two additional, thicker layers. An outer layer of elastic connective tissue allows the vessels to stretch and recoil. Between this layer and the epithelial cells is a middle layer of smooth muscle. By constricting or relaxing in response to signals from the brain, smooth muscle can narrow or widen the blood vessels, thereby regulating blood flow. Certain blood pressure medications work by blocking the chemicals involved in this system. This action widens the vessels, lowers blood pressure, and reduces the risk of heart attack and stroke (death of brain tissue due to lack of oxygen). Veins convey blood back to the heart at low velocity and pressure after the blood has passed through capillary beds. Veins (but not arteries) also have one-way valves that prevent backflow, ensuring that blood always moves toward the heart. ☑

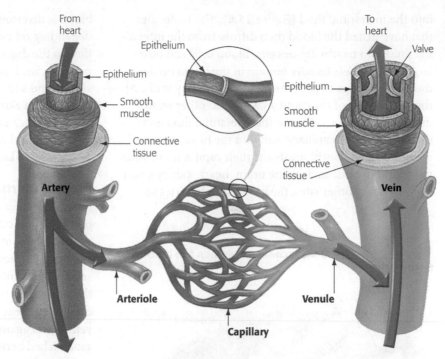

▲ Figure 23.7 **The structure of blood vessels.** Arteries branch into smaller vessels called arterioles, which in turn branch into capillaries. Chemical exchange between blood and interstitial fluid occurs across the thin walls of the capillaries. The capillaries converge into venules, which deliver blood to veins. All of these vessels are lined by a thin, smooth epithelium. Arteries and veins have additional layers of smooth muscle and connective tissue.

Blood Flow through Arteries

When you go to a doctor's office, you routinely have your blood pressure measured. **Blood pressure** is the force that blood exerts against the walls of your arteries. Created by the pumping of the heart, blood pressure pushes blood from the heart through the arteries and arterioles to the capillary beds. When the ventricles contract during systole, blood is forced into the arteries faster than it can flow into the arterioles. This stretches the elastic walls of the arteries. (You can feel this rhythmic stretching of your arteries when you take your pulse.) The elastic walls of the arteries snap back during diastole, maintaining enough pressure to sustain a constant flow into arterioles and capillaries. Thus, blood pressure is recorded as two numbers, such as 120/80 ("120 over 80"). The first number is blood pressure during systole (in millimeters of mercury, a standard pressure unit); the second number is the blood pressure that remains in the arteries during diastole.

Normal blood pressure falls within a range of values, but optimal blood pressure for adults is below 120 systolic and below 80 diastolic. Lower values are generally considered better, although very low blood pressure may lead to light-headedness and fainting. Blood pressure higher than normal may indicate a serious cardiovascular disorder.

High blood pressure, or **hypertension**, is persistent systolic blood pressure higher than 140 and/or diastolic blood pressure higher than 90. Hypertension affects nearly 1 in 3 adults (2/3 aged 65 and older) in the United States. It is sometimes called a "silent killer" because it often displays no outward symptoms for years while increasing the risks of heart disease, a heart attack, or a stroke. To control hypertension, you can eat a healthy diet, avoid smoking and excess alcohol, exercise regularly, and maintain a healthy weight. If lifestyle changes don't work, there are several medications that can help lower blood pressure.

Blood Flow through Capillary Beds

The most important function of the circulatory system is the chemical exchange between the blood and tissue cells within capillary beds. The walls of capillaries are thin and leaky. Consequently, as blood enters a capillary at the arterial end, blood pressure pushes fluid rich in O_2, nutrients, and other molecules out of the capillary and

☑ **CHECKPOINT**

How are capillary walls an example of structure fitting function?

Answer: The thinness of capillary walls helps the exchange of substances with cells.

into the interstitial fluid (Figure 23.8). The molecules that have exited the blood then diffuse from the interstitial fluid into nearby tissue cells. Blood cells and other large components usually remain in the blood because they are too large to pass through the capillary walls. At the venous end of the capillary, CO_2 and other wastes diffuse from tissue cells into the interstitial fluid and then through the capillary wall into the bloodstream.

Blood flows continuously through capillaries in your most vital organs, such as the brain, heart, kidneys, and liver. In many other sites, the blood supply varies as

blood is diverted from one part of the body to another, depending on need. After a meal, for instance, blood flow to the digestive tract increases. During strenuous exercise, blood is shunted away from the digestive system and supplied more generously to the skeletal muscles and skin. This is one reason why heavy exercise right after eating may cause indigestion or muscle cramping (and why you shouldn't swim too soon after eating—just like mom always said).

Blood Return through Veins

After molecules are exchanged between the blood and body cells, blood flows from the capillaries into small venules, then into larger veins, and finally to the inferior and superior venae cavae, the two large blood vessels that flow into the heart. By the time blood enters the veins, the pressure originating from the heart has dropped to near zero. The blood still moves through veins, even against the force of gravity, because veins are sandwiched between skeletal muscles (Figure 23.9). As these muscles contract (when you walk, for example), they squeeze the blood along.

You may have experienced firsthand the importance of muscle contraction in conveying blood through your veins if you have ever stood still for too long. After standing a while without contracting muscles, a person will start to become weak and dizzy and could even faint because gravity prevents blood from returning to the heart in sufficient amounts to supply the brain with oxygen.

Over time, leg veins may stretch and enlarge and the valves within them weaken. As a result, veins just under the skin can become visibly swollen, a condition called varicose veins. Health problems, such as cramping, blood clots, and open breaks in the skin, may result from varicose veins. ☑

▼ Figure 23.8 **Chemical exchange between the blood and tissue cells.**

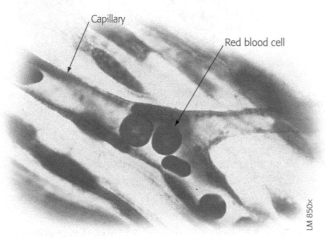

Capillary

Red blood cell

LM 850×

(a) **Capillaries.** Blood flowing through the circulatory system eventually reaches capillaries, the small vessels where exchange with cells actually takes place.

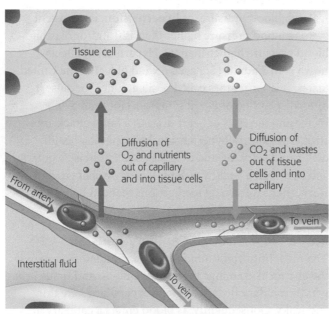

Tissue cell

Diffusion of O_2 and nutrients out of capillary and into tissue cells

Diffusion of CO_2 and wastes out of tissue cells and into capillary

From artery

To vein

To vein

Interstitial fluid

(b) **Chemical exchange.** Within the capillary beds, there is local exchange of molecules between the blood and interstitial fluid, which bathes the cells of tissues.

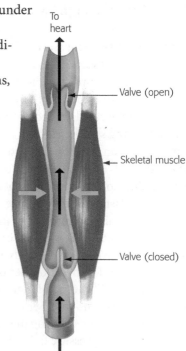

To heart

Valve (open)

Skeletal muscle

Valve (closed)

► Figure 23.9 **Blood flow in a vein.** The contraction of muscles surrounding veins squeezes blood toward the heart. Flaps of tissue in the veins act as one-way valves, preventing backflow.

Blood

Now that we have examined the structures and functions of the heart and blood vessels, let's focus on the composition of blood itself (Figure 23.10). The circulatory system of an adult human contains about 5 L of blood. Just over half this volume consists of a yellowish liquid called **plasma**. Plasma consists of water and dissolved salts, proteins, and various other molecules, such as nutrients, wastes, and hormones. Suspended within the plasma are three types of cellular elements: red blood cells, white blood cells, and platelets. Let's examine each of them in turn.

 **Structure/ Function** ## Red Blood Cells

Red blood cells, also called erythrocytes, are by far the most numerous type of blood cell. There are about 25 trillion of these cells in the average person's bloodstream. Their primary function is to carry oxygen from the lungs to the tissues.

Each red blood cell contains approximately 250 million molecules of **hemoglobin**, an iron-containing protein that transports oxygen. As red blood cells pass through the capillary beds of your lungs, oxygen diffuses into them and binds to the hemoglobin. This process is reversed in the capillaries of the systemic circuit, where the hemoglobin unloads its cargo of oxygen to the body's cells. How does the structure of a red blood cell support this function?

Human red blood cells are shaped like disks with indentations in the middle, increasing the surface area available for gas exchange. In addition to their shape, the red blood cells of mammals have another feature that enhances their oxygen-carrying capacity: They lack nuclei and other organelles, leaving more room to carry hemoglobin. Red blood cells are also very small relative to other cells, resulting in a surface-area-to-volume ratio that is efficient for gas exchange.

Adequate amounts of hemoglobin and red blood cells are essential to the normal functioning of the body. **Anemia** is a condition in which there is an abnormally low amount of hemoglobin or a low number of red blood cells. A person who is anemic feels constantly tired and run-down because the body cells do not get enough oxygen via the blood. In response, the kidneys produce a hormone called erythropoietin (EPO) that stimulates the bone marrow's production of oxygen-carrying red blood cells. Sometimes the body's response is not enough, however, and a synthetic version of EPO is used to boost red blood cell production in patients with anemia.

▼ Figure 23.10 **The composition of human blood.** Just over half the volume of blood is liquid plasma. Blood cells and platelets account for the remaining blood volume.

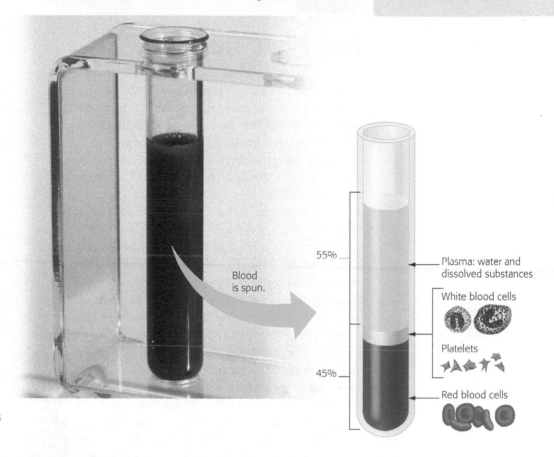

Blood is spun.

55%
45%

Plasma: water and dissolved substances
White blood cells
Platelets
Red blood cells

Blood Doping

Athletes sometimes abuse synthetic EPO to enhance their blood oxygen level, a practice referred to as blood doping, which is discussed in the Biology and Society section. As EPO increases the number of red blood cells in the bloodstream, more oxygen reaches working muscle cells and athletic endurance improves. Blood doping is banned in competitive sports, because it not only confers athletes with an unfair advantage but also is dangerous: If too many red blood cells are produced, blood flow becomes sluggish, tissues are deprived of oxygen, and death can occur.

Blood doping is difficult to detect because EPO is a hormone produced naturally by the body and because synthetic EPO is rapidly cleared from the bloodstream. One way that athletic commissions test for EPO abuse is by measuring the percentage of red blood cells in blood volume. High percentages and erratic spikes in these values are grounds for disqualification. Fortunately, there are other ways to increase athletic performance that are healthier than blood doping, as we'll see next. ☑

☑ **CHECKPOINT**
Which blood cells do athletes want to increase by blood doping?

Answer: red blood cells

Athletic Endurance THE PROCESS OF SCIENCE

Live High, Train Low?

Exercise physiologists have known for years that properly designed training regimens can vastly improve the performance of endurance athletes. Athletes have learned from **observation** that it is difficult to train at low altitude (near sea level) and then compete at high altitude. Often adjustments are made in competitions to compensate for the difference in difficulty. For example, a 1 minute adjustment is made for a 10 kilometer race in Boulder, Colorado (about 1 mile in elevation) compared to the same race at sea level. Running is more difficult at higher altitudes because the air is at a lower pressure and therefore "thinner," with fewer molecules of oxygen taken in per breath, making it more difficult for the body to supply the O_2 needed for muscles to function properly. But if an athlete stays at high altitude for weeks or months, the kidneys will compensate for the low O_2 level in the body by boosting production of EPO, which results in more red blood cells, which in turn results in more oxygen carried to muscle tissues.

For the last few decades, the International Olympic Committee has sponsored research that asks the **question**: Will athletes who live for a time at high altitude increase their performance when they train and compete at lower altitudes? One leading hypothesis can be summed up by the phrase "live high, train low." This **hypothesis** holds that living at high altitudes will boost red blood cell production, which can then improve stamina when training and competing at lower altitudes. One **experiment** published in 2006 involved 11 elite runners as test participants. Five of these athletes lived for 18 days in rooms that were gradually adjusted to simulate an altitude of 3,000 m (nearly

2 miles, about twice the elevation of Denver), while the other six athletes lived at 1,200 m. All the athletes trained at 1,200 m.

At the start of the study (day 0), the **results** of endurance tests showed that both groups were similar in their aerobic capacity, a measure of the body's ability to take in and use oxygen (**Figure 23.11**, first pair of bars). But by the end of the training regimen (day 18), the "live high, train low" group had a higher aerobic capacity (middle pair of bars) that gradually fell over two weeks (day 33, third pair of bars). These results, as well as those from many other similar studies, are providing a scientific basis for new training regimens aimed to improve the performance of today's endurance athletes.

▼ **Figure 23.11 Live high, train low.** The results of this study suggest that living at high altitude while training at low altitude can improve athletic performance, but the effect fades.

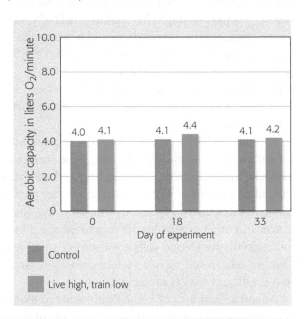

White Blood Cells and Defense

Red blood cells are just one of the cellular components of blood (**Figure 23.12**). The blood also contains **white blood cells**, or leukocytes. Unlike red blood cells, white blood cells contain nuclei and other organelles, and they are larger and lack hemoglobin. There are far fewer white blood cells than red blood cells, although, as a major component of your immune system, white blood cells temporarily increase when the body is combating an infection (see Chapter 24).

Platelets and Blood Clotting

Although you get cuts and scrapes, you do not bleed to death because your blood contains two components that help form clots: platelets and fibrinogen. **Platelets** are bits of membrane pinched off from larger cells in the bone marrow. Clotting begins when the lining of a blood vessel is damaged. Almost immediately, platelets form sticky clusters that adhere to and seal the damaged vessel. Platelets also release clotting factors, molecules that convert **fibrinogen**, a protein found in the plasma,

▼ **Figure 23.12 The three cellular components of blood.** Unlike red blood cells and platelets that carry out their functions in the blood, white blood cells exit the blood to fight local infections in body tissues.

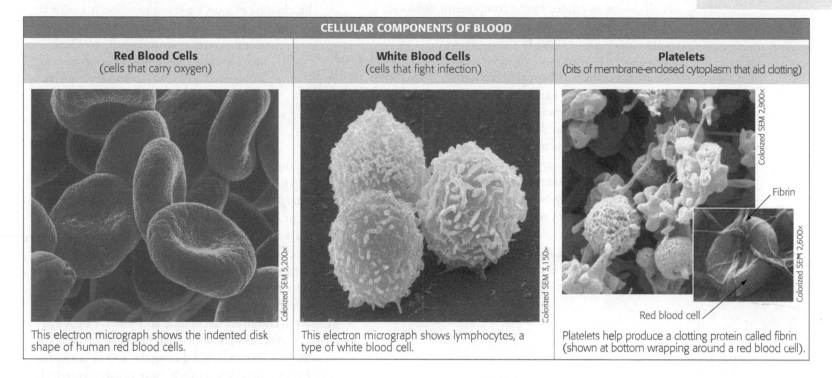

CELLULAR COMPONENTS OF BLOOD

Red Blood Cells (cells that carry oxygen)	White Blood Cells (cells that fight infection)	Platelets (bits of membrane-enclosed cytoplasm that aid clotting)

This electron micrograph shows the indented disk shape of human red blood cells.

This electron micrograph shows lymphocytes, a type of white blood cell.

Platelets help produce a clotting protein called fibrin (shown at bottom wrapping around a red blood cell).

Colorized SEM 5,200×

Colorized SEM 3,150×

Colorized SEM 2,900×

Colorized SEM 2,600×

Fibrin

Red blood cell

to a threadlike protein called **fibrin**. Molecules of fibrin form a dense network to create a patch (see the inset in Figure 23.12).

Too much or too little clotting can be life-threatening. An embolus is a blood clot that forms within a blood vessel and then dislodges from that point of origin and travels elsewhere in the body via the blood. If lodged in vessels of the lungs, an embolus affects gas exchange. An embolus that blocks blood flow to the heart may cause a heart attack, and an embolus in the brain may cause a stroke. The inability to form clots is also dangerous. In the disease hemophilia, a genetic mutation in a gene for a clotting factor results in excessive, sometimes fatal bleeding (see Figure 9.28).

Donating Blood Components

Excessive bleeding and certain diseases can be treated through blood transfusion, the process by which blood is intravenously (through veins) transferred from one person into another. Transfusion is only possible because of whole or partial blood donation.

When whole blood is donated, all the components—plasma, red blood cells, white blood cells, and platelets—are collected and then separated. White blood cells are discarded because of immune system reactions they cause in recipients, but the remaining components are saved and eventually transfused into patients being treated for cancer, accidents, burns, or

major surgery. A recipient must "match" a donor—for instance, a recipient with type A blood cannot receive type B blood (see Figure 9.20).

Partial blood donation is also possible. In a procedure called apheresis, only platelets or plasma is removed from a donor's blood; the remaining blood components are returned to the donor's circulation. Apheresis is used to collect stem cells, too. Stem cells that form the various blood cells normally reside inside the bone marrow. With a drug treatment, bone marrow stems cells can be coaxed into the bloodstream and then collected by apheresis for use in medical treatments.

Leukemia, a type of cancer that originates in the cells of bone marrow, can be treated with stem cells. The treatment involves completely destroying cancerous bone marrow and replacing it with healthy bone marrow stem cells from a matching donor, often a sibling. Matching bone marrow cell type between two unrelated individuals is complicated because of an extreme diversity in the proteins on the surface of bone marrow cells. In two-thirds of bone marrow replacement cases, the National Marrow Donor Program, Be the Match, successfully matches patients with unrelated donors.

Only about 10% of people eligible to donate components of the blood take action. Why? Most people who don't donate have "never thought about it." ✓

✓ CHECKPOINT

What is the purpose of apheresis?

Answer: to separate and retrieve only one component of blood, such as platelets or stem cells

Cardiovascular Disease

Blood clots in the lungs, hemophilia, and leukemia are all disorders related to the components of blood. Yet the cardiovascular system also includes a pump (heart) and a series of tubes (vessels). Diseases of the heart and blood vessels are collectively called **cardiovascular disease**, and they account for 1 in 3 deaths in the United States. We already discussed one form of cardiovascular disease—pacemaker malfunction that leads to erratic heart rhythms and sudden cardiac arrest. Let's turn our attention to disease of the **coronary arteries**, the vessels that supply oxygen-rich blood to the heart muscle (the dark red branches from the aorta in **Figure 23.13**).

Chronic disease of the coronary and other arteries throughout the body is called **atherosclerosis**. During the course of atherosclerosis, deposits called plaque

Half a million metal stents, not much wider than spaghetti, save lives every year.

develop in the inner walls of arteries (**Figure 23.14**). Arteries narrowed by plaque can easily form and trap clots. When a coronary artery becomes partially blocked by plaque, a person may feel occasional chest pain, a condition called angina. If the coronary artery becomes fully blocked, heart muscle cells quickly die (gray area in Figure 23.13). The coronary arteries are only about 4–5 mm in diameter, so just a tiny blockage—an amount of plaque smaller than a pea—can cause a catastrophic heart attack. Approximately one-third of heart attack victims die almost immediately. For those who survive, the ability of the damaged heart to pump blood may be seriously impaired for life because heart muscle cannot be replaced or repaired.

For people living with cardiovascular disease, there are treatments available (**Figure 23.15**). Certain drugs can lower the risk of developing clots. Angioplasty is the insertion of a tiny catheter with a balloon that is inflated to compress the plaque and widen clogged arteries. A stent, a small wire mesh tube that props open an artery, is often inserted during the angioplasty process. Bypass surgery is a much more drastic remedy. In this procedure, a vein is removed from a patient's leg and is sewn onto the heart, shunting blood around the clogged artery. Unfortunately, surgery of any kind treats only the disease symptoms, not the underlying cause; cardiovascular disease will return if risk factors are not minimized.

To some extent, the tendency to develop cardiovascular disease is inherited. However, three everyday behaviors significantly impact the risk of cardiovascular disease and heart attack. Smoking doubles the risk of heart attack and increases the severity if one

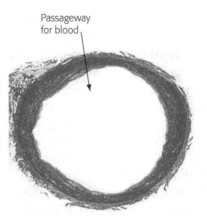

Aorta

Coronary artery (supplies oxygen to the heart muscle)

Dead muscle tissue

Blockage

▲ Figure 23.13 **Blockage of a coronary artery, resulting in a heart attack.** If one or more coronary arteries become blocked, the heart muscle cells that they feed will die from lack of oxygen. Such an event, called a heart attack, can lead to permanent damage of the heart muscle.

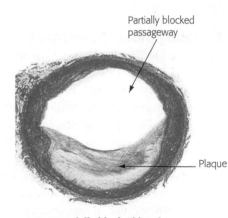

Passageway for blood

Partially blocked passageway

Plaque

Normal artery

Artery partially blocked by plaque

▲ Figure 23.14 **A normal artery and an artery showing atherosclerosis.**

does occur. Regular exercise (particularly "cardio" or "aerobic" workouts) can reduce the risk of heart disease by 50%. Eating a healthy diet high in fruits, vegetables, and whole grains can reduce the risk of developing atherosclerosis.

On the bright side, the death rate from cardiovascular disease in the United States has been cut in half during the past 50 years. Health education, early diagnosis, and reduction of risk factors, particularly smoking, are mostly responsible. ☑

☑ **CHECKPOINT**

Name three things you can do to lower your risk of cardiovascular disease.

Answer: eat a healthy diet, exercise regularly, and avoid smoking

▼ **Figure 23.15 Methods of repairing arteries blocked by plaque.**

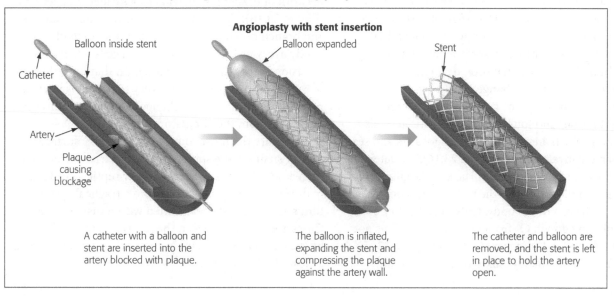

Angioplasty with stent insertion

A catheter with a balloon and stent are inserted into the artery blocked with plaque.

The balloon is inflated, expanding the stent and compressing the plaque against the artery wall.

The catheter and balloon are removed, and the stent is left in place to hold the artery open.

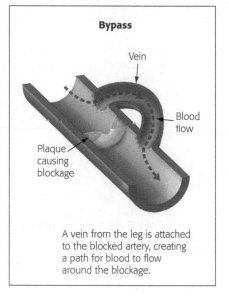

Bypass

A vein from the leg is attached to the blocked artery, creating a path for blood to flow around the blockage.

Unifying Concepts of Animal Respiration

Slowly breathe in and out, and try to imagine what is happening at the cellular level with each inhale and exhale. Recall that within cells, the process of cellular respiration (the topic of Chapter 6) uses O_2 and glucose to produce water, CO_2, and energy in the form of ATP. All working cells therefore require a steady supply of O_2 from the environment and must continuously dispose of CO_2. Thus, our need to breathe O_2 is directly related to one of the major themes in biology: energy transformation.

Your survival depends on close cooperation between the respiratory and circulatory systems. The **respiratory system** consists of several organs that facilitate exchange of O_2 and CO_2 between the environment and cells. (Note that in the context of a whole organism, the word *respiratory* refers to the process of breathing, not to cellular respiration.) As we've already learned, the circulatory system transports O_2 and CO_2 in the body to and from all your cells.

The part of an animal where O_2 from the environment diffuses into living cells and CO_2 diffuses out to the surrounding environment is called the **respiratory surface**. The respiratory surface usually has three major characteristics: It is covered with a single layer of living cells, it is thin, and it is moist. These characteristics allow rapid diffusion between the body and the environment. Additionally, there must be enough surface area to take up O_2 for every cell in the body.

Within the animal kingdom, a variety of respiratory surfaces have evolved. For some animals, such as sponges and flatworms, the plasma membrane of every cell in the body is close enough to the outside environment for gases to diffuse in and out. In many animals, however, the bulk of the body does not have direct access to the outer environment, and the respiratory surface is the place where the environment and the blood meet. The blood then transports gases to and from the rest of the body.

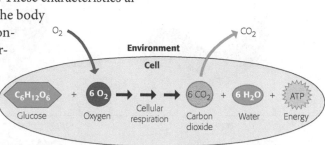

O_2 → **Environment** → CO_2

Cell

$C_6H_{12}O_6$ + $6 O_2$ → → → $6 CO_2$ + $6 H_2O$ + ATP

Glucose / Oxygen / Cellular respiration / Carbon dioxide / Water / Energy

✓ CHECKPOINT

1. Why do all human cells require a continuous supply of O_2 and disposal of CO_2?
2. In what basic way do insects carry out gas exchange differently from the way fishes and humans do?

Answers: 1. O_2 is required for cellular respiration, which generates ATP for cellular work and also produces CO_2 as a waste product. 2. The respiratory system of insects uses tracheae that bring O_2 directly to body cells and carry CO_2 directly away; fishes and humans have a circulatory system that transports O_2 from the respiratory surface to body cells and CO_2 back.

Some animals, such as leeches, earthworms, and frogs, use their entire outer skin as a respiratory surface (Figure 23.16). The skin is thin and moist, resulting in an exchange of O_2 and CO_2 between the environment and a dense net of capillaries just below the skin. Unfortunately, a consequence of thin skin is that water and toxins can diffuse through the skin, too. Some frogs are at risk because of toxins dissolved in the runoff water from farms. Pesticides and fertilizers easily diffuse through their thin skin and can result in reproductive problems.

For most animals, however, the outer surface either is impermeable to gases or lacks sufficient surface area to exchange gases for the whole body. In such animals, specialized regions of the body surface have extensively folded or branched tissues that provide a large respiratory surface area for gas exchange.

Gills are outfoldings of the body surface that are suspended in water and found in most aquatic animals, such as fishes, which are vertebrates, and lobsters and sea slugs, which are invertebrates (Figure 23.16). As water passes over the large respiratory surface of the gills, gases diffuse between the water and the blood. The blood then carries O_2 to the rest of the body. Gills must be very efficient to obtain enough O_2 from water because O_2 is not very soluble in water. (Dissolved O_2 makes up only 3–5% of water; in comparison, O_2 makes up 21% of the air.)

In most land-dwelling animals, the respiratory surfaces are folded into the body. The infolded surfaces are open to the air only through narrow tubes. This anatomical arrangement means that only those respiratory cells needed for gas exchange must be kept moist, rather than the entire body surface (as in earthworms or frogs). The two types of respiratory structures seen in land-dwelling animals are tracheal systems and lungs.

Insects breathe using a **tracheal system**, an extensive network of branching internal tubes called tracheae (Figure 23.16). Tracheae begin near the body's surface and branch down to narrower tubes that extend to nearly every cell. Gas exchange with cells occurs via diffusion across the moist epithelium that lines the tips of the tubes. Because almost every body cell is near the respiratory surface, gas exchange in insects requires no assistance from the circulatory system.

Lungs are the most common respiratory surface among snails, some spiders, and terrestrial vertebrates such as amphibians, birds and other reptiles, and mammals (Figure 23.16). In contrast to a tracheal system, **lungs** are localized organs lined with moist epithelium. Gases are carried between the lungs and the body cells by the circulatory system. In the next section, we'll take a closer look at human lungs. ✓

▼ Figure 23.16 **The diversity of respiratory organs.** Respiratory organs are external in some animals (skin and gills) and internal in other animals (tracheae and lungs).

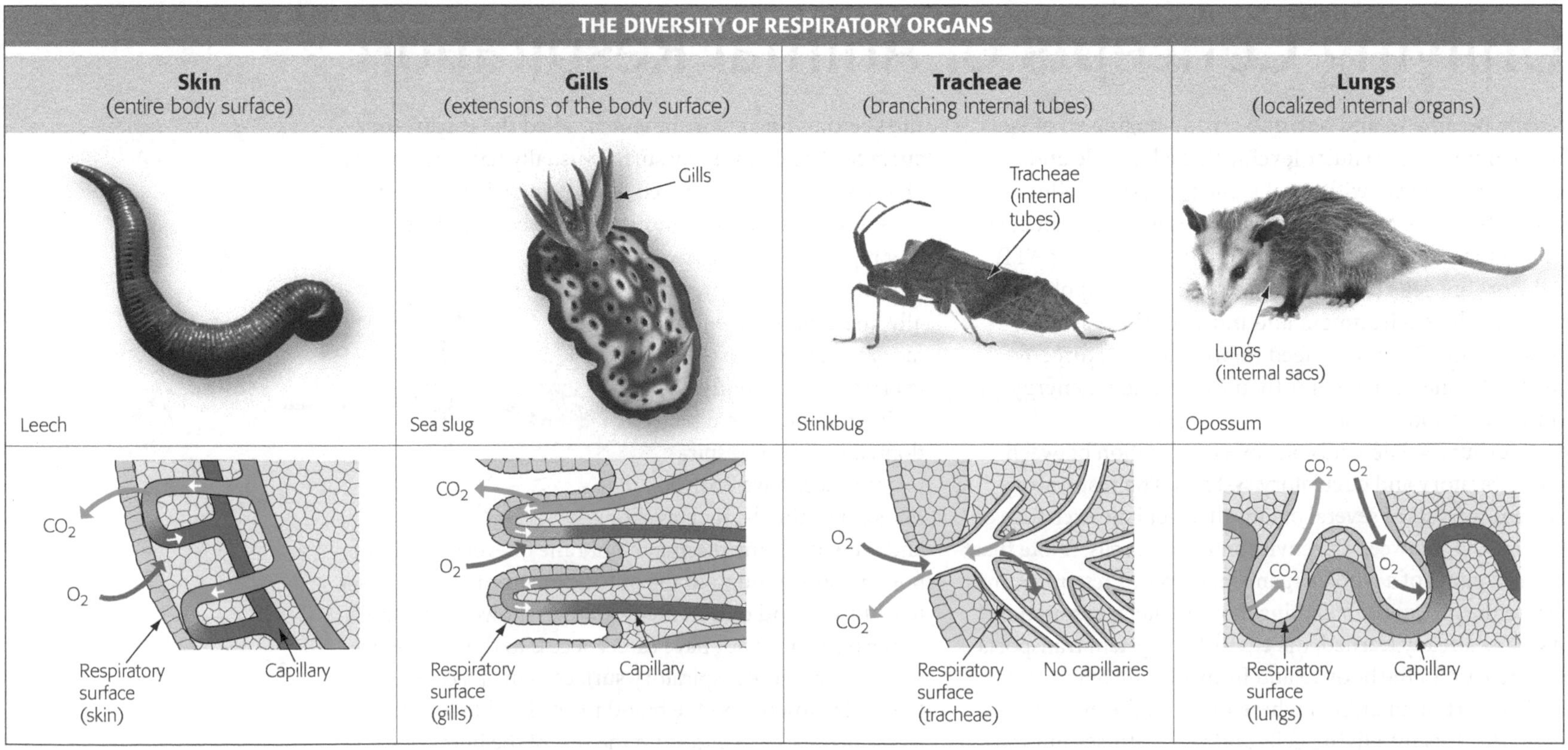

The Human Respiratory System

Figure 23.17 provides an overview of three phases of gas exchange in humans. **1** The first step of gas exchange is **breathing**, the ventilation of the lungs by alternate inhalation and exhalation. When an animal with lungs breathes, a large, moist internal surface is exposed to air. Oxygen diffuses across the cells lining the lungs and into the surrounding blood vessels. Simultaneously, CO_2 passes out of the blood vessels, into the lungs, and is exhaled. Most land animals require a lot of O_2. The inner tubes of the lungs are extensively branched, providing a large respiratory surface.

2 The second step in gas exchange is the transport of O_2 from the lungs to the rest of the body via the circulatory system. The blood also carries CO_2 from the tissues back to the lungs. **3** In the final step of gas exchange, O_2 diffuses from red blood cells into body cells. The delivered O_2 is used by the body cells to make ATP from food via the process of cellular respiration. This same process produces CO_2 as a waste product that diffuses from cells to the blood. The circulatory system transports the CO_2 back to the lungs, where it is exhaled. ☑

☑ CHECKPOINT

Which type of cellular transport is responsible for moving respiratory gases between the circulatory system and the rest of the body?

Answer: *diffusion*

▼ Figure 23.17 **Gas exchange.**

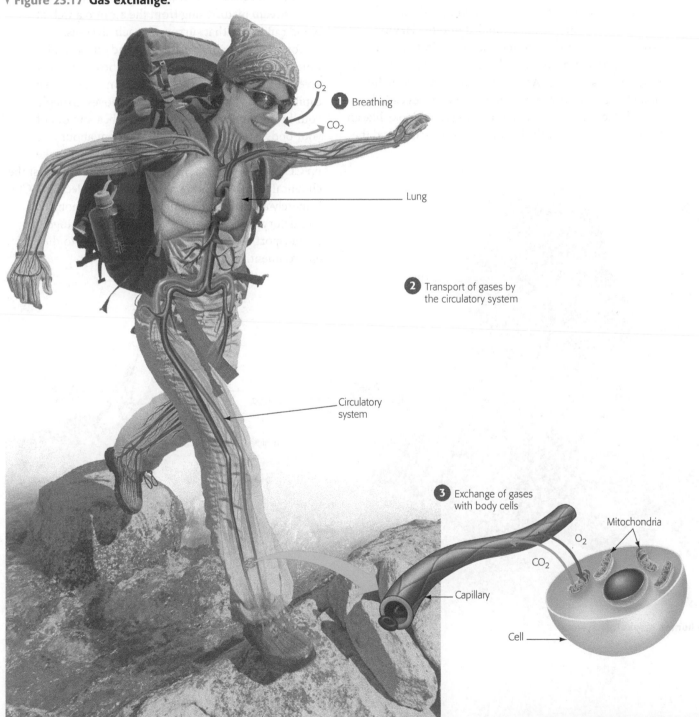

O_2

1 Breathing

CO_2

Lung

2 Transport of gases by the circulatory system

Circulatory system

3 Exchange of gases with body cells

Mitochondria

O_2

CO_2

Capillary

Cell

The Path of Air

Take a few deep breaths, and as you do, pay attention to the path air takes through your body. Let's get an overview of the respiratory system by following the flow of air into the lungs.

Figure 23.18a shows the human respiratory system. Air enters the respiratory system through the nostrils and mouth. In the nasal cavity, the air is filtered by hairs and mucus, warmed, humidified, and sampled by smell receptors. The air passes to the **pharynx**, where the digestive and respiratory systems meet. If a piece of food lodges in your pharynx, blocking passage of air to your lungs, you can choke and quickly die. The Heimlich maneuver, quick thrusts to the diaphragm to compress the lungs and force air upward through the trachea, can dislodge an obstruction and prevent a person from choking to death.

From the pharynx, air is inhaled into the **larynx** (voice box) and then into the **trachea** (windpipe). The trachea forks into two **bronchi** (singular, *bronchus*), one leading to each lung. Within the lungs, each bronchus branches repeatedly into finer and finer tubes called bronchioles. The system of branching tubes looks like an upside-down tree, with the trachea as the trunk and the

bronchioles as the smallest branches. Bronchitis is an illness in which these small tubes become inflamed. Bronchioles significantly narrow during an asthma attack, producing a wheezing sound as air is moved inward.

The bronchioles dead-end in grapelike clusters of air sacs called **alveoli** (singular, *alveolus*) **(Figure 23.18b)**. Each of your lungs contains millions of these tiny sacs, which provide about 50 times more surface area than your skin. The inner surface of each alveolus is lined with a layer of epithelial cells, where the exchange of gases actually takes place. The tiny alveoli are delicate and easily damaged, and after age 20 they are not replaced. Destruction of alveoli (usually by smoking, but also by involuntary exposure to air pollution) causes the lung disease emphysema. As the air we have been following reaches its final destination, O_2 enters the bloodstream by diffusing from the air into a web of blood capillaries that surrounds each alveolus.

You have probably breathed in and out several times while reading the paragraphs above. Your exhale reverses the process: CO_2 diffuses from blood in your capillaries into the alveoli and then moves through your bronchioles, bronchus, and trachea and out of your body. Stop and think for a moment about where this matter (CO_2) came from: The atoms that make up CO_2 originated in a meal you ate. As cells harvest the chemical energy of food (carbon-based molecules), CO_2 is merely a waste product of cellular respiration. The circulatory and respiratory systems function together to transport this waste from individual cells to the environment. ☑

► Figure 23.18 **The human respiratory system.**

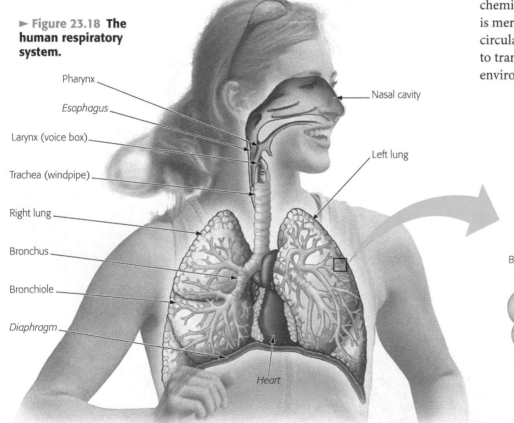

Pharynx
Esophagus
Larynx (voice box)
Trachea (windpipe)
Right lung
Bronchus
Bronchiole
Diaphragm
Nasal cavity
Left lung
Heart

(a) Overview of the human respiratory system

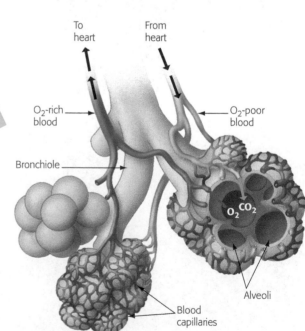

To heart
From heart
O_2-rich blood
O_2-poor blood
Bronchiole
O_2 CO_2
Alveoli
Blood capillaries

(b) The structure of alveoli. Capillaries surround air sacs called alveoli. These are the sites of gas exchange in the lungs.

Information Flow
The Brain's Control over Breathing

Figure 23.19 shows the changes that occur during breathing, the alternate inhaling and exhaling of air. When you place your hands on your rib cage and inhale, you can feel your ribs move upward and spread out as muscles between them contract. Meanwhile, your **diaphragm**, a sheet of muscle, moves downward, expanding the chest cavity. All of this increases the volume of the lungs, dropping the air pressure in the lungs to below the air pressure of the atmosphere. The result is that air rushes in through the mouth and nostrils from an area of higher pressure to an area of lower pressure, filling the lungs. Although it may seem that you actively suck in air when you inhale, air actually moves into your lungs passively after the air pressure in them drops. This type of ventilation is called **negative pressure breathing**. (In contrast, amphibians use muscles in the mouth and throat to push air into their lungs, a process called positive pressure breathing.)

During exhalation, the rib and diaphragm muscles relax, decreasing the volume of the chest cavity. This decreased volume increases the air pressure inside the lungs, forcing air to rush out of the respiratory system. Movement of the diaphragm is vital to normal breathing. That is why a punch to the diaphragm can "knock the wind out of you." It shocks the diaphragm muscle, prevents movement of the chest cavity, and stops you from taking a normal breath. Other times, sudden involuntary contractions of the diaphragm force air through the voice box, causing hiccups.

You can consciously speed up or slow down your breathing. You can even hold your breath (although, despite the claims of angry children, you would pass out and return to normal breathing before you turned blue). Usually, however, you aren't aware of breathing; you certainly aren't aware of it when you're asleep. What, then, controls your breathing?

Most of the time, nerves from breathing control centers in the brainstem maintain a respiratory rate of 10–14 inhalations per minute. This rate can vary, however, as when you exercise. **Figure 23.20** highlights the flow of information (from stimulus to response) in one respiratory control system. Levels of CO_2 in the blood affect breathing rate. ❶ During exercise, cellular respiration kicks into high gear, producing more ATP for your muscles and raising the amount of CO_2 in the blood. ❷ When the brain senses the higher CO_2 level (the stimulus), ❸ breathing control centers send information (nerve signals) to rib cage and diaphragm muscles that contract (the response), increasing the breathing rate and depth of ventilation. As a result, more CO_2 is eliminated in an exhale and more O_2 is provided to exercising muscles. ☑

▼ **Figure 23.19 How a human breathes.**

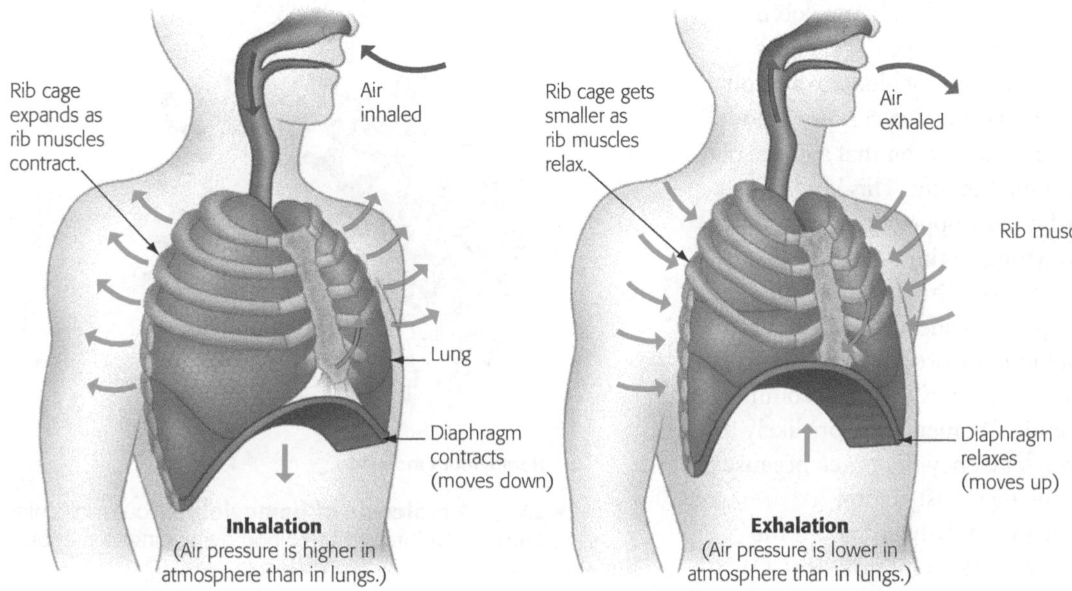

Rib cage expands as rib muscles contract.

Air inhaled

Lung

Diaphragm contracts (moves down)

Inhalation
(Air pressure is higher in atmosphere than in lungs.)

Rib cage gets smaller as rib muscles relax.

Air exhaled

Diaphragm relaxes (moves up)

Exhalation
(Air pressure is lower in atmosphere than in lungs.)

▼ **Figure 23.20 During exercise, control centers in the brain send signals to the body that increase breathing rate and depth.**

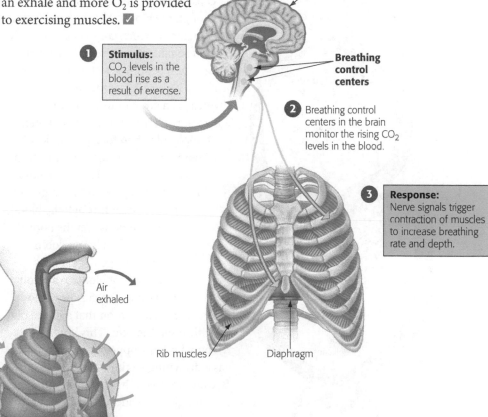

Brain

❶ **Stimulus:** CO_2 levels in the blood rise as a result of exercise.

Breathing control centers

❷ Breathing control centers in the brain monitor the rising CO_2 levels in the blood.

❸ **Response:** Nerve signals trigger contraction of muscles to increase breathing rate and depth.

Rib muscles

Diaphragm

545

The Role of Hemoglobin in Gas Transport

▼ Figure 23.21 Gas transport and exchange in the body. O_2-rich blood is sent from the lungs to capillaries in other body tissues via the heart. O_2 enters body tissue cells and CO_2 leaves, entering the blood. The O_2-poor blood is circulated back to the heart, then to the lungs.

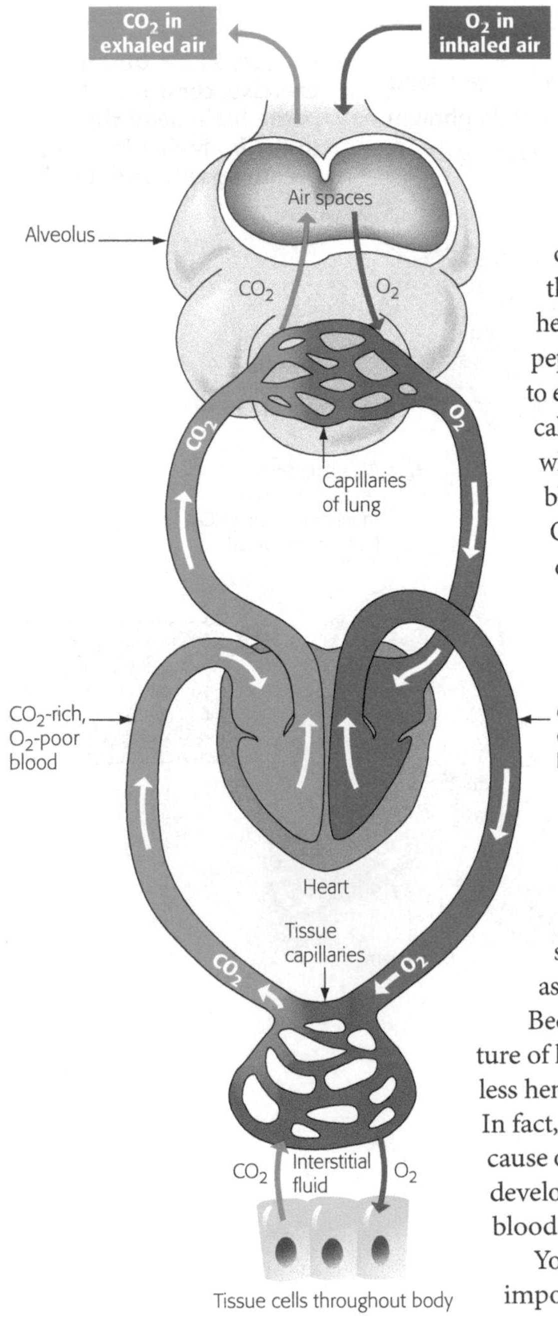

The human respiratory system takes O_2 into the body and expels CO_2, but it relies on the circulatory system to shuttle these gases between the lungs and the body's cells (Figure 23.21). In the lungs, O_2 diffuses from air spaces inside the alveoli into capillaries surrounding the alveoli. Oxygen moves out of the air and into the blood because air is richer in O_2. At the same time, CO_2 diffuses along its own gradient, from the blood out to the air in the lungs.

But there is a problem with this simple scheme: Oxygen does not readily dissolve in blood, so O_2 does not tend to move from the air into the blood on its own. Just as cargo is loaded onto a truck before it is transported along a highway, most O_2 must be loaded into red blood cells before it can be transported through the blood. The oxygen binds to hemoglobin, which consists of four polypeptide chains (Figure 23.22). Attached to each polypeptide is a chemical group called a heme ( ), at the center of which is an atom of iron (shown in black). Each iron atom can hold one O_2 molecule; therefore, one molecule of hemoglobin can carry a maximum of four molecules of O_2. Hemoglobin loads up on oxygen in the lungs, transports it through the blood, and unloads it at the body's cells. When hemoglobin binds oxygen, it changes the color of blood to a bright cherry red. Oxygen-poor blood is a dark maroon that appears blue through the skin. This is why a person who has stopped breathing—such as a drowning victim—turns blue.

Because iron is so important in the structure of hemoglobin, a shortage of iron causes less hemoglobin to be produced by the body. In fact, iron deficiency is the most common cause of anemia. Women are more likely to develop iron deficiency than men because of blood lost during menstruation.

You have probably heard about the importance of carbon monoxide (CO)

If hemoglobin in your blood could choose between binding to life-sustaining oxygen or poisonous carbon monoxide, it would pick the poison.

detectors in your home. CO is a colorless, odorless, poisonous gas that can bind to hemoglobin even more tightly than O_2 does. Breathing CO can therefore interfere with the delivery of O_2 to body cells, blocking cellular respiration and causing rapid death. CO is produced from the combustion of fuels, and if the gas accumulates in a sealed space, it can be deadly. Potential sources of CO poisoning in your home are faulty gas-powered appliances or use of grills or gas-powered vehicles inside enclosed spaces. Despite its potentially deadly effects, millions of Americans willingly inhale CO in the form of cigarette smoke.

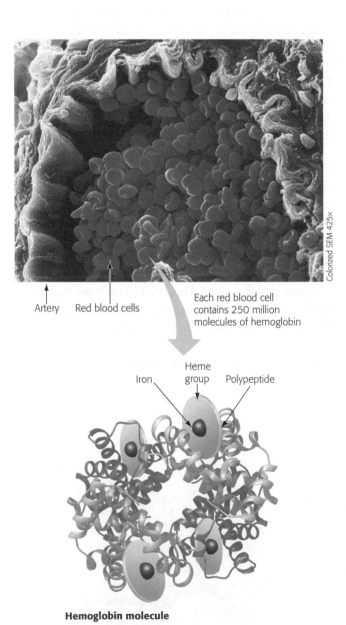

Colorized SEM 425x

Artery   Red blood cells

Each red blood cell contains 250 million molecules of hemoglobin

Iron   Heme group   Polypeptide

Hemoglobin molecule

▲ Figure 23.22 A molecule of hemoglobin. Four molecules of O_2 are transported through the blood by binding to the four iron atoms of one hemoglobin molecule.

Figure 23.21 labels

CO_2 in exhaled air

O_2 in inhaled air

Alveolus

Air spaces

CO_2 O_2

Capillaries of lung

CO_2

O_2

CO_2-rich, O_2-poor blood

O_2-rich, CO_2-poor blood

Heart

Tissue capillaries

CO_2 O_2

CO_2 Interstitial fluid O_2

Tissue cells throughout body

Smoking Damages the Structure and Function of the Lungs

Every breath you take exposes your respiratory tissues to potentially damaging chemicals. One of the worst sources of air pollution is cigarette smoke.

Smoking takes its toll on the respiratory system. A chronic cough and difficulty breathing are common problems for heavy smokers, and together these symptoms are known as chronic obstructive pulmonary disease (COPD). With COPD, both air flow and gas exchange are impaired. The epithelial tissue lining the bronchioles becomes irritated and swollen, and it produces more mucus than usual, blocking airflow to and from the alveoli. (Coughing occurs as a way to move the thick mucus.) The alveoli, where gas exchange occurs, are also damaged in COPD. The walls of the alveoli lose their elasticity, affecting their ability to expel air. Like walls between rooms being removed to form one big room, the alveolar walls are destroyed such that large air sacs form. With fewer alveoli and less surface area, gas exchange decreases.

Almost 20% of American adults smoke. Smoking and secondary exposure are responsible for about 1 in 5 deaths every year in the United States, more than all the deaths caused by accidents, alcohol and drug abuse, HIV, and murders *combined*. One in two American smokers will die from their habit. Smokers account for 80–90% of all cases of lung cancer, one of the deadliest forms of cancer (Figure 23.23). Lung cancer kills more Americans than any other form of cancer by a wide margin.

No lifestyle choice can have a more positive impact on your long-term health than not smoking. After quitting, it takes about 15 years for a former smoker's risk to even out with that of a nonsmoker. Even nonsmokers may be affected by tobacco, as studies have found that secondhand cigarette smoke is a substantial health hazard, particularly to young children. ☑

▼ Figure 23.23 **Healthy versus cancerous lungs.**

(a) Healthy lung (nonsmoker)　　**(b) Cancerous lung (smoker)**

☑ CHECKPOINT

How does smoking damage the lungs?

Answer: Smoking damages the epithelial tissue that lines airways, preventing air from flowing properly and destroying the delicate alveoli of the lungs where gas exchange occurs.

Athletic Endurance EVOLUTION CONNECTION

Evolving Endurance

In this chapter, we've used the topic of athletic endurance to illustrate the importance of the circulatory and respiratory systems. We've seen how conditioning and even artificially enhancing the blood can be used to boost athletic endurance by improving the ability of the circulatory and respiratory systems to deliver oxygen to muscles. But as it turns out, not everyone needs a boost to increase their stamina: Some people are born with more.

Tibet, a plateau within the Himalayan mountain range, is the highest region on Earth, with an average elevation of 16,000 feet—nearly 3 miles high. Many Tibetans live and work above 13,000 feet, where the amount of oxygen that reaches the blood is 40% less than at sea level (Figure 23.24). Without extensive conditioning, such altitudes render most people sick or even unconscious. Many Tibetans work as mountain guides yet rarely experience "mountain sickness." How have the Tibetan people come to be so tolerant of their surroundings?

A 2010 study by a team of Chinese biologists demonstrated that over the last several thousand years, Tibetans have evolved the ability to thrive at high altitude. By comparing the genomes of 50 Tibetans with the genomes of 40 nearby low-dwelling Chinese, the researchers found at least 30 genes that differed between the two populations. In each case, one version of the gene that is rare among the low-dwelling Chinese group had evolved to become common among Tibetans. Many of these genes are ones already known to contribute to the functioning of the circulatory and respiratory systems. Furthermore, scientists conducting this study were able to identify the mechanism of natural selection: At high altitude, people who lack such gene mutations have three times the infant mortality rate of people who harbor the mutations. There is a clear survival and reproductive advantage for those Tibetans who inherit the mutant genes. The Tibetan ability to work and thrive atop the Himalayans is a testament to their inborn endurance and another manifestation of biology's unifying theme of evolution.

▲ Figure 23.24 **A Tibetan mountaineer.** Tibetans are renowned for their ability to live and work at high altitudes.

Chapter Review

◼ SUMMARY OF KEY CONCEPTS

Unifying Concepts of Animal Circulation

The circulatory system facilitates the exchange of materials and energy between the cells of an organism and its environment. In an open circulatory system, circulating fluid is pumped through open-ended tubes and circulates freely among cells. In a closed circulatory system, the circulating fluid (blood) is confined within tubes.

The Human Cardiovascular System

The Path of Blood

Trace the path of blood in this figure, noting where blood is oxygen-poor and where it is oxygen-rich.

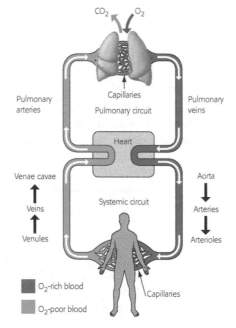

How the Heart Works

Review the structure of the heart by tracing the flow of blood in this diagram.

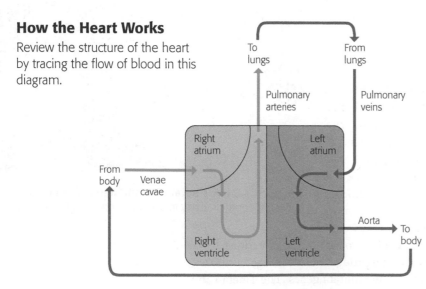

The cardiac cycle is composed of two phases: systole (contraction) and diastole (relaxation). The pacemaker, which sets the tempo of the heartbeat, generates electrical impulses that stimulate the atria and ventricles to contract.

Blood Vessels

Arteries carry blood away from the heart. Exchange between the blood and interstitial fluid occurs across the thin walls of capillaries. Valves in the veins and contractions of surrounding skeletal muscle keep blood moving back to the heart.

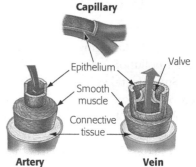

Blood

Red blood cells, platelets, plasma, and blood stem cells can be donated to patients in need.

CELLULAR COMPONENTS OF BLOOD		
Red Blood Cells (transport oxygen)	**White Blood Cells** (fight infections)	**Platelets** (allow for blood clotting)
SEM 2,750×	SEM 1,795×	SEM 1,825×

Cardiovascular Disease

Diseases of the heart and blood vessels—including heart attack and stroke—kill more Americans than any other type of disease. Atherosclerosis is the buildup of plaque in the inner walls of arteries.

Unifying Concepts of Animal Respiration

The respiratory system facilitates gas exchange. Various respiratory surfaces have evolved in animals. The respiratory surface is the part of the body where gas exchange takes place. This can include the entire body surface, gills, tracheae, or lungs.

The Human Respiratory System

The Path of Air

When you take a breath, air moves sequentially from the nostrils and/or mouth to the pharynx, larynx, trachea, bronchi (which enter the lungs), bronchioles, and finally the alveoli (the actual respiratory surfaces).

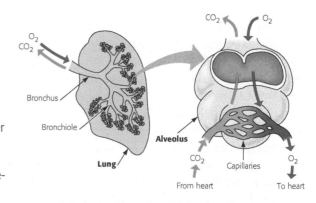

Information Flow: The Brain's Control over Breathing

During inhalation, the chest cavity expands and air pressure in the lungs decreases, causing air to rush into the lungs. During exhalation, air pressure in the lungs increases and air moves out of the lungs. The brain is influenced by CO_2 concentrations in the blood and sends signals to the muscles of the rib cage and diaphragm to control the rate and depth of ventilation.

The Role of Hemoglobin in Gas Transport

After O_2 enters the lungs, it binds to hemoglobin in red blood cells and is transported to body tissue cells by the circulatory system.

Smoking Damages the Structure and Function of the Lungs

Tobacco smoke damages the respiratory surfaces of the lungs, impairing alveolar function and air flow.

MasteringBiology®

For practice quizzes, BioFlix animations, MP3 tutorials, video tutors, and more study tools designed for this textbook, go to MasteringBiology®

SELF-QUIZ

1. What is the difference between an open circulatory system and a closed circulatory system?

2. Why is the following statement false? "All arteries carry oxygen-rich blood, while all veins carry oxygen-poor blood."

3. Match each chamber of the heart with its function.
   a. left atrium   1. receives blood from the body via the venae cavae
   b. right atrium   2. sends blood to the lungs via the pulmonary arteries
   c. left ventricle   3. receives blood from the lungs via the pulmonary veins
   d. right ventricle 4. sends blood to the body via the aorta

4. Some babies are born with a small hole in the wall of muscle that separates the left and right ventricles. How does this affect the oxygen content of the blood pumped out of the heart in the systemic circuit?

5. Oxygen is usually bound to _____ when it is transported through the body.

6. How does the structure of a human red blood cell support its function?

7. What happens to CO_2 output when a person exercises? Why?

8. What is the main difference between the respiratory organ of fishes and that of humans in terms of where they are located in the body?

9. In the human respiratory system, gas exchange occurs across the cells of the
   a. diaphragm.   c. bronchi.
   b. trachea.   d. alveoli.

10. When you hold your breath, which of the following changes in blood leads initially to the urge to breathe again?
    a. rising oxygen level
    b. falling oxygen level
    c. rising carbon dioxide level

11. Why do cigarette smokers cough more than most people do?
    a. The tar in cigarette smoke tends to make alveoli stick together. Coughing opens them.
    b. Coughing is the respiratory system's attempt to clear itself of the toxins found in smoke.
    c. Coughing stimulates blood flow to the lungs.

Answers to these questions can be found in Appendix: Self-Quiz Answers.

THE PROCESS OF SCIENCE

12. **Interpreting Data** Scientists in Brazil made a model of emphysema using mice, by injecting a chemical into the trachea of mice and waiting 28 days for the disease to develop. As a control, they injected other mice with a simple saline (salt) solution. The scientists measured the velocity at which all mice could run on a treadmill before injection and 28 days after injection. Control mice could run 18 miles/min at both time points. The emphysema-induced mice ran 19 miles/min before injection and 16 miles/min 28 days later. Draw a graph with the data presented in this description of the experiment. What conclusion do you draw from the data?

13. Eero Mäntyranta, an Olympic gold medalist cross-country skier from Finland, inherited a mutation such that he makes more red blood cells than the average human. Propose a hypothesis for what kind of mutation Mäntyranta could have inherited and how you could test your hypothesis.

14. Many athletes—runners in particular—have begun to wear compression socks, similar to those that people wear on airplanes in order to avoid blood clotting. How can these socks improve athletic performances?

BIOLOGY AND SOCIETY

15. Alcohol and cigarette pose a major threat to health due to their negative effects on the cardiovascular system. Health problems arising from the consumption of these legal drugs have a huge financial impact on the health services. By comparison, several illegal drugs, such as heroin, pose a much smaller burden on the economy of the industrialized countries. Why are alcohol and cigarette legal, whereas other drugs are not?

16. A 12-year-old girl will die from leukemia if she does not get a bone marrow transplant, but a matching donor cannot be found. A newborn's umbilical cord contains blood-forming stem cells, which can be harvested after birth and donated. The girl's parents decide to have another child in a final attempt to provide their daughter with a matching donor. Doctors report that having a "savior sibling" is not uncommon. In your opinion, is it acceptable to have a child in order to save another child's life? Why or why not?

24 The Body's Defenses

One of the things standing between you and cancer is a natural killer cell. ►

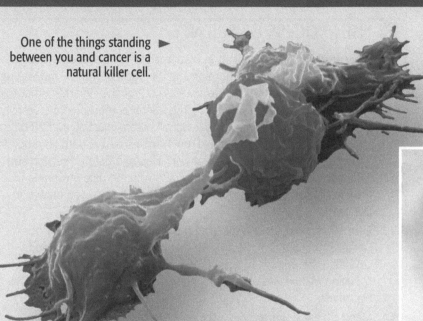

Depending on where you live, you could ▼ develop an allergy to meat.

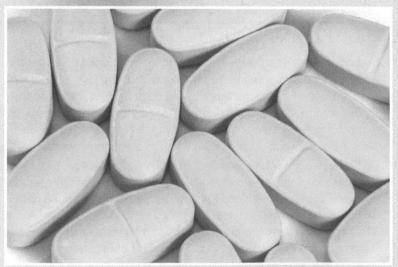

▲ A pill that prevents HIV is no longer an abstract concept: It's a reality.

Vaccines BIOLOGY AND SOCIETY

Herd Immunity

In the not so distant past, the United States had impressive control over the extremely contagious disease measles. A measles vaccine developed in the late 1950s and widely distributed thereafter eliminated the disease from the United States in the year 2000. But, in recent years, the virus that causes measles has been making a comeback in some regions of the country. Two recent cases illustrate the ease with which the virus is re-entering society. In 2013, 23 people developed measles in rural North Carolina, and in 2014, 68 measles cases occurred in an Ohio Amish community. In both outbreaks, individuals who had never been vaccinated against measles unknowingly brought the virus home to their families and communities after traveling abroad in countries where the disease is prevalent. A North Carolinian carried back the virus from India, and Amish missionaries returned from the Philippines with the virus.

Outbreaks of dangerous viral diseases, like measles, are rare in the United States because most people in the country are protected by vaccination. Those who can't be vaccinated, like infants and people with certain illnesses, are protected to a large

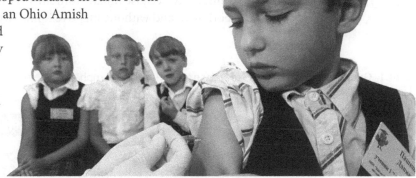

Vaccinating a community. When almost all members of a community are vaccinated, disease is much less likely to spread.

extent because others in the community who are vaccinated prevent the virus from taking hold. This community protection, referred to as "herd immunity," is the rationale behind state-mandated vaccinations for children in public schools and national health campaigns reminding you to get your flu vaccine each year. Still, the cases of measles in rural North Carolina and Amish Ohio remind us that even after successful elimination of a disease by widespread vaccination outbreaks can occur. According to the Centers for Disease Control and Prevention (CDC), herd immunity fails when 5% of a community is unvaccinated.

Like firefighters containing a forest fire before it rages out of control, a concerted public health effort in North Carolina and Ohio isolated individuals with measles and vaccinated the unvaccinated to stop the disease from spreading. One of the greatest medical advances of the 1900s was the development and distribution of vaccines against deadly diseases. Massive vaccination efforts have eliminated measles and other viral diseases such as polio and mumps from many countries, and the deadly viral disease smallpox has been eradicated worldwide because of heroic vaccination campaigns. The more people who are vaccinated against viral diseases, the safer it is for everyone.

Vaccinations work by priming your body's natural defenses, and in this chapter we'll examine how these defenses make up your body's immune system. We'll also learn how our knowledge of the immune system has been applied to improve human health, what happens when the immune system malfunctions, and where we are in our fight against AIDS (acquired immune deficiency syndrome).

An Overview of the Immune System

Almost everything in the environment—including the door handles you touch and even the air you breathe—teems with microbes, some of which can make you sick. Yet you're well most of the time, thanks to your **immune system**, your body's defense against infectious disease.

The human immune system is a collection of organs, tissues, and cells that together perform the vital function of safeguarding the body from a constant barrage of **pathogens**, disease-causing agents like viruses and bacteria. The protection provided by your immune system consists of two parts **(Figure 24.1)**. One part, **innate immunity**, doesn't change much from the time you are born, and its components attack pathogens indiscriminately. The other part, **adaptive immunity**, continually develops over your lifetime as it encounters and attacks specific pathogens. With its continual development, adaptive immunity ensures you will have better protection from a specific pathogen each time you encounter it. Scientists suspect that innate immunity evolved before adaptive immunity because innate immunity is found in animals both with and without adaptive immunity.

Innate immunity includes both external and internal defenses, which are, respectively, considered the first and second lines of defense. External innate defenses are on the frontline, preventing pathogens from getting deep inside the body, whereas internal innate defenses lie in wait, confronting pathogens that make it past external defenses. Adaptive immunity, the third line of defense, is strictly internal, deploying when innate immunity defenses fail to ward off a pathogen. Nearly all animals have innate immunity (the first and second lines of defense), but only vertebrates have adaptive immunity (the third line of defense).

Innate and adaptive immunities—the immune system—interact with and rely on the other systems in the body, but particularly the lymphatic system. The **lymphatic system** is a network of vessels, tissues, and organs where pathogens and cells involved in innate immunity and adaptive immunity interact with each other to carry out defensive actions. We'll discuss the lymphatic system and adaptive immunity in detail, but before we do, let's learn a bit more about innate immunity. ☑

☑ CHECKPOINT

What are the two parts of the immune system? Which part provides better protection the second time a specific pathogen is encountered?

Answer: innate immunity and adaptive immunity; adaptive immunity continually "adapts" to better fight specific pathogens each time they are encountered.

▼ **Figure 24.1 Overview of the body's defenses.** Note that the lymphatic system is involved in both innate and adaptive defenses.

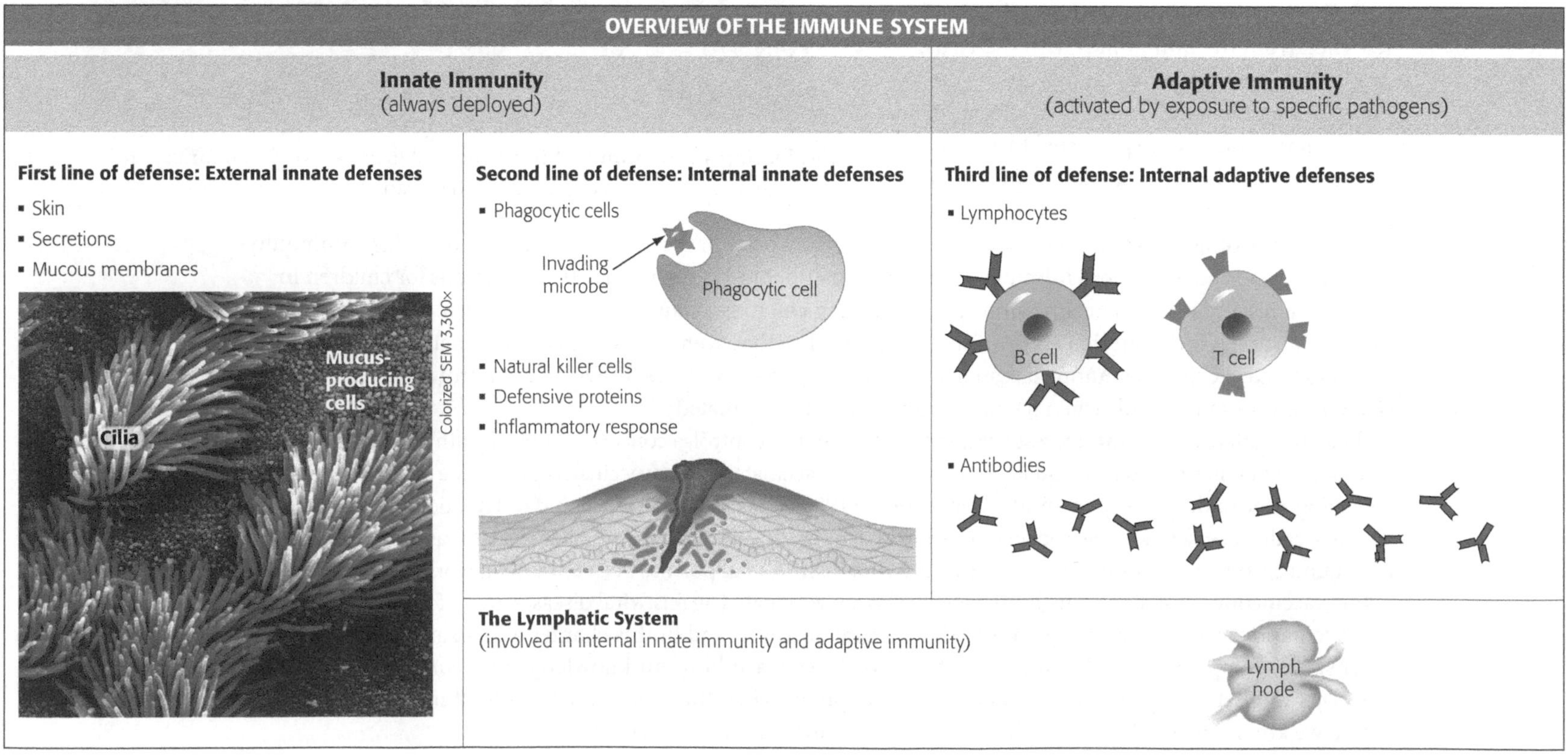

OVERVIEW OF THE IMMUNE SYSTEM

Innate Immunity (always deployed)

Adaptive Immunity (activated by exposure to specific pathogens)

First line of defense: External innate defenses
- Skin
- Secretions
- Mucous membranes

Cilia

Mucus-producing cells

Colonized SEM 3,300×

Second line of defense: Internal innate defenses
- Phagocytic cells

Invading microbe

Phagocytic cell

- Natural killer cells
- Defensive proteins
- Inflammatory response

Third line of defense: Internal adaptive defenses
- Lymphocytes

B cell   T cell

- Antibodies

The Lymphatic System (involved in internal innate immunity and adaptive immunity)

Lymph node

Innate Immunity

Innate immunity protects the body when a pathogen first attempts to infect the body. This protection is accomplished with two lines of defense. External innate defenses keep the pathogen from entering the body and do so in a variety of ways. Internal innate defenses are ready with immune cells and defensive proteins should a pathogen make it past the external innate defenses. Let's delve further into the functions of these two lines of defenses by looking at how external and internal innate defenses protect humans.

External Innate Defenses

For a pathogen to get inside your body, it must first get by your body's external innate defenses. These defenses form the frontline of your immune system because they *prevent* infection, as opposed to your body's other defenses, which fight an infection *after* it occurs.

Some barriers of external innate defenses block or filter out pathogens. Intact skin forms a tough outer layer that most bacteria and viruses cannot penetrate (**Figure 24.2a**). Nostril hairs filter many particles from the incoming air. Ear wax traps pathogens before they can travel too far down the ear canal. Organ systems that are open to the external environment (such as the respiratory and reproductive systems) are lined with cells that secrete mucus, a sticky fluid that traps bacteria, dust, and other particles (**Figure 24.2b**). Beating cilia extending from cells of the respiratory tract sweep mucus, including trapped pathogens, outward until the mucus is either swallowed or expelled when you sneeze, cough, or blow your nose.

External innate defenses also include chemical barriers in the form of antimicrobial secretions. Sweat, saliva, and tears contain enzymes that disrupt bacterial cell walls. The skin contains oils and acids that make it inhospitable to many microorganisms. The cells of the stomach produce acid that kills most of the bacteria we swallow before they can get past the digestive system's lining and enter the bloodstream.

Internal Innate Defenses

The internal innate defenses depend on white blood cells and defensive proteins. Two types of white blood cells contribute to your internal innate defenses: phagocytic cells and natural killer cells. **Phagocytic cells** (also called phagocytes) engulf foreign cells or molecules and debris from dead cells by phagocytosis, or "cellular eating" (see Figure 5.18). **Natural killer (NK) cells** recognize virus-infected and cancerous body cells. When contact is made, the NK cells release chemicals that kill the diseased cells. New therapies have been developed that utilize NK cells and their ability to infiltrate and kill cancerous tissue.

Along with white blood cells, a variety of defensive proteins are part of the internal innate defenses. Some defensive proteins prevent viral reproduction within body cells, others destroy pathogens directly, and still others trigger the inflammatory response.

What happens to your finger if you get a splinter and don't remove it? The next day, the area around the splinter is often red, swollen, painful, and warm to the touch.

> One of the things standing between you and cancer is a natural killer cell.

◄ **Figure 24.2 External innate defenses block or filter out pathogens**.

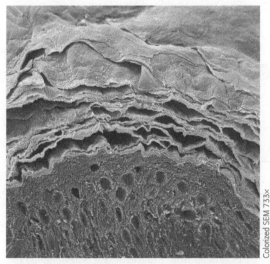

(a) Skin forms a protective barrier. The outermost layers of skin (brown) consist of dead cells that are continually shed and replaced by layers of living cells below (red).

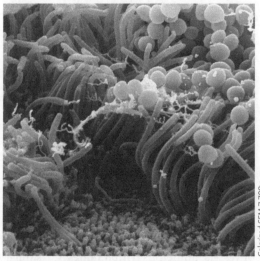

(b) Cilia on cells in the nasal cavity sweep mucus (blue) and trapped bacteria (yellow) out of the body.

These are signs of the **inflammatory response** to injury and infection, another example of an internal innate defense in action (**Figure 24.3**). When tissue becomes damaged (from cuts, scratches, bug bites, or burns, for example), ❶ the injured cells release chemicals that trigger various internal innate defenses. One such chemical signal, **histamine**, ❷ causes nearby blood vessels to dilate (widen) and leak fluid into the wounded tissue, a process called swelling. The excess fluid heals damaged tissue by diluting toxins in it, bringing it extra oxygen, and delivering platelets and clotting proteins to it that promote scabbing. The chemical signals also attract phagocytic cells. ❸ These cells engulf bacteria and the remains of body cells killed by bacteria or by the physical injury. The pus that often fills an infected injury consists of phagocytic cells, fluid that leaked from capillaries, and other tissue debris.

Damaged cells release chemical signals that increase blood flow to the damaged area, causing the wound to turn red and warm. (The word *inflammation* means "setting on fire.") In response to severe tissue damage or infection, damaged cells may release other chemicals that travel through the bloodstream to the brain, where they stimulate a fever that may discourage bacterial growth.

Anti-inflammatory drugs dampen the normal inflammatory response and help reduce swelling and fever. Ibuprofen, for example, inhibits dilation of blood vessels, which reduces swelling. This is an example of treating the symptoms of an illness (in this case, swelling) without addressing the underlying cause (which is the injury or infection).

All the defenses you've learned about so far are called *innate* because they're ready "off the rack"; that is, innate defenses are already deployed in the body and at the ready without any preparation. Soon you'll learn about the body's adaptive defenses—ones that are "tailored" to specific invaders. But first let's look at a system that contributes to both innate and adaptive defenses: the lymphatic system. ☑

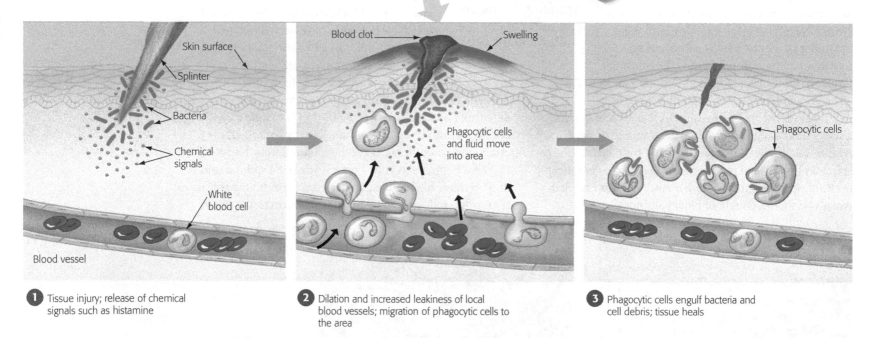

❶ Tissue injury; release of chemical signals such as histamine

❷ Dilation and increased leakiness of local blood vessels; migration of phagocytic cells to the area

❸ Phagocytic cells engulf bacteria and cell debris; tissue heals

▲ **Figure 24.3 The inflammatory response.** Whenever tissue is damaged, the body responds with a coordinated set of internal innate defenses called the inflammatory response.

The Lymphatic System

As you see in **Figure 24.4a**, the lymphatic system consists of a branching network of vessels, numerous **lymph nodes** (little round organs packed with white blood cells), and a few other organs, such as the spleen, appendix, and tonsils.

The lymphatic vessels carry fluid called **lymph**, which is similar to the interstitial fluid surrounding body cells. The two main functions of the lymphatic system are to return tissue fluid to the circulatory system and to fight infection.

Circulatory Function

As blood travels in the circulatory system, fluid exits the blood through small gaps between the cells of capillaries. The fluid then enters the interstitial space surrounding tissues **(Figure 24.4b)**. Nutrients and wastes for all cells are exchanged in this interstitial fluid. After nutrient and waste exchange, most of the fluid reenters the blood and the circulatory system through capillaries, but some remains in the tissue. This excess fluid flows into small lymphatic vessels. From the small lymphatic vessels, the fluid, now called *lymph*, drains into larger and larger lymphatic vessels. Eventually, lymph enters the circulatory system through two large lymphatic vessels that fuse with veins near the shoulders. If lymph doesn't drain well from tissues (a condition you have no doubt experienced), the tissue swells.

Immune Function

When your body is fighting an infection, the lymphatic system is the main battleground. Because lymphatic vessels penetrate nearly every tissue, lymph can pick up pathogens from infection sites just about anywhere in the body. As this fluid circulates, phagocytic cells inside lymphatic tissues and organs engulf the invaders. Lymph nodes are key sites where particular white blood cells called **lymphocytes** multiply during times of infection. Normally glands in your neck and armpits are about the size of a green pea, but with an infection, a swollen gland can be as big as an acorn for a few days as it expands with lymphocytes. The types of lymphocytes that multiply in the lymph nodes are essential to the immune system's third line of defenses, the adaptive defenses, which we'll discuss next. ☑

☑ CHECKPOINT
What are swollen glands?

Answer: lymph nodes where lymphocytes (white blood cells) are multiplying during an infection

▼ Figure 24.4 **The human lymphatic system.**

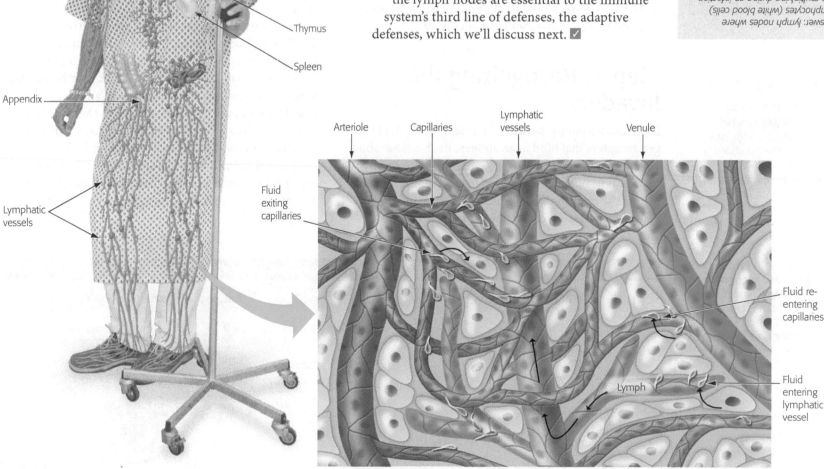

Tonsil

Lymph nodes

Lymphatic vessels entering veins

Thymus

Spleen

Appendix

Lymphatic vessels

(a) The organs and vessels of the lymphatic system.

Arteriole

Capillaries

Lymphatic vessels

Venule

Fluid exiting capillaries

Fluid re-entering capillaries

Lymph

Fluid entering lymphatic vessel

(b) Lymphatic vessels. Interstitial fluid that enters lymphatic vessels is called lymph.

Adaptive Immunity

The immune system's third line of defense, adaptive immunity, depends on two types of lymphocytes that recognize and respond to specific invading pathogens. Like all blood cells, lymphocytes originate from stem cells in bone marrow. **B cells** fully develop and become specialized in bone marrow, whereas immature **T cells** migrate via the blood to the thymus, a gland in the chest, where they mature and become specialized. Both B cells and T cells eventually make their way to the lymph nodes and other lymphatic organs and wait to encounter an invader.

Any molecule that elicits a response from a lymphocyte is called an **antigen**. Most antigens are molecules on the surfaces of viruses or foreign cells, such as cells of bacteria or parasitic worms. Any given pathogen will have many different antigens on its surface. Antigens also include toxins secreted from bacteria, molecules from mold spores, pollen, house dust, as well as molecules on cell surfaces of transplanted tissue.

Unlike innate immunity defenses, which are ready to fight pathogens at any time, the adaptive immunity defenses, specifically B cells and T cells, must be primed before they attack foreign molecules. In the next few sections, we look at how B cells and T cells recognize, respond to, and remember their enemy. ✓

Step 1: Recognizing the Invaders

Protruding from the surface of B cells and T cells are antigen receptors that bind to an antigen. Each cell has about 100,000 copies of an antigen receptor that detects only a single type of antigen. One cell may recognize an antigen on the mumps virus, for instance, whereas another detects an antigen on a tetanus-causing bacterium.

The antigen receptors of B cells and T cells have similarities in their structures, but they recognize antigens in different ways (**Figure 24.5**). Antigen receptors on B cells specialize in recognizing intact antigens that are on the surface of pathogens or circulating freely in body fluids. The unique shape of the antigen and the complementary antigen receptor on the B cell results in a lock-and-key fit that activates the B cell. In contrast, antigen receptors on T cells only recognize fragments of antigens, and the fragments must be displayed, or presented, on the surface of body cells by special proteins before T cells are activated.

The fragments of antigens that T cells recognize originate from pathogens that have entered a body cell. Enzymes inside the body cell break the pathogen into fragments. Like a retailer displaying a few pieces of merchandise in her shop window, the body cell "advertises" the antigen fragments on the outer surface of its own proteins (green "self proteins" in Figure 24.5). A T cell bearing a receptor with specificity for this antigen fragment binds to both the antigen and self protein. This three-part interaction among a self protein, an antigen fragment, and an antigen receptor is required for a T cell to function.

The immune system develops a great diversity of B cells and T cells—enough to recognize and bind to just about every possible antigen. A small population of each kind of lymphocyte lies in wait in your body, ready to recognize and respond to a specific antigen. Over your lifetime, only a tiny fraction of the various types of your B cells and T cells will ever be used, but all the variations are available if needed. It is as if the immune system maintains a huge standing army of soldiers, each made to recognize one particular kind of invader. The majority of soldiers, however, never encounter an invader. ✓

▶ **Figure 24.5 Two ways antigens are recognized by the adaptive defenses.** Note that both B cells and T cells bind to the antigen, but T cells additionally require that the antigen be bound to a self protein.

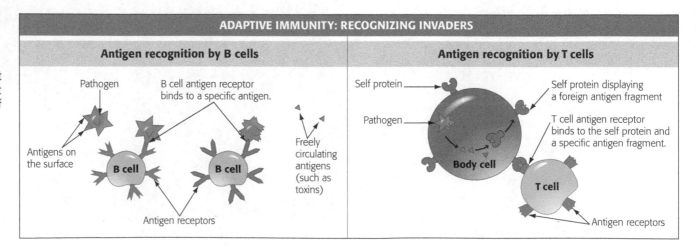

ADAPTIVE IMMUNITY: RECOGNIZING INVADERS

| Antigen recognition by B cells | Antigen recognition by T cells |

Antigen recognition by B cells: Pathogen; B cell antigen receptor binds to a specific antigen; Antigens on the surface; B cell; B cell; Freely circulating antigens (such as toxins); Antigen receptors

Antigen recognition by T cells: Self protein; Pathogen; Body cell; Self protein displaying a foreign antigen fragment; T cell antigen receptor binds to the self protein and a specific antigen fragment; T cell; Antigen receptors

Step 2: Cloning the Responders

Once a pathogen is inside the body it multiplies. The number of lymphocytes able to respond to this specific pathogen similarly rises. Let's now examine the process of lymphocyte proliferation, which is alike in both B and T cells.

Recall that the immune system maintains a vast collection of different kinds of lymphocytes. With so many kinds of B and T cells, how does the body marshal enough of the right kind of lymphocyte to fight a specific invading pathogen? The key is a process called **clonal selection**. The concept is simple. At first, an antigen activates only a tiny number of lymphocytes with specific antigen receptors. These "selected" lymphocytes then multiply through cell division, forming a clone of lymphocytes (a population of genetically identical B cells or T cells) with a specific antigen receptor.

Figure 24.6 illustrates how clonal selection of lymphocytes works, using B cells as an example. Note that a similar mechanism activates clonal selection for T cells. The top row shows three different B cells, each with its own specific type of antigen receptor embedded in its surface. ❶ Once a pathogen enters the body, antigens on its surface bind with a B cell that has complementary antigen receptors. Other lymphocytes without the appropriate binding sites are not affected. ❷ The binding activates the B cell—it grows, divides, and develops further. This produces clones of B cells specialized for defending against the very antigen that triggered the response. ❸ Some of the newly produced B cells are short-lived cells that have an immediate effect against the antigen and are therefore called **effector cells**. In this example with clonal selection of B cells, the effector cells secrete huge quantities of **antibodies**, defensive proteins that bind antigens. You'll soon learn how antibodies help destroy invaders.

During the first response to an antigen, called the **primary immune response**, it takes several days for clonal selection to produce effector cells. The primary immune response peaks about two to three weeks after the first exposure, at which point it starts to decline. Each effector cell lives only four or five days, and the primary immune response subsides as the effector cells die out.

Clonal selection also produces memory cells that will help fight subsequent exposures to a specific antigen. ❹ **Memory cells** are long-lived cells found in the lymph nodes, ready to attack should a "known" antigen infect the body again. If memory cells are exposed to a previously encountered antigen, they rapidly give rise to new effector cells and memory cells, a process known as the **secondary immune response**. Thus, clonal selection produces not only cells that will fight the first exposure to an antigen (effector cells) but also cells that will respond to future exposures (memory cells). ☑

▼ Figure 24.6 **Clonal selection during the first exposure to an antigen.** In this example of clonal selection, an antigen binds to a B cell and causes the cell's proliferation. A similar kind of clonal selection operates on T cells.

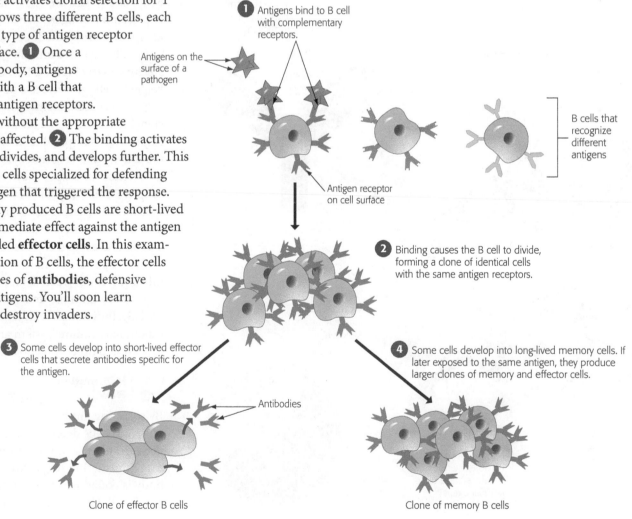

❶ Antigens bind to B cell with complementary receptors.

Antigens on the surface of a pathogen

B cells that recognize different antigens

Antigen receptor on cell surface

❷ Binding causes the B cell to divide, forming a clone of identical cells with the same antigen receptors.

❸ Some cells develop into short-lived effector cells that secrete antibodies specific for the antigen.

Antibodies

❹ Some cells develop into long-lived memory cells. If later exposed to the same antigen, they produce larger clones of memory and effector cells.

Clone of effector B cells

Clone of memory B cells

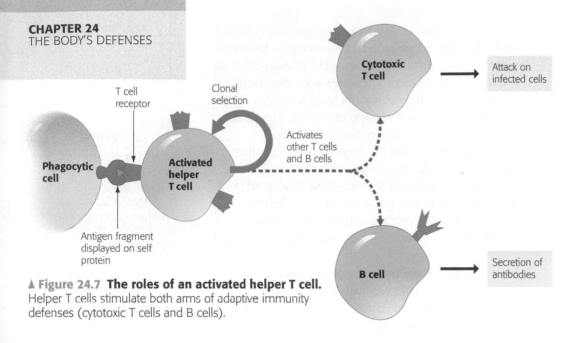

▲ **Figure 24.7 The roles of an activated helper T cell.** Helper T cells stimulate both arms of adaptive immunity defenses (cytotoxic T cells and B cells).

▼ **Figure 24.8 The B cell response.** Effector B cells secrete antibodies that bind antigens.

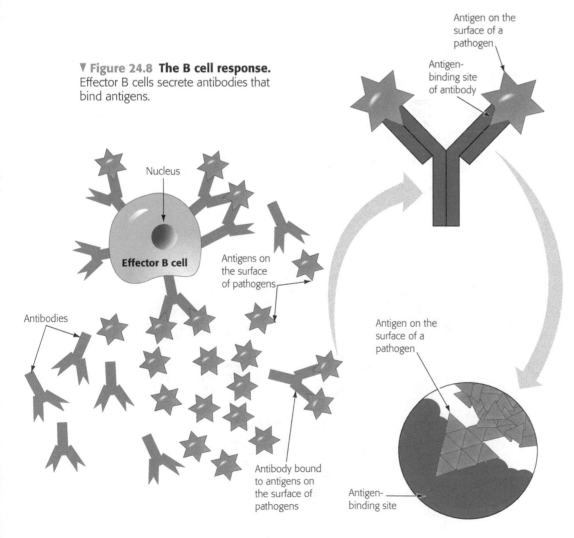

Step 3: Responding to Invaders

Now that you know how lymphocytes clone themselves, we'll look more closely at how B cells and T cells differ. While the antibody response from B cells helps to eliminate pathogens in the blood and lymph, the **cytotoxic T cell** destroys pathogens within body cells. Another type of T cell, the **helper T cell**, does not directly carry out attacks on pathogens but aids in stimulating both the B cells and the cytotoxic T cells in their responses. Without helper T cells, B cells and cytotoxic T cells could not initiate defenses against invaders.

The Helper T Cell Response

Like other T cells, each helper T cell only recognizes a specific antigen fragment displayed on self proteins **(Figure 24.7)**. Additionally, helper T cells are only activated when particular white blood cell types "advertise" the antigen. Phagocytes of the internal innate defenses that migrate into the lymphatic system are one of the cell types capable of presenting antigens to helper T cells—thus, phagocytes provide a direct link between innate immunity, adaptive immunity, and the lymphatic system (review Figure 24.1). Once activated, helper T cells give rise to a population of effector helper T cells and memory helper T cells through clonal selection. As we see in Figure 24.7, effector helper T cells respond to infection by stimulating the activity of B cells and cytotoxic T cells.

Because the helper T cell has a central role in adaptive immunity, the destruction of this cell type has devastating consequences. The human immunodeficiency virus (HIV) infects helper T cells, and as we'll explore later, HIV infection causes helper T cell numbers to decline significantly. If not treated, HIV infection results in acquired immune deficiency syndrome (AIDS). Individuals with AIDS lack a completely functional immune system and can die from exposure to other infectious agents.

Information Flow **The B Cell Response**

By secreting antibodies into the blood and lymph, B cells defend primarily against pathogens circulating in body fluids **(Figure 24.8)**. The information for making a specific antibody is coded within the DNA in the nucleus of a B cell. Ultimately, four polypeptides are synthesized and joined to form a single Y-shaped antibody protein that is then secreted. The tip of each "Y" forms a region, an antigen-binding site, that will recognize and bind to a specific antigen in a lock-and-key structure. Antibodies are secreted at a furious pace: One effector B cell can produce up to 2,000 antibodies per second.

How do secreted antibodies defend against pathogens circulating in body fluids? Antibodies may serve as physical barriers that prevent pathogens from entering body cells (Figure 24.9a). In this example, antibodies block the viral attachment proteins necessary for entering and infecting a body cell. Antibodies also aid in pathogen destruction. The binding of antibodies to antigens on the surface pathogens can also result in clumps that are easily engulfed and destroyed by circulating phagocytic cells (Figure 24.9b).

Because of their ability to tag specific molecules or cells, antibodies are useful in medicine. Herceptin, a genetically engineered antibody, treats certain cases of aggressive breast cancer. When Herceptin binds to a receptor present on some breast cancer cells, it interferes with the cells' ability to receive cell growth signals. With growth receptors blocked, cancer cells cannot grow and divide. In home pregnancy tests, antibodies are used to detect a hormone called human chorionic gonadotropin (HCG), present in the urine of pregnant women. When a testing strip is dipped into urine, antibodies on the strip bind to HCG, causing the strip to change color.

▼ Figure 24.9 **The binding of antibodies to antigens blocks or helps to destroy an invader.**

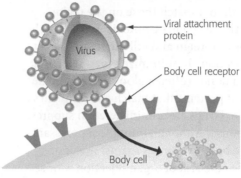

Virus entering a body cell

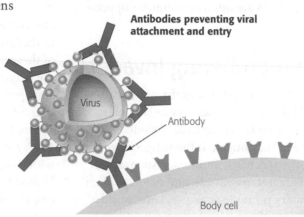

Antibodies preventing viral attachment and entry

(a) **Antibodies block a virus from entering a body cell.**

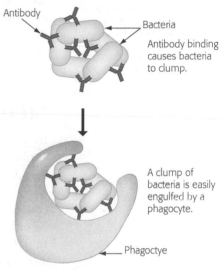

(b) **Antibodies enhance phagocytosis.**

The Cytotoxic T Cell Response

Whereas the B cell response defends against pathogens in body fluids, the cytotoxic T cell response defends against pathogens that have entered body cells. Cytotoxic T cells are the only T cells that actually kill infected cells. They identify infected body cells because foreign antigen fragments are "advertised," or bound to a self protein (see Figure 24.5). As shown in Figure 24.10, ❶ a cytotoxic T cell binds to an infected cell. The binding activates the T cell. ❷ The activated cell synthesizes and discharges several proteins that enter the infected cell, and trigger ❸ death of the cell.

▼ Figure 24.10 **The cytotoxic T cell response.**
Effector cytotoxic T cells kill infected body cells.

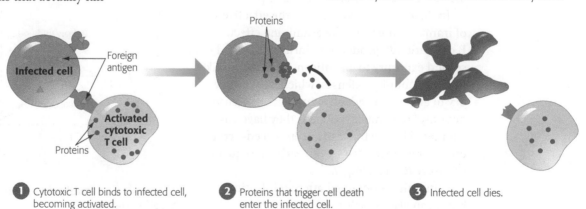

❶ Cytotoxic T cell binds to infected cell, becoming activated.

❷ Proteins that trigger cell death enter the infected cell.

❸ Infected cell dies.

559

The cytotoxic T cell response does not always work in a person's favor. Consider organ transplantation. When an organ is transplanted from a donor into a recipient, the newly transplanted cells contain self proteins that do not match those on the recipient's cells. Thus, the recipient's cytotoxic T cells tag the transplanted cells as foreign and kill them, ultimately causing organ rejection. To minimize organ rejection, doctors look for a donor (often a blood relative of the recipient) with self proteins that match the recipient's as closely as possible. Drugs are administered to suppress the immune response. Although immunosuppressant drugs weaken the body's defenses against infections, they greatly reduce the risk of organ rejection. Consequently, organ recipients are often on immunosuppressants for life. ☑

Step 4: Remembering Invaders

In contrast to short-lived effector cells that produce antibodies or kill infected cells, memory cells can last decades in the lymph nodes, ready to be activated by a second exposure to the antigen. If the antigen is encountered again, the secondary immune response will be more rapid, of greater magnitude, and of longer duration than the primary immune response. Let's consider a childhood infection. If you had chicken pox as a child, your body has memory cells ready to recognize the virus that causes the disease. If the virus that causes chicken pox enters your body, memory cells recognize the pathogen, multiply quickly, and produce many new B and T effector cells that destroy the invader—even before you feel ill. Because of your first childhood exposure, we say you have long-term "immunity" to the virus. The immune system's ability to remember prior infection provides long-term protection against many diseases. This type of protection was noted almost 2,400 years ago by the Greek historian Thucydides. He observed that individuals who had recovered from the plague could safely care for those who were sick or dying.

In **Figure 24.11**, you can see the effectiveness of immune "memory" by examining effector B cell antibody production during a first and second exposure to an antigen. Note that unlike innate immunity defenses, which have no memory of an antigen, adaptive immunity defenses not only "remember" antigens they have encountered before but also are enhanced—come on faster and stronger—with each subsequent exposure to that antigen.

Immunity is obtained after an infection, but it can also be achieved artificially by

vaccination (also called immunization). The immune system can be confronted with a **vaccine** composed of an inactive or otherwise harmless version of a pathogen. Vaccination induces the primary immune response that produces memory cells. An encounter with the pathogen later in life then elicits a rapid and strong secondary immune response. When a person has been successfully vaccinated, the immune system responds quickly and effectively to the pathogen. Such protection may last for life.

In the United States, most children receive a series of shots starting soon after birth. The shots include vaccinations against diphtheria/pertussis/tetanus (DPT), polio, hepatitis, chicken pox, and measles/mumps/rubella (MMR). As referenced in the Biology and Society section, widespread childhood vaccination in the United States has virtually eliminated several of these diseases, including polio and mumps. One of the major triumphs of modern vaccination involves smallpox, a potentially fatal viral infection that affected over 50 million people a year worldwide in the 1950s. A massive vaccination effort was so effective that in 1977, a man from Somalia was the last known patient with smallpox. The World Health Organization of the United Nations confidently declared smallpox to be completely eradicated in 1980.

So far, we have seen how lymphocytes recognize invaders and then remember them upon subsequent encounters. We have also explored how vaccines utilize the memory response of adaptive immunity to provide protection. Next, we'll look at one of history's most publicized vaccine studies to see how scientists determine the effectiveness of a vaccine. ☑

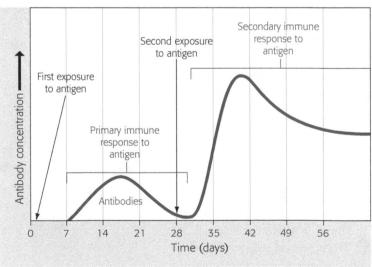

▲ Figure 24.11 **Antibody production during the two phases of the B cell response.**

How Do We Know Vaccines Work?

In 1954, as in previous years, a large number of polio cases were reported around the United States, and many thousands of people infected with poliovirus were experiencing degrees of paralysis in the muscles that control walking, swallowing, and breathing. Polio is a contagious disease that attacks the nervous system, and although paralysis is a temporary symptom of polio in some infected individuals, it can be a permanent outcome of the disease for others. However grim 1954 was in terms of polio cases, it was also a year of hope. A researcher, Doctor Jonas Salk, had developed a polio vaccine and was testing its effectiveness.

Salk's idea for the polio vaccine stemmed from two historic medical **observations**: (1) Once infected with a disease, the immune system protects a person from getting that same disease again (a phenomenon known as immune memory), and (2) infectious diseases are caused by microbes (not by "bad air," as polio was once thought to be transmitted). In the early 1950s, Salk set out to answer a **question**: Could a vaccine be developed that would prevent polio? His **hypothesis** was that an inactivated form of the polio virus could prime the immune system. By 1953, after developing a polio vaccine and trying it out on himself and his own family, Salk made the **prediction** that his polio vaccine would confer protection from the disease in the population at large.

To test his prediction, Salk and fellow researchers conducted a controlled **experiment**. Because paralysis from the disease is more likely to occur in children than in adults, the polio vaccine experiment was conducted on children. Approximately 400,000 children participated, with each child randomly assigned to either a control group or an experimental group.

Each group ultimately contained about 200,000 children. Children in the control group were injected with salt water (a placebo); children in the experimental group were injected with the polio vaccine (**Figure 24.12**). The experiment was "double-blind," meaning that none of the children, nor their parents, nor the researchers injecting the children knew to which group a child was assigned. Once all the children were injected, scientists waited to see how many of the children became infected with the poliovirus over a six-month period.

In 1955, the **results** of the experiment were revealed to the world via television, radio, and newspapers. There were significantly fewer cases of polio and of its paralytic symptoms in children who received the vaccine injection instead of the placebo (**Table 24.1**). These results were decisive: Salk's polio vaccine was effective. The vaccine was immediately licensed, and within a few years, the number of polio cases in the United States dropped by 85–90%. By 1979, polio had been eradicated in the United States.

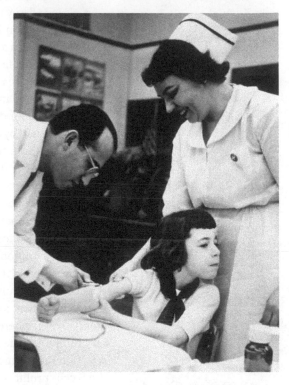

▲ Figure 24.12 **The polio vaccine.** Jonas Salk (1914–1995) is seen injecting a young girl with the polio vaccine that he developed.

Table 24.1	Results from the Salk Polio Vaccine Trials of 1954		
Study Group	**Number of children**	**Number of polio cases**	**Number of polio cases involving paralysis**
Control group (received placebo injection)	201,229	142	115
Experimental group (received polio vaccine)	200,745	57	33

Immune Disorders

If the intricate interplay between innate immunity and adaptive immunity goes awry, problems can arise that range from mild irritations to deadly diseases. In this section, we'll examine some of the consequences of a malfunctioning immune system.

Allergies

An **allergy** is an exaggerated sensitivity to an otherwise harmless antigen in the environment. Antigens that cause allergies are called **allergens**. Common allergens include

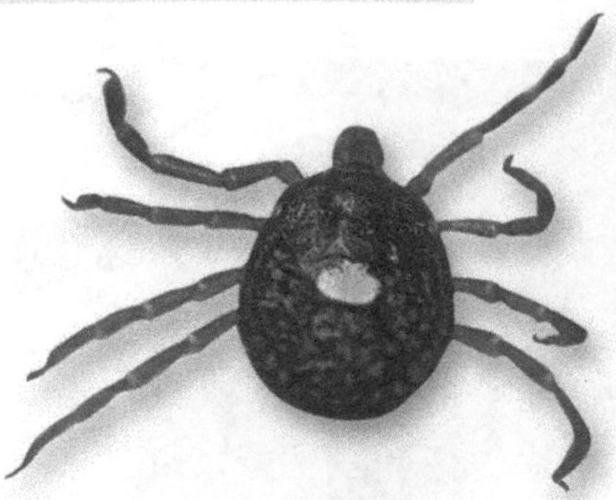

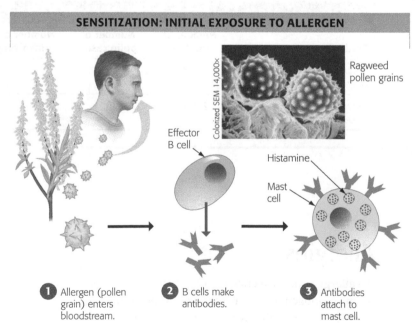

▲ **Figure 24.13 The lone star tick.** This tick is to blame for the rising numbers of meat allergies across the southern and eastern regions of the United States.

✓ **CHECKPOINT**

Why does it take at least two exposures to an allergen to trigger allergic symptoms?

Answer: The first time you are exposed to an allergen, it merely sensitizes you by causing the attachment of specific antibodies to mast cells. A second exposure causes the mast cells to release chemicals that trigger the symptoms of an allergic response.

protein molecules on pollen grains, on the feces of tiny mites that live in house dust, in animal dander (shed skin cells), and in various foods. New allergens are periodically discovered. For example, an allergy to meat, which was first described in 2009, is on the rise in the southern and eastern regions of the United States. Adults affected tend to awake in the middle of the night covered in hives hours after ingesting meat for dinner. The allergy appears to be caused by a tick bite, a common occurrence in southern and eastern states (Figure 24.13). Those who suffer from a meat allergy must avoid eating beef, pork, venison, and other mammalian meat.

Allergic symptoms typically occur very rapidly in response to tiny amounts of an allergen but can sometimes be delayed for hours, as is typical in meat allergy cases. The symptoms of an allergy result from a two-stage reaction outlined in Figure 24.14. The first stage, called sensitization, occurs when a person is first exposed to an allergen—pollen, for example. ❶ An allergen enters the bloodstream, where it binds to B cells with complementary receptors. The B cells are thereby tagged for clonal selection (as in Figure 24.6). ❷ During clonal selection, the B cells proliferate and then secrete large amounts of antibodies to the allergen. ❸ Some of the antibodies attach to receptor proteins on the surfaces of mast cells, body cells that produce histamine and other chemicals that trigger the inflammatory response.

Depending on where you live, you could develop an allergy to meat.

Allergy symptoms do not arise until the second stage, which begins when the allergen enters the body again. ❹ With the second and subsequent exposures, the allergen binds to antibodies on mast cells. In response to the binding, ❺ the mast cells release histamine, which triggers allergy symptoms. Histamine causes blood vessels to dilate and leak fluid, which in turn results in nasal irritation, itchy skin, and tears. Because allergens usually enter the body through the nose and throat, symptoms are often most prominent there. Antihistamines are drugs that interfere with histamine's action and give temporary relief from an allergy.

Allergic reactions range from seasonal nuisances to severe, life-threatening responses. Anaphylactic shock is an especially dangerous type of allergic reaction. Some people are extremely sensitive to certain allergens, such as the venom from a bee sting or allergens in peanuts or shellfish. Any contact with these allergens causes a sudden release of inflammatory chemicals. Blood vessels dilate abruptly, causing a rapid and potentially fatal drop in blood pressure, a condition called shock. Fortunately, anaphylactic shock can be counteracted with injections of the hormone epinephrine (Figure 24.15). ✓

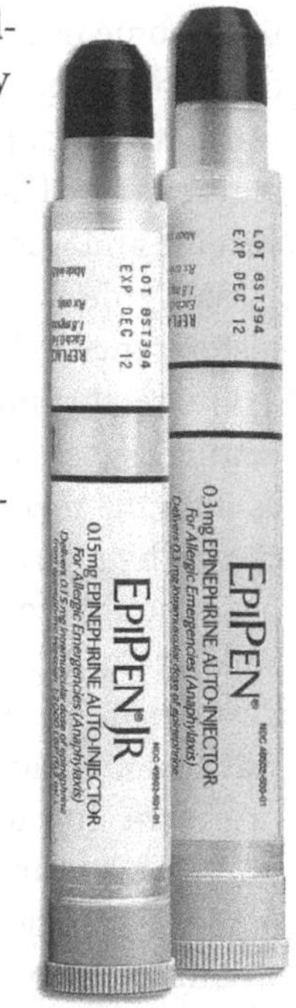

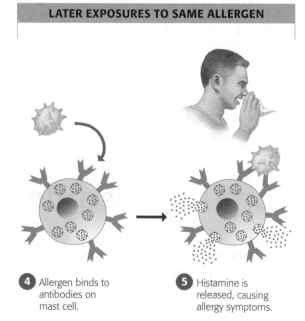

► **Figure 24.15 Single-use epinephrine syringes.** People who know they have severe allergies can carry single-use syringes that contain the hormone epinephrine (also called adrenaline). An injection can quickly stop a life-threatening allergic reaction.

▼ **Figure 24.14 How allergies develop.** The first exposure to an allergen causes sensitization. Subsequent exposures to the same allergen produce allergy symptoms.

SENSITIZATION: INITIAL EXPOSURE TO ALLERGEN	LATER EXPOSURES TO SAME ALLERGEN

Ragweed pollen grains

Colorized SEM 14,000×

Effector B cell

Histamine

Mast cell

❶ Allergen (pollen grain) enters bloodstream.

❷ B cells make antibodies.

❸ Antibodies attach to mast cell.

❹ Allergen binds to antibodies on mast cell.

❺ Histamine is released, causing allergy symptoms.

Autoimmune Diseases

Your immune system's ability to identify its own cells enables your body to battle foreign invaders without harm to itself. Like your Social Security number, which legally can't be used by anyone else, your cells have a particular set of self proteins that are unique. The unique proteins mark your cells as off-limits to attacks by your immune system. Pathogen invaders do not have your self proteins and thus are fair game in immune system assaults. Sometimes, however, the immune system falters. In **autoimmune diseases**, the immune system actually turns against the body's own molecules.

You've probably heard of the more common autoimmune diseases. In systemic lupus erythematosus (lupus), B cells make antibodies against many sorts of self molecules, even proteins and DNA released by the normal breakdown of body cells. Rheumatoid arthritis leads to damage and painful inflammation of the cartilage and bones of joints (**Figure 24.16**). In type 1 (insulin-dependent) diabetes, the insulin-producing cells of the pancreas are the targets of cytotoxic T cells. In multiple sclerosis (MS), T cells wrongly attack proteins in neurons (see Figure 27.2), often causing progressive muscle paralysis. Recent research suggests that Crohn's disease, a chronic inflammation of the digestive tract, may be caused by an autoimmune reaction against bacteria that normally inhabit the intestinal tract.

Immunodeficiency Diseases

In contrast to autoimmune diseases are a variety of defects called **immunodeficiency diseases**. Immunodeficient people lack one or more of the components of the immune system and are therefore susceptible to infections that would not ordinarily cause a problem. In the rare genetic disease called severe combined immunodeficiency (SCID), both B cells and T cells are absent or inactive. Because people with SCID are extremely vulnerable to even minor infections, they must live behind protective barriers or receive bone marrow transplants.

Immunodeficiency is not always an inborn condition; it may be acquired later in life. For instance, Hodgkin's disease, a type of cancer that affects lymphocytes, can depress the immune system. Radiation therapy and drug treatments used against many cancers can have the same effect. Next, we will look at perhaps the most dreaded acquired immunodeficiency disease of our time: AIDS. ☑

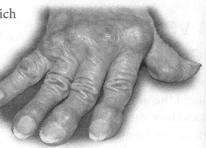
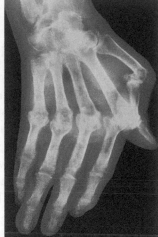

▼ **Figure 24.16 Rheumatoid arthritis.** The hands in this photo and X-ray are from a person with rheumatoid arthritis, an autoimmune disease in which the cartilage and bones of joints become damaged, resulting in malformation of the hands.

AIDS

Since the epidemic was first recognized in 1981, AIDS has killed more than 35 million people worldwide, and more than 35 million people are currently living with HIV, the virus that causes AIDS. The vast majority of HIV infections and AIDS deaths occur in sub-Saharan Africa and southeastern Asia, but approximately 1.1 million people in the United States are currently living with HIV. To put that number in perspective, consider that it is roughly equal to the population of Rhode Island.

HIV is deadly because it destroys the immune system, leaving the body defenseless against most invaders. HIV can infect a variety of cells, but it most often attacks helper T cells—the cells that activate cytotoxic T cells and B cells (see Figure 24.7). As HIV depletes the body of helper T cells, the adaptive immune defenses are severely impaired. If not treated with anti-HIV drugs, a final stage of HIV infection occurs, known as AIDS. The name AIDS refers to the fact that the disease is *acquired* through an infection that results in severe *immune deficiency*, with patients presenting a common set of symptoms, or a *syndrome*. (See Figure 10.31 to review the course of cell infection by HIV.) AIDS patients die from infectious agents or cancer that cannot be defended against without a functional immune system.

Drugs, vaccines, and education are areas of focus for prevention of HIV infection. Drug development has led to a drastic reduction in transmission rates of HIV from mother to child, and in 2012, the Federal Drug Administration (FDA) approved the first HIV prevention pill, for people who have a high risk of infection. The HIV pill is taken daily to block the virus from establishing an infection. A vaccine against HIV has been challenging to develop, and one does not yet exist. Currently, the most effective HIV/AIDS prevention is education. Safe sex behaviors, such as reducing promiscuity and using condoms, can save many lives.

A pill that prevents HIV is no longer an abstract concept: It's a reality.

☑ CHECKPOINT

1. What is an autoimmune disease?
2. What is an immunodeficiency disease?

Answers: 1. a disease in which cells of the immune system attack certain of the body's own cells 2. a disease in which one or more parts of the immune system are defective, resulting in an inability to fight infections

Vaccines EVOLUTION CONNECTION

Viral Evolution versus the Flu Vaccine

Throughout this chapter, we have returned to the topic of vaccines as a way to highlight the importance of the immune system. To understand how vaccines work or fail to work, we need to keep in mind biology's unifying theme of evolution. As viruses reproduce, mutations occur, generating new strains of a virus. These new strains are different enough that they can cause disease in individuals who had immunity to the ancestral strain. Some viruses mutate much faster than others; these viruses usually have genomes that consist of RNA rather than DNA. Both HIV and flu viruses have genomes made of RNA and, not surprisingly, both mutate rapidly. Let's explore the consequence of viral evolution on vaccine development by considering the flu virus, which is associated with 3,000–49,000 deaths in the United States in a given year.

Since most vaccines protect us for several years or even a lifetime, you might question whether you really need a flu vaccine every year. Because the flu virus mutates so rapidly, the antigens from the virus that our immune system recognizes one year often have changed significantly by the next year. Thus, even though memory cells in our immune system will protect us from flu strains that went around in previous years, our immune system doesn't have memory cells that will protect us from flu strains that are circulating currently or that will in future years. This is why scientists need to make new flu vaccine "recipes" every year.

Developing a flu vaccine is a challenge. Several major flu virus strains circulate in a given flu season, each strain mutating at a high rate. To examine which strains are likely to cause the next flu season's epidemic, scientists use a worldwide surveillance system, a computer program that gathers data on flu outbreaks from countries around the globe. From these data, scientists typically select the three most prevalent strains for use in a single vaccine. The timing of this choice is tricky. The selection of strains occurs between January and March, but it takes more than six months to produce and deliver a vaccine to healthcare providers **(Figure 24.17)**. In the interim, viral evolution continues, and the developed flu vaccine becomes less and less of a match to the flu virus strains that ultimately circulate. Thus, the vaccine given to us in November is an "educated guess," based on a strain that circulated several months earlier. When scientists fail to guess correctly which flu virus strains will become prevalent, a flu vaccine is likely to provide little or no protection against infection, and the number of flu related deaths will increase.

Scientists are working toward what they are calling a "universal flu vaccine" that would provide broad protection against several flu virus strains and would not be needed each year. For now, however, the battle continues, with medical science on one side and the constantly evolving flu viruses on the other side.

▼ **Figure 24.17 The flu vaccine production timeline.** A new flu vaccine "recipe" is made each year to keep up with the rapidly evolving flu virus.

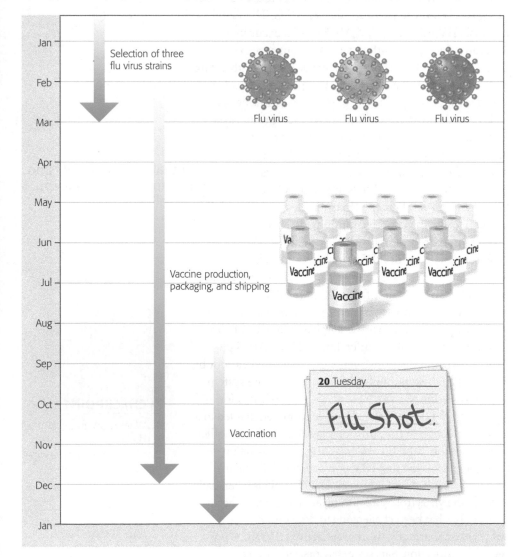

Chapter Review

SUMMARY OF KEY CONCEPTS

An Overview of the Immune System

The body's immune system protects it from pathogens. One part of the immune system, innate immunity, is present at birth, always deployed, and attacks pathogens nonspecifically. Adaptive immunity, the other part of the immune system, develops over a lifetime and recognizes, defends against, and remembers specific pathogens.

Innate Immunity

The human body contains two lines of defense (external barriers and internal defenses) that are innate, fully ready to respond before an invader has been encountered.

Innate Immunity			
External	**Internal**		
• Skin • Mucous membranes • Secretions	White blood cells	Defensive proteins	The inflammatory response
	• Phagocytic cells • Natural killer cells		• Chemical signals and phagocytic cells

The Lymphatic System

The lymphatic system maintains fluid balance in the body's tissues and is the main battleground for fighting infections.

Circulatory Function

During circulation, fluid from blood moves from capillaries to the interstitial space surrounding tissues. Cells exchange nutrients and wastes in interstitial fluid. Any residual fluid either resumes circulation in the circulatory system by reentering capillaries or drains into lymphatic vessels for circulation in the lymphatic system.

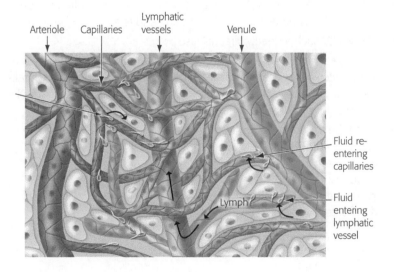

Arteriole Capillaries Lymphatic vessels Venule

Fluid re-entering capillaries

Lymph

Fluid entering lymphatic vessel

Immune Function

As fluid circulates through the lymphatic vessels, it carries pathogens to the immune cells of the lymphatic system. Pathogens swept past lymph nodes activate white blood cells that multiply and cause the lymph nodes to enlarge.

Adaptive Immunity

The adaptive defenses consist of a large collection of B and T lymphocytes that respond to specific invaders. Antigens are molecules that elicit responses from lymphocytes.

Step 1: Recognizing the Invaders

Antigen receptors on B cells and T cells recognize antigens that are circulating, on the surface of pathogens, or within body cells. The diversity of antigen receptors allows the adaptive immune system to recognize millions of antigens.

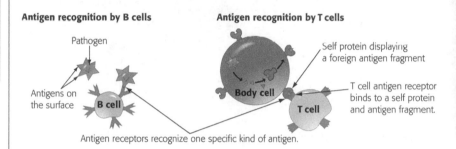

Antigen recognition by B cells

Pathogen

Antigens on the surface

B cell

Antigen recognition by T cells

Self protein displaying a foreign antigen fragment

Body cell

T cell antigen receptor binds to a self protein and antigen fragment.

T cell

Antigen receptors recognize one specific kind of antigen.

Step 2: Cloning the Responders

When an antigen enters the body, it activates only lymphocytes with complementary receptors, a process called clonal selection. Effector cells and memory cells are produced.

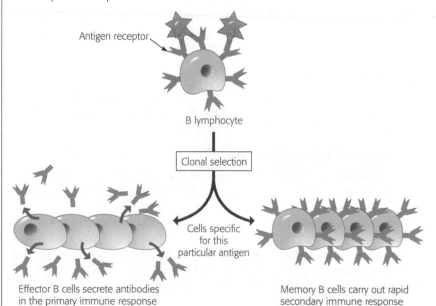

Antigen receptor

B lymphocyte

Clonal selection

Cells specific for this particular antigen

Effector B cells secrete antibodies in the primary immune response to the first exposure.

Memory B cells carry out rapid secondary immune response upon subsequent exposures.

Step 3: Responding to Invaders

Helper T cells help activate the two arms of adaptive immunity defenses: B cells secrete antibodies to help eliminate pathogens in the blood and lymph, and cytotoxic T cells destroy body cells that are infected.

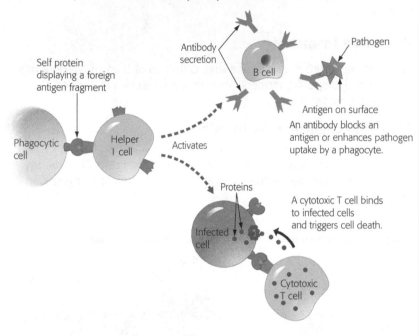

Self protein displaying a foreign antigen fragment

Phagocytic cell

Helper T cell

Activates

Antibody secretion

B cell

Pathogen

Antigen on surface

An antibody blocks an antigen or enhances pathogen uptake by a phagocyte.

Proteins

Infected cell

Cytotoxic T cell

A cytotoxic T cell binds to infected cells and triggers cell death.

Step 4: Remembering Invaders

Memory cells are activated by a second exposure to an antigen, and they initiate a faster and stronger secondary immune response.

Immune Disorders

Allergies

Allergies are abnormal sensitivities to otherwise harmless antigens, known as allergens. An allergic reaction produces inflammatory responses that result in uncomfortable and sometimes dangerous symptoms.

Autoimmune Diseases

The immune system normally reacts only against foreign molecules and cells, not against self (the body's own molecules). In autoimmune diseases, the system turns against some of the body's own molecules.

Immunodeficiency Diseases

In immunodeficiency diseases, immune components are lacking, and infections recur. Immunodeficiencies may arise through inborn genetic mutations or through disease.

AIDS

AIDS is a worldwide epidemic that kills millions of people each year. HIV, the AIDS virus, attacks helper T cells, crippling both the B cell and cytotoxic T cell responses. Safe sex practices could save many lives.

SELF-QUIZ

1. Molecules that elicit a response from lymphocytes are called _____. Proteins secreted by lymphocytes that bind to these molecules are _____.

2. Classify each of the following components of the immune system as a part of either innate immunity or adaptive immunity.
   a. cytotoxic T cells
   b. antigen receptors
   c. mucus membranes
   d. natural killer cells
   e. memory B cells

3. The immune system is capable of mounting specific responses to particular microorganisms because
   a. lymphocytes are able to change their antigen specificity as required to fight infection.
   b. the body is able to determine which type of B and T cells to make.
   c. the body contains an enormous diversity of lymphocytes, each with a specific kind of antigen receptor.
   d. the body is able to make different antigen receptors depending on the invading microorganism.

4. What makes a secondary immune response faster than a primary immune response?

5. Match each type of defensive cell with its function or description.
   a. lymphocyte
   b. cytotoxic T cell
   c. helper T cell
   d. phagocytic cell
   e. B cell
   f. memory cell

   1. attacks infected body cells
   2. secretes antibodies
   3. white blood cell that engulfs cells or molecules
   4. general name for B or T cell
   5. initiates the secondary immune response
   6. cell most commonly attacked by HIV

6. Explain how each of the following characteristics of the inflammatory response helps protect the body: swelling and fever.

7. Why is HIV such a deadly viral disease?

8. DiGeorge syndrome is a congenital disorder caused by the deletion of a short stretch of the chromosome 22. DiGeorge patients often lack T cells due to an underdeveloped thymus. Which type of infection is a particular concern in these patients?

9. When the immune system improperly turns against the body's own molecules, the result is
   a. an allergy.
   b. lupus.
   c. an autoimmune disease.
   d. progressive paralysis.

10. Once vaccinated, you have had a primary exposure to specific antigens. If you ever encounter these antigens again, your body will mount a rapid immune response. The cells that account for this rapid secondary response are called _____. The process that produces these long-lived cells is called _____.

Answers to these questions can be found in Appendix: Self-Quiz Answers.

THE PROCESS OF SCIENCE

11. **Interpreting Data** The graph below details food allergies of children in the United States over time. Determine whether each claim below is supported or not supported by the graph. Explain your answers.

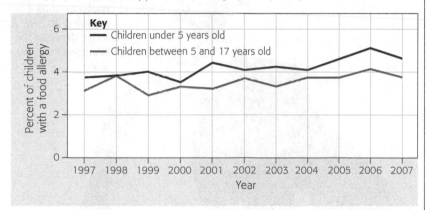

a. Food allergies in children are on the rise.

b. In a group of 100 children age 5 or younger, we would expect about 10 children to have a food allergy.

c. Food allergies are higher in younger children than in older children.

12. In the past, viral vaccines were often generated by isolating and enriching whole virus particles and then inactivating them using heat or reactive chemicals. The vaccines against polio and hepatitis A viruses are still generated in this way. With the advent of recombinant DNA techniques (see Chapter 12), scientists have started to generate vaccines that contain only fragments of viruses—hepatitis B and influenza viruses are examples of this. What are the advantages and disadvantages of both the methods?

BIOLOGY AND SOCIETY

13. Vaccination programs have greatly contributed to an extended human life span; however, controversies have accompanied these programs for several decades. Many people worry about the safety of some vaccines, and vaccination programs have been criticized by civil liberty campaigners, who oppose the idea of government infringing on people's freedom to choose what medications they take. In your opinion, should governments be given the right to enforce vaccination programs?

14. After the peak of HIV infections in the 1980s, the number of cases of new infections declined but leveled off at about 50,000 new cases in the United States per year. If your job was to use limited public funds to educate certain groups of people about HIV transmission, which groups might you focus on? (You will need to do some research to look at rates of infection based on age, geographic location, and other demographic identifiers.) Make a pamphlet that has two goals. First, ensure that the pamphlet states the problem by detailing statistics about HIV infections in the United States among different groups (use both words and bold graphics). Second, provide information on a program that has worked to educate a specific target group or describe a novel idea you have to educate a target group.

15. Ebola is a dangerous and often lethal viral disease spread from person to person through direct contact with blood or body secretions. The largest outbreak of Ebola began in West Africa in 2014. Many healthcare workers from the United States and Europe traveled to West Africa to help treat Ebola patients in that region. One healthcare worker, Kaci Hickox, returned to the United States and was told she was to be isolated, or quarantined, in a tent behind a hospital in New Jersey. She protested the precautionary quarantine and was released early. Ebola experts note that an infected person can only transmit the virus when they are showing symptoms of the disease. How long does a person exposed to Ebola need to be quarantined to be declared disease-free? Should healthcare workers exposed to Ebola who are not showing symptoms be forced into quarantine? Answer this last question from the perspective of a doctor and then again from the viewpoint of a government official.

25 Hormones

◄ Pay attention to your posture during a job interview and you just might decrease the hormone that triggers stress.

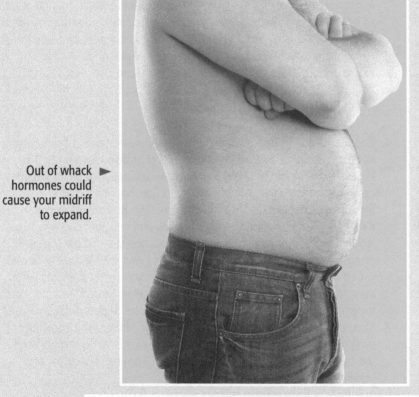

Out of whack hormones could cause your midriff to expand. ►

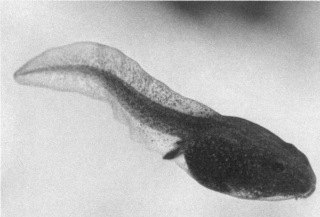

▲ To turn a male tadpole into a female frog, add a pinch of endocrine disruptor and wait.

568

Steroid Abuse BIOLOGY AND SOCIETY

Should Steroids Be Allowed in Sports?

Anabolic steroids, which are known to enhance athletic performance, were banned from Major League Baseball (MLB) in 2001. Nevertheless, the sport has been plagued with scandals of anabolic steroid abuse ever since. In 2013, 13 MLB players were suspended from playing for 50 games or more after being linked to a clinic that sold the banned substances. The severity of the punishments was unprecedented in the history of baseball. Among the suspended was the National League's Most Valuable Player in 2011, Ryan Braun of the Milwaukee Brewers.

Steroids are hormones, chemical signals that help regulate body functions, and everyone's body naturally produces them. Anabolic steroids, on the other hand, are synthetic hormones that mimic the male sex hormone testosterone. Large doses of synthetic testosterone can increase muscle mass and improve athletic performance, such as the ability of a baseball player to hit a home run. Athletes are reluctant to come clean about using anabolic steroids because doing so calls into question their natural abilities. When initially accused, Braun denied allegations of anabolic steroid use, but following his suspension, he publicly admitted to it.

Generally, sports fans are in agreement that athletes should push their game to the limit, but when it comes to the issue of banned steroids, sports enthusiasts are divided in their opinions. Banning anabolic steroids protects players from health risks, such as aggression, psychiatric disorders, and even kidney, liver, and cardiovascular damage. Fans opposed to the ban argue that sports officials should not be the ones deciding what is and isn't healthy for players—that is, individual athletes can make their own decisions about the safety of anabolic steroids. Those fans supporting the ban raise the issue of cheating. Because they increase muscle mass, anabolic steroids can enhance athletic performance, creating an unfair advantage. If the ban was lifted, athletes who opted *not* to use these substances would be at an unfair *disadvantage*.

Admitting to steroid use. In 2013, Ryan Braun of the Milwaukee Brewers publicly admitted to using anabolic steroids.

Although baseball has been perceived as a sport with widespread anabolic steroid abuse, the suspension of numerous high-profile baseball players is a signal to fans that MLB has substantially cleaned up the game. After serving his lengthy suspension, Braun returned to the Milwaukee Brewers in 2014 ready to play.

In this chapter, we further discuss synthetic hormones, such as anabolic steroids, and the physical effects of them, but we'll mainly explore natural hormones and the way they maintain homeostasis within the human body. We begin with an overview on the functions of natural hormones in all vertebrates, including the main system in which animal hormones function, the endocrine system. Then we'll focus specifically on the human endocrine system. Along the way, we'll consider many examples of the effects of hormonal imbalance.

Hormones: An Overview

The cells of your body are constantly communicating with one another. They do so through chemical and electrical signals, traveling by way of two major organ systems: the endocrine system and the nervous system. The **endocrine system** is a group of interacting glands and tissues throughout the animal body that produce and secrete chemicals to initiate and maintain body functions. Chemical signals called **hormones** are released by endocrine cells and carried to all locations in the body, usually via the blood. In the nervous system, the signals are primarily electrical and are transmitted via nerve cells called neurons (Chapter 27).

Figure 25.1 shows the release of hormone molecules from an endocrine cell. The circulatory system carries hormones throughout the body, but a hormone can only bind to a **target cell** with receptors for that specific hormone (bottom of Figure 25.1). Imagine target cells with receptors as locked doors and hormones as keys; each key only opens doors that have a matching

lock. Because hormones reach all parts of the body, the endocrine system is especially important in controlling whole-body activities. For example, hormones govern our metabolic rate, growth (including muscle development, the target of anabolic steroid abuse), maturation, and reproduction. Hormones trigger changes in target cells in different ways, depending on whether the hormone is water-soluble or lipid-soluble: Water-soluble hormones trigger responses without entering the cell, whereas lipid-soluble hormones trigger responses after entering the cell.

Water-soluble hormones cannot pass through the phospholipid bilayer of the plasma membrane, but they can bring about cellular changes without entering their target cells **(Figure 25.2)**. To start, ❶ a water-soluble hormone (•) binds to a specific receptor protein (⬤) in the plasma membrane of the target cell. The binding activates the receptor protein, which ❷ initiates a signal transduction pathway: a series of changes to molecules

► Figure 25.1 **Hormone secretion from an endocrine cell.** A cell within an endocrine gland (upper-right diagram) secretes hormone molecules. The hormone is carried via the circulatory system to all cells of the body, but it only affects target cells that have matching receptors (lower-right diagram).

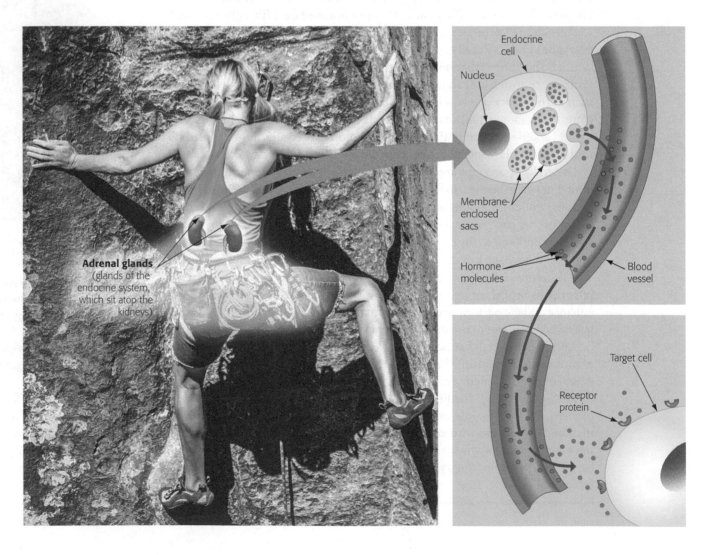

Adrenal glands (glands of the endocrine system, which sit atop the kidneys)

Endocrine cell

Nucleus

Membrane-enclosed sacs

Hormone molecules

Blood vessel

Target cell

Receptor protein

that converts a chemical message from outside the cell to a specific response inside the cell. ❸ The final relay molecule (⬤) activates a protein that either carries out a response in the cytoplasm (such as activating an enzyme) or affects gene regulation in the nucleus (such as turning on or off genes). This is a bit like an email string, where a message with an instruction is forwarded to one person and then another and so on until it reaches someone who actually takes action. An example of a water-soluble hormone is the protein erythropoietin (EPO), which regulates the production of red blood cells. Some athletes abuse EPO to increase their endurance.

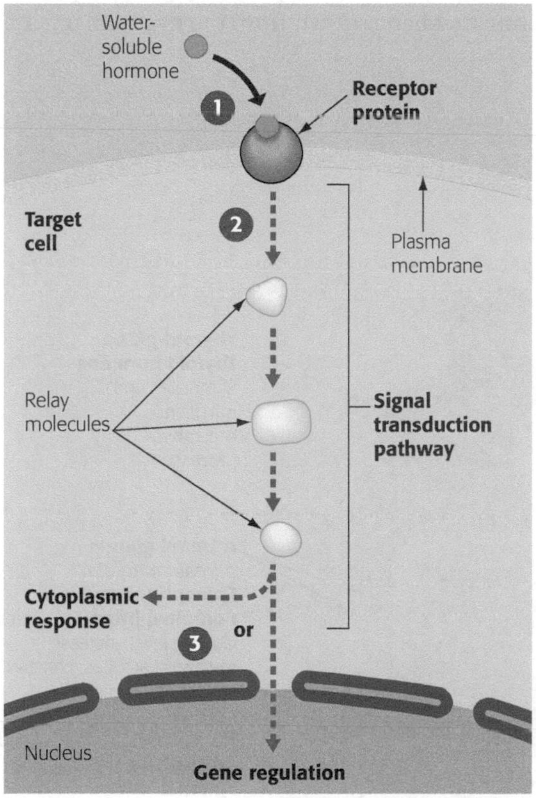

In contrast, lipid-soluble hormones pass through the phospholipid bilayer and trigger responses by binding to receptors inside the target cell. Steroid hormones—including natural ones such as the sex hormones (testosterone and estrogen) as well as artificial anabolic steroids—work in this manner. As shown in **Figure 25.3**, ❶ a lipid-soluble hormone (▼) enters a cell by diffusing through the plasma membrane. If the cell is a target cell, the hormone ❷ binds to a receptor protein (⬤) in the cytoplasm or nucleus. Rather than triggering a signal transduction pathway, the receptor itself carries the hormone's signal. ❸ The hormone-receptor complex attaches to specific sites on the cell's DNA in the nucleus. ❹ The binding to DNA turns specific genes on or off.

We've now completed an overview of how hormones work. The principles presented so far apply to all vertebrates (as well as many invertebrates), reminding us of one of the major themes in biology: evolution. Animals have many hormones in common, a result of shared ancestry. Yet over the course of evolution, the functions of a given hormone have diverged between species. For example, thyroid hormone plays a role in regulating metabolism in many vertebrates, yet also stimulates metamorphosis in frogs (the process in which tadpoles become adults). In the next section, we'll take a closer look at the endocrine system of one particular vertebrate: humans. ☑

☑ **CHECKPOINT**
What is the difference in the way that lipid-soluble and water-soluble hormones interact with their target cells?

Answer: *Lipid-soluble hormones bind to receptors inside the cell, whereas water-soluble hormones bind to receptors in the plasma membrane.*

▲ Figure 25.2 **A hormone that binds to a plasma membrane receptor.** Hormones that are water-soluble cannot cross the phospholipid bilayer but can bind to membrane receptors outside target cells and activate a signal transduction pathway.

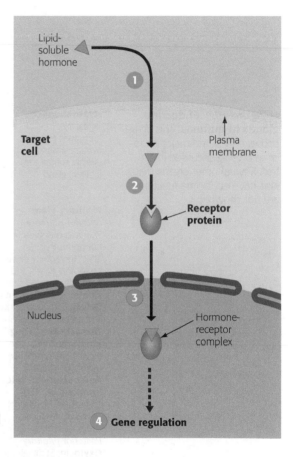

► Figure 25.3 **A hormone that binds to an intracellular receptor.** Hormones that are lipid-soluble can cross the phospholipid bilayer and bind to receptors inside target cells. Steroid hormones and other lipid-soluble hormones affect the cell by turning specific genes on and off.

The Human Endocrine System

Hormones are made and secreted mainly by organs called **endocrine glands**. The human endocrine system consists of about a dozen major glands. Some of these, such as the thyroid and pituitary glands, are endocrine specialists; that is, their primary function is to secrete hormones into the blood. Other organs, for example, the stomach and the pancreas of the digestive system, are primarily nonendocrine but have some cells that secrete hormones. The stomach releases ghrelin (the "hunger hormone"), which travels to the brain and stimulates appetite. The pancreas secretes insulin and glucagon, hormones that influence the level of glucose in the blood.

Hormones have a wide range of targets. Some hormones, like adrenaline, affect many tissues of the body. Others, such as glucagon from the pancreas, have only a few kinds of target cells (in the case of glucagon, liver and fat cells). In some instances, a hormone elicits different responses in different target cells, depending on the type of cell and its signal transduction pathway.

Figure 25.4 shows the locations of some of the major human endocrine glands. The rest of this chapter focuses on the hormones each of these glands secretes and how these hormones help the body maintain homeostasis. Keep in mind there are many more hormone-secreting structures and hormones that we will not discuss. ☑

Information Flow

The Hypothalamus and Pituitary Gland

The **hypothalamus**, a gland in the brain, acts as the main control center of the endocrine system (**Figure 25.5**). It receives information from nerves about the internal condition of the body and about the external environment and then responds by sending out appropriate nervous or endocrine signals (**Figure 25.6**). These signals directly control the **pituitary gland**, a pea-sized structure that hangs down from the hypothalamus.

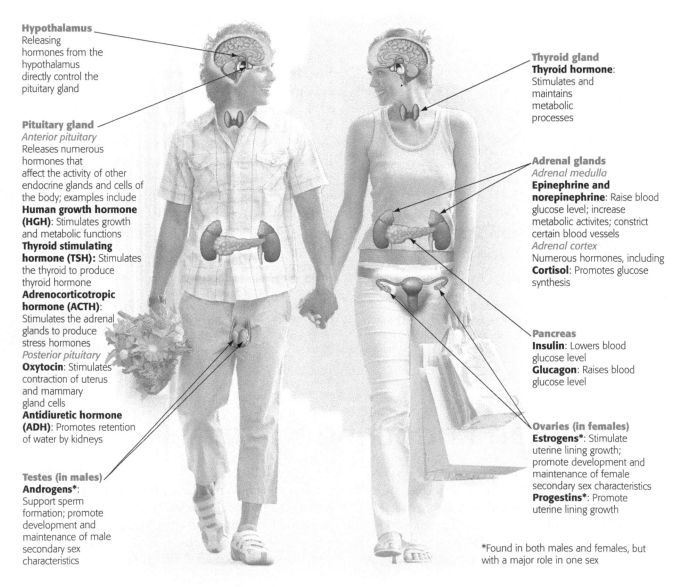

► Figure 25.4 **Endocrine glands in humans.** This figure shows only some major endocrine organs discussed in the text. Several other organs that also have endocrine functions are not shown.

Hypothalamus
Releasing hormones from the hypothalamus directly control the pituitary gland

Pituitary gland
Anterior pituitary
Releases numerous hormones that affect the activity of other endocrine glands and cells of the body; examples include
Human growth hormone (HGH): Stimulates growth and metabolic functions
Thyroid stimulating hormone (TSH): Stimulates the thyroid to produce thyroid hormone
Adrenocorticotropic hormone (ACTH): Stimulates the adrenal glands to produce stress hormones
Posterior pituitary
Oxytocin: Stimulates contraction of uterus and mammary gland cells
Antidiuretic hormone (ADH): Promotes retention of water by kidneys

Testes (in males)
Androgens*:
Support sperm formation; promote development and maintenance of male secondary sex characteristics

Thyroid gland
Thyroid hormone:
Stimulates and maintains metabolic processes

Adrenal glands
Adrenal medulla
Epinephrine and norepinephrine: Raise blood glucose level; increase metabolic activites; constrict certain blood vessels
Adrenal cortex
Numerous hormones, including
Cortisol: Promotes glucose synthesis

Pancreas
Insulin: Lowers blood glucose level
Glucagon: Raises blood glucose level

Ovaries (in females)
Estrogens*: Stimulate uterine lining growth; promote development and maintenance of female secondary sex characteristics
Progestins*: Promote uterine lining growth

*Found in both males and females, but with a major role in one sex

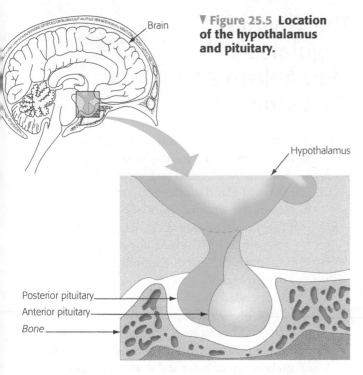

▼ **Figure 25.5 Location of the hypothalamus and pituitary.**

Brain

Hypothalamus

Posterior pituitary

Anterior pituitary

Bone

▼ **Figure 25.6 Master control exerted by the hypothalamus.** The hypothalamus receives information and via the pituitary gland controls activities that maintain homeostasis throughout the body.

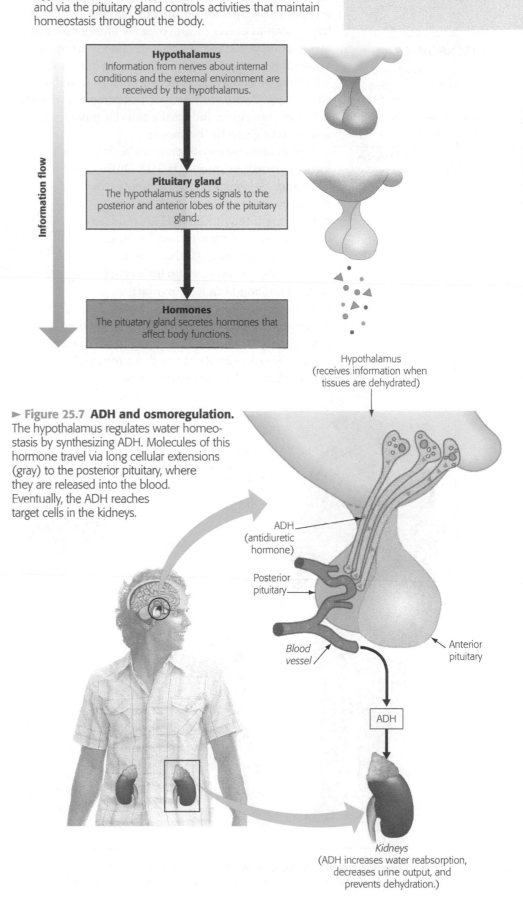

THE HUMAN ENDOCRINE SYSTEM

Hypothalamus
Information from nerves about internal conditions and the external environment are received by the hypothalamus.

Information flow

Pituitary gland
The hypothalamus sends signals to the posterior and anterior lobes of the pituitary gland.

Hormones
The pituatary gland secretes hormones that affect body functions.

Hypothalamus
(receives information when tissues are dehydrated)

▶ **Figure 25.7 ADH and osmoregulation.** The hypothalamus regulates water homeostasis by synthesizing ADH. Molecules of this hormone travel via long cellular extensions (gray) to the posterior pituitary, where they are released into the blood. Eventually, the ADH reaches target cells in the kidneys.

ADH (antidiuretic hormone)

Posterior pituitary

Blood vessel

Anterior pituitary

ADH

Kidneys
(ADH increases water reabsorption, decreases urine output, and prevents dehydration.)

In response to signals from the hypothalamus, the pituitary secretes hormones that influence numerous body functions. Like a chief executive officer (CEO) of a business directing managers who then instruct their subordinates, the hypothalamus (CEO) exerts master control over the endocrine system by using the pituitary (manager) to relay directives to other glands (subordinates).

The flow of information to and from the hypothalamus is illustrated when we consider how the endocrine system maintains homeostasis. Let's look at a specific example: the endocrine system interacting with the urinary system to maintain water balance in all of the body's tissues. As you see in **Figure 25.7**, when the hypothalamus receives information that the body tissues are dehydrated, it makes antidiuretic hormone (ADH) (▲), which is stored and released by the pituitary gland. ADH signals kidney cells to reabsorb more water so that less urine is produced and serious dehydration is prevented. As an individual rehydrates, the hypothalamus slows the release of ADH from the pituitary, and the kidneys increase the output of urine. As you will learn later in this chapter, the hypothalamus maintains homeostasis in many more ways, including regulating metabolism, growth, development, and reproduction.

As Figure 25.5 shows, the pituitary gland just under the hypothalamus consists of two distinct parts: a posterior lobe (green) and an anterior lobe (yellow). The **posterior pituitary** is actually an extension of the hypothalamus that stores and secretes hormones made in the hypothalamus. Two hormones are released by the posterior pituitary: ADH and oxytocin. We've already discussed ADH. Among other functions, oxytocin stimulates the contraction of the uterus during

☑ **CHECKPOINT**

Outline the flow of information in the endocrine system with respect to the hypothalamus and pituitary glands.

Answer: *The hypothalamus receives specific information about the body and directs the pituitary gland to secrete hormone signals in response.*

childbirth and causes the mammary glands to release milk. In contrast, the **anterior pituitary** synthesizes and secretes its own hormones directly into the blood. The hypothalamus exerts control over the anterior pituitary by secreting two kinds of hormones into short blood vessels that connect the glands: releasing hormones and inhibiting hormones. Releasing hormones stimulate the anterior pituitary to secrete specific hormones, whereas inhibiting hormones induce the anterior pituitary to stop secreting specific hormones.

One of several anterior pituitary secretions is **human growth hormone (HGH)**. During childhood and adolescence, HGH promotes the development and enlargement of all parts of the body. If too much HGH is produced in a very young person, usually because of a pituitary tumor, gigantism can result (**Figure 25.8**). In contrast, too little HGH during childhood can lead to dwarfism. Administering HGH to children with HGH deficiency can prevent dwarfism. HGH is sometimes abused by athletes attempting to build muscle mass. Abuse of HGH can be dangerous and may lead to joint swelling, diabetes, enlarged brow and jaw bones, and heart complications.

Next we will discuss information flow from the hypothalamus and anterior pituitary glands to the thyroid, testes, ovaries, and adrenal glands. Let's first examine the thyroid gland. ☑

> Out of whack hormones could cause your midriff to expand.

The Thyroid Regulates Metabolism and Development

You've probably had a routine physical and had your physician feel and press the front of your neck. Your doctor is checking the size and texture of your thyroid gland. The **thyroid gland** produces thyroid hormone in response to thyroid stimulating hormone (TSH) released from the anterior pituitary (**Figure 25.9**). **Thyroid hormone** affects many functions of the body including metabolic rate (the amount of energy expended over a period of time), heart rate, blood pressure, and tolerance to cold. Thyroid hormone plays additional crucial roles in children, affecting the development and maturation of bone and nerve cells. Homeostasis is maintained when thyroid hormone level in the blood hovers near a "set point." An insufficient level of thyroid hormone in the blood (hypothyroidism) or excess level (hyperthyroidism) can result in serious metabolic disorders. Fortunately, thyroid disorders can be treated easily.

Hypothyroidism (too little thyroid hormone) can result in weight gain, tiredness, a slowed heart rate, and intolerance to cold. As you see in Figure 25.9, thyroid

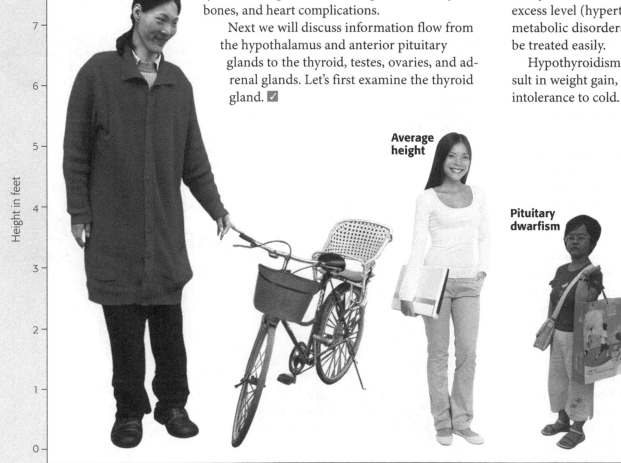

Gigantism

Average height

Pituitary dwarfism

Height in feet

▲ Figure 25.8 **Pituitary growth hormone disorders.** A pituitary tumor producing too much HGH causes gigantism (left). Dwarfism (right) has many causes, including low production of HGH during childhood.

▼ **Figure 25.9 Thyroid hormone production.** When stimulated by TSH, the thyroid produces thyroid hormone. Thyroid hormone level is regulated through negative feedback, as thyroid hormone inhibits the release of TSH.

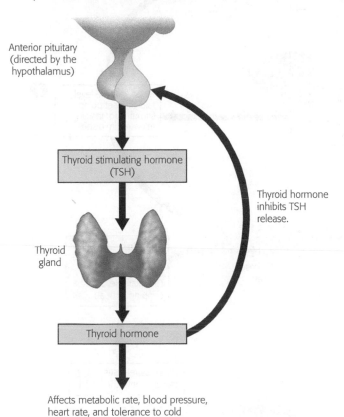

Anterior pituitary (directed by the hypothalamus)

Thyroid stimulating hormone (TSH)

Thyroid hormone inhibits TSH release.

Thyroid gland

Thyroid hormone

Affects metabolic rate, blood pressure, heart rate, and tolerance to cold

hormone inhibits the release of TSH, an example of negative feedback. When thyroid hormone is absent or low, negative feedback is absent, too, and the continual secretion of TSH stimulates the growth of the thyroid gland. Thus, minor swelling of the thyroid or even goiter (see Figure 2.3) may develop from hypothyroidism.

You may not realize it, but salt in your diet helps prevent hypothyroidism. It's not actually the salt, but an additive to salt, iodine, that is necessary for normal thyroid function. Just as pancakes can't be made without flour, thyroid hormone cannot be synthesized without iodine. Without iodine, and thus a lack of thyroid hormone, goiter eventually develops. Goiter still affects many people in less industrialized nations where iodine is not available in the diet.

Disease and environmental toxins are causes of hypothyroidism. Hashimoto's disease is an autoimmune disorder in which the body's immune system attacks cells of the thyroid, interfering with its ability to make thyroid hormone. Environmental toxins that adversely affect the endocrine system, termed **endocrine disruptors**, are commonly found in pesticides and plastic containers. For example, BPA (bisphenol A) is an endocrine disruptor found in some plastic water bottles. A growing

body of evidence links particular endocrine disruptors to thyroid disease in animals. Fortunately, no matter what the cause (known or unknown), pills containing synthetic thyroid hormone easily relieve symptoms of hypothyroidism.

Hyperthyroidism (overproduction of thyroid hormone) is characterized by weight loss, high blood pressure, rapid heart rate, and increased sensitivity to heat (profuse sweating). Treatment for hyperthyroidism takes advantage of the fact that the thyroid accumulates iodine. Patients drink a solution containing a low dose of radioactive iodine, which kills off enough cells to reduce thyroid output and relieve symptoms.

Whereas a low dose of radioactive iodine treats disease, a high dose of it can be deadly. Nuclear fallout often produces dangerous plumes of radioactive iodine that enter the air, water, and food supply. (This was a big concern after the earthquake and tsunami that caused the 2011 nuclear disaster in Fukushima, Japan.) When radioactive iodine accumulates in the thyroid, the risk of thyroid cancer increases, especially in young children. To help protect the thyroid, a stable form of iodine, potassium iodide (KI), is taken in pill form before or immediately after exposure to nuclear fallout. This treatment "fills up" the thyroid with stable iodine so that the thyroid cannot take up the harmful radioactive iodine. These pills are sometimes advertised to consumers as a necessity for "doomsday" preparation. ☑

The Pancreas Regulates Blood Glucose

The **pancreas** secretes two hormones, **insulin** and **glucagon**, that play important roles in managing the body's energy supplies. Insulin and glucagon help maintain a homeostatic balance between the amount of glucose available in the blood and the amount of glucose stored as the polymer glycogen in body cells. Just as you would deposit money in a bank if you had too much cash in your wallet, the liver stores excess glucose inside cells in the form of glycogen. And, like withdrawing money from your bank account when your wallet is empty, when blood glucose level is low, the liver breaks down the glycogen in cells to return glucose to the blood.

► Figure 25.10 **Glucose homeostasis.** This diagram traces the regulation of blood glucose level by insulin and glucagon, two antagonistic hormones released by the pancreas.

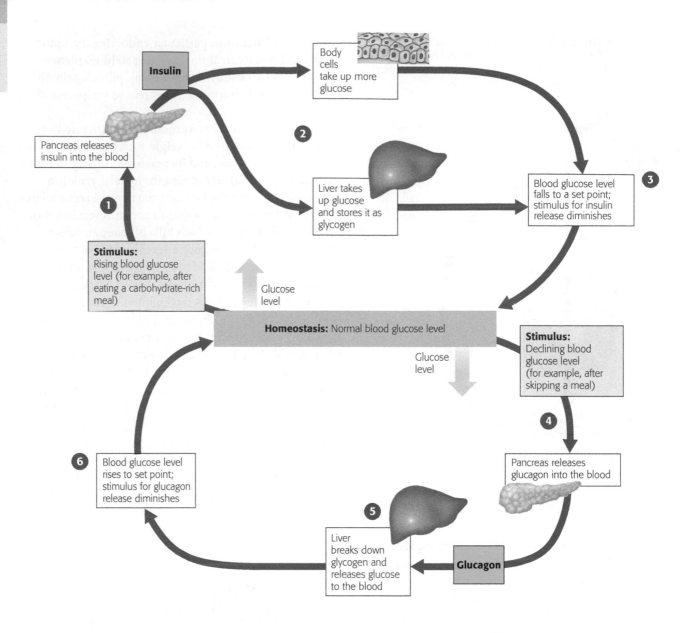

As shown in **Figure 25.10**, insulin and glucagon are **antagonistic hormones**, countering each other in a feedback circuit that precisely manages the level of glucose in the blood. By negative feedback, the concentration of glucose in the blood determines the relative amounts of insulin and glucagon secreted by pancreatic cells. ❶ Rising glucose concentration in the blood—as happens shortly after you eat a carbohydrate-rich meal, for example—stimulates the pancreas to secrete more insulin into the blood. ❷ Insulin binds to body cells with insulin receptors, stimulating them to take up more glucose from the blood. Liver and skeletal muscle cells take up glucose and use it to make glycogen, which they store. ❸ As a result, the blood glucose level falls to the set point, and the cells of the pancreas lose their stimulus to secrete insulin.

❹ When the blood glucose level dips below the set point, as it may between meals, pancreatic cells respond by secreting more glucagon. ❺ Glucagon helps mobilize stored fuel in liver cells, the target cells of glucagon. In response, liver cells break glycogen down into glucose and release it into the blood. ❻ Then, when the blood glucose level returns to the set point, the pancreas slows its secretion of glucagon.

Diabetes mellitus is a serious hormonal disease that affects about 1 in 12 Americans, and millions of them don't even know they are ill. If current trends continue, 1 in 3 Americans will be affected by 2050. In diabetes, body cells are unable to absorb glucose from the blood, either because there is not enough insulin in the blood (as in type 1, or insulin-dependent, diabetes) or because the target cells do not respond normally to the insulin in the blood (as in type 2, or non–insulin-dependent, diabetes). A third type of diabetes, called gestational diabetes, can affect any woman during pregnancy, even one who has never shown symptoms of diabetes before. Of those Americans who have diabetes, more than 90% have type 2.

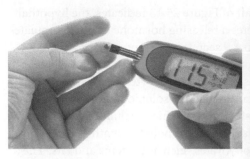

► **Figure 25.11 Measuring blood glucose.** People with diabetes use glucose meters (left) to measure the amount of glucose in the blood to determine if an insulin injection is needed. Insulin pumps (right) both measure glucose level and inject insulin as needed.

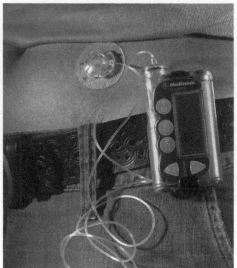

▼ **Figure 25.12 Rate of hospital admissions for hypoglycemia in low- and high-income patients.**

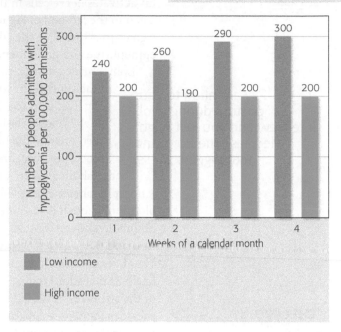

Number of people admitted with hypoglycemia per 100,000 admissions

240, 200, 260, 190, 290, 200, 300, 200

300
200
100
0

1   2   3   4
Weeks of a calendar month

■ Low income
■ High income

In a person with diabetes, a lack of insulin or the inability of cells to access it causes glucose to build up in the blood. Cells cannot obtain enough glucose from the blood, even though there is plenty of glucose in the blood. Starved for fuel, cells are forced to burn the body's supply of fats and proteins. Meanwhile, the digestive system continues to absorb glucose from food, causing the glucose concentration in the blood to become dangerously high. Diabetics measure their blood glucose routinely throughout the day **(Figure 25.11)**.

There are treatments for diabetes but no cure. Type 1 patients require regular injections of insulin. Type 2 diabetes is almost always associated with being overweight and underactive, although whether obesity causes diabetes (and if so, how) remains unknown. By controlling sugar intake and exercising and dieting to reduce weight, type 2 diabetes can often be managed without medications. Despite the treatments available for both types of diabetes, every year at least 200,000 Americans die from the disease or its complications, which include severe dehydration, cardiovascular and kidney disease, and nerve damage.

Low income appears to be a factor in developing type 2 diabetes, and low-income diabetics often suffer greater complications from the disease. Scientists from the University of California, San Francisco, hypothesized that diabetics with low income would have more health problems at the end of a calendar month (compared to the beginning), when they have little to no money for food. The scientists predicted they would find an increase in hypoglycemia (low blood sugar) at the end of each month. (Hypoglycemia is most commonly associated with diabetics who fast while taking glucose lowering medicine.) Tracking California hospital admissions from 2000 to 2008, the researchers measured the number of patients admitted for hypoglycemia and identified each patient as either "low" or

"high" (non-low) income. As you see in **Figure 25.12**, the scientists' hypothesis about low-income diabetics was supported: The data show a strong association between hypoglycemia and the end of the calendar month for low-income but not high-income patients. We can't know for certain if a lack of food caused the results, but it is a strong correlation. ☑

The Adrenal Glands Respond to Stress

Imagine this: You are home alone and you hear a door slam. How does your body react? You've probably felt your heart beat faster and your skin develop goose bumps when you've sensed danger or approached a stressful situation. These reactions are triggered by two "fight-or-flight" hormones: **epinephrine** (also called adrenaline) and **norepinephrine**. These two water-soluble hormones ensure a rapid, short-term response to stress that can be activated in seconds and last for minutes or hours.

The human body has two **adrenal glands,** one sitting atop each kidney. Each adrenal gland is actually two glands in one: a central portion called the **adrenal medulla** and an outer portion called the **adrenal cortex**. Though the cells they contain and the hormones they produce are different, both the adrenal medulla and the adrenal cortex secrete hormones that enable the body to respond to stress.

☑ CHECKPOINT

If someone with type 1 diabetes eats a big meal and does not take any medication, what will happen to that person's blood glucose level?

Answer: Insufficient insulin in the bloodstream will cause a high level of glucose to remain in the blood after a meal.

The "fight-or-fight" hormones are secreted by the adrenal medulla. Let's examine how they are controlled and the effects they have on the body. A stressful stimulus activates nerve cells in the hypothalamus. As indicated in the left half of **Figure 25.13**, these cells ❶ send signals that stimulate the adrenal medulla to ❷ secrete epinephrine and norepinephrine (•) into the blood.

Epinephrine and norepinephrine both contribute to the short-term stress response by stimulating liver cells to release glucose, making more fuel available for cellular work. The hormones also prepare the body for action by raising blood pressure, breathing rate, heart rate, and metabolic rate. In addition, epinephrine and norepinephrine change blood flow patterns to shuttle blood to where it is most needed: Blood vessels in the brain and skeletal muscles are widened, increasing alertness and the muscles' ability to react to stress, while blood vessels elsewhere are narrowed, reducing activities (such as digestion) that are not immediately involved in the stress response. The short-term stress response subsides rapidly—explaining why once you discover that the door slamming was only the wind, you feel calm within a few minutes. Positive stimuli, such as riding a roller coaster, can trigger the same response.

In contrast to epinephrine and norepinephrine (secreted by the adrenal medulla), hormones secreted by the adrenal cortex can provide a slower, longer-lasting response to stress that can last for hours or days.

> Pay attention to your posture during a job interview and you just might decrease the hormone that triggers stress.

☑ CHECKPOINT

Which hormones regulate the short-term stress response? What family of hormones regulates the long-term stress response?

Answer: epinephrine and norepinephrine; corticosteroids

As the right half of Figure 25.13 indicates, the hypothalamus ❸ secretes a releasing hormone (•) that stimulates the pituitary to ❹ secrete a hormone called ACTH (adrenocorticotropic hormone) (•). In turn, ACTH stimulates cells of the adrenal cortex to ❺ synthesize and secrete a family of lipid-soluble steroid hormones called **corticosteroids** (•), which includes the hormone cortisol. **Cortisol** helps promote the synthesis of glucose from noncarbohydrates, such as proteins and fats. The energy is needed when the body is coping with stressful situations. As cortisol increases energy for vital functions, it can suppress less immediate functions, such as the body's immune system. A very high level of cortisol in the blood can suppress the inflammatory response that occurs at infection sites. For this reason, physicians may use a similar molecule known as cortisone to treat inflammation from a variety of causes, such as skin reactions or injured joints. Cortisone injections are commonly used to relieve the pain of athletic injuries.

Can you affect the balance of your stress hormones? Researchers examined how hormones were affected when individuals posed for 2 minutes in positions of power. On average, 17 minutes after power posing, individuals had a decrease in the stress hormone cortisol and an increase in the sex hormone testosterone. Those in lower power poses for 2 minutes (like with arms folded across the chest) had opposite results. The researchers suggest: When interviewing for a job, power pose to stimulate the hormones that can help you feel confident and get you hired. ☑

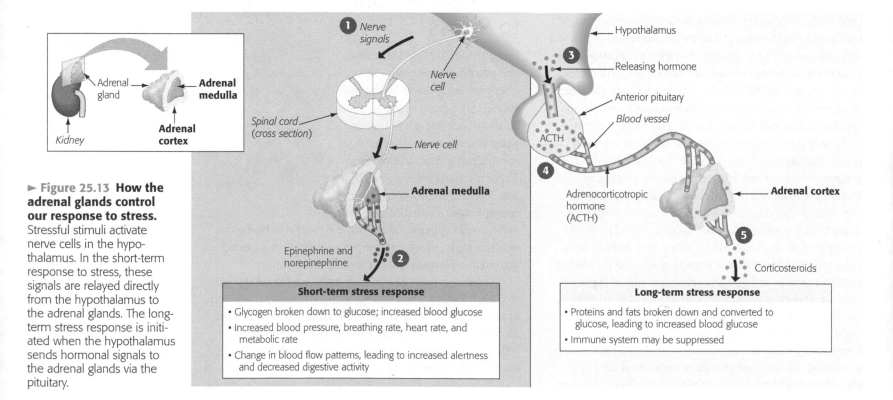

▶ **Figure 25.13 How the adrenal glands control our response to stress.** Stressful stimuli activate nerve cells in the hypothalamus. In the short-term response to stress, these signals are relayed directly from the hypothalamus to the adrenal glands. The long-term stress response is initiated when the hypothalamus sends hormonal signals to the adrenal glands via the pituitary.

Adrenal gland
Adrenal medulla
Adrenal cortex
Kidney

❶ Nerve signals
Nerve cell
Spinal cord (cross section)
Nerve cell
Adrenal medulla
Epinephrine and norepinephrine ❷

Short-term stress response
- Glycogen broken down to glucose; increased blood glucose
- Increased blood pressure, breathing rate, heart rate, and metabolic rate
- Change in blood flow patterns, leading to increased alertness and decreased digestive activity

Hypothalamus
❸ Releasing hormone
Anterior pituitary
Blood vessel
ACTH
❹ Adrenocorticotropic hormone (ACTH)
Adrenal cortex
❺ Corticosteroids

Long-term stress response
- Proteins and fats broken down and converted to glucose, leading to increased blood glucose
- Immune system may be suppressed

The Gonads Produce Sex Hormones

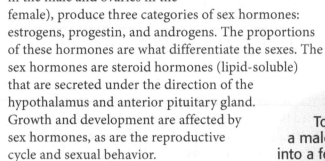

Do women make testosterone? Do men produce estrogen? Yes and yes. Many people don't realize that women and men produce the same sex hormones. The **gonads**, or sex glands (testes in the male and ovaries in the female), produce three categories of sex hormones: estrogens, progestin, and androgens. The proportions of these hormones are what differentiate the sexes. The sex hormones are steroid hormones (lipid-soluble) that are secreted under the direction of the hypothalamus and anterior pituitary gland. Growth and development are affected by sex hormones, as are the reproductive cycle and sexual behavior.

Estrogens and progestins are found in higher concentrations in women than in men. **Estrogens** maintain the female reproductive system and promote the development of such female features as breasts and wider hips. **Progestins**, such as progesterone, are primarily involved in preparing the uterus to support a developing embryo. Both hormones are critical to the female menstrual cycle (see Figure 26.9).

Men have high levels of **androgens** (the main one being testosterone), which stimulate the development and maintenance of the male reproductive system. Androgens produced by embryos during the seventh week of gestation stimulate development of a male rather than a female. During puberty, a high concentration of testosterone triggers the development of male characteristics, such as a lower-pitched voice, facial hair, and skeletal muscle growth.

As we learned, some endocrine disruptors affect the thyroid, but others wreak havoc on reproductive functions by altering the balance of sex hormones. In one study, scientists examined the effects of an endocrine disruptor on frogs. They raised male frogs in the absence (control) or presence of a chemical found in weed-killers called atrazine. When male frogs reached adulthood, they found that atrazine-exposed males were far less successful at mating behaviors and had a lower testosterone level. Furthermore, 10% underwent complete sex reversal—that is, they became females capable of producing eggs! ☑

> To turn a male tadpole into a female frog, add a pinch of endocrine disruptor and wait.

Structure/Function Mimicking Sex Hormones

Chemicals that affect the balance of sex hormones are not always detrimental. A drug used in breast cancer treatment decreases estrogen's effects, and synthetic hormones are useful for boosting function in individuals deficient in particular hormones. These chemicals have similarities in structure and function to the natural hormones they mimic.

Approximately 70% of all breast cancers rely on estrogen to grow and divide, so inhibiting estrogen binding or its effects is a successful way to treat many breast cancers. Tamoxifen is one drug that inhibits estrogen. Once in the body, tamoxifen is broken down into a molecule that binds to estrogen receptors (Figure 25.14). Although each molecule mimics the structure of estrogen enough to attach to its receptor, estrogen's function (stimulating cell division) is not mimicked. Tamoxifen is effective because it competes with estrogen. Similar to people fighting for a limited number of seats on a bus, the drug outcompetes

> ☑ **CHECKPOINT**
> If cells within a male embryo do not secrete testosterone during the proper time in development, what will be the consequence?
>
> Answer: *The fetus, and ultimately the baby that is born, will be genetically male but physically female.*

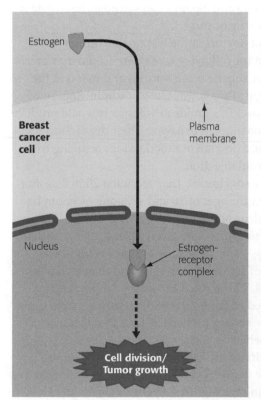
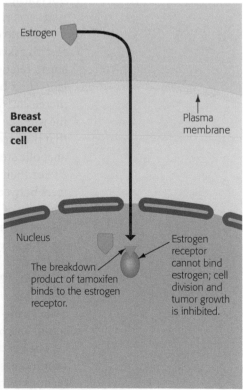

▲ **Figure 25.14 Breast cancer drug treatment.** Some breast cancers rely on estrogen to grow. The drug tamoxifen (and its breakdown products) binds to estrogen receptors in breast cells, preventing estrogen binding and cell division.

estrogen for a limited number of receptors. As long as the level of tamoxifen is high enough, the breast cancer cells will not receive a signal to divide, even in the presence of estrogen.

As individuals age, sex hormone levels decline. One treatment for the side effects of menopause (the time in a woman's life when estrogen production drops off) is synthetic estrogen. Anabolic steroids, artificial hormones that are similar in structure and function to testosterone, are sometimes used to treat the gradual decline of testosterone in men. Because anabolic steroids function to build muscle quickly, some athletes abuse these compounds (as we discussed in the Biology and Society section). What effects might steroid abuse have? We explore that question next. ☑

☑ **CHECKPOINT**

What is the consequence at the cellular level if the dosing (amount) of tamoxifen in a breast cancer patient is too low?

Answer: If the tamoxifen level is low, estrogen will bind to its receptor and readily stimulate cell division of the cancer cells.

Steroid Abuse THE PROCESS OF SCIENCE

Do 'Roids Cause Rage?

For years, the general public, media, and doctors made **observations** of an apparent link between abuse of anabolic steroids and violent mood swings, a phenomenon commonly referred to as "roid rage." But such anecdotal evidence—generalizations made from individual observations—does not clearly establish that abuse of anabolic steroids causes roid rage. Anecdotal evidence is not subject to analysis by the scientific method and therefore is not considered valid. So how can scientists test to see if roid rage actually occurs?

A group of doctors at the National Institutes of Health asked the following **question**: Could they measure a relationship between steroids and mood in the lab? Their **hypothesis** was that the administration of increasing doses of steroids to volunteers would produce measurable changes in behavior. Their **prediction** was that they could document mood swings resulting from anabolic steroid injection.

For their **experiment**, they recruited 20 male volunteers between the ages of 18 and 42, none of whom had a history of drug abuse or psychiatric problems. The research participants were given a placebo drug with no steroids (days 1–3), a low dose of steroids (days 4–6), a high dose of steroids (days 7–9), and then a placebo (days 10–12). Throughout the study, the participants were observed continuously and questioned three times a day using standard psychiatric tests that measure mood and behavior.

The **results** of the experiment revealed a significant increase in ratings for hostility, anger, and violent feelings during the high-dose period (Figure 25.15).

Even though the sample size was small, the researchers noted that their participants received doses well below the levels reportedly taken by some abusers. Also, the test participants used just a single drug (whereas some athletes abuse several at once), hinting that real-world abuse could produce even stronger mood swings. Together with many similar studies, the data suggest a link between the abuse of anabolic steroids and violent behavior.

▼ Figure 25.15 **The mood-altering effects of anabolic steroids.** Each set of bars represents one measure of mood. Each color represents one dosage state. Notice the increased scores during the high-dose period (the green bar in each set).

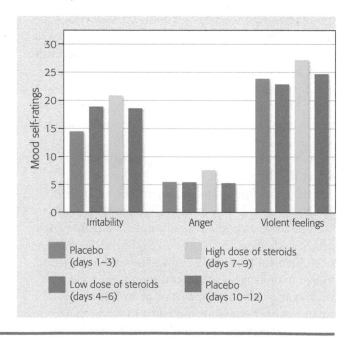

Steroid Abuse EVOLUTION CONNECTION

Steroids and Male Aggression

Among human males, the primary role of testosterone and other androgens is to promote the development and maintenance of male reproductive anatomy and secondary sexual characteristics (such as facial hair). Androgens play a similar role—promoting uniquely male characteristics—in many other species. In fact, research has established that the process of sex determination by androgens occurs in a highly similar manner in all vertebrates, suggesting that androgens had this role early in evolution.

Androgens have been shown to produce a wide variety of effects in different species. In many vertebrates, they are responsible for male vocalizations, such as the territorial songs of birds and the mating calls of frogs. Androgens are connected with aggressive behavior—and not just in people. For example, researchers have measured androgens in a type of cichlid fish called the Mozambique tilapia, a native of eastern Africa. The researchers found elevated androgen levels in males engaged in territorial battles; the victor tends to be the one with the higher level. And in male elephant seals, androgens promote the development of bodies weighing more than 2 tons and aggressive behavior toward other males. These males fight by slamming their bodies against each other (Figure 25.16). After a fight, one male will have established dominance over the other, earning the right to mate with many females, an obvious advantage in terms of Darwinian fitness. Thus, the idea that steroid abuse can lead to aggressive behavior may have an evolutionary basis.

The roles that androgens play among vertebrates illustrate two central aspects of life that result from Darwinian evolution: unity (a consistent effect, such as the development of the male gonads) and diversity (a variety of secondary effects, such as facial hair, male vocalization, and aggressive behavior). The universal nature of gonad development is a strong indication that androgen regulation was an early evolutionary adaptation among vertebrates. But over millions of years of evolution, the specific effects of androgens changed in response to varied environments.

▼ Figure 25.16 **Male elephant seals battling for dominance.**

Chapter Review

SUMMARY OF KEY CONCEPTS

Hormones: An Overview

Hormones are chemical signals carried by the circulatory system that communicate regulatory messages throughout the body. The endocrine system consists of a collection of hormone-secreting cells and is the body's main system for internal chemical regulation, particularly of whole-body activities such as growth, reproduction, and control of metabolic rate. Endocrine glands are the primary sites of hormone production and secretion. Changes in target cells are triggered either indirectly by water-soluble hormones or directly by lipid-soluble hormones.

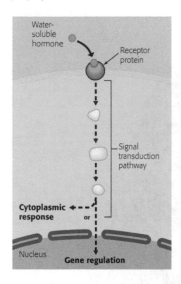
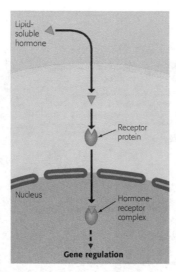

The Human Endocrine System

The human endocrine system consists of about a dozen glands that secrete several dozen hormones. These glands and hormones, such as those summarized in Figure 25.4, vary widely in their functions, means of regulation, and targets.

Information Flow: The Hypothalamus and Pituitary Gland

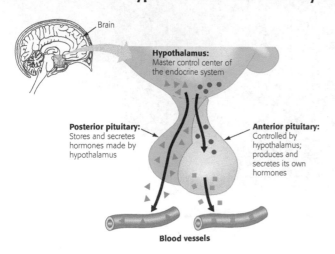

The Thyroid Regulates Metabolism and Development

Hormones from the thyroid gland regulate an animal's development and metabolism. Too little thyroid hormone in the blood (hypothyroidism) or too much (hyperthyroidism) can lead to metabolic disorders.

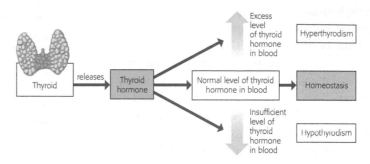

The Pancreas Regulates Blood Glucose

The pancreas secretes two antagonistic hormones, insulin and glucagon, that control the level of glucose in the blood. Insulin signals cells to take up glucose and the liver to store glucose. Glucagon causes the liver to release stored glucose into the blood. Diabetes mellitus results from a lack of insulin (type 1 diabetes) or a failure of cells to respond to it (type 2 diabetes).

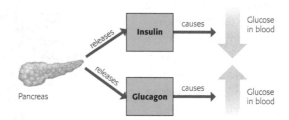

The Adrenal Glands Respond to Stress

Hormones from the adrenal glands help maintain homeostasis when the body is stressed. Nerve signals from the hypothalamus stimulate the adrenal medulla to secrete epinephrine and norepinephrine, which quickly trigger the fight-or-flight response. The long-term stress response is initiated when the hypothalamus sends hormonal signals to the pituitary. In response, ACTH from the anterior pituitary causes the adrenal cortex to secrete corticosteroids, which promote the synthesis of glucose.

The Gonads Produce Sex Hormones

Estrogens, progestins, and androgens are steroid sex hormones produced by the ovaries in females and the testes in males. Estrogens and progestins stimulate the development of female characteristics and maintain the female reproductive system. Androgens, such as testosterone, trigger the development of male characteristics. The secretion of sex hormones is controlled by the hypothalamus and pituitary gland.

Structure/Function: Mimicking Sex Hormones

Hormones that mimic the structure of sex hormones are useful in treating breast cancer as well as age-related decline of estrogen and testosterone.

MasteringBiology®

For practice quizzes, BioFlix animations, MP3 tutorials, video tutors, and more study tools designed for this textbook, go to MasteringBiology®

SELF-QUIZ

1. Which of the following statements concerning homeostasis is *not* true?
   a. It is the maintenance of constant internal conditions.
   b. It involves the regulation of body conditions, such as glucose level.
   c. It usually prevents major fluctuations in body conditions.
   d. Its maintenance is solely the responsibility of the endocrine system.

2. Hormones are made from all of the following classes of compounds *except*
   a. peptides.
   b. proteins.
   c. carbohydrates.
   d. amines.

3. Explain how the same hormone can have different effects on two different target cells and no effect on a third type of cell.

4. Which of the following has both endocrine and nonendocrine functions?
   a. pancreas
   b. salivary gland
   c. anterior pituitary gland
   d. skeletal muscle

5. Why does the consumption of alcohol lead to frequent urination?

6. A patient comes to a local health clinic with a large swelling in the neck that appears to be goiter. As the physician is taking the patient's history, the patient mentions not using iodized salt. The patient is overweight and has a slow metabolism. Does this patient have hyperthyroidism or hypothyroidism? How is this problem related to the lack of iodine in the diet?

7. The pancreas increases its output of insulin in response to
   a. an increase in blood glucose.
   b. a hormone secreted by the anterior pituitary.
   c. both of these choices.

8. Which one of the following statements is true?
   a. The anterior pituitary is composed of nervous tissue.
   b. The pituitary is the master control center of the entire endocrine system.
   c. Inhibiting hormones make the anterior pituitary secrete hormones.
   d. The pituitary secretes hormones that influence numerous body functions.

9. Name the gland that each of the following hormones is secreted from: cortisol, epinephrine, oxytocin, releasing hormones, estrogen, glucagon, and progesterone.

10. Testosterone belongs to a class of sex hormones called _____.

11. For each of the following situations, name the hormone(s) that is most likely responsible for the effect described.
   a. The first recognized autoimmune disease is Hashimoto's disease, which is characterized by fatigue, weight gain, a slowed heart rate, and several other problems. It is caused by a lack of _____ due to an inflammation of the _____.
   b. In order to generate energy for stressful situations, your body releases _____ that causes the breakdown of proteins and fats into sugars.
   c. A friend of yours, who is an avid bodybuilder, appears to have increased mood swings and outbursts of anger. You suspect that he is using _____.

Answers to these questions can be found in Appendix: Self-Quiz Answers.

THE PROCESS OF SCIENCE

12. **Interpreting Data**. Look back at Figure 25.12 and its associated description in the text. Determine whether each statement below is supported or not supported by the data and explain your reasoning.
   a. About 30% of low-income patients admitted to the hospital suffer from hypoglycemia in the last week of a month.
   b. The number of patients admitted with hypoglycemia was always higher in the low-income group compared to the high-income group.
   c. The difference between low-income and high-income patients more than doubles between week 1 and week 4.

13. Older men often develop a benign enlargement of breast tissue, a condition known as senile gynecomastia. This can be treated to some extent with changes in lifestyle (healthy and low-fat diet, reduced alcohol intake, exercise, etc.) and with cosmetic surgery. Which of the following hypotheses explains why older men develop this condition? (Several answers may be correct.)
   a. decreasing estrogen levels with increasing age
   b. decreasing testosterone levels with increasing age
   c. increased subcutaneous fat with increasing age
   d. accumulation of estrogen-like chemicals from cosmetic products

BIOLOGY AND SOCIETY

14. In Chapter 9, we discussed the role of an individual's chromosomal makeup in sex determination. In the current chapter, we have learnt that different doses of sex hormones determine sex-specific development and behavior, and that disruptions in the sex hormones' signaling pathways can cause morphological sex reversals, such that a genetically male fetus (XY) develops into a morphologically female baby or vice versa. Would you consider this baby a boy or a girl? What determines society's perception of sex and gender? How do these two categories differ?

15. Type 2 diabetes is becoming increasingly common in the United States. The primary risk factor for type 2 diabetes is a history of obesity. Statistics show that children are becoming obese at an alarmingly high rate, and this correlates with increased rates of type 2 diabetes. When young children become overweight, the problem is usually blamed on nutrition and exercise choices made by the parents. Why do you think today's parents have a harder time feeding their children nutritious meals and providing an exercise program than did parents of previous generations? What can be done to solve this problem? Additionally, diabetes is more common in low-income populations than in wealthy populations. What are possible reasons for this?

26 Reproduction and Development

At one time, you had ▶ gill slits and a tail.

◀ A switch from briefs to boxers solves many cases of infertility.

A 9-month ▶ pregnancy may seem long, but gestation in elephants drags on for 22 months.

High-Tech Babies BIOLOGY AND SOCIETY

New Ways of Making Babies

Throughout their 18-year marriage, Gary and Francia P. always wanted children. But after multiple miscarriages, the couple came to accept that they were infertile, unable to bear children naturally. Seeking help, they visited a fertility clinic, hoping to use technology to fulfill their dream of starting a family.

Two weeks later, Gary died suddenly of a heart attack. At the hospital, six hours after his death, Francia requested that a sample of Gary's sperm be removed. Months later, Francia returned to the fertility clinic. The sample obtained from her husband's body provided just enough sperm for one attempt at producing a child. Her eggs were surgically removed and combined with Gary's sperm. The resulting embryos were implanted into Francia's uterus, and one grew into a baby son, Jacob, born about one year after his father's death.

Because of his career as a captain in the Newark, New Jersey, police department, Gary's family was entitled to pension benefits that were tied to the size of his family. After Jacob's birth, Francia applied to the pension board for an increase in her widow's benefits. Should Francia be entitled to this increase? The pension board did not think so, ruling that a child born more than a year after a parent's death is not entitled to benefits from that parent.

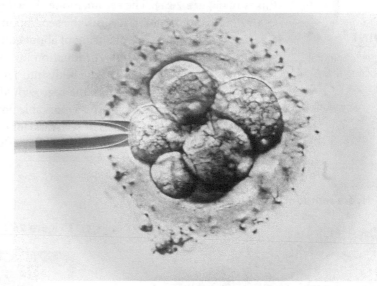

LM 40x

Fertilization in a test tube. This micrograph shows a technician manipulating a human embryo in preparation for implantation.

Francia filed a legal appeal. There were several previous cases where the courts determined that children born to women who used a deceased husband's sperm had full rights as heirs, but these cases involved men who had donated their sperm before death. Francia's case was thus entering new legal territory. As of 2014, the final outcome was still under consideration.

This case illustrates two important points. First, modern technologies allow us to circumvent the natural course of reproduction in many ways. Second, society's laws have not yet caught up to these new technologies. It is up to all of us to decide what is right in such circumstances. To quote the U.S. Court of Appeals in Philadelphia from a related case, "[We] cannot help but observe that this is, indeed, a new world."

In this chapter, we'll explore this new world, beginning with the anatomy and physiology of animal reproduction, paying particular attention to the reproductive structures of humans and how human babies develop from a single cell. Along the way, we'll also consider how modern health practices and technologies circumvent the natural process of sexual reproduction.

Unifying Concepts of Animal Reproduction

Although every individual animal has a relatively short life span, species last much longer because of **reproduction**, the creation of new individuals from existing ones. Animals reproduce in a great variety of ways, but there are two principal modes: asexual reproduction and sexual reproduction.

Asexual Reproduction

Asexual reproduction (reproduction without sex) is the creation of genetically identical offspring by a lone parent. Several types of asexual reproduction can be found in the animal kingdom. Many invertebrates reproduce asexually by **budding**, the outgrowth and eventual splitting off of a new individual from a parent. Hydras, freshwater relatives of jellies, reproduce this way **(Figure 26.1)**. The sea anemone shown in **Figure 26.2** is undergoing **fission**, the separation of a parent into two or more offspring of about equal size. Another type of asexual reproduction involves fragmentation, the breaking of the body into several pieces, followed by regeneration, the regrowth of lost body parts. Some organisms have remarkable powers of regeneration. In certain species of sea stars, for example, an arm plus a bit of the central body that has split off can give rise to a new sea star (see Figure 8.1). In some species of sea sponges, if a single sponge is pushed through a wire mesh, each of the resulting clumps of cells can grow into a new sponge. (These are examples of asexual reproduction as it occurs in nature. Other species have been the target of artificial asexual reproduction; see the discussion of cloning in Chapter 11.)

In nature, asexual reproduction has several potential advantages. Because it eliminates the need to find a mate, asexual reproduction allows the individual members of a species to perpetuate themselves even if they are isolated from one another. Asexual reproduction also allows organisms to multiply quickly, without spending time or energy producing sperm and eggs. If an individual is very well suited to its environment, asexual reproduction allows it to reproduce rapidly and exploit available resources.

A potential disadvantage of asexual reproduction is that it produces genetically uniform populations. Genetically similar individuals may thrive in a particular environment. But if the environment changes and becomes less favorable to survival (as a result of some natural disaster, say, or a new predator or pathogen), then all individuals may be affected equally, and the entire population may die out. ☑

▼ Figure 26.1 **Hydra reproducing by budding.**

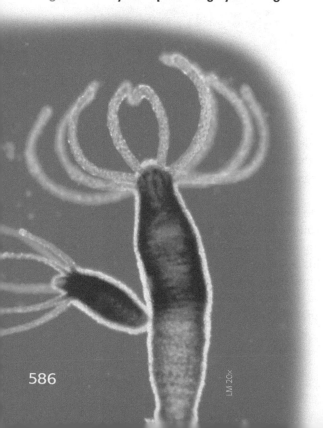

▼ Figure 26.2 **Sea anemone reproducing by fission.**

LM 20x

Sexual Reproduction

You know that you have a mix of traits from your mother and father. But you also have a genome distinct from every other human (unless you have an identical twin). You are the product of **sexual reproduction**, the creation of offspring by the fusion of two haploid sex cells called **gametes** to form a diploid **zygote**. The male gamete is the **sperm**, and the female gamete is the **egg**. The zygote and the new individual it develops into contain a unique combination of genes carried from the parents via the egg and sperm.

Unlike asexual reproduction, sexual reproduction increases genetic variability among offspring as a result of the huge variety of gametes produced by meiosis (see Chapter 8). Variation is the raw material of evolution by natural selection. When an environment changes, there is a better chance that some of the various offspring will survive and reproduce than if all the offspring were genetically identical.

Some animals can reproduce both asexually and sexually, benefiting from both modes. In **Figure 26.3**, you can see two sea anemones of the same species. The one on the left is reproducing asexually (via fission), while the one on the right is releasing eggs. Why would such dual reproductive capabilities be advantageous to an animal? Some animals reproduce asexually when food is ample and conditions are favorable. But when conditions change (becoming colder or drier, for example), these animals switch to sexual reproduction, producing a generation of genetically varied individuals that has a better potential to have some well-adapted individuals.

Most individual animals are a single sex, either male or female. But in some species, each individual is a **hermaphrodite**, meaning that it has both male and female reproductive systems. For hermaphrodites, any two individuals can mate. Some hermaphrodites (such as tapeworms, which can live as parasites in the human intestine) can fertilize their own eggs. Other species require a partner. For example, when earthworms mate, each individual donates and receives sperm.

The mechanics of fertilization play an important part in sexual reproduction. Many aquatic animals use external fertilization, in which the parents discharge their gametes into the water, where fertilization occurs (**Figure 26.4**). The female and male don't necessarily have to touch to mate. Among some species with external fertilization, individuals clustered in the same area release their gametes into the water simultaneously, a process called spawning.

In contrast to animals that reproduce externally, nearly all terrestrial animals reproduce by internal fertilization, in which a male deposits sperm within a female's body. This adaptation enables sperm to reach eggs despite a dry environment. In the next section, we'll examine the reproductive anatomy that allows one particular terrestrial animal to achieve sexual reproduction. ☑

☑ **CHECKPOINT**

What is the most important difference in the genetic makeup of the offspring resulting from sexual versus asexual reproduction?

Answer: Asexual reproduction produces genetically identical offspring, whereas sexual reproduction produces genetically diverse offspring.

▼ Figure 26.3 **Asexual (left) and sexual (right) reproduction in a starlet sea anemone.**

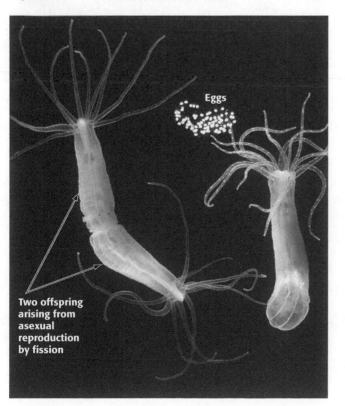

Eggs

Two offspring arising from asexual reproduction by fission

▼ Figure 26.4 **External fertilization in frogs.** Frogs release eggs and sperm (too small to be seen) into the water, where fertilization takes place. The embrace between the two frogs is a mating ritual that coordinates the simultaneous release of gametes.

Eggs

☑ **CHECKPOINT**

Arrange the following organs
of the male reproductive
system in the correct
sequence for the travel of
sperm: epididymis, testis,
urethra, vas deferens.

*Answer: testis, epididymis, vas
deferens, urethra*

Human Reproduction

Although we tend to focus on the anatomical differences between the human male and female reproductive systems, there are some important similarities. Both sexes have a pair of **gonads**, the organs that produce gametes. And both sexes have ducts that store and deliver the gametes as well as structures that allow mating. We'll now examine the anatomical features of human reproduction, beginning with the male anatomy.

Male Reproductive Anatomy

Figure 26.5 presents side and front views of the male reproductive system. The **penis** is the structure that transfers sperm from the male to the female during sexual intercourse. It contains three cylinders of erectile tissue (shown in blue), which consists of modified veins and capillaries that can fill with blood and cause an erection during sexual arousal. The penis consists of a shaft that supports a highly sensitive glans (head). A **prepuce**, or foreskin, covers the glans. The foreskin may be surgically removed, a procedure known as circumcision.

The **testes** (singular, *testis*), the male gonads, are located outside the abdominal cavity in a sac called the **scrotum**. A testis and a scrotum together are called a **testicle**. Because sperm do not develop optimally at body temperature, the scrotum promotes sperm formation by keeping sperm-forming cells about 2°C cooler. In cold conditions, muscles around the scrotum contract, pulling the testes toward the body, thereby maintaining the proper temperature.

From puberty into old age, the testes produce hundreds of millions of sperm each day. From the testes, sperm pass into a coiled tube called the epididymis. During **ejaculation**—expulsion of sperm-containing fluid from the penis—the sperm leave the epididymis and travel through a duct called the **vas deferens** (the target of a vasectomy). The **seminal vesicles** and **prostate gland** add fluid that nourishes the sperm and provides protection from the natural acidity of the vagina. The prostate gland enlarges in most men over age 40, and prostate cancer is the second most commonly diagnosed cancer in the United States. Each vas deferens, one from each of the two testes, empties into the **urethra**. The urethra conveys, at different times, both sperm and urine out through the penis. Ejaculation, caused by the contraction of muscles along the sperm ducts, releases about 5 mL (1 teaspoonful) of **semen**. Only 5% of semen consists of sperm (typically 200–500 million of them); the remaining 95% of semen is fluid secreted by the various glands. ☑

▼ **Figure 26.5 The male reproductive system.** The color blue highlights erectile tissue. Some nonreproductive structures are also labeled (in italics) to help keep you oriented.

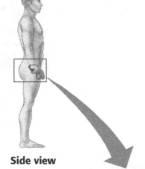

Side view

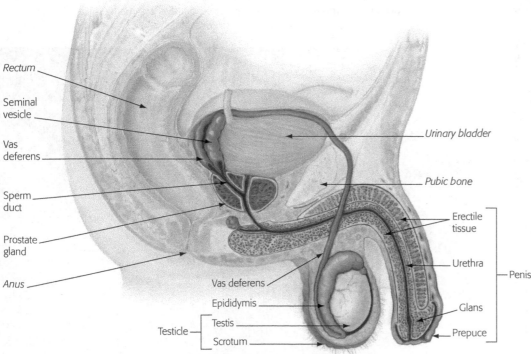

Rectum

Seminal vesicle

Vas deferens

Sperm duct

Prostate gland

Anus

Urinary bladder

Pubic bone

Erectile tissue

Urethra — Penis

Vas deferens

Epididymis

Testis — Testicle

Scrotum

Glans

Prepuce

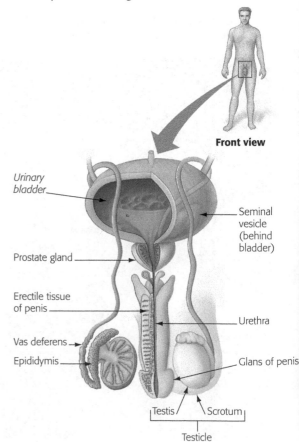

Front view

Urinary bladder

Prostate gland

Erectile tissue of penis

Vas deferens

Epididymis

Seminal vesicle (behind bladder)

Urethra

Glans of penis

Testis / Scrotum — Testicle

Female Reproductive Anatomy

Two views of the female reproductive system are shown in **Figure 26.6**. The outer features of the female reproductive anatomy are collectively called the **vulva**. The **vagina**, or birth canal, opens to the outside just behind the opening of the urethra, the tube through which urine is excreted. An outer pair of thick, fatty ridges, the **labia majora**, protects the entire genital region, and a pair of inner skin folds, the **labia minora**, borders the openings. A thin membrane called the **hymen** partly covers the vaginal opening until sexual intercourse or other vigorous physical activity ruptures it. The **clitoris** is an organ that engorges with blood and becomes erect during sexual arousal. Analogous to the male penis, the clitoris consists of a short shaft supporting a rounded glans, or head, covered by a small hood of skin called the prepuce. The clitoris has an enormous number of nerve endings and is very sensitive to touch.

The **ovaries** are the female gonads, the site of gamete production. Each ovary is about an inch long and has a bumpy surface. The bumps are **follicles**, each consisting of a single developing egg cell surrounded by cells that nourish and protect it. The follicles also produce estrogen, the female sex hormone (see Chapter 25).

A female is born with over a million follicles, but only several hundred will release egg cells during her reproductive years. Starting at puberty and continuing until menopause (the end of fertility, which usually occurs during middle age), a woman undergoes ovulation about every 28 days. During **ovulation**, one follicle (or rarely two or more) matures and ejects an immature egg cell. After ovulation, what remains of the follicle grows to form a solid mass called the **corpus luteum**, which secretes hormones during the reproductive cycle (as you'll see later in the chapter). The released egg enters an **oviduct** (also called a fallopian tube), where cilia sweep it toward the uterus, like a crowd surfer being carried hand-over-hand across a mosh pit. If sperm are present, fertilization may take place in the upper part of the oviduct. If the released egg is not fertilized, it is shed during menstruation, and a new follicle matures during the next cycle.

The **uterus** (also called the womb) is the actual site of pregnancy. The uterus is about the size and shape of an upside-down pear, but grows to several times that size during pregnancy. The uterus has a thick muscular wall lined with a blood-rich layer of tissue called the **endometrium**. An embryo implants in the endometrium and grows there. The term **embryo** is used for the stage in development from the first division of the zygote until body structures begin to appear, about the 9th week. From the 9th week until birth, a developing human is called a **fetus**.

The narrow neck at the bottom of the uterus is the **cervix**. It is recommended that a woman have a yearly Pap test in which cells are removed from around the cervix and examined under a microscope for signs of cervical cancer. Regular Pap smears greatly increase the chances of detecting cervical cancer early and treating it successfully. The cervix opens into the vagina. During intercourse, the vagina serves as a repository for sperm. ☑

☑ **CHECKPOINT**

In which organ of the human female reproductive system does fertilization occur? In which organ does pregnancy occur?

Answer: the oviduct; the uterus

▼ Figure 26.6 **The female reproductive system.** The color blue highlights erectile tissue. Some nonreproductive structures are also labeled (in italics) to help keep you oriented.

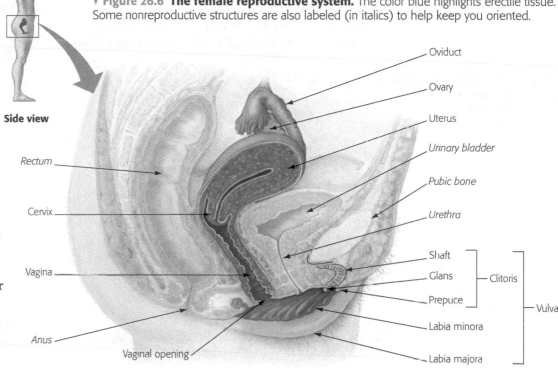

Side view

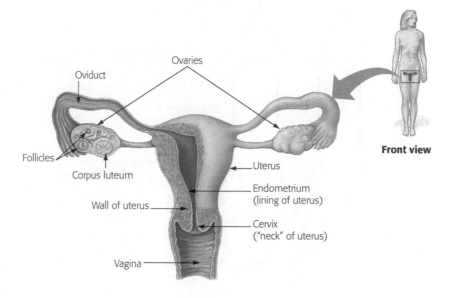

Front view

589

Gametogenesis

The production of gametes is called **gametogenesis**. Human gametes—sperm and egg—are haploid cells with 23 chromosomes that develop by meiosis from diploid cells with 46 chromosomes. (You may find it helpful to review the discussion of meiosis in Chapter 8, particularly Figures 8.13–8.15.) There are significant differences in gametogenesis between human males and females, so we'll examine the processes separately.

Spermatogenesis

The formation of sperm cells is called **spermatogenesis** (Figure 26.7). Sperm cells develop inside the testes in coiled tubes called the **seminiferous tubules**. Cells near the outer walls of the tubules multiply constantly by mitosis. Each day, about 3 million of them differentiate into primary spermatocytes, the cells that undergo meiosis.

Meiosis I of a primary spermatocyte produces two haploid secondary spermatocytes. Then meiosis II forms four cells, each with the haploid number of chromosomes. A sperm cell that develops from one of these haploid cells is gradually pushed toward the center of the seminiferous tubule. From there, it passes into the epididymis, where it matures, becomes motile, and is stored until ejaculation. In human males, spermatogenesis takes about 10 weeks. Because the pre-sperm cells continuously replenish themselves, there is a never-ending supply of spermatocytes, allowing males to produce sperm throughout their adult lives.

Oogenesis

Figure 26.8 summarizes **oogenesis**, the development of mature egg cells, also called ova (singular, *ovum*).

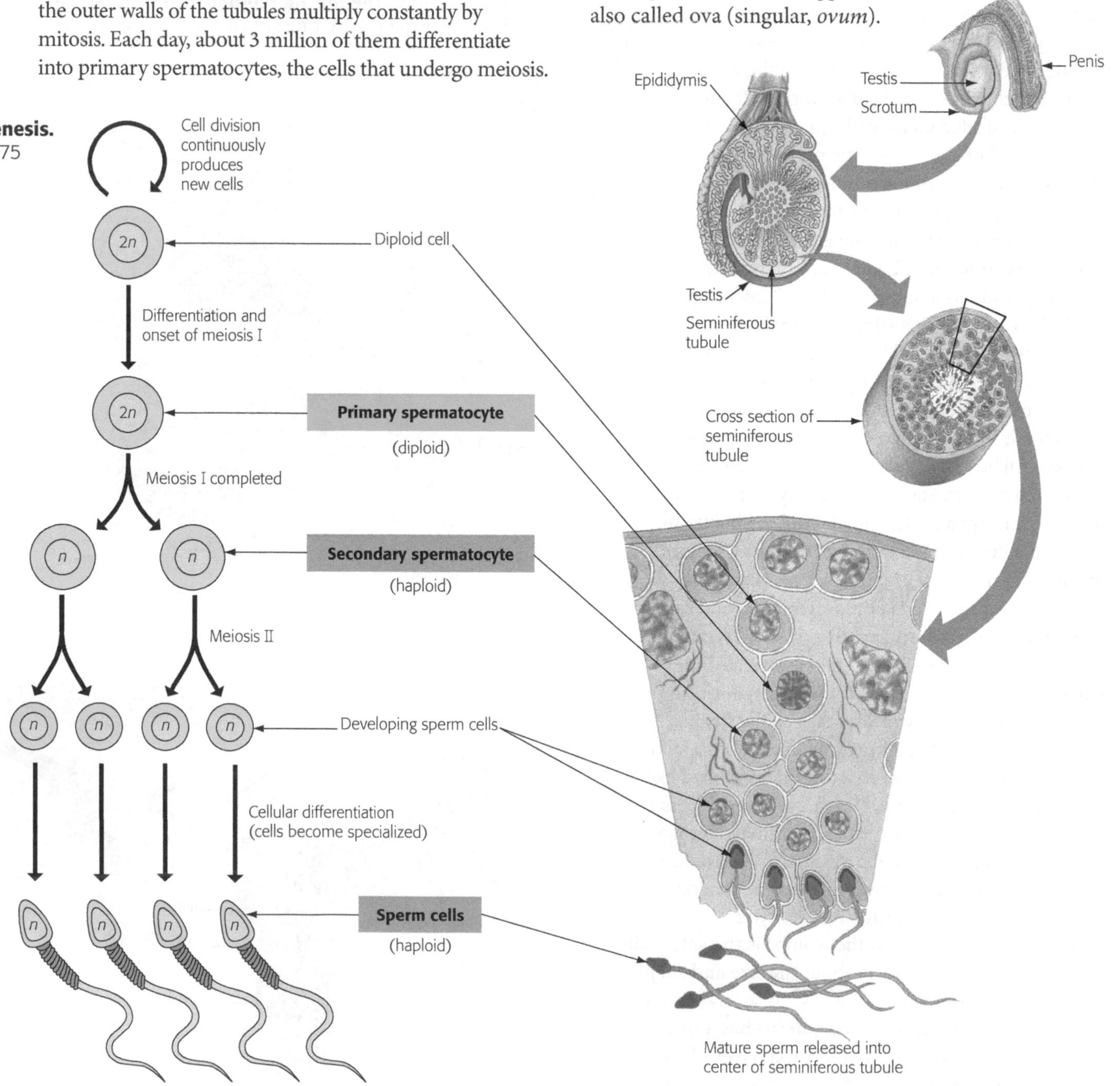

▶ Figure 26.7 **Spermatogenesis.** This process takes about 65–75 days in the human male.

At birth, each ovary contains many thousands of follicles. Each follicle contains a single dormant primary oocyte, a diploid cell that has paused its cell cycle in prophase of meiosis I.

A primary oocyte can be triggered to develop further by the hormone FSH (follicle-stimulating hormone). After puberty and until menopause, about every 28 days, FSH from the pituitary gland stimulates one of the dormant follicles to develop. The follicle enlarges, and the primary oocyte within it completes meiosis I and begins meiosis II. The division of the cytoplasm in meiosis I is unequal, with a single secondary oocyte receiving almost all of it. The smaller of the two daughter cells, called the first polar body, receives almost no cytoplasm.

About the time the secondary oocyte forms, the pituitary gland secretes LH (luteinizing hormone), which triggers ovulation. The ripening follicle bursts, releasing its secondary oocyte from the ovary. The ruptured follicle then develops into a corpus luteum. The secondary oocyte enters the oviduct, and if a sperm cell fuses with it, the secondary oocyte completes meiosis II. Meiosis II is also unequal, yielding a small polar body and a mature egg (ovum). The chromosomes of the egg can then fuse with the chromosomes of the sperm cell, producing a diploid

zygote. The polar bodies, which are quite small because they received almost no cytoplasm, degenerate. The zygote thus acquires nearly all the cytoplasm and the bulk of the nutrients and organelles contained in the original cell.

Both oogenesis and spermatogenesis produce haploid gametes, but there are several important differences between the two processes. One obvious difference is location: testes in the male and ovaries in the female. Furthermore, human males create new sperm every day from puberty through old age. Human females, on the other hand, create primary oocytes only during their fetal development. Another difference is that four gametes result from each diploid parent cell during spermatogenesis, whereas oogenesis results in only one gamete from each parent cell. There are also significant differences in the cells produced by meiosis: Sperm are small, move by means of a whiplike flagellum, and contain relatively few nutrients; eggs are large, not self-propelled, and well stocked with nutrients and organelles. Finally, spermatogenesis is completed before sperm leave the testis, whereas oogenesis cannot be completed without stimulation from a sperm cell. ☑

☑ CHECKPOINT

During gametogenesis in a human female, how many ova are produced from one primary oocyte? In a male, how many sperm arise from one primary spermatocyte?

Answer: one; four

▼ Figure 26.8 **Oogenesis and development of an ovarian follicle.** Notice that oogenesis starts before birth, but the final ovum does not form until fertilization.

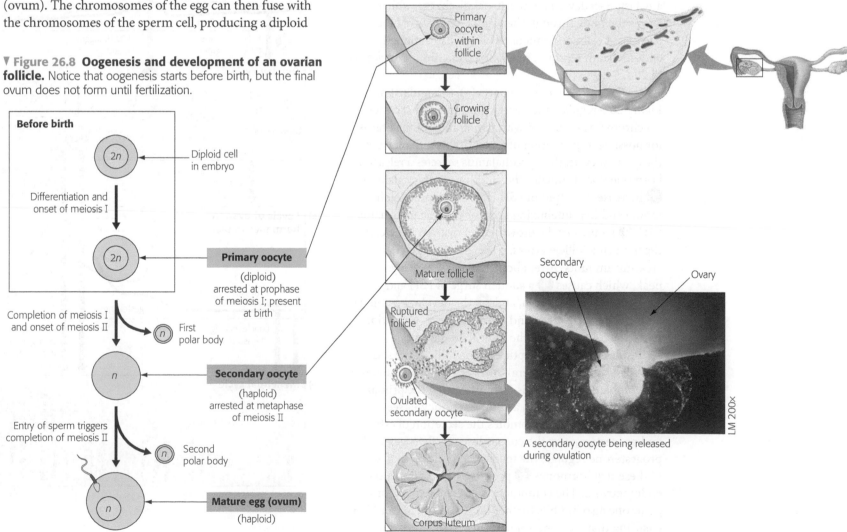

The Female Reproductive Cycle

Human females have a **reproductive cycle**, a recurring series of events that produces gametes, makes them available for fertilization, and prepares the body for pregnancy (**Figure 26.9**). The reproductive cycle repeats every 28 days, on average, but cycles from 20 to 40 days are not uncommon. The reproductive cycle is actually two cycles in one. The **ovarian cycle** (Figure 26.9c) controls the growth and release of an egg. During the **menstrual cycle** (Figure 26.9e), the uterus is prepared for possible implantation of an embryo. Hormonal messages coordinate the two cycles, keeping them synchronized with each other.

The menstrual cycle begins on the first day of a woman's period. **Menstruation** is uterine bleeding caused by the breakdown of the endometrium, the blood-rich inner lining of the uterus. If an embryo implants in the uterine wall, it will obtain nutrients from the endometrium, and the thickened lining will not be discharged. Menstruation is thus a sign that pregnancy has not occurred during the previous cycle. Menstruation usually lasts 3–5 days. The menstrual discharge, which leaves the body through the vagina, consists of blood, clusters of cells, and mucus. After menstruation, the endometrium regrows, reaching its maximum thickness in 20–25 days.

The hormones shown in parts (a), (b), and (d) of Figure 26.9 regulate the ovarian and menstrual cycles, synchronizing ovulation with preparation of the uterus for possible implantation of an embryo. At the start of the ovarian cycle, the hypothalamus secretes a releasing hormone that stimulates the anterior pituitary gland to ❶ increase its output of FSH (follicle-stimulating hormone) and LH (luteinizing hormone). True to its name, FSH ❷ stimulates the growth of an ovarian follicle. As the maturing follicle grows, it secretes estrogen in increasing amounts. After about 12 days, ❸ estrogen levels peak, which causes ❹ a sudden surge of FSH and LH. This stimulates ovulation, and ❺ the developing follicle within the ovary bursts and releases its egg. Ovulation takes place on day 14 of the typical 28-day cycle.

Besides promoting rupture of the follicle, the sudden surge of LH has several other effects. It stimulates the completion of meiosis I, transforming the primary oocyte in the follicle into a secondary oocyte (see Figure 26.8). LH also promotes the secretion of estrogen and progesterone by the corpus luteum. Estrogen and progesterone regulate the menstrual cycle. Rising levels of these two hormones ❻ promote thickening of the endometrium. The combination of estrogen and progesterone also inhibits further secretion of FSH and LH, ensuring that a second follicle does not mature during

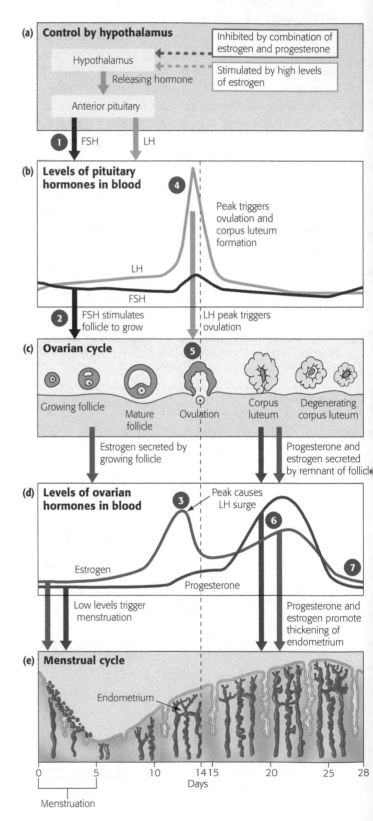

▼ **Figure 26.9 The reproductive cycle of the human female.** This figure shows how (c) the ovarian cycle and (e) the menstrual cycle are regulated by changing hormonal levels, represented in parts (a), (b), and (d). The time scale at the bottom of the figure applies to parts (b)–(e).

(a) **Control by hypothalamus**

Inhibited by combination of estrogen and progesterone

Stimulated by high levels of estrogen

Hypothalamus

Releasing hormone

Anterior pituitary

❶ FSH   LH

(b) **Levels of pituitary hormones in blood**

❹

Peak triggers ovulation and corpus luteum formation

LH

FSH

❷ FSH stimulates follicle to grow

LH peak triggers ovulation

(c) **Ovarian cycle**

❺

Growing follicle   Mature follicle   Ovulation   Corpus luteum   Degenerating corpus luteum

Estrogen secreted by growing follicle

Progesterone and estrogen secreted by remnant of follicle

(d) **Levels of ovarian hormones in blood**

❸   Peak causes LH surge

❻

Estrogen

Progesterone

❼

Low levels trigger menstruation

Progesterone and estrogen promote thickening of endometrium

(e) **Menstrual cycle**

Endometrium

0   5   10   14 15   20   25   28
Days

Menstruation

this cycle. Further degeneration of the corpus luteum causes the levels of estrogen and progesterone to fall off. Once these hormones ❼ fall below a critical level, the endometrium begins to shed, starting menstruation again. Now that estrogen and progesterone are no longer there to inhibit it, the pituitary secretes FSH and LH, and a new cycle begins.

The description of the reproductive cycle in this section assumes that fertilization has not occurred. If it does, the embryo implants into the endometrium and secretes a hormone called HCG (human chorionic gonadotropin). HCG maintains the corpus luteum, which continues to secrete progesterone and estrogen, keeping the endometrium intact. As you will see later, some forms of contraception work by mimicking the high levels of hormones that occur during pregnancy. Most home pregnancy tests work by detecting HCG that passes from a pregnant woman's blood to her urine. In males, HCG boosts testosterone production, so it can be abused as a performance-enhancing drug by athletes; its use is therefore banned by many sports organizations. ☑

☑ **CHECKPOINT**
What hormonal changes trigger the start of menstruation?

Answer: decreasing levels of estrogen and progesterone

Reproductive Health

Now that you've read about the anatomy and physiology of the human reproductive system, you can apply this knowledge to two issues of reproductive health: contraception and the transmission of disease.

Contraception

Contraception is the deliberate prevention of pregnancy. There are many forms of contraception that interfere with different steps in the process of becoming pregnant. Table 26.1 lists the most common methods of birth control and their failure rates when used correctly and when used in typical practice. Note that these two rates are often quite different, emphasizing the importance of learning to use contraception properly.

Complete abstinence (avoiding intercourse) is the only totally effective method of contraception, but other methods are effective to varying degrees. Sterilization, surgery that prevents sperm from reaching an egg, is very reliable.

A man may have a **vasectomy**, in which a doctor cuts a section out of each vas deferens to prevent sperm from reaching the urethra. A woman may have a **tubal ligation** (have her "tubes tied"). In this procedure, a doctor removes a short section from each oviduct, often tying (ligating) the remaining ends and thereby blocking the route of sperm to egg. Both forms of sterilization are free from side effects and relatively safe. They are meant to be permanent but can sometimes be surgically reversed. An **intrauterine device (IUD)** is a T-shaped device placed within the uterus by a healthcare provider. IUDs are safe and highly effective at preventing pregnancy for up to 12 years, but can be safely removed at any time.

The effectiveness of other methods of contraception depends on how they are used. Temporary abstinence, also called the **rhythm method** or **natural family planning**, depends on refraining from intercourse during the days around ovulation, when fertilization is most likely. In theory, the time of ovulation can be determined by monitoring

Table 26.1	Effectiveness of Some Common Contraceptive Methods	
	Pregnancies per 100 Women per Year	
Method	**Used Correctly**	**Typically**
None		85
Birth control pill	0.1	5
Vasectomy	0.1	n/a
Tubal ligation	0.5	n/a
IUD	0.2–0.8	n/a
Rhythm method	1–9	20
Withdrawal	4	19
Condom (male)	2	15
Diaphragm and spermicide	6	16
Spermicide alone	6	29

changes in body temperature and the composition of cervical mucus, but careful monitoring and record keeping are required. Additionally, the length of the reproductive cycle can vary from month to month, and sperm can survive for 3–5 days within the female reproductive tract, making natural family planning among the most unreliable methods of contraception in actual practice. Withdrawal of the penis from the vagina before ejaculation is also ineffective, because sperm may exit the penis before climax.

If used correctly, methods that physically block the union of sperm and egg can be quite effective. **Condoms** are sheaths, usually made of latex, that fit over the penis or within the vagina. A **diaphragm** is a dome-shaped rubber cap that covers the cervix. It requires a doctor's visit for proper fitting. To be more effective, condoms and diaphragms can be used in combination with **spermicides**, sperm-killing chemicals in the form of a jelly, cream, or foam; spermicides used alone are not reliable contraceptives.

Some of the most effective methods of contraception prevent the release of eggs altogether. Oral contraceptives, or **birth control pills**, come in several different forms that contain varying formulations of a synthetic estrogen and/or a synthetic progesterone. Steady intake of these hormones simulates their constant levels during pregnancy. In response to being "fooled" that pregnancy has occurred, the hypothalamus fails to send the signals that start development of an ovarian follicle. Ovulation ceases, preventing pregnancy. In addition to pills, various combinations of these hormones are also available as a shot (Depo-Provera), a ring inserted into the vagina, or a skin patch **(Figure 26.10)**. Some of these contraceptives maintain hormone levels for up to 3 months, so a user will have only four periods a year. Currently, there is no chemical contraceptive available that can prevent the production or release of sperm.

Certain drugs can prevent fertilization or implantation even after intercourse. **Morning-after pills** ("Plan B") are birth control pills that can be taken in high doses for emergency contraception. If taken within 3 days of intercourse, morning-after pills are about 75% effective at preventing pregnancy. Such treatments should only be used in emergency situations because they have significant side effects. If pregnancy has occurred, the drug RU-486 (mifepristone) can induce an abortion, the termination of a pregnancy in process. RU-486 must be taken within the first 7 weeks of pregnancy and requires a doctor's prescription and several visits to a medical facility.

Responsible sex involves more than preventing unwanted pregnancies. It also involves preventing the spread of sexually transmitted diseases, the focus of the next section. ✓

☑ CHECKPOINT

What methods of contraception are foolproof or nearly so?

Answer: abstinence and sterilization (vasectomy and tubal ligation)

▼ Figure 26.10 **A contraceptive skin patch containing birth control hormones.**

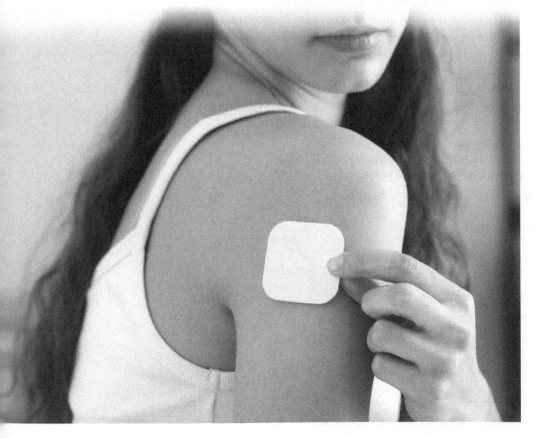

Sexually Transmitted Diseases

Sexually transmitted diseases (STDs), also referred to as sexually transmitted infections (STIs), are contagious diseases spread by sexual contact. **Table 26.2** lists the STDs most common in the United States, organized by the type of infectious agent. Notice that bacteria, viruses, protists, and fungi can all cause STDs.

The most common bacterial STD (with nearly a million new cases reported annually in the United States and over 90 million worldwide) is **chlamydia**. The primary symptoms are genital discharge and a burning sensation during urination. But chlamydia is frequently "silent"— half of infected men and three-quarters of infected women do not notice any symptoms. Long-term complications are rare among men, but up to 40% of infected women develop pelvic inflammatory disease (PID). The inflammation associated with PID may block the oviducts or scar the uterus, causing infertility. Fortunately, chlamydia can be easily treated with a single dose of an antibiotic. But early screening is required to catch the disease before any scarring occurs. Sexually active women are encouraged to be screened for chlamydia and other STDs annually.

In contrast to bacterial STDs, viral STDs are not curable. They can be controlled by medications, but their symptoms and the ability to infect others remain a possibility throughout a person's lifetime. One in five Americans is infected with **genital herpes**, caused by the herpes simplex virus type 2, a variant of the virus that causes cold sores. Most outbreaks heal within a few weeks, but the virus lies dormant within nearby nerve cells. Months or years later, it can reemerge, causing fresh sores that allow the virus to be spread to sexual partners. Abstinence during outbreaks, the use of condoms, and the use of antiviral medications that minimize symptoms can reduce the spread of herpes. But there is no cure, so infection lasts a lifetime.

AIDS, caused by HIV (discussed in Chapters 10 and 24), poses one of the greatest health challenges in the world today, particularly among the developing nations of Africa and Asia. But even within the United States, there are 56,000 new infections each year, one-third of which result from heterosexual contact.

Another sexually transmitted virus is the human papillomavirus (HPV). There are many known strains of HPV, some of which cause genital warts. Furthermore, HPV infection causes nearly all cases of cervical cancer. A vaccine against HPV is available, but it is only effective before infection. Therefore, it is recommended that preteen boys and girls (ages 11–12) be vaccinated via a series of three shots over 6 months before sexual activity begins.

Anyone who is sexually active should have regular medical exams, be tested for STDs, and seek immediate help if any suspicious symptoms appear—even if they are mild. STDs are most prevalent among teenagers and young adults; nearly two-thirds of infections occur among people under 25. The best way to avoid the spread of STDs is, of course, abstinence. Alternatively, latex condoms provide the best protection for "safe sex." ☑

☑ **CHECKPOINT**

What is the most important difference between STDs caused by viruses and STDs caused by bacteria in terms of their treatment?

Answer: Bacterial STDs can be cured with antibiotics; viral STDs are permanent.

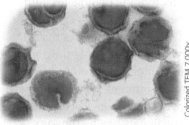
Chlamydia trachomatis
Colonized TEM 7,000×

Neisseria gonorrhoeae
Colonized TEM 17,000×

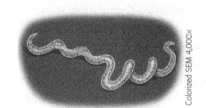
Treponema pallidum
Colonized SEM 4,000×

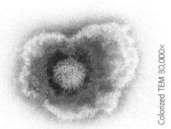
Herpes simplex virus
Colonized TEM 30,000×

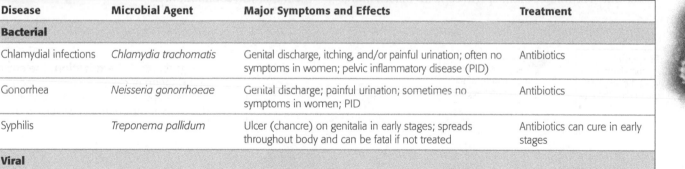

Table 26.2	STDs Common in the United States		
Disease	**Microbial Agent**	**Major Symptoms and Effects**	**Treatment**
Bacterial			
Chlamydial infections	*Chlamydia trachomatis*	Genital discharge, itching, and/or painful urination; often no symptoms in women; pelvic inflammatory disease (PID)	Antibiotics
Gonorrhea	*Neisseria gonorrhoeae*	Genital discharge; painful urination; sometimes no symptoms in women; PID	Antibiotics
Syphilis	*Treponema pallidum*	Ulcer (chancre) on genitalia in early stages; spreads throughout body and can be fatal if not treated	Antibiotics can cure in early stages
Viral			
Genital herpes	Herpes simplex virus type 2, occasionally type 1	Recurring symptoms: small blisters on genitalia, painful urination, skin inflammation; linked to cervical cancer, miscarriage, birth defects	Drugs can prevent recurrences, but there is no cure
Genital warts	Human papillomaviruses (HPV), multiple types	Painless growths on genitalia; HPV can cause cervical cancer	Removal by freezing; can be prevented by HPV vaccine
AIDS and HIV infection	HIV	Destruction of the immune system; see Chapter 24	Combination of drugs can slow progression
Protistan			
Trichomoniasis	*Trichomonas vaginalis*	Vaginal irritation, itching, and discharge; usually no symptoms in men	Antiprotist drugs
Fungal			
Candidiasis (yeast infections)	*Candida albicans*	Vaginal irritation, itching, and discharge; frequently acquired nonsexually	Antifungal drugs

Papillomaviruses
Colonized TEM 80,000×

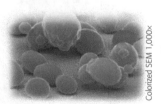
Trichomonas vaginalis
Colonized SEM 2,200×

Candida albicans
Colonized SEM 1,000×

Human Development

Embryonic development begins with **fertilization**, or conception, the union of a sperm and egg to form a zygote. In this section, we will examine the process of fertilization and the subsequent development of a human.

Fertilization by Sperm

 **Structure/Function**

Sexual intercourse releases hundreds of millions of sperm into the vagina, but only a few thousand sperm survive the several-hour trip to the egg in the oviduct. Of these sperm, only a single one can enter and fertilize the egg. All the other millions of sperm die.

Repeating a theme we've seen many times during our exploration of anatomy, the shape of sperm is related to what they do—in other words, form fits function. A mature human sperm has a streamlined shape that enables it to swim through fluids in the vagina, uterus, and oviduct **(Figure 26.11)**. The sperm's head contains a haploid nucleus and is tipped with a membrane-enclosed sac called the acrosome. The middle of the sperm contains mitochondria that use high-energy nutrients from the semen, especially the sugar fructose, to fuel movement of the flagellum.

Figure 26.12 traces one sperm through the events of fertilization. The sperm ❶ approaches and then ❷ contacts the jelly coat (red) that surrounds the egg. The acrosome in the sperm's head releases a cloud of enzymes that digest a hole in the jelly. This hole allows the sperm head to ❸ fuse its plasma membrane with that of the egg. Fusion of the two membranes makes it possible for the sperm nucleus to ❹ enter the cytoplasm of the egg. Fusion also triggers completion of meiosis II in the egg. Furthermore, contact of sperm with egg triggers a change in the egg's plasma membrane that makes it impenetrable to other sperm cells. This blockage ensures that the zygote contains only the diploid number of chromosomes. The chromosomes of the egg and sperm nuclei ❺ are eventually enclosed in a single diploid nucleus. In the new diploid zygote, the egg's metabolic machinery awakens from dormancy and gears up in preparation for the enormous growth and development that will soon follow. ☑

☑ CHECKPOINT

What is the function of a sperm's acrosome?

Answer: It releases enzymes that dissolve the jelly coat that surrounds the egg, allowing the sperm to penetrate the egg.

▼ **Figure 26.11 A human sperm cell.**

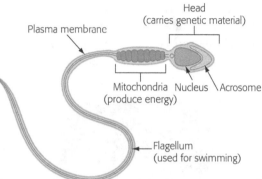

Head (carries genetic material)

Plasma membrane

Mitochondria (produce energy) Nucleus Acrosome

Flagellum (used for swimming)

◄ **Figure 26.12 Fertilization.** The entry of the sperm nucleus into the egg requires several steps. The electron micrograph (below) shows a human egg surrounded by sperm.

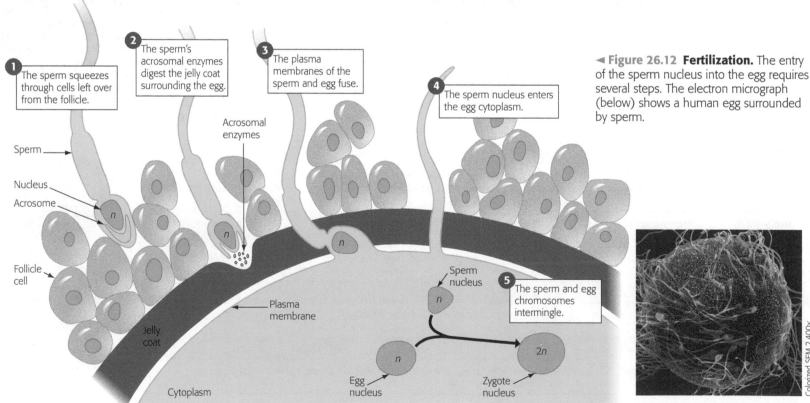

❶ The sperm squeezes through cells left over from the follicle.

❷ The sperm's acrosomal enzymes digest the jelly coat surrounding the egg.

❸ The plasma membranes of the sperm and egg fuse.

❹ The sperm nucleus enters the egg cytoplasm.

❺ The sperm and egg chromosomes intermingle.

Sperm

Nucleus

Acrosome

Follicle cell

Acrosomal enzymes

Plasma membrane

Jelly coat

Sperm nucleus

Egg nucleus

Zygote nucleus

Cytoplasm

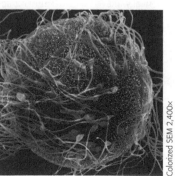

Colonized SEM 2,400x

Basic Concepts of Embryonic Development

A single-celled human zygote formed by fertilization is smaller than the period at the end of this sentence. From this humble start, the zygote develops into a full-fledged organism with trillions of cells organized into complex tissues and organs. Clearly, this process requires an astonishing amount of cell division and specialization. The key to development in all organisms is that each stage takes place in a highly organized fashion.

Development begins with cleavage, a series of rapid cell divisions that produces a multicellular ball. After the zygote divides for the first time, about 24 hours after fertilization, it is called an embryo. Rarely, and apparently at random, the two cells of the early embryo separate from each other. When this separation happens, each cell may "reset" and act as a zygote; the result is the development of identical (monozygotic) twins. (Nonidentical, or dizygotic, twins result from a completely different mechanism: Two separate eggs fuse with two separate sperm to produce two genetically unique zygotes that develop at the same time.)

During cleavage, DNA replication, mitosis, and cytokinesis occur rapidly, but the total amount of cytoplasm remains unchanged. As a result, the overall size of the embryo does not change; instead, each cell division partitions the embryo into twice as many smaller cells (Figure 26.13). Cleavage continues as the embryo moves down the oviduct toward the uterus. A central cavity begins to form in the embryo. About 6–7 days after fertilization, the embryo has reached the uterus as a fluid-filled hollow ball of about 100 cells called a blastocyst. Protruding into the central cavity on one side of the human blastocyst is a small clump of cells called the inner cell mass, which will eventually form the fetus.

Not all embryos are capable of completing development. Occasionally, an embryo does not travel down the oviduct to the uterus. The result is an ectopic pregnancy, one in which the embryo develops in the wrong location. Ectopic pregnancies are invariably fatal to the embryo and can be dangerous to the mother, requiring immediate medical attention. Many other embryos spontaneously stop developing as a result of chromosomal or developmental abnormalities. The result, called a spontaneous abortion or miscarriage, occurs in as many as one-third of all pregnancies, often before the woman is even aware she is pregnant.

For viable embryos, the second stage of development, gastrulation, begins about 9 days after conception (see the bottom of Figure 26.13). The cells of the early embryo begin an organized migration that produces the **gastrula**, an embryo with three main layers. The three layers produced in gastrulation are embryonic tissues called ectoderm,

endoderm, and mesoderm. The ectoderm eventually develops into the nervous system and outer layer of skin (epidermis). The endoderm becomes the innermost lining of the digestive system and organs such as the liver, pancreas, and thyroid. The mesoderm gives rise to most other organs and tissues, such as the heart, kidneys, and muscles.

At this stage, various cellular changes contribute to the formation of embryonic structures. All developmental processes depend on chemical signals passed between neighboring cells and cell layers, telling embryonic cells precisely what to do and when. For example, a chemical signal might switch on a set of genes whose expression makes the receiving cells differentiate into a specific type of cell or tissue. Other signals might cause a cell to change shape or migrate from one location to another within the developing embryo. Another key developmental process

▼ **Figure 26.13 Early development of an embryo.** Notice that each round of division during cleavage does not change the total size of the embryo. Instead, there are more cells, and each cell is smaller.

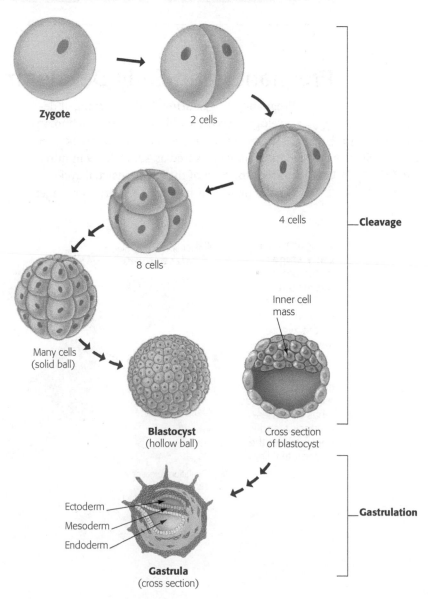

Zygote

2 cells

4 cells

8 cells

Many cells (solid ball)

Cleavage

Inner cell mass

Blastocyst (hollow ball)

Cross section of blastocyst

Gastrulation

Ectoderm
Mesoderm
Endoderm

Gastrula (cross section)

☑ **CHECKPOINT**

Place these steps of
development in order:
organ formation, fertilization,
gastrulation, cleavage.

Answer: fertilization, cleavage,
gastrulation, organ formation

is programmed cell death, which kills selected cells. For example, some genes encode proteins that kill certain cells in developing human hands and feet, separating the fingers and toes (**Figure 26.14**).

Further along the course of embryonic development, a sequence of chemical signals between cells leads to increasingly greater specialization as organs begin to take shape. The importance of these processes is underscored by birth defects that result from improper signaling. Spina bifida is a condition that results from the failure of a tube of ectoderm cells to properly close and form the spine during the first month of fetal development. Infants born with spina bifida often have permanent nerve damage that results in paralysis of the lower limbs. ☑

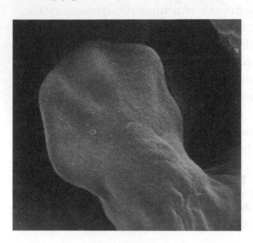
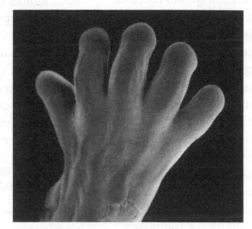

◄ **Figure 26.14 Programmed cell death in a developing human hand.** Proteins produced during human development destroy cells in the early hand (left), creating the spaces between fingers (right).

Pregnancy and Early Development

A 9-month pregnancy
may seem long, but
gestation in elephants
drags on for 22 months.

Pregnancy, or **gestation**, is the carrying of developing young within the uterus. In humans, pregnancy averages 266 days (38 weeks), but is usually measured as 40 weeks (9 months) from the start of the last menstrual cycle. Other mammals have much shorter or longer gestation periods (3 weeks in mice, 270 days in cows, and 22 months in elephants).

The early stages in human development are summarized in **Figure 26.15**. ❶ After an oocyte is released during ovulation, ❷ fertilization and ❸ cleavage take place. Fertilization and cleavage

► **Figure 26.15 Early stages of human development.** Fertilization takes place in the oviduct. As the zygote travels down the oviduct, cleavage starts. By the time the embryo reaches the uterus, it has become a blastocyst, and it implants into the endometrial lining.

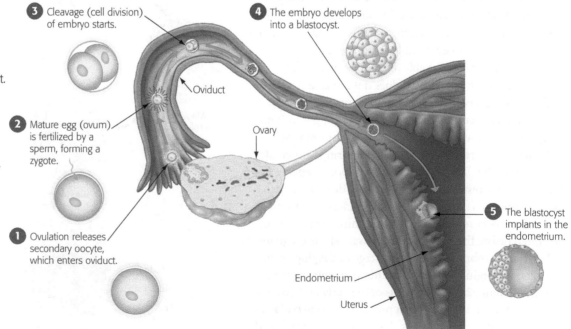

❸ Cleavage (cell division) of embryo starts.

❹ The embryo develops into a blastocyst.

Oviduct

❷ Mature egg (ovum) is fertilized by a sperm, forming a zygote.

Ovary

❶ Ovulation releases secondary oocyte, which enters oviduct.

❺ The blastocyst implants in the endometrium.

Endometrium

Uterus

take place in the oviduct. **4** About a week after conception, the embryo reaches the uterus and **5** implants itself into the endometrium. By this time, the embryo has become a blastocyst, with a central fluid-filled cavity and an inner cell mass (**Figure 26.16a**). These early embryonic cells are **stem cells**, with the potential to give rise to every type of cell in the body. (As discussed in Chapter 11, embryonic stem cells have great potential as medical tools.) The outer cell layer becomes part of the **placenta**, the organ that provides nourishment and oxygen to the embryo and helps dispose of its wastes (**Figure 26.16b**).

Figure 26.16c shows the embryo about a month after conception. Besides the growing embryo, there are now several pieces of life-support equipment: the amnion, the developing umbilical cord, and the placenta. The amnion is a fluid-filled sac that encloses and protects the embryo. The amnion usually breaks just before childbirth, and the amniotic fluid leaves the mother's body through her vagina. Many anxious couples are startled when the woman's "water breaks," and take this as a sign to get to the hospital. The **umbilical cord** serves as a lifeline between the embryo and the placenta.

The placenta develops chorionic villi, finger-like outgrowths containing embryonic blood vessels. These blood vessels are closely associated with blood vessels of the mother's endometrium. The chorionic villi absorb nutrients and oxygen from the mother's blood via diffusion and pass these substances to the embryo. The villi also diffuse wastes from the embryo to the mother's bloodstream. Although substances can pass back and forth via diffusion, the fetus has its own blood supply that does not mix with the mother's. A small sample of chorionic villus tissue can be removed for prenatal genetic testing (see Chapter 9).

The placenta provides for other needs of the embryo as well. For example, it allows protective antibodies to pass from the mother to the fetus. Unfortunately, harmful substances may also cross the placenta and harm the developing embyro. Most drugs—both prescription and nonprescription—can harm the developing embryo. Alcohol, the chemicals in tobacco smoke, and other drugs increase the risk of miscarriage and can cause developmental abnormalities. Additionally, viruses—the German measles virus and HIV, for example—can cross the placenta and cause disease. German measles can cause serious birth defects; if untreated, HIV-infected babies usually die of AIDS within a few years. ☑

▼ Figure 26.16 **A human embryo: the first month.**

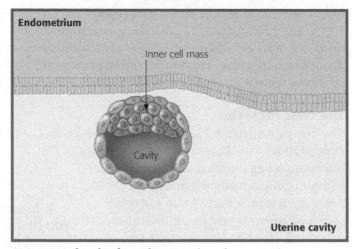

(a) **Day 6: Before implantation.** By 6 days after conception, the embryo has developed into a blastocyst.

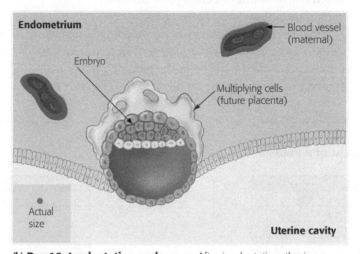

(b) **Day 10: Implantation under way.** After implantation, the inner cell mass of embryonic stem cells will develop into the fetus. The outer layer of trophoblast cells will become the embryo's contribution to the placenta.

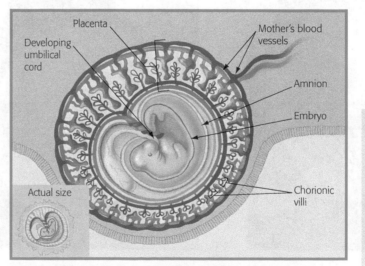

(c) **Day 31: The embryo and its life-support equipment**

January					
				0	Conception
	Day 6			Day 10	
				Day 31	

☑ CHECKPOINT

Why should a woman who is trying to get pregnant avoid drugs such as alcohol and nicotine?

Answer: Drugs can pass to the developing embryo from the mother's bloodstream and affect it before the mother even knows she is pregnant.

599

The Stages of Pregnancy

In this section, we use the series of photographs in **Figure 26.17** to illustrate the rest of human development as it takes place in the uterus. For convenience, we divide pregnancy (the period from conception to birth) into three **trimesters** of about 3 months each.

The First Trimester

The first photograph in Figure 26.17 shows a human embryo about 5 weeks after fertilization. In that brief time, a single cell has developed into a highly organized multicellular embryo about 7 mm (0.28 inch) long. Its brain and spinal cord have begun to take shape. It also has four stumpy limb buds, a short tail, and primitive gill-like structures. Overall, a month-old human embryo looks pretty much like a month-old embryo of any other vertebrate species (see Figure 13.8).

The next photograph shows a developing human, now called a fetus, about 9 weeks after fertilization. The large pinkish structure on the left is the placenta, which is attached to the fetus by the umbilical cord. The clear sac

At one time, you had gill slits and a tail.

around the fetus is the amnion. The fetus is about 5.5 cm (2.2 inches) long and has all of its organs and major body parts, including muscles and the bones of the back and ribs. The limb buds have become tiny arms and legs with fingers and toes. By the end of the first trimester, the fetus looks like a miniature human being, albeit one with an oversized head. By this time, the sex of the fetus can be determined by an ultrasound exam **(Figure 26.18)**.

The Second Trimester

The main developmental changes during the second and third trimesters involve an increase in size and general refinement of the human features—nothing as dramatic as the changes of the first trimester. A 14-week-old fetus is about 6 cm (2.4 inches) long.

At 20 weeks, well into the second trimester, the fetus is about 19 cm (7.6 inches) long, weighs about half a kilogram (1 pound), and has the face of an infant, complete with eyebrows and eyelashes. Its arms, legs, fingers, and toes have lengthened. It also has fingernails and toenails and is covered with fine hair. By this time, the fetal heartbeat is detectable with a stethoscope, and the mother can

▼ Figure 26.17 **The development of a human.**

January	February	March	April
0	35 days	63 days	98 days
Conception			

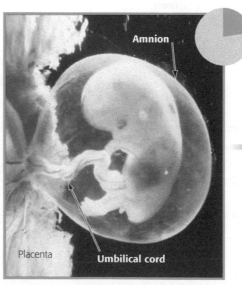

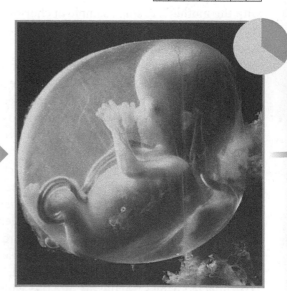

Gill pouches

Limb buds Tail

Amnion

Placenta Umbilical cord

5 weeks (35 days)   **9 weeks** (63 days)   **14 weeks** (98 days)

usually feel the fetus moving. Because of the limited space in the uterus, the fetus flexes forward into the so-called fetal position, with its head tucked against its knees. By the end of the second trimester, the fetus's eyes are open, its teeth are forming, and its bones have begun to harden.

The Third Trimester

The third trimester is a time of rapid growth as the fetus gains the strength it will need to survive outside the protective environment of the uterus. Babies born prematurely—as early as 24 weeks—may survive, but they require special medical care after birth. During the third trimester, the fetus's circulatory and respiratory systems undergo changes that will allow the newborn infant to breathe air. The fetus begins to gain the ability to maintain its own temperature, and its muscles thicken. It also loses much of its fine body hair, except on its head. The fetus usually rotates so that its head points down toward the cervix, and it becomes less active as it fills the space in the uterus. As the fetus grows and the uterus expands around it, the mother's abdominal organs may be squeezed, causing frequent urination, digestive troubles, and backaches. At birth, babies average about 50 cm (20 inches) in length and weigh 3–4 kg (6–8 pounds). ☑

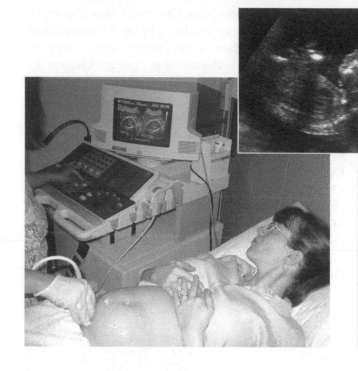

▼ **Figure 26.18 Ultrasound imaging.** An ultrasound image is produced when high-frequency sounds from a scanner held against a pregnant woman's abdomen bounce off the fetus. The inset image shows a fetus in the uterus at about 18 weeks.

☑ CHECKPOINT

As a human develops in the uterus, why are the first-trimester changes considered the most dramatic?

Answer: The development proceeds from a single-celled zygote to a miniature human. Subsequent changes are mainly growth and refinement of the structures developed during the first trimester.

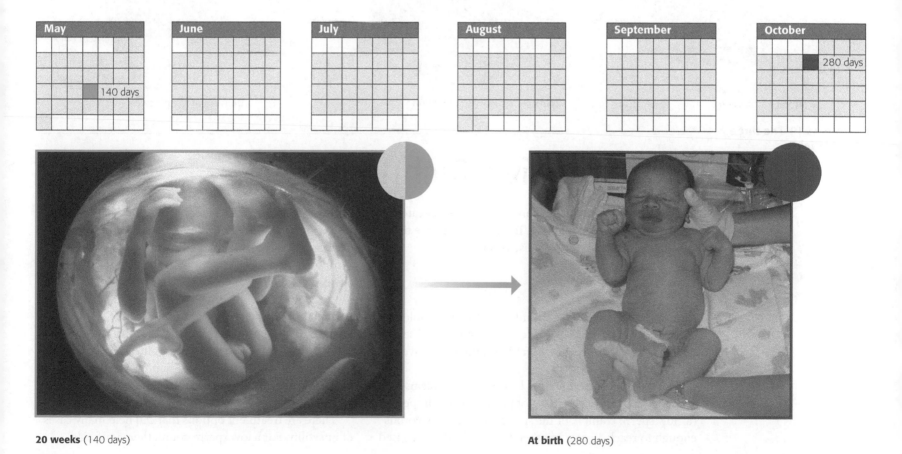

20 weeks (140 days)　　　　**At birth** (280 days)

Childbirth

The birth of a child is brought about by **labor**, a series of strong, rhythmic contractions of the uterus. Several hormones bring about labor. One of the most important is oxytocin, a hormone produced by the fetus's cells and, late in pregnancy, the mother's pituitary gland. Oxytocin stimulates muscles in the uterus to contract. Uterine contractions stimulate the release of more oxytocin, which in turn increases the contractions—an example of positive-feedback control. The result is climactic—the intense muscle contractions that propel a baby from the womb. If a mother is overdue, or if labor has continued for a long time, a doctor may inject oxytocin to promote a more rapid birth.

Figure 26.19 shows the three stages of labor. The first stage is dilation (shown on the left of the figure), the time from the onset of labor until the cervix dilates, or opens, to a width of about 10 cm (about 4 inches). Dilation is the longest stage of labor, typically lasting 6–12 hours, sometimes considerably longer. The period from full dilation of the cervix to delivery of the infant is called the expulsion stage (middle). Strong uterine contractions, lasting about 1 minute each, occur every 2–3 minutes, and the mother feels an increasing urge to push with her abdominal muscles. Within a period of 20 minutes to an hour or so, the infant is forced down and out of the uterus and vagina. The final stage is the delivery of the placenta ("afterbirth"), usually within 15 minutes after the birth of the baby (right). After the baby is born, the umbilical cord is clamped and cut. The stump of the cord generally remains for several weeks, then shrivels and falls off, leaving the belly button.

Hormones continue to be important after birth. Decreasing levels of progesterone and estrogen allow the uterus to start returning to its prepregnancy state. The pituitary hormone prolactin promotes milk production (called lactation) by the mammary glands. At first, a yellowish antibody-rich fluid called colostrum is secreted. After 2–3 days, normal milk production begins. ☑

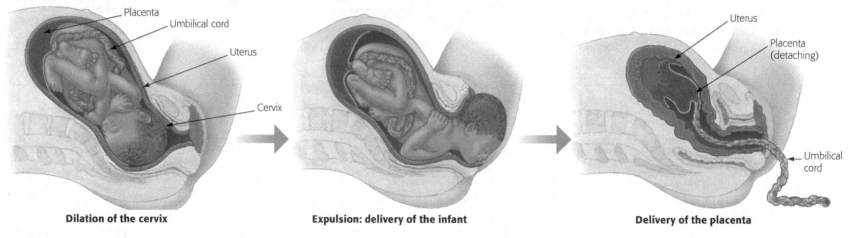

Dilation of the cervix | Expulsion: delivery of the infant | Delivery of the placenta

▲ Figure 26.19 **The three stages of labor.**

Reproductive Technologies

Many couples are unable to conceive children because of one or more abnormalities. Today, reproductive technologies can solve many of these problems.

Infertility

About 15% of couples who want children are unable to conceive, even after a year of regular, unprotected sex. Such a condition is called **infertility**, and it can have many causes.

In most cases, infertility can be traced to problems with the man. His testes may not produce enough sperm (a "low sperm count"), or the sperm may not be vigorous enough to reach an egg. In other cases, infertility is caused by **impotence**, also called erectile dysfunction, the inability to maintain an erection. To a certain degree, erectile dysfunction is a normal part of the male aging process. Temporary impotence can result from alcohol or drug use or from psychological problems. Permanent impotence can result from nervous system or circulatory problems.

Female infertility can result from a lack of eggs, a failure to ovulate, or blocked oviducts (often due to scarring resulting from a sexually transmitted disease). Other women are able to conceive but cannot support a growing embryo in the uterus. The resulting multiple miscarriages can take a heavy emotional toll.

There are treatment options that can help many cases of infertility. For a low sperm count, there is often a simple solution. Underproduction of sperm is frequently caused by a man's scrotum being too warm, so a switch of underwear from briefs (which hold the scrotum close to the body) to boxers may help. If that doesn't work, sperm can be collected, concentrated, and then injected into the woman's uterus via the vagina. Drug therapies (such as Viagra) and penile implants can treat impotence. If a man produces no functioning sperm, the couple may elect to use another man's sperm from a sperm bank.

If a woman has normal eggs that are not being released properly, hormone injections can induce ovulation; such treatments frequently result in multiple pregnancies (twins, triplets, or more). If a woman has no eggs of her own, they can be obtained from a donor for fertilization and injection into the uterus. Although sperm can be collected without any danger to the donor, the collection of eggs is a surgical procedure that involves pain and risk for the donating woman. For this reason, egg "donors" are typically paid several thousand dollars for their time and discomfort.

If a woman cannot maintain pregnancy, she and her partner may enter into a legal contract with a surrogate mother who agrees to carry the couple's child to birth. This method has worked for many couples, but serious

> A switch from briefs to boxers solves many cases of infertility

ethical and legal problems can arise. A number of states have laws restricting surrogate motherhood.

In Vitro Fertilization

A procedure performed in vitro ("in glass") happens under artificial laboratory conditions rather than within the living body (in vivo). **In vitro fertilization (IVF)**—sometimes called the creation of a "test-tube baby"—begins with the administration of drugs that promote the development of multiple eggs (instead of the one egg that typically occurs during each ovulation). The eggs are surgically removed and then fertilized with sperm in a petri dish, allowed to develop for several days, and then injected into a woman's uterus **(Figure 26.20)**. There, one or more embryos may successfully implant and continue development. The sperm and eggs, as well as the embryos created, can be used immediately or frozen for later use.

Since its introduction in 1978, over 5 million babies have been born after in vitro fertilization treatments. From the outset, doctors have been concerned that such babies might suffer health consequences due to their unusual conception. We'll examine that question next.

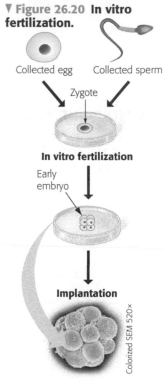

▼ **Figure 26.20 In vitro fertilization.**

Collected egg   Collected sperm

Zygote

In vitro fertilization

Early embryo

Implantation

Colorized SEM 520×

High-Tech Babies THE PROCESS OF SCIENCE

Are Babies Conceived via In Vitro Fertilization as Healthy as Babies Conceived Naturally?

Researchers in Sweden started with the **observation** that tens of thousands of Swedish children were conceived after in vitro fertilization, but no data existed on the long-term risk of neurological disorders in such children. This caused the researchers to **question** whether developmental disabilities, such as autism and low IQ, were more or less prevalent in IVF children than in non-IVF children. Their **hypothesis** was that the use of IVF is associated with an increased risk of developmental disabilities in the offspring. Their **prediction** was that rates of developmental disabilities would be higher among children born after IVF compared to those conceived naturally.

For their **experiment**, the researchers accessed a national database containing health data on over 2.5 million Swedish children born between 1982 and 2007. Of these, nearly 31,000 were born after IVF. The database also identified all children who had been diagnosed with

autistic disorder or abnormally low IQ (below 70). By adjusting for age, the researchers were able to compare rates of developmental disabilities in the IVF and non-IVF populations, expressed as the rate of disability per 100,000 person-years. The **results** are summarized in **Figure 26.21**. Although the rate of diagnosis for both disabilities was higher among IVF children, only the difference in the rates for low IQ was statistically significant (46.3 vs. 39.8 cases per 100,000 person-years). The difference in the rates of diagnosis of autism (19.0 vs. 15.6) was not statistically significant and could be due to random chance.

► **Figure 26.21 Comparing IVF and non-IVF children.** Although both autism and low IQ occurred more frequently in children born after IVF, only the difference in the data for a low IQ is statistically significant. Data from: S. Sandin et al., Autism and Mental Retardation Among Offspring Born After In Vitro Fertilization. *Journal of the American Medical Association* 310: 75–84 (2013).

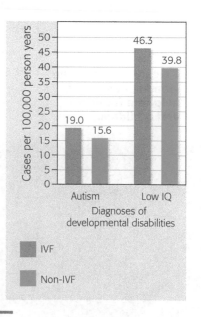

Cases per 100,000 person years

19.0   15.6   46.3   39.8

Autism   Low IQ

Diagnoses of developmental disabilities

■ IVF

■ Non-IVF

The Ethics of IVF

The study just described is part of a growing body of evidence that IVF babies have a slightly greater risk of birth defects than babies that are conceived naturally. Despite such risks and the high cost (around $10,000 for each attempt, whether it succeeds or not), IVF is now routinely performed at medical centers throughout the world and results in the birth of thousands of babies each year.

One major difference between natural and in vitro fertilization is that technology offers choices that nature does not. For example, sperm can be sorted based on whether they contain an X or Y chromosome, increasing the likelihood of creating a zygote of a particular sex. Furthermore, cells can be harmlessly removed from early embryos; they can then be tested for disease-causing (or other) genes. As analysis of the human genome progresses, the potential to screen for diseases and physical traits increases. Parents may thus have a lot of information they can use in choosing which embryos to implant.

The concept of parenthood has been greatly complicated by modern reproductive technologies. Mix-ups in fertility clinics have resulted in the implantation of embryos into the wrong women. When this happens, who has what rights? Divorcing couples have sued each other for custody of frozen embryos. An infertile couple could purchase eggs from a woman, sperm from a man, and then hire a surrogate to carry the baby. Who, then, are the baby's true parents? Each new reproductive technology raises moral and legal questions. Many of these have not yet been addressed, much less resolved, by society. All of us need to understand the science behind these complex issues so that we can make informed decisions as citizens and potential parents. ✓

High-Tech Babies EVOLUTION CONNECTION

The "Grandmother Hypothesis"

In this chapter, you have read how reproductive technologies can allow infertile couples to have babies. In recent years, technology has also made it possible for women to become pregnant after menopause, the cessation of ovulation and menstruation caused by changes in hormone levels. Menopause typically occurs after 500 menstrual cycles, usually around age 50, but in vitro fertilization has enabled some postmenopausal women to bear children up to age 70.

Given such efforts to artificially extend a woman's childbearing years, it may seem that menopause is an evolutionary disadvantage. Indeed, in most species, females retain their reproductive capacity throughout life. If evolutionary fitness is measured by the number of surviving, fertile offspring, why does menopause occur in humans? Evolutionary biologists have formed several intriguing hypotheses.

A larger brain is one of the hallmarks of human evolution (see Chapter 17), but the development of a large brain requires significant time and nutrients. Humans, unlike almost every other species, continue to depend on their mothers to provide food and care for many years after weaning. Natural selection may thus have favored reproductive adaptations that promote extended maternal care. Human mothers may bear fewer children than most other animals, but each child receives a significant investment of resources and therefore has an increased chance of survival.

Some researchers speculate that menopause actually increases a woman's evolutionary fitness. Perhaps losing the ability to become pregnant allows a woman to focus her energy on caring for the children she has, rather than producing more that might not survive. Furthermore, women who can no longer bear children themselves can help to raise their grandchildren (Figure 26.22). Because a woman shares one-half of her genes with each of her children and one-quarter of her genes with each grandchild, helping to raise two grandchildren can perpetuate as many genes to future generations as bearing one more child. Thus, even though menopause may cause a woman to have fewer children, this "grandmother hypothesis" suggests that it may actually increase the number of closely related children who will themselves reach maturity, ensuring the continuation of the postmenopausal grandmother's genes. Viewed this way, menopause in women may be another example of biology's unifying theme of evolution.

▼ **Figure 26.22 Human grandmothers often help care for their grandchildren.**

Chapter Review

SUMMARY OF KEY CONCEPTS

Unifying Concepts of Animal Reproduction

Asexual Reproduction

In asexual reproduction, one parent produces genetically identical offspring by budding, fission, or fragmentation followed by regeneration. Asexual reproduction enables a single individual to produce many offspring rapidly, but the resulting genetically identical population may be less able to survive environmental changes.

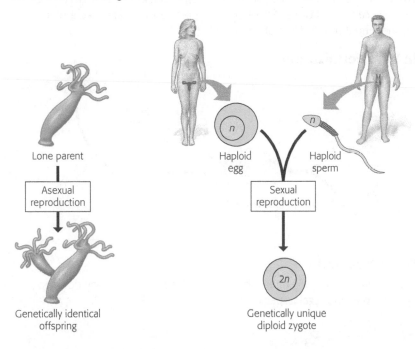

Sexual Reproduction

Sexual reproduction increases the variation among offspring, which may enhance reproductive success in changing environments.

Human Reproduction

The human reproductive system consists of a pair of gonads, ducts that carry gametes, and structures for sexual intercourse.

Male Reproductive Anatomy

Located in an external sac called the scrotum, a man's gonads (the testes) produce sperm, which are expelled through ducts during ejaculation. Several glands contribute to the formation of fluid that carries, nourishes, and protects sperm. This fluid and the sperm constitute semen.

Female Reproductive Anatomy

A woman's gonads (her ovaries) contain follicles that nurture eggs and produce sex hormones. Oviducts convey eggs to the uterus, where a fertilized egg develops into a fetus. The uterus opens into the vagina, which receives the penis during intercourse and forms the birth canal.

Gametogenesis

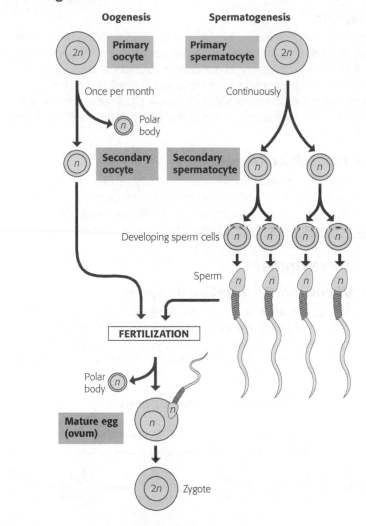

The Female Reproductive Cycle

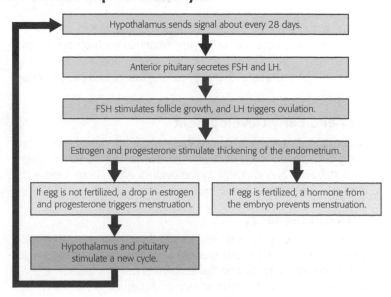

Reproductive Health

Contraception

Contraception is the deliberate prevention of pregnancy. Different forms of contraception block different steps in the process of becoming pregnant and have different levels of reliability. In addition to abstinence, particularly effective methods of contraception include sterilization (tubal ligation for women and vasectomy for men), methods that physically block sperm from reaching egg (condom or diaphragm in conjunction with a spermicide), and birth control pills.

Sexually Transmitted Diseases

STDs are contagious diseases that can be spread by sexual contact. Viral STDs (such as AIDS, genital herpes, and genital warts) cannot be cured, whereas STDs caused by bacteria, protists, and fungi are generally curable with drugs. Abstinence or the use of latex condoms can prevent STDs.

Human Development

Structure/Function: Fertilization by Sperm

During fertilization, the union of sperm and egg to form a zygote, a sperm releases enzymes that pierce the egg's membrane; then the sperm and egg plasma membranes fuse, and the two nuclei unite. Changes in the egg membrane prevent entry of additional sperm, and the fertilized egg is stimulated to develop into an embryo.

Basic Concepts of Embryonic Development

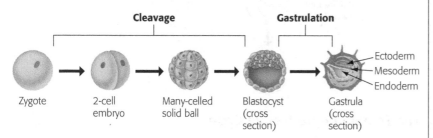

Cleavage			Gastrulation

Zygote → 2-cell embryo → Many-celled solid ball → Blastocyst (cross section) → Gastrula (cross section) — Ectoderm, Mesoderm, Endoderm

After gastrulation, the three embryonic tissue layers give rise to specific organ systems. Adjacent cells and cell layers influence each other's differentiation via chemical signals. Tissues and organs take shape in a developing embryo as a result of cell shape changes, cell migration, and programmed cell death.

Pregnancy and Early Development

Pregnancy, measured as 40 weeks from the start of the last menstrual cycle, is the carrying of the developing young in the uterus. Human development begins with fertilization and cleavage in the oviduct. After about 1 week, the embryo implants itself in the uterine wall.

The Stages of Pregnancy

Human embryonic development is divided into three trimesters of about 3 months each. The most rapid changes occur during the first trimester. By 9 weeks, all organs are formed, and the embryo is called a fetus. The second and third trimesters are times of growth and preparation for birth.

Childbirth

Hormonal changes induce birth. The cervix dilates, the baby is expelled by strong muscular contractions, and the placenta follows. Prolactin then stimulates milk production.

Reproductive Technologies

Infertility

Infertility, the inability to have children after one year of trying, is most often due to problems in the man, such as underproduction of sperm. Female infertility can arise from a lack of eggs or a failure to ovulate. Technologies can help treat many forms of infertility.

In Vitro Fertilization

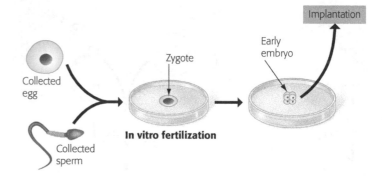

Collected egg — Collected sperm — **In vitro fertilization** — Zygote — Early embryo — Implantation

The Ethics of IVF

In vitro fertilization (IVF) offers choices that the natural pathway does not, raising moral and legal issues.

MasteringBiology®

For practice quizzes, BioFlix animations, MP3 tutorials, video tutors, and more study tools designed for this textbook, go to MasteringBiology®

SELF-QUIZ

1. Some animals, such as rotifers and aphids, are able to alternate between sexual and asexual reproduction. Under what conditions might it be advantageous to reproduce asexually? Sexually?

2. A fertility specialist determines that a patient is producing a normal amount of sperm, but the semen does not contain enough of the fluids needed to nourish the sperm. Which male reproductive system anatomical structures are most likely responsible for this problem?

3. Match each reproductive structure with its description:
   a. uterus
   b. vas deferens
   c. oviduct
   d. ovary
   e. endometrium
   f. testis

   1. female gonad
   2. site of spermatogenesis
   3. site of fertilization
   4. site of gestation
   5. lining of uterus
   6. sperm duct

4. A woman has had several miscarriages. Her doctor suspects that a hormonal insufficiency has been causing the lining of the uterus to break down as it does during menstruation, terminating her pregnancies. Treatment with which of the following might help her remain pregnant?

  a. oxytocin

  b. HCG

  c. follicle-stimulating hormone

  d. prolactin

5. The menstrual cycle is regulated by multiple feed-forward signals and one negative feedback loop. What constitutes the negative feedback loop?

6. What advantage do abstinence and condoms have over other forms of contraception?

7. Which one of the following is the earliest event in the process of fertilization?

  a. The sperm nucleus enters the cytoplasm of the egg.

  b. The sperm head plasma membrane fuses with the egg plasma membrane.

  c. Enzymes from the acrosome are released.

  d. Sperm contact the jelly coat around the egg.

8. The liver, pancreas, and lining of the digestive tract come from

  a. ectoderm.

  b. endoderm.

  c. mesoderm.

  d. neural crest cells.

9. A pregnant woman is two weeks past her due date, and the doctor decides to induce her labor. The natural hormone that would be needed is _____.

10. The procedure that creates a "test-tube baby" is _____. What does this term literally mean?

Answers to these questions can be found in Appendix: Self-Quiz Answers.

THE PROCESS OF SCIENCE

11. Early embryos of vertebrates look rather similar to each other, with rudimentary gills on either side of the head and a tail. They become increasingly different from each other as development proceeds. What can we learn from these early similarities? The German biologist and philosopher Ernst Haeckel suggested that embryogenesis recapitulates the evolutionary/phylogenetic development of each species. Do you think this is a correct interpretation?

12. A physician is trying to determine the cause of a male patient's infertility. Based on what you know about the male reproductive system, what would be the most logical things to test for?

13. **Interpreting Data** How effective are condoms at preventing the spread of HIV? Although this question is obviously very important to public health, it is difficult to study for many reasons: Animal models can't be used because HIV infects only humans, researchers cannot perform controlled studies, and research subjects may be hard to identify. A 2007 study was able to overcome some of these issues by studying HIV-negative partners of HIV-positive individuals. Such couples were divided into those who reported always using condoms and those who reported never doing so. The "always" group was found to have an HIV transmission rate of 1.1 new infections per 100 person-years. The "never" group had an HIV transmission rate of 5.8 infections per 100 person-years. What does it mean for the data to be reported in "person-years" and why is this important? What would be wrong about simply comparing the number of people who became infected in each group?

BIOLOGY AND SOCIETY

14. Some oppose abortion on the grounds that an individual human life is generated at the moment of conception. However, others argue that the unborn fetus in the first trimester (when most abortions take place) is entirely dependent on the mother and cannot be considered an individual human being. If you look at the development of a human embryo, at which point does individual life begin in your opinion? Explain.

15. Some infertile couples hire a surrogate mother to have their biological child. When couples enter into an agreement with a surrogate mother, all parties sign a contract. But there have been cases where a surrogate mother signed a contract but then decided not to give up the baby to the couple who hired her. Do you think a surrogate mother should have the right to change her mind and keep the baby, even though it is not genetically hers? In some cases, the surrogate will also donate her eggs to be fertilized by the male's sperm, making the baby genetically hers. Should she have a right to change her mind after she has signed a contract and the baby is born? How should judges decide these cases? Can you suggest any laws or regulations for such situations?

16. When a couple uses in vitro fertilization to produce a baby, they are faced with some novel choices. Typically, more embryos are produced than will be used during any one procedure. Thus, a subset of prepared embryos has to be chosen for implantation. How should parents decide which embryos to use? Should they have the right to choose embryos based on the presence or absence of disease-causing genes? What about the sex of the embryo? Should parents be able to choose embryos for implantation based on any criteria? How would you distinguish acceptable criteria from unacceptable ones? Do you think such options should be legislated?

27 Nervous, Sensory, and Locomotor Systems

Why the Nervous, Sensory, and Locomotor Systems Matter

It's undeniable: Sports-related head injuries and neurological disorders are linked.

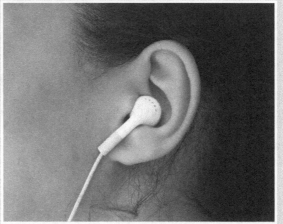

It's tempting to turn up the volume on your media player, but if you do, you're risking permanent hearing loss.

Forensic scientists may check a murder victim for muscle stiffness to estimate the time of death.

Extrahuman Senses BIOLOGY AND SOCIETY

The Seismic Sense of Elephants

We humans have keen senses, but many animals perceive the world in ways we cannot. Some bats use echolocation, emitting high-pitched sounds (above the range of human hearing) and measuring the echoes produced. Many fishes hunt prey by electroreception, using organs in their tails to detect changes in the electric fields. Other animals—including birds, fishes, turtles, and bees—navigate using magnetoreception, the ability to sense Earth's magnetic field. Elephants, the largest land animals on Earth, communicate using infrasounds, low-rumbling vocal calls that can't be heard by human ears.

Under ideal conditions, infrasounds can travel for 5 miles through the air, allowing elephants to communicate over long distances. Elephants are warned of danger when they hear certain infrasounds, and males listen for specific infrasounds made by females when they are ready to mate. Exceptional hearing, however, may not be the only extra-human sense that these animals possess.

Wildlife researchers have observed that a group of elephants in unison and for no apparent reason will often freeze. Most will freeze with their trunk in the air and all four feet planted, but a few will hold still with one foot lifted in the air or their trunk laid along the ground. Applying sensitive listening equipment, similar to the instruments used to detect earthquakes, wildlife researchers discovered that the elephants are responding to a sense that humans lack: the ability to detect seismic waves, which are low-frequency vibrations in the ground. Recent research has suggested that elephants use specialized pressure-sensing nerve endings in their feet and trunk to sense seismic waves.

Feeling the waves. Migrating elephants may communicate with other members of the herd through seismic waves.

Seismic detection may be an evolutionary adaptation that allows elephants to migrate, meet, and mate over great distances of the African savannah. Perhaps the elephants are sending specific signals to each other; after all, a typical elephant weighs 6 tons or more, so a purposeful foot stomp can produce a sizable ground wave. Or maybe the seismic activity is merely a by-product of elephant activities such as running. Vibrations produced from a running herd might warn of nearby threats. Research into the senses of these remarkable creatures remains an active and fascinating area of biology.

Although as a human you lack some of the senses that other animals have, you are able to perceive the world and to process your perceptions. You accomplish this using your brain, which is the hub of an intricate network of structures that detect, integrate, and respond to stimuli from the environment. By examining the structure and function of the nervous, sensory, and motor systems, we will explore how stimuli are translated into responses within the body. Let's begin by focusing on the processing center: the nervous system.

An Overview of Animal Nervous Systems

The **nervous system** forms a communication and coordination network throughout an animal's body. Nervous systems are the most intricately organized data processing systems on Earth. Your brain, for instance, contains an estimated 100 billion **neurons**, nerve cells that carry electrical signals from one part of the body to another. Each neuron may communicate with thousands of others, forming networks that enable us to move, perceive our surroundings, learn, and remember. In this section, we'll focus on features that are common to the nervous systems of most animals. We begin with the structure and function of neurons.

Neurons

Neurons can vary in structure, but most of them share some common features because of their common function: to receive and send signals. **Figure 27.1** depicts one kind of neuron, the kind that carries signals from your spinal cord to your skeletal muscles. The numerous short and highly branched extensions, called **dendrites**, *receive* incoming messages from other neurons and convey this information toward the cell body. The cell body houses the nucleus and other organelles. A different type of extension, a single long fiber called an **axon**, or nerve fiber, *transmits* signals toward a receiving cell, such as a muscle cell. Some nerve fibers, for example, the ones that reach from your spinal cord to muscle cells in your feet, can be over a meter long. Notice that the axon ends in a cluster of branches. A typical axon has hundreds or thousands of these branches, and each branch ends with a **synaptic terminal**, which relays signals to a receiving cell.

You drop a heavy object toward your foot, and without thinking you jerk your foot out of harm's way. What allows for such a fast response? Electrical transmission occurs at over 330 mph in our bodies—that's over 100 mph faster than a NASCAR racer—which means that a command from your brain can make you move your foot in just a few milliseconds. In humans and many other vertebrates, the speed of nerve signals can be attributed to the **myelin sheath**, a chain of bead-like supporting cells that wrap around and insulate nerves. Nerve signals are only generated at gaps in the myelin sheath and are transmitted from one gap to the next. Now that we have covered the basic structure of neurons, let's take a closer look at how neurons are organized in the nervous system. ☑

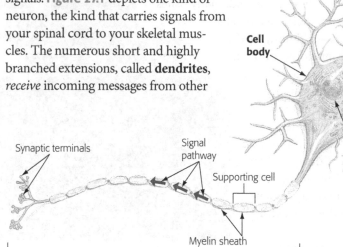

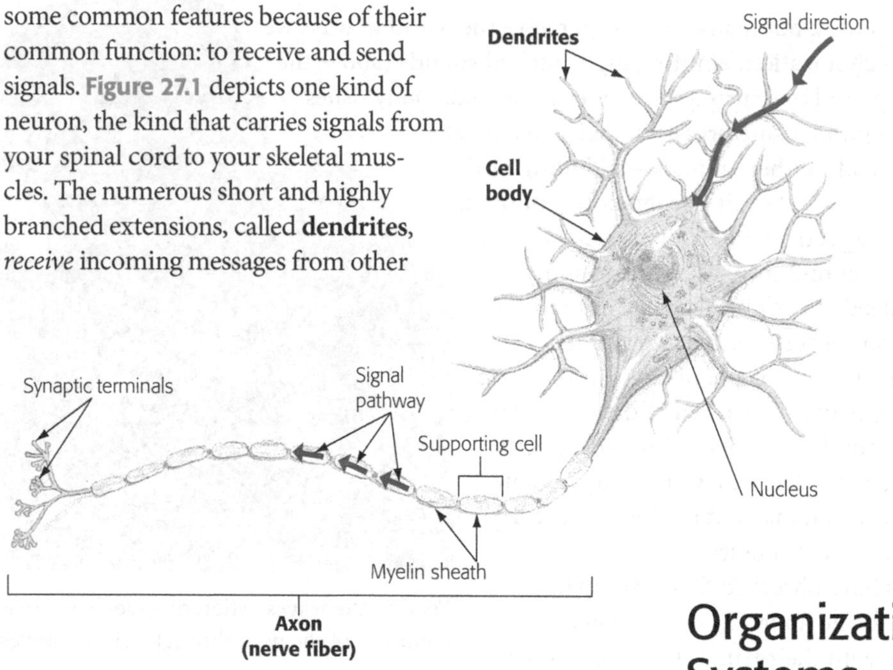

▶ **Figure 27.1 Structure of a neuron.** This neuron, typical of a neuron that signals muscles to contract, shows the flow of an electrical signal from a dendrite → cell body → axon.

Dendrites

Signal direction

Cell body

Signal pathway

Supporting cell

Nucleus

Synaptic terminals

Myelin sheath

Axon (nerve fiber)

Organization of Nervous Systems

The nervous systems of most animals have two main anatomical divisions. The **central nervous system (CNS)** consists of the brain and spinal cord. The CNS is the control center of the nervous system, interpreting stimuli and sending out responses. The **peripheral nervous system (PNS)** is made up mostly of **nerves** (bundles of axons) in the rest of the body that carry signals into and out of the CNS.

Figure 27.2 highlights the three interconnected functions of the nervous system and the three types of neurons that carry out these functions. **Sensory input** (shown in blue) is the process of sending signals from external or internal stimuli to the CNS. Examples of stimuli include an increase or decrease in light, a soft or loud sound, or a change in the pH of the body fluids. **Sensory neurons** are nerve cells often located in the

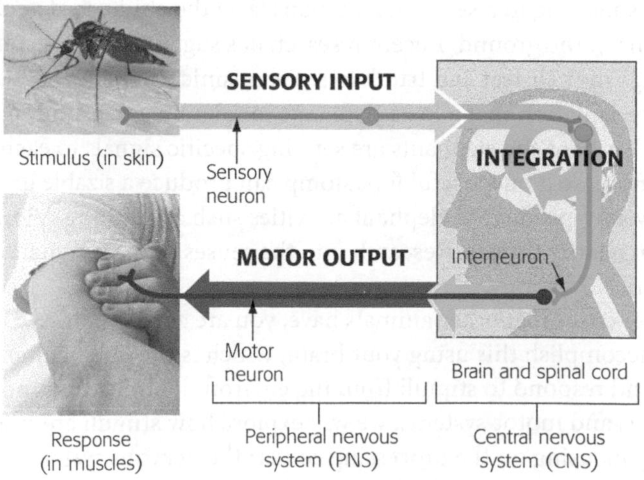

▶ **Figure 27.2 Organization of a nervous system.** As a simplified version of the nervous system, this figure shows only one neuron of each functional type, but body activity actually involves many. A simplified version of this figure will keep you oriented throughout the chapter.

SENSORY INPUT

Stimulus (in skin)

Sensory neuron

INTEGRATION

MOTOR OUTPUT

Interneuron

Motor neuron

Brain and spinal cord

Response (in muscles)

Peripheral nervous system (PNS)

Central nervous system (CNS)

PNS that convey information about a stimulus to the CNS. In the figure, a mosquito bite causes sensory neurons in the skin to send signals of pain to the spinal cord and then the brain. During **integration** (green), sensory signals are interpreted and appropriate responses are planned. The nerve cells that integrate the sensory input, **interneurons**, are typically located in the CNS. Most neurons in the brain are interneurons. Integration leads to **motor output** (purple), the sending of signals from the CNS to the PNS that result in the body's response. **Motor neurons** are nerve cells that carry output signals to muscles and glands of the body. When a mosquito bites you (sensory input), interneurons in your brain process the information (integration) to produce a response by your muscles (motor output), resulting in you slapping the mosquito away. With the basic functions and types of neurons of the nervous system covered, let's turn our attention to the signals sent by neurons. ☑

Sending a Signal through a Neuron

To understand nerve signals, we must first study a resting neuron, one that is not transmitting a signal. A resting neuron has potential energy that exists in the form of an electrical charge difference across the neuron's plasma membrane. The inside of the cell is negatively charged relative to the outside. Because opposite charges are pulled toward each other, the membrane stores energy by holding opposite charges apart, much like a battery. This difference in charge (voltage) that exists across the plasma membrane of a resting neuron is called the **resting potential**.

How does the membrane keep ions apart in a resting neuron? The hydrophobic interior of the membrane's phospholipid bilayer doesn't let ions pass through (Chapter 4). However, the membrane has protein channels and pumps that can allow positive ions across. By selectively allowing the outflow of positive ions, the membrane's interior becomes negative, and a charge separation is created in a resting neuron.

The Action Potential

Turn on a flashlight, and you use potential energy stored in a battery to create light. In a similar way, stimulating a neuron's plasma membrane can trigger the use of the membrane's potential energy to generate an electrical nerve signal. A stimulus of sufficient strength can trigger an **action potential**, a nerve signal that carries information along a neuron. The signal is actually a self-propagating change in the voltage across the plasma membrane.

Figure 27.3 shows the series of events involved in generating an action potential. As you'll see, two different sets of ion channels are involved in this process. ❶ At first, the membrane is at its resting potential, positively charged outside and negatively charged inside. ❷ A stimulus triggers the opening of a few of the first set of ion channels in the membrane (represented by the blue arrows), allowing a few positive ions to enter the neuron. This tiny change makes the inside surface of the membrane slightly less negative than before. If the stimulus is strong enough, a sufficient number of channels open to reach the **threshold**, the minimum change in a membrane's voltage that must occur to trigger the action potential. ❸ Once threshold is reached, more of these channels open and more positive ions rapidly rush in. As a result, the interior of this region of the cell becomes positively charged with respect to the outside. ❹ This electrical change triggers the closing of the first set of channels. Meanwhile, a second set of channels opens (green arrows), allowing other positive ions to diffuse rapidly out and returning the membrane to its resting potential. Within a living neuron, this whole process takes just a few milliseconds, meaning a neuron can produce hundreds of nerve signals in a second.

Propagation of the Signal

An action potential is a localized event—a change from the resting potential at a specific place along the neuron. A nerve signal starts as an action potential generated near a neuron's cell body. To function as a long-distance signal, this local change must be propagated, that is, passed along the length of the neuron. Propagation of the signal is like tipping the first domino in a standing row: The first domino does not tip over all the other dominos in the row; rather, its fall causes each domino in the row to hit and knock down the next.

☑ **CHECKPOINT**

Arrange the following neurons into the correct sequence for information flow, from stimulus to response: interneuron, sensory neuron, motor neuron.

Answer: sensory neuron, interneuron, motor neuron

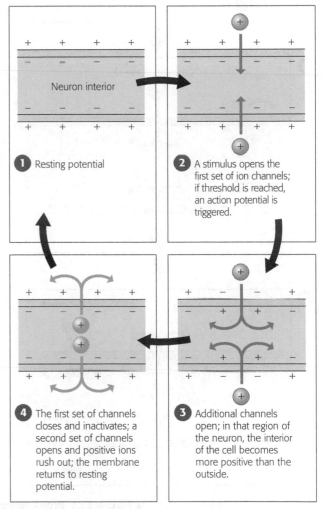

❶ Resting potential

❷ A stimulus opens the first set of ion channels; if threshold is reached, an action potential is triggered.

Neuron interior

❸ Additional channels open; in that region of the neuron, the interior of the cell becomes more positive than the outside.

❹ The first set of channels closes and inactivates; a second set of channels opens and positive ions rush out; the membrane returns to resting potential.

▲ Figure 27.3 **Generation of an action potential.**

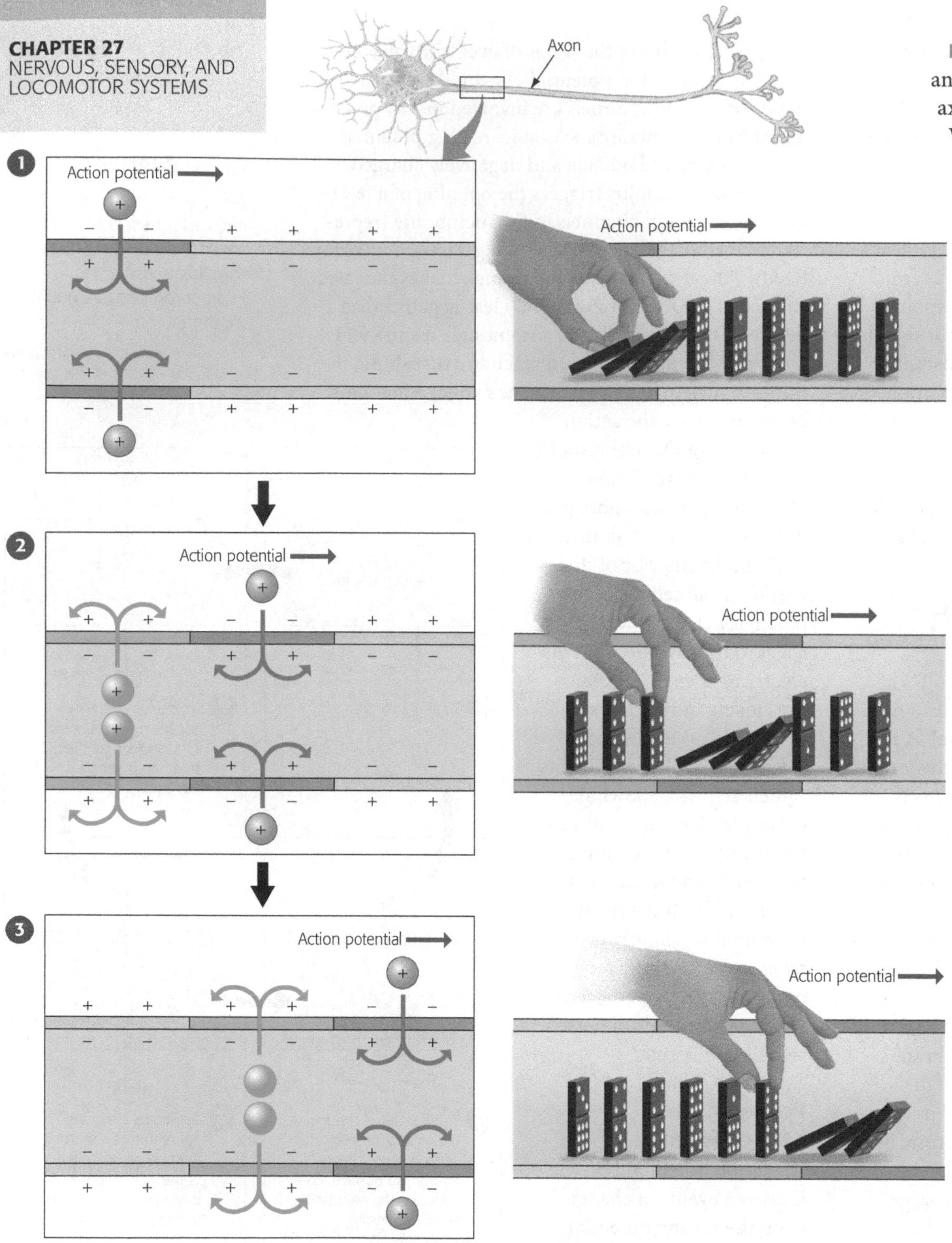

Axon

❶

❷

❸

Action potential →

Figure 27.4 shows the changes that occur as an action potential moves down a neuron's axon (from left to right in the figure). ❶ When the ion channels are open (indicated by a blue line in the illustration), positive ions diffuse inward (blue arrows), and an action potential is generated. The blue arrows show local spreading (in both directions) of the electrical changes caused by the positive ions entering the neuron. These changes trigger the opening of ion channels in only one direction along the membrane of the axon (to the right in the illustration). ❷ As a result, the initial action potential spreads in that direction (the middle of the illustration). ❸ In the same way, the action potential eventually spreads further down the axon (the far right in the figure). Thus, at each point along the axon, the action potential triggers changes in the adjacent region that result in propagation of the action potential in one direction along the neuron.

If local electrical changes spread in both directions, why does the action potential only move in one direction? Let's revisit the domino analogy by looking at the right side of Figure 27.4. Dominoes that have fallen need to be set upright before they can be knocked down again. Similarly, an action potential cannot be generated in regions where resting potential is being restored (green in figure). Once the resting potential is reestablished (yellow), an action potential can be generated again.

An action potential is an all-or-nothing event. Just as either a light tap or forceful flick of your finger knocks over just one domino, an action potential is the same no matter how weak or strong the stimulus is that initiates it (as long as threshold is reached). How, then, do action potentials relay different intensities of stimuli, such as soft and loud noises? Varying intensities can be achieved by sending more action potentials in a given amount of time. For example, in the neurons connecting your ears to your brain, loud sounds generate more action potentials per second than quiet sounds. ☑

▲ **Figure 27.4 Propagation of an action potential along an axon.** For quick behavioral responses, action potentials must travel fast along an axon. Wider nerves conduct action potentials quicker than narrower nerves. Action potentials will travel fastest if an axon is myelinated, because opening and closing of ion channels only occur at the gaps in the myelin sheath.

☑ CHECKPOINT

In what way is an action potential an example of positive feedback?

Answer: Positive feedback occurs when the result of a process further stimulates that process. The opening of channels caused by stimulation of the neuron changes the membrane potential, and this change causes more channels to open.

Passing a Signal from a Neuron to a Receiving Cell

What happens when an action potential arrives at the end of the neuron? To continue conveying information, the signal must be passed to another cell. This occurs at a **synapse**, a relay point between a neuron and a receiving cell. The receiving cell can be another neuron or another cell type such as a muscle cell or a hormone-secreting cell.

Synapses come in two varieties: electrical and chemical. In an electrical synapse, electric currents pass directly from one neuron to the next and provide very rapid responses. These synapses are common in invertebrates and are found in some organs of vertebrates, where they help maintain rhythmic muscle contractions. However, most synapses in the nervous system of vertebrates are chemical synapses. Chemicals provide more diverse and subtle responses than electrical synapses. Let's see how chemical synapses function.

Information Flow

Chemical Synapses

Chemical synapses have a narrow gap, called the **synaptic cleft**, separating the synaptic terminal of the sending neuron from the receiving cell **(Figure 27.5)**. When the action potential (an electrical signal) reaches the end of the sending neuron, it is converted to a chemical signal consisting of molecules of neurotransmitter. A **neurotransmitter** is a chemical "messenger" that carries information from a nerve cell to a receiving cell. Once a neurotransmitter conveys a chemical signal from the sending neuron, an action potential may then be generated in the receiving cell.

Let's follow the events that occur at a synapse between two neurons in Figure 27.5. ① An action potential (red arrow) arrives at the synaptic terminal. ② The action potential causes vesicles filled with neurotransmitter ( ) to fuse with the plasma membrane of the sending neuron and ③ release their neurotransmitter molecules into the synaptic cleft. ④ The neurotransmitter molecules diffuse across the cleft and bind to complementary receptors on ion channel proteins in the receiving neuron's plasma membrane. ⑤ The binding of neurotransmitter to receptors opens the ion channels. With the channels open, ions can diffuse into the receiving neuron. In many cases, this triggers a new action potential; in other cases, a neurotransmitter inhibits the generation of an action potential. ⑥ The neurotransmitter is broken down or transported back into the sending neuron (known as reuptake), causing the receiving cell's ion channels in the plasma membrane to close. Step 6 ensures that the neurotransmitter's effect on the receiving neuron is brief and precise.

▼ Figure 27.5 **Neuron communication at a synaptic cleft.**

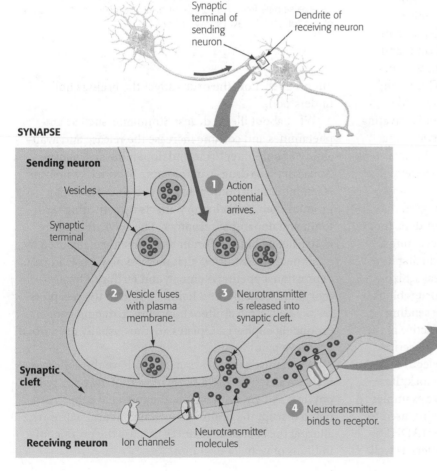

Synaptic terminal of sending neuron

Dendrite of receiving neuron

SYNAPSE

Sending neuron

Vesicles

Synaptic terminal

① Action potential arrives.

② Vesicle fuses with plasma membrane.

③ Neurotransmitter is released into synaptic cleft.

Synaptic cleft

④ Neurotransmitter binds to receptor.

Receiving neuron　Ion channels　Neurotransmitter molecules

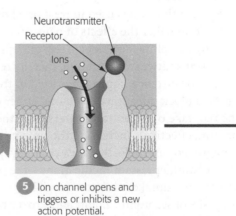

Neurotransmitter

Receptor

Ions

⑤ Ion channel opens and triggers or inhibits a new action potential.

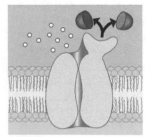

⑥ Ion channel closes. Neurotransmitter is either broken down, as shown, or transported back into the sending neuron.

Neurotransmitters result in many different actions, and a single neurotransmitter, such as dopamine, can play a role in pleasure, attention, memory, sleep, movement, and learning. What if a gene involved in manufacturing a neurotransmitter or the neurotransmitter's receptor is defective? The result might be too much or too little of a chemical message at a synapse. Some types of depression have been linked to reduced levels of neurotransmitters, specifically norepinephrine and serotonin. The degenerative illness Parkinson's disease is associated with a lack of the neurotransmitter dopamine, while an excess of dopamine is linked to schizophrenia. Although genetic studies have identified defective genes in a subset of individuals with these types of disorders, how other genes and environmental factors contribute to one's overall risk for these disorders is still unclear.

Chemical synapses can process extremely complex information. A neuron may receive input from hundreds of other neurons via thousands of synaptic terminals (Figure 27.6). The inputs can be highly varied because each sending neuron may secrete a different quantity or kind of neurotransmitter. These factors account for the nervous system's ability to process huge amounts of complex stimuli and formulate appropriate responses. ☑

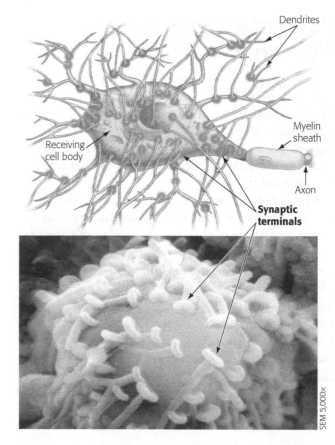

▲ **Figure 27.6 A neuron's multiple synaptic inputs.** As shown in this drawing and micrograph, a single neuron may receive signals from hundreds of other neurons via thousands of synaptic terminals (shown in orange).

Drugs and the Brain

Many common substances, such as caffeine, nicotine, and alcohol, affect the actions of neurotransmitters at the brain's trillions of synapses. For example, caffeine, found in coffee, chocolate, and many soft drinks and energy drinks, counters the effects of neurotransmitters that normally suppress nerve signals. This is why caffeine tends to stimulate you and keep you awake. Nicotine acts as a stimulant by binding to and activating receptors for a neurotransmitter called acetylcholine. Alcohol is a strong depressant; its effect on the nervous system is complex, affecting the function of several neurotransmitters.

Prescription drugs used to treat psychological disorders often alter the effects of neurotransmitters, too. A drug may physically block a receptor, preventing a neurotransmitter from binding, thereby reducing its effect. For instance, some drugs used to treat schizophrenia block dopamine receptors. Other drugs block the reuptake of neurotransmitters from the sending neuron, effectively increasing the time it is active at the synapse. For example, the most commonly prescribed class of antidepressant medications—the selective serotonin reuptake inhibitors (or SSRIs)—block the reuptake of serotonin. Zoloft and Prozac are examples of SSRIs. Drugs like methylphenidate (Ritalin), used to treat attention deficit hyperactivity disorder (ADHD), likely inhibit the reuptake of neurotransmitters as well, but precisely how the drugs affect the brain is not yet understood.

What about illegal drugs? Stimulants such as amphetamines and cocaine increase the release and availability of norepinephrine and dopamine at synapses. Abuse of these drugs can therefore produce symptoms resembling schizophrenia. LSD and mescaline may produce their hallucinatory effects by activating serotonin and dopamine receptors. The active ingredient in marijuana binds to brain receptors normally used by other neurotransmitters that seem to play a role in pain, depression, appetite, memory, and fertility. Opiates—morphine, codeine, and heroin—bind to the receptors for the neurotransmitter endorphin, reducing pain and producing euphoria. Opiates are thus usually prescribed to relieve pain.

While the drugs mentioned here have the ability to increase alertness and a sense of well-being or to reduce physical and emotional pain, they also have the potential to disrupt the brain's finely tuned neural pathways, altering the chemical balances that are the product of millions of years of evolution. ☑

The Human Nervous System:
A Closer Look

To this point, we have concentrated on cellular mechanisms that are fundamental to nearly all animal nervous systems. Although there is remarkable uniformity throughout the animal kingdom in the way nerve cells function, there is great variety in how nervous systems as a whole are organized. For the rest of this chapter, we'll focus on vertebrate nervous systems, with particular emphasis on the human nervous system.

Vertebrate nervous systems are diverse in both structure and level of sophistication—the brains of dolphins and humans are much more complex than the brains of frogs and fishes, for instance. However, some features are common to the nervous systems of all vertebrates. All have a nervous system that is concentrated at the head end, and all have a central nervous system (CNS) distinct from a peripheral nervous system (PNS) **(Figure 27.7)**.

The Central Nervous System

INTEGRATION

Similar to how you receive and send messages through your email account, the CNS interprets signals coming from peripheral nerves and sends out appropriate responses through other peripheral nerves. The CNS consists of two parts in vertebrates: the brain and spinal cord. The **brain** functions as the master control center for the nervous system, integrating data from the sense organs and sending out commands from motor control centers. The **spinal cord** is a bundle of nerve fibers running through the bony spinal column, acting as a communication conduit between the brain and the rest of the body. Millions of nerve fibers within the spinal cord convey sensory information (such as pain) from the body to the brain, while other fibers carry motor information (such as movement) from the brain to the muscles and organs.

Injury to the CNS is often permanent. Although protected by the bony spinal column, a traumatic blow to the spine can crush delicate nerve bundles. Such trauma may cause quadriplegia (paralysis from the neck down) or paraplegia (paralysis of the lower half of the body), depending where on the spinal column the injury occurs. In the United States, males suffer the majority of spinal cord injuries (more than 80%), which are caused primarily by car accidents, falls, and gun shot wounds. Spinal cord injuries are often permanent because the neurons of the CNS, unlike those in many other body tissues, cannot be repaired. Trauma to the brain can have a variety of effects—such as personality changes, coma, or death—depending on what part of the brain is affected.

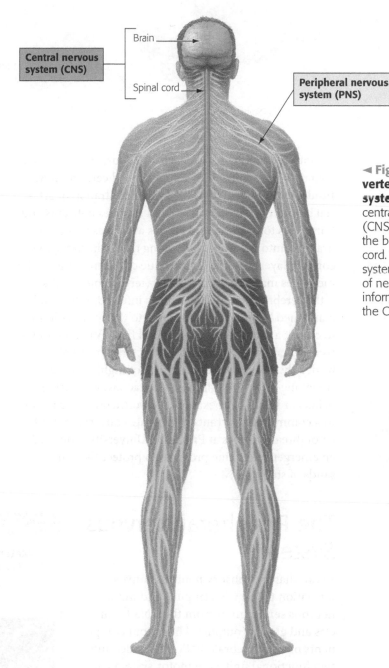

Brain

Central nervous system (CNS)

Spinal cord

Peripheral nervous system (PNS)

◄ **Figure 27.7 A vertebrate nervous system (back view).** The central nervous system (CNS) of humans includes the brain and the spinal cord. The peripheral nervous system (PNS) consists of nerves that convey information to and from the CNS.

615

► Figure 27.8 **Fluid-filled spaces of the vertebrate CNS.** Spaces in the brain and spinal cord are filled with cerebrospinal fluid, which helps to cushion and supply nutrients to the CNS.

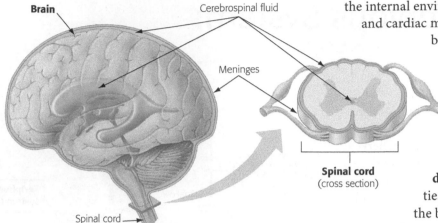

Brain — Cerebrospinal fluid — Meninges — Spinal cord (cross section) — Spinal cord

CHECKPOINT

If a doctor declares that a patient has sustained central nervous system injuries, what structures in the CNS might be damaged?

Answer: the brain and/or spinal cord

Both the brain and spinal cord are surrounded by and contain spaces filled with a liquid called **cerebrospinal fluid**, which is formed in the brain by filtration of arterial blood (Figure 27.8). The fluid supplies nutrients and hormones to the CNS and carries away wastes, before draining into the veins. Protecting the brain and spinal cord are layers of connective tissue called **meninges**. The meninges may become inflamed (termed meningitis) if the cerebrospinal fluid becomes infected. Infection is confirmed by inserting a needle into the spinal column to collect a sample of cerebrospinal fluid. Viral meningitis is generally not harmful, but bacterial meningitis can be deadly. Bacterial meningitis is not highly contagious, spreading only through direct contact, when sharing drinks or kissing, for example. Nevertheless, an infection in a community warrants immediate concern. In 2014, an outbreak of cases at Princeton University prompted an emergency vaccine program to protect their thousands of students. ✓

the internal environment by controlling smooth and cardiac muscles and the glands of several body systems. This control is generally involuntary. Let's look more closely at the autonomic nervous system.

The autonomic nervous system contains two sets of neurons with opposing effects on most body organs. One set, called the **parasympathetic division**, primes the body for activities that gain and conserve energy for the body ("rest and digest"). These effects include stimulating the digestive organs, decreasing the heart rate, and narrowing the bronchi, which correlates with a decreased breathing rate. The other set of neurons in the autonomic nervous system, the **sympathetic division**, tends to have the opposite effect, preparing the body for intense, energy-consuming activities, such as fighting or fleeing ("fight or flight"). When this division is stimulated, the digestive organs are inhibited, the bronchi dilate so that more air can pass through, and the adrenal glands secrete the hormones epinephrine (also called adrenaline) and norepinephrine. Relaxation and the fight-or-flight response are opposite extremes. Your body usually operates somewhere in between, with most of your organs receiving both sympathetic and parasympathetic signals.

The motor system and autonomic nervous system constitute lower levels of the nervous system's hierarchy. In the next section, we'll take a closer look at the highest level of the hierarchy, the brain. ✓

The Peripheral Nervous System

Recall that the sensory neurons convey information to the CNS (input), and motor neurons send signals from the CNS to muscles and glands (output). The output components of the vertebrate PNS are divided into two functional units: the motor system and the autonomic nervous system (Figure 27.9). Neurons of the **motor system** carry signals to skeletal muscles, mainly in response to external stimuli. When you lift an object, for instance, these neurons carry commands that make your arms move. The control of skeletal muscles is usually voluntary. The **autonomic nervous system** regulates

CHECKPOINT

How would a drug that inhibits the parasympathetic nervous system affect your heart rate?

Answer: Your heart rate would increase.

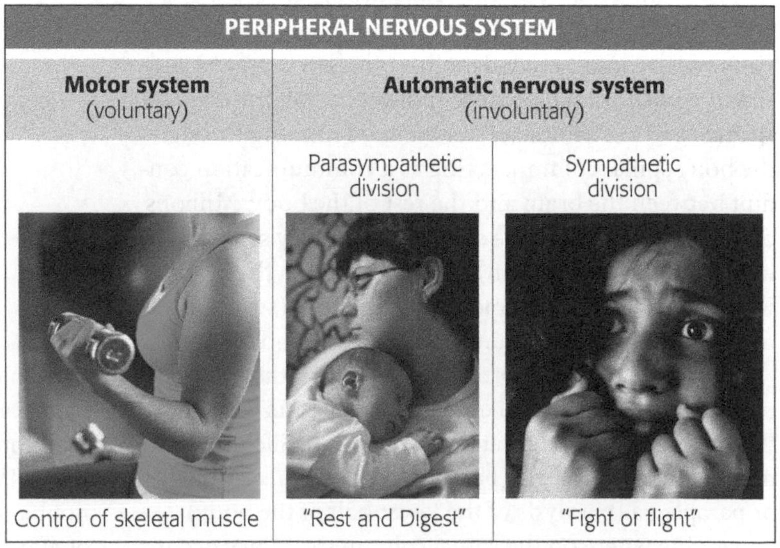

PERIPHERAL NERVOUS SYSTEM		
Motor system (voluntary)	**Automatic nervous system** (involuntary)	
	Parasympathetic division	Sympathetic division
Control of skeletal muscle	"Rest and Digest"	"Fight or flight"

▲ Figure 27.9 **Functional divisions of the vertebrate PNS.**

The Human Brain

Composed of up to 100 billion intricately organized neurons, with a much larger number of supporting cells, the human brain is more powerful than the most sophisticated computer. **Figure 27.10** highlights some major parts of the brain and associated functions. We'll focus on four major areas of the brain: the brainstem, the cerebellum, the hypothalamus, and the cerebrum.

The **brainstem** is the core of the brain, and it receives and sends information to other major regions. It regulates some very vital and basic functions, including breathing and consciousness. Thus individuals with severe brainstem damage may lose the ability to breathe on their own and may become comatose. The brainstem is particularly vulnerable when other parts of the brain are injured, because as an injured part of the brain swells, it pushes downward and compresses the brainstem. (The brain has very little room to swell outward because of the hard skull.) This compression of the brainstem in large part explains why a patient's health may deteriorate hours and even days after an initial brain injury. Structures of the brainstem, including the medulla oblongata, the pons, and the midbrain, are shown in Figure 27.10 along with descriptions of the functions associated with each.

Individuals who have damage to their cerebellum often have trouble with coordinated movements, such as balance or hand-eye coordination; for example, they can follow a tennis ball with their eyes but are not able to move the racket toward the ball. From this example, you might have inferred that the **cerebellum** serves as a planning center for body movements. The cerebellum receives input—for instance, visual cues about a moving tennis ball and body awareness of arm position—and uses the input to coordinate movement and balance—such as swinging a racket toward a tennis ball.

A complex integrating center, the **hypothalamus** regulates the autonomic nervous system and controls the secretion of hormones from the pituitary gland and other organs. The functions of the hypothalamus affect body temperature, blood pressure, hunger, thirst, sex drive, and the fight-or-flight response. The hypothalamus also helps us experience emotions, such as rage and pleasure. The pleasure center of the hypothalamus is strongly affected by certain addictive drugs. Another area of the hypothalamus functions as an internal timekeeper that maintains our daily biological rhythms, such as cycles of sleepiness and hunger.

The most complex integrating center, and undoubtedly the image you conjure up when you think of the brain, is the cerebrum. The **cerebrum** is the largest and most sophisticated part of our brain. The outer layer of the cerebrum, the **cerebral cortex**, is a thin, highly folded ("wrinkled") layer of tissue that accounts for over 80% of the total brain mass yet is thinner than the width of a pencil. The intricate neural circuitry of the cerebral cortex helps produce our most distinctive human traits: reasoning and mathematical abilities, language skills, imagination, artistic talent, and personality traits. Integrating information it receives from our senses, the cerebral cortex creates our sensory perceptions—what we are actually aware of when we see, hear, smell, taste, or touch. The cerebral cortex also regulates our voluntary movements. Let's now take a closer look at the structure and function of the cerebral cortex. ☑

☑ **CHECKPOINT**

Which brain structure—the cerebellum or the cerebrum—contains sophisticated thinking centers?

Answer: the cerebrum

▼ Figure 27.10 **Some major parts of the human brain.**

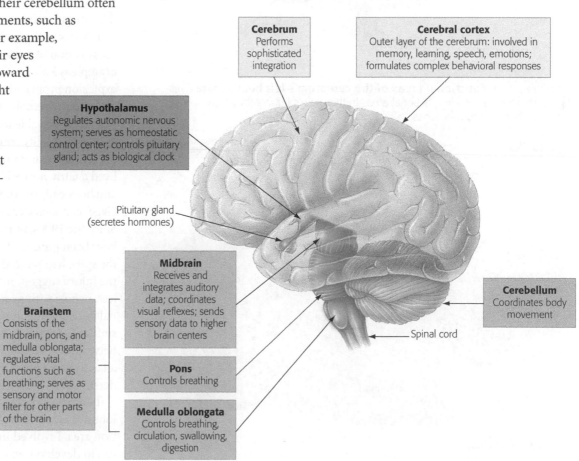

Cerebrum
Performs sophisticated integration

Cerebral cortex
Outer layer of the cerebrum: involved in memory, learning, speech, emotions; formulates complex behavioral responses

Hypothalamus
Regulates autonomic nervous system; serves as homeostatic control center; controls pituitary gland; acts as biological clock

Pituitary gland (secretes hormones)

Midbrain
Receives and integrates auditory data; coordinates visual reflexes; sends sensory data to higher brain centers

Brainstem
Consists of the midbrain, pons, and medulla oblongata; regulates vital functions such as breathing; serves as sensory and motor filter for other parts of the brain

Pons
Controls breathing

Medulla oblongata
Controls breathing, circulation, swallowing, digestion

Cerebellum
Coordinates body movement

Spinal cord

Structure/ Function   The Cerebral Cortex

The cerebral cortex is divided into right and left hemispheres (**Figure 27.11**). Each half receives information from and controls the movement of the opposite side of the body. The **corpus callosum**, a thick cable of nerve fibers, bridges the two sides and enables them to process information together. The corpus callosum offers a good illustration of structural differences among

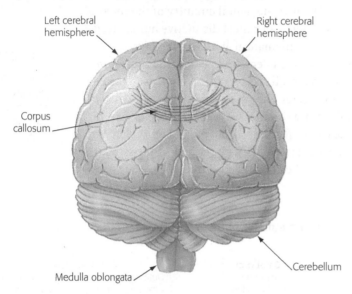

▼ **Figure 27.11 A rear view of the brain.** The large cerebrum (yellow) consists of left and right cerebral hemispheres connected by a thick band of nerves called the corpus callosum.

▼ **Figure 27.12 Functional areas of the cerebrum's left hemisphere.** This figure identifies the main functional areas in the brain's left cerebral hemisphere, which is divided into four lobes.

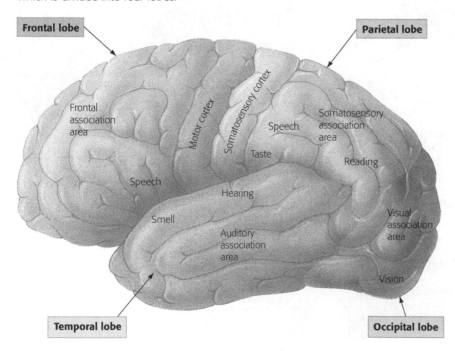

human brains. For example, the corpus callosum is often reduced in size and sometimes even absent in autistic individuals. One notable individual who was missing a corpus callosum was Kim Peek, the inspiration for the 1988 Oscar-winning movie *Rain Man*. Kim, known to many as "Kimputer," could catalog and recall an enormous amount of knowledge. How reductions in the corpus callosum affect the "wiring" in the rest of the brain is still unclear to neurobiologists.

Both the right and left hemispheres of the cerebral cortex have four major lobes, each named for a nearby skull bone: the frontal, parietal, temporal, and occipital lobes (represented by different colors in **Figure 27.12**). Researchers have identified a number of functional areas within each lobe: Primary sensory areas receive and process sensory information (taste, hearing, etc.), association areas integrate the information, and motor areas transmit instructions to other parts of the CNS and eventually the body. When you walk toward a person you know on the street, a visual area processes the information about the person's face, an association area plans for how you'll greet them, and the motor areas send information to muscles that cause you to smile and wave. Note that these association areas are the sites of higher mental activities—roughly, what we call thinking. Extremely complex interactions among several association areas of different lobes are involved in some activities you might take for granted—such as recognizing a familiar face.

Injuries to the brain provide scientists with clues about the function of certain association areas. A well-known example is Phineas Gage (**Figure 27.13**). In 1848, an explosion propelled a 3-foot-long railroad spike through his head. Incredibly, after recovery from the trauma, it appeared he was able to think normally. However, changes in his personality soon appeared, including a propensity for meanness and impulsiveness. Historians believe Gage lived a fairly normal life for 12 years after his accident and before his death in 1860. A few years after his death, Gage's remains were exhumed and his skull preserved. Since the 1990s, numerous computer models of his injury have been produced. Although it was once thought that the spike had pierced both frontal lobes, new models of the injury suggest only the left side was affected. Nonetheless, people with frontal lobe injuries often exhibit irrational decision making and difficulty processing emotions, similar to what was first seen with Phineas Gage. The link between the frontal lobes and personality continues to be an active area of neuroscience research, especially in understanding degenerative neurological disorders.

In a typical brain, the association areas of the frontal lobe are some of the last areas to mature. One association area, involved in decision making, likely continues to develop into our 20s. This might account for

The railroad spike removed from Gage's brain

The railroad spike entered Gage's left cheek and exited through his skull.

◀ **Figure 27.13 Phineas Gage's accident.** Phineas Gage holds the 3-foot railroad spike that punctured his brain in 1848 (left). On the right is one of numerous computer simulations of his accident, depicting frontal lobe damage. Such an injury could account for Gage's personality changes after the accident.

the riskier decisions made by teens and young adults compared to older adults. Because neuroscientists have described how the brain typically develops, comparisons can define atypical brain maturation.

A delay in the maturation of the brain could explain a common brain disorder. Attention deficit hyperactivity disorder (ADHD) is conservatively estimated to affect 1 in 20 school-age children. The disorder is characterized by a lack of focus, impulsive behavior, and restlessness. In one study, scientists examined brains of over 400 youths, some with ADHD and others with no signs of the disorder. They concluded that identical regions of the brain thicken in the same order, yet the maturation of the brain is delayed by approximately 3 years in youths with ADHD.

Advanced imaging technologies of the brain show variations in its maturation rate, its activity during certain cognitive tasks, and even in its anatomy. By studying the differences among our brains, scientists are beginning to unravel the mysteries that underlie not only ADHD but also autism, dyslexia, schizophrenia, and many other disorders. ☑

Neuroplasticity

Evidence from brain surgery patients indicates that our brain's wiring is not necessarily fixed. One of the most radical surgical alterations of the brain is hemispherectomy, the removal of almost one-half of the brain **(Figure 27.14)**. This procedure is performed to alleviate severe seizures that originate from one of the brain's hemispheres. Incredibly, with just half a brain, hemispherectomy patients recover quickly, often leaving the hospital within a few weeks. Although the side of the body opposite the surgery always has partial paralysis,

hemispherectomy patients have undiminished intellectual capacities. Higher brain functions that previously originated from the missing half of the brain begin to be controlled by the opposite side. Recovery after hemispherectomy is a striking example of the brain's remarkable **neuroplasticity**, its ability to reorganize its neural connections.

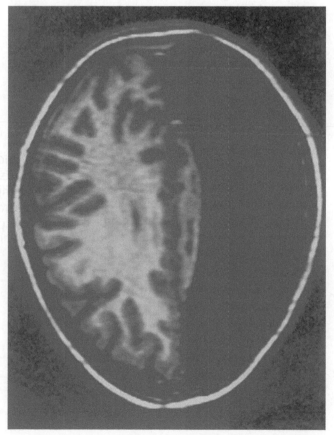

◀ **Figure 27.14 Hemispherectomy.** This top-down X-ray view shows the skull and brain of a hemispherectomy patient after surgery.

Neurological Disorders

Despite its neuroplasticity, there are conditions caused by injury or disease that result in permanent decreased brain function. Neurological disorders, diseases of the nervous system, often involve the brain. These disorders may cause portions of the brain to deteriorate, or they may alter activity of certain brain regions. Let's look at a few examples.

Alzheimer's disease is a form of mental deterioration, or dementia, characterized by confusion, memory loss, and a variety of other symptoms. One particularly common characteristic of Alzheimer's disease is the loss of neurons in the temporal lobe. Its incidence is usually age related, rising from about 12% at age 65 to almost 50% at age 85. The disease is progressive; patients gradually become less able to function. There are also personality changes, almost always for the worse. Patients often lose their ability to recognize people, even family members, and may treat them with suspicion and hostility.

Another disorder that causes structural changes to the brain is **chronic traumatic encephalopathy (CTE)**. Frequently with CTE, the frontal lobe and temporal lobes show significant deterioration. CTE is caused by repeated brain trauma, particularly concussions, and numerous deceased football players, including well-known NFL linebacker Junior Seau, have been diagnosed with CTE posthumously. The link between injury and brain damage is so strong that the NFL has agreed to pay a million dollars or more to each of its thousands of former players who have brain dysfunction. It's still unclear how forceful a hit to the head must be to cause CTE to develop or how many head injuries it takes to affect progression of the disease. Like Alzheimer's disease, early symptoms include personality changes. Depression, loss of impulse control, and, eventually, memory loss and dementia are prominent features of CTE.

Rather than causing large structural changes, other neurological disorders alter the activity of certain brain regions. A prominent area of research is determining how the brain's physiology changes during depression **(Figure 27.15)**. Nearly 20 million American adults are affected by depression. Two broad forms of depressive illness have been identified: major depression and bipolar disorder. People with **major depression** may experience sadness, loss of interest in pleasurable activities, changes in body weight and sleep patterns, and suicidal thoughts. Major depression is extreme and persistent, leaving the sufferer unable to live a normal life. **Bipolar disorder**, or manic-depressive disorder, involves extreme mood swings. The manic phase is characterized by high self-esteem, increased energy, a flood of thoughts and ideas, as well as behaviors that often involve increased risk taking. The depressive phase is marked by sleep disturbances, feelings of worthlessness, and decreased ability to experience pleasure.

Why do networks of neurons in the brain malfunction and disrupt our personalities, memories, and emotions? This continues to be a challenging and intriguing question. How the brain receives information from the environment via the sense organs is another engaging area of biological research. We examine the senses next. ☑

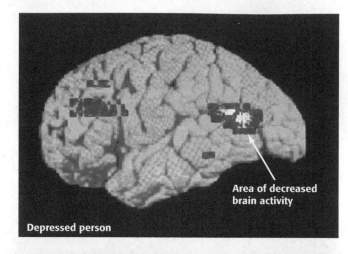

▼ **Figure 27.15 Brain activity in a depressed person and a healthy person.** The purple and pink colors in these computer images indicate areas of low brain activity. Notice that the brain of the depressed person shows decreased activity in certain areas of the brain.

Area of decreased brain activity

Depressed person

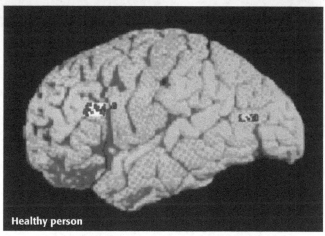

Healthy person

It's undeniable: Sports-related head injuries and neurological disorders are linked.

The Senses

In this section, we'll focus on the human body's sensory structures. To begin, we'll examine how these structures gather information and pass it on to the central nervous system (CNS). After that, we'll take a closer look at two of the human senses, vision and hearing.

Sensory Input

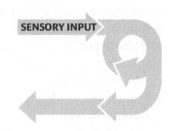

SENSORY INPUT

You're asleep, when a fire alarm sounds and lights begin flashing. Sensory receptors in your ears and eyes relay information about the alarm to your CNS, and you get up and move outside. Sensory receptors, such as those that helped you react to the fire alarm, are tuned to the condition of both the external world and internal organs. Detecting stimuli such as light, sounds, cold, heat, and touch, a sensory receptor's job is completed when it sends information to the CNS by triggering action potentials.

What exactly do we mean when we say that a sensory receptor detects a stimulus? In stimulus detection, the receptor cell converts one type of signal (the stimulus) to an electrical signal. This conversion of the signal, called sensory transduction, occurs as a change in the membrane potential of the receptor cell.

Converting a stimulus to an electrical signal

Figure 27.16 shows sensory receptors in a taste bud detecting sugar molecules. ❶ When the sugar molecules (🔵) first come into contact with the taste bud, they bind to membrane receptors of the sensory receptor cells. ❷ This binding triggers a signal transduction pathway that ❸ causes some ion channels in the membrane to close and others to open. Changes in the flow of ions (⊕) alter the membrane potential. This change in membrane potential is called the **receptor potential**. In contrast to action potentials, which are all-or-none phenomena, receptor potentials vary in intensity; the stronger the stimulus (the more sugar present), the stronger the receptor potential.

Once a receptor cell converts a stimulus to a receptor potential, this potential usually results in signals being sent to the CNS. In Figure 27.16, ❹ each receptor cell forms a synapse with a sensory neuron. When there are enough sugar molecules, a strong receptor potential is triggered. This receptor potential makes the receptor cell release enough neurotransmitter (🔵) to increase the rate (frequency) of action potential generation in

the sensory neuron (more action potentials in a given time). The brain interprets the intensity of the stimulus from the rate at which it receives action potentials. It gains additional information about stimulus intensity by keeping track of how many sensory neurons it receives signals from.

There is an important qualification to what we have just said about stimulus intensity. Have you ever noticed how an odor that is strong at first seems to fade with time, even when you know the smell is still there? Or how the water in a pool is shockingly cold when you first jump in, but then you get used to it? This effect is called **sensory adaptation**, the tendency of some

▼ Figure 27.16 **Converting a chemical stimulus to an electrical signal in a human taste bud.**

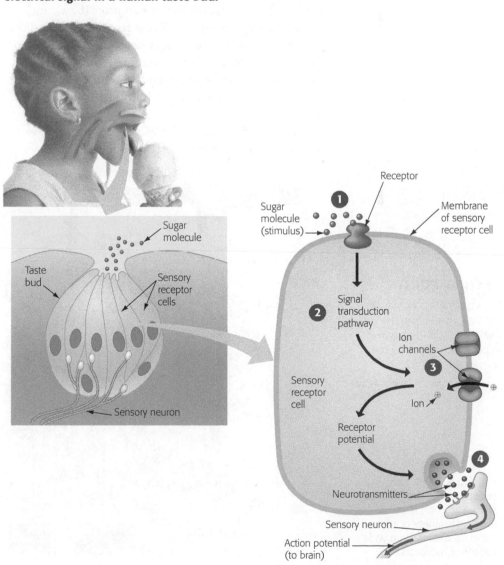

☑ **CHECKPOINT**

What happens to the action potentials the brain receives when you go from a club with loud music to the outside?

Answer: The rate of action potentials decreases.

sensory receptors to become less sensitive when they are stimulated repeatedly. When receptors become less sensitive, they trigger fewer action potentials, causing the brain to receive fewer stimuli. Sensory adaptation keeps the body from continuously reacting to normal background stimuli. Without it, our nervous system would become overloaded with unnecessary information. ☑

Types of Sensory Receptors

Based on the type of signals to which they respond, we can group sensory receptors into five general categories: pain receptors, thermoreceptors (sensors for heat and cold), mechanoreceptors (sensors for touch, pressure, motion, and sound), chemoreceptors (sensors for chemicals), and electromagnetic receptors (sensors for energy such as light and electricity). These five types of receptors work in various combinations to produce the five human senses: hearing, smell, taste, touch, and vision.

Figure 27.17, showing a section of human skin, reveals why the surface of our body is sensitive to such a variety of stimuli. Our skin contains pain receptors (red label in the figure), thermoreceptors (blue), and mechanoreceptors (green). These receptors are modified dendrites of individual sensory neurons. A single receptor is capable of recognizing a stimulus and responding to it by sending an action potential to the CNS. In other words, each receptor serves as both a receptor cell and a sensory neuron.

All parts of the human body except the brain have pain receptors. Pain is important because it often indicates danger and usually makes an animal withdraw to safety. Pain can also make us aware of injury or disease. **Pain receptors** may respond to excessive heat or pressure or to chemicals released from damaged or inflamed tissues. Aspirin and ibuprofen reduce pain by inhibiting molecules that stimulate pain receptors.

Thermoreceptors in the skin detect either heat or cold. Other temperature sensors located deep in the body monitor the temperature of the blood. The body's major thermostat is the hypothalamus. Receiving action potentials both from surface sensors and from deep sensors, the hypothalamus keeps a mammal's or bird's body temperature within a narrow range (see Figure 21.14).

Mechanoreceptors are highly diverse. Different types are stimulated by different forms of mechanical energy, such as touch and pressure, stretching, motion, and sound. All these forces produce their effects by bending or stretching the plasma membrane of a receptor cell. When the membrane changes shape, it becomes more permeable to positive ions, and the mechanical energy of the stimulus is transduced into a receptor potential. One example of a mechanoreceptor is the touch receptor at the base of a cat's whisker. These receptors are extremely sensitive and enable the animal to detect objects by touch in the dark. The elephants you learned about at the beginning of the chapter provide another example; they detect seismic waves through mechanoreceptors in their feet and trunk.

Chemoreceptors include the sensory cells in your nose and taste buds, which are attuned to chemicals in the external environment, as well as some internal receptors that detect chemicals in your body's internal environment. Internal chemoreceptors include sensors in your arteries that monitor your blood, with some sensors detecting changes in pH and others detecting changes in O_2 concentration. In all types of chemoreceptors, a receptor cell develops receptor potentials in response to chemicals dissolved in fluid such as blood or saliva.

Electromagnetic receptors are sensitive to energy of various wavelengths, which takes such forms as magnetism and light. For example, **photoreceptors** detect the electromagnetic energy of light. In the next section, we'll focus on one specific organ that uses photoreceptors: the human eye. ☑

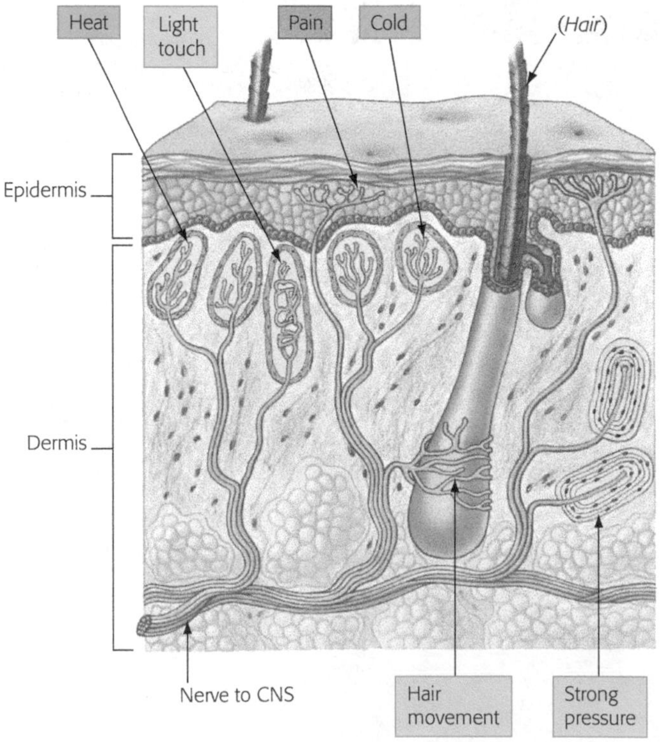

Vision

The human eye is a remarkable sense organ, able to detect a multitude of colors, form images of objects both near and far, and respond to minute amounts of light energy. In this section, you'll learn about the structure of the human eye and how it processes images. You'll also learn why vision problems occur and how they can be corrected.

☑ **CHECKPOINT**

Fish can detect vibrations and movements in the water around them; what type of sensory receptor allows this sense?

Answer: mechanoreceptors

► **Figure 27.17 Sensory receptors in the human skin.** Your skin is sensitive to a wide variety of stimuli because it contains a wide variety of receptors.

Heat Light touch Pain Cold (Hair)

Epidermis

Dermis

Nerve to CNS Hair movement Strong pressure

Lighting the Retina

To understand how photoreceptors function in vision, we first need to follow light's path through the eye to the **retina**—the layer of tissue that contains these photoreceptors **(Figure 27.18)**. Although the outer layer of the human eyeball is covered in a whitish layer of connective tissue (the sclera or "white of your eye"), it becomes transparent at the front of the eye. This curved, transparent "window" that lets light in and helps to focus it is called the **cornea**. Through the cornea and a fluid-filled chamber, light reaches a pigmented structure called the **iris**. (Your iris is what you describe when you tell someone your eye color.) Light travels through the iris at its center through a circular opening called the **pupil**. Muscles of the iris regulate the size of the pupil, controlling the amount of light that enters further into the eye. Once past the pupil, light passes through the transparent, disklike **lens**, through a jellylike fluid, and on to the retina. The retina contains the photoreceptor cells that sense light.

Before we consider the photoreceptors of the retina in more detail, let's think about how the parts of the eye we discussed change to aid our vision in different situations. Move your eyes from this text to an object far away and then back again. What just happened? You had to change your focus. This is accomplished through a change in the curvature of the lens, which focuses light on the retina by bending light rays **(Figure 27.19)**. When your eyes focus on a close object, muscles controlling the lens contract, the ligaments that suspend the lens slacken, and with this reduced tension, the elastic lens becomes thicker and rounder. When your eyes focus on a distant object, muscles controlling the lens relax, putting tension on ligaments and flattening the lens.

Besides seeing objects near and far, you often challenge your vision in bright light and dim light. How do your eyes adjust to varying light? Look in a mirror and shine a bright light near your face, and you'll see that your pupils become smaller. The reverse happens as you darken the room. Because of our common ancestors, the pupils of other animals behave similarly in bright and dim light. One difference is that some nocturnal animals, such as cats, have pupils that are slits, not circular like ours. Slits allow some nocturnally active species to quickly close out light that is too bright during the day. Cats also have larger lenses and more light-sensitive retinas than humans.

Now that we have seen how light passes through the eye and how the eye adjusts to control the amount of light coming into it, let's see how light stimulates the cells of the retina. We turn our attention now to photoreceptors of the retina. ☑

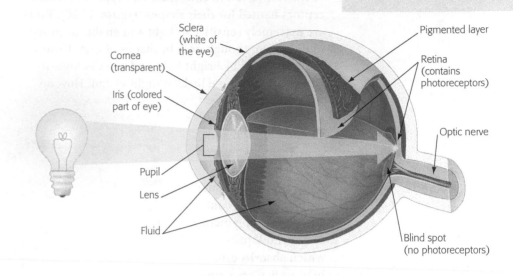

▼ **Figure 27.18 The path of light to the retina.** Once light passes through the eye, it is focused on photoreceptor cells in the retina.

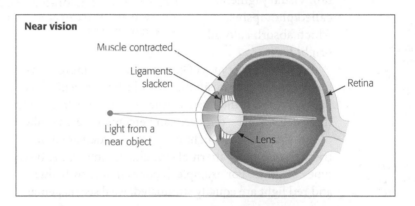

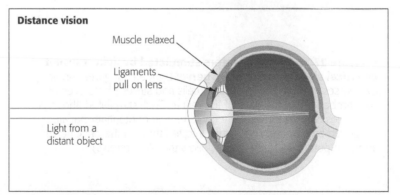

▲ **Figure 27.19 How the lens of the eye focuses light.** When viewing a nearby object (top), the lens becomes thicker and rounder and focuses the image on the retina. When viewing a distant object (bottom), the lens becomes flattened, and the image is again focused on the retina.

☑ CHECKPOINT

By age 80, half of Americans have cataracts (the lens becomes cloudy). How do cataracts affect vision?

Answer: Light does not pass through a lens with a cataract as well, decreasing the lens's ability to focus light on the retina and thus the person's ability to see clearly.

Photoreceptors

As light passes through the front of the human eye to the retina, it will encounter two types of photoreceptors named for their shapes (**Figure 27.20**). **Rods** are extremely sensitive to light and enable us to see in dim light, though only in shades of gray. **Cones** are stimulated by bright light and can distinguish color but contribute little to night vision. How do rods and cones detect light? As Figure 27.20 shows, each rod and cone includes an array of disks containing light-absorbing visual pigments. Rods contain a visual pigment called rhodopsin, which absorbs dim light well. Cones contain visual pigments called photopsins, which absorb colored, bright light.

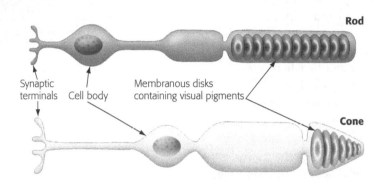

Rod

Cone

Synaptic terminals   Cell body   Membranous disks containing visual pigments

▲ **Figure 27.20 Photoreceptor cells.** Your eyes contain rods, which allow you to see in dim light, and cones that allow you to perceive color in bright light.

We have three types of cones, each containing a different type of photopsin that absorbs blue, green, or red light. Just as a printer can synthesize all colors from only three color toners, our eyes can perceive a great number of colors because the light from each particular color triggers a unique pattern of stimulation among the three types of cones. For example, if cones that absorb blue and red light are equally stimulated, we'll see magenta.

Colorblindness results from defects in genes producing proteins involved in one or more types of cones. Daytime visual acuity is sharpest for vertebrates with a high density of cones. Some birds, such as hawks, have ten times more cones than we do, enabling them to spot small prey from high altitudes.

Like all receptor cells, rods and cones convert a stimulus to an electrical signal. In **Figure 27.21**, we see light reaching these photoreceptors on the right side of the illustration. As they absorb light, rhodopsin and photopsin change chemically, and the change alters the permeability of the cell's membrane. The resulting receptor potentials trigger a complex integration process that begins in the retina. Other retinal neurons start to integrate the receptor potentials and in doing so produce action potentials. These action potentials (arrows in the figure) travel along the **optic nerve**, which connects the retina with the brain. The optic nerve carries the partly integrated information into the brain. Three-dimensional perceptions (what we actually see) result from further integration in several processing centers of the cerebral cortex. ☑

► **Figure 27.21 Photoreceptors stimulated by light transmit electrical signals to the optic nerve.** Light entering the human eye passes through many layers of cells and strikes photoreceptor cells embedded in the back of the retina. Once stimulated, the rods and cones convert the light energy to receptor potentials that are integrated by neurons and communicated through the optic nerve as action potentials to the brain (follow the black arrows).

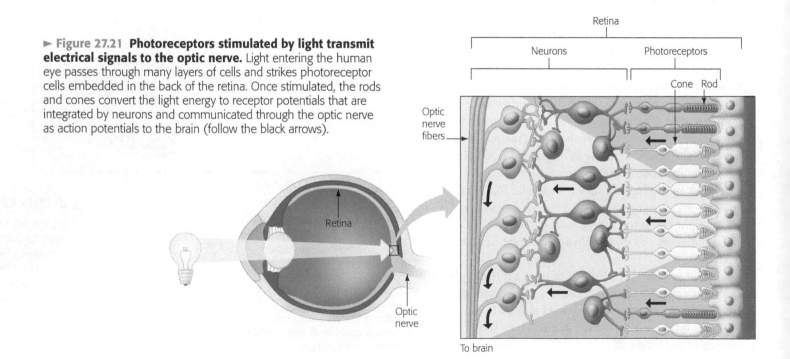

Retina

Neurons   Photoreceptors

Cone   Rod

Optic nerve fibers

Retina

Optic nerve

To brain

Evolution — Our Eyes Are Not Perfect

All vertebrates have eyes with a single lens, as do some invertebrates such as squid. Yet the single-lens eye arose independently in these two groups; the eyes originated from two different tissues that acquired new functions through natural selection. When existing structures acquire new functions, imperfections are sometimes apparent. For example, in humans, there are no photoreceptor cells in the part of the retina where the optic nerve passes through the back of the eye, known as the "blind spot" (see its location in Figure 27.18). Light that is focused on the blind spot cannot be detected. You can find your own blind spot using **Figure 27.22**. Despite the blind spot, having two eyes with overlapping fields of view enables us to perceive uninterrupted images. Would our eyes function better if the optic nerve didn't pass through the retina and instead went behind the photoreceptors? Probably. The single lens eye found in squids is structured this way, and it has no blind spot. Yet our quite functional eyes demonstrate that anatomical structures need not be perfect.

Vision Problems and Corrections

Three of the most common visual problems are nearsightedness, farsightedness, and astigmatism. All three are focusing problems, easily corrected with artificial lenses. People with **nearsightedness** cannot focus well on distant objects, although they can see well at short distances. A nearsighted eyeball **(Figure 27.23a)** is longer than normal, and it focuses distant objects in front of the retina instead of on it. **Farsightedness** occurs when the eyeball is shorter than normal, causing the lens to focus images behind the retina **(Figure 27.23b)**. Farsighted people see distant objects normally, but they can't focus on close objects. Both nearsightedness and farsightedness can be corrected by glasses or contact lenses that help focus light at the correct point on the retina.

Astigmatism is blurred vision caused by a misshapen lens or cornea. Any such distortion makes light rays converge unevenly and not focus at any one point on the retina. Lenses that correct astigmatism are asymmetrical in a way that compensates for the asymmetry in the eye.

Surgical procedures are an option for treating vision disorders. In a laser-assisted surgery known as LASIK, the cornea is reshaped to change its focusing ability. Close to 1 million LASIK procedures are performed each year to correct a variety of vision problems. ☑

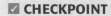

☑ **CHECKPOINT**

Glasses can correct what kind of structural defects of the eye?

Answer: a short eyeball (farsightedness), a long eyeball (nearsightedness), or a misshaped cornea or lens (astigmatism)

▲ **Figure 27.22 Detecting the blind spot.** Hold your right eye closed. With your left eye, look at the ✚ in the image above. Starting about two feet away from the page, slowly bring your head closer while looking at the ✚. You should notice that, at a specific distance from the page, the ● disappears. The ● is in the blind spot of your left eye when this happens. If you look at the ● directly, it will reappear.

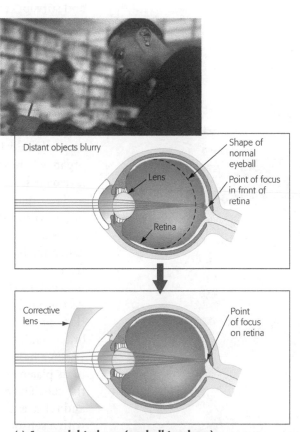

(a) A nearsighted eye (eyeball too long). Nearsightedness is corrected by lenses that are thinner toward the middle.

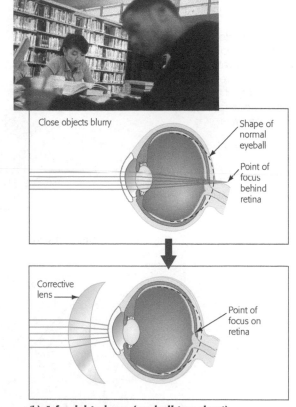

(b) A farsighted eye (eyeball too short). Farsightedness is corrected by lenses that are thicker in the middle.

▲ **Figure 27.23 A nearsighted eye and a farsighted eye.** Corrective lenses help vision problems by focusing the image exactly on the retina.

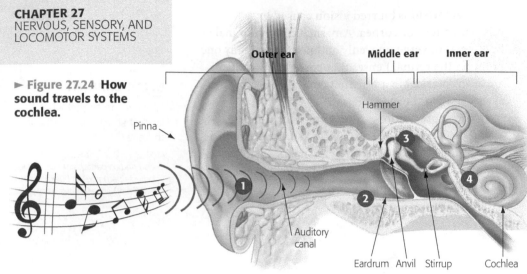

► Figure 27.24 **How sound travels to the cochlea.**

► Figure 27.25 **The organ of Corti.** The organ of Corti is located within a fluid-filled canal inside the cochlea. Sounds cause vibrations of the basilar membrane, forcing hairlike projections on hair cells (mechanoreceptors) to bend. Hair cells convert the stimuli to receptor potentials that are communicated through the auditory nerve as action potentials to the brain.

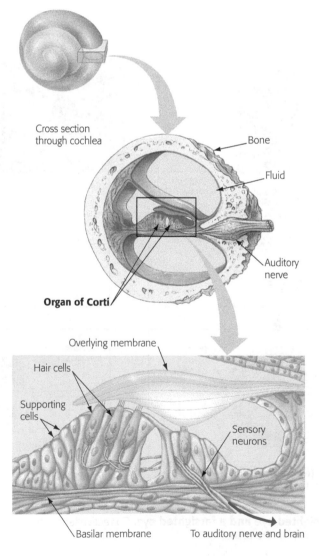

Hearing

As we learned in the Biology and Society section, elephants can sense very low frequency sounds that are not audible to humans. Although the human ear is less sensitive than that of elephants and many other animals, it is capable of hearing a great range of auditory signals. In this section, we'll examine the structure and function of the human ear.

When you hear a sound, such as the alert of a text message, the sound travels inside your ear to sensory receptors that produce electrical signals, which are relayed to the brain. **Figure 27.24** shows how sound travels through the ear to where the receptors are located. ❶ The bendable structure we commonly refer to as our ear, the pinna, acts like a funnel focusing sound waves into the auditory canal. The pinna and auditory canal are considered the **outer ear**. ❷ The sound waves then reach a sheet of tissue called the **eardrum**, which separates the outer ear from the middle ear. The sound waves cause the eardrum to vibrate. ❸ From the eardrum, the vibrations are concentrated as they pass through the hammer, anvil, and stirrup—three bones of the **middle ear**. These are the smallest bones of the human body—so small, all three together could fit on a shirt button. ❹ The stirrup transmits the vibrations to a sheet of tissue between the middle ear and the third part of the ear, the **inner ear**. As the bones vibrate, they push on the membrane to generate vibrations in the fluid that fills the inner ear. A component of the inner ear is the **cochlea**, a coiled, fluid-filled structure containing our hearing organ, where sensory receptors are located and sound is detected.

The organ of Corti is the name of our hearing organ within the cochlea (**Figure 27.25**). Receptor cells known as hair cells are anchored on a structure called the basilar membrane. The vibrations traveling from the middle ear produce pressure waves within the cochlea that cause the basilar membrane to vibrate. As the basilar membrane vibrates, hairlike projections on the hair cells bend against an overlying membrane. Hair cells are an example of mechanoreceptors. When a hair cell's projections are bent, ion channels in its plasma membrane open, and positive ions enter the cell. The hair cell develops a receptor potential and releases neurotransmitter molecules at its synapse with a sensory neuron. In turn, the sensory neuron sends action potentials through the auditory nerve to the brain. Your brain interprets the sound and plans for action—if it's a text message alert, you'll likely grab your phone.

Of course, whether you hear your text alert or not depends on your phone's volume settings. Loudness is the perception of the intensity (amplitude) of sound waves. More intense sound waves cause stronger displacement of the fluid in the inner ear. The more intense displacement of the fluid the louder the perceived sound, and vice versa (Figure 27.26). You can distinguish particular tones (pitches) because various frequencies of sound waves cause different hair cells along the organ of Corti in the inner ear to bend and vibrate. Thus, shifts in fluid and vibration of particular hair cells in the inner ear allow you to discriminate sounds of different volumes and pitches from your phone or anything else in your environment.

Deafness, the inability to hear, can be present at birth or acquired. Mutations in genes known to play a role in hearing may be inherited and can cause a child to be born with defective ear structures. The inability to hear may also result from middle-ear infections, a ruptured eardrum, or stiffening of the middle-ear bones (a common age-related problem). Frequent or prolonged exposure to sounds of more than 90 decibels can damage or destroy hair cells, which are never replaced. Earplugs can provide protection.

It's tempting to turn up the volume on your media player, but if you do, you're risking permanent hearing loss.

However, many people routinely listen to music in earphones exceeding 90 decibels. Deafness is often progressive and permanent. One treatment for hearing loss is a device called a cochlear implant, which bypasses defective structures and directly stimulates the auditory nerve. ☑

☑ CHECKPOINT

How does the ear convert sound waves in the air to pressure waves in the fluid of the inner ear?

Answer: Sound waves in the air cause the eardrum to vibrate. The small bones attached to the inside of the eardrum transmit vibrations to the fluid of the inner ear, producing pressure waves.

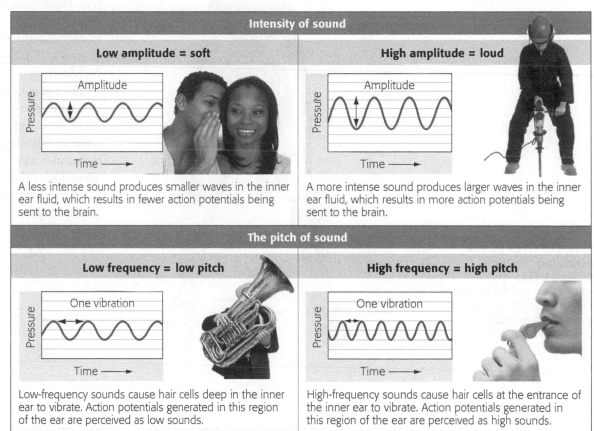

▼ Figure 27.26 **The characteristics of sound affect how we perceive volume and pitch.**

Intensity of sound

Low amplitude = soft

A less intense sound produces smaller waves in the inner ear fluid, which results in fewer action potentials being sent to the brain.

High amplitude = loud

A more intense sound produces larger waves in the inner ear fluid, which results in more action potentials being sent to the brain.

The pitch of sound

Low frequency = low pitch

Low-frequency sounds cause hair cells deep in the inner ear to vibrate. Action potentials generated in this region of the ear are perceived as low sounds.

High frequency = high pitch

High-frequency sounds cause hair cells at the entrance of the inner ear to vibrate. Action potentials generated in this region of the ear are perceived as high sounds.

Locomotor Systems

So far, we've seen how the body senses the environment via sensory receptors and integrates this information within the central nervous system. In this section, we'll consider how the body couples stimulus to locomotor response (movement), thereby completing the neural circuit.

MOTOR OUTPUT

Movement is one of the most distinctive features of animals. Whether an animal walks or runs on two, four, six legs, or more, swims, crawls, flies, or sits, an interplay of organ systems is responsible for its movement. Controlled by the nervous system, the muscular system exerts the force that actually makes an animal or its parts move. But the force exerted by muscles produces movement only when it is applied against a firm structure. In vertebrates, this structure is the skeletal system.

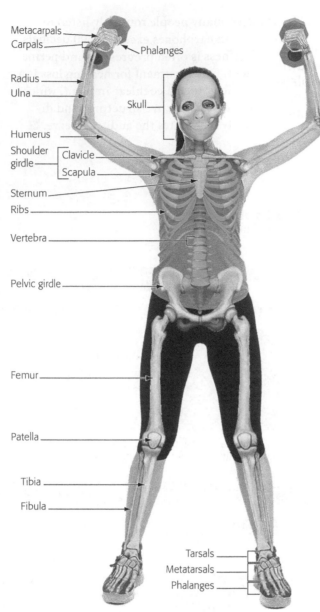

Metacarpals
Carpals
Phalanges
Radius
Ulna
Skull
Humerus
Shoulder girdle — Clavicle
Scapula
Sternum
Ribs
Vertebra
Pelvic girdle
Femur
Patella
Tibia
Fibula
Tarsals
Metatarsals
Phalanges

► **Figure 27.27 The human endoskeleton.** The bones supporting the trunk are shown in green, and the bones supporting the limbs are seen as yellow. Cartilage (blue) provides flexibility at various points.

The Skeletal System

The **skeletal system** provides support, protection, and anchoring. If not for the support provided by the skeleton, most land animals would sag from their own weight. An animal's soft organs, such as the brain, heart, and lungs, are largely protected by the skeleton. The anchoring of muscle on the skeleton allows bones to move. In this section, we'll discuss the human skeleton: its organization, its structure, and some of the problems that may arise with it.

Organization of the Human Skeleton

Humans, like all vertebrates, have an **endoskeleton**—hard supporting elements situated among soft tissues. The human endoskeleton is a combination of cartilage and the 206 bones that make up the skeletal system **(Figure 27.27)**; the cartilage provides flexibility in certain areas.

The bones of the skeleton are held together at movable joints by strong fibrous tissues called **ligaments**. Much of the versatility of our skeleton comes from three types of movable joints **(Figure 27.28)**. Humans have **ball-and-socket joints** in the shoulder and the hip. These joints enable us to rotate our arms and legs and move them in several directions. In the elbow, a **hinge joint** permits movement in a single direction, such as when we do a bicep curl. A **pivot joint** enables us to rotate our forearms at the elbow and to turn our heads side to side (as if saying "no").

▼ **Figure 27.28 Three kinds of joints.**

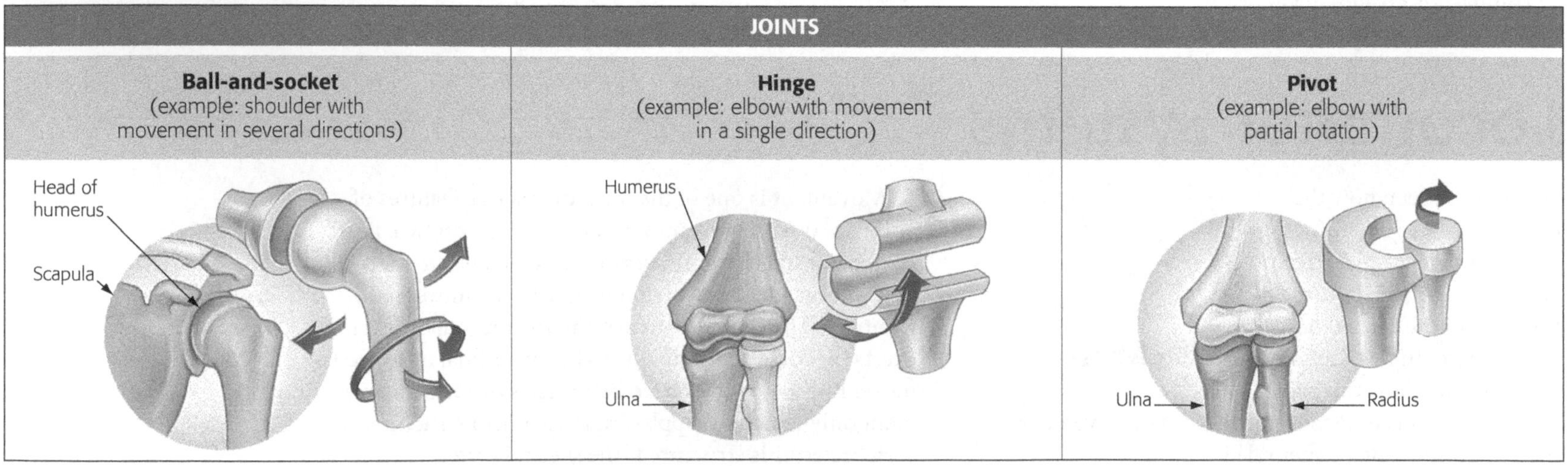

JOINTS		
Ball-and-socket (example: shoulder with movement in several directions)	**Hinge** (example: elbow with movement in a single direction)	**Pivot** (example: elbow with partial rotation)
Head of humerus / Scapula	Humerus / Ulna	Ulna / Radius

The Structure of Bones

Although they may appear dry and dead, bones are complex organs consisting of several kinds of living tissues. You can get a sense of some of a bone's complexity from the cutaway view of the human humerus (the upper arm bone) shown in **Figure 27.29**. A sheet of fibrous connective tissue, shown in pink (most visible in the enlargement on the lower right), covers most of the outside surface. This tissue can form new bone during normal growth or in the event of a fracture. At either end of the bone is a thin sheet of cartilage (blue), also living tissue that cushions joints, protecting the ends of bones as they move against each other. The bone itself contains cells that secrete a surrounding material, or matrix (see Figure 21.4). Like all living tissues, bone tissues require nutrients and oxygen. Blood vessels course through channels in the bone, transporting nutrients and regulatory hormones to its cells and waste materials from them.

Notice that the shaft of the long bone shown in Figure 27.29 surrounds a central cavity. The central cavity contains yellow bone marrow, which is mostly stored fat brought into the bone by the blood. The ends of the bone have cavities that contain red bone marrow in certain bones (not shown in the figure), a specialized tissue that produces blood cells (see Chapter 23). ☑

▼ Figure 27.29 **The structure of an arm bone.**

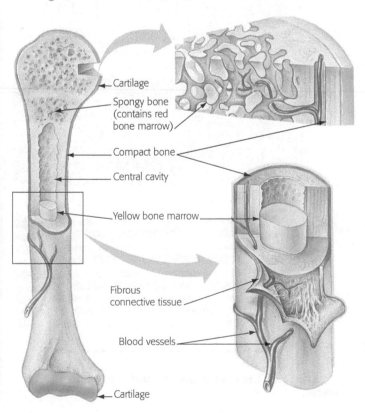

- Cartilage
- Spongy bone (contains red bone marrow)
- Compact bone
- Central cavity
- Yellow bone marrow
- Fibrous connective tissue
- Blood vessels
- Cartilage

Skeletal Diseases and Injuries

Your skeleton is quite strong and provides you with reliable support, but it is susceptible to injury and disease. Bones are rigid but not inflexible; they will bend in response to external forces. However, as many of us know from personal experience, the skeletal system has its limits. If a force is applied that exceeds a bone's capacity to bend, the result is a broken bone, or fracture. The average American will break two bones during his or her lifetime, most commonly the forearm or, for people over 75, the hip.

Treatment of a fracture involves two steps: putting the bone back into its natural shape and then immobilizing it until the body's natural bone-building cells can repair the fracture. A splint or a cast is usually sufficient to protect the area, prevent movement, and promote healing. In more severe cases, a fracture can only be repaired surgically by inserting plates, rods, and/or screws that hold the broken pieces together (**Figure 27.30**).

Arthritis—inflammation of the joints—affects one out of every seven people in the United States. The most common form of arthritis occurs as a result of aging: The joints become stiff and sore and often swell as the cartilage between the bones wears down. Sometimes the bones thicken at the joints, restricting movement. This form of arthritis is irreversible but not crippling in most cases, and moderate exercise, rest, and over-the-counter pain medications usually relieve most symptoms.

Rheumatoid arthritis is a debilitating autoimmune disease. The joints become highly inflamed, and their tissues may be destroyed by the body's immune system (see Figure 24.16). Rheumatoid arthritis usually begins between ages 40 and 50 and affects more women than men. Anti-inflammatory drugs relieve symptoms, but there is no cure.

Osteoporosis is another serious bone disorder. It is most common in women after menopause: Estrogen contributes to normal bone maintenance, and with lowered production of the hormone, bones may become thinner, more porous, and more easily broken. Insufficient exercise, smoking, diabetes mellitus, and an inadequate intake of protein and calcium may also contribute to the disease. Treatments include calcium and vitamin supplements, hormone replacement therapy, and drugs that slow bone loss or increase bone formation. Prevention of osteoporosis begins with sufficient calcium intake while bones are still increasing in density (up until about age 35). Weight-bearing exercise (walking, jogging, lifting weights) builds bone mass and is beneficial to bone health throughout life. ☑

☑ **CHECKPOINT**

Why is the saying "dry as a bone" inaccurate?

Answer: Living bones contain a blood supply and marrow.

▼ **Figure 27.30 Broken bone.** The X-ray shows a fractured tibia (shin bone) after insertion of a plate and screws to repair it.

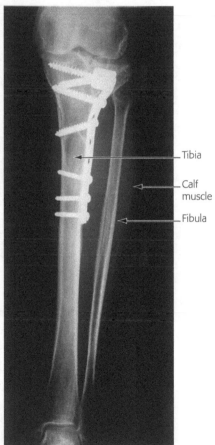

- Tibia
- Calf muscle
- Fibula

☑ **CHECKPOINT**

What two things can a young person do to help prevent osteoporosis?

Answer: ingest plenty of calcium and engage in weight-bearing exercise regularly.

The Muscular System

Now that you've learned about the skeletal system, we can focus on the **muscular system**, which is made up of all the skeletal muscles in the body.

Skeletal muscles are attached to the skeleton and produce voluntary body movement by interacting with it. The two other muscle types of the body, smooth muscle and cardiac muscle, contract involuntarily and are found inside organs (see Figure 21.5).

Figure 27.31 illustrates how skeletal muscles interact with bones to raise and lower the human forearm. As shown in the figure, strong fibrous tissues called **tendons** connect muscles to bones. For instance, the triceps muscle is attached to bones of the shoulder and also to bones in the forearm by tendons.

The ability to move the forearm in opposite directions requires that two muscles work as an antagonistic pair—that is, the two muscles must perform opposite tasks.

Notice in Figure 27.31 that when one muscle is contracted, the antagonistic (opposite) muscle is relaxed.

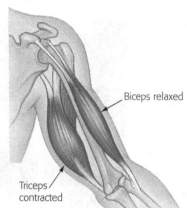

▲ Figure 27.31 **Antagonistic action of muscles in the human arm.** Contraction of the biceps shortens the muscle and pulls up the forearm (top). Contraction of the triceps pulls the forearm down.

The Cellular Basis of Muscle Contraction

As shown at the top of **Figure 27.32**, a muscle consists of bundles of parallel muscle fibers. Each muscle fiber is a single, long cell with many nuclei. The bulging muscles of weight lifters are a result of larger sized muscle fibers. Notice that each muscle fiber is itself a bundle of smaller **myofibrils**. A myofibril consists of repeating units called **sarcomeres**. Functionally, the sarcomere is the contractile apparatus in a myofibril—the muscle fiber's fundamental unit of action.

The micrograph and the diagram at the bottom of Figure 27.32 reveal the structure of a sarcomere in more detail. A myofibril is composed of regular arrangements of two kinds of filaments: **thick filaments** made primarily from the protein myosin (red in the diagram) and **thin filaments** made primarily from the protein actin (blue).

Forensic scientists may check a murder victim for muscle stiffness to estimate the time of death.

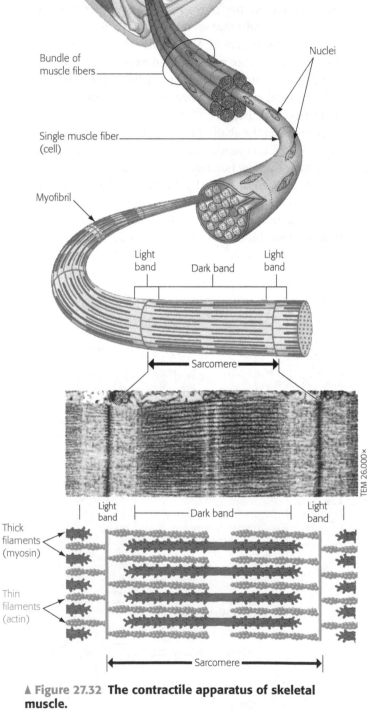

▲ Figure 27.32 **The contractile apparatus of skeletal muscle.**

► Figure 27.33 **The sliding-filament model of muscle contraction.** Notice that the lengths of the thick and thin filaments do not change; the two types of filaments merely slide past each other to overlap.

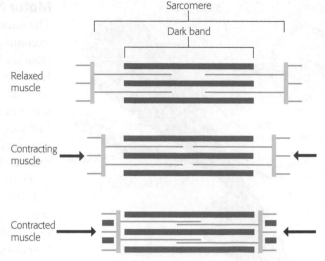

Sarcomere

Dark band

Relaxed muscle

Contracting muscle

Contracted muscle

LOCOMOTOR SYSTEMS

✓ **CHECKPOINT**

Based on what muscles require for contraction, what cell organelle would you expect to find in abundance within muscle fibers?

Answer: mitochondria (to perform cellular respiration and produce the needed ATP for contraction)

A sarcomere contracts (shortens) when its thin filaments slide past its thick filaments. The model describing this process is known as the sliding-filament model of muscle contraction. **Figure 27.33** is a simplified diagram of this model. Notice in the contracting sarcomere that the thin filaments (blue) have moved toward the middle of the sarcomere. When the muscle is fully contracted, the thin filaments overlap in the middle of the sarcomere. Contraction only shortens the sarcomere; it does not change the lengths of the thick and thin filaments.

If you have ever trained for a sport, you'll know that you need to eat more calories the more you exercise. The chemical energy stored in the food you eat is converted to adenosine triphosphate (ATP) energy, and it is the ATP that your working muscles require. Why? Without ATP, thin filaments cannot slide to contract the muscle. **Figure 27.34** indicates how sliding is thought to work. ❶ ATP binds to a myosin head (🥄), causing the head to detach from a binding site on actin (⬤). ❷ The myosin head gains energy from the breakdown of ATP and changes to a high-energy position. ❸ The energized myosin head binds to an exposed binding site on actin. ❹ In what is called the power stroke, the myosin head bends back to its low-energy position, pulling the thin filament toward the center of the sarcomere. ❺ After the power stroke, the whole process repeats. As long as sufficient ATP is present, the process continues until the signal to contract stops.

What if ATP is no longer present, such as when an animal dies? Without ATP, myosin heads are "locked" to actin molecules, as in step 4 of Figure 27.34. The result is rigor mortis (muscle stiffness) that sets in not long after an animal dies. You may have noticed rigor mortis in dead animals on roadsides. Rigor mortis typically reverses within a few days because as an animal decays so does its muscle fibers. Since rigor mortis is a temporary condition, forensic scientists investigating a murder can sometimes use the presence of rigor mortis to approximate time of death. ✓

► Figure 27.34 **The mechanism of filament sliding.** Though we show only one myosin head in this figure, a typical thick filament has about 350 heads, each of which can bind and unbind to a thin filament about five times per second.

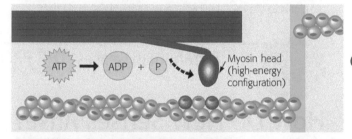

Thick filament (myosin)

Thin filament (actin)

ATP

Myosin head (low-energy configuration)

❶ ATP binds to a myosin head, which is then released from an actin filament.

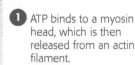

ATP → ADP + P

Myosin head (high-energy configuration)

❷ The breakdown of ATP cocks the myosin head.

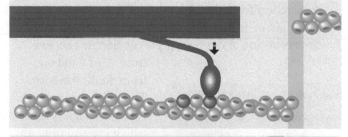

❸ The myosin head attaches to an actin binding site.

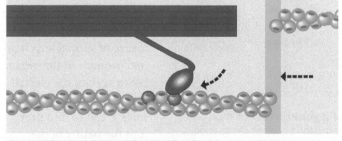

❹ The power stroke slides the actin (thin) filament toward the center of the sarcomere.

❺ As long as ATP is available, the process can be repeated until the muscle is fully contracted.

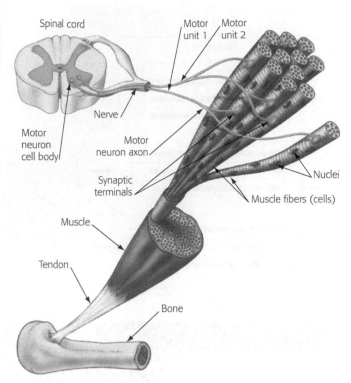

▲ **Figure 27.35 The relationship between motor neurons and muscle fibers.** A motor unit consists of a motor neuron and the one or more muscle fibers that it stimulates. When a motor neuron receives signals from the CNS, it passes them via synaptic terminals to the muscle fibers, causing the fibers to contract.

Motor Neurons: Control of Muscle Contraction

The sarcomeres of a muscle fiber are stimulated to contract by motor neurons. Responding to a signal sent from the brain via the spinal cord a motor neuron releases a neurotransmitter, which causes the muscle fiber to contract. A typical motor neuron can stimulate more than one muscle fiber because each neuron has many branches. In the example shown in **Figure 27.35**, you can see two orange **motor units**, each consisting of a neuron and all the muscle fibers it controls (two or three, in this case).

The organization of individual neurons and muscle cells into motor units is the key to the action of whole muscles. Each motor neuron may serve just one or up to several hundred fibers scattered throughout a muscle. Stimulation of the muscle by a single motor neuron produces only a weak contraction. More forceful contractions result when additional motor units are activated. Thus, depending on how many motor units your brain commands to contract, you can apply a small amount of force to lift a fork or considerably more to lift a barbell. In muscles requiring precise control, such as those controlling eye movements, a motor neuron may control only one fiber.

We have focused on how skeletal muscles produce voluntary movement of the skeleton. Let's now look at how an ancestral muscle could have given rise to an electric sense in certain fishes.

Extrahuman Senses THE PROCESS OF SCIENCE

How Do New Senses Arise?

Researchers from the University of Texas investigated the origin of an electric sense in fish by starting with **observations** about two species of electric fish, one from Africa and one from South America. These fishes can activate an electric organ, which evolved from an ancestral muscle. Electric fields are generated because of special ion channel proteins in the organ. Both species use electrical emissions to communicate and to find prey, but the nature of the electrical discharges varies considerably between the two lineages. This led the researchers to **question** whether different ion channel proteins had evolved in the two electric species.

The researchers formed the **hypothesis** that the ion channel genes of the two electric species had mutated in unique ways. In their **experiment**, they determined the DNA sequence of the genes in the two electric fishes, as well as in one closely related but nonelectric South American fish. **Figure 27.36** summarizes their **results**. In the common ancestor of all three of these species, a single ion channel gene had duplicated, and the gene copies later mutated into two forms, referred to as *a* and *b*. In the nonelectric fish, both of these genes function in muscles, much as they do in many other vertebrates. In the two species of electric fish, however, gene *a* had mutated into forms that allow the fishes to produce a current in the electric organ. Slight differences in how gene *a* mutated account for the differences in how each species generates electrical signals. Gene *b*, however, functions in the muscles of the electric fishes as it does in nonelectric fishes. This research demonstrates how a new sense can arise from an old one by gene duplication followed by mutation and natural selection.

Ion channel genes	South American species		African species
	Nonelectric fish	**Electric fish**	**Electric fish**
Gene *a*	Functions in muscles	Functions in electric organ (mutant variant 1)	Functions in electric organ (mutant variant 2)
Gene *b*	Functions in muscles	Functions in muscles	Functions in muscles

▲ **Figure 27.36 Duplication of a gene.** In the common ancestor of three species of fish, the gene for a muscle ion channel protein duplicated and then mutated, producing genes *a* and *b*. In a nonelectric fish species, both genes function in muscle. In the electric fish, gene *a* functions in the electric organ, while gene *b* functions in muscle.

Stimulus and Response: Putting It All Together

Car keys are tossed to a woman, she looks up, and she catches them in her hand (Figure 27.37). We can now appreciate this simple response in terms of sensory input, nervous system integration, and locomotor response.

The woman obtains information about the keys coming near her with sensory structures. Photoreceptors in her eyes track the keys. Sensory neurons convey this information to her brain as a series of action potentials (sensory input). Within her brain, information about the angle and speed of the keys is interpreted (integration). In response, motor neurons signal multiple muscles in her shoulder, arm, and fingers to perform very specific actions (motor output). She opens her hand, the keys land, and she closes her hand around them. Underlying this simple event is the nervous system directing muscles to respond to information that sensory receptors have gathered about the environment.

We have now completed the unit on animal structure and function. Our overriding theme has been that the evolution of new structures allows adaptations for new functions; that is, the structural adaptations of a cell, tissue, organ, or organ system determine the job it can

▼ Figure 27.37 **The nervous, sensory, and locomotor systems in action.**

perform. We'll see the structure-function theme emerge again in the context of adaptations to the environment when we take up the study of plants in the next unit.

Extrahuman Senses EVOLUTION CONNECTION

Seeing UV

In the Biology and Society section and elsewhere in this chapter, you have read about animal senses, some not present in humans. Through natural selection, new adaptations can arise from mutations in existing genes, for example, the gene mutation that led to the evolution of electroreception in some fishes. This principle is also clearly evident in the evolution of another nonhuman sense: ultraviolet (UV) vision.

Many birds can see beyond what we humans can see (Figure 27.38). Ultraviolet vision—the ability to see light of wavelengths shorter than those in our visual spectrum (see Figure 7.4)—is believed to be important in social communication and food gathering. As you learned in this chapter, visual information is transmitted to the brain after pigment molecules in the eye absorb light. Researchers studying vision in birds found that a single amino acid change in the pigment protein rhodopsin converted it to a UV-detecting form.

This is yet another example of a large change—an important extension of a sense—that can be traced to a small change: a single mutation. In nature, if a mutation such as this one confers a survival advantage (the ability to more easily locate food, for example), then natural selection would tend to favor individuals with the mutation. Over time, the new adaptation would eventually become the norm.

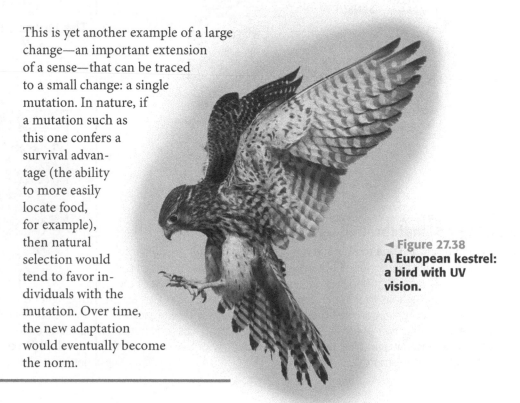

◄ Figure 27.38
A European kestrel: a bird with UV vision.

Chapter Review

■ SUMMARY OF KEY CONCEPTS

An Overview of Animal Nervous Systems

The nervous system is a communication and coordination network composed of neurons, specialized nerve cells capable of carrying signals throughout the body.

Neurons

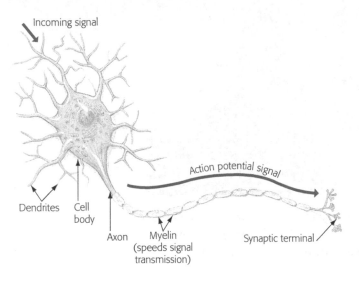

Organization of Nervous Systems

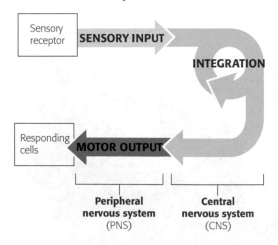

Sending a Signal through a Neuron

At rest, a neuron's plasma membrane has a resting potential caused by the membrane's ability to maintain a positive charge on its outer surface opposing a negative charge on its inner surface. A stimulus alters a portion of the membrane, allowing ions to enter and exit the neuron and creating an action potential. Action potentials are self-propagated in a one-way chain reaction along a neuron. An action potential is an all-or-none event; its size is not affected by differences in stimulus strength. The frequency of action potentials does change with the strength of the stimulus.

Passing a Signal from a Neuron to a Receiving Cell

Signals are transmitted beyond an individual neuron at relay points called synapses. In an electrical synapse, electric current passes directly from one neuron to the next. At chemical synapses, the sending cell secretes a chemical signal, a neurotransmitter, which crosses the synaptic cleft (a gap between the cells) and binds to a specific receptor on the surface of the receiving cell. A cell may receive different signals from many neurons. Many drugs act at synapses by increasing or decreasing the normal effect of neurotransmitters.

The Human Nervous System: A Closer Look

The Central Nervous System and the Peripheral Nervous System

The central nervous system interprets signals coming from peripheral nerves and sends out responses through peripheral nerves. Outgoing responses of the peripheral nervous system lead to both voluntary and involuntary movements in the body.

NERVOUS SYSTEM			
Central Nervous System (CNS)		**Peripheral Nervous System** (PNS)	
Brain	Spinal cord: nerve bundle that communicates with body	Motor system: voluntary control over muscles	Autonomic nervous system: involuntary control over organs ▪ Parasympathetic division: rest and digest ▪ Sympathetic division: fight or flight

The Human Brain

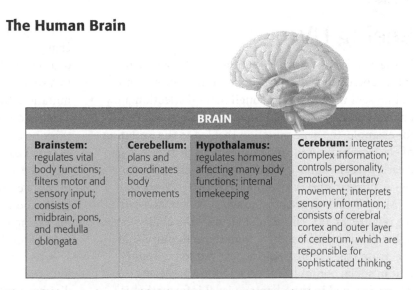

BRAIN			
Brainstem: regulates vital body functions; filters motor and sensory input; consists of midbrain, pons, and medulla oblongata	**Cerebellum:** plans and coordinates body movements	**Hypothalamus:** regulates hormones affecting many body functions; internal timekeeping	**Cerebrum:** integrates complex information; controls personality, emotion, voluntary movement; interprets sensory information; consists of cerebral cortex and outer layer of cerebrum, which are responsible for sophisticated thinking

The right hemisphere and left hemisphere of the cerebral cortex are bridged by the corpus callosum, a thick band of nerves. Both hemispheres of the cerebral cortex have four major lobes: the frontal, parietal, temporal, and occipital. Within each lobe are association areas where sensory information is interpreted and appropriate responses are planned.

The Senses

Sensory Input

Converting a stimulus to an electrical signal:

The strength of the stimulus alters the rate of action potential transmission. Some sensory neurons tend to become less sensitive when stimulated repeatedly, a phenomenon known as sensory adaptation. Pain receptors sense stimuli that may indicate tissue damage; thermoreceptors detect heat or cold; mechanoreceptors respond to mechanical energy (such as touch, pressure, or sound); chemoreceptors respond to chemicals in the external environment or body fluids; and electromagnetic receptors respond to electricity, magnetism, or light (photoreceptors).

Vision

In the human eye, the cornea and lens focus light on the retina. The human lens changes shape to bring objects at different distances into sharp focus. The retina contains two types of photoreceptor cells called rods and cones. Rods contain the visual pigment rhodopsin and function in dim light. Cones contain photopsin, which enables us to see color in full light. Nearsightedness, farsightedness, and astigmatism result from focusing problems in the lens. Corrective lenses bend the light rays to compensate.

Hearing

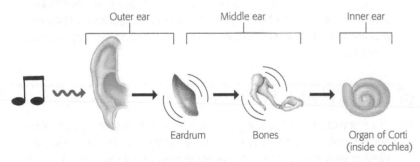

The waves generated in the cochlear fluid move hair cells (mechanoreceptors) of the organ of Corti against an overlying membrane. Bending of the hair cells triggers receptor potentials, sending nerve signals to the brain. Deafness can be present at birth or caused by infections, injury, or overexposure to loud noises.

Locomotor Systems

Movement is one of the most distinctive features of animals. In the process of locomotor output, the nervous system issues commands to the muscular system, and the muscular system exerts force against the skeleton.

The Skeletal System

The skeletal system functions in support, movement, and the protection of internal organs. The human endoskeleton is composed of cartilage and bone. Movable joints provide flexibility. A bone is a living organ containing several kinds of tissues. It is covered with a connective tissue membrane. Cartilage at the ends of the bone cushions the joints. The human skeleton is versatile, but it is also subject to problems, such as arthritis and osteoporosis. Broken bones can be realigned and immobilized; bone cells then build new bone and repair the break.

The Muscular System

Skeletal muscles pull on bones to produce movements. Antagonistic pairs of muscles produce opposite movements. Each skeletal muscle cell, or fiber, contains bundles of myofibrils. Each myofibril contains bundles of overlapping thick (myosin) and thin (actin) protein filaments. Repeating units of thick and thin filaments, called sarcomeres, are the muscle fiber's contractile units. The sliding-filament model explains the molecular process of muscle contraction. Using ATP, the myosin heads of the thick filaments attach to binding sites on the actin molecules and pull the thin filaments toward the center of the sarcomere. Motor neurons carry action potentials that initiate muscle contraction. A neuron can branch to contact a number of muscle fibers; the neuron and the muscle fibers it controls constitute a motor unit. The strength of a muscle contraction depends on the number of motor units activated.

Stimulus and Response: Putting It All Together

An animal's nervous system connects sensations derived from environmental stimuli to responses carried out by its muscles.

MasteringBiology®

For practice quizzes, BioFlix animations, MP3 tutorials, video tutors, and more study tools designed for this textbook, go to MasteringBiology®

SELF-QUIZ

1. The cells that carry electric signals from one part of the body to another in the nervous system are
   a. dendrites.
   b. neurons.
   c. effectors.
   d. synapses.

2. What is the function of the myelin sheath?

3. Your nervous system can be divided into two broad subsystems: the _____ and the _____.

4. A neuron can relay a more intense signal by _____.
   a. sending a stronger action potential
   b. sending more action potentials in a given amount of time
   c. sending an action potential that travels more quickly

5. After lunch at your friend's place on a Sunday, you sit back, relax, and chat with your friend. After a while, you notice that your breathing has become slow, you feel sleepy, and you have problems keeping your eyes open. What is causing this reaction? Which part of the nervous system is involved in this response?

6. A victim of a severe head injury can live for years in a nonresponsive state in which the cerebral cortex is not functioning but the person is still alive and performing metabolic functions. Based on your knowledge of brain structure and function, how can this be possible? Make sure your answer indicates which brain structures might be keeping the person alive.

7. How is an action potential different from a receptor potential?

8. For each of the following senses in humans, identify the type of receptor: seeing, tasting, hearing, smelling.

9. Mr. Johnson is becoming slightly deaf. To test his hearing, his doctor holds a vibrating tuning fork tightly against the back of Mr. Johnson's skull. This sends vibrations through the bones of the skull, setting the fluid in the cochlea in motion. Mr. Johnson can hear the tuning fork this way, but not when it is held away from his skull a few inches from his ear. Where is Mr. Johnson's hearing problem located? (Explain your answer.)
    a. in the auditory nerve leading to the brain
    b. in the hair cells in the cochlea
    c. in the bones of the middle ear
    d. in the fluid of the cochlea

10. Where we have ball-and-socket joints (shoulders and hips), horses have hinge joints. How does this affect the movements they can perform?

11. One of the early symptoms of amyotrophic lateral sclerosis (ALS), often referred to as Lou Gehrig's disease, is skeletal muscle weakness. As the disease progresses, a patient with ALS loses voluntary movement. ALS is not a muscular disease; it is a degenerative neurological disorder in which motor neurons that cause skeletal muscles to contract are lost. Explain what is happening to a person with ALS who can still move his or her muscles but whose muscle contraction is weak.

Answers to these questions can be found in Appendix: Self-Quiz Answers.

THE PROCESS OF SCIENCE

12. Studies of families with patients suffering from neuropsychiatric disorders, such as autism, depression, dyslexia, and schizophrenia, have revealed that the risk of developing one of these diseases is significantly high in the siblings of the patients (for example, an estimated 30% compared to the 1.5% average risk across the general population for autism spectrum disorders). What does this signify? Propose a hypothesis to explain the cause of these diseases.

13. **Interpreting Data** Marine mammals, such as seals, whales, and dolphins, have adapted to their watery environment in many ways, including a respiratory system that allows for long underwater dives and a language used to communicate through water. Another adaption studied by researchers involves bottlenose dolphins. These dolphins spend approximately 33% of a 24-hour day with one eye open and the other closed, alternating which eye is open and which is closed at intervals of approximately 1 hour. During this time, the animals continuously swim and intermittently rise to the surface of the water to breathe air. Look below at data collected on brain activity in bottlenose dolphins. Why do the dolphins swim with one eye open and the other eye closed—that is, what are they doing? How does their brain activity relate to what is happening with their eyes? Why might this be a successful adaptation?

Location	Brain activity at first measurement	Brain activity at second measurement (after 1 hour)
Left hemisphere		
Right hemisphere		

Key

Low-frequency waves characteristic of sleep

High-frequency waves characteristic of wakefulness

14. Sensory organs tend to come in pairs. We have two eyes and two ears. Similarly, a planarian worm has two eyecups, a rattlesnake has two infrared receptors, and a butterfly has two antennae. Propose a testable hypothesis that could explain the advantage of having two eyes or ears instead of one.

BIOLOGY AND SOCIETY

15. In a foreword to Dale Carnegie's book *How to Win Friends and Influence People*, the writer Lowell Thomas falsely cites the psychologist William James, saying that "the average man develops only 10% of his latent mental ability". Since then the idea that most humans only use 10% of their brains has been widely perpetuated. What do you think about this claim?

16. Victoria Falls in Zimbabwe is a big tourist attraction, and many tourists like to fly over the falls by helicopter. As many as 20 helicopters may fly over the falls and surrounding areas at the same time. The low rumbling sounds of the helicopters can cause elephant stampedes. On behalf of an environmental conservation group, construct a poster to educate tourists as they enter the region. Explain why helicopter noise pollution is particularly detrimental to elephants and how a behavior change in elephants could affect other animals. Suggest other (more elephant-friendly) ways to enjoy the falls.

Unit 6
Plant Structure and Function

The Life of a Flowering Plant

Why Plant Structures Matter

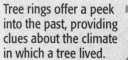

Tree rings offer a peek into the past, providing clues about the climate in which a tree lived.

◄ Blame your seasonal allergies on angiosperm sex.

▲ All seedless oranges are the product of asexual reproduction.

▲ Angiosperms feed the world: Just ten species account for 90% of humanity's calories.

Agriculture BIOLOGY AND SOCIETY

The Buzz on Coffee Plants

From the dawn of civilization to the frontiers of genetic engineering, human progress has always depended on expanding our use of plants—for food, fuel, clothing, and countless other trappings of modern life.

To illustrate, think about a particular crop, one that may help get you through a long day of classes: coffee. Today, coffee is one of the most important agricultural products in the world. The average American (man, woman, and child) drinks an eye-opening 28 gallons of coffee each year!

Ground coffee is made from the seeds of plants belonging to several species in the genus *Coffea*. Historians believe that coffee was discovered in Ethiopia during the 1200s. Locals may have noticed that the plants provided an energizing effect on animals that ate its fruit. African farmers eventually learned to roast, grind, and filter the coffee beans in water. From Africa, coffee cultivation and consumption spread to the Middle East. By the 1600s, coffee houses had appeared throughout Europe. And by the 1700s, coffee plants imported to the Americas were an important crop. Today, coffee is grown in at least 50 tropical countries around the world. A small but growing percentage of commercial coffee is "Fair Trade Certified." To receive such a designation, farmers agree to provide humane working conditions for their employees and to abide by agricultural practices that promote a healthy and sustainable environment. In exchange, the farmers are guaranteed a fair price for their crop.

Coffee, an important agricultural crop. People have been cultivating, drying, roasting, and drinking coffee for hundreds of years.

Clearly, we have benefited from our agricultural relationship with coffee. But ask yourself this: What is the basis of coffee's popularity? No doubt we love the jolt we get from the caffeine, which is produced in coffee seeds and stimulates the nervous system. Caffeine acts as a toxin to many herbivores (plant-eaters) and therefore serves as a form of self-defense for the plant. Your morning buzz may therefore be a by-product of an evolutionary adaptation that helps protect the coffee plant from being eaten by predators.

Plants are vital to the well-being of not just humans but the entire biosphere. Above and below the ground, plants provide shelter, food, and breeding areas for animals, fungi, and microorganisms. Because angiosperms—the flowering plants—make up more than 90% of the plant kingdom, we focus primarily on them in this unit. (If you'd like to learn about other groups of plants—such as mosses, ferns, and cone-bearing gymnosperms—see Chapter 16.) In this chapter, we begin by examining angiosperm structure, first at the level of the whole plant and then at the microscopic level of tissues and cells. Then we'll see how plant structures function in growth and reproduction.

The Structure and Function of a Flowering Plant

Coffea arabica

Angiosperms have dominated the land for more than 100 million years, and about 250,000 species of flowering plants exist today. Most of our foods come from 100 or so species of domesticated flowering plants. Among these foods are roots, such as beets and carrots; the fruits of trees and vines, such as apples, nuts, berries, and squashes; the fruits and seeds of legumes, such as peas, peanuts, and beans; and grains, the fruits of grasses such as wheat, rice, and corn. (Notice that the term *coffee bean* is not strictly correct, because the coffee plant is not a bean-producing legume; it is *coffee seeds* that we roast and consume.)

Angiosperms feed the world: Just ten species account for 90% of humanity's calories.

☑ CHECKPOINT

How do the terms *monocot* and *eudicot* relate to an anatomical difference between these two groups of plants?

Answer: A monocot embryo has a single (mono- means "one") cotyledon (seed leaf), and a eudicot embryo has two (di- means "two").

Monocots and Eudicots

On the basis of several structural differences, botanists (biologists who study plants) classify most angiosperms into two groups: monocots and eudicots **(Table 28.1)**. The names of the groups refer to embryonic structures called **cotyledons**, or seed leaves, the first leaves to emerge from a growing seedling. A **monocot** embryo has one seed leaf; a **eudicot** embryo has two seed leaves.

Monocots include orchids, palms, and lilies, as well as grains and other grasses. Most monocots have leaves with parallel veins. Monocot stems have vascular tissues (internal tissues that transport water and nutrients) organized into bundles that are arranged in a scattered pattern. The flowers of most monocots have petals and other parts in multiples of three. Monocot roots form a fibrous system—a mat of threads—that spreads out just below the soil surface. With most of their roots in the top few centimeters of soil, monocots, especially grasses, make excellent ground covers and can help reduce soil erosion.

Most flowering plants are eudicots, including many food crops (such as nearly all of our fruits and vegetables), the majority of ornamental plants, and most shrubs and trees (except for the gymnosperms). Coffee, for example, grows as a eudicot shrub. Eudicot leaves have a multibranched network of veins, and eudicot stems have vascular bundles arranged in a ring. Eudicot flowers usually have petals and other floral parts in multiples of four or five. The large, main root of a eudicot, known as a taproot, goes deep into the soil, as you know if you've ever tried to pull up a dandelion. ☑

Table 28.1	Comparing Monocots and Eudicots				
	Seed Leaves	**Leaf Veins**	**Stems**	**Flowers**	**Roots**
Monocots (Examples: grasses, grains, orchids, palms)	One cotyledon	Veins usually parallel	Vascular bundles in scattered arrangement	Floral parts usually in multiples of three	Fibrous root system
Eudicots (Examples: nearly all fleshy fruits, vegetables, shrubs, and broad-leaved trees)	Two cotyledons	Veins usually branched	Vascular bundles arranged in ring	Floral parts usually in multiples of four or five	Taproot usually present

Structure/Function

Roots, Stems, and Leaves

As we see throughout our study of biology, a close look at a structure often reveals its function. Conversely, function provides insight into the "logic" of a structure. In the sections that follow, we'll take a detailed look at the correlation between plant structure and function.

Plants, like most animals, have organs made up of different tissues, and these tissues in turn are made up of one or more types of cells. An **organ** consists of several types of tissues that together carry out a particular function. In looking at plant structure, we'll begin with organs—roots, stems, and leaves—because they are the most familiar plant structures.

Plants (along with fungi) first occupied land around 470 million years ago (see Chapter 16). Among the evolutionary adaptations that made it possible for plants to thrive on land were structures for absorbing water and minerals from the soil, a large light-collecting surface, the ability to take in carbon dioxide from the air for photosynthesis, and adaptations for surviving dry conditions. In a land plant, the roots (the belowground structures) and the shoots (the aboveground structures, including stems, leaves, and flowers) perform all these vital functions (**Figure 28.1**). Neither roots nor shoots can survive without the other. Lacking chloroplasts and living in the dark, roots would starve without sugar and other organic nutrients transported from photosynthetic stems and leaves. Conversely, stems and leaves depend on the water and minerals absorbed by roots.

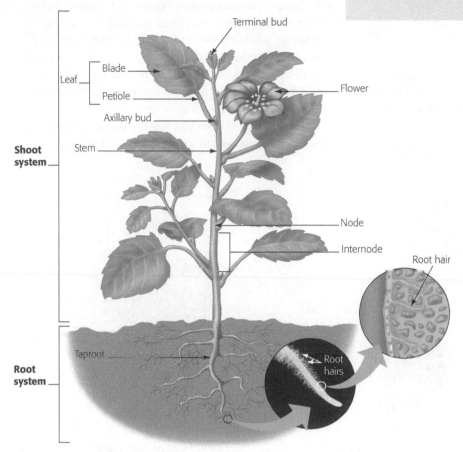

▲ Figure 28.1 **The structure of a flowering plant.**

Roots

The **root** is an organ that anchors a plant in the soil, absorbs and transports minerals and water, and stores food. All of a plant's roots make up its **root system**. The fibrous root system of a monocot provides broad exposure to soil water and minerals as well as firm anchorage. In contrast, the root system of a eudicot includes a vertical taproot with many small secondary roots growing outward. Near root tips are **root hairs**, tiny projections that greatly increase the surface area, providing an extensive outer layer for absorption (Figure 28.1, far right). Each root hair is an outgrowth of a cell on the surface of the root. It is difficult to move an established plant without injury because transplantation often damages the plant's delicate root hairs.

Large taproots, such as those found in carrots, turnips, and sugar beets, store food as starch or sucrose (**Figure 28.2a**). The plants use these carbohydrates during periods of growth and when they are producing flowers and fruit. Other types of modified roots include buttress roots that extend above the ground and look like the supporting beams of a building (**Figure 28.2b**).

▼ Figure 28.2 **Modified roots.**

(a) **The root of a sugar beet stores carbohydrates.**

(b) **The buttress roots of this *Canarium* tree in Madagascar strengthten its trunk.**

Stems

The **shoot system** of a plant is made up of stems, leaves, and flowers. As indicated in Figure 28.1, **stems** generally grow above the ground and support the flowers and leaves, allowing them to maximize their exposure to sunlight. In the case of a tree, the stems are the trunk and all the branches. A stem has **nodes**, the points at which leaves are attached, and **internodes**, the portions of the stem between nodes.

The two types of buds you saw in Figure 28.1 are undeveloped shoots. When a stem is growing in length,

the **terminal bud**, at the tip of the stem, has developing leaves and a compact series of nodes and internodes. The **axillary buds**, one in each of the crooks formed by a leaf and the stem, are usually dormant. In many plants, the terminal bud produces hormones that inhibit the growth of the axillary buds, a phenomenon called **apical dominance**. By concentrating the plant's resources on growing taller, apical dominance is an evolutionary adaptation that increases the plant's exposure to light. Branching is also important for increasing the exposure of the shoot system to the environment, and under certain conditions, the axillary buds begin growing and developing into branches. As any home gardener knows, removing the terminal bud by pruning a fruit tree or "pinching back" a houseplant will make the plant bushier (**Figure 28.3**).

Stems can take many forms. Strawberry plants have horizontal stems, or runners, that grow along the surface of the ground. A runner is a means of asexual reproduction; as shown in **Figure 28.4a**, a new plant can emerge from it. This is why strawberries, if left unchecked, can rapidly fill your garden. Stems that we commonly eat include asparagus. Other kinds of stems are often mistaken for roots. If you dig up an iris or ginger plant, you'll see large, brownish, rootlike structures near the soil surface; these are actually horizontal underground stems called **rhizomes** (**Figure 28.4b**). Rhizomes store food; because they have buds, they can also form new plants. About every three years, gardeners can dig up iris rhizomes, split them, and plant the partial rhizomes to get multiple identical plants. A potato plant has rhizomes ending in enlarged structures called **tubers** (the potatoes we eat), where food is stored in the form of starch (**Figure 28.4c**). The "eyes" of a potato are axillary buds, which can grow into new plants, allowing potatoes to be easily propagated.

▼ **Figure 28.3 Apical dominance and the effect of pruning on a rosemary plant.**

Terminal bud

The terminal bud of this rosemary plant produces hormones that inhibit growth of the axillary buds, an effect called apical dominance. The result is a tall, thin plant.

Terminal bud removed

This rosemary plant has been pruned, which decreases apical dominance and increases branching and leaf production. The result is an increase in the useful yield of leaves.

▼ **Figure 28.4 Modified stems.**

Runner

Rhizome

Root

Taproot

Rhizome

Tuber at end of rhizome

(a) Strawberry plants

(b) Ginger plant

(c) Potato plant

Leaves

Leaves are the primary sites of photosynthesis in most plants. Leaves also exchange gases with the atmosphere, dissipate heat, and provide defenses against herbivores and pathogens. A typical leaf consists of a flattened **blade** and a stalk, or **petiole**, which joins the leaf to the stem (see Figure 28.1).

Plant leaves are highly varied. **Figure 28.5** displays one example of each leaf type, but keep in mind that leaves come in a great variety of sizes and shapes. Grasses and most other mono-cots have long leaves without petioles. Some eudicots have enormous petioles that contain a lot of water and stored food, such as the edible stalks of celery. A modified leaf called a tendril can help plants such as sweet peas or grapes climb up their supports (**Figure 28.6a**). And the spines of the cactus (**Figure 28.6b**) are modified leaf parts that may protect the plant from predatory animals. In many cactus species, the main photosynthetic organ is the large green stem, which also stores water.

So far, we have examined plants as we see them with the unaided eye. In the next section, we begin to dissect a plant and explore its microscopic organization. ☑

LEAF ARRANGEMENT		
Simple	**Compound**	**Doubly Compound**
Petiole / Axillary bud	Petiole / Axillary bud	Leaflet / Petiole / Axillary bud
A single individual blade	One blade consisting of many leaflets (which themselves lack axillary buds)	Each leaflet divided into smaller leaflets

▲ **Figure 28.5 Arrangement of leaves.** You can distinguish simple leaves from compound leaves by looking for axillary buds: Each leaf has only one axillary bud, where the petiole attaches to the stem; leaflets of compound leaves do not have axillary buds.

☑ CHECKPOINT

Name the organ you are eating when you eat each of the following foods: asparagus, lettuce, carrot, potato, turnip.

Answer: stem, leaf, root, stem, root

▼ **Figure 28.6 Modified leaves.**

(a) Tendrils of a gourd plant

Tendril

(b) Spines of a cactus

Plant Tissues and Tissue Systems

As in animals, the organs of plants contain tissues. Each **tissue** is a group of cells that together perform a specialized function. For example, plants have two types of vascular tissue that form an internal plumbing system within the plant body: **Xylem** tissue conveys water and dissolved minerals upward from the roots to the stems and leaves, while **phloem** tissue transports sugars from leaves or storage tissues to other parts of the plant.

A **tissue system** consists of one or more tissues organized into a functional unit within a plant. Each plant organ—root, stem, or leaf—is made up of three tissue systems: the dermal, vascular, and ground tissue systems. Each tissue system is continuous throughout the entire plant body, but the systems are arranged differently in leaves, stems, and roots **(Figure 28.7)**.

The **dermal tissue system** (brown in Figure 28.7) forms an outer protective covering. Like our skin, it forms a first line of defense against physical damage and infectious organisms. On leaves and on most stems, dermal cells secrete a waxy coating called the **cuticle**, which helps prevent water loss—a key adaptation that allowed plants to move onto land. The **vascular tissue system** (purple) provides support and long-distance transport throughout the plant; xylem and phloem are part of this system. The **ground tissue system** (yellow) accounts for most of the bulk of a plant; it has diverse functions, including photosynthesis, storage, and support.

Figure 28.8, a cross section of a eudicot root, illustrates the three tissue systems. The epidermis (the outer layer of the dermal tissue system) is a single layer of tightly packed cells covering the entire root. Water and minerals from the soil enter the plant through these cells. Some of the young epidermal cells grow outward and form root hairs (see Figure 28.1). In the center of the root, the vascular tissue system forms a cylinder, with xylem cells radiating from the center like the spokes of a wheel and phloem cells filling in the wedges between the spokes. The ground tissue system of the root forms the **cortex**, where cells store food and take up water and minerals. The innermost layer of cortex is the **endodermis**, a thin cylinder one cell thick. The endodermis is a selective barrier that regulates the passage of substances between the cortex and the vascular tissue.

All plant stems have vascular tissue systems arranged in numerous vascular bundles. The location and arrangement of these bundles differ between monocots and eudicots (see Figure 28.7). In eudicots, another part of the ground tissue system, known as the **pith**, fills the center of the stem and often stores food.

▶ **Figure 28.7 The three tissue systems of a plant.** The roots, stems, and leaves of a plant are made from three tissue systems (the dermal tissue system, vascular tissue system, and ground tissue system) that are continuous throughout the plant body.

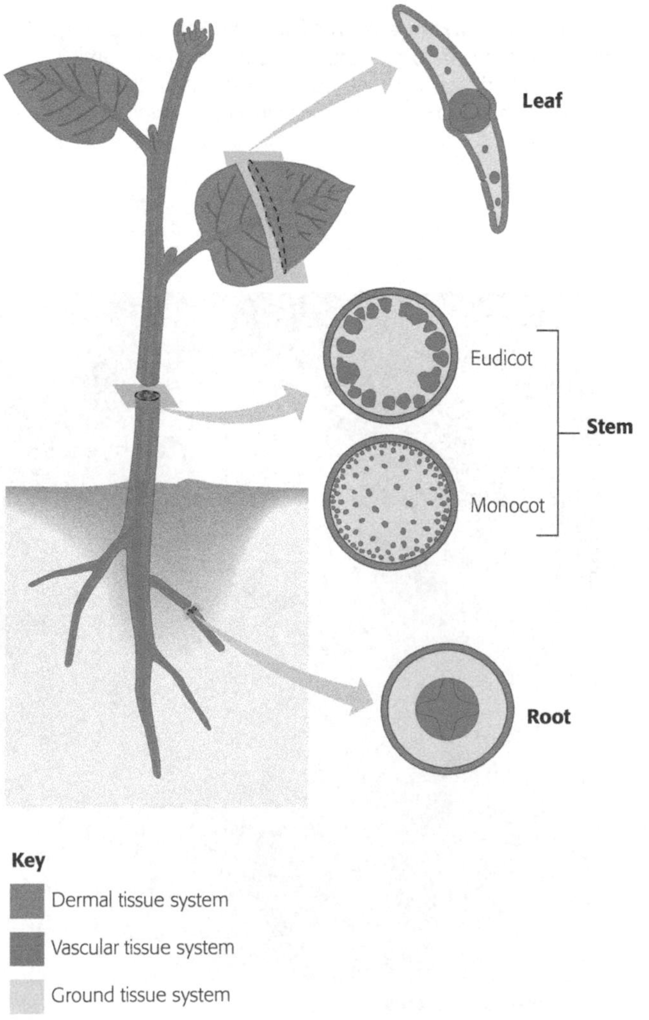

Leaf

Eudicot

Stem

Monocot

Root

Key

Dermal tissue system

Vascular tissue system

Ground tissue system

▼ **Figure 28.8 Tissues in a eudicot root.** This colored micrograph is a cross section of a young buttercup root. You can see the single-layered epidermis, the cylindrical vascular tissue with xylem and phloem, and the ground tissue that makes up the bulk of the root.

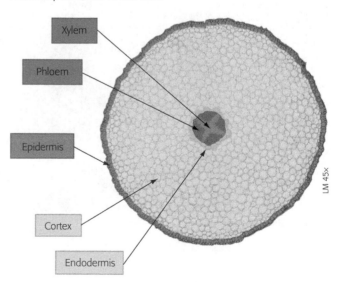

Xylem

Phloem

Epidermis

Cortex

Endodermis

LM 45×

Figure 28.9 illustrates the arrangement of the three tissue systems found in a typical eudicot leaf. The epidermis contains **stomata** (singular, *stoma*), which are tiny pores between two specialized cells called guard cells. The **guard cells** regulate the opening and closing of stomata (see Figure 29.10), allowing gas exchange between the surrounding air and photosynthetic cells inside the leaf.

The main site of photosynthesis is the **mesophyll**, which is the ground tissue of a leaf. Mesophyll consists mainly of photosynthetic cells containing chloroplasts (green in Figure 28.9).

The leaf's vascular tissue is made up of a network of veins. As you can see in Figure 28.9, each vein is a vascular bundle composed of xylem and phloem. The veins are in close contact with the leaf's photosynthetic tissues. This ensures that those tissues are supplied with water and minerals from the soil and that sugars made in the leaves are transported throughout the plant. ☑

☑ **CHECKPOINT**

Which tissue system is the main site of photosynthesis? Which tissue system covers and protects the outside of the plant?

Answer: ground; dermal

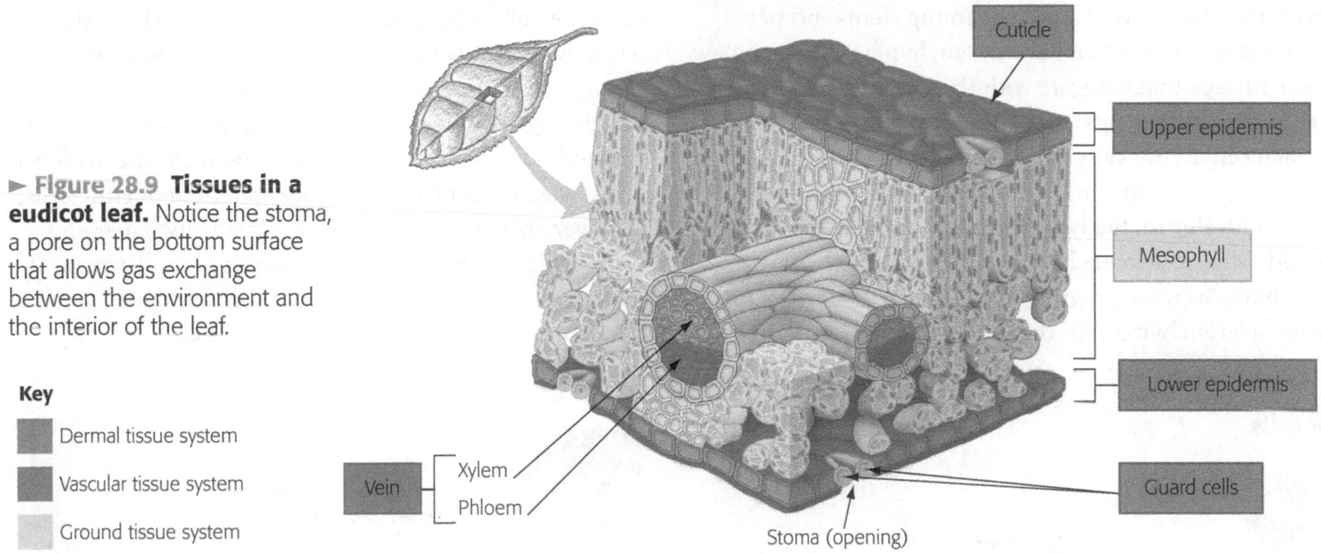

▶ **Figure 28.9 Tissues in a eudicot leaf.** Notice the stoma, a pore on the bottom surface that allows gas exchange between the environment and the interior of the leaf.

Cuticle
Upper epidermis
Mesophyll
Lower epidermis
Guard cells
Vein — Xylem / Phloem
Stoma (opening)

Key
- Dermal tissue system
- Vascular tissue system
- Ground tissue system

Plant Cells

So far, we have examined plant structure at the level of organs and tissues. Now we'll zoom in even closer to examine cells.

In addition to features shared with other eukaryotic cells (see Figure 4.3), most plant cells have three unique structures, as shown in **Figure 28.10**: **chloroplasts**, the sites of photosynthesis; a large **central vacuole** containing fluid that helps maintain the cell's firmness (also referred to as turgor); and a protective **cell wall** that surrounds the plasma membrane.

Plant cell walls consist largely of the structural carbohydrate cellulose. Some plant cells, especially those that provide structural support, have a two-layered cell wall; a primary cell wall grows first, and then a thicker, more rigid secondary cell wall is deposited between the plasma membrane and the primary wall. Plasmodesmata are open channels in adjacent cell walls through which cytoplasm and various substances can flow from cell to cell.

▼ **Figure 28.10 The structure of a plant cell.** The colored boxes highlight the unique features of plant cells. The enlargement shows two adjoining cells.

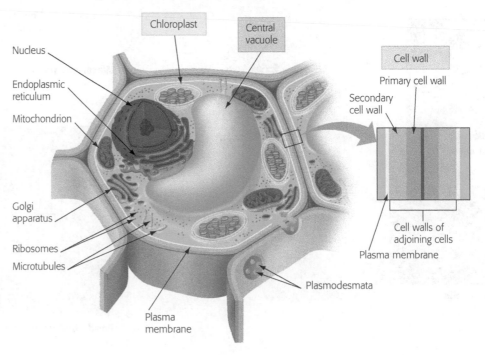

Chloroplast
Central vacuole
Nucleus
Endoplasmic reticulum
Mitochondrion
Golgi apparatus
Ribosomes
Microtubules
Plasma membrane
Cell wall
Primary cell wall
Secondary cell wall
Cell walls of adjoining cells
Plasma membrane
Plasmodesmata

Figure 28.11 presents an overview of plant cell types. **Parenchyma cells** are the most abundant type of cell in most plants. They have only primary cell walls, which are often thin. Parenchyma cells perform a variety of functions, including food storage and photosynthesis. Most parenchyma cells can divide and develop into other types of plant cells, which they may do during repair of an injury.

Collenchyma cells resemble parenchyma cells in lacking secondary walls, but they have unevenly thickened primary walls. These cells provide support in parts of the plant that are actively growing. Young stems and petioles, for example, often have collenchyma cells just below their surface that elongate with the growing stem (these cells form the "strings" of a celery stalk, for example).

Sclerenchyma cells have thick secondary cell walls (colored pale yellow in Figure 28.11) usually strengthened with **lignin**, the main chemical component of wood. Mature sclerenchyma cells develop only in regions that have stopped growing in length. After they mature, most sclerenchyma cells die, leaving behind a "skeleton"

that supports the plant, much as steel beams support a building. Sclerenchyma cells make up some commercially important plant products, such as the highly versatile hemp fiber used to make rope and clothing.

The xylem tissue of angiosperms contains two types of water-conducting cells: tracheids and vessel elements. Both have rigid, lignin-containing secondary cell walls.

Tracheids are long, thin cells with tapered ends, while **vessel elements** are wider, shorter, and less tapered. Water-conducting cells are arranged in chains with overlapping ends that form a system of water-carrying tubes. The tubes are hollow because water-conducting cells are dead when mature; only their cell walls remain. Water passes through pits in the walls of tracheids and vessel elements and through openings in the end walls of vessel elements.

Food-conducting cells within phloem are also arranged end to end, forming tubes. Unlike water-conducting cells, however, these cells remain alive at maturity. Their end walls, which are perforated with large plasmodesmata, allow sugars, other compounds, and some minerals to move between adjacent food-conducting cells. ☑

☑ CHECKPOINT

Which of the cell types shown in Figure 28.11 has the potential to give rise to all the others?

Answer: parenchyma

▼ **Figure 28.11 Types of plant cells.**

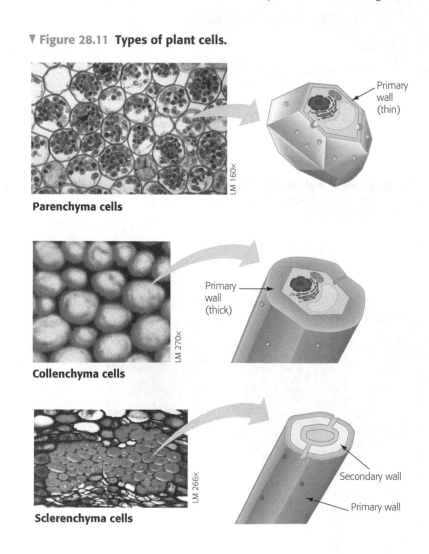

Parenchyma cells

Primary wall (thin)

Collenchyma cells

Primary wall (thick)

Sclerenchyma cells

Secondary wall

Primary wall

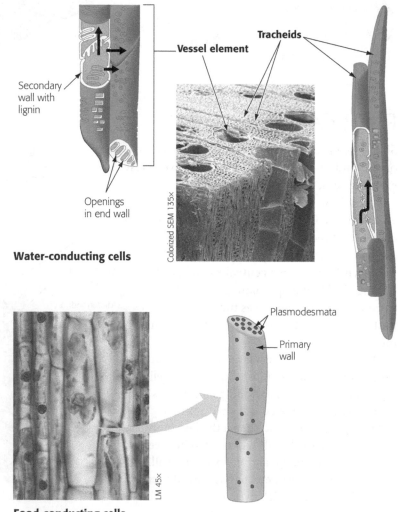

Tracheids

Vessel element

Secondary wall with lignin

Openings in end wall

Water-conducting cells

Plasmodesmata

Primary wall

Food-conducting cells

Plant Growth

The growth of a plant differs from that of an animal in a fundamental way. Most animals cease growing after reaching a certain size. Most plants, in contrast, continue to grow for as long as they live. But this does not mean that plants are immortal. In fact, different types of plants have very different life spans (Figure 28.12). Plants called **annuals** emerge from seed, mature, reproduce, and die in a single year or growing season. Our most important food crops—wheat, corn, and rice, for example—are annuals. **Biennials** live for two years; flowering and seed production usually occur during the second year. Carrots are biennials, but we usually harvest them in their first year and so miss seeing their flowers. Plants that live and reproduce for many years, including trees, shrubs (such as the coffee plant), many herbaceous plants and flowering bulbs (such as tulips), and some grasses, are known as **perennials**. Some perennials have life spans that extend well beyond the longest-lived animals; for example, a fig tree planted in Sri Lanka in the year 288 B.C. is still growing today, over 2,300 years later.

Primary Growth: Lengthening

Growth in all plants is made possible by tissues called meristems. A **meristem** consists of undifferentiated (unspecialized) cells that divide when conditions permit, generating new cells and tissues. Meristems at the tips of roots and in the buds of shoots are called **apical meristems**. Cell division in apical meristems produces new cells that enable a plant to grow in length, a process called **primary growth** (Figure 28.13).

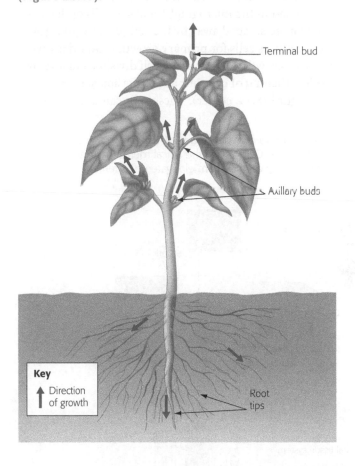

◄ **Figure 28.13 Primary growth at apical meristems.** Meristems are tissues responsible for growth in all plants. The apical meristems are located at the root tips and in the buds of shoots.

Key
↑ Direction of growth

Terminal bud

Axillary buds

Root tips

▼ Figure 28.12 **Plants with varying life spans.**

PLANT LIFE SPANS		
Annuals (live for just one growing season)	**Biennials** (live for two growing seasons)	**Perennials** (live for many growing seasons)
Rice	Wild chervil (also called cow parsley)	Rose

☑ **CHECKPOINT**

What two basic cellular
mechanisms account for
primary growth?

Answer: cell division and cell
lengthening

Figure 28.14 shows a slice through a growing onion root. Primary growth enables roots to push through the soil. (A very similar process results in the upward growth of shoots.) At the tip of the root is the **root cap**, a thimble-like cone of cells that protects the delicate, actively dividing cells of the apical meristem. The root's apical meristem (marked with the orange oval in the art and circled in the micrograph) replaces cells of the root cap that are scraped away by the soil (downward arrow) and produces cells for primary growth (upward arrow). Primary growth is achieved by cell division and also by the lengthening of cells just above the apical meristem (see Figure 28.14, center). These cells can undergo a

tenfold increase in length, mainly by taking up water. The elongation of these cells is what forces a root down through the soil. The elongating cells begin to differentiate, forming the epidermis, cortex, and vascular tissue (see Figure 28.14, top). Cells of this last type eventually differentiate into vascular tissues called primary xylem and primary phloem. ☑

Secondary Growth: Thickening

In addition to lengthwise primary growth, the stems and roots of many eudicot species (but few monocots) also thicken by a process called **secondary growth** (Figure 28.15). Such thickening is most evident in the woody plants—trees, shrubs, and vines—whose stems last from year to year and consist mainly of thick layers of mature, mostly dead xylem tissue, called **wood**.

Secondary growth involves cell division in two meristems we have not yet discussed: the vascular cambium and the cork cambium. The **vascular cambium** (blue-green in Figure 28.16) is a cylinder of actively dividing cells between the primary xylem and primary phloem, as you can see in the pie-shaped section at the left of the figure. Secondary growth (red arrows) adds cells on either side of the vascular cambium.

The middle and right drawings in Figure 28.16 show the results of secondary growth. In the middle drawing, the vascular cambium has given rise to two new tissues: secondary phloem to its exterior and secondary xylem to its interior. Yearly production of a new layer of secondary xylem accounts for most of the growth in thickness of a perennial plant.

▼ Figure 28.14 **Close-up of primary growth in a root tip.** Primary growth is achieved through two mechanisms. Cells within the apical meristem actively divide. In addition, cells behind the apical meristem elongate up to tenfold.

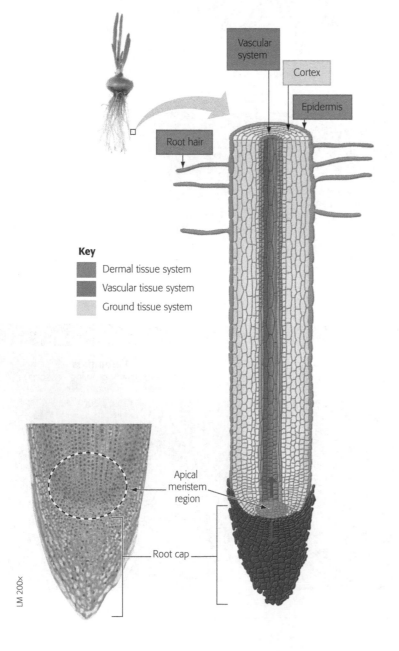

Vascular system

Cortex

Epidermis

Root hair

Key

■ Dermal tissue system

■ Vascular tissue system

■ Ground tissue system

Apical meristem region

Root cap

LM 200x

▼ Figure 28.15 **Secondary growth.** Trees (such as these live oaks) thicken from year to year due to secondary growth.

▼ Figure 28.16 **Secondary growth of a woody stem.** The branches of most woody plants are made up of tissues of varying ages, with the youngest regions near the tip and the older regions nearer the trunk. The cross section at the left shows a region of the stem that is just beginning secondary growth. The middle and right cross sections show progressively older regions.

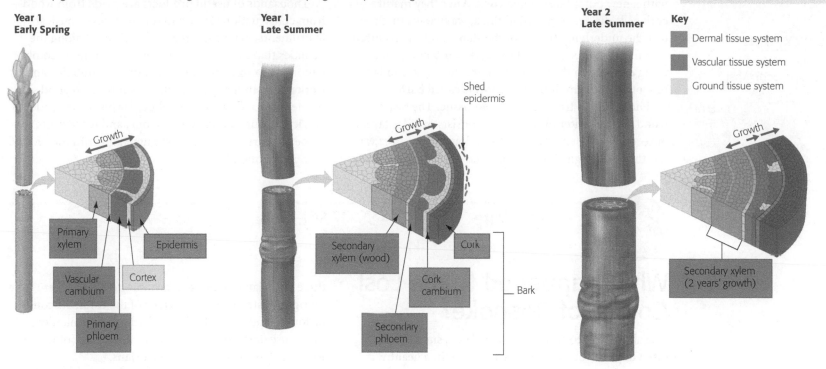

Annual growth rings result from the layering of secondary xylem (Figure 28.17). The layers are visible as rings because of uneven activity of the vascular cambium during the year. In woody plants that live in temperate regions, such as most of the United States, the vascular cambium becomes dormant each year during winter, and secondary growth is interrupted. When secondary growth resumes in the spring, a cylinder of early wood forms. Made up of the first new xylem cells to develop, early wood cells are usually larger in diameter and have thinner walls than those produced later in summer. The boundary between the large cells of early wood and the smaller cells of the late wood produced during the previous growing season is usually a distinct ring visible in cross sections of tree trunks and roots. As you've probably heard, a tree's age can be estimated by counting

> Tree rings offer a peek into the past, providing clues about the climate in which a tree lived.

its annual rings. The rings may have varying thicknesses, reflecting the amount of seasonal growth in different years and therefore climate conditions. In fact, the pattern of growth rings in older trees is one source of evidence for recent global climate change. As we'll see in the Process of Science section, tree ring data can even provide insight into subjects outside the scope of science.

▼ Figure 28.17 **Anatomy of a tree trunk.** On the left, you see a cross section of a locust trunk, with several decades of growth rings visible. On the right, the various layers of a mature trunk are separated for easier viewing.

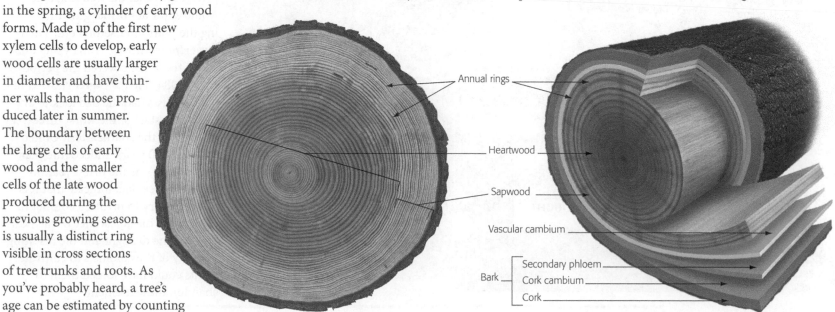

☑ **CHECKPOINT**

What type of plant tissue makes up wood? What is bark?

Answer: secondary xylem; all tissues exterior to the vascular cambium: secondary phloem, cork cambium, and cork.

The epidermis and cortex make up the young stem's external covering (see Figure 28.16). When secondary growth begins, the epidermis is shed and is replaced with a new outer layer called **cork**. After they mature, cork cells die, leaving behind thick, waxy walls that protect the underlying tissues of the stem. Cork is produced by a meristem tissue called the cork cambium. Everything external to the vascular cambium (the secondary phloem, cork cambium, and cork) is called **bark**.

The bulk of a tree trunk is dead tissue. The heartwood, in the center of the trunk, consists of older layers of secondary xylem. These cells no longer transport water; they are clogged with resins and other compounds that make the heartwood very dense and resistant to rotting. The lighter-colored sapwood consists of younger secondary xylem that does conduct water.

Thousands of useful products are made from wood—from construction lumber to fine furniture, musical instruments, paper, and many chemicals. Among the qualities that make wood so useful are a unique combination of strength, hardness, lightness, durability, and workability. In many cases, there is simply no good substitute for wood. A wooden oboe, for instance, produces far richer sounds than a plastic one, and fence posts made of heartwood often last much longer in the ground than metal ones. ☑

Agriculture THE PROCESS OF SCIENCE

What Happened to the Lost Colony of Roanoke?

The ability of people to settle new lands successfully often depends on their ability to maintain a healthy system of agriculture. On July 22, 1587, a group of 121 English settlers landed on Roanoke Island, in present-day North Carolina (**Figure 28.18**). Their supply ship returned to England later that year, leaving 115 colonists behind. When the next ship returned three years later, the settlement was deserted. There were no signs of struggle, yet no trace of the colonists could be found, and all the buildings had been dismantled. The cause of the disappearance of the "Lost Colony" of Roanoke has long intrigued historians.

In 1998, a group of biologists made the **observation** that a large number of very old bald cypress trees (*Taxodium distichum*) around Roanoke Island could provide a reliable record of the climate over the past 800 years. This led them to **question** whether the tree ring data could provide insight into the colony's mysterious disappearance. Their **hypothesis** was that the loss of the colony corresponded to a period of drought, leading to the **prediction** that the tree ring data would show abnormal growth during the years after settlement. Their **experiment** involved analyzing dozens of trees that covered the period 1185–1984. Their **results** showed that the colonists had the bad luck to arrive at the start of the worst three-year drought in the southeastern United States in 800 years. Indeed, the year they settled was the driest year on record. These data indicate that it is not necessary to invoke a conflict to explain the disappearance of the Lost Colony; the drought and the resulting inability of the colonists to cultivate plants could have forced them to abandon the settlement.

► **Figure 28.18 The Lost Colony of Roanoke.** Notice the very narrow growth rings corresponding to the years 1587–1590.

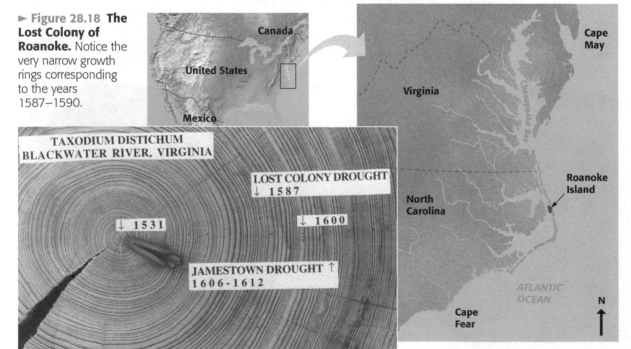

The Life Cycle of a Flowering Plant

The life cycle of an organism is the sequence of stages from the adults of one generation to the adults of the next. In completing their life cycle, many flowering plants can reproduce both asexually and sexually, and both modes have played an important role in the evolutionary adaptation of plant populations to their environments. **Figure 28.19** shows three examples of **asexual reproduction**, the creation of offspring derived from a single parent without fertilization. Through asexual reproduction, a single plant can produce many offspring quickly and efficiently. This ability of plants is useful in agriculture because it allows farmers to grow a large crop of identical plants from a parent with desirable traits. For example, because navel oranges lack seeds, they can only be propagated asexually; every navel orange tree growing today is a clone of a single tree discovered in Brazil around the year 1820. As in animals, **sexual reproduction** in plants involves **fertilization**, the union of gametes from two parents to produce genetically distinct offspring. This section discusses the sexual life cycle of flowering plants, starting with a look at the flower.

▼ Figure 28.19 **Asexual reproduction in plants.**

Garlic. This garlic bulb is actually an underground stem that functions in storage. A single large bulb fragments into several cloves. Each clove can give rise to a separate plant.

Holly trees. Each of the small trees is a sprout from the roots of a single holly tree. Eventually, one or more of these root sprouts may take the place of its parent.

Creosote bushes. This ring of plants is a clone of creosote bushes in the Mojave Desert in southern California. All these bushes came from generations of asexual reproduction by roots. The original plant probably occupied the center of the ring.

> All seedless oranges are the product of asexual reproduction.

neck (the style) with a sticky stigma at its tip. The **stigma** is the landing platform for pollen grains, acting like flypaper to which pollen sticks. The base of the carpel is the **ovary**. Within the ovary are reproductive structures called **ovules**, each containing one developing egg and the cells that support it. The term *pistil* is sometimes used to refer to a single carpel or a group of fused carpels. ☑

The Flower

In angiosperms, the structure specific to sexual reproduction is a shoot called the **flower**. The main parts of a flower—the sepals, petals, stamens, and carpels—are modified leaves (**Figure 28.20**; also see Figure 28.1). The **sepals** enclose and protect the flower bud. The **petals** are often colorful and fragrant, which may serve to advertise the flower to insects and other pollinators.

The flower's reproductive organs are the stamens and carpel. A **stamen** consists of a stalk (filament) tipped by an **anther**. Within the anther are sacs where meiosis occurs and pollen grains develop. Pollen grains house the cells that develop into sperm. A **carpel** has a long, slender

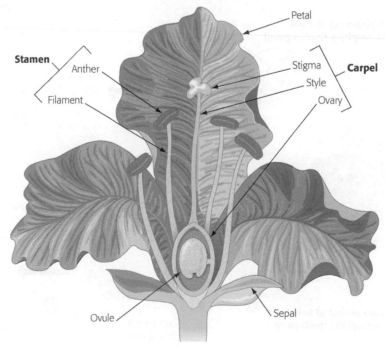

◄ Figure 28.20 **The structure of a flower.** The flower is the structure used by angiosperms for sexual reproduction.

Labels: Stamen, Anther, Filament, Petal, Stigma, Style, Ovary, Carpel, Ovule, Sepal

Overview of the Flowering Plant Life Cycle

The life cycle of a sexually reproducing angiosperm is shown in **Figure 28.21**. After fertilization, the ovule of a flower matures into a seed containing the embryo.

Meanwhile, the ovary develops into a fruit, which protects the seed and aids in dispersing it. Completing the life cycle, the seed **germinates** (begins to grow), the embryo develops into a seedling, and the seedling grows into a mature plant.

Let's look a bit closer at this process from a genetic point of view. The life cycles of all plants include a haploid generation (when each cell has a single set of chromosomes, abbreviated n) and a diploid generation (when each cells has two sets of chromosomes, abbreviated $2n$; see Figure 16.10). The roots, stems, leaves, and most of the reproductive structures of angiosperms are diploid. The diploid plant body, called a **sporophyte**, produces the anthers and ovules in which cells undergo meiosis to produce haploid cells called spores. Each spore then divides via mitosis and becomes a multicellular **gametophyte**, the plant's haploid generation. The gametophyte produces gametes by mitosis. Fertilization occurs when the male and female gametes (sperm and egg, respectively) unite, producing a diploid zygote. The life cycle is completed when the zygote divides by mitosis and develops into a new sporophyte. In the rest of this section, we'll examine each stage in the angiosperm sexual life cycle in more detail.

▼ **Figure 28.21 The life cycle of an angiosperm.**

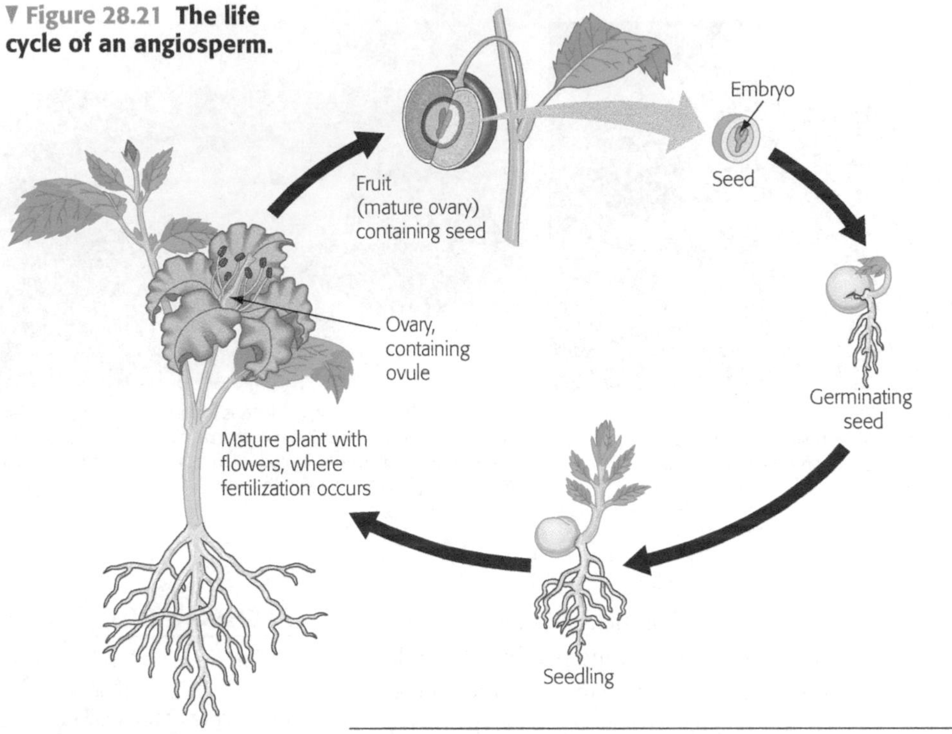

Embryo

Seed

Fruit
(mature ovary)
containing seed

Ovary,
containing
ovule

Germinating
seed

Mature plant with
flowers, where
fertilization occurs

Seedling

Pollination and Fertilization

Fertilization requires gametes, which are produced by gametophytes. The male gametophyte is called the **pollen grain**, which is essential for pollination and produces sperm. The female gametophyte is a multicellular structure called the **embryo sac**, which produces the egg. As shown at the top of **Figure 28.22**, ❶ cells within a flower's anthers ❷ undergo meiosis to form four haploid spores. ❸ Each spore then divides by mitosis into two haploid cells. A thick wall forms around these cells, and the resulting pollen grain is ready for release from the anther.

Moving to the bottom of the figure, we see that ❶ within an ovule, a central cell enlarges and ❷ undergoes meiosis, producing four haploid spores. Three of the spores usually degenerate, but the surviving one enlarges and ❸ divides by mitosis, producing the embryo sac. The sac contains a large central cell with two haploid nuclei. One of its other cells is the haploid egg, ready to be fertilized.

▼ **Figure 28.22 Development of gametophytes in an angiosperm.** The male gametophyte (pollen grain) develops within the anther (top). The female gametophyte (embryo sac) develops within the ovule (bottom). In most species, the ovary of a flower contains several ovules, but only one is shown here.

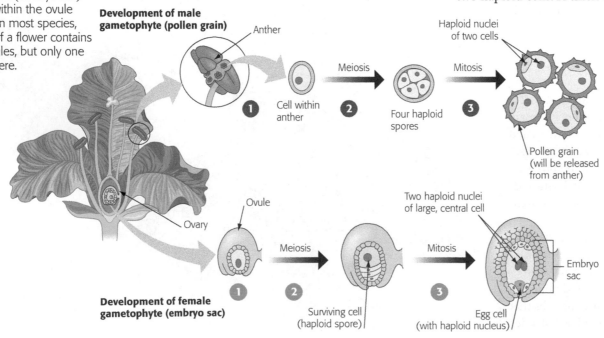

Development of male gametophyte (pollen grain)

Anther

Haploid nuclei
of two cells

❶ Cell within anther
Meiosis
❷ Four haploid spores
Mitosis
❸ Pollen grain (will be released from anther)

Ovule

Two haploid nuclei
of large, central cell

Ovary

Embryo sac

Development of female gametophyte (embryo sac)

❶
Meiosis
❷ Surviving cell (haploid spore)
Mitosis
❸ Egg cell (with haploid nucleus)

How does fertilization occur? As shown in **Figure 28.23**, the first step **1** is **pollination**, the delivery of pollen grains from anther to stigma. Many angiosperms are dependent on insects, birds, or other animals to transfer their pollen (a fact we will discuss further in the Evolution Connection section). But the pollen of some plants, such as grasses, is wind-borne—as anyone with pollen allergies knows!

2 After pollination, the pollen grain germinates on the stigma. It divides by mitosis, forming **3** two haploid sperm that travel to the ovule through a pollen tube that grows out from the pollen. **4** One sperm fertilizes the egg, forming the diploid zygote. The other sperm contributes its haploid nucleus to the large diploid central cell of the embryo sac. This cell, now with a triploid ($3n$) nucleus, will give rise to a food-storing tissue called **endosperm**. The formation of both a zygote and a fertilized central cell with a triploid nucleus is called **double fertilization**. This occurs only in plants, mainly in angiosperms.

Blame your seasonal allergies on angiosperm sex.

▼ **Figure 28.23 Pollination and double fertilization.**

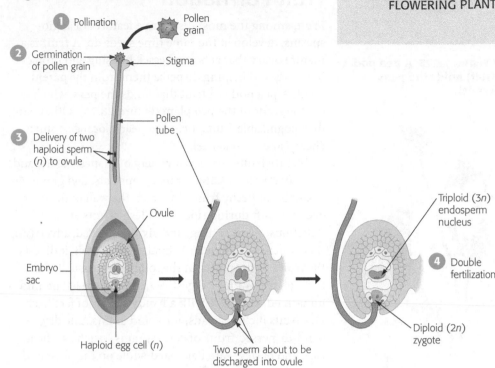

1 Pollination

Pollen grain

2 Germination of pollen grain

Stigma

3 Delivery of two haploid sperm (n) to ovule

Pollen tube

Ovule

Embryo sac

Haploid egg cell (n)

Two sperm about to be discharged into ovule

Triploid ($3n$) endosperm nucleus

4 Double fertilization

Diploid ($2n$) zygote

Seed Formation

After fertilization, the ovule, containing the zygote and the triploid central cell, begins developing into a seed (**Figure 28.24**). The zygote divides via mitosis into a ball of cells that becomes the embryo. Meanwhile, the triploid cell divides and develops into the endosperm. The endosperm is like a lunchbox packed with food for the developing embryo. Coconut "meat" and the white fluffy part of popcorn are familiar examples of endosperm. As cotyledons develop, they absorb nutrients from the endosperm. The result of embryonic development in the ovule is a mature seed (see Figure 28.24, bottom). A **seed** is a plant embryo and endosperm packaged within a tough protective covering called a seed coat. (When you grind coffee, you can't see the embryo contained within the coffee bean—which is actually a seed—because it was destroyed by the drying and roasting process.)

At this point, the embryo stops developing and the seed becomes dormant; growth and development are suspended until the seed germinates. Seeds as old as 2,000 years (excavated from archaeological sites) can still sprout and form a new plant. Seed dormancy is an important evolutionary adaptation. It allows time for seed dispersal and increases the chance that a new generation of plants will begin growing only when environmental conditions, such as temperature and moisture, favor their survival. ☑

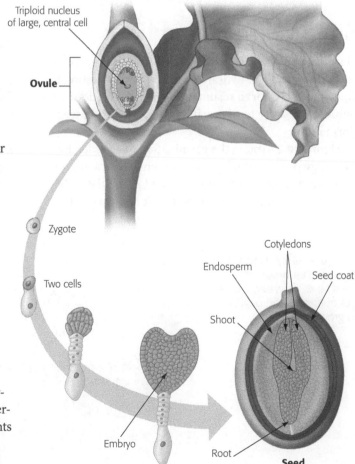

Triploid nucleus of large, central cell

Ovule

Zygote

Two cells

Embryo

Cotyledons

Endosperm

Seed coat

Shoot

Root

Seed

◄ **Figure 28.24 Development of a seed.** The plant in this drawing is a eudicot; a monocot would have only one cotyledon.

☑ **CHECKPOINT**

Of these human structures—sperm, egg, ovary, embryo—which one is analogous to a plant seed?

Answer: embryo (a seed contains a plant embryo suspended during growth)

▼ **Figure 28.25 A pea pod (fruit) holds the peas (seeds).**

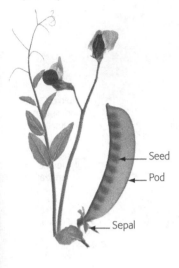

Seed

Pod

Sepal

Fruit Formation

Fruit, among the most distinctive features of angiosperms, develop at the same time seeds do. A **fruit** is a mature ovary that acts as a vessel, housing and protecting seeds and helping disperse them from the parent plant. A pea pod is a fruit that holds the peas, which are the seeds of the pea plant (**Figure 28.25**). Other easily recognizable fruits include a peach, orange, tomato, cherry, and corn kernel.

Mature fruits come in a variety of shapes, colors, and textures (**Figure 28.26**). Oranges, apricots, and grapes are examples of fleshy fruits, in which the wall of the ovary becomes soft during ripening. Fleshy fruits are usually nutritious, sweet tasting, and vividly colored, advertising their ripeness. When an animal eats the fruit, it digests the fruit's fleshy part, but the tough seeds usually pass through the animal's digestive tract and are left behind unharmed, complete with a handy supply of fertilizer. Dry fruits include beans, nuts, and grains. The dry, wind-dispersed fruits of cereal grains—such as wheat, rice, and barley—are harvested while on the plant and serve as staple foods for people.

▼ **Figure 28.26 A variety of fruits.**

Fleshy fruits

Dry fruits

Seed Germination

Germination of a seed usually begins when the seed takes up water. The hydrated seed expands, bursting its coat, and the embryo resumes the growth and development that were temporarily suspended during seed dormancy.

Figure 28.27 traces the germination of a garden bean. The embryonic root of the bean emerges first and grows downward from the germinating seed. Next, the embryonic shoot emerges, and a hook forms near its tip. The hook protects the delicate shoot tip by holding it downward as it pushes through the abrasive soil. As the shoot breaks through the soil surface, exposure to light stimulates the hook to straighten, and the tip is lifted. The first foliage leaves then expand from the shoot tip and begin making food for the plant by photosynthesis.

A germinating seed is fragile. In the wild, only a small fraction of seedlings endure long enough to reproduce. The great numbers of seeds produced by most plants compensate for the odds against each seedling, increasing the chances of reproductive success. ☑

☑ **CHECKPOINT**

Is the garden bean shown in Figure 28.27 a monocot or a eudicot?

Answer: It's a eudicot. (You can tell because it has two cotyledons.)

▶ **Figure 28.27 Germination in a garden bean.** After a seed takes up water, it bursts out of its coat and begins to grow.

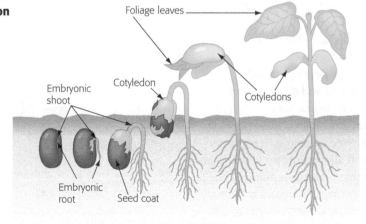

Foliage leaves

Cotyledon

Cotyledons

Embryonic shoot

Embryonic root

Seed coat

Agriculture EVOLUTION CONNECTION

The Problem of the Disappearing Bees

Throughout this chapter, we've discussed how flowering plants and land animals have had mutually beneficial relationships throughout their evolutionary history. You've learned that most angiosperms depend on insects, birds, or mammals for pollination and seed dispersal. And most land animals depend on angiosperms for food and shelter. Such mutual dependencies tend to improve the reproductive success of both the plants and the animals and are thus favored by natural selection.

The flowers of many angiosperms attract pollinators that rely entirely on the flowers' nectar (a sugary fluid) and pollen for food. Evolutionary adaptations for advertising the presence of nectar include flower color and fragrance, which are keyed to pollinators' senses of sight and smell. For example, many flowers pollinated by birds are red or pink, colors to which bird eyes are especially sensitive. Flowers may also have markings that attract pollinators, leading them past pollen-bearing organs on their way to gathering nectar (Figure 28.28). For example, flowers pollinated by bees often have markings that reflect ultraviolet light. Such markings are invisible

to us, but vivid to bees. Many other animals—including hummingbirds, butterflies, and fruit bats—have similar relationships with other flower species. To a large extent, flowering plants are as diverse and successful as they are today because of their close connections with animals.

The importance of such interactions has been underscored by concern over colony collapse disorder. In late 2006, U.S. beekeepers noticed a sudden and drastic die-off in their bee colonies. The next year, similar problems were reported in Europe. Agricultural scientists fear that a shortage of bees could have devastating consequences on many food crops that require bees for pollination, such as almonds, berries, and fruits. Bees pollinate $15 billion worth of crops each year, and an estimated one-third of our food supply relies on bees. The cause of the die-off has not yet been explained, although several possible causes have been proposed, including pesticides, infection by mites, nutritional deficiencies, and various pathogens. This much is certain: Colony collapse disorder is yet another reminder of how much we all depend on plants for our survival.

▼ Figure 28.28 **A busy bee.**

Chapter Review

▌SUMMARY OF KEY CONCEPTS

The Structure and Function of a Flowering Plant

Flowering plants, or angiosperms, account for nearly 90% of the plant kingdom.

Monocots and Eudicots

Angiosperms can be grouped into two categories based on the number of cotyledons (seed leaves) found in the embryo and other structural differences.

Plant Structures					
	Seed leaves	Leaf veins	Vascular bundles	Floral parts	Roots
Monocots	One cotyledon	Parallel	Scattered	Multiples of three	Fibrous
Eudicots	Two cotyledons	Branched	Ring	Multiples of four or five	Taproot

Structure/Function: Roots, Stems, and Leaves

A plant body consists of a root system and a shoot system, each dependent on the other.

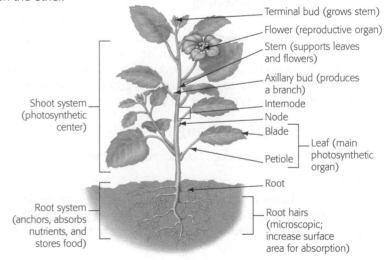

Shoot system (photosynthetic center)

Root system (anchors, absorbs nutrients, and stores food)

Terminal bud (grows stem)
Flower (reproductive organ)
Stem (supports leaves and flowers)
Axillary bud (produces a branch)
Internode
Node
Blade
Leaf (main photosynthetic organ)
Petiole
Root
Root hairs (microscopic; increase surface area for absorption)

A growing shoot often lengthens only at the terminal bud; its axillary buds are dormant. This condition is called apical dominance and allows the plant to grow tall quickly.

Plant Tissues and Tissue Systems

Roots, stems, and leaves are made up of tissues organized into three tissue systems: the dermal, vascular, and ground tissue systems.

The Three Tissues of Plant Organs		
Dermal Tissue System	**Vascular Tissue System**	**Ground Tissue System**
• Functions in protection • Includes the epidermis, an outer layer that covers and protects the plant • Stomata (surrounded by guard cells): pores that regulate gas exchange between the leaf and environment	• Functions in support and transport • Xylem: transports water and dissolved minerals • Phloem: transports sugars	• Functions mainly in storage • Includes leaf mesophyll, where most photosynthesis occurs

Plant Cells

Plant cells are distinguished by the presence of chloroplasts, a large central vacuole, and cell walls. There are several types of plant cells: parenchyma cells, collenchyma cells, sclerenchyma cells, water-conducting cells (tracheids and vessel elements), and food-conducting cells.

Plant Growth

Most plants continue to grow as long as they are alive. Some plants grow, reproduce, and die in one year (annuals); some live for two years (biennials); and some live for many years (perennials).

Primary Growth: Lengthening

Plant growth originates in meristems, areas of undifferentiated, dividing cells. Apical meristems are located at the tips of roots and in the terminal and axillary buds of shoots. These meristems initiate lengthwise growth by producing new cells. A root or shoot lengthens further as the new cells elongate. This process of cell division and elongation is called primary growth. The new cells eventually differentiate into specialized tissues.

Secondary Growth: Thickening

An increase in a plant's girth, called secondary growth, arises from cell division in a cylindrical meristem called the vascular cambium. The vascular cambium produces layers of secondary xylem, or wood, next to its inner surface. Outside the vascular cambium is bark, which consists of secondary phloem, a meristem called the cork cambium, and cork cells. The outer layers of bark are sloughed off as the plant thickens.

The Life Cycle of a Flowering Plant

Many flowering plants can reproduce both asexually and sexually.

The Flower

In angiosperms, the flower is the structure specific to sexual reproduction. The stamen contains the anther, in which pollen grains are produced. The carpel contains the stigma, which receives pollen, and the ovary, which contains the ovule.

Overview of the Flowering Plant Life Cycle

The plant life cycle alternates between diploid ($2n$) and haploid (n) generations. The spores in the anthers give rise to male gametophytes, or pollen grains, each of which produces two sperm. A spore in an ovule produces the female gametophyte, called an embryo sac. Each embryo sac contains an egg cell.

Pollination and Fertilization

After pollen lands on the stigma (pollination), the sperm travels through a pollen tube into the ovule. One sperm combines with the egg, and the other combines with a diploid cell. This process is called double fertilization.

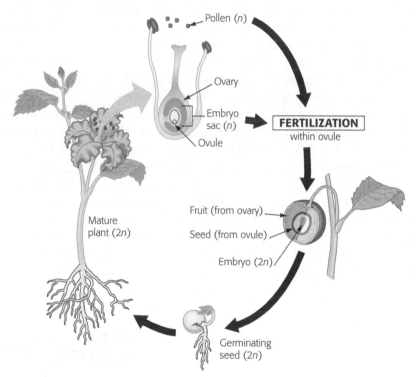

Seed Formation

After fertilization, the ovule becomes a seed, and the fertilized egg within it divides to become an embryo. The other fertilized cell (now triploid) develops into the endosperm, which stores food for the embryo. A tough seed coat protects the embryo and endosperm. The embryo develops cotyledons (seed leaves), which absorb food from the endosperm.

Fruit Formation

While the ovule becomes a seed, the ovary develops into a fruit, which helps protect and disperse the seeds.

Seed Germination

A seed starts to germinate when it takes up water, expands, and bursts its seed coat. The embryo resumes growth, an embryonic root emerges, and a shoot pushes upward and expands its leaves.

MasteringBiology®

For practice quizzes, BioFlix animations, MP3 tutorials, video tutors, and more study tools designed for this textbook, go to MasteringBiology®

SELF-QUIZ

1. While walking in the woods, you encounter an unfamiliar flowering plant. Which of the following plant features would help you determine whether that plant is a monocot or a eudicot? (Choose all that are appropriate.)
   a. size of the plant
   b. number of seed leaves
   c. shape of its root system
   d. number of petals in its flowers
   e. arrangement of vascular bundles in its stem
   f. whether or not it produces seed-bearing cones

2. Your friend plants a new herb (which is a eudicot and a perennial). During the first year, she allows the plants to grow naturally. Her plants grow tall and relatively spindly during the growing season, and when it comes time to harvest, she doesn't have a large yield. What phenomenon is responsible for her poor yield? What could your friend do to increase her yield next year?

3. Match each flower structure with its function.
   a. pollen grain   1. attracts pollinators
   b. ovule   2. develops into seed
   c. anther   3. protects an unopened flower
   d. ovary   4. produces sperm
   e. sepal   5. produces pollen
   f. petal   6. contains ovules

4. Many plant species can reproduce both sexually and asexually. Which mode of reproduction would generally be more advantageous in a location where the composition of the soil is changing rapidly? Why?

5. What part of a plant are you eating when you consume each of the following?
   a. tomato   d. sweet potato
   b. apple   e. black bean
   c. spinach

6. In the angiosperm life cycle, which of the following processes is directly dependent on meiosis?
   a. production of gametophytes   c. production of spores
   b. production of gametes   d. all of the above

7. In angiosperms, each pollen grain produces two sperm. What do these sperm do?
   a. Each one fertilizes a separate egg cell.
   b. One fertilizes an egg, and the other fertilizes the fruit.
   c. One fertilizes an egg, and the other is kept in reserve.
   d. One fertilizes an egg, and the other fertilizes a cell that develops into stored food.

8. Which of the following is a multicellular mass that nourishes the embryo until it becomes a self-supporting seedling?
   a. a fruit   c. the endosperm
   b. a seed coat   d. a cotyledon

Answers to these questions can be found in Appendix: Self-Quiz Answers.

THE PROCESS OF SCIENCE

9. Dendrochronology—the study of growth rings in trees—is used as a tool in studying climate change. How do you think tree ring data can be used to draw conclusions about the climate? How would you calibrate those studies?

10. **Interpreting Data** Botanists observed that the number of "teeth" along the edges of red maple (*Acer rubrum*) leaves varies depending on where the tree grows. The botanists wondered whether these differences were due to climate. To find out, red maple seeds were collected from four sites (from north to south, Ontario, Pennsylvania, South Carolina, and Florida) and grown in two sites (Rhode Island and Florida). After a few years, leaves were collected, and the average number of teeth per leaf area were calculated:

Site of origin	Number of teeth per cm² of leaf area	
	Grown in RI	Grown in FL
Ontario	3.9	3.2
PA	3.0	3.5
SC	2.3	1.9
FL	2.1	0.9

Based on these data, does the number of teeth vary from north to south? Do the data suggest that these differences are due to climate or genetic differences?

BIOLOGY AND SOCIETY

11. The look of a fruit or a vegetable adds to its market value as consumers tend to prefer well-grown, perfectly shaped, and polished products. Large quantities of fruits and vegetables are often thrown away by supermarkets because they do not meet certain aesthetic standards. To what extent do you think we should be guided by the look of these products?

12. When buying coffee, many consumers will pay more for coffee that is "Fair Trade Certified" because they believe that doing so will help family farmers in developing nations. A Fair Trade label means that the coffee was grown and harvested in accordance with a set of standards overseen by an independent nongovernmental regulatory agency. There are different certificates with different standards, but in general the Fair Trade Certified designation is meant to ensure that the coffee was grown by farmers in a cooperative that avoids child labor, does not use certain herbicides and pesticides, and is paid at least a fair minimum price for the product. Do these seem like worthwhile criteria to you? Would you be willing to pay more for coffee that meets these standards? How much more—10% more? Double? If you do pay a premium for such coffee, do you think you can be reasonably sure that it accomplishes the Fair Trade goals?

Why Plant Functions Matter

Because of a gaseous hormone, one overripe apple can spoil the whole bunch. ▶

▲ Without soil bacteria, we'd all starve.

▲ Compost happens: Discarded kitchen scraps can become a valuable addition to any garden.

The Interdependence of Organisms BIOLOGY AND SOCIETY

Planting Hope in the Wake of Disaster

On March 11, 2011, a magnitude 9.0 earthquake originating beneath the Pacific Ocean produced a 100-foot tsunami that swept 6 miles into northeastern Japan. One of the five strongest earthquakes ever recorded, the disaster had far-reaching consequences, killing more than 15,000 people and damaging or destroying at least 125,000 buildings. The surge of ocean water from the tsunami flooded the Fukushima Nuclear Power Plant, resulting in a reactor core meltdown. Fukushima joined the 1986 Chernobyl disaster in Ukraine as one of the worst nuclear accidents the world has ever seen.

The cleanup effort in Japan will take decades. In addition to bulldozers and wrecking crews, help is coming from a natural source: plants. The use of plants to help clean up polluted soil and groundwater is called phytoremediation. Plants can efficiently move minerals and other compounds from the soil into the plant body. Usually, the absorbed compounds are nutrients that help the plants grow. But many species of plants, including sunflowers, are capable of absorbing large amounts of radioactive isotopes and heavy metals. In the years after the disaster, millions of sunflowers were planted in the Fukushima Prefecture to help restore the area. This effort was modeled in part after similar work carried out in New Orleans in the aftermath of Hurricane Katrina, work that continues to this day.

Cleansing plants. Throughout Japan, people are planting sunflowers to take advantage of their natural ability to absorb radioactive elements.

Because cleanup crews have to remove only a few cubic meters of plant material rather than thousands of cubic meters of contaminated soil, phytoremediation can be a cost-effective and time-efficient process. An added benefit: Using plants to clean the soil can detoxify an area without dramatically disturbing the landscape. And sunflowers provide an uplifting message to the community, bringing beauty and life back to an area weary with grief and loss.

By using plants to clean up toxic wastes, we are benefiting from millions of years of plant evolution. Although plants are unable to move about in search of food, they have evolved an amazing ability to pull water and nutrients out of the soil and air. In this chapter, we'll study the physiology of the working plant to see how plants obtain essential nutrients and how they transport them throughout their roots, stems, and leaves. Then we'll examine other aspects of the life of a working plant, including the crucial roles played by hormones and how certain plant activities are affected by stimuli from the environment. Throughout the discussion, we'll explore the theme of how different forms of life—such as plants and animals—depend on each other for survival.

How Plants Acquire and Transport Nutrients

▼ Figure 29.1 **The uptake of nutrients by a plant.** A plant absorbs the CO_2 it needs from the air and minerals and H_2O from the soil.

Minerals (inorganic ions)

Watch a plant grow from a tiny seed and you can't help wondering where all the mass comes from. About 96% of a plant's dry weight is organic (carbon-containing) material synthesized from inorganic nutrients extracted from the plant's surroundings. Plants obtain carbon dioxide (CO_2) from the air and water (H_2O) and **minerals** (inorganic ions) from the soil (Figure 29.1). From the CO_2 and H_2O, plants produce sugars via photosynthesis. These sugars, combined with minerals, are used to construct all the other organic materials a plant needs (Figure 29.2). Plants, like people, require a balanced diet, so let's start by discussing plant nutrition.

▼ Figure 29.2 **A banyan tree is a giant product of photosynthesis.** Plants use sugars created by photosynthesis to produce all the organic materials they need. The giant trunk and long branches of this Hawaiian banyan tree consist largely of molecules derived from sugars.

Plant Nutrition

A chemical element is considered an **essential element** if a plant must obtain it from its environment to complete its life cycle. Seventeen elements are essential to all plants, and a few others are essential to some plants. Of the 17 essential elements, 9 are called **macronutrients** because plants require relatively large amounts of them. Elements that plants need in extremely small amounts are called **micronutrients**.

Macronutrients

Six of the nine macronutrients—carbon, oxygen, hydrogen, nitrogen, sulfur, and phosphorus—make up almost 98% of a plant's dry weight. The other three macronutrients—calcium, potassium, and magnesium—make up another 1.5%.

What does a plant do with macronutrients? Carbon, oxygen, and hydrogen are the basic ingredients of a plant's organic compounds. Nitrogen is a component

of all nucleic acids and proteins, as well as ATP and chlorophyll (the key molecule in photosynthesis). Sulfur is a component of most proteins. Phosphorus is a major component of nucleic acids, phospholipids, and ATP. The other macronutrients, although present in smaller amounts, play similarly important roles. For example, magnesium is an essential component of chlorophyll.

Micronutrients

Eight micronutrients—iron, chlorine, copper, manganese, zinc, molybdenum, boron, and nickel—make up the remaining 0.5% of a plant's dry weight. A plant recycles the atoms of micronutrients over and over, so it needs them in only minute quantities. Yet a deficiency of any micronutrient can kill a plant.

Fertilizers

The quality of soil, especially the availability of nutrients, affects the health of plants and, for plants we consume, the quality of our own nutrition ("You are what you eat!"). A shortage of nitrogen is the most common nutritional problem for plants. Nitrogen-deficient crop plants may produce grain, but the grain will have a lower nutritional value, and its nutrient deficiencies will be passed on to livestock or human consumers.

Fertilizers are compounds that are applied to the soil to promote plant growth. There are two basic types of fertilizers: inorganic and organic. Inorganic fertilizers contain simple inorganic minerals, such as mined limestone (rich in calcium) or phosphate rock. Inorganic fertilizers come in a wide variety of formulations, but most emphasize the "N-P-K" ratio, the relative amounts of the three nutrients most often deficient in depleted soils: nitrogen (N), phosphorus (P), and potassium (K). An all-purpose fertilizer, for example, might be "5-5-5," with 5% of each of these vital nutrients **(Figure 29.3)**.

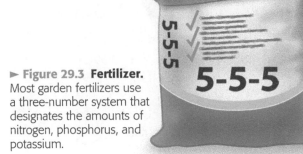

► **Figure 29.3 Fertilizer.** Most garden fertilizers use a three-number system that designates the amounts of nitrogen, phosphorus, and potassium.

Organic fertilizers are composed of chemically complex organic matter such as **compost**, a soil-like mixture of decomposed organic matter. Many gardeners maintain a compost pile or bin to which they add leaves, grass clippings, yard waste, and kitchen scraps (avoiding meat, fat, and bones). Over time, the vegetable matter is broken down by microbes, fungi, and animals **(Figure 29.4)**. Occasional turning and watering speed the process. The compost, which is highly valued by gardeners for its nutrient-rich composition, can be applied to outdoor gardens or indoor pots.

Compost and other organic fertilizers are the key to **organic farming**. Organic farmers follow guidelines that promote sustainable agriculture—a system of farming that protects biological diversity, maintains and replenishes soil quality (as by crop rotation), manages pests with no (or few) synthetic pesticides, avoids genetically modified organisms, conserves water, and uses no (or few) synthetic fertilizers. In the United States, organic farming is one of the fastest-growing segments of agriculture, but organic farms still account for less than 1% of total agricultural acreage. ☑

▼ **Figure 29.4 Compost.** Steam is produced by the metabolic activity of naturally occurring organisms that break down the organic matter of rotting vegetation.

Compost happens: Discarded kitchen scraps can become a valuable addition to any garden.

☑ **CHECKPOINT**

What inorganic nutrient do plants acquire from the air? What nutrients do plants extract from the soil?

Answer: carbon dioxide; water and minerals (inorganic ions)

▼ **Figure 29.5 Root hairs, which enable efficient absorption.** Root hairs, clearly visible in this radish seedling, are extensions of epidermal cells on the outer surface of the root.

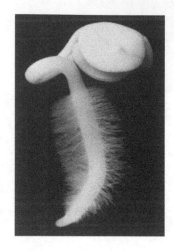

From the Soil into the Roots

Plant roots have a remarkable capacity for absorbing water and essential nutrients from soil because of their root hairs, extensions of epidermal cells that dramatically increase the surface area available for absorption (**Figure 29.5**). For example, the root hairs of a single sunflower plant, if laid end to end, could stretch many miles. Root hairs provide a huge surface area in contact with nutrient-containing soil.

All substances that enter a plant root are dissolved in water. To be transported, water and solutes must move from the soil through the epidermis and cortex of the root and then into the water-conducting xylem tissue in the root's central cylinder (see Figure 28.8). To reach the xylem, the solution must pass through the plasma membranes of root cells. Because these membranes are selectively permeable, only certain solutes reach the xylem. This selectivity helps regulate the mineral composition of a plant's vascular system.

Many plants gain significant surface area for absorption through symbiotic associations with fungi. Together, the plant's roots and the associated fungus form a mutually beneficial (mutualistic) structure called a **mycorrhiza** (plural, *mycorrhizae*) (**Figure 29.6**). Fungal filaments around the roots absorb water and minerals much more rapidly than can the roots alone. Some of the water and minerals taken up by the fungus are transferred to the plant, while the plant's photosynthetic products nourish the fungus.

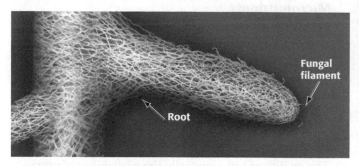

Fungal filament

Root

▲ **Figure 29.6 A mycorrhiza on a small tree root.**

Energy Transformations
The Role of Bacteria in Nitrogen Nutrition

Air penetrates the soil around roots, but plants cannot use the form of nitrogen found in air—gaseous N_2. To be useful, N_2 must first be converted to ammonium (NH_4^+) or nitrate (NO_3^-). Such conversions illustrate the importance of transformations of energy and matter in living systems. In this instance, vital elements within the ecosystem are recycled, converted from one form to another. A plant cannot perform this vital transformation, but bacteria can. Although we often think of bacteria as troublesome, it is important to note that we utterly depend on bacteria: Without them, our crop plants could not obtain sufficient nitrogen, and we would be unable to feed ourselves.

> Without soil bacteria, we'd all starve.

Soil Bacteria and Nitrogen

Figure 29.7 shows three types of soil bacteria that play essential roles in supplying plants with nitrogen. Nitrogen-fixing bacteria (⬤) convert atmospheric N_2 to ammonium, a process called **nitrogen fixation**. Ammonifying bacteria (◯) add to the soil's supply of ammonium by decomposing organic matter. Nitrifying bacteria (⬭) convert soil ammonium to nitrate. Plants take up most of their nitrogen in this form. They then convert nitrate back to ammonium, for use in building organic molecules. This process is a bit like a manufacturer that assembles appliances from pre-made parts; the bacteria are the suppliers that convert raw materials into parts needed by the manufacturer (the plant).

▼ **Figure 29.7 The role of bacteria in supplying nitrogen to plants.**

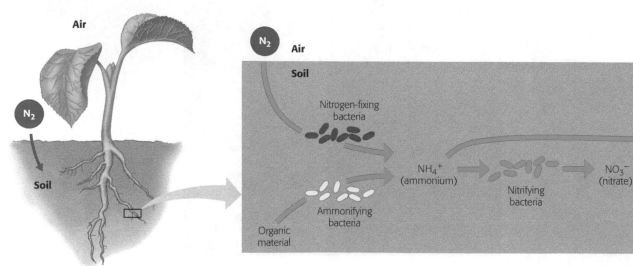

Air

N_2

Soil

N_2

Air

Soil

Nitrogen-fixing bacteria

Ammonifying bacteria

Organic material

NH_4^+ (ammonium)

Nitrifying bacteria

NO_3^- (nitrate)

NH_4^+

Root

Root Nodule Bacteria and Nitrogen

Some plant families, including legumes—peas, beans, peanuts, and many other plants that produce their seeds in pods—have their own built-in source of ammonium: nitrogen-fixing bacteria that live in swellings called root nodules (Figure 29.8). Within these nodules, plant cells have been "infected" by nitrogen-fixing bacteria that reside inside cytoplasmic vesicles. The symbiotic relationship between a plant and its nitrogen-fixing bacteria is mutually beneficial ("You scratch my back; I'll scratch yours"). The bacteria have enzymes that help transform nitrogen; the plant, in turn, provides the bacteria with other nutrients. To continue the manufacturer analogy, the root nodules of a legume are like a "company town" where the manufacturer provides living space for its suppliers.

When conditions are favorable, root nodule bacteria fix so much nitrogen that the nodules secrete excess ammonia, which improves the soil. This is one reason farmers practice crop rotation, one year planting a non-legume such as corn and the next year planting a legume such as soybeans. The legume crops may be plowed under to decompose into "green manure," reducing the need for fertilizer. ☑

☑ **CHECKPOINT**

Why might a pollutant that kills soil bacteria result in nitrogen deficiency in plants?

Answer: because certain soil bacteria make nitrogen available to plants in forms they can use

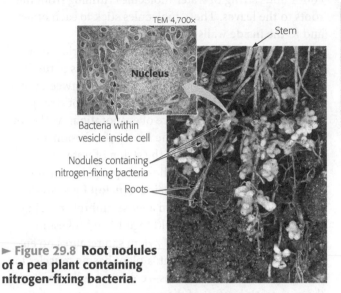

TEM 4,700×

Stem

Nucleus

Bacteria within
vesicle inside cell

Nodules containing
nitrogen-fixing bacteria

Roots

▶ Figure 29.8 **Root nodules of a pea plant containing nitrogen-fixing bacteria.**

The Transport of Water

As a plant grows upward toward sunlight, it needs an increasing supply of resources from the soil. Imagine hauling a 5-gallon bucket of water up five flights of stairs every 20 minutes, all day long. To thrive, a typical tree must transport that quantity of water from its roots to the rest of the plant.

In a mature plant, water-conducting cells of the xylem are arranged end to end to form very thin vertical tubes (see Figure 28.11). A solution of water and inorganic nutrients, called **xylem sap**, flows through these tubes all the way from a plant's roots to the tips of its leaves.

The Ascent of Xylem Sap

What force moves xylem sap up against the downward pull of gravity? Xylem sap is pulled upward by **transpiration**, the loss of water from the leaves of a plant by evaporation (Figure 29.9). Most transpiration occurs through the stomata (pores) on the undersides of leaves. The stomata open into air spaces, which are filled with water molecules that have evaporated from the surrounding cells. ❶ Water vapor diffuses out of the stomata because the surrounding air is usually drier than the inside of the leaf.

Transpiration can pull xylem sap up a tree because of two special properties of water: adhesion and cohesion. **Adhesion** is the sticking together of molecules of different kinds. ❷ Water molecules tend to adhere to cellulose molecules in the walls of xylem cells. **Cohesion** is the sticking together of molecules

▼ Figure 29.9 **How transpiration pulls water up the xylem of a tree.**

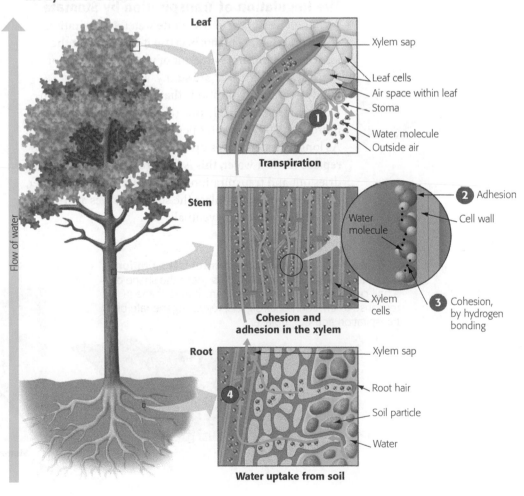

Flow of water

Leaf

Xylem sap
Leaf cells
Air space within leaf
Stoma
❶
Water molecule
Outside air

Transpiration

Stem

❷ Adhesion
Water molecule
Cell wall
❸ Cohesion, by hydrogen bonding
Xylem cells

Cohesion and adhesion in the xylem

Root

❹
Xylem sap
Root hair
Soil particle
Water

Water uptake from soil

☑ CHECKPOINT

1. Water moves from the roots to the tips of the leaves without any expenditure of energy by the plant. What is this process called? What two forces between molecules are responsible?

2. What is the evolutionary advantage of stomata that close at night?

Answers: 1. transpiration; cohesion and adhesion 2. Closed stomata prevent water loss when photosynthesis cannot occur.

of the same kind. ❸ In the case of water, hydrogen bonds make the H_2O molecules stick to one another (see Figure 2.8). ❹ Together, adhesion and cohesion create a continuous string of water molecules running from the roots to the leaves. These molecules stick to each other and to the inside walls of the xylem tubes.

Because the air outside the leaf is much drier than the moist interior of the leaf, water molecules naturally diffuse outward. Cohesion (the attraction between water molecules) resists this pulling force, but is not strong enough to overcome the force of evaporation. As the top water molecule breaks off, the forces of cohesion and transpiration pull the rest of the string of water molecules upward. Biologists call this explanation for the ascent of xylem sap the **cohesion-tension hypothesis**: Transpiration exerts a pull on a tense, unbroken string of water molecules that is held together by cohesion.

Note that the transport of xylem sap requires no energy expenditure by the plant; a plant has no muscles for pumping. Physical properties—cohesion, adhesion, and the evaporating effect of the sun—move water and dissolved minerals from a plant's roots to its shoots. Transpiration is thus a highly efficient means of moving a lot of water upward through the body of a plant.

The Regulation of Transpiration by Stomata

Although necessary to distribute water, transpiration also works against plants because it can result in the loss of an astonishing amount of water. Transpiration is greatest on days that are sunny, warm, dry, and windy—weather conditions that promote evaporation. An average-sized maple tree (about 20 m high) can lose more than 200 L of water per hour on a hot summer day. As long as water moves up from the soil fast enough to replace the lost water, this isn't a problem. But if the soil dries out and transpiration exceeds the delivery of water, the leaves will wilt. Unless the soil and leaves are rehydrated, the plant will eventually die.

The leaf stomata, which can open and close, are evolutionary adaptations that help plants adjust their transpiration rates to changing environmental conditions. In many plants, stomata are open during the day, which allows CO_2 to enter the leaf from the atmosphere and thus keep photosynthesis going when sunlight is available. However, the plant also loses water through transpiration. At night, when there is no light for photosynthesis and therefore no need for CO_2, many plants close their stomata, saving water. Stomata may also close during the day if a plant is losing water too fast. The opening and closing of stomata are controlled by the changing shape of the two guard cells flanking each stoma **(Figure 29.10)**. ☑

The Transport of Sugars

A plant's survival depends not only on the transport of water and minerals absorbed from the soil but also on the transport of the sugars it makes by photosynthesis. This is the main function of phloem.

Phloem consists of living food-conducting cells arranged end to end into long tubes. Through perforations in the end walls of these cells, a sugary liquid called **phloem sap** moves freely from one cell to the next **(Figure 29.11)**. The main solute in phloem sap is usually the disaccharide sugar sucrose (table sugar). Phloem sap also contains inorganic ions, amino acids, and hormones in transit from one part of the plant to another.

In contrast to xylem sap, which only flows upward from the roots, phloem sap moves throughout the plant in various directions. A location in a plant where sugar is being produced (either by photosynthesis or by the breakdown of stored starch) is called a **sugar source**. A receiving location in the plant, where the sugar will be stored or consumed, is called a **sugar sink**. Sugar moves within food-conducting tubes of phloem from a source, such as a leaf, to a sink, such as a root or a fruit.

▶ **Figure 29.10 How stomata control transpiration via guard cells.** A pair of guard cells (each the shape of a half-circle) surrounding a stoma can change shape in response to environmental signals, regulating the rate of transpiration and CO_2 uptake.

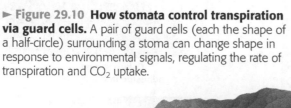

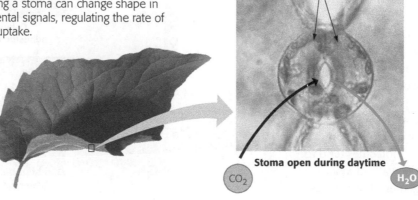

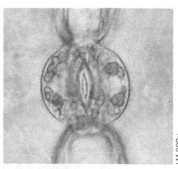

Guard cells

Stoma open during daytime

Stoma closed at night

CO_2

H_2O

LM 900x

What causes phloem sap to flow from a source to a sink? **Figure 29.12** uses a beet plant to illustrate a widely accepted model called the **pressure-flow mechanism**. At the sugar source (the beet leaves), ➊ sugar is loaded from a photosynthetic cell into a phloem tube via active transport. Sugar loading at the source end raises the solute (sugar) concentration inside the phloem tube. ➋ The high solute concentration draws water into the tube by osmosis, usually from the xylem. The flow of water from the xylem into the phloem raises the water pressure at the source end of the phloem tube.

At the sugar sink (the beet root), both sugar and water leave the phloem tube. ➌ As sugar leaves the phloem, lowering the sugar concentration at the sink end, ➍ water moves by osmosis back into the xylem. The exit of water lowers the water pressure in the tube. The increase of water pressure at the source end of the phloem tube and the reduction of water pressure at the sink end cause phloem sap to flow from source to sink.

The pressure-flow mechanism explains why phloem sap always flows from a sugar source to a sugar sink, regardless of their locations in the plant. The same plant structure may act as a sugar source at one time of year and a sugar sink at another. For example, beet plants transport sugars to their roots during the summer and then draw sugars out of those roots during springtime growth. Imagine a series of warehouses that can ship or receive goods as needed; goods are shipped from warehouses where they are plentiful (sugar sources) to warehouses where they are needed (sugar sinks), but a particular warehouse may change roles depending on supply and demand.

We now have a broad picture of how a plant absorbs substances from the soil and transports materials from one part of its body to another: Water and inorganic ions enter from the soil and are pulled upward through xylem by transpiration. Carbon dioxide enters leaves through the stomata and is incorporated into sugars, which are distributed by phloem. Pressure flow drives the phloem sap from leaves and storage sites to other parts of the plant, where the sugars are used or stored. In the next section, we'll consider various ways that a working plant uses these resources. ☑

☑ **CHECKPOINT**

In the pressure-flow mechanism, phloem sap always flows from a _____ to a _____.

Answer: sugar source; sugar sink

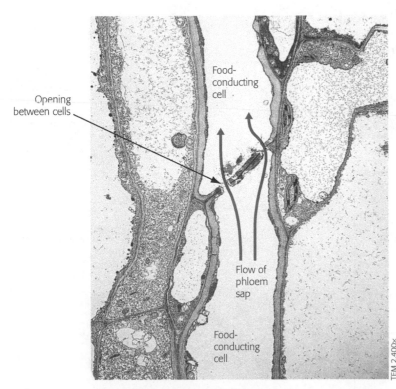

▲ **Figure 29.11 Food-conducting cells of phloem.** The main function of phloem is to transport sugars through living food-conducting cells arranged into tubes. Phloem sap moves through openings between the cells.

Opening between cells

Food-conducting cell

Flow of phloem sap

Food-conducting cell

TEM 2,400×

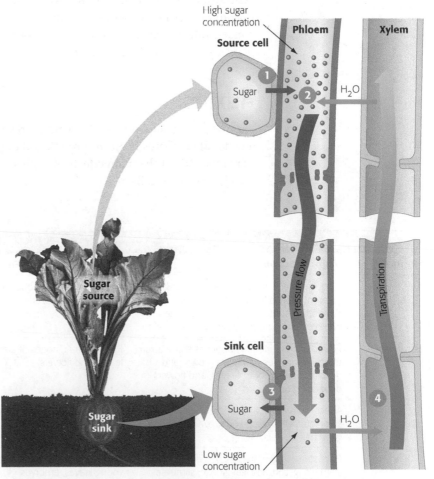

High sugar concentration

Source cell

Phloem

Xylem

Sugar

H_2O

Pressure flow

Transpiration

Sink cell

Sugar source

Sugar sink

Sugar

H_2O

Low sugar concentration

▲ **Figure 29.12 Pressure flow in plant phloem.** The red dots represent sugar molecules; a concentration gradient decreasing from top to bottom creates a pressure flow (wide red arrow). The wide blue arrow represents water pressure due to transpiration in the xylem sap; water pressure decreases from bottom to top. Xylem tubes transport the water back from sink to source.

Plant Hormones

A **hormone** is a chemical signal produced in one part of an organism and transported to other parts, where it acts on target cells to change their functioning (see Chapter 25 for a discussion of animal hormones). Plants produce hormones in very small amounts, but even a tiny amount of any of these chemicals can have profound effects on growth and development. Plant hormones control growth and development by affecting the division, elongation, and differentiation of cells.

Biologists have identified five major types of plant hormones (Table 29.1). Each type can produce a variety of effects, depending on the developmental stage of the plant and the hormone's concentration and site of action. In most situations, a plant hormone does not act alone. Instead, it is the relative concentration of two or more hormones that controls a plant's growth and development.

▲ Figure 29.13 **Phototropism: growing toward light.**

Phototropism and Cell Elongation

Phototropism is the directional growth of a plant shoot in response to light. Such growth directs both seedlings and mature plants toward the sunlight they use for photosynthesis (Figure 29.13). How does a plant grow in a particular direction? Microscopic observations of growing plants reveal the cellular mechanism that underlies phototropism (Figure 29.14). Cells on the side of a stem that gets the least amount of light are larger—actually, they have elongated faster—than those on the brighter side, causing the shoot to bend toward the light. If a seedling is illuminated uniformly from all sides or if it is kept in the dark, the cells all elongate at a similar rate and the seedling grows straight upward.

What causes plant cells on the dark side of a shoot to grow faster than those on the bright side? Our present understanding of this phenomenon emerged from a series of classic experiments conducted by two scientists with a very familiar name, as we see next.

Auxins

Auxins are a group of related hormones responsible for a wide range of effects on the growth and development of plants. One of the main effects of auxins is to promote the elongation of cells.

Table 29.1	Major Types of Plant Hormones	
Hormone	**Major Functions**	**Where Produced or Found in Plant**
Auxins	Stimulate stem elongation; affect root growth, differentiation, and branching; stimulate development of fruit, apical dominance, phototropism, and gravitropism	Meristems of apical buds, young leaves, embryos within seeds
Ethylene	Promotes fruit ripening; opposes some auxin effects; promotes or inhibits growth and development of roots, leaves, and flowers, depending on species	Ripening fruit, nodes of stems, aging leaves and flowers
Cytokinins	Affect root growth and differentiation; stimulate cell division and growth; stimulate germination; delay aging	Made in roots and transported to other organs
Gibberellins	Promote seed germination, bud development, stem elongation, and leaf growth; stimulate flowering and fruit development; affect root growth and differentiation	Meristems of apical buds and roots, young leaves, embryos
Abscisic acid	Inhibits growth; closes stomata when water is scarce; helps maintain dormancy	Leaves, stems, roots, green fruit

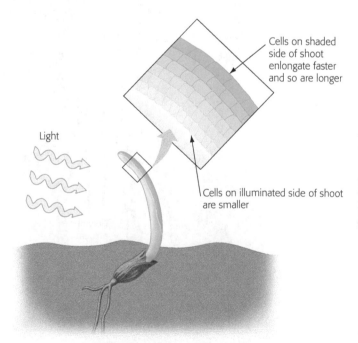

Cells on shaded side of shoot enlongate faster and so are longer

Light

Cells on illuminated side of shoot are smaller

▲ Figure 29.14 **Cell elongation causes phototropism.** A grass seedling curves toward light coming from one side because of the faster elongation of cells on the shaded side.

The Interdependence of Organisms THE PROCESS OF SCIENCE

Do Chemical Signals Affect Plant Growth?

In the late 1800s, Charles Darwin and his son Francis performed some of the earliest experiments on phototropism (Figure 29.15). They began with the **observation** that grass seedlings bent toward light only if the tips of their shoots were present. This led them to **question** whether the tips of the seedlings produced some kind of chemical growth signal. Their **hypothesis** was that plant tips produced a growth signal in response to light. To test this hypothesis, they made a **prediction**: Removing a shoot tip or blocking its access to light would prevent phototropism. Several of their **experiments** are summarized in Figure 29.15.

The Darwins' **results** showed that removing the tip of a grass shoot did prevent growth toward light. The shoot also remained straight when they placed an opaque cap on its tip. However, the shoot curved normally when they placed a transparent cap on its tip or an opaque shield around its base. The Darwins concluded that the tip of the shoot was responsible for sensing light. They also recognized that the growth response, the bending of the shoot, occurred below the tip. Therefore, they speculated that some unknown growth signal was transmitted downward from the tip to the growing region of the shoot.

Building on the Darwins' results, botanists have since discovered the mechanisms of many other plant hormones. This understanding is one of the cornerstones of modern agriculture, allowing people to control the growth of plants in a variety of ways.

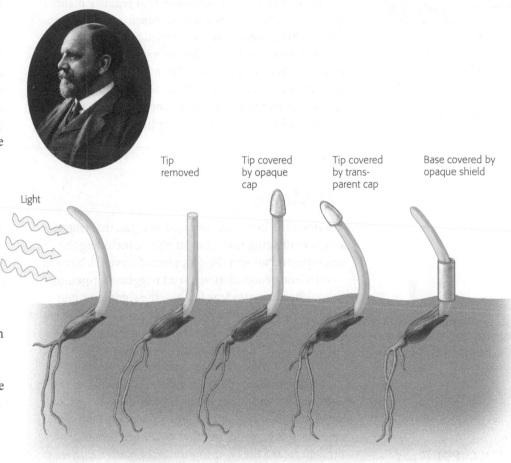

▲ Figure 29.15 **The Darwins' classic experiments on phototropism.** This art summarizes experiments conducted by Charles Darwin and his son Francis (pictured above). Together, these results suggested that a substance produced by the tip of a seedling controls phototropism.

The Action of Auxins

The hormones responsible for the phototropism observed by the Darwins are auxins produced by the apical meristem at the tip of a shoot. Illuminating one side of a shoot results in a concentration of auxin in the cells on the dark side. This uneven distribution of auxins makes cells on the dark side elongate, causing the shoot to bend.

Auxins promote cell elongation in stems only when they are present within a certain concentration range. Interestingly, this same concentration range inhibits cell elongation in roots. Above this range, auxins inhibit cell elongation in stems; below this range, they cause root cells to elongate. This complex set of effects reinforces two points: (1) Different concentrations of a hormone may have different effects in the same target cell, and (2) a particular concentration of a hormone may have different effects on different target cells.

In addition to causing stems and roots to lengthen, auxins promote growth in stem diameter. Furthermore, auxins made in developing seeds promote the growth of fruit. Farmers sometimes produce seedless tomatoes, cucumbers, and eggplants by spraying unpollinated plants with synthetic auxins.

667

The discovery of plant hormones has led to many advances in agriculture. The use of synthetic plant hormones allows more food to be produced at lower cost. For example, one of the most widely used herbicides, or weed killers, is a synthetic auxin that disrupts the normal balance of hormones that regulate plant growth. Because dicots are more sensitive than monocots to this herbicide, it can be used to selectively remove dandelions and other unwanted broadleaf dicot weeds from a monocot lawn or field of grain. By applying herbicides to cropland, a farmer can reduce the amount of plowing required to control weeds, thus reducing soil erosion, fuel consumption, and labor costs.

At the same time, there is growing concern that the heavy use of synthetic chemicals in food production may pose environmental and health hazards. Also, many consumers are concerned that foods produced using synthetic hormones may not be as tasty or nutritious as those raised naturally. This highlights an important question about modern agricultural techniques: Should we continue to produce cheap, plentiful food using synthetic chemicals and tolerate the potential problems, or should we put more of our agricultural effort into organic farming, recognizing that foods may be less plentiful and more expensive as a result? This is something you need to consider as you make choices about the foods you buy and eat. ☑

Ethylene

Ethylene is a hormone, released as a gas, that triggers a variety of aging responses in plants, including the ripening of fruit and the dropping of leaves. A burst of ethylene production in a fruit triggers its ripening.

Because of a gaseous hormone, one overripe apple can spoil the whole bunch.

Because ethylene is a gas, the signal to ripen spreads from fruit to fruit: One bad apple really can spoil the barrel! You can make fruit ripen faster if you store it with already ripened fruit in a bag so that the ethylene gas accumulates **(Figure 29.16)**. On a commercial scale, many kinds of fruit—such as tomatoes—are often picked green and then ripened in huge storage bins into which ethylene gas is piped. In other cases, growers take measures to retard the ripening action of natural ethylene. Stored apples are often flushed with CO_2, which inhibits the action of ethylene. In this way, apples picked in autumn can be stored for sale the following summer.

The loss of leaves in autumn is affected by ethylene. Leaf drop is triggered by environmental stimuli, including the shortening days and cooler temperatures of autumn. These stimuli cause a change in the balance of ethylene and auxins that weakens the cell walls in a layer of cells at the base of the leaf stalk. The weight of the leaf, often assisted by wind, causes this layer to split, releasing the leaf. ☑

▼ **Figure 29.16 Effect of ethylene on the ripening of bananas.** Two unripe bananas were stored for the same amount of time in bags. The banana on top was stored alone; the bottom banana was stored with an ethylene-releasing peach.

Banana alone

Banana plus ethylene-releasing peach

Cytokinins

Cytokinins are a group of closely related hormones that act as growth regulators. These regulators promote cell division, or cytokinesis. Cytokinins are produced in actively growing tissues, particularly in roots, embryos, and fruits.

Cytokinins stimulate the growth of axillary buds, making a plant grow more branches and become bushy.

Christmas tree growers sometimes use cytokinins to produce attractive branching. Cytokinins entering the shoot system from the roots counter the inhibitory effects of auxins coming down from the terminal buds. The complex growth patterns of most plants probably result from the relative concentrations of auxins and cytokinins.

Gibberellins

Gibberellins are another group of growth-regulating hormones. Roots and young leaves are major sites of gibberellin production. One of the main effects of gibberellins is to stimulate cell elongation and cell division in stems and leaves. This action generally enhances that of auxins. Also in combination with auxins, gibberellins can influence fruit development, and gibberellin-auxin sprays can make apples, currants, and eggplants develop without pollination and seed production. Gibberellins are used commercially to make seedless grapes grow larger and farther apart in a cluster **(Figure 29.17)**.

Gibberellins are also important in seed germination in many plants. Many seeds that require special environmental conditions to germinate, such as exposure to light or cold, will germinate when sprayed with gibberellins. In nature, gibberellins in seeds are probably the link between environmental cues and the metabolic processes that reactivate growth of the embryo after a period of dormancy. For example, when water becomes available to a grass seed, it causes the embryo in the seed to release gibberellins, which promote germination by mobilizing nutrients stored in the seed.

Abscisic Acid

Unlike the growth-stimulating hormones we have studied so far—auxin, cytokinins, and gibberellins—**abscisic acid** *slows* growth. One of the times in a plant's life when it is advantageous to suspend growth is at the onset of seed dormancy (inactivity of the seed). What prevents a seed dispersed in autumn from germinating immediately, only to be killed by cold temperatures in winter? Seed dormancy is an evolutionary adaptation that ensures that a seed will germinate only when the conditions of light, temperature, and moisture are appropriate for growth.

Many types of dormant seeds will germinate only when abscisic acid is removed or inactivated. The seeds of some plants—such as the perennials rosemary and rhubarb—require prolonged exposure to cold during the winter to inactivate abscisic acid and trigger seed germination in the spring. The seeds of some desert plants remain dormant for years (even decades) until a sufficient downpour washes out the abscisic acid, allowing them to germinate only when water is available **(Figure 29.18)**. ABA often counteracts the actions of growth hormones. Therefore, for many plants, the ratio of abscisic acid (which inhibits seed germination) to gibberellins (which promote seed germination) determines whether the seed will germinate or remain dormant. ☑

▼ Figure 29.17 **Effect of gibberellins on grapes.** The right cluster of grapes shows the effect of gibberellin treatment: larger grapes farther apart in the cluster.

▼ Figure 29.18 **The effect of abscisic acid removal on seed dormancy.** These wildflowers grew in Anza-Borrego Desert State Park in southern California just after a hard rain washed abscisic acid out of the seeds, allowing them to germinate.

Untreated

Treated with gibberellins

Response to Stimuli

In this section, we'll consider how a plant responds to physical stimuli from the environment—including light, touch, and gravity.

Tropisms

Tropisms are directed growth responses that cause parts of a plant to grow toward or away from a stimulus **(Figure 29.19)**. Phototropism, as you saw in Figure 29.13, is a particularly important example, but there are other directed growth responses. For example, **thigmotropism** is a growth in response to touch, as when a pea tendril contacts a string or wire and coils around it for support. The tendril in the center photo in Figure 29.19 grew straight until it touched the support. Contact then caused different growth rates in cells on opposite sides of the tendril, making the tendril coil around the wire.

Gravitropism is the directional growth of a plant organ in response to gravity: Shoots grow upward and roots grow downward, regardless of the orientation of the seed. Auxins play an important role in gravitropism. The seedling shown on the right in Figure 29.19 responded to gravity by redistributing auxins to the lower side of its horizontally growing root. A high concentration of auxins inhibits the growth of root cells. As growth of the lower side of the root was slowed, cells on the upper side continued to elongate normally, and the root curved downward.

Photoperiod

In addition to providing energy for photosynthesis and directing growth, light helps regulate a plant's life cycle. Flowering, seed germination, and the onset and end of dormancy are stages in plant development that usually occur at specific times of the year. The environmental stimulus that plants most often use to detect the time of year is called **photoperiod**, the relative lengths of day and night.

Plants whose flowering is triggered by photoperiod fall into two groups: long-night plants and short-night plants **(Figure 29.20)**. (Originally, botanists referred to these plants as short- and long-day plants, but we now know that it is night length, not day length, that affects a plant's flowering.) Long-night plants, such as chrysanthemum and poinsettia, generally flower in fall or winter, when nights lengthen. Short-night plants, such as lettuce, iris, and many grains, usually flower in spring or summer, when nights are briefer. Some plants, such as dandelions, are night-neutral; their flowering is unaffected by photoperiod. Florists apply knowledge of the photoperiod of particular plants to bring us flowers out of season. The blooming of chrysanthemums, for instance, can be stalled until spring by interrupting each long night with a flash of light, thus turning one long night into two short nights. ☑

☑ **CHECKPOINT**

1. Some species of ivy grow along tree trunks, always staying in close contact with the host tree. This is an example of what kind of growth response?
2. A particular long-night plant won't flower in the spring. You try to induce flowering by using a short, dark interruption to split a long-day period of spring into two short-day periods. What result do you predict?

Answers: 1. thigmotropism 2. The plants still won't flower, because it is actually night length, not day length, that controls flowering.

▼ **Figure 29.19 Tropisms.**

TROPISMS
(directed growth responses)

Phototropism

These seedlings bending toward the light are young cilantro plants (a common herb).

Thigmotropism

Thigmotropism, growth in response to touch, allows vining plants to wrap around support structures. In this photo, the tendril of a pea plant coils around a wire fence.

Gravitropism

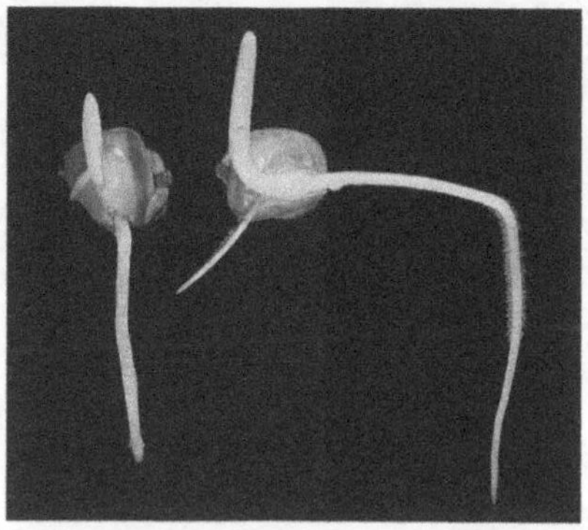

These seedlings were both germinated in the dark. The one on the left was left alone. The one on the right was turned on its side two days after germination so that the shoot and root were horizontal. When this photo was taken, the shoot had turned back upward and the root had turned down.

▼ Figure 29.20 **Photoperiod and flowering.** Photoperiod, the relative lengths of day and night, can trigger flowering.

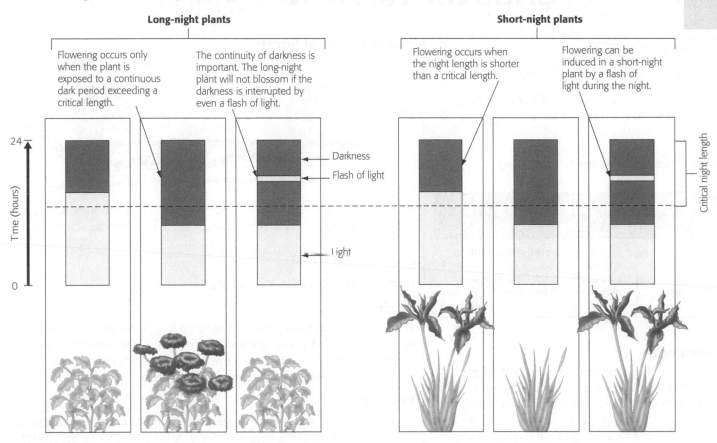

Long-night plants

Flowering occurs only when the plant is exposed to a continuous dark period exceeding a critical length.

The continuity of darkness is important. The long-night plant will not blossom if the darkness is interrupted by even a flash of light.

Short-night plants

Flowering occurs when the night length is shorter than a critical length.

Flowering can be induced in a short-night plant by a flash of light during the night.

Darkness

Flash of light

Light

Critical night length

Time (hours)

24

0

The Interdependence of Organisms EVOLUTION CONNECTION

Plants, Bugs, and People

As you learned in the last chapter, most flowering plants depend on insects or other animals for pollination and seed dispersal. In this chapter, you saw that plants rely on organisms from two other kingdoms to help acquire nutrients: soil bacteria and the fungi of mycorrhizae. Indeed, the mycorrhizal connection may have altered the entire course of evolution by helping make possible the colonization of land. Today, nearly all land animals depend on plants (or animals that eat plants) for food.

Both the cover of this book and **Figure 29.21** show thorn bugs (*Umbonia crassicornis*), a particularly striking member of the family Membracidae, a group of insects commonly called treehoppers. Within subtropical regions of North America, thorn bugs feed and lay eggs within the stems of various plant species including shrubs, grasses, and hardwood trees. Thorn bugs may form mutualistic relationships with various other species, including ants, wasps, and geckos. In turn, thorn bugs are prey for many

species of birds and lizards; they also have to contend with parasitic fungi that feed on the digested sap that they secrete. And thorn bugs are of concern to humans because some of the plants they feed on are important agricultural crops, such as avocadoes.

Thus, we close this chapter—and this book— illustrating one of biology's overarching themes: the interrelatedness of organisms. Animals depend on other animals and plants, which in turn depend on animals, fungi, and protists. And all eukaryotic life depends on prokaryotes. This interdependency reminds us that it is impossible to separate ourselves from all of the living creatures that share the biosphere. We hope that *Campbell Essential Biology with Physiology* has given you a new appreciation of your place in the living world.

▼ Figure 29.21 **The interdependence of organisms.** These thorn bugs emphasize how animals, plants, fungi—indeed all life— are interrelated.

Chapter Review

SUMMARY OF KEY CONCEPTS

How Plants Acquire and Transport Nutrients

As a plant grows, its leaves absorb CO_2 from the air, and its roots absorb water and minerals (inorganic ions) from the soil.

Plant Nutrition

A plant must obtain the nutrients it requires from its surroundings.

Nutrients	
Macronutrients	**Micronutrients**
• Needed in large amounts • Used to build organic molecules • Carbon, oxygen, hydrogen, nitrogen, sulfur, phosphorus, calcium, potassium, magnesium	• Reusable, so needed in much smaller amounts • Iron, chlorine, copper, manganese, zinc, molybdenum, boron, nickel

From the Soil into the Roots

Root hairs greatly increase a root's absorbing surface. Water and solutes move through the root's epidermis and cortex and then into the xylem (water-conducting tissue) for transport upward. Relationships with other organisms help plants obtain nutrients. Many plants form mycorrhizae, mutually beneficial associations with fungi. A network of fungal threads increases a plant's absorption of water and mineral nutrients, and the fungus receives sugars from the plant.

Energy Transformations: The Role of Bacteria in Nitrogen Nutrition

Most plants depend on bacteria to convert nitrogen to a form that is usable by the plant. Most plants depend on multiple types of bacteria to convert atmospheric N_2 (which is abundant but unusable) into forms of nitrogen that can be absorbed and used. Legumes and certain other plants have nodules in their roots that house vesicles filled with nitrogen-fixing bacteria.

The Transport of Water

Water (and the solutes dissolved in it) moves upward from the roots within xylem tissue due to transpiration.

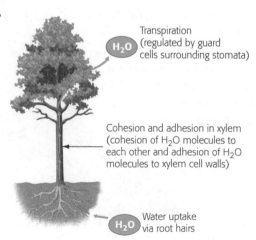

Transpiration (regulated by guard cells surrounding stomata)

H_2O

Cohesion and adhesion in xylem (cohesion of H_2O molecules to each other and adhesion of H_2O molecules to xylem cell walls)

H_2O Water uptake via root hairs

The Transport of Sugars

Within phloem, a pressure-flow mechanism transports sugar molecules.

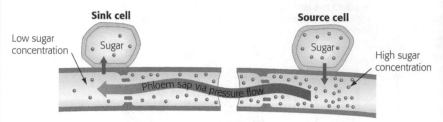

Sink cell
Low sugar concentration
Sugar

Source cell
Sugar
High sugar concentration

Phloem sap via pressure flow

An increase in sugar concentration and water pressure at the sugar source and a decrease at the sugar sink cause phloem sap to flow from source to sink.

Plant Hormones

Hormones coordinate the activities of plants.

Auxins

Phototropism, directional growth in response to light, results from faster cell growth on the shaded side of the shoot. This effect is regulated by auxins. Plants produce auxins in the apical meristems at the tips of shoots. At different concentrations, auxins stimulate or inhibit elongation.

Ethylene

As a fruit ages, it gives off ethylene gas, which hastens ripening. A changing ratio of auxins to ethylene, triggered mainly by longer nights, causes the loss of leaves from certain trees.

Cytokinins

Cytokinins, produced by growing roots, embryos, and fruits, are hormones that promote cell division. The ratio of auxins to cytokinins helps coordinate the growth of roots and shoots.

Gibberellins

Gibberellins stimulate the elongation of stems and leaves and the development of fruits. Gibberellins released from embryos function in some of the early events of seed germination. Auxins and gibberellins are used in agriculture to produce seedless fruits.

Abscisic Acid

Abscisic acid inhibits the germination of seeds. Seeds of many plants remain dormant until their abscisic acid is inactivated or washed away.

Response to Stimuli

Tropisms

Tropisms are growth responses that make a plant grow toward or away from a stimulus. Important tropisms include phototropism, thigmotropism (a response to touch), and gravitropism (a response to gravity).

Photoperiod

Plants mark the seasons by measuring photoperiod, the relative lengths of night and day. The timing of flowering is one of the seasonal responses to photoperiod. Long-night plants flower when nights exceed a certain critical length; short-night plants flower when nights are shorter than a critical length.

MasteringBiology®

For practice quizzes, BioFlix animations, MP3 tutorials, video tutors, and more study tools designed for this textbook, go to MasteringBiology®

SELF-QUIZ

1. Certain fungi cause diseases in plants. There are a variety of antifungal sprays that can be used to control this problem. Some gardeners constantly spray their plants with fungicides, even when no signs of disease are evident. How might this be disadvantageous to the plant?

2. Which of the following best describes the role of root nodules in plants such as peas and beans?
   a. They house nitrogen-fixing bacteria.
   b. They help absorb nutrients from the soil.
   c. They are plant adaptations to kill bacteria that destroy root hairs.
   d. They store nutrients.

3. Plants transport two types of sap. _____ sap, a solution of mostly water and sucrose, is transported from sites of sugar production to other parts of the plant. _____ sap, a solution of mostly water and inorganic ions, is transported from the roots to the rest of the plant body.

4. Contrast cohesion and adhesion, and describe the role of each in the ascent of xylem sap.

5. Which of the following best describes the condition of a twig if the auxin produced by its apical meristem is conveyed equally down all sides?
   a. It will elongate.
   b. It will branch near its tip.
   c. It will produce a flower.
   d. It will bend to one side.

6. The status of axillary buds—dormant or growing—depends on the relative concentrations of _____ moving down from the shoot tip and _____ moving up from the roots.

7. Why do many plants become bushier if you pinch off their terminal buds?

8. Match each of the following hormones to its primary role in a plant.
   a. auxins
   b. ethylene
   c. cytokinins
   d. gibberellins
   e. abscisic acid

   1. at different concentrations, stimulate or inhibit the elongation of shoots and roots
   2. produced by roots, promote cell division
   3. inhibits seed germination
   4. hastens fruit ripening
   5. promote fruit development

9. Match each of the following terms to its meaning.
   a. cohesion-tension hypothesis
   b. pressure-flow mechanism
   c. phototropism
   d. photoperiod

   1. growth response to light
   2. relative lengths of night and day
   3. solute redistribution by differential osmosis
   4. transpirational pull on a cohesive water column

Answers to these questions can be found in Appendix: Self-Quiz Answers.

THE PROCESS OF SCIENCE

10. A potted plant in your house looks a bit droopy. Having read this book, you know that the cell walls of plants, which contribute to a plant's stability, are rich in cellulose, a polysaccharide. So, when watering the plant, will it be a good idea to add sugar to the water to help the plant regenerate?

11. **Interpreting Data** A botanist from Harvard University hypothesized that garlic mustard, an invasive plant found in forests of the northeastern United States, suppresses the growth of trees by harming mycorrhizal fungi in nearby soil. To test this hypothesis, soil samples from invaded and uninvaded woodlands were collected, and half the samples were sterilized (to kill any fungi present). Two species of maple tree seedlings were planted, and the growth was calculated (as a percentage increase in the mass of the plant). The graph below presents the data. Do these data support the hypothesis that garlic mustard retards tree growth by harming soil fungi?

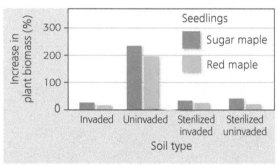

Source K. A. Stinson et al., Invasive plant suppresses the growth of native tree seedlings by disrupting belowground mutualisms, *Public Library of Science: Biology* 4(5): e140 (2006).

BIOLOGY AND SOCIETY

12. Bioremediation is the use of living organisms to detoxify and restore polluted ecosystems. Bioremediation can involve different types of organisms, such as bacteria, protists, and plants. What advantages do you think bioremediation presents over other means of cleaning up polluted sites? If your neighborhood became polluted by a natural disaster, would you prefer bioremediation to more traditional means of cleanup? What if it cost extra or took longer?

13. Students of biology often tend to find the botany modules of their courses less engaging than the zoology modules. After reading this book, how do you feel about the botanical versus the zoological aspects of biology? In your opinion, which one is more important in influencing our quality of life?

Appendix A Metric Conversion Table

Measurement	Unit and Abbreviation	Metric Equivalent	Approximate Metric-to-English Conversion Factor	Approximate English-to-Metric Conversion Factor
Length	1 kilometer (km)	= 1,000 (10^3) meters	1 km = 0.6 mile	1 mile = 1.6 km
	1 meter (m)	= 100 (10^2) centimeters	1 m = 1.1 yards	1 yard = 0.9 m
		= 1,000 millimeters	1 m = 3.3 feet	1 foot = 0.3 m
			1 m = 39.4 inches	
	1 centimeter (cm)	= 0.01(10^{-2}) meter	1 cm = 0.4 inch	1 foot = 30.5 cm
				1 inch = 2.5 cm
	1 millimeter (mm)	= 0.001 (10^{-3}) meter	1 mm = 0.04 inch	
	1 micrometer (μm)	= 10^{-6} meter (10^{-3} mm)		
	1 nanometer (nm)	= 10^{-9} meter (10^{-3} μm)		
	1 angstrom (Å)	= 10^{-10} meter (10^{-4} μm)		
Area	1 hectare (ha)	= 10,000 square meters	1 ha = 2.5 acres	1 acre = 0.4 ha
	1 square meter (m²)	= 10,000 square centimeters	1 m² = 1.2 square yards	1 square yard = 0.8 m²
			1 m² = 10.8 square feet	1 square foot = 0.09 m²
	1 square centimeter (cm²)	= 100 square millimeters	1 cm² = 0.16 square inch	1 square inch = 6.5 cm²
Mass	1 metric ton (t)	= 1,000 kilograms	1 t = 1.1 tons	1 ton = 0.91 t
	1 kilogram (kg)	= 1,000 grams	1 kg = 2.2 pounds	1 pound = 0.45 kg
	1 gram (g)	= 1,000 milligrams	1 g = 0.04 ounce	1 ounce = 28.35 g
			1 g = 15.4 grains	
	1 milligram (mg)	= 10^{-3} gram	1 mg = 0.02 grain	
	1 microgram (μg)	= 10^{-6} gram		
Volume *(solids)*	1 cubic meter (m³)	= 1,000,000 cubic centimeters	1 m³ = 1.3 cubic yards	1 cubic yard = 0.8 m³
			1 m³ = 35.3 cubic feet	1 cubic foot = 0.03 m³
	1 cubic centimeter (cm³ or cc)	= 10^{-6} cubic meter	1 cm³ = 0.06 cubic inch	1 cubic inch = 16.4 cm³
	1 cubic millimeter (mm³)	= 10^{-9} cubic meter (10^{-3} cubic centimeter)		
Volume *(liquids and gases)*	1 kiloliter (kL or kl)	= 1,000 liters	1 kL = 264.2 gallons	1 gallon = 3.79 L
	1 liter (L)	= 1,000 milliliters	1 L = 0.26 gallon	1 quart = 0.95 L
			1 L = 1.06 quarts	
	1 milliliter (mL or ml)	= 10^{-3} liter	1 mL = 0.03 fluid ounce	1 quart = 946 mL
		= 1 cubic centimeter	1 mL = approx. ¼ teaspoon	1 pint = 473 mL
			1 mL = approx. 15–16 drops	1 fluid ounce = 29.6 mL
	1 microliter (μL or μl)	= 10^{-6} liter (10^{-3} milliliters)		1 teaspoon = approx. 5 mL
Time	1 second (s)	= $\frac{1}{60}$ minute		
	1 millisecond (ms)	= 10^{-3} second		
Temperature	Degrees Celsius (°C)		°F = $\frac{9}{5}$°C + 32	°C = $\frac{5}{9}$(°F − 32)

Appendix B The Periodic Table

Atomic number (number of protons) · Element symbol · Atomic mass (number of protons plus number of neutrons averaged over all isotopes)

Metals · Metalloids · Nonmetals

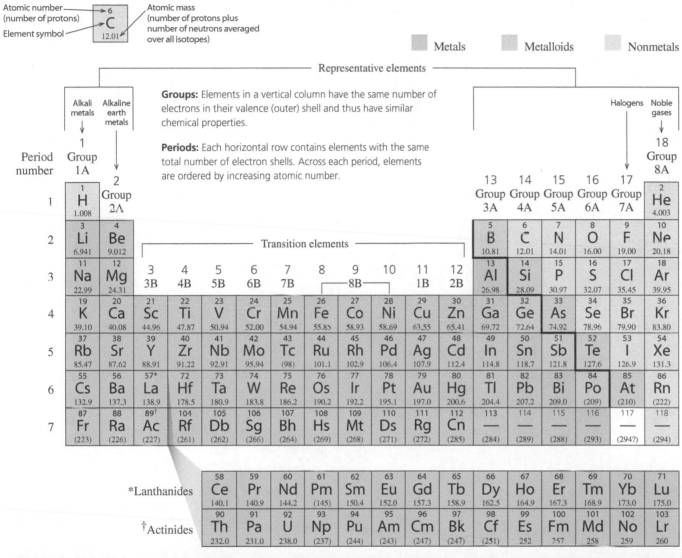

Groups: Elements in a vertical column have the same number of electrons in their valence (outer) shell and thus have similar chemical properties.

Periods: Each horizontal row contains elements with the same total number of electron shells. Across each period, elements are ordered by increasing atomic number.

Name (Symbol)	Atomic Number	Name (Symbol)	Atomic Number	Name (Symbol)	Atomic Number	Name (Symbol)	Atomic Number	Name (Symbol)	Atomic Number
Actinium (Ac)	89	Copernicium (Cn)	112	Iridium (Ir)	77	Palladium (Pd)	46	Sodium (Na)	11
Aluminum (Al)	13	Copper (Cu)	29	Iron (Fe)	26	Phosphorus (P)	15	Strontium (Sr)	38
Americium (Am)	95	Curium (Cm)	96	Krypton (Kr)	36	Platinum (Pt)	78	Sulfur (S)	16
Antimony (Sb)	51	Darmstadtium (Ds)	110	Lanthanum (La)	57	Plutonium (Pu)	94	Tantalum (Ta)	73
Argon (Ar)	18	Dubnium (Db)	105	Lawrencium (Lr)	103	Polonium (Po)	84	Technetium (Tc)	43
Arsenic (As)	33	Dysprosium (Dy)	66	Lead (Pb)	82	Potassium (K)	19	Tellurium (Te)	52
Astatine (At)	85	Einsteinium (Es)	99	Lithium (Li)	3	Praseodymium (Pr)	59	Terbium (Tb)	65
Barium (Ba)	56	Erbium (Er)	68	Lutetium (Lu)	71	Promethium (Pm)	61	Thallium (Tl)	81
Berkelium (Bk)	97	Europium (Eu)	63	Magnesium (Mg)	12	Protactinium (Pa)	91	Thorium (Th)	90
Beryllium (Be)	4	Fermium (Fm)	100	Manganese (Mn)	25	Radium (Ra)	88	Thulium (Tm)	69
Bismuth (Bi)	83	Fluorine (F)	9	Meitnerium (Mt)	109	Radon (Rn)	86	Tin (Sn)	50
Bohrium (Bh)	107	Francium (Fr)	87	Mendelevium (Md)	101	Rhenium (Re)	75	Titanium (Ti)	22
Boron (B)	5	Gadolinium (Gd)	64	Mercury (Hg)	80	Rhodium (Rh)	45	Tungsten (W)	74
Bromine (Br)	35	Gallium (Ga)	31	Molybdenum (Mo)	42	Roentgenium (Rg)	111	Uranium (U)	92
Cadmium (Cd)	48	Germanium (Ge)	32	Neodymium (Nd)	60	Rubidium (Rb)	37	Vanadium (V)	23
Calcium (Ca)	20	Gold (Au)	79	Neon (Ne)	10	Ruthenium (Ru)	44	Xenon (Xe)	54
Californium (Cf)	98	Hafnium (Hf)	72	Neptunium (Np)	93	Rutherfordium (Rf)	104	Ytterbium (Yb)	70
Carbon (C)	6	Hassium (Hs)	108	Nickel (Ni)	28	Samarium (Sm)	62	Yttrium (Y)	39
Cerium (Ce)	58	Helium (He)	2	Niobium (Nb)	41	Scandium (Sc)	21	Zinc (Zn)	30
Cesium (Cs)	55	Holmium (Ho)	67	Nitrogen (N)	7	Seaborgium (Sg)	106	Zirconium (Zr)	40
Chlorine (Cl)	17	Hydrogen (H)	1	Nobelium (No)	102	Selenium (Se)	34		
Chromium (Cr)	24	Indium (In)	49	Osmium (Os)	76	Silicon (Si)	14		
Cobalt (Co)	27	Iodine (I)	53	Oxygen (O)	8	Silver (Ag)	47		

677

Appendix C Credits

PHOTO CREDITS

UNIT OPENERS: Unit 1 top left Vaclav Volrab/ Shutterstock; **unit 1 left** Stone/Getty Images; **unit 2 top left** Nhpa/Age Fotostock; **unit 3 top left** Gay Bumgarner/Alamy; **unit 3 left** Eric Isselee/Shutterstock; **unit 4 top left** Jean-Paul Ferrero/Mary Evans Picture Library Ltd/Age Fotostock; **unit 4 left** Masyanya/Shutterstock; **unit 6 top right** John Coletti/Getty Images; **unit 6 left** Vilor/Shutterstock.

CHAPTER 1: Chapter opening photo right David Malan/Getty Images; **chapter opening photo bottom** Katherine Dickey; **chapter opening photo top** Eric J. Simon; **p. 37 right** Eric J. Simon; **1.1 left** M. I. Walker/ Science Source; **1.1 center** SPL/Science Source; **1.1 right** Educational Images Ltd./Custom Medical Stock; **1.2 left** Michael Nichols/National Geographic/Getty Images; **1.2 right** Tim Ridley/Dorling Kindersley; **1.3** Adrian Sherratt/ Alamy; **1.4a p. 40** xtock/Shutterstock; **1.4b p. 40** Lorne Chapman/Alamy; **1.4c p. 40** Trevor Kelly/Shutterstock; **1.4d p. 40** Eric J. Simon; **1.4e p. 41** Cathy Keife/Fotolia; **1.4f p. 41** Eric J. Simon; **1.4g p. 41** Redchanka/Shutterstock; **1.5** NASA/JPL California Institute of Technology; **1.6** Edwin Verin/Shutterstock; **1.7 top to bottom** Science Source; Neil Fletcher/Dorling Kindersley; Kristin Piljay/Lonely Planet Images/Getty Images; Stockbyte/Getty Images; Dr. D. P. Wilson/Science Source; **1.10 bottom** Michael Nolan/Robert Harding; **1.10 top left** Mike Hayward/Alamy; **1.10 top right** Public domain/Wikipedia; **1.11 top left** The Natural History Museum/Alamy; **1.13 left to right** Africa Studio/ Shutterstock; Pablo Paul/Alamy; Foodcollection/Alamy; Barbro Bergfeldt/Shutterstock; Volff/Fotolia; Sarycheva Olesia/Shutterstock; **1.14 left to right** Eric Baccega/Nature Picture Library; Erik Lam/Shutterstock; **1.15** Martin Dohrn/ Royal College of Surgeons/Science Source; **1.16** Susumu Nishinaga/Science Source; **1.18** Ian Hooton/Science Source; **1.19** Sergey Novikov/Shutterstock; **1.20 top to bottom** NASA Goddard Institute for Space Studies and Surface Temperature Analysis; Prasit Chansareekorn/Moment Collection/Getty Images; Dave G. Houser/Alamy; Biology Pics/Science Source.

CHAPTER 2: Chapter opening photo top Kevin Grant/ Fotolia; **chapter opening photo bottom** PM Images/ Ocean/Corbis; **chapter opening photo right** Ji Zhou/ Shutterstock; **p. 23 right** Vaclav Volrab/Shutterstock **p. 58 top** Geoff Dann/Dorling Kindersley; **p. 58 bottom** Clive Streeter/Dorling Kindersley; **2.3 woman left** Ivan Polunin/ Photoshot; **montage clockwise from bottom left** Jiri Hera/ Fotolia; Anne Dowie/Pearson Education; Howard Shooter/ Dorling Kindersley; Alessio Cola/Shutterstock; Ulkastudio/ Shutterstock; **2.5** Raguet H/Age Fotostock; **2.9** William Allum/Shutterstock; **2.10 left** Stephen Alvarez/National Geographic/Getty Images; **2.10 right** Andrew Syred/ Science Source; **2.11** Alasdair Thomson/Getty Images; **2.12** Ohn Leyba/The Denver Post/Getty Images; **2.13** Nexus 7/ Shutterstock; **2.14** Kristin Piljay/Pearson Education; **p. 65 left** Doug Allan/Science Source; **2.15 top to bottom** Beth Van Trees/Shutterstock; Steve Gschmeissner/Science Source; VR Photos/Shutterstock; Terekhov igor/Shutterstock; Jsemeniuk/ E+/Getty Images; **2.16** Luiz A. Rocha/Shutterstock.

CHAPTER 3: Chapter opening photo top left Nikola Bilic/ Shutterstock; **chapter opening photo top right** Monkey Business/Fotolia; **chapter opening photo bottom right** Brian P. Hogan; **chapter opening photo bottom left** Eric J. Simon; **p. 71 right** Sean Justice/Cardinal/Corbis; **p. 71 top left** Sean Justice/Cardinal/Corbis; **3.3 left** James Marvin Phelps/Shutterstock; **3.3 right** Ilolab/Fotolia; **p. 73 center** Gavran333/Fotolia; **3.5** StudioSmart/Shutterstock; **3.8 bottom** Kristin Piljay/Pearson Education; **3.8 top** Stieberszabolcs/Fotolia; **3.9 top** Science Source; **3.9 center** Dr. Lloyd M. Beidler; **3.9 bottom** Biophoto Associates/ Science Source; **3.10** Martyn F. Chillmaid/SPL/Science Source; **3.12 left box clockwise from left** Thomas M Perkins/Shutterstock; Hannamariah/Shutterstock; Valentin Mosichev/Shutterstock; Kellis/Fotolia; **3.12 right box clockwise from left** Kayros Studio/Fotolia; Multiart/ Shutterstock; Maksim Shebeko/Fotolia; Subbotina Anna/Fotolia; Vladm/Shutterstock; **3.13 left** EcoPrint/ Shutterstock; **3.13 right** Stockbyte/Getty Images; **3.14 top left** Brian Cassella/Rapport Press/Newscom; **3.14 top right** Matthew Cavanaugh/epa/Corbis Wire/Corbis; **3.14 bottom left** Gabriel Bouys/AFP/Getty Images; **3.14 bottom right** Shannon Stapleton/Reuters; **3.15 left to right** GlowImages/ Alamy; Geoff Dann/Dorling Kindersley; Dave King/Dorling Kindersley; Tim Parmenter/Dorling Kindersley; Starsstudio/ Fotolia; Steve Gschmeissner/Science Source; Kristin Piljay/ Pearson Education; **3.19b** Ingram Publishing/Vetta/Getty Images; **3.20 top to bottom** Science Source/Science Source; Michael Patrick O'Neill/Alamy; Stefan Wackerhagen/ imageBROKER/Newscom; **3.27** Ralph Morse/Contributor/ Time Life Pictures/Getty Images.

CHAPTER 4: Chapter opening photo top left Asfloro/ Fotolia; **chapter opening photo top right** Lily/Fotolia; **chapter opening photo bottom left** Jennifer Waters/ Science Source; **chapter opening photo bottom right** Viktor Fischer/Alamy; **p. 89 right** SPL/Science Source; **p. 89 top left** SPL/Science Source; **p. 90 top left** Tmc_photos/ Fotolia; **4.5** Science Source; **4.6 left** Biophoto Associates/ Science Source; **4.6 right** Don W. Fawcett/Science Source; **4.9** MedImage/Science Source; **4.13 left** SPL/Science Source; **4.14 right** Daniel S. Friend; **4.15a** Michael Abbey/Science Source; **4.15b** Dr. Jeremy Burgess/Science Source; **4.17** Biology Pics/Science Source; **4.18** Daniel S. Friend; **4.19a** Dr. Torsten Wittmann/Science Source; **4.19b** Roland Birke/ Photolibrary/Getty Images; **4.20a** Eye of Science/Science Source; **4.20b** Science Source; **4.20c** Charles Daghlian/ Science Source.

CHAPTER 5: Chapter opening photo bottom left Jdwfoto/ Fotolia; **chapter opening photo bottom right** William87/ Fotolia; **chapter opening photo top left** Dmitrimaruta/ Fotolia; **chapter opening photo top right** Eric J. Simon; **p. 109 right** Don W. Fawcett/Science Source; **p. 109 top left** Don W. Fawcett/Science Source; **5.1** Stephen Simpson/ Photolibrary/Getty Images; **5.3a** George Doyle/Stockbyte/ Getty Images; **5.3b** Monkey Business/Fotolia; **5.19** Peter B. Armstrong.

CHAPTER 6: Chapter opening photo right National Motor Museum/Motoring Picture Library/Alamy; **chapter opening photo top** Allison Herreid/Shutterstock; **chapter opening photo bottom** Nickola_che/Fotolia; **p. 125 right** Philippe Psaila/Science Source; **p. 125 top left** Philippe Psaila/Science Source; **6.1** Eric J. Simon; **6.2 left to right** Dudarev Mikhail/Shutterstock; Eric Isselée/Shutterstock; **6.3** Dmitrimaruta/Fotolia; **6.12** Alex Staroseltsev/Shutterstock; **6.13** Maridav/Shutterstock; **6.16** Kristin Piljay/Alamy.

CHAPTER 7: Chapter opening photo top right NASA; **chapter opening photo top left** Sunny studio/Shutterstock; **chapter opening photo bottom** Mangostock/Shutterstock; **p. 141 right** Martin Bond/Science Source; **p. 141 top left** Martin Bond/Science Source; **7.1 left to right** Neil Fletcher/ Dorling Kindersley; Jeff Rotman/Nature Picture Library; Susan M. Barns, Ph.D; **7.2 top center** John Fielding/Dorling Kindersley; **7.2 top right** John Durham/Science Source; **7.2 center left** Biophoto Associates/Science Source; **7.6** Joyce/Fotolia; **7.7** JulietPhotography/Fotolia; **7.8b** Photos by LQ/Alamy; **7.14** Pascal Goetgheluck/Science Source.

CHAPTER 8: Chapter opening photo top left Franco Banfi/ Science Source; **chapter opening photo bottom left** Brian Jackson/Fotolia; **chapter opening photo bottom right** Zephyr/Science Source; **chapter opening photo top right** Andrew Syred/Science Source; **p. 155 right** Nhpa/Age Fotostock; **p. 155 top left** Nhpa/Age Fotostock; **8.1 p. 156 left to right** Dr. Torsten Wittmann/Science Source; Dr. Yorgos Nikas/Science Source; **8.1 p. 157 left to right** Biophoto Associates/Science Source; Image Quest Marine; John Beedle/Getty Images; **8.2 top to bottom** Eric Isselee/ Shutterstock; Christian Musat/Shutterstock; Michaeljung/ Fotolia; Eric Isselee/Shutterstock; Milton H. Gallardo; **8.3** Ed Reschke/Getty Images; **8.4** Biophoto Associates/ Science Source; **8.7** Conly L. Rieder, Ph.D; **8.8a** Don W. Fawcett/Science Source; **8.8b** Kent Wood/Science Source; **8.10** Sarahwolfephotography/Getty Images; **8.11** CNRI/ Science Source; **8.12** Iofoto/Shutterstock; **8.14** Ed Reschke/ Getty Images; **8.17** David M. Phillips/Science Source; **8.19** Dr. David Mark Welch; **8.22 bottom** CNRI/Science Source; **8.22 top** Lauren Shear/Science Source; **8.23** Nhpa/Superstock.

CHAPTER 9: Chapter opening photo top left Dora Zett/ Shutterstock; **chapter opening photo bottom** ICHIRO/ Getty Images; **chapter opening photo top right** Classic Image/Alamy; **p. 179 right** Eric J. Simon; **p. 179 top left** Eric J. Simon; **9.1** Science Source; **9.4** Patrick Lynch/ Alamy; **9.5** James King-Holmes/Science Source; **9.8** Martin Shields/Science Source; **9.9 left to right** Tracy Morgan/ Dorling Kindersley; Tracy Morgan/Dorling Kindersley; Eric Isselee/Shutterstock; Victoria Rak/Shutterstock; **9.10 left to right** Tracy Morgan/Dorling Kindersley; Eric Isselee/ Shutterstock; **9.12 top left to right** James Woodson/Getty Images; Ostill/Shutterstock; Oleksii Sergieiev/Fotolia; **9.12 bottom left to right** Blend Images - KidStock/ Getty Images; Image Source/Getty Images; Blend Images/ Shutterstock; **9.13 left to right** Jupiterimages/Stockbyte/ Getty Images; Diego Cervo/Fotolia; **Table 9.1 top to bottom** David Terrazas Morales/Corbis; Editorial Image, LLC/ Alamy; Eye of Science/Science Source; Science Source; **9.16 left to right** Cynoclub/Fotolia; CallallooAlexis/ Fotolia; **p. 190 center left** Eric J. Simon; **p. 191 bottom** Astier/BSIP SA/Alamy; **9.20** Mauro Fermariello/Science

Source; **9.21** Oliver Meckes & Nicole Ottawa/Science Source; **9.23** Eric J. Simon; **p. 195 center** Szabolcs Szekeres/Fotolia; **9.25 top left to right** In Green/Shutterstock; Rido/Shutterstock; **9.25 bottom left to right** Dave King/Dorling Kindersley; Jo Foord/Dorling Kindersley; **9.25 right** Andrew Syred/Science Source; **9.28** Archive Pics/Alamy; **9.29 left to right** Jerry Young/Dorling Kindersley; Gelpi/Fotolia; Dave King/Dorling Kindersley; Dave King/Dorling Kindersley; Tracy Morgan/Dorling Kindersley; Dave King/Dorling Kindersley; Jerry Young/Dorling Kindersley; Dave King/Dorling Kindersley; Dave King/Dorling Kindersley; Shutterstock; Tracy Morgan/Dorling Kindersley.

CHAPTER 10: Chapter opening photo bottom right Mopic/Alamy; **chapter opening photo top right** Oregon Health Sciences University, ho/AP Images; **chapter opening photo top left** Science Source; **chapter opening photo bottom left** Marcobarone/Fotolia; **p. 205 right** Stringer Russia/Kazbek Basaev/Reuters; **p. 205 top left** Stringer Russia/Kazbek Basaev/Reuters; **10.3 left to right** Barrington Brown/Science Source; Library of Congress; **10.11** Jay Cheng/Reuters; **10.22** Michael Follan - Mgfotouk.com/Getty Images; **10.23** Mixa/Alamy; **10.24** Oliver Meckes/Science Source; **10.25** Russell Kightley/Science Source; **10.26** N. Thomas/Science Source; **10.28** Hazel Appleton, Health Protection Agency Centre for Infections/Science Source; **10.29** Jeff Zelevansky JAZ/Reuters; **10.31** NIBSC/Science Photo Library/Science Source; **10.32** Will & Deni McIntyre/Science Source.

CHAPTER 11: Chapter opening photo top left Alila Medical Media/Shutterstock; **chapter opening photo right** Keren Su/Corbis; **chapter opening photo bottom left** Chubykin Arkady/Shutterstock; **p. 231 right** David McCarthy/Science Source; **p. 231 top left** David McCarthy/Science Source; **11.1 left to right** Steve Gschmeissner/Science Source; Steve Gschmeissner/Science Photo Library/Alamy; Ed Reschke/Getty Images; **11.4** Iuliia Lodia/Fotolia; **11.9 top to bottom** F. Rudolf Turner; F. Rudolf Turner; **11.10** American Association for the Advancement of Science; **11.11** Videowokart/Shutterstock; **p. 239 top center** Joseph T. Collins/Science Source; **11.13a** Courtesy of the Roslin Institute, Edinburgh; **11.13b** University of Missouri; **11.13c left to right** Pasqualino Loi; Robert Lanza; Robert Lanza/F. Rudolf Turner; **11.15** Mauro Fermariello/Science Source; **11.15 bottom right inset** Craig Hammell/Cord Blood Registry; **11.18 bottom** ERproductions Ltd/Blend Images/Alamy; **11.18 top** Simon Fraser/Royal Victoria Infirmary, Newcastle upon Tyne/Science Source; **11.19** Geo Martinez/Shutterstock; **11.21** Alastair Grant/AP Images; **11.22** CNRI/Science Source.

CHAPTER 12: Chapter opening photo top right Alexander Raths/Shutterstock; **chapter opening photo bottom** Stocksnapper/Shutterstock; **chapter opening photo top left** Gerd Guenther/Science Source; **p. 251 right** Tek Image/Science Source; **p. 251 top left** Tek Image/Science Source; **12.1** AP Images; **12.2** Huntington Potter/University of South Florida College of Medicine; **12.2 inset** Prof. S. Cohen/Science Source; **12.5** Andrew Brookes/National Physical Laboratory/Science Source; **12.6** Eric Carr/Alamy; **12.7** Volker Steger/

Science Source; **12.8** Inga Spence/Alamy; **12.9** Christopher Gable and Sally Gable/Dorling Kindersley; **12.9 inset** U.S. Department of Agriculture (USDA); **12.10 top to bottom** International Rice Research Institute; Fotosearch RM/Age Fotostock; **12.13** Applied Biosystems, Inc; **12.16** Steve Helber/AP Images; **12.17** Fine Art Images/Heritage Image Partnership Ltd/Alamy; **12.18** Volker Steger/Science Source; **12.19 bottom** David Parker/Science Photo Library/Science Source; **12.19 top** Edyta Pawlowska/Shutterstock; **12.22** Scott Camazine/Science Source; **12.23** James King-Holmes/Science Source; **12.24** 68/Ben Edwards/Ocean/Corbis; **12.25** Alex Milan Tracy/NurPhoto/Sipa U/Newscom; **12.26** Isak55/Shutterstock; **12.27** Image Point Fr/Shutterstock; **12.28** Public domain; **12.29** Daily Mail/Rex/Alamy.

CHAPTER 13: Chapter opening photo top left victoriaKh/Shutterstock; **chapter opening photo top center** BW Folsom/Shutterstock; **chapter opening photo top right** Parameswaran Pillai Karunakaran/FLPA; **chapter opening photo bottom right** John Bryant/Getty Images; **chapter opening photo center left** W. Perry Conway/Ramble/Corbis; **chapter opening photo bottom left** G. Newman Lowrance/AP Images; **p. 277 right** Gay Bumgarner/Alamy; **p. 243 top left** Gay Bumgarner/Alamy; **13.1** Aditya Singh/Moment/Getty Images; **13.2 left to right** Chris Pole/Shutterstock; Sabena Jane Blackbird/Alamy; **13.3 left** Science Source; **13.3 right** Classic Image/Alamy; **13.4a** Celso Diniz/Shutterstock; **13.4b** Tim Laman/National Geographic/Getty Images; **13.5 top left** Francois Gohier/Science Source; **13.5 center** Francois Gohier/Science Source; **13.5 top right** Science Source; **13.5 bottom left** Pixtal/SuperStock; **13.5 bottom right** Vostok Sarl/Mammuthus; **13.8 left to right** Dr. Keith Wheeler/Science Source; Lennart Nilsson/Tidningarnas Telelgrambyra AB Nilsson/TT Nyhetsbyrun; **13.10 top to bottom** Eric Isselee/Shutterstock; Tom Reichner/Shutterstock; Eric Isselee/Shutterstock; Christian Musat/Shutterstock; **13.11** Laura Jesse; **13.12** Zoonar/Poelzer/Age Fotostock; **13.13** Philip Wallick/Spirit/Corbis; **13.14** Edmund D. Brodie III; **13.15** Adam Jones/The Image Bank/Getty Images; **13.16** Andy Levin/Science Source; **13.18** Pearson Education; **13.21** Steve Bloom Images/Alamy; **13.22** Planetpix/Alamy; **13.23** Heather Angel/Natural Visions/Alamy; **13.24** Mariko Yuki/Shutterstock; **13.26a** Reinhard/ARCO/Nature Picture Library; **13.26b** John Cancalosi/Age Fotostock; **13.27** Centers for Disease Control and Prevention

CHAPTER 14: Chapter opening photo top left Keneva Photography/Shutterstock; **chapter opening photo right** Peter Augustin/Photodisc/Getty Images; **chapter opening photo bottom left** Oberhaeuser/Caro/Alamy; **chapter opening photo bottom right** GL Archive/Alamy; **p. 303 right** Kevin Schafer/Lithium/Age Fotostock; **p. 303 top left** Kevin Schafer/Lithium/Age Fotostock; **14.1 left to right** Cathleen A Clapper/Shutterstock; Peter Scoones/Nature Picture Library; **14.2 Clockwise from far left** Rolf Nussbaumer Photography/Alamy; David Kjaer/Nature Picture Library; Jupiterimages/Stockbyte/Getty images; Photos.com; Phil Date/Shutterstock; Robert Kneschke/Shutterstock; Comstock/Stockbyte/Getty Images; Comstock Photos/Fotosearch; **14.4 p. 306 left to right** USDA/

APHIS Animal and Plant Health Inspection Service; Brian Kentosh; McDonald/Photoshot Holdings Ltd.; Michelsohn Moses; **14.4 p. 307 left to right** J & C Sohns/Tier und Naturfotografie/Age Fotostock; Asami Takahiro; Danita Delimont/Gallo Images/Age Fotostock; **14.5 left** Brown Charles W; **14.5 center top to bottom** Dogist/Shutterstock; Alistair Duncan/Dorling Kindersley; Dorling Kindersley/Dorling Kindersley; **14.5 right** Okuno Kazutoshi; **14.6** Morey Milbradt/Getty Images; **14.6 top left inset** John Shaw/Photoshot; **14.6 top right inset** Clement Vezin/Fotolia; **14.8** Michelle Gilders/Alamy; **14.9** Loraart8/Shutterstock; **14.11 top to bottom** Interfoto/Alamy; Mary Plage/Oxford Scientific/Getty Images; Jim Clare/Nature Picture Library; **14.12** Robert Glusic/Corbis; **14.14** Kyodo/Xinhua/Photoshot/Newscom; **14.16 left to right** Jurgen & Christine Sohns/FLPA; Eugene Sergeev/Shutterstock; Kamonrat/Shutterstock; **14.17 right** Mark Pilkington/Geological Survey of Canada/SPL/Science Source; **14.18** Juniors Bildarchiv GmbH/Alamy; **14.19** Jean Kern; **14.20** Chris Hellier/Science Source; **14.21 left to right** Image Quest Marine; Christophe Courteau/Science Source; Reinhard Dirscherl/Alamy; Image Quest Marine; Image Quest Marine; **p. 320 top right** FloridaStock/Shutterstock.com; **14.23** National Geographic Image Collection/Alamy; **14.27** Arco Images GmbH/Alamy.

CHAPTER 15: Chapter opening photo bottom right vlabo/Shutterstock; **chapter opening photo bottom left** Harry Vorsteher/Cultura Creative (RF)/Alamy; **chapter opening photo bottom center** Odilon Dimier/PhotoAlto sas/Alamy; **p. 327 right** Steve Gschmeissner/Science Source; **p. 293 top left** Steve Gschmeissner/Science Source; **15.2** Mark Garlick/Science Photo Library/Corbis; **15.5** George Luther; **15.6 left to right** Jupiter Images; Dr. Tony Brain and David Parker/Science Photo Library/Science Source; **15.7 left to right** Scimat/Science Source; Niaid/CDC/Science Source; CNRI/SPL/Science Source; **15.8a** David M. Phillips/Science Source; **15.8b** Susan M. Barns; **15.8c** Heide Schulz/Max-Planck Institut für Marine Mikrobiologie; **15.9** Science Photo Library - Steve Gschmeissner/Brand X Pictures/Getty Images; **15.10** Eye of Science/Science Source; **15.11a** Sinclair Stammers/Science Source; **15.11b** Dr. Gary Gaugler/Science Source; **15.12** Image Quest Marine; **15.13 inset** Arco Images/Huetter, C./Alamy; **15.13** Nigel Cattlin/Alamy; **15.15** SIPA USA/SIPA/Newscom; **15.16** Eric J. Simon; **15.17 left** Jim West/Alamy; **15.18 right** SPL/Science Source; **15.19 left to right** Centers for Disease Control and Prevention (CDC); Scott Camazine/Science Source; Dariusz Majgier/Shutterstock; David M. Phillips/Science Source; **15.22a** Carol Buchanan/F1online/Age Fotostock; **15.22b** Oliver Meckes/Science Source; **15.22c** blickwinkel/Alamy; **15.23 p. 342 left to right** Eye of Science/Science Source; David M. Phillips/The Population Council/Science Source; Biophoto Associates/Science Source; **15.23 p. 343 left to right** Claude Carre/Science Source; Dr. Masamichi Aikawa; Michael Abbey/Science Source; **15.24** The Hidden Forest, www.hiddenforest.co.nz; **15.25 right top to bottom** Courtesy of Matt Springer, Stanford University; Courtesy of Robert Kay, MRC Cambridge; Courtesy of Robert Kay, MRC Cambridge; **15.26a** Eye of Science/Science Source;

15.26b Steve Gschmeissner/Science Source; **15.26c** Manfred Kage/Science Source; **15.27 left to right** Marevision/Age Fotostock; Marevision/Age Fotostock; David Hall/Science Source; **15.28** Lucidio Studio, Inc/ Moment/Getty Images.

CHAPTER 16: Chapter opening photo bottom left Francesco de marco/Shutterstock; **chapter opening photo top left** Neale Clarke/Robert Harding; **chapter opening photo bottom right** Scenics & Science/Alamy; **chapter opening photo top right** Webphotographeer/E+/ Getty Images; **p. 349 right** Marco Mayer/Shutterstock; **p. 349 top left** Marco Mayer/Shutterstock; **16.2** Science Source; **16.3** Steve Gorton/Dorling Kindersley; **16.4** Kent Graham; **16.5** Linda E. Graham; **16.7 left to right** Photolibrary/Getty Images; James Randklev/Photographer's Choice RF/; V. J. Matthew/Shutterstock; Dale Wagler/ Shutterstock; **16.8** Duncan Shaw/Science Source; **16.9** John Serrao/Science Source; **16.11 center right** Jon Bilous/ Shutterstock; **16.11 top right inset** Biophoto Associates/ Science Source; **16.11 bottom left inset** Ed Reschke/ Photolibrary/Getty Images; **16.11 bottom right inset** Art Fleury; **16.12** Field Museum Library/Premium Archive/ Getty Images; **16.13** Danilo Donadoni/Marka/Age Fotostock; **16.15 left to right** Stephen P. Parker/Science Source; Morales/Age Fotostock; Gunter Marx/Alamy; **16.16** Gene Cox/Science Source; **16.18 left to right** Jean Dickey; Tyler Boyes/Shutterstock; Christopher Marin/ Shutterstock; Jean Dickey; **16.20 left to right** Jean Dickey; Scott Camazine/Science Source; Sonny Tumbelaka/ AFP Creative/Getty Images; **16.21** Prill/Shutterstock; **Table 16.1 top to bottom** Steve Gorton/Dorling Kindersley; Dionisvera/Fotolia; Radu Razvan/Shutterstock; Colin Keates/Courtesy of the Natural History Museum, London/ Dorling Kindersley; Sally Scott/Shutterstock; Alle/ Shutterstock; **16.22 bottom left** Jean Dickey; **16.22 center** Stan Rohrer/Alamy; **16.22 center right** Science Source; **16.22 bottom left** Pabkov/Shutterstock; **16.22 bottom center** VEM/Science Source; **16.22 bottom right** Astrid & Hanns-Frieder Michler/Science Source; **16.23 top to bottom** Jupiterimages/Photos.com/360/Getty Images; Blickwinkel/ Alamy; **16.24a** Bedrich Grunzweig/Science Source; **16.24b** Nigel Cattlin/Science Source; **16.25** North Wind Picture Archives; **16.26** Mikeledray/Shutterstock; Imagebroker.net/ SuperStock; Will Heap/Dorling Kindersley; **16.27** Christine Case; **16.28 left** Jean Dickey; **16.28 right** Eye of Science/ Science Source.

CHAPTER 17: Chapter opening photo top left Image Source/Getty Images; **chapter opening photo bottom right** David Gomez/E+/Getty Images; **chapter opening photo right** Vishnevskiy Vasily/Shutterstock; **chapter opening photo bottom center** Nunosilvaphotography/Shutterstock; **chapter opening photo bottom left** S. Plailly/E. Daynes/ Science Source; **p. 371 right** Tim Wiencis/Splash News/ Newscom; **p. 371 top left** Tim Wiencis/Splash News/ Newscom; **17.1** Gunter Ziesler/Photolibrary/Getty Images; **17.4 left to right** Sinclair Stammers/Science Source; Sinclair Stammers/Science Source; **17.5 left to right** Publiphoto/ Science Source; LorraineHudgins/Shutterstock; **17.9** Image Quest Marine; **17.10 top left to right** Sue Daly/

Nature Picture Library; Michael Klenetsky/Shutterstock; Lebendkulturen.de/Shutterstock; **17.10 bottom** Pavlo Vakhrushev/Fotolia; **17.13 left to right** Georgette Douwman/Nature Picture Library; Image Quest Marine; Christophe Courteau/Natural Picture Library; Marevision/ Age Fotostock; Reinhard Dirscherl/Alamy; **17.14 bottom** Geoff Brightling/Gary Stabb - modelmaker/Dorling Kindersley; **17.14 center right** CMB/Age Fotostock; **17.14 bottom right** Eye of Science/Science Source; **17.15 left to right** Daphne Keller; Ralph Keller/Photoshot Holdings Ltd.; F1online digitale Bildagentur GmbH/Alamy; Wolfgang Poelzer/Wa/Age Fotostock; **17.17a** Steve Gschmeissner/ Science Source; **17.17b** Eye of Science/Science Source; **17.17c** Sebastian Kaulitzki/Shutterstock; **17.18 top to bottom** Mark Kostich/Getty Images; Maximilian Weinzierl/ Alamy; Jean Dickey; **17.19** Dave King/Dorling Kindersley; **17.20 top left to right** Herbert Hopfensperger/Age Fotostock; Dave King/Dorling Kindersley; Andrew Syred/ Science Source; **17.20 bottom left to right** Mark Kostich/ Getty Images; Larry West/Science Source; **17.21 bottom right** Nancy Sefton/Science Source; **17.21 top** Dave King/ Dorling Kindersley; **17.21 center left to right** Maximilian Weinzierl/Alamy; Tom McHugh/Science Source; Nature's Images/Science Source; **17.22 left to right** Jean Dickey; Tom McHugh/Science Source; **17.23** Radius Images/ Alamy; **17.24 counterclockwise from top left** NH/ Shutterstock; Doug Lemke/Shutterstock; lkpro/Shutterstock; Jean Dickey; Jean Dickey; Cordier Huguet/Age Fotostock; Jean Dickey; **17.24 center** Stuart Wilson/Science Source; **17.25 left to right** Thomas Kitchin & Victoria Hurst/Design Pics Inc./Alamy; Thomas Kitchin & Victoria Hurst/Design Pics Inc./Alamy; Thomas Kitchin & Victoria Hurst/Design Pics Inc./Alamy; Thomas Kitchin & Victoria Hurst/Design Pics Inc./Alamy; Thomas Kitchin & Victoria Hurst/Design Pics Inc./Alamy; **17.25 bottom** Keith Dannemiller/Alamy; **17.26 top left** Image Quest Marine; **17.26 inset** Andrew J. Martinez/Science Source; **17.26 top right** Jose B. Ruiz/Nature Picture Library; **17.26 bottom left to right** tbkmedia.de/Alamy; Image Quest Marine; Image Quest Marine; **17.27** Colin Keates/ Courtesy of the Natural History Museum/Dorling Kindersley; **17.29 left** Heather Angel/Natural Visions/ Alamy; **17.29 right** Image Quest Marine; **17.31a** Tom McHugh/Science Source; **17.31b** F Hecker/Blickwinkel/ Age Fotostock; **17.31b inset** A Hartl/Blickwinkel/Age Fotostock; **17.31c** George Grall/National Geographic/ Getty Images; **17.31d** Christian Vinces/Shutterstock; **17.32a** LeChatMachine/Fotolia; **17.32b left** Gary Meszaros/Science Source; **17.32b right** Bill Brooks/Alamy; **17.32c left** Tom McHugh/Science Source; **17.32c right** Jack Goldfarb/Age Fotostock; **17.34 p. 392 left to right** Encyclopaedia Britannica/Alamy; Sylvain Cordier/Getty Images; **17.34 p. 393 left to right** Jerry Young/Dorling Kindersley; Dlillc/Corbis; Miguel Periera/Courtesy of the Instituto Fundacion Miguel Lillo, Argentina/Dorling Kindersley; **17.35** Adam Jones/Getty Images; **17.36 left to right** Jean-Philippe Varin/Science Source; Rebecca Jackrel/ Age Fotostock; Barry Lewis/Alamy; **17.38 clockwise from top left** Creativ Studio Heinem/Age Fotostock; Siegfried Grassegger/Age Fotostock; P. Wegner/Age Fotostock;

Arco Images Gmbh/Tuns/Alamy; Juan Carlos Munoz/Age Fotostock; Anup Shah/Nature Picture Library; Ingo Arndt/ Nature Picture Library; Lexan/Fotolia; John Kelly/Getty Images; **17.40b** John Reader/SPL/Science Source; **17.41** Achmad Ibrahim/AP Images; **17.43** Hemis.fr/Superstock; **17.44** Kablonk/Superstock; **17.45** Stefan Espenhahn/Image Broker/Age Fotostock.

CHAPTER 18: Chapter opening photo bottom Balazs Kovacs Images/Shutterstock; **chapter opening photo top right** Erni/Shutterstock; **chapter opening photo top left** Olegusk/Shutterstock; **chapter opening photo right** Vince Clements/Shutterstock; **p. 407 right** Jean-Paul Ferrero/ Mary Evans Picture Library Ltd/Age Fotostock; **p. 373 top left** Jean-Paul Ferrero/Mary Evans Picture Library Ltd/ Age Fotostock; **18.1** Philippe Psaila/Science Source; **18.2** Jay Directo/AFP/Getty Images/Newscom; **18.3a** Barry Mansell/Nature Picture Library; **10.3b** Sue Flood/Alamy; **18.3c** Juniors Bildarchiv/Age Fotostock; **18.3d** Jeremy Woodhouse/Getty Images; **18.4** David Wall/Alamy; **18.5** NASA Earth Observing System; **18.6 right inset** Image Quest Marine; **18.6** Verena Tunnicliffe/AFP/Newscom; **18.7** Wayne Lynch/All Canada Photos/Getty Images; **18.8a** Jean Dickey; **18.8b** Age Fotostock/Superstock; **18.9** Jean Dickey; **18.11 left to right** Outdoorsman/Shutterstock; Tier und Naturfotografie/J & C Sohns/Age Fotostock; **18.12** Ed Reschke/Photolibrary/Getty Images; **18.13** Robert Stainforth/ Alamy; **18.14** WorldSat International/Science Source; **18.16** Ishbukar Yalilfatar/Shutterstock; **18.17** Kevin Schafer/ Alamy; **18.18** Jean Dickey; **18.20** Digital Vision/Photodisc/ Getty Images; **18.21** Ron Watts/All Canada Photos/Getty Images; **18.22** George McCarthy/Nature Picture Library; **18.29** Age Fotostock/Superstock; **18.30** Eric J. Simon; **18.31** Age Fotostock/Juan Carlos Munoz/Age Fotostock; **18.32** The California/Chaparral Institute; **18.33** Mark Coffey/All Canada Photos/Superstock; **18.34** Joe Sohm/Visions of America, LLC/Alamy; **18.35** Jorma Luhta/Nature Picture Library; **18.36** Paul Nicklen/National Geographic/Getty Images; **18.37** Gordon Wiltsie/National Geographic/Getty Images; **18.39 top to bottom** UNEP/GRID-Arenda; UNEP/GRID-Arendal; **18.40** Crack Palinggi/Reuters; **18.41a** UNEP/GRID-Arenda; **18.41b** USGS; **18.42** Jim West/Alamy; **18.43** Kamira/ Shutterstock; **18.47** Chris Martin Bahr/Science Source; **18.48** Design Pics/Kip Evans/Newscom; **18.49** Drake Fleege/Alamy; **18.50** University of North Carolina; **18.51** Chris Cheadle/All Canada Photos/Getty Images; **18.52a** William E. Bradshaw; **18.52b** Christopher Wood/Shutterstock.

CHAPTER 19: Chapter opening photo bottom right Johan Swanepoel/Shutterstock; **chapter opening photo top left** James Watt/Getty Images; **chapter opening photo bottom left** Gemenacom/Fotolia; **chapter opening photo top right** rob245/Fotolia; **p. 437 right** Karen Doody/ Stocktrek Images/Getty Images; **p. 437 top left** Karen Doody/Stocktrek Images/Getty Images; **19.1a** Henry, P/ Arco Images/Age Fotostock; **19.1b** Jose Azel/Aurora Photos/ Corbis; **19.1c** Sadd/FLPA; **19.2** Wave Royalty Free/Design Pics Inc/Alamy; **19.3** WorldFoto/Alamy; **19.4 left to right** Roger Phillips/Dorling Kindersley; Jane Burton/Dorling Kindersley; Yuri Arcurs/Shutterstock; **p. 441 center** Prill/ Shutterstock; **p. 441 right** Anke van Wyk/Shutterstock;

19.6 Bikeriderlondon/Shutterstock; **19.7 left to right** Joshua Lewis/Shutterstock; Wizdata/Shutterstock; **19.8** Marcin Perkowski/Shutterstock; **19.9** Design Pics/Super Stock; **19.10** Meul/ARCO/Nature Picture Library; **19.11 bottom right** Alan & Sandy Carey/Science Source; **19.12 left to right** William Leaman/Alamy; USDA Forest Service; Gilbert S. Grant/Science Source; **19.13** Dan Burton/Nature Picture Library; **19.14** Ed Reschke/Getty Images; **19.15** St. Petersburg Times/Tampa Bay Times/Zuma/Newscom; **19.16** Chris Johns/National Geographic/Getty Images; **19.17** Jean Dickey; **19.18** Science Photo Library/Alamy; **19.19** Nigel Cattlin/Alamy; **19.20** imageBROKER/Alamy; **19.25 left to right** Horizons WWP/Alamy; Maskot/Getty Images; **19.26** Franzfoto.com/Alamy.

CHAPTER 20: Chapter opening photo bottom left Philippe Psaila/Science Source; **chapter opening photo center left** Ming-Hsiang Chuang/Shutterstock; **chapter opening photo center right** Joe Tucciarone/Science Source; **chapter opening photo center** Jag_cz/Shutterstock; **chapter opening photo center** Frannyanne/Shutterstock; **p. 459 center right** Amar and Isabelle Guillen - Guillen Photo LLC/Alamy; **p. 459 top left** Amar and Isabelle Guillen - Guillen Photo LLC/Alamy; **20.1** James King-Holmes/Science Source; **20.2 left** Gerard Lacz/Age Fotostock; **20.2 right** Mark Carwardine/Getty Images; **20.3** Andre Seale/Age Fotostock; **20.4** American Folklife Center, Library of Congress; **20.5** Gerald Herbert/AP Images; **20.6** Richard D. Estes/Science Source; **20.7a** Jim Zipp/Science Source **20.7b** Tim Zurowski/All Canada Photos/Superstock; **20.8 top** M. I. Walker/Science Source; **20.8 bottom** M. I. Walker/Science Source; **20.9** Jurgen Freund/Nature Picture Library; **20.10** Eric Lemar/Shutterstock; **20.11** P. Wegner/ARCO/Age Fotostock; **20.12 left** Robert Hamilton/Alamy; **20.12 right** Barry Mansell/Nature Picture Library; **20.13 left** Dante Fenolio/Science Source; **20.13 right** Peter J. Mayne; **20.14 top left** Bildagentur-online/TH Foto-Werbung/Science Source; **20.14 top right** Luca Invernizzi Tetto/Age Fotostock; **20.14 bottom** ImageState/Alamy; **20.16** audaxl/Shutterstock; **20.21** Mark Conlin/Vwpics/Visual&Written SL/Alamy; **20.22** Todd Sieling/Corvus Consulting; **20.23** HiloFoto/Getty Images; **20.24** AdstockRF/Universal Images Group Limited/Alamy; **20.30** Wing-Chi Poon; **20.35** Michael Marten/Science Source; **20.36 top** NASA/Goddard Space Flight Center; **20.36 bottom left** NASA Goddard Space Flight Center; **20.36 bottom right** NASA Goddard Space Flight Center; **20.38 bottom** Matthew Dixon/Shutterstock; **20.39** Alan Sirulnikoff/Science Source; **20.40** R. O. Bierregaard, Jr., Biology Department, University of North Carolina, Charlotte; **20.41** United States Department of Agriculture; **20.42** South Florida Water Management District; **20.43 left** Andrew_Shurtleff/The Daily Progress/AP Images; **20.43 right** The Fresno Bee/ZUMA Press, Inc/Alamy; **20.44 top right** Juliet Shrimpton/Age Fotostock; **20.44 bottom left** Matt Jeppson/Shutterstock; **20.44 bottom right** Image Source/Getty Images; **p. 484 Biodiversity table left top to bottom** James King-Holmes/Science Source; Gerard Lacz/Age Fotostock; Andre Seale/Age Fotostock; **p. 484 Interactions table left top to bottom** Jim Zipp/Science Source; Tim Zurowski/All

Canada Photos/Superstock; Jurgen Freund/Nature Picture Library; **p. 484 Interactions table right top to bottom** Outdoor-Archiv/Kukulenz/Alamy; Jean Dickey; Renaud Visage/Getty Images.

CHAPTER 21: Chapter opening photo bottom Everett Collection/Newscom; **chapter opening photo left** drbimages/Getty Images; **chapter opening photo right** Africa Studio/Shutterstock; **p. 489 center right** Stephen Coburn/Shutterstock; **p. 489 top left** Stephen Coburn/Shutterstock; **21.1** Andy Crawford/Dorling Kindersley; **21.2a** Eric Isselée/Fotolia; **21.2b** Sheldon Janice; **21.2c** Thomas Deerinck, NCMIR/Science Source; **21.3** Paul Burns/Blend Images/Alamy; **21.4 center** Nina Zanetti/Pearson Education; **21.4 center right** Nina Zanetti/Pearson Education; **21.4 center right** Dr. Gopal Murti/Science Source; **21.4 bottom right** Biophoto Associates/Science Source; **21.4 bottom center** Chuck Brown/Science Source; **21.4 bottom left** Nina Zanetti; **21.4 left** Rubberball/Mike Kemp/Getty Images; **21.5a left** Nina Zanetti; **21.5b top right** Manfred Kage/Science Source; **21.5c bottom right** Michael Abbey/Science Source; **21.5 center** Blend Images/Alamy; **21.6 left** James Cavallini/Science Source; **21.7** Blend Images/Alamy; **21.8 p. 462 top left** Mast3r/Shutterstock; **21.8 p. 462 bottom left** Andy Crawford/Dorling Kindersley; **21.8 p. 462 center** Sean Justice/Getty Images; **21.8 p. 462 top right** Douglas Pulsipher/Alamy; **21.8 p. 462 bottom right** Tim Tadder/Fancy/Corbis; **21.8 p. 463 bottom right** moodboard/Corbis; **21.8 p. 463 top left** Cardinal/Corbis; **21.8 p. 463 bottom left** Goodshoot/Corbis; **21.8 p. 463 top right** Photodisc/Getty Images; **21.8 p. 463 center** wavebreakmedia/Shutterstock; **21.9** James W. Evarts/Science Source; **21.11** Science Source; **21.12** Jane Burton/Dorling Kindersley; **21.16** anakondasp/Fotolia; **21.17 left** Tim Tadder/Fancy/Corbis; **21.19** Tek Image/Science Photo Library/Corbis; **21.20 left** Volodymyr Goinyk/Shutterstock; **21.20 right** Regien Paassen/Shutterstock; **p. 505 right** John Michael Evan Potter/Shutterstock; **p. 505 center** Ger Bosma/Alamy.

CHAPTER 22: Chapter opening photo bottom ffolas/Shutterstock; **chapter opening photo left** LorenzoArcobasso/Shutterstock; **chapter opening photo right** Eric J. Simon; **p. 509 right** Erik Isakson/Getty Images; **p. 509 top left** Erik Isakson/Getty Images; **22.1 left to right** Arco Images GmbH/Alamy; Bence Mate/Nature Picture Library; Reggie Casagrande/Getty Images; **22.2** Colorblind/Cardinal/Corbis; **22.5** Geo Martinez/Shutterstock; **22.17 top left** Lindsay Noechel/Shutterstock; **22.17 top right** Roger Phillips/Dorling Kindersley; **22.17 bottom left to right** Clive Streeter/Dorling Kindersley; Susanna Price/Dorling Kindersley; David Murray and Jules Selmes/Dorling Kindersley; **p. 521** Rebvt/Shutterstock; **22.18** Kristin Piljay; **22.19** Stephen Morrison/epa/Corbis Wire/Corbis; **22.21** Jackson Laboratory; **22.23** manley099/Getty Images.

CHAPTER 23: Chapter opening photo top E+/Getty Images; **chapter opening photo bottom** BioMedical/Shutterstock; **chapter opening photo right** Thomas Deerinck, NCMIR/Science Source; **p. 529 right** Stefan Schurr/Shutterstock; **p. 529 top left** Stefan Schurr/

Shutterstock; **23.1 bottom left to right** EcoView/Fotolia; Crisod/Fotolia; Hotshotsworldwide/Fotolia; **23.6b** Gustoimages/Science Source; **23.8a** Lennart Nilsson/Tidningarnas Telelgrambyra AB; **23.10** Martyn F. Chillmaid/Science Source; **23.11** Photodisc/Getty Images; **23.12 left to right** Susumu Nishinaga/Science Source; SPL/Science Source; Scimat/Science Source; **23.12 right inset** The Gillette Co; **23.14 left** Biophoto Associates/Science Source; **23.14 right** Biophoto Associates/Science Source; **23.16 left to right** Arco/H Reinhard/Age Fotostock; Matthew Oldfield, Scubazoo/Science Source; Eric Isselée/Fotolia; Tony Wear/Shutterstock; **23.17** Image broker/Superstock; **23.22** Steve Gschmeissner/Science Source; **23.23** Medical-on-Line/Alamy; **23.24** Pep Roig/Alamy.

CHAPTER 24: Chapter opening photo top left Eye of Science/Science Source; **chapter opening photo right** ilolab/Shutterstock; **chapter opening photo bottom left** Martin Lízal/Shutterstock; **p. 551 center right** RIA Novosti/Science Source; **24.1** Charles Daghlian/Science Source; **24.2 left** Juergen Berger/Science Source; **24.2 right** Eye of Science/Science Source; **24.3 center** Getty Images; **24.4a** wavebreakmedia/Shutterstock; **24.12** Science Source; **24.13** epantha/Fotolia; **24.14** Ralph C. Eagle Jr./Science Source; **24.15** Dey L. P. Pharmaceuticals; **24.16 left** SPL/Science Source; **24.16 right** Salisbury District Hospital/Science Source.

CHAPTER 25: Chapter opening photo top left Halfpoint/Fotolia; **chapter opening photo top right** Warren Goldswain/Shutterstock; **chapter opening photo bottom left** FLPA/Alamy; **chapter opening photo bottom right** Michiel de Wit/Shutterstock; **p. 569 right** Alan Diaz/Brewers-Marlins/AP Images; **p. 569 top left** Alan Diaz/Brewers-Marlins/AP Images; **25.1** Bob Ingelhart/Getty Images; **25.4** Cardinal/Corbis; **25.8 left** Nir Elias/Reuters; **25.8 center** Maridav/Fotolia; **25.8 right** Huihe - CNImaging/Newscom; **p. 575** Mega Pixel/Shutterstock; **25.10 top left** Dmitry Lobanov/Shutterstock; **25.10 top center** ZUMA Press/Newscom; **25.16 bottom center** Wayne Lynch/Age Fotostock.

CHAPTER 26: Chapter opening photo bottom left Hill Street Studios/Blend Images/Corbis; **chapter opening photo bottom right** Duncan Noakes/Fotolia; **chapter opening photo top left** StampCollection/Alamy; **chapter opening photo top right** 4x6/Getty Images; **p. 585 right** ZEPHYR/Getty Images; **p. 585 top left** ZEPHYR/Getty Images; **26.1** Biophoto Associates/Science Source; **26.2** Mary Jo Adams; **26.3** Adam Reitzel, John R. Finnerty, Boston University; **26.4** Lighthouse\UIG\Age Fotostock; **26.8** C Edelman/Science Source; **26.10** Image Point Fr/Shutterstock; **Table 26.2 clockwise from top left p. 595** David M. Phillips/Science Source; Kwangshin Kim/Science Source; Chris Bjornberg/Science Source; Hazel Appleton/Science Source; Dr. Linda M. Stannard, University of Cape Town/Science Source; Eye of Science/Science Source; Eye of Science/Science Source; **26.12** David M. Phillips/Science Source; **26.14 left** David Barlow; **26.14 right** David Barlow; **26.17 p. 600 left** Lennart Nilsson/Tidningarnas Telelgrambyra AB

Nilsson/Tidningarnas Telelgrambyra; **26.17 p. 600 center** Lennart Nilsson/Tidningarnas Telelgrambyra AB Nilsson/Tidningarnas Telelgrambyra AB; **26.17 p. 600 right** Lennart Nilsson/Tidningarnas Telelgrambyra AB; **26.17 p. 601 left** Lennart Nilsson/Tidningarnas Telelgrambyra AB Nilsson/Tidningarnas Telelgrambyra AB; **26.17 p. 601 right** Eric J. Simon; **26.18** Eric J. Simon; **26.18 inset** Trevor Smith/Alamy; **26.20** Dr. Yorgos Nikas/Science Source; **26.22** Don Hammond/Design Pics Inc./Alamy.

CHAPTER 27: Chapter opening photo top right Mimagephotography/Shutterstock; **chapter opening photo left** John Green/Cal Sport Media/Alamy; **chapter opening photo bottom right** Peter Macdiarmid/Getty Images; **p. 609 right** Eric J. Simon; **p. 609 top left** Eric J. Simon; **27.2 top** claffra/Shutterstock; **27.2 bottom** Alice Day/Shutterstock; **27.6** Edwin R. Lewis; **27.9 left to right** Dragon Images/Shutterstock; Eternalfeelings/Shutterstock; Zurijeta/Shutterstock; **27.13 center** Patrick Landmann/Science Source; **27.13 left** Public domain; **27.14** Johns Hopkins University School of Education; **27.15** WDCN/Univ. College London/Science Source; **27.16** Fuse/Getty Images; **27.23** moodboard/Corbis; **27.26 top left** Andrey_Popov/Shutterstock; **27.26 top right** Djtaylor/Fotolia; **27.26 bottom left** Tereshchenko Dmitry/Shutterstock; **27.26 bottom right** Rissy Story/Shutterstock; **27.27** mast3r/Shutterstock; **27.30** Puwadol Jaturawutthichai/Shutterstock; **27.32 top right** Franzini Armstrong; **27.32 bottom right** Clara Franzini-Armstrong; **27.38** Frank Greenaway/Dorling Kindersley.

CHAPTER 28: Chapter opening photo bottom Pablo Corral V/Corbis/Glow Images; **chapter opening photo top left** Fekete Tibor/Shutterstock; **chapter opening photo right** Tim UR/Shutterstock; **chapter opening photo center** Serhiy Kobyakov/Fotolia; **p. 639 right** John Coletti/Getty Images; **p. 640 top left** Eric J. Simon; **28.2a** Nigel Cattlin/Alamy; **28.2b** Philip Lee Harvey/Age Fotostock; **28.3 left** Smit/Shutterstock; **28.3 right** Bluemagenta/Shotshop GmbH/Alamy; **28.4a** Nhpa/Superstock; **28.4b** Donald Gregory Clever; **28.4c** Datacraft/Age Fotostock; **28.6a** Yogesh S. More/Age Fotostock; **28.6b** Ana Abadía/AGE Fotostock; **28.11 left top to bottom** Ed Reschke/Getty Images; Graham Kent; Graham Kent; **28.11 right top** Andrew Syred/Science Source; **28.11 right bottom** Biophoto Associates/Science Source; **28.12 left to right** Swapan/Fotolia; Bob Gibbons/ardea/Age Fotostock; M. Wolf/Shutterstock; **28.14** Biology Pics/Science Source; **28.15** Ray Hendley/Age Fotostock; **28.17** Don Mason/Flame/Corbis; **28.18** David W. Stahle; **28.19 left** Tim Hill/Alamy; **28.19 center** Duncan Smith/Science Source; **28.19 right** Dan Suzio/Science Source; **28.25** Jopelka/Shutterstock; **28.26 top** Ilya Genkin/Shutterstock; **28.26 bottom** Ever/Shutterstock; **28.27** Nigel Cattlin/Alamy; **28.28** Volodymyr Pylypchuk/Shutterstock.

CHAPTER 29: Chapter opening photo bottom Marina Lohrbach/Shutterstock; **chapter opening photo right** Candy1812/Fotolia; **chapter opening photo top** Ljupco Smokovski/Fotolia; **p. 659 right** katsumi.takahashi/Getty Images; **p. 659 top left** katsumi.takahashi/Getty Images; **29.2** Eric J. Simon; **29.4** Paul Rapson/Science Source; **29.5** Brian Capon; **29.8 left** Science Source; **29.8 right** Science Source; **29.10 left to right** Frank Greenaway/Dorling Kindersley; Jeremy Bur/Science Source; Jeremy Bur/Science Source; **29.11** Professor Ray F. Evert; **29.12** Matt Walford/Age Fotostock; **29.13** Martin Shields/Alamy; **29.15** Apic/Getty Images; **29.16** Kristin Piljay/Pearson Education/Pearson Science; **29.17** Fred Jensen; **29.18** Sam Antonio Photography/Getty Images; **29.19 left to right** Maryann Frazier/Science Source; Scott Camazine/Science Source; Michael Evans.

ILLUSTRATION AND TEXT CREDITS

CHAPTER 1: 20: Data from Clifton, P. M., Keogh, J. B., and Noakes, M. (2004), "Trans Fatty Acids in Adipose Tissue and the Food Supply Are Associated with Myocardial Infarction," *J. Nutr.* 134: 874–79.

CHAPTER 3: 3.19: Based on Protein Databank: http://www.pdb.org/pdb/explore/explore.do?structureId=1a00; **3.20:** Based on *The Core* module 8.11 and *Campbell Biology* 10e Fig. 19.10.

CHAPTER 5: 5.3: Data from S. E. Gebhardt and R. G. Thomas, *Nutritive Values of Foods* (USDA, 2002); S. A. Plowman and D. L. Smith, *Exercise Physiology for Health, Fitness and Performance,* 2nd edition. Copyright 2003 Pearson Education Inc. Publishing as Pearson Benjamin Cummings.

CHAPTER 7: 7.5: Adapted from Richard and David Walker, *Energy, Plants and Man,* fig. 4.1, p. 103. Oxygraphics. Copyright Richard Walker. Used courtesy of Richard Walker, http://www.oxygraphics.co.uk; **7.12:** Adapted from Richard and David Walker, *Energy, Plants and Man,* fig. 4.1, p. 103. Oxygraphics. Copyright Richard Walker. Used courtesy of Richard Walker, http://www.oxygraphics.co.uk.

CHAPTER 10: 10.33: CDC, http://www.cdc.gov/westnile/statsMaps/finalMapsData/index.html; **page 226:** Text quotation by Joshua Lederberg, from Barbara J. Culliton, "Emerging Viruses, Emerging Threat," *Science, 247,* p. 313, 1/19/1990.

CHAPTER 11: Table 11.1: Data from "Cancer Facts and Figures 2014" (American Cancer Society Inc.)

CHAPTER 12: page 270: MARYLAND v. KING CERTIORARI TO THE COURT OF APPEALS OF MARYLAND No. 12–207. Argued February 26, 2013—Decided June 3, 2013. SUPREME COURT OF THE UNITED STATES.

CHAPTER 13: 13.10: Data from Phylogenetic Relationships among Cetartiodactyls Based on Insertions of Short and Long Interspersed Elements: Hippopotamuses Are the Closest Extant Relatives of Whales. Authors: Masato Nikaido, Alejandro P. Rooney and Norihiro Okada. *Proceedings of the National Academy of Sciences of the United States of America,* Vol. 96, No. 18 (Aug. 31, 1999), pp. 10295–10266. Copyright (1999) National Academy of Sciences, U.S.A. Reproduced by permission.

CHAPTER 14: page 304: Charles Darwin, *The Voyage of the Beagle* (Auckland: Floating Press, 1839); **14.13:** Adapted from "Active Volcanoes and Plate Tectonics, 'Hot Spots' and the 'Ring of Fire'" by Lyn Topinka, U.S. Geological Survey website, January 2, 2003; **page 323:** Charles Darwin in *The Origin of Species* (London: Murray, 1859).

CHAPTER 15: 15.14: Gut Microbiota from Twins Discordant for Obesity Modulate Metabolism in Mice, Vanessa K. Ridaura et al., *Science* 341 (2013); DOI: 10.1126/science.1241214. 9th Ed., © 2007. Reprinted and electronically reproduced by permission of Pearson education, Inc., Upper Saddle River, New Jersey; **15.21:** V. K. Ridaura et al., "Gut Microbiota from Twins Discordant for Obesity Modulate Metabolism in Mice," *Science,* 341.

CHAPTER 16: Table 16.1: Data from Randy Moore et al., *Botany,* 2nd ed. Dubuque, IA: Brown, 1998, Table 2.2, p. 71; **page 369:** Data from Lewis Ziskaa., et al. "Recent Warming by Latitude Associated with Increased Length of Ragweed Pollen Season in Central North America," *Proceedings of the National Academy of Sciences,* 108: 4248–4251 (2011).

CHAPTER 17: 17.39: Drawn from photos of fossils: *A. ramidus* adapted from www.age-of-the-sage.org/evolution/ardi_fossilized_skeleton.html. *H. neanderthalensis* adapted from *The Human Evolution Coloring Book. P. boisei* drawn from a photo by David Bill.

CHAPTER 18: 18.25: J. H. Withgott and S. R. Brennan, *Environment: The Science Behind the Stories,* 3rd Ed., © 2008. Reprinted and electronically reproduced by permission of Pearson Education, Inc., Upper Saddle River, New Jersey; **18.44:** NASA.gov website GISS Surface Temperature Analysis (global map generator), http://data.giss.nasa.gov/gistemp/maps/ (settings for generating this specific map are shown below the map); **18.45:** Based on *Climate Change 2013: The Physical Science Basis.* Working Group I Contribution to the Fourth Assessment Report of the Intergovernmental Panel on Climate Change; **page 435:** Based on data from http://www.ncdc.noaa.gov/land-based-station-data/climate-normals/1981-2010-normals-data.

CHAPTER 19: Table 19.1: Data from Centers for Disease Control and Prevention website; **19.8a:** Data from P. Arcese et al., "Stability, Regulation and the Determination of Abundance in an Insular Song Sparrow Population," *Ecology,* 73: 805–882 (1992); **19.8b:** Data from T. W Anderson, "Predator Responses, Prey Refuges, and Density Dependent Mortality of a Marine Fish," *Ecology* 82: 245–257 (2001); **19.13:** Data from Fisheries and Oceans, Canada, 1999; **19.21:** Data from United Nations, "The World at 6 Billion," 2007; **Table 19.3:** Data from Population Reference Bureau; **19.22:** Data from U.S. Census Bureau; **19.23:** Data from U.S. Census Bureau; **19.24:** Data from Living Planet Report, 2012: "Biodiversity, Biocapacity and Better Choices," World Wildlife Fund (2012); **page 457:** Data from "Transitions in World Population," *Population Bulletin,* 59: 1 (2004).

CHAPTER 20: page 486: Data from J. M. Teal, "Energy Flow in the Salt Marsh Ecosystem of Georgia," *Ecology,* 43:614–624 (1962); **20.43:** © Pearson Education, Inc.

CHAPTER 21: 21.15: Adapted from Fig. 2 in V. H. Hutchison et al., "Thermoregulation in a Brooding Female Indian Python, *Python molurus bivittatus,*" *Science,* 151: 694–695, Fig. 2. Copyright © 1966. Reprinted with permission from AAAS.

CHAPTER 22: Table 22.1: Data from S. E. Gebhardt and R. G. Thomas, *Nutritive Values of Foods* (USDA, 2002); S. A. Plowman and D. L. Smith, *Exercise Physiology for Health, Fitness, and Performance,* 2nd ed. Copyright © 2003 Pearson Education, Inc. publishing as Benjamin Cummings; **22.20:** Data adapted from *Clinical Guidelines on the Identification, Evaluation, and Treatment of Overweight and Obesity in Adults: The Evidence Report.* Data found at the National Heart, Lung, and Blood Institute: http://www.nhlbi.nih.gov/guidelines/obesity/bmi_tbl.htm.

CHAPTER 23: 23.11: "Eighteen Days of 'Living High, Training Low' Stimulate Erythropoiesis and Enhance Aerobic Performance in Elite Middle-Distance Runners," *Journal of Applied Physiology.* Julien V. Brugniaux, Laurent Schmitt, Paul Robach, Gérard Nicolet, Jean-Pierre Marie-Claude Chorvot, Jérémy Cornolo, Niels V. Olsen and Jean-Paul Richalet Fouillot, Stéphane Moutereau, Françoise Lasne, Vincent Pialoux, Philippe Saas, *J. Appl. Physiol.* 100: 203–211, 2006. First published 22 September 2005; doi:10.1152/japplphysiol.00808.2005; **23.22:** Based on the illustration of hemoglobin structure by Irving Geis. Image from Irving Geis Collection, HHMI. Rights owned by Howard Hughes Medical Institute (HHMI).

CHAPTER 24: 24.11: Data from: F. M. Burnet et al., *The Production of Antibodies: A Review and Theoretical Discussion.* Monographs from the Walter and Eliza Hall Institute of Research in Pathology and Laboratory Medicine, Number One, Melbourne: Macmillan and Company Limited (1941); **page 567:** http://www.cdc.gov/nchs/data/databriefs/db10.htm.

CHAPTER 25: 25.12: Based on Hilary K. Seligman, Ann F. Bolger, David Guzman, Andrea López and Kirsten Bibbins-Domingo, "Exhaustion of Food Budgets at Month's End and Hospital Admissions for Hypoglycemia," *Health Affairs,* 33: 116–123 (2014): doi: 10.1377/hlthaff.2013.0096; **25.15:** Data from "Neuropsychiatric Effects of Anabolic Steroids in Male Normal Volunteers," Tung-Ping Su, MD; Michael Pagliaro, RN; Peter J. Schmidt, MD; David Pickar, MD; Owen Wolkowitz, MD; David R. Rubinow, MD, *JAMA* 269(21): 2760–2764 (1993). doi:10.1001/jama.1993.

CHAPTER 26: page 585: U.S. Court of Appeals in Philadelphia; **Table 26.1:** Data from R. Hatcher et al., *Contraceptive Technology: 1990–1992,* p. 168 (New York: Irvington, 1990); **26.21:** Data from "Autism and Mental Retardation Among Offspring Born After In Vitro Fertilization," Sven Sandin, MSc; Karl-Gösta Nygren, PhD; Anastasia Iliadou, PhD; Christina M. Hultman, PhD; Abraham Reichenberg, PhD. *JAMA* 310(1): 75–84 (2013).

CHAPTER 27: 27.5: W. M. Becker, *World of the Cell,* © 1986. Reproduced and electronically reproduced by permission of Pearson Education, Inc., Upper Saddle River, New Jersey; **page 636:** Based on "Sleep in Marine Animals" by L. M. Mukhametov, from *Sleep Mechanisms,* edited by Alexander A. Borberly and J. L. Valatx (Springer, 1984).

CHAPTER 29: Page 673: Based on Kristina A. Stinson, Stuart A. Campbell, Jeff R. Powell, Benjamin E. Wolfe, Ragan M. Callaway, Giles C. Thelen, Steven G. Hallett, Daniel Prati, John N. Klironomos, "Invasive Plant Suppresses the Growth of Native Tree Seedlings by Disrupting Belowground Mutualisms," *PLoS Biology,* 4(5): e140 (2006). doi:10.1371/journal.pbio.0040140.

Appendix D Self-Quiz Answers

CHAPTER 1

1. a
2. atom, molecule, cell, tissue, organ, organism, population, ecosystem, biosphere; the cell
3. Photosynthesis cycles nutrients by converting the carbon in carbon dioxide to sugar, which is then consumed by other organisms. Additionally, the oxygen in water is released as oxygen gas. Photosynthesis contributes to energy flow by converting sunlight to chemical energy, which is then also consumed by other organisms, and by producing heat.
4. a4, b1, c3, d2
5. On average, those individuals with heritable traits best suited to the local environment produce the greatest number of offspring that survive and reproduce. This increases the frequency of those traits in the population over time. The result is the accumulation of evolutionary adaptations.
6. c
7. c
8. artificial
9. a3, b2, c1, d4

CHAPTER 2

1. electrons; neutrons
2. protons; neutrons
3. Sodium-23 has an atomic number of 11 and a mass number of 23. The radioactive isotope, sodium-22, has an atomic number of 11 and a mass number of 22.
4. Organisms incorporate radioactive isotopes of an element into their molecules just as they do nonradioactive isotopes, and researchers can detect the presence of the radioactive isotopes.
5. Each carbon atom has only three covalent bonds instead of the required four.
6. The positively charged hydrogen regions would repel each other.
7. d
8. c
9. The positive and negative poles cause adjacent water molecules to become attracted to each other, forming hydrogen bonds. The properties of water such as

cohesion, temperature regulation, and water's ability to act as a solvent all arise from this atomic "stickiness."
10. Because a nonpolar molecule cannot form hydrogen bonds, it would not have the properties that allow water to act as the basis of life, such as the ability to dissolve substances and water's cohesive properties.
11. The cola is an aqueous solution, with water as the solvent, sugar as the main solute, and the CO_2 making the solution acidic.

CHAPTER 3

1. Isomers have different structures, or shapes, and the shape of a molecule usually helps determine the way it functions in the body.
2. dehydration reactions; water
3. hydrolysis
4. d
5. $C_6H_{12}O_6 + C_6H_{12}O_6 \rightarrow C_{12}H_{22}O_{11} + H_2O$
6. fatty acid; glycerol; triglyceride
7. c
8. c
9. If the change does not affect the shape of the protein in any way, then that change would not affect the function of the protein.
10. Hydrophobic amino acids are most likely to be found within the interior of a protein, far from the watery environment.
11. a
12. nucleotide
13. Both DNA and RNA are polynucleotides; both have the same phosphate group along the backbone; and both use A, C, and G bases. But DNA uses T while RNA uses U as a base; the sugar differs between them; and DNA is usually double-stranded, while RNA is usually single-stranded.
14. Structurally, a gene is a long stretch of DNA. Functionally, a gene contains the information needed to produce a protein.

CHAPTER 4

1. d
2. A membrane is fluid because its components are not locked into

place. A membrane is mosaic because it contains a variety of suspended proteins.
3. endomembrane system
4. smooth
5. rough ER, Golgi apparatus, plasma membrane
6. Both organelles use membranes to organize enzymes and both provide energy to the cell. But chloroplasts capture energy from sunlight during photosynthesis, whereas mitochondria release energy from glucose during cellular respiration. Chloroplasts are only in photosynthetic plants and protists, whereas mitochondria are in almost all eukaryotic cells.
7. a3, b1, c5, d2, e4
8. nucleus, nuclear pores, ribosomes, rough ER, Golgi apparatus
9. Both are appendages that aid in movement and that extend from the surface of a cell. Cells with flagella typically have one long flagellum that propels the cell in a whiplike motion; cilia are usually shorter, are more numerous, and beat in a coordinated fashion.

CHAPTER 5

1. You convert the chemical energy from food to the kinetic energy of your upward climb. At the top of the stairs, some of the energy has been stored as potential energy because of your higher elevation. The rest has been converted to heat.
2. kinetic; potential
3. 3,733 g (or 3.733 kg); remember that 1 Calorie on a food label equals 1,000 Calories of heat energy.
4. The three phosphate groups store chemical energy, a form of potential energy. The release of a phosphate group makes some of this potential energy available to cells to perform work.
5. Hydrolases are enzymes that participate in hydrolysis reactions, breaking down large molecules into the smaller molecules that make them up. Enzymes often have names that end in -ase, so a hydrolase is an enzyme that performs hydrolysis reactions.
6. An inhibitor's binding to another site on the enzyme can cause the

enzyme's active site to change shape.
7. b
8. *Hypertonic* and *hypotonic* are relative terms. A solution that is hypertonic to tap water could be hypotonic to seawater. When using these terms, you must provide a comparison, as in "The solution is hypertonic to the cell's cytoplasm."
9. Passive transport moves atoms or molecules along their concentration gradient (from higher to lower concentration), and active transport moves them against their concentration gradient.
10. c

CHAPTER 6

1. d
2. Plants produce organic molecules by photosynthesis. Consumers must acquire organic material by consuming it rather than making it.
3. In breathing, your lungs exchange CO_2 and O_2 between your body and the atmosphere. In cellular respiration, your cells consume the O_2 in extracting energy from food and release CO_2 as a waste product.
4. the citric acid cycle
5. NAD^+
6. The majority of the energy provided by cellular respiration is generated during the electron transport chain. Shutting down that pathway will deprive cells of energy very quickly.
7. b
8. Glycolysis
9. a
10. Because fermentation supplies only 2 ATP per glucose molecule compared with about 32 from cellular respiration, the yeast will have to consume 16 times as much glucose to produce the same amount of ATP.

CHAPTER 7

1. thylakoids; stroma
2. Because NADPH and ATP are produced by the light reactions on the stroma side, they are more readily available for the Calvin cycle, which consumes the NADPH and ATP and occurs in the stroma.
3. b

4. "Photo" refers to the light required for photosynthesis to proceed, and "synthesis" refers to the fact that it makes sugar. Putting it together, the word "photosynthesis" means "to make using light."

5. Since chlorophyll *b* filters out blue light, it has got the complementary color yellow.

6. carbohydrates

7. c

8. The reactions of the Calvin cycle require the outputs of the light reactions (ATP and NADPH).

9. c

CHAPTER 8

1. c

2. They have identical genes (DNA).

3. They are in the form of very long, thin strands.

4. d

5. prophase and telophase

6. a. 1, 1; b. 1, 2; c. 2, 4; d. $2n$, n; e. individually, by homologous pair; f. identical, unique; g. repair, growth, asexual reproduction; gamete formation

7. 39

8. Prophase II or metaphase II; it cannot be during meiosis I because then you would see an even number of chromosomes; it cannot be during a later stage of meiosis II because then you would see the sister chromatids separated.

9. benign; malignant

10. 16

11. Nondisjunction would create just as many gametes with an extra copy of chromosome 3 or 16, but extra copies of chromosome 3 or 16 are probably fatal.

CHAPTER 9

1. genotype; phenotype

2. Statement **a.** is the law of independent assortment; statement **b.** is the law of segregation.

3. c

4. b

5. b

6. d

7. d

8. Rudy must be $X^D Y^0$. Carla must be $X^D X^d$ (because she had a son with the disease). There is a ¼ chance that their second child will be a male with the disease.

9. Height appears to result from polygenic inheritance, like human skin color. See Figure 9.22.

10. The brown allele appears to be dominant, the white allele recessive. The brown parent appears to be homozygous dominant, *BB*, and the white mouse is homozygous recessive, *bb*. The F_1 mice are all heterozygous, *Bb*. If two of the F_1 mice are mated, ¾ of the F_2 mice will be brown.

11. The best way to find out whether a brown F_2 mouse is homozygous dominant or heterozygous is to do a testcross: Mate the brown mouse with a white mouse. If the brown mouse is homozygous, all the offspring will be brown. If the brown mouse is heterozygous, you would expect half the offspring to be brown and half to be white.

12. Freckles are dominant, so Tim and Jan must both be heterozygous. There is a ¾ chance that they will produce a child with freckles and a ¼ chance that they will produce a child without freckles. The probability that the next two children will have freckles is $\frac{3}{4} \times \frac{3}{4} = \frac{9}{16}$.

13. 50%

14. Because it is the father's sperm (that could carry either an X or a Y chromosome) that determines the sex of the offspring, not the mother's egg (which always carries an X).

15. The mother is a heterozygous carrier, and the father is normal. See the brown boxed area at the bottom of Figure 9.28 for a pedigree. One-quarter of their children will be boys suffering from hemophilia; ¼ will be female carriers.

16. For a woman to be colorblind, she must inherit X chromosomes bearing the colorblindness allele from both parents. Her father has only one X chromosome, which he passes on to all his daughters, so he must be colorblind. A male only needs to inherit the colorblindness allele from a carrier mother; both his parents are usually phenotypically normal.

17. The genotype of the black short-haired parent rabbit is *BBSS*. The genotype of the brown long-haired parent is *bbss*. The F_1 rabbits will

all be black and short-haired, *BbSs*. The F_2 rabbits will be black short-haired, black long-haired, brown short-haired, and brown long-haired, in a proportion of 9:3:3:1.

CHAPTER 10

1. polynucleotides; nucleotides

2. sugar (deoxyribose), phosphate, nitrogenous base

3. c

4. Each daughter DNA molecule will have half the radioactivity of the parent molecule, because one polynucleotide from the original parental DNA molecule winds up in each daughter DNA molecule.

5. UGG; ACC; TGG; no

6. A gene is the polynucleotide sequence with information for making one polypeptide. Each codon—a triplet of bases in DNA or RNA—codes for one amino acid. Transcription occurs when RNA polymerase produces mRNA using one strand of DNA as a template. A ribosome is the site of translation, or polypeptide synthesis, and tRNA molecules serve as interpreters of the genetic code. Each tRNA molecule has an amino acid attached at one end and a three-base anticodon at the other end. Beginning at the start codon, mRNA moves relative to the ribosome a codon at a time. A tRNA with a complementary anticodon pairs with each codon, adding its amino acid to the polypeptide chain. The amino acids are linked by peptide bonds. Translation stops at a stop codon, and the finished polypeptide is released. The polypeptide folds to form a functional protein, sometimes in combination with other polypeptides.

7. a3, b3, c1, d2, e2 and 3

8. b

9. d

10. The genetic material of these viruses is RNA, which is replicated inside the infected cell by special enzymes encoded by the virus. The viral genome (or its complement) serves as mRNA for the synthesis of viral proteins.

11. reverse transcriptase; The process of reverse transcription occurs only in infections by

RNA-containing retroviruses like HIV. Cells do not require reverse transcriptase (their RNA molecules do not undergo reverse transcription), so reverse transcriptase can be knocked out without harming the human host.

CHAPTER 11

1. c

2. operon

3. b

4. a

5. The cells exhibit different patterns of gene expression.

6. No, the mitochondrial DNA is derived from the acceptor cell.

7. nuclear transplantation

8. The production of genetically identical animals for experimentation, the production of organs in pigs for transplant into humans, and restocking populations of endangered animals are a few possible uses of reproductive cloning.

9. b

10. embryonic tissue (ES cells), umbilical cord blood, and bone marrow (adult stem cells)

11. Proto-oncogenes are normal genes involved in the control of the cell cycle. Mutation or viruses can cause them to be converted to oncogenes, or cancer-causing genes. Proto-oncogenes are necessary for normal control of cell division.

12. Master control genes, called homeotic genes, regulate many other genes during development.

CHAPTER 12

1. b

2. DNA ligase

3. Such an enzyme creates DNA fragments with "sticky ends," single-stranded regions whose unpaired bases can hydrogen-bond to the complementary sticky ends of other fragments created by the same enzyme.

4. PCR

5. Different people tend to have different numbers of repeats at each STR site. DNA fragments prepared from the STR sites of different people will thus have different lengths, causing them to migrate to different locations on a gel.

6. b
7. b
8. Chop the genome into fragments using restriction enzymes, clone and sequence each fragment, and reassemble the short sequences into a continuous sequence for every chromosome.
9. c, b, a, d

CHAPTER 13
1. species, genus, family, order, class, phylum, kingdom, domain
2. c
3. Lyell and other geologists presented evidence for the gradual change of geologic features over millions of years. Darwin applied this idea to suggest that species evolve through the slow accumulation of small changes over long periods of time.
4. *Bb*: 0.42; *BB*: 0.49; *bb*: 0.09
5. The fitness of an individual (or of a particular genotype) is measured by the relative number of alleles that it contributes to the gene pool of the next generation compared with the contribution of others. Thus the number of fertile offspring produced determines an individual's fitness.
6. b
7. b
8. Both effects result in populations small enough for significant sampling error in the gene pool for the first few generations. A bottleneck event reduces the size of an existing population in a given location. The founder effect occurs when a new, small population colonizes a new territory.
9. disruptive selection
10. b

CHAPTER 14
1. Microevolution is a change in the gene pool of a population, often associated with adaptation. Speciation is an evolutionary process in which one species splits into two or more species. Macroevolution is evolutionary change above the species level, for example, the origin of evolutionary novelty and new taxonomic groups and the impact of mass extinctions on the diversity of life and its subsequent recovery. Macroevolution is marked by major

changes in the history of life, and these changes are often noticeable enough to be evident in the fossil record.
2. b
3. prezygotic: a, b, c, e; postzygotic: d
4. because a small gene pool is more likely to be changed substantially by genetic drift and natural selection
5. exaptations
6. b
7. d
8. 1.4
9. Homologies reflected a shared evolutionary history, while analogies do not. Analogies result from convergent evolution.
10. Eukarya

CHAPTER 15
1. g, a, d, f, c, b, e
2. d, c, a, b, e
3. DNA polymerase is a protein, which must be transcribed from a gene. But a DNA gene requires DNA polymerase to be replicated. This creates a paradox about which came first—DNA or protein. But RNA can act as both an information storage molecule and an enzyme, suggesting that dual-role RNA may have preceded both DNA and proteins.
4. Exotoxins are poisons secreted by pathogenic bacteria; endotoxins are components of the outer membrane of pathogenic bacteria.
5. a large array of nutrients from the host's intestinal tract
6. They can form endospores.
7. Prokaryotes in soil or water decompose the organic matter in leaves and other plant and animal remains, returning the elements to the environment in inorganic form. Prokaryotes in a sewage treatment facility decompose the organic matter in sewage, converting it to an inorganic form.
8. No, cyanobacteria also perform photosynthesis.
9. b
10. a

CHAPTER 16
1. cuticle
2. flowers
3. a. sporophyte b. cones; angiosperms c. fruit
4. a

5. c
6. a fern
7. Because the plants do not lose their leaves during autumn and winter, the leaves are already fully developed for photosynthesis when the short growing season begins in spring.
8. vascular plant
9. a
10. Green fruits are harder to be spotted and thus less likely to be eaten than other fruits.
11. cell walls; cellulose; chitin
12. A fungus digests its food externally by secreting digestive juices into the food and then absorbing the small nutrients that result from digestion. In contrast, humans and most other animals ingest relatively large pieces of food and digest the food within their bodies.

CHAPTER 17
1. c
2. arthropod
3. Insects
4. a
5. chordata; notochord; cartilage disks between your vertebrae
6. bipedalism
7. a
8. *Australopithecus* species, *Homo habilis, Homo erectus, Homo sapiens*
9. a4, b5, c1, d2, e3

CHAPTER 18
1. organismal ecology, population ecology, community ecology, ecosystem ecology
2. light, water temperature, chemicals added
3. acclimation
4. a
5. a. desert; b. grassland; c. tropical rain forest; d. temperate broadleaf forest; e. coniferous forest; f. tundra
6. chaparral
7. permafrost, very cold winters, and high winds
8. agriculture
9. Carbon dioxide and other gases in the atmosphere absorb heat energy radiating from Earth and reflect it back toward Earth. This is called the greenhouse effect. As the carbon dioxide concentration in the atmosphere increases, more heat is retained, causing global warming.
10. a

11. populations of organisms that have high genetic variability and short life spans

CHAPTER 19
1. the number of people and the land area in which they live
2. III; I
3. a. The *x*-axis is time; the *y*-axis is the number of individuals; the red curve represents exponential growth; the blue curve represents logistic growth.
   b. carrying capacity
   c. In exponential growth, the size of the population increases more and more rapidly. In logistic growth, the population grows fastest when it is about one-half the carrying capacity.
   d. exponential growth curve, though the worldwide growth rate is slowing
4. c
5. opportunistic
6. b
7. a

CHAPTER 20
1. habitat destruction
2. d
3. c
4. a2, b5, c5, d3 or d4, e1
5. because the pesticides become concentrated in their prey
6. succession
7. Only 10% of the energy trapped by photosynthesis is turned into biomass by the plant, and only 10% of that energy is turned into the meat of a grazing animal. Therefore, grain-fed beef provides only about 1% of the energy captured by photosynthesis.
8. Many nutrients come from the soil, but carbon comes from the air.
9. endemic
10. d

CHAPTER 21
1. A tissue that lines an organ (such as the throat) is epithelial. Because epithelium is continually renewing, it is at a much higher risk of cell division errors, a cause of cancer, compared to other tissues, such as smooth muscle.
2. A very thin or flat organism may have all of its body cells in contact with the environment and therefore

would not have a circulatory system. In contrast, a thicker organism will require a circulatory system to transport substances to all body cells.

3. b
4. digestive; urinary
5. c
6. The pigs roll in mud to cool off. Similar to how sweat drying from your skin cools you off, mud drying from the pig's skin cools it off. (Pigs have few sweat glands.)
7. water, urea, ions, drugs, other solutes; pregnancy, illicit drug use, urinary tract infection, diabetes
8. d
9. Most of the water is reabsorbed into the blood; the rest is excreted in the urine.
10. Both the temperature in a room and your body temperature are maintained through negative feedback. In a room, the thermostat senses low temperature and sends signals to turn a heater on; in your body, the brain's control center senses low temperature and sends signals that result in shivering and blood vessel constriction near the skin. When the temperature in a room rises above a set point, the thermostat switches the heater off. When your body temperature gets too high, the brain sends signals to dilate blood vessels at the skin and activate sweat glands.
11. Having large amounts of brown fat is an anatomical response, and increased food intake is a behavioral response.

CHAPTER 22

1. Large food items could not be digested, it would be difficult to eliminate wastes, and it would be difficult to nourish the many cells that compose a large animal.
2. Yes. Absorption is the uptake of nutrients into body cells. Without it, you would never gain any calories (or nutritional value) from your food.
3. Digestion; absorption
4. because peristalsis pushes the food along
5. proteins; pepsin

6. bacteria (specifically, *Helicobacter pylori*); by using antibiotics and bismuth
7. c
8. The antibiotics kill the bacteria that synthesize vitamin K in the colon.
9. because cellular respiration requires oxygen to break down food
10. b

CHAPTER 23

1. In an open circulatory system, vessels have open ends and the circulating fluid flows directly around the body cells. In a closed circulatory system, the circulating fluid stays within a closed system of vessels.
2. Arteries carry oxygen-rich blood to the body, but they carry oxygen-poor blood to the lungs. Veins carry oxygen-poor blood from the body to the heart, but they carry oxygen-rich blood from the lungs to the heart. Arteries are defined as vessels that carry blood away from the heart; veins carry blood to the heart.
3. a3, b1, c4, d2
4. Oxygen levels are reduced, as oxygen-depleted blood mixes with oxygen-rich blood.
5. hemoglobin
6. The shape (disk-like) and size (small) provide human red blood cells a large surface area relative to volume, making them efficient at their main function: gas exchange. Also, the lack of nuclei and other organelles leaves more room to pack hemoglobin into the cells.
7. Working muscles (as in exercise) require a great deal of ATP and thus produce more CO_2 than resting muscles by cellular respiration. As a result, an exercising person needs to breathe more quickly to expel an overabundance of CO_2 (and to bring in extra O_2) than a person at rest.
8. Gills (the respiratory organ of fishes) extend outward from the body into the surrounding environment, whereas lungs (the respiratory organ of humans) are internal.
9. d
10. c
11. b

CHAPTER 24

1. antigens; antibodies
2. innate: c, d; adaptive: a, b, e
3. c
4. It takes several days for a clone of effector cells to be formed during the primary response. In a secondary response, memory cells respond more quickly.
5. a4, b1, c6, d3, e2, f5
6. Swelling (excess fluid in the tissue) dilutes toxins, delivers more oxygen, and promotes scabbing; fever may inhibit bacterial growth.
7. HIV destroys helper T cells, thereby impairing both the B cell and cytotoxic T cell immune response.
8. infections by intracellular pathogens
9. c
10. memory cells; clonal selection

CHAPTER 25

1. d
2. c
3. The first two cells could have membrane receptors for that hormone, but they might trigger different signal transduction pathways. The third cell could lack membrane receptors for that particular hormone.
4. a
5. Alcohol slows the release of antidiuretic hormone (ADH) from the pituitary. Decreased levels of ADH result in decreased levels of water reabsorption by the kidneys, increasing urine output.
6. The patient has hypothyroidism. The thyroid cannot produce its metabolism-regulating hormone, thyroid hormone, without iodine. Without thyroid hormone, there is no negative feedback on TSH, resulting in the growth of the thyroid gland.
7. a
8. d
9. cortisol—adrenal cortex; epinephrine—adrenal medulla; oxytocin—posterior pituitary; releasing hormones—hypothalamus; estrogen—ovaries and testes;

glucagon—pancreas; progesterone—ovaries and testes
10. androgens
11. a. thyroid hormone; thyroid gland
    b. cortisol
    c. anabolic steroids

CHAPTER 26

1. Asexual reproduction is favorable in a stable environment with many resources. Sexual reproduction may enhance success in a changing environment by producing genetically variable offspring.
2. the seminal vesicles and/or the prostate
3. a4, b6, c3, d1, e5, f2
4. b
5. the combined release of estrogen and progesterone from the corpus luteum, which inhibits FSH and LH release from the anterior pituitary gland
6. They also help prevent the spread of sexually transmitted diseases.
7. d
8. b
9. oxytocin
10. in vitro fertilization; fertilization "in glass"

CHAPTER 27

1. b
2. It speeds up transmission of signals along an axon.
3. central nervous system; peripheral nervous system
4. b
5. activation of the parasympathetic division of the peripheral central nervous system
6. The cerebral cortex deals with sensory information, while structures in the brainstem maintain vital functions. If a person suffers a severe head injury, the cerebral cortex may be damaged while the brainstem is still able to function, leaving the person unresponsive but alive.
7. An action potential (used by neurons) is an all-or-none situation; it doesn't vary in intensity. A receptor potential (used by sensory receptor cells) conveys information about the strength of the stimulus.

8. photoreceptor, chemoreceptor, mechanoreceptor, chemoreceptor

9. c (He can hear the tuning fork against his skull, so the cochlea, nerve, and brain are all functioning properly. Apparently, sounds are not being transmitted to the cochlea; therefore, the bones of the middle ear are the problem.)

10. Hinge joints restrict the movements of the legs to a single direction, making horse legs less flexible than human legs.

11. A strong muscle contraction requires that many muscle fibers scattered throughout the muscle are stimulated. In patients with ALS, as individual motor neurons die, potentially hundreds of muscles fibers (that make up the motor unit) are no longer activated to contract.

With fewer motor units being stimulated, the contractions become weaker.

CHAPTER 28

1. b, c, d, e

2. When most plants grow naturally, they exhibit apical dominance, which inhibits the outgrowth of the axillary buds. If your friend were to pinch back the terminal buds, there would be increased growth from the axillary buds. This would result in more branches and leaves, which would increase next year's yield.

3. a4, b2, c5, d6, e3, f1

4. sexual, because it generates genetic variation among the offspring, enhancing the potential for adaptation to a changing environment

5. a. fruit (ripened ovary); b. fruit; c. leaf; d. root; e. seed

6. c

7. d

8. c

CHAPTER 29

1. Excessive amounts of fungicides could destroy mycorrhizae, symbiotic associations of fungi and plant root hairs. The fungal filaments provide lots of surface area for absorption of water and nutrients. Destroying the mycorrhizae could cause a water or nutrient deficiency in the plant.

2. a

3. Phloem; Xylem

4. Cohesion is the sticking together of identical molecules—water molecules in the case of xylem sap. Adhesion is the sticking together

of different kinds of molecules, as in the adhesion of water to the cellulose of xylem walls. Cohesion enables transpiration to pull xylem sap up without the water in the vessels separating; adhesion helps to support the sap against the downward pull of gravity.

5. a

6. auxins; cytokinins

7. The terminal bud produces auxins, which counter the effects of cytokinins from the roots and inhibit the growth of axillary buds. If the terminal bud is removed, the cytokinins predominate, and branching occurs at the axillary buds.

8. a1; b4; c2; d5; e3

9. a4; b3; c1; d2

Glossary

A

abiotic factor (ā'bī-ot'-ik)
A nonliving component of an ecosystem, such as air, water, light, minerals, or temperature.

abiotic reservoir (ā'bī-ot'-ik)
The part of an ecosystem where a chemical, such as carbon or nitrogen, accumulates or is stockpiled outside of living organisms.

ABO blood groups
Genetically determined classes of human blood that are based on the presence or absence of carbohydrates A and B on the surface of red blood cells. The ABO blood group phenotypes, also called blood types, are A, B, AB, and O.

abscisic acid (ab-sis'-ik)
A plant hormone that inhibits cell division and promotes dormancy. Abscisic acid interacts with gibberellins in regulating seed germination.

absorption
The uptake of small nutrient molecules by an organism's own body. In animals, absorption is the third main stage of food processing, following digestion; in fungi, it is acquisition of nutrients from the surrounding medium.

acclimation (ak'-li-mā'-shun)
Physiological adjustment that occurs gradually, though still reversibly, in response to an environmental change.

acid
A substance that increases the hydrogen ion (H^+) concentration in a solution.

action potential
A self-propagating change in the voltage across the plasma membrane of a neuron; a nerve signal.

activation energy
The amount of energy that reactants must absorb before a chemical reaction will start. An enzyme lowers the activation energy of a chemical reaction, allowing it to proceed faster.

activator
A protein that switches on a gene or group of genes by binding to DNA.

active site
The part of an enzyme molecule where a substrate molecule attaches—typically, a pocket or groove on the enzyme's surface.

active transport
The movement of a substance across a biological membrane against its concentration gradient, aided by specific transport proteins and requiring the input of energy (often as ATP).

adaptive immunity
The part of the immune system's defenses that is found only in vertebrates and that must be activated by exposure to a specific invader.

adenine (A) (ad'-uh-nēn)
A double-ring nitrogenous base found in DNA and RNA.

adhesion
The attraction between different kinds of molecules.

adipose tissue
A type of connective tissue in which the cells contain fat.

ADP Adenosine diphosphate (a-den'-ō-sēn dī-fos'-fāt)
A molecule composed of adenosine and two phosphate groups. The molecule ATP is made by combining a molecule of ADP with a third phosphate in an energy-consuming reaction.

adrenal cortex (uh-drē'-nul)
The outer portion of an adrenal gland, controlled by ACTH from the anterior pituitary; secretes hormones called corticosteroids.

adrenal gland (uh-drē'-nul)
One of a pair of endocrine glands, located atop each kidney in mammals, composed of an outer cortex and a central medulla.

adrenal medulla (uh-drē'-nul muh-dul'-uh)
The central portion of an adrenal gland, controlled by nerve signals; secretes the fight-or-flight hormones epinephrine and norepinephrine.

adult stem cell
A cell present in adult tissues that generates replacements for nondividing differentiated cells.

aerobic (ār-ō'-bik)
Containing or requiring molecular oxygen (O_2).

age structure
The relative number of individuals of each age in a population.

AIDS
Acquired immunodeficiency syndrome; the late stages of HIV infection, characterized by a reduced number of T cells; usually results in death caused by infections that would be defeated by a properly functioning immune system.

alga (al'-guh)
(plural, **algae**) An informal term that describes a great variety of photosynthetic protists, including unicellular, colonial, and multicellular forms. Prokaryotes that are photosynthetic autotrophs are also regarded as algae.

alimentary canal (al'-uh-men'-tuh-rē)
A digestive tube running between a mouth and an anus; also called a digestive tract.

allele (uh-lē'-ul)
An alternative version of a gene.

allergen (al'-er-jen)
An otherwise harmless antigen that causes an allergic reaction.

allergy
An exaggerated sensitivity to an antigen. Symptoms are triggered by histamines released from mast cells.

allopatric speciation
The formation of a new species in populations that are geographically isolated from one another. *See also* sympatric speciation.

alternation of generations
A life cycle in which there is both a multicellular diploid form, the sporophyte, and a multicellular haploid form, the gametophyte; a characteristic of plants and multicellular green algae.

alternative RNA splicing
A type of regulation at the RNA-processing level in which different mRNA molecules are produced from the same primary transcript, depending on which RNA segments are treated as exons and which as introns.

alveolus (al-vē'-oh-lus)
(plural, **alveoli**) One of millions of tiny sacs within the vertebrate lungs where gas exchange occurs.

Alzheimer's disease
A form of mental deterioration, or dementia, characterized by confusion and memory loss.

amino acid (uh-mēn'-ō)
An organic molecule containing a carboxyl group, an amino group, a hydrogen atom, and a variable side chain (also called a radical group or R group); serves as the monomer of proteins.

amniote
Member of a clade of tetrapods that has an amniotic egg containing specialized membranes that protect the embryo. Amniotes include mammals and reptiles (including birds).

amniotic egg (am'-nē-ot'-ik)
A shelled egg in which an embryo develops within a fluid-filled amniotic sac and is nourished by yolk. Produced by reptiles (including birds) and egg-laying mammals, it enables them to complete their life cycles on dry land.

amoeba (uh-mē'-buh)
A general term for a protozoan (animal-like protist) characterized by great structural flexibility and the presence of pseudopodia.

amphibian
Member of a class of vertebrate animals that includes frogs and salamanders.

anaerobic (an'-ār-ō'-bik)
Lacking or not requiring molecular oxygen (O_2).

analogy
The similarity between two species that is due to convergent evolution rather than to descent from a common ancestor with the same trait.

anaphase
The third stage of mitosis, beginning when sister chromatids separate from each other and ending when a complete set of daughter chromosomes has arrived at each of the two poles of the cell.

anatomy
The study of the structure of an organism and its parts.

androgen (an'-drō-jen)
A lipid-soluble sex hormone secreted by the gonads that promotes the development and maintenance of the male reproductive system and male body features.

anemia (uh-nē'-mē-ah)
A condition in which an abnormally low amount of hemoglobin or a low number of red blood cells results in the body cells not receiving enough oxygen.

angiosperm (an'-jē-ō-sperm)
A flowering plant, which forms seeds inside a protective chamber called an ovary.

animal
A eukaryotic, multicellular, heterotrophic organism that obtains nutrients by ingestion.

annelid (an'-uh-lid)
A segmented worm. Annelids include earthworms, polychaetes, and leeches.

annual
A plant that completes its life cycle in a single year or growing season.

anorexia nervosa
An eating disorder that results in self-starvation due to an intense fear of gaining weight, even when the person is underweight.

antagonistic hormones
Two hormones that have opposite effects.

anterior pituitary (puh-tū'-uh-tār-ē)
An endocrine gland, adjacent to the hypothalamus and the posterior pituitary, that synthesizes and secretes several hormones, including some that control the activity of other endocrine glands.

anther
A sac in which pollen grains develop, located at the tip of a flower's stamen.

anthropoid (an'-thruh-poyd)
Member of a primate group made up of the apes (gibbons, orangutans, gorillas, chimpanzees, and bonobos), monkeys, and humans.

antibody (an'-tih-bod'-ē)
A protein that is secreted by a B cell and attaches to a specific kind of antigen, helping counter its effects.

anticodon (an'-tī-kō'-don)
On a tRNA molecule, a specific sequence of three nucleotides that is complementary to a codon triplet on mRNA.

antigen (an'-tuh-jen)
Any molecule that elicits a response from a lymphocyte.

anus
The digestive system opening through which undigested materials are expelled.

aphotic zone (ā-fō'-tik)
The region of an aquatic ecosystem beneath the photic zone, where light levels are too low for photosynthesis to take place.

apical dominance (ā'-pik-ul)
In a plant, the hormonal inhibition of axillary buds by a terminal bud.

apical meristem (ā'-pik-ul mer'-uh-stem)
A meristem at the tip of a plant root or in the terminal or axillary bud of a shoot.

apicomplexan (ap'-ē-kom-pleks'-un)
A type of parasitic protozoan (animal-like protist). Some apicomplexans cause serious human disease.

appendix (uh-pen'-dix)
A small, finger-like extension near the union of the small and large intestines.

aqueous solution (ā'-kwē-us)
A solution in which water is the solvent.

arachnid
Member of a major arthropod group that includes spiders, scorpions, ticks, and mites.

Archaea (ar-kē'-uh)
One of two prokaryotic domains of life, the other being Bacteria.

archaean (ar-kē'-uhn)
(plural, **archaea**) An organism that is a member of the domain Archaea.

artery
A vessel that carries blood away from the heart to other parts of the body.

arthritis (ar-thrī'-tis)
A skeletal disorder characterized by inflamed joints and deterioration of the cartilage between bones.

arthropod (ar'-thruh-pod)
Member of the most diverse phylum in the animal kingdom; includes the horseshoe crab, arachnids (for example, spiders, ticks, scorpions, and mites), crustaceans (for example, crayfish, lobsters, crabs, and barnacles), millipedes, centipedes, and insects. Arthropods are characterized by a chitinous exoskeleton, molting, jointed appendages, and a body formed of distinct groups of segments.

artificial selection
The selective breeding of domesticated plants and animals to promote the occurrence of desirable traits in the offspring.

asexual reproduction
The creation of genetically identical offspring by a single parent, without the participation of gametes (sperm and egg).

astigmatism (uh-stig'-muh-tizm)
Blurred vision caused by a misshapen lens or cornea.

atherosclerosis (ath'-uh-rō'-skluh-rō'-sis)
A cardiovascular disease in which a buildup of fatty deposits called plaque develops on the inner walls of the arteries, narrowing the passageways through which blood can flow.

atom
The smallest unit of matter that retains the properties of an element.

atomic mass
The total mass of an atom.

atomic number
The number of protons in each atom of a particular element. Elements are ordered by atomic number in the periodic table of the elements.

ATP Adenosine triphosphate
(a-den'-ō-sēn trī-fos'-fāt)
A molecule composed of adenosine and three phosphate groups; the main energy source for cells. A molecule of ATP can be broken down to a molecule of ADP (adenosine diphosphate) and a free phosphate; this reaction releases energy that can be used for cellular work.

ATP synthase
A protein cluster, found in a cellular membrane (including the inner membrane of mitochondria, the thylakoid membrane of chloroplasts, and the plasma membrane of prokaryotes), that uses the energy of a hydrogen ion concentration gradient to make ATP from ADP. An ATP synthase provides a port through which hydrogen ions (H+) diffuse.

atrium (ā'-trē-um)
(plural, **atria**) A heart chamber that receives blood from the body or lungs via veins.

autoimmune disease
An immunological disorder in which the immune system improperly attacks the body's own molecules.

autonomic nervous system
(ot'-ō-nom'-ik)
A component of the peripheral nervous system of vertebrates that regulates the internal environment. The autonomic nervous system is made up of the sympathetic and parasympathetic subdivisions.

The autonomic nervous system is primarily under involuntary control.

autosome
A chromosome not directly involved in determining the sex of an organism; in mammals, for example, any chromosome other than X or Y.

autotroph (ot'-ō-trōf)
An organism that makes its own food from inorganic ingredients, thereby sustaining itself without eating other organisms or their molecules. Plants, algae, and photosynthetic bacteria are autotrophs.

auxin (ok'-sin)
A term that refers to a group of related plant hormones that have a variety of effects, including cell elongation, root formation, secondary growth, and fruit growth.

axillary bud (ak'-sil-ār-ē)
An embryonic shoot present in the angle formed by a leaf and stem.

axon (ak'-son)
A neuron fiber that extends from the cell body and conducts signals to another neuron or to a muscle or gland cell.

B

B cell
A type of lymphocyte that matures in the bone marrow and later produces antibodies to help combat pathogens circulating in body fluid.

bacillus (buh-sil'-us)
(plural, **bacilli**) A rod-shaped prokaryotic cell.

Bacteria
One of two prokaryotic domains of life, the other being Archaea.

bacteriophage (bak-tēr'-ē-ō-fāj)
A virus that infects bacteria; also called a phage.

bacterium
(plural, **bacteria**) An organism that is a member of the domain Bacteria.

ball-and-socket joint
A joint that allows rotation and movement in several directions. Examples in humans are the shoulder and hip joints.

bark
All the tissues external to the vascular cambium in a plant with secondary

growth. Bark is made up of secondary phloem, cork cambium, and cork.

basal metabolic rate (BMR)
The number of kilocalories a resting animal requires to fuel its essential body processes for a given time.

base
A substance that decreases the hydrogen ion (H^+) concentration in a solution.

benign tumor
An abnormal mass of cells that remains at its original site in the body.

benthic realm
A seafloor or the bottom of a freshwater lake, pond, river, or stream. The benthic realm is occupied by communities of organisms known as benthos.

biennial
A plant that completes its life cycle in two years.

bilateral symmetry
An arrangement of body parts such that an organism can be divided equally by a single cut passing longitudinally through it. A bilaterally symmetric organism has mirror-image right and left sides.

bile
A solution of salts secreted by the liver that emulsifies fats, breaking them into small droplets, and aids in their digestion.

binary fission
A means of asexual reproduction in which a parent organism, often a single cell, divides into two individuals of about equal size.

binomial
A two-part Latinized name of a species; for example, *Homo sapiens*.

biocapacity
Earth's capacity to produce the resources such as food, water, and fuel consumed by humans and to absorb human-generated waste.

biodiversity
The variety of living things; includes genetic diversity, species diversity, and ecosystem diversity.

biodiversity hot spot
A small geographic area that contains a large number of threatened or endangered species and an exceptional concentration of endemic species (those found nowhere else).

biofilm
A surface-coating cooperative colony of prokaryotes.

biogeochemical cycle
Any of the various chemical circuits occurring in an ecosystem, involving both biotic and abiotic components of the ecosystem.

biogeography
The study of the past and present distribution of organisms.

bioinformatics
A scientific field of study that uses mathematics to develop methods for organizing and analyzing large sets of biological data.

biological control
The intentional release of a natural enemy to attack a pest population.

biological magnification
The accumulation of persistent chemicals in the living tissues of consumers in food chains.

biological species concept
The definition of a species as a population or group of populations whose members have the potential in nature to interbreed and produce fertile offspring.

biology
The scientific study of life.

biomass
The amount, or mass, of living organic material in an ecosystem.

biome (*bī'-ōm*)
A major terrestrial or aquatic life zone, characterized by vegetation type in terrestrial biomes and the physical environment in aquatic biomes.

biophilia
The human desire to affiliate with other life in its many forms.

bioremediation
The use of living organisms to detoxify and restore polluted and degraded ecosystems.

biosphere
The global ecosystem; the entire portion of Earth inhabited by life; all of life and where it lives.

biotechnology
The manipulation of living organisms to perform useful tasks.

biotic factor (*bī-ot'-ik*)
A living component of a biological community; any organism that is part of an individual's environment.

bipolar disorder
Depressive mental illness characterized by extreme mood swings; also called manic-depressive disorder.

bird
Member of a group of reptiles with feathers and adaptations for flight.

birth control pill
A chemical contraceptive that inhibits ovulation, retards follicular development, or alters a woman's cervical mucus to prevent sperm from entering the uterus.

bivalve
Member of a group of molluscs that includes clams, mussels, scallops, and oysters.

blade
The flattened portion of a typical leaf.

blastula (*blas'-tyū-luh*)
An embryonic stage that marks the end of cleavage during animal development; a hollow ball of cells in many species.

blood
A type of connective tissue with a fluid matrix called plasma in which blood cells are suspended.

blood pressure
The force that blood exerts against the walls of the blood vessels.

body cavity
A fluid-filled space separating the digestive tract from the outer body wall.

body mass index (BMI)
A ratio of weight to height used as a measure of obesity.

body segmentation
Subdivision of an animal's body into a series of repeated parts called segments.

bone
A type of connective tissue consisting of living cells held in a rigid matrix of collagen fibers embedded in calcium salts.

bony fish
A fish that has a stiff skeleton reinforced by calcium salts.

bottleneck effect
Genetic drift resulting from a drastic reduction in population size. Typically, the surviving population is no longer genetically representative of the parent population.

brain
The master control center of the central nervous system, which is involved in regulating and controlling body activity and interpreting information from the senses.

brainstem
A functional unit of the vertebrate brain, composed of the midbrain, medulla oblongata, and pons; receives and sends information to other brain regions; regulates basic vital functions like breathing and consciousness.

breathing
The ventilation of the lungs by alternate inhalation and exhalation, supplying a lung or gill with O_2-rich air or water and expelling CO_2-rich air or water.

bronchus (*brong'-kus*)
(plural, **bronchi**) One of a pair of breathing tubes that branch from the trachea into the lungs.

bryophyte (*brī'-uh-fīt*)
A type of plant that lacks xylem and phloem; a nonvascular plant. Bryophytes include mosses and their close relatives.

budding
A means of asexual reproduction in which a new individual splits off after developing from an outgrowth of a parent.

buffer
A chemical substance that resists changes in pH by accepting hydrogen ions from or donating hydrogen ions to solutions.

bulimia
An eating disorder characterized by episodes of binge eating followed by purging through induced vomiting, abuse of laxatives, or excessive exercise.

C

calorie
The amount of energy that raises the temperature of 1 g of water by 1°C. Commonly reported as Calories, which are kilocalories (1,000 calories).

Calvin cycle
The second of two stages of photosynthesis; a cyclic series of chemical reactions that occur in

the stroma of a chloroplast, using the carbon in CO_2 and the ATP and NADPH produced by the light reactions to make the energy-rich sugar molecule G3P, which is later used to produce glucose.

cancer
A malignant growth or tumor caused by abnormal and uncontrolled cell division.

cap
Extra nucleotides added to the beginning of an RNA transcript in the nucleus of a eukaryotic cell.

capillary (*kap'-il-ar-ē*)
A microscopic blood vessel that conveys blood between an artery and a vein or between an arteriole and a venule; enables the exchange of nutrients and dissolved gases between the blood and interstitial fluid.

carbohydrate (*kar'-bō-hī'-drāt*)
A biological molecule consisting of a simple sugar (a monosaccharide), two monosaccharides joined into a double sugar (a disaccharide), or a chain of monosaccharides (a polysaccharide).

carbon fixation
The initial incorporation of carbon from CO_2 into organic compounds by autotrophic organisms such as photosynthetic plants, algae, or bacteria.

carbon footprint
The amount of greenhouse gas emitted as a result of the actions of a person, nation, or other entity.

carcinogen (*kar-sin'-uh-jin*)
A cancer-causing agent, either high-energy radiation (such as X-rays or UV light) or a chemical.

cardiac cycle (*kar'-dē-ak*)
The alternating contractions and relaxations of the heart.

cardiac muscle (*kar'-dē-ak*)
Striated muscle that forms the contractile tissue of the heart.

cardiovascular disease
(*kar'-dē-ō-vas'-kyū-ler*)
A set of diseases of the heart and blood vessels.

cardiovascular system
(*kar'-dē-ō-vas'-kyū-ler*)
A closed circulatory system, found in vertebrates, with a heart and a branching network of arteries, capillaries, and veins.

carnivore
An animal that mainly eats other animals. *See also* herbivore; omnivore.

carpel (*kar'-pul*)
The egg-producing part of a flower, consisting of a stalk with an ovary at the base and a stigma, which traps pollen, at the tip.

carrier
An individual who is heterozygous for a recessively inherited disorder and who therefore does not show symptoms of that disorder.

carrying capacity
The maximum population size that a particular environment can sustain.

cartilage (*kar'-ti-lij*)
A type of connective tissue consisting of living cells embedded in a rubbery matrix with collagen fibers.

cartilaginous fish (*kar-ti-laj'-uh-nus*)
A fish that has a flexible skeleton made of cartilage.

cell cycle
An ordered sequence of events (including interphase and the mitotic phase) that extends from the time a eukaryotic cell is first formed from a dividing parent cell until its own division into two cells.

cell cycle control system
A cyclically operating set of proteins that triggers and coordinates events in the eukaryotic cell cycle.

cell division
The reproduction of a cell.

cell plate
A membranous disk that forms across the midline of a dividing plant cell. During cytokinesis, the cell plate grows outward, accumulating more cell wall material and eventually fusing into a new cell wall.

cell theory
The theory that all living things are composed of cells and that all cells come from earlier cells.

cell wall
A protective layer external to the plasma membrane in plant cells, bacteria, fungi, and some protists; protects the cell and helps maintain its shape.

cellular respiration
The aerobic harvesting of energy from food molecules; the energy-releasing chemical breakdown of food

molecules, such as glucose, and the storage of potential energy in a form that cells can use to perform work; involves glycolysis, the citric acid cycle, the electron transport chain, and chemiosmosis.

cellulose (*sel'-yū-lōs*)
A large polysaccharide composed of many glucose monomers linked into cable-like fibrils that provide structural support in plant cell walls. Because cellulose cannot be digested by animals, it acts as fiber, or roughage, in the diet.

centipede
A carnivorous terrestrial arthropod that has one pair of long legs for each of its numerous body segments, with the front pair modified as poisonous claws.

central nervous system (CNS)
The integration and command center of the nervous system; the brain and, in vertebrates, the spinal cord.

central vacuole (*vak'-yū-ōl*)
A membrane-enclosed sac occupying most of the interior of a mature plant cell, having diverse roles in reproduction, growth, and development.

centromere (*sen'-trō-mer*)
The region of a chromosome where two sister chromatids are joined and where spindle microtubules attach during mitosis and meiosis. The centromere divides at the onset of anaphase during mitosis and anaphase II of meiosis.

cephalopod
Member of a group of molluscs that includes squids and octopuses.

cerebellum (*sār'-ruh-bel'-um*)
Part of the vertebrate brain; mainly a planning center that interacts closely with the cerebrum in coordinating body movement.

cerebral cortex (*suh-rē'-brul kor'-teks*)
A highly folded layer of tissue that forms the surface of the cerebrum. In humans, it contains integrating centers for higher brain functions, such as reasoning, speech, language, and imagination.

cerebrospinal fluid (*suh-rē'-brō-spī'-nul*)
Fluid that surrounds, cushions, and nourishes the brain and spinal cord and protects them from infection.

cerebrum (*suh-rē'-brum*)
The largest, most sophisticated, and most dominant part of the vertebrate brain, made up of right and left cerebral hemispheres. The cerebrum contains the cerebral cortex.

cervix (*ser'-viks*)
The narrow neck at the bottom of the uterus, which opens into the vagina.

chaparral (*shap-uh-ral'*)
A terrestrial biome limited to coastal regions where cold ocean currents circulate offshore, creating mild, rainy winters and long, hot, dry summers; also known as the Mediterranean biome. Chaparral vegetation is adapted to fire.

character
A heritable feature that varies among individuals within a population, such as flower color in pea plants or eye color in humans.

charophyte (*kār'-uh-fīt'*)
A member of the green algal group that shares features with land plants. Charophytes are considered the closest relatives of land plants; modern charophytes and modern plants likely evolved from a common ancestor.

chemical bond
An attraction between two atoms resulting from a sharing of outer-shell electrons or the presence of opposite charges on the atoms.

chemical cycling
The use and reuse of chemical elements such as carbon within an ecosystem.

chemical energy
Energy stored in the chemical bonds of molecules; a form of potential energy.

chemical reaction
A process leading to chemical changes in matter, involving the making and/or breaking of chemical bonds. A chemical reaction involves rearranging atoms, but no atoms are created or destroyed.

chemoreceptor (*kē'-mō-rē-sep'-ter*)
A sensory receptor that detects chemical changes within the body or a specific kind of chemical in the external environment; chemoreceptors are involved in our senses of taste and smell.

chemotherapy (kē'-mō-ther'-uh-pē)
Treatment for cancer in which drugs are administered to disrupt cell division of the cancer cells.

chlamydia
A common sexually transmitted disease caused by a bacterial infection. Its primary symptoms are genital discharge and burning during urination. It is easily treatable with antibiotics.

chlorophyll (klor'-ō-fil)
A light-absorbing pigment in chloroplasts that plays a central role in converting solar energy to chemical energy.

chlorophyll a (klor'-ō-fil ā)
A green pigment in chloroplasts that participates directly in the light reactions.

chloroplast (klō'-rō-plast)
An organelle found in plants and photosynthetic protists. Enclosed by two membranes, a chloroplast absorbs sunlight and uses it to power the synthesis of organic food molecules (sugars).

chordate (kōr'-dāt)
An animal that at some point during its development has a dorsal, hollow nerve cord, a notochord, pharyngeal slits, and a post-anal tail. Chordates include lancelets, tunicates, and vertebrates.

chromatin (krō'-muh-tin)
The combination of DNA and proteins that constitutes chromosomes; often used to refer to the diffuse, very extended form taken by the chromosomes when a eukaryotic cell is not dividing.

chromosome (krō'-muh-sōm)
A gene-carrying structure found in the nucleus of a eukaryotic cell and most visible when compacted during mitosis and meiosis; also, the main gene-carrying structure of a prokaryotic cell. Each chromosome consists of one very long threadlike DNA molecule and associated proteins. *See also* chromatin.

chromosome theory of inheritance
A basic principle in biology stating that genes are located on chromosomes and that the behavior of chromosomes during meiosis accounts for inheritance patterns.

chronic traumatic encephalopathy (CTE)
A degenerative disease leading to deterioration of the brain, caused by brain injuries such as concussion.

chyme (kīm)
A mixture of recently swallowed food and gastric juice.

ciliate (sil'-ē-it)
A type of protozoan (animal-like protist) that moves and feeds by means of cilia.

cilium (sil'-ē-um)
(plural, **cilia**) A short appendage that propels some protists through the water and moves fluids across the surface of many tissue cells in animals.

circulatory system
The organ system that transports materials such as nutrients, O_2, and hormones to body cells and transports CO_2 and other wastes from body cells.

citric acid cycle
The metabolic cycle that is fueled by acetyl CoA formed after glycolysis in cellular respiration. Chemical reactions in the cycle complete the metabolic breakdown of glucose molecules to carbon dioxide. The cycle occurs in the matrix of mitochondria and supplies most of the NADH molecules that carry energy to the electron transport chains. Also referred to as the Krebs cycle.

clade
An ancestral species and all its descendants—a distinctive branch in the tree of life.

cladistics (kluh-dis'-tiks)
The study of evolutionary history; specifically, an approach to systematics in which organisms are grouped by common ancestry.

class
In classification, the taxonomic category above order.

cleavage
The process of cytokinesis in animal cells, characterized by pinching of the plasma membrane.

clitoris (klit'-uh-ris)
An organ in the female that engorges with blood and becomes erect during sexual arousal.

clonal selection (klōn'-ul)
The production of a population of genetically identical lymphocytes (white blood cells) that recognize and attack the specific antigen that stimulated their proliferation. Clonal selection is the mechanism that underlies the immune system's specificity and memory of antigens.

clone
As a verb, to produce genetically identical copies of a cell, organism, or DNA molecule. As a noun, the collection of cells, organisms, or molecules resulting from cloning; also (colloquially), a single organism that is genetically identical to another because it arose from the cloning of a somatic cell.

closed circulatory system
A circulatory system in which blood is confined to vessels and is kept separate from the interstitial fluid.

cnidarian (nī-dār'-ē-an)
An animal characterized by cnidocytes, radial symmetry, a gastrovascular cavity, and a polyp or medusa body form. Cnidarians include hydras, jellies, sea anemones, and corals.

coccus (kok'-us)
(plural, **cocci**) A spherical prokaryotic cell.

cochlea (kok'-lē-uh)
A coiled tube in the inner ear that contains the hearing organ, the organ of Corti.

codominant
Expressing two different alleles of a gene in a heterozygote.

codon (kō'-don)
A three-nucleotide sequence in mRNA that specifies a particular amino acid or polypeptide termination signal; the basic unit of the genetic code.

cohesion (kō-hē'-zhun)
The attraction between molecules of the same kind.

cohesion-tension hypothesis
The transport mechanism of xylem sap whereby transpiration exerts a pull that is relayed downward along a string of water molecules held together by cohesion and helped upward by adhesion.

collenchyma cell (kuh-leng'-kuh-muh)
In plants, a type of cell with a thick primary wall and no secondary wall, functioning mainly in supporting growing parts.

colon (kō'-lun)
Most of the length of the large intestine; the tubular portion of the vertebrate alimentary canal between the small intestine and the rectum; functions mainly in water absorption and the formation of feces.

community
All the organisms inhabiting and potentially interacting in a particular area; an assemblage of populations of different species.

community ecology
The study of how interactions between species affect community structure and organization.

competitive exclusion principle
The concept that populations of two species cannot coexist in a community if their niches are nearly identical. Using resources more efficiently and having a reproductive advantage, one of the populations will eventually outcompete and eliminate the other.

complementary DNA (cDNA)
A DNA molecule made in vitro using mRNA as a template and the enzyme reverse transcriptase. A cDNA molecule therefore corresponds to a gene but lacks the introns present in the DNA of the genome.

complete digestive tract
A digestive tube with two openings, a mouth and an anus.

compost
A soil-like mixture of decomposed organic matter used to fertilize plants.

compound
A substance containing two or more elements in a fixed ratio; for example, table salt (NaCl) consists of one atom of the element sodium (Na) for every atom of chlorine (Cl).

concentration gradient
An increase or decrease in the density of a chemical substance within a given region. Cells often maintain concentration gradients of hydrogen ions across their membranes. When a gradient exists, the ions or other chemical substances involved tend

to move from where they are more concentrated to where they are less concentrated.

condom
A flexible sheath, usually made of thin rubber or latex, designed to cover the penis during sexual intercourse for contraceptive purposes or as a means of preventing sexually transmitted diseases.

cone
(1) In vertebrates, a photoreceptor cell in the retina, stimulated by bright light and enabling color vision. (2) In conifers, a reproductive structure bearing pollen or ovules.

conifer (*kon'-uh-fer*)
A gymnosperm, or naked-seed plant, most of which produce cones.

coniferous forest (*kō-nif'-rus*)
A terrestrial biome characterized by conifers, cone-bearing evergreen trees.

connective tissue
Tissue consisting of cells held in an abundant extracellular matrix.

conservation biology
A goal-oriented science that seeks to understand and counter the loss of biodiversity.

conservation of energy
The principle that energy can be neither created nor destroyed.

consumer
An organism that obtains its food by eating plants or by eating animals that have eaten plants.

contraception
The deliberate prevention of pregnancy.

convergent evolution
The evolution of similar features in different evolutionary lineages, which can result from living in very similar environments.

coral reef
Tropical marine biome characterized by hard skeletal structures secreted primarily by the resident cnidarians.

cork
The outermost protective layer of a plant's bark, produced by the cork cambium.

cornea (*kor'-nē-uh*)
The transparent front portion of the white connective tissue that admits light into the vertebrate eye.

coronary artery (*kōr'-uh-nār-ē*)
A large blood vessel that conveys blood from the aorta to the tissues of the heart.

corpus callosum (*kor'-pus kuh-lō'-sum*)
A thick band of nerve fibers that connects the right and left cerebral hemispheres, enabling them to process information together.

corpus luteum (*kor'-pus lū'-tē-um*)
A small body of endocrine tissue that develops from an ovarian follicle after ovulation. The corpus luteum secretes progesterone and estrogen during pregnancy.

cortex
In plants, the ground tissue system of a root, which stores food and absorbs water and minerals that have passed through the epidermis. *See also* adrenal cortex; cerebral cortex.

corticosteroid
One of a family of steroid hormones, synthesized and secreted by the adrenal cortex in response to stress.

cortisol
A lipid-soluble hormone produced by the adrenal cortex that promotes the synthesis of glucose from noncarbohydrates, such as proteins and fats, under stressful situations.

cotyledon (*kot'-uh-lē'-don*)
The first leaf that appears on an embryo of a flowering plant; a seed leaf. Monocot embryos have one cotyledon; eudicot embryos have two.

covalent bond (*kō-vā'-lent*)
An attraction between atoms that share one or more pairs of electrons.

cross
The cross-fertilization of two different varieties of an organism or of two different species; also called hybridization.

crossing over
The exchange of segments between chromatids of homologous chromosomes during prophase I of meiosis.

crustacean
Member of a major arthropod group that includes lobsters, crayfish, crabs, shrimps, and barnacles.

cryptic coloration
Adaptive coloration that makes an organism difficult to spot against its background.

cuticle (*kyū'-tuh-kul*)
(1) In animals, a tough, nonliving outer layer of the skin. (2) In plants, a waxy coating on the surface of stems and leaves that helps retain water.

cystic fibrosis (CF)
A human genetic disorder caused by a recessive allele and characterized by an excessive secretion of mucus and consequent vulnerability to infection; fatal if untreated.

cytokinesis (*sī'-tō-kuh-nē'-sis*)
The division of the cytoplasm to form two separate daughter cells. Cytokinesis usually occurs during telophase of mitosis, and the two processes (mitosis and cytokinesis) make up the mitotic (M) phase of the cell cycle.

cytokinin (*sī'-tō-kī'-nin*)
Any of a family of plant hormones that promote cell division, retard aging in flowers and fruits, and may counter the effects of auxins in regulating plant growth and development.

cytoplasm (*sī'-tō-plaz'-um*)
Everything within a eukaryotic cell inside the plasma membrane and outside the nucleus; consists of a semifluid medium (cytosol) and organelles; can also refer to the interior of a prokaryotic cell.

cytosine (C) (*sī'-tuh-sēn*)
A single-ring nitrogenous base found in DNA and RNA.

cytoskeleton
A meshwork of fine fibers in the cytoplasm of a eukaryotic cell; includes microfilaments, intermediate filaments, and microtubules.

cytosol (*sī'-tuh-sol*)
The fluid part of the cytoplasm, in which organelles are suspended.

cytotoxic T cell (*sī'-tō-tok'-sik*)
A type of lymphocyte that directly attacks body cells infected by pathogens.

D

data
Recorded verifiable observations.

decomposer
An organism that secretes enzymes that digest molecules in organic material and convert them to inorganic form.

dehydration reaction (*dē-hī-drā'-shun*)
A chemical process in which a polymer forms when monomers are linked by the removal of water molecules. One molecule of water is removed for each pair of monomers linked. The atoms in the water molecule are provided by the two monomers involved in the reaction. A dehydration reaction is essentially the reverse of a hydrolysis reaction.

dendrite (*den'-drīt*)
A short, branched neuron fiber that receives signals and conveys them from its tip inward, toward the rest of the neuron.

density-dependent factor
A limiting factor whose effects intensify with increasing population density.

density-independent factor
A limiting factor whose occurrence and effects are not related to population density.

dermal tissue system
In plants, the tissue system that forms an outer protective covering.

desert
A terrestrial biome characterized by low and unpredictable rainfall (less than 30 cm per year).

detritivore (*di-trī'-tuh-vor*)
An organism that consumes dead organic matter (detritus).

detritus (*di-trī'-tus*)
Dead organic matter.

diabetes mellitus
(*dī'-uh-bē'-tēz mel'-uh-tus*)
A human hormonal disease in which body cells cannot absorb enough glucose from the blood and become energy starved. Body fats and proteins are then consumed for their energy. Type 1 diabetes results when the pancreas does not produce insulin. Type 2 diabetes results when body cells fail to respond to insulin. A third type of diabetes, called gestational diabetes, can affect any pregnant woman, even one who has never shown symptoms of diabetes before.

diaphragm (*dī'-uh-fram*)
(1) The sheet of muscle separating the chest cavity from the abdominal cavity in mammals. Its contraction expands the chest cavity, and its relaxation reduces it. (2) A dome-shaped rubber cap that covers a

woman's cervix, serving as a method of contraception.

diastole (dī-as'-tuh-lē)
The stage of the cardiac cycle in which the heart muscle is relaxed, allowing the chambers to fill with blood. *See also* systole.

diatom (dī'-uh-tom)
A unicellular photosynthetic alga with a unique glassy cell wall containing silica.

diffusion
The spontaneous movement of particles of any kind down a concentration gradient, that is, movement of particles from where they are more concentrated to where they are less concentrated.

digestion
The mechanical and chemical breakdown of food into molecules small enough for the body to absorb; the second stage of food processing, following ingestion.

digestive tract
A digestive compartment with two openings: a mouth for the entrance of food and an anus for the exit of undigested wastes. Most animals, including humans, have a digestive tract, also called an alimentary canal.

dihybrid cross (dī'-hī'-brid)
A mating of individuals differing at two genetic loci.

dinoflagellate (dī'-nō-flaj'-uh-let)
A unicellular photosynthetic alga with two flagella situated in perpendicular grooves in cellulose plates covering the cell.

diploid (dip'-loid)
Containing two sets of chromosomes (pairs of homologous chromosomes) in each cell, one set inherited from each parent; referring to a 2n cell.

directional selection
Natural selection that acts in favor of the individuals at one end of a phenotypic range.

disaccharide (dī-sak'-uh-rīd)
A sugar molecule consisting of two monosaccharides (simple sugars) linked by a dehydration reaction.

discovery science
The process of scientific inquiry that focuses on using observations to describe nature. *See also* hypothesis, science.

disruptive selection
Natural selection that favors extreme over intermediate phenotypes.

disturbance
In an ecological sense, a force that damages a biological community, at least temporarily, by destroying organisms and altering the availability of resources needed by organisms in the community. Disturbances such as fires and storms play a pivotal role in structuring many biological communities.

DNA Deoxyribonucleic acid
(dē-ok'-sē-rī'-bo-nu-klō'-ik)
The genetic material that organisms inherit from their parents; a double-stranded helical macromolecule consisting of nucleotide monomers with deoxyribose sugar, a phosphate group, and the nitrogenous bases adenine (A), cytosine (C), guanine (G), and thymine (T). *See also* gene.

DNA ligase (lī'-gās)
An enzyme, essential for DNA replication, that creates new chemical bonds between adjacent DNA nucleotides; used in genetic engineering to paste a specific piece of DNA containing a gene of interest into a bacterial plasmid or other vector.

DNA microarray
A glass slide containing thousands of different kinds of single-stranded DNA fragments arranged in an array (grid). Tiny amounts of DNA fragments, representing different genes, are attached to the glass slide. These fragments are tested for hybridization with various samples of cDNA molecules, thereby measuring the expression of thousands of genes at one time.

DNA polymerase (puh-lim'-er-ās)
An enzyme that assembles DNA nucleotides into polynucleotides using a preexisting strand of DNA as a template.

DNA profiling
A procedure that analyzes an individual's unique collection of genetic markers using PCR and gel electrophoresis. DNA profiling can be used to determine whether two samples of genetic material came from the same individual.

DNA sequencing
Determining the complete nucleotide sequence of a gene or DNA segment.

domain
A taxonomic category above the kingdom level. The three domains of life are Archaea, Bacteria, and Eukarya.

dominant allele
In a heterozygote, the allele that determines the phenotype with respect to a particular gene; the dominant version of a gene is usually represented with a capital italic letter (e.g., *F*).

dorsal, hollow nerve cord
One of the four hallmarks of chordates; the chordate brain and spinal cord.

double circulation system
A circulation scheme where blood flows through the heart by way of two separate loops, the pulmonary circuit and the systemic circuit.

double fertilization
In flowering plants, the formation of both a zygote and a cell with a triploid nucleus, which develops into the endosperm.

double helix
The form assumed by DNA in living cells, referring to its two adjacent polynucleotide strands wound into a spiral shape.

Down syndrome
A human genetic disorder resulting from a condition called trisomy 21, the presence of an extra chromosome 21; characterized by heart and respiratory defects and varying degrees of developmental disability.

duodenum (dū-ō-dē'-num)
The first portion of the vertebrate small intestine after the stomach, where chyme from the stomach is mixed with bile and digestive enzymes.

E

eardrum
A sheet of connective tissue separating the outer ear from the middle ear. The eardrum vibrates when stimulated by sound waves and passes the vibrations to the middle ear.

earthworm
A type of annelid, or segmented worm, that extracts nutrients from soil.

echinoderm (ih-kī'-nuh-derm)
Member of a group of slow-moving or stationary marine animals characterized by a rough or spiny skin, a water vascular system, typically an endoskeleton, and radial symmetry in adults. Echinoderms include sea stars, sea urchins, and sand dollars.

ecological footprint
An estimate of the amount of land required to provide the resources, such as food, water, fuel, and housing, that an individual or a nation consumes and to absorb the waste it generates.

ecological niche
The sum of a species' use of the biotic and abiotic resources in its environment.

ecological succession
The process of biological community change resulting from disturbance; transition in the species composition of a biological community, often following a flood, fire, or volcanic eruption. *See also* primary succession; secondary succession.

ecology
The scientific study of the interactions between organisms and their environments.

ecosystem (ē'-kō-sis-tem)
All the organisms in a given area, along with the nonliving (abiotic) factors with which they interact; a biological community and its physical environment.

ecosystem ecology
The study of energy flow and the cycling of chemicals among the various biotic and abiotic factors in an ecosystem.

ecosystem service
Function performed by an ecosystem that directly or indirectly benefits people.

ectotherm (ek'-tō-therm)
An animal that warms itself mainly by absorbing heat from its surroundings.

effector cell
A short-lived lymphocyte that has an immediate effect against a specific pathogen.

egg
A female gamete.

ejaculation (*ih-jak'-yū-lā'-shun*)
The expulsion of sperm-containing fluid (semen) from the penis.

electromagnetic receptor
A sensory receptor that detects energy of different wavelengths, such as magnetism and light.

electromagnetic spectrum
The full range of radiation, from the very short wavelengths of gamma rays to the very long wavelengths of radio signals.

electron
A subatomic particle with a single unit of negative electrical charge. One or more electrons move around the nucleus of an atom.

electron transport
A reaction in which one or more electrons are transferred to carrier molecules. A series of such reactions, called an electron transport chain, can release the energy stored in high-energy molecules such as glucose. *See also* electron transport chain.

electron transport chain
A series of electron carrier molecules that shuttle electrons during the final stage of cellular respiration, ultimately using the energy of the electrons to make ATP; located in the inner membrane of mitochondria, the thylakoid membrane of chloroplasts, and the plasma membrane of prokaryotes.

element
A substance that cannot be broken down into other substances by chemical means. Scientists recognize 92 chemical elements that occur naturally and several more that have been created in the laboratory.

elimination
The passing of undigested material out of the digestive compartment; the fourth stage of food processing, following absorption.

embryo (*em'-brē-ō*)
A stage in the development of a multicellular organism. In humans, the stage in the development of offspring from the first division of the zygote until body structures begin to appear, about the 9th week of pregnancy.

embryo sac
The female gametophyte contained in the ovule of a flowering plant.

embryonic stem cell (ES cell)
Any of the cells in the early animal embryo that differentiate during development to give rise to all the kinds of specialized cells in the body.

emerging virus
A virus that has appeared suddenly or has recently come to the attention of medical scientists.

endangered species
As defined in the U.S. Endangered Species Act, a species that is in danger of extinction throughout all or a significant portion of its range.

endemic species
A species whose distribution is limited to a specific geographic area.

endocrine disruptor (*en'-dō-krin*)
A chemical that has adverse effects on vertebrates by interfering with chemical signaling within the endocrine system.

endocrine gland (*en'-dō-krin*)
A gland that synthesizes hormone molecules and secretes them directly into the bloodstream.

endocrine system (*en'-dō-krin*)
The body's main system for internal chemical regulation, consisting of all hormone-secreting cells; cooperates with the nervous system in regulating body functions and maintaining homeostasis.

endocytosis (*en'-dō-sī-tō'-sis*)
The movement of materials from the external environment into the cytoplasm of a cell via vesicles or vacuoles.

endodermis
The innermost layer (a one-cell-thick cylinder) of the cortex of a plant root. The endodermis forms a selective barrier, determining which substances pass from the cortex into the vascular tissue.

endomembrane system
A network of organelles that partitions the cytoplasm of eukaryotic cells into functional compartments. Some of the organelles are structurally connected to each other, whereas others are structurally separate but functionally connected by the traffic of vesicles among them.

endometrium (*en'-dō-mē'-trē-um*)
The inner lining of the uterus in mammals, richly supplied with blood vessels that provide the maternal part of the placenta and nourish the developing embryo.

endoplasmic reticulum (ER) (*reh-tik'-yuh-lum*)
An extensive membranous network in a eukaryotic cell, continuous with the outer nuclear membrane and composed of ribosome-studded (rough) and ribosome-free (smooth) regions. *See also* rough ER; smooth ER.

endoskeleton
A hard interior skeleton located within the soft tissues of an animal; found in all vertebrates and a few invertebrates (such as echinoderms).

endosperm
In flowering plants, a nutrient-rich mass formed by the union of a sperm cell with the diploid central cell of the embryo sac during double fertilization; provides nourishment to the developing embryo in the seed.

endospore
A thick-coated, protective cell produced within a prokaryotic cell exposed to harsh conditions.

endosymbiosis
A relationship in which one organism lives inside the cell or cells of a host organism.

endotherm
An animal that derives most of its body heat from its own metabolism.

endotoxin
A poisonous component of the outer membrane of certain bacteria.

energy
The capacity to cause change, or to move matter in a direction it would not move if left alone.

energy flow
The passage of energy through the components of an ecosystem.

enhancer
A eukaryotic DNA sequence that helps stimulate the transcription of a gene at some distance from it. An enhancer functions by means of a transcription factor called an activator, which binds to it and then to the rest of the transcription apparatus.

entropy (*en'-truh-pē*)
A measure of disorder, or randomness. One form of disorder is heat, which is random molecular motion.

enzyme (*en'-zīm*)
A protein that serves as a biological catalyst, changing the rate of a chemical reaction without itself being changed in the process.

enzyme inhibitor (*en'-zīm*)
A chemical that interferes with an enzyme's activity by changing the enzyme's shape, either by plugging up the active site or by binding to another site on the enzyme.

epigenetic inheritance
Inheritance of traits transmitted by mechanisms not directly involving the nucleotide sequence of a genome; frequently involves chemical modification of DNA bases and/or histone proteins.

epinephrine (*ep'-uh-nef'-rin*)
A water-soluble hormone (also called adrenaline) that is secreted by the adrenal medulla and that prepares body organs for "fight or flight"; also serves as a neurotransmitter.

epithelial tissue (*ep'-uh-thē'-lē-ul*)
A sheet of tightly packed cells lining organs and cavities; also called epithelium.

epithelium (*ep'-uh-thē'-lē-um*)
(plural, **epithelia**) *See* epithelial tissue.

equilibrial life history (*ē-kwi-lib'-rē-ul*)
The pattern of reaching sexual maturity slowly and producing few offspring but caring for the young; often seen in long-lived, large-bodied species.

esophagus (*ih-sof'-uh-gus*)
The channel through which food passes in a digestive tube, connecting the pharynx to the stomach.

essential amino acid
Any amino acid that an animal cannot synthesize itself and must obtain from food. Eight amino acids are essential for human adults and nine for human babies.

essential element
In plants, a chemical element that a plant must acquire from its

environment to complete its life cycle (to grow from a seed and produce another generation of seeds).

essential fatty acid
An unsaturated fatty acid that an animal needs but cannot make.

essential nutrient
A substance that an organism must absorb in preassembled form because it cannot synthesize the nutrient from any other material. A dietary shortage of any essential nutrient causes disease. Humans require vitamins, minerals, essential amino acids, and essential fatty acids.

estrogen (*es'-trō-jen*)
One of several chemically similar lipid-soluble sex hormones secreted by the gonads. Estrogen maintains the female reproductive system and promotes the development of female body features.

estuary (*es'-chuh-wār-ē*)
The area where a freshwater stream or river merges with seawater.

ethylene (*eth'-uh-lēn*)
A gas that functions as a hormone in plants, triggering aging responses such as fruit ripening and leaf drop.

eudicot (*yū-dī'-kot*)
Member of a group consisting of the vast majority of flowering plants that have two cotyledons (embryonic seed leaves).

Eukarya (*yū-kār'-yuh*)
The domain of eukaryotes, organisms made up of eukaryotic cells; includes all of the protists, plants, fungi, and animals.

eukaryote (*yū-kār'-ē-ōt*)
An organism characterized by eukaryotic cells. *See also* eukaryotic cell.

eukaryotic cell (*yū-kār'-ē-ot'-ik*)
A type of cell that has a membrane-enclosed nucleus and other membrane-enclosed organelles. All organisms except bacteria and archaea (including protists, plants, fungi, and animals) are composed of eukaryotic cells.

eutherian (*yū-thēr'-ē-un*)
See placental mammal.

evaporative cooling
A property of water whereby a body becomes cooler as water evaporates from it.

evo-devo
Evolutionary developmental biology, which studies the evolution of developmental processes in multicellular organisms.

evolution
Descent with modification; genetic change in a population or species over generations; the heritable changes that have produced Earth's diversity of organisms.

evolutionary adaptation
Modification resulting from natural selection that suits organisms to their environment.

evolutionary tree
A branching diagram that reflects a hypothesis about evolutionary relationships between groups of organisms.

excretion (*ek-skrē'-shun*)
The disposal of nitrogen-containing metabolic wastes.

exocytosis (*ek'-sō-sī-tō'-sis*)
The movement of materials out of the cytoplasm of a cell via membranous vesicles or vacuoles.

exon (*ek'-son*)
In eukaryotes, a coding portion of a gene. *See also* intron.

exoskeleton
A hard, external skeleton that protects an animal and provides points of attachment for muscles.

exotoxin
A poisonous protein secreted by certain bacteria.

exponential population growth
A model that describes the expansion of a population in an ideal, unlimited environment.

extracellular matrix
The meshwork that surrounds animal cells, consisting of a web of protein and polysaccharide fibers embedded in a liquid, jelly, or solid.

F

F₁ generation
The offspring of two parental (P generation) individuals. F_1 stands for first filial.

F₂ generation
The offspring of the F_1 generation. F_2 stands for second filial.

facilitated diffusion
The passage of a substance across a biological membrane down its concentration gradient aided by specific transport proteins.

family
In classification, the taxonomic category above genus.

farsightedness
An inability to focus on close objects; occurs when the eyeball is shorter than normal and the focal point of the lens is behind the retina.

fat
A large lipid molecule made from an alcohol called glycerol and three fatty acids; a triglyceride. Most fats function as energy-storage molecules.

feces
The wastes of the digestive tube expelled through the anus.

fermentation
The anaerobic harvest of energy from food by some cells. Different pathways of fermentation can produce different end products, including ethanol and lactic acid.

fern
Any of a group of seedless vascular plants.

fertilization
The union of a haploid sperm cell with a haploid egg cell, producing a zygote.

fertilizer
A compound applied to the soil to promote plant growth.

fetus (*fē'-tus*)
A developing human from the 9th week of pregnancy until birth. The fetus has all the major structures of an adult.

fever
An abnormally high internal body temperature, usually the result of an infection.

fibrin (*fī'-brin*)
The activated form of the blood-clotting protein fibrinogen, which aggregates into threads that form the fabric of a blood clot.

fibrinogen (*fī-brin'-uh-jen*)
The plasma protein that is activated to form a clot when a blood vessel is injured.

fibrous connective tissue
A dense tissue with large numbers of collagen fibers organized into parallel bundles. This is the dominant tissue in tendons and ligaments.

filtrate
Fluid extracted by the excretory system from the blood or body cavity. The excretory system produces urine from the filtrate after extracting valuable solutes from it and concentrating it.

filtration
In the vertebrate kidney, the extraction of water and small solutes, including metabolic wastes, from the blood by the nephrons.

fission
A means of asexual reproduction whereby a parent separates into two or more genetically identical individuals of about equal size.

flagellate (*flaj'-uh-lit*)
A protozoan (animal-like protist) that moves by means of one or more flagella.

flagellum (*fluh-jel'-um*)
(plural, **flagella**) A long appendage that propels protists through the water and moves fluids across the surface of many tissue cells in animals. A cell may have one or more flagella.

flatworm
A bilateral animal with a thin, flat body form, a gastrovascular cavity with a single opening, and no body cavity. Flatworms include planarians, flukes, and tapeworms.

flower
In an angiosperm, a short shoot with four sets of modified leaves, bearing structures that function in sexual reproduction.

fluid mosaic
A description of membrane structure, depicting a cellular membrane as a mosaic of diverse protein molecules suspended in a fluid bilayer of phospholipid molecules.

follicle (*fol'-uh-kul*)
A cluster of cells surrounding, protecting, and nourishing a developing egg cell in the ovary. The follicle also secretes estrogen.

food chain
The sequence of food transfers between the trophic levels of a community, beginning with the producers.

food web

A network of interconnecting food chains.

foram

A marine protozoan (animal-like protist) that secretes a shell and extends pseudopodia through pores in its shell.

forensics

The scientific analysis of evidence for crime scene investigations and other legal proceedings.

fossil

A preserved imprint or remains of an organism that lived in the past.

fossil fuel

An energy deposit formed from the fossilized remains of long-dead plants and animals.

fossil record

The ordered sequence of fossils as they appear in rock layers, marking the passing of geologic time.

founder effect

The genetic drift resulting from the establishment of a new, small population whose gene pool represents only a sample of the genetic variation present in the parent population.

fruit

A ripened, thickened ovary of a flower, which protects dormant seeds and aids in their dispersal.

functional group

A group of atoms that form the chemically reactive part of an organic molecule. A particular functional group usually behaves similarly in different chemical reactions.

fungus

(plural, **fungi**) A heterotrophic eukaryote that digests its food externally and absorbs the resulting small nutrient molecules. Most fungi consist of a netlike mass of filaments called hyphae. Molds, mushrooms, and yeasts are examples of fungi.

G

gallbladder

An organ that stores bile and releases it as needed into the small intestine.

gamete (*gam'-ēt*)

A sex cell; a haploid egg or sperm. The union of two gametes of opposite sex (fertilization) produces a zygote.

gametogenesis (*guh-mē'-tō-gen-e-sis*)

The formation of gametes within the gonads.

gametophyte (*guh-mē'-tō -fīt*)

The multicellular haploid form in the life cycle of organisms undergoing alternation of generations; results from a union of spores and mitotically produces haploid gametes that unite and grow into the sporophyte generation.

gastric juice

The collection of fluids secreted by the epithelium lining the stomach.

gastropod

Member of the largest group of molluscs, including snails and slugs.

gastrovascular cavity

A digestive compartment with a single opening that serves as both the entrance for food and the exit for undigested wastes; may also function in circulation, body support, and gas exchange. Jellies and hydras are examples of animals with a gastrovascular cavity.

gastrula (*gas'-trū-luh*)

An embryonic stage in animal development. Most animals have a gastrula made up of three layers of cells: ectoderm, endoderm, and mesoderm.

gel electrophoresis

(*jel e-lek'-trō-fōr-ē'-sis*)

A technique for sorting macromolecules. A mixture of molecules is placed on a gel between a positively charged electrode and a negatively charged one; negatively charged molecules migrate toward the positive electrode. The molecules separate in the gel according to their rates of migration.

gene

A unit of inheritance in DNA (or RNA, in some viruses) consisting of a specific nucleotide sequence that programs the amino acid sequence of a polypeptide. Most of the genes of a eukaryote are located in its chromosomal DNA; a few are carried by the DNA of mitochondria and chloroplasts.

gene cloning

The production of multiple copies of a gene.

gene expression

The process whereby genetic information flows from genes to proteins; the flow of genetic information from the genotype to the phenotype: DNA → RNA → protein.

gene flow

The gain or loss of alleles from a population by the movement of individuals or gametes into or out of the population.

gene pool

All the alleles for all the genes in a population at any one time.

gene regulation

The turning on and off of specific genes within a living organism.

genetic code

The set of rules giving the correspondence between nucleotide triplets (codons) in mRNA and amino acids in protein.

genetic drift

A change in the gene pool of a population due to chance.

genetic engineering

The direct manipulation of genes for practical purposes.

genetically modified (GM) organism

An organism that has acquired one or more genes by artificial means. If the gene is from another organism, typically of another species, the recombinant organism is also known as a transgenic organism.

genetics

The scientific study of heredity (inheritance).

genital herpes

A common sexually transmitted disease caused by a virus. The primary symptom is sores on the genitalia. Although outbreaks can be controlled with medication, genital herpes is incurable.

genome

The genetic material of an organism or virus; the complete complement of an organism's or virus's genes along with its noncoding nucleic acid sequences.

genomics

The study of whole sets of genes and their interactions.

genotype (*jē'-nō-tīp*)

The genetic makeup of an organism.

genus (*jē'-nus*)

(plural, **genera**) In classification, the taxonomic category above species; the first part of a species' binomial; for example, *Homo*.

geologic time scale

A time scale established by geologists that reflects a consistent sequence of geologic periods, grouped into four divisions: Precambrian, Paleozoic, Mesozoic, and Cenozoic.

germinate

To initiate growth, as in a plant seed or a plant or fungal spore.

gestation (*jes-tā'-shun*)

Pregnancy; the state of carrying developing young within the female reproductive tract.

gibberellin (*jib'-uh-rel'-in*)

Any of a family of plant hormones that trigger the germination of seeds and interact with auxins in regulating growth and fruit development.

gill

An extension of the body surface of an aquatic animal, specialized for gas exchange and/or suspension feeding.

glucagon (*glū'-kuh-gon*)

A water-soluble hormone secreted by the pancreas that raises the level of glucose in the blood.

glycogen (*glī'-kō-jen*)

A complex, extensively branched polysaccharide made up of many glucose monomers; serves as a temporary energy-storage molecule in liver and muscle cells.

glycolysis (*glī-kol'-uh-sis*)

The multistep chemical breakdown of a molecule of glucose into two molecules of pyruvic acid; the first stage of cellular respiration in all organisms; occurs in the cytoplasmic fluid.

Golgi apparatus (*gol'-jē*)

An organelle in eukaryotic cells consisting of stacks of membranous sacs that modify, store, and ship products of the endoplasmic reticulum.

gonad

An animal sex organ that produces gametes; an ovary or a testis.

granum (*gran'-um*)

(plural, **grana**) A stack of hollow disks formed of thylakoid membrane in a chloroplast. Grana are the sites where light energy is trapped by chlorophyll and converted to chemical energy during the light reactions of photosynthesis.

gravitropism (*grav'-uh-trō'-pizm*)
A plant's directional growth in response to gravity.

green alga
One of a group of photosynthetic protists that includes unicellular, colonial, and multicellular species. Green algae are the photosynthetic protists most closely related to plants.

greenhouse effect
The warming of the atmosphere caused by CO_2, CH_4, and other gases that absorb heat radiation and slow its escape from Earth's surface.

greenhouse gas
Any of the gases in the atmosphere that absorb heat radiation, including CO_2, methane, water vapor, and synthetic chlorofluorocarbons.

ground tissue system
A tissue that makes up the bulk of a plant, filling the space between the epidermis and the vascular tissue system. The ground tissue system fulfills a variety of functions, including storage, photosynthesis, and support.

growth factor
A protein secreted by certain body cells that stimulates other cells to divide.

guanine (G) (*gwa'-nēn*)
A double-ring nitrogenous base found in DNA and RNA.

guard cell
A specialized epidermal cell in plants that regulates the size of a stoma, allowing gas exchange between the surrounding air and the photosynthetic cells in the leaf.

gymnosperm (*jim'-nō-sperm*)
A naked-seed plant. Its seed is said to be naked because it is not enclosed in an ovary.

H

habitat
A place where an organism lives; a specific environment in which an organism lives.

haploid
Containing a single set of chromosomes; referring to an *n* cell.

Hardy-Weinberg equilibrium
The condition describing a nonevolving population (one that is in genetic equilibrium).

heart
(1) The chambered muscular organ in vertebrates that pumps blood received from the veins into the arteries, thereby maintaining the flow of blood through the entire circulatory system. (2) A similarly functioning structure in invertebrates.

heart rate
The number of heartbeats per minute.

heat
The amount of kinetic energy contained in the movement of the atoms and molecules in a body of matter. Heat is energy in its most random form.

helper T cell
A type of lymphocyte that helps activate cytotoxic T cells to attack infected body cells and helps stimulate B cells to produce antibodies.

hemoglobin (*hē'-muh-glō-bin*)
An iron-containing protein in red blood cells that reversibly binds O_2 and transports it to body tissues.

herbivore
An animal that eats mainly plants and/or algae. *See also* carnivore; omnivore.

herbivory
The consumption of plant parts or algae by an animal.

heredity
The transmission of traits from one generation to the next.

hermaphrodite (*her-maf'-rō-dīt*)
An individual that has both female and male reproductive systems, producing both sperm and eggs.

heterotroph (*het'-er-ō-trōf*)
An organism that cannot make its own organic food molecules from inorganic ingredients and must obtain them by consuming other organisms or their organic products; a consumer (such as an animal) or a decomposer (such as a fungus) in a food chain.

heterozygous (*het'-er-ō-zī'-gus*)
Having two different alleles for a given gene.

hinge joint
A joint that allows movement in only one direction. In humans, examples include the elbow and knee.

histamine (*his'-tuh-mēn*)
A chemical alarm signal released by injured cells that causes blood vessels to dilate during an inflammatory response.

histone (*his'-tōn*)
A small protein molecule associated with DNA and important in DNA packing in the eukaryotic chromosome.

HIV
Human immunodeficiency virus; the retrovirus that attacks the human immune system and causes AIDS.

homeostasis (*hō'-mē-ō-stā'-sis*)
The steady state of body functioning; the tendency to maintain relatively constant conditions in the internal environment even when the external environment changes.

homeotic gene (*hō'-mē-ot'-ik*)
A master control gene that determines the identity of a body structure of a developing organism, presumably by controlling the developmental fate of groups of cells. (In plants, such genes are called organ identity genes.)

hominin (*hah'-mi-nin*)
Any anthropoid on the human branch of the evolutionary tree, more closely related to humans than to chimpanzees.

homologous chromosomes (*hō-mol'-uh-gus*)
The two chromosomes that make up a matched pair in a diploid cell. Homologous chromosomes are of the same length, centromere position, and staining pattern and possess genes for the same characteristics at corresponding loci. One homologous chromosome is inherited from the organism's father, the other from the mother.

homology (*hō-mol'-uh-jē*)
Similarity in characteristics resulting from a shared ancestry.

homozygous (*hō'-mō-zī'-gus*)
Having two identical alleles for a given gene.

hormone
In multicellular organisms, a regulatory chemical that travels in body fluids from its production site to other sites, where target cells respond to the regulatory signal.

host
An organism that is exploited by a parasite or pathogen.

human gene therapy
A recombinant DNA procedure intended to treat disease by altering an afflicted person's genes.

Human Genome Project
An international collaborative effort that sequenced the DNA of the entire human genome.

human growth hormone (HGH)
A protein hormone, secreted by the anterior pituitary, that promotes development and growth and stimulates metabolism.

hybrid
The offspring of parents of two different species or of two different varieties of one species; the offspring of two parents that differ in one or more inherited traits; an individual that is heterozygous for one or more pairs of genes.

hydrogen bond
A type of weak chemical bond formed when a partially positive hydrogen atom from one polar molecule is attracted to the partially negative atom in another molecule (or in another part of the same molecule).

hydrogenation
The artificial process of converting unsaturated fats to saturated fats by adding hydrogen.

hydrolysis (*hī-drol'-uh-sis*)
A chemical process in which macromolecules are broken down by the chemical addition of water molecules to the bonds linking their monomers; an essential part of digestion. A hydrolysis reaction is essentially the opposite of a dehydration reaction.

hydrophilic (*hī'-drō-fil'-ik*)
"Water-loving"; pertaining to polar, or charged, molecules (or parts of molecules), which are soluble in water.

hydrophobic (hī'-drō-fō'-bik)
"Water-fearing"; pertaining to nonpolar molecules (or parts of molecules), which do not dissolve in water.

hymen
A thin membrane that partly covers the vaginal opening in the human female and is ruptured by sexual intercourse or other vigorous activity.

hypertension
Abnormally high blood pressure consisting of a persistent systolic blood pressure higher than 140 mm HG and/or diastolic blood pressure higher than 90 mm HG. This condition can lead to a variety of serious cardiovascular disorders.

hypertonic
In comparing two solutions, referring to the one with the greater concentration of solutes.

hypha (hī'-fuh)
(plural, **hyphae**) One of many filaments making up the body of a fungus.

hypothalamus (hī'-pō-thal'-uh-mus)
The main control center of the endocrine system, located in the vertebrate brain. The hypothalamus functions in maintaining homeostasis, especially in coordinating the endocrine and nervous systems. It synthesizes hormones secreted by the posterior pituitary and regulates the secretion of hormones by the anterior pituitary.

hypothesis (hī-poth'-uh-sis)
(plural, **hypotheses**) A tentative explanation that a scientist proposes for a specific phenomenon that has been observed.

hypotonic
In comparing two solutions, referring to the one with the lower concentration of solutes.

I

immune system
The body's system of defenses against infectious disease.

immunodeficiency disease
An immunological disorder in which the body lacks one or more components of the immune system, making a person susceptible to infectious agents that would not ordinarily cause a problem.

impotence
The inability to maintain an erection; also called erectile dysfunction.

in vitro fertilization (IVF) (vē'-tro)
Uniting sperm and egg in a laboratory container, followed by the placement of a resulting early embryo into the mother's uterus.

incomplete dominance
A type of inheritance in which the phenotype of a heterozygote (*Aa*) is intermediate between the phenotypes of the two types of homozygotes (*AA* and *aa*).

induced fit
The interaction between a substrate molecule and the active site of an enzyme, which changes shape slightly to embrace the substrate and catalyze the reaction.

infertility
The inability to conceive after one year of regular, unprotected sexual intercourse.

inflammatory response
An example of an internal innate defense involving the release of histamine and other chemical alarm signals, which trigger increased blood flow, a local increase in white blood cells, and fluid leakage from the blood. The results include redness, heat, and swelling in the affected tissues.

ingestion
The act of eating; the first stage of food processing.

innate immunity
The part of the immune system's defenses that is always present in its final form and ready to act.

inner ear
One of three main regions of the human ear; includes the cochlea, which contains the organ of Corti, the hearing organ.

insect
An arthropod that usually has three body segments (head, thorax, and abdomen), three pairs of legs, and one or two pairs of wings.

insulin
A water-soluble hormone, secreted by the pancreas, that lowers the level of glucose in the blood.

integration
The interpretation of sensory signals and the formulation of responses within the central nervous system.

interneuron (in'-ter-nūr'-on)
A nerve cell, entirely within the central nervous system, that integrates sensory signals and may relay command signals to motor neurons.

internode
The portion of a plant stem between two nodes.

interphase
The phase in the eukaryotic cell cycle when the cell is not actually dividing. During interphase, cellular metabolic activity is high, chromosomes and organelles are duplicated, and cell size may increase. Interphase accounts for 90% of the cell cycle. *See also* mitosis.

interspecific competition
Competition between populations of two or more species that require similar limited resources.

interspecific interaction
Any interaction between members of different species.

interstitial fluid (in'-ter-stish'-ul)
An aqueous solution that surrounds body cells and through which materials pass back and forth between the blood and the body tissues.

intertidal zone (in'-ter-tīd'-ul)
A shallow zone where the waters of an estuary or ocean meet land.

intraspecific competition
Competition between individuals of the same species for the same limited resources.

intrauterine device (IUD)
A T-shaped device that, when placed within the uterus, acts as female contraception.

intron (in'-tron)
In eukaryotes, a nonexpressed (noncoding) portion of a gene that is excised from the RNA transcript. *See also* exon.

invasive species
A non-native species that has spread far beyond the original point of introduction and causes environmental or economic damage by colonizing and dominating suitable habitats.

invertebrate
An animal that does not have a backbone.

ion (ī'-on)
An atom or molecule that has gained or lost one or more electrons, thus acquiring an electrical charge.

ionic bond (ī-on'-ik)
An attraction between two ions with opposite electrical charges. The electrical attraction of the opposite charges holds the ions together.

iris
The colored part of the vertebrate eye; muscles of the iris regulate the amount of light that enters farther into the eye.

isomer (ī'-sō-mer)
One of two or more molecules with the same molecular formula but different structures and thus different properties.

isotonic (ī-sō-ton'-ik)
Having the same solute concentration as another solution.

isotope (ī'-sō-tōp)
A variant form of an atom. Different isotopes of an element have the same number of protons and electrons but different numbers of neutrons.

K

karyotype (kār'-ē-ō-tīp)
A display of micrographs of the metaphase chromosomes of a cell, arranged by size and centromere position.

keystone species
A species whose impact on its community is much larger than its biomass or abundance indicates.

kilocalorie (kcal)
A quantity of heat equal to 1,000 calories. When used to measure the energy content of food, it is usually called a "Calorie."

kinetic energy (kuh-net'-ik)
Energy of motion. Moving matter performs work by transferring its motion to other matter, such as leg muscles pushing bicycle pedals.

kingdom
In classification, the broad taxonomic category above phylum.

L

labia majora (*lā'-bē-uh muh-jor'-uh*)
A pair of outer thickened folds of skin that protect the female genital region.

labia minora (*lā'-bē-uh mi-nor'-uh*)
A pair of inner folds of skin bordering and protecting the female genital region.

labor
A series of strong, rhythmic uterine contractions that expels a baby out of the uterus and vagina during childbirth.

lancelet
One of a group of bladelike invertebrate chordates.

landscape
A regional assemblage of interacting ecosystems.

landscape ecology
The application of ecological principles to the study of land-use patterns; the scientific study of the biodiversity of interacting ecosystems.

large intestine
The tubular portion of the vertebrate alimentary canal between the small intestine and the anus. *See also* colon.

larva
An immature individual that looks different from the adult animal.

larynx (*lār'-inks*)
The voice box, containing the vocal cords.

lateral line system
A row of sensory organs along each side of a fish's body. Sensitive to changes in water pressure, it enables a fish to detect minor vibrations in the water.

law of independent assortment
A general rule of inheritance, first proposed by Gregor Mendel, that states that when gametes form during meiosis, each pair of alleles for a particular character segregates (separates) independently of each other pair.

law of segregation
A general rule of inheritance, first proposed by Gregor Mendel, that states that the two alleles in a pair segregate (separate) into different gametes during meiosis.

leaf
The main site of photosynthesis in a plant; consists of a flattened blade and a stalk (petiole) that joins the leaf to the stem.

leech
A type of annelid, or segmented worm, that typically lives in fresh water.

lens
The disklike structure in an eye that focuses light rays onto the retina.

leukemia (*lū-kē'-mē-ah*)
Cancer of the white blood cells (leukocytes), characterized by excessive production of these cells, resulting in an abnormally high number in the blood.

lichen (*lī'-ken*)
A symbiotic association between a fungus and an alga or between a fungus and a cyanobacterium.

life
Defined by the set of common characteristics that distinguish living organisms from nonliving matter, including such properties and processes as order, regulation, growth and development, energy use, response to the environment, reproduction, and the capacity to evolve over time.

life cycle
The entire sequence of stages in the life of an organism, from the adults of one generation to the adults of the next.

life history
The traits that affect an organism's schedule of reproduction and survival.

life table
A listing of survivals and deaths in a population in a particular time period and predictions of how long, on average, an individual of a given age will live.

ligament
A type of fibrous connective tissue that joins bones together at joints.

light reactions
The first of two stages in photosynthesis; the steps in which solar energy is absorbed and converted to chemical energy in the form of ATP and NADPH. The light reactions power the sugar-producing Calvin cycle but produce no sugar themselves.

lignin (*lig'-nin*)
A chemical that hardens the cell walls of plants. Lignin makes up most of what we call wood.

limiting factor
An environmental factor that restricts the number of individuals that can occupy a particular habitat, thus holding population growth in check.

linked genes
Genes located close enough together on a chromosome that they are usually inherited together.

lipid
An organic compound consisting mainly of carbon and hydrogen atoms linked by nonpolar covalent bonds and therefore mostly hydrophobic and insoluble in water. Lipids include fats, waxes, phospholipids, and steroids.

liver
The largest organ in the vertebrate body. The liver performs diverse functions, such as producing bile, preparing nitrogenous wastes for disposal, and detoxifying poisonous chemicals in the blood.

lobe-finned fish
A bony fish with strong, muscular fins supported by bones. *See also* ray-finned fish.

locus
(plural, **loci**) The particular site where a gene is found on a chromosome. Homologous chromosomes have corresponding gene loci.

logistic population growth
A model that describes population growth that decreases as population size approaches carrying capacity.

loose connective tissue
The most widespread connective tissue in the vertebrate body. It binds epithelia to underlying tissues.

lung
An internal sac, lined with moist epithelium, where gases are exchanged between inhaled air and the blood.

lymph
A fluid similar to interstitial fluid that circulates in the lymphatic system.

lymph node
A small organ that is located along a lymph vessel and that filters lymph.

lymphatic system (*lim-fat'-ik*)
The organ system through which lymph circulates; includes lymph vessels, lymph nodes, and several other organs. The lymphatic system helps remove toxins and pathogens from the blood and interstitial fluid and returns fluid and solutes from the interstitial fluid to the circulatory system.

lymphocyte (*lim'-fuh-sīt*)
A type of white blood cell that carries out adaptive defenses—recognizing and responding to specific invading pathogens. There are two types of lymphocytes: B cells and T cells. *See also* B cell; T cell.

lysogenic cycle (*lī-sō-jen'-ik*)
A bacteriophage reproductive cycle in which the viral genome is incorporated into the bacterial host chromosome as a prophage. New phages are not produced, and the host cell is not killed or lysed unless the viral genome leaves the host chromosome.

lysosome (*lī'-sō-sōm*)
A digestive organelle in eukaryotic cells; contains enzymes that digest the cell's food and wastes.

lytic cycle (*lit'-ik*)
A viral reproductive cycle resulting in the release of new viruses by lysis (breaking open) of the host cell.

M

macroevolution
Evolutionary change above the species level. Examples of macroevolutionary change include the origin of a new group of organisms through a series of speciation events and the impact of mass extinctions on the diversity of life and its subsequent recovery.

macromolecule
A giant molecule formed by joining smaller molecules. Examples of macromolecules include proteins, polysaccharides, and nucleic acids.

macronutrient
A chemical element that an organism must obtain in relatively large amounts.

major depression
Depressive mental illness characterized by persistent sadness and loss of interest in pleasurable activities.

malignant tumor
An abnormal tissue mass that spreads into neighboring tissue and to other parts of the body; a cancerous tumor.

malnutrition
The absence of one or more essential nutrients from the diet.

mammal
Member of a class of endothermic amniotes that possesses mammary glands and hair.

mantle
In molluscs, the outgrowth of the body surface that drapes over the animal. The mantle produces the shell and forms the mantle cavity.

marsupial (*mar-sū'-pē-ul*)
A pouched mammal, such as a kangaroo, opossum, or koala. Marsupials give birth to embryonic offspring that complete development while housed in a pouch and attached to nipples on the mother's abdomen.

mass
A measure of the amount of matter in an object.

mass number
The sum of the number of protons and neutrons in an atom's nucleus.

matter
Anything that occupies space and has mass.

mechanoreceptor
(*mek'-uh-nō-ri-sep'-ter*)
A sensory receptor that detects physical changes in the environment, associated with pressure, touch, stretch, motion, and sound.

medusa (*med-ū'-suh*)
(plural, **medusae**) One of two types of cnidarian body forms; a floating, umbrella-like body form; also called a jelly.

meiosis (*mī-ō'-sis*)
In a sexually reproducing organism, the process of cell division that produces haploid gametes from diploid cells within the reproductive organs.

memory cell
A long-lived lymphocyte that responds to subsequent exposures to a specific pathogen. A memory cell is formed during the primary immune response and is activated by exposure to the same antigen that triggered its formation. When activated, a memory cell forms large clones of effector cells and memory cells that mount the secondary immune response.

meninges (*muh-nin'-jēz*)
Layers of connective tissue that enwrap and protect the brain and spinal cord.

menstrual cycle (*men'-strū-ul*)
The hormonally synchronized cyclic buildup and breakdown of the endometrium of the uterus in preparation for a possible implantation of an embryo.

menstruation (*men'-strū-ā'-shun*)
Uterine bleeding resulting from the breakdown of the endometrium during a menstrual cycle.

meristem (*mār'-eh-stem*)
Plant tissue consisting of undifferentiated cells that divide and generate new cells and tissues.

mesophyll (*mes'-ō-fil*)
The green tissue in the interior of a leaf; a leaf's ground tissue system, the main site of photosynthesis.

messenger RNA (mRNA)
The type of ribonucleic acid that encodes genetic information from DNA and conveys it to ribosomes, where the information is translated into amino acid sequences.

metabolic rate
Energy expended by the body per unit time.

metabolism (*muh-tab'-uh-liz-um*)
The total of all the chemical reactions in an organism.

metamorphosis (*met'-uh-mōr'-fuh-sis*)
The transformation of a larva into an adult.

metaphase (*met'-eh-fāz*)
The second stage of mitosis. During metaphase, the centromeres of all the cell's duplicated chromosomes are lined up along the center line of the cell.

metastasis (*muh-tas'-tuh-sis*)
The spread of cancer cells beyond their original site.

microbiota
The community of microorganisms that live in and on the body of an animal.

microevolution
A change in a population's gene pool over a succession of generations.

micronutrient
A chemical element that an organism needs in very small amounts.

microtubule
The thickest of the three main kinds of fibers making up the cytoskeleton of a eukaryotic cell; a straight, hollow tube made of globular proteins called tubulins. Microtubules form the basis of the structure and movement of cilia and flagella.

middle ear
One of three main regions of the human ear; a chamber containing three small bones (the hammer, anvil, and stirrup) that convey vibrations from the eardrum to the inner ear.

millipede
A terrestrial arthropod that has two pairs of short legs for each of its numerous body segments and that eats decaying plant matter.

mineral
In nutrition, an inorganic chemical element (other than carbon, hydrogen, oxygen, or nitrogen) that an organism requires for proper body functioning.

mitochondrion (*mī'-tō-kon'-drē-on*)
(plural, **mitochondria**) An organelle in eukaryotic cells where cellular respiration occurs. Enclosed by two concentric membranes, it is where most of the cell's ATP is made.

mitosis (*mī-tō'-sis*)
The division of a single nucleus into two genetically identical daughter nuclei. Mitosis and cytokinesis make up the mitotic (M) phase of the cell cycle.

mitotic (M) phase
The phase of the cell cycle when mitosis divides the nucleus and distributes its chromosomes to the daughter nuclei and cytokinesis divides the cytoplasm, producing two daughter cells.

mitotic spindle
A spindle-shaped structure formed of microtubules and associated proteins that is involved in the movement of chromosomes during mitosis and meiosis. (A spindle is shaped roughly like a football.)

molecular biology
The study of the molecular basis of heredity.

molecule
A group of two or more atoms held together by covalent bonds.

mollusc (*mol'-lusk*)
A soft-bodied animal characterized by a muscular foot, mantle, mantle cavity, and radula. Molluscs include gastropods (snails and slugs), bivalves (clams, oysters, and scallops), and cephalopods (squids and octopuses).

monocot (*mon'-uh-kot*)
A flowering plant whose embryos have a single seed leaf, or cotyledon.

monohybrid cross
A mating of individuals differing at one genetic locus.

monomer (*mon'-uh-mer*)
A chemical subunit that serves as a building block of a polymer.

monosaccharide (*mon'-uh-sak'-uh-rīd*)
The smallest kind of sugar molecule; a single-unit sugar; also known as a simple sugar.

monotreme (*mon'-uh-trēm*)
An egg-laying mammal, such as the duck-billed platypus.

morning-after pill (MAP)
A birth control pill taken within 3 days of unprotected intercourse to prevent fertilization or implantation.

moss
Any of a group of seedless nonvascular plants.

motor neuron
A nerve cell that conveys command signals from the central nervous system to receiving cells, such as muscle cells or gland cells.

motor output
The process of sending signals from the central nervous system to the peripheral nervous system that results in a response, such as movement.

motor system
A component of the peripheral nervous system of vertebrates composed of neurons that carry signals to skeletal muscles, mainly in response to external stimuli. The motor system is primarily under voluntary control.

motor unit
A motor neuron and all the muscle fibers it controls.

mouth

See oral cavity.

movement corridor

A series of small clumps or a narrow strip of quality habitat (usable by organisms) that connects otherwise isolated patches of quality habitat.

muscle tissue

Tissue consisting of long muscle cells that are capable of contracting when stimulated by nerve impulses. *See also* skeletal muscle; cardiac muscle; smooth muscle.

muscular system

All the skeletal muscles in the body. (Cardiac muscle and smooth muscle are components of other organ systems.)

mutagen (*myū'-tuh-jen*)

A chemical or physical agent that interacts with DNA and causes a mutation.

mutation

A change in the nucleotide sequence of DNA; a major source of genetic diversity.

mutualism

An interspecific interaction in which both partners benefit.

mycelium (*mī-sē'-lē-um*)

(plural, **mycelia**) The densely branched network of hyphae in a fungus.

mycorrhiza (*mī'-kō-rī'-zuh*)

(plural, **mycorrhizae**) A mutually beneficial symbiotic association of a plant root and fungus.

myelin sheath (*mī'-uh-lin*)

A chain of bead-like supporting cells that insulate the axon of a nerve cell in vertebrates. This insulation helps to speed electrical transmission along the axon.

myofibril (*mī'-ō-fī'-bril*)

A contractile unit in a muscle cell (fiber) made up of many sarcomeres. Longitudinal bundles of myofibrils make up a muscle fiber.

N

NADH

An electron carrier (a molecule that carries electrons) involved in cellular respiration and photosynthesis. NADH carries electrons from glucose and other fuel molecules and deposits

them at the top of an electron transport chain. NADH is generated during glycolysis and the citric acid cycle.

NADPH

An electron carrier (a molecule that carries electrons) involved in photosynthesis. Light drives electrons from chlorophyll to NADP$^+$, forming NADPH, which provides the high-energy electrons for the reduction of carbon dioxide to sugar in the Calvin cycle.

natural family planning

See rhythm method.

natural killer (NK) cell

A white blood cell that attacks cancer cells and infected body cells as part of internal innate defenses.

natural selection

A process in which organisms with certain inherited characteristics are more likely to survive and reproduce than are organisms with other characteristics; unequal reproductive success.

nearsightedness

An inability to focus on distant objects; occurs when the eyeball is longer than normal and the lens focuses distant objects in front of the retina.

negative feedback

A control mechanism in which a chemical reaction, metabolic pathway, or hormone-secreting gland is inhibited by the products of the reaction, pathway, or gland. As the concentration of the products builds up, the product molecules themselves inhibit the process that produced them.

negative pressure breathing

A breathing system in which lower air pressure in the lungs causes air to enter the lungs.

nematode (*nēm'-uh-tōd*)

See roundworm.

nephron

The tubular excretory unit and associated blood vessels of the vertebrate kidney. The nephron extracts filtrate from the blood and refines it into urine.

nerve

A communication line made up of cable-like bundles of neuron fibers (axons) tightly wrapped in connective tissue.

nervous system

The organ system that forms a communication and coordination network throughout an animal's body.

nervous tissue

Tissue made up of neurons and supportive cells.

neuron (*nūr'-on*)

A nerve cell; the fundamental structural and functional unit of the nervous system, specialized for carrying signals from one location in the body to another.

neuroplasticity

The brain's ability to reorganize neural connections throughout life, caused by normal development, changes in behavior and environment, and injury.

neurotransmitter

A chemical messenger that carries information from a transmitting neuron to a receiving cell, which is another neuron, muscle cell, or gland cell.

neutron

An electrically neutral particle (a particle having no electrical charge), found in the nucleus of an atom.

nitrogen fixation

The conversion of atmospheric nitrogen (N_2) to ammonia (NH_3). NH_3 then picks up another H$^+$ to become NH_4^+ (ammonium), which plants can absorb and use.

node

The point of attachment of a leaf on a stem.

nondisjunction

An accident of meiosis or mitosis in which a pair of homologous chromosomes or a pair of sister chromatids fails to separate at anaphase.

norepinephrine (*nor'-ep-uh-nef'-rin*)

A water-soluble hormone (also called noradrenaline) that is secreted by the adrenal medulla and that prepares body organs for "fight or flight"; also serves as a neurotransmitter.

notochord (*nō'-tuh-kord*)

A flexible, cartilage-like, longitudinal rod located between the digestive tract and nerve cord in chordate animals, present only in embryos in many species.

nuclear envelope

A double membrane, perforated with pores, that encloses the nucleus

and separates it from the rest of the eukaryotic cell.

nuclear transplantation

A technique in which the nucleus of one cell is placed into another cell that already has a nucleus or in which the nucleus has been previously destroyed. The cell is then stimulated to grow, producing an embryo that is a genetic copy of the nucleus donor.

nucleic acid (*nū-klā'-ik*)

A polymer consisting of many nucleotide monomers; serves as a blueprint for proteins and, through the actions of proteins, for all cellular structures and activities. The two types of nucleic acids are DNA and RNA.

nucleoid

A non–membrane-enclosed region in a prokaryotic cell where the DNA is concentrated.

nucleolus (*nū-klē'-ō-lus*)

A structure within the nucleus of a eukaryotic cell where ribosomal RNA is made and assembled with proteins to make ribosomal subunits; consists of parts of the chromatin DNA, RNA transcribed from the DNA, and proteins imported from the cytoplasm.

nucleosome (*nū'-klē-ō-sōm*)

The bead-like unit of DNA packing in a eukaryotic cell; consists of DNA wound around a protein core made up of eight histone molecules.

nucleotide (*nū'-klē-ō-tīd*)

An organic monomer consisting of a five-carbon sugar covalently bonded to a nitrogenous base and a phosphate group. Nucleotides are the building blocks of nucleic acids, including DNA and RNA.

nucleus

(plural, **nuclei**) (1) An atom's central core, containing protons and neutrons. (2) The genetic control center of a eukaryotic cell.

O

obesity

An excessively high body mass index, a ratio of weight to height.

omnivore

An animal that eats both plants and animals. *See also* carnivore; herbivore.

oncogene (on'-kō-jēn)
A cancer-causing gene; usually contributes to malignancy by abnormally enhancing the amount or activity of a growth factor made by the cell.

oogenesis (ō'-uh-jen'-uh-sis)
The formation of egg cells within the ovaries.

open circulatory system
A circulatory system in which the circulating fluid is pumped through open-ended vessels and out among the body cells. In an animal with an open circulatory system, the circulating fluid and interstitial fluid are the same.

open system
Any system that exchanges chemicals and energy with its surroundings. All organisms are open systems.

operator
In prokaryotic DNA, a sequence of nucleotides near the start of an operon to which an active repressor can attach. The binding of repressor prevents RNA polymerase from attaching to the promoter and transcribing the genes of the operon.

operculum (ō-per'-kyū-lum)
(plural, **opercula**) A protective flap on each side of a bony fish's head that covers a chamber housing the gills.

operon (op'-er-on)
A unit of genetic regulation common in prokaryotes; a cluster of genes with related functions, along with the promoter and operator that control their transcription.

opportunistic life history
The pattern of reproducing when young and producing many offspring that receive little or no parental care; often seen in short-lived, small-bodied species.

optic nerve
A nerve that arises from the retina in each eye and carries visual information to the brain.

oral cavity
An opening through which food is taken into an animal's body; also known as the mouth.

order
In classification, the taxonomic category above family.

organ
A structure consisting of two or more tissues that coordinate to perform specific functions.

organ system
A group of organs that work together in performing vital body functions.

organelle (ōr-guh-nel')
A membrane-enclosed structure with a specialized function within a eukaryotic cell.

organic compound
A chemical compound containing the element carbon.

organic farming
A method of farming intended to promote environmental sustainability through such practices as crop rotation, water conservation, and avoidance of synthetic fertilizers, pesticides, and genetically modified organisms.

organism
An individual living thing, such as a bacterium, fungus, protist, plant, or animal.

organismal ecology
The study of the evolutionary adaptations that enable individual organisms to meet the challenges posed by their abiotic environments.

osmoregulation
The control of the gain or loss of water and dissolved solutes in an organism.

osmosis (oz-mō'-sis)
The diffusion of water across a selectively permeable membrane.

osteoporosis (os'-tē-ō-puh-rō'-sis)
A skeletal disorder characterized by thinning, porous, and easily broken bones. Osteoporosis is common among women after menopause and is often related to low estrogen levels.

outer ear
One of three main regions of the ear in humans and some other animals. The outer ear is made up of the auditory canal and the pinna.

ovarian cycle (ō-vār'-ē-un)
Hormonally synchronized cyclic events in the mammalian ovary, culminating in ovulation.

ovary
(1) In animals, the female gonad, which produces egg cells and reproductive hormones. (2) In flowering plants, the base of a carpel in which the egg-containing ovules develop.

oviduct (ō'-vuh-dukt)
The tube that conveys egg cells away from an ovary; also called a fallopian tube.

ovulation (ah'-vyū-lā'-shun)
The release of an egg cell from an ovarian follicle.

ovule (ō'-vyūl)
In a seed plant, a reproductive structure that contains the female gametophyte and the developing egg. An ovule develops into a seed.

P

P generation
The parent individuals from which offspring are derived in studies of inheritance. P stands for parental.

pacemaker
A region of cardiac muscle located in the wall of the right atrium that maintains the heart's pumping rhythm (heartbeat) by setting the rate at which the heart contracts.

paedomorphosis (pē'-duh-mōr'-fuh-sis)
The retention in the adult of features that were juvenile in ancestral species.

pain receptor
A sensory receptor that detects painful stimuli.

paleontologist
A scientist who studies fossils.

pancreas (pan'-krē-us)
A gland with dual functions: The nonendocrine portion secretes digestive enzymes and an alkaline solution into the small intestine via a duct; the endocrine portion secretes the hormones insulin and glucagon into the blood.

parasite
An organism that lives in or on another organism (the host) from which it obtains nourishment; an organism that benefits at the expense of another organism, which is harmed in the process.

parasympathetic division
One of two sets of neurons in the autonomic nervous system; generally promotes body activities that gain and conserve energy ("rest and digest"). *See also* sympathetic division.

parenchyma cell (puh-reng'-kuh-muh)
In plants, an abundant and relatively unspecialized type of cell with a thin primary wall and no secondary wall; functions in photosynthesis, food storage, and aerobic respiration and may differentiate into other cell types.

passive transport
The diffusion of a substance across a biological membrane without any input of energy.

pathogen
A disease-causing virus or organism.

pedigree
A family tree representing the occurrence of heritable traits in parents and offspring across a number of generations.

pelagic realm (puh-laj'-ik)
The open-water region of an ocean.

penis
The structure in male mammals that functions in sexual intercourse.

pepsin
An enzyme in gastric juice that begins the breakdown of proteins via hydrolysis reactions.

peptide bond
The covalent linkage between two amino acid units in a polypeptide, formed by a dehydration reaction between two amino acids.

perennial (puh-ren'-ē-ul)
A plant that lives for many years.

periodic table of the elements
A table listing all of the chemical elements (both natural and human-made) ordered by atomic number (the number of protons in the nucleus of a single atom of that element).

peripheral nervous system (PNS)
The network of nerves carrying signals into and out of the central nervous system.

peristalsis (par'-uh-stal'-sis)
Rhythmic waves of contraction of smooth muscles. Peristalsis propels food through a digestive tube and also enables many animals, such as earthworms, to crawl.

permafrost
Continuously frozen subsoil found in the arctic tundra.

petal
A modified leaf of a flowering plant. Petals are the often colorful parts of a

flower that advertise it to insects and other pollinators.

petiole (*pet'-ē-ōl*)
The stalk of a leaf, which joins the leaf to a node of the stem.

pH scale
A measure of the relative acidity of a solution, ranging in value from 0 (most acidic) to 14 (most basic).

phage (*fāj*)
See bacteriophage.

phagocytic cell (*fag'-ō-si-tik*)
A white blood cell that engulfs bacteria, foreign proteins, and the remains of dead body cells as part of internal innate defenses.

phagocytosis (*fag'-ō-sī-tō'-sis*)
Cellular "eating"; a type of endocytosis whereby a cell engulfs large molecules, other cells, or particles into its cytoplasm.

pharyngeal slit (*fuh-rin'-jē-ul*)
A gill structure in the pharynx, found in chordate embryos and some adult chordates.

pharynx (*far'-inks*)
The organ in a digestive tract that receives food from the oral cavity; in terrestrial vertebrates, the throat region where the air and food passages cross.

phenotype (*fē'-nō-tīp*)
The expressed traits of an organism.

phloem (*flō'-um*)
The portion of a plant's vascular system that conveys sugars, nutrients, and hormones throughout a plant. Phloem is made up of live food-conducting cells.

phloem sap
The solution of sugars, other nutrients, and hormones conveyed throughout a plant via phloem tissue.

phospholipid (*fos'-fō-lip'-id*)
A molecule that is a part of the inner bilayer of biological membranes, having a hydrophilic head and a hydrophobic tail.

phospholipid bilayer (*fos'-fō-lip'-id*)
A double layer of phospholipid molecules (each molecule consisting of a phosphate group bonded to two fatty acids) that is the primary component of all cellular membranes.

photic zone (*fō'-tik*)
Shallow water near the shore or the upper layer of water away from the shore; region of an aquatic ecosystem where sufficient light is available for photosynthesis.

photon (*fō'-ton*)
A fixed quantity of light energy. The shorter the wavelength of light, the greater the energy of a photon.

photoperiod
The length of the day relative to the length of the night; an environmental stimulus that plants use to detect the time of year.

photoreceptor
A type of electromagnetic receptor that detects light.

photosynthesis (*fō'-tō-sin'-thuh-sis*)
The process by which plants, algae, and some bacteria transform light energy to chemical energy stored in the bonds of sugars. This process requires an input of carbon dioxide (CO_2) and water (H_2O) and produces oxygen gas (O_2) as a waste product.

photosystem
A light-harvesting unit of a chloroplast's thylakoid membrane; consists of several hundred molecules, a reaction-center chlorophyll, and a primary electron acceptor.

phototropism (*fō'-tō-trō'-pizm*)
The directional growth of a plant shoot in response to light.

phylogenetic tree (*fī'-lō-juh-net'-ik*)
A branching diagram that represents a hypothesis about evolutionary relationships between organisms.

phylum (*fī'-lum*)
(plural, **phyla**) In classification, the taxonomic category above class. Members of a phylum all have a similar general body plan.

physiology (*fi'-zē-ol'-uh-jē*)
The study of the function of an organism's structural equipment.

phytoplankton
Photosynthetic organisms, mostly microscopic, that drift near the surfaces of ponds, lakes, and oceans.

pith
Part of the ground tissue system of a eudicot plant. Pith fills the center of a stem and may store food.

pituitary gland (*puh-tū'-uh-tār'-ē*)
An endocrine gland at the base of the hypothalamus; consists of a posterior lobe, which stores and releases two hormones produced by the hypothalamus, and an anterior lobe, which produces and secretes many hormones that regulate diverse body functions.

pivot joint
A joint that allows partial rotation. Examples in humans are the elbow and neck (turning head side to side).

placenta (*pluh-sen'-tuh*)
In most mammals, the organ that provides nutrients and oxygen to the embryo and helps dispose of its metabolic wastes. The placenta is formed from embryonic tissue and the mother's endometrial blood vessels.

placental mammal (*pluh-sen'-tul*)
Mammal whose young complete their embryonic development in the uterus, nourished via the mother's blood vessels in the placenta; also called a eutherian.

plant
A multicellular eukaryote that carries out photosynthesis and has a set of structural and reproductive terrestrial adaptations, including a multicellular, dependent embryo.

plasma
The yellowish liquid of the blood in which the blood cells are suspended.

plasma membrane
The thin double layer of lipids and proteins that sets a cell off from its surroundings and acts as a selective barrier to the passage of ions and molecules into and out of the cell; consists of a phospholipid bilayer in which proteins are embedded.

plasmid
A small ring of self-replicating DNA separate from the larger chromosome(s). Plasmids are most frequently derived from bacteria.

platelet
A piece of cytoplasm from a large cell in the bone marrow; a blood-clotting element.

plate tectonics (*tek-tahn'-iks*)
The theory that the continents are part of great plates of Earth's crust that float on the hot, underlying portion of the mantle. Movements in the mantle cause the continents to move slowly over time.

pleiotropy (*plī'-uh-trō-pē*)
The control of more than one phenotypic character by a single gene.

polar ice
A terrestrial biome that includes regions of extremely cold temperature and low precipitation located at high latitudes north of the arctic tundra and in Antarctica.

polar molecule
A molecule containing an uneven distribution of charge due to the presence of polar covalent bonds (bonds having opposite charges on opposite ends). A polar molecule will have a slightly positive pole (end) and a slightly negative pole.

pollen grain
In a seed plant, the male gametophyte that develops within the anther of a stamen. It houses cells that will develop into sperm.

pollination
In seed plants, the delivery, by wind or animals, of pollen from the male (pollen-producing) parts of a plant to the stigma of a carpel on the female part of a plant.

polychaete (*pahl'-ē-kēt*)
A type of annelid, or segmented worm, that typically lives on the seafloor.

polygenic inheritance (*pol'-ē-jen'-ik*)
The additive effect of two or more genes on a single phenotypic character.

polymer (*pol'-uh-mer*)
A large molecule consisting of many identical or similar molecular units, called monomers, covalently joined together in a chain.

polymerase chain reaction (PCR) (*puh-lim'-uh-rās*)
A technique used to obtain many copies of a DNA molecule or many copies of part of a DNA molecule. A small amount of DNA mixed with the enzyme DNA polymerase, DNA nucleotides, and a few other ingredients replicates repeatedly in a test tube.

polynucleotide (*pol'-ē-nū'-klē-ō-tīd*)
A polymer made up of many nucleotides covalently bonded together.

polyp (*pol'-ip*)
One of two types of cnidarian body forms; a stationary (sedentary), columnar, hydra-like body.

polypeptide
A chain of amino acids linked by peptide bonds.

polyploidy
Having more than two complete sets of chromosomes as a result of an accident of cell division.

polysaccharide (*pol'-ē-sak'-uh-rīd*)
A carbohydrate polymer consisting of many monosaccharides (simple sugars) linked by covalent bonds.

population
A group of interacting individuals belonging to one species and living in the same geographic area at the same time.

population density
The number of individuals of a species per unit area or volume of the habitat.

population ecology
The study of how members of a population interact with their environment, focusing on factors that influence population density and growth.

population momentum
In a population in which fertility (the number of live births over a woman's lifetime) averages two children (replacement rate), the continuation of population growth as girls reach their reproductive years.

positive feedback
A control mechanism in which the products of a process stimulate the process that produced them.

post-anal tail
A tail posterior to the anus, found in chordate embryos and most adult chordates.

posterior pituitary (*puh-tū'-uh-ter-ē*)
An extension of the hypothalamus composed of nervous tissue that secretes hormones made in the hypothalamus; a temporary storage site for hypothalamic hormones.

postzygotic barrier (*pōst'-zī-got'-ik*)
A reproductive barrier that operates if interspecies mating occurs and forms hybrid zygotes.

potential energy
Stored energy; the energy that an object has due to its location and/or arrangement. Water behind a dam and chemical bonds both possess potential energy.

predation
An interaction between species in which one species, the predator, kills and eats the other, the prey.

prepuce (*prē'-pyūs*)
A fold of skin covering the head of the clitoris or penis.

pressure-flow mechanism
The method by which phloem sap is transported through a plant from a sugar source, where sugars are produced, to a sugar sink, where sugars are used. The mechanism relies on building up water pressure at the source end of the phloem tube and reducing water pressure at the sink end.

prezygotic barrier (*prē'-zī-got'-ik*)
A reproductive barrier that impedes mating between species or hinders fertilization of eggs if members of different species attempt to mate.

primary consumer
An organism that eats only autotrophs; an herbivore.

primary growth
Growth in the length of a plant root or shoot produced by an apical meristem.

primary immune response
The initial immune response to an antigen, which includes production of effector cells that respond to the antigen within a few days and memory cells that will respond to future exposure to the antigen.

primary production
The amount of solar energy converted to the chemical energy stored in organic compounds by autotrophs in an ecosystem during a given time period.

primary succession
A type of ecological succession in which a biological community begins in an area without soil. *See also* secondary succession.

primate
Member of the mammalian group that includes lorises, bush babies, lemurs, tarsiers, monkeys, apes, and humans.

primer
A short stretch of nucleic acid bound by complementary base pairing to a DNA sequence and elongated with DNA nucleotides. During PCR, primers flank the desired sequence to be copied.

prion (*prī'on*)
An infectious form of protein that may multiply by converting related proteins to more prions. Prions cause several related diseases in different animals, including scrapie in sheep, mad cow disease, and Creutzfeldt-Jakob disease in humans.

producer
An organism that makes organic food molecules from carbon dioxide, water, and other inorganic raw materials: a plant, alga, or autotrophic bacterium; the trophic level that supports all others in a food chain or food web.

product
An ending material in a chemical reaction.

progestin (*prō-jes'-tin*)
One of a family of lipid-soluble sex hormones, including progesterone, produced by the mammalian ovary. Progestins prepare the uterus for pregnancy.

prokaryote (*prō-kār'-ē-ōt*)
An organism characterized by prokaryotic cells. *See also* prokaryotic cell.

prokaryotic cell (*prō-kār'-ē-ot'-ik*)
A type of cell lacking a nucleus and other membrane-bound organelles. Prokaryotic cells are found only among organisms of the domains Bacteria and Archaea.

promoter
A specific nucleotide sequence in DNA, located at the start of a gene, that is the binding site for RNA polymerase and the place where transcription begins.

prophage (*prō'-fāj*)
Phage DNA that has inserted into the DNA of a prokaryotic chromosome.

prophase
The first stage of mitosis. During prophase, duplicated chromosomes condense to form structures visible with a light microscope, and the mitotic spindle forms and begins moving the chromosomes toward the center of the cell.

prostate gland (*pros'-tāt*)
A gland in human males that secretes an acid-neutralizing component of semen.

protein
A biological polymer constructed from hundreds to thousands of amino acid monomers. Proteins perform many functions within living cells, including providing structure, transport, and acting as enzymes.

proteomics
The systematic study of the full protein sets (proteomes) encoded by genomes.

protist (*prō'-tist*)
Any eukaryote that is not a plant, animal, or fungus.

proton
A subatomic particle with a single unit of positive electrical charge, found in the nucleus of an atom.

proto-oncogene (*prō'-tō-on'-kō-jēn*)
A normal gene that can be converted to a cancer-causing gene.

protozoan (*prō'-tō-zō'-un*)
A protist that lives primarily by ingesting food; a heterotrophic, animal-like protist.

provirus
Viral DNA that inserts into a host genome.

pseudopodium (*sū'-dō-pō'-dē-um*)
(plural, **pseudopodia**) A temporary extension of an amoeboid cell. Pseudopodia function in moving cells and engulfing food.

pulmonary circuit
One of two main blood circuits in terrestrial vertebrates; conveys blood between the heart and the lungs. *See also* systemic circuit.

pulse
The rhythmic stretching of the arteries caused by the pressure of blood forced through the arteries by contractions of the ventricles during systole.

Punnett square
A diagram used in the study of inheritance to show the results of random fertilization.

pupil
The opening in the iris that admits light into the interior of the vertebrate eye. Muscles in the iris regulate its size.

pyramid of production

A diagram depicting the cumulative loss of energy with each transfer in a food chain.

Q

quaternary consumer (*kwot'-er-nār-ē*)
An organism that eats tertiary consumers.

R

radial symmetry

An arrangement of the body parts of an organism like pieces of a pie around an imaginary central axis. Any slice passing longitudinally through a radially symmetric organism's central axis divides the organism into mirror-image halves.

radiation therapy

Treatment for cancer in which parts of the body that have cancerous tumors are exposed to high-energy radiation to disrupt cell division of the cancer cells.

radioactive isotope

An isotope whose nucleus decays spontaneously, giving off particles and energy.

radiometric dating

A method for determining the age of fossils and rocks from the ratio of a radioactive isotope to the nonradioactive isotope(s) of the same element in the sample.

radula (*rad'-yū-luh*)
A file-like organ found in many molluscs, typically used to scrape up or shred food.

ray-finned fish

A bony fish in which fins are webs of skin supported by thin, flexible skeletal rays. All but one living species of bony fishes are ray-fins. *See also* lobe-finned fish.

reabsorption

In the vertebrate kidney, the reclaiming of water and valuable solutes from the filtrate.

reactant

A starting material in a chemical reaction.

receptor potential

The change in membrane potential of a sensory receptor cell that results when a stimulus is converted to an electrical signal.

recessive allele

In heterozygotes, the allele that has no noticeable effect on the phenotype; the recessive version of a gene is usually represented with a lowercase italic letter (e.g., *f*).

recombinant DNA

A DNA molecule carrying genes derived from two or more sources, often from different species.

rectum

The terminal portion of the large intestine where feces are stored until they are eliminated.

red blood cell

A blood cell containing hemoglobin, which transports O_2; also called an erythrocyte.

regeneration

The regrowth of body parts from pieces of an organism.

relative abundance

The proportional representation of a species in a biological community; one component of species diversity.

relative fitness

The contribution an individual makes to the gene pool of the next generation relative to the contribution of other individuals in the population.

repetitive DNA

Nucleotide sequences that are present in many copies in the DNA of a genome. The repeated sequences may be long or short and may be located next to each other or dispersed in the DNA.

repressor

A protein that blocks the transcription of a gene or operon.

reproduction

The creation of new individuals from previous ones.

reproductive barrier

Anything that prevents individuals of closely related species from interbreeding, even when populations of the two species live together.

reproductive cloning

Using a body cell from a multicellular organism to make one or more genetically identical individuals.

reproductive cycle

In females, a recurring series of events that produces gametes, makes them available for fertilization, and prepares the body for pregnancy.

reptile

Member of the clade of amniotes that includes snakes, lizards, turtles, crocodiles, alligators, birds, and a number of extinct groups (including most of the dinosaurs).

respiratory surface

The part of an animal where gases are exchanged with the environment.

respiratory system

The organ system that functions in exchanging gases with the environment, taking in O_2 and disposing of CO_2.

resting potential

The voltage across the plasma membrane of a resting neuron, positively charged outside and negatively charged inside.

restoration ecology

A field of ecology that develops methods of returning degraded ecosystems to their natural state.

restriction enzyme

A bacterial enzyme that cuts up foreign DNA at one very specific nucleotide sequence. Restriction enzymes are used in DNA technology to cut DNA molecules in reproducible ways.

restriction fragment

A molecule of DNA produced from a longer DNA molecule cut up by a restriction enzyme.

restriction site

A specific sequence on a DNA strand that is recognized and cut by a restriction enzyme.

retina (*ret'-uh-nuh*)
The light-sensitive layer in an eye, made up of photoreceptor cells and sensory neurons.

retrovirus

An RNA virus that reproduces by means of a DNA molecule. It reverse-transcribes its RNA into DNA, inserts the DNA into a cellular chromosome, and then transcribes more copies of the RNA from the viral DNA. HIV and a number of cancer-causing viruses are retroviruses.

reverse transcriptase (*tran-skrip'-tās*)
An enzyme that catalyzes the synthesis of DNA on an RNA template.

rheumatoid arthritis (*ar-thrī'-tis*)
An autoimmune disease in which the joints become highly inflamed.

rhizome (*rī'-zōm*)
A horizontal stem that grows below the ground.

rhythm method

A form of contraception that relies on refraining from sexual intercourse when conception is most likely to occur; also called natural family planning.

ribosomal RNA (rRNA)
(*rī'-buh-sōm'-ul*)
The type of ribonucleic acid that, together with proteins, makes up ribosomes.

ribosome (*rī'-buh-sōm*)
A cellular structure consisting of RNA and protein organized into two subunits and functioning as the site of protein synthesis in the cytoplasm. The ribosomal subunits are constructed in the nucleolus and then transported to the cytoplasm where they act.

RNA Ribonucleic acid (*rī'-bō-nū-klā'-ik*)
A type of nucleic acid consisting of nucleotide monomers, with a ribose sugar, a phosphate group, and the nitrogenous bases adenine (A), cytosine (C), guanine (G), and uracil (U); usually single-stranded; functions in protein synthesis and as the genome of some viruses.

RNA polymerase (*puh-lim'-uh-rās*)
An enzyme that links together the growing chain of RNA nucleotides during transcription, using a DNA strand as a template.

RNA splicing

The removal of introns and joining of exons in eukaryotic RNA, forming an mRNA molecule with a continuous coding sequence; occurs before mRNA leaves the nucleus.

rod

A photoreceptor cell in the vertebrate retina, enabling vision in dim light (but only in shades of gray).

root

The underground organ of a plant. Roots anchor the plant in the soil, absorb and transport minerals and water, and store food.

root cap

A cone of cells at the tip of a plant root that protects the root's apical meristem.

root hair

An outgrowth of an epidermal cell on a root, which increases the root's absorptive surface area.

root system

All of a plant's roots, which anchor it in the soil, absorb and transport minerals and water, and store food.

rough ER (rough endoplasmic reticulum) (reh-tik'-yuh-lum)

A network of interconnected membranous sacs in a eukaryotic cell's cytoplasm. Rough ER membranes are studded with ribosomes that make membrane proteins and secretory proteins. The rough ER constructs membrane from phospholipids and proteins.

roundworm

An animal characterized by a cylindrical, wormlike body form and a complete digestive tract; also called a nematode.

rule of multiplication

A rule stating that the probability of a compound event is the product of the separate probabilities of the independent events.

S

salivary gland

A gland associated with the oral cavity that secretes substances that lubricate food and begin the process of chemical digestion.

sarcomere (sar'-kō-mēr)

The fundamental unit of muscle contraction, composed of thin filaments and thick filaments; thin filaments slide past thick filaments to shorten (contract) the sarcomere.

saturated

Pertaining to fats and fatty acids whose hydrocarbon chains contain the maximum number of hydrogens and therefore have no double covalent bonds. Because of their straight, flat shape, saturated fats and fatty acids tend to be solid at room temperature.

savanna

A terrestrial biome dominated by grasses and scattered trees. The temperature is warm year-round. Frequent fires and seasonal drought are significant abiotic factors.

scavenger

An animal that feeds on the carcasses of dead animals.

science

Any method of learning about the natural world that follows the scientific method. See also discovery science; hypothesis.

scientific method

Scientific investigation involving the observation of phenomena, the formulation of a hypothesis concerning the phenomena, experimentation to demonstrate the truth or falseness of the hypothesis, and results that validate or modify the hypothesis.

sclerenchyma cell (skli-reng'-kuh-muh)

In plants, a supportive type of cell with a rigid secondary wall hardened with lignin.

scrotum

A pouch of skin outside the abdomen that houses a testis. The scrotum functions in cooling the sperm, thereby keeping them viable.

seaweed

A large, multicellular marine alga.

secondary consumer

An organism that eats primary consumers.

secondary growth

An increase in a plant's girth, involving cell division in the vascular cambium and cork cambium.

secondary immune response

The immune response elicited by memory cells upon exposure to a previously encountered antigen. The secondary immune response is more rapid, of greater magnitude, and of longer duration than the primary immune response.

secondary succession

A type of ecological succession that occurs where a disturbance has destroyed an existing biological community but left the soil intact. See also primary succession.

secretion

In the vertebrate kidney, the transport of certain substances, such as some ions and drugs, from the blood into the filtrate.

seed

A plant embryo packaged with a food supply within a protective covering.

semen (sē'-mun)

The sperm-containing fluid that is ejaculated by the male during orgasm.

seminal vesicle (sem'-uh-nul ves'-uh-kul)

A gland in males that secretes a fluid component of semen that lubricates and nourishes sperm.

seminiferous tubule (sem'-uh-nif'-uh-rus)

A coiled sperm-producing tube in a testis.

sensory adaptation

The tendency of some sensory receptors to become less sensitive when they are stimulated repeatedly.

sensory input

The transmission of signals from sensory receptors to integration centers in the central nervous system.

sensory neuron

A nerve cell that receives information from sensory receptors and conveys signals into the central nervous system.

sepal (sē'-pul)

A modified leaf of a flowering plant. A whorl of sepals encloses and protects the flower bud before it opens.

sex chromosome

A chromosome that determines whether an individual is male or female; in mammals, for example, the X or Y chromosome.

sex-linked gene

A gene located on a sex chromosome.

sexual dimorphism

Distinction in appearance based on secondary sexual characteristics, noticeable differences not directly associated with reproduction or survival.

sexual reproduction

The creation of genetically distinct offspring by the fusion of two haploid sex cells (gametes: sperm and egg), forming a diploid zygote.

sexual selection

A form of natural selection in which individuals with certain characteristics are more likely than other individuals to obtain mates.

sexually transmitted disease (STD)

A contagious disease spread by sexual contact. Also called a sexually transmitted infection (STI).

shoot

The aerial organ of a plant, consisting of stem and leaves. Leaves are the main photosynthetic structures of most plants.

shoot system

All of a plant's stems, leaves, and reproductive structures.

short tandem repeat (STR)

DNA consisting of tandem (in a row) repeats of a short sequence of nucleotides.

signal transduction pathway

A series of molecular changes that converts a signal received on a target cell's surface to a specific response inside the cell.

silencer

A eukaryotic DNA sequence that inhibits the start of gene transcription; may act analogously to an enhancer, binding a repressor.

single circulation system

A circulation scheme where blood flows through the heart by way of a single closed loop, or circuit.

sister chromatid (krō'-muh-tid)

One of the two identical parts of a duplicated chromosome. While joined, two sister chromatids make up one chromosome; chromatids are eventually separated during mitosis or meiosis II.

skeletal muscle

Striated muscle attached to the skeleton. The contraction of striated muscle produces voluntary movements of the body.

skeletal system

The organ system that provides body support, protects body organs such as the brain, heart, and lungs, and anchors the muscles.

slime mold

A multicellular protist related to amoebas.

small intestine
The longest section of the alimentary canal; the principal site of the enzymatic breakdown of food molecules and the absorption of nutrients.

smooth ER (smooth endoplasmic reticulum) (reh-tik'-yuh-lum)
A network of interconnected membranous tubules in a eukaryotic cell's cytoplasm. Smooth ER lacks ribosomes. Enzymes embedded in the smooth ER membrane function in the synthesis of certain kinds of molecules, such as lipids.

smooth muscle
Muscle made up of cells without striations, found in the walls of organs such as the digestive tract, urinary bladder, and arteries.

solute (sol'-yūt)
A substance that is dissolved in a liquid (which is called the solvent) to form a solution.

solution
A liquid consisting of a homogeneous mixture of two or more substances: a dissolving agent, the solvent, and a substance that is dissolved, the solute.

solvent
The dissolving agent in a solution. Water is the most versatile known solvent.

somatic cell (sō-mat'-ik)
Any cell in a multicellular organism except a sperm or egg cell or a cell that develops into a sperm or egg; a body cell.

speciation (spē'-sē-ā'-shun)
An evolutionary process in which one species splits into two or more species.

species
A group of populations whose members possess similar anatomical characteristics and have the ability to interbreed. See also biological species concept.

species diversity
The variety of species that make up a biological community; the number and relative abundance of species in a biological community.

species richness
The total number of different species in a community; one component of species diversity.

sperm
A male gamete.

spermatogenesis (sper-mat'-ō-jen'-uh-sis)
The formation of sperm cells in the testis.

spermicide
A sperm-killing chemical, in the form of a cream, jelly, or foam, that works with a barrier device as a method of contraception.

spinal cord
In vertebrates, a jellylike bundle of nerve fibers located within the vertebral column. The spinal cord and the brain together make up the central nervous system.

sponge
An aquatic stationary animal characterized by a highly porous body, choanocytes (specialized cells used for suspension feeding), and no true tissues.

spore
(1) In plants and algae, a haploid cell that can develop into a multicellular haploid individual, the gametophyte, without fusing with another cell. (2) In fungi, a haploid cell that germinates to produce a mycelium.

sporophyte (spōr'-uh-fīt)
The multicellular diploid form in the life cycle of organisms undergoing alternation of generations; results from a union of gametes and meiotically produces haploid spores that grow into the gametophyte generation.

stabilizing selection
Natural selection that favors intermediate variants by acting against extreme phenotypes.

stamen (stā'-men)
A pollen-producing part of a flower, consisting of a stalk (filament) and an anther.

starch
A storage polysaccharide found in the roots of plants and certain other cells; a polymer of glucose.

start codon (kō'-don)
On mRNA, the specific three-nucleotide sequence (AUG) to which an initiator tRNA molecule binds, starting translation of genetic information.

stem
That part of a plant's shoot system that generally grows above the ground and supports the leaves and reproductive structures.

stem cell
A relatively unspecialized cell that can give rise to one or more types of specialized cells. See also embryonic stem cell (ES cell); adult stem cell.

steroid (stir'-oyd)
A type of lipid with a carbon skeleton in the form of four fused rings: three 6-sided rings and one 5-sided ring. Examples are cholesterol, testosterone, and estrogen.

stigma (stig'-muh)
(plural, **stigmata**) The sticky tip of a flower's carpel that traps pollen.

stoma (stō'-muh)
(plural, **stomata**) A pore surrounded by guard cells in the epidermis of a leaf. When stomata are open, CO_2 enters the leaf, and water and O_2 exit. A plant conserves water when its stomata are closed.

stomach
A pouch-like organ in a digestive tube that stores and churns food.

stop codon (kō'-don)
In mRNA, one of three triplets (UAG, UAA, UGA) that signal gene translation to stop.

STR analysis
A method of DNA profiling that compares the lengths of STR sequences at specific sites in the genome.

stroma (strō'-muh)
A thick fluid enclosed by the inner membrane of a chloroplast. Sugars are made in the stroma by the enzymes of the Calvin cycle.

substrate
(1) A specific substance (reactant) on which an enzyme acts. Each enzyme recognizes only the specific substrate of the reaction it catalyzes. (2) A surface in or on which an organism lives.

sugar-phosphate backbone
The alternating chain of sugar and phosphate to which DNA and RNA nitrogenous bases are attached.

sugar sink
A plant organ that is a net consumer or storer of sugar. Growing roots,

shoot tips, stems, and fruits are sugar sinks supplied by phloem.

sugar source
A plant organ in which sugar is being produced by either photosynthesis or the breakdown of starch. Mature leaves are the primary sugar sources of plants.

survivorship curve
A plot of the number of individuals that are still alive at each age in the maximum life span; one way to represent the age-specific death rate.

sustainability
The goal of developing, managing, and conserving Earth's resources in ways that meet the needs of people today without compromising the ability of future generations to meet their needs.

sustainable development
Development that meets the needs of people today without limiting the ability of future generations to meet their needs.

swim bladder
A gas-filled internal sac that helps bony fishes maintain buoyancy.

symbiosis (sim'-bē-ō'-sis)
An interaction between organisms of different species in which one species, the symbiont, lives in or on another species, the host.

sympathetic division
One of two sets of neurons in the autonomic nervous system; generally prepares the body for energy-consuming activities ("fight or flight"). See also parasympathetic division.

sympatric speciation
The formation of a new species in populations that live in the same geographic area. See also allopatric speciation.

synapse (sin'-aps)
A junction, or relay point, between two neurons or between a neuron and a muscle or gland cell. Electrical and chemical signals are relayed from one cell to another at a synapse.

synaptic cleft (sin-ap'-tik)
A narrow gap at a synapse separating the synaptic terminal of a transmitting neuron from a receiving cell, which is another neuron, a muscle cell, or a gland cell.

synaptic terminal
The bulb-like structure at the tip of a transmitting neuron's axon, where signals are sent to another neuron or to a muscle or gland cell.

systematics
A discipline of biology that focuses on classifying organisms and determining their evolutionary relationships.

systemic circuit
One of two main blood circuits in terrestrial vertebrates; conveys blood between the heart and the rest of the body. *See also* pulmonary circuit.

systole (*sis'-tō-lē*)
The contraction stage of the cardiac cycle, when the heart chambers actively pump blood. *See also* diastole.

T

T cell
A type of lymphocyte that matures in the thymus. *See also* helper T cell and cytotoxic T cell.

taiga (*tī'-guh*)
The northern coniferous forest, characterized by long, snowy winters and short, wet summers. Taiga extends across North America and Eurasia, to the southern border of the arctic tundra; it is also found just below alpine tundra on mountainsides in temperate zones.

tail
Extra nucleotides added at the end of an RNA transcript in the nucleus of a eukaryotic cell.

target cell
A cell that responds to a regulatory signal, such as a hormone.

taxonomy
The branch of biology concerned with identifying, naming, and classifying species.

telophase
The fourth and final stage of mitosis, during which daughter nuclei form at the two poles of a cell. Telophase usually occurs together with cytokinesis.

temperate broadleaf forest
A terrestrial biome located throughout midlatitude regions where there is sufficient moisture to

support the growth of large, broadleaf deciduous trees.

temperate grassland
A terrestrial biome located in the temperate zone and characterized by low rainfall and nonwoody vegetation. Tree growth is hindered by occasional fires and periodic severe drought.

temperate rain forest
A coniferous forest of coastal North America (from Alaska to Oregon) supported by warm, moist air from the Pacific Ocean.

temperate zones
Latitudes between the tropics and the Arctic Circle in the north and the Antarctic Circle in the south; regions with milder climates than the tropics or polar regions.

tendon
Fibrous connective tissue connecting a muscle to a bone.

terminal bud
Embryonic tissue at the tip of a shoot, made up of developing leaves and a compact series of nodes and internodes.

terminator
A special sequence of nucleotides in DNA that marks the end of a gene. It signals RNA polymerase to release the newly made RNA molecule, which then departs from the gene.

tertiary consumer (*ter'-shē-ār-ē*)
An organism that eats secondary consumers.

testcross
The mating between an individual of unknown genotype for a particular character and an individual that is homozygous recessive for that same character.

testicle
A structural component of the male reproductive system consisting of a testis and scrotum.

testis
(plural, **testes**) The male gonad in an animal. The testis produces sperm and, in many species, reproductive hormones.

tetrapod
A vertebrate with four limbs. Tetrapods include mammals, amphibians, and reptiles (including birds).

theory
A widely accepted explanatory idea that is broader in scope than a hypothesis, generates new hypotheses, and is supported by a large body of evidence.

therapeutic cloning
The cloning of human cells by nuclear transplantation for therapeutic purposes, such as the replacement of body cells that have been irreversibly damaged by disease or injury. *See also* nuclear transplantation; reproductive cloning.

thermoreceptor
A sensory receptor that detects heat or cold.

thermoregulation
The maintenance of internal temperature within a range that allows cells to function efficiently.

thick filament
The thicker of the two types of filaments that make up a sarcomere, consisting of the protein myosin.

thigmotropism (*thig'-mō-trō'-pizm*)
Growth of a plant in response to touch.

thin filament
The thinner of the two types of filaments that make up a sarcomere, consisting of the protein actin.

threatened species
As defined in the U.S. Endangered Species Act, a species that is likely to become endangered in the foreseeable future throughout all or a significant portion of its range.

three-domain system
A system of taxonomic classification based on three basic groups: Bacteria, Archaea, and Eukarya.

threshold
The minimum change in a membrane's voltage that must occur to generate a nerve signal (action potential).

thylakoid (*thī'-luh-koyd*)
One of a number of disk-shaped membranous sacs inside a chloroplast. Thylakoid membranes contain chlorophyll and the enzymes of the light reactions of photosynthesis. A stack of thylakoids is called a granum.

thymine (T) (*thī'-mēn*)
A single-ring nitrogenous base found in DNA.

thyroid gland (*thī'-royd*)
An endocrine gland, located in the neck, that secretes hormones that increase oxygen consumption and metabolic rate and help regulate development and maturation.

thyroid hormone (*thī'-royd*)
A hormone secreted by the thyroid gland that affects many functions of the body, including metabolic rate, heart rate, blood pressure, and tolerance to cold.

tissue
An integrated group of similar cells that performs a specific function within a multicellular organism.

tissue system
An organized collection of plant tissues. The organs of plants (such as roots, stems, and leaves) are formed from the dermal, vascular, and ground tissue systems.

tongue
A muscular organ of the mouth that helps to taste and swallow food.

trace element
An element that is essential for the survival of an organism but is needed in only minute quantities. Examples of trace elements needed by people include iron and zinc.

trachea (*trā'-kē-uh*)
(plural, **tracheae**) (1) The windpipe; the portion of the respiratory tube between the larynx and the bronchi. (2) One of many tiny tubes that branch throughout an insect's body, enabling gas exchange between outside air and body cells.

tracheal system
In insects, an extensive network of branching internal tubes called tracheae, used in respiration.

tracheid (*trā'-kē-id*)
A tapered, porous, water-conducting, supportive cell in plants. Chains of tracheids or vessel elements make up the water-conducting, supportive tubes in xylem.

trait
A variant of a character found within a population, such as purple flowers in pea plants or blue eyes in people.

trans fats
An unsaturated fatty acid produced by the partial hydrogenation of vegetable oils and present in

hardened vegetable oils, most margarines, many commercial baked foods, and many fried foods.

transcription

The synthesis of RNA on a DNA template.

transcription factor

In the eukaryotic cell, a protein that functions in initiating or regulating transcription. Transcription factors bind to DNA or to other proteins that bind to DNA.

transfer RNA (tRNA)

A type of ribonucleic acid that functions as an interpreter in translation. Each tRNA molecule has a specific anticodon, picks up a specific amino acid, and conveys the amino acid to the appropriate codon on mRNA.

transgenic organism

An organism that contains genes from another organism, typically of another species.

translation

The synthesis of a polypeptide using the genetic information encoded in an mRNA molecule. There is a change of "language" from nucleotides to amino acids. *See also* genetic code.

transpiration

The evaporative loss of water from a plant. Transpiration is used to transport water from the roots to the tips of shoots.

transport vesicle

A tiny membranous sphere in a cell's cytoplasm carrying molecules produced by the cell. The vesicle buds from the endoplasmic reticulum or Golgi apparatus and eventually fuses with another organelle or the plasma membrane, releasing its contents.

triglyceride (*trī-glis'-uh-rīd*)

A dietary fat that consists of a molecule of glycerol linked to three molecules of fatty acids.

trimester

In human development, one of three 3-month-long periods of pregnancy.

trisomy 21

See Down syndrome.

trophic structure (*trō'-fik*)

The feeding relationships among the various species in a community.

tropical forest

A terrestrial biome characterized by warm temperatures year-round.

tropics

The region between the Tropic of Cancer and the Tropic of Capricorn; latitudes between 23.5° north and south.

tropism (*trō'-pizm*)

A growth response that makes a plant grow toward or away from a stimulus.

tubal ligation

A means of sterilization in which a segment is removed from each of a woman's two oviducts (fallopian tubes) and the remaining ends are tied shut, thereby blocking the route of sperm to egg.

tuber

An enlargement at the end of a rhizome, in which food is stored.

tubule

A thin tube within the internal structure of the human kidney.

tumor

An abnormal mass of cells that forms within otherwise normal tissue.

tumor-suppressor gene

A gene whose product inhibits cell division, thereby preventing uncontrolled cell growth.

tundra

A terrestrial biome characterized by bitterly cold temperatures. Plant life is limited to dwarf woody shrubs, grasses, mosses, and lichens. Arctic tundra has permanently frozen subsoil (permafrost); alpine tundra, found at high elevations, lacks permafrost.

tunicate

One of a group of stationary invertebrate chordates.

U

umbilical cord

A structure containing arteries and veins that connects a developing embryo to the placenta of the mother.

unsaturated

Pertaining to fats and fatty acids whose hydrocarbon chains lack the maximum number of hydrogen atoms and therefore have one or more double covalent bonds. Because of their bent shape, unsaturated fats and

fatty acids tend to stay liquid at room temperature.

uracil (U) (*yū'-ruh-sil*)

A single-ring nitrogenous base found in RNA.

ureter (*yū-rē'-ter* or *yū'-reh-ter*)

A duct that conveys urine from the kidney to the urinary bladder.

urethra (*yū-rē'-thruh*)

A duct that conveys urine from the urinary bladder to outside the body. In the male, the urethra also conveys semen out of the body during ejaculation.

urinary bladder

The pouch where urine is stored before elimination.

urine

Concentrated filtrate produced by the kidneys and excreted via the bladder.

uterus (*yū'-ter-us*)

In the reproductive system of a mammalian female, the organ where the development of young occurs; the womb.

V

vaccination (*vak'-suh-nā'-shun*)

A procedure that presents the immune system with a harmless version of a pathogen, thereby stimulating adaptive immunity when the pathogen itself is encountered.

vaccine (*vak-sēn'*)

A harmless version or piece of a pathogen (a disease-causing virus or organism) used to stimulate a host organism's immune system to mount a long-term defense against the pathogen.

vacuole (*vak'-ū-ōl*)

A membrane-enclosed sac, part of the endomembrane system of a eukaryotic cell, having diverse functions.

vagina (*vuh-jī'-nuh*)

Part of the female reproductive system between the uterus and the outside opening; the birth canal in mammals. The vagina accommodates the male's penis and receives sperm during copulation.

vas deferens (*vas def'-er-enz*)

(plural, **vasa deferentia**) Part of the male reproductive system that

conveys sperm away from the testis; the sperm duct; in humans, the tube that conveys sperm between the epididymis and the common duct that leads to the urethra.

vascular cambium

(*vas'-kyū-ler kam'-bē-um*)

In plants, a cylinder of meristem tissue found between the primary xylem and phloem. During secondary growth, the vascular cambium produces secondary xylem and phloem.

vascular tissue

Plant tissue consisting of cells joined into tubes that transport water and nutrients throughout the plant body. Xylem and phloem make up vascular tissue.

vascular tissue system

A system formed by xylem and phloem throughout a plant, serving as a long-distance transport system for water and nutrients, respectively.

vasectomy (*vuh-sek'-tuh-mē*)

Surgical removal of a section of the two sperm ducts (vasa deferentia) to prevent sperm from reaching the urethra; a means of sterilization in the male.

vector

A piece of DNA, usually a plasmid or a viral genome, that is used to move genes from one cell to another.

vein

(1) In animals, a vessel that returns blood to the heart. (2) In plants, a vascular bundle in a leaf, composed of xylem and phloem.

ventricle (*ven'-truh-kul*)

A heart chamber that pumps blood out of the heart to the body or lungs via arteries.

vertebrate (*ver'-tuh-brāt*)

A chordate animal with a backbone. Vertebrates include lampreys, cartilaginous fishes, bony fishes, amphibians, reptiles (including birds), and mammals.

vesicle

A membranous sac in the cytoplasm of a eukaryotic cell.

vessel element

A short, open-ended, water-conducting, supportive cell in plants. Chains of vessel elements or tracheids make up the water-conducting, supportive tubes in xylem.

vestigial structure (*ve-sti'-gē-al*)
A structure of marginal, if any, importance to an organism. Vestigial structures are historical remnants of structures that had important functions in ancestors.

virus
A microscopic particle capable of infecting cells of living organisms and inserting its genetic material. Viruses have a very simple structure and are generally not considered to be alive because they do not display all of the characteristics associated with life.

vitamin
An organic nutrient that an organism requires in very small quantities. Vitamins generally function as coenzymes.

vulva
The outer features of the female reproductive anatomy.

W

warning coloration
The bright color pattern, often yellow, red, or orange in combination with black, of animals that have effective chemical defenses.

water vascular system
In echinoderms, a radially arranged system of water-filled canals that branch into extensions called tube feet. The system provides movement and circulates water, facilitating gas exchange and waste disposal.

wavelength
The distance between crests of adjacent waves, such as those of the electromagnetic spectrum including light.

wetland
An ecosystem intermediate between an aquatic ecosystem and a terrestrial ecosystem. Wetland soil is saturated with water permanently or periodically.

white blood cell
A blood cell that functions in defending the body against infections; also called a leukocyte.

whole-genome shotgun method
A method for determining the DNA sequence of an entire genome by cutting it into small fragments, sequencing each fragment, and then placing the fragments in the proper order.

wild-type trait
The trait most commonly found in nature.

wood
Secondary xylem of a plant.

X

X chromosome inactivation
In female mammals, the inactivation of one X chromosome in each somatic cell. Once X inactivation occurs in a given cell (during embryonic development), all descendants of that cell will have the same copy of the X chromosome inactivated.

xylem (*zī'-lum*)
The portion of a plant's vascular system that provides support and conveys water and inorganic nutrients from the roots to the rest of the plant. Xylem consists mainly of vessel elements and/or tracheids, water-conducting cells.

xylem sap
The solution of inorganic nutrients conveyed in xylem tissue from a plant's roots to its shoots.

Z

zooplankton
In aquatic environments, free-floating animals, including many microscopic ones.

zygote (*zī'-gōt*)
The fertilized egg, which is diploid, that results from the union of haploid gametes (sperm and egg) during fertilization.

Index

Page numbers with *f* indicate figure, *t* indicate table, and those in bold indicate page where defined as key term.

A

Abalones, eye complexity of, 319*f*
Abdomen, arthropod, 381
Abiotic factors
  of biosphere, 410
  in ecology, **409**
  in ecosystem ecology, 474–76
Abiotic reservoirs, **474**–76
ABO blood groups, **193**
Abortion, RU-486 and, 594
Abscisic acid, 666*t*, 669
Absorption, **363**
  of nutrients, 516–17
  of nutrients, in plants, 660–65
Abstinence, sexual, 593, 595
Acclimation, **412**
Acetic acid, 131
Acetylcholine, 614
Acetyl CoA, 131, 134
Achondroplasia (dwarfism), 189*t*, 190, 190*f*
Acidification, ocean, 417, 429
Acidity, in food digestion, 56, 65
Acid precipitation, 406, 425
Acid reflux, 515
Acids, 56, **65**–66
  bases, pH, and, 65–66
  in human gastric juice, 514–15
Acquired defenses. *See* Adaptive defenses
Acrosome, 596
ACTH (adrenocorticotropic hormone), 572*f*, 578
Actin, 630
Actinomycetes, 334*f*
Action potentials, **611**–12, 611*f*, 612*f*, 621–22
  in nervous system signal transmission, 611–12
  in vision, 624
Activation energy, enzymes and, **114**, 114*f*
Activators, DNA, **235**
Active site, enzyme, **116**
Active transport, **120**, 120*f*
Adam's apples, 514
Adaptations, evolutionary. *See* Evolutionary
  adaptations
Adaptive coloration, predation and, 413*f*, 464–65.
  See also Camouflage
Adaptive defenses, 552–55, 560
Addiction
  hypothalamus pleasure center and, 617
  smooth ER and, 99
Adaptive immunity, **552**, 556*f*
Adélie penguins, effect of global climate change on, 407

Adenine (A), 83, 207
Adenosine, 113
Adenoviruses, 220*f*
ADH. *See* Antidiuretic hormone
ADHD (attention deficit hyperactivity disorder),
  614, 619
Adhesion, **663**–64
Adipose tissue, 77, **493**, 493*f*
ADP (adenosine diphosphate), **113**
Adrenal cortex, 572*f*, **577**
Adrenal glands, 572*f*, **577**–78*f*
  endocrine system functions of, 577–78
  stress and, 577–78
Adrenaline, 572
Adrenal medulla, 572*f*, **577**
Adrenocorticotropic hormone. *See* ACTH
Adult stem cells, therapeutic cloning
  using, **242**
Aerobic capacity
  athletic conditioning and, 125
  high-altitude athletic training for, 538
Aerobic processes, **128**
Aerobic vs. anaerobic lifestyles
  evolution of glycolysis and, 128–34
  muscle fatigue as lactic acid accumulation from
    fermentation, 135
African violets, 157*f*
Afterbirth, 602
Age structure, population, 439
Aggressive behavior, androgens and male, 581
Agriculture
  angiosperms in, 360
  auxins in, 666–68
  biological control of pests in, 448
  cheatgrass as invasive species and, 447*f*
  climate change and, 431
  coffee plants and, 639
  crop rotation and, 663
  deforestation for, 361, 426
  effects of bee colony collapse disorder
    on, 655
  genetically modified organisms in, 256–57
  gibberellins and, 669
  integrated pest management in, 450
  loss of plant diversity through deforestation,
    423, 426
  nitrogen and, 662
  organic farming and sustainable, 661
  sustainable, 427
  synthetic auxins in, 667–68
  in temperate grassland biomes, 422
  thorn bugs and, 671

AIDS (acquired immunodeficiency syndrome),
  224–25, 224*f*, 595
  HIV and, 563
  search for vaccine for, 563
  as sexually transmitted disease, 595
Air pollution
  declining biodiversity and, 426
  effects of, on human lungs, 544
  tobacco smoke as, 547
Albinism, 189*t*
Alcohol
  as a depressant, 614
  effects of, on fetal development, 599
  effects of, on human brain, 614
  effects of, on neurotransmitters, 614
Alcoholic fermentation, 136–37
Algae, **341**
  adaptations to terrestrial life, 352
  brown, 345*f*
  in freshwater biomes, 427
  green, 344, 345*f*, 352
  in marine biomes, 416
  as photosynthetic protists, 341
  red, 345*f*
Alimentary canal, 513, 516*f*, 518*f*
Alleles, **182**
  ABO blood groups as example of codominant, 193
  biological diversity and, 302, 320
  dominant vs. recessive, 182
  frequencies in a populations, 293–98
  gene pools of, 291–93
  human genetic disorders and, 188–91
  law of segregation and, 182–83, 186*f*
  multiple, and codominance, 193
  sickle-cell disease as example of pleiotropy
    by, 194
Allergens, **561**
Allergic reactions, 562
Allergies, **561**–62, 653
  development of, 562, 562*f*
  to meat, 550, 562
Allopatric speciation, **308**–09
Alternation of generations, in plants and multicellular
  green algae, **354**, 356*f*
Alternative RNA splicing, **236**, 236*f*
Altitude
  athletic training and, 538
  human adaptations and, 547
  vegetation types and, 418*f*
Alvarez, Luis and Walter, 316
Alveoli, **544**, 547
Alzheimer's disease, 60, 189*t*